AF785185

Genesis and Metallogeny of Non-ferrous and Precious Metal Deposits

Genesis and Metallogeny of Non-ferrous and Precious Metal Deposits

Guest Editors

Yunsheng Ren
Qun Yang

MDPI

Basel • Beijing • Wuhan • Barcelona • Belgrade • Novi Sad • Cluj • Manchester

Guest Editors
Yunsheng Ren
College of Earth Sciences
Jilin University
Changchun
China

Qun Yang
College of Earth Sciences
Jilin University
Changchun
China

Editorial Office
MDPI AG
Grosspeteranlage 5
4052 Basel, Switzerland

This is a reprint of the Special Issue, published open access by the journal *Minerals* (ISSN 2075-163X), freely accessible at: https://www.mdpi.com/journal/minerals/special_issues/NFPMD.

For citation purposes, cite each article independently as indicated on the article page online and as indicated below:

Lastname, A.A.; Lastname, B.B. Article Title. *Journal Name* **Year**, *Volume Number*, Page Range.

ISBN 978-3-7258-5099-0 (Hbk)
ISBN 978-3-7258-5100-3 (PDF)
https://doi.org/10.3390/books978-3-7258-5100-3

Contents

Preface

This Special Issue plans to focus on four types of research:

(1) Systematic papers on global deposits discovered in the past decade, which use mineralogical analysis and isotope tracing to reveal the sources of ore-forming fluids and the tectonic ore-controlling laws;

(2) Groundbreaking review articles that integrate deep geophysical data with mineral deposit evidence to innovate traditional metallogenic theories, aiming to explain new types of deposits such as the enrichment of "critical rare, rare earth, and scattered" metals;

(3) Case studies on the application of advanced technologies, analyzing post-mineralization modifications and ultramicroscopic mineral inclusions;

(4) Research on regional metallogenic regularities, which establishes a metallogenic pedigree of deep-Paleoproterozoic deposits by combining tectonic evolution data from orogenic belts such as the Central Asian Orogenic Belt, providing support for "deep exploration and blind ore prospecting".

The launch of the third volume is driven by dual needs: disciplinary development and resource security. On the one hand, the first two volumes have covered traditional models such as porphyry–skarn systems, while current research urgently needs to respond to cutting-edge directions like "deep-earth metallogeny" and "non-traditional deposits". On the other hand, given China's over 70% external dependence for key minerals such as copper, cobalt, and lithium, this Special Issue will focus on metallogenic belt regularities along the "Belt and Road" to serve overseas exploration strategies. Meanwhile, by integrating geophysics and environmental science, it will promote interdisciplinary research on "mineralization–environment" coupling.

By including breakthrough achievements in theories, technologies, and regional regularities, this volume aims to build a complete chain of "metallogenic theory–technical methods–exploration applications". Academically, it will enhance the citation influence in top journals such as *Economic Geology*; practically, it will provide scientific support for China's "Deep Earth Resource Exploration Program" and "Strategic Mineral Prospecting Campaign". Ultimately, it will serve as a key link connecting basic research and global resource exploration practices.

Yunsheng Ren and Qun Yang
Guest Editors

minerals

MDPI

Editorial

Editorial for Special Issue "Genesis and Metallogeny of Non-Ferrous and Precious Metal Deposits"

Sitong Chen [1], Qun Yang [1,*], Yunsheng Ren [2], Chan Li [1], Haozhe Li [1] and Xiaolei Peng [1]

1 College of Earth Sciences, Jilin University, Changchun 130061, China; 19834922485@163.com (S.C.); chanli1112@163.com (C.L.); haozhe0304@163.com (H.L.); pengxl@jlu.edu.cn (X.P.)
2 School of Earth Science, Institute of Disaster Prevention, Beijing 101601, China; renys@jlu.edu.cn
* Correspondence: yangqun@jlu.edu.cn

1. Introduction

In recent decades, an increasing number of non-ferrous and precious metal deposits have been discovered all over the world. Concurrently, new and advanced analytical techniques in deposit research have been developed, such as isotope dating of ore minerals by U-Pb, Ar-Ar, and Re-Os methods.

This focus of this Special Issue includes ore properties and genesis, ore-controlling tectonic–magmatic events, geochronology, tectonic setting, and metallogenic models.

2. Highlights from the Special Issue

Yu et al. (2023) [1] conducted a comprehensive analysis of fluid inclusions and H–O–S–Pb isotopes of the Songjianghe gold deposit in Jilin Province, and determined that fluid immiscibility triggered by rapid pressure drop is the main mechanism for the precipitation of gold–polymetallic sulfides. The deposit formed during the hiatus of Paleo-Pacific plate subduction, and it is aged between 157 and 156 Ma (Late Jurassic).

Sun et al. (2022) [2] studied the Huoluotai porphyry Cu (Mo) deposit in the northern Greater Khingan Range, showing that sulfur is derived from magma, while lead has a crust–mantle mixed source. A key finding is that mineralization is dominated by medium–low temperature fluid boiling, which challenges the traditional porphyry metallogenic model. This mineralization system formed at approximately 149 Ma, correlating with the closure event of the Mongol-Okhotsk Ocean in the Late Mesozoic.

Chen et al. (2022) [3] defined the Dongbuzi deposit in the Liaodong Peninsula as a fault-controlled, low-sulfidation epithermal gold deposit hosted in Cretaceous trachyte porphyry. Fluid cooling and dilution drove gold precipitation. Regarding isotope analysis, H–O isotopes indicate meteoric water while C–O isotopes point to a mantle/magmatic source of carbon. Moreover, S–Pb isotopes indicates a mixed source from trachyteporphyry and from host volcanic rocks. The deposit formed during the Early Cretaceous lithospheric extension in the eastern North China Craton.

Wang et al. (2022) [4] traced the magmatic–hydrothermal origin of magnetite and pyrite in the Dongping deposit on the northern margin of the North China Craton through in situ electron probe microanalysis (EPMA) and laser ablation inductively coupled-plasma mass spectrometry (LA-ICP-MS). Magnetite formed in a low-temperature hydrothermal environment. Gold enrichment is related to fluid cooling and boiling, supporting a deep-source magmatic–hydrothermal model associated with Mesozoic tectono-magmatic events.

Zhao et al. (2022) [5] first reported the occurrence of Paleoproterozoic (approximately 1851 Ma) gold mineralization in the Lijiabaozi deposit of the Liaodong Peninsula. S–Pb

 https://doi.org/10.3390/min15080810

isotopes indicate that ore-forming materials are mainly derived from Paleoproterozoic sedimentary strata. Mineralization is formed by regional metamorphic fluids through water–rock reactions, and its precipitation is triggered by changes in temperature/pressure.

Zhang et al. (2022) [6] combined monazite U-Pb dating, in situ trace elements of sulfides, and sulfur isotope analysis to confirm that there are two stages of orogenic gold mineralization events (Early Paleozoic and Mesozoic) in the Wulonggou gold ore field of the East Kunlun Orogenic Belt, verifying that the deposit is a composite mineralization system related to multi-stage tectono-magmatic activities.

Li et al. (2022) [7] classified the Toudaochuan gold deposit in Jilin Province as a mesothermal magmatic–hydrothermal deposit controlled by NE-trending faults, with three stages of mineralization. Studies on fluid inclusions and C–H–O–S–Pb isotopes show that fluid immiscibility caused by CO_2 escape and pH increase leads to gold precipitation. S–Pb isotopes indicate a crust–mantle mixed source, and the ore bodies are closely related to granitic porphyry dykes.

3. Conclusions

Based on studies of gold deposits in Northeast China and adjacent regions [1–7] in this Special Issue, we propose focusing future research on hydrodynamic mechanisms controlling gold precipitation (particularly immiscibility/boiling), spatiotemporal coupling of multi-epoch hydrothermal events, deep magma–hydrothermal system connections, regional mineralization models, and integrated 3D "tectonics–magma–fluids" dynamic systems.

Author Contributions: Conceptualization, Q.Y., Y.R. and X.P.; methodology, H.L.; software, C.L.; validation, S.C., C.L. and H.L.; formal analysis, S.C.; investigation, S.C.; resources, X.P.; data curation, Q.Y.; writing—original draft preparation, S.C.; writing—review and editing, S.C.; visualization, C.L.; supervision, Y.R.; project administration, Q.Y.; funding acquisition, Q.Y. and X.P. All authors have read and agreed to the published version of the manuscript.

Funding: This work was financially supported by the Natural Science Foundation of Jilin Province (20230101097JC), Jilin Provincial Department of Education Science Foundation (JJKH20241294KJ), National Natural Science Foundation of China (42202070), and China Postdoctoral Science Foundation (2022M721305).

Acknowledgments: The Guest Editors of this Special Issue would like to express their gratitude to all authors for their high-quality contributions and profound insights into the research of non-ferrous and precious metal deposits. We also extend our thanks to this journal for its constructive comments, which have significantly enhanced the clarity and rigor of the papers. Additionally, we are sincerely grateful to the editorial team for their support throughout the publication process. Finally, we acknowledge the broader scientific community, whose ongoing research continues to provide new understandings of the complex genesis of various types of non-ferrous and precious metal deposits.

Conflicts of Interest: The authors declare no conflicts of interest.

References

1. Yu, Q.; Wang, K.; Zhang, X.; Sun, Q.; Bai, W.; Ma, C.; Xiao, Y. Fluid Evolution and Ore Genesis of the Songjianghe Au Deposit in Eastern Jilin Province, NE China: Constraints from Fluid Inclusions and H-O-S-Pb Isotope Systematics. *Minerals* **2023**, *13*, 652. [CrossRef]
2. Sun, Y.; Li, B.; Chen, X.; Meng, F.; Ding, Q.; Qian, Y.; Wang, L. Fluid Inclusions and C–H–O–S–Pb Isotopes of the Huoluotai Porphyry Cu (Mo) Deposit in the Northern Great Xing'an Range, NE China: Implications for Ore Genesis. *Minerals* **2022**, *12*, 1072. [CrossRef]
3. Chen, C.; Wu, T.; Sha, D.; Li, D.; Yang, Z.; Zhang, J.; Shang, Q. Genesis of the Dongpuzi Gold Deposit in the Liaodong Peninsula, NE China: Constraints from Geology, Fluid Inclusion, and C–H–O–S–Pb Isotopes. *Minerals* **2022**, *12*, 1008. [CrossRef]

4. Wang, C.; Yu, J.; Ren, Y.; Zhao, J.; Sun, Z. Ore Genesis of the Dongping Gold Deposit in the Northern Margin of North China Craton: Constraints from In-Situ Major, Trace Elemental Analysis of Magnetite and Pyrite. *Minerals* **2022**, *12*, 978. [CrossRef]
5. Zhao, Y.; Zhang, P.; Yang, H.; Kou, L.; Bi, Z.; Yang, Z.; Chen, J. Paleoproterozoic Mineralization of the Lijiapuzi Gold Deposit in the Liaodong Peninsula, NE China: Constraints from 40Ar-39Ar Age, S-Pb Isotopes, and In Situ Analyses. *Minerals* **2022**, *12*, 971. [CrossRef]
6. Zhang, Z.; Zeng, Q.; Pan, T.; Xie, H.; Wei, Z.; Fan, H.; Wu, J.; Yang, K.; Li, X.; Liang, G. Two Epochs of Mineralization of Orogenic Gold Deposit in the East Kunlun Orogenic Belt: Constraints from Monazite U–Pb Age, In Situ Sulfide Trace Elements and Sulfur Isotopes in Wulonggou Gold Field. *Minerals* **2022**, *12*, 968. [CrossRef]
7. Li, J.; Ren, Y.; Yang, Q.; Sun, X. Ore Genesis of the Toudaochuan Gold Deposit in Central Jilin Province, NE China: Constraints from Fluid Inclusions and C–H–O–S–Pb Isotopes. *Minerals* **2022**, *12*, 964. [CrossRef]

minerals

MDPI

Article

Geochemical and Isotopic Compositions of Fluorites from the Yama Fluorite Deposit in the Qilian Orogen in Northwest China, and Their Metallogenic Implications

He Jiao [1], Guo-Biao Huang [1], Wei Ma [1], Qiang-Qiang Cui [1], Wei-Hu Wang [1], Qing-Feng Ding [2,*], Xuan Zhou [2] and Rui-Zhe Wu [2]

1 Qinghai Provincial Key Laboratory of Salt Lake Resources Exploration and Research in Qaidam Basin, Qaidam Integrated Geological Exploration Institute of Qinghai Province, Golmud 816099, China; jiaofushun666@163.com (H.J.); hsy20200205@163.com (G.-B.H.); 13369796086@163.com (W.M.); zailushang1215@163.com (Q.-Q.C.); hu861025.cool@163.com (W.-H.W.)

2 College of Earth Sciences, Jilin University, Changchun 130061, China; xuanzhou21@mails.jlu.edu.cn (X.Z.); rzwu22@mails.jlu.edu.cn (R.-Z.W.)

* Correspondence: dingqf@jlu.edu.cn

Abstract: The Yama area is characterized by numerous large-scale fluorite–quartz veins that are located along faults within the widespread Late Devonian–Late Silurian syenogranites in the Tataleng granitic batholith, Qilian Orogen, Northwest China. These fluorite–quartz veins contribute to an important fluorite reserve, but their ore genesis remains unresolved so far. In this study, trace elements, rare earth elements (REEs), and hydrogen, oxygen, and strontium isotopic compositions of fluorites are analyzed. The studied fluorite samples have similar chondrite-normalized REEs, including Y patterns, with relatively strong enrichment in heavy REEs, negative Eu anomalies, strongly positive Y anomalies, and comparably invariable Y/Ho ratios of 41.43–73.79, suggesting a unique hydrothermal genesis. The relatively variable values of δD and $\delta^{18}O$ are $-77.4‰$ to $-102.4‰$ and $-12.7‰$ to $-4.3‰$, respectively, close to the meteoric water line. These fluorites yield relatively invariable analytical $^{87}Sr/^{86}Sr$ ratios of 0.749089−0.756628 (except for an anomalously high ratio), and their calculated initial $^{87}Sr/^{86}Sr$ ratios, based on the ore-forming ages provided, are apparently higher than the calculated initial $^{87}Sr/^{86}Sr$ ratios of syenogranite wall rocks. Collectively, the geochemistry of trace elements, REEs, and stable isotopes (H, O, and Sr) suggests that the ore-forming fluids were of meteoric origin and that the Sr sources were directly derived from the ore-forming fluids themselves rather than syenogranite wall rocks. Finally, it was considered that the Yama fluorite deposit is a fault-controlled hydrothermal vein-type deposit which was possibly related to the evolution of the Paleo-Tethys Ocean in the Permian–Triassic.

Keywords: H–O isotope; Sr isotope; trace and rare earth elements; ore genesis; fluorites

1. Introduction

The Qilian Orogen (QLO) is traditionally subdivided from north to south into the North Qilian Belt (NQB), the Central Qilian Block (CQB), and the South Qilian Belt (SQB) (Figure 1a) [1–6]. It is generally believed that the QLO has developed as a result of the closure of the Proto-Tethys Ocean [1,2,6–10]. Because numerous granitic rocks in the QLO are thought to be associated with subducted, collisional, or post-collisional magmatism that was connected to the Proto-Tethys Ocean, these rocks have been studied in the past [1,4,6,10–20]. Following the widespread Early Paleozoic granitic magmatism, numerous hydrothermal mineralizations of W, Mo, Nb, Ta, Cu, Pb, Zn, Au, and F have been developed throughout the NQB and CQB [6,21–23]. However, in the SQB, there are apparently less granitic-magmatism-related mineralizations than those in the NQB and CQB. Currently, the SQB has only a limited number of significant mineral deposits, including a few gold deposits in the Danghe Nanshan mountain in the northern margin of the SQB [6,21], a few W deposits in the Tianjun area in the central segment of the SQB [6,23], and the Yama fluorite deposit and the Yansenhada (i.e., Yamatu) quartz vein deposit in the Tataleng granitic batholith in the western segment of the SQB [6,10,23,24]. Any detailed reports of geological and geochemical research on the mineralization in the SQB are lacking, leaving the ore genesis unresolved for the time being. Specifically, there are no reports on the geological and geochemical characteristics, the ore-forming age, and the ore genesis of the Yama fluorite deposit. Furthermore, in order to assess more accurately the potential for mineralization in the granitic rocks located in and around the Yama fluorite deposit within the Tataleng granitic batholith, it is necessary to conduct detailed geological and geochemical research on this fluorite mineralization.

This paper presents the trace elements and REE abundance, as well as the H, O, and Sr isotopic compositions of fluorites from various types of fluorite ores in the Yama fluorite deposit within the Tataleng granitic batholith in the SQB. The aim is to investigate the ore-forming fluid origin, metallogenic source, and ore genesis of the fluorite–quartz veins in the Yama area as well as the potential fluorite mineralization of syenogranites within the Tataleng granitic batholith.

2. Geological Background and Geology of the Yama Fluorite Deposit

2.1. Geological Background

The Qinghai–Tibet Plateau's northeastern edge hosts the QLO, which is bordered by the Alxa Block to the north and the Quanji massif (QM) to the south (Figure 1a) [4,6]. The QLO, the northernmost orogenic collage of the Tethyan domain, has an excellent record of the Early to Middle Paleozoic subduction–accretion that resulted from the convergence of the Alax and Qaidam blocks [1,6,9,25]. The North Qilian and South Qilian suture zones normally split the OLO into three tectonic belts, i.e., the NQB, CQB, and SQB (Figure 1a) [1,3,4,6,7,9,10,26].

The NQB is widely recognized as a typical cold oceanic suture zone resulting from the closure of the Early Paleozoic North Qilian Ocean along the North Qilian suture zone [1,2,4,6,18,27]. It hosts Early Paleozoic ophiolite sequences (560–450 Ma), high-pressure metamorphic belts, island-arc-related volcanic rocks (510–450 Ma), and granitic intrusions or plutons (510–420 Ma), as well as Silurian flysch formations, Devonian molasses, and Carboniferous to Triassic sedimentary cover sequences [2–4,6,20,27,28]. The CQB has been previously considered as either a detached fragment of the Yangtze Craton [3,6,20,29], a fragment rifted from the Alxa Block [2,4,6,30,31], or a fragment of an accretionary orogen [1,6]. Within the CQB, numerous Precambrian rocks, Cambrian to Ordovician volcanic sequences, Paleozoic sedimentary sequences, and Early Paleozoic granites have been identified [4,6,28,31]. The SQB is generally considered to be an accretionary belt comprising microcontinents, ophiolite complexes, and Early Paleozoic arc-related magmatic rocks [1,5,6,20,25]. The Early Paleozoic arc-related magmatic rocks in the SQB are mainly composed of dominant granitic rocks including several large-scale granitic batholiths, such as the Tataleng granitic batholith [6,10,11,20,32–37], the Delingha pluton [18], and the Huangyuan pluton [38]. These granitic rocks from the Early Paleozoic era are typically believed to have formed due to subduction and collisional events related to the South Qilian Ocean, which was a southern branch of the Proto-Tethys Ocean in the Qilian Orogen [14,17–20]. The Yama fluorite deposit is situated in the eastern section of the extensive Tataleng granitic batholith (Figure 2).

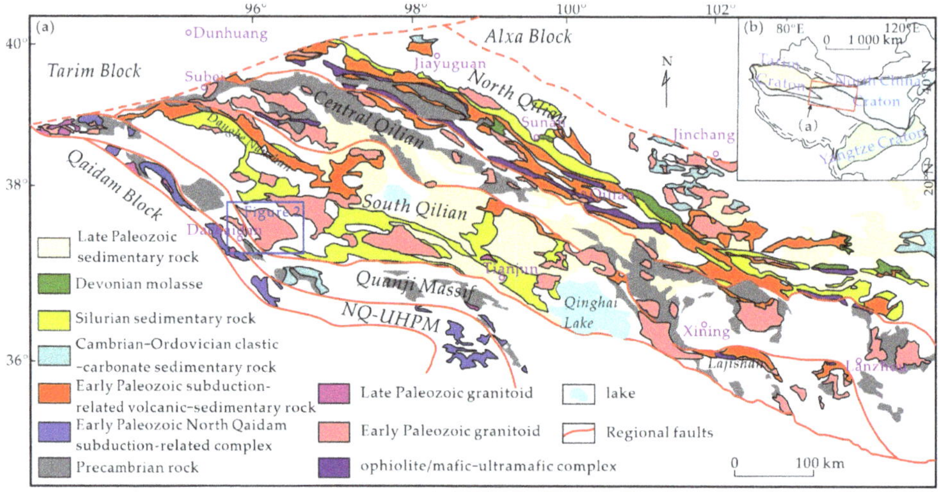

Figure 1. (**a**) Regional geological map of the Qilian Orogen (revised after [4]). NQ–UHPM (North Qaidam–ultrahigh-pressure metamorphic belt). (**b**) Schematic map showing major tectonic units in China (simplified after [39]).

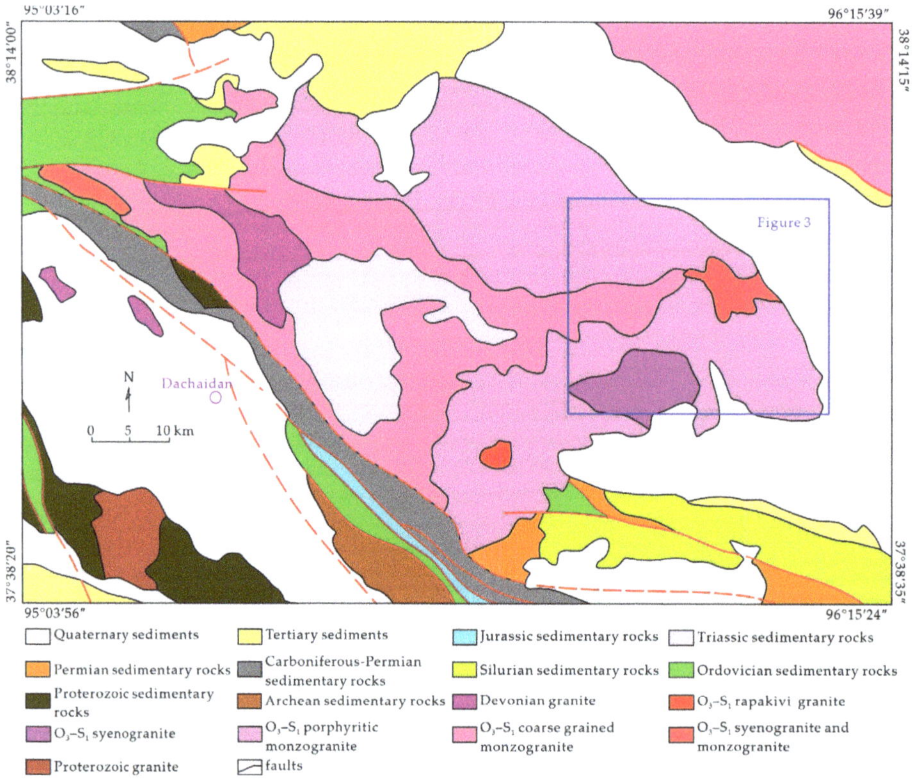

Figure 2. Geological map of the Tataleng granite batholith in the Southern Qilian Belt revised by [24]. O_3–S_1: the Late Ordovician to Early Silurian.

2.2. Geology of the Yama Fluorite Deposit

The Tataleng granitic batholith is situated in the northwestern part of the SQB (Figure 2) [10,20]. Quaternary sediments cover the northern, eastern, and southeastern regions of the Tataleng granitic batholith, while its southwest edge is adjacent to northwest-striking ductile faults with the Carboniferous to Permian sedimentary rocks (Figure 2). The Tataleng granitic batholith underwent a long crystallization period, ranging from the Late Ordovician to the Early Silurian (O_3–S_1) [10,20,24,32,34,35,37,40,41]. The central part of the Tataleng granitic batholith is covered by Triassic sedimentary rocks (Figure 2) [20].

The Yama fluorite deposit is situated in the eastern section of the Tataleng granitic batholith. Throughout the study region, the O_3–S_1 granitic rock outcrops are widely exposed, with the exception of the southeast part which is extensively covered by massive Quaternary sediments (Figure 3) [10,24]. Numerous northeast-striking faults and minor northwest-striking faults have developed in the O_3–S_1 syenogranites. Numerous fluorite-free quartz veins and fluorite–quartz veins were emplaced along these faults within the O_3–S_1 syenogranite in the Yama area (Figure 3) [10,24]. Fluorites are of high economic significance due to their abundance in numerous large-scale fluorite–quartz, of which 81 fluorite ore bodies or fluorite mineralized occurrences, ranging from No. M1 to No. M81, has been delineated in the Yama area. Among 81 fluorite ore bodies or fluorite mineralized occurrences, 76 fluorite ore bodies contribute to an important, inferred 3,555,500 tons fluorite reserve at 34.53% grade in the Yama fluorite deposit [24], in which the No. M23 fluorite ore body yield inferred a 1,844,400 tons fluorite reserve at 37.38% grade. Most of the fluorite ore bodies are located in the Yama ore district (Figure 4a), except ore body M33 which is located in the Wulandawu ore district (Figure 4b). The Yama porphyritic syenogranite, located in the central part of the Yama area, and the Wulandawu syenogranite, located in the northwest of the Yama area, make up the wall rocks of the fluorite–quartz veins in the Yama fluorite deposit (Figure 3) [10].

Based on the geological relationship (Figure 5), it can be observed that generally the fluorite-free quartz veins were formed earlier than the fluorite–quartz veins, and then a few quartz–calcite veinlets can be seen cutting the fluorite–quartz veins. Additionally, the paragenetic sequences for the Yama fluorite mineralization can be divided into two major stages, i.e., the pre-ore quartz stage and the ore-forming hydrothermal stage, and the latter stage could be subdivided into the quartz–fluorite epoch and the quartz–calcite epoch (Figure 6). Fluorite-free quartz veins, fluorite–quartz veins, and a few quartz–calcite veins were also formed in the pre-ore stage, the quartz–fluorite epoch, and the quartz–calcite epoch, respectively.

Fluorite ore bodies are commonly located within fluorite–quartz veins, which strike predominantly north–northwest (NNW), with minor northeast (NE) or north (Figure 4), and they dip predominantly in the NE (Figure 5). Rare alterations can be observed close to fluorite–quartz veins within the syenogranite wall rocks. The fluorite ore bodies are usually continuous in striking length from 67 m to 1200 m and are generally invariable in width from 0.72 to 7.3 m. A few of them have branches and variable widths in single veins (Figures 4 and 5). Fluorite–quartz veins are generally characterized by vein structures (Figure 7a,d,g), and some narrow quartz–fluorite veins intruded into brecciated syenogranite wall rocks, showing veinlets, stringer structures (Figure 7d), and/or pectinate (where the quartz is vertical to the margin of veins, Figure 7g–i); herein, we named these fluorite ores as Type 1. Type 1 ores commonly consist of fluorite and quartz, with minor quantities of sericite and calcite (Figure 7). However, some important fluorite–quartz veins may also enclose certain variable sizes of wall rock breccias, showing breccia structures within the veins (Figure 7j,m,n); herein, we named these fluorite ores as Type 2. Breccias in Type 2 ores should be derived from variable numbers and sizes of syenogranite wall rocks (Figure 7j,m), and the fine-grained fluorites and quartz which commonly grow around syenogranite breccias show crusty or colloform structures (Figure 7l,n,o), indicating such fluorites might be crystalized earlier than those in Type 2 ores during a low-temperature mineralization process.

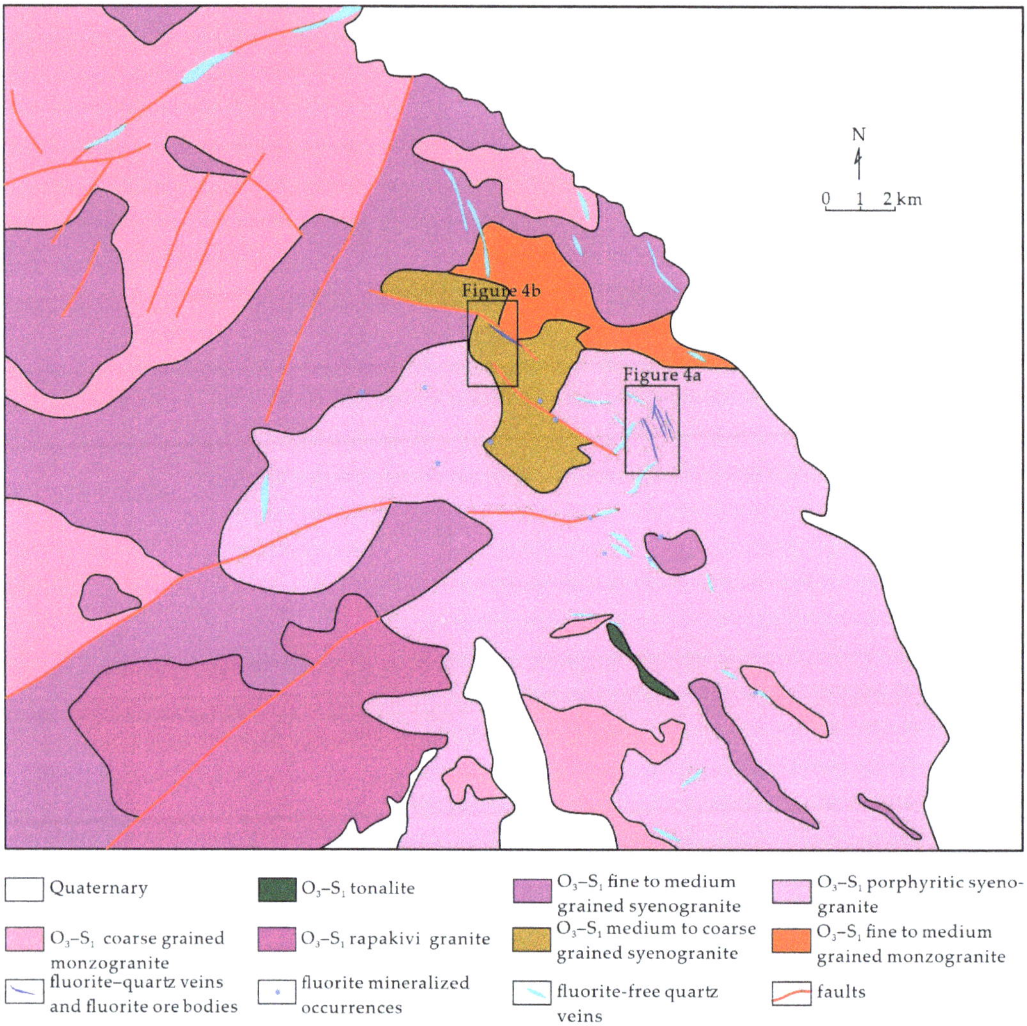

Figure 3. Geological map of the Yama fluorite deposit in the Eastern segment of the Tataleng granitic batholith (revised after [24]). O_3–S_1: the Late Ordovician to Early Silurian.

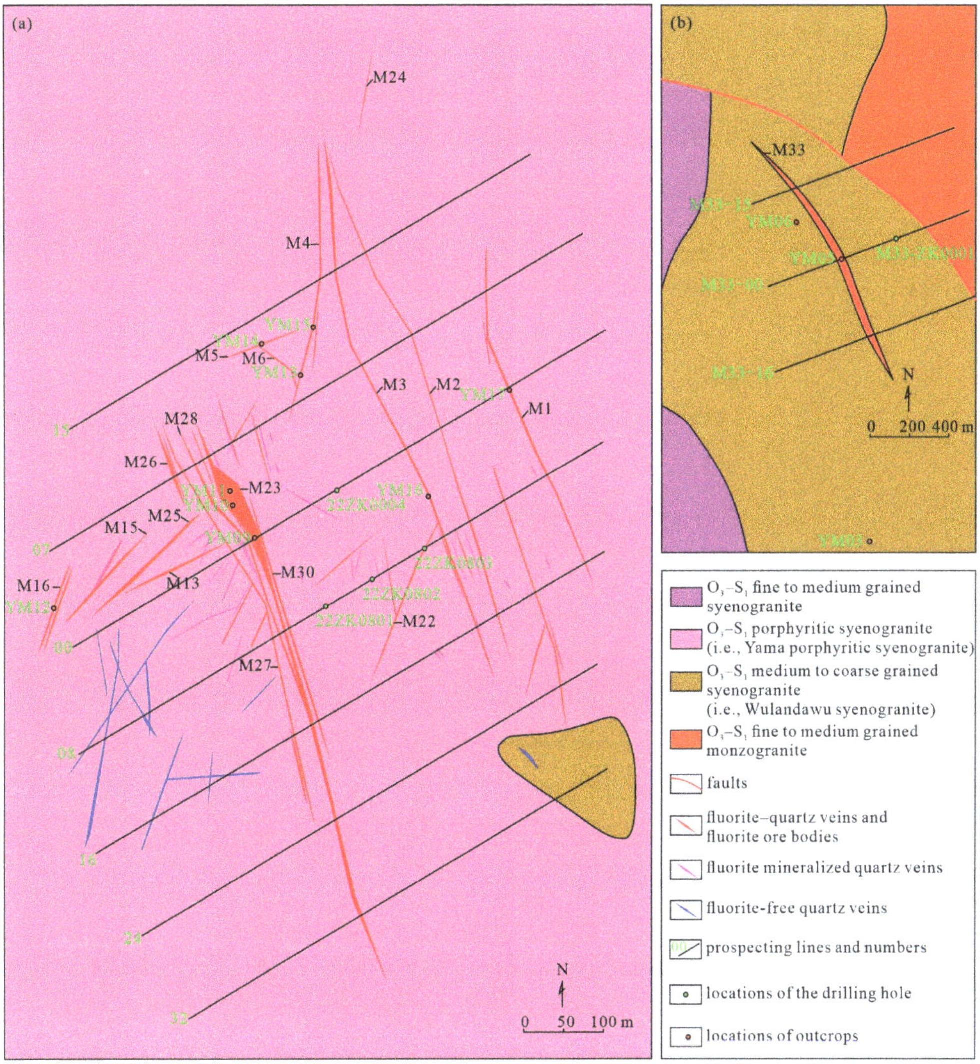

Figure 4. Geological map of the Yama fluorite deposit (simplified after [24]). (**a**) The Yama ore district and (**b**) the Wulandawu district.

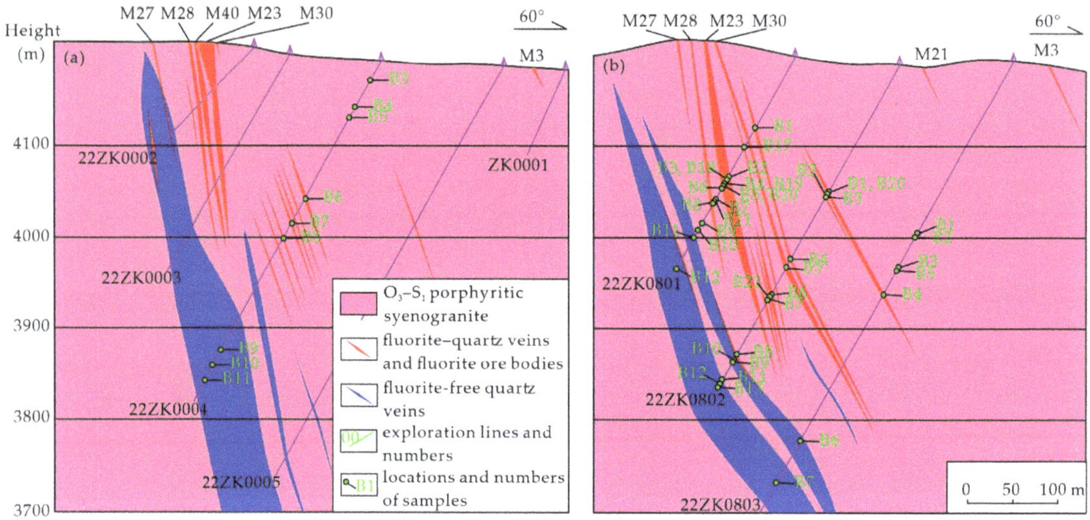

Figure 5. Typical sections of the Yama fluorite deposit (simplified after [24]). (**a**) Prospecting line 00 and (**b**) prospecting line 08.

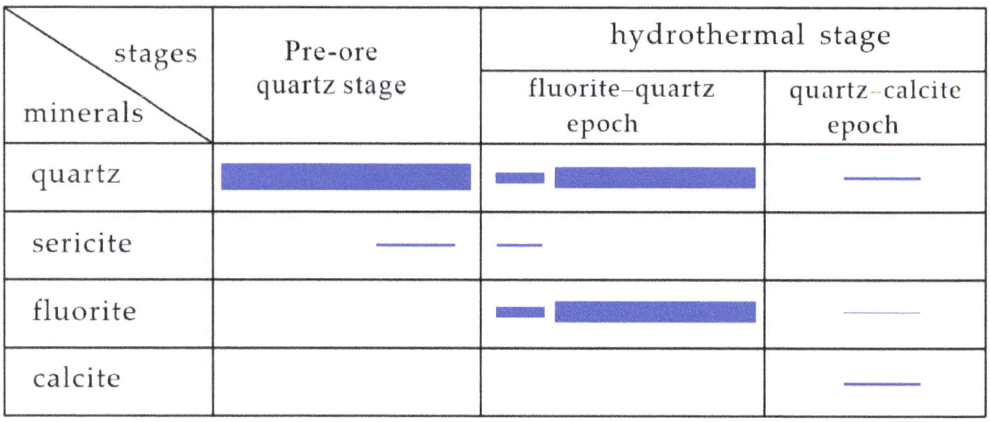

Figure 6. Paragenetic sequences of minerals in different ore-forming stages in the Yama fluorite deposit. The width of the blue line means the relative mineral abundance, and the length of the blue line means the relative forming period.

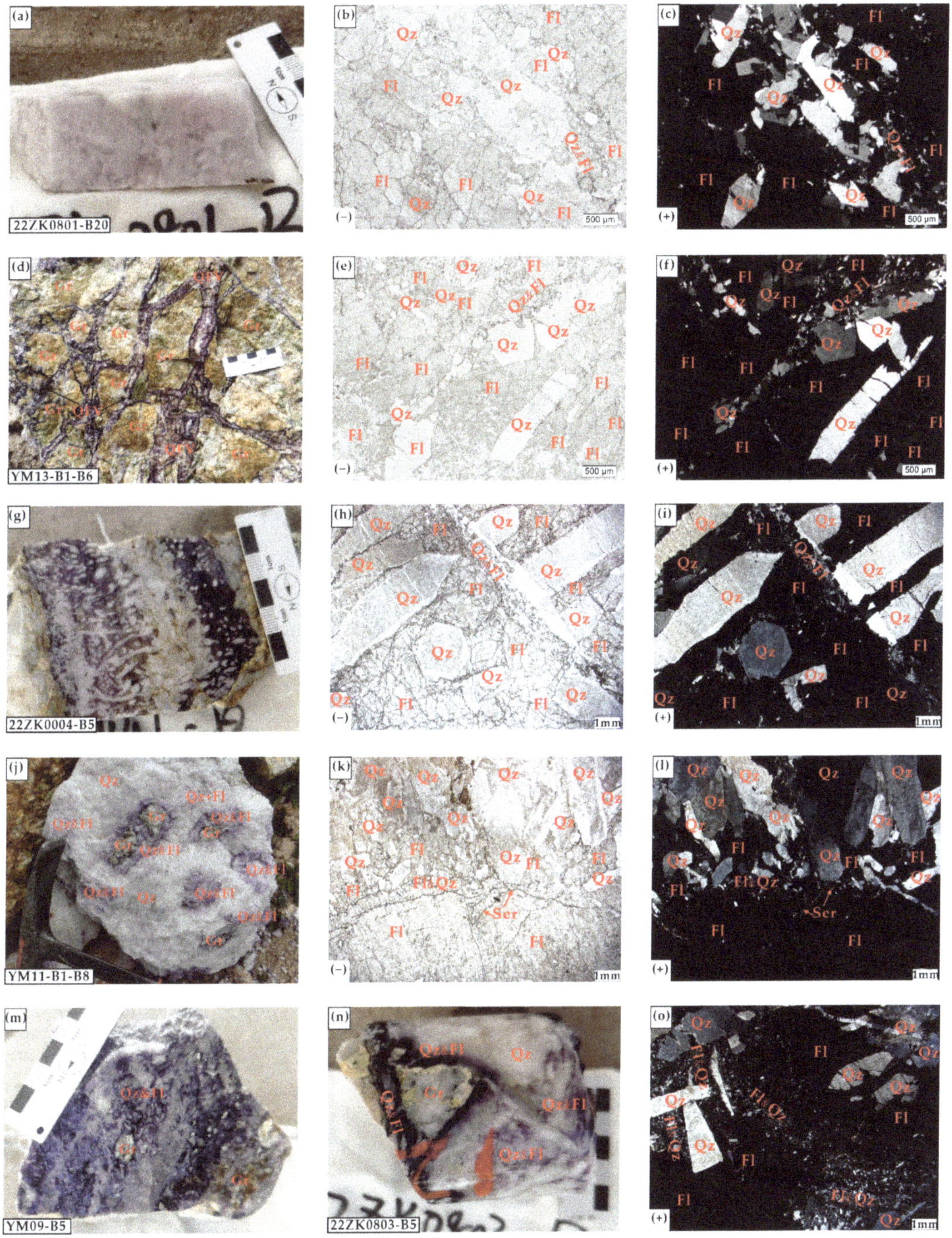

Figure 7. Photographs and photomicrographs showing the morphology and textural features of the fluorite ores in the Yama fluorite deposit. (**a**) Photograph and (**b**,**c**) photomicrograph of sample 22ZK0801-B20 of Type 1 ore; (**d**) photograph and (**e**,**f**) photomicrograph of sample YM13-B1-B6 of

Type 1 ore; (**g**) photograph and (**h**,**i**) photomicrograph of sample 22ZK0004-B5 of Type 1 ore; (**j**) photograph and (**k**,**l**) photomicrograph of sample YM11-B1-B8 of Type 2 ore; (**m**) photograph of sample YM09-B5 of Type 2 ore; (**n**) photograph and (**o**) photomicrograph of sample 22ZK0803-B5 of Type 2 ore. Qz, quartz; Fl, fluorite; Ser, sericite; Gr, syenogranite; (−), plane-polarized light; (+), cross-polarized light.

3. Sampling and Analytical Methods

3.1. Sampling

Seven samples from two types of typical fluorite–quartz veins were collected from drilling holes (22ZK0802, 22ZK0803, and 22ZK0004) in the Yama area (Figures 4a and 5). The other eleven fluorite–quartz vein samples were collected from outcrops (locations YM05, YM09–YM11, YM13, and YM15) in the Yama fluorite deposit (Figure 4a). Table 1 provides detailed descriptions of the samples. In the processing of the selected samples, crushing, conventional magnetic, and heavy liquid separation methods were used to extract quartz for H and O isotopes, fluorites for H, O, and Sr isotopes, as well as for trace element analyses.

Table 1. Sample list in the Yama fluorite deposit.

No.	Sample Name	Rock Type	Ore Type	Mineral Assembly	Location
1	22ZK0801-B2	fluorite–quartz vein	Type 1	quartz, fluorite	90.5 m at the core of drilling hole 22ZK0801
2	22ZK0801-B6	breccia-bearing fluorite–quartz vein	Type 2	quartz, fluorite, syenogranite breccia	161.7 m at the core of drilling hole 22ZK0801
3	22ZK0801-B17	fluorite–quartz vein	Type 1	quartz, fluorite	149.4 m at the core of drilling hole 22ZK0801
4	22ZK0802-B21	fluorite–quartz vein	Type 1	quartz, fluorite	253.6 m at the core of drilling hole 22ZK0802
5	22ZK0803-B1	breccia-bearing fluorite–quartz vein	Type 2	quartz, fluorite, syenogranite breccia	202.8 m at the core of drilling hole 22ZK0803
6	22ZK0803-B3	pectinal fluorite–quartz vein	Type 1	quartz, fluorite, minor syenogranite breccia	251.0 m at the core of drilling hole 22ZK0803
7	22ZK0004-B6	pectinal fluorite–quartz vein	Type 2	quartz, fluorite	171.6 m at the core of drilling hole 22ZK0004
8	YM05-B3	fluorite–quartz vein	Type 1	quartz, fluorite, minor sericite	96°01′03″ E 37°58′50″ N
9	YM05-B7	fluorite–quartz vein	Type 1	quartz, fluorite, minor sericite	96°01′03″ E 37°58′50″ N
10	YM09-B2	breccia-bearing banded fluorite–quartz vein	Type 2	quartz, fluorite	96°03′58″ E 37°57′27″ N
11	YM09-B6	breccia-bearing banded fluorite–quartz vein	Type 2	quartz, fluorite, minor sericite	96°03′58″ E 37°57′27″ N
12	YM10-B2	fluorite–quartz vein	Type 2	quartz, fluorite	96°03′57″ E 37°57′31″ N
13	YM11-B1	breccia-bearing fluorite–quartz vein	Type 2	quartz, fluorite, syenogranite breccia	96°03′57″ E 37°57′31″ N
14	YM11-B2	breccia-bearing crusty fluorite–quartz vein	Type 2	quartz, fluorite, minor sericite, syenogranite breccia	96°03′57″ E 37°57′31″ N
15	YM11-B4	breccia-bearing crusty fluorite–quartz vein	Type 2	quartz, fluorite, minor sericite, syenogranite breccia	96°03′57″ E 37°57′31″ N

Table 1. *Cont.*

No.	Sample Name	Rock Type	Ore Type	Mineral Assembly	Location
16	YM11-B6	breccia-bearing crusty fluorite–quartz vein	Type 2	quartz, fluorite, minor sericite, syenogranite breccia	96°03′57″ E 37°57′31″ N
17	YM13-B2	pectinal fluorite–quartz vein	Type 1	quartz, fluorite, minor sericite	96°04′00″ E 37°57′40″ N
18	YM15-B1	pectinal fluorite–quartz vein	Type 1	quartz, fluorite	96°04′02″ E 37°57′43″ N
19	22ZK0802-B16	Yama porphyritic syenogranite	wall rock	alkali feldspar (perthite), plagioclase, quartz, muscovite, and biotite, with a few accessory minerals including zircon and epidote	58.6 m at the core of drilling hole 22ZK0802
20	22ZK0802-B18	Yama porphyritic syenogranite			138.0 m at the core of drilling hole 22ZK0802
21	22ZK0803-B8	Yama porphyritic syenogranite			36.3 m at the core of drilling hole 22ZK0803
22	YM03-B1	Wulandawu syenogranite	wall rock	alkali feldspar (perthite), plagioclase, quartz, biotite, and muscovite, with a few accessory minerals including tourmaline, zircon, and epidote	96°01′29″ E 37°58′03″ N
23	YM06-B1	Wulandawu syenogranite			96°01′04″ E 37°58′52″ N
24	M33-ZK0001-B1	Wulandawu syenogranite			18.0 m at the core of drilling hole 22ZK0803

To compare the Sr isotope of syenogranite wall rocks, six syenogranite samples were collected from drilling holes (22ZK0802, 22ZK0803, and M33-ZK0001-B1) and outcrops (locations YM03 and YM06) (Figure 4). Sample descriptions for the selected syenogranite wall rocks are listed in Table 1 and for detailed descriptions of the samples we refer to our previously published paper [10].

3.2. Trace Element and REE Analyses of Fluorites

Trace element contents (excluding REEs) for samples were determined using inductively coupled plasma–atomic emission spectroscopy (ICP-AES) (Agilent 5110 ICP–AES, Santa Clara, CA, USA) and inductively coupled plasma–mass spectrometry (ICP–MS) (Agilent 7900 ICP–MS, Santa Clara, CA, USA) at ALS Chemex (Guangzhou, China) Co., Ltd. Perchloric, nitric, hydrofluoric, and hydrochloric acids were used to digest a prepared sample. The residue was leached with dilute hydrochloric acid and diluted to volume. After that, ICP–AES and ICP–MS were used to analyze it. REEs for samples were also determined using ICP–MS (Agilent 7900 ICP–MS, Santa Clara, CA, USA) at ALS Chemex (Guangzhou, China) Co., Ltd. A prepared sample was added to lithium borate flux, mixed well, and fused in a furnace at 1025 °C. The resulting melt was then cooled and dissolved in an acidic mixture containing nitric, hydrochloric, and hydrofluoric acids. Finally, this solution was analyzed via ICP–MS. The precision for trace elements was better than 10%, which was controlled by the system settings, and the detailed analytical procedure was also referred to in [6].

3.3. H, O Isotope Analyses of Fluorites and Quartz

H and O isotope analyses were carried out at the Analytical Laboratory of the Beijing Research Institute of Uranium Geology, China National Nuclear Corporation (BRIUG). The O isotope analysis of quartz was carried out with the bromine pentafluoride method [42]. Water was liberated from 5 to 10 mg of quartz by reaction with BrF_5 and converted to CO_2

on a platinum-coated carbon rod. The $\delta^{18}O$ determinations were made with a Thermo Scientific Delta V Advantage mass spectrometer. Because quartz is an O-bearing mineral, the O isotope of fluid inclusions in quartz, i.e., the $\delta^{18}O_{H2O}$ values, should be calculated by using the isotopic fractionation equation of O between quartz and water as shown in Equation (1) as follows [43]:

$$1000\ln\alpha = 3.38 \times 10^6/T^2 - 3.4 \quad (1)$$

1000lnα is commonly equal to the value of $\delta^{18}O_{quartz}$ minus $\delta^{18}O_{H2O}$.

By mechanically crushing 5–10 mg of quartz, the H isotopic ratios of decrepitated quartz-hosted fluid inclusions were measured. Samples were heated to 1400 °C in order to release water from fluid inclusions after being degassed for 4 h under vacuum at 105 °C. After that, the released water was captured, reduced to H_2 by glassy carbon, and then analyzed for δD values using a Thermo Scientific MAT-253 mass spectrometer.

Through the mechanical crushing of 0.5 g of fluorites, H and O isotopic ratios of decrepitated fluorite-hosted fluid inclusions were measured. Water in fluid inclusions within fluorite samples was first released by heating at 550 °C; then, the released water was trapped, reduced to H_2 by glassy carbon at 1400 °C, and then analyzed for δD values using a Thermo Scientific MAT-253 mass spectrometer. Water was liberated from fluorites by reaction with BrF_5 and converted to CO_2 on a platinum-coated carbon rod. The $\delta^{18}O$ determinations were performed on a Thermo Scientific MAT-253 mass spectrometer.

The results are reported relative to the Vienna Standard Mean Ocean Water (V-SMOW), and the analytical precisions are ±0.1‰ for $\delta^{18}O$ and ±1‰ for δD. The detailed analytical procedures were described in [44].

3.4. Sr Isotope Analyses of Fluorites

At Beijing Createch Testing Technology Co., Ltd., (Beijing, China) Sr isotope analyses were performed. Every chemical preparation was carried out on Class 100 work benches within a Class 1000 clean laboratory. A 15 mL Savillex™ PFA screw-top beaker was filled with about 150 mg of sample powder. After adding concentrated HNO_3 and HF (1 mL and 2 mL, respectively) to the samples, the sealed beakers were heated for one week at 150 °C on a hotplate. Following digestion, acids were evaporated on the hotplate and residues were dissolved in 1 mL of concentrated HCl. After that, the same procedures were repeated 3 times. Finally, the samples were dissolved in 1.5 mL of 3.5 M HNO_3 for Sr purification. Sr-Spec (Triskem, 100–150 μm) resin was used to separate and purify Sr from samples. Isotopic compositions of Sr analyses were carried out using a Thermo Fisher Scientific Neptune Plus MC-ICP-MS. $^{87}Sr/^{86}Sr$ ratios were corrected for instrumental mass fractionation using the exponential fractionation law and assuming $^{88}Sr/^{86}Sr = 8.375209$. The analytical methods were modified from those in [45].

Fifteen fluorite and six syenogranite samples were analyzed for Sr isotopic compositions; furthermore, the Sr isotope international standard NBS 987 was repeatedly tested for accuracy monitoring, yielding an average $^{87}Sr/^{86}Sr$ ratio of 0.710249 ± 14 (2SD, n = 21) during the analytical procedure of fluorites and $^{87}Sr/^{86}Sr$ ratios = 0.710247 ± 13 (2SD, n = 24) during the analytical procedure of syenogranites. Initial $^{87}Sr/^{86}Sr$ ratios (i.e., $(^{87}Sr/^{86}Sr)_i$) were calculated based on rock-forming ages for syenogranite wall rocks dated by [10], and on the possible ore-forming ages of fluorites.

4. Results

4.1. Trace Elements and REEs of Fluorites

The trace element and REE data for fluorites are listed in Table 2. Most trace and REE elements show concentrations of <10 ppm. Relatively high concentrations (a few to several tens of ppm) were observed for Rb (1.8–21.5 ppm), Sr (15.2–64.5 ppm), Y (2.2–107 ppm), and Ba (1.8–13.3 ppm), whereas very low concentrations (usually <1 ppm) were found for Fe, Co, W, Pb, Th, U and some of REEs, such as Eu, Tb, Tm, and Lu, and some of them were even below the detection limits, such as Mg, Mn, Ni, Zn, Nb, Cd, Mo, Sn, Sb, Hf, and Ta.

Table 2. Trace elements and REEs of fluorites in the Yama fluorite deposit.

Sample Name	22ZK0801-B2	22ZK0802-B21	22ZK0803-B3	YM05-B3	YM05-B7	22ZK0004-B6	YM09-B2	YM09-B6	YM10-B2	YM11-B2	YM11-B4	YM11-B6	YM13-B2	YM15-B1
Types	Type 1	Type 1	Type 1	Type 1	Type 1	Type 2	Type 2	Type 2	Type 2	Type 2	Type 2	Type 2	Type 1	Type 1
Li	2.60	3.60	1.90	5.20	5.20	2.80	0.50	2.30	7.00	0.60	0.70	0.60	1.80	1.70
Be	0.1	0.4	0.1	0.4	0.5	0.2	0.8	0.6	7.5	<0.05	<0.05	0.1	0.21	0.13
Mg	<0.01	0.01	0.01	0.01	0.01	<0.01	<0.01	0.01	0.01	<0.01	<0.01	<0.01	0.02	0.01
Sc	0.1	0.1	0.1	0.1	0.1	0.1	0.1	0.1	0.1	0.1	0.1	0.1	0.1	0.1
Ti	<0.005	<0.005	<0.005	<0.005	<0.005	<0.005	<0.005	<0.005	<0.005	<0.005	<0.005	<0.005	<0.005	<0.005
V	1	1	1	1	2	1	1	1	1	<1	1	<1	1	1
Cr	1	1	1	1	1	<1	<1	<1	1	1	<1	1	<1	<1
Mn	<5	<5	<5	<5	<5	<5	<5	<5	<5	<5	<5	<5	<5	<5
Fe	0.01	0.01	0.01	0.04	0.04	0.01	0.01	<0.01	0.01	0.01	<0.01	0.01	0.02	0.01
Co	0.7	0.6	0.7	0.6	0.7	0.7	0.7	0.7	0.6	0.8	0.8	0.8	0.8	0.8
Ni	<0.2	<0.2	<0.2	<0.2	<0.2	<0.2	<0.2	<0.2	<0.2	<0.2	<0.2	<0.2	<0.2	<0.2
Cu	0.8	0.5	0.6	0.6	1.2	0.8	0.8	0.8	0.6	0.8	0.6	0.8	0.8	0.5
Zn	<2	<2	<2	<2	<2	<2	<2	<2	<2	<2	<2	<2	<2	<2
Ga	0.2	0.3	0.2	1.3	1.6	0.2	0.3	0.2	0.5	0.2	0.1	0.2	0.25	0.23
Rb	2.6	4.6	4.4	20.0	21.5	5.3	3.5	2.8	6.9	2.3	1.8	3.3	4.1	3
Sr	20.1	18.3	17.8	64.5	54.3	21.2	15.2	15.2	17	23.1	22.2	22	21.9	26.7
Y	11.2	14.0	29.5	107.0	64.1	5.6	7.5	3.2	8.2	2.9	2.2	2.7	53.9	5.9
Zr	3.00	<2	<2	3.00	<2	<2	3.00	4.00	6.00	3.00	<2	<2	<2	3
Nb	<0.1	<0.1	<0.1	0.10	0.10	<0.1	0.10	0.10	<0.1	0.10	<0.1	<0.1	<0.1	<0.1
Mo	<0.05	<0.05	0.09	0.06	<0.05	<0.05	<0.05	0.05	0.42	<0.05	<0.05	<0.05	0.07	0.42
Cd	<0.02	<0.02	<0.02	<0.02	<0.02	<0.02	<0.02	<0.02	<0.02	<0.02	<0.02	<0.02	<0.02	<0.02
Sn	<0.2	<0.2	<0.2	<0.2	<0.2	<0.2	<0.2	<0.2	<0.2	<0.2	<0.2	<0.2	<0.2	0.2
Sb	<0.05	<0.05	<0.05	0.07	0.06	<0.05	0.3	<0.05	0.08	<0.05	<0.05	<0.05	<0.05	<0.05
Cs	0.2	0.3	0.2	1.0	1.1	0.2	0.1	0.1	0.3	0.1	0.1	0.1	0.2	0.21
Ba	2.3	7.0	2.1	2.1	2.4	5.6	2.6	3.5	1.8	4.3	13.3	8.1	6.9	6.3
Ca	47.00	46.70	>50	46.20	46.70	>50	49.10	48.50	45.50	>50	>50	>50	>50	49.8
La	0.8	1.4	7.6	1.4	1.1	0.4	0.6	0.3	0.2	1.2	0.3	0.4	1.2	1.2
Ce	1.60	2.70	16.20	3.10	2.40	0.90	1.00	0.40	0.20	2.00	1.30	1.20	3	1.7
Pr	0.18	0.30	1.86	0.51	0.39	0.10	0.15	0.06	0.03	0.19	0.06	0.09	0.37	0.3
Nd	0.80	1.20	6.80	2.30	1.80	0.40	0.60	0.20	0.20	0.70	0.20	0.30	1.5	1.1
Sm	0.32	0.32	1.77	1.21	0.93	0.13	0.19	0.08	0.08	0.10	0.09	0.10	0.66	0.25
Eu	0.03	0.05	0.14	0.10	0.08	<0.02	0.02	<0.02	<0.02	0.02	<0.02	<0.02	0.09	0.04
Gd	0.63	0.62	2.30	3.10	2.07	0.24	0.35	0.15	0.23	0.23	0.15	0.17	1.65	0.41
Tb	0.13	0.13	0.42	0.71	0.49	0.06	0.08	0.03	0.05	0.05	0.03	0.04	0.37	0.07
Dy	1.02	1.04	2.91	5.88	3.96	0.44	0.63	0.24	0.44	0.31	0.24	0.29	3.39	0.5
Ho	0.24	0.26	0.66	1.45	1.00	0.11	0.15	0.06	0.14	0.07	0.05	0.06	0.94	0.12
Er	0.77	0.90	2.12	4.33	3.00	0.32	0.47	0.19	0.47	0.19	0.16	0.19	3.54	0.34

Table 2. *Cont.*

Sample Name	22ZK0801-B2	22ZK0802-B21	22ZK0803-B3	YM05-B3	YM05-B7	22ZK0004-B6	YM09-B2	YM09-B6	YM10-B2	YM11-B2	YM11-B4	YM11-B6	YM13-B2	YM15-B1
Tm	0.1	0.2	0.4	0.7	0.5	0.1	0.1	0.0	0.1	0.0	0.0	0.1	0.59	0.04
Yb	0.78	1.09	2.33	4.80	3.58	0.32	0.51	0.28	0.62	0.21	0.23	0.23	4.39	0.33
Lu	0.1	0.2	0.4	0.8	0.7	0.1	0.1	0.1	0.1	0.1	0.1	0.0	0.75	0.05
Hf	0.1	<0.1	<0.1	<0.1	<0.1	<0.1	<0.1	<0.1	<0.1	<0.1	<0.1	<0.1	<0.1	<0.1
Ta	<0.05	<0.05	<0.05	<0.05	<0.05	<0.05	<0.05	<0.05	<0.05	<0.05	<0.05	<0.05	<0.05	<0.05
W	0.1	0.2	0.1	0.2	0.4	0.1	0.1	0.1	0.1	<0.1	<0.1	<0.1	0.3	0.1
Pb	1.00	<0.5	<0.5	0.70	0.70	<0.5	<0.5	1.30	6.00	<0.5	0.60	<0.5	26.3	<0.5
Bi	2.35	0.02	0.10	0.25	0.11	0.74	0.05	8.31	15.15	2.64	1.87	1.55	0.16	0.01
Th	0.31	0.16	0.34	0.54	0.30	0.21	0.43	0.27	<0.05	0.11	0.06	0.08	2.77	1.13
U	0.15	0.21	1.04	0.47	0.63	0.23	0.14	0.21	0.29	0.28	0.19	0.19	0.30	0.28
∑REEs	7.56	10.36	45.86	30.42	21.96	3.53	4.91	2.07	2.87	5.35	2.89	3.16	22.44	6.45
LREEs/HREEs	0.97	1.36	2.99	0.40	0.44		1.09			3.69			0.44	2.47
$(La/Yb)_N$	0.69	3.34	1.89	8.43	4.24	1.29	3.03	0.77	1.02	0.75	0.85	1.03	7.62	1.20
$(La/Sm)_N$	0.82	0.60	0.79	0.77	0.70	0.86	0.78	0.54	0.45	0.93	0.66	0.79	0.49	0.95
Eu/Eu*	0.20	0.34	0.21	0.16	0.18		0.24			0.40			0.26	0.38
Y/Y*	1.64	1.95	1.54	2.65	2.33	1.84	1.77	1.93	2.39	1.43	1.45	1.48	2.19	1.74

Note: Units for Ca, Mg, Fe, and Ti are wt. %, and units for other trace elements are ppm. ∑REEs—total REE contents, LREEs—light rare earth elements, including La, Ce, Pr, Nd, Sm, and Eu, HREEs—heavy rare earth elements, including Gd, Tb, Dy, Ho, Er, Tm, Yb, and Lu, $Eu/Eu^* = Eu_N/\sqrt{[Sm_N \times Gd_N]}$, $Y/Y^* = (Y_N)/\sqrt{[Dy_N \times Ho_N]}$. REE normalization values were obtained from [46]. The Y normalization value was obtained from [47].

The total REE contents (∑REEs) in studied fluorites are relatively low (<2.09–45.86 ppm), i.e., 2.89–30.42 ppm for the fluorites in Type 1 ores, and 2.09–45.89 ppm for the fluorites in Type 2 ores. The Y concentrations for all types of fluorites are relatively high, i.e., 5.90–107 ppm for the fluorites in Type 1 ores and 2.20–29.50 ppm for the fluorites in Type 2 ores. In general, Type 1 ores have slightly higher total REE contents and Y contents than Type 2 ores (Table 2). In chondrite-normalized REE, including Y, pattern (namely a REY pattern) diagrams (Y inserted between Dy and Ho), the fluorite samples have roughly similar chondrite-normalized REY patterns, basically indicating a similar source and ore genesis. They are also characterized by relatively strong enrichment in heavy rare earth elements (HREEs), showing negative Eu anomalies (Eu/Eu* = 0.16–0.40, Table 2), strongly positive Y anomalies (Y/Y* = 1.43–2.39, Table 2), and relatively high $(La/Yb)_N$ values of 0.69–8.43 (Figure 8), although fluorites in Type 1 ores show higher total REE contents (average = 14.58 ppm) than those in Type 2 ores (average = 9.69 ppm). However, both types of fluorite ores show slightly different chondrite-normalized REY patterns compared to the average of syenogranite wall rocks (which show a flat HREE pattern and are slightly enriched in LREEs) relative to HREEs as described in our previously published paper [10].

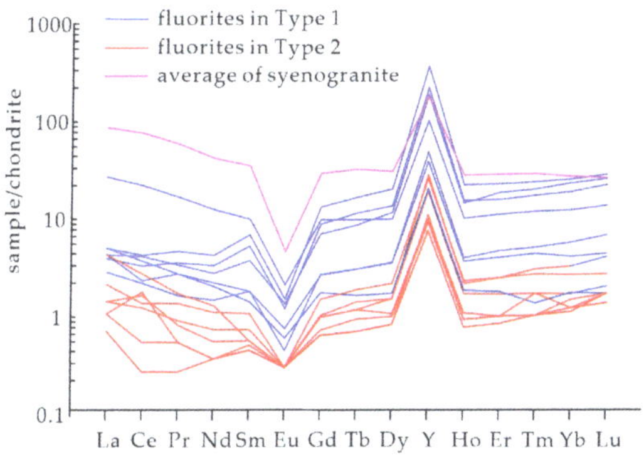

Figure 8. Chondrite-normalized REE and Y element patterns of fluorites in the Yama fluorite deposit. REE normalization values were obtained from [46]. The Y normalization value was obtained from [47].

4.2. H and O Isotopic Compositions of Quartz

The Yama fluorite deposit contains fluorites with varying δD and δ^{18}O values from −77.4‰ to −102.4‰ (average = −90.1‰) and from −12.7‰ to −4.3‰ (average = −8.0‰), respectively (Table 3). These values indicate the initial isotope compositions of the ore-forming fluids in the Yama fluorite deposit. The δD value ranges of two types of fluorite ores overlap each other, but Type 2 samples have slightly heavier δ^{18}O values than Type 1 (Figure 9).

Table 3. H and O isotope data for quartz and fluorites in the Yama fluorite deposit.

No.	Sample Name	Rock Type	Mineral	$\delta D_{V\text{-}SMOW}$ (‰)	$\delta^{18}O_{V\text{-}SMOW}$ (‰)	$\delta^{18}O_{H2O}$ (‰) *
1	22ZK0801-B2	Type 1	quartz	−82.1	−3.6	−17.2
2	22ZK0801-B17	Type 1	quartz	−85.3	−2.8	−16.4
3	22ZK0802-B21	Type 1	quartz	−80.6	−3.6	−17.2
4	22ZK0803-B3	Type 1	quartz	−87.3	−4.2	−17.8
5	22ZK0004-B6	Type 2	quartz	−83.6	−1.4	−15.0
6	YM09-B2	Type 2	quartz	−86.0	−3.4	−17.0
7	YM10-B2	Type 2	quartz	−83.6	−2.7	−16.3
8	YM11-B1	Type 2	quartz	−91.2	−4.5	−18.1
9	YM11-B4	Type 2	quartz	−86.1	−4.2	−17.8
10	YM11-B6	Type 2	quartz	−82.9	−4.4	−18.0

Table 3. *Cont.*

No.	Sample Name	Rock Type	Mineral	$\delta D_{V\text{-}SMOW}$ (‰)	$\delta^{18}O_{V\text{-}SMOW}$ (‰)	$\delta^{18}O_{H2O}$ (‰) *
11	YM13-B2	Type 1	quartz	−81.0	−2.3	−15.9
12	YM15-B1	Type 1	quartz	−79.6	−2.6	−16.2
13	22ZK0801-B2	Type 1	fluorite	−84.7	−12.7	
14	22ZK0801-B6	Type 2	fluorite	−88.8	−9.1	
15	22ZK0802-B21	Type 1	fluorite	−89.3	−10.2	
16	22ZK0803-B1	Type 2	fluorite	−92.1	−11.7	
17	22ZK0803-B3	Type 1	fluorite	−87.0	−9.8	
18	YM05-B3	Type 1	fluorite	−77.4	−9.1	
19	22ZK0004-B6	Type 2	fluorite	−83.3	−10.7	
20	YM09-B2	Type 2	fluorite	−83.5	−4.6	
21	YM09-B6	Type 2	fluorite	−92.6	−4.4	
22	YM10-B2	Type 2	fluorite	−97.1	−4.3	
23	YM11-B2	Type 2	fluorite	−98.2	−5.0	
24	YM11-B4	Type 2	fluorite	−94.6	−5.7	
25	YM11-B6	Type 2	fluorite	−102.4	−7.8	
26	YM13-B2	Type 1	fluorite	−94.6	−6.5	
27	YM15-B1	Type 1	fluorite	−86.5	−8.8	

* The calculated $\delta^{18}O_{H2O}$ values of the fluids are based on an average homogenization temperature of quartz of 173 °C using Equation (1) [43].

The Yama fluorite deposit contains quartz in different types of fluorite ores with slightly invariable δD values ranging from −91.2‰ to −79.6‰ (average = −84.1‰). The values are within the range of δD values of fluorite (Figure 9). The quartz in different types of fluorite ores in this deposit also has invariable $\delta^{18}O$ values ranging from −4.5‰ to −1.4‰ (average = −3.3‰). Calculating the precise $\delta^{18}O_{H2O}$ values of fluids that were in equilibrium with the quartz is challenging due to the large variation in temperatures obtained from fluid inclusions (e.g., [48]). However, for a temperature range of 200–500 °C, Equation (1) provides a good approximation of the isotopic fractionation of O between quartz and water [43]. However, the homogenization temperatures of quartz in fluorite–quartz veins ranged from 135 °C to 209 °C (average = 173 °C) in the Yama fluorite deposit (unpublished data). Therefore, if using 173 °C as the trapping temperature, the calculated $\delta^{18}O_{H2O}$ values of the fluids for fluid inclusions from the quartz in fluorite ores were from −18.0‰ to −15.0‰ (Table 3), which are totally different from the analytical O isotopic data from −12.7‰ to −4.3‰ of fluorite (Figure 9).

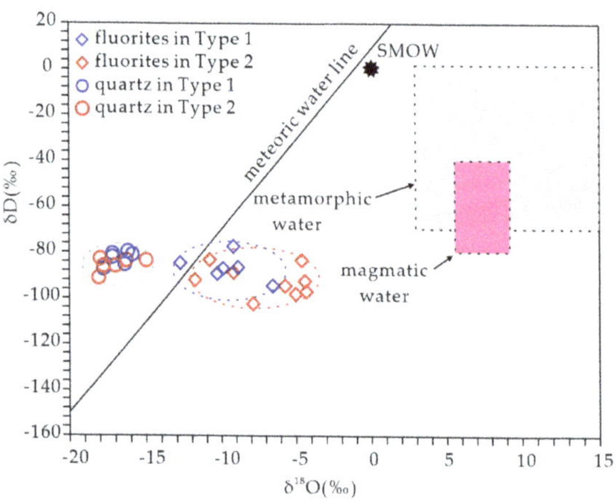

Figure 9. Plot of δD versus $\delta^{18}O$ diagram in the Yama fluorite deposit. Magmatic and metamorphic water fields after [49].

4.3. Sr Isotopic Compositions of Fluorites

Table 4 lists the Sr isotopic compositions of fluorites. The samples for the Yama porphyritic syenogranite and the Wulandawu syenogranite have variable analytical $^{87}Sr/^{86}Sr$ ratios of 0.832625−0.880566 and 1.056637−1.191950, respectively. Based on the average age of their rock-forming ages of 444.3 Ma and 441.2 Ma [10], their calculated initial $^{87}Sr/^{86}Sr$ ratios (i.e., $(^{87}Sr/^{86}Sr)_i$) are 0.660469−0.678689 and 0.668841−0.697535, respectively.

Table 4. Sr isotopic compositions of fluorites and syenogranites from the Yama fluorite deposit.

Sample Name	Rock/Mineral Types	Rb [a] (ppm)	Sr [a] (ppm)	$^{87}Rb/^{86}Sr$	$^{87}Sr/^{86}Sr$	±2σ	$(^{87}Sr/^{86}Sr)I$ [b]	$(^{87}Sr/^{86}Sr)I$ [c]
22ZK0801-B2	Type 1	2.6	20.1	0.374295	0.753393	0.000007	0.751031	
22ZK0801-B6	Type 2	2.6	20.1	0.374295	0.753517	0.000008	0.751155	
22ZK0802-B21	Type 1	4.6	18.3	0.727351	0.755463	0.000008	0.750873	
22ZK0803-B1	Type 2	4.4	17.8	0.715269	0.754075	0.000007	0.749562	
22ZK0803-B3	Type 1	4.4	17.8	0.715269	0.755551	0.000007	0.751037	
YM05-B3	Type 1	20	64.5	0.897237	0.828185	0.000008	0.822523	
22ZK0004-B6	Type 2	5.3	21.2	0.723398	0.756299	0.000006	0.751734	
YM09-B2	Type 2	3.5	15.2	0.666287	0.756628	0.000006	0.752424	
YM09-B6	Type 1	2.8	15.2	0.533030	0.753935	0.000009	0.750571	
YM10-B2	Type 2	6.9	17	1.174457	0.754550	0.000007	0.747139	
YM11-B2	Type 2	2.3	23.1	0.288106	0.752564	0.000006	0.750746	
YM11-B4	Type 2	1.8	22.2	0.234615	0.751325	0.000008	0.749845	
YM11-B6	Type 2	3.3	22	0.434039	0.753245	0.000006	0.750506	
YM13-B2	Type 1	4.1	21.9	0.541722	0.749089	0.000007	0.745671	
YM15-B1	Type 1	3	26.7	0.325122	0.753748	0.000006	0.751696	
22ZK0802-B16	Yama syenogranite	374	37.6	28.781987	0.854448	0.000009		0.672287
22ZK0802-B18	Yama syenogranite	439	46.7	27.200986	0.832625	0.000008		0.660469
22ZK0803-B8	Yama syenogranite	377	34.2	31.897178	0.880566	0.000007		0.678689
YM03-B1	Wulandawu syenogranite	477	20.5	67.328902	1.091983	0.000010		0.668841
YM06-B1	Wulandawu syenogranite	443	23.7	54.086938	1.056637	0.000009		0.716716
M33-ZK0001-B1	Wulandawu syenogranite	522	19.2	78.669482	1.191950	0.000013		0.697535

[a] Contents determined by ICP-MS (fluorites from Table 2, and syenogranites from [10]). [b] Recalculation of $(^{87}Sr/^{86}Sr)i$ ratios to 443 Ma for fluorites (ages obtained from [10]). [c] Recalculation of $(^{87}Sr/^{86}Sr)i$ ratios to 444.3 Ma for the Yama porphyritic syenogranite, and to 441.2 Ma for the Wulandawu syenogranite (ages obtained from [10]), using $\lambda^{87}Rb = 1.42 \times 10^{-11}$ year^{-1} [50].

Both types of fluorites have variable $^{87}Sr/^{86}Sr$ ratios ranging from 0.749089 to 0.828185, with an average of 0.758771. However, except for an anomalously high ratio of 0.828185 for the sample YM05-B3, the other analytical $^{87}Sr/^{86}Sr$ ratios are focused on a relatively invariable range of 0.749089–0.756628, with an average of 0.753813. As we do not know the exact ore-forming age of fluorite ore bodies in the Yama fluorite deposit, it should be considered to be either the same or later than the rock-forming ages of the syenogranite wall rocks, as provided by [10]. If the ore-forming age is equal to the average formation age of 443 Ma of syenogranite wall rocks provided in [10], the initial $^{87}Sr/^{86}Sr$ ratios for fluorites can be calculated as ranging from 0.745671 to 0.822523, with an average of 0.755101 (Figure 10). If we discarded the anomalously high ratio of 0.822523 for the sample YM05-B3, the calculated $(^{87}Sr/^{86}Sr)_i$ ratios of 0.745671 to 0.751696 for fluorites in Type 1 ores are very similar to the calculated $(^{87}Sr/^{86}Sr)_i$ ratios of 0.747139 to 0.752424 for fluorites in Type 2 ores, and both $(^{87}Sr/^{86}Sr)_i$ ratios of fluorites are relatively invariable, ranging from 0.745671 to 0.752424.

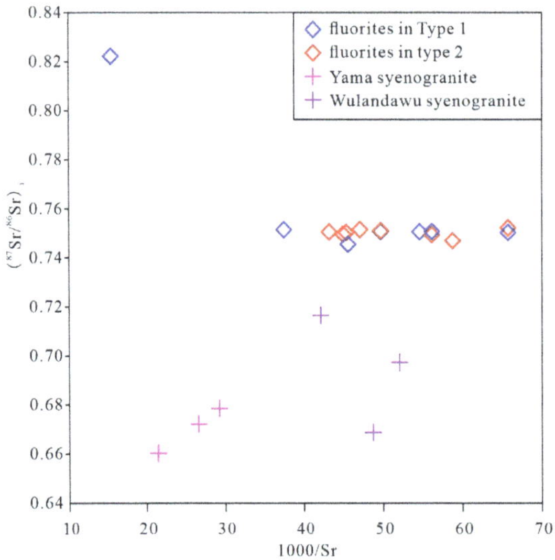

Figure 10. Plot of calculated ($^{87}Sr/^{86}Sr$)i ratio versus 1000/Sr for fluorites and syenogranite wall rocks from the Yama fluorite deposit. All isotopic data for fluorites are corrected at the average rock-forming age (443 Ma) of syenogranite wall rocks [10]. All isotopic data for syenogranites are corrected at their own rock-forming age, i.e., 444.3 Ma for the Yama porphyritic syenogranite and 441.2 Ma for the Wulandawu syenogranite [10].

5. Discussion

5.1. Origin of Ore-Forming Fluids for Fluorite Mineralization

A latitude effect on δD values has been proposed for fluorite deposits in Eastern China [51]. The δD value of these deposits decreases with increasing latitude, with implications of a significant origin from meteoric water. Figure 9 shows the hydrogen and oxygen isotopic compositions of fluorites in the Yama fluorite deposit. Both types of fluorites display relatively unvaried δD values and variable $\delta^{18}O_{H2O}$ values, and they roughly overlap with each other. The plot of the data is close to the meteoric water line with relatively lower δD values than the typical composition of magmatic or metamorphic water, indicating that the ore-forming fluids might predominantly be meteoric water. Their relatively constant δD values might indicate there is no apparent H isotope exchange between the ore-forming water and wall rocks.

The $\delta^{18}O_{H2O}$ values of different types of fluorites have a more variable range; thus, samples show a horizontal trend in $\delta^{18}O_{H2O}$ values ("O isotope shift") extending from the meteoric water line, which might result from possible fluid–rock interactions either before or during fluorite mineralization (e.g., [52]). Certainly, mixing with minor other deep water also cannot be totally precluded. Because quartz–fluorite veins were dominantly filled in the brittle faults within syenogranite wall rocks, rare fluid–rock interactions happened during fluorite mineralization, which is proven by the different initial Sr isotopic compositions between fluorites and syenogranite wall rocks and rare alterations close to fluorite–quartz veins within the syenogranite wall rocks. Therefore, the $\delta^{18}O_{H2O}$ values of fluorite samples showing an O isotope shift extending from the meteoric water line may have resulted from possible deep fluid–rock interactions before fluorite mineralization. The fluorites in Type 2 ores have slightly heavier and variable $\delta^{18}O_{H2O}$ values than those in Type 1 ores, possibly because they contain some syenogranite breccias within such breccia-bearing fluorite–quartz veins. There would also possibly be a slight exchange between ore-forming fluids and syenogranite breccias during fluorite mineralization.

Compared to fluorites, the quartz in different fluorite ores has a similar δD value range for fluid inclusions to those of fluorites (Figure 9); however, their calculated $\delta^{18}O_{VSMOW}$ values from −18.0‰ to −15.0‰ for fluid inclusions from the quartz in the Yama fluorite deposit are totally different from the analytical $\delta^{18}O_{H2O}$ values from −12.7‰ to −4.3‰ of different types of fluorites. We have argued that the reasons might be as follows: firstly, the trapping temperature (173 °C) is lower than the temperature range of 200–500 °C for Equation (1), which is used as a good approximation of isotopic fractionation of O between quartz and water [43]; secondly, bulk fluids extracted from quartz, which are a mixture of primary, pseudosecondary, and secondary inclusions, usually reflect H and O isotopic compositions of important secondary fluid inclusions formed in the presence of meteoric water during the uplift of the deposits (e.g., [53–55]). Therefore, the H and O isotopic compositions for the ore-forming fluids should be represented by the data obtained from studied fluorites rather than quartz, in particular from those fluorites which usually contain more cleavages than quartz and cannot trap the considerable secondary fluid inclusions. These are consistent with our microscopic observations.

In conclusion, the origin of ore-forming fluids for the Yama fluorite mineralization might predominantly originate from paleo-meteoric water, but a mixture with certain deep-water sources (e.g., magmatic water or metamorphic water) cannot be totally precluded.

5.2. Source of Ore-Forming Materials for Fluorite Mineralization

Because strontium and calcium have similar geochemical natures, fluorites are therefore usually characterized by enriched Sr and depleted Nd and could keep their initial Sr isotopic compositions when they are formed. Two types of fluorites typically have invariable analytical $^{87}Sr/^{86}Sr$ ratios of 0.749089−0.756628 with an average of 0.753813, except for the anomalously high ratio. In addition, the corresponding calculated $(^{87}Sr/^{86}Sr)_i$ ratios for both types of fluorites, which range from 0.745671 to 0.752424 with an average of 0.750285 based on the average formation age of 443 Ma of syenogranite wall rocks [10], are apparently higher than the calculated $(^{87}Sr/^{86}Sr)_i$ ratios of 0.660469−0.697535 of the syenogranite wall rocks (Figure 10). If a younger ore-forming age than the 443 Ma is used, then the calculated $(^{87}Sr/^{86}Sr)_i$ ratios should be larger than the range of 0.745671−0.752424 with the corrected age of 443 Ma. Furthermore, they will be larger than the calculated $(^{87}Sr/^{86}Sr)_i$ ratios of 0.660469−0.697535 of the syenogranite wall rocks. Therefore, whatever the ore-forming age which was the same or younger than 443 Ma, the calculated $(^{87}Sr/^{86}Sr)_i$ ratios of fluorites should be much greater than the calculated $(^{87}Sr/^{86}Sr)_i$ ratios of the syenogranite wall rocks. This indicates that the Sr source should be predominantly derived from the ore-forming fluids themselves rather than from the syenogranite wall rocks, as do other ore-forming materials. This argument has further been proven by the geological fact that the fluorite–quartz veins were formed by the low-temperature-filling mineralization rather than by an apparent water–rock reaction between the ore-forming fluids and syenogranite wall rocks.

5.3. Implication for the Genesis of Fluorite Mineralization

The Tb/Ca ratio can reveal the assimilation of Ca from the surrounding rocks or the adsorption of REEs during the crystallization of fluorite [56]. The Tb/La ratio can be employed to evaluate both the fractionation degree of REEs in fluorite and the reactivation of fluorite [57,58]. Therefore, the Tb/Ca versus Tb/La diagram was used to discriminate fluorite genetic types, i.e., pegmatitic, hydrothermal, and sedimentary genesis [56]. In this study, four fluorites in Type 2 and two samples in Type 1 have a Ca content of >50% (the detection limit) by LA-ICP-MS trace element analyses; therefore, it is approximated that n(Ca) is equal to the theoretical value (51.3328%) in CaF_2 (e.g., [59]). Meanwhile, the Tb/Ca ratios are calculated for such samples. In this way, fluorites in Type 1 ores in the Yama fluorite deposit predominantly plot in the hydrothermal field on the Tb/Ca versus the Tb/La diagram (Figure 11a); meanwhile, fluorites in Type 2 ores predominantly plot in the sedimentary field in the Tb/Ca versus the Tb/La diagram (Figure 11a). Moreover, as the partitioning behavior of Dy into fluorite is similar to that of Tb [56], Dy can be substituted for Tb in the Tb/Ca versus the Tb/La discrimination diagram [60]. Similarly, in the Dy/Ca vs. Dy/La plot (Figure 11b), fluorites in Type 1 ores also predominantly plot in the hydrothermal field, and fluorites in Type 2 ores predominantly plot in the sedimentary field. Therefore, it is reasonably considered that the crusty structures for fluorite in Type 2 ores might not only indicate the low-temperature structure but also the slow speed of crystallization of ore-forming fluids while they infilled in the ore-controlling faults. The process of such low-temperature and slow-speed crystallization of the ore-forming fluids might be similar to a chemical sedimentary process; therefore, they plot in the sedimentary fields in the Tb/Ca vs. Tb/La plot (Figure 11a) or the Dy/Ca vs. Dy/La plot (Figure 11b). However, with geological observations in the field, fluorite−quartz veins in the study area were all infilled along the faults within the syenogranite wall rocks; thus, indicating that all types of fluorites within fluorite−quartz veins should be a hydrothermal filling genesis. In the Tb/Ca vs. Tb/La plot (Figure 11a), fluorites in the Type 1 ore also show a crystallization trend, indicating normal hydrothermal crystallization. The slightly increased Tb/La atom ratios (Figure 11) and the increased ∑REE contents (Table 2) in fluorites from Type 2 to Type 1 ores suggest that REEs in the ore-forming fluid were gradually fractionated. This is consistent with the fact that fine-grained fluorites showing crusty or colloform structures in Type 2 ores were crystallized earlier than other fluorites within fluorite−quartz veins (Figures 6 and 7), and the more perfect crystal structure of fluorite might be more suitable for the entry of REEs (e.g., [61]).

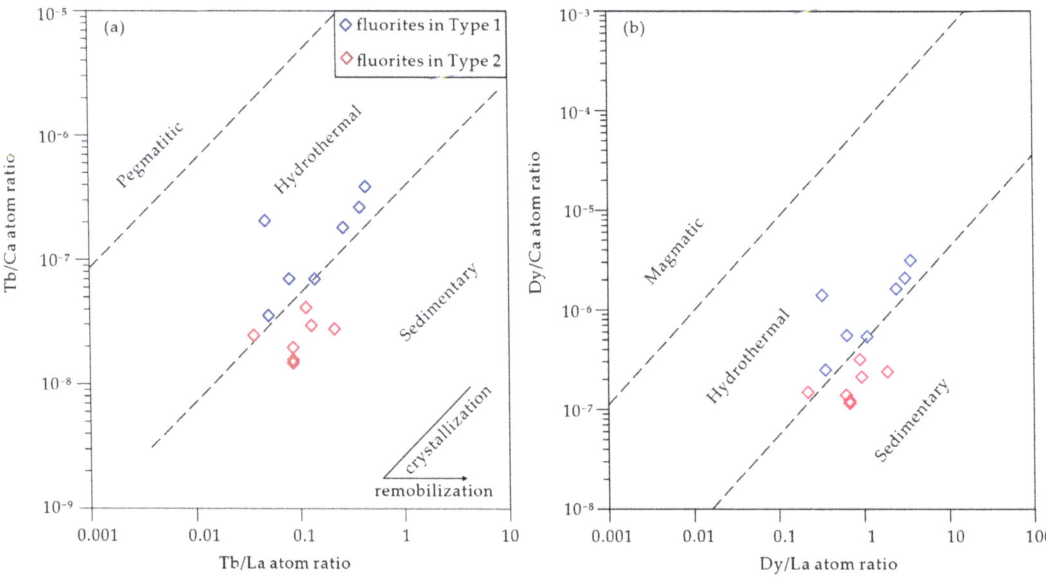

Figure 11. (**a**) Plot of Tb/Ca versus Tb/La ratios for fluorites in the Yama fluorite deposit (the base map was modified from [56]); (**b**) plot of Dy/Ca versus Dy/La for fluorites in the Yama fluorite deposit (the base map was modified from [60]).

Moreover, the Y/Ho ratios of fluorite can be used to determine the origin of the fluid because the Y and Ho have comparable ionic radii and potentials [58], and the Y/Ho ratios of fluorite from the same origin are usually constant [62]. In this study, the Y contents of fluorite in the Yama fluorite deposit are basically positively correlated with ∑REE contents (Figure 12a), and the Y/Ho ratios of fluorite from both types of fluorite ores have a comparably invariable range of 41.43–73.79. Apparently, fluorites from the Yama fluorite deposit or their origins of ore-forming fluids follow a recrystallization trend, indicating an identical fluid origin (Figure 12b; e.g., [58]). Furthermore, hydrothermal fluorites are identified by their high Y/Ho ratios (average between 35 and 250; e.g., [58]). The Y/Ho ratio range of 41.43–73.79 for the studied fluorite samples is much higher than the chondrite Y/Ho ratio [63], PAAS Y/Ho ratio [47], and most igneous rocks [58], and is located within the Y/Ho value range of hydrothermal fluorites (Figure 12c) [58], suggesting that the fluorites in the Yama fluorite deposit might be derived from a hydrothermal source rather than the surrounding syenogranites. Moreover, fluorites in the study area showing positive Y anomalies also suggested that Y enrichment depends on the presence of Y–F complexes (e.g., [64]), which have a higher stability when compared with Ho–F complexes [65]. Therefore, Y preferentially stays within the fluorite formed from F-rich fluids.

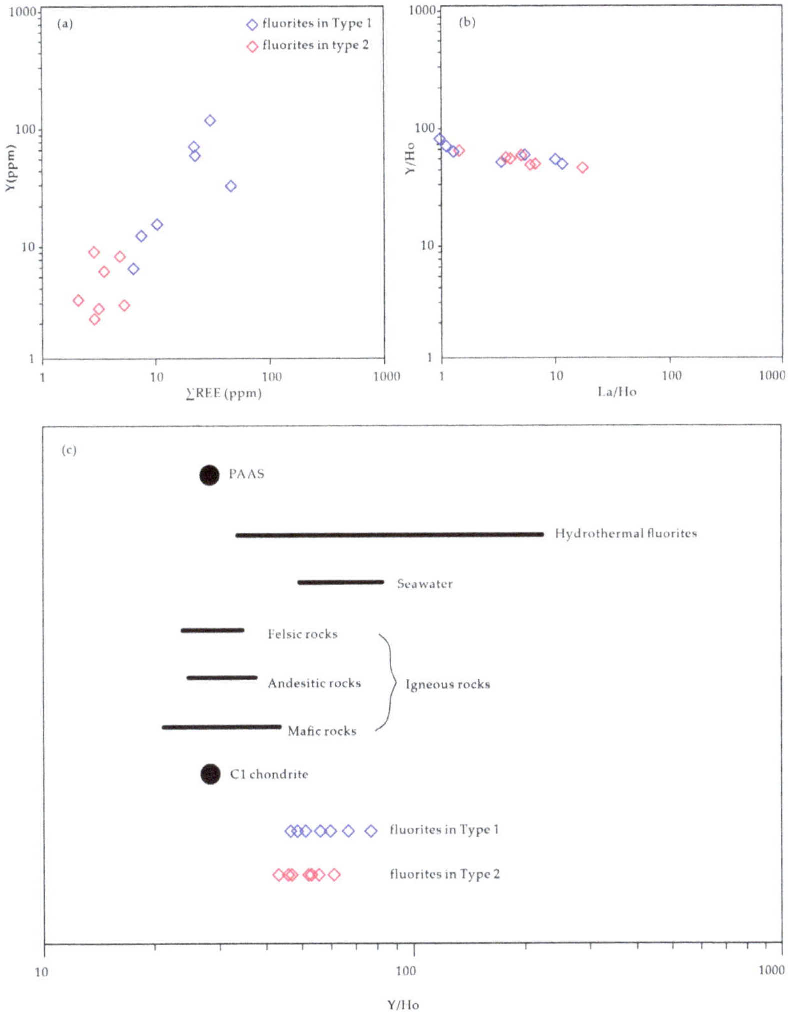

Figure 12. Plot of (**a**) Y versus ∑REEs and (**b**) Y/Ho versus La/Ho for fluorites from the Yama deposit; (**c**) comparison of the Y/Ho ratios of fluorites from the Yama deposit. Igneous rocks and hydrothermal fluorite are also plotted for comparison (after [58]). PAAS data were obtained from [47], and C1 chondrite data were obtained from [63].

The valence state of Eu in fluorite and the Eu anomalies relative to other REEs can reveal the redox state of the ore-forming fluid [58]. In the reducing environment, some Eu takes on the form of Eu^{2+}, and the Eu^{2+} ionic radius is large and cannot easily replace Ca^{2+}, thus forming a negative Eu anomaly [66]. There were weak to moderate negative Eu anomalies (Eu/Eu* = 0.16–0.40) of fluorite in the Yama fluorite deposit; therefore, Eu^{2+} should exist in the fluorite in the Yama fluorite deposit and indicate a reducing environment. These negative Eu anomalies indicate the low O fugacity conditions of the hydrothermal fluids (e.g., [67–69]).

To obtain a precise ore-forming age for fluorite mineralization in the Yama fluorite deposit, we attempted to carry out the Sm-Nd isotope analysis of five fluorite samples. However, we did not obtain reliable Sm-Nd isotope data for our samples, as well as an isochron age. Fluorite mineralization is a type of hydrothermal filling of fluorite–quartz veins in the Yama fluorite deposit, and no coeval hydrothermal alterations can be observed close to the quartz–fluorite veins within the syenogranite wall rocks. According to the above geological fact, it was indicated that there is very weak fluid–rock interactions between ore-forming meteoric fluid and surrounding syenogranites during fluorite and quartz filling. Therefore, the ore-forming age for fluorite mineralization should be later than the forming age of syenogranite wall rocks (i.e., mean age = 443 Ma [10]).

Commonly, the Late Ordovician–Early Silurian syenogranites in the study area were considered to be emplaced in a post-collisional setting which was related to the closure of the South Qilian Ocean (a branch of the Proto-Tethys Ocean) between the Qaidam Block and SQB [10]. Since the occurrence of such numerous Early Paleozoic granitic emplacements, subsequent dominant tectonic and magmatic activities within the SQB had happened predominantly in the Permian–Triassic and were considered to be closely related to the evolution of the Paleo-Tethys Ocean between the QM and the SQB [6]. Therefore, we assume that the meteoric water under the Yama area was possibly heated and convected in the Permian–Triassic, and then might be turned into a type of reduced, low-temperature, F-bearing hydrothermal fluid which might extract F from the deep surrounding rocks. The late, large-scale brittle faults within the Late Ordovician–Early Silurian syenogranites in the Yama area would provide a wide space for migration pathways and the rapid filling of F-bearing ore-forming fluids. Finally, quartz veins and quartz–fluorite veins filled along faults within the syenogranite wall rocks in the Yama area. In essence, the Yama fluorite deposit is a fault-controlled hydrothermal vein-type deposit (e.g., [70,71]), and might be also classified as a Qiquanba type of meso-epithermal fluorite deposit in China, as proposed by [72]. The Late Ordovician–Early Silurian syenogranites as well as other granites in the Tataleng granitic batholith are just wall rocks for fluorite mineralization; thus, potential fluorite mineralization should be searched for within the subsidiary brittle faults within granites rather than in the granitic bodies themselves.

6. Conclusions

Petrographic observation; H, O, and Sr isotope analyses; and trace element and REE analyses were conducted on fluorites within the quartz–fluorite veins in the Yama fluorite deposit in the Tataleng granitic batholith in the SQB. Based on these analyses, the following conclusions can be drawn:

(1) The paragenetic sequences for the Yama fluorite mineralization can be divided into two major stages, i.e., the pre-ore quartz stage, and the hydrothermal stage, and the latter stage can be subdivided into the quartz–fluorite epoch and the quartz–calcite epoch. Fluorite-free quartz veins, fluorite–quartz veins, and a few quartz–calcite veins were formed in the pre-ore stage, the quartz–fluorite epoch, and the quartz–calcite epoch, respectively. Typical fluorite–quartz veins consist of Type 1 fluorite ores showing vein structures, veinlet, stringer structures, and/or pectinate, and Type 2 ores showing breccia structures within veins.

(2) Fluorite samples from fluorite–quartz veins have roughly similar REY patterns with relatively strong enrichment in HREEs, negative Eu anomalies (Eu/Eu* = 0.16–0.40),

strongly positive Y anomalies (Y/Y* = 1.43–2.39), relatively high $(La/Yb)_N$ values of 0.69–8.43, and comparably invariable range Y/Ho ratios (41.43–73.79), suggesting a reduced hydrothermal genesis.

(3) Fluorites in two types of fluorite ore have relatively variable δD and $\delta^{18}O$ values of −77.4‰ to −102.4‰ and −12.7‰ to −4.3‰, respectively, and the δD value ranges of two types of fluorite ores roughly overlap each other; however, Type 2 samples show slightly heavier $\delta^{18}O$ values than Type 1, suggesting the origin of the ore-forming fluid for the Yama fluorite mineralization might predominantly originate from paleo-meteoric water, and a mixture with certain deep water sources (e.g., magmatic water or metamorphic water) cannot also be totally precluded.

(4) Two types of fluorites have a relatively invariable range of 0.749089−0.756628 except an anomalously high ratio of 0.828185, and the syenogranite wall rocks have variable analytical $^{87}Sr/^{86}Sr$ ratios of 0.832625−1.191950. The calculated initial $^{87}Sr/^{86}Sr$ ratios of fluorites based on given ore-forming ages less than 443 Ma are apparently higher than the initial $^{87}Sr/^{86}Sr$ ratios of syenogranites, indicating the Sr source should be predominantly derived from the ore-forming fluids themselves rather than from the syenogranite wall rocks.

(5) Combining the geological characteristics, trace element and REE geochemistry, and the isotope (H, O, and Sr) geochemistry, it was suggested that the Yama fluorite deposit can be classified as a fault-controlled hydrothermal vein-type deposit, and ore-forming water and materials might be directly derived from the evolutionary meteoric water rather than syenogranite wall rocks. The late large-scale faults within the Late Ordovician–Early Silurian syenogranites in the Yama area would provide a wide space for migration pathways and rapid filling of F-bearing ore-forming fluids. Finally, quartz veins and quartz–fluorite veins are filled alongside them, and weak hydrothermal alterations can be observed close to the veins within syenogranite wall rocks.

Author Contributions: Author Contributions: Conceptualization, Q.-F.D.; methodology, H.J., X.Z. and R.-Z.W.; software, X.Z. and R.-Z.W.; validation, H.J., Q.-F.D. and R.-Z.W.; formal analysis, H.J., X.Z., R.-Z.W. and Q.-F.D.; investigation, H.J., R.-Z.W., X.Z., G.-B.H., W.M., Q.-Q.C., W.-H.W. and Q.-F.D.; resources, H.J., G.-B.H., Q.-Q.C. and Q.-F.D.; data curation, H.J. and Q.-F.D.; writing—original draft preparation, H.J. and Q.-F.D.; writing—review and editing, H.J. and Q.-F.D.; visualization, H.J., X.Z. and Q.-F.D.; supervision, Q.-F.D.; project administration, H.J. and Q.-F.D.; funding acquisition, H.J. and Q.-F.D. All authors have read and agreed to the published version of the manuscript.

Funding: This research was funded by the First Group of Geological Survey Project in 2024 from the Bureau of Geological Exploration & Development of Qinghai Province, grant number: File (2023) 50.

Data Availability Statement: The datasets generated during and/or analyzed during the current study are available in full within the published article.

Acknowledgments: We are grateful to the analytical Laboratories at ALS Chemex (Guangzhou, China) Co., Ltd., BRIUG, and BCTT for their help with the element and stable isotope analyses. We also thank the three reviewers for their thoughtful reviews.

Conflicts of Interest: The authors declare no conflict of interest.

References

1. Xiao, W.; Windley, B.F.; Yong, Y.; Yan, Z.; Yuan, C.; Liu, C.; Li, J. Early Paleozoic to Devonian multiple-accretionary model for the Qilian Shan, NW China. *J. Asian Earth Sci.* **2009**, *35*, 323–333. [CrossRef]
2. Song, S.; Niu, Y.; Su, L.; Xia, X. Tectonics of the North Qilian orogen, NW China. *Gondwana Res.* **2013**, *23*, 1378–1401. [CrossRef]
3. Song, S.; Niu, Y.; Su, L.; Zhang, C.; Zhang, L. Continental orogenesis from ocean subduction, continent collision/subduction, to orogen collapse, and orogen recycling: The example of the North Qaidam UHPM belt, NW China. *Earth-Sci. Rev.* **2014**, *129*, 59–84. [CrossRef]
4. Wang, C.; Li, R.; Smithies, R.H.; Li, M.; Peng, Y.; Chen, F.; He, S. Early Paleozoic felsic magmatic evolution of the western Central Qilian belt, Northwestern China, and constraints on convergent margin processes. *Gondwana Res.* **2017**, *41*, 301–324. [CrossRef]

5. Yan, Z.; Fu, C.; Aitchison, J.C.; Buckman, S.; Niu, M.; Cao, B.; Sun, Y.; Guo, X.; Wang, Z.; Zhou, R. Retro-foreland Basin Development in Response to Proto-Tethyan Ocean Closure, NE Tibet Plateau. *Tectonics* **2019**, *38*, 4229–4248. [CrossRef]
6. Ding, Q.F.; Pan, T.; Zhou, X.; Cheng, L.; Li, S.P.; Han, J.; Qian, Y.; Gao, Y.; Shang, M.Y. Geochronology, petrogenesis, and tectonic significance of the granites in the Chaqiabeishan area of the Quanji Massif, northwestern China. *Geol. J.* **2022**, *57*, 1241–1261. [CrossRef]
7. Song, S.; Yang, L.; Zhang, Y.; Niu, Y.; Wang, C.; Su, L.; Gao, Y. Qi-Qin Accretionary Belt in Central China Orogen: Accretion by trench jam of oceanic plateau and formation of intra-oceanic arc in the Early Paleozoic Qin-Qi-Kun Ocean. *Sci. Bull.* **2017**, *62*, 1035–1038. [CrossRef]
8. Sun, Y.; Niu, M.; Li, X.; Wu, Q.; Cai, Q.; Yuan, X.; Li, C. Petrogenesis and tectonic implications from the Ayishan Group in the South Qilian Belt, NW China. *Geol. J.* **2020**, *55*, 6860–6877. [CrossRef]
9. He, X.; Fu, C.; Yan, Z.; Wang, B.; Niu, M.; Li, X. Petrogenesis and tectonic implication of the lower Silurian high-Sr/Y subvolcanic rocks from the South Qilian suture zone in the Qilian Orogen, NW China. *Geol. Mag.* **2021**, *158*, 1383–1402. [CrossRef]
10. Jiao, H.; Lu, H.; Huang, G.; Cui, Q.; Ding, Q.; Zhou, X.; Wu, R. Geochronology, petrogenesis, and mineralization potential of the syenogranites in the Yama fluorite deposit, Tataleng granitic batholith, Qilian Orogen, NW China. *Can. J. Earth Sci.* **2023**, *60*, 1647–1673. [CrossRef]
11. Wu, C.L.; Yang, J.S.; Wooden, J.L.; Liou, J.G.; Li, H.B.; Meng, F.C.; Persing, H.; Meibom, A. Zircon SHRIMP dating of granite from Qaidamshan, NW China. *Chin. Sci. Bull.* **2002**, *47*, 418–422. [CrossRef]
12. Tseng, C.; Yang, H.; Yang, H.; Liu, D.; Wu, C.; Cheng, C.; Chen, C.; Ker, C. Continuity of the North Qilian and North Qinling orogenic belts, Central Orogenic System of China: Evidence from newly discovered Paleozoic adakitic rocks. *Gondwana Res.* **2009**, *16*, 285–293. [CrossRef]
13. Chen, Y.; Song, S.; Niu, Y.; Wei, C. Melting of continental crust during subduction initiation: A case study from the Chaidanuo peraluminous granite in the North Qilian suture zone. *Geochim. Cosmochim. Acta* **2014**, *132*, 311–336. [CrossRef]
14. Yang, H.; Zhang, H.; Luo, B.; Zhang, J.; Xiong, Z.; Guo, L.; Pan, F. Early Paleozoic intrusive rocks from the eastern Qilian orogen, NE Tibetan Plateau: Petrogenesis and tectonic significance. *Lithos* **2015**, *224–225*, 13–31. [CrossRef]
15. Yu, S.; Zhang, J.; Qin, H.; Sun, D.; Zhao, X.; Cong, F.; Li, Y. Petrogenesis of the early Paleozoic low-Mg and high-Mg adakitic rocks in the North Qilian orogenic belt, NW China: Implications for transition from crustal thickening to extension thinning. *J. Asian Earth Sci.* **2015**, *107*, 122–139. [CrossRef]
16. Tung, K.; Yang, H.; Yang, H.; Smith, A.; Liu, D.; Zhang, J.; Wu, C.; Shau, Y.; Wen, D.; Tseng, C. Magma sources and petrogenesis of the early–middle Paleozoic backarc granitoids from the central part of the Qilian block, NW China. *Gondwana Res.* **2016**, *38*, 197–219. [CrossRef]
17. Yang, H.; Zhang, H.; Xiao, W.; Tao, L.; Gao, Z.; Luo, B.; Zhang, L. Multiple Early Paleozoic granitoids from the southeastern Qilian orogen, NW China: Magma responses to slab roll-back and break-off. *Lithos* **2021**, *380–381*, 105910. [CrossRef]
18. Li, J.; Xia, Y.; Jiang, H.; Zhang, X.; Zhang, X.; Wang, Y.; Lei, T.; Liu, Y. Magma hybridization and crust-mantle interaction revealed by mineralogical and geochemical footprints: A case study from the South Qilian Accretionary Belt at the northern Tibetan Plateau. *Int. Geol. Rev.* **2021**, *64*, 2284–2305. [CrossRef]
19. Zhao, J.; Wu, B.; Zhang, X.; Chen, W.; Ma, X. Petrogenesis of Early Paleozoic adakitic granitoids in the eastern Qilian Block, northwest China: Implications for the South Qilian Ocean subduction. *Mineral. Petrol.* **2021**, *115*, 687–708. [CrossRef]
20. Zhu, X.; Chen, D.; Ren, Y.; Yang, M.; Wang, C. Late Silurian mafic and felsic magmatism in the South Qilian Belt, northern Tibet Plateau: Response to slab breakoff. *Lithos* **2021**, *380*, 105860. [CrossRef]
21. Zhang, B.W. Study on Metalliferous Deposits Mineralization in Southern Qilian Orogenic Belt, Qinghai Province. Ph.D. Thesis, Jilin University, Changchun, China, 2010.
22. Li, S.J. Geodynamic Evolution of Qilian Orogenic Belt and Metallogenesis of Endogenous Metals. Ph.D. Thesis, Jilin University, Changchun, China, 2011.
23. Li, S.J.; Zeng, X.P.; Wang, F.C.; Tong, H.K.; Yang, Y.Q.; Zhang, D.X.; Kang, B.; Shi, H.Y.; Zheng, Z.H. Main metallogenic series and prospecting potential in Qinghai province. *J. Jilin Univ. (Earth Sci. Ed.)* **2022**, *52*, 1419–1445. [CrossRef]
24. QIGEIQP. *Interim Report of Geological Survey for Fluorite Mineralizations in Yama Area in Delingha City, Qinghai Province*; Qaidam Integrated Geological Exploration Institute of Qinghai Province (QIGEIQP): Golmud, China, 2023.
25. Yan, Z.; Fu, C.; Aitchison, J.C.; Niu, M.; Buckman, S.; Cao, B. Early Cambrian Muli arc–ophiolite complex: A relic of the Proto-Tethys oceanic lithosphere in the Qilian Orogen, NW China. *Int. J. Earth Sci. (Geol. Rundsch.)* **2019**, *108*, 1147–1164. [CrossRef]
26. Fu, C.; Yan, Z.; Wang, Z.; Buckman, S.; Aitchison, J.C.; Niu, M.; Cao, B.; Guo, X.; Li, X.; Li, Y.; et al. Lajishankou Ophiolite Complex: Implications for Paleozoic Multiple Accretionary and Collisional Events in the South Qilian Belt. *Tectonics* **2018**, *37*, 1321–1346. [CrossRef]
27. Kang, H.; Chen, Y.; Li, D. The nature and history of the South Qilian orogenic belt: Constraints from compositions of rivers' sediments and their detrital zircon U-Pb geochronology, Lu-Hf isotopic compositions. *Geol. J.* **2020**, *55*, 712–727. [CrossRef]
28. Wang, D.; Wang, T.; Yan, J.; Lin, X. Petrogenesis and tectonic implications of early Palaeozoic igneous rocks of the western South Qilian Belt, central China. *Int. Geol. Rev.* **2018**, *60*, 844–864. [CrossRef]
29. Tung, K.; Yang, H.; Yang, H.; Liu, D.; Zhang, J.; Wan, Y.; Tseng, C. SHRIMP U-Pb geochronology of the zircons from the Precambrian basement of the Qilian Block and its geological significances. *Chin. Sci. Bull.* **2007**, *52*, 2687–2701. [CrossRef]

30. Xu, Z.Q.; Xu, H.F.; Zhang, J.X.; Li, H.B.; Zhu, Z.Z.; Qu, J.C.; Chen, D.N.; Chen, J.L.; Yang, K.C. The Zoulang Nanshan Caledonian subduction complex in the Northern Qilian Mountains and its dynamics1. *Acta Geol. Sin.* **1994**, *7*, 241.
31. Feng, Y.M.; He, S.P. *Geotectonics and Orogeny of the Qilian Mountains*; Geological Publish House: Beijing, China, 1996; p. 721.
32. Wu, C.L.; Yang, J.S.; Wooden, J.; Liou, J.G.; Li, H.B.; Meng, F.C.; Persing, H.; Meibom, A. Zircon SHRIMP dating for the Qaidam moutain granite. *Sci. Bull.* **2001**, *46*, 1743–1747. (In Chinese)
33. Hu, G.N.; Wang, R.; Wang, X.X.; Sun, Y.G.; Lin, W.; Shen, W.H. Mineralogical characteristics and petrological significance of the Tataleng Rapakivi granite in the Northern margin of the Qaidam basin. *J. Earth Sci. Environ.* **2008**, *30*, 337–345. (In Chinese with English Abstract)
34. Zhou, B.; Zheng, Y.Y.; Xu, R.K.; Wang, Y.K.; Zhang, X.P.; Wu, L. LA-ICP-MS zircon U-Pb dating and Hf isotope geochemical characteristics of Qaidamshan intrusive body. *Geol. Bull. China* **2013**, *32*, 1027–1034. (In Chinese with English Abstract)
35. Zhu, X.H.; Wang, H.L.; Yang, M. Zircon U–Pb age of the monzogranite from the middle segment of the Qaidam Mountain composite granite on the South margin of the Qilian mountain. *Geol. China* **2016**, *43*, 751–767. (In Chinese with English Abstract)
36. He, X.Y.; Wang, Y.; Guo, R.H. Petrology, geochemistry, and zircon U-Pb geochronology of the Chaidamu mountain granite. *Geol. Rev.* **2019**, *65*, 55–56. (In Chinese)
37. He, X.Y.; Yang, X.K.; Wang, Y.; Guo, R.H.; Liao, Y.Y.; Fan, F.H. Petrology, geochemistry, and zircon U-Pb geochronology of the Chaidamushan granite from the southern margin of Qilianshan. *Acta Geol. Sin.* **2020**, *94*, 1248–1263. (In Chinese with English Abstract)
38. Huang, H.; Niu, Y.; Nowell, G.; Zhao, Z.; Yu, X.; Mo, X. The nature and history of the Qilian Block in the context of the development of the Greater Tibetan Plateau. *Gondwana Res.* **2015**, *28*, 209–224. [CrossRef]
39. Wang, Q.; Dong, Y.; Pan, Y.; Liao, F.; Guo, X. Early Paleozoic Granulite-Facies Metamorphism and Magmatism in the Northern Wulan Terrane of the Quanji Massif: Implications for the Evolution of the Proto-Tethys Ocean in Northwestern China. *J. Earth Sci.* **2018**, *29*, 1081–1101. [CrossRef]
40. Wu, C.L.; Wooden, J.L.; Mazadab, F.K.; Mattinson, C.; Gao, Y.H.; Wu, S.P.; Chen, Q.L. Zircon SHRIMP U-Pb dating of granites from the Da Qaidam area in the north margin of Qaidam basin, NW China. *Acta Petrol. Sin.* **2007**, *23*, 1861–1875. (In Chinese with English abstract)
41. Lu, X.X.; Sun, Y.G.; Zhang, X.T.; Xiao, Q.H.; Wang, X.X.; Wei, X.D.; Gu, D.M. The SHRIMP age of Tataleng Rapakivi granite at the north margin of Qaidam basin. *Acta Geol. Sin.* **2007**, *81*, 626–634. (In Chinese with English Abstract)
42. Clayton, R.N.; Mayeda, T.K. The use of bromine pentafluoride in the extraction of oxygen from oxides and silicates for isotopic analysis. *Geochim. Cosmochim. Acta* **1963**, *27*, 43–52. [CrossRef]
43. Clayton, R.N.; O'Neil, J.R.; Mayeda, T.K. Oxygen isotope exchange between quartz and water. *J. Geophys. Res.* **1972**, *77*, 3057–3067. [CrossRef]
44. Liu, H.B.; Jin, G.S.; Li, J.J.; Han, J.; Zhang, J.F.; Zhang, J.; Zhong, F.W.; Guo, D.Q. Determination of stable isotope composition in uranium geological samples. *World Nucl. Geosci.* **2013**, *30*, 174–179. (In Chinese with English Abstract)
45. Yang, Y.; Zhang, H.; Chu, Z.; Xie, L.; Wu, F. Combined chemical separation of Lu, Hf, Rb, Sr, Sm and Nd from a single rock digest and precise and accurate isotope determinations of Lu–Hf, Rb–Sr and Sm–Nd isotope systems using Multi-Collector ICP-MS and TIMS. *Int. J. Mass Spectrom.* **2010**, *290*, 120–126. [CrossRef]
46. Boynton, W.V. *Geochemistry of the Rare Earth Elements: Meteorites Studies*; Elsevier: New York, NY, USA, 1984; pp. 63–114.
47. Taylor, S.R.; McLennan, S.M. *The Continental Crust: Its Composition and Evolution*; Blackwell: Oxford, UK, 1985; p. 312.
48. Goldfarb, R.J.; Ayuso, R.; Miller, M.L.; Ebert, S.W.; Marsh, E.E.; Petsel, S.A.; Miller, L.D.; Bradley, D.C.; Johnson, C.; McClelland, W. The Late Cretaceous Donlin Creek gold deposit, Southwestern Alaska: Controls on epizonal ore formation. *Econ. Geol.* **2004**, *99*, 643–671. [CrossRef]
49. Sheppard, S. Characterization and isotopic variations in natural waters. *Rev. Mineral. Geochem.* **1986**, *16*, 165–183. [CrossRef]
50. Steiger, R.H.; Jäger, E. Subcommission on geochronology: Convention on the use of decay constants in geo- and cosmochronology. *Earth Planet. Sci. Lett.* **1977**, *36*, 359–362. [CrossRef]
51. Pei, Q.; Li, C.; Zhang, S.; Zou, H.; Liang, Y.; Wang, L.; Li, S.; Cao, H. Vein-type fluorite mineralization of the Linxi district in the Great Xing'an Range, Northeast China: Insights from geochronology, mineral geochemistry, fluid inclusion and stable isotope systematics. *Ore Geol. Rev.* **2022**, *142*, 104708. [CrossRef]
52. Craig, H. The isotopic composition of water and carbon in geothermal areas. In *Nuclear Geology on Geothermal Areas*; Tongiorgi, E., Ed.; CNR Laboratorio di Geologia Nuclear: Pisa, Italy, 1963; pp. 17–53.
53. Goldfarb, R.J.; Snee, L.W.; Miller, L.D.; Newberry, R.J. Rapid dewatering of the crust deduced from ages of mesothermal gold deposits. *Nature* **1991**, *354*, 296–298. [CrossRef]
54. Jiang, S.H.; Nie, F.J.; Hu, P.; Lai, X.R.; Liu, Y.F. Mayum: An orogenic gold deposit in Tibet, China. *Ore Geol. Rev.* **2009**, *36*, 160–173. [CrossRef]
55. McCuaig, C.T.; Kerrich, R. P—T—T—Deformation—Fluid characteristics of lode gold deposits: Evidence from alteration systematics. *Ore Geol. Rev.* **1998**, *12*, 381–453. [CrossRef]
56. Möller, P.; Parekh, P.P.; Schneider, H.J. The ApplicaUon of Tb/Ca-Tb/La Abundance Ralios to Problems of Fluorspar Genesis. *Miner. Depos.* **1976**, *11*, 111–116. [CrossRef]
57. Barbieri, M.; Tolomeo, L.; Voltaggio, M. Yttrium, lanthanum and Manganese geochemistry in fluorite deposits from Sardinia (Italy). *Chem. Geol.* **1983**, *40*, 43–50. [CrossRef]

58. Bau, M.; Dulski, P. Comparative study of yttrium and rare-earth element behaviours in fluorine-rich hydrothermal fluids. *Contrib. Mineral. Petrol.* **1995**, *119*, 213–223. [CrossRef]
59. Ye, X.; Bai, F. Spectral characteristics, rare earth elements, and ore-forming fluid constrains on the origin of fluorite deposit in Nanlishu, Jilin Province, China. *Minerals* **2022**, *12*, 1195. [CrossRef]
60. Singh, T.; Upadhyay, D.; Patel, A.K.; Mishra, B. High MREE-HREE solubility in a carbonatite-derived hydrothermal fluid: Evidence from fluorite-hosted fluid inclusions in the Amba Dongar carbonatite complex, India. *Chem. Geol.* **2022**, *613*, 121162. [CrossRef]
61. Liu, B.; Wu, Q.; Li, H.; Wu, J.; Cao, J.; Jiang, J.; Liang, W. Fault-controlled fluid evolution in the Xitian W-Sn-Pb-Zn-fluorite mineralization system (South China): Insights from fluorite texture, geochemistry and geochronology. *Ore Geol. Rev.* **2020**, *116*, 103233. [CrossRef]
62. Irber, W. The lanthanide tetrad effect and its correlation with K/Rb, Eu/Eu∗, Sr/Eu, Y/Ho, and Zr/Hf of evolving peraluminous granite suites. *Geochim. Cosmochim. Acta* **1999**, *63*, 489–508. [CrossRef]
63. Anders, E.; Grevesse, N. Abundances of the elements: Meteoritic and solar. *Geochim. Cosmochim. Acta* **1989**, *53*, 197–214. [CrossRef]
64. Möller, P. Europium anomalies in hydrothermal minerals. Kinetic versus thermodynamic interpretation. In Proceedings of the Ninth Quadrennial IAGOD Symposium, Beijing, China, 12–18 August 1994; Schweizerbart: Stuttgart, Germany, 1998; pp. 239–246.
65. Bau, M. Controls on the fractionation of isovalent trace elements in magmatic and aqueous systems: Evidence from Y/Ho, Zr/Hf, and lanthanide tetrad effect. *Contrib. Mineral. Petrol.* **1996**, *123*, 323–333. [CrossRef]
66. Möller, P. REE fractionation in hydrothermal fluorite and calcite. In *Source, Transport and Deposition of Metals*, 1st ed.; Pagel, M., Leroy, J.L., Eds.; Balkema: Rotterdam, The Netherlands, 1991; pp. 91–94.
67. Sasmaz, A.; Yavuz, F.; Sagiroglu, A.; Akgul, B. Geochemical patterns of the Akdagmadeni (Yozgat, Central Turkey) fluorite deposits and implications. *J. Asian Earth Sci.* **2005**, *24*, 469–479. [CrossRef]
68. Sasmaz, A.; Kryuchenko, N.; Zhovinsky, E.; Suyarko, V.; Konakci, N.; Akgul, B. Major, trace and rare earth element (REE) geochemistry of different colored fluorites in the Bobrynets region, Ukraine. *Ore Geol. Rev.* **2018**, *102*, 338–350. [CrossRef]
69. Duan, Z.; Jiang, S.; Su, H.; Zhu, X.; Zou, T.; Cheng, X. Trace and Rare Earth Elements, and Sr Isotopic Compositions of Fluorite from the Shihuiyao Rare Metal Deposit, Inner Mongolia: Implication for Its Origin. *Minerals* **2020**, *10*, 882. [CrossRef]
70. Zhao, Y.; Pei, Q.; Zhang, S.; Guo, G.; Li, J.; Liu, W.; Hu, X.; Song, K.; Wang, L. Formation timing and genesis of Madiu fluorite deposit in East Qinling, China: Constraints from fluid inclusion, geochemistry, and H–O–Sr–Nd isotopes. *Geol. J.* **2020**, *55*, 2532–2549. [CrossRef]
71. Pei, Q.; Zhang, S.; Hayashi, K.; Wang, L.; Cao, H.; Zhao, Y.; Hu, X.; Song, K.; Chao, W. Nature and Genesis of the Xiaobeigou Fluorite Deposit, Inner Mongolia, Northeast China: Evidence from Fluid Inclusions and Stable Isotopes. *Resour. Geol.* **2019**, *69*, 148–166. [CrossRef]
72. Han, B.; Shang, P.; Gao, Y.; Jiao, S.; Yao, C.; Zou, H.; Li, M.; Wang, L.; Zheng, H. Fluorite deposits in China: Geological features, metallogenic regularity, and research progress. *China Geol.* **2020**, *3*, 473–489.

minerals

MDPI

Article

Age and Tectonic Setting of Layered Lead–Zinc Ore Bodies in the Xiaohongshilazi Deposit: Constraints from Geochronology and Geochemistry of the Volcanic Rocks in Central Jilin Province, NE China

Qun Yang [1], Qingqing Shang [1,*], Yunsheng Ren [2] and Zhongjie Yang [3]

1 College of Earth Sciences, Jilin University, Changchun 130061, China; yangqun@jlu.edu.cn
2 School of Earth Science, Institute of Disaster Prevention, Langfang 065201, China; renys@jlu.edu.cn
3 Mudanjiang Natural Resources Comprehensive Survey Center, China Geological Survey, Mudanjiang 157000, China; yangzj870227@163.com
* Correspondence: shangqq@jlu.edu.cn

Abstract: The newly discovered Xiaohongshilazi deposit located in Panshi City, central Jilin Province, NE China, is a medium-scale Pb–Zn–(Ag) deposit. The Pb–Zn–(Ag) orebodies are divided into layered and vein-type orebodies, which have different ore geneses. The layered Pb–Zn orebodies are mainly hosted within and spatially controlled by the volcanic rocks. To constrain the age and tectonic setting of the layered Pb–Zn mineralization, we completed laser-ablation–ICP–MS zircon U–Pb dating and whole-rock major and trace element analyses of the ore-bearing volcanic rocks. The dacite samples were confirmed as belonging to the Daheshen Formation and were the main ore-bearing volcanic rocks for the layered orebodies. They yielded concordia U–Pb ages of 278.1 ± 1.8 Ma and 278.3 ± 1.8 Ma, respectively, indicating that the volcanic rocks from the Daheshen Formation and related layered Pb–Zn mineralization were formed in the early Permian. The andesite and rhyolite located above the layered orebodies yielded concordia U–Pb ages of 225.0 ± 1.1 Ma, 225.3 ± 1.5 Ma, and 224.7 ± 1.2 Ma, respectively; these substances are considered to be of the Sihetun Formation and were first reported in the area. The dacite samples associated with layered Pb–Zn mineralization were high in SiO_2 (62.54–65.02 wt.%), enriched in LREEs and LILEs (e.g., Rb, Ba, and K), and showed depletion in HFSEs (e.g., P and Ti). It showed slightly negative Eu anomalies (δEu = 0.60–0.65) and negative Nb anomalies, with Th/Nb (1.12–1.21) and La/Nb (2.8–4.7) ratios, presenting subduction-related arc magma affinity formed in an active continental margin setting. In agreement with previous studies on zircon Hf isotopes ($\varepsilon_{Hf}(t)$ = +0.23~ +10.60) of the volcanic rocks from the Daheshen Formation, we infer that they were derived from the partial melting of the depleted lower crust. In conclusion, mineralization characteristics, geochronological data, geochemical features, and regional tectonic evolution suggest that two Pb–Zn–(Ag) mineralization stages from the Xiaohongshilazi deposit occurred: the layered VMS-type Pb–Zn mineralization associated with the marine volcanic rocks from the early Permian Daheshen Formation, which was induced by the subduction of the Paleo-Asian oceanic plate beneath the northern margin of the North China Craton, and the vein-type Pb–Zn–(Ag) mineralization caused by the subduction of the Paleo-Pacific Plate in the early Jurassic. Considering this, along with the mineralization characteristics of the same-type polymetallic deposits in this region, we propose that the early Permian marine volcanic rocks have great prospecting potential for the VMS-type Pb–Zn polymetallic deposits.

Keywords: LA-ICP-MS zircon U–Pb dating; geochemistry of volcanic rocks; layered Pb–Zn orebodies; Xiaohongshilazi deposit; central Jilin Province

1. Introduction

Central Jilin Province, Northeast China, located on the intersection between the eastern segments of the northern margin of North China Craton (NCC) and Xing'an Mongolian

Orogenic Belt (XMOB), experienced the evolution, superposition, and transition of the Paleo-Asian Ocean and Paleo-Pacific Ocean tectonic domains from the late Paleozoic to the early Mesozoic [1–5]. Numerous porphyry Mo deposits, mesothermal hydrothermal lode gold deposits, and magmatic Cu–Ni sulfide deposits have been discovered in this region, and they are temporally and spatially related to the widespread Mesozoic intrusions [6–13]. However, the deposits related to Paleozoic magmatism are relatively scarce and poorly studied in the area. In recent years, Pb–Zn polymetallic deposits (such as the Xiaohongshilazi Pb–Zn deposit, Hongtaiping Pb–Zn–Cu deposit, Dongfengnanshan Pb–Zn–Cu deposit, etc.) occurring in Paleozoic marine volcanic rocks have been discovered along the Solonker–Xra Moron–Changchun–Yanji suture zone [14–17]. The discovery of these deposits can contribute to the study of Paleozoic mineralization in Northeast China and provide a new direction for the exploration of polymetallic deposits in this area, and also has great scientific significance for the study of the evolution and metallogenesis of the Paleo-Asian Ocean.

The newly discovered Xiaohongshilazi deposit is located ~30 km east of Panshi City in central Jilin Province, NE China. It is a medium-scale Pb–Zn–(Ag) deposit containing reserves of 34,968 t Pb, 100,150 t Zn, and 158 t Ag at average grades of 1.39% Zn, 0.486% Pb, and 21.94 g/t Ag [18]. Until recently, studies of the Xiaohongshilazi deposit were still focused on the vein-type orebodies [13,19,20], but the studies on the layered Pb–Zn orebodies were scarce. Based on current research, the vein-type Pb–Zn–(Ag) mineralization is related to an early Mesozoic (~200 Ma) tectono-magmatic–hydrothermal event induced by the initial subduction of the Paleo-Pacific Plate [18–20]. The layered Pb–Zn orebodies have the characteristics of VMS-type mineralization, according to the limited studies of the ore-forming conditions [21,22]. However, significant scientific studies on the layered Pb–Zn orebodies have not been reported so far. Firstly, the layered orebodies are interbedded with the marine volcanic rocks, and clearly cut by the late vein-type orebodies. Nevertheless, the mineralization timing of the layered Pb–Zn ore remains unknown due to the lack of clarity on the age of the ore-bearing volcanic rocks. Secondly, there are two distinct viewpoints about the tectonic setting in this region in the late Paleozoic: (1) the continental rift, taphrogenic trough, and post-orogenic extension after the final closure of the Paleo-Asian Ocean [23–26]; (2) the subduction of the Paleo-Asian oceanic plate beneath the northern margin of NCC [5,8,12,27–32]. It is therefore uncertain whether the layered Pb–Zn mineralization of the Xiaohongshilazi deposit was formed in an extensional setting after the final closure of the Paleo-Asian Ocean or formed in the subduction of the Paleo-Asian oceanic plate. Solving these scientific problems will not only improve understanding of the mineralization process of the Xiaohongshilazi deposit, but will also be important for the study of Paleozoic mineralization in this region and the evolution of the ancient Asian Ocean.

As a result, our focus is on the characteristics of layered Pb–Zn mineralization in the Xiaohongshilazi Pb–Zn–(Ag) deposit. We conducted laser-ablation–ICP–MS zircon U–Pb dating and collected whole-rock major and trace element data from the ore-bearing volcanic rocks to determine the age and tectonic setting of the layered Pb–Zn mineralization. This study holds importance for the research and exploration of similar deposits in this region.

2. Geological Background and Deposit Geology

Located at the junction between the XMOB and the northern margin of NCC, central Jilin Province successively experienced two distinct tectonic evolution stages during the Paleozoic to Cenozoic era [3,5] (Figure 1A). The Paleozoic tectonic evolution was controlled mainly by the progressive subduction of the Paleo-Asian oceanic plate beneath the northern margin of NCC, and the Mesozoic–Cenozoic tectonic evolution was dominated by the oblique subduction of the Paleo-Pacific oceanic plate beneath the Eurasian continent [11,12,30–35]. More than 10 large Mesozoic deposits, including porphyry Mo deposits (Daheishan, Jidetun, and Fu'anbu), lode gold deposits (Toudaochuan and Cuyu), and magmatic Cu–Ni sulfide deposits (Hongqiling), have been discovered in this region.

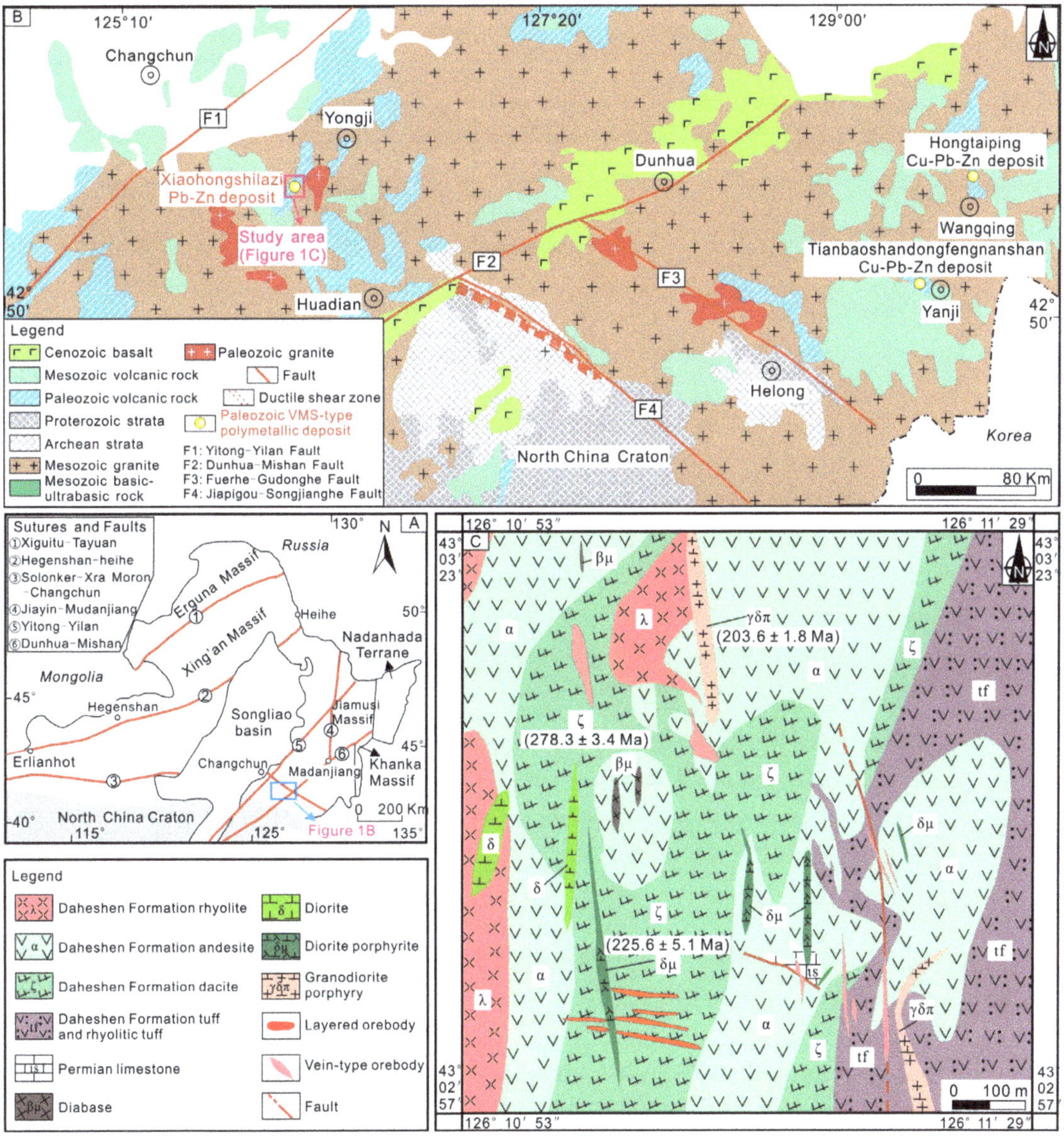

Figure 1. (**A**) Simplified tectonic map of NE China (modified from Wu et al. [2]); (**B**) Regional geological map of the central Jilin Province (modified from Yang [13]). (**C**) Geological map of Xiaohongshilazi Pb–Zn–(Ag) deposit (modified from Yang et al. [20]).

The Xiaohongshilazi deposit in central Jilin Province is located in the southeast segment of the intersection between the Yilan–Yitong fault and the Solonker–Xra Moron–Changchun suture (Figure 1B). The mainly exposed strata in the area are Daheshen Formation volcanic rocks (Figure 1C). The Daheshen Formation is composed of marine intermediate-acidic volcanic rocks, terrigenous clastic rocks, and carbonate rocks, which are locally mixed with a small number of continental clastic rocks. In addition, there are a large number of fusulina fossils, mainly Monodiexodina; plant fossils, mainly Cardioneura;

corals; brachiopods; and bryozoan fossils [13], indicating a marine environment. The volcanic rocks in the Xiaohongshilazi deposit are composed of andesitic volcanic breccia, andesite, dacite, rhyolite, andesitic tuff, dacitic tuff, rhyolitic tuff, etc. (Figure 2A). The micropetrographic characteristics of the volcanic rocks from the Xiaohongshilazi deposit are listed in Table 1 and shown in Figure 3. Dacite interbedded with the layered Pb–Zn orebodies is the primary ore-bearing volcanic rock Figures 2A,B and 4A.

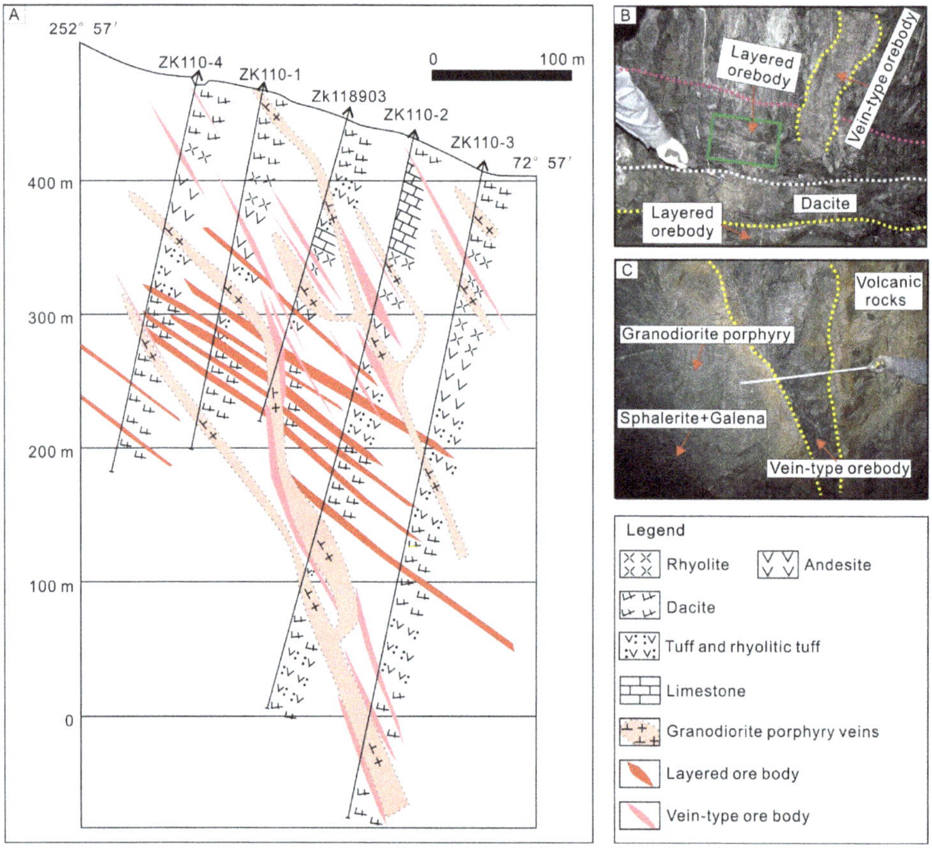

Figure 2. (**A**) Profile map of exploration line from the Xiaohongshilazi Pb–Zn–(Ag) deposit (modified from Chang [18]). (**B**) Vein-type Pb–Zn–(Ag) orebodies cut the layered orebodies and the volcanic rocks; (**C**) vein-type Pb–Zn–(Ag) orebodies occur around the granodiorite porphyry, and both develop along fissures and fractures and cut the volcanic rocks.

Table 1. Characteristics of the volcanic rocks in the Xiaohongshilazi ore district.

Sample No.	Lithology	Location	Texture/Structure	Phenocrysts	Matrix
X370-9	Andesite	370 m-depth of Xifeng mine	Porphyritic texture massive structure	Phenocrysts account for 45% of the rock, and consist of plagioclase (~35%, 2–7.5 mm) and amphibole (~10%, 0.5–4 mm, partly altered to chlorite)	Matrix is primarily pilotaxitic texture and dominated by plagioclase (~0.2 mm) and minor amphibole (~0.2 mm)
7XH-3	Andesite	Above the layered orebodies of Dongfeng mine	Porphyritic texture massive structure	Phenocrysts account for 40% of the rock, and consist of plagioclase (0.25–0.6 mm, partly altered to calcite)	Matrix is primarily pilotaxitic texture and dominated by plagioclase (0.1–0.2 mm) and minor amphibole (~0.1 mm)

Table 1. *Cont.*

Sample No.	Lithology	Location	Texture/Structure	Phenocrysts	Matrix
7XH-4	Rhyolite	Above the layered orebodies of Dongfeng mine	Porphyritic texture rhyolitic structure	Phenocrysts account for 25% of the rock, and consist of quartz (~20%, 0.2–0.5 mm,) and minor sanidine (5%, 0.3–0.4 mm)	Matrix is primarily cryptocrystalline texture and dominated by quartz (~30%), K-feldspar (~25%) and plagioclase
7XH-8-1	Dacite	Interbedding with the layered orebodies of Dongfeng mine	Porphyritic texture massive structure	Phenocrysts account for 55% of the rock, and consist of quartz (15%, ~0.2 mm) and plagioclase (40%, 0.5–2 mm) with minor amphibole	Matrix is primarily microcrystalline texture and dominated by plagioclase and quartz (10%, 0.03–0.1 mm)
7XH-8-2	Dacite	Interbedding with the layered orebodies of Xifeng mine	Porphyritic texture massive structure	Phenocrysts account for 40% of the rock, and consist of quartz (10%, 0.2–0.25 mm), plagioclase (30%, 0.3–2.2 mm) with minor amphibole	Matrix is primarily microcrystalline texture and dominated by plagioclase and quartz (15%, 0.01–0.1 mm)

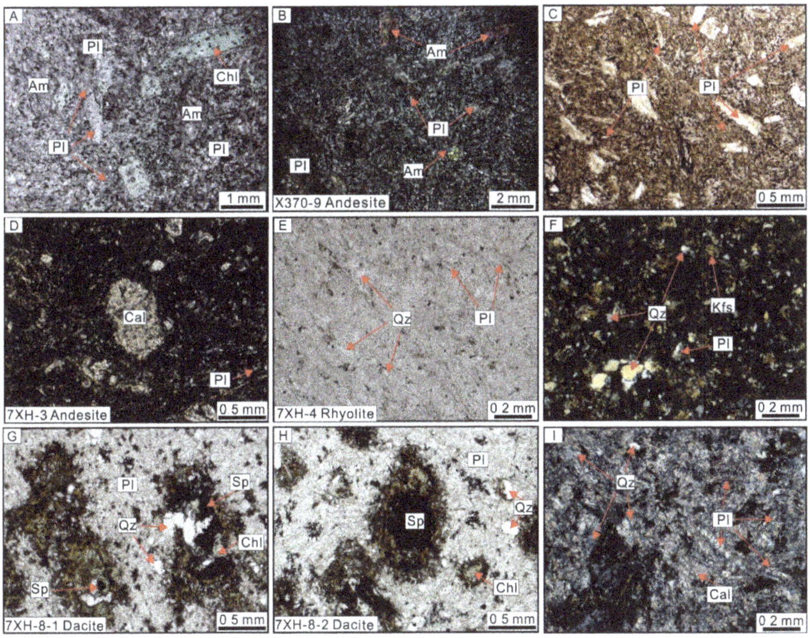

Figure 3. Photomicrographs of the volcanic rocks from Xiaohongshilazi deposit. (**A**,**B**) Andesite (X370-9; A under polarized light; B under cross-polarized light); (**C**,**D**) andesite (7XH-3; C under polarized light; D under cross-polarized light); (**E**,**F**) rhyolite (7XH-4; E under polarized light; F under cross-polarized light); (**G**) dacite (7XH-8-1; polarized light); (**H**,**I**) dacite (7XH-8-2; H under polarized light; I under cross-polarized light). Abbreviations: Sp: sphalerite; Am: amphibole; Pl: plagioclase; Kfs: K-feldspar; Qz: quartz; Chl: chlorite; Cal: calcite.

According to the cross-cutting relationships of the faults, the structures in the ore district can be classified into two groups: (1) the major N–S-trending faults, which controlled the distribution of vein-type Pb–Zn–(Ag) mineralization and intrusions; and (2) the post-mineralization NE–SW-trending faults, which cut the vein-type orebody and intrusions (Figure 1C). The intrusions in the ore district, including diorite porphyry and granodiorite porphyry, with minor diorite and diabase dikes, are commonly N–S-trending and parallel to the vein-type Pb–Zn–(Ag) orebodies (Figure 1C). The granodiorite porphyry (203.6 ± 1.8 Ma, U–Pb zircon) within the internal Pb–Zn mineralization (Figure 4F–G) cut the diorite porphyry (225.6 ± 5.1 Ma, U–Pb zircon), and was closely associated with the vein-type Pb–Zn–(Ag) mineralization of the Xiaohongshilazi deposit (195 ± 17 Ma; sulfide Rb–Sr dating) [20].

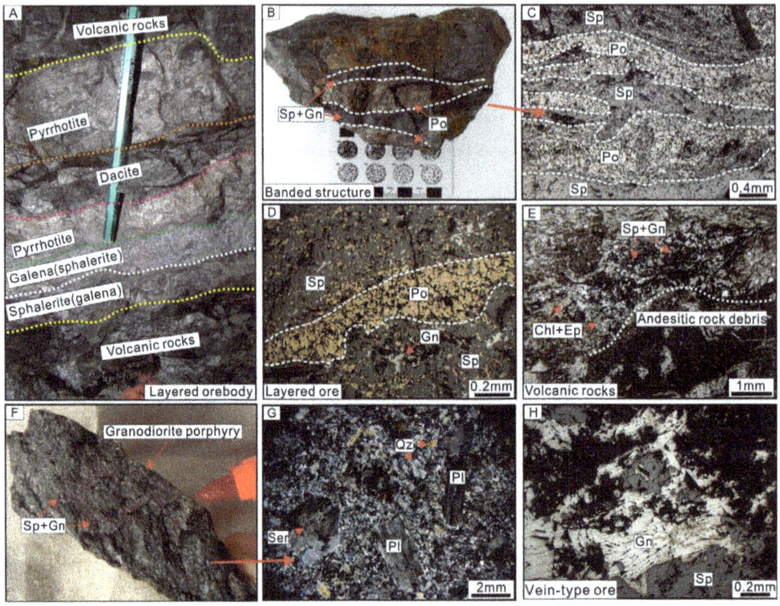

Figure 4. Representative photographs showing the texture and structure of orebodies in the Xiaohongshilazi Pb–Zn–(Ag) deposit. (**A**) Layered Pb–Zn orebodies, displaying a banded structure and interbedding with the volcanic rocks; (**B**) banded structure of the layered ore; (**C**) pyrrhotite interbedded with sphalerite in the layered ore (reflected light); (**D**) pyrrhotite interbedded with sphalerite and minor galena in the layered ore (reflected light); (**E**) chloritization and epidotization associated with layered Pb–Zn mineralization in the volcaniclastic rock (polarized light); (**F**) veinlet Pb–Zn mineralization in the granodiorite porphyry; (**G**) granodiorite porphyry (cross-polarized light); (**H**) sphalerite replaced by galena during the major stage of the vein-type mineralization (reflected light). Abbreviations: Sp: sphalerite; Gn: galena; Po: pyrrhotite; Chl: chlorite; Ep: epidote; Qz: quartz; Pl: plagioclase; Ser: sericite.

Forty-one Pb–Zn–(Ag) orebodies in the ore district were categorized into layered orebodies (n = 19) and vein-type orebodies (n = 22). The layered Pb–Zn orebodies that were mainly hosted within and spatially dominated by volcanic rocks were cut by vein-type Pb–Zn–(Ag) orebodies (Figure 2), which had lengths of 100–400 m, depths of 110–280 m, and thicknesses of 1–3 m in the ore district [18,19,22]. In addition, the ore fabrics, mineral assemblages, and wall-rock alteration showed obvious differences between the layered and vein-type ores. Layered ore occurred mainly as a banded structure (Figure 4A–D), and the vein-type ore displayed a vein structure (Figure 2C). The metallic minerals in layered ore were mainly composed of sphalerite, galena, pyrrhotite (Figure 4A–D), pyrite, and minor chalcopyrite, and those in the vein-type ore dominantly consisted of galena, sphalerite (Figure 4F,H), pyrite, and silver. In contrast with the metallic minerals of layered ore, pyrrhotite and chalcopyrite were not found in the vein-type ore. Moreover, the layered ore had more sphalerite, while the vein-type ore had more galena. Accompanied by Pb–Zn mineralization in the layered ore, most of the crystal and rock fabric in the volcanic rocks, including biotite, amphibole, plagioclase, and andesitic rock fabric, had been altered to chlorite, epidote, and calcite minerals (Figure 4E). The wall-rock alteration related to the layered Pb–Zn mineralization would have been controlled by the reaction between the marine volcanic rocks and hot brine, because the layered orebodies were interbedded with the marine volcanic rocks (Figure 4A). The volcanic rocks and the layered orebodies were cut by the vein-type Pb–Zn–(Ag) mineralization along fractures and fissures that mainly occurred at the edge of the granodiorite porphyry (Figure 2B,C). In the granodiorite porphyry, the primary plagioclase, biotite, and amphibole had been altered to become hydrothermal quartz, chlorite, epidote, sericite, and calcite (Figure 4G), which was related to vein-type Pb–Zn–(Ag) mineralization and most likely dominated by the reaction between ore-bearing hydrothermal fluids and rock-forming minerals of the granodiorite porphyry. The silicification related to vein-type Pb–Zn–(Ag) mineralization, forming abundant quartz-polymetallic sulfide veins, was not found in the layered Pb–Zn mineralization, indicating that the ore geneses of the layered and vein-type orebodies were different.

3. Sample and Analytical Methods

3.1. LA–ICP–MS Zircon U–Pb Dating

More than 100 zircon grains from each volcanic rock in the Xiaohongshilazi ore district (Table 1), were extracted using magnetic and heavy liquid separation techniques, and then handpicked under a binocular microscope at the Integrity Geological Service Corporation in Langfang City, Heibei Province, China. The handpicked zircon grains were mounted in the epoxy resin disk and polished to about half their thickness for exposing crystal cores. Photomicrographs of the zircon grains were taken under reflected and transmitted light. Cathodoluminescence (CL) images of the zircon grains were obtained using a JEOL scanning electron microscope, and used to exhibit their internal structures and determine the spot locations for zircon U–Pb isotope analyses. The spot locations should be selected as far as possible on the growth zoning of zircon grains without inclusions and cracks (Figure 5).

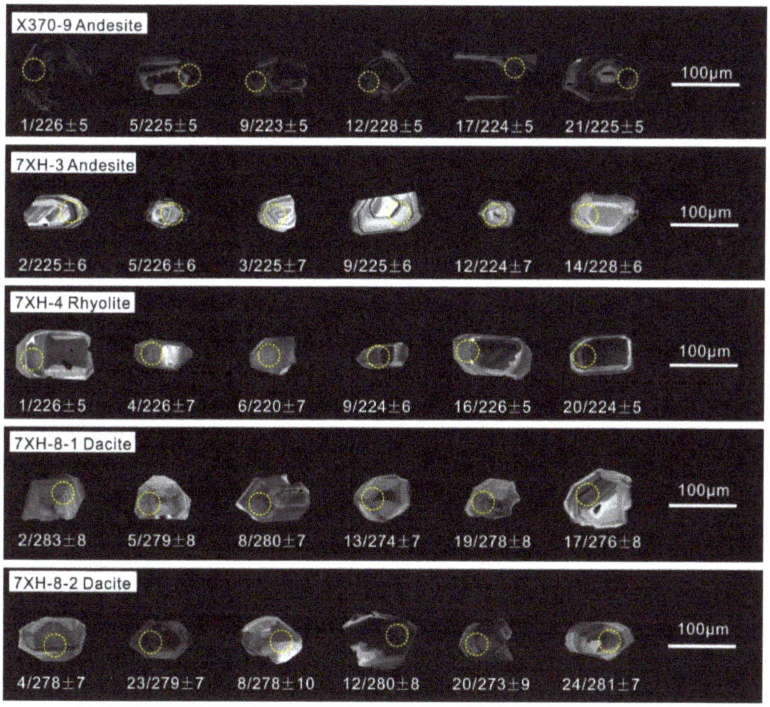

Figure 5. Cathodoluminescence (CL) images of representative zircon grains from the andesite (X370-9 and 7XH-3), rhyolite (7XH-4), and dacite (7XH-8-1 and 7XH-8-2) of the Xiaohongshilazi deposit. Yellow circles denote the locations for U–Pb dating.

Laser ablation (LA)–ICP–MS zircon U–Pb dating for more than 20 zircon grains from each volcanic rock was completed at the Key Laboratory of Mineral Resources Evaluation in Northeast Asia, Ministry of Land and Resources of China, Changchun, Jilin Province, China. The instrument for LA–ICP–MS zircon U–Pb analyses couples a quadrupole ICP–MS (Agilient 75,00c) and 193-nm ArF Excimer laser (COMPexPro 102, Coherent, DE, Saxonburg, Pennsylvania) with the automatic positioning system. The diameter of the laser spot was 32 μm. One zircon standard (91,500) and National Institute of Science and Technology (NIST) 610 reference standard were analyzed after each set of five unknown analyses. The external zircon standard (91,500) was used to correct for isotope ratio fractionation. The NIST610 reference standard was used in the calculations of element concentrations and Si was used as an internal standard. Uncertainties on isotope ratios and ages are presented as $\pm 2\sigma$. The analytical process was described in detail by Hou et al. [36]. The isotope data were calculated using the Glitter 4.0 [37]. Concordia and weighted-mean age diagrams were produced using Isoplot 3.0 [38]. Corrections for common lead were carried out by the method of Andersen [39].

3.2. *Major and Trace Element Concentrations*

Removing the weathered surfaces, 10 fresh rock samples from the rhyolite (7XH-4) and dacite (7XH-8-1 and 7XH-8-2) were crushed, cleaned repeatedly with deionized water, pulverized, and ground to 200 mesh using an agate mill. Major and trace elements analyses were completed at the State Key Laboratory of Ore Deposit Geochemistry, Institute of Geochemistry, Chinese Academy of Sciences (IGCAS), Guiyang, China. Major and trace element concentrations were determined by the PANalytical Axios-advance X-ray fluorescence spectrometer (XRF) and Perkin-Elmer-ELAN 6000 inductively coupled plasma Mass Spectrometer (ICP–MS), respectively. The analysis error is less than 5%. Detailed experimental operation process and method were described by Qi and Zhou [40].

4. Analytical Results

4.1. *Zircon U–Pb Age*

LA–ICP–MS zircon U–Pb ages for the andesite (X370-9 and 7XH-3), rhyolite (7XH-4) and dacite (7XH-8-1 and 7XH-8-2) are listed in Supplementary Table S1 and shown in Figure 6. All zircon grains from the five volcanic rock samples were euhedral–subhedral in shape and transparent, and mostly displayed oscillatory growth zoning in CL images (Figure 5) and high Th/U ratios (0.34–0.87; Supplementary Table S1), indicating their magmatic origin [41–43].

For sample X370-9, except for one analytical spot showing a $^{206}Pb/^{238}U$ age of 279 ± 6 Ma, 20 analytical spots yielded $^{206}Pb/^{238}U$ ages ranging from 223 ± 5 Ma to 228 ± 5 Ma, with a concordia U–Pb age of 225.0 ± 1.1 Ma (Figure 6A) and a weighted mean age of 225.0 ± 2.2 Ma (MSWD = 0.055, n = 20; Figure 6B). The $^{206}Pb/^{238}U$ ages of 18 analytical spots from sample 7XH-3 varied from 222 ± 9 Ma to 228 ± 6 Ma, with a concordia U–Pb age and a weighted mean age of 225.3 ± 1.5 Ma (Figure 6C) and 225.5 ± 2.9 Ma (MSWD = 0.032, n =18; Figure 6D), respectively. Twenty-five analytical spots from sample 7XH-4 yielded $^{206}Pb/^{238}U$ ages ranging from 220± 7 Ma to 226 ± 7 Ma, and a concordia U–Pb age of 224.7 ± 1.2 Ma (Figure 6E) and a weighted mean age of 224.7 ± 2.3 Ma (MSWD = 0.063, n =25; Figure 6F). From sample 7XH-8-1, the $^{206}Pb/^{238}U$ ages of 20 analytical spots ranged from 274 ± 7 Ma to 284 ± 7 Ma, yielding a concordia U–Pb age of 278.1 ± 1.8 Ma (Figure 6G) and a weighted mean age of 278.3 ± 3.4 Ma (MSWD = 0.16, n = 20; Figure 6H). Twenty-four analytical spots from sample 7XH-8-2 yielded $^{206}Pb/^{238}U$ ages varying from 273 ± 9 Ma to 282 ± 13 Ma, which gave a concordia U–Pb age of 278.3 ± 1.8 Ma (Figure 6I) and a weighted mean age of 278.3 ± 3.4 Ma (MSWD = 0.063, n = 24; Figure 6J). The older zircon grain (279 ± 6 Ma) from sample X370-9 was interpreted to be the crystallization age of inherited or captured zircon entrained by the magma.

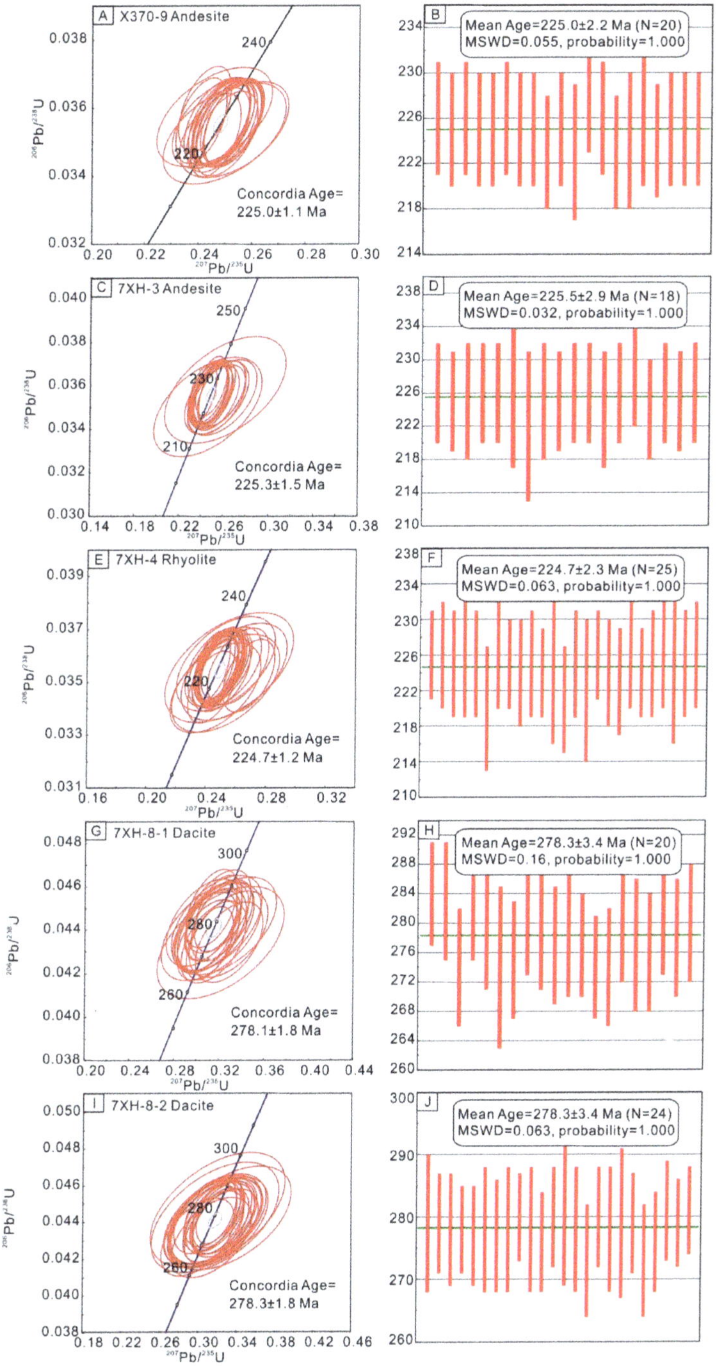

Figure 6. LA–ICP–MS zircon U–Pb diagrams of the concordia age and weighted mean age for the andesite ((**A**,**B**) X370-9; (**C**,**D**) 7XH-3), rhyolite ((**E**,**F**) 7XH-4), and dacite ((**G**,**H**) 7XH-8-1; (**I**,**J**) 7XH-8-2) from the Xiaohongshilazi deposit.

4.2. Major and Trace Element

The whole-rock major and trace element concentrations of the rhyolite (7XH-4) and dacite (7XH-8-1 and 7XH-8-2) from the Xiaohongshilazi deposit are listed in Supplementary Table S2. The rhyolite had high SiO_2 (73.70–76.44 wt.%) and $Na_2O + K_2O$ (7.31–7.64 wt.%), and contained a Al_2O_3 content of 12.76–13.09 wt.%, a TFe_2O_3 content of 1.54–2.89 wt.%, a CaO content of 0.22–0.47 wt.%, a TiO_2 content of 0.12–0.20 wt.%, and a MgO content of 0.68–0.87 wt.%, with Mg# [Mg# = 100 × $Mg^{2+}/(Mg^{2+} + TFe^{2+})$] values varying from 31.62 to 50.51. Compared with the rhyolite, the dacite possessed relatively low SiO_2 ratios ranging from 62.54 to 65.02 wt.% and $Na_2O + K_2O$ varying from 5.23 to 5.45 wt.%, but relatively high Al_2O_3 (15.45–15.71 wt.%), TFe_2O_3(7.97–9.16 wt.%), CaO (0.82–1.25 wt.%), and TiO_2 (0.56–0.68 wt.%), and similar MgO (0.76–0.81 wt.%), with lower Mg# values ranging from 14.88 to 16.12. Their A/CNK ratios varied from 1.30 to 1.36 and ranged from 1.66 to 1.82, respectively, which is indicative of peraluminous rocks.

In the chondrite-normalized rare earth element (REE) patterns (Figure 7A), the rhyolite and dacite samples from the Xiaohongshilazi deposit had different degrees of enrichment in light rare-earth elements (LREEs), relative to heavy rare-earth elements (HREEs). The dacite samples [LREE/HREE = 9.63–13.80, $(La/Yb)_N$ = 9.02–15.55] were more fractionated than the rhyolite samples [LREE/HREE =5.37–6.71, $(La/Yb)_N$ = 3.94–5.15]. The rhyolite had obviously negative Eu anomalies (δEu = 0.04–0.13), while the dacite showed slightly negative Eu anomalies (δEu = 0.60–0.65). In the primitive mantle-normalized geochemical patterns (Figure 7B), both of them were enriched in large-ion lithophile elements (LILEs, e.g., Rb, Ba, and K) with obviously negative Nb, Sr, P, and Ti anomalies. The rhyolite, relative to the dacite, displayed more intense depletion in high-field-strength elements (HFSEs, e.g., P and Ti) and more enrichment in Zr and Hf. The rhyolite and dacite samples were divided into the subalkaline series, and mainly plotted into the rhyolite and dacite regions, respectively, on the Nb/Y vs. Zr/TiO_2*0.0001 diagram (Figure 8A). They were classified as high-K calc-alkaline and shoshonite series, respectively, on Co vs. Th and SiO_2 vs. K_2O diagrams (Figure 8B,C).

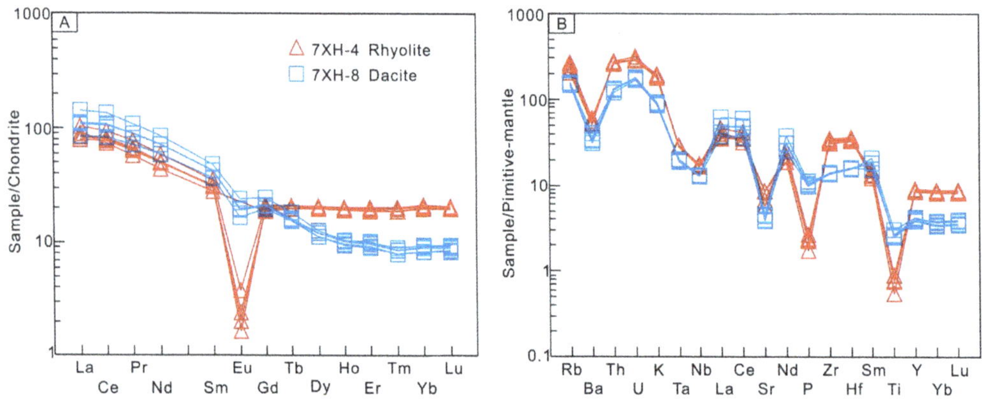

Figure 7. (**A**) Chondrite-normalized rare earth element (REE) patterns and (**B**) primitive mantle-normalized geochemical patterns for the rhyolite and dacite from the Xiaohongsilazi deposit. Chondrite and primitive-mantle values are from Boynton [44] and Sun and McDinough [45], respectively.

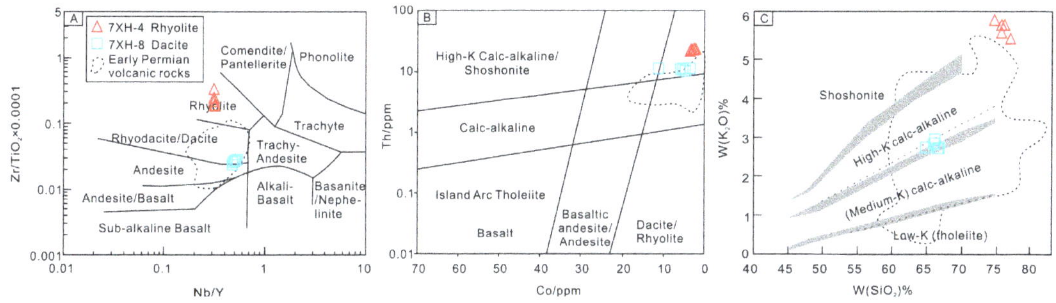

Figure 8. (**A**) Nb/Y vs. $Zr/TiO_2 \times 0.0001$ diagram (modified from Winchester and Floyd [46]); (**B**) Co vs. Th diagram (modified from Hastie et al. [47]); (**C**) SiO_2 vs. K_2O diagram (modified from Peccerillo and Taylor [48]). The data of early Permian volcanic rocks in the area are from Hou [16], Cao [49], and Yu et al. [50].

5. Discussion

5.1. Age of Magmatic Hydrothermal Events and Layered Pb–Zn Mineralization

The dacite samples from Dongfeng and Xifeng mine of Xiaohongshilazi ore district yielded concordia U–Pb ages of 278.1 ± 1.8 Ma and 278.3 ± 1.8 Ma, respectively, indicating that the dacite formed during the early Permian. The andesite and rhyolite samples yielded concordia U–Pb ages of 225.0 ± 1.1 Ma, 225.3 ± 1.5 Ma and 224.7 ± 1.2 Ma, respectively, and were identical within error to the age of diorite porphyry from the Xiaohongshilazi deposit (225.3 ± 5.1 Ma) (Yang et al., 2020). This would suggest that the intermediate-silicic magmatic hydrothermal event in the area occurred during late Triassic time. In addition, one zircon grain from the andesite furnished a $^{206}Pb/^{238}U$ age of 279 ± 6 Ma comparable to that of the dacite samples, indicating that the magmatic zircon from the early Permian was inherited or captured. The ages of the dacite samples (~278.3 Ma) were obviously older than the andesite and rhyolite samples (~225 Ma). These data combined with the distribution of strata in different epoch and their rock types in central Jilin Province [13,51] meant that the dacite samples (~278.3 Ma) could be classified as Daheshen Formation, and the andesite and rhyolite samples (~225 Ma) could be confirmed as Sihetun Formation. The volcanic rocks from the Sihetun Formation were first reported in this area, which provided new geological information for regional geological mapping of central Jilin Province.

As mentioned above, the layered Pb–Zn orebodies were mainly hosted within and controlled by Daheshen Formation and Sihetun Formation volcanic rocks. Nevertheless, the volcanic rocks from the Daheshen Formation were spatially, temporally, and genetically related to the layered Pb–Zn mineralization, which is evidenced by the following: (1) Dacite samples of the Daheshen Formation from Dongfeng and Xifeng mine in the Xiaohongshilazi ore district were all interbedded with the layered orebodies (Figures 2B and 4A). (2) The volcanic rocks from the Daheshen Formation and the layered orebodies were overlain by the volcanic rocks from the Sihetun Formation (~225Ma), and both of them were cut by the vein-type Pb–Zn–(Ag) orebodies and the granodiorite porphyry (~200Ma) (Figure 2A) [20]. (3) No Pb–Zn mineralization was found in the rhyolite and andesite from the Sihetun Formation (Figure 3A–F). (4) In addition to the Xiaohongshalazi deposit, several newly discovered VMS-type polymetallic deposits in this region all occurred in the marine volcanic rocks from the early Permian, such as the Hongtaiping Pb–Zn–Cu deposit, the Dongfengnanshan Pb–Zn–Cu deposit, etc. The zircon U-Pb ages of dacite samples from the Daheshen Formation in the Xiaohongshilazi deposit was consistent with the sphalerite Rb–Sr isochron age of the layered ore from Hongtaiping deposit [52]. In summary, the volcanic rocks from the Daheshen Formation have a closely genetic relationship with the layered Pb–Zn mineralization, suggesting that the layered Pb–Zn mineralization of the Xiaohongshilazi deposit occurred during the early Permian (~278 Ma).

In light of the research of the vein-type Pb–Zn–(Ag) mineralization in the Xiaohongshilazi deposit (~200 Ma) [20], it can be concluded that there were three stages of tectono-magmatic–hydrothermal events that occurred in the Xiaohongshilazi area from the early Permian to the early Jurassic: (1) the early Permian (~278 Ma) forming of the Daheshen Formation marine volcanic rocks related to the layered Pb–Zn mineralization; (2) the late Triassic (~225 Ma) forming of the Sihetun Formation volcanic rocks and the diorite porphyry dikes; (3) the early Jurassic (~200 Ma) forming of the granodiorite porphyry associated with the vein-type Pb–Zn–(Ag) mineralization.

5.2. Tectonic Setting and Associated Mineralization

The dacite from the Daheshen Formation genetically related to the layered Pb–Zn mineralization was geochemically distinguished by high SiO_2, high K, and high Al_2O_3 composition, and was identified as a peraluminous high-K calc-alkaline rock (σ < 3.3 wt.%, Figure 8B,C). The dacite was characterized by an enrichment in LREEs and LILEs (e.g., Rb, Ba, and K), and a depletion in HFSEs (e.g., P and Ti), suggesting that this dacite was formed in a subduction-related tectonic environment as island arc rocks. The Th/Nb (1.12–1.21) and La/Nb (2.8–4.7) ratios (Supplementary Table S2) presented arc magmas that could have been contributed by crustal source contents [53–55]. Meanwhile, the dacite had obviously negative Nb anomalies and slightly negative Eu anomalies (δEu = 0.60–0.65), differing markedly from mantle-derived rocks with strong negative Eu anomalies, indicating that the magma was derived from the crust [56]. The Mg# values (14.88–16.12) of the dacite were lower than those of mantle-derived rocks (Mg# > 40), further implying that it was formed by the partial melting of basaltic lower crust [57]. Moreover, the zircon $\varepsilon_{Hf}(t)$ values of the volcanic rocks from the Daheshen Formation were all positive ($\varepsilon_{Hf}(t)$ = + 0.23~ + 10.60), and Hf two-stage model ages (T_{DM2}) ranged from 1384 Ma to 662 Ma [49,50], suggesting that the original magma was derived from the partial melting of the depleted basaltic lower crust newly accreted during the Meso-Neoproterozoic, which is generally consistent with the Proterozoic accretionary event in this region [58]. In summary, we conclude that the volcanic rocks from the Daheshen Formation most likely originated from the partial melting of depleted lower crust.

The rhyolite from the Sihetun Formation was classified as a peraluminous shoshonitic volcanic rock (Figure 8B,C) with high SiO_2 and high-K (Na_2O/K_2O = 0.28–0.34), and was characterized by an intense enrichment in LREEs and LILEs and a depletion in HFSEs, as well as obviously negative Eu anomalies (δEu = 0.04–0.13) and negative Nb, Sr, P, and Ti anomalies. These geochemical features were similar to A-type rhyolites, and implied that the rhyolitic magma was likely derived from the partial melting of the lower crust, accompanied by the crystallization differentiation of minerals, such as plagioclase, apatite, sphene, etc. [59–61].

The dacite samples are plotted within the field of the volcanic arc granites on the Yb vs. Ta and Y + Nb vs. Rb diagrams (Figure 9A,B), mainly near the region of the active continental margin (continental arc) on the Ta/Yb vs. Th/Yb diagram (Figure 9C). The dacite samples displayed LILE-enrichment and HFSE-depletion, as well as obvious Ta–Nb–Sr–P–Ti troughs, which provide a scenario for a subduction setting similar to the active continental margin. Meanwhile, the tectonic setting of Hongtaiping and Dongfengnanshan VMS-type Pb–Zn–Cu deposits, which had the same ore genesis as the layered Pb–Zn mineralization of the Xiaohongshilazi deposit, indicated an active continental margin [15,50]. Furthermore, a large amount of geochronology data, geochemistry characteristics, and regional tectonic evolution indicated that this region was under an active continental margin arc setting in the early Permian, which related to the southward subduction of the Paleo-Asian oceanic plate beneath the North China Craton [16,17,27–32,49,50,62]. We propose, therefore, the dacite and associated layered Pb–Zn mineralization formed in the subduction setting of the Paleo-Asian oceanic plate in the early Permian (~278 Ma; Figure 10A). The rhyolite samples were plotted close to the field of the within-plate granites and post-orogenic setting in the discrimination diagrams of Yb vs. Ta diagram (Figure 9A), Y + Nb vs. Rb diagram (Figure 9B) and R1 vs. R2 diagram (Figure 9D), respectively. Moreover, the rhyolite displayed an affinity to A-type rhyolites in its geochemical composition, proving that it was formed in the post-collisional extensional setting after the final closure of the Paleo-Asian Ocean. In addition, the existence of bimodal volcanic rocks and A-type rhyolites in the eastern Heilongjiang–Jilin provinces during the late Triassic [5], as well as Hongqiling magmatic Cu–Ni sulfide deposit in the XMOB (~223 Ma) [8,63], further indicate that this region evolved into a period of post-collisional extensional setting involving lithosphere delamination in the late Triassic (Figure 10C). The granodiorite porphyry associated with the vein-type Pb–Zn–(Ag) mineralization was emplaced in the initial subduction stage of the Paleo-Pacific plate in the early Jurassic (~200 Ma; Figure 10D) [20].

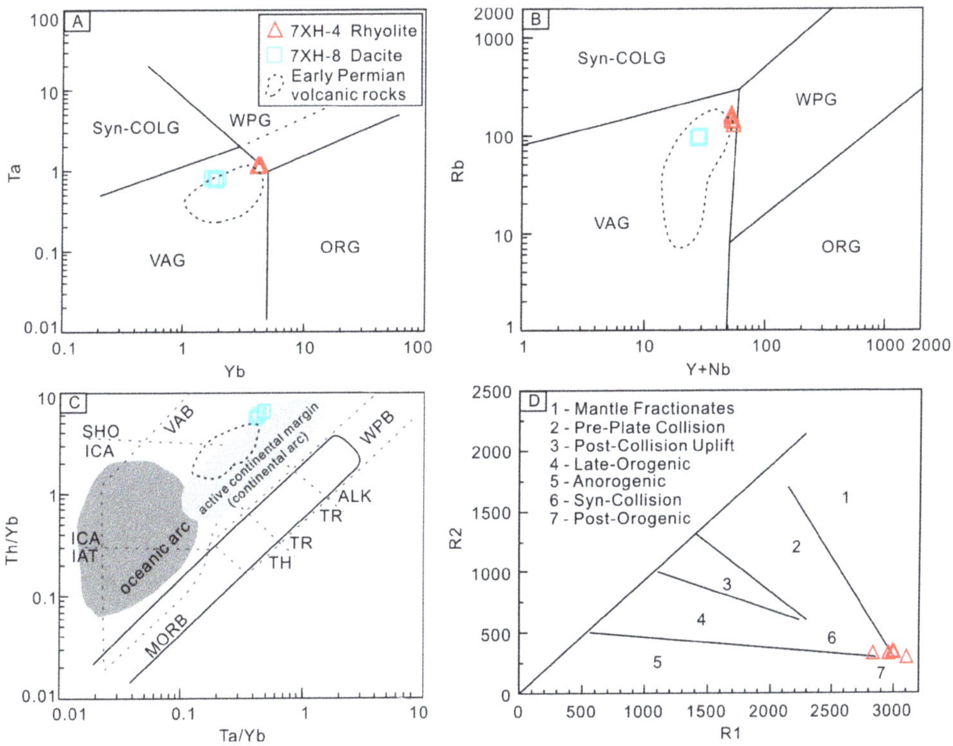

Figure 9. (**A**) Yb vs. Ta diagram (modified from Pearce et al. [64]); (**B**) Y + Nb vs. Rb diagram (modified from Pearce et al. [64]); (**C**) Ta/Yb vs. Th/Yb diagram (modified from Pearce et al. [65]); (**D**) R1 vs. R2 (modified from Batchelor and Bowden [66]); R1 = 4Si−11(Na + K)−2(Fe + Ti) and R2 = 6Ca + 2 Mg + Al). The data of early Permian volcanic rocks in the area are from Hou [16], Cao [49], and Yu et al. [50]. Abbreviations: VAG: volcanic arc granites; Syn-COLG: syn-collisional granites; WPG: within-plate granites; ORG: ocean ridge granites; SHO: shoshonitic series; ICA: island calc-alkaline series; IAT: island arc tholeiites; VAB: volcanic arc basalts; WPB: within-plate basalts; MORB: mid-ocean ridge basalts; ALK: alkalic series; TR: transitional series; TH: tholeiitic series.

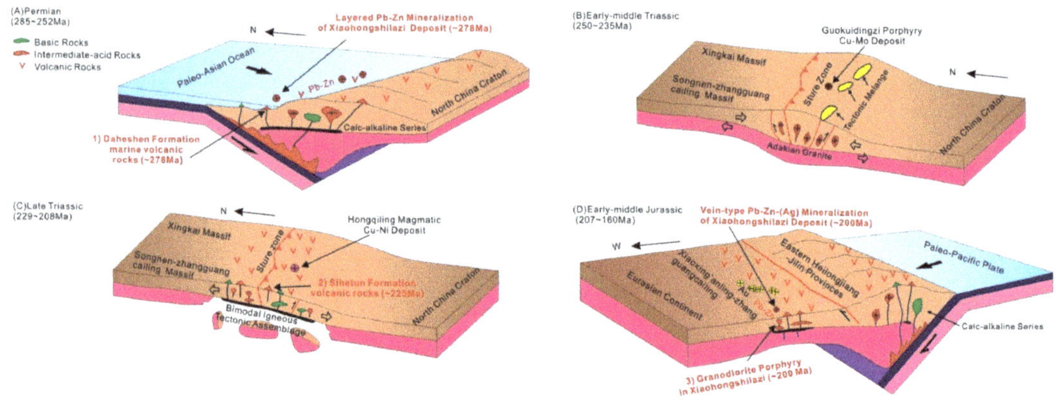

Figure 10. Cartoon block diagrams showing the tectonic evolution and associated mineralization in this region (modified from Yang [13]).

5.3. *Implication for the Late Paleozoic Pb–Zn Polymetallic Mineralization*

Located on the intersection between the Central Asian Orogenic Belt and the eastern segments of NCC, the central Jilin Province successively experienced the evolution and transition of the Paleo-Asian Ocean and the Paleo-Pacific Ocean metallogenic domains [1–5,30–35]. In the Permian, the sustained subduction of the Paleo-Asian oceanic plate beneath the North China Craton formed the marine volcanic rocks and associated VMS-type polymetallic mineralization (Figure 10A). This area experienced the transition from the collision and closure of the Paleo-Asian Ocean to post-collision extension during the Triassic (Figure 10B,C). Large numbers of Mesozoic polymetallic deposits have been found, including porphyry Cu–Mo deposits (e.g., Guokuidingzi) and magmatic Cu–Ni sulfide deposits (e.g., Hongqiling) [8,58,63]. In the early–middle Jurassic, the regional tectonic setting transformed into the subduction of the Paleo-Pacific oceanic plate beneath the Eurasian continent (Figure 10D), forming mesothermal hydrothermal lode gold deposits (e.g., Cuyu), skarn-type gold deposits (e.g., Guanma), and hydrothermal vein-type Pb–Zn–(Ag) deposits (e.g., Xiaohongshilazi). Nevertheless, few Paleozoic deposits have been discovered. The newly discovered late Paleozoic Pb–Zn polymetallic deposits in Northeast China, including the layered polymetallic mineralization of the Xiaohongshilazi, Hongtaiping, and Dongfengnanshan, are along the Solonker–Xra Moron–Changchun–Yanji suture zone (Figure 1B) and were formed in the active continental margin setting induced by the subduction of the Paleo-Asian Ocean (Figure 10A). The discovery of these deposits is not only helpful to the study of Paleozoic mineralization in this area, but also provides theoretical support for the exploration of the same-type polymetallic deposits and the study of the evolution and associated metallogenesis of the Paleo-Asian Ocean in Northeast China.

At present, the discovered VMS-type mineralization generally occurred in the marine volcanic rocks during the Permian, and no Mesozoic VMS-type deposits have been found in the eastern Heilongjiang–Jilin Provinces, which may be related to the local extension or rifting tectonics under the subduction of the Paleo-Asian oceanic plate in this region during the Paleozoic. Early Permian marine volcanic rocks are the main ore-bearing strata of VMS-type layered mineralization in this area, and they have extremely high zinc background anomaly values [13]. Therefore, the early Permian marine volcanic rocks have large prospecting potential and could be important prospecting indicators for VMS-type polymetallic deposits in this region. In particular, several lead–zinc polymetallic deposits have been found in the marine volcanic rocks of the early Permian Miaoling Formation and Daheshen Formation (Figure 1B). These include the layered Pb–Zn–Cu mineralization of the Hongtaiping and Dongfengnanshan VMS-type deposits in the Yanbian area, eastern

Jilin Province, and the layered Pb–Zn mineralization of the Xiaohongshilazi deposit in Panshi City, central Jilin Province [15,52]. All of these occurred in the early Permian marine volcanic rocks, as seen in Figure 1B. These early Permian marine volcanic rocks related to the Paleo-Asian Ocean evolution are mainly distributed in central (Panshi, Yongji) and eastern (Wangqing, Longjing, Kaishantun) Jilin Province. They represent important lead–zinc polymetallic mineralization areas and are prospecting indicators for VMS-type deposits.

6. Conclusions

(1) The layered Pb–Zn mineralization was interbedded with the volcanic rocks from the Daheshen Formation. It manifested mainly as a banded structure and had more sphalerite than the vein-type mineralization. The dacite samples from the Daheshen Formation yielded concordia ages of 278.1 ± 1.8 Ma and 278.3 ± 1.8 Ma, respectively, and were spatially, temporally, and genetically related to the layered Pb–Zn mineralization, indicating that they were formed during the early Permian.
(2) The andesite and rhyolite from the Sihetun Formation yielded ages of 225.0 ± 1.1 Ma, 225.3 ± 1.5 Ma, and 224.7 ± 1.2 Ma, respectively, and were unrelated to the layered Pb–Zn mineralization and first reported in this region.
(3) The Daheshen Formation and Sihetun Formation volcanic rocks were formed in the active continental margin arc setting induced by the subduction of the Paleo-Asian oceanic plate during the early Permian and the post-collisional extensional setting after the final closure of the Paleo-Asian Ocean during the late Triassic, respectively.
(4) Three tectono-magmatic events occurred in the Xiaohongshilazi area during the early Permian (278 Ma), the late Triassic (225 Ma), and the early Jurassic (200 Ma). The layered Pb–Zn mineralization was formed during the early Permian (278 Ma), and the vein-type Pb–Zn–(Ag) mineralization was formed during the early Jurassic (200 Ma).
(5) Early Permian marine volcanic rocks, closely related to the layered Pb–Zn mineralization of VMS-type polymetallic deposits, are important prospecting indicators for similar polymetallic deposits in this region.

Supplementary Materials: The following supporting information can be downloaded at: https://www.mdpi.com/article/10.3390/min13111371/s1, Table S1: LA–ICP–MS zircon U–Pb dating data of the volcanic rocks in the Xiaohongshilazi deposit; Table S2: Whole-rock major and trace element data of the rhyolite and dacite in the Xiaohongsilazi deposit.

Author Contributions: Conceptualization: Q.Y., Q.S. and Z.Y.; field investigation: Q.Y. and Y.R.; experimental analysis: Q.Y. and Z.Y.; software: Q.S.; validation: Q.Y. and Q.S.; resources: Y.R. and Q.Y.; data curation: Q.Y.; writing—original draft preparation: Q.S.; writing—review and editing: Q.Y.; visualization: Q.S.; funding acquisition: Q.Y. and Y.R. All authors have read and agreed to the published version of the manuscript.

Funding: This work was financially supported by the Natural Science Foundation of Jilin Province (no. 20230101089JC) and China Postdoctoral Science Foundation (no. 2022M721305), and the Dynamic Evaluation Project of Gold Resource Potential in Eastern Jilin-Heilongjiang Area (DD20230373).

Data Availability Statement: Not applicable.

Acknowledgments: We would like to thank the leaders and geologists of the Xiaohongshilazi ore district for their support of our fieldwork.

Conflicts of Interest: The authors declare no conflict of interest.

References

1. Wu, F.Y.; Sun, D.Y.; Li, H.M.; Jahn, B.M.; Wilde, S. A-type granites in northeastern China: Age and geochemical constraints on their petrogenesis. *Chem. Geol.* **2002**, *187*, 143–173. [CrossRef]
2. Wu, F.Y.; Sun, D.Y.; Ge, W.C.; Zhang, Y.B.; Grant, M.L.; Wilde, S.A.; Jahn, B.M. Geochronology of the Phanerozoic granitoids in northeastern China. *J. Asian Earth Sci.* **2011**, *41*, 1–30. [CrossRef]
3. Zeng, Q.; Liu, J.; Chu, S.; Wang, Y.; Sun, Y.; Duan, X.; Zhou, L. Mesozoic molybdenum deposits in the East Xingmeng orogenic belt, northeast China: Characteristics and tectonic setting. *Int. Geol. Rev.* **2012**, *54*, 1843–1869. [CrossRef]

4. Zeng, Q.; Liu, J.; Qin, K.; Fan, H.; Chu, S.; Wang, Y.; Zhou, L. Types, characteristics, and time-space distribution of molybdenum deposits in China. *Int. Geol. Rev.* **2013**, *55*, 1311–1358. [CrossRef]
5. Xu, W.L.; Pei, F.P.; Wang, F.; Meng, E.; Ji, W.Q.; Yang, D.B.; Wang, W. Spatial-temporal relationships of Mesozoic volcanic rocks in NE China: Constraints on tectonic overprinting and transformations between multiple tectonic systems. *J. Asian Earth Sci.* **2013**, *74*, 167–193. [CrossRef]
6. Wang, H.; Ren, Y.S.; Zhao, H.L.; Ju, N.; Qu, W.J. Re–Os dating of molybdenite from the Liushengdian Molybdenum deposit in Antu Area of Jilin Province and its geological significance. *Acta Geosci. Sin.* **2011**, *32*, 707–715. (In Chinese with English Abstract)
7. Zhang, Y.; Sun, J.G.; Chen, Y.J.; Zhao, K.Q. Re–Os and U–Pb geochronology of porphyry Mo deposits in central Jilin Province: Mo ore-forming stages in northeast China. *Int. Geol. Rev.* **2013**, *14*, 1763–1785. [CrossRef]
8. Han, C.M.; Xiao, W.J.; Zhao, G.C.; Sun, B.X.; Ao, S.J.; Zhang, J.E.; Wan, B.; Wang, Z.M.; Ding, J.X.; Qu, W.J.; et al. Re–Os isotopic age of the Hongqiling Cu–Ni sulfide deposit in Jilin Province, NE China and its geological significance. *Resour. Geol.* **2014**, *64*, 247–261. [CrossRef]
9. Deng, J.; Yuan, W.; Carranza, E.J.M.; Yang, L.; Wang, C.; Yang, L.; Hao, N. Geochronology and thermochronometry of the Jiapigou gold belt, northeastern China: New evidence for multiple episodes of mineralization. *J. Asian Earth Sci.* **2014**, *89*, 10–27. [CrossRef]
10. Zhang, Y.; Xing, S.W.; Song, Q.H.; Wang, Y.; Yu, Z.T.; Du, X.H.; Ma, Y.B.; Zhang, Z.J. Re–Os and U–Pb geochronology of porphyry and skarn types copper deposits in Jilin Province, NE China. *Resour. Geol.* **2015**, *65*, 394–404. [CrossRef]
11. Zhou, L.L.; Zeng, Q.D.; Liu, J.M.; Friis, H.; Zhang, Z.L.; Duan, X.X. Ore genesis and fluid evolution of the Daheishan giant porphyry molybdenum deposit, NE China. *J. Asian Earth Sci.* **2015**, *97*, 486–505. [CrossRef]
12. Wang, Z.G.; Wang, K.Y.; Wan, D.; Yassa, K.; Yang, T.N.; Liang, Y.H. Metallogenic age and hydrothermal evolution of the Jidetun Mo deposit in central Jilin Province, northeast China: Evidence from fluid inclusions, isotope systematics, and geochronology. *Ore Geol. Rev.* **2017**, *89*, 731–751. [CrossRef]
13. Yang, Q. Late Paleozoic to Early Mesozoic Copper-Gold Polymetallic Metallogeny and Metallogenic Regularity in Central Jilin Province, Northeast China. Ph.D. Thesis, Jilin University, Changchun, China, 2020. (In Chinese with English Abstract)
14. Ren, Y.S.; Ju, N.; Chen, C.; Yang, Q.; Shang, Q.Q.; Lai, K. Late Paleozoic VMS-type deposit and its tectonic significance in eastern Jilin Province, NE China. *Acta Mineral. Sin.* **2015**, *35*, 493. (In Chinese)
15. Lu, S.Y.; Ren, Y.S.; Yang, Q.; Sun, Z.M.; Hao, Y.J.; Sun, X.H. Ore genesis for stratiform ore bodies of the Dongfengnanshan copper polymetallic deposit in the Yanbian Area, NE China: Constraints from LA-ICP-MS in situ trace elements and sulfide S–Pb isotopes. *Acta Geol. Sin. (Engl. Ed.)* **2019**, *93*, 1591–1606. [CrossRef]
16. Hou, H.N. Tectonic Setting and Metallogenic Significance of Late Paleozoic Volcanic Rock Series in the Yanbian Area, NE China. Ph.D. Thesis, Jilin University, Changchun, China, 2020. (In Chinese with English Abstract)
17. Wang, Z.G.; Wan, D.; Wang, K.Y.; Konare, Y.; Liang, Y.H. Isotope systematics and fluid inclusion studies of the Hongtaiping Cu–Pb–Zn deposit in Yanbian, NE China: Implications for ore genesis. *Geol. J.* **2020**, *55*, 6912–6935. [CrossRef]
18. Chang, J.J. Study on Geological Characteristics and Genesis of Xiaohongshilazi Lead–Znic Deposit, Panshi, Jilin Province. M.S. Thesis, Jilin University, Changchun, China, 2016. (In Chinese with English Abstract)
19. Li, Y. Ore Genesis and Tectonic Setting of Xiaohongshilazi Lead-Zinc Deposit in Panshi Area, Jilin Province. M.S. Thesis, Jilin University, Changchun, China, 2017. (In Chinese with English Abstract)
20. Yang, Q.; Ren, Y.S.; Li, Y.; Hao, Y.J.; Li, J.M.; Shang, Q.Q.; Sun, X.H. Age and tectonic setting of mesothermal magmatic hydrothermal vein-type Pb–Zn–(Ag) mineralization in the Xiaohongshilazi deposit, central Jilin Province, NE China. *Res. Geol.* **2020**, *70*, 70–88. [CrossRef]
21. Zhu, Y.Y.; Zhang, J.Q.; Cao, J.F. Geological features and prospecting criteria of the Xiaohongshilazi silver lead-zinc deposit. *Jilin Geol.* **2012**, *31*, 55–59, (In Chinese with English Abstract)
22. Cao, J.F.; Sun, Z.Y.; Xue, X.G. The characteristics and significance of the Pb and Zn massive sulfide deposits in Xiaohongshilazi, Panshi City, Jilin Province. *J. Changchun Inst. Tech. (Nat. Sci. Ed.)* **2012**, *13*, 90–94. (In Chinese with English Abstract)
23. Zhang, X.Z.; Zhou, J.B.; Chi, X.G.; Wang, C.W.; Hu, D.Q. Late Paleozoic tectonic-sedimentation and petroleum resources in Northeast China. *J. Jilin Univ. (Earth Sci. Ed.)* **2008**, *38*, 719–725. (In Chinese with English Abstract)
24. Peng, X.D.; Zhang, M.S.; Li, X.M. The evolution of Paleozoic tectonic paleogeography in Jilin and Heilongjiang orogenic belt. *World Geol.* **1999**, *18*, 24–28. (In Chinese with English Abstract)
25. Peng, Y.J.; Zhao, C.B. The evolution of the Paleo Jihei orogenic belt and accretion of the continental crust. *Jilin Geol.* **2001**, *20*, 1–9. (In Chinese with English Abstract)
26. Zhang, X.Z.; Ma, Y.X.; Chi, X.G.; Zhang, F.X.; Sun, Y.W.; Guo, Y.; Zeng, Z. Discussion on Phanerozoic tectonic evolution in Northeast China and eastern Inner Mongolia. *J. Jilin Univ. (Earth Sci. Ed.)* **2012**, *42*, 1269–1285. (In Chinese with English Abstract)
27. Jia, D.C.; Hu, R.Z.; Lu, Y.; Qiu, X.L. Collision belt between the Khanka block and the North China block in the Yanbian Region, Northeast China. *J. Asian Earth Sci.* **2004**, *23*, 211–219.
28. Li, J.Y. Permian geodynamic setting of Northeast China and adjacent regions: Closure of the Pale-Asian Ocean and subduction of the Paleo-Pacific Plate. *J. Asian Earth Sci.* **2006**, *26*, 207–224. [CrossRef]
29. Lin, W.; Faure, M.; Nomade, S.; Shang, Q.H.; Renne, P.R. Permian-Triassic amalgamation of Asia: Insights from Northeast China sutures and their place in the final collision of North China and Siberia. *Comptes Rendus Geosci.* **2008**, *340*, 190–201. [CrossRef]

30. Cao, H.H.; Xu, W.L.; Pei, F.P.; Wang, Z.W.; Wang, F.; Wang, Z.J. Zircon U–Pb geochronology and petrogenesis of the Late Paleozoic–Early Mesozoic intrusive rocks in the eastern segment of the northern margin of the North China Block. *Lithos* **2013**, *170–171*, 191–207. [CrossRef]
31. Chen, B.; Jahn, B.M.; Tian, W. Evolution of the Solonker suture zone: Constraints from zircon U–Pb ages, Hf isotopic ratios and whole-rock Nd-Sr isotope compositions of subduction and collision-related magmas and forearc sediments. *J. Asian Earth Sci.* **2009**, *34*, 245–257. [CrossRef]
32. Zhou, J.B.; Wilde, S.A. The crustal accretion history and tectonic evolution of the NE China segment of the Central Asian Orogenic Belt. *Gondwana Res.* **2013**, *23*, 1365–1377. [CrossRef]
33. Sengör, A.M.C.; Natal'in, B.A. Paleotectonics of Asia: Fragments of a synthesis. In *The Tectonic Evolution of Asia*; Yin, A., Harrison, M., Eds.; Cambridge University Press: London, UK, 1996; pp. 486–640.
34. Zhang, Y.B.; Wu, F.Y.; Wilde, S.A.; Zhai, M.G.; Lu, X.P.; Zhang, X.F. Geochronology and tectonic implications of the "Proterozoic" Seluohe Group at the Northern margin of the North China Craton. *Int. Geol. Rev.* **2008**, *50*, 135–153. [CrossRef]
35. Wang, Z.J. Late Paleozoic—Triassic Tectonic Evolution of Eastern Segment of the Southern Margin of the Xing'an–Mongolia Orogenic Belt: Evidence from Detrital Zircon U-Pb Geochronology and Igneous Rock Associations. Ph.D. Thesis, Jilin University, Changchun, China, 2016. (In Chinese with English Abstract)
36. Hou, K.J.; Li, Y.H.; Tian, Y.R. In situ U—Pb zircon dating using, laser ablation–multiion counting-ICP–MS. *Miner. Deposits* **2009**, *28*, 481–492. (In Chinese with English Abstract)
37. Griffin, W.L.; Powell, W.J.; Pearson, N.J.; O'Reilly, S.Y. GLITTER: Data reduction software for laser ablation ICP–MS. In *Laser Ablation–ICP–MS in the Earth Sciences: Current Practices and Outstanding Issues*; Mineralogical Association Canada Short Course; Sylvester, P., Ed.; Mineralogical Association Canada: Quebec City, QC, Canada, 2008; Volume 40, pp. 308–311.
38. Ludwig, K.R. User's manual for Isoplot 3.0: A geochronological toolkit for Microsoft Excel. *Berkeley Geochrono. Cent. Spec. Publ.* **2003**, *4*, 1–70.
39. Anderson, T. Correction of common lead in U–Pb analyses that do not report ^{204}Pb. *Chem. Geol.* **2002**, *192*, 59–79. [CrossRef]
40. Qi, L.; Zhou, M.F. Platinum-group elemental and Sr-Nd-Os isotopic geochemistry of Permian Emeishan flood basalts in Guizhou Province, SW China. *Chem. Geol.* **2008**, *248*, 83–103. [CrossRef]
41. Belousova, E.A.; Griffin, W.L.; O'Reilly, S.Y. Igneous zircon: Trace element composition as an indicator of source rock type. *Contrib. Mineral. Petrol.* **2002**, *143*, 602–622. [CrossRef]
42. Hoskin, P.W.O.; Schaltegger, U. The composition of zircon and igneous and metamorphic petrogenesis. *Rev. Mineral. Geochem.* **2003**, *53*, 27–62. [CrossRef]
43. Qiu, K.F.; Marsh, E.; Yu, H.C.; Pfaff, K.; Gulbransen, C.; Gou, Z.Y.; Li, N. Fluid and metal sources of the Wenquan porphyry molybdenum deposit, Western Qinling, NW China. *Ore Geol. Rev.* **2017**, *86*, 459–473.
44. Boynton, W.V. Geochemistry of the rare earth elements: Meteorite studies. In *Rare Earth Element Geochemistry*; Henderson, P., Ed.; Elsevier: Amsterdam, The Netherlands, 1984; pp. 63–114.
45. Sun, S.S.; McDonough, W.F. Chemical and isotopic systematics of oceanic basalts: Implications for mantle composition and processes. *Geol. Soc.* **1989**, *42*, 313–345.
46. Winchester, J.A.; Floyd, P.A. Geochemical discrimination of different magma series and their differentiation products using immobile elements. *Chem. Geol.* **1977**, *20*, 325–343. [CrossRef]
47. Hastie, A.R.; Kerr, A.C.; Pearce, J.A.; Mitchell, S.F. Classification of altered volcanic island arc rocks using immobile trace elements: Development of the Th–Co discrimination diagram. *J. Petrol.* **2007**, *48*, 2341–2357. [CrossRef]
48. Peccerillo, A.; Taylor, A.R. Geochemistry of Eocene calc-alkaline volcanic rocks from the Kastamonu area, Northern Turkey. *Contrib. Miner. Petrol.* **1976**, *58*, 63–81.
49. Cao, H.H. Geochronology and Geochemistry of the Late Paleozoic-Early Mesozoic Igneous Rocks in the Eastern Segment of the Northern Margin of the North China Block. Ph.D. Thesis, Jilin University, Changchun, China, 2013. (In Chinese with English Abstract)
50. Yu, Q.; Ge, W.C.; Yang, H.; Zhao, G.C.; Zhang, Y.L.; Su, L. Petrogenesis of late Paleozoic volcanic rocks from the Daheshen Formation in central Jilin Province, NE China, and its tectonic implications: Constraints from geochronology, geochemistry and Sr-Nd-Hf isotopes. *Lithos* **2014**, *192/195*, 116–131.
51. Li, D.J. *Stratigraphy (Lithostratic) of Jilin Province*; China University of Geosciences Press: Wuhan, China, 1997; pp. 1–324. (In Chinese)
52. Lu, S.Y.; Ren, Y.S.; Hao, Y.J.; Hou, H.H.; Yang, Q. Two-phase mineralization of Hongtaiping Cu polymetallic deposit in Yanbian area (NE China): Evidence from sulfde Rb–Sr dating and in-situ trace element analysis. *Ore Geol. Rev.* **2021**, *137*, 104295.
53. Gill, J.B. Early geochemical evolution of an oceanic island arc and backarc: Fiji and the South Fiji Basin. *J. Geol.* **1987**, *95*, 589–615. [CrossRef]
54. Nurlu, N.; Akinci, A.C.; Kohút, M. Geochemistry and geochronology of Middle Eocene subduction-related felsic volcanics in the Misis-Andirin Complex of the Southeast Anatolian Orogenic Belt. *Geosci. J.* **2023**. [CrossRef]
55. Sünger, A.; Akyıldız, M.; Nurlu, N. Significance of the Felsic Sub-Volcanics in Development of Related Au–Ag Deposits at Maden Village (Bolkar Mountain Region), South Turkey: Constraints from Whole–Rock Geochemistry, Geochronology and Pb-S Isotopes. *Geotectonics* **2022**, *56*, 361–381.

56. Baker, M.B.; Hirschmann, M.M.; Ghiorso, M.S.; Stolper, E.M. Compositions of near-solidus peridotite melts from experiments and thermodynamic calculations. *Nature* **1995**, *375*, 308–311. [CrossRef]
57. Kelemen, P.B.; Hanghj, K.; Greene, A.R. One view of the geochemistry of subduction-related magmatic arcs, with an emphasis on primitive andesite and lower crust. *Trea. Geochem.* **2007**, 3, 1–70.
58. Rollinson, H.R. *Using Geochemical Data: Evaluation, Presentation, Interpretation*; Longman Publishing Group: London, UK, 1993; pp. 1–384.
59. Rapp, R.P.; Watson, E.B. Dehydration melting of metabasalt at 8 ~ 32 kbar: Implications for continental growth and crust-mantle recycling. *J. Petrol.* **1995**, *36*, 891–931. [CrossRef]
60. Yang, Q.; Ren, Y.S.; Huizenga, J.M.; Li, J.M.; Wang, E.D.; Wang, X.; Wang, B. Geological signifcance of Early Triassic porphyry Cu mineralization in the eastern Xar Moron–Changchun Metallogenic Belt, northeast China: A case study of the newly-discovered Guokuidingzi Cu deposit. *Ore Geol. Rev.* **2021**, *133*, 104092.
61. Collins, W.J.; Beams, S.D.; White, A.J.R.; Chappell, B.W. Nature and origin of A-type granites with particular reference to southeastern Australia. *Contrib. Mineral. Petrol.* **1982**, *80*, 189–200.
62. Whalen, J.B.; Currie, K.L.; Chappell, B.W. A-type granites: Geochemical characteristics, discrimination and petrogenesis. *Contrib. Mineral. Petrol.* **1987**, *95*, 407–419. [CrossRef]
63. Bonin, B. A-type granites and related rocks: Evolution of a concept, problems and prospects. *Lithos* **2007**, *97*, 1–29. [CrossRef]
64. Lu, S.Y.; Ren, Y.S.; Yang, Q.; Hao, Y.J.; Zhao, X. Petrogenesis and tectonic implication of the Hongtaiping high-mg diorite in the Wangqing area, NE China: Constraints from geochronology, geochemistry and Hf isotopes. *Minerals* **2022**, *12*, 1002. [CrossRef]
65. Lü, L.S.; Mao, J.W.; Li, H.B.; Pirajno, F.; Zhang, Z.H.; Zhou, Z.H. Pyrrhotite Re-Os and SHRIMP zircon U–Pb dating of the Hongqiling Ni-Cu sulfide deposits in Northeast China. *Ore Geol. Rev.* **2011**, *43*, 106–119.
66. Pearce, J.A.; Harris, N.B.W.; Tindle, A.G. Trace element discrimination diagrams for the tectonic interpretation of granitic rocks. *Petrol.* **1984**, *25*, 956–983. [CrossRef]

minerals

MDPI

Article

Fluid Evolution and Ore Genesis of the Songjianghe Au Deposit in Eastern Jilin Province, NE China: Constraints from Fluid Inclusions and H-O-S-Pb Isotope Systematics

Qi Yu [1,2], Keyong Wang [1,*], Xuebing Zhang [3], Qingfei Sun [1], Wenqiang Bai [4], Chao Ma [2] and Yongchun Xiao [2]

1 College of Earth Sciences, Jilin University, Changchun 130061, China; yuqi2018611011@163.com (Q.Y.)
2 Geological Survey Institute of Jilin Province, Changchun 130102, China
3 College of Geology and Mining Engineering, Xinjiang University, Urumqi 830047, China
4 Team 603 of Jilin Nonferrous Metal Geological Exploration Bureau, Yanji 133300, China
* Correspondence: wangky@jlu.edu.cn

Abstract: The medium-sized Songjianghe Au deposit is located in the southeastern part of the Jiapigou-Haigou gold belt (JHGB) in central eastern Jilin Province, NE China. The gold mineralization is primarily characterized by disseminated-style ores and hosted in the low-/medium-grade metamorphic rocks of the Seluohe Group. The ore bodies are governed by NNW-striking brittle-ductile structures and spatially correlated with silicic and sericitic alterations. Four alteration/mineralization stages have been distinguished: (I) Quartz-pyrrhotite-pyrite, (II) quartz-polymetallic sulfides, (III) quartz-pyrite, and (IV) quartz-calcite. The fluid inclusion (FI) assemblage in quartz from Stage I comprises C1-type, C2-type, C3-type, and VL-type FIs, with total homogenization temperatures (Th-total) of 292.8 to 405.6 °C and salinities of 2.8 to 9.3 wt% NaCl eqv. Quartz from Stage II (main ore stage) developed C2-, C3-, and VL-type FIs, with a Th-total of 278.5 to 338.9 °C and salinities of 2.8 to 8.1 wt% NaCl eqv. Stage III is characterized by coexisting C3- and VL-type FIs in quartz, with a Th-total of 215.9 to 307.3 °C and salinities of 2.4 to 7.2 wt% NaCl eqv. Only VL-type FIs are observed in Stage IV, with a Th-total of 189.5 to 240.4 °C and salinities of 3.7 to 5.7 wt% NaCl eqv. The Laser Raman spectroscopic results demonstrated minor CH_4 in the C-type FIs from Stages I and II. The results suggest that ore fluids may have evolved from a medium-high temperature, low-salinity immiscible CO_2-NaCl-$H_2O \pm CH_4$ system to a low temperature, low-salinity homogeneous NaCl-H_2O system. Fluid immiscibility caused by the rapid drop in pressure may have been the main trigger for gold-polymetallic sulfide precipitation. The Songjianghe Au deposit may have been formed under 352–448 °C and 850–1380 bar pressure, based on the isochore intersection for Stage II fluid inclusions. The H-O isotopic compositions (Stage I: $\delta^{18}O_{fluid}$ = 5.6 to 5.8‰, δD = −96.2 to −95.7‰; Stage II: $\delta^{18}O_{fluid}$ = 3.7 to 4.2‰, δD = −98.7 to −89.8‰; Stage III: $\delta^{18}O_{fluid}$ = 1.2 to 1.4‰, δD = −103.5 to −101.2‰) indicate that the hydrothermal fluids are dominated by magmatic water in the early stages (Stages I and II) and mixed with meteoric water since Stage III. The pyrite S-Pb isotope data ($\delta^{34}S$: −2.91 to 3.40‰; $^{206}Pb/^{204}Pb$: 16.3270 to 16.4874; $^{207}Pb/^{204}Pb$: 15.2258 to 15.3489; $^{208}Pb/^{204}Pb$: 36.6088 to 36.7174), combined with Pb isotopic compositions of the intrusive rocks and wall rocks (the Seluohe Group) in the ore district, indicate that the ore-forming materials at Songjianghe are predominantly from a magmatic source and may have been affected by the contamination of the Seluohe Group. In accordance with the features of ore geology, ore-forming fluids and metals, and geodynamic setting, the Songjianghe Au deposit belongs to a mesothermal magmatic hydrothermal vein gold deposit, which formed in the intermittent stage of Paleo-Pacific plate subduction during the Late Jurassic.

Keywords: fluid evolution; ore genesis; H-O-S-Pb isotopes; Songjianghe Au deposit; NE China

1. Introduction

The Jiapigou-Haigou gold belt (JHGB), located in the tectonic transition between the northeastern North China Craton (NCC) margin and the eastern Central Asian Orogenic Belt (CAOB), is a key gold province with an explored gold reserve of >200 metric tonnes (t) in NE China [1–3]. The region has undergone prolonged tectonic evolution with the Paleo-Asian Ocean closure and the Paleo-Pacific subduction, forming a series of hydrothermal gold deposits related to CO_2-enriched fluids (e.g., Erdaogou, Xiaobeigou, Bajiazi, Xiaxitai, Sandaocha, and Haigou). Extensive studies have been undertaken to unravel the geological setting, metallogenic ages, and mineralization systematics of these gold deposits [4–11]. However, the origin of ore-forming fluids has long been a controversial topic: (1) Predominantly metamorphic water with later meteoric water infusion [6,12]; (2) a mixed source from magmatic water and meteoric water [13–15]; and (3) mantle-derived fluids [8,16,17]. Meanwhile, diverging insights also exist on the source of metals: (1) Extracted from metamorphic country rocks [18–20] or (2) a mixture of crust and mantle [17,21]. Consequently, in many cases, both the origin of ore fluids and metal source are indistinct or under dispute, which hinders the understanding of the ore-forming process and the mechanism of gold deposition. Considering the aforementioned distinctions, the genetic type of these gold deposits is a subject of continuous discussion and with conflicting conclusions: (1) A greenstone-type gold deposit [12,18,20], (2) a magmatic hydrothermal vein-type gold deposit [4,22,23], or (3) an orogenic gold deposit [6,21,24].

The Songjianghe Au deposit is sited in the southeastern JHGB, hosting a 9 metric tonne (t) gold reserve @ 3.02 g/t. Compared with other major gold deposits, the Songjianghe Au deposit has received less attention. Prior studies on the Songjianghe Au deposit primarily addressed ore geology [25,26], ore fluids [27–29], geochronology, and the geodynamic setting [30,31]. These studies proposed that the Songjianghe Au deposit was formed in the intermittent stage that followed the first subduction of the Paleo-Pacific plate during the Late Jurassic (157–156 Ma [31]) and gold precipitation was primarily caused by fluid immiscibility [27–29]. Nevertheless, the links between the evolution of ore-forming fluids and gold mineralization have not been convincing. Meanwhile, the sources of ore fluids and ore-forming materials are still obscure, resulting in unclear ore precipitation processes. In addition, the ores in the Songjianghe Au deposit are predominately of the disseminated style [25,26], in contrast to the quartz vein-type ores that describe the majority of gold deposits (e.g., Erdaogou, Xiaobeigou, Bajiazi, and Haigou) in the JHGB [17,20,32,33].

In this study, we give detailed descriptions of the geology and mineralogy, as well as a systematic study of the fluid inclusions and H-O stable isotopes, to provide distinct constraints on the property, origin, and evolution of the ore fluids and the mineralization mechanisms. Additionally, the S-Pb isotopic data are reported to reveal the source of the ore-forming materials. In accordance with the research results generated, together with the previous studies, the genetic type is classified, and a metallogenic model is established for the Songjianghe Au deposit. It is expected that these results can provide new insights into regional metallogeny in the Jiapigou-Haigou gold metallogenic belt.

2. Geological Setting

The Central Asian Orogenic Belt (CAOB), extending over 5500 km from west to east, evolved through complex tectonics of the Neoproterozoic-Middle Triassic Paleo-Asian Ocean closure, the Mesozoic Mongolia-Okhotsk Ocean closure, and the Paleo-Pacific subduction [34–42]. The CAOB separates the Siberia Craton in the north from the Tarim-North China Craton in the south (Figure 1a). NE China, the adjacent Russia Far East, and eastern Mongolia are composed primarily of micro-continental masses, including (from NW to SE) the Erguna, Xing'an, Songliao, and Jiamusi–Khanka blocks, and the Sikhote–Alin accretionary complex (Nadanhada terrane), bounded by the Xinlin-Xiguitu, Hegenshan-Heihe, Mudanjiang-Yanji, and Primoria faults, respectively [43,44] (Figure 1b).

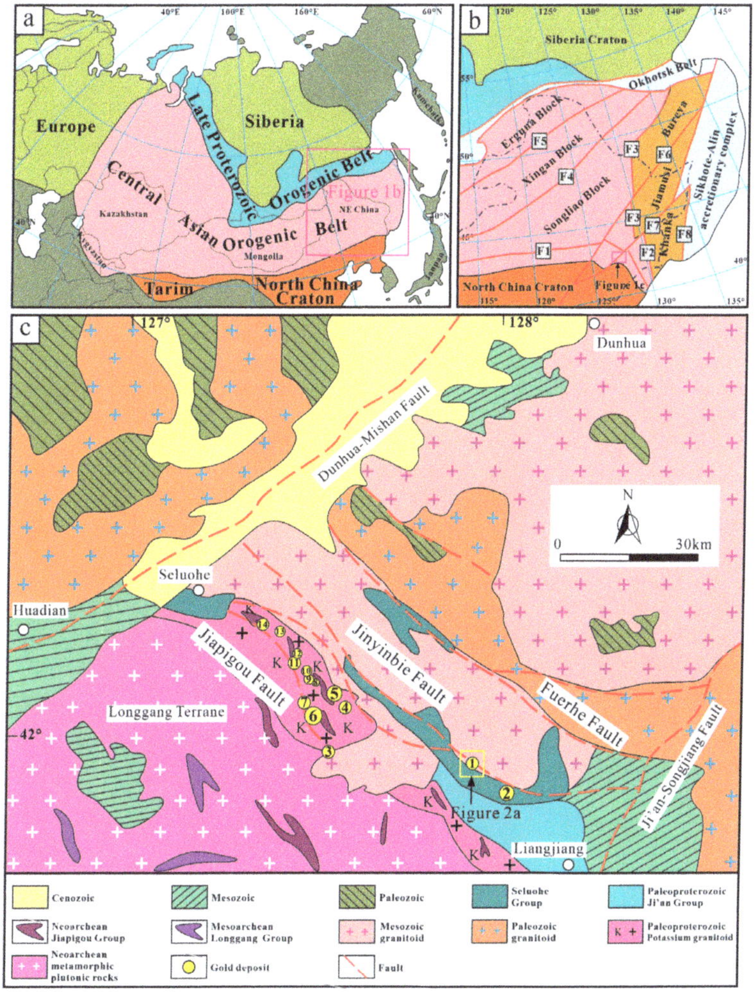

Figure 1. (**a**) Tectonic map of the Central Asian Orogenic Belt, showing the location of NE China (modified after [45,46]). (**b**) Tectonic sub-divisions of NE China and its adjacent regions, showing the location of the Jiapigou-Haigou gold belt (modified after [47,48]). F1: Solonker-Xar Moron-Changchun zone; F2: Yanji Fault; F3: Mudanjiang Fault; F4: Hegenshan-Heihe Fault; F5: Xinlin-Xiguitu Fault; F6: Yilan-Yitong Fault; F7: Dunhua-Mishan Fault; F8: Primoria Fault. (**c**) Simplified regional geological map of the Jiapigou-Haigou gold belt, showing the locations of representative gold deposits (modified after [9,31]), including (number in the figure) 1. Songjianghe; 2. Haigou; 3. Liupiye; 4. Bajiazi; 5. Xiaxitai; 6. Erdaogou; 7. Miaoling; 8. Sidaocha; 9. Sandaocha; 10. Daxiangou; 11. Xiaobeigou; 12. Renaogou; 13. Caiqiangzi; 14. Banmiaozi.

The JHGB is located in the tectonic transition between the northeastern NCC margin (locally named the Longgang terrane) and the eastern CAOB (Figure 1b,c). The formation, accretion, and solidification of the continental nucleus, as well as its cracking and subsequent cratonization, are primary tectonic evolution processes during Precambrian [2,49]. The Early Paleozoic witnessed the beginning of the closure of the Paleo-Asian Ocean, which had completely closed before the Middle Triassic [43,50]. The subduction of the

Paleo-Pacific plate beneath Eurasia is the most significant tectonic event in the JHGB during the Mesozoic [7,17].

The regional strata comprise Archean medium-/high-grade metamorphic rocks, Proterozoic low-/medium-grade rocks, Paleozoic low metamorphic/sedimentary rocks, Mesozoic volcanic/sedimentary rocks, and Cenozoic sediments (Figure 1c). The Archean metamorphic basement composed of the Longgang and Jiapigou Groups is exposed in the Longgang terrane on the southwestern side of the JHGB (Figure 1c). The Mesoarchean Longgang Group may have experienced medium-/high-grade metamorphism and primarily consists of felsic gneiss, granulite, plagioclase amphibolite, and supracrustal rocks [51]. The Neoarchean Jiapigou Group comprises characteristically biotite plagioclase gneiss, plagioclase amphibolite, hornblende plagioclase gneiss, and minor magnetite quartzite [21,52], which is the main host rock of many gold deposits (e.g., Bajiazi, Xiaxitai, and Sandaocha). The Paleoproterozoic Ji'an Group composed of plagioclase amphibolite, biotite granulite, biotite schist, and marble was formed in the early rifting stage [53]. The Seluohe Group is discontinuously exposed along the Jinyinbie and Fuerhe faults (Figure 1c), which experienced low-/medium-grade metamorphism. As the host rock of the Songjianghe Au deposit, the Seluohe Group has long been considered a Meso-Neoproterozoic metamorphic suite [54,55], yet more recent studies have shown that the Seluohe Group may actually represent a complex of large tectonic slices that primarily comprise metamorphic strata and mylonitized granites of different ages and origins [31,56–58]. The Paleozoic strata consist primarily of the Qinglongcun Group and the Hulan Group, which are characterized by low metamorphic volcano-sedimentary rocks. The Late Triassic greenschist facies retrograde metamorphism of the Hulan Group represents a post-collisional extension setting after the closure of the Paleo-Asian Ocean [43,59]. The Mesozoic strata crop out in the northern and southeastern JHGB (Figure 1c), primarily comprising terrestrial sedimentary and intermediate-acid pyroclastic rocks.

The oldest granitoid, a TTG (tonalite-trondhjemite-granodiorite) suite, and a set of potassium-rich granitoids are exposed in the central and northern Longgang terrane, respectively (Figure 1c), which are characterized by sodic to potassic compositional evolution [60]. These two granitoid suites are likely the product of late Neoarchean to early Paleoproterozoic magmatic events under an active continental margin environment [52,61]. The Paleozoic intrusive rocks are exposed along NW-striking faults (Figure 1c), primarily comprising late Hercynian biotite granite and granodiorite, followed by monzogranite and quartz diorite, which were formed in the process of the Paleo-Asian Ocean closure [44,62]. The Mesozoic intrusive rocks are characteristically composed of large batholiths of granodiorite, monzogranite, and quartz diorite. These intrusive rocks were considered to be formed in the tectonic settings of post-collisional extension after the Paleo-Asian Ocean closure and/or the multiple subductions of the Paleo-Pacific plate [14,63,64].

Several fault zones developed in the region, including the NE-striking Dunhua-Mishan and Ji'an-Songjiang faults, as well as the NW-striking Fuerhe, Jinyinbie, and Jiapigou faults (Figure 1c). The Dunhua-Mishan fault is considered to be a branch of the Tancheng-Lujiang fault zone [10,65,66], whilst the NW-striking Jiapigou, Jinyinbie, and Fuerhe faults are the subsidiary structures of the Dunhua-Mishan fault [21,67], which experienced brittle deformation superimposed on brittle-ductile deformation [68]. These subsidiary faults, characterized by dextral shear [17,69], control the distribution of gold deposits in the study area [7,17] (Figure 1c). The ultra-crustal Ji'an-Songjiang fault locally controls the Mesozoic magma activities and the distribution of Early Jurassic and Cretaceous strata [1,70].

3. Deposit Geology

The Songjianghe Au deposit is located in the southeastern part of the Jiapigou-Haigou gold belt (JHGB). The only exposed stratigraphic unit at Songjianghe is the Seluohe Group (NE-dipping with 40–55° dip angle), which is also the ore host (Figure 2a). The Seluohe Group at Songjianghe was formed in the Late Neoarchean, primarily composed of chlorite-epidote-actinolite schist, biotite-plagioclase schist, amphibolite-plagioclase schist, muscovite-biotite schist, granitic gneiss, meta-sandstone, and marble, and experienced stages of multistage metamorphism and deformation. The protoliths of the Seluohe Group may have primarily consisted of intermediate-basic and acidic volcanic and sedimentary rocks [31].

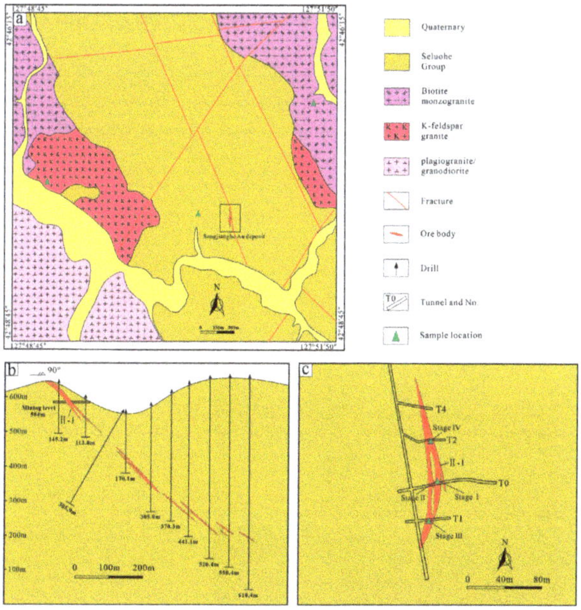

Figure 2. (**a**) Simplified geological map of the Songjianghe Au deposit, showing the Pb isotope sample locations in the pluton and wallrocks; (**b**) geologic profile along Exploration Line 0, showing No. II-1 orebody; (**c**) geological plan of mining level (584 m), showing the sampling locations for ores of the various mineralization stages.

There are three sets of major structures, which are (from old to young) EW-, NNW-, and NE-striking fractures (Figure 2a). The EW-striking (N-dipping with 40–80° dip angle) fractures are cut by the NNW- and NE-striking fractures, exposed on the north side of the ore district. The orebodies are controlled by the NNW-striking fractures (E or NE-dipping at 40 to 50°), which are characterized by the early-stage ductile shearing superimposed by the late-stage brittle-ductile fractures. The NE-striking fractures marked by SE-dipping with a 60–70° dip angle are characteristically transtensional, which crosscut the NNW-striking ore-controlling structures.

The local intrusive rocks consist primarily of plagiogranite and granodiorite (171 ± 5 Ma [71]), K-feldspar granite (169.6~173 Ma [25,30]), and biotite monzogranite (168.2 ± 2 Ma [72]) (Figure 2a). These Middle Jurassic igneous rocks were considered to have formed in the tectonic setting of the island arc or active continental margin under the subduction of the Paleo-Pacific plate [30,71,72].

A total of 10 orebodies have been delineated at Songjianghe, with a gold reserve of 9 t (@ 3.02 g/t). The largest orebody (II-1: 240 m long, 600 m deep, and 0.8–21.6 m thick) is lensoidal and developed along the NNW-striking structures (Figure 2b,c). The ores are primarily characterized by disseminated or veinlet types and minor quartz vein types. The ore textures primarily include anhedral granular, metasomatism, metasomatic relict, and minor euhedral-subhedral granular. Metallic minerals include characteristically pyrite, pyrrhotite, chalcopyrite, galena, sphalerite, and native gold. Gangue minerals primarily include quartz, calcite, mica, chlorite, and epidote.

Based on field observations, ore texture/structure, mineral assemblage, paragenetic sequence, and vein crosscutting relations, four alteration/mineralization stages were recognized (Figure 3). Stage I (quartz-pyrrhotite-pyrite) is characterized by medium-fine-grained milky white quartz with a euhedral-subhedral texture, and euhedral-subhedral fine-medium-grained pyrrhotite and pyrite (Figure 3a,e,f). Stage II (quartz-polymetallic sulfides) is the main Au ore stage, characterized by pyrite, chalcopyrite, galena, sphalerite, native gold, and quartz (Figure 3a–c,g–k). The quartz is usually smoke gray in color with subhedral-anhedral grains. The metal sulfides are mostly anhedral grains, and a few are euhedral-subhedral (Figure 3g–k). Pyrite, the predominant metal sulfide (~90 vol%), can be locally observed in coexistence with native gold, yet most native gold is characteristically hosted in the intergranular spaces and fissures of quartz (Figure 3j,k). Stage III is marked by the coexistence of quartz and pyrite (Figure 3c,l). It is visible that Stage II quartz-polymetallic sulfides were cut by Stage III quartz-pyrite veins (Figure 3c). Stage IV represents the end of hydrothermal activity, characterized by quartz-calcite veins (Figure 3d). The ore and gangue mineral paragenetic sequence is summarized in Figure 4. The Wall rock alteration zoning (from orebody to wallrock) includes sericite-silicic → weak potassic-biotite → epidote-chlorite → carbonate [31].

Figure 3. Photographs and photomicrographs of the Songjianghe Au ores at various stages: (**a**) Stage I quartz-pyrrhotite-pyrite cut by Stage II quartz-polymetallic sulfides vein. (**b**) Stage II quartz-polymetallic sulfide veins. (**c**) Stage II quartz-polymetallic sulfides were cut by Stage III quartz-pyrite vein. (**d**) Stage IV quartz-calcite vein. (**e**,**f**) Stage I euhedral-subhedral pyrite and pyrrhotite in quartz. (**g**–**i**) Stage II pyrite-chalcopyrite-galena-sphalerite assemblage. (**j**,**k**) Stage II native gold in coexistence with pyrite and in intergranular spaces and fissures of quartz. (**l**) Stage III quartz and pyrite. Abbreviations: Cal—calcite; Ccp—chalcopyrite; Gl—native gold; Gn—galena; Po—pyrrhotite; Py—pyrite; Qz—quartz; Sp—sphalerite.

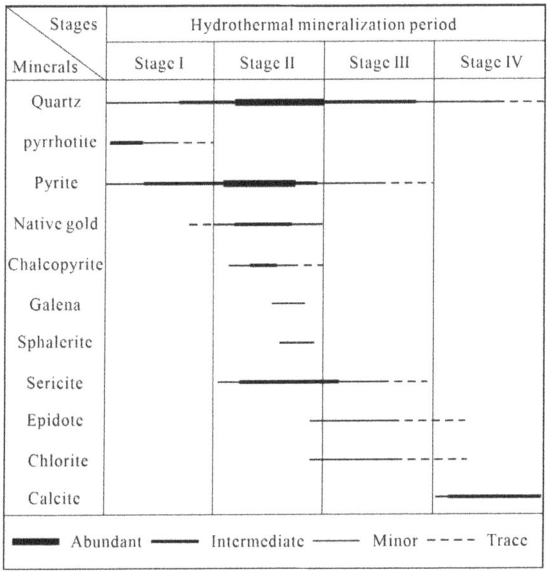

Figure 4. Mineral paragenetic sequence of the Songjianghe Au deposit.

4. Sampling and Analytical Methods

4.1. Fluid Inclusion Analysis

In order to determine the ore-fluid evolution and physicochemical conditions of mineralization, 20 quartz samples (Stage I to IV) were selected from the II-1 orebody of the 584 m mining level (Figure 2c) to make doubly polished thin sections (~150 μm thick) for fluid inclusion (FI) petrographic and microthermometric analyses. Following the detailed fluid inclusion assemblage (FIA) definition by [73], representative primary FIAs were selected for microthermometric measurements on a Linkam THMS-600 heating–freezing stage (with a −195 to +600 °C temperature range, mounted on a Carl Zeiss Axiolab microscope), at the Key Laboratory of Geological Fluid in Jilin University (China). The equipment was calibrated by the triple point of CO_2 (−56.6 °C) and the freezing point of pure water (0.0 °C) to ensure analysis accuracy (±0.1 °C (<0 °C) and ±2 °C (>200 °C)). The heating and cooling rates are 1–5 °C/min, which decreased to 0.1 °C/min when approaching phase transitions. Representative single FI compositions were analyzed on a Renishaw System-1000 Laser Raman Spectrometer (UK) with a 514 nm laser source. The spectral scanning range was set to between 1200 and 4000 cm^{-1}, with a 60 s accumulation time for each scan.

4.2. Isotope Analysis

Nine quartz samples from representative ores (Stage I to III) were selected from the II-1 orebody of the 584 m mining level (Figure 2c) for hydrogen-oxygen isotope analyses to constrain the ore-fluid origin. The ore samples were examined and identified under a microscope to make sure they were representative of various hydrothermal stages. The quartz samples were crushed to 40–60 mesh, and then handpicked under a binocular microscope to obtain >98% purity. Hydrogen-oxygen isotope composition analyses were completed in the laboratory of the Beijing Research Institute of Uranium Geology and Beijing GeoAnalysis Co. Ltd. (Beijing, China). The quartz was reacted with BrF_5 in a vacuum at a high temperature to extract oxygen, which was then converted to CO_2 gas on a burning Pt-coated carbon rod [74]. The oxygen isotope composition of the obtained CO_2 gas was analyzed on a MAT-253EM mass spectrometer. The same samples were heated at 150 °C for 3 h in a vacuum, and then heated to >500 °C in an induction furnace to release the inclusion water, which was converted to H_2 by reacting with zinc powder at 800 °C. Finally, the hydrogen isotope composition was analyzed on a MAT-253EM mass spectrometer. The oxygen and hydrogen isotope ratios are presented in standard δ notation (‰) relative to the Standard Mean Ocean Water (SMOW). The analysis accuracy is ±0.2‰ for oxygen isotope and ±2‰ for hydrogen isotope.

Twenty-two pyrite samples (Stage I and II) were selected from the II-1 orebody of the 584 m mining level (Figure 2c) for sulfur isotope analyses, which were conducted at Beijing GeoAnalysis Co., Ltd. Sulfur isotope compositions were determined by conventional combustion and LA-MC-ICP-MS in situ S isotope analysis. SO_2 was liberated for S isotope analysis using the conventional extraction technique of [75], and then measured with a Delta V Plus mass spectrometer. Results are reported as $\delta^{34}S$ relative to Vienna Canyon Diablo Troilite (V-CDT) with a precision of ±0.2‰. The in situ S isotope was analyzed using an NWR 193 laser ablation system connected to Neptune Plus MC-ICP-MS. The single-spot ablation mode was used with a large spot size (44 μm) and low pulse frequency (2 Hz). Approximately 100 laser pulses were completed in a single analysis [76]. Details of the S isotope analytical methods follow those described by [77,78].

Five pyrite samples (Stage II) were selected from the II-1 orebody of the 584 m mining level (Figure 2c) for lead isotope analyses. In addition, lead isotope analyses of the wallrock, K-feldspar granite, and monzogranite (Figure 2a) were also carried out. All these analyses were conducted at Beijing GeoAnalysis Co., Ltd. Furthermore, 10–50 mg powder samples were washed with acetone and distilled water to remove surface contamination before air drying at 60 °C. The cleaned samples were dissolved in HF + HNO_3 at 150 °C for 7 days. The conventional cation exchange technique was used to purify the lead, and then placed in 0.5 mol/L HBr-HCl solution for Pb separation. The lead isotope analyses were conducted on an ISOPROBE-T thermal ionization mass spectrometer (TIMS) with an analysis accuracy of better than 0.005%. In addition, Pb isotope ratios were calibrated by the NBS SRM 981 standard. For details on lead isotope analysis methods, refer to [79].

5. Results

5.1. Fluid Inclusion

5.1.1. Fluid Inclusion Petrography

On the basis of the criteria of [73], primary FIs in isolation or random clusters were selected for the analyses, whereas secondary FIs along micro-fractures or penetrating the crystal boundary were not selected [80]. Based on the FI phase composition at room temperature (25 °C) and the phase transition characteristics during heating or freezing, two types of primary FIs are identified: CO_2 (C-type) and liquid-rich two-phase (VL-type) FIs. C-type FIs usually have one to two components, i.e., CO_2 ± liquid H_2O. Depending on the CO_2/H_2O ratio, C-type FIs can be divided into three sub-types: CO_2-pure (C1-type, entirely CO_2), CO_2-rich (C2-type, CO_2/H_2O ratio > 1), and CO_2-bearing (C3-type, CO_2/H_2O ratio < 1).

C1-type FIs are usually composed of two phases (liquid CO_2 + vapor CO_2) or a single liquid CO_2 phase at room temperature (25 °C), and the latter would appear as vapor CO_2 when cooled to <10 °C. The volume range of vapor CO_2 is 5% to 20% at room temperature (25 °C). These FIs are usually rounded/elliptical with a size of 8–12 μm and primarily distributed in groups (Figure 5(a-4); Table 1).

C2-type FIs usually contain three phases (liquid CO_2 + vapor CO_2 + liquid H_2O). The volume range of the CO_2 phase (liquid CO_2 + vapor CO_2) is over 50% (primarily 70–90%, i.e., CO_2/H_2O ratio > 1), with the vapor CO_2 phase accounting for 85 to 90% (a few 30–35%). These FIs are commonly elliptical with a size of 6–15 μm. Most of these FIs are isolated, while a few are distributed in groups (Figure 5(a-5,b-1,b-2,b-4); Table 1).

C3-type FIs are composed of three phases (liquid CO_2 + vapor CO_2 + liquid H_2O). The volume range of the CO_2 phase (liquid CO_2 + vapor CO_2) is < 50% (primarily 35–40%, i.e., CO_2/H_2O ratio < 1), with the vapor CO_2 phase accounting for 65 to 90% (a few with 35–45%). These FIs are commonly elliptical or irregular-shaped with a size of 5–20 μm.

Most of these FIs are isolated, and a few are clustered (Figure 5(a-1–a-3,a-5,b-3,b-4,c-1,c-2); Table 1).

VL-type FIs have two phases (vapor H_2O + liquid H_2O) at room temperature (25 °C), with vapor H_2O accounting for <50 vol.% (primarily 15–30 vol.%, minor 5–10 vol.%). These FIs are commonly elliptical, irregular-, or occasionally negative-crystal-shaped, with a size of 5–22 μm, and most of them occur in clusters, with a few isolated FIs (Figure 5(a-1,a-5,b-1,b-4,c-1,c-3,d-1,d-2); Table 1).

Table 1. Microthermometric data of fluid inclusion assemblages from different mineralization stages.

Mineralization Stages	Host Minerals	Types (Number)	Size (μm)	Tm-CO_2 (°C)	Tm-Cla (°C)	Th-CO_2 (°C)	Tm-Ice (°C)	Th-Total (°C)	Salinity (wt.% NaCl eqv)
Stage I	quartz	C1-type (10)	8–12	−60.8 to −60.0		17.6 to 30.0			
		C2-type (22)	6–15	−61.0 to −56.9	6.5 to 8.6	17.5 to 30.2		305.0 to 386.4	2.8 to 6.6
		C3-type (27)	6–20	−62.2 to −59.5	6.9 to 8.5	16.2 to 30.9		327.6 to 405.6	3.0 to 5.9
		VL-type (31)	5–20				−6.1 to −3.9	292.8 to 368.4	6.3 to 9.3
Stage II	quartz	C2-type (10)	8–12	−60.1 to −57.5	7.6 to 8.3	25.4 to 30.5		301.5 to 337.6	3.4 to 4.7
		C3-type (21)	5–15	−60.0 to −56.9	7.4 to 8.6	21.8 to 30.7		301.9 to 338.6	2.8 to 5.1
		VL-type (23)	6–22				−5.2 to −3.6	278.5 to 338.9	5.8 to 8.1
Stage III	quartz	C3-type (20)	5–15	−57.7 to −56.7	7.9 to 8.8	11.8 to 28.0		225.4 to 307.3	2.4 to 4.1
		VL-type (24)	5–15				−4.5 to −2.8	215.9 to 297.3	4.6 to 7.2
Stage IV	quartz	VL-type (19)	5–15				−3.5 to −2.2	189.5 to 240.4	3.7 to 5.7

Note: Tm-CO_2 = melting temperatures of solid CO_2; Tm-cla = melting temperatures of CO_2-H_2O clathrate; Th-CO_2 = homogenization temperatures of the CO_2 phase; Tm-ice = temperatures of ice melting; Th-total = total homogenization temperatures.

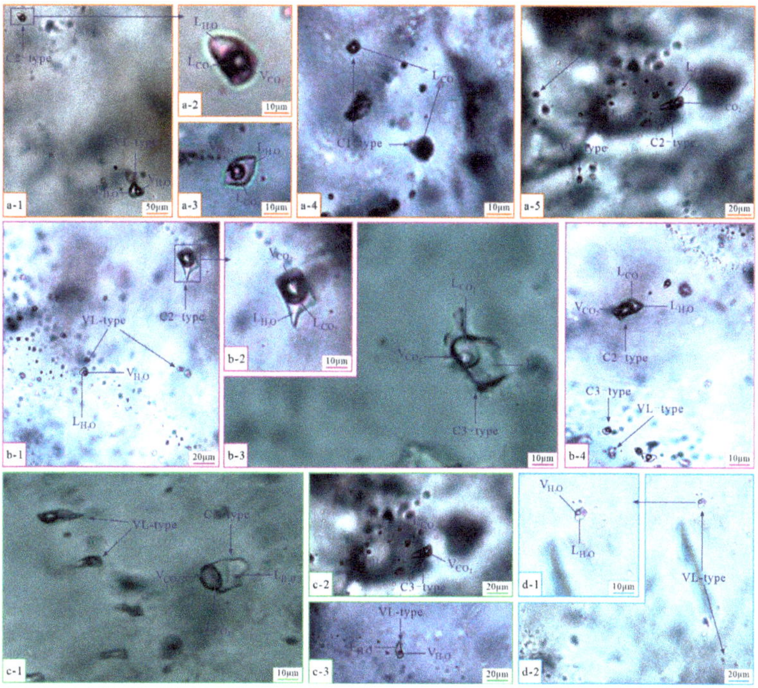

Figure 5. Photomicrographs of representative primary fluid inclusion assemblages from the Songjianghe Au deposit. (**a-1–a-5**) C1-, C2-, C3-, and VL-type FIA in Stage I quartz; (**b-1–b-4**) C2-, C3-, and VL-type FIA in Stage II quartz; (**c-1–c-3**) C3- and VL-type FIA in Stage III quartz; (**d-1,d-2**) VL-type FIs in Stage IV quartz. Abbreviations: L = liquid, V = vapor.

5.1.2. Fluid Inclusion Microthermometry

The FI microthermometric data are presented in Table 1 and illustrated in Figure 6. The FI salinities (expressed as wt.% NaCl eqv) were calculated according to the ice-melting temperature (Tm-ice) of VL-type FIs [81]. For C2- and C3-type FIs, the salinities were calculated based on the CO_2-H_2O clathrate melting temperatures (Tm-cla) [73].

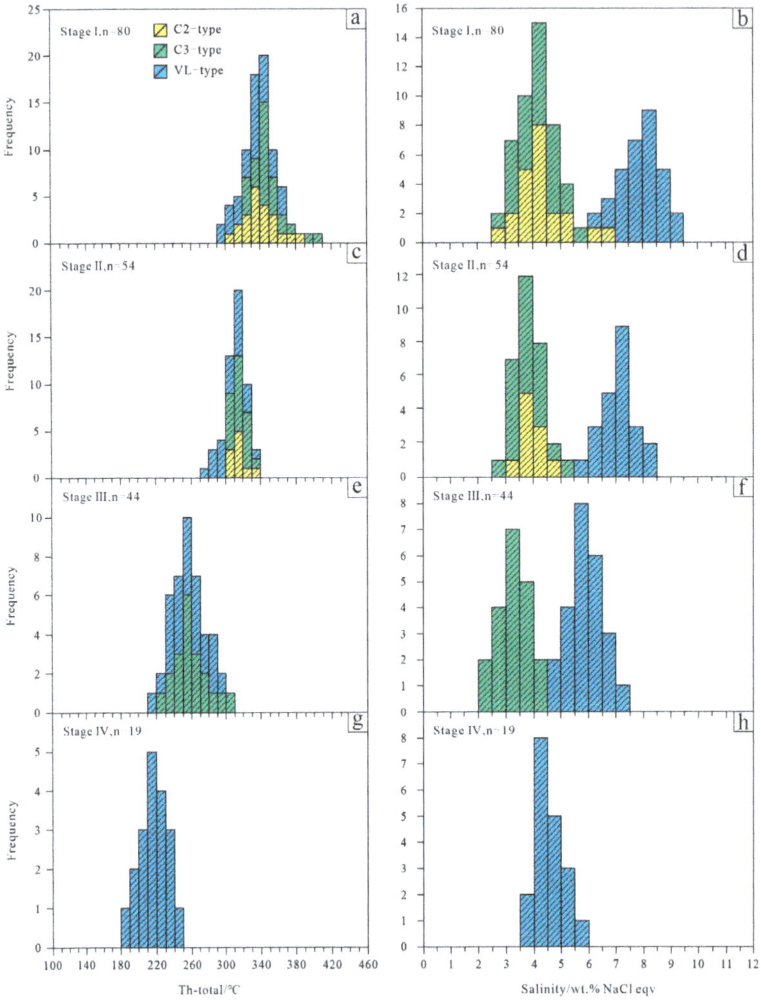

Figure 6. Histogram of fluid inclusion total homogenization temperature (Th-total) and salinity of the different hydrothermal stages at Songjianghe.

Stage I: C- (C1-, C2-, and C3-) and VL-type FIs were identified (Figure 5(a-1–a-5)). C-type FIs have similar solid CO_2 (Tm-CO_2) melting temperatures (−62.2 to 56.9 °C) and homogenization temperatures to the CO_2 phase (Th-CO_2 = 16.2 to 30.9 °C). The CO_2 phase of C1- and C2-type FIs homogenized to the liquid CO_2 phase, while the CO_2 phase of C3-type FIs homogenized to the liquid or vapor CO_2 phase. The C2- and C3-type FIs yielded similar Tm-cla (6.5 to 8.6 °C), with corresponding salinities of 2.8 to 6.6 wt.% NaCl eqv (Figure 6). The total homogenization temperatures (Th-total) of C2- and C3-type type FIs are 305.0 to 405.6 °C (primarily 330–350 °C) during heating (Figure 6). It should be noted that the C2-type FIs are totally homogenized to the liquid CO_2 phase, while the C3-type FIs are totally homogenized to the liquid H_2O phase. VL-type FIs homogenized to the liquid H_2O phase with a Th-total of 292.8 to 368.4 °C (primarily 330 to 350 °C) (Figure 6).

Tm-ice of VL-type FIs is −6.1 to −3.9 °C, with calculated salinities of 6.3 to 9.3 wt.% NaCl eqv (Figure 6).

Stage II: FIs are predominantly C2-, C3-, and VL-type (Figure 5(b-1–b-4)). C2- and C3-type FIs have similar Tm-CO_2 (−60.1 to −56.9 °C) and Th-CO_2 (21.8 to 30.7 °C). The CO_2 phase of C2-type FIs homogenized to the liquid CO_2 phase, while the CO_2 phase of C3-type FIs homogenized to the liquid or vapor CO_2 phase. Tm-cla of C2- and C3-type FIs are 7.4 to 8.6, with corresponding salinities of 2.8 to 5.1 wt.% NaCl eqv (Figure 6). C2-type FIs are totally homogenized to the liquid CO_2 phase at 301.5 to 337.6 °C, while C3-type FIs are totally homogenized to the liquid H_2O phase at 301.9 to 338.6 °C. The Th-total of C2- and C3-type FIs cluster around 310 to 320 °C (Figure 6). VL-type FIs homogenized to the liquid H_2O phase at 278.5 to 338.9 °C (primarily 310 to 320 °C) (Figure 6). Tm-ice of VL-type FIs is −5.2 to −3.6 °C, with corresponding salinities of 5.8 to 8.1 wt.% NaCl eqv (Figure 6).

Stage III: C3- and VL-type FIs were identified (Figure 5(c-1–c-3)). C3-type FIs have Tm-CO_2 of −59.8 to −56.7 °C and Th-CO_2 of 11.8 to 28.0 °C. The CO_2 phase of C3-type FIs homogenized to the liquid or vapor CO_2 phase. Tm-cla values of C3-type FIs are 7.9 to 8.8 °C, with corresponding salinities of 2.4 to 4.1 wt.% NaCl eqv (Figure 6). C3-type FIs are totally homogenized to the liquid H_2O phase at 225.4 to 307.3 °C (primarily 250 to 260 °C) (Figure 6). VL-type FIs homogenized to liquid H_2O phase at 215.9 to 297.3 °C (primarily 250 to 260 °C). These FIs have Tm-ice of −4.5 to −2.8 °C, with calculated salinities of 4.6 to 7.2 wt.% NaCl eqv (Figure 6).

Stage IV: Only VL-type FIs were identified (Figure 5(d-1,d-2)). The Tm-ice of these FIs is −3.5 to −2.2 °C, with corresponding salinities of 3.7 to 5.7 wt.% NaCl eqv (Figure 6). They homogenized into the liquid H_2O phase at 189.5 to 240.4 °C (primarily 210 to 230 °C) (Figure 6).

5.1.3. Laser Raman Spectroscopy

Representative primary fluid inclusions were selected for Laser Raman analysis. Stage I to II C-type FIs are generally CO_2-rich with minor CH_4 (Figure 7a–d), whereas only CO_2 was detected in Stage III C3-type FIs (Figure 7e). Stage IV VL-type FIs contain only H_2O (Figure 7f).

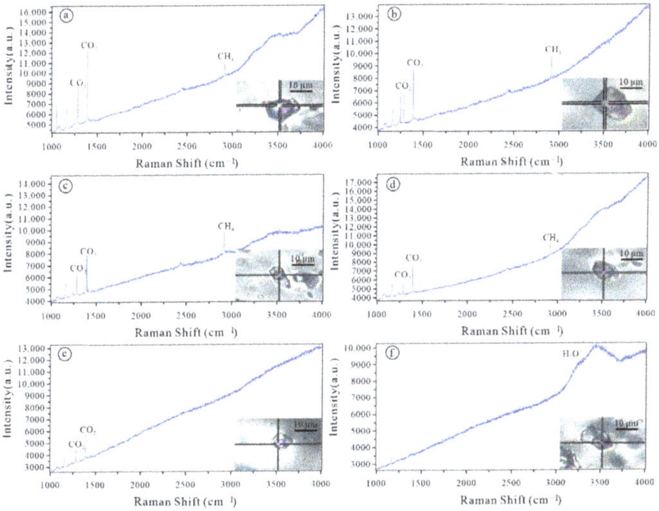

Figure 7. Laser Raman spectra for representative fluid inclusions (FIs) from the different hydrothermal stages at Songjianghe: (**a**,**b**) C1-type and C2-type FIs in Stage I quartz. (**c**,**d**) C3-type and C2-type FIs in Stage II quartz. (**e**) C3-type FIs in Stage III quartz. (**f**) VL-type FIs in Stage IV quartz.

5.2. Hydrogen-Oxygen Isotope Compositions

The H-O isotope data are presented in Table 2 and illustrated in Figure 8. Stage I $\delta D_{V\text{-}SMOW}$ and δ^{18}Oquartz values are −96.2 to −95.7‰ and 11.1 to 11.3‰, respectively. Stage II $\delta D_{V\text{-}SMOW}$ and δ^{18}Oquartz values are −98.7 to −89.8‰ and 10.1 to 10.6‰, respectively, whilst Stage III $\delta D_{V\text{-}SMOW}$ and δ^{18}Oquartz values are −103.5 to −101.2‰ and 9.8 to 9.9‰, respectively. Based on the host mineral-fluid fractionation equations [82] and the FI Th-total of Stages I to III, the calculated $\delta^{18}O_{fluid}$ values are 5.6 to 5.8‰ (Stage I), 3.7 to 4.2‰ (Stage II), and 1.2 to 1.4‰ (Stage III).

Table 2. Hydrogen and oxygen isotope data of fluid inclusions in quartz from different mineralization stages.

Sample No.	Mineralization Stages	Mineral	δDV-SMOW (‰)	δ18Oquartz (‰)	T (°C)	$\delta^{18}O_{fluid}$ (‰)
SJH-HO-1	Stage I	quartz	−95.7	11.1	341	5.6
SJH-HO-2		quartz	−96.2	11.3	341	5.8
SJH-HO-3	Stage II	quartz	−98.7	10.6	312	4.2
SJH-HO-4		quartz	−95.5	10.5	312	4.1
SJH-HO-5		quartz	−93.9	10.1	312	3.7
SJH-HO-6		quartz	−89.8	10.6	312	4.2
SJH-HO-7	Stage III	quartz	−101.2	9.9	258	1.4
SJH-HO-8		quartz	−103.5	9.8	258	1.2
SJH-HO-9		quartz	−102.1	9.9	258	1.4

Note: $\delta^{18}O_{fluid} = \delta^{18}O_{V\text{-}SMOW} - 3.38 \times 10^6/(T + 273)^2 + 3.40$ [82], T = average total homogenization temperatures (Th-total) of fluid inclusions.

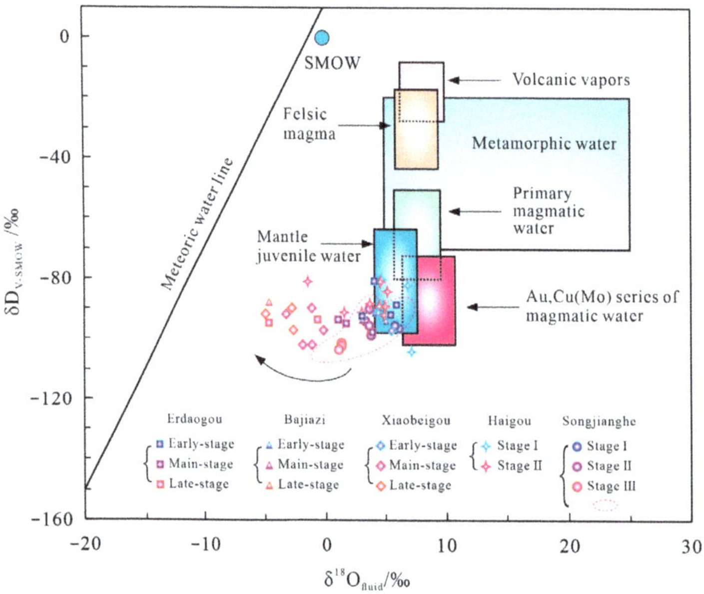

Figure 8. δDV-SMOW vs. δ18O$_{fluid}$ diagram for the different-stage hydrothermal fluids at Songjianghe (base map after [83]). The field of mantle juvenile water is from [84]. The field of Au, Cu (Mo) series of magmatic water is from [85]. The volcanic vapor and felsic magma fields are from [86]. The published data of δDV-SMOW and δ18O$_{fluid}$ compositions for the Erdaogou, Xiaobeigou, Bajiazi, and Haigou gold deposits in the JHGB are from [6,8,10,17,20,33,87].

5.3. Sulfur-Lead Isotope Compositions

Sulfur isotope compositions of Stages I and II (main-ore stage) pyrites are presented in Table 3 and illustrated in Figure 9. Seven pyrite samples from Stage I show $\delta^{34}S$ values varying from −2.91 to 0.42‰ (average = −1.04‰) and fifteen pyrite samples from Stage II have $\delta^{34}S$ values varying from −0.96 to 3.40‰ (average = 1.19‰). The samples as a whole yielded a relatively narrow $\delta^{34}S$ range (−2.91 to 3.40‰, mean 0.48‰) with a cluster around −1 to 1‰ (Figure 9a).

Table 3. Sulfur isotope compositions of the Songjianghe Au deposit.

Sample No.	Mineralization Stages	Mineral	$\delta^{34}S_{V-CDT}$ (‰)
SJH-S-1	Stage I	pyrite	−2.91
SJH-S-2	Stage I		−1.87
SJH-S-3	Stage I		−1.33
SJH-S-4	Stage I		−1.65
SJH-S-5	Stage I		0.42
SJH-S-6	Stage I		−0.15
SJH-S-7	Stage I		0.22
SJH-S-8	Stage II		1.39
SJH-S-9	Stage II		−0.84
SJH-S-10	Stage II		0.76
SJH-S-11	Stage II		−0.96
SJH-S-12	Stage II		3.40
SJH-S-13	Stage II		−0.33
SJH-S-14	Stage II		3.08
SJH-S-15	Stage II		1.55
SJH-S-16	Stage II		0.92
SJH-S-17	Stage II		2.42
SJH-S-18	Stage II		2.38
SJH-S-19	Stage II		2.76
SJH-S-20	Stage II		1.04
SJH-S-21	Stage II		−0.60
SJH-S-22	Stage II		0.90

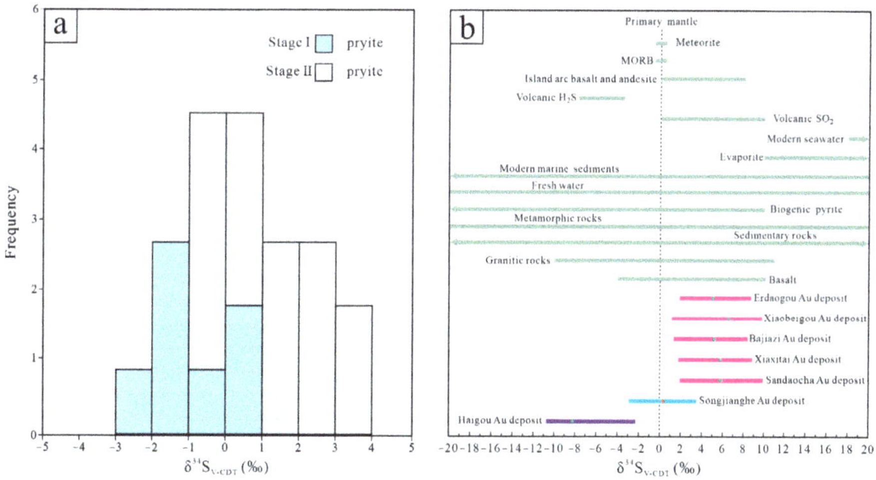

Figure 9. (**a**) $\delta^{34}S$ histogram from main-ore Stage II pyrite for the Songjianghe Au deposit; (**b**) $\delta^{34}S$ ranges for major Au deposits in the Jiapigou-Haigou gold belt (modified after [2,8,10,17,20,88–90].

Lead isotope compositions of stage II pyrite, Middle Jurassic monzogranite and K-feldspar granite, and the Seluohe Group biotite-plagioclase schist are presented in Table 4 and illustrated in Figure 10. The monzogranite ($^{206}Pb/^{204}Pb$ = 17.1720 to 17.3080, $^{207}Pb/^{204}Pb$ = 15.3720 to 15.4060, $^{208}Pb/^{204}Pb$ = 37.2280 to 37.3910) and K-feldspar granite ($^{206}Pb/^{204}Pb$ = 17.4630 to 17.5480, $^{207}Pb/^{204}Pb$ = 15.4080 to 15.4440, $^{208}Pb/^{204}Pb$ = 37.5280 to 37.6810) have relatively high Pb isotopic ratios, whilst the biotite-plagioclase schist ($^{206}Pb/^{204}Pb$ = 16.4550 to 16.8380, $^{207}Pb/^{204}Pb$ = 15.2050 to 15.2920, $^{208}Pb/^{204}Pb$ = 36.5910 to 36.9210) and Stage II pyrite ($^{206}Pb/^{204}Pb$ = 16.3270 to 16.4874, $^{207}Pb/^{204}Pb$ = 15.2258 to 15.3489, $^{208}Pb/^{204}Pb$ = 36.6088 to 36.7174) display low Pb isotopic ratios.

Table 4. Lead isotope compositions of ores and wallrocks from the Songjianghe Au deposit.

Sample No.	Sample Description	Test Object	$^{208}Pb/^{204}Pb$	$^{207}Pb/^{204}Pb$	$^{206}Pb/^{204}Pb$
SJH-Pb-1	monzogranite	whole rock	37.2820	15.3930	17.1720
SJH-Pb-2	monzogranite	whole rock	37.2280	15.3720	17.1760
SJH-Pb-3	monzogranite	whole rock	37.3910	15.4060	17.3080
SJH-Pb-4	K-feldspar granite	whole rock	37.5280	15.4080	17.4630
SJH-Pb-5	K-feldspar granite	whole rock	37.5980	15.4250	17.4970
SJH-Pb-6	K-feldspar granite	whole rock	37.6810	15.4440	17.5480
SJH-Pb-7	biotite-plagioclase schist	whole rock	36.9210	15.2920	16.8380
SJH-Pb-8	biotite-plagioclase schist	whole rock	36.5910	15.2140	16.4550
SJH-Pb-9	biotite-plagioclase schist	whole rock	36.6000	15.2050	16.4980
SJH-Pb-10	Stage II	pyrite	36.6265	15.2258	16.4874
SJH-Pb-11	Stage II	pyrite	36.6088	15.2655	16.4205
SJH-Pb-12	Stage II	pyrite	36.6153	15.2250	16.3270
SJH-Pb-13	Stage II	pyrite	36.6878	15.3489	16.4388
SJH-Pb-14	Stage II	pyrite	36.7174	15.3300	16.4230

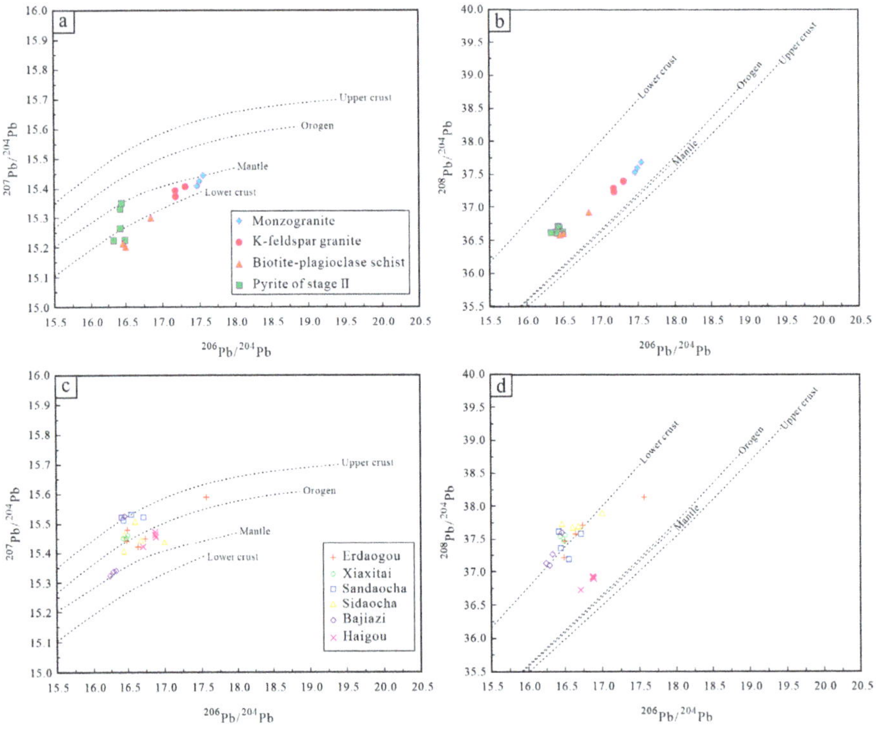

Figure 10. (**a**,**b**) Lead isotope compositions of the monzogranite, K-feldspar granite, Seluohe Group biotite-plagioclase schist, and the main ore-stage pyrite at Songjianghe. The average growth lines for the mantle, orogen, upper crust, and lower crust are from [91]. (**c**,**d**) The published data of Lead isotope compositions for the Erdaogou, Xiaxitai, Sandaocha, Sidaocha, Bajiazi, and Haigou gold deposits in the JHGB are from [20,89,92,93].

6. Discussion

6.1. Property and Origin of the Ore-Forming Fluids

Stage I: There are four types of primary FIs (C1-, C2-, C3-, and VL-type) coexisting in the quartz crystals (Figure 5(a1–a5)). These FIs have similar total homogenization temperatures (Th-total, 292.8 to 405.6 °C; Figure 6a) but diversified salinity (2.8 to 9.3 wt% NaCl eqv; Figure 6b) and different homogenization paths (in Section 5.1.2). The C- (C1-, C2-, and C3-) type FIs exhibit various CO_2/H_2O ratios (in Section 5.1.1). Moreover, the C-type FIs have similar homogenization temperatures of the CO_2 phase (Th-CO_2 = 16.2 to 30.9 °C). Thus, there has been considerable evidence that the ore fluids from Stage I were immiscible when the FIs were captured [94–96]. The melting temperatures of solid CO_2 (Tm-CO_2) of C-type FIs (−62.2 to 56.9 °C) are lower than the CO_2 triple point (−56.6 °C), implying the presence of a certain amount of CH_4 and/or N_2. The Laser Raman spectroscopic results demonstrated minor CH_4 in the C-type FIs, which is in accordance with the above inference (Figure 7a,b). The Stage I $\delta D_{V\text{-}SMOW}$ and $\delta^{18}O_{fluid}$ values have similar distribution ranges and plot into the magmatic domain (Figure 8). Additionally, the oxygen isotopic data ($\delta^{18}O_{fluid}$ = 5.6 to 5.8‰) are also in accordance with those of magmatic water (5.5‰ to 9.5‰), as proposed by [97]. The hydrogen-oxygen isotopic composition from the early stage at Songjianghe is comparable to other gold deposits (e.g., Erdaogou, Xiaobeigou, Bajiazi, and Haigou) in the JHGB (Figure 8), indicating a potentially similar source of the ore-forming fluids, which are distinctly inconsistent with the metamorphic fluids [8,17,33] (Figure 8). Taken together, Stage I ore fluids belong to a medium-high-temperature, low-salinity immiscible CO_2-NaCl-H_2O ± CH_4 system and are primarily derived from magmatic water.

Stage II: The coexisting C2-, C3-, and VL-type FIs have similar Th-totals (278.5 to 338.9 °C; Figure 6c), yet they have different salinity values (2.8 to 8.1 wt% NaCl eqv; Figure 6d) and homogenization paths (in Section 5.1.2). The C2- and C3-type FIs are characterized by various CO_2/H_2O ratios (in Section 5.1.1). In addition, the Th-CO_2 of C2- and C3-types have similar distribution ranges (21.8 to 30.7 °C). These features suggest that the ore fluids from Stage II are immiscible [94]. Tm-CO_2 of C-type (C2- and C3-type) FIs (−60.1 to −56.9 °C) and Laser Raman spectroscopic results (Figure 7c,d) demonstrated minor CH_4 in the fluid inclusions. Although $\delta^{18}O_{fluid}$ values have slightly drifted toward meteoric water relative to Stage I, $\delta D_{V\text{-}SMOW}$ values remain almost unchanged (Figure 8). In general, the fluid–rock interaction is a common process at temperatures of 280–380 °C in hydrothermal gold deposits [98]. Meanwhile, in contrast to the quartz vein-type ore, the ore fluids of the altered type are characteristically dominated by magmatic fluids and less admixed meteoric water [99], as supported by the hydrogen-oxygen isotope data of quartz vein-type gold deposits (e.g., Erdaogou, Xiaobeigou, Bajiazi, and Haigou) within the JHGB (Figure 8). Therefore, the slight variation in $\delta^{18}O_{fluid}$ values is interpreted as an exchange of material between the ore fluids and wallrocks, rather than being caused by the input of meteoric water. In combination, Stage II ore fluids belong to a moderate-temperature, low-salinity immiscible CO_2-NaCl-H_2O ± CH_4 system derived from magmatic water.

Stage III: The identified C3- and VL-type FIs commonly occur in groups in the same quartz grain and have a similar Th-total (215.9 to 307.3 °C; Figure 6e), suggesting that they were captured almost simultaneously. These FIs do not constitute an immiscible FIA and cannot represent a fluid-immiscible assemblage. This is because C3-type FIs have the same CO_2/H_2O ratio (in Section 5.1.1), and the vapor/liquid ratios of VL-type FIs are similar. The $\delta D_{V\text{-}SMOW}$ and $\delta^{18}O_{fluid}$ values fall between the fields of meteoric water and magmatic water (Figure 8), indicating meteoric water incursion. The volatiles (CO_2 and CH_4) of FIs from Stage I to III decreased apparently, which may have been attributed to decompression, crystallization-induced degassing, and/or meteoric water incursion [100]. Hence, Stage III ore fluids belong to a medium-low-temperature, low-salinity homogeneous CO_2-NaCl-H_2O system, which was sourced from the mixing of magmatic water and meteoric water.

Stage IV: Only VL-type FIs are identified, which commonly occur in groups in the same quartz grain. These FIs have similar vapor–liquid ratios (5 to 15%), Th-totals (189.5

to 240.4 °C), and salinity values (3.7 to 5.7 wt.% NaCl eqv), which decreased further in comparison to Stage III equivalents (Figure 6g,h). This suggests that the VL-type FIs are almost captured simultaneously and that the ore fluids were transformed into a low-temperature, low-salinity homogeneous NaCl-H_2O system accompanied by more meteoric water mixing.

6.2. Fluid Evolution and Gold Mineralization

The characteristics of FIs from Stages I and II demonstrate that fluid immiscibility occurred prior and synchronous to gold mineralization in the Songjianghe Au deposit, and the initial ore-forming fluids were derived from the exsolution of magma. Along with the continuous crystallization and evolution of the magma, the volatile-rich (CO_2, CH_4, and H_2O) magmatic fluids ascended to shallower depths owing to the fracturing and/or hydrofracturing of country rocks [73]. Subsequently, the rapid pressure drop caused by fault failure and fluid drainage triggered the fluid immiscibility and phase separation [101] (Stage I; Figure 11).

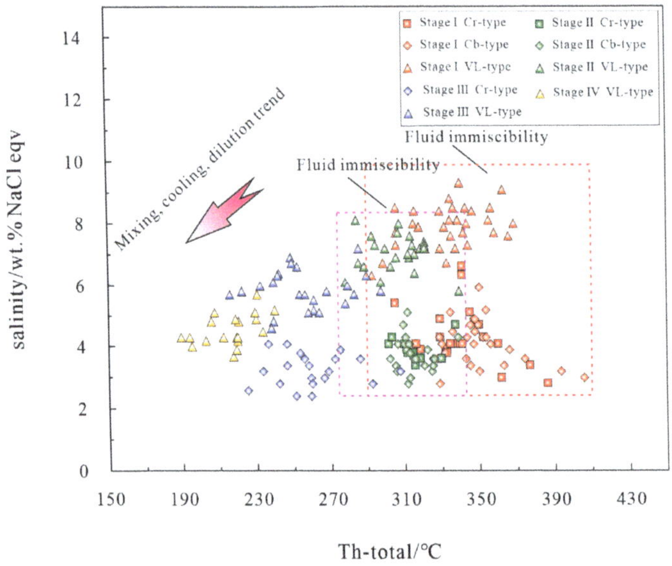

Figure 11. Total homogenization temperature (Th-total) vs. salinity plot for ore-stage fluid inclusions assemblages (FIAs) from the Songjianghe Au deposit.

In hydrothermal fluids, gold bisulfide complexes ($Au(HS)_2^-$ and/or $Au(HS)^0$) are the primary ligands for gold migration at a pressure of 200 MPa and temperatures of 200–400 °C [102,103]. Hence, gold bisulfide was the most probable ligand for gold migration in the Songjianghe Au deposit. Additionally, the primitive ore fluids were enriched in CO_2, which may be conducive to the transport of gold bisulfide complexes by buffering the pH of the solution [104]. Stage II (main ore stage) FIs document the continuous occurrence of fluid immiscibility (Figure 11). Relative to Stage I, the CO_2 component of FIs in Stage II is significantly reduced, and salinity drops (Figures 6a–d and 11), revealing the escape of volatile components due to fluid immiscibility. The release of volatiles can result in a decrease in oxygen fugacity, an increase in pH, and a drop in the pressure of the ore fluids, which destabilized the gold bisulfide complexes and accelerated the gold precipitation [80,105]. Synchronously, the HS^- generated by the decomposition of gold bisulfide complexes was combined with metal cations to form polymetallic sulfides (e.g., pyrite, chalcopyrite, galena, and sphalerite) and released H^+ [106].

Following a large amount of volatile components' escape and metal precipitation, the hydrothermal fluid entered a relatively open condition and meteoric water began to admix (Stage III; Figures 8 and 11). C3-type FIs have the same proportion of aqueous solution and CO_2, and the vapor–liquid ratios of VL-type FIs are similar, revealing that fluid immiscibility essentially ended. As meteoric water was continuously introduced, the ore fluids gradually cooled, diluted, and tended to resemble meteoric water (Figure 11), which signaled the decline and conclusion of the gold mineralization process with the production of calcite and quartz (Stage IV; Figure 4d).

6.3. Fluid Trapping Temperature-Pressure Conditions

In general, it is difficult to accurately estimate the mineralization pressure, unless the mineralization occurred under the exact temperature or immiscible/boiling conditions when the FIs were trapped [94–96]. As end-members trapped in immiscible CO_2-NaCl-H_2O $\pm$ CH_4 system fluids, the trapping temperature–pressure conditions of C2- and VL-type FIs could approximately represent the mineralization temperature and pressure. For Stage I and II, the Th-total of some VL-type FIs is slightly lower than that of C2-type FIs, which can be interpreted as a result of multiple fluid immiscibilities with fluctuating pressure during mineralization [107,108]. Therefore, we applied the isochore intersection method to constrain the trapping temperature–pressure conditions of C2- and VL-type FIs (with similar Th-totals) for Stages I and II in the same thin sheet. The volume range of the CO_2 phase for C2-type FIs is over 50% (primarily 80–90%), and VL-type FIs have two phases (vapor H_2O + liquid H_2O) with vapor H_2O accounting for < 50 vol.% (primarily 20–30 vol.%). The isochores of C2- and VL-type FIs were plotted using MacFlincor software [95]. As shown in Figure 12, the trapping temperature and pressure for Stage I and II are 395–525 °C, 1046–1778 bar, and 352–448 °C, 850–1380 bar, respectively. The Songjianghe Au deposit was likely formed at a higher pressure (850–1380 bar) than other major gold deposits characterized by the quartz vein-type in the JHGB (520 to 1030 bar; e.g., Bajiazi, Xiaxitai, and Sandaocha [2]), suggesting that the gold mineralization characterized by disseminated-style ores may have occurred at a greater mineralization depth with a small-scale and relatively close structural system. Meanwhile, field investigations also show that the ore-controlling structural system of the Songjianghe Au deposit is characterized by narrower brittle-ductile fractures, whilst other major quartz vein-type gold deposits in the JHGB (e.g., Jiapigoubenqu, Erdaogou, Sandaocha, Xiaobeigou, and Banmiaozi [8,17]) are controlled primarily by relatively wider ductile-brittle fractures.

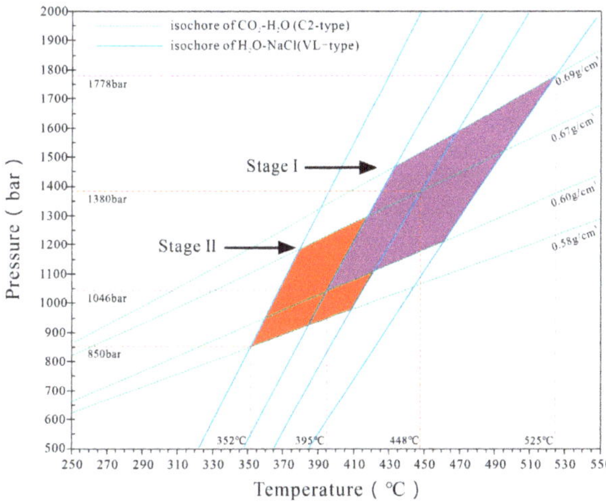

Figure 12. Trapping temperature–pressure diagram for Stage II C2-type and VL-type fluid inclusions from the Songjianghe Au deposit.

6.4. Source of Ore-Forming Materials

There are four stable isotopes of sulfur in nature, with a relative abundance of ^{32}S 95.02 %, ^{33}S 0.75 %, ^{34}S 4.21 %, and ^{36}S 0.02 % [109]. Geochemical studies generally consider only the most widely distributed and abundant $^{34}S/^{32}S$ ratio, expressed as $\delta^{34}S$. It is generally accepted that different $\delta^{34}S$ values can be derived from three common sulfur sources [110]: (1) Mantle or magmatic source ($\delta^{34}S = 0 \pm 3‰$); (2) seawater sulfates ($\delta^{34}S$ = approximately + 20‰ but varies greatly); and (3) reduced sulfur from sedimentary rocks (negative $\delta^{34}S$, generally $< -15‰$). In hydrothermal systems, where H_2S is the dominant sulfur species in the fluids or the fluid redox state is below the SO_2/H_2S boundary, the following relation exists as $\delta^{34}S$ (sulfide) = $\delta^{34}S$ (fluid) [111]. The sulfur speciation at Songjianghe is dominated by sulfides (especially pyrite, ~90 vol%) with no obvious sulfate development. Thus, the $\delta^{34}S$ values of pyrite could represent the $\delta^{34}S$ values of the hydrothermal fluid system [112]. There are major differences in sulfur isotopes between the Songjianghe gold deposit and other major gold deposits in the JHGB (Figure 9b): The latter (e.g., the Erdaogou, Xiaobeigou, Bajiazi, Xiaxitai, and Sandaocha deposits; Figure 1c) have relatively wide-range and positive $\delta^{34}S$ values (1 to 10‰), which clusters around 1 to 3‰ and 5 to 9‰ [2,87]. The low $\delta^{34}S$ values (1 to 3‰) exhibit a magmatic sulfur isotope signature [110]. The high $\delta^{34}S$ values (5 to 9‰) imply that the deep-sourced sulfur may have mixed with shallow crustal sulfur during the hydrothermal fluid ascent [17,21,87]. Sulfides from the Haigou gold deposit (southeast of Songjianghe; Figure 1c) have negative $\delta^{34}S$ values (−10.9 to −2.3‰ [90]), which may be caused by the following factors: (1) The early mineralization stage may have been oxidizing, causing the fractionation of ^{34}S, as suggested by the extensive development of magnetite and hematite in the early stage [113]; (2) fluid reactions with the low-$\delta^{34}S$ wallrocks (−10.9 to −10.5‰ [90]). In this study, Stage II $\delta^{34}S$ values (−0.96–3.40‰) are slightly higher than those of Stage I (−2.91–0.42‰), which may be caused by crustal components added during hydrothermal fluid evolution [114]. However, the samples as a whole yielded a relatively narrow $\delta^{34}S$ range (−2.91–3.40‰, mean 0.48‰; Figure 9), suggesting a typical magmatic sulfur source ($0 \pm 3‰$ [112]). Compared to other major gold deposits (e.g., the Erdaogou, Xiaobeigou, Bajiazi, Xiaxitai, and Sandaocha deposits) in the JHGB, the Songjianghe Au deposit has a single sulfur source and relatively lower $\delta^{34}S$ values. Wang and Yan [115] and Wang et al. [116] proposed that pyrite $\delta^{34}S$ values are relatively higher in the gold deposits hosted in large and open structural systems, and lower in those hosted in small-scale and relatively close structural systems. Hence, the Songjianghe Au deposit may have been formed in a small-scale and relatively close structural system, as supported by the trapping temperature–pressure conditions and field investigations.

The relative atomic mass of Pb is large, and the relative mass difference between different Pb isotopes is limited. As a result, Pb not only struggles to experience isotope fractionation when leaching from the ore source layer but also generally has no changes in composition during transfer into the ore-forming hydrothermal fluid and subsequent migration [117]. Therefore, lead isotope compositions of sulfides are widely used to constrain the source of lead [118,119]. The pyrite Pb isotope compositions from Stage II projected primarily toward the field between the mantle and lower crust evolution curve (Figure 10a). Meanwhile, they also exhibit a consistent range between $^{206}Pb/^{204}Pb$ and $^{208}Pb/^{204}Pb$ (Figure 10b). These characteristics indicate that the primitive Pb at Songjianghe may have sourced from a low-radiogenic lower crust and mantle lead reservoir. Although the Pb isotope data of monzogranite and K-feldspar granite in the Songjianghe Au deposit also fall between the field between the mantle and lower crust evolution curve, its relatively high-radiogenic member is distinctly different from those of Stage II pyrite (Figure 10a,b). The Pb isotope compositions from the Seluohe Group biotite-plagioclase schist fall close to the lower crust evolution curve (Figure 10a) and are similar to the ore Pb (Figure 10b), therefore the involvement of Pb from wallrock cannot be ruled out. The Pb isotope data from other major gold deposits (e.g., Erdaogou, Xiaxitai, Sandaocha, Sidaocha, Bajiazi, and Haigou) in the JHGB fall mostly between the mantle and lower crust (Figure 10c,d). Additionally, the Pb isotope compositions of these gold deposits show broad correlations, which suggest a two-component mixing of the lower crust and mantle components (Figure 10c,d), inconsistent with the Songjianghe Au deposit. These features indicate that the crustal-source Pb contributes significantly less to the Songjianghe Au deposit than other major gold deposits in JHGB. In conclusion, we propose that the Pb in the Songjianghe Au deposit was primarily derived from a magmatic source (mantle and lower crust) and may have been affected by the Seluohe Group wallrock contamination due to the fluid–rock interaction.

6.5. Ore Genesis

The Songjianghe Au mineralization characterized by disseminated-style ores is predominantly hosted in the metamorphic rocks of the Seluohe Group. The distribution of the ore body, spatially related to the sericitization-silicification zone, is primarily controlled by the NNW-striking structures, which were characterized by the early-stage ductile shearing superimposed by the late-stage brittle-ductile fractures.

Previous studies suggest that the Songjianghe Au deposit shares several characteristics with orogenic gold deposits, such as the mineralization type, composition of metallic minerals, characteristic of structural ore-control, and wallrock alteration [25,27], whereas there are apparent differences between the Songjianghe Au deposit and the typical orogenic gold deposits hosted in Precambrian medium-low grade metamorphic rocks [114]. The ore-forming fluids of orogenic gold deposits are usually derived from the progressive metamorphism process under the conditions of greenschist and amphibolite facies [120], whilst the ore fluids at Songjianghe are dominated by magmatic water with later meteoric water input. Additionally, the metamorphic age (2521–2506 Ma) of the ore-bearing wallrocks (the Seluohe Group) is much older than the mineralization age (157–156 Ma) [31]. In addition, there was no visible orogenic metamorphic event in the study region during the Late Jurassic [121,122].

The JHGB is located in the tectonic transition between the northeastern NCC margin and the eastern CAOB, witnessing the formation and extinction of the Paleo-Asian Ocean and the subduction of the Paleo-Pacific plate beneath Eurasia [17,50]. As mentioned in the geological setting, the greenschist facies retrograde metamorphism of the Hulan Group indicates that the Paleo-Asian Ocean closed before the Late Triassic [43,59]. In the Early-Middle Jurassic, the variation in composition polarity of igneous rocks from the continental margin to the intracontinental region revealed the beginning of subduction of the Paleo-Pacific plate in NE China [121–123]. In addition, the widely developed late Early Jurassic to early Middle Jurassic (178–168 Ma) medium-acid intrusions in the JHGB show characteristically geochemical signatures of island arcs or active continental margins, suggesting the occurrence of Paleo-Pacific plate subduction [2,30,49,71]. Abundant gold mineralization events occurred during this period in the JHGB (e.g., Xiaxitai, Sidaocha, Liupiye, and Xiaobeigou [2,4,7]). However, during the Late Middle Jurassic to Early Cretaceous (165–140 Ma), there has been a significant intermittence of magmatic activity in eastern Jilin Province and Songliao Basin, which is interpreted as the Paleo-Pacific plate experiencing an intermittent stage of subduction and/or a strike-slip tectonic regime existing between the continental margin of Northeast Asia and the Paleo-Pacific plate [121,122]. Simultaneously, under the effects of the Paleo-Pacific plate subduction in the Early-Middle Jurassic, the thickened lithospheric mantle from NE China was delaminated due to gravity instability, accompanied by lithospheric thinning [124–126] (Figure 13). According to the sericite $^{40}Ar/^{39}Ar$ dating, the Au mineralization at Songjianghe took place during the Late Jurassic (157–156 Ma [31]). The delamination and lithospheric thinning may have led to asthenosphere upwelling, resulting in the partial melting of the upper mantle and lower crust to generate original magma carrying abundant metals and volatile components [125,126] (Figure 13). With the evolution and exsolution of the magma, the initial ore-forming fluids were generated (Figure 13). As the ore fluids ascended to the shallower zone, the rapid pressure drop triggered the fluid immiscibility and phase separation, which contributes to the decrease in oxygen fugacity, the increase in pH, and the decomposition of gold bisulfide complexes. These processes led to significant gold-polymetallic sulfide precipitation (primarily in Stage II). The H-O isotope compositions suggest that the ore-forming fluids are dominated by magmatic water (Stage I and II) and mixed with meteoric water in the later stage (Stage III). As meteoric water was continuously introduced, the ore fluids gradually cooled, diluted, and tended to resemble meteoric water, which suggested the decline and termination of the gold mineralization process (Stage IV). The S-Pb isotope data indicate that ore-forming materials at Songjianghe are characterized by deep magmatic sources and may be affected by the contamination of Seluohe Group wallrocks due to the fluid–rock interaction. Consequently, the Songjianghe Au deposit belongs to a mesothermal magmatic hydrothermal vein gold deposit, which formed in the intermittent stage of Paleo-Pacific plate subduction during the Late Jurassic.

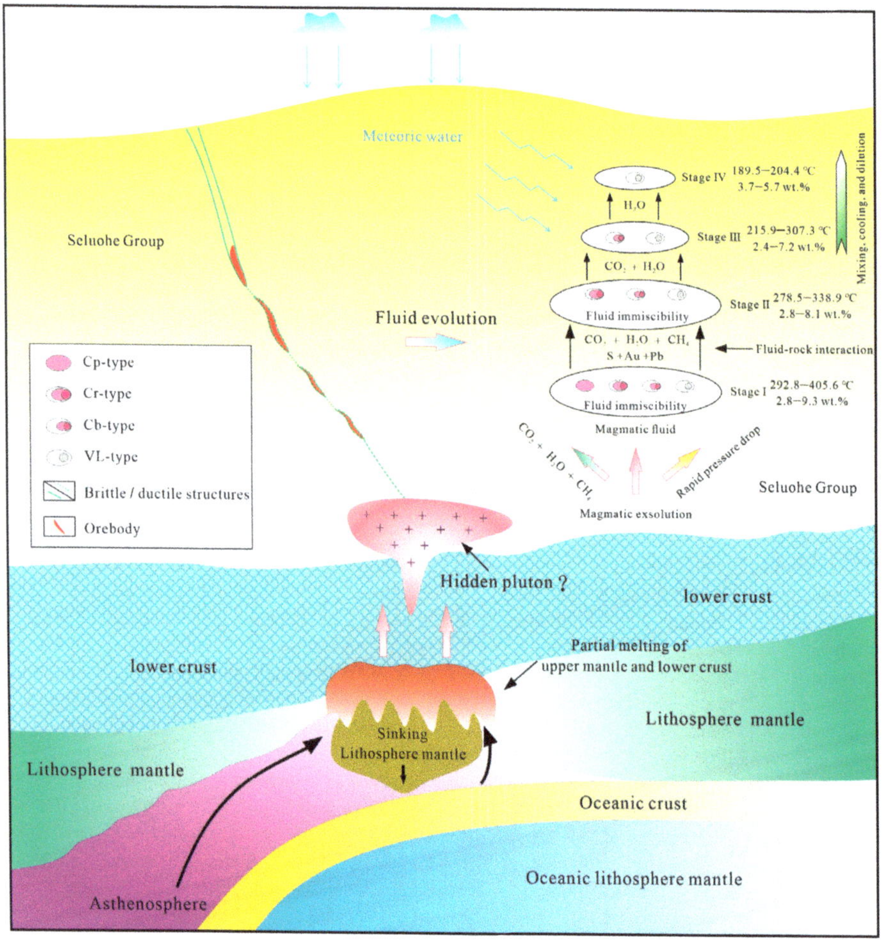

Figure 13. Schematic ore genetic model for the Songjianghe Au deposit.

7. Conclusions

1. The Songjianghe Au deposit is a magmatic hydrothermal gold deposit formed in the intermittent stage of Paleo-Pacific plate subduction during the Late Jurassic. Four alteration/mineralization stages were recognized in this gold deposit: (I) A quartz-pyrrhotite-pyrite stage; (II) a quartz-polymetallic sulfides stage; (III) a quartz-pyrite stage; and (IV) a quartz-calcite-minor pyrite stage.

2. The ore-forming fluids have evolved from a medium-high-temperature, low-salinity immiscible CO_2-NaCl-H_2O ± CH_4 system to a low-temperature, low-salinity homogeneous NaCl-H_2O system.
3. The H-O isotope data suggest that the ore-forming fluids were dominated by magmatic water (Stage I and II) and intruded by meteoric water since Stage III.
4. Fluid immiscibility caused by the rapid drop in pressure may have been the main trigger for gold-polymetallic sulfide precipitation in the Songjianghe Au deposit.
5. The S-Pb isotope data indicate that the ore-forming materials have predominantly derived from a magmatic source and may have been affected by the contamination of the Seluohe Group.

Author Contributions: Conceptualization, Q.Y. and K.W.; investigation, Q.Y., K.W., W.B., C.M. and Y.X.; funding acquisition, K.W.; project administration, K.W. and Q.S.; writing—original draft preparation, Q.Y.; writing—review and editing, Q.Y., K.W., X.Z. and Q.S. All authors have read and agreed to the published version of the manuscript.

Funding: This research was supported by the Opening Foundation of Key Laboratory of Mineral Resources Evaluation in Northeast Asia, Ministry of Land and Resources of China (DBY-ZZ-18-12, DBY-KY-18-08).

Data Availability Statement: The experimental data used to support the conclusions of this study are included within the article.

Acknowledgments: We would like to thank Gang Zhao and Hong Wei from the Songjianghe gold mine for helping with the field sampling. In addition, we wish to thank Shuo Zhao (Chinese Academy of Geological Sciences) for providing valuable advice on the manuscript, and to the anonymous reviewers and editors for their insightful comments.

Conflicts of Interest: The authors declare no conflict of interest.

References

1. Li, L. Research on Ore-Forming Fluids of Gold Deposits in Jiapigou-Haigou Gold Belt, Jilin Province and Deep-Seated Metallogenic Assessment. Ph.D. Thesis, Jilin University, Changchun, China, 2016. (In Chinese)
2. Zhang, X.T. Research on Geology, Geochemistry and Metallogenesis of the Gold Deposits of the Jiapigou Ore Field in the Continental Margin of Northeast China. Ph.D. Thesis, Jilin University, Changchun, China, 2018. (In Chinese)
3. Liu, S.F.; Wang, Q.F.; Groves, D.I.; Wang, Z.M.; Yang, L.; Wu, Z.L.; Yu, Z.Y.; Huang, P.F.; Deng, J. Adoption of a mineral system model in successful deep exploration at Erdaogou, China's deepest gold mine, on the northeastern margin of the North China Craton. *Ore Geol. Rev.* **2021**, *131*, 104060. [CrossRef]
4. Sun, Z.S.; Feng, Y.M. Main minerogenetic epoch, determine and exploratory direction of Jiapigou gold deposit, Jilin. *Acta Geosci. Sin.* **1997**, *18*, 367–372. (In Chinese)
5. Zhang, S.; Zeng, Q.D.; Liu, J.M.; Wang, Z.C.; Gao, J.H. The characteristics and its geological significance of fluid inclusions in Haigou Lode gold deposit, Jilin Province. *Acta Petrol. Sin.* **2011**, *27*, 1287–1298. (In Chinese)
6. Yang, L.Y.; Yang, L.Q.; Yuan, W.M.; Zhang, C.; Zhao, K.; Yu, H.J. Origin and evolution of ore fluid for orogenic gold traced by D-O isotopes: A case from the Jiapigou gold belt, China. *Acta Petrol. Sin.* **2013**, *29*, 4025–4035. (In Chinese)
7. Deng, J.; Yuan, W.M.; Carranza, E.J.M.; Yang, L.Q.; Wang, C.M.; Yang, L.Y.; Hao, N.N. Geochronology and thermochronometry of the Jiapigou gold belt, northeastern China: New evidence for multiple episodes of mineralization. *J. Asian Earth Sci.* **2014**, *89*, 10–27. [CrossRef]
8. Li, L.; Sun, J.G.; Men, L.J.; Chai, P. Origin and evolution of the ore-forming fluids of the Erdaogou and Xiaobeigou gold deposits, Jiapigou gold province, NE China. *J. Asian Earth Sci.* **2016**, *129*, 170–190. [CrossRef]
9. Li, L.; Sun, J.; Men, L.; Chai, P. Genesis of the Haigou gold deposit, Jilin Province, NE China: Evidence from fluid inclusions, $^{40}Ar/^{39}Ar$ geochronology and isotopes. *Geol. J.* **2017**, *52*, 992–1008. [CrossRef]
10. Liu, J.; Li, T.G.; Duan, C. Rb-Sr Isochron Dating and Isotopic Geochemistry Characteristics of the Bajiazi Large Gold Deposit, Jilin Province, China. *Acta Geol. Sin.* **2018**, *92*, 1432–1446, (In Chinese with English Abstract).
11. Zhai, D.G.; Williams-Jones, A.E.; Liu, J.J.; Selby, D.; Li, C.; Huang, X.W.; Qi, L.; Guo, D.H. Evaluating the Use of the Molybdenite Re-Os Chronometer in Dating Gold Mineralization: Evidence from the Haigou Deposit, Northeastern China. *Econ. Geol.* **2019**, *114*, 897–915. [CrossRef]
12. Li, J.J.; Shen, B.F.; Mao, D.B.; Li, S.B.; Zhou, H.F. Metallogenic epochs of the Jiapigou gold deposit, Jilin Province, China. *Acta Geol. Sin.* **1996**, *70*, 335–341. (In Chinese with English Abstract)
13. Dai, J.Z.; Wang, K.Y.; Cheng, X.M. Geochemical features of ore-forming fluids in the Jiapigou gold belt, Jilin Province. *Acta Petrol. Sin.* **2007**, *23*, 2198–2206. (In Chinese with English Abstract)

14. Huang, Z.X.; Yuan, W.M.; Wang, C.M.; Liu, X.W.; Xu, X.T.; Yang, L.Y. Metallogenic epoch of the Jiapigou gold belt, Jilin Province, China. Constrains from rare earth element, fluid inclusion geochemistry and geochronology. *J. Earth Syst. Sci.* **2012**, *121*, 1401–1420. [CrossRef]
15. Feng, Y.Y. A Study on the Genesis of Bajiazi Gold Ore in Jiapigou, Jilin Province. Master's Thesis, Jilin University, Changchun, China, 2017. (In Chinese)
16. Sun, Z.S.; Feng, B.Z.; Qing, X. The stable isotopic geology and exploration direction of the Jiaopigou gold ore deposit, Jilin Province. *J. Chang. Univ. Sci. Technol.* **1998**, *28*, 142–147. (In Chinese)
17. Zeng, Q.D.; Wang, Z.C.; He, H.Y.; Wang, Y.B.; Liu, J.M. Multiple isotope composition (S, Pb, H, O, He, and Ar) and genetic implications for gold deposits in the Jiapigou gold belt, Northeast China. *Miner. Depos.* **2014**, *49*, 145–164. [CrossRef]
18. Shen, B.F.; Peng, X.L.; Luo, H.; Mao, D.B. Archean greenstone belts in China. *Acta Geol. Sin.* **1993**, *67*, 208–220. (In Chinese with English Abstract)
19. Xu, X.T. On the Magmatic Rocks and Ore-Forming Fluid in the Jiapigou Gold Belt, Jilin Province. Master's Thesis, China University of Geosciences, Beijing, China, 2010. (In Chinese)
20. Huang, Z.X. Tectonics–Fluids–Mineralization System of Jiapigou Gold Belt in Jilin Province, China. Ph.D. Thesis, China University of Geosciences, Beijing, China, 2012. (In Chinese)
21. Miao, L.C.; Qiu, Y.M.; Fan, W.M.; Zhang, F.Q.; Zhai, M.G. Geology, geochronology, and tectonic setting of the Jiapigou gold deposits, southern Jilin Province, China. *Ore Geol. Rev.* **2005**, *26*, 137–165. [CrossRef]
22. Yu, H.J. On the Geology Geochemistry and Genetic Model Study of the Jiapigou Ore Deposit of Jilin Province. Master's Thesis, China University of Geosciences, Beijing, China, 2009. (In Chinese)
23. Zeng, Q.D.; Wang, Z.C.; Zhang, S.; Wang, Y.B.; Yang, Y.H.; Liu, J.M. The ore-forming epoch of the Jiapigou gold belt, NE China: Evidences from the zircon LA–ICP–MS U–Pb dating of the intrusive rocks. *Acta Geol. Sin.* **2014**, *88*, 1029–1030. [CrossRef]
24. Li, B.L.; Chen, G.J.; Song, Z.W. Discussion on minerogenetic epoch of the gold deposit in Jiapigou of Jilin Province. *Glob. Geol.* **2004**, *23*, 354–359. (In Chinese with English Abstract)
25. Li, B. Study on Geological Characteristics and Genesis of Songjianghe Gold Deposit, Dunhua City, Jilin Province. Master's Thesis, Jilin University, Changchun, China, 2015. (In Chinese)
26. Yang, Y. Geological Characteristics and Prospecting Direction of Songjianghe Gold Deposit in Dunhua City, Jilin Province. Master's Thesis, Jilin University, Changchun, China, 2019. (In Chinese)
27. Jin, H.Y.; Yu, Z.T.; Wu, J.X. Characteristics of fluid inclusions and genesis of Songjianghe gold deposit in Jilin Province. *Glob. Geol.* **2014**, *33*, 76–85. (In Chinese with English Abstract)
28. Qin, Z.; Sun, D.Y.; Zhang, J.L.; Li, L.W.; Wang, Z.Y.; Li, X.J. Characteristics of fluid inclusions and metallogenic physicochemical conditions of Songjianghe Gold Deposit in Dunhua, Jilin Province. *Gold* **2019**, *40*, 14–18+22. (In Chinese with English Abstract)
29. Liu, Y.M.; Sun, J.G.; Li, X.M.; Song, Q.H.; Wang, Q.H.; Zhang, X.T. Geology, fluid inclusion and genesis of Songjianghe gold deposit in Dunhua City, Jilin Province. *Glob. Geol.* **2021**, *40*, 273–287. (In Chinese with English Abstract)
30. Lu, Q.; Sun, J.G.; An, J.H.; Han, J.L.; Chu, X.L. Petrogenesis and Forming Environment of Mesozoic Porphyritic Granite in Songjianghe Region of Dunhua, Jilin Province: Evidences from Element Geochemistry, Hf Isotope and U-Pb. *J. Jilin Univ.* **2019**, *49*, 673–689. (In Chinese with English Abstract)
31. Zhang, X.T.; Sun, J.G.; Yu, Z.T.; Song, Q.H. LA-ICP-MS zircon U–Pb and sericite $^{40}Ar/^{39}Ar$ ages of the Songjianghe gold deposit in southeastern Jilin Province, Northeast China, and their geological significance. *Can. J. Earth Sci.* **2019**, *56*, 607–628. [CrossRef]
32. Hou, S.H.; Sang, H.Q. $^{40}Ar/^{39}Ar$ fast neutron activation dating method for sericite from Liupiye gold deposit, Jilin Province. *Miner. Depos.* **2004**, *23*, 502–508. (In Chinese with English Abstract)
33. Zeng, Q.D.; He, H.Y.; Zhu, R.X.; Zhang, S.; Wang, Y.B.; Su, F. Origin of ore-forming fluids of the Haigou gold deposit in the eastern Central Asian Orogenic belt, NE China: Constraints from H-O-He-Ar isotopes. *J. Asian Earth Sci.* **2017**, *144*, 384–397. [CrossRef]
34. Şengör, A.M.C.; Natal'in, B.A.; Burtman, V.S. Evolution of the Altaid tectonic collage and Paleozoic crustal growth in Eurasia. *Nature* **1993**, *364*, 299–307. [CrossRef]
35. Jahn, B.M.; Wu, F.Y.; Chen, B. Massive granitoid generation in central Asia: Nd isotopic evidence and implication for continental growth in the Phanerozoic. *Episodes* **2000**, *23*, 82–92. [CrossRef]
36. Windley, B.F.; Alexeiev, D.; Xiao, W.J.; Kröner, A.; Badarch, G. Tectonic model for accretion of the Central Asian Orogenic Belt. *J. Geol. Soc.* **2007**, *164*, 31–47. [CrossRef]
37. Guo, F.; Fan, W.M.; Gao, X.F.; Li, C.W.; Miao, L.C.; Zhao, L.; Li, H.X. Sr–Nd–Pb isotope mapping of Mesozoic igneous rocks in NE China: Constraints on tectonic framework and Phanerozoic crustal growth. *Lithos* **2010**, *120*, 563–578. [CrossRef]
38. Kravchinsky, V.A.; Cogn, J.P.; Harbert, W.P.; Kuzmin, M.I. Evolution of the Mongol-Okhotsk Ocean as constrained by new palaeomagnetic data from the Mongol-Okhotsk suture zone, Siberia. *Geophys. J. R. Astron. Soc.* **2010**, *148*, 34–57. [CrossRef]
39. Cao, H.H.; Xu, W.L.; Pei, F.P.; Wang, Z.W.; Wang, F.; Wang, Z.J. Zircon U–Pb geochronology and petrogenesis of the Late Paleozoic–Early Mesozoic intrusive rocks in the eastern segment of the northern margin of the North China Block. *Lithos* **2013**, *170–171*, 191–207. [CrossRef]
40. Yu, J.J.; Wang, F.; Xu, W.L.; Gao, F.H.; Tang, J. Late Permian tectonic evolution at the southeastern margin of the Songnen–Zhangguangcai range massif, NE China: Constraints from geochronology and geochemistry of granitoids. *Gondwana Res.* **2013**, *24*, 635–647. [CrossRef]

41. Wilde, S.A. Final amalgamation of the Central Asian orogenic belt in NE China: Paleo-Asian Ocean closure versus Paleo-Pacific plate subduction—A review of the evidence. *Tectonophysics* **2015**, *662*, 345–362. [CrossRef]
42. Zhou, J.B.; Wilde, S.A.; Zhao, G.C.; Han, J. Nature and assembly of microcontinental blocks within the Paleo-Asian Ocean. *Earth-Sci. Rev.* **2018**, *186*, 76–93. [CrossRef]
43. Wu, F.Y.; Zhao, G.C.; Sun, D.Y.; Wilde, S.A.; Zhang, G.L. The Hulan Group: Its role in the evolution of the Central Asian Orogenic Belt of NE China. *J. Asian Earth Sci.* **2007**, *30*, 542–556. [CrossRef]
44. Wu, F.Y.; Sun, D.Y.; Ge, W.C.; Zhang, Y.B.; Grant, M.L.; Wilde, S.A.; Jahn, B.M. Geochronology of the Phanerozoic granitoids in northeastern China. *J. Asian Earth Sci.* **2011**, *41*, 1–30. [CrossRef]
45. Huang, H.; Zhang, Z.C.; Santosh, M.; Zhang, D.Y.; Wang, T. Petrogenesis of the Early Permian volcanic rocks in the Chinese South Tianshan: Implications for crustal growth in the Central Asian Orogenic Belt. *Lithos* **2015**, *228–229*, 23–42. [CrossRef]
46. Bo, H.; Zhang, Z.; Santosh, M.; Cheng, Z.; Xie, Q.; Kong, W.; Liu, B.; Qin, J.; Li, B. Petrogenesis of an Early Permian bimodal intermediate-felsic suite in the East Junggar in Central Asian Orogenic Belt and tectonic implications. *Geol. J.* **2020**, *56*, 547–571. [CrossRef]
47. Wilde, S.A.; Zhang, X.Z.; Wu, F.Y. Extension of a newly-identified 500 Ma metamorphic terrain in northeast China: Further U–Pb SHRIMP dating of the Mashan Complex, Heilongjiang Province, China. *Tectonophysics* **2000**, *328*, 115–130. [CrossRef]
48. Zhou, J.B.; Wilde, S.A. The crustal accretion history and tectonic evolution of the NE China segment of the Central Asian Orogenic Belt. *Gondwana Res.* **2013**, *23*, 1365–1377. [CrossRef]
49. Han, J.L. Metallogenesis and Geodynamic Setting of the Typical Gold Deposits in the Liuhe Area of the Huadian City, Jilin Province. Ph.D. Thesis, Jilin University, Changchun, China, 2019. (In Chinese).
50. Zhou, J.B.; Wang, B.; Wilde, S.A.; Zhao, G.C.; Cao, J.L.; Zheng, C.Q.; Zeng, W.S. Geochemistry and U-Pb Zircon Dating of the Toudaoqiao Blueschists in the Great Xing'an Range, Northeast China, and Tectonic Implications. *J. Asian Earth Sci.* **2015**, *97*, 197–210. [CrossRef]
51. Zhou, X.D.; Zhu, D.Y.; Sun, Y.J. The Precambrian strata of Jilin Province. *Jilin Geol.* **2006**, *25*, 1–10. (In Chinese)
52. Guo, B.R.; Liu, S.W.; Zhang, J.; Wang, W.; Fu, J.H.; Wang, M.J. Neoarchean Andean-type active continental margin in the northeastern North China Craton: Geochemical and geochronological evidence from metavolcanic rocks in the Jiapigou granite-greenstone belt, Southern Jilin Province. *Precambr. Res.* **2016**, *285*, 147–169. [CrossRef]
53. Qin, Y.; Chen, D.D.; Liang, Y.H.; Zou, C.M.; Bai, L.A. Geochronology of Ji'an group in Tonghua area, Southern Jilin province. *Earth Sci. -J. China Univ. Geosci.* **2014**, *39*, 1587–1599. (In Chinese with English Abstract)
54. Liu, Z.H.; Liu, Y.Q.; Feng, B.Z. The establishment and tectonic evolution of Proterozoic Orogenic Belt in the north margin of North China Plate. *J. Chang. Univ. Sci. Technol.* **2000**, *30*, 110–114. (In Chinese with English Abstract)
55. Chen, Y.J.; Peng, Y.J.; Lu, X.P.; Liu, Y.W. Proterozoic tectonic slices along the northern margin of North China plate. *J. Jilin Univ.* **2002**, *32*, 134–139. (In Chinese with English Abstract)
56. Li, C.D.; Zhang, F.Q.; Miao, L.C.; Jie, H.Q.; Hua, Y.Q.; Xu, W.Y. Reconsideration of the Seluohe Group in Seluohe area, Jilin Province. *J. Jilin Univ.* **2007**, *37*, 841–847. (In Chinese)
57. Li, C.D.; Zhang, F.Q.; Miao, L.C.; Du, Y.L.; Hua, Y.Q.; Xu, W.Y.; Kang, S.M. Zircon SHRIMP U–Pb geochronology of the Zhangsangou Formation complex in the northeastern margin of North China Block, and its geological significance. *Acta Geol. Sin.* **2009**, *83*, 642–650. (In Chinese)
58. Zhang, Y.B.; Wu, F.Y.; Wilde, S.A.; Zhai, M.G.; Lu, X.P.; Zhang, H.F. Geochronology and Tectonic Implications of the "Proterozoic" Seluohe Group at the Northern Margin of the North China Craton. *Int. Geol. Rev.* **2008**, *50*, 135–153. [CrossRef]
59. Liu, Z.H.; Wang, C.; Song, J.; Gao, X.; Sun, L.N. ^{40}Ar-^{39}Ar dating and its tectonic significance of the Hulan Group at the northern margin of the North China Plate. *Acta Petrol. Sin.* **2016**, *32*, 2757–2764. (In Chinese)
60. Guo, B.R.; Liu, S.W.; Santosh, M.; Wang, W. Neoarchean arc magmatism and crustal growth in the north-eastern North China Craton: Evidence from granitoid gneisses in the Southern Jilin Province. *Precambrian Res.* **2017**, *303*, 30–53. [CrossRef]
61. Guo, B.R.; Liu, S.W.; Chen, X.; Wang, W.; Guo, R.R.; Yan, M. K–rich granitoid magmatism at the Archean–Proterozoic transition in southern Jilin: Insights into the Neoarchean crustal evolution of the northeastern part of the North China Craton. *Gondwana Res.* **2018**, *58*, 87–104. [CrossRef]
62. Liu, Y.S.; Gao, S.; Hu, Z.C.; Gao, C.G.; Zong, K.Q.; Wang, D.B. Continental and oceanic crust recycling-induced melt-peridotite interactions in the Trans-North China Orogen: U-Pb dating, Hf isotopes and trace elements in zircons of mantle xenoliths. *J. Petrol.* **2010**, *51*, 537–571. [CrossRef]
63. Zhang, Y.B.; Wu, F.Y.; Li, H.M.; Lu, X.P.; Sun, D.Y.; Zhou, H.Y. Single grain zircon U-Pb ages of the Huangniling granite in Jilin Province. *Acta Petrol. Sin.* **2002**, *18*, 475–481. (In Chinese)
64. Liu, S.; Hu, R.Z.; Feng, C.X.; Feng, G.Y.; Yu, X.F.; Li, C.; Jia, D.C.; Qi, Y.Q.; Wang, T. Zircon U–Pb age, Hf isotopic compositions of Dapuchaihe adakites in eastern Jilin Province and their significance. *Acta Petrol. Sin.* **2009**, *25*, 3153–3464. (In Chinese)
65. Xu, J.W.; Ma, G.F. Review of ten years (1981–1991) of research on the Tancheng–Lujiang fault zone. *Geol. Rev.* **1992**, *38*, 316–323. (In Chinese with English Abstract)
66. Wang, F.; Xu, W.; Ge, W.; Yang, H.; Pei, F.; Wu, W. The offset distance of the Dunhua-Mishan Fault; Constraints from Paleozoic-Mesozoic magmatism within the Songnen-Zhangguangcai Range, Jiamusi, and Khanka massifs. *Acta Petrol. Sin.* **2016**, *32*, 1129–1140. (In Chinese with English Abstract)

67. Xie, H.Y.; Shen, Y.C.; Jiao, X.D.; Meng, Q.Y. Discussion some important geological problems of Jiapigou gold belt, Jilin Province. *Chin. J. Geol.* **2000**, *35*, 111–120, (In Chinese with English Abstract).
68. Deng, J.; Yang, L.Q.; Gao, B.F.; Sun, Z.S.; Guo, C.Y.; Wang, Q.F.; Wang, J.P. Fluid evolution and metallogenic dynamics during tectonic regime transition: Example from the Jiapigou gold belt in northeast China. *Resour. Geol.* **2009**, *59*, 140–152. [CrossRef]
69. Wu, F.Y.; Li, J.W.; Evans, K.; Koenig, A.E.; Li, Z.K.; O'Brien, H.; Lahaye, Y.; Rempel, K.; Hu, S.Y.; Zhang, Z.P.; et al. Ore-forming processes of the Daqiao epizonal orogenic gold deposit, west Qinling orogen, China: Constraints from textures, trace elements, and sulfur isotopes of pyrite and marcasite, and raman spectroscopy of carbonaceous material. *Econ. Geol.* **2018**, *113*, 1093–1132. [CrossRef]
70. Qiu, C. Metallogenic Epoch and Geodynamic Background of the Haigou Gold Deposit, Jilin Province. Master's Thesis, Jilin University, Changchun, China, 2017. (In Chinese)
71. Zhang, Y.B.; Wu, F.Y.; Wilde, S.A.; Zhai, M.G.; Lu, X.P.; Sun, D.Y. Zircon U–Pb ages and tectonic implications of 'Early Paleozoic' granitoids at Yanbian, Jilin Province, northeast China. *Isl. Arc* **2004**, *13*, 484–505. [CrossRef]
72. Wang, G.Z.; Li, B.L.; Peng, B.; Ning, C.Q.; Yang, W.L. Songjianghe biotite monzonitic granite zircon U–Pb geochronology and geochemical significance. *Glob. Geol.* **2016**, *19*, 133–143.
73. Roedder, E. *Fluid Inclusions: An Introduction to Studies of All Types of Fluid Inclusions, Gas, Liquid, or Melt, Trapped in Materials from Earth and Space, and Their Application to the Understanding of Geologic Processes*; Mineralogical Society of America: Chantilly, VA, USA, 1984; Volume 12, pp. 11–304.
74. Clayton, R.N.; Mayeda, T.K. The use of bromine pentafluoride in the extraction of oxygen from oxides and silicates for isotopic analysis. *Geochim. Cosmochim. Acta* **1963**, *27*, 43–52. [CrossRef]
75. Robinson, B.W.; Kusakabe, M. Quantitative preparation of sulfur dioxide, for $^{34}S/^{32}S$ analyses, from sulfides by combustion with cuprous oxide. *Anal. Chem.* **1975**, *47*, 1179–1181. [CrossRef]
76. Wu, S.S.; Zhang, X.B.; Zhang, Y.; Huang, G.H.; Zhang, T.Z.; Lai, C.K. In-situ trace element and S isotope compositions of sulfides from the Tonggou Cu deposit, Eastern Tianshan Orogen: Implication for ore-forming process. *Ore Geol. Rev.* **2022**, *149*, 105085. [CrossRef]
77. Chen, L.; Chen, K.; Bao, Z.; Liang, P.; Sun, T.; Yuan, H. Preparation of standards for in situ sulfur isotope measurement in sulfides using femtosecond laser ablation MC-ICP-MS. *J. Anal. At. Spectrom.* **2017**, *32*, 107–116. [CrossRef]
78. Fan, H.R.; Li, X.H.; Zuo, Y.B.; Chen, L.; Liu, S.; Hu, F.F.; Feng, K. In-situ LA-(MC)-ICPMS and (Nano) SIMS trace elements and sulfur isotope analyses on sulfides and application to confine metallogenic process of ore deposit. *Acta Petrol. Sin.* **2018**, *34*, 3479–3496. (In Chinese)
79. Sun, Q.F.; Wang, K.Y.; Wang, Y.C.; Cai, W.Y.; Ma, X.L. Superimposing mineralization in the Zhengguang Au-Zn deposit, NE China: Evidence from pyrite Re-Os geochronology, fluid inclusion, and H-O-S-Pb isotopes. *Ore Geol. Rev.* **2021**, *137*, 104307. [CrossRef]
80. Lu, H.Z.; Fan, H.R.; Ni, P.; Ou, G.X.; Shen, K.; Zhang, W.H. *Fluid Inclusion*; Science Press: Beijing, China, 2004; pp. 193–200. (In Chinese)
81. Hall, D.L.; Sterner, S.M.; Bodnar, R.J. Freezing point depression of NaCl–KCl–H_2O solution. *Econ. Geol.* **1988**, *83*, 197–202. [CrossRef]
82. Clayton, R.N.; O'Neil, J.R.; Mayeda, T.K. Oxygen isotope exchange between quartz and water. *J. Geophys. Res.* **1972**, *77*, 3057–3067. [CrossRef]
83. Taylor, H.P. The application of oxygen and hydrogen isotope studies to problems of hydrothermal alteration and ore deposition. *Econ. Geol.* **1974**, *69*, 843–883. [CrossRef]
84. Craig, H.; Lupton, J.E. Lupton primordial neon, helium and hydrogen in oceanic basalts. *Earth Planet. Sci. Lett.* **1976**, *31*, 369–385. [CrossRef]
85. Zhang, L.G. *Geological Appliance for the Stable Isotope*; Shanxi Science and Technology Publishing House: Xi'an, China, 1985; pp. 54–61. (In Chinese)
86. Hedenquist, J.W.; Lowenstern, J.B. The role of magmas in the formation of hydrothermal ore deposits. *Nature* **1994**, *370*, 519–527. [CrossRef]
87. Wutiepu, W. Multi-Information Metallogenic Prediction of Gold Deposits in Jiapigou Metallogenic Belt, Jilin Province. Ph.D. Thesis, Jilin University, Changchun, China, 2019. (In Chinese)
88. Li, S.R.; Santosh, M. Metallogeny and craton destruction: Records from the North China Craton. *Ore Geol. Rev.* **2014**, *56*, 376–414. [CrossRef]
89. Gao, S. Properties and Mineralizing Significance of CO_2-Rich Fluid in Haigou Gold Deposit, Jilin Province. Master's Thesis, Jilin University, Changchun, China, 2016. (In Chinese)
90. Chen, Y.L.; Li, H.; Zheng, C.Y.; Maoayipu, N.; Khan, R. Genesis and ore prospecting model of Haigou gold deposit in Jilin Province: Evidence from LA-ICP-MS trace element and sulfur isotopic analysis of pyrite. *J. Cent. South Univ.* **2021**, *52*, 2990–3002. (In Chinese)
91. Zartman, R.E.; Doe, B.R. Plumbotectonics-the model. *Tectonophysics* **1981**, *75*, 135–162. [CrossRef]
92. Liu, Y.Q. Isotope geochemistry and metallogenic regularity of the Haigou gold deposit in Jilin Province. *Miner. Depos.* **1991**, *10*, 131–142. (In Chinese)

93. Yang, L.Y. The Geochemistry and Genesis of Jiapigou Gold Deposit in Jilin Province. Master's Thesis, China University of Geosciences, Beijing, China, 2011. (In Chinese)
94. Roedder, E.; Bodnar, R.J. Geologic pressure determinations from fluid inclusion studies. Annu. Rev. *Earth Planet. Sci.* **1980**, *8*, 263–301. [CrossRef]
95. Brown, P.E.; Hagemann, S.G. MacFlincor and its application to fluid in Archean lode-gold deposits. *Geochim. Cosmochim. Acta* **1995**, *59*, 3943–3952. [CrossRef]
96. Bouzari, F.; Clark, A.H. Prograde evolution and geothermal affinities of a major porphyry copper deposit: The Cerro Colorado hypogene protore, I region, northern Chile. *Econ. Geol.* **2006**, *101*, 95–134. [CrossRef]
97. Ohmoto, H. Stable isotope geochemistry of ore deposits. *Rev. Mineral. Geochem.* **1986**, *16*, 491–559.
98. Heinrich, C.A. The physical and chemical evolution of low-salinity magmatic fluids at the porphyry to epithermal transition: A thermodynamic study. *Miner. Depos.* **2005**, *39*, 864–889. [CrossRef]
99. Mao, J.W.; Wang, Y.T.; Li, H.M.; Pirajno, F.; Zhang, C.Q.; Wang, R.T. The relationship of mantle-derived fluids to gold metallogenesis in the Jiaodong Peninsula: Evidence from D-O-C-S isotope systematics. *Ore Geol. Rev.* **2008**, *33*, 361–381. [CrossRef]
100. Lowenstern, J.B. Carbon dioxide in magmas and implications for hydrothermal systems. *Miner. Depos.* **2001**, *36*, 490–502. [CrossRef]
101. Sibson, R.H.; Robert, F.; Poulsen, H. High angle reverse faults fluid pressure cycling and mesothermal gold quartz deposits. *Geology* **1988**, *16*, 551–555. [CrossRef]
102. Gammons, C.H.; Williams-Jones, A.E.; Yu, Y. New data on the stability of gold (I) chloride complexes at 300 °C. *Mineral. Mag.* **1994**, *58*, 309–310. [CrossRef]
103. Benning, L.G.; Seward, T.M. Hydrosulfide complexing of Au (I) in hydrothermal solutions from 150 to 400 °C and 500–1500 bar. *Geochim. Cosmochim. Acta* **1996**, *60*, 1849–1871. [CrossRef]
104. Phillips, G.N.; Evans, K.A. Role of CO_2 in the formation of gold deposits. *Nature* **2004**, *429*, 860–863. [CrossRef]
105. Guha, J.; Lu, H.Z.; Dube, B.; Robert, F.; Gagnon, M. Fluid characteristics of vein and altered wall rock in Archean mesothermal gold deposits. *Econ. Geol.* **1991**, *86*, 667–684. [CrossRef]
106. Reed, M.H.; Spycher, N.F. Boiling, cooling, and oxidation in epithermal system: A numerical modeling approach. *Rev. Econ. Geol.* **1985**, *1*, 249–272.
107. Robert, F.; Kelly, W.C. Ore-forming fluids in Archean gold-bearing quartz veins at the Sigma Mine, Abitibi greenstone belt, Quebec, Canada. *Econ. Geol.* **1987**, *82*, 1464–1482. [CrossRef]
108. Tbrahim, M.S.; Kyser, T.K. Fluid inclusion and isotope systematics of the high-T Proterozoic Star Lake Lode gold deposit, Northern Soskatchewan, Canada. *Econ. Geol.* **1991**, *86*, 1468–1490. [CrossRef]
109. Macnamara, J.; Thode, H.G. Comparison of the isotopic constitution of terrestrial and meteoritic sulfur. *Phys. Rev.* **1950**, *78*, 307–308. [CrossRef]
110. Chaussidon, M.; Lorand, J.P. Sulphur isotope composition of orogenic spinel lherzolite massifs from Ariege (North-Eastern Pyrenees, France): An ion microprobe study. *Geochim. Cosmochim. Acta* **1990**, *54*, 2835–2846. [CrossRef]
111. Kelley, W.C.; Rye, R.O. Geologic, fluid inclusion, and stable isotope studies of the Tin-Tungsten deposits of Panasqueira, Portugal. *Econ. Geol.* **1979**, *74*, 1721–1819. [CrossRef]
112. Ohmoto, H.; Rye, R.O. Isotopes of sulfur and carbon. In *Geochemistry of Hydrothermal Ore Deposits*; Barnes, H.L., Ed.; Wiley: New York, NY, USA, 1979; pp. 509–567.
113. Wang, Z.C.; Liu, J.M.; Liu, H.T.; Zeng, Q.D.; Zhang, S.; Wang, Y.B. Complexity and uncertainty of tracing fluid sources by means of H-O, C, S, N isotope systems: A case study of orogenic lode gold deposits. *Acta Petrol. Mineral.* **2010**, *29*, 577–590. (In Chinese)
114. Goldfarb, R.J.; Groves, D.I.; Gardoll, S. Orogenic gold and geologic time: A synthesis. *Ore Geol. Rev.* **2001**, *18*, 1–75. [CrossRef]
115. Wang, T.J.; Yan, F. Ore fluid evolution and ore prediction of magmatic hydrothermal gold deposits in Jiaodong area. *J. Geol. Explor.* **2002**, *17*, 169–174. (In Chinese)
116. Wang, Y.W.; Zhu, F.S.; Gong, R.T. Tectonic isotope geochemistry: Further study on sulphur isotope of Jiaodong gold concentration area. *Gold* **2002**, *23*, 1–16. (In Chinese)
117. Wu, K.X.; Hu, R.Z.; Bi, X.W.; Peng, J.T.; Tang, Q.L. Ore lead isotopes as a tracer for ore-forming material sources: A review. *Geol. Geochem.* **2002**, *30*, 73–81. (In Chinese)
118. Goldfarb, R.J.; Ayuso, R.; Miller, M.L.; Ebert, S.W.; Marsh, E.E.; Petsel, S.A.; Miller, L.D.; Bradley, D.C.; Johnson, C.; McClelland, W. The Late Cretaceous Donlin Creek gold deposit, Southwestern Alaska: Controls on epizonal ore formation. *Econ. Geol.* **2004**, *99*, 643–671. [CrossRef]
119. Mortensen, J.K.; Hall, B.V.; Bissig, T.; Friedman, R.M.; Danielson, T.; Oliver, J.; Rhys, D.A.; Ross, K.V.; Gabites, J.E. Age and paleotectonic setting of volcanogenic massive sulfide deposits in the Guerrero Terrane of Central Mexico: Constraints from U–Pb age and Pb isotope studies. *Econ. Geol.* **2008**, *103*, 117–140. [CrossRef]
120. Wen, B.J.; Fan, H.R.; Hu, F.F.; Liu, X.; Yang, K.F.; Sun, Z.F.; Sun, Z.F. Fluid evolution and ore genesis of the giant Sanshandao gold deposit, Jiaodong gold province, China: Constrains from geology, fluid inclusions and H-O-S-He-Ar isotopic compositions. *J. Geochem. Explor.* **2016**, *171*, 96–112. [CrossRef]
121. Xu, W.L.; Wang, F.; Pei, F.P.; Meng, E.; Tang, J.; Xu, M.J.; Wang, W. Mesozoic tectonic regimes and regional ore-forming background in NE China: Constraints from spatial and temporal variations of Mesozoic volcanic rock associations. *Acta Petrol. Sin.* **2013**, *29*, 339–353. (In Chinese)

122. Tang, J.; Xu, W.L.; Wang, F.; Ge, W.C. Subduction History of the Paleo-Pacific Slab beneath Eurasian Continent: Mesozoic-Paleogene Magmatic Records in Northeast Asia. *Sci. China Earth Sci.* **2018**, *61*, 527–559. [CrossRef]
123. Wang, F.; Xu, Y.G.; Xu, W.L.; Yang, L.; Wu, W.; Sun, C.Y. Early Jurassic calc-alkaline magmatism in northeast China: Magmatic response to subduction of the Paleo-Pacific Plate beneath the Eurasian continent. *J. Asian Earth Sci.* **2017**, *143*, 249–268. [CrossRef]
124. Wu, F.Y.; Sun, D.Y. The Mesozoic magmatism and lithospheric thinning in eastern China. *J. Chang. Univ. Sci. Technol.* **1999**, *29*, 313–318. (In Chinese)
125. Wu, F.Y.; Sun, D.Y.; Zhang, G.L.; Ren, X.W. Deep geodynamics of Yanshan movement. *Geol. J. China Univ.* **2000**, *6*, 379–388. (In Chinese)
126. Wu, F.Y.; Ge, W.C.; Sun, D.Y.; Guo, C.L. Discussions on the lithospheric thinning in eastern China. *Earth Sci. Front.* **2003**, *10*, 51–60. (In Chinese)

minerals MDPI

Article

Automated Hyperparameter Optimization of Gradient Boosting Decision Tree Approach for Gold Mineral Prospectivity Mapping in the Xiong'ershan Area

Mingjing Fan [1,2], Keyan Xiao [1,*], Li Sun [1,*], Shuai Zhang [3,*] and Yang Xu [1,4]

[1] MNR Key Laboratory of Metallogeny and Mineral Resource Assessment, Institute of Mineral Resources, Chinese Academy of Geological Sciences, Beijing 100037, China
[2] Institute of Geological Survey, China University of Geosciences, Wuhan 430074, China
[3] China Aero Geophysical Survey and Remote Sensing Center for Natural Resources, Beijing 100083, China
[4] Institute of Earth Science, China University of Geosciences, Beijing 100083, China
* Correspondence: kyanxiao@sohu.com (K.X.); sunli0727@163.com (L.S.); zhangs01@mail.cgs.gov.cn (S.Z.); Tel.: +86-010-68993164 (S.Z.)

Abstract: The weak classifier ensemble algorithms based on the decision tree model, mainly include bagging (e.g., fandom forest-RF) and boosting (e.g., gradient boosting decision tree, eXtreme gradient boosting), the former reduces the variance for the overall generalization error reduction while the latter focuses on reducing the overall bias to that end. Because of its straightforward idea, it is prevalent in MPM (mineral prospectivity mapping). However, an inevitable problem in the application of such methods is the hyperparameters tuning which is a laborious and time-consuming task. The selection of hyperparameters suitable for a specific task is worth investigating. In this paper, a tree Parzen estimator-based GBDT (gradient boosting decision tree) model (TPE-GBDT) was introduced for hyperparameters tuning (e.g., loss criterion, n_estimators, learning_rate, max_features, subsample, max_depth, min_impurity_decrease). Then, the geological data of the gold deposit in the Xiong 'ershan area was used to create training data for MPM and to compare the TPE-GBDT and random search-GBDT training results. Results showed that the TPE-GBDT model can obtain higher accuracy than random search-GBDT in a shorter time for the same parameter space, which proves that this algorithm is superior to random search in principle and more suitable for complex hyperparametric tuning. Subsequently, the validation measures, five-fold cross-validation, confusion matrix and success rate curves were employed to evaluate the overall performance of the hyperparameter optimization models. The results showed good scores for the predictive models. Finally, according to the maximum Youden index as the threshold to divide metallogenic potential areas and non-prospective areas, the high metallogenic prospect area (accounts for 10.22% of the total study area) derived by the TPE-GBDT model contained > 90% of the known deposits and provided a preferred range for future exploration work.

Keywords: mineral prospectivity mapping; machine learning; hyperparameter optimization; gradient boosting decision tree

1. Introduction

Mineral prospectivity mapping (MPM) can guide us to conduct deep and peripheral ore prospecting in chosen study area. From the spatial dimension, it can be divided into two-dimensional MPM and three-dimensional MPM [1,2]. Two-dimensional MPM can facilitate the regional-scale prospecting and delineation of the prospective area [3–8], while three-dimensional MPM can be used to guide the delineation of the deep metallogenic target area at the deposit scale [9–16]. In terms of methods, MPM can be divided into knowledge-driven MPM and data-driven MPM. Knowledge-driven MPM is mainly based on geological expert experience for statistical analysis methods [17–24], and data-driven MPM is mainly based on big data methods (i.e., machine learning or deep learning algorithms) [25–35].

No matter which MPM is performed, at either the regional scale or at the deposit scale, MPM is carried out by way of data-driven or knowledge-driven approaches, geological exploration data are essential. The quality of exploration data collected matters a great deal for the accuracy of MPM. However, the geoscientific data available are generally limited, and there are a few target/positive samples (ore spots) in the real world. Moreover, machine learning and deep learning algorithms are data-hungry [36–41]. Therefore, there are some inevitable problems for the application of big data algorithms based on geological data (such as with the data structure, extreme imbalance of positive (prospect) and negative samples (non-prospect) etc.) [42].

In resolving these problems, the current research mainly focused on data augmentation and algorithm optimization, among which the data augmentation mainly included: multi-sample data augmentation (e.g., under-sampling, over-sampling, synthetic sampling) [43–51], single-sample data augmentation (e.g., image up-and-down, left-right inversion and center inversion, etc.) [52–57], and unsupervised data enhancement (e.g., GAN, AutoML) [58–61]. The improved algorithms mainly included design cost sensitive functions and used ensemble learning, such as cost-sensitive neural networks [62,63], balanced fuzzy support vector machine algorithms [64–68], a CBP-SVM algorithm based on a hybrid model [69], the RSBoost algorithm [70–72], anomaly detection algorithms (e.g., isolation forest) [73–76], differential Siamese convolutional neural network [77–81], and the MLA (machine learning algorithms) based on the decision tree model (e.g., GBDT, weighted random forest) [82–84]. Among them, the tree model was suitable for few-shot data training [85–87].

The MLA based on the tree model has high computational efficiency and strong model discrimination ability with a simple principle. It is one of the few "white box models" in MLA [88–91]. Moreover, in terms of the function of the tree model itself, the decision tree can be used for classification and regression, which also produce additional criteria such as importance of features and continuous variable box indexes. In ensemble learning, the tree model is the most commonly used primary classifier [92]. These advantages make the tree model one of the most important in MLA. The GBDT model is an essential boosting algorithm, based on a decision tree, that utilizes the strategy of multi-model integration to fit the residuals for reducing the deviation and variance of the model [93,94]. Nevertheless, it is also faced with an important problem: hyperparametric optimization (HPO). However, there are few studies on HPO with GBDT. Many researchers tend to set default parameters values for the model, leading to weak end results [95,96]. The optimization of hyperparameters, for certain case studies, has a great impact on the effectiveness of the algorithm, and it plays a crucial role in the accuracy and generalization ability of the final generated model. There are a large number of hyperparameters (e.g., n_estimation, learning_rate, max_features, subsample, etc.) in the GBDT, with different parameters having different effects on the GBDT model. Therefore, the selection of the hyperparameter combination is important to achieve the optimal model.

With regard to its algorithms, we aspire to eventually realize the perfect automation of all processes. The discipline specializing in automation machine learning is called AutoML, and hyperparameter automatic optimization is the most mature, in-depth, and well-known direction [97–99]. Theoretically, when the computational power and data are sufficient, the performance of HPO should exceed that of human beings. HPO can reduce human workload, and the results obtained by HPO are more likely to be reproduced than searched, so HPO can greatly improve the reproducibility and fairness of scientific research [100–103]. Contemporary HPO algorithms can be mainly divided into grid-based search (grid), Bayesian optimization (Bayesian), gradient-based optimization (gradient-based), and population-based optimization (evolutionary algorithm, genetic algorithm, etc.), among which the grid search and Bayes-based optimization are the most popular [104–110]. These HPO methods have great effect and significance on optimizing complex ensemble algorithms.

In this paper, a knowledge-driven synthetic sampling method was proposed for dealing with imbalanced data, which made the data form more suitable for MLA. The GBDT algorithm based on the decision tree model for few-shot geological data was selected. The random grid search and the Bayesian optimization based on the TPE algorithm were used to study the hyperparametric of the GBDT model, taking the gold deposit in Xiong 'ershan area as an example, in order to verify the difference of MPM based on two algorithms for HPO with few shot geological data. The study can be used as a reference for the research of MLA based on HPO in the MPM.

2. Methodology

2.1. GBDT Algorithm

The gradient boosting decision tree (GBDT) is a representative boosting algorithm, which is the cornerstone of XGBoost, LightGBM and other tree model algorithms. It is also one of the most widely used MLAs in the industry and the most stable one in practice. When it was first proposed, GBDT was written as gradient boosting machine (GBM), which integrated bagging and boosting ideas and can accept all types of weak estimators as input. After the weak estimator was defined as a decision tree, it was slowly renamed the gradient boosting tree.

Inspired by the boosting algorithm, GBDT naturally contains three boosting elements: loss function $L(x, y)$, weak estimators $f(x)$ and comprehensive integration results $H(x)$ [111]. Yet, some improvements have since been made: (1) The weak estimator output type of GBDT is no longer consistent with the ensemble algorithm. For AdaBoost or random forest (RF) algorithms, weak estimators are regressions when ensemble algorithms perform regression tasks. When ensemble algorithms perform classification tasks, weak estimators are classifiers; no matter whether the GBDT is performing regression or classification tasks as a whole, the weak estimator must be a regression. It outputs specific classification results by sigmoid or softmax functions, but the actual weak estimator must be a regression; (2) The loss function has been extended to any differentiable function in the mathematical principle, and is no longer limited to a fixed or single loss function; (3) Before each weak estimator is established, the sample weight is not modified, but the residual error is fitted to affect the structure of the subsequent weak estimator; (4) It joins the idea of random sampling from RF, allowing samples and features to be sampled before each tree is built to increase the independence between weak estimators (thus allowing for out-of-bag datasets to be used to verify the establishment results of each weak evaluator). It can further increase the stability of boosting algorithm.

As is well known, MPM is considered a classification problem in the MLA, and is categorized as either a prospect or non-prospect in the study region. So, we should know the GBDT binary classification algorithm flow [112]:

First, initialize weak classifier as:

$$H_0 = \log \frac{P(Y = 1|x)}{1 - P(Y = 1|x)} \tag{1}$$

where the $P(Y = 1|x)$ is the proportion of positive (as y = 1) in train datasets, using prior information to initialize the learner.

Second, the loss function is defined, and the GBDT binary classification algorithm uses a logarithmic loss function:

$$L(\theta) = -y_i \log \widehat{y}_i - (1 - y_i) \log(1 - \widehat{y}_i) \tag{2}$$

where the $\hat{y}_i$ is the result of logistic regression prediction $H(x)$.

$$H(x) = \frac{1}{1 + e^{-f(x)}} \tag{3}$$

where the $f(x)$ are the weak estimators. Flowing Formula (3) into Formula (2) as:

$$L(y_i, f(x_i)) = y_i \log\left(1 + e^{-f(x_i)}\right) + (1 - y_i)\left[f(x_i) + \log\left(1 + e^{-f(x_i)}\right)\right] \tag{4}$$

Then, calculation of negative gradient of loss function (pseudo residual):

$$r_{m,i} = -\left|\frac{\partial L(y_i, f(x_i))}{\partial f(x_i)}\right|_{f(x) = f_{m-1}(x)} = y_i - \frac{1}{1 + e^{-f(x)}} = y_i - \widehat{y}_i \tag{5}$$

Subsequently, calculate the best residual fitting value of each leaf node as:

$$C_{m,j} = \operatorname{argmin} \sum_{x_i \in R_{m,j}} L(y_i, f_{m-1}(x_i) + C) \tag{6}$$

where the m is the mth trees. In order to solve the $C_{m,j}$ value, we introduce the second-order Taylor expansion as:

$$C_{m,j} = \frac{\sum_{x_i \in R_{m,j}} r_{m,i}}{\sum_{x_i \in R_{m,j}} (y_i - r_{m,i})(1 - y_i + r_{m,i})} \tag{7}$$

Meanwhile, update strong learners as:

$$H_m = H_{m-1}(x) + \sum_{j=1}^{J_m} C_{m,j} I(x \in R_{m,j}) \tag{8}$$

Finally, the $H_M(x)$ as:

$$H_M(x) = H_m + \sum_{m=1}^{M} \sum_{j=1}^{J_m} C_{m,j} I(x \in R_{m,j}) \tag{9}$$

In summary, the GBDT binary classification algorithm uses multiple CART regression trees to fit the log probability with the positive label (as y = 1), and the loss function uses the logarithmic loss, but it is necessary to replace the predicted value $\hat{y}_i$ (the probability predicted as y = 1) with the logarithmic probability $H(x)$ predicted by the regression tree, and then fit the residuals each round. After the final classifier output is obtained, the probability that the prediction label is positive (as y = 1) can be obtained by the sigmoid function.

2.2. *Hyperparameter Optimization Method*

Model optimization is one of the most difficult challenges in MLA implementation. Parameter adjustment is the core of model optimization, but the process is complex and cumbersome. Especially, there are a large number of hyperparameters in the MLA and deep learning algorithms (DLA), which not only makes the method extremely flexible but also affects the algorithm performance through the combination hyperparameters. Therefore, the application and selection of an HPO method that can automatically obtain hyperparameters is crucial. In this paper, the stochastic network search (RandomizedSearchCV) and a tree-structured Parzen estimator approach (TPF) based on the Bayesian optimization algorithm was used for HPO with a GBDT model.

2.2.1. Random Grid Search Optimization

The random grid search method is the advanced version of the grid optimization algorithm. Based on the original grid search optimization, it abandons the global hyper-parameter space that was used in the original search, and instead randomly selects some parameter combinations to construct the hyperparameters subspace, and searches only in the subspace [113]. This greatly reduces the search space, the number of parameter groups that need to be enumerated and compared, and shortens the overall search time. Additionally, the minimum loss obtained by random grid search is very close to the minimum loss obtained by enumeration grid search. It improves the computing speed without reducing the search accuracy. In random search, random sampling is not put back, so there will be no problem for extracting the same set of parameters twice. We can assign a fixed amount of computation to the random grid search, and when all computation is consumed, the random grid search is completed.

2.2.2. Tree Parzen Estimator in Bayesian Optimization

Bayesian optimization is a parameter adjustment method with a prior process. It is the SOTA in the current hyperparameter optimization and is the most advanced optimization framework. It can be used in various fields of AutoML, and it can also be used in advanced fields such as neural network architecture search for NAS and meta-learning [114]. The mathematical process of Bayesian optimization mainly includes the following processes:

(1) Define the objective function $f(x)$ to be estimated and the definition domain of x;
(2) Take out the values on finite nx, and solve the $f(x)$ of these x (solve the observed values);
(3) According to the limited observations, the function is estimated (this assumption is called the prior knowledge in Bayesian Optimization), and the target value (maximum or minimum) on the estimated $f(\hat{x})$ is obtained;
(4) Define a rule to determine the next observation point to be calculated.

Continue to cycle in steps (2)–(4) until the target value on the assumption distribution reaches, or all computing resources are used up (e.g., up to m observations or up to t minutes allowed to run).

With the tree Parzen estimator (TPE), a different idea is used to model the probability distribution. According to Bayesian theorem:

$$P(Y|X) = \frac{p(X|Y)P(Y)}{P(X)} \tag{10}$$

TPE divides $P(X|Y)$ as:

$$P(X|Y) = \begin{cases} l(x), Y < Y^* \\ g(x), Y > Y^* \end{cases} \tag{11}$$

In other words, the different distribution of TPE for observation points on both sides of the threshold Y^* can be regarded as the hyperparametric probability distribution of a good and a bad score. The threshold Y^* is determined by the hyperparametric γ, that is the quantile of Y. Through the above division, we can obtain:

$$P(X) = \int_R P((X|Y)P(Y)dy = \gamma l(x) + (1-\gamma)g(x) \tag{12}$$

Then, bring the Formula (12) into the EI formula as:

$$\mathrm{EI}_{\mathrm{Y}^*} = \int_{-\infty}^{+\infty} \max(\mathrm{Y}^* - \mathrm{Y}, 0) \mathrm{P}_{\mathrm{M}} \left(\mathrm{Y} \middle| \mathrm{X}\right) dy = \frac{\int_{-\infty}^{\mathrm{Y}^*} (\mathrm{Y}^* - \mathrm{Y}) \mathrm{P}(\mathrm{Y}) d(y)}{\gamma + (1 - \gamma) \frac{g(x)}{l(x)}}. \tag{13}$$

Thus, $\mathrm{EI}_{\mathrm{Y}^*}(x) \propto \left(\gamma + (1-\gamma)\frac{g(x)}{l(x)}\right)^{-1}$, where the $\mathrm{EI}_{\mathrm{Y}^*}$ is proportional to the reciprocal of the denominator, and when γ is determined, the denominator value only depends on the ratio $\frac{g(x)}{l(x)}$ of the two segments of x , and the physical meaning of this ratio is the probability of x being a good score to the probability of x being a bad score. Therefore, the result is used to find the max x to obtain the ratio maximum.

2.3. GBDT Modeling

GBDT, as the representative boosting algorithm, has a large number of parameters which can be roughly divided into five categories, such as parameters based on an the iterative process; parameters of a weak evaluator structure; parameters for an early stop; training data parameters of weak evaluator; and others (Table 1). Among them, the number of iterations and the parameters of the weak evaluator have a great influence on the GBDT model. For example, then estimators parameters determine the number of iterations, the learning rate parameters that affect the overall learning efficiency of the algorithm, and the max depth, min impurity decrease parameters are used to prune the tree model to reduce the complexity of the model. Faced with so many parameters, quickly selecting the appropriate parameters to optimize the model is a challenge. This paper selected eight parameters that greatly impact the GBDT model (Table 2), and carried out parameter optimization based on the Sklean package in Python.

Table 1. Categories of GBDT hyperparameters.

Type	Parameter
boosting	n_estimators, learning_rate, loss, alpha, init
Weak evaluator structure	criterion, max_depth, min_samples_split, min_samples_leaf, min_weight_fraction_leaf, max_leaf_nodes, min_impurity_decrease
early stop	validation_fraction, n_iter_no_change, tol, n_estimators_
Weak evaluator training data	subsample, max_features, random_state
others	ccp_alpha, warm_start

Table 2. GBDT hyperparameters selected and default values in Sklearn.

Parameter	Function of Parameters	Default Value
loss	loss function	"deviance"
criterion	Impurity measurement index of weak estimate when branching	"friedman_mse"
n_estimators	The Actual number of iterations	100
learning_rate	Weighted summation process affecting weak estimator results	0.1
max_features	Maximum number of features considered in constructing optimal CART tree model	None
subsample	The Proportion of random samples released from the full dataset before each CART tree is built	1.0
max_depth	Maximum allowable depth of weak estimator	3
min_impurity_decrease	The minimum reduction in impurity is allowed when the weak evaluator branches	0.0

The basic process of hyperparametric optimization is as follows: First, the objective function is defined. We used the Formulation (2) as the objective function. Second, the search space is determined. For the GBDT model, most parameters have a fixed range, so we chose to explore the unbounded parameters. Generally, a large space will be set initially, and during the iterative optimization classification model, the parameter space range and dimension are gradually reduced many times. Finally, the function is optimized, and the iterative optimization model is trained.

3. Study area and Geological Data

3.1. Geological Setting

The Xiong'ershan area is located in the southern margin of the North China Craton (NCC) and the eastern segment of the Qinling orogenic belt, which is an important part of Huaxiong block (Figure 1A,B). The strata in this area are mainly composed of the Neoarchean-Paleoproterozoic Taihua Group, the Paleoproterozoic-Mesoproterozoic Xiong'er Group, the Mesoproterozoic Guankou Group, the Luanchuan Group, and finally Cenozoic formations. The Taihua Group is composed of Late Archean-Early Proterozoic metamorphic rocks (e.g., amphibolite gneiss, amphibolite, leptynite, granulite), exposed along the central and western in the southern margin of the NCC. It is one of the metamorphic crystalline basements in NCC [115–117]. Xiong'er Group is the product of important magmatic events after the stability of the NCC. It comprises intermediate-acid volcanic strata distributed in the southern margin of the NCC. The main lithologies are andesites, rhyolites, dacites, etc. [118,119]. The Guandaokou Group is mainly composed of a series of shallow marine terrigenous clastic-carbonate rock (e.g., dolomite and quartz sandstone) formations located in the southern part of the Xiong'ershan area, overlain by the Xiong'er group in an unconformable contact relationship. The Luanchuan Group is distributed in the south of Luanchuan, and is mainly a set of shallow metamorphic clastic and carbonate rocks (e.g., sandstone, mudstone, limestone, etc.) (Figure 1C) [120–122].

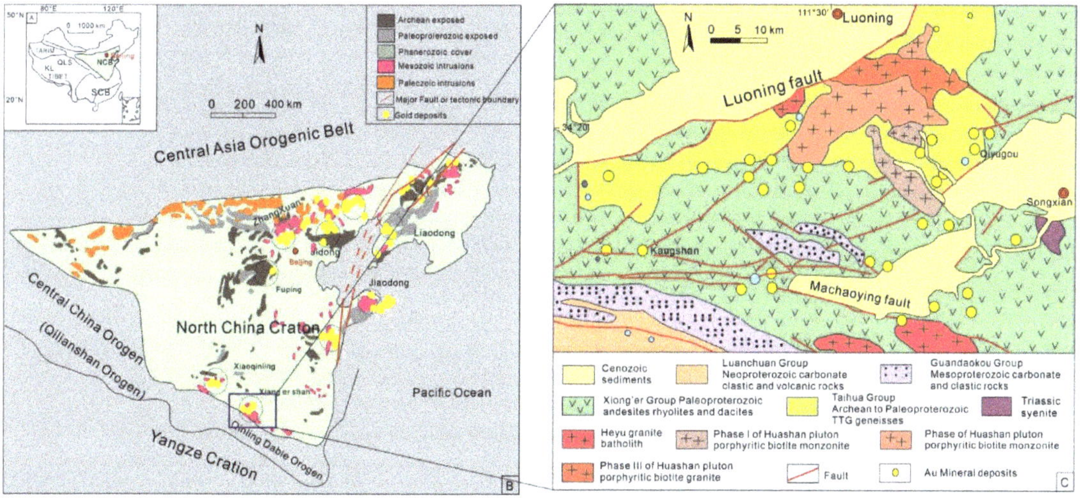

Figure 1. Simplified geological maps of the Xiong'ershan, Henan province in China: (**A**) map showing the location of the NCC; (**B**) the location of the Xiong'ershan ore clusters within the CNN (modified after [123–125]); and (**C**) regional geologic map of Xiong'erhsan district and distribution of gold deposits in the region (modified after [126–128]).

The regional structure is mainly composed of three EW trending deep regional faults (e.g., Luonan-Luanchuan fault, Machaoying fault and Sanmenxia-Baofeng fault). A series of NE and NW trending faults are distributed among the three main faults. The Machaoying fault zone experienced at least five deformation cycles and seven generations of tectonic events, which is the main ore-guiding-hosting structure in this area [129].

The magmatic activity is relatively frequent. According to the age of magmatic activity, the characteristics of magmatic rocks and the corresponding geodynamic background are roughly divided into three stages: (1) The first cycle mainly occurred in the Late Archean-Early Proterozoic, during which a large number of intermediate-acid volcanic rocks and TGG granites were produced, and the Taihua Group metamorphic basement was formed by metamorphism in the later period. (2) The second cycle mainly occurred in the Middle-Late Proterozoic, forming the Xiong'er Group of volcanic strata, which is discordant to the Taihua Group. (3) The third magmatic cycle took place mainly in the Mesozoic, being the most important magmatic activity in the Xiong'ershan area. This magmatic activity is also an important tectonic-magmatic thermal event that formed a large number of deposits in the central and eastern China (Figure 2a), hosted in the Wuzhangshan, Huashan, Heyu and other granite batholiths [130,131].

3.2. Geological Exploration Datasets

The used datasets were: (1) the geological map of Xiong'ershan district at 1:50,000 scale provided by the Geological Survey in the Henan province; (2) geochemical data derived from 1156 stream sediment samples at the scale of 1:200,000 with the density of 2 Km $*$ 2 Km; and (3) gravity anomaly data at scale of 1:200,000.

The metallogenic model needed to be translated into predictor layers for better support to the MPM of the Xiong'ershan area. This section describes the methods used to generate the predictor layers based on the exploration criteria.

3.2.1. Source

The source of ore-forming material provides the basis for the deposit formation [132,133]. A large number of studies on fluid inclusions and stable isotopes (O, H, C) in the Xiong 'ershan area, show that the main metallogenic mechanism of gold deposits is fluid boiling, and the source of ore-forming fluids is mainly deep source materials [134–138]. The large-scale tectonic-magmatic-metallogenetic thermal event in the Early Cretaceous is characterized by the widespread exposure of faults and granitic batholiths in the study area (Figure 2). Granite batholith provides a heat source for the formation of gold and molybdenum deposits; therefore, the contemporaneous granite base in the region can be used as an important prediction factor. Reasonable estimation of the influence distance between granite rock mass and gold deposits indicates the positioning function for metallogenic prediction. Here, we analyzed the buffer distance layer by the distribution histogram (Figure 3) as one of the prediction factors. It showed the relationship between the granite rock mass (GRB) and ore-occurrence (Figure 4p). The granite intrusions usually exhibit low and gentle gravity anomalies owing to their low density. Figure 4l shows that the low gravity is roughly consistent with granitic intrusions, and the outline of several huge granitic intrusions can be recognized. The residual gravity anomaly (RGA) can highlight concealed granite geological bodies. We adopted the RGA as another prediction factor for MPM in the Xiong'ershan area.

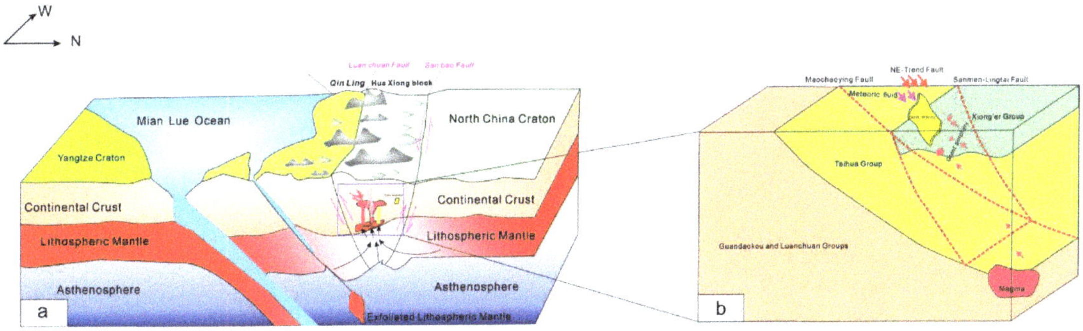

Figure 2. Metallogenic model map of Xiong'er shan gold district: (**a**) Late Mesozoic tectonic-magmatic evolution scenario in Xiong'er shan Area; and (**b**) metallogenic model for the Au polymetallic deposits in Xiong'ershan Area (Modified after [139,140]).

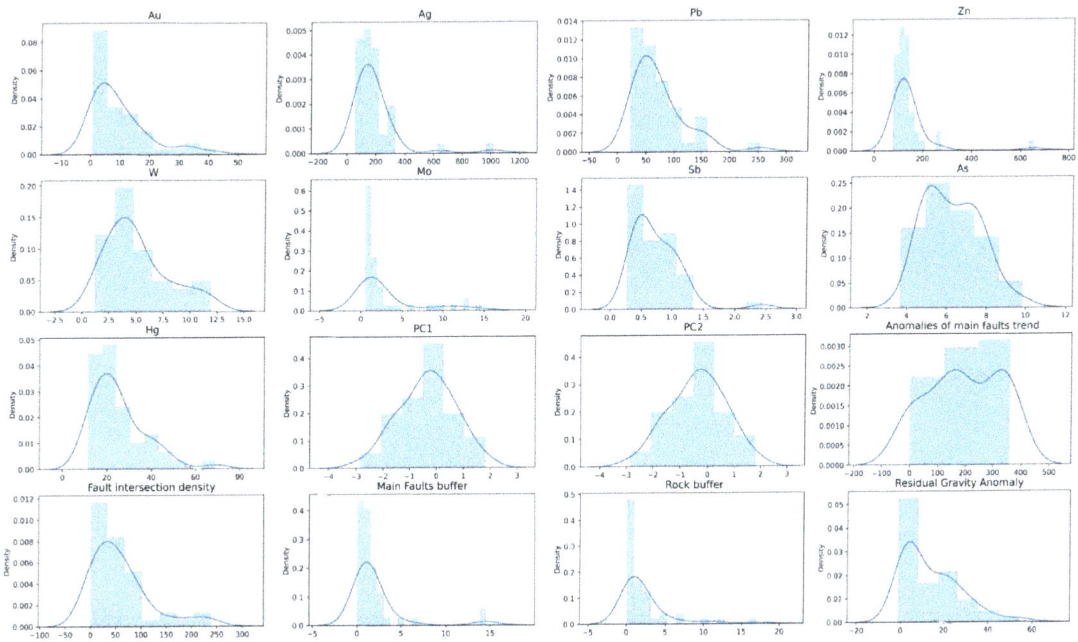

Figure 3. Distribution Histogram for Geological exploration dataset in the Xiong'ershan area.

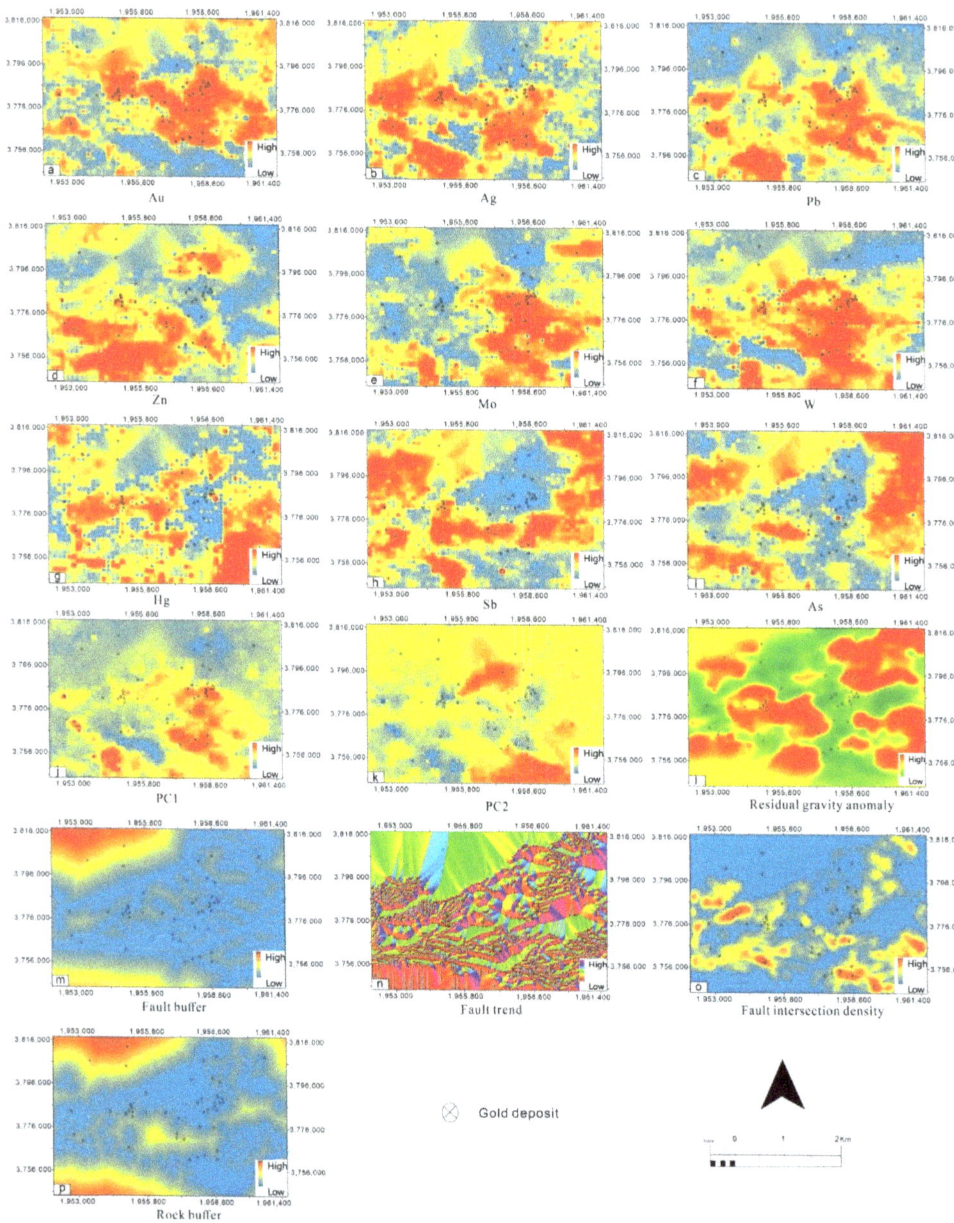

Figure 4. Evidence layers with major gold deposits: (**a–k**) clr-transformed Au, Ag, Pb, Zn, Mo, W, Hg, Sb, As, PC1, PC2; and (**l–p**) with gold deposits RGA, FB, FT, FID, GRB.

3.2.2. Transport and Deposition

Most gold deposits are hosted in faults. The ore-bearing faults in the study area are mainly NE and NWW faults, and a few are NW faults (i.e., Mao Chaoying fault). The Mao Chaoying fault zone is a deep fault that acted as pathways for the upward flow of deep-seated fluids (Figure 2b). The thick ore bodies (such as Qianhe and Hongzhuang gold deposits) are common in the intersection of NWW and NE faults in the Xiong'ershan area. Therefore, the fault buffer zone (FB), the European direction of the fault (FT), and the fault intersection density (FID) were selected as the prediction factor layers (Figure 4m–o).

Stream sediment geochemical data can reveal regional to district-scale patterns associated with Au mineralization in the study area. We adopted the geochemical anomalies of Au, Ag, Pb, Zn, As, Sb, Hg, W and Mo, associated with fault-controlled mineralization on a regional scale (Figure 4a–k). In addition, considering the compositional nature of geochemical data (i.e., to address the constant sum problem existing in compositional data), we applied centered log-ratio transformation (clr) prior principal component analysis (PCA) [141–143]. Figure 5 shows the positive loading for Zn, Sb, Pb, Mo, Hg, Au and Ag elements, and the negative loading for As in the first principal component (PC1), where the Ag and Pb had a positive high factor load. This indicated that the high-value area of PC1 is more closely related to Ag, and Pb metallogenetic elements. Where Sb, Au, As, and Ag had a positive loading for the second principal component (PC2), the As and Sb as the front elements had a positive high factor load, and the high-value area was consistent with the low-value of PC2. This indicated that the low-value area of PC2 had great metallogenic potential.

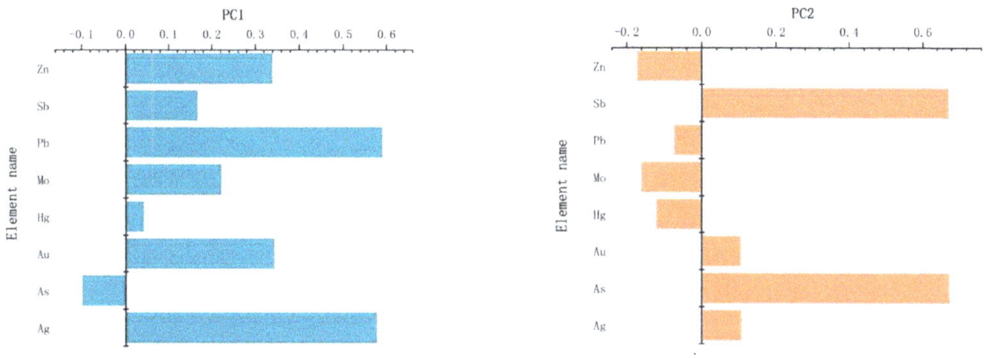

Figure 5. Loading chart for the first and second principal components in the Xiong'er shan area.

Therefore, the distribution histogram was used for the statistical analysis of the quantitative prediction factor as shown in Figure 3, along with the sixteen evidence maps derived by the prediction factor layers, which have been processed using a grid with a pixel size 130 m × 130 m (Figure 4).

3.2.3. Training and Validation Data

In this paper, 45 gold deposits developed in the study area were used as the training set, and each gold deposit includes 16 prediction variables (such as Au, Ag, Pb, Zn, Mo, W, Hg, Sb, As, PC1, PC2, RGA, FB, FT, FID, RB). The classification for MLA has a common basic assumption: the number of positive and negative datasets (samples) should be balanced [144,145]. If the positive and negative datasets are extremely imbalanced, the predicted results may be biased towards the majority classes with a large number of datasets [146,147]. Mineralization is a rare event, resulting in an insufficient number of training samples by MLA, and the number of deposits (positive samples) and non-deposits (negative samples) is not equal. In this paper, we used the synthetic minority over-sampling technique (SMOTE) based on geological knowledge constraints to balance data. The specific process is as follows: (1) based on the data of 45 known Au deposits (including 16 features) in the Xiong'ershan area and constrained by the optimal fault buffer radius of 3 km (Figure 3), yielded 945 SMOTE-augmented positive samples in the optimal ore-controlling adjacent area; (2) A total of 900 SMOTE-augmented negative samples were randomly selected from outside the 3 km threshold range of the non-anomalous area; (3) The knowledge-driven SMOTE datasets (generated by the (1) and (2)) are called MS (with the 945 deposits/positive and the 940 non-deposits/negative datasets) as the first datasets. The second datasets (denoted hereafter as OS datasets) were composited with 90 datasets (with the original known 45 deposits, 45 non-deposits are randomly selected non-mineralized locations)l (4) Then, we used the sixteen derived evidence maps, and the optimal threshold was used to show the Kernel density estimation curve (Figure 3) needed to accomplish the metallogenic predictionl (5) Finally, the dataset (with OS and MS) was divided into a test set and a train set (ratio of 2:8), and the test set was used to evaluate the accuracy and robustness of the model.

4. Results and Discussion

4.1. Parameters Optimization

Based on the Section 2.3 hyperparametric optimization process, the initially selected parameter space and the final reduced parameter space are shown in Table 3.

Table 3. GBDT parameter space for training datasets.

Parameter	Initial Parameter Space	Final Parameter Space
loss	["deviance","exponential"]	["deviance","exponential"]
criterion	["friedman_mse", "squared_error"]	["friedman_mse", "squared_error"]
n_estimators	(25,200,25)	(55,200,1)
learning_rate	(0.1,2.1,0.1)	(0.05,1,0.005)
max_features	(4,20,2)	(1,16,1)
subsample	(0.1,0.8,0.1)	(0.5,1.0,0.05)
max_depth	(2,30,2)	(10,35,1)
min_impurity_decrease	(0,5,1)	(0,5,0.1)

In order to compare and verify the performance of the two hyperparameter optimization algorithms on the final classification model, we plotted the kernel density estimates curve to understand the distribution of hyperparameters in both algorithms. Figure 6a–h shows that the distribution density curves of two hyperparameter optimization algorithms are consistent in the same domain space. It indicates that the hyperparameter distribution of the two algorithms is basically consistent. However, TPE tends to concentrate near the high-density area (placing more probabilities), resulting in the minimum loss in cross-validation. Figure 6i shows that the parameters criterion is mainly "friedman_mse"in random search and TPE-GBDT model. For pruning parameters "max_depth", "min_impurity_decrease" TPE has relatively lower values. Figure 7 shows that the value of the loss function is related to the spatial relationship based on the iterative process and the weak estimators. The random search reached the minimum loss function value at 77 iterations, while TPE reached the minimum loss function value at 54 iterations. Figure 6g also verified that in the same domain control interval, the training time of TPE was lower than that of the random search. The aforementioned results indirectly prove that the core idea of TPE is to spend more time evaluating promising hyperparametric values (i.e., the minimum loss function value). The loss value decreases with the increase of "learning rate" and "max_feature" (Figure 7b,d). The subsample and the "max_depth" shows a low loss value between 0.8–0.95, 15–25, respectively (Figure 7a,c). There is no obvious monotonic relation between the "min_impurity_decrease" parameter and loss value in the Xiong'ershan dataset. Compared with random search and TPE, the latter has a less objective function evaluation and better generalization performance on test sets. The optimal hyperparameters determined by the minimum loss function are shown in Table 4.

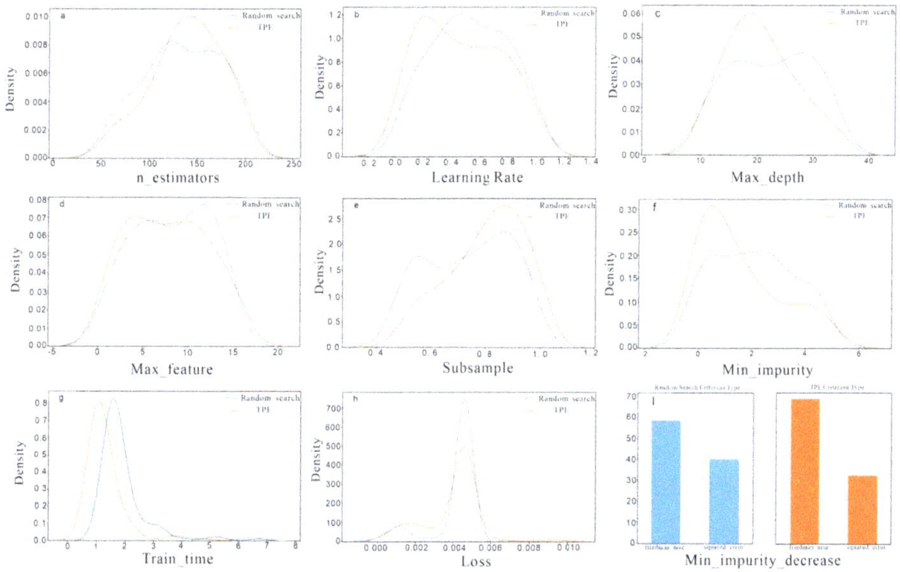

Figure 6. The kernel density map of parameter optimization based on random search and TPE on the test sets (iteration = 100).(**a–i**) the parameter of n_estimators, learning rate, max_depth, max_feature, subsample, min_impurity, train_time, loss, min_impurity_decrease.

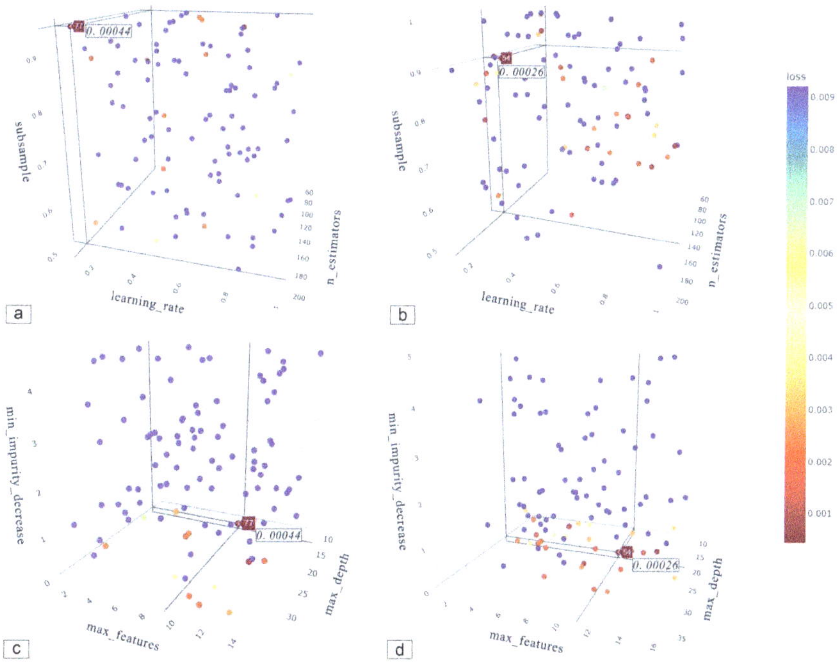

Figure 7. 3D visualization of the random search-GBDT model and TPE-GBDT model results: (**a**) random search-GBDT based on boosting hyperparameters; (**b**) TPE-GBDT based on boosting hyperparameters; (**c**) random search-GBDT based on weak learner hyperparameters; (**d**) and TPE-GBDT based on weak learner hyperparameters.

Table 4. The Optimized parameters for the GBDT model.

Parameter	Parameter Value Based on the Random Search	Parameter Value Based on the TPE
loss	"deviance"	"deviance"
criterion	"friedman_mse"	"friedman_mse"
n_estimators	186	69
learning_rate	0.09	0.7
max_features	9	12
subsample	0.95	0.8
max_depth	12	10
min_impurity_decrease	0.1	0.1

4.2. Performance Evaluation

Model evaluation plays a vital role in machine learning models. It helps to find the best model to represent our data. Different types of models (regression model and classification model) have different evaluation indexes. The evaluation indexes of the regression model include mean absolute error (MAE), mean square error (MSE), root mean square error (RMSE), and Coefficient of determination (R2) etc. The evaluation methods commonly used for classification models include k-fold cross validation to verify the model fitting effect, accuracy, recall rate and AUC index to evaluate the model generalization ability [148–151]. In our study, we selected the five-fold cross-validation, confusion matrix and the AUC as the performance evaluation for the GBDT model.

The five-fold cross-validation refers to roughly dividing the datasets into five parts, one of which is reserved for the validation model; the other samples are used for training. Cross-validation is repeated five times, with each sub-sample verified iterately. The results of an average of five times or other combination methods are used to finally obtain a single estimation. In this way, the model is more accurate. The GBDT model was trained according to the hyperparameters in Table 4, and the average accuracy of the five-fold cross-validation results obtained based on random search and TPE on the MS test set was 0.963 and 0.966, and on the OS test set was 0.764 and 0.786, respectively. This showed that the augmented datasets of the MS and the parameter adjustment improved the accuracy of the GBDT model (detail shows in Table 5). The results of the random search and TPE hyperparameter optimization were compared and analyzed after the five-fold cross-validation. Figure 8 shows that the results based on TPE had the best anti-overfitting effect for the GBDT model with the higher accuracy compared with the test sets.

Table 5. The result of five-fold cross-validation for GBDT model.

Model	Train Accuracy	Test Accuracy	Five-Fold Cross-Validation Time	Datasets
GBDT	0.981	0.940	0.55 s	MS
GBDT-Random	1.000	0.963	3.16 s	
GBDT-TPE	1.000	0.966	0.71 s	
GBDT	1.000	0.754	0.07 s	OS
GBDT-Random	1.000	0.764	0.33 s	
GBDT-TPE	1.000	0.786	0.09 s	

The confusion matrix is a situation analysis table that summarizes the prediction results of the classification model in MLA. In the form of the matrix, records in datasets are summarized according to two criteria: the real category and the prediction category. The rows of the matrix represent the real value, and the columns of the matrix represent the predicted value.

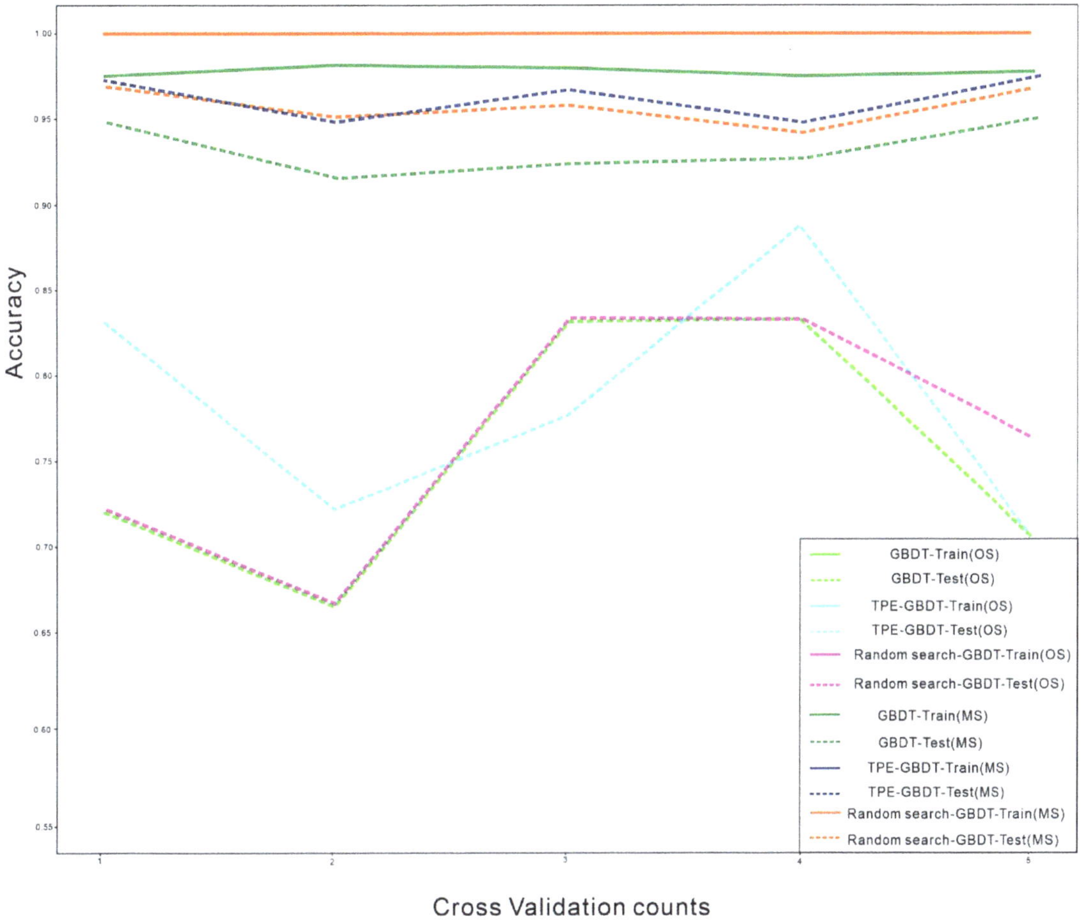

Figure 8. The Accuracy of five-fold cross-validation based on the GBDT model.

This mainly includes the real value being negative that is composed of the TP (true positive) and FN (false negative), and the real value being positive that is composed of the TN (true negative) and FP (false positive). The model's generalization ability was analyzed and verified by the test sets (20% of the total training data), and it selected 369 data from the MS and 18 from OS datasets. Figure 9a,b shows that the random search-GBDT model and TPE-GBDT model have the same performance when the real value is negative with the MS datasets, as TN = 165, FP = 12, and a slightly different performance when the real value is positive, as FN = 3, TP = 189 in the TPE-GBDT model, that is, three of the positive samples were incorrectly predicted in this model, while the four of positive samples were incorrectly predicted in the Random search-GBDT model. Figure 9c,d shows that with the OS datasets, the random search-GBDT model and TPE-GBDT model had the same TP and FP. However, the FN with the TPE-GBDT model was lower than the random search-GBDT model, and the TN with the TPE-GBDT model was the opposite. According to the confusion matrix value, the secondary indicators (i.e., accuracy, recall, specificity) can be calculated to evaluate the model. Table 6 shows that Accuracy = 0.9593, recall = 0.984, and Precision = 0.941 with the TPE-GBDT model with the MS datasets. The values were slightly higher than the random search-GBDT model, indicating that the TPE-GBDT model performed better than the random search-GBDT model.

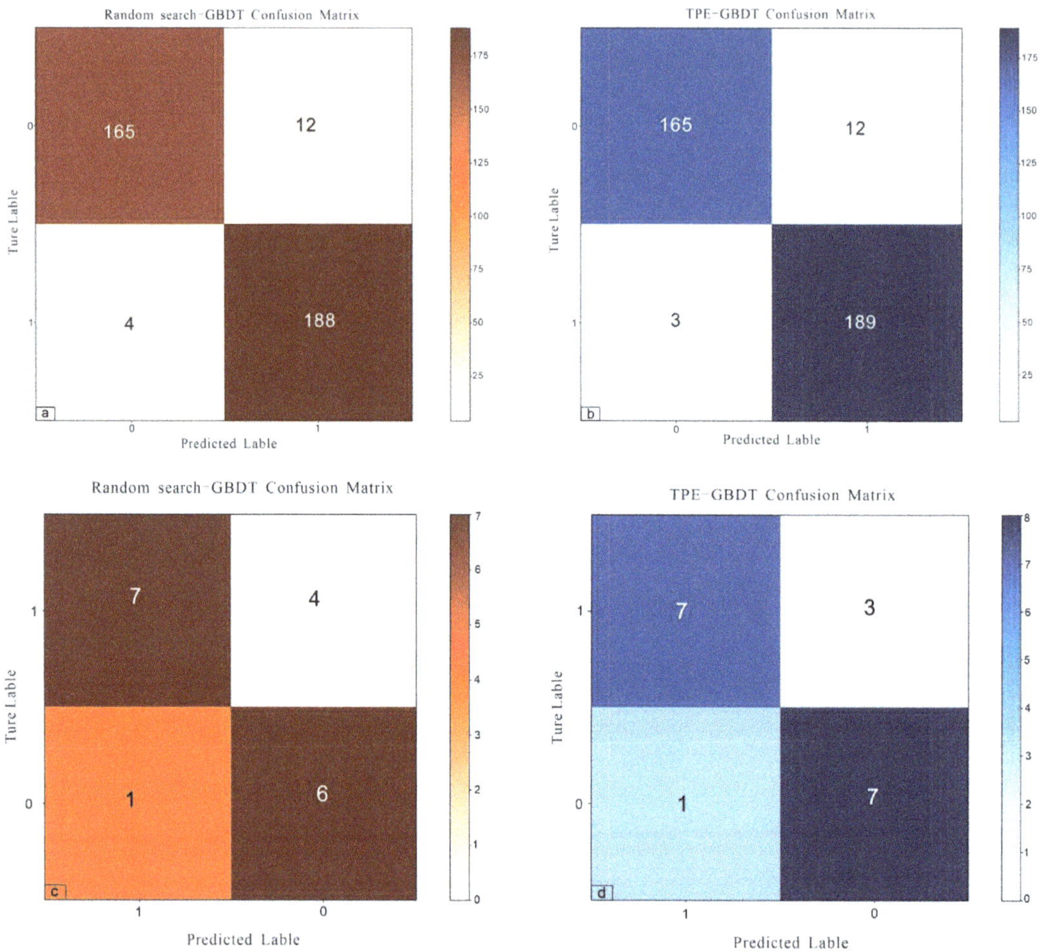

Figure 9. The confusion matrix of the: (**a**) random search-GBDT; (**b**) TPE-GBDT model with the MS datasets; (**c**) random search-GBDT; and (**d**) TPE-GBDT model with the OS datasets.

Table 6. The secondary indicators of the GBDT model base on the confusion matrix.

Model	Accuracy	Recall	Precision	Specificity	Datasets
GBDT-Random	0.9566	0.9791	0.940	0.9322	MS
GBDT-TPE	0.9593	0.9844	0.941	0.9322	
GBDT-Random	0.777	0.636	0.875	0.857	OS
GBDT-TPE	0.833	0.727	0.875	0.875	

Another indicator to evaluate the performance of overall classification of the model is the ROC curve and the AUC value. The ROC (short for Receiver Operating Characteristic) is a curve in two-dimensional plane space. The AUC is the calculation result of the area under the ROC curve, which is a specific value. For any model, the closer the ROC curve is to the upper left, the larger the area is under the ROC curve, and the better the classification performance of the model. Figure 10 shows that ROC curve with the MS datasets all in the upper left corner; the AUC values of the TPE-GBDT model and random search-GBDT model were 0.985 and 0.981, respectively.

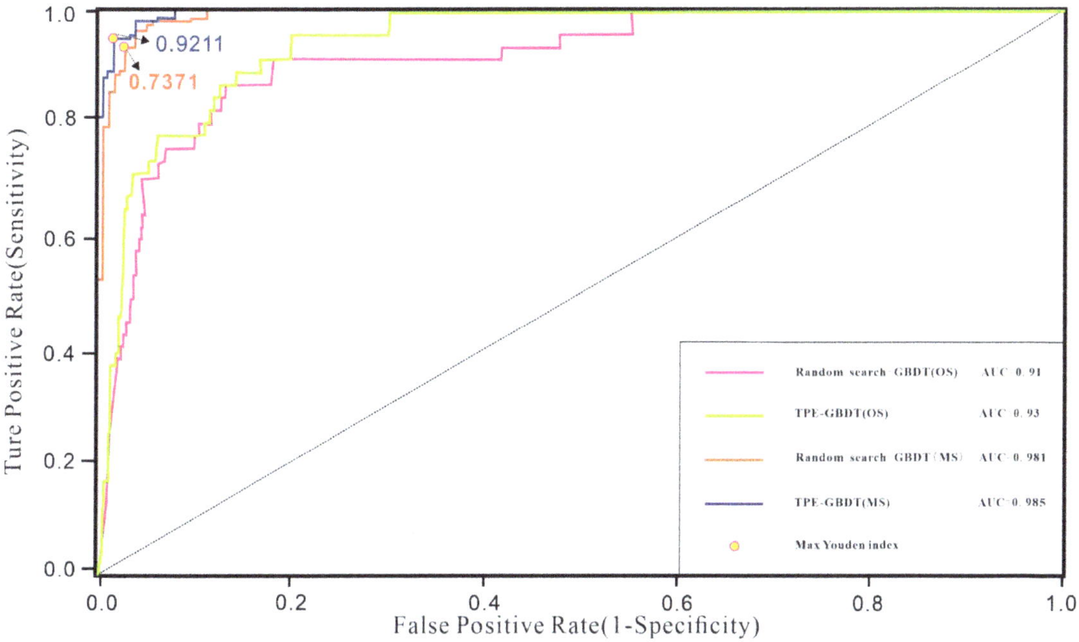

Figure 10. ROC curves of Random search-GBDT model and TPE-GBDT model.

4.3. Mapping of Mineral Prospectivity

The MPM of the Xiong'ershan area using ArcGIS software was finally obtained through the above hyperparameter determination and model evaluation. The evaluation of prospectivity models with continuous probabilities (ranging from 0–1) of random search-GBDT (Figure 11a) and TPE-GBDT (Figure 11b) was conducted by measuring the correlation between the prospectivity values and known mineral occurrences, and for each model, the success-rate curves were plotted. The success-rate curve is a capture efficiency curve that indicates the relationship between the probability distribution and Au deposit locations. Figure 12 shows the proportion of the known gold deposits according to different percentages of prospective areas. It can be observed that the random search-GBDT and TPE-GBDT start from a similar success-rate curve, although the slope of the success-rate curve is steeper in the TPE-GBDT model, indicating that the TPE improved the performance of the GBDT predictive modeling.

Then combined with the fact that the high probability part of the Random search-GBDT and TPE-GBDT models occupy 25% of the total study area. However thisthis 25% study area contains more than 93% of the known gold deposits (Figure 12). Finally, in order to delineate high-favorable targets in the study area. The cut-off values of Youden index in the ROC curve were adopted to discretize the TPE-GBDT and random search-GBDT predictive models, with the thresholds of 0.9211 and 0.7371, divided the prospectivity map of TPE-GBDT and Random search models into high potential (favorable) and low potential (non-favorable) areas, respectively. The highly favorable areas in the TPE-GBDT model (Figure 13b) captured 91% of known Au deposits within only 10.22% of the Xiong'er shan area, while the highly favorable areas of the Random search-GBDT model (Figure 13a) contained the same percentage of known Au occurrences but within larger areas (16.84%). Results suggested that the TPE-GBDT model is more consistent with the actual geological conditions, and more suitable for the next step of the exploration work.

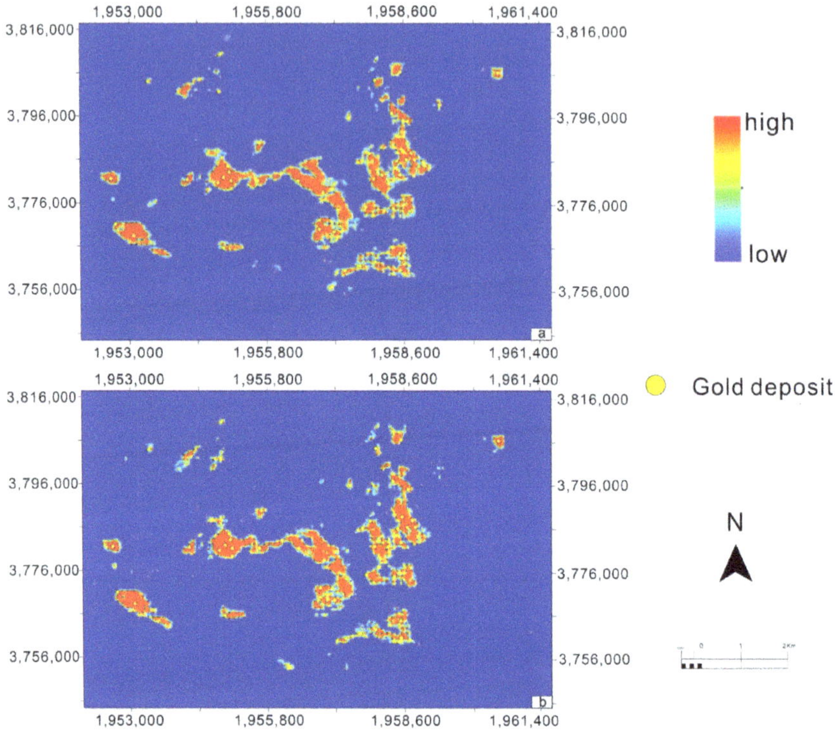

Figure 11. Predictive models of mineral prospectivity derived by: (**a**) random search-GBDT; and (**b**) TPE-GBDT models.

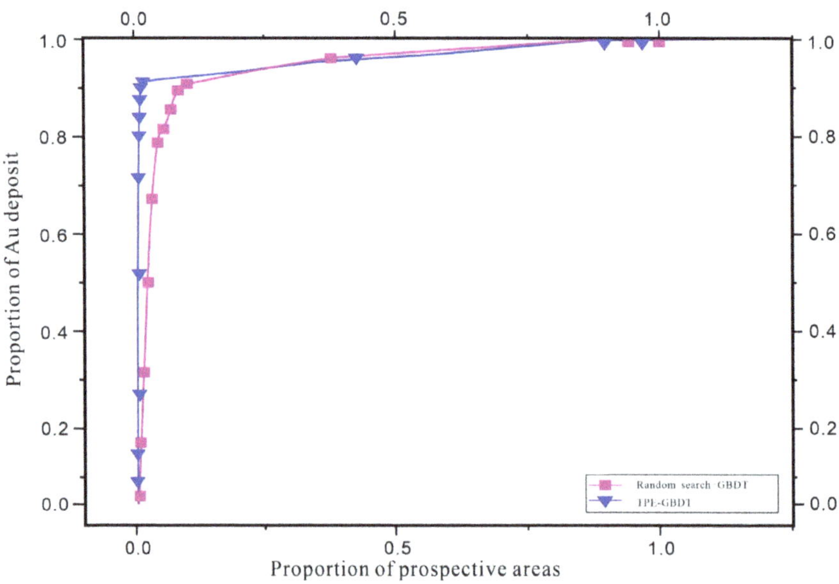

Figure 12. Performance of the Random search-GBDT and TPE-GBDT models of mineral prospectivity measured by success rate curve.

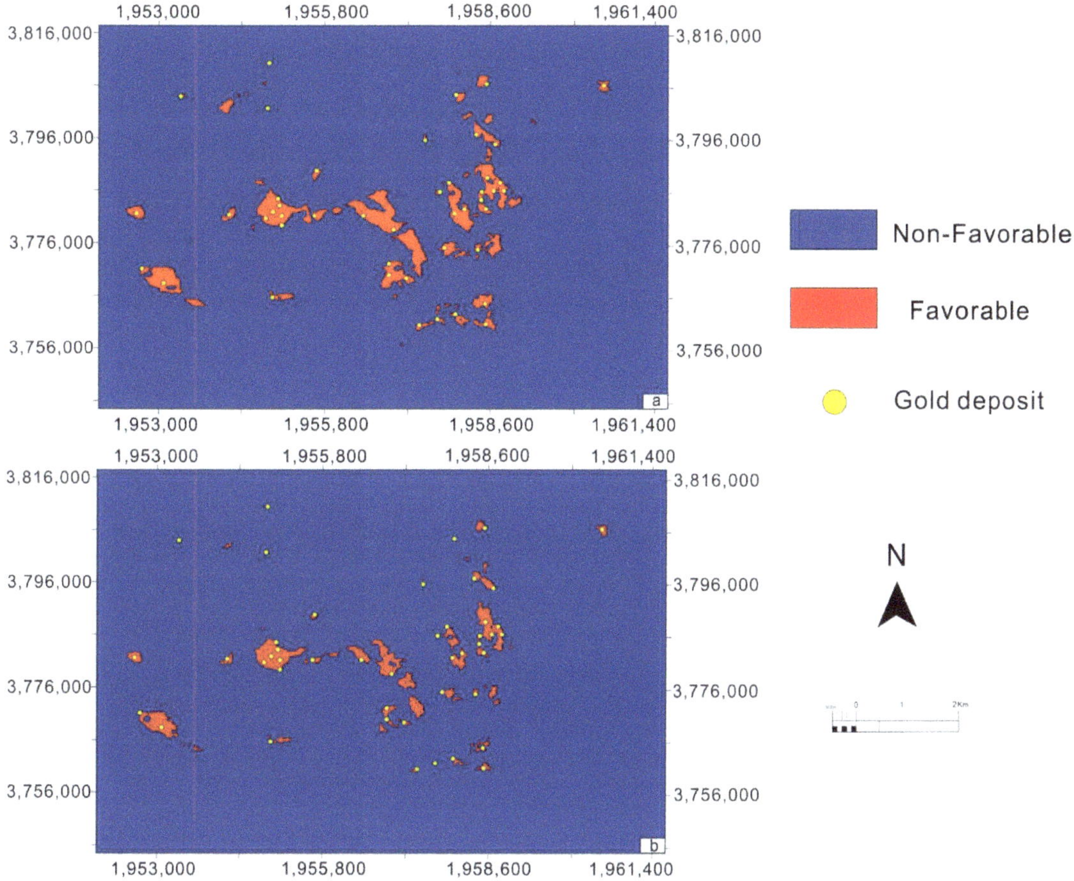

Figure 13. Predictive maps of: (**a**) random search-GBDT; and (**b**) TPE-GBDT models, showing favorable and non-favorable areas by a threshold value of the Max Youden index.

5. Conclusions

In this study, we proposed a new method of TPE-GBDT generating MPM, which solved the problem of hyperparametric optimization in MLA. To some extent, it improved the accuracy of the MPM. The summary includes the following points:

(1) In view of the data imbalance of geological datasets, we used the knowledge-driven SMOTE method for data augmentation (as the MS datasets), which reached the balance of positive and negative sets, and then compared and analyzed these with the OS datasets (original 45 ore occurrence and the randomly selected 45 non-ore occurrence datasets), showing that it improved the precision and increased the interpretability of the model;

(2) The comprehensive AUC value and the accuracy of the models indicated that the GBDT models suit small data (insufficient known mineral deposits) training. The proposed random search-GBDT model and the TPE-GBDT model were able to adjust the GBDT for HPO automatically, and the AUC value was higher than that in the conventional GBDT, indicating that the HPO increased GBDT model accuracy;

(3) The spatial distribution of the random search-GBDT model and the TPE-GBDT model predicted results was consistent. A comparison with known Au deposits indicated that the TPE-GBDT model required less training time and had a more reasonable probability distribution than the random search-GBDT model. Thus, TPE-GBDT modeling

can further reduce the uncertainty of predictions and enhance the predictive accuracy of mineral exploration.

Author Contributions: M.F.: Writing original draft, Data curation, Investigation, Methodology. K.X.: Supervision, Funding acquisition. L.S.: Data curation, Funding acquisition. S.Z.: Writing—review & editing. Y.X.: Visualization, Investigation. All authors have read and agreed to the published version of the manuscript.

Funding: This research was supported by the National Key R&D Program of China (Grant No. 2016YFC0600504).

Data Availability Statement: Not applicable.

Conflicts of Interest: The authors declare no conflict of interest.

References

1. Payne, C.E.; Cunningham, F.; Peters, K.J.; Nielsen, S.; Puccioni, E.; Wildman, C.; Partington, G.A. From 2D to 3D: Prospectivity modelling in the Taupo volcanic zone, New Zealand. *Ore Geol. Rev.* **2015**, *71*, 558–577. [CrossRef]
2. Zhang, Z.; Zhang, J.; Wang, G.; Carranza, E.J.M.; Pang, Z.; Wang, H. From 2D to 3D modeling of mineral prospectivity using multi-source geoscience datasets, Wulong Gold District, China. *Nat. Resour. Res.* **2020**, *29*, 345–364. [CrossRef]
3. Joly, A.; Porwal, A.; McCuaig, T.C.; Chudasama, B.; Dentith, M.C.; Aitken, A.R. Mineral systems approach applied to GIS-based 2D-prospectivity modelling of geological regions: Insights from Western Australia. *Ore Geol. Rev.* **2015**, *71*, 673–702. [CrossRef]
4. Yousefi, M.; Carranza, E.J.M. Prediction–area (P–A) plot and C–A fractal analysis to classify and evaluate evidential maps for mineral prospectivity modeling. *Comput. Geosci.* **2015**, *79*, 69–81. [CrossRef]
5. Jiang, W.; Korsch, R.J.; Doublier, M.P.; Duan, J.; Costelloe, R. Mapping deep electrical conductivity structure in the mount isa region, northern australia: Implications for mineral prospectivity. *J. Geophys. Res. Solid Earth* **2019**, *124*, 10655–10671. [CrossRef]
6. Li, T.; Zuo, R.; Xiong, Y.; Peng, Y. Random-drop data augmentation of deep convolutional neural network for mineral prospectivity mapping. *Nat. Resour. Res.* **2021**, *30*, 27–38. [CrossRef]
7. Yang, N.; Zhang, Z.; Yang, J.; Hong, Z.; Shi, J. A convolutional neural network of GoogLeNet applied in mineral prospectivity prediction based on multi-source geoinformation. *Nat. Resour. Res.* **2021**, *30*, 3905–3923. [CrossRef]
8. Wang, Z.; Zuo, R. Mineral prospectivity mapping using a joint singularity-based weighting method and long short-term memory network. *Comput. Geosci.* **2022**, *158*, 104974. [CrossRef]
9. Xiao, K.; Li, N.; Porwal, A.; Holden, E.J.; Bagas, L.; Lu, Y. GIS-based 3D prospectivity mapping: A case study of Jiama copper-polymetallic deposit in Tibet, China. *Ore Geol. Rev.* **2015**, *71*, 611–632. [CrossRef]
10. Li, X.; Yuan, F.; Zhang, M.; Jia, C.; Jowitt, S.M.; Ord, A.; Zheng, T.; Hu, X.; Li, Y. Three-dimensional mineral prospectivity modeling for targeting of concealed mineralization within the Zhonggu iron orefield, Ningwu Basin, China. *Ore Geol. Rev.* **2015**, *71*, 633–654. [CrossRef]
11. Li, X.; Yuan, F.; Zhang, M.; Jowitt, S.M.; Ord, A.; Zhou, T.; Dai, W. 3D computational simulation-based mineral prospectivity modeling for exploration for concealed Fe–Cu skarn-type mineralization within the Yueshan orefield, Anqing district, Anhui Province, China. *Ore Geol. Rev.* **2019**, *105*, 1–17. [CrossRef]
12. Xiang, J.; Xiao, K.; Carranza, E.J.M.; Chen, J.; Li, S. 3D mineral prospectivity mapping with random forests: A case study of Tongling, Anhui, China. *Nat. Resour. Res.* **2020**, *29*, 395–414. [CrossRef]
13. Mao, X.; Zhang, W.; Liu, Z.; Ren, J.; Bayless, R.C.; Deng, H. 3D mineral prospectivity modeling for the low-sulfidation epithermal gold deposit: A case study of the axi gold deposit, western Tianshan, NW China. *Minerals* **2020**, *10*, 233. [CrossRef]
14. Qin, Y.; Liu, L.; Wu, W. Machine learning-based 3D modeling of mineral prospectivity mapping in the Anqing Orefield, Eastern China. *Nat. Resour. Res.* **2021**, *30*, 3099–3120. [CrossRef]
15. Mohammadpour, M.; Bahroudi, A.; Abedi, M. Three dimensional mineral prospectivity modeling by evidential belief functions, a case study from Kahang porphyry Cu deposit. *J. Afr. Earth Sci.* **2021**, *174*, 104098. [CrossRef]
16. Xiao, K.; Xiang, J.; Fan, M.; Xu, Y. 3D mineral prospectivity mapping based on deep metallogenic prediction theory: A case study of the Lala Copper Mine, Sichuan, China. *J. Earth Sci.* **2021**, *32*, 348–357. [CrossRef]
17. Porwal, A.; Carranza, E.J.M.; Hale, M. Knowledge-driven and data-driven fuzzy models for predictive mineral potential mapping. *Nat. Resour. Res.* **2003**, *12*, 1–25. [CrossRef]
18. Carranza, E.J.M.; Van Ruitenbeek, F.J.A.; Hecker, C.; van der Meijde, M.; van der Meer, F.D. Knowledge-guided data-driven evidential belief modeling of mineral prospectivity in Cabo de Gata, SE Spain. *Int. J. Appl. Earth Obs. Geoinf.* **2008**, *10*, 374–387. [CrossRef]
19. Abedi, M.; Torabi, S.A.; Norouzi, G.H.; Hamzeh, M. ELECTRE III: A knowledge-driven method for integration of geophysical data with geological and geochemical data in mineral prospectivity mapping. *J. Appl. Geophys.* **2012**, *87*, 9–18. [CrossRef]
20. Harris, J.R.; Grunsky, E.; Behnia, P.; Corrigan, D. Data-and knowledge-driven mineral prospectivity maps for Canada's North. *Ore Geol. Rev.* **2015**, *71*, 788–803. [CrossRef]

21. Hosseini, S.A.; Abedi, M. Data envelopment analysis: A knowledge-driven method for mineral prospectivity mapping. *Comput. Geosci.* **2015**, *82*, 111–119. [CrossRef]
22. Abedi, M.; Kashani, S.B.M.; Norouzi, G.H.; Yousefi, M. A deposit scale mineral prospectivity analysis: A comparison of various knowledge-driven approaches for porphyry copper targeting in Seridune, Iran. *J. Afr. Earth Sci.* **2017**, *128*, 127–146. [CrossRef]
23. Skirrow, R.G.; Murr, J.; Schofield, A.; Huston, D.L.; van der Wielen, S.; Czarnota, K.; Coghlan, R.; Highet, L.M.; Connolly, D.; Doublier, M.; et al. Mapping iron oxide Cu-Au (IOCG) mineral potential in Australia using a knowledge-driven mineral systems-based approach. *Ore Geol. Rev.* **2019**, *113*, 103011. [CrossRef]
24. Daviran, M.; Parsa, M.; Maghsoudi, A.; Ghezelbash, R. Quantifying uncertainties linked to the diversity of mathematical frameworks in knowledge-driven mineral prospectivity mapping. *Nat. Resour. Res.* **2022**, *31*, 2271–2287. [CrossRef]
25. Carranza, E.J.M.; Hale, M.; Faassen, C. Selection of coherent deposit-type locations and their application in data-driven mineral prospectivity mapping. *Ore Geol. Rev.* **2008**, *33*, 536–558. [CrossRef]
26. Carranza, E.J.M. Objective selection of suitable unit cell size in data-driven modeling of mineral prospectivity. *Comput. Geosci.* **2009**, *35*, 2032–2046. [CrossRef]
27. Carranza, E.J.M.; Laborte, A.G. Data-driven predictive mapping of gold prospectivity, Baguio district, Philippines: Application of Random Forests algorithm. *Ore Geol. Rev.* **2015**, *71*, 777–787. [CrossRef]
28. Carranza, E.J.M.; Laborte, A.G. Data-driven predictive modeling of mineral prospectivity using random forests: A case study in Catanduanes Island (Philippines). *Nat. Resour. Res.* **2016**, *25*, 35–50. [CrossRef]
29. Yousefi, M.; Nykänen, V. Data-driven logistic-based weighting of geochemical and geological evidence layers in mineral prospectivity mapping. *J. Geochem. Explor.* **2016**, *164*, 94–106. [CrossRef]
30. Yousefi, M.; Carranza, E.J.M. Data-driven index overlay and Boolean logic mineral prospectivity modeling in greenfields exploration. *Nat. Resour. Res.* **2016**, *25*, 3–18. [CrossRef]
31. McKay, G.; Harris, J.R. Comparison of the data-driven random forests model and a knowledge-driven method for mineral prospectivity mapping: A case study for gold deposits around the Huritz Group and Nueltin Suite, Nunavut, Canada. *Nat. Resour. Res.* **2016**, *25*, 125–143. [CrossRef]
32. Chen, Y.; Wu, W. Isolation forest as an alternative data-driven mineral prospectivity mapping method with a higher data-processing efficiency. *Nat. Resour. Res.* **2019**, *28*, 31–46. [CrossRef]
33. Sun, T.; Li, H.; Wu, K.; Chen, F.; Zhu, Z.; Hu, Z. Data-driven predictive modelling of mineral prospectivity using machine learning and deep learning methods: A case study from southern Jiangxi Province, China. *Minerals* **2020**, *10*, 102. [CrossRef]
34. Zhang, S.; Carranza, E.J.M.; Wei, H.; Xiao, K.; Yang, F.; Xiang, J.; Zhang, S.; Xu, Y. Data-driven Mineral Prospectivity Mapping by Joint Application of Unsupervised Convolutional Auto-encoder Network and Supervised Convolutional Neural Network. *Nat. Resour. Res.* **2021**, *30*, 1011–1031. [CrossRef]
35. Parsa, M.; Carranza, E.J.M. Modulating the impacts of stochastic uncertainties linked to deposit locations in data-driven predictive mapping of mineral prospectivity. *Nat. Resour. Res.* **2021**, *30*, 3081–3097. [CrossRef]
36. Bacardit, J.; Llorà, X. Large scale data mining using genetics-based machine learning. In Proceedings of the 11th Annual Conference Companion on Genetic and Evolutionary Computation Conference, Montréal, Canada, 8–12 July 2009; pp. 3381–3412.
37. Najafabadi, M.M.; Villanustre, F.; Khoshgoftaar, T.M.; Seliya, N.; Wald, R.; Muharemagic, E. Deep learning applications and challenges in big data analytics. *J. Big Data* **2015**, *2*, 1. [CrossRef]
38. Angermueller, C.; Pärnamaa, T.; Parts, L.; Stegle, O. Deep learning for computational biology. *Mol. Syst. Biol.* **2016**, *12*, 878. [CrossRef]
39. Sun, C.; Shrivastava, A.; Singh, S.; Gupta, A. Revisiting unreasonable effectiveness of data in deep learning era. In Proceedings of the IEEE International Conference on Computer Vision, Venice, Italy, 22–29 October 2017; pp. 843–852.
40. Rusk, N. Deep learning. *Nat. Methods* **2016**, *13*, 35. [CrossRef]
41. Nguyen, G.; Dlugolinsky, S.; Bobák, M.; Tran, V.; López García, Á.; Heredia, I.; Malík, P.; Hluchý, L. Machine learning and deep learning frameworks and libraries for large-scale data mining: A survey. *Artif. Intell. Rev.* **2019**, *52*, 77–124. [CrossRef]
42. Zuo, R. Geodata science-based mineral prospectivity mapping: A review. *Nat. Resour. Res.* **2020**, *29*, 3415–3424. [CrossRef]
43. Chawla, N.V.; Bowyer, K.W.; Hall, L.O.; Kegelmeyer, W.P. SMOTE: Synthetic minority over-sampling technique. *J. Artif. Intell. Res.* **2002**, *16*, 321–357. [CrossRef]
44. Han, H.; Wang, W.Y.; Mao, B.H. Borderline-SMOTE: A new over-sampling method in imbalanced data sets learning. In Proceedings of the International Conference on Intelligent Computing, Hefei China, 23–26 August 2005; Springer: Berlin/Heidelberg, Germany, 2005; pp. 878–887.
45. Bunkhumpornpat, C.; Sinapiromsaran, K.; Lursinsap, C. Safe-level-smote: Safe-level-synthetic minority over-sampling technique for handling the class imbalanced problem. In Proceedings of the Pacific-Asia Conference on Knowledge Discovery and Data Mining, Bangkok, Thailand, 27–30 April 2009; Springer: Berlin/Heidelberg, Germany, 2009; pp. 475–482.
46. Bunkhumpornpat, C.; Sinapiromsaran, K.; Lursinsap, C. DBSMOTE: Density-based synthetic minority over-sampling technique. *Appl. Intell.* **2012**, *36*, 664–684. [CrossRef]
47. Guzmán-Ponce, A.; Sánchez, J.S.; Valdovinos, R.M.; Marcial-Romero, J.R. DBIG-US: A two-stage under-sampling algorithm to face the class imbalance problem. *Expert Syst. Appl.* **2021**, *168*, 114301. [CrossRef]
48. Soltanzadeh, P.; Hashemzadeh, M. RCSMOTE: Range-Controlled synthetic minority over-sampling technique for handling the class imbalance problem. *Inf. Sci.* **2021**, *542*, 92–111. [CrossRef]

49. Peng, C.Y.; Park, Y.J. A New Hybrid Under-sampling Approach to Imbalanced Classification Problems. *Appl. Artif. Intell.* **2022**, *36*, 1975393. [CrossRef]
50. Lenka, S.R.; Bisoy, S.K.; Priyadarshini, R.; Nayak, B. Representative-based cluster undersampling technique for imbalanced credit scoring datasets. In *Innovations in Computational Intelligence and Computer Vision*; Springer: Singapore, 2022; pp. 119–129.
51. Amirruddin, A.D.; Muharam, F.M.; Ismail, M.H.; Tan, N.P.; Ismail, M.F. Synthetic Minority Over-sampling TEchnique (SMOTE) and Logistic Model Tree (LMT)-Adaptive Boosting algorithms for classifying imbalanced datasets of nutrient and chlorophyll sufficiency levels of oil palm (Elaeis guineensis) using spectroradiometers and unmanned aerial vehicles. *Comput. Electron. Agric.* **2022**, *193*, 106646.
52. Perez, L.; Wang, J. The effectiveness of data augmentation in image classification using deep learning. *arXiv* **2017**, arXiv:1712.04621. *Preprint*.
53. Abbaszadeh, M.; Soltani-Mohammadi, S.; Ahmed, A.N. Optimization of support vector machine parameters in modeling of Iju deposit mineralization and alteration zones using particle swarm optimization algorithm and grid search method. *Comput. Geosci.* **2022**, *165*, 105140. [CrossRef]
54. Shorten, C.; Khoshgoftaar, T.M. A survey on image data augmentation for deep learning. *J. Big Data* **2019**, *6*, 60. [CrossRef]
55. Inoue, H. Data augmentation by pairing samples for images classification. *arXiv* **2018**, arXiv:1801.02929; Preprint.
56. Jackson, P.T.; Abarghouei, A.A.; Bonner, S.; Breckon, T.P.; Obara, B. Style augmentation: Data augmentation via style randomization. *CVPR Workshops* **2019**, *6*, 10–11.
57. Raj, R.; Mathew, J.; Kannath, S.K.; Rajan, J. Crossover based technique for data augmentation. *Comput. Methods Programs Biomed.* **2022**, *218*, 106716. [CrossRef] [PubMed]
58. Radford, A.; Metz, L.; Chintala, S. Unsupervised representation learning with deep convolutional generative adversarial networks. *arXiv* **2015**, arXiv:1511.06434. *Preprint*.
59. Creswell, A.; White, T.; Dumoulin, V.; Arulkumaran, K.; Sengupta, B.; Bharath, A.A. Generative adversarial networks: An overview. *IEEE Signal Process. Mag.* **2018**, *35*, 53–65. [CrossRef]
60. Cubuk, E.D.; Zoph, B.; Mane, D.; Vasudevan, V.; Le, Q.V. Autoaugment: Learning augmentation policies from data. *arXiv* **2018**, arXiv:1805.09501. *Preprint*.
61. Lim, S.; Kim, I.; Kim, T.; Kim, C.; Kim, S. Fast autoaugment. *Adv. Neural Inf. Process. Syst.* **2019**, *32*, 1–11.
62. Xiong, Y.; Zuo, R. Effects of misclassification costs on mapping mineral prospectivity. *Ore Geol. Rev.* **2017**, *82*, 1–9. [CrossRef]
63. Xiong, Y.; Zuo, R. GIS-based rare events logistic regression for mineral prospectivity mapping. *Comput. Geosci.* **2018**, *111*, 18–25. [CrossRef]
64. Lin, C.F.; Wang, S.D. Fuzzy support vector machines. *IEEE Trans. Neural Netw.* **2002**, *13*, 464–471.
65. Min, R.; Cheng, H.D. Effective image retrieval using dominant color descriptor and fuzzy support vector machine. *Pattern Recognit.* **2009**, *42*, 147–157. [CrossRef]
66. Batuwita, R.; Palade, V. FSVM-CIL: Fuzzy support vector machines for class imbalance learning. *IEEE Trans. Fuzzy Syst.* **2010**, *18*, 558–571. [CrossRef]
67. Yu, H.; Sun, C.; Yang, X.; Zheng, S.; Zou, H. Fuzzy support vector machine with relative density information for classifying imbalanced data. *IEEE Trans. Fuzzy Syst.* **2019**, *27*, 2353–2367. [CrossRef]
68. Maldonado, S.; López, J.; Vairetti, C. Time-weighted Fuzzy Support Vector Machines for classification in changing environments. *Inf. Sci.* **2021**, *559*, 97–110. [CrossRef]
69. Yu, X.; Zhang, X. Imbalanced data classification algorithm based on hybrid model. In Proceedings of the International Conference on Machine Learning and Cybernetics, Xi'an, China, 15–17 July 2012; IEEE: New York, NY, USA, 2012; Volume 2, pp. 735–740.
70. Zhang, M.; Wu, M. Efficient super greedy boosting for classification. In Proceedings of the 2020 10th Institute of Electrical and Electronics Engineers International Conference on Cyber Technology in Automation, Control, and Intelligent Systems (CYBER), Xi'an, China, 10–13 October 2020; IEEE: New York, NY, USA, 2020; pp. 192–197.
71. Ding, J.; Wang, S.; Jia, L.; You, J.; Jiang, Y. Spark-based Ensemble Learning for Imbalanced Data Classification. *Int. J. Perform. Eng.* **2018**, *14*, 955. [CrossRef]
72. Wang, Y.; Liao, X.; Lin, S. Rescaled boosting in classification. *IEEE Trans. Neural Netw. Learn. Syst.* **2019**, *30*, 2598–2610. [CrossRef]
73. Lim, S.K.; Loo, Y.; Tran, N.T.; Cheung, N.M.; Roig, G.; Elovici, Y. Doping: Generative data augmentation for unsupervised anomaly detection with gan. In Proceedings of the 2018 IEEE International Conference on Data Mining (ICDM), Singapore, 17–20 November 2018; IEEE: New York, NY, USA, 2018; pp. 1122–1127.
74. Wen, Q.; Sun, L.; Yang, F.; Song, X.; Gao, J.; Wang, X.; Xu, H. Time series data augmentation for deep learning: A survey. *arXiv* **2002**, arXiv:2002.12478. *Preprint*.
75. Al Olaimat, M.; Lee, D.; Kim, Y.; Kim, J.; Kim, J. A learning-based data augmentation for network anomaly detection. In Proceedings of the 2020 29th International Conference on Computer Communications and Networks (ICCCN), Honolulu, HI, USA, 3–6 August 2020; IEEE: New York, NY, USA, 2020; pp. 1–10.
76. Sinha, A.; Ayush, K.; Song, J.; Uzkent, B.; Jin, H.; Ermon, S. Negative data augmentation. *arXiv* **2021**, arXiv:2102.05113. *Preprint*.
77. Chen, R.T.; Rubanova, Y.; Bettencourt, J.; Duvenaud, D.K. Neural ordinary differential equations. In *Advances in Neural Information Processing Systems*; NeurIPS: New Orleans, LA, USA, 2018; p. 31.

78. Song, L.; Gong, D.; Li, Z.; Liu, C.; Liu, W. Occlusion robust face recognition based on mask learning with pairwise differential siamese network. In Proceedings of the IEEE/CVF International Conference on Computer Vision, Seould, Republic of Korea, 27 October–2 November 2019; pp. 773–782.
79. Meldo, A.A.; Utkin, L.V. A new approach to differential lung diagnosis with ct scans based on the siamese neural network. *J. Phys.* **2019**, *1236*, 12–58. [CrossRef]
80. Ruthotto, L.; Haber, E. Deep neural networks motivated by partial differential equations. *J. Math. Imaging Vis.* **2020**, *62*, 352–364. [CrossRef]
81. Soleymani, S.; Chaudhary, B.; Dabouei, A.; Dawson, J.; Nasrabadi, N.M. Differential morphed face detection using deep siamese networks. In *International Conference on Pattern Recognition*; Springer: Cham, Switzerland, 2021; pp. 560–572.
82. Booth, A.; Gerding, E.; McGroarty, F. Automated trading with performance weighted random forests and seasonality. *Expert Syst. Appl.* **2014**, *41*, 3651–3661. [CrossRef]
83. Li, H.B.; Wang, W.; Ding, H.W.; Dong, J. Trees weighting random forest method for classifying high-dimensional noisy data. In Proceedings of the 2010 IEEE 7th International Conference on e-Business Engineering, Shanghai, China, 10–12 November 2010; IEEE: New York, NY, USA, 2010; pp. 160–163.
84. Gajowniczek, K.; Grzegorczyk, I.; Ząbkowski, T.; Bajaj, C. Weighted random forests to improve arrhythmia classification. *Electronics* **2020**, *9*, 99. [CrossRef] [PubMed]
85. Chowdhury, M.M.U.; Hammond, F.; Konowicz, G.; Xin, C.; Wu, H.; Li, J. A few-shot deep learning approach for improved intrusion detection. In Proceedings of the 2017 IEEE 8th Annual Ubiquitous Computing, Electronics and Mobile Communication Conference (UEMCON), New York, NY, USA, 19–21 October 2017; IEEE: New York, NY, USA, 2017; pp. 456–462.
86. Wang, A.; Zhang, Y.; Wu, H.; Jiang, K.; Wang, M. Few-shot learning based balanced distribution adaptation for heterogeneous defect prediction. *IEEE Access* **2020**, *8*, 32989–33001. [CrossRef]
87. Zhang, B.; Jiang, H.; Li, X.; Feng, S.; Ye, Y.; Ye, R. MetaDT: Meta Decision Tree for Interpretable Few-Shot Learning. *arXiv* **2022**, arXiv:2203.01482. *Preprint*.
88. Bishop, C.M. Model-based machine learning. *Philosophical Trans. R. Soc. A Math. Phys. Eng. Sci.* **2013**, *371*, 20120222. [CrossRef]
89. Singh, A.; Thakur, N.; Sharma, A. A review of supervised machine learning algorithms. In Proceedings of the 2016 3rd International Conference on Computing for Sustainable Global Development (INDIACom), New Delhi, India, 16–18 March 2016; IEEE: New York, NY, USA, 2016; pp. 1310–1315.
90. Kern, C.; Klausch, T.; Kreuter, F. Tree-based machine learning methods for survey research. *Surv. Res. Methods* **2019**, *13*, 73.
91. Mahesh, B. Machine learning algorithms-a review. *Int. J. Sci. Res.* **2020**, *9*, 381–386.
92. Charbuty, B.; Abdulazeez, A. Classification based on decision tree algorithm for machine learning. *J. Appl. Sci. Technol. Trends* **2021**, *2*, 20–28. [CrossRef]
93. Flores, V.; Keith, B.; Leiva, C. Using artificial intelligence techniques to improve the prediction of copper recovery by leaching. *J. Sens.* **2020**, *2020*, 2454875. [CrossRef]
94. Zou, Y.; Chen, Y.; Deng, H. Gradient boosting decision tree for lithology identification with well logs: A case study of zhaoxian gold deposit, shandong peninsula, China. *Nat. Resour. Res.* **2021**, *30*, 3197–3217. [CrossRef]
95. Kotthoff, L.; Thornton, C.; Hoos, H.H.; Hutter, F.; Leyton-Brown, K. Auto-WEKA: Automatic model selection and hyperparameter optimization in WEKA. In *Automated Machine Learning*; Springer: Cham, Switzerland, 2019; pp. 81–95.
96. Wong, J.; Manderson, T.; Abrahamowicz, M.; Buckeridge, D.L.; Tamblyn, R. Can hyperparameter tuning improve the performance of a super learner? A case study. *Epidemiology* **2019**, *30*, 521. [CrossRef]
97. Rafique, D.; Velasco, L. Machine Learning for Network Automation: Overview, Architecture, and Applications [Invited Tutorial]. *J. Opt. Commun. Netw.* **2018**, *10*, D126–D143. [CrossRef]
98. Marshall, I.J.; Wallace, B.C. Toward systematic review automation: A practical guide to using machine learning tools in research synthesis. *Syst. Rev.* **2019**, *8*, 163. [CrossRef] [PubMed]
99. Wang, W.; Siau, K. Artificial intelligence, machine learning, automation, robotics, future of work and future of humanity: A review and research agenda. *J. Database Manag.* **2019**, *30*, 61–79. [CrossRef]
100. Bergstra, J.; Bardenet, R.; Bengio, Y.; Kégl, B. Algorithms for hyper-parameter optimization. *Adv. Neural Inf. Process. Syst.24.* **2011**, 2546–2554.
101. Feurer, M.; Hutter, F. Hyperparameter optimization. In *Automated Machine Learning*; Springer: Cham, Switzerland, 2019; pp. 3–33.
102. Akiba, T.; Sano, S.; Yanase, T.; Ohta, T.; Koyama, M. Optuna: A next-generation hyperparameter optimization framework. In Proceedings of the 25th ACM SIGKDD International Conference on Knowledge Discovery & Data Mining, Anchorage, AK, USA, 4–8 August 2019; pp. 2623–2631.
103. Yang, L.; Shami, A. On hyperparameter optimization of machine learning algorithms: Theory and practice. *Neurocomputing* **2020**, *415*, 295–316. [CrossRef]
104. Bergstra, J.; Bengio, Y. Random search for hyper-parameter optimization. *J. Mach. Learn. Res.* **2012**, *13*, 281–305.
105. Snoek, J.; Larochelle, H.; Adams, R.P. Practical bayesian optimization of machine learning algorithms. *Adv. Neural Inf. Process. Syst.* **2012**, *25*, 1–9.
106. Wu, J.; Chen, X.Y.; Zhang, H.; Xiong, L.D.; Lei, H.; Deng, S.H. Hyperparameter optimization for machine learning models based on Bayesian optimization. *J. Electron. Sci. Technol.* **2019**, *17*, 26–40.

107. Maclaurin, D.; Duvenaud, D.; Adams, R. Gradient-based hyperparameter optimization through reversible learning. In Proceedings of the International Conference on Machine Learning, Lille, France, 6–11 July 2015; pp. 2113–2122.
108. Nalçakan, Y.; Ensari, T. Decision of neural networks hyperparameters with a population-based algorithm. In Proceedings of the International Conference on Machine Learning, Optimization, and Data Science, Volterra, Italy, 13–16 September 2018; Springer: Cham, Switzerland, 2018; pp. 276–281.
109. Bakhteev, O.Y.; Strijov, V.V. Comprehensive analysis of gradient-based hyperparameter optimization algorithms. *Ann. Oper. Res.* **2019**, *289*, 51–65. [CrossRef]
110. Li, W.; Wang, T.; Ng, W.W.Y. Population-Based Hyperparameter Tuning With Multitask Collaboration. *IEEE Trans. Neural Networks Learn. Syst.* **2021**. [CrossRef]
111. Friedman, J.H. Greedy function approximation: A gradient boosting machine. *Ann. Stat.* **2001**, *29*, 1189–1232. [CrossRef]
112. Ll, M.; Baxter, J. *Boosting Algorithms as Gradient Descent in Function Space*; NIPS: New Orleans, LA, USA, 1999.
113. Bhat, P.C.; Prosper, H.B.; Sekmen, S.; Stewart, C. Optimizing event selection with the random grid search. *Comput. Phys. Commun.* **2018**, *228*, 245–257. [CrossRef]
114. Nguyen, D.A.; Kong, J.; Wang, H.; Menzel, S.; Sendhoff, B.; Kononova, A.V.; Bäck, T. Improved automated cash optimization with tree parzen estimators for class imbalance problems. In Proceedings of the 2021 IEEE 8th international conference on data science and advanced analytics (DSAA), Porto, Portugal, 6–9 October 2021; IEEE: New York, NY, USA, 2021; pp. 1–9.
115. Zhai, M.G.; Peng, P. Paleoproterozoic tectonic events in North China Craton. *Acta Petrol. Sin.* **2007**, *11*, 2665–2682, (In Chinese with English abstract).
116. Wan, Y.S.; Dong, C.Y.; Ren, P. A Review of the Temporal and Spatial Distribution, Composition and Evolution of Archean TTG Rocks in the North China Craton. *Acta Petrol. Sin.* **2017**, *33*, 1405–1419.
117. Jia, X.; Zhai, M.; Xiao, W.; Li, L.; Ratheesh-Kumar, R.; Wu, J.; Liu, Y. Mesoarchean to Paleoproterozoic crustal evolution of the Taihua Complex in the southern North China Craton. *Precambrian Res.* **2019**, *337*, 105451. [CrossRef]
118. Zhao, G.; He, Y.; Sun, M. The Xiong'er volcanic belt at the southern margin of the North China Craton: Petrographic and geochemical evidence for its outboard position in the Paleo-Mesoproterozoic Columbia Supercontinent. *Gondwana Res.* **2009**, *16*, 170–181. [CrossRef]
119. He, Y.; Zhao, G.; Sun, M. Geochemical and Isotopic Study of the Xiong'er Volcanic Rocks at the Southern Margin of the North China Craton: Petrogenesis and Tectonic Implications. *J. Geol.* **2010**, *118*, 417–433. [CrossRef]
120. Wang, C.; He, X.; Carranza, E.J.M.; Cui, C. Paleoproterozoic volcanic rocks in the southern margin of the North China Craton, central China: Implications for the Columbia supercontinent. *Geosci. Front.* **2019**, *10*, 1543–1560. [CrossRef]
121. Li, Y.F. *The Temporal-Spital Evolution of Mesozoid Granitoids in the Xiong'ershan Area and Their Relationships to Molybdenum-Gold Mineralization*; China University of Geosciences: Beijing, China, 2005; pp. 1–122, (In Chinese with English abstract).
122. Wenxiang, X.; Fang, P.; Guangjin, B. *Rock Strata in Henan Province*; China University of Geosciences: Wuhan, China, 1997; pp. 1–209, (In Chinese with English abstract).
123. Hu, X.-K.; Tang, L.; Zhang, S.-T.; Santosh, M.; Spencer, C.J.; Zhao, Y.; Cao, H.-W.; Pei, Q.-M. In situ trace element and sulfur isotope of pyrite constrain ore genesis in the Shapoling molybdenum deposit, East Qinling Orogen, China. *Ore Geol. Rev.* **2018**, *105*, 123–136. [CrossRef]
124. Zhai, M.-G.; Santosh, M. The early Precambrian odyssey of the North China Craton: A synoptic overview. *Gondwana Res.* **2011**, *20*, 6–25. [CrossRef]
125. Li, S.-R.; Santosh, M. Geodynamics of heterogeneous gold mineralization in the North China Craton and its relationship to lithospheric destruction. *Gondwana Res.* **2017**, *50*, 267–292. [CrossRef]
126. Li, L.; Li, C.; Li, Q.; Yuan, M.-W.; Zhang, J.-Q.; Li, S.-R.; Santosh, M.; Shen, J.-F.; Zhang, H.-F. Indicators of decratonic gold mineralization in the North China Craton. *Earth Sci. Rev.* **2022**, *228*, 103995. [CrossRef]
127. Mao, J.; Goldfarb, R.J.; Zhang, Z.; Xu, W.; Qiu, Y.; Deng, J. Gold deposits in the Xiaoqinling–Xiong'ershan region, Qinling Mountains, central China. *Miner. Depos.* **2002**, *37*, 306–325. [CrossRef]
128. Cao, M.; Yao, J.; Deng, X.; Yang, F.; Mao, G.; Mathur, R. Diverse and multistage Mo, Au, Ag–Pb–Zn and Cu deposits in the Xiong'er Terrane, East Qinling: From Triassic Cu mineralization. *Ore Geol. Rev.* **2017**, *81*, 565–574. [CrossRef]
129. Deng, J.; Gong, Q.; Wang, C.; Carranza, E.J.M.; Santosh, M. Sequence of Late Jurassic–Early Cretaceous magmatic–hydrothermal events in the Xiong'ershan region, Central China: An overview with new zircon U–Pb geochronology data on quartz porphyries. *J. Asian Earth Sci.* **2014**, *79*, 161–172. [CrossRef]
130. Yan, J.S.; Wang, M.S.; Yang, J.C. Tectonic evolution of the Machaoying fault zone in western Henan and its relationship with Au-polymetallic mineralization. *Reg. Geol. China* **2000**, *19*, 166–171, (In Chinese with English abstract).
131. Kefei, T. *Characteristics, Genesis, and Geodynamic Setting of Representative Gold Deposits in the Xiong'ershan District, Southern Margin of the North China Craton*; China University of Geosciences: Wuhan, China, 2014; pp. 1–131, (In Chinese with English abstract).
132. Tang, L.; Zhang, S.T.; Yang, F.; Santosh, M.; Li, J.J.; Kim, S.W.; Hu, X.K.; Zhao, Y.; Cao, H.W. Triassic alkaline magmatism and mineralization in the Xiong'ershan area, East Qinling, China. *Geol. J.* **2019**, *54*, 143–156. [CrossRef]
133. McCuaig, T.C.; Hronsky, J.M. *The Mineral System Concept: The Key to Exploration Targeting*; Society of Economic Geologists, Inc.: Littleton, CO, USA, 2014; Volume 18, pp. 153–175.
134. McCuaig, T.C.; Beresford, S.; Hronsky, J. Translating the mineral systems approach into an effective exploration targeting system. *Ore Geol. Rev.* **2010**, *38*, 128–138. [CrossRef]

135. Ni, S.J.; Li, C.Y.; Zhang, C.; Gao, R.D.; Liu, C.F. Contribution of Meso-basic dykerocks to gold deposits—An example from gold deposits in Xiaoqinling area. *J. Chengdu Inst. Technol.* **1994**, *21*, 70–78.
136. Li, S.M.; Huang, J.J.; Wang, X.S.; Zhai, L.Q. *The Geology of Xiaoqinling Gold Deposits and Metallogenetic Prospecting*; Beijing, Geological Publishing House: Beijing, China, 1996; pp. 1–250, (In Chinese with English abstract).
137. Xu, J.H.; Xie, Y.L.; Liu, J.M.; Zhu, H.P. Trace elements in fluid inclusions of Wenyu-Dongchuang gold deposits in the Xiaoqinling area, China. *Geol. Prospect.* **2004**, *40*, 1–6, (In Chinese with English abstract).
138. Wang, T.H.; Xie, G.Q.; Ye, A.W.; Li, Z.Y. Material sources of gold deposits in Xiaoqinling–Xiong'ershan area of Western Henan Province as well as the relationship between gold deposits and intermediate-basic dykes. *Acta Geosci. Sin.* **2009**, *30*, 27–38.
139. Yanjing, C.; Jigu, F.; Bing, L. Classification of genetic types and series of gold deposits. *Adv. Earth Sci.* **1992**, *3*, 73–79, (In Chinese with English abstract).
140. Chen, Y.J.; Santosh, M. Triassic tectonics and mineral systems in the Qinling Orogen, central China. *Geol. J.* **2014**, *49*, 338–358. [CrossRef]
141. Deng, X.H.; Chen, Y.J.; Santosh, M.; Yao, J.M.; Sun, Y.L. Re–Os and Sr–Nd–Pb isotope constraints on source of fluids in the Zhifang Mo deposit, Qinling Orogen, China. *Gondwana Res.* **2016**, *30*, 132–143. [CrossRef]
142. Aitchison, J. The statistical analysis of compositional data. *J. R. Stat. Soc. Ser. B Methodol.* **1982**, *44*, 139–160. [CrossRef]
143. Van den Boogaart, K.G.; Tolosana-Delgado, R. *Analyzing Compositional Data with R*; Springer: Berlin, Germany, 2013; Volume 122, pp. 1–200.
144. Galletti, A.; Maratea, A. Numerical stability analysis of the centered log-ratio transformation. In Proceedings of the 2016 12th International Conference on Signal-Image Technology & Internet-Based Systems (SITIS), Naples, Italy, 28 November–1 December 2016; IEEE: New York, NY, USA, 2016; pp. 713–716.
145. Brodersen, K.H.; Ong, C.S.; Stephan, K.E.; Buhmann, J.M. The balanced accuracy and its posterior distribution. In Proceedings of the 2010 20th International Conference on Pattern Recognition, Istanbul, Turkey, 23–26 August 2010; IEEE: New York, NY, USA, 2010; pp. 3121–3124.
146. Wei, Q.; Dunbrack, R.L., Jr. The role of balanced training and testing data sets for binary classifiers in bioinformatics. *PLoS ONE* **2013**, *8*, e67863. [CrossRef]
147. Velez, D.R.; White, B.C.; Motsinger, A.A.; Bush, W.S.; Ritchie, M.D.; Williams, S.M.; Moore, J.H. A balanced accuracy function for epistasis modeling in imbalanced datasets using multifactor dimensionality reduction. *Genet. Epidemiol.* **2007**, *31*, 306–315. [CrossRef]
148. Weng, C.G.; Poon, J. A new evaluation measure for imbalanced datasets. In Proceedings of the 7th Australasian Data Mining Conference, Glenelg, Australia, 27–28 November 2008; Volume 87, pp. 27–32.
149. Chuang, C.L.; Chang, P.C.; Lin, R.H. An efficiency data envelopment analysis model reinforced by classification and regression tree for hospital performance evaluation. *J. Med. Syst.* **2011**, *35*, 1075–1083. [CrossRef]
150. Gu, Q.; Zhu, L.; Cai, Z. Evaluation measures of the classification performance of imbalanced data sets. In Proceedings of the International Symposium on Intelligence Computation and Applications, Guangzhou, China, 20–21 November 2009; Springer: Berlin/Heidelberg, Germany, 2009; pp. 461–471.
151. Wong, T.T. Performance evaluation of classification algorithms by k-fold and leave-one-out cross validation. *Pattern Recognit.* **2015**, *48*, 2839–2846. [CrossRef]

minerals

MDPI

Article

Geology, Mineralogy, and Age of Li-Bearing Pegmatites: Case Study of Tochka Deposit (East Kazakhstan)

Natalya A. Zimanovskaya [1], Tatyana A. Oitseva [1,2,*], Sergey V. Khromykh [3], Alexey V. Travin [3], Ainel Y. Bissatova [1,2], Irina Yu. Annikova [3] and Saltanat S. Aitbayeva [1]

[1] Faculty of Earth Sciences, D. Serikbayev East Kazakhstan Technical University, 19, Serikbayev str., Ust-Kamenogorsk 070000, Kazakhstan
[2] Geos LLP, 83, Protozanov str., Ust-Kamenogorsk 070000, Kazakhstan
[3] V.S. Sobolev Institute of Geology and Mineralogy Siberian Branch of the Russian Academy of Sciences, 3 Koptyug ave., 630090 Novosibirsk, Russia
* Correspondence: tatiana.oitseva@gmail.com; Tel.: +7-705-150-76-53

Abstract: New geological, mineralogical, geochemical, and geochronological data have been obtained for Li-bearing pegmatites from the Tochka deposit located within the Karagoin–Saryozek zone in East Kazakhstan. Earlier, the exploration works in this zone were carried out to detect only Ta and Sn mineralization, but other ores (including Li) were not considered. The estimation of lithium resources in pegmatites from the area was methodologically imperfect. Previously, it was believed that the formation of rare-metal pegmatite veins was associated with Late Carboniferous Na-granites. The obtained geological observation confirms that the ore-bearing rare-metal pegmatites at the Tochka deposits cut the Late Carboniferous Na-granites and do not cut the Early Permian Kalba granites. The associations of the accessory minerals in host hornfels, Na-granites, and rare-metal pegmatites are different and the accessory minerals in pegmatites are similar to the accessory minerals in the Kalba granites. Geochemical data show that the behavior of rare elements (Ba, Th, HFSE, and REE) and the levels of accumulation of rare metals prove that pegmatites are similar to the product of the differentiation of the granitic magmas of the Kalba complex. The $^{40}Ar/^{39}Ar$ muscovite age of the Tochka pegmatites (~292 Ma) fits the age range of the Kalba granite complex. Based on the main principles of the generation of rare-metal pegmatites, the Tochka pegmatites formed during the fluid–magmatic fractionation of magma in large granitic reservoirs of the Kalba complex. The Karagoin–Saryozek zone—located between several large granite massifs of the Kalba complex where host rocks play a role as a roof—may be very promising for rare-metal pegmatite mineralization.

Keywords: rare-metal pegmatite; granite; lithium resources; development of Li pegmatite deposits; Ar-Ar dating; East Kazakhstan

1. Introduction

The Central Asian Orogenic Belt (also known as the Altaids) accommodates voluminous Phanerozoic granitoid batholiths [1–9]. Granitic magmatism hosts Rb, Be, Mo, and W metals [10–13], including especially valuable Li, Cs, Ta, Nb, and Sn. Many rare-metal pegmatite bodies (first of all the lithium–cesium–tantalum (LCT) pegmatite family) are spatially and genetically associated with large granitoid intrusions and often occur as fields or swarms of veins, commonly along the margins or above the roofs of plutons [11,12,14–17].

Rare-metal pegmatite mineralization (LCT family) related to granitic magmatism is abundant in East Kazakhstan, which is part of a Late Paleozoic collisional suture between Siberia and Kazakhstan [18,19]. Magmatic activity in the region was the most active in the Early Permian [20–22] and produced, among others, the large Kalba and Zharma granitoid batholiths [21,22]. Large-scale magmatism occurred in a setting of post-collisional extension and was maintained by heat from the Tarim plume [23–27]. The area—with multiple and diverse juxtaposed igneous complexes—stores extremely rich resources of base, noble, and rare metals [28–33]. Rare-metal deposits cluster mainly within the Kalba–Narym belt bound by the West Kalba and Irtysh shear zones (Figure 1). A great part of rare-metal mineralization of pegmatite, albitite–greisen, greisen–quartz vein, and hydrothermal types is hosted by Early Permian granitoids of the Kalba batholith, including spodumene pegmatites bearing economic grades of Ta, Nb, Be, Li, Cs, and Sn located in the Asubulak ore district along intrusion margins in the Kalba granite complex [14]. Some sites of rare-metal mineralization in the area are located outside the Kalba complex: F-rich granite porphyry (ongonites) dikes in the northwestern end of the Kalba–Narym belt [34,35] and the Novoakhmirovo porphyry granite stock [36,37].

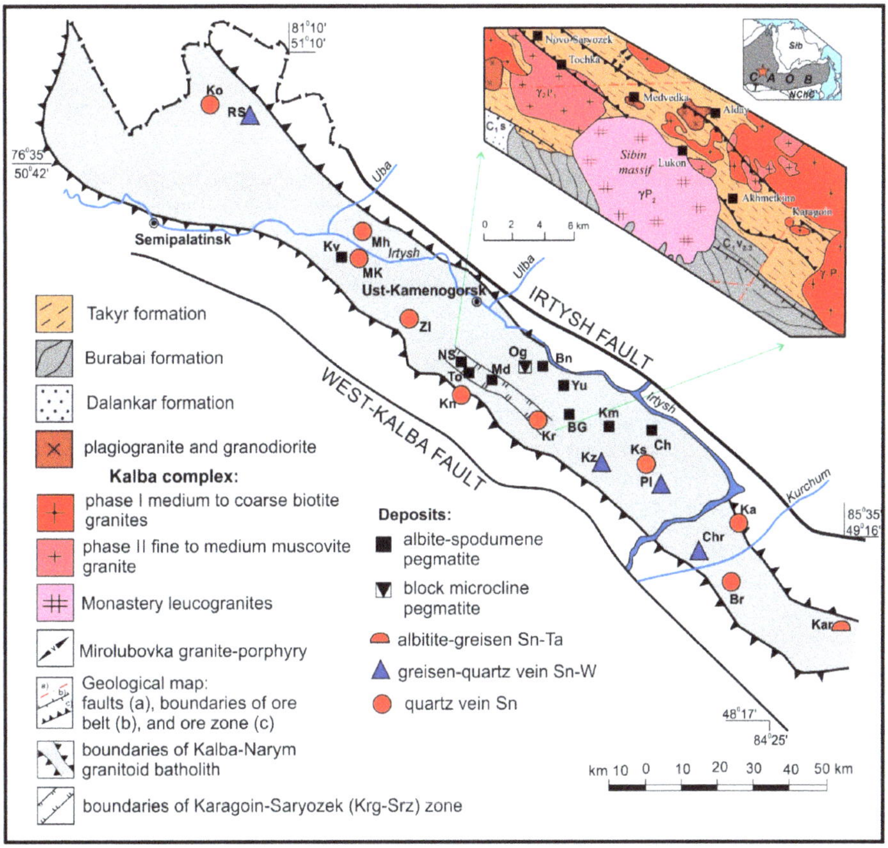

Figure 1. Location map of rare-metal deposits and occurrences in the Kalba–Narym belt [30]. Abbreviations stand for names of mineral deposits: Ko—Korosteli; RS—Rzhavaya Sopka; Mh—Mokhnatukha; MK—Malaya Kainda;, Kv—Kvartz; Zl—Zelenoye; Kn—Kanaika; NS—Novo-Saryozek; Md—Medvedka; To—Tochka; Kr—Karagoin; Og—Ognevka; Bn—Bakennoye; Yu—Yubileinoye; BG—Belaya Gora; Kz—Kozlovskoye; Km—Komarovskoye; Ks—Komsomolskoye; Ch—Chebuntay; Pl—Palatsy; Chr—Cherdoyak; Ka—Kasatka; Br—Burabay; Kar—Karasu. Sibiny pluton.

Some other rare-metal occurrences have no immediate linkage with large granitic plutons, although are located within their vicinity. They are, namely, the Tochka, Medvedka, Akhmetkino, Lukon, Alday, and other swarms of spodumene pegmatite veins (Figure 1) between the plutons within the Karagoin–Saryozek ore zone [38,39]. From 1955 to 1994, exploration works in this zone were carried out to detect Ta and Sn mineralization, but other ores (including Li) were not considered. The estimation of lithium resources in pegmatites from the area was methodologically imperfect. Pegmatite veins were sampled on 200 m × 150–200 m or 400 m × 200 m grids; however, this spacing is insufficient to pick ore intersections in the short veins, which are highly variable in morphology and composition, and to evaluate the volume of potentially metalliferous albite-spodumene material in individual veins and over the deposit as a whole. Average reported Li_2O contents [38,40–42] were underestimated, because Li grades were averaged over barren and albite–spodumene pegmatites taken together. Moreover, laboratory works, mainly by flame photometry and poorly sensitive semi-quantitative spectral analyses, failed to furnish appropriate information.

Thus, it is relevant to consider the mineralogical and geochemical characteristics of deposits in the Karagoin–Saryozek zone as promising for Li mineralization. We report geological, mineralogical, geochemical, and geochronological data from the Tochka deposit of rare-metal spodumene pegmatites, with implications for the age of mineralization and its relation to the granitoids.

2. Geological Background

The Karagoin–Saryozek ore potential zone—60 km long and 4–6 km wide—strikes in the NW direction along the margins of intrusions within the Kalba–Narym batholith. The batholith comprises the large Shubarshoky, Dvoryansky, Eshkulmes, and Sibiny plutons composed of granites and leucogranites, as well small NW aplitic intrusions and granite porphyry dikes between them, which may belong to a zone of hidden supra–plutonic granitic bodies.

The Tochka deposit is located in the northwestern part of the Karagoin–Saryozek zone, near Bayash-Utepov Village. It is a field of rare-metal pegmatite veins, more than 3 km long and 300 m wide, at the northeastern margin of the Shubarshoky pluton (Figure 2). The area accommodates two small intrusions that consist of granites of two types: (1) the intrusion of Na-rich biotite granites on the east that belong to the Late Carboniferous Kunush complex and have a 299 ± 2 Ma age [43]; (2) the intrusion of K-Na biotite granites that are an apophysis of the Shubarshoky pluton assigned to the Early Permian Kalba complex with ages from 296 to 286 Ma [6,23,24]. The Tochka pegmatite veins crosscut the Kunush Na-granites but do not intrude the Kalba granites (Figure 2).

As it was found out previously [25,27–29,33], the ore bodies of the deposit are mainly localized in deformed and metamorphosed black shales of the Takyr Fm. (D_3–C_1) and in Na-granites (Bariernaya vein, etc.).

Pegmatite veins, 3–5 m thick and 250–300 m long, strike in the NW direction (315–330°) concordantly with the black shale sequence and dip mainly to the NE at 70–85°; later quartz veins are 0.2–1.0 m thick. The pegmatites are of microcline, microcline–albite, albite, and albite–spodumene types. No zonation is observed within the pegmatite veins; each vein is represented by any one mineralogical type. However, in the contact zones of some veins, there are medium-grained or fine-grained quartz–microcline–muscovite greisens. All pegmatite veins are subparallel and have almost no crossing relationships. Thus, it is impossible to unambiguously establish the sequence of intrusion of different pegmatite types; most likely they were introduced synchronously.

The pegmatite veins within the field make up the central and southern zones displaced 380 m one relative to another by sublatitudinal faults. Within the field, three pegmatite formations of the northwestern direction are distinguished, located in a vertical section one under the other: the Eastern Fm. (hanging side), Central Fm., and Western Fm. (recumbent). The Central Fm. is the most promising, composed mainly of albite–spodumene pegmatites, which form two bundles of closely spaced veins, with a total thickness of about 23 m. Each bundle consists of 3–5 closely spaced bodies 170–200 m long and it is considered as a single ore body. The pegmatite veins are exposed on the surface (veins 13, 14, etc.) and exhibit nest-like clusters of spodumene crystals more than 1 m in size, most often in the vein interior. According to drilling evidence, economic albite–spodumene mineralization is unevenly distributed and is traceable to depths of 200–300 m or more [38].

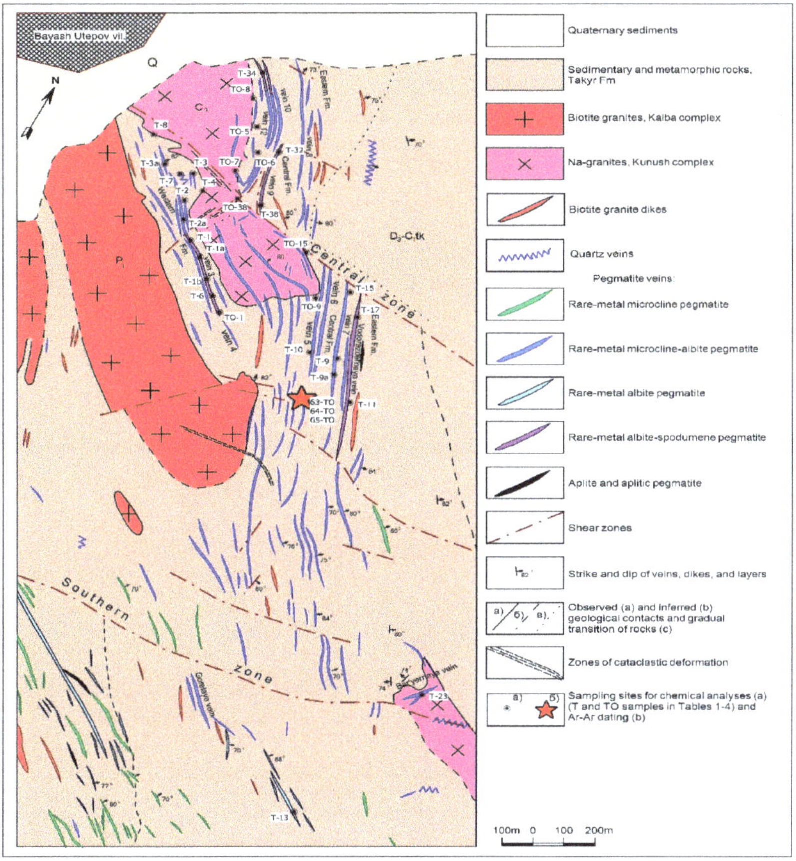

Figure 2. Local geology of the Tochka deposit, from Kashcheev et al. [38].

3. Materials and Methods

The study included field and analytical work and was made with reference to the available published evidence and survey reports of different years [38,40–42,44]. Fifty samples of rare-metal pegmatites and country rocks were collected for analyses. Large-volume samples of pegmatites weighing up to 10 kg were taken. Chemical analyses were carried out for bulk rocks and, in some cases, for the spodumene monomineral fraction. Additionally, to assess the effect of pegmatite veins on the host rocks, the host rocks samples (hornfelses and Na-granites) were taken and analyzed.

The samples were dried at a temperature of 105 °C and then ground to a fraction of less than 71 mkm. Before geochemical measurements, the samples were transferred into a solution using polyacid decomposition. During acid decomposition, standard samples of the composition of aqueous solutions of elements were used as calibration solutions. Calibration solutions were prepared by diluting standard samples with a solution of nitric acid with a molar concentration of 1 mol/dm^3, a mass concentration of 2% or 5%; hydrochloric acid solution with a mass concentration of 10% or water. A solution of nitric acid with a mass concentration of 5% was used as a background solution.

The analyses were performed at the VERITAS Laboratory of the D. Serikbaev East Kazakhstan Technical University (Ust'-Kamenogorsk, Kazakhstan). The compositions of rocks and minerals were determined, respectively, by mass spectrometry with inductively coupled plasma (ICP-MS) on an Agilent 7500cx (Agilent Technologies, Santa Clara, CA,

USA) spectrometer and by electron probe microanalysis (EPMA), under standard operation conditions, on a Cameca MS-46 analyzer (Cameca, Gennevilliers, France) that allowed detection and precise measurements of seventy-three elements (Au, Ag, Pt, Cd, In, Ir, Y, Cd, REE, U, etc.). Scanning electron microscopy and energy dispersive spectrometry (SEM-EDS), with a Jeol-100C microscope with a Kevex-Ray detector and a Jeol ISM-6390 LV microscope (JEOL, Tokyo, Japan) with an Oxford INCA Energy system, were applied to study micrometer inclusions of metallic and gangue minerals and to analyze impurity elements (Au, Ag, Sb, As, Pt, In, U, etc.).

The $^{40}Ar/^{39}Ar$ dating of muscovite was carried out using the stepwise heating technique [45,46]. The samples were irradiated in a nuclear BWR-reactor at the Tomsk Polytechnic University (Tomsk). A standard biotite sample MSA-11 (OSO No. 129-88) with an age of 311.0 ± 1.5 Ma, validated against international standard samples Bern-4M (muscovite) and LP-6 (biotite), was used as the fluence monitor. The ages of the Bern 4M and LP-6 standard samples were assumed to be 18.51 and 128.1 Ma, respectively. The Ar isotopic composition in the irradiated samples was measured using a Noble Gas 5400 (Micromass) mass spectrometer [45,46]. The separation of gas fractions and the isotopic analysis of argon were carried out in a temperature range from 500 to 1200 °C. The raw data were corrected for procedural blank mass discrimination by analysis of atmospheric Ar, decay of radioactive ^{37}Ar and ^{39}Ar isotopes produced by irradiation, and interferences of ^{36}Ar, ^{39}Ar, and ^{40}Ar produced from ^{40}Ca, ^{42}Ca, and ^{40}K, respectively. The corrections were applied using the following coefficients: $(^{39}Ar/^{37}Ar)_{Ca} = 0.001279 \pm 0.000061$, $(^{36}Ar/^{37}Ar)_{Ca} = 0.000613 \pm 0.000084$, $(^{40}Ar/^{39}Ar)_K = 0.0191 \pm 0.0018$. The amount of ^{40}Ar in the blank did not exceed 0.5 ng.

4. Results

4.1. Mineralogy

4.1.1. Host Rocks

Rare-metal mineralization in the area is hosted by the Late Devonian–Early Carboniferous (D_3–C_1) Takyr Fm. of carbonaceous shales, fine sandstones, and mudrocks. The rocks at the contacts with pegmatite veins include hornfelsed shales (Figure 3a) and muscovitized hornfels with ≤0.5 cm flaky and platy muscovite (Figure 3b,d). The shales bear signatures of contorted bedding with small thin folds delineated by 0.5–1.0 cm thick veins (Figure 3c) [47,48].

The hornfels rocks enclose micrometer grains of ilmenite, magnetite, zircon, monazite, and xenotime (Figure 4a–d). A pyrite microcrystal with inconspicuous faces was also revealed (Figure 4e). In addition, a micro-inclusion of cassiterite was found in a microcrack (Figure 4f), reflecting the overlay of ore-bearing fluids on hornfelses. The composition of accessory minerals from hornfels are given in Supplementary Table S1. Ilmenite contains Mn (from 0.69% to 2.07%); monazite contains La (from 2.37% to 5.49%), Nd (from 2.77% to 6.02%), Th (from 1.12% to 1.79%); xenotime contains 3.2% of Dy.

4.1.2. Na-Granites

Na-granites are heavily cracked on the surface and enclose xenoliths of the Takyr Fm. (Figure 5a). They are deformed and metamorphosed in fault zones with chloritization of biotite, quartzitization, etc. Core samples are light grey, inequigranular, with distinct grains of white plagioclase (Figure 5b). Dikes are composed of quartz porphyry and quartz albitophyre (Figure 5c,d). Of special interest is a dike of quartz albitophyre with spherulites and streaks of fine albite enclosing acicular schorl tourmaline (Figure 5d), which indicates quite high concentrations of boron in granitic melts.

Figure 3. Takyr Fm. shales and hornfels from the Tochka area: black shale (**a**), muscovitized hornfels (**b**), folded hornfels with quartz veinlets (**c**), quartz–biotite–muscovite (I) and quartz–feldspar–muscovite (II) hornfels and pegmatite (**d**).

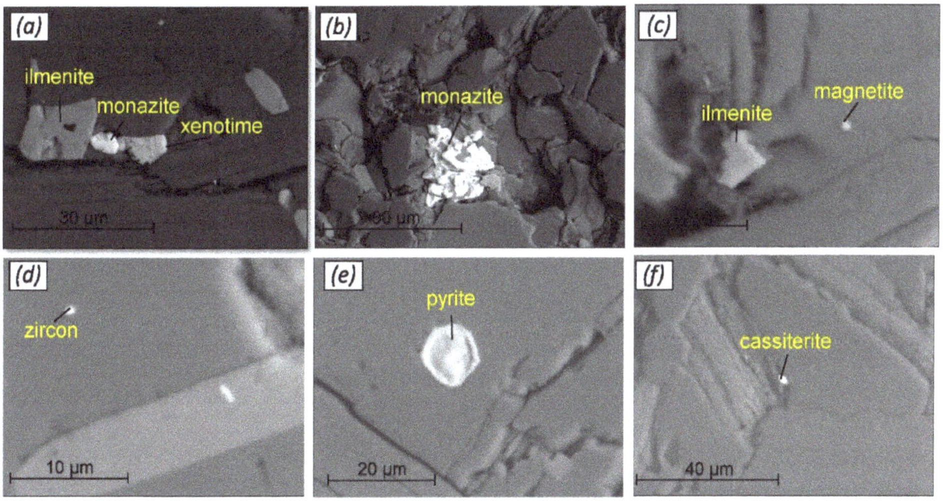

Figure 4. Micrometer accessory minerals in hornfels from the Tochka area (vein 2, sample TO-11, TO-5). (**a**) Nests of ilmenite, monazite, and xenotime; (**b**) a nest of monazite; (**c**) ilmenite and magnetite; (**d**) sub-micrometer zircon in hornfels and tourmaline; (**e**) oval pyrite; (**f**) cassiterite in a crack.

Figure 5. Na-granites from the Tochka area. (**a**) Outcrop of Na-granites with hornfels xenoliths; (**b**) core samples of medium-grained biotite (Bt) Na-granite; (**c**) quartz porphyry dike with albite and quartz phenocrysts; (**d**) quartz albitophyre dike with albite (Ab) spherulites that enclose tourmaline (Tur) (schorl).

Scanning electron microscopy of Na-granite revealed accessory minerals: zircon, titanite, monazite, sphalerite and apatite (Figure 6a,b), tungstite ($WO_3 \cdot H_2O$) (Figure 6c), as well as a gersdorffite (NiAsS) (Figure 6d) with an admixture of Co (5.24 wt.%) and Fe (3.54 wt.%). The composition of accessory minerals from hornfels are given in Supplementary Table S2. Apatite contains 0.38 wt.% of La, 0.89 wt.% of Ce, and 0.38 wt.% of Nd; monazite contains 5.66 wt.% of La, 13.12 wt.% of Ce, and 6.06 wt.% of Nd.

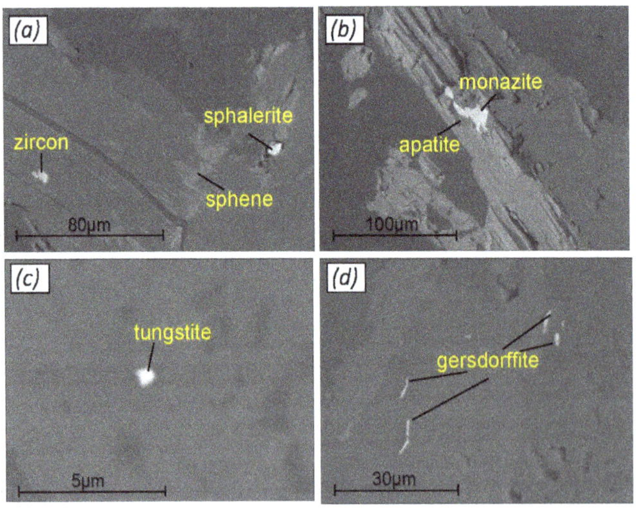

Figure 6. Micrometer accessory minerals in Na-granite (samples TO-4 and TO-10). (**a**) Zircon (spectrum 1) and sphalerite (spectrum 2) in a matrix defect; (**b**) nests of monazite in apatite; (**c**) tungstite; (**d**) gersdorffite streaks.

4.1.3. Pegmatites

The systematics of rare-metal pegmatites are given in the works of Aytaliev Zh.A., Smirnov I.A., Shcherba G.N., Shavlo S.G., Ginzburg A.I., Solodov N.A., Kuznetsov V.I., Rodionov G.G., Dyachkov B.A., and other researchers [48–53].

Traditionally, based on the model of formation of rare-metal pegmatites of the Kalba region, graphic and oligoclase–microcline (ore-free), microcline block (with beryl and columbite), microcline–albite, albite, and albite–spodumene with rich complex ores (Ta, Nb, Li, Cs, Be, Sn) occurred. During their staged formation from early to late, an increasing concentration of rare-metal mineralization occurred with the formation of unique minerals (tantalite–columbite, lepidolite, pollucite, spodumene, petalite, amblygonite, colored and polychrome tourmalines, etc.).

Pegmatites from three type veins: microcline–albite, albite, and albite–spodumene pegmatites were sampled. In addition, the albite–muscovite greisens from contact zones of microcline–albite pegmatite veins were studied.

Microcline–albite pegmatites occur mainly as quartz–albite–muscovite veins with tantalite–columbite, beryl, cassiterite, and sporadic spodumene crystals. They are of several varieties: coarse-grained weakly albitized quartz–microcline, greisenized albite–muscovite, mainly quartz–albite with nests of iron phosphate and manganese, and albite pegmatites (Figure 7) [54,55].

Figure 7. Pegmatites from the Tochka area. (**a**) Coarse-grained weakly albitized quartz–microcline pegmatite; (**b**) albitized pegmatite with nests of iron phosphate and manganese (dark inclusions); (**c**) albite pegmatite with disseminated metallic minerals.

Albite–muscovite greisens bear REE phases of monazite and xenotime (Figure 8a–c), as well as acicular zircon (Figure 8d).

Albite pegmatites with spodumene enclose fluid-derived lumpy cassiterite (Figure 9a) and submicrometer native tin (Figure 9b), a Ni phase, presumably bunsenite (NiO?) discovered for the first time (Figure 9c), and rutile micro-inclusions (Figure 9d). Another sample of albitized pegmatite with cymatolite and garnet bears rare-metal phases of columbite, tantalite–columbite, and cassiterite (Figure 9a), as well as micrometer grains of accessory apatite, zircon, and pyrolusite (Figure 9e). In this respect, cymatolite, which is a pseudomorph after spodumene, can be considered an indicator mineral of Li-bearing albite–spodumene pegmatites. The SEM images (Figure 9f) reveal an ultramicroscopic-zoned grain of radioactive zircon with a disseminated white U mineral (9.7 wt.% Zr and 11.2 wt.% U).

Albite–spodumene pegmatites provide economic Li resources, mainly in spodumene, occurring as flat prismatic crystals. Spodumene crystals (up to 25%–50%) are often oriented across the veins (Figure 10) or cluster in nests, where they are overgrown with fibrous cymatolite and garnet crystals (Figure 10).

The composition of accessory minerals from hornfels are given in Supplementary Table S3. Apatite contains 0.42% of Mn and 0.38% of Nd; monazite contains La (from 1.93% to 2.92%) and Nd (from 2.64% to 3.85%); xenotime contains 1.41% of Dy; pyrolusite contains 0.60% of Cu; rutile contains 1.1% of Nb; columbite contains Mn (from 1.54% to 4.27%).

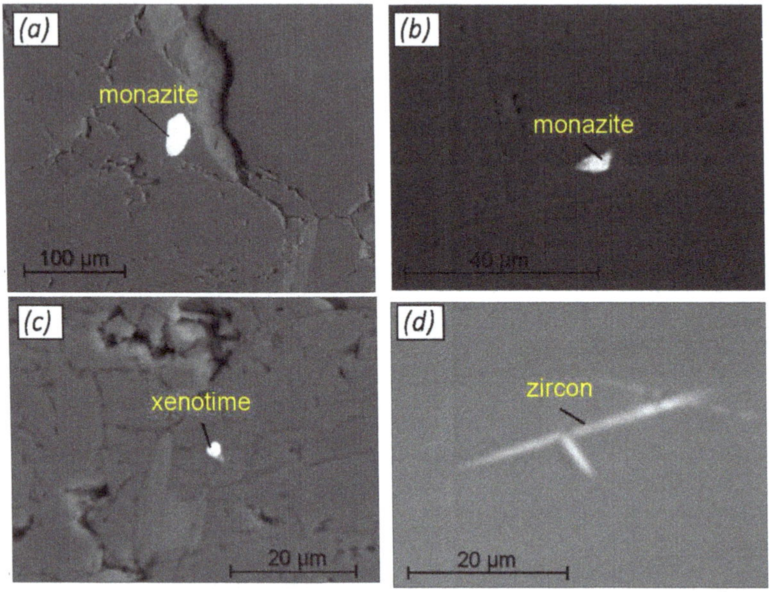

Figure 8. Inclusions of fluid-derived minerals in albite–muscovite greisens (sample TO-9). (**a**,**b**) Monazite crystals (**a**) and lumps (**b**); (**c**) droplet-shaped xenotime; (**d**) zircon in a microcrack.

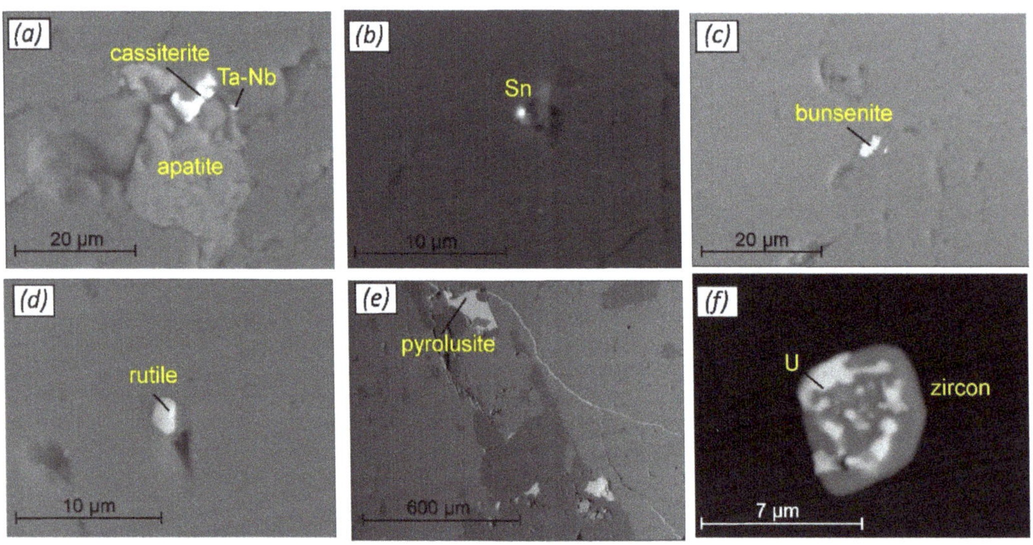

Figure 9. Micrometer mineral inclusions in albite pegmatites with spodumene (sample TO-13, TO-12). (**a**) Cassiterite and tantalite–columbite in apatite; (**b**) native tin; (**c**) Ni phase (bunsenite?); (**d**) rutile; (**e**) Mn phase (pyrolusite); (**f**) zircon with disseminated U phase (uraninite).

Figure 10. Albite–spodumene pegmatites from the Tochka area. (**a**) Nest of albite–spodumene pegmatite; (**b**) albite–spodumene pegmatite; (**c**) nest of flat spodumene crystals; (**d**) albitized pegmatite with cymatolite nests and garnet crystals.

4.2. Geochemistry

Trace elements in rare-metal pegmatites and their hornfels and Na-granite hosts were analyzed by ICP-MS. The results of the analyses are given in Supplementary Table S4. Unfortunately, we have no analyses from the intrusion of Kalba granites located within the studied area (see Figure 2). Thus, we had to take the data on the composition of Kalba granites from early publications [24,47]. These granites are from Shubarshoky, Dvoryansky, Sibiny, and Asubulak plutons located near the Karagoin–Saryozek zone. The rocks of these plutons are Bt granites and Bt($\pm$Ms) leucogranites. Their compositions are also given in Supplementary Table S4.

Hornfels are similar to each other, according to the contents of the main elements. They have higher concentrations of Ti, Fe, Mg, and P and lower concentrations of Na than all granites and pegmatites. Pegmatites and Na-granites show similar concentrations of the main elements; they contain less Ti, Fe, Na, and K than the Kalba granites (Figure 11).

Hornfels can be divided into two groups, according to the distribution of rare metal. Hornfels without muscovite (analyses 1 and 2 in Table S4) have generally low rare-metal concentrations. The Na-granites show a similar behavior to rare metals (Figure 12).

However, muscovite hornfelses (analyses 3 and 4 in Table S4) show elevated concentrations of Li, Be, Rb, Cs, Sn, and W. The behavior of rare metals in pegmatites is generally close to the Kalba granites, although the pegmatites contain higher concentrations of Li, Rb, Nb, and Ta (Figure 12).

The trace element (Ba, Th, HFSE, REE) behavior in the hornfelses is similar to the Kalba granites, whereas Na-granites and pegmatites show lower concentrations of these elements relative to the Kalba granites (Figure 13).

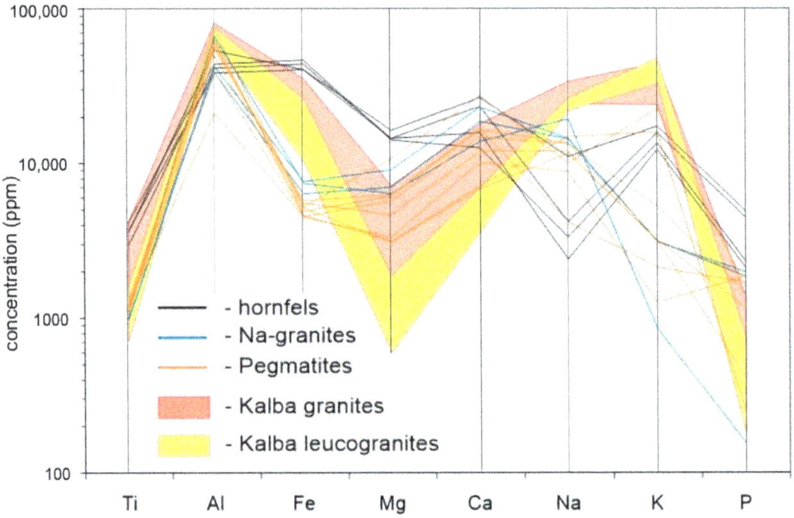

Figure 11. Distribution of main elements (concentrations in ppm) in hornfelses, Na-granites, and pegmatites of the Tochka deposit in comparison with the Kalba granites and leucogranites. The composition of the Kalba granites and leucogranites are taken from [24,47].

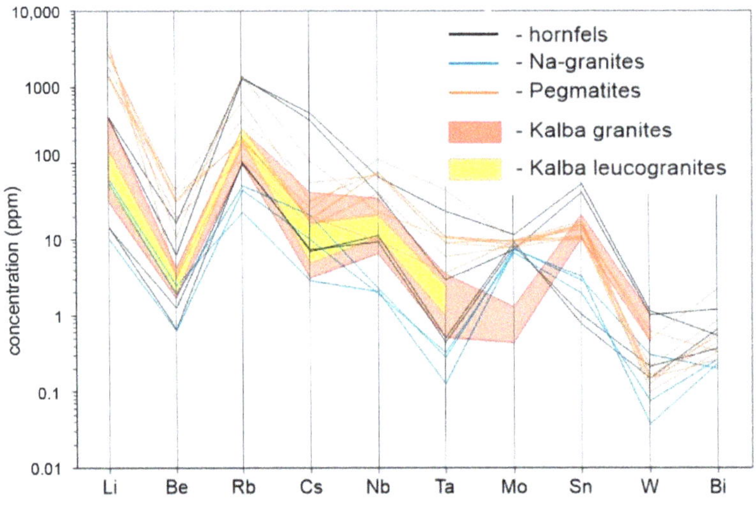

Figure 12. Distribution of rare metals (concentrations in ppm) in hornfelses, Na-granites, and pegmatites of the Tochka deposit in comparison with the Kalba granites and leucogranites. The composition of the Kalba granites and leucogranites are taken from [24,47].

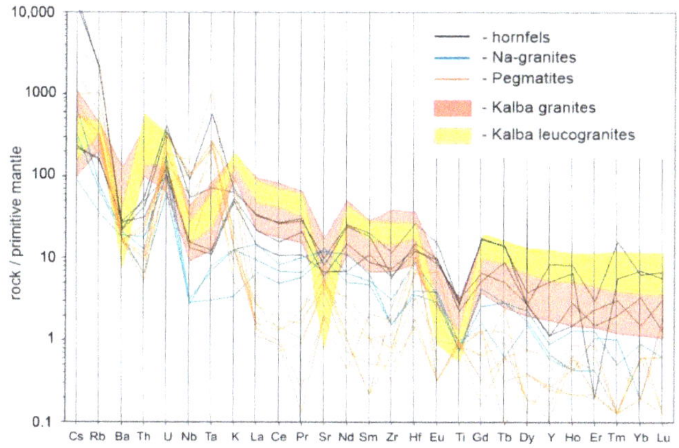

Figure 13. Distribution (spider diagram) of trace elements of rare metals in hornfelses, Na-granites, and pegmatites of the Tochka deposit in comparison with the Kalba granites and leucogranites. The composition of the Kalba granites and leucogranites are taken from [24,47]. The concentrations are normalized to the primitive mantle [56].

4.3. Ar-Ar Age

The age of the Tochka rare-metal pegmatites was estimated in muscovite monofractions from two albite pegmatites (samples 63-TO and 64-TO) from the one vein and from the microcline–albite pegmatite with muscovite (sample 65-TO) from a nearby vein. The measurement results are given in Table S5. All obtained $^{40}Ar/^{39}Ar$ spectra (Figure 14) show distinct plateaus corresponding to 96.8% to 99.3% cumulative fraction of ^{39}Ar released. The resulting plateau ages are 293 ± 4 Ma for 63-TO, 292 ± 4 Ma for 64-TO (albite pegmatite samples), and 290 ± 4 Ma for 65-TO (microcline–albite pegmatite with muscovite). The latter date is slightly younger, but all three determinations are generally similar within the measurement error. Thus, the Tochka pegmatites formed about 292 Ma ago (total age over three samples).

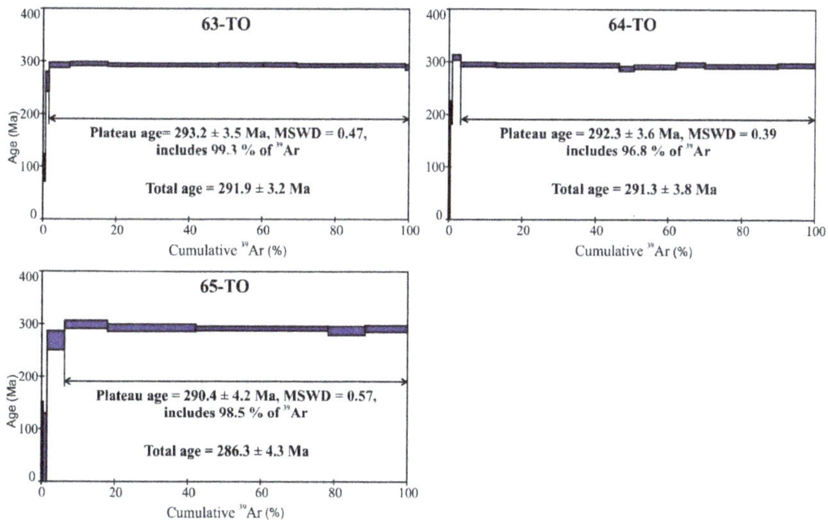

Figure 14. $^{40}Ar/^{39}Ar$ age spectra of muscovite from the Tochka pegmatite.

5. Discussion

Granite magmatism in the Kalba–Narym zone, according to recent studies, occurred in the Early Permian in a post-collision geodynamic setting as a result of the effect of Tarim mantle plume on the lithosphere during its extension [6,22,24,43,57,58]. It was determined that the granites of the Kunush complex have Na specialization, similar to I-type granites. It is believed that they occurred during the melting of lower crustal metabasite substrates [43, 58]. Granites of the Kalba complex have a K-Na specialization, similar to S-type granites but, according to some geochemical characteristics, similar to A-type granites. It is believed that they occurred during the melting of medium-crustal substrates of mixed composition (mainly metapelites with a small proportion of metabasites) [24].

The origin of the Tochka rare-metal pegmatites was previously attributed to the Na-granite magmatism of the Kunush complex [38,40–42], proceeding their spatial proximity to and intrusion into a Na-granite body. There was a point of view that pegmatites emplaced within the field of Na-granites, possibly because the ascending pegmatitic magma followed the same paths as the older Na-granite magma [59]. However, our results show the invalidity of this model.

Firstly, there are the field observations and the age data. It has been established that, at the Tochka deposit, pegmatite veins cut only the Na-granites of the Kunsh complex and do not cut the granites of the Kalba complex. In addition, the Kalba granites do not cut the pegmatite veins and the arrangement of the pegmatite veins and the Kalba granites' body is sub-according. This is geological evidence that the Kalba granites and the rare-metal pegmatites intruded synchronously. This is confirmed by the results of the Ar-Ar dating of pegmatites from the Tochka deposit (293–290 Ma). They are younger than the 299 $\pm$ 2 Ma Na-granites of the Kunush complex (U-Pb dating of zircons [43]) but fall within the 296–286 Ma interval for the Kalba granites (U-Pb dating of zircons [6,24]). Moreover, the age of the Tochka rare-metal pegmatites is close to the age (approximately between 291 and 286 Ma) of rare-metal pegmatites from the Asubulak ore district located about 40 km east [14].

Secondly, there are the composition data. Mineralogical data show that the typical accessory minerals for gneisses—ilmenite, magnetite, pyrite, zircon, monazite, and xenotime—are present in hornfels. In Na-granites, the typical accessory minerals for granites—apatite, zircon, monazite, and titanite—are present. Pegmatites also contain accessory apatite, zircon, monazite, and xenotime, but these are common minerals for all granitoid rocks. In addition, apatite in pegmatites contains an admixture of Mn. Other accessory minerals in pegmatites are not similar to accessory minerals in hornfelses and Na-granites; these are rutile, cassiterite, sphalerite, pyrolusite, uraninite, and columbite. The Kalba granites and leucogranites, according to published data, contain accessory cassiterite, rutile, columbite, scheelite, and wolframite [47,59–61]. Thus, the mineral associations of pegmatites are more similar to the mineral associations of the Kalba granites than to those in hornfels or Na-granites. Micro-inclusions of cassiterite in cracks in hornfels and micro-inclusions of tungstite in cracks in Na-granites can be explained by the influence of fluids separated from pegmatite magmas on them.

The composition of muscovite hornfelses near the pegmatite veins shows enrichment in rare metals. This indicates that the pegmatites are a source of rare-metal mineralization, but the Na-granites are not. Thus, the location of rare-metal pegmatites among Na-granites at the Tochka deposit seems to be random.

Geochemical data show that the concentrations of Ti, Fe, Na, K, REE, and HFSE in pegmatites are lower than in the Kalba granites (Figures 11 and 13). The concentrations of some elements, such as Fe, Ca, K, Ba, Th, Zr, Hf, and HREE, in pegmatites and Na-granites are at the same level. However, this similarity is not proof of the genetic relationship between pegmatites and Na-granites but, on the contrary, refutes it. It is generally accepted that pegmatitic magmas are formed during the evolution of granitic magmas, often in large chambers [62–66]. During this, the components compatible with silicate minerals have lower concentrations in pegmatites; that is, if pegmatites were associated with Na-granites, the concentrations of the main elements—REE and HFSE—in them would be even lower than in Na-granites. Since this is not the case, the concentrations of these elements in pegmatites allow us to conclude that they are the result of the differentiation of the magmas of the Kalba granites but not the Na-granites. This is also confirmed by the increased concentrations of rare metals (those components that accumulate in residual magmas during differentiation) in pegmatites compared with the Kalba granites (Figure 12).

These conclusions are confirmed by the results of a study of rare-metal pegmatites in the Asubulak ore district [14,47,66]. There are numerous veins of rare-metal pegmatites that cut granites of the first phase but are not cut by granites of the second intrusive phase. According to geological and geochronological data, the Kalba rare-metal pegmatites are related to granites of the Kalba complex, which may be a source of ore material (considering their relative enrichment in F, Li, Ta, and Be). Thus, the rare-metal pegmatite mineralization may have a magmatic origin and may associate with granitic magma differentiation processes in large chambers. It is also important that the pegmatite fields of the Asubulak ore district were formed in the upper (near-roof) part of the granite massif [56,66].

In this respect the Karagoin–Saryozek zone—located between several large granite massifs of the Kalba complex where host rocks play a role as a roof—may be very promising for rare-metal pegmatite mineralization. Specifically, the source of the Shubarshoky pluton may be such a reservoir of fractionating magma that eventually produced the Tochka pegmatites. The new results mean positive prospects for the Karagoin–Saryozek zone as a potentially rich source of Li in albite–spodumene pegmatites.

It should be noted that the earlier assessment of the resources of the Tochka deposit showed concentrations of Li_2O (0.219–1.132 wt.%), even according to the flame photometry. The low-average values in the reports [38,40–42] apparently represent rather the bulk composition of mixed-mineral assemblages in pegmatites and are thus poorly reliable. The present advanced analytical tools, such as mass spectrometry with inductively coupled plasma (ICP-MS) or atomic emission spectrometry (AES), reveal higher Li grades of the Tochka albite–spodumene pegmatites (0.22% to 1.66%–2.87% Li_2O) that are comparable with the estimates for the economic operations, such as Yubileinoye (0.306%), Bakennoye (0.119%), and Akhmetkino (0.76%) in the Kalba–Narym zone, as well as Greenbushes mine in Australia (up to 2.9%), Vishnevskoye in Russia (1.06%), and other pegmatite deposits [11,65,66].

6. Conclusions

The studies in the area of the Tochka Li deposit in East Kazakhstan have provided updates for local geology, mineralogy, geochemistry, and ages of Li-pegmatites and country rocks. The conducted mineralogical, geochemical, and geochronological studies made it possible to link the formation of pegmatite veins of the Tochka deposit not with Na-granites (as previously thought) but with the Kalba granites. A similar relationship of rare-metal pegmatites with the Kalba granites was shown earlier for the deposits of the Asubulak ore region. Considering that the Karagoin–Saryozek zone is located between large granitic plutons of the Kalba batholith, a significant potential for the discovery of rare-metal deposits is probably concentrated within it.

Supplementary Materials: The following supporting information can be downloaded: https://www.mdpi.com/article/10.3390/min12121478/s1, Table S1. The composition of accessory minerals from hornfels, wt. %; Table S2 The composition of accessory minerals from Na-granites, wt. %; Table S3. The composition of accessory minerals from pegmatites, wt. %; Table S4. The composition of rocks (concentrations in ppm) from Tochka deposit (authors data) and from Kalba batolith (data from [24,47]. Table S5. 40Ar-39Ar dating results of muscovite from pegmatite samples of the Tochka deposit.

Author Contributions: T.A.O., N.A.Z., A.Y.B., A.V.T. and S.V.K.: idea and subject development; N.A.Z., T.A.O. and S.S.A.: sampling, preconditioning of samples, and analytical work; A.V.T. and I.Y.A.: ^{40}Ar/^{39}Ar dating; N.A.Z., A.V.T. and S.V.K.: data handling; A.Y.B. and T.A.O.: illustrations; T.A.O., N.A.Z., A.Y.B., A.V.T., S.V.K. and I.Y.A.: manuscript writing and editing. All authors have read and agreed to the published version of the manuscript.

Funding: The work was supported by the Committee for Science of the Ministry of Science and Education of Kazakhstan (registration No. AP08052371). The research of geological structure was carried out on a government assignment to the V.S. Sobolev Institute of Geology and Mineralogy (Novosibirsk). ^{40}Ar/^{39}Ar dating was supported by grant 22-17-00038 from the Russian Science Foundation. The comparison of pegmatites' and Kalba granites' and leucogranites' compositions was supported by grant 21-17-00175 from the Russian Science Foundation.

Data Availability Statement: The data presented in this study are available on request from the corresponding author.

Acknowledgments: We wish to thank analysts from the VERITAS Engineering Laboratory and from the V.S. Sobolev Institute of Geology and Mineralogy (Novosibirsk) for the analytical work. Thanks are extended to Tatiana Perepelova for aid in manuscript preparation.

Conflicts of Interest: The authors declare no conflict of interest.

References

1. Jahn, B.-M.; Wu, F.; Chen, B. Granitoids of the Central Asian Orogonic Belt and continental growth in the Phanerozoic. *Earth Environ. Sci. Trans. R. Soc. Edinb.* **2000**, *91*, 181–193. [CrossRef]
2. Kozakov, I.K.; Kovach, V.P.; Bibikova, E.V.; Kirnozova, T.I.; Zagornaya, N.Y.; Plotkina, Y.V.; Podkovyrov, V.N. Age and sources of granitoids in the junction zone of the Caledonides and Hercynides in southwestern Mongolia: Geodynamic implications. *Petrology* **2007**, *15*, 126–150. [CrossRef]
3. Zhang, S.H.; Zhao, Y.; Kröner, A.; Liu, X.M.; Xie, L.W.; Chen, F.K. Early Permian plutons from the northern North China Block: Constraints on continental arc evolution and convergent margin magmatism related to the Central Asian Orogenic Belt. *Int. J. Earth Sci.* **2009**, *98*, 1441–1467. [CrossRef]
4. Yarmolyuk, V.V.; Kuzmin, M.I.; Kozlovsky, A.M. Late paleozoic-Early Mesozoic within-plate magmatism in North Asia: Traps, rifts, giant batholiths, and the geodynamics of their origin. *Petrology* **2013**, *21*, 101–126. [CrossRef]
5. Tsygankov, A.A. Late Paleozoic granitoids in western Transbaikalia: Sequence of formation, sources of magmas, and geodynamics. *Russ. Geol. Geophys.* **2014**, *55*, 153–176. [CrossRef]
6. Khromykh, S.V.; Tsygankov, A.A.; Kotler, P.D.; Navozov, O.V.; Kruk, N.N.; Vladimirov, A.G.; Travin, A.V.; Yudin, D.S.; Burmakina, G.N.; Khubanov, V.B.; et al. Late Paleozoic granitoid magmatism of Eastern Kazakhstan and Western Transbaikalia: Plume model test. *Russ. Geol. Geophys.* **2016**, *57*, 773–789. [CrossRef]
7. Tsygankov, A.A.; Burmakina, G.N.; Khubanov, V.B.; Buyantuev, M.D. Geodynamics of Late Paleozoic batholith-forming processes in western Transbaikalia. *Petrology* **2017**, *25*, 396–418. [CrossRef]
8. Şengör, A.M.C.; Natal'In, B.A.; Burtman, V.S. Evolution of the Altaid tectonic collage and Palaeozoic crustal growth in Eurasia. *Nature* **1993**, *364*, 299–307. [CrossRef]
9. Şengör, A.C.; Sunal, G.; Natal'In, B.A.; van der Voo, R. The Altaids: A review of twenty-five years of knowledge accumulation. *Earth Sci. Rev.* **2022**, *228*, 104013. [CrossRef]
10. Yarmolyuk, V.V.; Kuzmin, M.I. Late Paleozoic and Early Mesozoic rare-metal magmatism of Central Asia: Stages, provinces, and formation settings. *Geol. Ore Depos.* **2012**, *54*, 313–333. [CrossRef]
11. Zagorsky, V.; Vladimirov, A.G.; Makagon, V.M.; Kuznetsova, L.G.; Smirnov, S.Z.; D'yachkov, B.A.; Annikova, I.; Shokalsky, S.P.; Uvarov, A.N. Large fields of spodumene pegmatites in the settings of rifting and postcollisional shear-pull-apart dislocations of continental lithosphere. *Russ. Geol. Geophys.* **2014**, *55*, 237–251. [CrossRef]
12. Kuznetsova, L.G. Interaction of crustal and mantle materials, sources of trace elements during the formation and evolution of Early Paleozoic Li-rich granite-pegmatite systems in southeastern Tuva. *Russ. Geol. Geophys.* **2018**, *59*, 1660–1678. [CrossRef]
13. Antipin, V.S.; Kushch, L.V.; Odgerel, D.; Belozerova, O.Y. Early Mesozoic rare-metal granites and metasomatites of Mongolia: Mineral and geochemical features and hosted ore mineralization (Baga Gazriin Chuluu pluton). *Russ. Geol. Geophys.* **2021**, *62*, 1061–1073. [CrossRef]
14. Khromykh, S.V.; Oitseva, T.A.; Kotler, P.D.; D'yachkov, B.A.; Smirnov, S.Z.; Travin, A.V.; Vladimirov, A.G.; Sokolova, E.N.; Kuzmina, O.N.; Mizernaya, M.A.; et al. Rare-Metal Pegmatite Deposits of the Kalba Region, Eastern Kazakhstan: Age, Composition and Petrogenetic Implications. *Minerals* **2020**, *10*, 1017. [CrossRef]
15. Yarmolyuk, V.V.; Kozlovsky, A.M.; Travin, A.V.; Kirnozova, T.I.; Fugzan, M.M.; Kozakov, I.K.; Plotkina, Y.V.; Eenzhin, G.; Oyunchimeng, T.; Sviridova, O.E. Duration and geodynamic nature of Giant Central Asian Batholiths: Geological and geochronological stadies of the Khangai batholith. *Strat. Geol. Correl.* **2016**, *27*, 73–94. [CrossRef]

16. Cerny, P. Rare-element granitic pegmatites. Part I: Anatomy and internal evolution of pegmatite deposits. *Geosci. Can.* **1991**, *18*, 49–67.
17. Cerny, P.; Ercit, S. The classification of granitic pegmatites revisited. *Can. Mineral.* **2005**, *43*, 2005–2026. [CrossRef]
18. Zonenshain, L.P.; Kuzmin, M.I.; Natapov, L.M. *Geology of the USSR: A Plate Tectonic Synthesis*; Geodynamic Series 21; American Geophysical Union: Washington, DC, USA, 1990; p. 242.
19. Vladimirov, A.G.; Kruk, N.N.; Rudnev, S.N.; Khromykh, S.V. Geodynamics and granitoid magmatism of collision orogens. *Russ. Geol. Geophys.* **2003**, *44*, 1321–1338.
20. Khromykh, S.V.; Vladimirov, A.G.; Izokh, A.E.; Travin, A.V.; Prokop'ev, I.R.; Azimbaev, E.; Lobanov, S.S. Petrology and geochemistry of gabbro and picrites from the Altai collisional system of Hercynides: Evidence for the activity of the Tarim plume. *Russ. Geol. Geophys.* **2013**, *54*, 1288–1304. [CrossRef]
21. Khromykh, S.V.; Tsygankov, A.A.; Burmakina, G.N.; Kotler, P.D.; Sokolova, E.N. Mantle-crust interaction in petrogenesis of gabbro-granite association in Preobrazhenka intrusion, Eastern Kazakhstan. *Petrology* **2018**, *26*, 368–388. [CrossRef]
22. Khromykh, S.V. Basic and associated granitoid magmatism and geodynamic evolution of the Altai accretion–collision system (Eastern Kazakhstan). *Russ. Geol. Geophys.* **2022**, *63*, 279–299. [CrossRef]
23. Kotler, P.D.; Khromykh, S.V.; Vladimirov, A.G.; Travin, A.V.; Kruk, N.N.; Murzintsev, N.G.; Navozov, O.V.; Karavaeva, G.S. New data on the age and geodynamic interpretation of the Kalba-Narym granitic batholith, Eastern Kazakhstan. *Dokl. Earth Sci.* **2015**, *462*, 565–569. [CrossRef]
24. Kotler, P.D.; Khromykh, S.V.; Kruk, N.N.; Sun, M.; Li, P.; Khubanov, V.B.; Semenova, D.V.; Vladimirov, A.G. Granitoids of the Kalba batholith, Eastern Kazakhstan: U–Pb zircon age, petrogenesis and tectonic implications. *Lithos* **2021**, *388–389*, 106056. [CrossRef]
25. Vladimirov, A.G.; Kruk, N.N.; Khromykh, S.V.; Polyansky, O.P.; Chervov, V.V.; Vladimirov, V.G.; Travin, A.V.; Babin, G.A.; Kuibida, M.L.; Khomyakov, V.D. Permian magmatism and lithospheric deformation in the Altai caused by crustal and mantle thermal processes. *Russ. Geol. Geophys.* **2008**, *49*, 468–479. [CrossRef]
26. Khromykh, S.V.; Kotler, P.D.; Semenova, D.V. Geochemistry, age, and geodynamic settings of formation of the Saur gab-bro–granitoid intrusive series (Eastern Kazakhstan). *Geosfernye Issled. Geosph. Res.* **2019**, *2*, 6–26. [CrossRef]
27. Khromykh, S.V.; Vishnevskiy, A.V.; Kotler, P.D.; Antsiferova, T.N.; Semenova, D.V.; Kulikova, A.V. The Kalba batholith dyke swarms (Eastern Kazakhstan): Mafic magmas effect on granite formation. *Lithos* **2022**, *426–427*, 106801. [CrossRef]
28. D'yachkov, B.A.; Titov, D.V.; Sapargaliev, E.M. Ore belts of the Greater Altai and their ore resource potential. *Geol. Ore Depos.* **2009**, *51*, 197–211. [CrossRef]
29. D'yachkov, B.; Aitbayeva, S.; Mizernaya, M.; Amralinova, B.; Bissatova, A. New data on non-traditional types of East Kazakhstan rare metal ore. *Sci. Bull. Nat. Min. Univ.* **2020**, *4*, 11–16. [CrossRef]
30. D'yachkov, B.A.; Bissatova, A.Y.; Mizernaya, M.A.; Khromykh, S.V.; Oitseva, T.A.; Kuzmina, O.N.; Zimanovskaya, N.A.; Aitbayeva, S.S. Mineralogical Tracers of Gold and Rare-Metal Mineralization in Eastern Kazakhstan. *Minerals* **2021**, *11*, 253. [CrossRef]
31. D'yachkov, B.A.; Mizernaya, M.A.; Khromykh, S.V.; Bissatova, A.Y.; Oitseva, T.A.; Miroshnikova, A.P.; Frolova, O.V.; Kuzmina, O.N.; Zimanovskaya, N.A.; Pyatkova, A.P.; et al. Geological history of the Great Altai: Implications for mineral exploration. *Minerals* **2022**, *12*, 744. [CrossRef]
32. Kalinin, Y.A.; Kovalev, K.R.; Serdyukov, A.N.; Gladkov, A.S.; Sukhorukov, V.P.; Naumov, E.A.; Travin, A.V.; Semenova, D.V.; Serebryakov, E.V.; Greku, E.D. Age constraints and metallogenic prediction of gold deposits in the Akzhal-Boko-Ashalin ore zone (Altai accretion-collision system). *Geodyn. Tectonophys.* **2021**, *12*, 392–408. [CrossRef]
33. Suiekpayev, Y.S.; Sapargaliyev, Y.M.; Dolgopolova, A.V.; Pirajno, F.; Seltmann, R.; Khromykh, S.V.; Bekenova, G.K.; Kotler, P.D.; Kravchenko, M.M.; Azelhanov, A.Z. Mineralogy, geochemistry and U-Pb zircon age of the Karaotkel Ti-Zr placer de-posit, Eastern Kazakhstan and its genetic link to the Karaotkel-Preobrazhenka intrusion. *Ore Geol. Rev.* **2021**, *131*, 104015. [CrossRef]
34. Khromykh, S.V.; Sokolova, E.N.; Smirnov, S.Z.; Travin, A.V.; Annikova, I.Y. Geochemistry and age of rare-metal dyke belts in Eastern Kazakhstan. *Dokl. Earth Sci.* **2014**, *459*, 1587–1591. [CrossRef]
35. Sokolova, E.N.; Smirnov, S.Z.; Khromykh, S.V. Conditions of crystallization, composition, and sources of rare-metal magmas forming ongonites in the Kalba-Narym zone, Eastern Kazakhstan. *Petrology* **2016**, *24*, 153–177. [CrossRef]
36. Annikova, I.Y.; Vladimirov, A.G.; Smirnov, S.Z.; Oitseva, T.A.; Mikheev, E.I.; Jes, E.N.; Travin, A.V.; D'yachkov, B.A.; Maslov, V.I.; Gertner, I.F. Geology and mineralogy of the Novo-Akhmirovskoe deposit of lithium topaz-zinnwaldite granites (East Kazakhstan). *Lithosphere* **2019**, *19*, 304–326. [CrossRef]
37. Murzintsev, N.G.; Annikova, I.Y.; Travin, A.V.; Vladimirov, A.G.; Dyachkov, B.A.; Maslov, V.I.; Oitseva, T.A.; Gavryushkina, O.A. Thermochronology and mathematical modeling of the formation dynamics of rare-metal-granite deposits of the Altai collision system. *Geodyn. Tectonophys.* **2019**, *10*, 375–404. [CrossRef]
38. Osipova, G.F.; Tupicyn, A.V.; Ksijashko, P.L.; Rassolov, S.M. *Exploration in the Area of Tochka Deposit, 1988–1990*; Technical Report; Vostkazgeologiya Production Geological Association: Ust-Kamenogorsk, Kazakhstan, 1990. (In Russian)
39. Shcherba, G.N.; Bespaev, X.A.; Dyachkov, B.A.; Mysnik, A.M.; Ganzhenko, G.D.; Sapargaliev, E.M. *Big Altai: Geology and Metallogeny*; Book 2. Metallogeny; RIO VAK RK: Almaty, Kazakhstan, 2000; p. 400.
40. Kashcheev, V.F.; Butko, A.R.; Distryanov, V.M.; Sinishin, P.I. *Report on Results of Exploration Works in Karagoyn-Saryozek Zone in 1956*; Vostkazgeologiya Production Geological Association: Ust-Kamenogorsk, Kazakhstan, 1957; 190p. (In Russian)

41. Ermolin, V.T. *Report on Detailed Exploration Works by Asubulak GEP in 1974-1975 (Apogranitny, Medvedka, Tochka Sites)*; Vostkazgeologiya Production Geological Association: Ust-Kamenogorsk, Kazakhstan, 1976; 136p. (In Russian)
42. Bayeva, I.V. *Report on Exploration Works in Medvedka-Tochka ore Field in 1990–1992*; Vostkazgeologiya Production Geological Association: Ust-Kamenogorsk, Kazakhstan, 1992; 134p. (In Russian)
43. Kuibida, M.L.; Kruk, N.N.; Vladimirov, A.G.; Polyanskii, N.V.; Nikolaeva, I.V. U-Pb isotopic age, composition, and sources of the plagiogranites of the Kalba range, Eastern Kazakhstan. *Dokl. Earth Sci.* **2009**, *424*, 72–76. [CrossRef]
44. Oitseva, T.A.; D'yachkov, B.A.; Kuzmina, O.N.; Bissatova, A.Y.; Ageyeva, O.V. LI-bearing pegmatites of the Kalba-Narym metallogenic zone (East Kazakhstan): Mineral potential and exploration criteria. *News Acad. Sci. Repub. Kazakhstan* **2022**, *1*, 83–90. [CrossRef]
45. Travin, A.V.; Yudin, D.S.; Vladimirov, A.G.; Khromykh, S.V.; Volkova, N.I.; Mekhonoshin, A.S.; Kolotilina, T.B. Thermochronology of the Chernorud granulite zone, Ol'khon Region, Western Baikal area. *Geochem. Int.* **2009**, *47*, 1107–1124. [CrossRef]
46. Baksi, A.K.; Archibald, D.A.; Farrar, E. Intercalibration of 40Ar–39Ar dating standards. *Chem. Geol.* **1996**, *129*, 307–324. [CrossRef]
47. Oitseva, T.A.; D'yachkov, B.A.; Vladimirov, A.G.; Kuzmina, O.N.; Ageeva, O.V. New data on the substantial composition of Kalba rare metal deposits. *IOP Conf. Ser. Earth Environ. Sci.* **2017**, *110*, 012018. [CrossRef]
48. Aytaliev, Z.A.; Brach, P.A.; Bryushkov, V.I.; Ergaliev, A.E.; Shavlo, S.G.; Shcherba, G.N. *Promising Areas for Tin and Rare Metals in the Kalba-Narym Ore district*; AN KazSSR: Alma-Ata, Kazakhstan, 1954; 232p. (In Russian)
49. D'yachkov, B.A. *Intrusive Magmatism and Metallogeny of the Eastern Kalba*; Nedra: Moskow, Russia, 1972; 212p. (In Russian)
50. D'yachkov, B.A. *Genetic Types of Rare Metal Deposits of the Kalba-Narym Belt*; EKSTU: Ust-Kamenogorsk, Kazakhstan, 2012; 130p. (In Russian)
51. Solodov, N.A. *Internal Structure and Geochemistry of Rare-Metal Granite Pegmatites*; AN SSSR: Moskow, Russia, 1962; 233p. (In Russian)
52. Shavlo, S.G. *Pegmatites and Hydrothermalites of the Kalba Range*; Science: Alma-Ata, Kazakhstan, 1958; 325p. (In Russian)
53. Shcherba, G.N.; Kudryashov, A.V.; Senchilo, N.P. *Rare Metal Mineralization of Kazakhstan*; Science: Alma-Ata, Kazakhstan, 1988; 224p. (In Russian)
54. Levashova, E.V.; Skublov, S.G.; Oitseva, T.A.; Dyachkov, B.A.; Li, X.-H.; Li, Q.-L.; Shatova, N.V.; Shatov, V.V. First Age and Geochemical Data on Zircon from Riebeckite Granites of the Verkhnee Espe Rare Earth–Rare Metal Deposit, East Kazakhstan. *Geochem. Int.* **2022**, *60*, 1–15. [CrossRef]
55. Boynton, W.V. Cosmochemistry of the rare earth elements: Meteorite studies. In *Rare Earth Element Geochemistry*; Henderson, P., Ed.; Elsevier: Amsterdam, The Netherlands, 1984; pp. 63–114.
56. Sun, S.S.; McDonough, W.F. Chemical and isotopic systematics of oceanic basalts: Implications for mantle composition and processes. *Geol. Soc. Spec. Publ.* **1989**, *42*, 313–345. [CrossRef]
57. Khromykh, S.V.; Kotler, P.D.; Izokh, A.E.; Kruk, N.N. A review of Early Permian (300–270 Ma) magmatism in Eastern Kazakhstan and implications for plate tectonics and plume interplay. *Geodyn. Tectonophys.* **2019**, *10*, 79–99. [CrossRef]
58. Kuibida, M.L.; Dyachkov, B.A.; Vladimirov, A.G.; Kruk, N.N.; Khromykh, S.V.; Kotler, P.D.; Rudnev, S.N.; Kruk, E.A.; Kuibida, Y.V.; Oitseva, T. Contrasting granitic magmatism of the Kalba fold belt (East Kazakhstan): Evidence for late Paleozoic postorogenic events. *J. Asian Earth Sci.* **2019**, *175*, 178–198. [CrossRef]
59. Lopatnikov, V.V.; Izokh, E.P.; Ermolov, P.V.; Ponomareva, A.P.; Stepanov, A.S. *Magmatism and Metallogeny of the Kalba-Narym zone, Eastern Kazakhstan*; Nauka: Moskow, Russia, 1982; 248p. (In Russian)
60. Ponomareva, Y.V.; Yermolov, P.V.; Portnov, V.S. Age and geodynamic of the Irtysh shear zone. In *Naukovyi Visnyk Natsionalnoho Hirnychoho Universytetu*; National Mining University: Dnipro, Ukraine, 2018; Volume 6, pp. 13–19. [CrossRef]
61. Dyachkov, B.A.; Oitseva, T.A.; Frolova, O.V.; Mataibaeva, I.Y.; Kuzmina, O.N. Geotectonic position and ore potential of intrusive-dyke belts in East Kazakhstan. In Proceedings of the 17th International multidisciplinary scientific geoconference, Albena, Bulgaria, 27 June–6 July 2017; pp. 239–246. [CrossRef]
62. Tkachev, A.V. Evolution of metallogeny of granitic pegmatite associated with orogens throughout geological time. *Geol. Soc. Spec. Publ.* **2011**, *350*, 7–23. [CrossRef]
63. Tkachev, A.V.; Rundqvist, D.V.; Vishnevskaya, N.A. Metallogeny of lithium through geological time. *Russ. J. Earth Sci.* **2018**, *18*, 1–13. [CrossRef]
64. London, D. Ore-forming processes within granitic pegmatites. *Ore Geol. Rev.* **2018**, *101*, 349–383. [CrossRef]
65. Dittrich, T.; Seifert, T.; Schulz, B.; Hagemann, S.; Gerdes, A.; Pfänder, J. *Archean Rare-Metal Pegmatites in Zimbabwe and Western Australia*; Springer Briefs in World Mineral Deposits; Springer: Cham, Switzerland, 2019; 125p.
66. D'yachkov, B.A.; Mayorova, N.P.; Shcherba, G.N.; Abdrakhmanov, K.A. *Granitoids and Ore-Bearing Mineral Assemblages in the Kalba-Narym Belt (Rudny Altai)*; Gylym: Almaty, Kazakhstan, 1994; 208p. (In Russian)

minerals

MDPI

Article

Magma Evolution and Constraints on the Graphite Mineralization Hosted by the Huangyangshan Alkaline Granite Suite in the East Junggar of Xinjiang Province: Evidence from In Situ Analyses of Silicate Minerals

Xinhao Sun [1], Yunsheng Ren [2,3,*], Jingmou Li [3], Mengjia Huang [4], Zhenjun Sun [2] and Zuowu Li [5]

1 Chengdu Center of China Geological Survey, Chengdu 610081, China
2 Hebei Key Laboratory of Earthquake Dynamics, Institute of Disaster Prevention, Sanhe 065201, China
3 School of Earth Sciences and Resources, China University of Geosciences, Beijing 100083, China
4 Sichuan Institute of Land Science and Technology, Chengdu 610045, China
5 Xinjiang Branch of China National Geological Exploration Center of Building Materials Industry, Urumqi 830092, China
* Correspondence: rys@cidp.edu.cn

Abstract: The Huangyangshan super-large graphite deposit, located in the East Junggar area of the Xinjiang Province, is hosted in and has closely temporal, spatial, and genetic relationships with the Huangyangshan alkaline granites. There are such silicate minerals as amphibole, biotite, pyroxene, and plagioclase occurring in the graphite-bearing granites. The integration of the electron microprobe analysis (EMPA) and laser ablation-inductively coupled plasma-mass spectrometry (LA-ICP-MS) enabled us to reveal the physicochemical conditions and evolution process, as well as the relationship of alkaline magmatism with graphite mineralization. The results show that the amphiboles generally have low Al and high Ti, K, Si, and Fe contents, as well as similar rare-earth elements (REEs) patterns and trace element distribution patterns to granites with significantly negative Eu anomalies. In the analyzed samples, primary biotite belongs to Fe-biotite and has characteristics of high Si and Fe and low Al and Mg contents. In the graphite orbicules, the pyroxene phenocrysts develop multiple zonal structures and are characterized by high Si and low Ca and Fe contents. The dominant plagioclase phenocrysts in the graphite orbicules are oligoclase and andesine, with normal and occasionally oscillatory zoning. The calculated crystallization temperature of the pyroxene, amphibole, and primary biotite in graphite orbicules are 840–1012 °C, 681–761 °C, and 658–720 °C, respectively, corresponding with their crystallization order. The pressure and depth calculation results of the amphibole, representing those of the magmatism, are 157–220 Mpa and 5.95–8.32 km, respectively. Both amphibole and biotite crystallized in a reducing environment with extremely low oxygen fugacity. The elemental compositions of these silicates indicate that the Huangyangshan pluton experienced significant mixing of mafic mantle-derived magma and felsic crust-derived magma. The cores of graphite orbicules were formed in a relatively earlier magmatic stage, while the granites and their dioritic enclaves were formed in a later magmatic stage. During magmatism, the mixing of mantle-derived basic magma had an important influence on the evolution and differentiation of the melts. According to the coexisting sulfides with graphite and compositional difference of amphibole and biotite in the granites and graphite ores, the graphite mineralization might be triggered by a magma mixing process.

Keywords: mineral elemental composition; EPMA; crystallization conditions; magma mixing; Huangyangshan graphite deposit; NW China

1. Introduction

Generally, there are three main mineralization processes that graphite crystallizes, namely, organic matter transforming by high-grade metamorphism in regional metamorphic graphite deposit [1–3], conversion of organic matter in contact with pyrometamorphism between igneous intrusions (mostly felsic) and coal-bearing sedimentary rocks [4–6], and carbon transforming by certain reactions in carbonaceous fluids or melts in fluid-deposited graphite [7–10]. The fluid-deposited graphite is the most complicated type due to the variability of ore-hosting rocks, melts/fluids properties, and ore-forming mechanisms [8,11–13].

In China, the first fluid-deposited graphite deposit, the Sujiquan occurring in the Kalamaili area in Xinjiang Province, was distinguished in the 1960s. In the last ten years, more crystalline graphite deposits have been discovered in this region such as the Huangyangshan super-large deposit. Then, the graphite deposits in the Eastern Junggar have attracted more attention. Some scholars have reported the ages and properties of the Huangyangshan pluton and concluded that belongs to the A-type granites formed in the Late Carboniferous [14–16]. Besides, the ore genesis of the Huangyangshan deposit has been involved in some references on the basis of fluid inclusion analysis laser Raman and graphite C isotopic analyses, as well as S and Pb isotopic analyses of associating sulfides in the graphite ores [16,17]. The C isotopic composition of the Huangyangshan graphite showed organic characteristics [16], whereas the S and Pb isotopic composition of associating sulfides showed a magmatic origin [17]. As a result, there are two major genetic viewpoints on the ore-forming mechanism of the Huangyangshan and the Sujiquan graphite deposits. One suggests that graphite is directly crystallized from magma [17–19]. The other holds that the graphite deposited from the hydrothermal process in the late stage of magma evolution [10,16]. The critical factors for the two controversial points of view are the accurate understanding on the magma evolution process and the relationship between graphite mineralization and the magmatic evolution of the ore-hosting granites.

Some magmatic minerals such as amphibole, biotite, pyroxene, and feldspar are generally used to reflect the magma composition and physicochemical condition when they precipitated according to their geochemical behavior [20–26]. As the main minerals in granitic rocks, they all have broad stability field to precipitate during the whole magmatic process and are capable of recording the temperature, pressure, and oxygen fugacity [27,28]. Our present study focuses mainly on the amphibole, biotite, pyroxene and plagioclase textures and chemical compositions in the graphite-hosting granites, the pluton microgranular dioritic enclaves, and orbicular graphite ores from the Huangyangshan deposit. We present in situ major and trace element analyses in order to reveal the physicochemical conditions of different magmatic stages and their implication on the mineralization of the investigated graphite. These studies are presumably helpful for better understanding the magma evolution process and unravel the genetic relationship between graphite and magmatism, which would have an impact on the mineral exploration for graphite deposits in China and elsewhere in the world.

2. Geological Setting and Graphite Mineralization

2.1. Regional Geology

The northern Xinjiang area, tectonically located in the southern part of the Central Asian Orogenic Belt (CAOB; Figure 1a), is split into the Altay, Junggar, and Tianshan blocks from north to south (Figure 1b). The East Junggar orogenic belt (EJOB) is situated in the northeast of the Junggar terrane and comprises the Durat–Baytag arc, the Aermantai ophiolite belt, the Yemaquan arc, the Kalamaili ophiolite belt, and the Harlik–Dananhu arc from north to south (Figure 1b); [29–32]. Based on the available geochronological data for this area, the Junggar ocean separating the Junggar and Altai block was formed at least from early Ordovician, and gradually closed during the late Paleozoic [33–38]. The Cambrain Aermantai ophiolite and early Devonian Kalamaili ophiolite witnessed the tectonic evolution process in the Paleozoic [39–42]. After its closure, the EJOB transferred into the post-collisional orogenic stage when the Huangyangshan alkaline pluton were emplaced, which is the main concern of the present study [19,43,44].

The Kalamaili area is located in the central part of the EJOB and is controlled by the NWW-trending Kalamaili fault system, which hosts typical ophiolite and several metallic deposits (Figure 1c). The exposed strata in the area mainly comprise small-scale Silurian sediments, Permian volcanogenic sediments, and a thick pile of Devonian and Carboniferous volcano-sedimentary rocks. Some Mesozoic and Cenozoic strata are distributed in the eastern and southern parts of this area. All these sediments are controlled by the Beitashan, Kubusu, Qingshui–Sujiquan, and Kalamaili faults and are deposited in a forearc basin-type sedimentary system dominated by active continental margin sediments [45–47]. The magmatic activities were characterized by late Paleozoic post-collisional granitic magmatism according to the geochronology and geochemical data from previous studies [14,32,48,49]. They produced many alkaline granitic plutons including the Laoyaquan, Beilekuduke, Yemaquan, Huangyangshan, and Yebushan plutons, as well as multiple Au, Cu, Sn, and graphite mineralization (Figure 1c); [44,50–53].

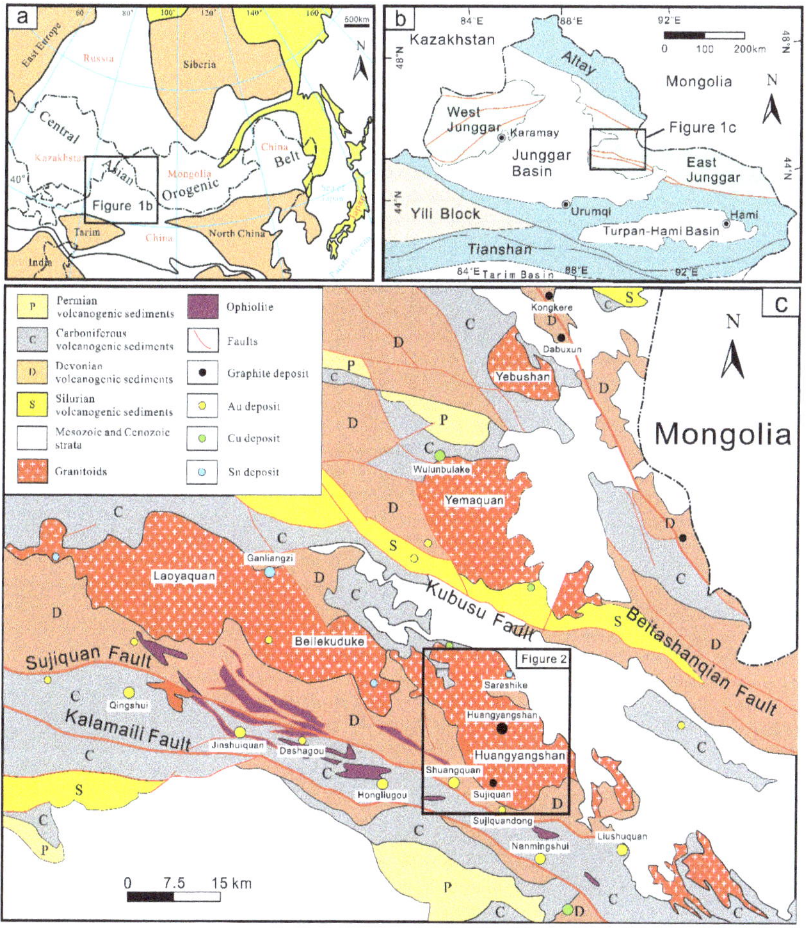

Figure 1. (**a**) Sketch map of the Central Asian Orogenic Belt (after [54]); (**b**) Geological map of the northern Xinjiang Province (after [38]); (**c**) Simplified geological map of the Kalamaili area, Eastern Junggar (modified after [19]).

2.2. Huangyangshan Pluton

The Huangyangshan pluton, located in the northeastern Kalamaili fault, is a typical A-type alkaline granitic pluton within the Kalamaili felsic intrusions of alkaline affinity. It roughly shows a NW–SE trend and crops out as a nearly elliptical shape on the surface, generating a concentrically zoned domain as a result of multiple phases of magmatism (Figure 2a). It intruded into the Devonian strata and shows sharp contact combined with small-scale contact metamorphism, indicating shallow depth for emplacement [14]. According to the grain sizes and main types of dark minerals, this pluton is divided into five lithofacies: medium-grained arfvedsonite granite, from medium- to fine-grained arfvedsonite granite, from medium- to fine-grained amphibole granite, medium-grained biotite granite, and fine-grained biotite granite. Their crystallizing ages show a slightly younger trend from north to south (~320–300 Ma) [16,19]. In our study, we focus on the arfvedsonite and amphibole granites, which occur near orebodies, consisting of K-feldspar (~60%), quartz (~25%), arfvedsonite (~3%–5%), hornblende (~5%–8%), plagioclase (<3%), and some accessory minerals such as zircon, titanite, ilmenite, and apatite.

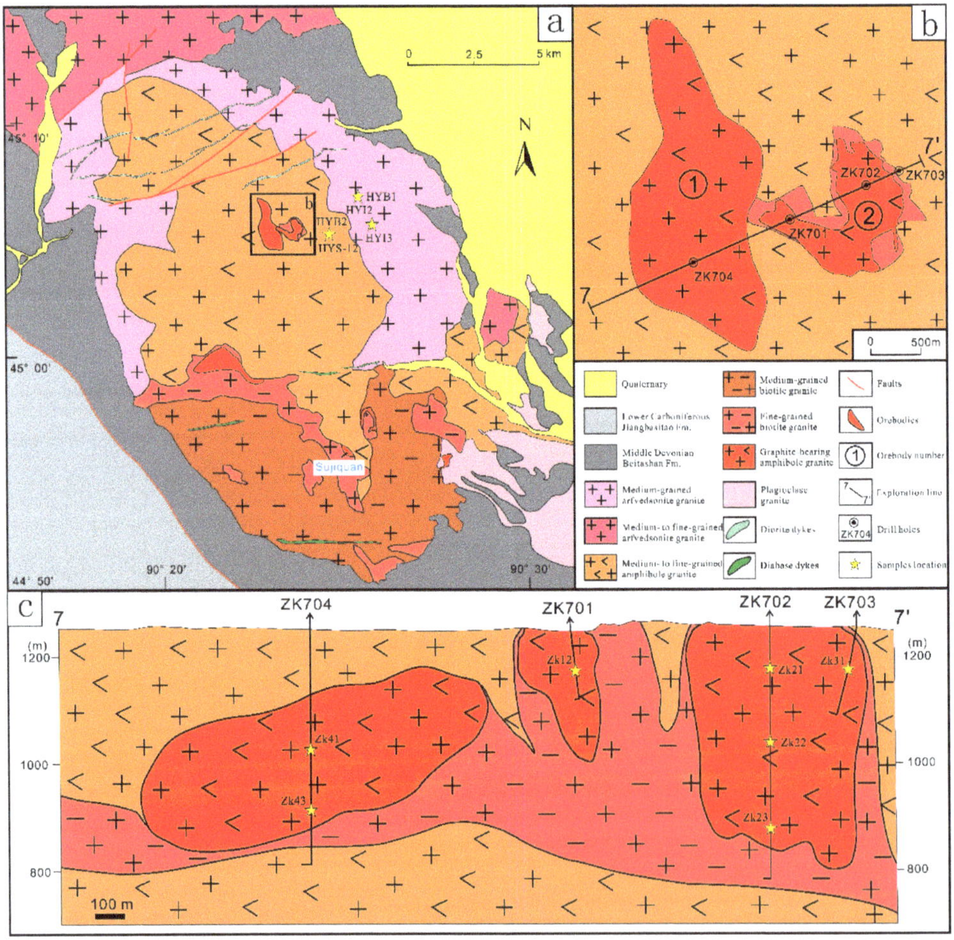

Figure 2. (**a**) Geological map of the Huangyangshan pluton (after [15]). (**b**) Simplified map of the No.1 and 2 ore bodies of the Huangyangshan graphite deposit. (**c**) Cross-section along line 7–7′ in Figure 2b (modified after [17]).

One of the most remarkable features of the Huangyangshan granites is a great abundance of microgranular dioritic enclaves. They are arranged randomly in different sizes and develop black chilled margins (Figure 3a,b). These enclaves are mostly diorite in composition, consisting of plagioclase (~45%), amphibole (mainly richterite, ~30%), quartz (~10%), biotite (<5%), and K-feldspar (~5%). Their grain sizes (~0.2–0.5 mm) are significantly smaller than those of the host granites (~5 mm). Some gabbro and diorite porphyrite dykes are occurred inside the Huangyangshan granites and their strikes are nearly east–west, which have clear boundaries with the granites on the surface. The diorite porphyrite dykes are mainly distributed in from medium- to fine-grained amphibole granite in the north, while the gabbro dykes are mainly distributed in medium-grained biotite granite in the south of the pluton (Figure 2a).

Figure 3. Photographs (**a**–**c**) and photomicrographs (**d**–**o**) of representative granite and graphite ore samples of the Huangyangshan pluton. The pink points represent some measuring locations. (**a**) A dark-colored ME in arfvedsonite granite. (**b**) Enclaves of different sizes in amphibole granite. (**c**) Zoning

graphite orbicules consisting of graphite shell, light-colored core, and fine-grained dark inner core. (**d**) Subhedral arfvedsonite phenocryst with ilmenite and titanite inclusions in arfvedsonite granite (−). (**e**) Contact zone between granite and ME containing subhedral richterite phenocryst and fine-grained richterite (−). (**f**) Altered amphibole phenocryst and biotite at the edge in enclaves (−). (**g**) Amphibole and its chloritization in amphibole granite. (**h**) Euhedral–anhedral hornblendes in the light-colored core of graphite orbicules (−). (**i**) Fine-grained hornblendes and distributed carbon materials in the dark inner core of graphite orbicules (−). (**j**) Altered amphibole, biotite, and chlorite in the granite collected on the surface (−). (**k**) Fine-grained biotite coexisting with richterite in ME (−). (**l**) Euhedral biotite and hornblendes in the light-colored core of graphite orbicules (−). (**m**) A pyroxene phenocryst besides cluster graphite aggregate (−). (**n**) Corona texture of pyroxene, amphibole, and biotite (BSE). (**o**) A plagioclase phenocryst without any alteration and inclusions in light-colored cores (+). Gr = graphite, Arf = arfvedsonite, Ilm = ilmenite, Ttn = titanite, Kfs = K-feldspar, Rct = richterite, Qz = quartz, Pl = plagioclase, Bt = biotite, Amp = amphibole, CM = carbon materials, Chl = chlorite, Mag = magnetite, Po = pyrrhotite, and Px = pyroxene.

2.3. *Graphite Mineralization*

The graphite deposit hosted in the Huangyangshan pluton is a large-scale deposit with over 78 million tons predicted graphite reserves [55]. All the orebodies occur as lenticular, saddle, or irregular shapes in the middle and southern part of the pluton (Figure 2a). Among them, the No. 1 lenticular and No. 2 saddle orebodies are the largest in scale, with a predicted resource of more than 70 million tons (Figure 2b) [15,55]. The graphite ore is often characterized by the unique spherical structure and coexist with metal sulfides such as pyrrhotite and chalcopyrite (Figure 3c). The graphite orbicules are abundant and randomly oriented in the orebodies. Most of them are composed of a dark graphite shell and light color core forming typical "zonal structures". Some of them include two cores in a dark and a light color, respectively (Figure 3c). The cores, as well as the host granites, have obvious different mineral compositions. The light-colored cores contain euhedral to anhedral amphiboles in brown or green color (Figure 3h,k). The dark-colored cores consist of fine-grained amphiboles, plagioclases, graphite, and some pyroxene phenocrysts (Figure 3i,m).

3. Sampling and Analytical Methods

Twelve samples were collected either from the surface outcrops or the drill cores, including the granites, enclaves, and graphite ore in the Huangyangshan graphite deposit (Figure 2). The HYI and HYS are the slightly weathered samples collected on surface from the arfvedsonite and amphibole granites, respectively. The HYB are the dark enclaves in arfvedsonite and amphibole granites. From ZK12 to ZK43 represents the fresh ore samples collected from the drill cores in No. 1 and No. 2 orebodies. All the samples were polished into thin sections for petrographic observation, electron microprobe analysis (EMPA), and LA-ICP-MS analysis.

Major elements compositions of amphibole, biotite, and plagioclase were analyzed using a JEOL JXA-8230 electron probe microanalyzer hosted at the State Key Laboratory of geological process and mineral resources, China University of Geosciences (Wuhan). The operating conditions were 15 kV accelerating voltage, 20 nA probe current, and 2–5 μm beam diameter. The natural minerals and oxides standards were used for calibrating the elemental X-ray intensity. A modified ZAF matrix-correction was applied during the data reduction. The analytical precision for SiO_2, MgO, FeO^T, Al_2O_3, TiO_2, MnO, Na_2O, K_2O, and Cr_2O_3 are all below 0.01 wt%. The analytical errors are generally less than 2%.

The trace element analysis of the amphiboles was performed using the LA-ICP-MS method at the State Key Laboratory of Ore Deposit Geochemistry, IGCAS. The laser sampling was performed using an ASI RESOLution-LR-S155 laser microprobe equipped with a Coherent Compex-Pro 193 nm ArF excimer laser. An Agilent 7700x ICP-MS instrument was used to acquire the ion-signal intensities. Helium (350 mL/min) was applied as a carrier gas. The ablated aerosol was mixed with Ar (900 mL/min) as a transport gas, before exiting the cell. Each analysis incorporated a background acquisition of approximately 30 s (gas blank) followed by 60 s of data acquisition from the sample. The calibration line for each element was constructed by analyzing three USGS reference glasses (BHVO-2G, BIR-1G and BCR-2G) with Si as an internal standard element. The spot size and frequency of the laser were set to 40 μm and 7 Hz, respectively. An Excel-based software, ICPMS-DataCal, (10.2, Wuhan, China) was used to perform the offline selection and integration of background and analyzed signals, time–drift correction, and quantitative calibration for the trace element analysis [56].

4. Results

4.1. Mineral Petrography

4.1.1. Amphibole

The amphibole in the Huangyangshan pluton can be divided into five types: Type-I is typical arfvedsonite in the arfvedsonite granite and occurs pervasively as phenocrysts (size up to 3 mm). It is euhedral and deep green in color (Figure 3d). There always occur some inclusions inside such as ilmenite and titanite. Type-II occurs as phenocryst at the contact zone between the arfvedsonite granite and enclaves (Figure 3e). It is from deep to light green in color, from euhedral to subhedral, has a relatively smaller size (size in 1–2 mm) than Type-I, and has no inclusions inside. Type-III amphibole is fine-grained (size in 0.2–0.5 mm) occurring in enclaves (Figure 3f). It is green-colored and mostly in a columnar or acicular shape. The common alterations such as chloritization, epidotization, and kaolinization are mainly due to the weathering on the surface. Type-IV is the subhedral-anhedral and yellow-green amphibole developing in the amphibole granite. It is mainly short columnar or granular in shape and develops chloritization at the edge (Figure 3g). Type-V amphibole developing in the cores of the graphite orbicules is enclosed by graphite. It is from brown to green in color, mostly from euhedral to subhedral, and almost fresh (Figure 3h). The interior of the graphite orbicule is diorite, whereas the exterior is granitic. The dark-colored cores in graphite orbicules contain sparse graphite grains, whereas the graphite shells are packet aggregates (Figure 3i), possibly showing the growing process of graphite.

4.1.2. Biotite

The biotite contents in both granites and enclaves are significantly lower than the amphibole. They can be divided into two types according to the alteration degree. Altered biotite mostly occurs in enclaves and granites metasomatizing arfvedsonite and/or developing chloritization (Figure 3f,j). It mainly has a rough surface, reabsorbed structure, and symbiosis with secondary minerals. Some relatively fresh biotites in enclaves are fine-grained (<0.1 mm) in light brown color and coexist with amphibole (Figure 3k). Fresh biotite all occurs in the graphite orbicule core. It is commonly euhedral in flake, dark-brown, and coexists with Type-IV amphibole, plagioclase, apatite, and some pyroxene (Figure 3l).

4.1.3. Pyroxene

The pyroxene occurs only in the graphite orbicules and is mostly colorless and in a euhedral granular shape. Most of them are phenocrysts and do not coexist with or surrounded by graphite, indicating their formation was earlier. A small amount of pyrrhotite and graphite can be found in the cracks inside some pyroxene (Figure 3m). Some other pyroxene phenocrysts are wrapped by amphibole and biotite in a corona structure (Figure 3n).

4.1.4. Plagioclase

The Huangyangshan granites barely contains plagioclase resulted from the high alkalinity. The plagioclase content of the dioritic enclaves is more than 40% but the grains are too small (<0.5 mm) to observe oscillatory zones. Therefore, we mainly focused on the relatively coarse-grained plagioclase in graphite orbicules and their host granites collected in the drill cores. The plagioclase in granites are commonly 0.5–3 mm in size and distributed in clusters, with some accessory mineral inclusions developed inside such as apatite and zircon. The plagioclase content in the graphite orbicules is significantly higher than that in the granites. They are mostly euhedral and hardly contain fine-grained amphibole and biotite as inclusions (Figure 3o). Most plagioclases in the granites are coarse-grained and develop normal zoning, while those in the graphite orbicules are medium- to fine-grained and develop inapparent normal and oscillatory zoning (Figure 3o).

4.2. Mineral Chemistry

4.2.1. Amphibole

The amphiboles in the Huangyangshan pluton and graphite ores are characterized by low Al_2O_3 (commonly < 6%, Figure 4), extremely high SiO_2 (mostly > 45%, Figure 4a), FeO^T (>20%, Figure 4d), and relatively high TiO_2 (mostly > 0.8%), K_2O (0.7%–1.9%, Figure 4f). These results are shown in Table S1. The MgO, CaO, and Na_2O contents of the five types of amphibole are significantly different (Figure 4b,c,e). According to the major element compositions of the amphiboles, their structural formulae on a 23-cations anhydrous basis after adjusting the ferrous iron charge balance were calculated following the procedure of Leake et al. [57] The results show that Type-I amphibole has a high FeO^T (32.54%–34.03%), Na_2O (6.37%–7.03%) contents, and low MgO (0.1%–0.24%), CaO (1.5%–2.5%) contents, belonging to sodic amphiboles [$Na_B \geq 1.5$, $(Na + K)_A \geq 0.5$)]. It is classified as typical arfvedsonite in Figure 5a. Type-II and -III amphiboles belong to sodic–calcic amphiboles [$Ca_B < 1.5$; $(Na + K)_A \geq 0.5$] and have a relatively high Na_2O (3.57%–4.99%) and low CaO (4.73%–7.93%) content. They are significantly different in MgO (<1% and >8%, respectively) and FeO (32.82%–33.56% and 18.87%–26.65%, respectively) contents, being consistent with the characteristics that the enclaves are neutral (rich in magnesium) and the granites are acidic (rich in iron). They are both plotted in the field of ferrorichterite in Figure 5b. Type-IV and -V amphiboles both have high CaO (9.37%–10.72%), medium MgO (4.97%–8.44%), and low Na_2O (1.61%–2.07%) content, showing calcic-amphiboles characteristics [$Ca_B \geq 1.5$; $(Na + K)_A < 0.5$]. They both are classified as ferrohornblende in Figure 5c. The main difference between them is that Type-IV has a low $Mg/(Mg + Fe^{2+})$ value and most measuring points do not contain Fe^{3+} (Table S1). There is no obvious difference between the amphiboles in light-colored and dark-colored cores.

The trace element compositions of 17 amphibole points in this study collected from enclaves and graphite orbicules and 7 arfvedsonite points reported by Ye et al. [58] are shown in Table S2. They have relatively consistent REE patterns that show significantly negative Eu anomalies and inapparent fractionation of the REEs (Figure 6a). Type-I arfvedsonite in granite and Type-II and -III richterites in enclaves have almost the same patterns with a slight HREE enrichment of Er–Tm–Yb–Lu compared with whole-rock REE patterns [19], except for one of the HYB-2 points (ΣREE = 52.67 ppm). Type-IV and -V amphiboles in amphibole granite and graphite orbicules have similar REE patterns, showing a significant enrichment in the REE content (ΣREE = 515.3–569.29 and 887.64–2126.49 ppm, respectively). The REE content of the Type-V amphibole is obviously higher than that from the Type-I to -IV amphiboles, indicating that the cores of graphite have not experienced the crystallization process of REE-rich minerals similar to the granites. The primitive mantle-normalized multi-element variation diagram patterns (Figure 6b) suggest all types of amphiboles have similar distribution characteristics to the granite with stronger Ba and Sr negative anomalies. Compared to the other four types, Type-V amphibole shows a slightly stronger La–Ce–Sm–Nd positive and Pb–Zr negative anomalies.

Note that one point in the enclaves (HYB-2) shows altered amphibole characteristics with low Nb (0.48 ppm) and REE (ΣREE = 52.67) contents and significantly different patterns in Figure 6b, hence the data were eliminated in the statistics. From calcic to sodic amphiboles, the Li content has an upward trend (average of 36, 485 and 1627 ppm, respectively), whereas the Sc–Y contents show a downward trend. The V–Cr contents also have a decreasing trend from calcic to sodic–calcic amphiboles due to the lack of sodic amphiboles data.

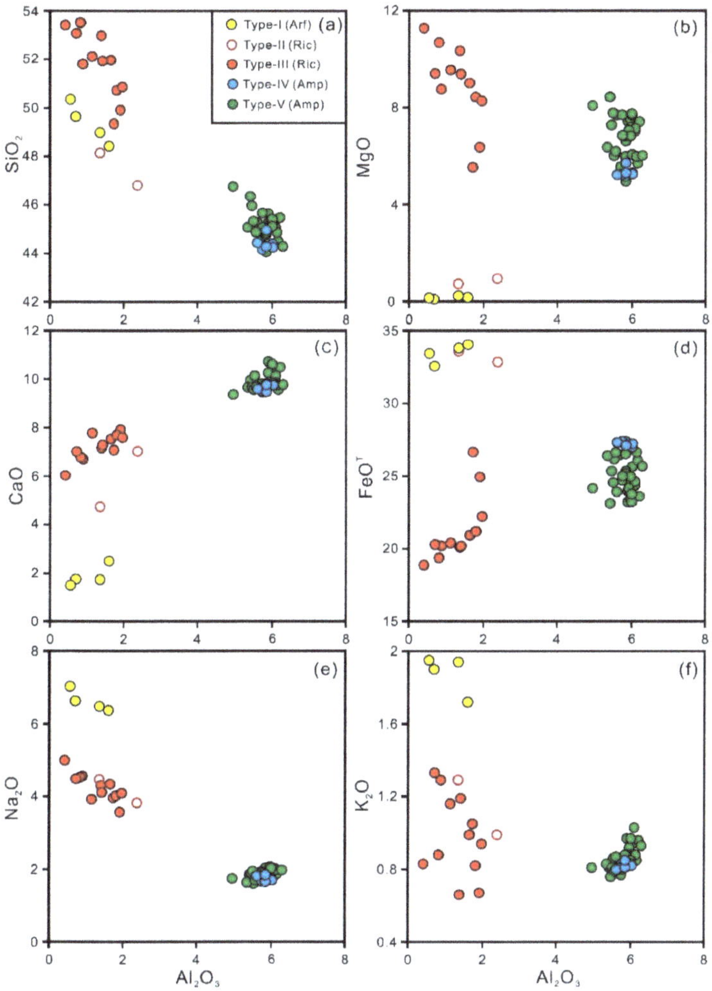

Figure 4. Diagrams of SiO_2 (**a**), MgO (**b**), CaO (**c**), FeO^T (**d**), Na_2O (**e**), and K_2O (**f**) vs. Al_2O_3 of amphibole in the Huangyangshan deposit in wt%.

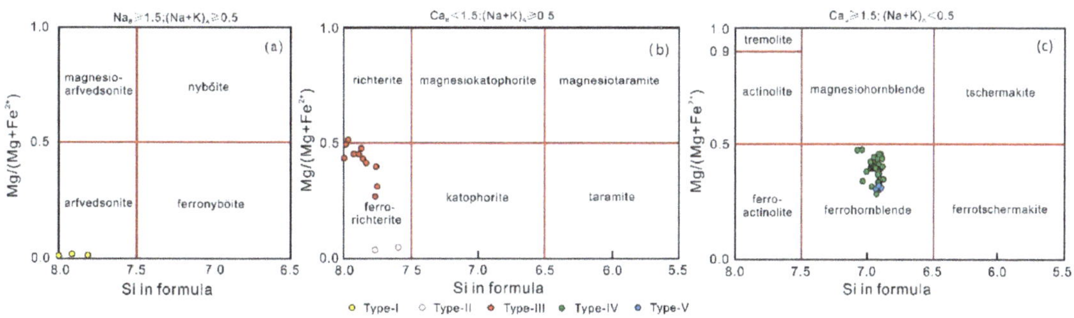

Figure 5. Classification diagrams of (**a**) sodic, (**b**) sodic–calcic, and (**c**) calcic amphiboles (after [57]). Symbols are as in Figure 4.

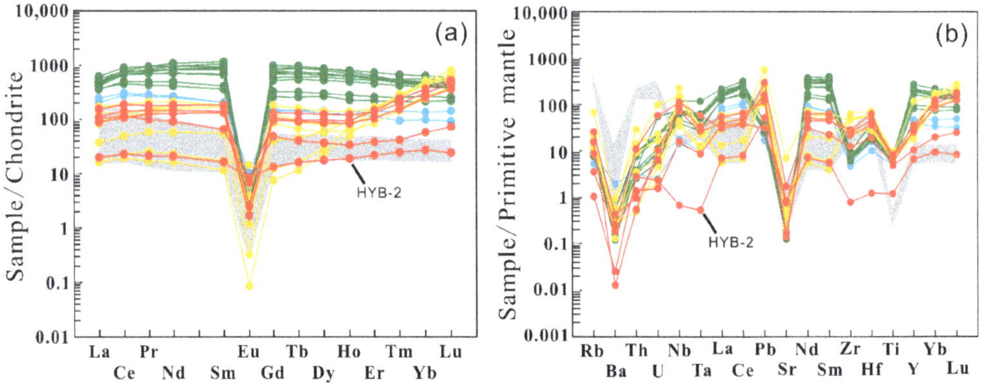

Figure 6. Chondrite-normalized REE patterns (**a**) and primitive mantle-normalized trace element variation patterns (**b**) of amphiboles and granites in the Huangyangshan deposit. The normalizing values for the REEs and trace elements are from Sun and McDonough [59]. The yellow, red, blue, green lines and the shadow parts represent the trace element data of Type-I, Type-II and -III, Type-IV, Type-V and granites, respectively. The granites data are from Sun et al. [19].

4.2.2. Biotite

The major element compositions of the biotite are shown in Table S3. The cation number and structural formulae of biotite were calculated by 22 oxygen atoms as a unit. We obtained only three available biotite EPMA points in dioritic enclaves, two of which are altered and one is relatively fresh. They have a relatively high SiO_2 (40.53%–42.32%) and MgO (9.02%–11.48%) and low Al_2O_3 (8.75%–9.51%) and FeO (19.6%–21.87%) contents; 3 and 12 valid data were obtained from amphibole granite and graphite orbicules, respectively, which have similar compositional characteristics. They have relatively low SiO_2 (35.57%–38.51%) and MgO (4.62%–10.12%), and high Al_2O_3 (11.12%–12.95%) and FeO (23.47%–30.17%) contents. The altered and fresh biotite samples are mainly different in TiO_2 content (0.99%–1.59% and 2.09%–4.46%, respectively) and are plotted in primary biotites and re-equilibrated biotites fields, respectively (Figure 7a). In the biotite classification plot (Figure 7b), the fresh biotite is classified as an Fe-biotite, whereas the altered biotite is classified as a Mg-biotite, indicating the alteration led to the loss of Fe or Mg enrichment.

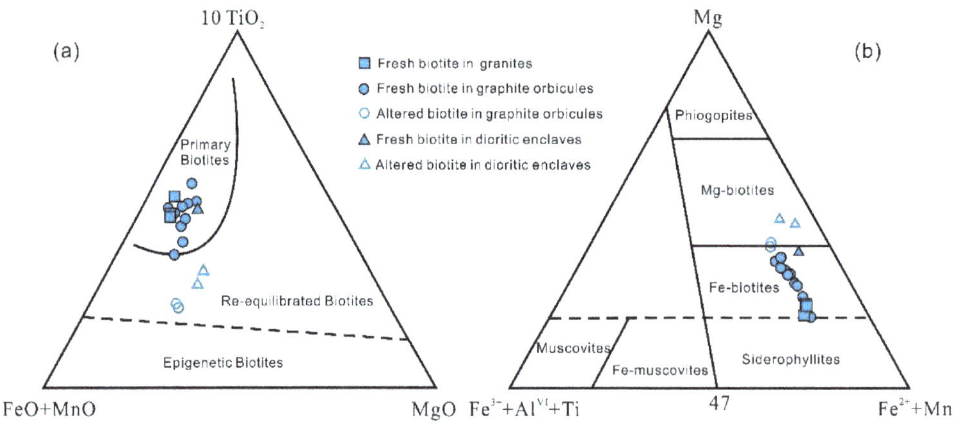

Figure 7. (**a**) Ternary 10 × TiO_2–(FeO + MnO)–MgO discrimination diagram of primary, re-equilibrated, and epigenetic biotite (after [60]). (**b**) Ternary Mg–(Fe^{2+} + Mn)–(Al^{VI} + Fe^{3+} + Ti) classification diagram of biotite (after [61]).

4.2.3. Pyroxene

Due to the low content of pyroxene in the Huangyangshan pluton, 16 measuring points of the 6 pyroxene phenocrysts in the graphite orbicules were analyzed in this study. According to their major element compositions (Table S4), 6 pyroxene phenocrysts were classified into 4 clinopyroxene phenocrysts with high calcic pyroxene endmembers (Wo) and 2 orthopyroxene phenocrysts with low Wo (Figure 8). The analyzed clinopyroxenes are mainly subhedral and fine-grained, which were found in granites, graphite orbicule shells, and deep-colored cores. Their enstatite endmembers (En) are similar (18.26–27.64) and have low Al_2O_3 (0.32%–1.12%), MgO (5.98%–9.26%), FeO (17.24%–24.26%), and high CaO (12.09%–20.43%) contents. The orthopyroxenes only occurred in graphite orbicules in the shape of coarse phenocrysts. They have extremely lower CaO (0.36%–1.69%), higher FeO (25.49%–36.33%), MgO (7.36%–19.96%), and similar Al_2O_3 (0.33%–1.01%) contents compared to the clinopyroxenes.

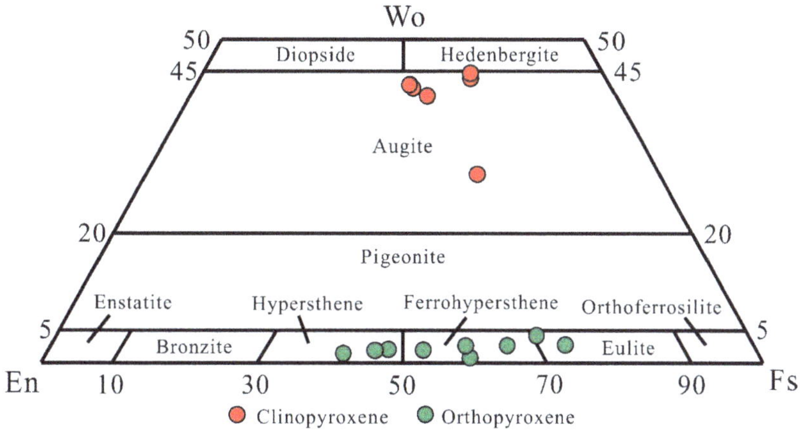

Figure 8. Ternary Calcic pyroxene (Wo)—Enstatite (En)—Orthoferrosilite (Fs) classification diagram of pyroxene (after [62]).

Some orthopyroxene phenocrysts develop typical zonal structures, including orthopyroxene, amphibole, and biotite from the core to the edge (Figure 3n). In the backscattered electron photograph (Figure 9a), an orthopyroxene has high protrusion and developed cleavage at the core, but low protrusion and undeveloped cleavage at the edge, indicating the orthopyroxene may have undergone metasomatic replacement and belongs to transitional phase minerals. From the core to the edge, the Fe content decreases, whereas the Mg content increases, showing the characteristics of reversely zoned structure. However, if it continues outward, the Fe content increases and the Mg content decreases (Figure 9b), suggesting that magma have experienced multiple mixing during the formation of pyroxene.

4.2.4. Plagioclase

The plagioclase grains in both granites and graphite orbicules have similar Al_2O_3 contents, ranging between 22.57% and 27.67% (Table S5). The former has relatively higher SiO_2 content than the latter, ranging in 59.51%–67.35% (63.48% on average) and 58.06%–66.81% (61.01% on average), respectively. The early plagioclase phenocryst in Sample ZK33 has a lower Si content (58.06%–65.36%, 59.89% on average), showing that the magma had a tendency of increasing acidity from early to late. Most plagioclases develop normal zoning characterized by dark low-An rims (30 > An > 50, oligoclase) and light high-An cores (10 > An > 30, andesine; Figure 10a–c), also suggesting the increasing magma alkalinity and acidity along with its evolution [20]. One plagioclase phenocryst enclosed by the graphite shows apparent occasional oscillatory zoning and has low a SiO_2, high An, and relatively small An spans (An = 32–42; Figure 10d), indicating a relatively stable neutral–basic environment in which it crystallized.

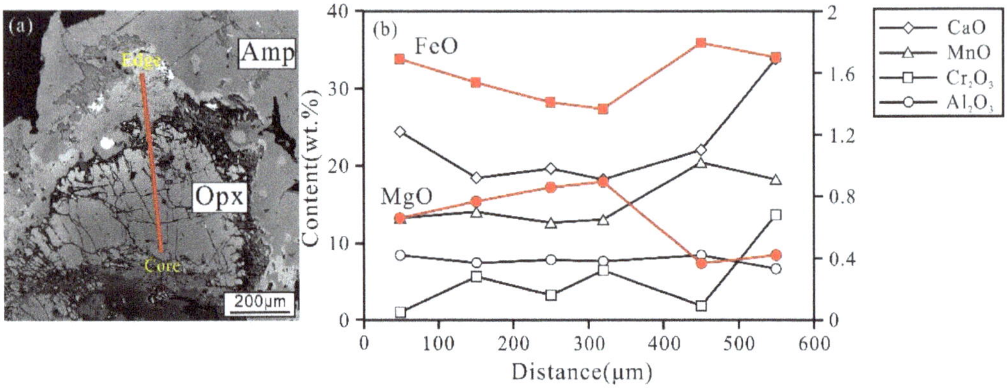

Figure 9. (**a**) BSE images of an orthopyroxene phenocryst in graphite orbicule. (**b**) Compositional variations along the profile of the Huangyangshan orthopyroxene phenocryst (based on the data from Table S4).

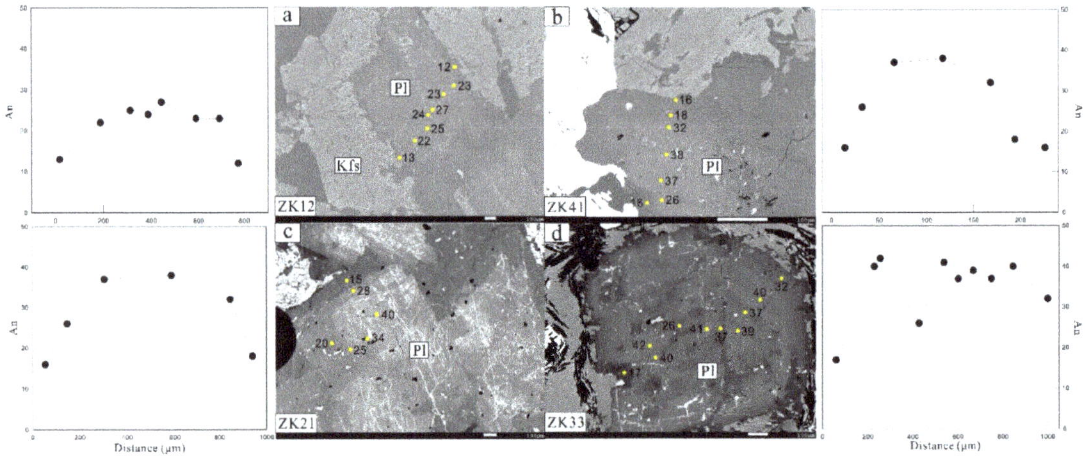

Figure 10. BSE images illustrating the petrographic characteristics of plagioclases in the granites and the graphite orbicules and their An (anorthite composition) variation trends. The numbers represent the An in plagioclase. (**a**,**b**) Plagioclase in granites showing high-An cores and low-An rims. (**c**) Plagioclase in the light-colored core of graphite orbicules having andesine core and oligoclase rim. (**d**) A plagioclase phenocryst surrounded by graphite having relatively high An and occasional oscillatory zoning.

5. Discussion

5.1. Mineral Crystallization Conditions

Previous studies have shown that amphibole, biotite, pyroxene, and plagioclase can be used to estimate their crystallization conditions, such as temperature, pressure, and oxidation, and their applicable objects of amphiboles mainly constrain in calcic amphiboles with relatively high Ca contents [21,23,24]. Ye et al. [58] tried to estimate the temperature and pressure of the sodic amphibole in the Huangyangshan pluton using multiple formulas. However, the pressure calculation result was not reliable for the sodic amphiboles.

The accurate temperature calculation result was more likely the closure temperature for the element exchange to reach equilibrium other than the crystallization temperatures. Through amphibole-only thermometers, an estimated temperature result with errors was obtained. In this study, only Type-IV and -V belong to calcic amphiboles. The Type-IV amphibole occurs in the amphibole alkaline granite with fine granularity and chloritization at prepheries. Type-V amphibole is mainly found in the cores of graphite orbicules with an even grain size and straight edges. It is commonly euhedral and rarely has obvious ductile deformation and directional arrangement, with the Si formula number less than 7.3 a.p.f.u. (6.88 a.p.f.u.–7.08 a.p.f.u.). All this evidence suggests a magmatic origin of Type-IV and -V amphiboles [63,64]. They hardly develop oscillatory zones, indicating their crystallized process can record the evolution of magma from melt to consolidation [65]. In this paper, therefore, the quantitative estimation of these two types of amphibole are mainly carried out. Then, the characteristics of the magma system were deduced in combination with the relevant parameters of biotite and pyroxene.

5.1.1. Pressure

Most amphibole thermobarometers are only applicable for amphiboles with relatively high Al and Mg/(Mg + Fe^{2+}) values [21–23,66]. However, Type-IV and -V amphiboles have significantly low Al (0.89 a.p.f.u–1.15 a.p.f.u) and Mg/(Mg + Fe^{2+}) (0.285 a.f.p.u–0.476 a.p.f.u) values. For the calcic ferrohornblendes in the Huangyangshan deposit, therefore, we need a more universal estimation method.

After experimental simulation, Mutch et al. [67] comprehensively considered the effects of fluid, temperature, and oxygen fugacity on pressure and proposed a new calculation formula. It is widely used in the composition of symbiotic amphibole (Mg/(Mg + Fe^{2+}) = 0.32 a.p.f.u.–0.81 a.p.f.u.) and feldspar (An = 15–80), but the rock is required to have a low variance mineral assemblage: amphibole + plagioclase + biotite + quartz + alkali feldspar + ilmenite/titanite + magnetite + apatite. Most composition parameters of the amphiboles in this study meet the above conditions and the mineral assemblages are very consistent. Based on this method, the estimated pressure of Type-IV and -V amphiboles is 157–220 Mpa and the corresponding depth is 5.95–8.32 km (with a crustal density of 2700 kg/m^3). This might represent the magmatic emplacement depth at the time of crystallization.

According to the Al^T-based biotite geobarometer proposed by Uchida et al. [68], the crystallized pressure of biotite in the graphite orbicules are mostly near 0, showing quite a difference with the results of the amphibole pressure. Kang et al. [69] pointed out that the calculation of granite diagenetic pressure has the following characteristics: If there exists an amphibole + biotite mineral assembly in the granite and the amphiboles are crystallized well, the calculated pressure of the amphibole will be reliable, whereas the biotite will not be reliable. If no amphibole exists in the granite, there will also be a large error in the biotite pressure results. The results of the biotite pressure will be reliable only if the amphibole crystallized poorly and the biotite crystallized well. Therefore, this study tends to use the amphibole pressure and depth as the formation pressure and depth of the graphite orbicule cores.

5.1.2. Temperature

The cores of the graphite orbicules are characterized by dioritic lithoface rich in plagioclase and calcic amphiboles, meeting the requirements of some geothermometers [70–72]. Using the thermometer proposed by Putirka [72] based on the correlation between the Si, Fe, Mg, Ti, Na, known pressure, and temperature of amphibole, we obtained the crystallization temperatures of 681–761 °C for Type-IV and -V amphiboles, with an average of 730 °C. Their low crystallized temperature condition suggests a late stage of magmatic evolution. Given that there is no thermobarometer for sodic and sodic–calcic amphibole, it can be inferred that the formation conditions roughly through the comparison of its compositional parameters. Generally, the Al content of the amphibole is closely related to the emplacement depth and low Al indicates the shallow emplacement depth [65]. The crystallization temperature has a strong negative correlation with the Si content, but negative correlations with Al, Na, and Ti contents [72]. Type-I arfvedsonite always contains some ilmenite and titanite inclusions, suggesting its high Ti content. From Type-I to -III amphiboles have significantly higher Si, Na, and lower Al than Type-IV (Figure 4). In alkaline rocks, the Na is much related to the amphibole compositions and magma components and cannot be used to estimate the temperature. Therefore, we inferred that the cores of the graphite orbicules were formed in a relatively high temperature and pressure condition compared with the granite and enclaves and may have crystallized in an earlier stage of magmatism. This is extremely consistent with the estimated temperature of the sodic amphibole (about 650 °C) and its late crystallization stage as reported by Ye et al. [58].

Under high temperature and high pressure, the content of Ti in biotite is closely related to the temperature change, which is usually used as the basis for estimating temperature [73,74]. Generally, biotite is Al-rich in peraluminous granite (S-type), Mg-rich in calc-alkaline granite (I-type), and Fe-rich in alkaline granite (A-type). The biotite in the graphite orbicules has extremely high Fe content (FeO = 23.47%–30.17%), showing typical A-type granite features. In comparison, the biotite in enclaves has a lower Fe content (FeO = 19.6%–21.87%) and higher Mg content (MgO = 9.02%–11.48%). These characteristics are similar to the compositions of amphiboles, indicating they crystallized at similar conditions. Based on the relation between the Ti content and crystallized temperature of the biotite in metapelites, Henry et al. [74] proposed a Ti-based empirical biotite geothermometer. It was later proved to be equally applicable to granite [75]. The crystallization temperature of the unaltered biotite in the enclaves is similar to that in the graphite orbicules, which yielded from 658 to 720 °C (686 °C on average). It is slightly lower than Type-IV and -V amphiboles, suggesting they crystallized slightly later than the amphiboles.

The thermometer for pyroxene is applicable when the pyroxene and the melt are in equilibrium. However, the pyroxenes in the studied graphite ore all occurred as phenocrysts, which are not contemporaneous products with the amphibole alkaline granite. Some other scholars put forward the estimation method of pyroxene thermometer through experiments and thermodynamic principles, which is suitable for most rocks [76,77], and requires the equilibrium parameter of symbiotic clinopyroxene and orthopyroxene (K_D) = 1.09 ± 0.14. Through calculation, there are five dipyroxene assemblages with K_D between 0.98 and 1.16, satisfying this condition. According to the formula, the crystallization temperatures of pyroxenes are 807–899 °C and 840–1012 °C, which are significantly higher than the crystallization temperatures of amphibole and biotite. This is also consistent with the pyroxene–amphibole–biotite crystallization order supported by the petrographic study.

5.1.3. Oxygen Fugacity

It is known that the oxygen fugacity of magma is positively correlated with the Mg content of amphibole and negatively correlated with the Fe content [78,79]. In addition, with the enhancement of reducibility, Fe^{3+} tends to change to Fe^{2+}. As a result, the Fe# (Fe/(Fe + Mg)) and ($Fe^{2+}/Fe^{2+} + Fe^{3+}$) values of amphibole can reflect the oxygen fugacity of magma. The higher these values, the lower the oxygen fugacity [78–80]. The results (Figure 11) show that Type-I and -II amphiboles have extremely high Fe# (0.951 a.p.f.u.–0.995 a.p.f.u.) and almost no Mg, indicating a very reductive condition in low oxygen fugacity. Type-III amphibole has relatively low Fe# (0.484 a.p.f.u.–0.730 a.p.f.u.) and mainly falls into the high oxygen fugacity field. Type-IV and -V amphiboles mostly have moderate Fe# (0.606 a.p.f.u.–0.741 a.p.f.u.) and are plotted in the field of intermediate oxygen fugacity. However, In Types I–IV amphiboles, all Fe elements exist in the form of Fe^{2+}, indicating an environment with very low oxygen fugacity. Besides, the ($Fe^{2+}/Fe^{2+} + Fe^{3+}$) value of the Type-V amphibole ranges from 0.12 a.p.f.u. to 0.34 a.p.f.u. Such values are indicative of a range of oxygen fugacity above the QFM buffer [81], suggesting the crystallization under from low to moderate conditions.

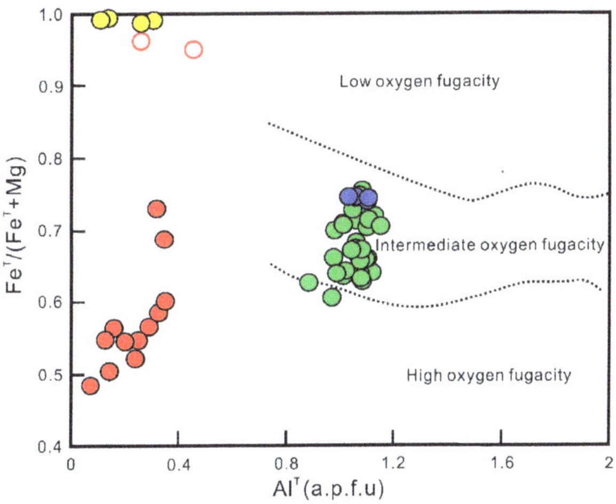

Figure 11. Fe# (Fe^T/Fe^T + Mg) versus Al^T of the amphiboles from the Huangyangshan deposit (after [80]). Symbols are as in Figure 4.

In mineralogy, Type-I arfvedsonite is an important petrogenetic indicator of alkaline magma, which forms in the reducing condition. The Sayashk hydrothermal Sn deposit occurring in the arfvedsonite granite in this area also reflects the low magmatic oxygen fugacity [82]. A large number of metal sulfides and oxides such as pyrrhotite, chalcopyrite, and ilmenite are developed in the graphite orbicules with rarely magnetite content, which also reflects the characteristics of relatively low oxygen fugacity. However, magnetite is also rare in the enclaves and the Fe^{3+} content is very low in the component parameters (Table S1). It does not show high oxygen fugacity characteristics, which is obviously ambiguous with the high oxygen fugacity of Type-III amphiboles shown in Figure 11. Amphibole is the main carrier of Mg in magma. During magma evolution, Mg is a refractory element that tends to reside in residual melt, whereas Fe readily migrates into differentiated minerals [83]. During magma mixing, the intermediate-basic magma end member with higher Mg and Fe contents evolved as the original melt. As a result, the Mg tends to remain in the amphiboles in enclaves, showing lower Fe# characteristics. The Fe tends to migrate into felsic magma to form minerals with higher Fe, such as arfvedsonite and ilmenite. Therefore, we used the characteristics of $Fe^{3+}/(Fe^{2+} + Fe^{3+})$ as the criterion. The enclaves were formed in a lower oxygen fugacity environment similar to the arfvedsonite granite.

Similar to those in amphiboles, the Fe^{2+}, Fe^{3+}, and Mg contents of biotite can also be used to estimate the oxygen fugacity of the magma and fluids [73]. In the ternary Fe^{2+}-Fe^{3+}-Mg plot (Figure 12a), the biotite in graphite orbicules is distributed near the NNO evolution line showing low oxygen fugacity, indicating it formed in a relatively reducing environment, which is consistent with the estimated results of amphibole. However, the biotite sample in the enclaves was plotted below the QFM evolution line due to its very low Fe^{3+} value, showing extremely low oxygen fugacity. In the Logf(O_2)–T diagram under P_{H2O} = 207.0 Mpa (Figure 12b), most samples are plotted near the NNO evolution line including the biotite in enclaves. Considering that the Fe^{3+}/Fe^{2+} value is more sensitive to oxygen fugacity but Figure 12b only considered the influence of the Fe# value, we tended to conclude that the enclaves might be formed under more reductive conditions based on Figure 12a.

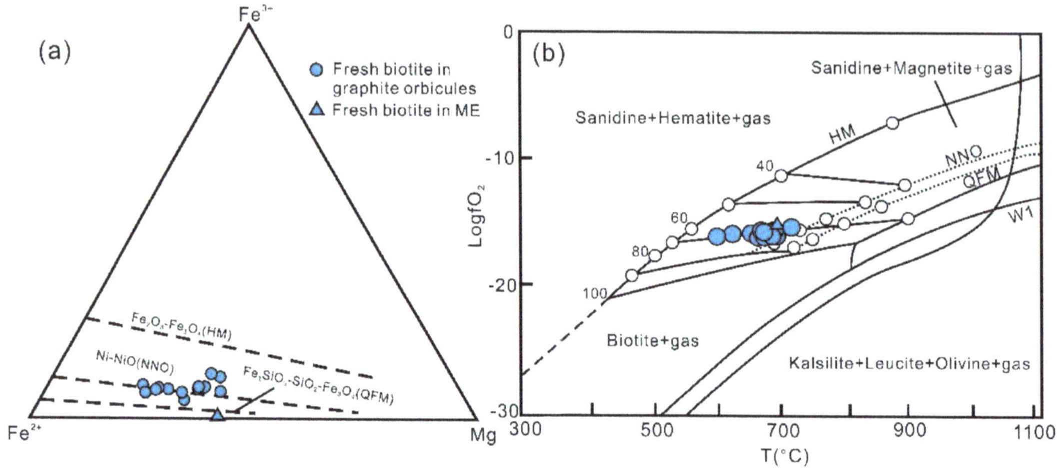

Figure 12. (**a**) Ternary Fe^{3+}-Fe^{2+}-Mg plot for the Huangyangshan biotite (after [73]). (**b**) Temperature (T) vs. Log(fO_2) for the Huangyangshan biotite (after [73]).

5.2. Magma Mixing and Evolution

The mixing of mafic and felsic magma is considered to be one of the most important mechanisms for forming enclaves characterized by the igneous textural features, finer grain size, and chilled margins in granites [84,85]. Up to now, some researchers have proved that the dark inclusions in the Huangyangshan pluton suggest magma mixing resulted from the injection of less differentiated mantle-derived intermediate–basic magma into the highly differentiated alkaline granitic magma [14,19,58]. The present study provides more evidence as follows: (1) The Type-II richterite phenocrysts at the contact between the enclaves and granites have Ca, Na, and K contents similar to the Type-III fine-grained richterite in enclaves (Figure 4c,e,f) but Mg and Fe contents and a crystalline form similar to the Type-I arfvedsonite phenocrysts in granites (Figure 4b,d), suggesting clear transitional characteristics. (2) Although the cores of graphite orbicules are characterized by neutral rocks with high contents of plagioclase and mafic minerals, the biotites in them have the feature of typical A-type granite (high Fe content). (3) The columnar and acicular shapes of Type-III amphibole indicate a rapid quenching possibly during fluctuations of temperature or water content of the magma [55,86]. (4) Amphiboles in dioritic-mafic rocks are commonly rich in Ca and Mg, rarely appearing Na-rich amphiboles. However, the type of amphiboles developed in dioritic enclaves is Na-rich richterite. (5) In the same magmatic system, the stronger the acidity of the rock, the higher the content of the REE. However, REE contents of Type-II, -III, and -V amphiboles in enclaves and graphite orbicules are significantly higher than Type-I amphiboles in granites (Figure 6a), indicating their different magma sources. (6) The pyroxene phenocrysts wrapped by amphibole and biotite indicates that they may be the residue of mafic magma crystallization in the early stage of magmatic evolution. (7) The reversely zoned structure developed in the pyroxenes indicate that the magma may have experienced partial mixing of basic mantle-derived magma during its evolution and then mixed with acidic crust-derived magma (Figure 9). (8) Plagioclases in orebody matrix have obvious oligoclase to andesine cores and albite rims, indicating the trend of magma transition from neutral to acidic. This partly reflects the contamination by a felsic magma.

The cores of graphite orbicules and the host granites have obvious different mineral compositions, indicating they formed in different conditions. The type of amphibole in the granites, enclaves, and graphite orbicules is sodic, calcic–sodic, and calcic amphibole, respectively, indicating that their forming environment or parent magma was obviously different. In Si-saturated alkaline magma, the evolution sequence of amphiboles is generally from katophorite to winchite and then to ferrorichterite from early to late. Alkaline types such as arfvedsonite and riebeckite are formed at the latest stage of crystallization [87]. With the evolution of magma, the quartz content and alkalinity of amphiboles will increase. The existence of arfvedsonite in granites indicates that the magma has reached to the latest stage. The Type-III ferrorichterite contains higher SiO_2 but lower $(Na + K)_A$ contents than Type-I arfvedsonite. The geochemical characteristics of minerals and whole-rock show that enclaves inherited the features of host A-type alkaline granitic magma [14,19], suggesting that the enclaves have experienced sufficient magma intermingling during their formation. Although showing the same dioritic rock characteristics dominated by plagioclase and amphiboles, the Type-III ferrorichterite in enclaves has more characteristics of crust source and alkaline magma compared with Type-V ferrohornblende in graphite orbicules, indicating the enclaves were formed in the environment with a higher magma mixing degree or in the magma with higher replenishment of acid magma. In addition, the graphite orbicules contain euhedral, coarse-grained amphibole, plagioclase, and clinopyroxene, whereas the enclaves consist of fine-grained, from subhedral to anhedral ferrorichterite, and plagioclase, indicating that the cores of the graphite orbicules formed in a deep environment of higher temperature and pressure. This is also consistent with the estimated amphibole temperature and pressure mentioned above. In summary, it can be concluded that the granites and the enclaves were formed in the latest magmatic evolution stage at the same time. The crystallization temperature and pressure of cores of graphite orbicules are higher and the characteristics of alkaline magma are weaker, indicating that they crystallized relatively earlier.

The rebalancing process below the solidus will not significantly change the geochemical characteristics of the amphiboles. Therefore, its composition, especially the REEs and incompatible elements, can be used to explore the properties and petrogenesis of the parent magma. Different types of amphibole in the Huangyangshan deposit all have extremely negative Eu anomalies, which is consistent with the whole-rock compositions (Figure 6a). It suggests that the magma experienced strong plagioclase crystallization and differentiation when amphibole crystallization, which is similar to the characteristics of acid rocks originated from mantle-derived magma [88]. Whereas the enclaves have relatively weak negative Eu anomalies compared with these amphiboles, indicating that these amphiboles crystallized later than plagioclase. The Ce is more incompatible than Pb [60], so the lg(CE/Pb) value of the amphibole can reflect the degree of crustal contamination. From Type-V to Type-I amphiboles, the lg(CE/Pb) value decreases (Figure 13), suggesting the proportion of crust-derived acidic magma increased gradually. Besides, Sc, V, and Cr are transition metals that are usually enriched in mafic, untra-mafic rocks, and depleted in acidic and alkaline rocks [89]. The Sc, V, and Cr contents of Type-IV and -V amphiboles are obviously higher than those of Type-I to -III amphiboles (Table S1), further showing the melt differentiation was weaker during the formation of cores of graphite orbicules. Therefore, the mixing of mantle-derived mafic magma has an important influence on the evolution and differentiation of magma.

One of the critical issues is that the dioritic cores surrounded by a large amount of graphite and the dioritic enclaves without graphite come from the same intermediate–basic parental magma or not. Generally, the REEs are lower in mafic and untra-mafic rocks and higher in acidic and alkaline rocks [60]. According to the above discussion, the parental magma of enclaves and granites had experienced sufficient differentiation and mixing. If the parent magma was from the same source, the amphiboles in the cores of graphite orbicules were supposed to have lower REEs. However, the REE contents of Type-IV and -V amphiboles are much higher than the others (Figure 6a). In addition, some early crystallized coarse-grained clinopyroxene and plagioclase phenocrysts are developed in the cores of graphite orbicules, whereas only fine-grained amphibole, plagioclase, and a small amount of biotite occurred in the enclaves. The mineral assemblage of amphibole and some biotite is a common cumulate result of andesitic melt [90,91]. The typical cumulate rocks formed by basaltic melt are gabbro and the main minerals are clinopyroxene and plagioclase [92]. Therefore, the parental magma of graphite orbicules cores may came from basaltic melt, while the enclaves were more likely derived from andesitic melt. Thirdly, some magmatic metal sulfides (pyrrhotite, chalcopyrite, and pentlandite assemblages) are developed along with graphite. The Type-V amphibole in the graphite orbicules has a higher oxygen fugacity (higher Fe^{3+} content) and more mantle-derived components than Type-III richterite in enclaves (Figure 11) and the biotite also has higher oxygen fugacity (Figure 12). Therefore, it can be concluded that the enclaves and the cores of graphite orbicules may not originated from the same parental magma. The basic parent magma that formed graphite orbicules might be characterized by rich metal elements, rich volatiles and high oxygen fugacity that were easier to crystallize sulfides. This may be one of the reasons for the crystallization of graphite.

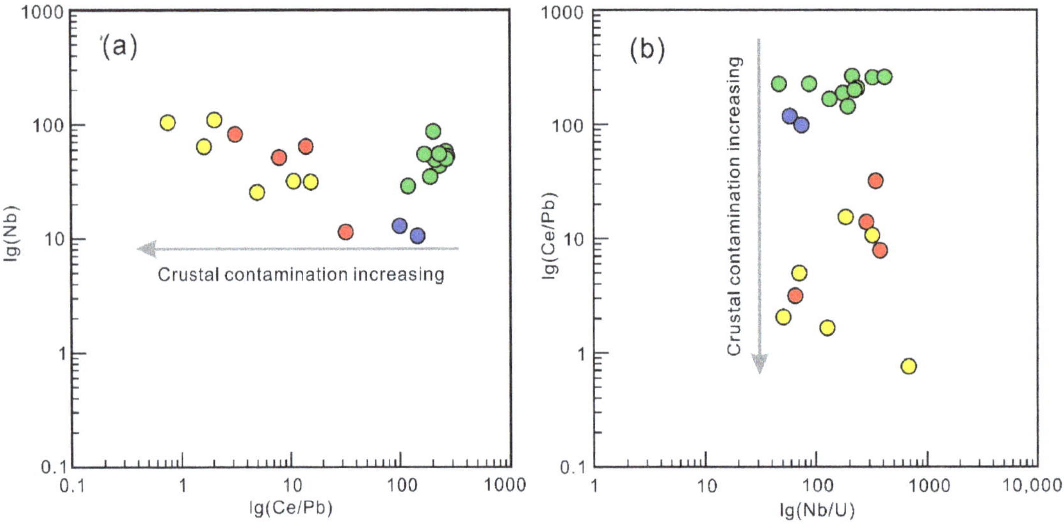

Figure 13. Plots of lg(Nb) vs. lg(Ce/Pb) (**a**) and lg(Ce/Pb) vs. lg(Nb/U) (**b**) of amphiboles from the Huangyangshan deposit. Symbols are as in Figure 4.

5.3. Implications for the Graphite Mineralization

During the evolution of magma, pyroxene, amphibole, and biotite in the Huangyangshan pluton crystallized successively with estimated crystallization temperatures of 840–1012 °C, 681–761 °C, and 658–720 °C, respectively. Their sources also evolved from the mantle source (pyroxene) to the crust–mantle mixed source (amphibole) and then to the crust source (biotite), suggesting the temperature changing trend and evolution process of magma. The early crystallized pyroxenes wrapped in the graphite orbicules were formed in an intermediate–basic magmatic environment. According to petrographic observation, the graphite is hardly direct contacts the pyroxenes, indicating that no graphite deposition occurred before the high-temperature mafic magma endmembers injected into the acid magma chamber. For melt or hydrothermal systems, graphite is usually deposited under relatively reducing conditions or precipitated and crystallized due to the reduction in the system [93,94]. The Huangyangshan pluton has significantly low oxygen fugacity and both amphibole and biotite were formed in a low oxygen fugacity environment. The reduction process of magma could be triggered by the mixing of mafic magma with the high oxygen fugacity and alkaline felsic magma.

At present, Raman graphite geothermometers have been widely used in order to determine the metamorphic grade and crystallization temperature of graphite [95–98]. Sun et al. [17] reported the crystallization temperature of the Huangyangshan graphite calculated by Raman geothermometers to be 750–800°C. It is higher than the temperature of amphibole and biotite and lower than that of pyroxene, which is also consistent with the petrographic observation. Sun et al. [17] also concluded the graphite crystallization temperature is similar to that of magmatic graphite deposit around the world. Therefore, it can be inferred that the Huangyangshan graphite might be crystallized at the end of magma mixing and before entering the low oxygen fugacity environment of the magma condition.

In a low oxygen fugacity environment, carbon mainly occurs in the form of methane [11]. There are two main processes for the formation of graphite from methane: oxidation ($CH_4 + O_2 \rightarrow C + 2H_2O$) and decomposition ($CH_4 \rightarrow C + 2H_2$). They both require the mixing of oxidizing melt/fluid to drastically change the physical and chemical conditions of the system [26]. However, no evidence of foreign melts/fluids mixing in the late magmatic stage has been found so far, so the mineralization ability of the late magma needs to be further confirmed. Nonetheless, the genetic relationship between the graphite mineralization and magma mixing can still be identified by the coexisted sulfides with graphite and compositional differences of amphibole and biotite in the granites and graphite ores. Combined with the crystallization temperature of graphite, therefore, graphite mineralization might be occurred from the late stage of magma mixing to the late stage of magmatism.

6. Conclusions

The amphibole in the Huangyangshan graphite deposit can be divided into five types. It generally has low Al and high Ti, K, Si, and Fe contents and has REE patterns with negative Eu anomalies and element distribution patterns similar to granites. The primary biotite in the granites, enclaves, and graphite orbicules belongs to Fe-biotite, which has the characteristics of high Si and Fe and low Al and Mg contents. The pyroxene phenocrysts in graphite orbicules develop significant zonal structures and are characterized by high Si and low Ca and Fe contents. The plagioclase phenocrysts are mainly oligoclase and andesine, with positive and oscillating zonal structures.

The crystallization temperature of pyroxene, Type-V amphibole, and primary biotite were 840–1012 °C, 681–761 °C, and 658–720 °C, respectively, suggesting their crystallization order. The magma pressure and depth were 157–220 Mpa and 5.95–8.32 km, respectively. Both the amphibole and biotite crystallized in a reducing environment with an extremely low oxygen fugacity.

The element compositions of amphibole, biotite, pyroxene, and plagioclase suggest that the Huangyangshan pluton has experienced significant mixing of mafic mantle-derived magma and felsic crust-derived magma. The graphite mineralization might be occurred from the late stage of magma mixing to the late stage of magmatism.

Supplementary Materials: The following supporting information can be downloaded at: https://www.mdpi.com/article/10.3390/min12111458/s1, Table S1: EPMA results of amphibole in the Huangyangshan pluton; Table S2: LA-ICP-MS results of amphibole; Table S3: EPMA results of biotite; Table S4: EPMA results of pyroxene; Table S5: EPMA results of plagioclase.

Author Contributions: Conceptualization, X.S.; Data curation, X.S., J.L. and M.H.; Formal analysis, X.S.; Funding acquisition, Y.R.; Investigation, X.S., Y.R., J.L., Z.S. and Z.L.; Methodology, X.S.; Project administration, Y.R.; Software, X.S. and M.H.; Supervision, Y.R.; Writing—original draft, X.S.; Writing—review and editing, Y.R. and J.L. All authors have read and agreed to the published version of the manuscript.

Funding: This research was funded by the Geological Survey project of the China Geological Survey (No. DD20190159).

Data Availability Statement: All data generated or used in the study are contained in the submitted article.

Acknowledgments: We sincerely appreciate Xiaolin Zhang and Chunjiang Wu from Xinjiang Branch of China National Geological Exploration Center of Building Materials Industry and Chengyang Wang from Institute of Disaster Prevention for their support of our fieldwork.

Conflicts of Interest: The authors declare no conflict of interest.

References

1. Wada, H.; Tomita, T.; Matsuura, K.; Iuchi, K.; Ito, M.; Morikiyo, T. Graphitization of carbonaceous matter during metamorphism with references to carbonate and politic rocks of contact and regional metamorphisms, Japan. *Contrib. Mineral. Petrol.* **1995**, *118*, 217–228. [CrossRef]
2. Manoel, T.N.; Leite, J.A.D. On the origin of the Neoproterozoic Peresopolis graphite deposit, Paraguay Belt, Brazil. *J. S. Am. Earth Sci.* **2018**, *84*, 104–112. [CrossRef]
3. Zhong, Y.; Ma, X.D.; Li, H.K.; Zhai, M.G. Revisit and comparative analysis of the typical graphite deposits in the Paleoproterozoic khondalite series, western North China Craton: Implications for genesis, depositional environment and prospecting potential. *Ore Geol. Rev.* **2019**, *109*, 370–380. [CrossRef]
4. Papineau, D.; De Gregorio, B.T.; Cody, G.D. Ancient graphite in the Eoarchean quartz-pyroxene rocks from Akilia in southern West Greenland I: Petrographic and spectroscopic characterization. *Geochim. Cosmochim. Acta* **2010**, *74*, 5862–5883. [CrossRef]
5. Wang, D.H.; Tang, J.X.; Ying, L.J.; Lin, B.; Ding, S. Hornfels feature in the Jiama ore deposit, Tibet and its significance on deep prospecting. *Acta Petrol. Sin.* **2011**, *27*, 2103–2108. (In Chinese)
6. Jiang, G.Z.; Li, Y.K.; Wang, A.J.; Yang, X.; Yang, B.; Ma, L. Genetic features of Dawudian graphite deposit in Urad Middle Banner, Inner Mongolia. *Earth Sci. Front.* **2017**, *24*, 306–316. (In Chinese)
7. Doroshkevich, A.G.; Wall, F.; Ripp, G.S. Magmatic graphite in dolomite carbonatite at Pogranichnoe, North Transbaikalia, Russia. *Contrib. Mineral. Petrol.* **2007**, *153*, 339–353. [CrossRef]
8. Huizenga, J.M. Thermodynamic modeling of a cooling C-O-H fluid-graphite system: Implications for hydrothermal graphite precipitation. *Miner. Deposita* **2011**, *46*, 23–33. [CrossRef]
9. Luque, F.J.; Ortega, L.; Barrenechea, J.F.; Huizenga, J.M.; Millward, D. Key factors controlling massive graphite deposition in volcanic settings: An example of a self-organized critical system. *J. Geol. Soc.* **2012**, *169*, 269–277. [CrossRef]
10. Li, C.; Wang, D.H.; Zhao, H.; Pei, H.X.; Li, X.W.; Zhou, L.M.; Du, A.D.; Qu, W.J. Minerogenetic regularity of graphite deposits in China. *Mineral. Depos.* **2015**, *34*, 1223–1236. (In Chinese)
11. Luque, F.J.; Pasteris, J.D.; Wopenka, B.; Rodas, M.; Barrenechea, J.F. Natural fluid-deposited graphite: Mineralogical characteristics and mechanisms of formation. *Am. J. Sci.* **1998**, *298*, 471–498. [CrossRef]
12. Lowenstern, J.B. Carbon dioxide in magmas and implications for hydrothermal systems. *Miner. Deposita* **2001**, *36*, 490–502. [CrossRef]
13. Luque, F.J.; Crespo, E.; Barrenechea, J.F. Carbon isotopes of graphite: Implications on fluid history. *Geosci. Front.* **2012**, *3*, 197–207. [CrossRef]
14. Yang, G.X.; Li, Y.J.; Wu, H.G.; Zhong, X.; Yang, B.K.; Yang, C.X.; Yan, J.; Si, G.H. Geochronological and geochemical constrains on petrogenesis of the Huangyangshan A-type granite from the East Junggar, Xinjiang, NW China. *J. Asian Earth Sci.* **2011**, *40*, 722–736. [CrossRef]
15. Bai, J.K.; Chen, J.L.; Peng, S.X. Geochronology and geochemistry of ore-bearing intrusions from Huangyangshan magmatic hydrothermal graphite deposit in Qitai County, Xinjiang. *Acta Petrol. Sin.* **2018**, *34*, 2327–2340. (In Chinese)
16. Ai, J.; Lu, X.B.; Li, Z.W.; Wu, Y.L. Genesis of the graphite orbicules in the Huangyangshan graphite deposit, Xinjiang, China: Evidence from geochemical, isotopic and fluid inclusion data. *Ore Geol. Rev.* **2020**, *122*, 103505. [CrossRef]
17. Sun, X.H.; Ren, Y.S.; Sun, Z.J.; Li, Z.W.; Wang, C.Y. Characteristics, ore sources and genesis of Huangyangshan graphite deposit in Qitai County, Xinjiang. *Acta Petrol. Sin.* **2021**, *37*, 1867–1882. (In Chinese)
18. Muo, R.J.; Liu, S.B.; Huang, C.R.; Zhang, G.R.; Tan, G.M.; Wang, B.X.; Xiao, X.Z. *Geology of Graphite Deposits in China*; China Architecture and Building Press: Beijing, China, 1989; p. 320. (In Chinese)
19. Sun, X.H.; Ren, Y.S.; Sun, Z.J.; Wang, C.Y.; Li, Z.W. Geochronology and geochemical properties of the large-scale graphite mineralization associated with the Huangyangshan alkaline pluton, Eastern Junggar, Xinjiang, NW China. *Geochemistry* **2021**, *81*, 125820. [CrossRef]
20. Tsuchiyama, A.; Takahashi, E. Melting kinetics of a plagioclase feldspar. *Contrib. Min. Petrol.* **1983**, *84*, 345–354. [CrossRef]
21. Hammarstrom, J.M.; Zen, E. Aluminum in hornblende: An empirical igneous geobarometer. *Am. Mineral.* **1986**, *71*, 1297–1313.
22. Schmidt, M.W. Amphibole composition in tonalite as a function of pressure: An experimental calibration of the Al-in-hornblende barometer. *Contrib. Mineral. Petrol.* **1992**, *110*, 304–310. [CrossRef]
23. Ridolfi, F.; Renzulli, A.; Puerini, M. Stability and chemical equilibrium of amphibole in calc-alkaline magmas: An overview, new thermobarometric formulations and application to subduction-related volcanoes. *Contrib. Mineral. Petrol.* **2010**, *160*, 45–66. [CrossRef]
24. Molina, J.F.; Moreno, J.A.; Castro, A.; Rodriguez, C.; Fershtater, G.B. Calcic amphibole thermobarometry in metamorphic and igneous rocks: New calibrations based on plagioclase/amphibole Al-Si partitioning and amphibole/liquid Mg partitioning. *Lithos* **2015**, *232*, 286–305. [CrossRef]
25. Li, J.X.; Qin, K.Z.; Li, G.M.; Evans, N.J.; Zhao, J.X.; Yue, Y.H.; Xie, J. Volatile variations in magmas related to porphyry Cu-Au deposits: Insights from amphibole geochemistry, Duolong district, central Tibet. *Ore Geol. Rev.* **2018**, *95*, 649–662. [CrossRef]
26. Zhang, L.; Chen, Z.Y.; Wang, F.Y.; Zhou, T.F. Whole-rock and biotite geochemistry of granites from the Miao'ershan batholith, South China: Implications for the sources of granite-hosted uranium ores. *Ore Geol. Rev.* **2021**, *129*, 103930. [CrossRef]
27. Krawczynski, M.J.; Grove, T.L.; Behrens, H. Amphibole stability in primitive arc magmas: Effects of temperature, H_2O content, and oxygen fugacity. *Contrib. Mineral. Petrol.* **2012**, *164*, 317–339. [CrossRef]

28. Li, W.K.; Yang, Z.M.; Cao, K.; Lu, Y.J.; Sun, M.Y. Redox-controlled generation of the giant porphyry Cu–Au deposit at Pulang, southwest China. *Contrib. Mineral. Petrol.* **2019**, *174*, 12. [CrossRef]
29. Li, J.Y.; Xiao, X.C.; Tang, Y.Q.; Zhao, M. Main characteristics of late Paleozoic plate tectonics in the southern part of east Junggar, Xinjiang. *Geol. Rev.* **1986**, *36*, 305–316. (In Chinese)
30. Li, J.Y.; Yang, T.N.; Li, Y.P.; Zhu, Z.X. Geological features of the Karamaili faulting belt, eastern Junggar region, Xinjiang, China and its constraints on the reconstruction of Late Paleozoic ocean-continental frame work of the Central Asian region. *Geol. Bull. China* **2009**, *28*, 1817–1826. (In Chinese)
31. Long, X.P.; Yuan, C.; Sun, M.; Safonova, I.; Xiao, W.J.; Wang, Y.J. Geochemistry and U–Pb detrital zircon dating of Paleozoic graywackes in East Junggar, NW China: Insights into subduction–accretion processes in the southern Central Asian Orogenic Belt. *Gondwana Res.* **2012**, *21*, 637–653. [CrossRef]
32. Han, Y.G.; Zhao, G.C. Final amalgamation of the Tianshan and Junggar orogenic collage in the southwestern Central Asian Orogenic Belt: Constraints on the closure of the Paleo-Asian Ocean. *Earth-Sci. Rev.* **2017**, *186*, 129–152. [CrossRef]
33. Xiao, W.J.; Han, C.M.; Yuan, C.; Sun, M.; Lin, S.; Chen, H.; Li, Z.; Li, J.; Sun, S. Middle Cambrian to Permian subduction-related accretionary orogenesis of Northern Xinjiang, NW China: Implications for the tectonic evolution of central Asia. *J. Asian Earth Sci.* **2008**, *32*, 102–117. [CrossRef]
34. Xiao, W.J.; Windley, B.F.; Yuan, C.; Sun, M.; Han, C.M.; Lin, S.F.; Chen, H.L.; Yan, Q.R.; Liu, D.Y.; Qin, K.Z.; et al. Paleozoic multiple subduction-accretion processes of the southern Altaids. *Am. J. Sci.* **2009**, *309*, 221–270. [CrossRef]
35. Yuan, C.; Sun, M.; Wilde, S.; Xiao, W.J.; Xu, Y.G.; Long, X.P.; Zhao, G.C. Postcollisional plutons in the Balikun area, East Chinese Tianshan: Evolving magmatism in response to extension and slab break-off. *Lithos* **2010**, *119*, 269–288. [CrossRef]
36. Zhang, Y.Y.; Pe-Piper, G.; Piper, D.J.W.; Guo, Z.J. Early Carboniferous collision of the Kalamaili orogenic belt, North Xinjiang, and its implications: Evidence from molasse deposits. *Geol. Soc. Am. Bull.* **2013**, *125*, 932–944. [CrossRef]
37. Xiao, W.J.; Windley, B.; Sun, S.; Li, J.L.; Huang, B.C.; Han, C.M.; Yuan, C.; Sun, M.; Chen, H.L. A tale of amalgamation of three Permo-Triassic collage systems in Central Asia: Oroclines, sutures, and terminal accretion. *Annu. Rev. Earth Planet. Sci.* **2015**, *43*, 477–507. [CrossRef]
38. Xu, Q.Q.; Zhao, L.; Niu, B.G.; Zheng, R.G.; Yang, Y.Q.; Liu, J.H. Early Paleozoic arc magmatism in the Kalamaili orogenic belt, Northern Xinjiang, NW China: Implications for the tectonic evolution of the East Junggar terrane. *J. Asian Earth Sci.* **2020**, *194*, 104072. [CrossRef]
39. Li, J.Y. Main characteristics and emplacement processes of the East Junggar Ophiolites, Xinjiang, China. *Acta Petrol. Sin.* **1995**, *11*, 71–84. (In Chinese)
40. Liu, X.J.; Xu, J.F.; Hou, Q.Y.; Bai, Z.H.; Lei, M. Geochemical characteristics of Karamaili ophiolite in east Junggar, Xinjiang: Products of ridge subduction. *Acta Petrol. Sin.* **2007**, *23*, 1591–1602. (In Chinese)
41. Huang, G.; Niu, G.Z.; Wang, X.L.; Guo, J.; Yu, F. Formation and emplacement age of Karamaili ophiolite: LA-ICP-MS zircon U-Pb age evidence from the diabase and tuff in eastern Junggar, Xinjiang. *Geol. Bull. China* **2012**, *31*, 1267–1278. (In Chinese)
42. Xu, X.W.; Jiang, N.; Li, X.H.; Wu, C.; Qu, X.; Zhou, G.; Dong, L.H. Spatial–temporal framework for the closure of the Junggar Ocean in central Asia: New SIMS zircon U–Pb ages of the ophiolitic mélange and collisional igneous rocks in the Zhifang area, East Junggar. *J. Asian Earth Sci.* **2015**, *111*, 470–491. [CrossRef]
43. Mao, Q.G.; Xiao, W.J.; Han, C.M.; Yuan, C.; Sun, M. Late Paleozoic southward accretionary polarity of the eastern Junggar orogenic belt: Insight from the Dajiashan and other A-type granites. *Acta Petrol. Sin.* **2008**, *24*, 733–742. (In Chinese)
44. Gan, L.; Tang, H.F.; Han, Y.J. Geochronology and geochemical characteristics of the Yemaquan granitic pluton in East Junggar, Xinjiang. *Acta Petrol. Sin.* **2010**, *26*, 2374–2388. (In Chinese)
45. Li, Y.P.; Li, J.Y.; Sun, G.H.; Zhu, Z.X.; Yang, Z.Q. Basement of Junggar basin: Evidence from detrial zircons in sandstone of previous Devonian Kalamaili formation. *Acta Petrol. Sin.* **2007**, *23*, 1577–1590. (In Chinese)
46. Guo, Z.J. A review on the Paleozoic tectonic evolution of northern Xinjiang and a discussion on the important role of geological maps in tectonic study. *Geol. Bull. China* **2012**, *31*, 1054–1060. (In Chinese)
47. Cai, X.F.; Fan, G.M.; Liao, Q.A.; Wang, F.M.; Wu, L.Y. Features of lower carboniferous intra-plate sedimentation in the Kalamai region, East Junggar, Xinjiang, China. *Bull. Miner. Petrol. Geochem.* **2015**, *34*, 1287–1302. (In Chinese)
48. Lin, J.F.; Yu, H.X.; Yu, X.Q.; Di, Y.J.; Tian, J.T. Zircon SHRIMP U–Pb dating and geological implication of the Sabei alkali-rich granite from Eastern Junggar of Xinjiang, NW China. *Acta Petrol. Sin.* **2007**, *23*, 1876–1884. (In Chinese)
49. Su, Y.P.; Tang, H.F.; Cong, F. Zircon U–Pb age and petrogenesis of the Huangyangshan alkaline granite body in East Junggar, Xinjiang. *Acta Mineral. Sin.* **2008**, *28*, 117–126. (In Chinese)
50. Han, B.F.; Ji, J.Q.; Song, B.; Chen, L.H.; Zhang, L. Late Paleozoic vertical growth of continental crust around the Junggar Basin, Xinjiang, China (Part I): Timing of postcollisional plutonism. *Acta Petrol. Sin.* **2006**, *22*, 1077–1086. (In Chinese)
51. Tang, H.F.; Qu, W.J.; Su, Y.P.; Hou, G.S.; Du, A.D.; Cong, F. Genetic connection of Sareshike tin deposit with the alkaline A-type granites of Sabei body in Xinjiang: Constraint from isotopic ages. *Acta Petrol. Sin.* **2007**, *23*, 1989–1997. (In Chinese)
52. Xu, B. Research on Tectonic-Magma Ore-Controlling Regularity and Prospecting Forecasting of Gold Metallogenic Belt in Kalamaili Area, Xinjiang. Ph.D. Thesis, China University of Geosciences (Beijing), Beijing, China, 2010. (In Chinese).
53. Liu, W.; Liu, X.J.; Liu, L.J. Underplating generated A- and I-type granitoids of the East Junggar from the lower and the upper oceanic crust with mixing of mafic magma: Insights from integrated zircon U–Pb ages, petrography, geochemistry and Nd–Sr–Hf isotopes. *Lithos* **2013**, *179*, 293–319. [CrossRef]

54. Safonova, I.Y.; Santosh, M. Accretionary complexes in the Asia–Pacific region: Tracing archives of ocean plate stratigraphy and tracking mantle plumes. *Gondwana Res.* **2014**, *25*, 126–158. [CrossRef]
55. Li, Z.W.; Zhang, X.L.; Zhang, L.; Chen, Z.G.; Chen, J.Y.; Wu, Y.L.; Wen, X.D. Geological characteristics and resource potential of Huangyangshan crystalline graphite deposit in Qitai County, Xinjiang. *China Non-Met. Miner. Indus.* **2019**, *S1*, 25–34. (In Chinese)
56. Liu, Y.S.; Hu, Z.C.; Gao, S.; Günther, D.; Xu, J.; Gao, C.G.; Chen, H.H. In situ analysis of major and trace elements of anhydrous minerals by LA–ICP–MS without applying an internal standard. *Chem. Geol.* **2008**, *257*, 34–43. [CrossRef]
57. Leake, B.E.; Woolley, A.R.; Arps, C.E.S.; Birch, W.D.; Gilbert, M.C.; Grice, J.D.; Hawthorne, F.C.; Kato, A.; Kisch, H.J.; Krivovichev, V.G.; et al. Nomenclature of amphiboles; report of the subcommittee on amphiboles of the international mineralogical association, commission on new minerals and mineral names. *Can. Mineral.* **1997**, *35*, 219–246.
58. Ye, C.Y.; Feng, Y.G.; Lei, R.X.; Yang, G.X. Compositional Variation of Amphiboles During Magma Mixing: A Case Study of Huangyangshan A-Type Granite in Kalamaili Metallogenic Belt, East Junggar, China. *Front. Earth Sci.* **2021**, *9*, 650014. [CrossRef]
59. Sun, S.S.; McDonough, W.F. Chemical and isotopic systematics of oceanic basalts: Implications for mantle composition and processes. *Geol. Soc. Lond. Spec. Publ.* **1989**, *42*, 313–345. [CrossRef]
60. Nachit, H.; Razafimahefa, N.; Stussi, J.M.; Carron, J.P. Composition chimique des biotites et typologie magmatique des granitoides. *C.R. Hebd. Séances Acad. Sci.* **1985**, *301*, 813–818.
61. Foster, M.D. Interpretation of the composition of trioctahedral micas. *U.S. Geol. Surv. Prof. Pap.* **1960**, *10*, 70–89.
62. Morimoto, N. Nomenclature of Pyroxenes. *Mineral. Petrol.* **1988**, *39*, 55–76. [CrossRef]
63. Leake, B.E. On Aluminous and Edenitic Hornblendes. *Mineral. Mag.* **1971**, *38*, 389. [CrossRef]
64. Ma, X.X.; Shi, B.; Xiong, F.H.; Li, H.B. Magma mixing of the Quxu batholith in the Gangdese belt, southern Tibet: Evidence from microstructure of hornblende in microgranular enclaves. *Acta Petrol. Sin.* **2020**, *36*, 3063–3080. (In Chinese)
65. Hu, R.G.; Liu, C.Q.; Feng, Z.H.; Zhao, Y.L.; Cai, Y.F.; Liu, Y.Z.; Xia, Z.P. The composition of amphibole in northern Guangxi and its implications for the magma evolution of the Xinzhai amphibole granite. *Acta Petrol. Mineral.* **2020**, *39*, 703–717. (In Chinese)
66. Hollister, L.S.; Grissom, G.C.; Peters, E.K.; Stowell, H.H.; Sisson, V.B. Confirmation of the empirical correlation of Al in hornblende with pressure of solidification of calc-alkaline plutons. *Am. Mineral.* **1987**, *72*, 231–239.
67. Mutch, E.J.F.; Blundy, J.D.; Tattitch, B.C.; Cooper, F.J.; Brooker, R.A. An experimental study of amphibole stability in low-pressure granitic magmas and a revised Al-in-hornblende geobarometer. *Contrib. Mineral. Petrol.* **2016**, *171*, 85. [CrossRef]
68. Uchida, E.; Endo, S.; Makino, M. Relationship between solidication depth of granitic rocks and formation of hydrothermal ore deposits. *Resour. Geol.* **2007**, *57*, 47–56. [CrossRef]
69. Kang, Z.Q.; Feng, Z.H.; Wang, R. Reliability comparison of Al-in-hornblende and biotite barometer: A case study of Guposhan-Huashan granite in North Guangxi. *J. Guilin Univ. Technol.* **1994**, *30*, 474–479. (In Chinese)
70. Holland, T.J.B.; Blundy, J. Non-ideal interactions in calcic amphiboles and their bearing on amphibole-plagioclase thermometry. *Contrib. Mineral. Petrol.* **1994**, *116*, 433–447. [CrossRef]
71. Stein, E.; Dietl, C. Hornblende thermobarometry of granitoids from the Central Odenwald (Germany) and their implications for the geotectonic development of the Odenwald. *Mineral. Petrol.* **2001**, *72*, 185–207. [CrossRef]
72. Putirka, K. Amphibole thermometers and barometers for igneous systems and some implications for eruption mechanisms of felsic magmas at arc volcanoes. *Am. Mineral.* **2016**, *101*, 841–858. [CrossRef]
73. Wones, D.R.; Eugster, H.P. Stability of biotite–experiment, theory, and application. *Am. Mineral.* **1965**, *50*, 1228–1272.
74. Henry, D.J.; Guidotti, C.V.; Thomson, J.A. The Ti-saturation surface for low to medium pressure metapelitic biotite: Implications for geothermometry and Ti-substitution mechanisms. *Am. Mineral.* **2005**, *90*, 316–328. [CrossRef]
75. Sarjoughian, F.; Kananian, A.; Ahmadian, J.; Murata, M. Chemical composition of biotite from the Kuh-e Dom pluton, Central Iran: Implication for granitoid magmatism and related Cu-Au mineralization. *Arabian J. Geosci.* **2015**, *8*, 1521–1533. [CrossRef]
76. Wood, B.J.; Banno, S. Garnet–orthopyroxene and orthopyroxene–clinopyroxene relationships in simple and complex systems. *Contrib. Mineral. Petrol.* **1973**, *42*, 109–124. [CrossRef]
77. Wells, P.R.A. Pyroxene thermometry in simple and complex systems. *Contrib. Mineral. Petrol.* **1977**, *62*, 129–139. [CrossRef]
78. Wones, D.R. Mafic minerals as indicators of intensive variables in granitic magmas. *Min. Geol.* **1981**, *31*, 9–22.
79. Anderson, J.L.; Smith, D.R. The effects of temperature and fO_2 on the Al-in hornblende barometer. *Am. Mineral.* **1995**, *80*, 549–559. [CrossRef]
80. Papoutsa, A.; Pe-piper, G. Geochemical variation of amphiboles in A-type granites as an indicator of complex magmatic systems: Wentworth pluton, Nova Scotia, Canada. *Chem. Geol.* **2014**, *384*, 120–134. [CrossRef]
81. Clowe, C.A.; Popp, R.K.; Fritz, S.J. Experimental investigation of oxygen fugacity on ferric–ferrous ratios and unit-cell parameters of four natural clinoamphiboles. *Am. Mineral.* **1988**, *73*, 487–499.
82. Sun, Z.J.; Liu, G.H.; Ren, Y.S.; Chen, X.; Sun, X.H.; Wang, C.Y.; Li, Z.W. Age, genesis and tectonic setting of the Sayashk tin deposit in the East Junggar region: Constraints from Lu–Hf isotopes, zircon U–Pb and molybdenite Re–Os dating. *Minerals* **2022**, *12*, 1063. [CrossRef]
83. Best, M.G. *Igneous and Metamorphic Petrology*; W.H. Freeman and Company: San Francisco, CA, USA, 1982; p. 630.
84. Shaw, S.E.; Flood, R.H. Zircon Hf isotopic evidence for mixing of crustal and silicic mantle-derived magmas in a zoned granite pluton, eastern Australia. *J. Petrol.* **2009**, *50*, 147–168. [CrossRef]

85. Zhang, J.Y.; Peng, T.P.; Fan, W.M.; Zhao, G.C.; Dong, X.H.; Gao, J.F.; Peng, B.X.; Wei, C.; Xia, X.P.; Chen, L.L.; et al. Petrogenesis of the Early Cretaceous granitoids and its mafic enclaves in the Northern Tengchong Terrane, southernmargin of the Tibetan Plateau and its tectonic implications. *Lithos* **2018**, *318–319*, 283–298. [CrossRef]
86. Vernon, R.H. Crystallization and hybridism in microgranitoid enclave magmas: Microstructural evidence. *J. Geophys. Res. Solid Earth* **1990**, *95*, 17849–17859. [CrossRef]
87. Giret, A.; Bonin, B.; Leger, J.M. Amphibole compositional trends in the oversaturated and undersaturated alkaline plutonic ring-complexes. *Can. Minral.* **1980**, *18*, 481–495.
88. Rollinson, H.R. *Using Geochemical Data: Evaluation, Presentation, Interpretation*; Longman Scientific and Technical Press: Harlow, UK, 1993; p. 352.
89. Lai, S.C.; Yi, H.S.; Liu, C.Y.; O'Reilly, S.Y. Trace element geochemistry and classification of amphiboles of the Cenozoic high-potassium calc-alkaline volcanic rock series from north Qiangtang, Qinghai-Tibetan plateau. *Acta Petro. Sin.* **2002**, *18*, 17–24. (In Chinese)
90. Chen, S.; Niu, Y.L.; Sun, W.L.; Zhang, Y.; Li, J.Y.; Guo, P.Y.; Sun, P. On the origin of mafic magmatic enclaves (MMEs) in syn-collisional granitoids: Evidence from the Baojishan pluton in the North Qilian Orogen, China. *Mineral. Petrol.* **2015**, *109*, 577–596. [CrossRef]
91. Shao, F.L.; Niu, Y.L.; Liu, Y.; Chen, S.; Kong, J.J.; Duan, M. Petrogenesis of Triassic granitoids in the East Kunlun Orogenic Belt, northern Tibetain Plateau and their tectonic implications. *Lithos* **2017**, *282–283*, 33–44. [CrossRef]
92. Niu, Y.L.; Zhao, Z.D.; Zhu, D.C.; Mo, X.X. Continental collision zones are primary sites for net continental crust growth–a testable hypothesis. *Earth Sci. Rev.* **2013**, *127*, 96–110. [CrossRef]
93. Kribek, B.; Sykorova, I.; Machovic, V.; Laufek, F. Graphitization of organic matter and fluid-deposited graphite in Palaeoproterozoic (Birimian) black shales of the Kaya-Goren greenstone belt (Burkina Faso, West Africa). *J. Metamorph. Geol.* **2008**, *26*, 937–958. [CrossRef]
94. Luque, F.J.; Ortega, L.; Barrenechea, J.F.; Millward, D.; Beyssac, O.; Huizenga, J.M. Deposition of highly crystalline graphite from moderate-temperature fluids. *Geology* **2009**, *3*, 275–278. [CrossRef]
95. Beyssac, O.; Goffé, B.; Chopin, C.; Rouzaud, J.N. Raman spectra of carbonaceous material in metasediments: A new geothermometer. *J. Metamorph. Geol.* **2002**, *20*, 859–871. [CrossRef]
96. Aoya, M.; Kouketsu, Y.; Endo, S.; Shimizu, H.; Mizukami, T.; Nakamura, D.; Wallis, S. Extending the applicability of the Raman carbonaceous-material geothermometer using data from contact metamorphic rocks. *J. Metamorph. Geol.* **2010**, *28*, 895–914. [CrossRef]
97. Ammar, M.R.; Rouzaud, J.N. How to obtain a reliable structural characterization of polished graphitized carbons by Raman microspectroscopy. *J. Raman Spectrosc.* **2012**, *43*, 207–211. [CrossRef]
98. Al-Ani, T.; Ahtola, T.; Cutts, K.; Torppa, T. Metamorphic evolution of graphite in the Paleoproterozoic Savo Schist Belt (SSB), Central Finland: Constraints from geothermetric modeling. *Ore Geol. Rev.* **2022**, *141*, 104672. [CrossRef]

minerals

MDPI

Article

Genesis of Metal Sulfides and Its Significance on Graphite Mineralization in the Huangyangshan Graphite Deposit, East Junggar, Xinjiang Province, China

Yunsheng Ren [1,2], Jingmou Li [2,*], Xinhao Sun [3], Zuowu Li [4] and Zhenjun Sun [1]

1 School of Earth Science, Institute of Disaster Prevention, Langfang 065201, China
2 School of Earth Sciences and Resources, China University of Geosciences, Beijing 100083, China
3 Chengdu Center of China Geological Survey, Chengdu 610081, China
4 Xinjiang Branch of China National Geological Exploration Center of Building Materials Industry, Urumqi 830092, China
* Correspondence: lijmmy2019@163.com

Abstract: The Huangyangshan super-large graphite deposit is located in the Qitai area of East Junggar in Xinjiang Province, China. This deposit is well known for its distinguishing properties, including the alkaline granite complex that hosts the graphite ore, the dominantly orbicular structure developed in the graphite ore, and the association of graphite with metal sulfides in the orbicular ore. This study aims to determine the genetic relationship between graphite and metal sulfides in order to better understand the graphite mineralization process of the Huangyangshan deposit. The methods applied in the study include X-ray micro-CT scanning and scanning electron microscopy (SEM) analyses of the orbicular graphite ore and in situ inductive laser ablation-coupled plasma mass spectrometry (LA-ICP-MS) trace element analyses of the pyrrhotite and chalcopyrite associated with the graphite. The analytical results show that the graphite ore is composed of crystalline graphite, K-feldspar, albite, quartz, biotite, amphibole, and metal sulfides. The metal sulfides in the orbicular ore include pyrite, pyrrhotite, pentlandite, and chalcopyrite. According to the color, crystalline shape, texture, and occurrence, pyrrhotite can be classified into four types (I, II, III, and IV), and chalcopyrite into two types (I and II), of which types I, II, and III pyrrhotite and type I chalcopyrite have a close genetic relationship with graphite. The granular types (I, II, and III) of pyrrhotite are enriched in Co, Ni, Se, Ge, and Te and are depleted in As, Sb, Ag, and Au; they also have a high value of Co/Ni, indicating that these types of pyrrhotite have a magmatic origin. Low values of Co/Ni suggest that type IV pyrrhotite has a hydrothermal origin. The similar contents of Co and Ni and the values of Co/Ni compared with the chalcopyrite from the magmatic Co–Ni sulfide deposits imply that type I chalcopyrite has a magmatic origin. In summary, the metal sulfides of the Huangyangshan deposit are genetically related to graphite mineralization and formed predominantly by magmatic processes.

Keywords: in situ LA–ICP–MS trace element analysis; CT and SEM scanning; magmatic origin; orbicular graphite ore; Huangyangshan graphite deposit; East Junggar

1. Introduction

The Huangyangshan graphite deposit was discovered in the early 21st century in the Karamaili area of East Junggar, Xinjiang Province. This deposit is the first nonmetamorphic super-large crystallized graphite deposit to be discovered in China and contains probable graphite resources of >70 Mt. As a result, the deposit has attracted extensive research and prospecting attention from geologists [1,2]. The Huangyangshan deposit, together with the adjacent Sujiquan graphite deposit discovered in the 1960s, has some distinguishing characteristics. For instance, it occurs within the Huangyangshan alkaline granite complex, is characterized by an orbicular ore structure, and contains metal sulfides associated with graphite in the ore. In recent years, more and more geochronological and petrogeochemical data for the ore-hosting Huangyangshan granite complex have shown that it is an A-type granite that was emplaced during the late Carboniferous period [2–4]. In order to determine the genesis of the Huangyangshan deposit and the original source of the carbon, studies of the graphite using X-ray powder diffraction (XRD), laser Raman spectroscopy, and carbon isotope analysis have led to the common view that the graphite mineralization in this area was sourced from the organic carbon in sedimentary strata at the periphery of the Huangyangshan pluton [5,6]. In contrast to the consensus regarding the carbon source, the genetic type and graphite mineralization mechanism of the Huangyangshan and adjacent Sujiquan graphite deposits remains uncertain. Two main hypotheses have been proposed: (1) graphite in the Huangyangshan and Sujiquan deposits crystallized during the magmatic period and are, therefore, of a magmatic origin [1,7–9]; (2) the graphite is of postmagmatic hydrothermal origin [2,5,10,11].

Available petrographic and ore microscopy data have shown that metal sulfides in the Huangyangshan graphite ore have a close spatial relationship with graphite, implying that the investigation of the metal sulfides should help to determine the mechanism of graphite mineralization. Aside from the S and Pb isotope data for the pyrrhotite samples, which indicate that the ore-forming material was derived from the mantle, as reported by Ai et al. [12] and Sun et al. [6], the nature and origin of the metal sulfides, including their occurrence and temporal and intergrowth relationships with the graphite and mineralization mechanism remain unclear. In this study, micro-CT scanning and scanning electron microscope (SEM) analyses were conducted on the graphite ore, and in situ trace element analysis using inductive laser-ablation coupled plasma mass spectrometry (LA–ICP–MS) was performed on the chalcopyrite and pyrrhotite. Integrating the results of these analyses, together with information on the ore-forming geological conditions and deposit geology, allow the genetic type and mineralization mechanism of the metal sulfides, as well as their genetic relationship with graphite mineralization in the Huangyangshan deposit, to be established.

2. Geological Background and Deposit Geology

2.1. Regional Geology

The East Junggar Orogenic Belt (EJOB) is located in the northeastern Junggar basin and comprises the Dulate arc, Armantai ophiolite belt, Yemaquan arc, Karamaili ophiolite belt, and the Halic-Dananhu arc from north to south [13–16]. The available data have suggested that the formation and evolution of the EJOB are related to the closure of the Junggar Ocean between the Junggar and Altai plates during the Paleozoic [17–19]. Geochronological data indicate that the Junggar Ocean formed during the early Ordovician and closed during the late Paleozoic, which caused the EJOB to evolve into a postcollision orogenic regime [17–21]. The Karamaili area, located in the southern EJOB, is a well-known tectonic–magmatic–metallogenic belt controlled by the deep NWW-trending Karamaili fault (Figure 1). Besides the scattered Silurian and Permian volcanic rocks and small amounts Mesozoic and Cenozoic sedimentary rocks, voluminous Devonian and Carboniferous volcanic sedimentary strata dominate the geology of the Karamaili area. The sedimentary rocks in this area were derived from active continental margin sediments that were deposited in a fore-arc basin setting, with the deposition also being influenced by various regional faults, including the Beitashan, Kubusu, Qingshui-Sujiquan, and Karamaili faults [22–24]. Alkaline granite complexes that were identified in the Karamaili area included the Laoyaquan, Beilekuduke, Yemaquan, Huangyangshan, and Yebushan complexes and were formed by postcollision alkaline magmatic activity during the Carboniferous [3,16,25,26]. Some of the gold, copper, tin, and graphite deposits in this area are considered to be products of these Carboniferous alkaline magmatic events [27–30].

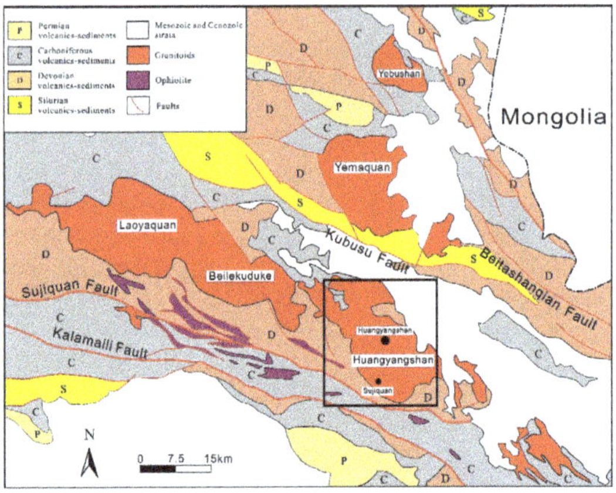

Figure 1. Simplified geological map of the Kalamaili area, Eastern Junggar (modified after [4]).

2.2. *Huangyangshan Complex*

The Huangyangshan graphite deposit is located in the northern segment of the Karamaili fault and within the Huangyangshan alkaline granite complex. The graphite-hosting Huangyangshan pluton is part of the Karamaili alkaline graphite belt, which extends NW–SE. In accordance with and owing to the multiple injections of magma, the Huangyangshan complex is characterized by a series of concentric rings (Figure 2a). According to mineral grain sizes and dark-colored mineral types, the Huangyangshan granite complex can be classified into five petrographic facies (five periods), from the north (early) to south (late), medium-grained arfvedsonite granite, medium–fine-grained arfvedsonite granite, medium–fine-grained hornblende granite, medium-grained biotite granite, and fine-grained biotite granite. All of the discovered graphite ore bodies have a close temporal–spatial relationship with the medium–fine-grained hornblende granite, medium-grained biotite granite, and fine-grained biotite granite (Figure 2a).

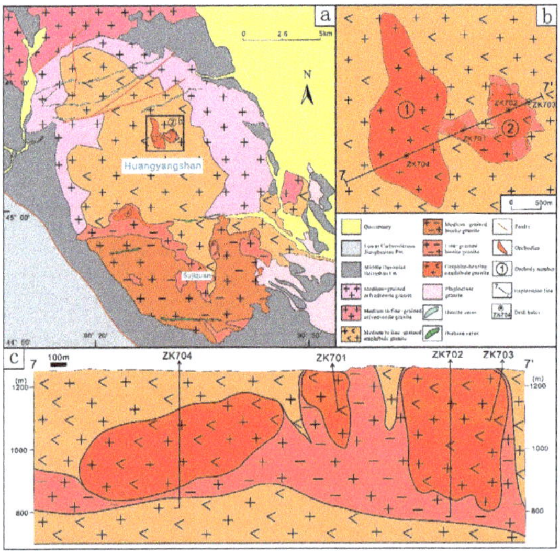

Figure 2. (**a**) Geological map of the Huangyangshan pluton; (**b**) simplified map of the No. 1 and 2 ore bodies of the Huangyangshan graphite deposit (after [2]); (**c**) cross-section along line 7–7′ in (**b**) (modified after [6]).

2.3. Characteristics of Graphite Mineralization

According to the graphite mineral reserves, three of the eight graphite ore bodies (i.e., No. 1, 2, and 4) are large-scale. Ore bodies No. 1 and 2 occur at the contact between the medium–fine-grained amphibole granite and fine-grained biotite granite, and the No. 4 ore body occurs within the fine-grained biotite granite (Figure 2a,b). The concealed No. 1 ore body, which is hosted by the graphite-bearing medium–fine-grained amphibole granite, is of lentiform appearance, extends N–S, and occurs as a small, gently dipping rock bed (Figure 2b,c). The No. 2 opencast ore body extends E–W and is almost entirely mineralized. This ore body displays a saddle shape at the surface (Figure 2b) and is steep-standing in the deep, showing that the No. 2 ore body is nearly upright and has a semiclosed cylindrical form (Figure 2c). The semiconcealed No. 4 ore body is located in the southwestern part of the Huangyangshan complex, where it extends nearly E–W and shows a spindle-shaped form, being thick in the north and thin in the south (Figure 2a).

The Huangyangshan deposit contains primary graphite ores, some of which are weathered. In the weathered ores, the graphite retains its original shape and is associated with malachite (Figure 3a). The primary graphite ores collected from the drill holes and prospecting trenches show an orbicular structure characterized by an inner core and an outer circle. The inner core is composed of felsic minerals and dark minerals, whereas the outer circle is composed of graphite and graphite-associated metal sulfides (Figure 3b).

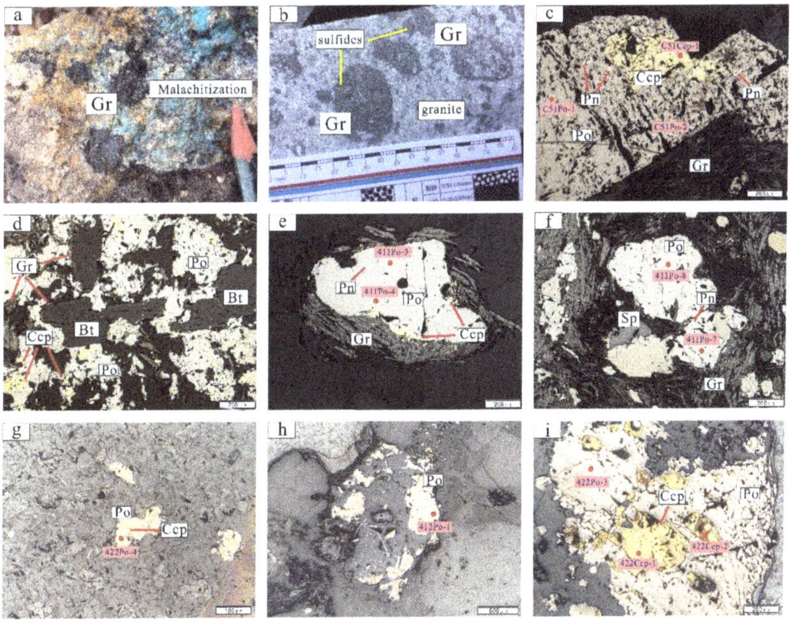

Figure 3. Characteristics and intergrowth of the graphite and metal sulfides in the orbicular graphite ore; (**a**) spherical graphite associated with malachitization on the surface; (**b**) metal sulfides surrounded by spherical and irregular graphite in drill holes; (**c**) nodular pyrrhotite containing chalcopyrite and pentlandite in the graphite (reflected light); (**d**) coarse-grained pyrrhotite coexisted with euhedral biotite (reflected light); (**e**) flake graphite growing along the edge of granular pyrrhotite (reflected light); (**f**) xenomorphic granular light yellow and brown pyrrhotite (reflected light); (**g**) fine-grained xenomorphic pyrrhotite in the graphite orbicular ore (reflected light); (**h**) metasomatic structure of pyrrhotite and sphalerite (reflected light); (**i**) copper yellow and brown chalcopyrite in graphite orbicular ore (reflected light). Gr—graphite, Po—pyrrhotite, Ccp—chalcopyrite, Pn—pentlandite, Bt—biotite Sp—sphalerite.

3. Samples and Analytical Methods

3.1. Sample Descriptions

The orbicular ore samples analyzed in this study were collected from the No. 704 drill holes of the No. 1 ore body (Figure 2c). These samples are composed mainly of graphite, metal sulfides, and felsic minerals, such as quartz and plagioclase. Samples for X-ray micro-CT scanning and SEM analysis were taken from a small drill core measuring ~16 mm in diameter, drilled from an 8 cm × 20 cm section of the drill core from the No.704 drill hole. Sulfide-rich ores were selected and produced 7 polished probe thin sections for the in situ LA–ICP–MS trace element analysis of the sulfides.

3.2. Analytical Methods

X-ray micro-CT scanning analysis, SEM rock mineral quantitative analysis (quantitative evaluation of minerals by SEM, referred to as "Qemscan"), and micro-scanning image-stitching analysis (Modular Automated Processing System, referred to as "MAPS") were conducted at the Beijing Institute of Digital China, Ltd., using a NanoVoxel-3502E X-ray three-dimensional microscope (Sanying precision instrument Co. Ltd., Tianjin, China), and a Quanta FEG 450 SEM and QEMSCAN 650F mineral quantitative analyzer (FEI, Hillsboro, OR, USA). The operating conditions for micro-CT scanning included an electric current of 20 mA, a voltage of 150 kV, and a scan duration of 90–120 min. After scanning, the obtained slice image data were imported into the 3D space using 3D reconstruction software. Each coordinate point was filled with data recorded from the X-ray CT scan, with reference to a scan resolution of 7.1999 μm. The CT images were denoised by median filtering method, and the 3D digital drill core images (Figure 4) were obtained by image enhancement and filtering. The scanned drill core portions were cut from the middle part and polished into thin sections (~5 mm thick). The thin-sections were mechanically polished, argon-ion polished, and surface carbon sprayed after being sealed with epoxy resin for QEMSCAN analysis. The principle of QEMSCAN analysis is to judge the element type of the scanning point material through the unique X-ray energy generated on the sample surface and calculate the mineral distribution information according to the mineral and element composition database so as to obtain the mineral distribution characteristics on the sample surface. Analyses were performed at an accelerating voltage of 20 kV, an emission current of 10 μA, and a working distance of 15 mm. The resolution for quantitative mineral analysis varied from 1 to 50 μm. In images obtained from the QEMSCAN analysis (Figure 5a), the elemental species and composition are determined mainly by X-ray energy, and different mineral types are represented by different colors. The area fraction can be obtained from the statistics of the number of pixel points of each mineral component in the image. Then, combined with the density of the different minerals, their mass fraction can be calculated. MAPS analysis was performed with a resolution of 250 nm and a field of view of 16 mm × 16 mm. The MAPS images were spliced from 1300 high-precision back-scattering electron microscope images.

In-situ LA–ICP–MS trace element analyses were conducted at the State Key Laboratory of Ore Deposit Geochemistry, Institute of Geochemistry, Chinese Academy of Sciences (IGCAS), Guizhou Province, China. Laser sampling was performed using an ASI RESOLution-LR-S155 laser microprobe equipped with a Coherent Compex-Pro 193 nm ArF excimer laser. An Agilent 7700x ICP–MS instrument was used to acquire ion-signal intensities. Helium was used as a carrier gas at a flux of 350 mL/min. Before exiting the cell, the ablated aerosol was mixed with Ar (900 mL/min) as a transport gas. Each analysis involved a background acquisition of ~30 s (gas blank) followed by 60 s of data acquisition from the sample. Ablation was performed using a uniform spot diameter of 26 μm pit size, 5 Hz laser pulse frequency, and 3 J/cm^2 energy fluence. The contents of S and Fe were calibrated using pure pyrite as an internal standard. The USGS reference glasses GSE-1G and GSD-1G were used to calibrate the elemental contents. STDGL3 was used to determine the elemental compositions of chalcophile and siderophile. Sulfide standard MASS-1 was analyzed as an unknown sample to check the accuracy of analyses.

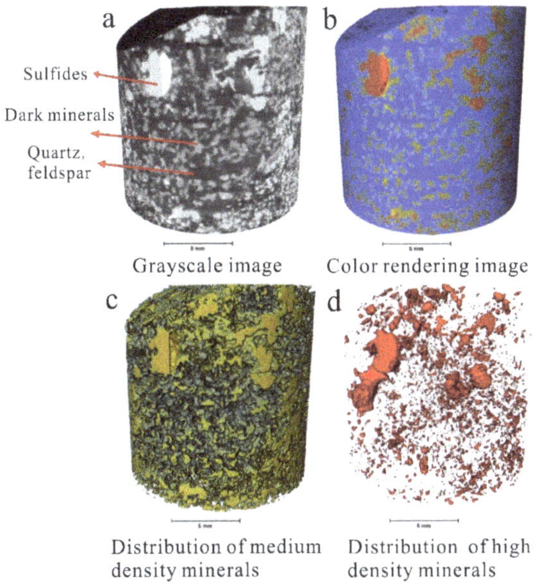

Figure 4. 3D sketch images of the drill core from the Huangyangshan graphite deposit; (**a**) grayscale image; (**b**) color rendering image; (**c**) distribution of medium density minerals; (**d**) distribution of high density minerals.

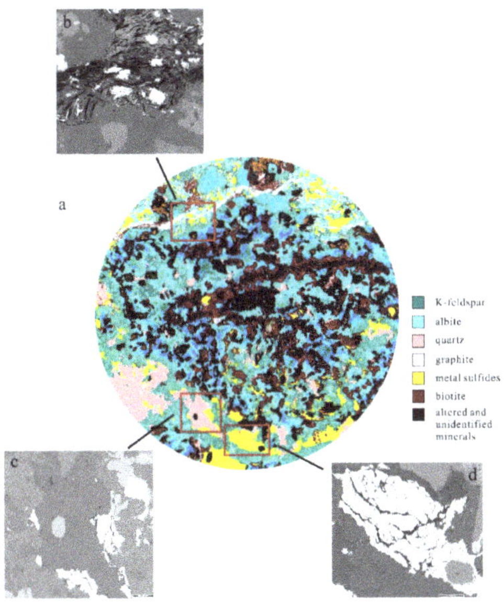

Figure 5. (**a**) Scan results of QEMSCAN and MAPS test; (**b**) the flake graphite was mainly distributed as bundled and banded aggregates, cutting through and metasomatizing sulfides; (**c**) the quartz, mainly occurring as xenomorphic granular metasomatized sulfides and contained a small amount of fine-grained disseminated graphite; (**d**) the sulfide aggregates cut and metasomatized by quartz contained fine-grained graphite bands.

4. Analytical Results

4.1. Metal Sulfides in Orbicular Ores

The metal sulfides associated with graphite in the orbicular ore (Figure 3b) include subhedral–xenomorphic pyrrhotite and xenomorphic chalcopyrite and minor pyrite, pentlandite, and sphalerite. These sulfides are largely enclosed or cut by the graphite, meaning that they were crystallized before the graphite. Of these metal sulfides, pyrrhotite has the highest content and the most extensive distribution in the Huangyangshan deposit, hosting mostly chalcopyrite and pentlandite and occurring in nodular aggregates composed of subhedral coarse-grained crystals (Figure 3c). The phenomenon whereby both the pyrrhotite and flake graphite are intruded by the biotite indicates they both formed earlier than the biotite (Figure 3d). In general, the pyrrhotite content has a positive relationship with the graphite content, and the flake graphite is commonly distributed along the edges of granular pyrrhotite (Figure 3e). The chalcopyrite and pentlandite are enclosed within the pyrrhotite. The contact boundaries of chalcopyrite and pentlandite with pyrrhotite are straight, suggesting that they are associated with pyrrhotite. (Figure 3c). The particle size of the pentlandite is small, mostly <0.1 mm (Figure 3e). The sulfides and graphite cut through the biotite that crystallized during the late magmatic stage (Figure 3d). Type IV pyrrhotite and type II chalcopyrite occur in silicate mineral cores lacking graphite, or replaced types I–III pyrrhotite and type I chalcopyrite (Figure 3h,i)

According to its color, crystalline shape, texture, and occurrence, the pyrrhotite associated with graphite in the Huangyangshan deposit can be grouped into four types. Type I pyrrhotite comprises coarse xenomorphic (Figure 3c) or medium–fine-grained xenomorphic granular grains that are included in the graphite (Figure 3e,f) and are cream-yellow in color, which encloses the associated chalcopyrite and pentlandite. Type II pyrrhotite consists of xenomorphic granular and light-yellow and brown grains (Figure 3f), which accompanies the type I pyrrhotite and encloses no other metal sulfides. Type III pyrrhotite is composed of fine-grained xenomorphic grains occurring in the core of the orbicular graphite (Figure 3g), which enclose small amounts of the chalcopyrite grains and is associated with fine-grained disseminated graphite. Type IV pyrrhotite comprises grains that are distributed within the granite intrusion and metasomatized along the edges of felsic mineral grains (Figure 3h).

The chalcopyrite in the orbicular graphite ore of the Huangyangshan deposit can be classified into two types: Type I chalcopyrite is composed of xenomorphic–granular grains and is enclosed within the pyrrhotite with a metasomatic texture (Figure 3c,i). Type II chalcopyrite consists of brown grains and is enclosed within the type I chalcopyrite (Figure 3i).

Type IV pyrrhotite and type II chalcopyrite occur in silicate mineral cores lacking graphite or replace types I–III pyrrhotite and type I chalcopyrite (Figure 3h,i).

4.2. X-ray Micro-CT Scanning and SEM Analysis

The 3D digital drill core image obtained by image enhancement and filtering is shown in Figure 4a. According to its density variation, the white high-density sulfides, gray medium-density dark minerals (and graphite), and black low-density quartz and feldspar can be clearly distinguished. Figure 4b was obtained after color rendering, in which the blue, yellow, and red represent low-, medium-, (Figure 4c), and high-density (Figure 4d) material areas, respectively.

The calculated percentages of various mineral contents for the studied section are given in Table 1. The MAPS images of the three representative horizons (graphite vein, quartz, and sulfide) of the section are shown in Figure 5b–d. The plagioclase content in the ore is highest (31.03%), followed by K-feldspar (13.39%) and dark minerals (14.26%), with a quartz content of only 12.02%, which is largely lower than that of the ore-bearing granite (>50% [3])

Table 1. Contents of the main minerals in the orbicular graphite ore analyzed by QEMSCAN testing.

Minerals	Area Fraction (%)	Mass Fraction (%)
Albite	19.53	21.62
K-feldspar	12.29	13.39
Graphite	11.82	8.55
Dark minerals (mainly biotite)	11.09	14.26
Quartz	10.78	12.02
Sulfide	6.47	12.67
Oligoclase, andesine	7.98	9.41
Altered and unidentified minerals	14.58	8.07

4.3. In Situ LA-ICP-MS Trace Element of Sulfides

The results of LA–ICP–MS trace element analysis for 40 analytical spots in the four types of pyrrhotite and 13 analytical spots in the two types of chalcopyrite are given in Tables 2 and 3 respectively. A comparison of the element compositions is presented in Figure 6.

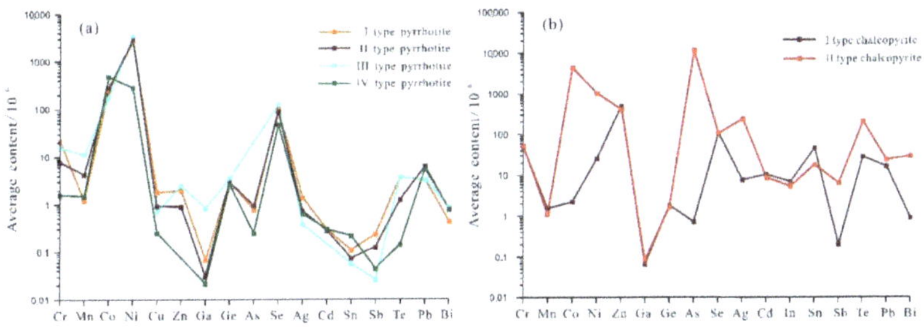

Figure 6. Comparison of the trace element compositions of pyrrhotite and chalcopyrite in the Huangyangshan deposit; (**a**) the trace element compositions of four types pyrrhotite; (**b**) the trace element compositions of two types chalcopyrite.

In type I pyrrhotite: element contents are Cr = 0.19 to 173.15 ppm (mean of 21.77 ppm), Mn = 0.03 to 11.18 ppm (mean of 1.24 ppm), Co = 123 to 357 ppm (mean of 251 ppm), Ni = 1998 to 3466 ppm (mean of 2769 ppm), Cu 0.11 to 16.94 ppm (mean of 1.82 ppm), Zn = 0.30 to 11.21 ppm (mean of 1.98 ppm), Ga = 0.01 to 0.09 ppm (mean of 0.07 ppm), Ge = 1.57 to 4.99 ppm (mean of 3.0 ppm), As = 0.04 to 2.17 ppm (mean of 0.78 ppm), Se = 66.44 to 129.47 ppm (mean of 99.52 ppm), Ag = 0.03 to 21.57 ppm (mean of 1.4 ppm), Cd = 0.01 to 0.64 ppm (mean of 0.30 ppm), Sn = 0.03 to 0.32 ppm (mean of 0.11 ppm), Sb = 0.03 to 1.09 ppm (mean of 0.32 ppm), Te = 0.06 to 20.60 ppm (mean of 3.92 ppm), Pb = 0.09 to 22.91 ppm (mean of 3.27 ppm), and Bi = 0.06 to 2.05 ppm (mean of 0.43 ppm) (Table 2).

In type II pyrrhotite: element contents are Cr = 0.35 to 23.07 ppm (mean of 7.82 ppm), Mn = 0.06 to 24.70 ppm (mean of 4.30 ppm), Co = 225 to 362 ppm (mean of 275 ppm), Ni = 2595 to 3233 ppm (mean of 2884 ppm), Cu = 0.08 to 2.23 ppm (mean of 0.94 ppm), Zn = 0.09 to 1.93 ppm (mean of 0.91 ppm), Ga = 0.03 to 0.06 ppm (mean of 0.03 ppm), Ge = 2.65 to 3.71 ppm (mean of 2.95 ppm), As = 0.37 to 1.33 ppm (mean of 0.96 ppm), Se = 74.03 to 102.85 ppm (mean of 89.09 ppm), Ag = 0.27 to 2.04 ppm (mean of 0.74 ppm), Cd = 0.03 to 0.65 ppm (mean of 0.29 ppm), Sn = 0.02 to 0.12 ppm (mean of 0.07 ppm), Sb = 0.04 to 0.34 ppm (mean of 0.13 ppm), Te = 0.45 to 2.37 ppm (mean of 1.25 ppm), Pb = 0.37 to 24.45 ppm (mean of 6.23 ppm), and Bi = 0.05 to 2.90 ppm (mean of 0.77 ppm) (Table 2).

Table 2. Trace element analysis results of pyrrhotite in Huangyangshan deposit (ppm).

Number	Type	Cr	Mn	Co	Ni	Cu	Zn	Ga	Ge	As	Se	Ag	Cd	Sn	Sb	Te	Pb	Bi
422Po-1		2.14	1.31	357	3109	/	0.69	/	4.36	/	117.02	0.5	/	0.07	/	1.81	1.72	0.29
412Po-2		2.02	0.15	336	2764	0.34	0.83	/	3.24	0.68	86.13	0.66	0.09	0.16	0.03	/	3.1	1.15
433Po-2		3.23	1.25	232	2482	0.27	1.64	0.03	2.64	0.04	100.04	0.57	0.01	0.12	0.14	1.07	0.81	0.19
423Po-2		1.78	0.34	227	3178	0.11	/	0.03	2.07	/	92.68	/	0.5	0.04	0.11	1.81	1.45	0.13
423Po-10		7.76	0.46	243	3172	0.3	1.07	0	2.23	0.41	80.59	0.3	0.54	/	/	0.5	1.72	0.19
423Po-13		75.07	0.76	238	3216	1.85	2.96	/	1.59	0.27	97.3	0.63	0.14	/	/	1.33	1.71	0.46
411Po-3		154.43	1.21	292	2534	0.85	2.07	/	3.76	/	122.89	0.42	0.26	0.07	0.61	0.25	0.36	0.06
411Po-13		0.99	0.41	256	2835	0.77	0.44	0.04	4.99	/	101.23	0.2	/	/	/	0.59	1.83	0.16
421Po-1		2.86	0.68	261	3466	0.5	0.44	0.03	3.96	0.41	100.24	0.42	/	/	/	/	0.85	/
421Po-7		173.15	0.46	224	2987	4.5	1.21	/	3.25	0.81	106.22	0.03	/	0.11	/	2.21	1.26	0.11
423Po-1		3.75	0.03	197	2885	0.11	/	/	2.53	/	95.01	0.09	/	0.23	0.02	0.06	2.2	0.4
ZK43Po-1	type I	0.19	0.37	249	2398	0.71	3.57	/	3.77	0.73	94.34	1.77	/	0.05	/	4.46	4.81	0.71
411Po-4	yellow	1.53	0.21	276	2356	0.31	1.69	0.04	2.99	/	101.06	0.2	0.42	/	/	1	0.6	0.34
411Po-11	pyrrhotite	70.53	4.9	281	3065	16.94	/	0.01	4.47	1.76	66.44	2.09	/	/	/	/	1.72	0.22
422Po-3		0.39	0.83	288	2962	0.25	/	0.07	2.23	0.26	108.56	0.3	0.09	/	/	0.31	0.78	/
421Po-2		2.49	0.47	302	3448	0.6	1.04	0	2.7	0.04	109.55	0.23	/	/	/	0.68	0.79	0.25
421Po-8		71.84	/	285	2910	0.15	3.5	/	2.95	0.99	83.3	/	0.25	0.03	/	/	0.86	0.14
433Po-1		4.31	0.83	239	2562	1.12	0.3	/	2.93	0.17	103.54	0.07	0.21	/	0.13	1.18	0.09	0.08
423Po-3		0.8	0.81	257	2964	0.8	0.5	0.04	1.57	0.51	129.47	/	0.1	0.04	0.15	/	0.97	0.08
ZK43Po-2		4.42	0.84	298	2883	2.37	/	/	3.05	/	100.09	2.34	/	/	/	10.03	11.59	0.76
ZK45Po-1		12.37	/	155	2117	/	/	0.02	3.39	1.45	89.63	0.12	0.12	0.13	0.25	2.15	0.91	0.38
ZK45Po-2		1.57	2.06	123	2240	1.24	/	/	2.88	1.72	81.89	1.03	0.64	0.32		5.77	3.77	0.53
423Po-4		1.28	0.9	247	3014	0.55	/	0.09	2.16	/	105.27	0.21	/	0.09	0.02	/	1.79	0.15
ZK41Po-2		4.69	0.4	195	2241	1.54	0.96	/	2.89	2.17	102.89	0.15	0.46	0.03	/	2.66	1.22	0.36
411Po-7		0.29	0.23	338	2779	2.52	/	0.09	2.28	/	89.7	1.47	/	/	/	10.17	10.93	1
C51Po-1		1.89	0.27	191	2215	2.9	/	0	3.61	/	100.88	21.57	0.56	0.17	0.02	20.6	7.91	0.5
C51Po-2	type I	13.66	11.18	153	1998	6.6	11.21	0.25		/	100.6	0.36	0.33	0.04	1.09	2.1	4.9	0.64
411Po-9	yellow	11.2	2.19	265	2707	0.89	/	0.38	2.93	/	112.86	0.12	0.46	0.25	/	/	1.21	0.39
421Po-3	pyrrhotite	0.74	0.03	262	2807	0.11	1.5	0.05	2.67	/	106.63	0.67	/	0.03	0.31	15.6	22.91	2.05
average		21.77	1.24	251	2769	1.82	1.98	0.07	3	0.78	99.52	1.4	0.3	0.11	0.24	3.92	3.27	0.43
411Po-8		14.09	0.08	298	2595	0.73	1.2	0.06	3.71	/	83.71	0.27	/	0.02	/	/	0.79	0.33
412Po-5	type II	0.35	0.06	362	2662	/	0.58	/	2.7	/	101.11	0.49	0.03	/	0.07	0.7	0.37	0.14
422Po-2	brown	23.07	/	348	2852	0.96	0.73	0.03	3.2	1.33	98.18	/	0.37	0.02	0.05	2.37	0.53	0.05
423Po-7	pyrrhotite	3.08	0.24	237	2991	0.68	1.93	/	2.66	0.98	102.85	0.43	/	0.12	0.05	0.45	4.08	0.54
423Po-8		6.34	24.7	225	2944	/	/	/	3.11	0.37	85.22	0.35	0.1	0.12	/	/	15.02	1.45

Table 2. *Cont.*

Number	Type	Cr	Mn	Co	Ni	Cu	Zn	Ga	Ge	As	Se	Ag	Cd	Sn	Sb	Te	Pb	Bi
423Po-9		4.2	0.31	227	2946	2.23	0.95	/	2.81	1.21	79.99	2.04	/	0.12	0.34	1.62	24.45	2.9
423Po-11		3.6	3.87	240	3233	0.08	0.09	/	2.65	/	87.59	/	/	0.04	0.2	/	1.23	0.26
423Po-12		/	0.87	267	2852	/	/	0	2.77	0.89	74.03	0.85	0.65	/	0.04	1.12	3.39	0.51
average		7.82	4.3	275	2884	0.94	0.91	0.03	2.95	0.96	89.09	0.74	0.29	0.07	0.13	1.25	6.23	0.77
422Po-4	type III Pyrrhotitein fine-grained core	6.67	17.94	181	4173	1.03	2.52	0.85	4.03	/	130	0.13	/	0.06	/	0.78	4.33	0.6
422Po-5		25.04	4.48	166	2691	0.37	/	/	3.16	/	124	0.65	/	/	0.03	6.8	2.33	1.15
average		15.85	11.21	174	3432	0.7	2.52	0.85	3.6	/	127	0.39	/	0.06	0.03	3.79	3.33	0.87
412Po-1	type IV	1.61	1.52	480	283	0.25	/	0.02	2.66	0.25	47.58	0.64	0.3	0.22	0.04	0.15	5.49	0.81

Table 3. Trace element analysis results of chalcopyrite in Huangyangshan deposit (ppm).

Number	Type	Cr	Mn	Co	Ni	Zn	Ga	Ge	As	Se	Ag	Cd	In	Sn	Sb	Te	Pb	Bi
411Ccp-1	type I copper yellow chalcopyrite	29.08	/	1.73	32.92	323	0.13	2.58	1.41	133.85	2.77	6.71	12.61	115.79	/	3.93	1.01	0.12
422Ccp-1		60.68	0.07	1.74	8.66	363	0.08	1.24	/	95.69	3.23	6.48	4.96	7.82	/	/	1.16	0.03
421Ccp-1		336.9	/	0.81	6.46	680	0.01	2.15	0.22	78.79	15.46	10.03	5.59	2.58	/	15.82	15.77	0.75
423Ccp-1		7.49	0.34	3.63	14.93	559	/	1.19	/	95	8.39	9.46	5.04	25.26	0.02	37.66	48.72	0.82
423Ccp-2		5.48	/	1.4	9.14	528	0.04	2.31	/	108.36	8.81	10.5	5.93	8.68	0.2	7.58	3.63	0.22
423Ccp-4		5	0.15	1.8	22.12	725	/	1.66	/	133.34	3.95	13.93	5.76	6.06	/	7.19	27.99	0.23
ZK46Ccp-1		13.87	9.36	1.63	13.5	445	0.09	1.81	/	93.15	6.84	9.71	7.43	36.2	0.38	36.2	8.67	1.98
ZK46Ccp-1		13.72	0.44	6	69.66	581	0.05	2.21	/	133.24	11.17	17.34	7.71	46.54	0.17	58.19	13.63	2.05
C51Ccp-1		0.53	0.21	1.85	43.26	546	0.07	2.34	/	124.52	7.79	10.31	5.85	87.3	/	54.98	11.72	1.03
411Ccp-2		27.87	0.34	1.74	35.8	453	0.03	0.93	0.53	75.63	6.53	8.44	6.5	117.85	/	29.24	25.51	1.45
average		50.06	1.56	2.23	25.64	520	0.06	1.84	0.72	107.16	7.49	10.29	6.74	45.41	0.19	27.87	15.78	0.87
422Ccp-4	type II brown chalcopyrite	50.43	0.56	4881	1098	435	0.08	0.87	10,784	83	266	7.99	4.74	17	7.05	203	24.06	31.47
422Ccp-5		57.89	/	4352	1138	465	0.05	2.51	12,572	116	244	9.37	5.1	18.91	6.1	213	22.96	29.11
412Ccp-1		51.02	1.68	4084	923	355	0.12	1.63	11,490	121	217	8.24	5.94	17.1	5.63	210	23.65	26.12
average		53.11	1.12	4439	1053	419	0.08	1.67	11,615	107	242	8.53	5.26	17.67	6.26	209	23.56	28.9

Detection limit (10^{-6}): Cr 0.34, Mn 0.39, Fe 24.13, Co 0.03, Ni 0.07, Cu 0.35, Zn 1.57, Ga 0.01, Ge 0.71, As 0.11, Se 0.91, Ag 0.13, Cd 0.04, In 0.01, Sn 0.08, Sb 0.26, Te 0.14, Pb 0.04, Bi 0.02.

In type III pyrrhotite: element contents are Cr = 6.67 to 25.04 ppm (mean of 15.85 ppm), Mn = 4.48 to 17.94 ppm (mean of 11.21 ppm), Co = 166 to 181 ppm (mean of 174 ppm), Ni = 2691 to 4173 ppm (mean of 3432 ppm), Cu = 0.37 to 1.03 ppm (mean of 0.7 ppm), Zn = 2.52 ppm, Ga = 0.85 ppm, Ge = 3.16 to 4.03 ppm (mean of 3.6 ppm), As = below detection limit (b.d.l.), Se = 124 to 130 ppm (mean of 127 ppm), Ag = 0.13 to 0.65 ppm (mean of 0.39 ppm), Cd = b.d.l., Sn = 0.60 ppm, Sb = 0.03 ppm, Te = 0.78 to 6.80 ppm (mean of 3.79 ppm), Pb = 2.33 to 4.33 ppm (mean of 3.33 ppm), and Bi = 0.6 to 1.15 ppm (mean of 0.87 ppm) (Table 2).

In type IV pyrrhotite: element contents are Cr = 1.61 ppm, Mn = 1.52 ppm, Co = 283 ppm, Ni = 480 ppm, Cu = 0.25 ppm, Zn = b.d.l., Ga = 0.02 ppm, Ge = 2.66 ppm, As = b.d.l., Se = 47.58 ppm, Ag = 0.64 ppm, Cd = 0.3 ppm, Sn = 0.22 ppm, Sb = 0.04 ppm, Te = 0.15 ppm, Pb = 5.49 ppm, and Bi = 0.81 ppm (Table 2).

In summary, the four types of pyrrhotite are enriched in Co, Ni, Se, Ge, Cr, Te, and Pb and depleted in Ga, As, Cd, Sn, and Sb. The contents of Co, Ge, and Se in the different types of pyrrhotite are similar and high. All of the spot analyses in types I, II, and III of the pyrrhotite have similar and high Ni contents compared with the low content of type IV pyrrhotite. The Cr, Mn, Cu, Zn, Ag, Sb, Te, and Pb content in the different types of pyrrhotites vary widely (Figure 6a).

In the type I chalcopyrite, the element contents are Cr = 0.53–336.9 ppm (mean of 50.06 ppm), Mn = 0.07–9.36 ppm (mean of 1.56 ppm), Co = 0.81–6.00 ppm (mean of 2.23 ppm), Ni = 6.46–69.66 ppm (mean of 25.64 ppm), Zn = 323–725 ppm (mean of 520 ppm), Ga = 0.01–0.13 ppm (mean of 0.06 ppm), Ge = 0.93–2.58 ppm (mean of 1.84 ppm), As = 0.22–1.41 ppm (mean of 0.72 ppm), Se = 76–134 ppm (mean of 107 ppm), Ag content of 2.77–15.46 ppm (mean of 7.49 ppm), Cd = 6.48–17.34 ppm (mean of 10.29 ppm), In = 4.96–12.61 ppm (mean of 6.74 ppm), Sn = 2.58–117.85 ppm (mean of 45.41 ppm), Sb = 0.02–0.38 ppm (mean of 0.19 ppm), Te = 3.93–58.19 ppm (mean of 27.87 ppm), Pb = 1.01–48.72 ppm (mean of 15.78 ppm), and Bi = 0.03–2.05 ppm (mean of 0.87 ppm) (Table 3).

In the type II chalcopyrite, the element contents are Cr = 50.43–57.89 ppm (mean of 53.11 ppm), Mn = 0.56–1.68 ppm (mean of 1.12 ppm), Co = 4084–4881 ppm (mean of 4439 ppm), Ni = 923–1138 ppm (mean of 1053 ppm), Zn = 356–465 ppm (mean of 419 ppm), Ga = 0.05–0.12 ppm (mean of 0.08 ppm), Ge = 0.87–2.51 ppm (mean of 1.67 ppm), As = 10,784–12,572 ppm (mean of 11,616 ppm), Se = 83–121 ppm (mean of 107 ppm), Ag = 217–266 ppm (mean of 242 ppm), Cd = 7.99–9.37 ppm (mean of 8.53 ppm), In = 4.74–5.94 ppm (mean of 5.26 ppm), Sn = 17.00–18.91 ppm (mean of 17.67 ppm), Sb = 5.63–7.05 ppm (mean of 6.26 ppm), Te = 203–213 ppm (mean of 209 ppm), Pb = 22.96–24.06 ppm (mean of 23.56 ppm), and Bi = 26.12–31.47 ppm (mean of 28.9 ppm) (Table 3).

In summary, the Co, Ni, As, and Ag content in the type I chalcopyrite are substantially lower than those in the type II chalcopyrite, with a maximum difference of 2–4 orders of magnitude. The Sb, Te, and Bi content in the type I chalcopyrite are also lower than those in the type II chalcopyrite but the differences are smaller. The contents of the remaining elements (i.e., Cr, Mn, Zn, Ga, Ge, Se, Cd, In, and Pb) are similar in the two types of chalcopyrite (Figure 6b).

5. Discussion

5.1. Mineral Characteristics of Graphite Ores

The X-ray micro-CT scanning results show (Figure 4) that the drill cores of the graphite ores are composed mainly of low-density minerals, such as quartz and feldspar. The medium-density dark minerals, such as amphibole and biotite, are distributed mostly among the quartz and feldspar grains (Figure 4c). There is a high content of high-density metal sulfides, and these occur as massive aggregations or are disseminated in the ore (Figure 4d).

In the QEMSCAN and MAPS images, the distribution and content characteristics of various minerals, as well as the shapes and intersecting relationships of the fine-grained minerals, can be clearly identified by their mineral element components (Figure 5). According to the mass fraction values (Table 1), the plagioclase content in the ore is highest (31.03%), followed by K-feldspar (13.39%) and dark minerals (14.26%), with a quartz content of only 12.02%, which is largely lower than that of the ore-bearing granite (>50%), suggesting that the crystallization conditions of the graphite are different from those of the ore-bearing granite. The scanning results of the QEMSCAN and MAPS analyses reveal that the groundmass in the orbicular ore is mainly albite and dark minerals (such as biotite and amphibole) and that the flake graphite is distributed predominantly as bundles and banded aggregates that cut through the sulfides (Figure 5b). The Quartz occurs as xenomorphic grains that have metasomatized sulfides and contain small amounts of fine-grained disseminated graphite (Figure 5c). Some coarse-grained nodular sulfide aggregates in the analyzed section are cut by the quartz. These sulfide aggregates contain densely distributed fine-grained graphite bands (Figure 5d), but the graphite appears mainly in the form of "scalelike", with relatively low degrees of crystallization. In addition, most of the flaky graphite only coexists with the sulfide and albite, not with the quartz and K-feldspar (Figure 5a), indicating that the formation of the flake graphite was closely related to the albite and sulfide. In addition, the dark minerals show high abundance in the middle of the analyzed section in association with andesine and oligoclase but not with quartz, indicating that the crystallization conditions of these dark minerals are different from those of the ore-bearing granite, which is consistent with the mineral mass fraction results (Table 1).

5.2. The Origin of Sulfide

The compositional and structural characteristics of the studied sulfides from the Huangyangshan deposit reveal important information about the genesis of these minerals; (1) the sulfides are dominated by pyrrhotite, with small amounts of associated chalcopyrite and pentlandite, whereas typical hydrothermal sulfides, such as pyrite and sphalerite, are rare; (2) the contact boundaries of most of the metal sulfides are straight (Figure 3c,f) and do not display metasomatic features; (3) pentlandite is distributed within the pyrrhotite and chalcopyrite (Figure 3c,e). These characteristics are highly similar to those of magmatic Cu–Ni sulfide deposits, indicating the magmatic origin of the metal sulfides [31,32].

The crystalline shapes and textures of the sulfides from the Huangyangshan deposit are rather consistent and lack recrystallization, regrowth, and superimposed reformation by metamorphism. The sulfides are highly homogeneous in the MAPS image (Figure 5d), indicating that the trace element components contained in these sulfides did not undergo pronounced element migration and, therefore, represent the primary component characteristics when the sulfides formed [33]. An elemental spectral line diagram (Figure 7) reveals that the main elements in the pyrrhotite are Fe, S, Ni, and Co, with Ni and Co displaying stable spectral line variations without obvious crests and troughs (Figure 7a). The chalcopyrite contains Cu, Fe, S, and Zn predominantly, and the spectral lines of these elements are also stable (Figure 7b), suggesting that these elements occur as isomorphic forms rather than inclusions [34], implying the formation occurred in a stable environment. Therefore, the trace element compositions of the pyrrhotite and chalcopyrite samples from the Huangyangshan graphite deposit can be used to interpret their genesis and formation conditions, as described, in turn, below.

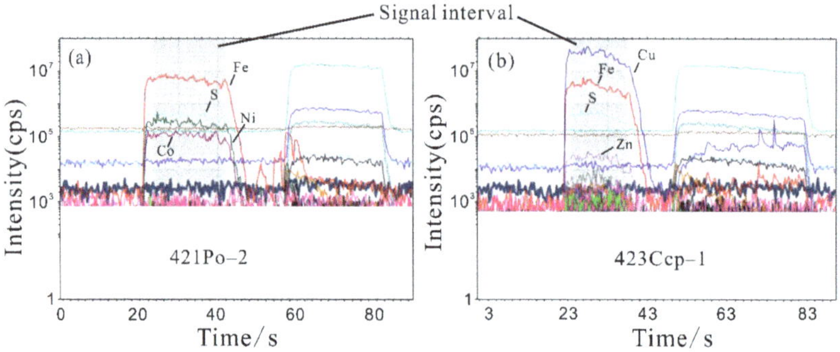

Figure 7. Examples of the time-resolved LA–ICP–MS depth profiles for the pyrrhotite and chalcopyrite from the Huangyangshan deposit. (**a**) Ablation profile for type I pyrrhotite from spot 421Po–2. (**b**) Ablation profile for type I chalcopyrite from spot 423Ccp–1.

5.2.1. Pyrrhotite

Previous studies have shown that Co and Ni replace Fe in pyrrhotite as isomorphic forms, and pyrrhotite is a major carrier mineral of Co and Ni. The contents of Co and Ni in pyrrhotite are predominantly affected by the physicochemical conditions applied during their formation, meaning that these elements can be used to infer the genetic type of pyrrhotite [35–37]. During magmatic evolution, Ni and Co are preferentially enriched in the early- and late-crystallized minerals, respectively [38]. Therefore, the Co/Ni ratio of magmatic-origin pyrrhotite tends to be lower than that of hydrothermal-origin pyrrhotite. The contents of Co and Ni are similar between types I–III of the pyrrhotite from the Huangyangshan deposit, and most of the Co/Ni values range from 0.07 to 0.12 (Table 2). All these values plot within the range of magmatic Cu–Ni sulfide deposits in a genetic discrimination map (Figure 8), indicating magmatic origin for types I–III of the pyrrhotite. Type IV pyrrhotite has higher Co and lower Ni contents compared with the other three types, with a mean Co/Ni ratio of 1.69, and it plots within the range of magmatic–hydrothermal deposits, indicating that this pyrrhotite might have formed during late hydrothermal metasomatic processes.

In general, Pb, Zn, Ge, and Te are enriched in hydrothermal pyrrhotite as micro inclusions [38,39]. The contents of these elements in the pyrrhotite from the graphite ore of the Huangyangshan deposit are extremely low, mostly below the detection limit, suggesting that these pyrrhotites were formed in a relatively stable, closed, and high-temperature magmatic environment. In addition, all of the analyzed pyrrhotites have low contents of low-temperature elements, such as As, Sb, Ag, and Au, which is in marked contrast to porphyry-type, skarn-type, and hydrothermal-vein–type deposits [40,41]. In summary, types I–III nodular and xenomorphic–granular pyrrhotite are of the same magmatic origin, whereas the type IV pyrrhotite in granitic groundmass is of hydrothermal origin, as discussed above.

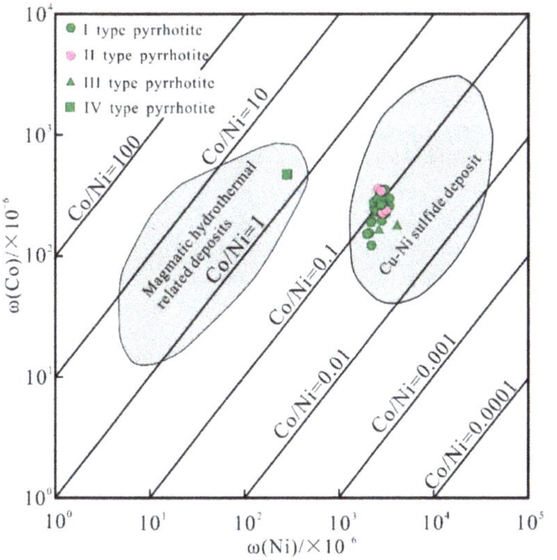

Figure 8. Co–Ni genetic discrimination diagram of the pyrrhotite from the Huangyangshan deposit (modified after [37]).

5.2.2. Chalcopyrite

Previous studies of the trace-element characteristics of chalcopyrite have proposed that the Co/Ni ratio is an important indicator reflecting the origin of chalcopyrite [37,42]. The trace element analytical results for chalcopyrite from the Huangyangshan deposit in this study reveal that type I chalcopyrite has low Co and Ni contents and low Co/Ni ratios, ranging from 0.04 to 0.24, whereas type II chalcopyrite contains abundant Co and Ni and has Co/Ni values ranging from 3.82 to 4.44. Previous studies have reported low Ni contents in chalcopyrite from hydrothermal deposits (averaging 1–5 ppm), with Co/Ni values >1, whereas the Ni contents of chalcopyrite from magmatic Cu–Ni sulfide deposits fluctuate widely (10–10^4 ppm), with Co/Ni values of <1 [37,43,44]. Therefore, the Co/Ni values of type I chalcopyrite from the Huangyangshan deposit are consistent with those of chalcopyrite from magmatic Cu–Ni sulfide deposits. The low Ni contents in type I chalcopyrite may have resulted from the low Ni content in the ore-forming melt itself or from Ni enrichment in the pre-existing pyrrhotite and pentlandite before the chalcopyrite was dissolved. Type II chalcopyrite from the Huangyangshan deposit has high Ni and very high Co contents in comparison with type I chalcopyrite, which is consistent with the characteristics of late hydrothermal chalcopyrite. The high contents of Ni in the type II chalcopyrite may have been caused by hydrothermal enrichment. Type II chalcopyrite also has substantially higher As and Ag contents relative to type I chalcopyrite. These two elements generally exist in chalcopyrite through the substitution of Cu by Ag and of S by As [45,46], which usually takes place at a low temperature. Therefore, the high contents of As and Ag suggest that the type II chalcopyrite underwent low-temperature replacement. In addition, low-temperature elements, such as Sb, Te, and Bi are enriched in type II chalcopyrite relative to type I, which further suggests a low-temperature environment for the formation of type II chalcopyrite. In summary, we believed that the type I chalcopyrite was of magmatic origin, coexisting with types I–III pyrrhotite, while type II chalcopyrite was of hydrothermal origin.

Types I–III pyrrhotite have high contents of Se (>80 ppm), Ge (>2.5 ppm) and low contents of Zn (mostly <1.5 ppm), Cd (mostly <0.5 ppm), and Sn (<0.5 ppm). The I-type chalcopyrite has high contents of Zn (mostly >500 ppm) and Se (mostly >100 ppm) and low contents of Sb (mostly <0.3 ppm) and Te (more than <20 ppm). These properties are similar to the trace element characteristics of Cu–Ni sulfide deposits in other regions, such as Jinchuan (China), Creighton (Canada), and Bushveld (South Africa), but they are clearly distinct from the hydrothermal deposits of skarn and porphyry types [47–51].

5.3. Implications of Metal Sulfides for Graphite Mineralization

The genetic type of the Huangyangshan graphite deposit has been debated since its discovery. The occurrence of the graphite ore in alkaline granite, the organic carbon source of the graphite (carbon isotopic compositions: −20.9 to −20.5‰), and the cutting of transparent minerals by flake graphite has led some previous studies to classify the Huangyangshan deposit as magmatic–hydrothermal in origin [2,10,12]. However, the results of the present study reveal that the metal sulfides closely associated with graphite are of magmatic origin, indicating the occurrence of a magmatic mineralization event related to mafic magmatism in the deposit.

Various lines of evidence in this study show that the formation of the Huangyangshan graphite deposit is closely related to the development of sulfides: (1) the graphite and metal sulfides occur mostly along the edges of the orbicular structures, whereas ores lacking graphite and granitic groundmass do not contain sulfides (Figure 3b); (2) the banded graphite coexists with the granular sulfides (Figure 5b) (3) the microscopy revealed that the sulfides and graphite cut through the biotite that crystallized during the late magmatic stage (Figure 3d); (4) hydrothermal-origin type IV pyrrhotite and type II chalcopyrite occur in the silicate mineral cores lacking graphite or replaced types I–III pyrrhotite and type I chalcopyrite (Figure 3h,i); (5) the Os isotope composition of the graphite indicates that its carbon was sourced at least partly from mantle-derived magma, which is the same as the sulfides in the deposit [6]; (6) the crystallization temperature of the graphite was about 750–800 °C, which is approximately similar to the crystallization temperature of high-temperature pyrrhotite and chalcopyrite from a magmatic system [12,52]. The above evidence shows that the graphite in the Huangyangshan deposit should have the same magmatic origin as sulfide. Moreover, the comparison studies showed that the intergrowth relationship between the graphite and sulfides, the orbicular mineralization characteristics of the graphite, the mineral assemblage of the sulfides, and the crystallization temperature of graphite from the Huangyangshan deposit were highly similar to the mineralization characteristics of the graphite hosted in the ultramafic rocks of the Ronda (Spain) and Beni-Bousera (Morocco) deposits [53]. These similar characteristics also indicate that the graphite in the Huangyangshan deposit is of magmatic origin.

In addition, Luque et al. [54] considered that metal sulfides have a catalytic effect on carbon's crystallization to graphite, for which pyrrhotite has the strongest catalytic effect. Graphite is commonly associated with sulfides in the Huangyangshan deposit (Figure 3e), which is consistent with the effect of catalysis on graphite mineralization. Based on all the above data, therefore, it can be proposed that the graphite in the Huangyangshan deposit has the same magmatic origin as the sulfide associated with graphite, and this sulfide acts as a catalyst to promote graphite mineralization.

6. Conclusions

In the Huangyangshan deposit of East Junggar, China, the graphite ores show an orbicular structure characterized by an inner core and an outer circle. The inner core is composed mainly of felsic minerals and dark minerals, whereas the outer circle is composed of graphite and associated metal sulfides, including pyrrhotite, chalcopyrite, and pentlandite. The results of the micro-CT and section-scanning electron microscopy of the drill core showed that the graphite ore has higher plagioclase and dark mineral content and substantially lower quartz content than the granitic wall rock. Flake graphite occurs mostly as bundles and banded aggregates, cutting through the sulfides. The formation of flake graphite is closely related to the formation of albite and sulfide.

Four types of pyrrhotite and two types of chalcopyrite were identified, with types I–III pyrrhotite and type I chalcopyrite being genetically related to graphite mineralization. Types I–III pyrrhotite are associated with graphite and have similar Co and Ni contents and low Co/Ni values, as well as generally low Au, Ag, Pb, Zn, As, Sb, Ge, and Te contents, suggesting a magmatic origin. Type IV pyrrhotite has higher Co and lower Ni contents compared with the other three types, suggesting a hydrothermal origin. Type I chalcopyrite has low Co and Ni contents and low Co/Ni values, implying a magmatic origin. Type II chalcopyrite has high Ni and very high Co contents and high Co/Ni values, as well as high low-temperature elements such as Sb, Te, and Bi, implying a hydrothermal origin. Graphite has the same origin as types I–III of pyrrhotite and type I of chalcopyrite, and the metal sulfides play a catalytic role, leading to graphite crystallization.

Author Contributions: Conceptualization, Y.R. and X.S.; formal analysis, X.S.; investigation, J.L., Y.R., Z.L., Z.S. and X.S.; data curation, X.S. and J.L.; writing—original draft preparation, Y.R.; writing—review and editing, J.L., Y.R., and X.S.; visualization, X.S.; supervision, Y.R.; project administration, Y.R.; funding acquisition, Y.R. All authors have read and agreed to the published version of the manuscript.

Funding: This research was funded by the Geological survey project of the China Geological Survey (No. DD20190159).

Data Availability Statement: All data generated or used in the study are contained in the submitted article.

Acknowledgments: We sincerely appreciate the geologists from the Xinjiang Branch of China National Geological Exploration Center of Building Materials Industry for their support of our fieldwork. We are also deeply grateful to anonymous reviewers for their reviews and constructive suggestions.

Conflicts of Interest: The authors declare no conflict of interest.

References

1. Zhang, X.L.; Fan, W.J.; Li, Z.W.; Chen, Z.G.; Chen, J.Y. The discovery of a superlarge magmatic graphite deposit in Huangyangshan area, Qitai County, Xinjiang. *Geol. China* **2017**, *44*, 1033–1034, (In Chinese with English Abstract).
2. Bai, J.K.; Chen, J.L.; Peng, S.X. Geochronology and geochemistry of ore-bearing intrusions from Huangyangshan magmatic hydrothermal graphite deposit in Qitai County, Xinjiang. *Acta Petrol. Sin.* **2018**, *34*, 2327–2340, (In Chinese with English Abstract).
3. Yang, G.X.; Li, Y.J.; Wu, H.E.; Zhong, X.; Yang, B.K.; Yang, C.X.; Yan, J.; Si, G.H. Geochronological and geochemical constrains on petrogenesis of the Huangyangshan A-type granite from the East Junggar, Xinjiang, NW China. *J. Asian Earth Sci.* **2011**, *40*, 722–736. [CrossRef]
4. Sun, X.H.; Ren, Y.S.; Sun, Z.J.; Wang, C.Y.; Li, Z.W. Geochronology and geochemical properties of the large-scale graphite mineralization associated with the Huangyangshan alkaline pluton, Eastern Junggar, Xinjiang, NW China. *Geochemistry* **2021**, *81*, 125820. [CrossRef]
5. Mo, R.J.; Liu, S.B.; Huang, C.R. *Geology of Graphite Deposits in China*; China Architecture & Building Press: Beijing, China, 1989; pp. 66–85. (In Chinese)
6. Zhang, G.X.; Hu, A.Q.; Zhang, H.B.; Zhang, Q.F.; Shen, Y.L. Caebon isotopic evidence for the origin of the spherical graphite in a granite-hosted graphite deposit, Sujiquan, Xinjiang, China. *Geochimica* **1996**, *25*, 379–386, (In Chinese with English Abstract).
7. Yang, G.X.; Li, Y.J.; Wu, H.E.; Si, G.H.; Zhang, Y.Z.; Jin, Z. A tentative discussion on the genesis of Huangyangshan granite body in Kalamaili Orogen, East Junggar. *Acta Geosci. Sin.* **2010**, *31*, 170–182, (In Chinese with English Abstract).
8. Li, C.; Wang, D.H.; Zhao, H.; Pei, H.X.; Li, X.W.; Zhou, L.M.; Du, A.D.; Qu, W.J. Minerogenetic regularity of graphite deposits in China. *Miner. Deposit.* **2015**, *34*, 1223–1236, (In Chinese with English Abstract).
9. Bai, J.K.; Chen, J.L.; Peng, S.X. Charcteristics and metallogeny regulation of graphite resources in Xinjiang. *Acta Geol. Sin.* **2017**, *91*, 2828–2840, (In Chinese with English Abstract).
10. Ai, J.; Lu, X.B.; Li, Z.W.; Wu, Y.L. Genesis of the graphite orbicules in the Huangyangshan graphite deposit, Xinjiang, China: Evidence from geochemical, isotopic and fluid inclusion data. *Ore Geol. Rev.* **2020**, *122*, 103505. [CrossRef]
11. Ai, J. Metallogenic Mechanism and Model of Huangyangshan Magmatic Fluid-Type Graphite Deposit in Xinjiang. Ph.D. Thesis, China University of Geosciences (Wuhan), Wuhan, China, 2020. (In Chinese with English Abstract).
12. Sun, X.H.; Ren, Y.S.; Sun, Z.J.; Wang, C.Y. Characteristics, ore sources and genesis of Huangyangshan graphite deposit in Qitai County, Xinjiang. *Acta Petrol. Sin.* **2021**, *37*, 1867–1882, (In Chinese with English Abstract).
13. Li, J.Y.; Xiao, X.C.; Tang, Y.Q.; Zhao, M. Main characteristics of late Paleozoic plate tectonics in the southern part of east Junggar, Xinjiang. *Geol. Rev.* **1986**, *36*, 305–316, (In Chinese with English abstract).
14. Li, J.Y.; Yang, T.N.; Li, Y.P.; Zhu, Z.X. Geological features of the Karamaili faulting belt, eastern Junggar region, Xinjiang, China and its constraints on the reconstruction of Late Paleozoic ocean-continental frame work of the Central Asian region. *Geol. Bull. China* **2009**, *28*, 1817–1826, (In Chinese with English abstract).

15. Long, X.P.; Yuan, C.; Sun, M.; Safonova, I.; Xiao, W.J.; Wang, Y.J. Geochemistry and U–Pb detrital zircon dating of Paleozoic graywackes in East Junggar, NW China: Insights into subduction–accretion processes in the southern Central Asian Orogenic Belt. *Gondwana Res.* **2012**, *21*, 637–653. [CrossRef]
16. Han, Y.G.; Zhao, G.C. Final amalgamation of the Tianshan and Junggar orogenic collage in the southwestern Central Asian Orogenic Belt: Constraints on the closure of the Paleo-Asian Ocean. *Earth-Sci. Rev.* **2017**, *186*, 129–152. [CrossRef]
17. Xiao, W.J.; Han, C.M.; Yuan, C.; Sun, M.; Lin, S.; Chen, H.; Li, Z.; Li, J.; Sun, S. Middle Cambrian to Permian subduction-related accretionary orogenesis of Northern Xinjiang, NW China: Implications for the tectonic evolution of central Asia. *J. Asian Earth Sci.* **2008**, *32*, 102–117. [CrossRef]
18. Xiao, W.J.; Windley, B.; Sun, S.; Li, J.L.; Huang, B.C.; Han, C.M.; Yuan, C.; Sun, M.; Chen, H.L. A tale of amalgamation of three Permo-Triassic collage systems in Central Asia: Oroclines, sutures, and terminal accretion. *Annu. Rev. Earth Planet. Sci.* **2015**, *43*, 477–507. [CrossRef]
19. Yuan, C.; Sun, M.; Wilde, S.; Xiao, W.J.; Xu, Y.G.; Long, X.P.; Zhao, G.C. Postcollisional plutons in the Balikun area, East Chinese Tianshan: Evolving magmatism in response to extension and slab break-off. *Lithos* **2010**, *119*, 269–288. [CrossRef]
20. Zhang, Y.Y.; Pe-Piper, G.; Piper, D.J.W.; Guo, Z.J. Early Carboniferous collision of the Kalamaili orogenic belt, North Xinjiang, and its implications: Evidence from molasse deposits. *Geol. Soc. Am. Bull.* **2013**, *125*, 932–944. [CrossRef]
21. Xu, Q.Q.; Zhao, L.; Niu, B.G.; Zheng, R.G.; Yang, Y.Q.; Liu, J.H. Early Paleozoic arc magmatism in the Kalamaili orogenic belt, Northern Xinjiang, NW China: Implications for the tectonic evolution of the East Junggar terrane. *J. Asian Earth Sci.* **2020**, *194*, 104072. [CrossRef]
22. Li, Y.P.; Li, J.Y.; Sun, G.H.; Zhu, Z.X.; Yang, Z.Q. Basement of Junggar basin: Evidence from detrial zircons in sandstone of previous Devonian Kalamaili formation. *Acta Petrol. Sin.* **2007**, *23*, 1577–1590, (In Chinese with English abstract).
23. Guo, Z.J. A review on the Paleozoic tectonic evolution of northern Xinjiang and a discussion on the important role of geological maps in tectonic study. *Geol. Bull. China* **2012**, *31*, 1054–1060, (In Chinese with English abstract).
24. Cai, X.F.; Fan, G.M.; Liao, Q.A.; Wang, F.M.; Wu, L.Y. Features of lower carboniferous intra-plate sedimentation in the Kalamai region, East Junggar, Xinjiang, China. *Bull. Miner. Petrol. Geochem.* **2015**, *34*, 1287–1302, (In Chinese with English abstract).
25. Lin, J.F.; Yu, H.X.; Yu, X.Q.; Di, Y.J.; Tian, J.T. Zircon SHRIMP U–Pb dating and geological implication of the Sabei alkali-rich granite from Eastern Junggar of Xinjiang, NW China. *Acta Petrol. Sin.* **2007**, *23*, 1876–1884, (In Chinese with English abstract).
26. Su, Y.P.; Tang, H.F.; Cong, F. Zircon U–Pb age and petrogenesis of the Huangyangshan alkaline granite body in East Junggar, Xinjiang. *Acta Miner. Sin.* **2008**, *28*, 117–126, (In Chinese with English abstract).
27. Han, B.F.; Ji, J.Q.; Song, B.; Chen, L.H.; Zhang, L. Late Paleozoic vertical growth of continental crust around the Junggar Basin, Xinjiang, China (Part I): Timing of postcollisional plutonism. *Acta Petrol. Sin.* **2006**, *22*, 1077–1086, (In Chinese with English abstract).
28. Tang, H.F.; Qu, W.J.; Su, Y.P.; Hou, G.S.; Du, A.D.; Cong, F. Genetic connection of Sareshike tin deposit with the alkaline A-type granites of Sabei body in Xinjiang: Constraint from isotopic ages. *Acta Petrol. Sin.* **2007**, *23*, 1989–1997, (In Chinese with English abstract).
29. Xu, B. Research on Tectonic-Magma ore-Controlling Regularity and Prospecting Forecasting of Gold Metallogenic Belt in Kalamaili Area, Xinjiang. Ph.D. Thesis, China University of Geosciences (Beijing), Beijing, China, 2007.
30. Gan, L.; Tang, H.F.; Han, Y.J. Geochronology and geochemical characteristics of the Yemaquan granitic pluton in East Junggar, Xinjiang. *Acta Petrol. Sin.* **2010**, *26*, 2374–2388, (In Chinese with English abstract).
31. Chen, G.Y.; Sun, D.S.; Yin, H.A. *Genetic Mineralogy and Prospecting Mineralogy*; Chongqing Publishing Group: Chongqing, China, 1987. (In Chinese)
32. Li, S.R. *Crystallography and Mineralogy*; Geology Press: Beijing, China, 2008. (In Chinese)
33. Zhang, Z.L.; Geng, X.X.; Yang, F.Q.; Zhang, Z.X.; Xu, Q.F. Gold enrichment and the Au–Te association in the Huangtan Au–Cu–Zn volcanogenic massive sulfide deposit, Kalatag, East Tianshan, China: An in-situ LA–ICP–MS analysis. *Ore Geol. Rev.* **2021**, *138*, 104407. [CrossRef]
34. Yang, Z.K.; Yang, Y.; Zhang, Z.K.; Lin, B.; He, J.; Zhang, Z.B.; Gao, F.T.; Tang, X.Q.; Tang, P.; Qi, J.; et al. Geochemistry of pyrrhotite in the Jiama deposit, Tibet and its relationship with gold enrichment and precipitation. *Geol. China* **2020**, *49*, 1198–1213, (In Chinese with English abstract).
35. Loftus-Hills, G.; Solomon, M. Cobalt, Nickel and selenium in sulphides as indicators of ore genesis. *Miner. Depos.* **1967**, *2*, 228–242. [CrossRef]
36. Bralia, A.; Sabatini, G.; Troja, F. A revaluation of the Co/Ni ratio in pyrite as geochemical tool in ore genesis problems. *Miner. Depos.* **1979**, *14*, 353–374. [CrossRef]
37. Chen, D.F. Characteristics of metallic minerals in some copper-nickel sulfide deposits of China. *Acta Petrol. Miner.* **1995**, *14*, 345–354, (In Chinese with English abstract).
38. Liu, Y.J. *Elemental Geochemistry*; Geology Press: Beijing, China, 1984; pp. 1–548. (In Chinese)
39. Zhao, Z.H. *Principle of Trace Element Geochemistry*; Science Press: Beijing, China, 1997; pp. 1–238. (In Chinese)
40. Steadman, J.A.; Large, R.R. Synsedimentary, diagenetic, and metamorphic pyrite, pyrrhotite, and marcasite at the Homestake BIF-hosted gold deposit, South Dakota, USA: Insights on Au-As ore genesis from textural and LA-ICP-MS trace element studies. *Econ. Geol.* **2016**, *111*, 1731–1752. [CrossRef]

41. Liu, W.S.; Zhao, R.Y.; Zhang, X.; Jiang, J.C.; Chen, Y.C.; Wang, D.H.; Ying, L.J.; Liu, Z.Q. The EPMA and LA-ICP-MS in situ geochemical features of pyrrhotite and pyrite in Dabaoshan Cu-polymetallic deposit, North Guangdong province, and their constraint on genetic mechanism. *Acta Geosci. Sin.* **2019**, *40*, 291–306, (in Chinese with English abstract).
42. Bajwah, Z.U.; Seccombe, P.K.; Offler, R. Trace element distribution, Co:Ni ratios and genesis of the Big Cadia iron-copper deposit, New South Wales, Australia. *Miner. Depos.* **1987**, *22*, 292–300. [CrossRef]
43. Wang, Q.L.; Zhang, J.Y.; Yan, D.T.; Min, H.; Liu, S.; Li, C. Genesis of ore deposits indicated by trace elements of chalcopyrite. *Bull. Geol. Sc. Tech.* **2021**, 1–19, (In Chinese with English abstract). Available online: https://kns.cnki.net/kcms/detail/42.1904.P.20211009.0839.012.html (accessed on 20 September 2022).
44. Sun, Y.C. Mineralogy of Chalcopyrite from the Typical Tungsten and Lead-Zinc Deposits in Southern Jiangxi. Master's Thesis, China University of Geosciences (Beijing), Beijing, China, 2017. (In Chinese with English abstract).
45. Butler, I.B.; Nesbitt, R.W. Trace element distributions in the chalcopyrite wall of a black smoker chimney: Insights from laser ablation inductively coupled plasma mass spectrometry (LA-ICP-MS). *Earth Planet Sci. Lett.* **1999**, *167*, 335–345. [CrossRef]
46. Reich, M.; Palacios, C.; Barra, F.; Chryssoulis, S. "Invisible" silver in chalcopyrite and bornite from the Mantos Blancos Cu deposit, northern Chile. *Eur. J. Mineral.* **2013**, *25*, 453–460. [CrossRef]
47. Merkle, R. The systematics of Co, Ni, and Cu contents in pyrrhotite from the upper zone of the western Bushveld Complex, South Africa. *S. Afr. J. Geol.* **2008**, *111*, 187–200. [CrossRef]
48. Cook, N.J.; Ciobanu, C.L.; Pring, A.; Skinner, W.; Danyushevsky, L.; Shimizu, M.; Saini-Eidukat, B.; and Melcher, F. Trace and minor elements in sphalerite: A LA–ICP–MS study. *Geochim. Cosmochim. Acta* **2009**, *73*, 4761–4791. [CrossRef]
49. Dare, S.A.S.; Barnes, S.J.; and Prichard, H.M. The distribution of platinum group elements and other chalcophile elements among sulfides from the Creighton Ni–Cu–PGE sulfide deposit, Sudbury, Canada, and the origin of Pd in pentlandite. *Miner. Depos.* **2010**, *45*, 765–793. [CrossRef]
50. George, L.L.; Cook, N.J.; Crowe, B.B.P.; Ciobanu, C.L. Trace elements in hydrothermal chalcopyrite. *Miner. Mag.* **2018**, *82*, 59–88. [CrossRef]
51. Lu, S.Y.; Ren, Y.S.; Yang, Q.; Sun, Z.M.; Hao, Y.J.; Sun, X.H. Ore genesis for stratiform ore bodies of the Dongfengnanshan copper polymetallic deposit, NE China: Constrains from LA-ICP-MS in situ trace elements and sulfide S-Pb isotopes. *Acta Geol. Sin.-Engl.* **2019**, *93*, 1591–1606. [CrossRef]
52. Durazzo, A.; Taylor, L.A. Exsolution in the mss-pentlandite system: Textural and genetic implications for Ni-sulfide ores. *Mineral. Depos.* **1982**, *17*, 313–332. [CrossRef]
53. Luque, F.J.; Crespo-Feo, E.; Barrenechea, J.F.; Ortega, L. Carbon isotopes of graphite: Implications on fluid history. *Geosci. Front.* **2012**, *3*, 197–207. [CrossRef]
54. Crespo, E.; Luque, F.J.; Rodas, M.; Wada, H.; Gervilla, F. Graphite-sulfide deposits in Ronda and Beni Bousera peridotites (Spain and Morocco) and the origin of carbon in mantle-derived rocks. *Gondwana Res.* **2006**, *9*, 279–290. [CrossRef]

minerals

MDPI

Article

Microscopic Characteristics and Formation Mechanism of Effective Reservoirs in the Xihu Depression, China: The Important Role of the Poikilotopic Calcite Cements in Tide-Dominated Delta Systems

Songxu Zhang [1], Keqiang Wu [2], Na Liu [1,*], Xiaolei Peng [1] and Ying Chen [3]

[1] College of Earth Sciences, Jilin University, Changchun 130061, China
[2] Hainan Branch, CNOOC China Limited, Haikou 570000, China
[3] CNOOC Research Institute, Beijing 100028, China
* Correspondence: liuna_jlu@126.com or liuna0620@jlu.edu.cn; Tel.: +86-138-4409-0821

Abstract: The Xihu depression is an offshore sag located on the East China Sea Shelf Basin, which is currently one of the major oil and gas basins along the coast of China. In this study, an integrated approach using thin sections, scanning electron microscopy (SEM), X-ray diffraction (XRD), cathodoluminescence (CL), high-resolution 3D CT core scanning and stable isotope analysis was applied to examine the diagenetic evolution and investigate the microscopic characteristics and formation mechanisms associated with effective reservoirs. Four types were distinguished: upper conventional reservoirs (UC reservoirs), lower conventional reservoirs (LC reservoirs), "bottom calcium" low-permeability reservoirs ("bottom calcium" reservoirs) and "MI clay" low-permeability reservoirs ("MI clay" reservoirs). Poikilotopic calcite cements play an important role in the diagenetic alterations and reservoir quality evolution, precipitating during early eogenesis, provided a framework that retards the adverse impacts of UC reservoirs by compaction. Conversely, in LC reservoirs, with limited poikilotopic calcite, secondary porosity is mostly due to the dissolution of feldspar or unstable rock fragments. UC reservoirs normally develop in the middle of tidal channels and in subaqueous distributary channels, with the base of the sand-body being extensively cemented by carbonate cements, such as late calcite, Fe-calcite and dolomite, which formed the "bottom calcium" reservoir. Combined evidence from petrographic and geochemical analyses suggests that calcite precipitates from diagenetic fluids of mixed marine and meteoric waters, with additional external sources from calcareous siltstones and bioclasts. The carbon sources of calcite mostly originate from the dissolution of carbonates clacts or bioclasts within sandstone beds or adjacent silty mudstones, while dolomite cements have an isotopic composition that is more comparable to the generation of biogenic methane. This study demonstrates how poikilotopic calcite, developed in tide-dominated delta systems, affects the vertical heterogeneity. The results can be used to improve the reservoir evolution model of tide-dominated delta systems and provide a basic understanding for researchers conducting reservoir studies of similar sedimentary systems. Our results can act as a geological basis for further oil and gas exploration.

Keywords: poikilotopic calcite; microscopic characteristics; reservoir quality; tide-dominated delta system; Pinghu Formation; Xihu depression

1. Introduction

The heterogeneity of sandstone reservoirs, which significantly affects hydrocarbon accumulation and producibility, is controlled by the geometry and internal structures of sand bodies, provenance, grain size, sorting and by the modes and distribution of diagenetic processes [1–8].

Calcite cementation, as a significant factor that controls the sandstone reservoir quality through reductions in porosity and permeability, has been the focus of reservoir studies [9–15]. The negative correlation between the calcite cement content and porosity suggests that pervasive calcite cementation is the dominant diagenetic factor affecting porosity reduction. For example, in the Flemish Pass Basin sandstone reservoir, over 75% original porosity loss was shown to be primarily due to calcite cementation [16]. However, the calcite precipitated during eogenesis could provide a framework that may reduce adverse impacts due to compaction overburden and retain porosity until the degradation of organic matter at greater depths [15–17]. Finally, the dissolution of calcite will become a key process that can be used to improve the reservoir quality through deep burial.

Recent studies have documented early-formed calcite cement and examined its origin and diagenetic environment by petrographic, microthermometric, fluid-inclusion gas ratios and geochemical (stable isotopes, major and rare earth elements) analyses. Such research has been conducted in offshore Norway [9], the Hibernia oil field [18], the Jeanne d'Basin [17], the Carnarvon Basin [19], the states of Wyoming and Utah, USA [20], the southeast Benue Trough [21], the St. George Group of Western Newfoundland [22], the Betic Cordillera [23], the Powder River Basin [14] and the Flemish Pass Basin [16].

The Xihu depression is one of the important petroleum locations in the coastal area of China. It is located in the eastern depression of the East China Sea Shelf Basin. It provides a powerful tool to predict the distribution of diagenetic alterations controlling quality and heterogeneity. Previous authors have conducted significant research and analyses on the physical reservoir properties and main factors controlling the Huagang Formation in the central inversion structural belt [24–32]. The most recent drilling results conducted in Xihu Sag indicate that coal, carbonaceous mudstone and mudstone in Pinghu Formation are essential source rocks [33,34], and the sand-body of the Pinghu Formation is theoretically rich in coalbed methane and shale gas. The current investigation mostly focuses on hydrocarbon generation [35,36], paleo-pressure evolution [37], geochemical characteristics and the sedimentary control of coal-bearing source rocks from the Pinghu Formation [38].

It has been confirmed by drilling results that the reservoirs in the Pinghu Formation are also strongly heterogeneous and are characterized by a complex fractal geometry and a range of pore sizes. Calcite has been reported to be the most prevalent cement [24], but no major petrographic and diagenetic studies have yet investigated the reservoir sandstones. Therefore, research on the reservoir characteristics, spatial differences and the intrinsic relationships between the reservoir quality and diagenesis evolution will provide an efficient approach for better understanding and predicting porosity distribution, which may therefore assist in future hydrocarbon exploration and production. The main objectives of the study were: (1) to reconstruct the paragenetic sequence and identify the effective reservoir characteristics of sandstone, and (2) to elucidate the formation mechanism of effective reservoirs, especially the influence of carbonate cements on porosity development and distribution.

2. Geological Setting

The East China Sea Shelf Basin (ECSSB) is located in the offshore eastern area (Figure 2) with a general NNE trend [39]. The basin was mainly formed from the Late Cretaceous to Pliocene and has an area of 26.7×10^4 km^2. Seven main secondary tectonic units developed, including the Yangte depression, Taipei depression and East Zhejiang depression, and several secondary depressions or uplifts developed in each depression [40–43]. The Xihu depression is adjacent to the Hupijiao Uplifts in the north, Diaoyubei Depression to the south, Haijiao and Yushandong Uplift to west, and the Diaoyudao Uplift to the east (Figure 2). The Xihu depression has undergone several stages of tectonic movement and can be divided into five zones. From west to east, there is the Western slope belt (WSB), Western subsag (WSS), Central anticline belt (CAB), Eastern subsag (ESS), and eastern fault zone (EF). The western slope can be further divided into Hangzhou, Pinghu and the Tiantai slope belt from north to south. The WSB and WS are important gas-producing regions in the central and western parts of the Xihu depression.

The structural evolution of the Pinghu slope belt has experienced a Late Cretaceous-Eocene rifting stage and an Oligocene-present post-rifting stage. Compressional tectonic events, including Yuquan, Huagang and Longjing movements, occurred in the depression stage. The Paleocene, Baoshi and Pinghu formations were deposited in the rifting stage. During the Oligocene-Middle Miocene, extension across bounding faults slowed significantly. The Huagang group, Longjing group, Yuquan group, and Liulang group were deposited in this stage. After the very intense Longjing movement, which was the most widespread movement, Xihu Sag entered a regional subsidence stage, and the Santan group and the Donghai group were deposited [24,43,44]. From the Paleocene to Quaternary, sedimentary succession in the Xihu depression typically comprised clastic rocks reaching a maximum thickness of nearly 15,000 m [45,46] (Figure 1).

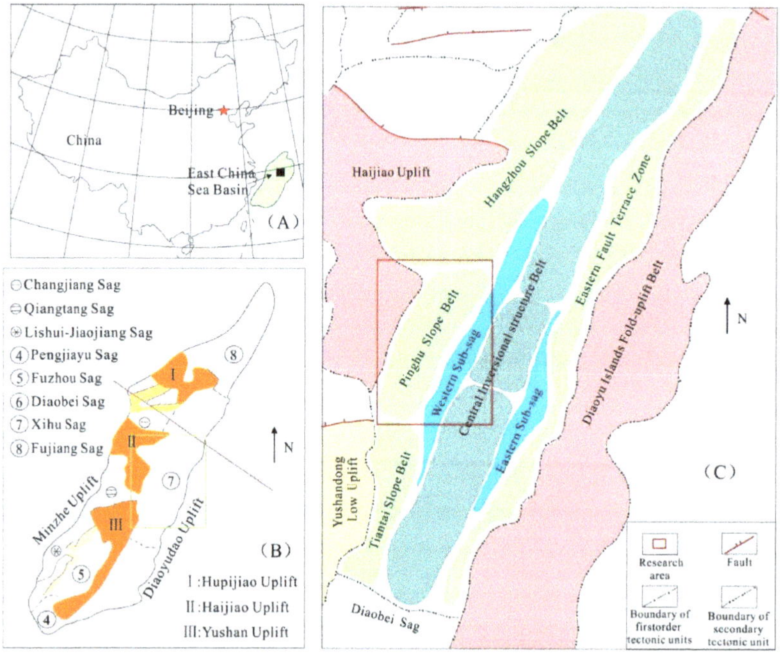

Figure 1. Schematic stratigraphic column of the Xihu Sag (modified from [24,31]).

The Eocene Pinghu Formation is a set of marine-continental transitional facies consisting of delta and tidal flats, which are characterized by sandstone, mudstone and coal interbed [49–51], The Pinghu Formation can be divided into three 3rd-order depositional sequences (SQ1–SQ3) (Figure 1). During SQ1, the sedimentary environment was a delta that was influenced by NE-SW-trending fault. During SQ2, it was a tidal flat and was greatly affected by tides. During SQ3, it was delta and tidal flat that were influenced by tides under wide backgrounds and gentle slopes [50].

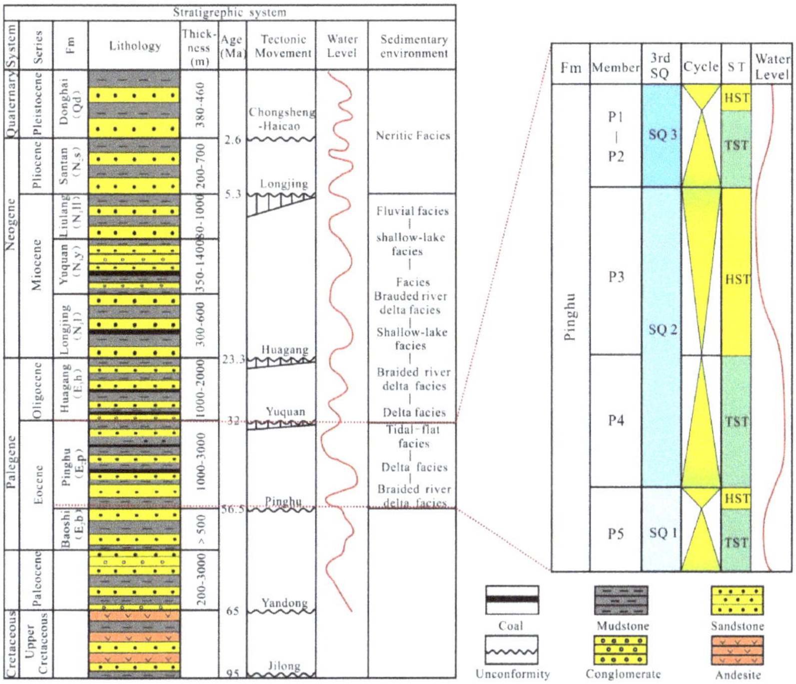

Figure 2. Structural characteristics and geographical location of the study area. (**A**) Geographical location of the East China Sea Basin (modified from [47]). (**B**) Geographical location of the Xihu depression (modified from [48]). (**C**) Tectonic division of the Xihu Depression.

3. Samples and Methods

This study focused on the Eocene Pinghu Formation (from P5 to P1) sandstones in the Xihu depression. Rock composition and section samples, reservoir porosity and permeability data from a total of 640 samples from the drill cores of 15 wells in the Pinghu slope belt were collected from the Shanghai Branch of CNOOC Ltd. Due to the constraints of the collected data, 24 core samples were selected from 9 wells and prepared for thin sections with depths ranging from 3445 to 4615 m.

Statistical analyses were carried out on sandstone framework compositions and the main cement components. The sandstone framework was quantified by counting 300–400 points per thin section, and the main composition of cement was marked by 10 views of straight lines with crosswire under a polarizing microscope as the reference. These uncovered thin sections were also stained with Alizazin Red-S and potassium ferricyanide solutions to recognize the carbonate cement types [52]. Ten selected samples were studied under cathodoluminescence (CL) with a Cambridge Image Technology instrument CL8200 and MK4 model. Cast thin sections were prepared on another 24 sections and filled with blue epoxy in order to identify the pores. Ten core samples were analyzed by scanning electron microscopy (SEM) in which they were observed under a scanning electron microscope at an acceleration voltage of 20 keV.

The whole rock mineral compositions of the samples were analyzed by X-ray diffraction at the Key Laboratory of Mineral Resources Evaluation in Northeast Asia Ministry of Natural Resources using DX-2700X ray diffraction, and the analysis was based on JY/009–1996 transtarget polycrystalline X-ray diffraction.

Analyses on carbon and oxygen stable isotope compositions were performed on 13 selected core samples using the MAT253 isotopic ratio mass spectrometer in the Institute of Geology and Geophysics, Chinese Academy of Sciences. The samples were selected on the basis that the carbonate minerals constituted in these 13 samples dominated in mesodiagenesis according to the petrographic observation. Whole powder samples were reacted with 100% orthophosphoric acid at 25 °C for 6 h and 48 h to extract CO_2 [53]. The different reaction rates of the different carbonate minerals are responsible for the reaction time selection. Thus, CO_2 from the 6 h reaction was prepared for the C and O isotopic analysis for calcite, and CO_2 from the 48 h reaction was used for dolomite, respectively. The final isotope data are presented per mil with respect to the PDB standards for carbon (C) and oxygen (O).

High-resolution 3D CT core scanning was performed on three selected samples with significant different microscopic characteristics, to carry out three-dimensional (3D) visualization and to finely quantify the characterization of rock microstructures and pore connectivity. The scanning was conducted at the Institute of Geology, Chinese Academy of Geological Sciences with a NiKon XTG225ST Industrial CT.

4. Results

4.1. Sandstone Petrology

The sandstone is mainly fine to very fine-grained feldspathic litharenite and lithic arkose (Figures 3 and 4) and some medium to coarse-grained sandstones are also found in P4 and P5. The most common detrital grains are quartz (30%~96% of the rock volume, average 68.95%), feldspar (2%~34%, average 12.55%) and rock fragments (2%~36% average 18.5%) (Table 1). The feldspar population is dominated by K-feldspar but also includes plagioclase. The rock fragments are mainly composed of granite, metamorphic and pyroclastic rock. The sorting degree of sandstone is medium to good, and the roundness is medium to poor, appearing as mainly subangular to subspherical grains.

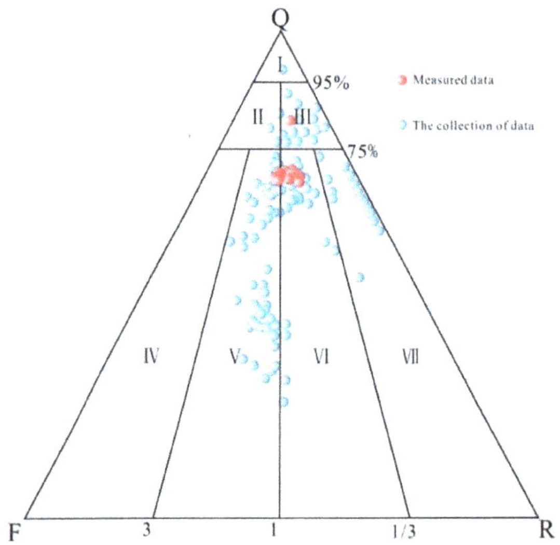

Figure 3. QFR triangular diagram of sandstone petrography showing the composition of the sandstones. Q, quartz; F, feldspar; R, rock fragments. I, quartz arenite; II, subarkose; III, sublitharenite; IV, arkose; V, lithic arkose; VI, feldspathic litharenite; VII, litharenite.

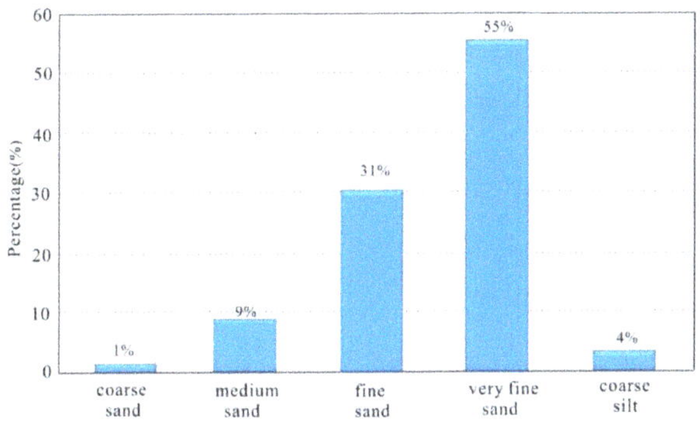

Figure 4. Distribution of grain size of sandstones (collected from Shanghai Branch of CNOOC Ltd., in Pinghu Formation).

Table 1. XRD data of Pinghu Formation.

Well	Depth (m)	Sedimentary Microfacies	Q (%)	Kp (%)	Pl (%)	Cc (%)	Dol (%)	Bit (%)	Clay (%)
A1	4202.46	Subaqueous distributary channel	61	12	23	0	0	1	3
A1	4196.54	Subaqueous distributary channel	82	7	10	0	0	0	1
A1	4202.77	Subaqueous distributary channel	82	2	3	2	0	0	1
A1	4197.37	Subaqueous distributary channel	96	2	0	1	0	0	1
A2	4183.87	Tidal channel	60	6	4	0	30	0	0
A2	4183.62	Tidal channel	30	0	5	0	60	0	5
A2	4186.50	Tidal channel	54	6	28	0	1	1	10
A3	3445.29	Tidal channel	30	15	12	5	3	2	33
A3	3446.57	Tidal channel	70	4	20	0	0	2	4
A3	3447.27	Tidal channel	51	29	5	0	15	0	0
A3	3443.67	Tidal channel	67	14	16	1	0	0	2
A5	4088.16	Mixed flat	60	2	23	10	2	1	2
A5	4089.26	Mixed flat	38	11	23	18	3	1	6
A5	4087.06	Tidal channel	60	6	22	7	1	1	3
A7	4348.82	Subaqueous distributary channel	40	5	29	14	2	2	8
A7	4057.18	Subaqueous distributary channel	70	25	0	0	0	0	5
A7	4055.64	Subaqueous distributary channel	67	30	0	0	0	0	3
A8	4343.57	Subaqueous distributary channel	40	2	15	40	0	0	3
A8	4340.74	Subaqueous distributary channel	49	4	24	9	1	1	12
A8	4342.54	Subaqueous distributary channel	68	2	22	8	0	0	0
A9	4220.00	Subaqueous distributary channel	50	5	10	12	14	0	7
A9	4610.80	Subaqueous distributary channel	65	30	0	0	0	0	5
A9	4612.10	Subaqueous distributary channel	58	10	15	3	1	1	12
A9	4615.00	Subaqueous distributary channel	62	16	14	6	0	0	1

Q—quartz; Kp—K—fledspar; Pl—plagioclases; Cc—calcite; Dol—dolomite; Bit—biotite; Clay—MI clay.

① values performed from Jilin university; ② values performed from Key of Laboratory Mineral Resources Evaluation in Northeast Asia of the Ministry of Natural Resources.

Blue symbols represent results collected from the Shanghai Branch of CNOOC Ltd. (from 13 wells) and red symbols are visual estimates from thin section using a conventional optical microscope (from 7 wells).

There are also grains that are typical of pseudomatrix and infiltrated clay minerals. Pseudomatrix consists of originally plastic particles (such as mudstone fragments) that have been plastically deformed or even broken under compaction, leading to a change in their original shape and the filling of pores between rigid particles, making them appear as interstitial clay materials (Figure 5a). Pseudomatrix mainly develops at the bottom of tidal flat deposits with a maximum content of 15%. Mechanically infiltrated clay (MI clay) was found to be mainly attached to the surface of clastic particles (Figure 5b). Sandstone with MI clay was mainly found in low-energy zones or in the transition zones from high energy to low energy, such as the subaqueous distributary bays and the top of subaqueous distributary channels in delta-front sediments.

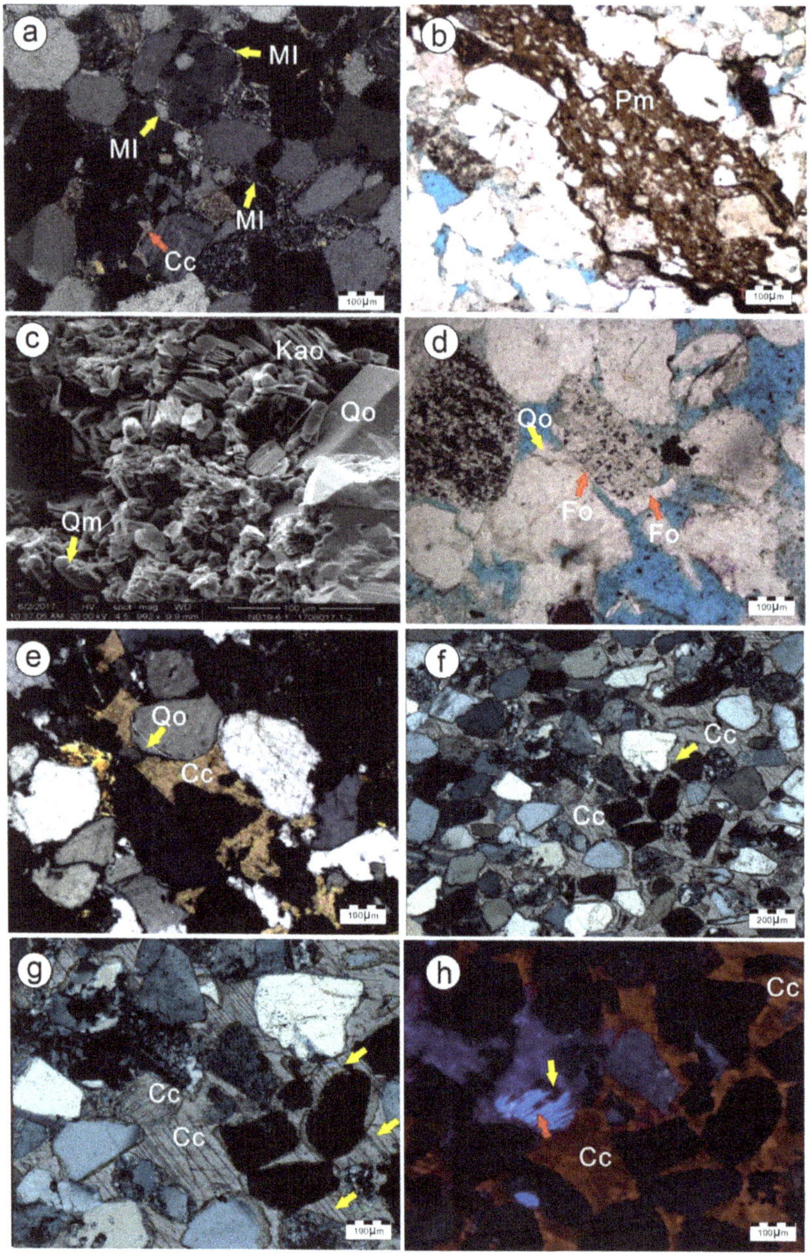

Figure 5. Photomicrographs of petrographic features. (**a**) MI clay fills pores with effective compaction, well A7, 4569 m, P3, thin-section micrograph, crossed polars; (**b**) pseudomatrix (Pm) is plastically deformed under compaction, and changes the original shape, filling the pores between rigid particles, well A8, 3443.47 m, P3, thin-section micrograph, single polars; (**c**) kaolinite (Kao), microcrystalline quartz (Qm) and quartz overgrowth (Qo) fill the pores, well A5, 4183.5 m, P3; SEM; (**d**) feldspar overgrowth (Fo) appears as a thin fringe on the edge of feldspar grains, the growth of quartz overgrowth (Qo) was limited by feldspar overgrowth (Fo), well A8, 3440.57 m, P3, thin-section micrograph, crossed polars; (**e**) quartz overgrowth (Qo) developed as syntaxial overgrowth around the detrital quartz grain, calcite (Cc) fills the residual pores, well A2, 4471.5 m, P2, thin-section micrograph,

crossed polars; (**f**) precipitation of calcite (Cc) cements completely filling intergranular pore areas and drastically reduced the porosity, framework grains exhibit contact relations ranging from point to line, indicating that compaction was not strong, well A1, 4202.77 m, P3,thin-section micrograph, crossed polars; (**g**) large calcite crystals present in the form of poikilotopic cement surrounding the detrital grains of the sandstones, twinning planes of calcite cements (marked by yellow arrows) occurring uninterruptedly on either side of the grains, well A1, 4202.77 m, P3,thin-section micrograph, crossed polars; (**h**) calcite cements, reddish in CL, fills the pores and replaced blue luminescent feldspar (yellow arrow), feldspar (red arrow), well A1, 4202.77 m, P3, CL. MI-mechanically infiltrated clay; Pm-pseudomatrix; Kao-kaolinite; Qm-microcrystalline quartz; Qo- quartz overgrowth; Cc-calcite cement.

4.2. Authigenic Minerals

Authigenic minerals include quartz, feldspar, kaolinite, illite-smectite, calcite, Fe-calcite and dolomite, which were observed under the optical microscope and with cathodoluminescence and SEM.

Quartz cements occur as overgrowth and microcrystalline pore-filling cements. The content of quartz overgrowth in the area was 0%–3%, with an average value of 0.84%. Microcrystalline quartz is associated with quartz overgrowth and kaolinite, and its formation time may be similar to that of quartz overgrowth (Figure 5c).

Feldspar overgrowth was found to represent <1% of the rock volume on average, appearing as a thin fringe on the edge of feldspar grains. Feldspar overgrowth enlargement occupies the growth positions of grains and restricts the development of quartz overgrowth (Figure 5d).

Kaolinite cements appear as book-like aggregates (Figure 5c) that mostly develop in places where feldspar has extensively dissolved. Quartz overgrowths, microcrystalline quartz and kaolinite are genetically related to dissolution of feldspar and are inferred to be para-simultaneous.

Carbonate minerals mainly include calcite, Fe-calcite and dolomite. The content of carbonate cements varies from top to bottom: that is, the upper part (P1–P3) is higher than the lower part (P4–P5) (Figure 6). The calcite cement content was found to range from 1% to 30% with an average of 5.6%. It appears pink after staining with alizarin red-S (Figure 5g), which distinguishes it from other carbonate cements. Calcite appears as pore-filling cements or replaces feldspar (Figure 5h). Two generations of calcite cements were distinguished: pokilitopic calcite, which completely fills intergranular pore areas and drastically reduces the porosity (Figure 5f,g); and the second calcite (Calcite II) which precipitates in the limited intergranular pores (Figure 5e). In addition, the poikilotopic calcite is normally well developed in P2 and P3 Formations.

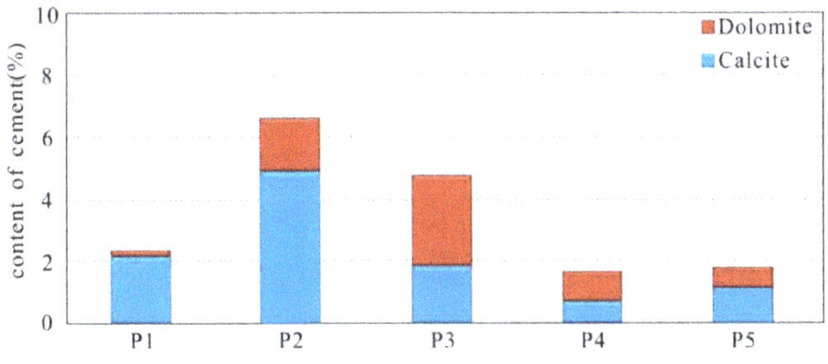

Figure 6. Content of carbonate cements in Pinghu Formation.

Under the CL system, calcite cements can be classified into two generations. Calcite I is dull to non-luminescent, appearing around the edges of debris (Figure 7e), while Calcite II is bright red, filling the residual pores left by calcite I (Figure 7e).

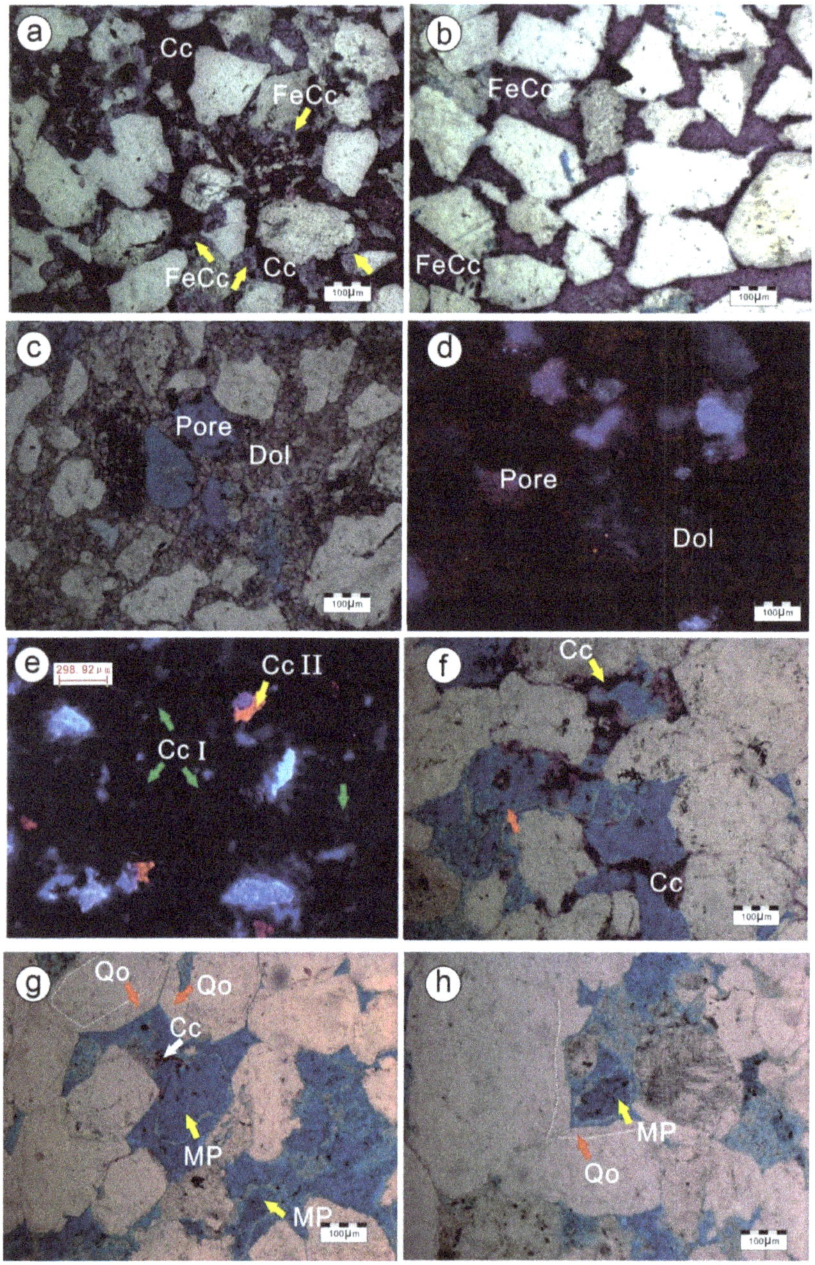

Figure 7. Photomicrographs of characteristic petrography and conventional reservoir. (**a**) calcite (Cc) engulfed by Fe-calcite (FeCc), well A9, 4220 m, P3, stained thin-section micrograph, single polars; (**b**) Fe-calcite (stained as purple red) fills pores forms "bottom calcium", the grain contacts are mainly point contacts indicating that compaction was not strong, well A7, 4202.7 m, P3, stained thin-section micrograph, single polars; (**c**) euhedral rhomb dolomite fills pores, note the development of secondary intergranular pores, well A4, 4183.8 m, thin-section micrograph, single polars; (**d**) euhedral rhomb dolomite among grains, well A2, 4183.87, P3, CL; (**e**) two generation of calcite cements with different luminescence characteristics, Calcite I (CcI) is dull to non-luminescent (green arrow), appearing around the edge of debris, Calcite II (CcII) is bright orange (yellow arrow), filling the residual pores,

well A5, 4183.8 m, P3, CL; (**f**) the extensive developed of secondary intergranular and intragranular pores (red arrow), with some calcite cements stained as red, well A6, 4203 m, P3, stained thin-section micrograph, single polars; (**g**) the grains were in point to long contact, with extensive development of intergranular pores, partial dissolution of feldspar grains create oversize secondary porosity, few calcite (stained as red) can be seen in the pores, well A6, 4203 m, P3, thin-section micrograph, single polars; (**h**) the grains were in long and concave–convex grain contacts; note the development of moldic pore, well A6, 4203 m, P5, thin-section micrograph, single polars. FeCc- Fe-calcite; Cc- calcite; Dol- dolomite; MP-moldic pore; Qo-quartz overgrowth; F-feldspar.

Dolomite and Fe-calcite are late carbonate cements and were found to have average contents of 3% and 5%, respectively. They are mostly distributed between clastic particles with point and long contacts (Figure 7b,c). Fe-calcite usually grows at the edge of calcite (Figure 7a), suggesting that Fe-calcite grows after calcite cement. Dolomite fills the spaces between particles, developing euhedral crystal faces (Figure 7c,d). In the study area, no mineral was shown to have formed after the Fe-calcite and dolomite cements.

4.3. Dissolution

Two secondary pore development zones developed longitudinally: the upper dissolution zone (3200–3500 m) and the lower dissolution zone (4000–4400 m). This large-scale dissolution has also affected the framework grains and accessory minerals and therefore, altered the original composition of the sandstones.

4.4. Paragenesis

Figure 8 summarizes the paragenetic sequence of the sandstones based on petrographic relationships established from the petrographic examination as follows: (1) pervasive precipitation of early poikilotopic calcite cements in high-energy zones; (2) precipitation of feldspar overgrowth; (3) major feldspar dissolution followed by quartz cement and kaolinite precipitation; and (4) precipitation of Calcite II, Fe-calcite, and dolomite.

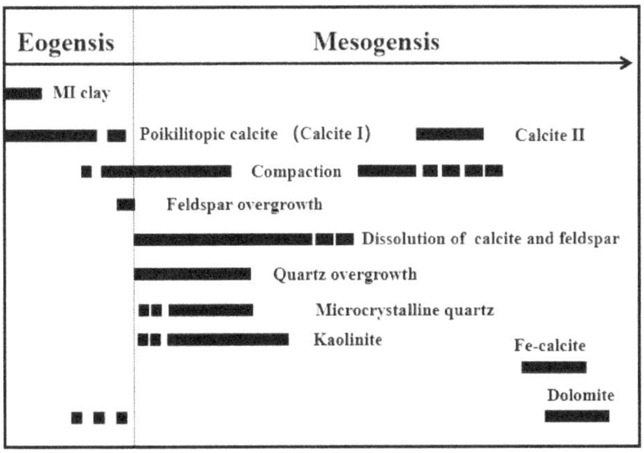

Figure 8. MI clay minerals and paragenetic sequence of the sandstones of the Pinghu Formation based on petrographic relationships.

4.5. *Effective Reservoir Characteristics*

Effective reservoirs include conventional reservoirs (k > 10 md) and low-permeability reservoirs (k = 0.2–10 md) [54]. The microscopic characteristics and 3D CT scanning of conventional reservoirs and low-permeability reservoirs were comprehensively analyzed to provide evidence for further definition of the formation mechanisms of effective reservoirs.

4.5.1. Microscopic Characteristics

Conventional reservoirs with relatively high porosity and permeability mainly develop in tidal channels and subaqueous distributary channels and can be divided into two types, upper conventional reservoirs (UC reservoirs) and lower conventional reservoirs (LC reservoirs), based on their different microscopic characteristics. UC reservoirs with fine-grained sandstones mainly developed in P1 to P3, while LC reservoirs with mainly medium and coarse grains developed in P4 to P5. Compared with the UC reservoirs, LC reservoirs have significantly higher contents of lithic grains and lower contents of feldspar (Figure 9).

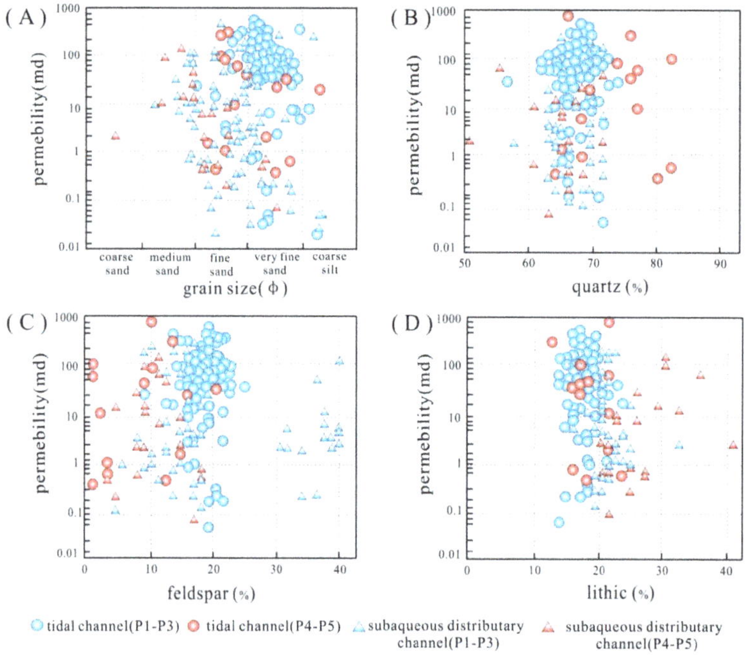

Figure 9. Cross plots of permeability versus (**A**) grains size; (**B**) quartz; (**C**) feldspar; (**D**) lithic.

The detrital grains in UC reservoirs are mainly fine-grained, presenting point to long grain contacts (Figure 7g), while detrital grains in LC reservoirs are mainly medium and coarse, showing long to concave–convex grain contacts (Figure 7h). Extensive dissolution occurs in both types of reservoirs.

Low-permeability reservoirs can also be divided into two types: "bottom calcium" low-permeability reservoirs ("bottom calcium" reservoirs) and "MI clay" low-permeability reservoirs ("MI clay" reservoirs).

"Bottom calcium" is used to describe the carbonate minerals that mainly develop at the base of the permeable layer at the interphase between the sand layer and impermeable layer. These carbonate minerals completely fill the intergranular pore and enclose several grains, simultaneously reducing the porosity drastically. "Bottom calcium" reservoirs are normally characterized by low permeability and high porosity with poikilotopic carbonate and euhedral rhomb dolomite cements filling in the intergranular pores (Figures 5g and 7a–f). They mainly develop at the bottom of tidal channels and subaqueous distributary channels. Poikilotopic carbonate cements include calcite and Fe-calcite. The debris grains are mainly point contacts. This indicates that without early effective compaction, cementation is the main factor that reduces the porosity and permeability.

"MI clay" reservoirs with low permeability and porosity have mainly developed in the low-energy zones and the transition zone from high energy to low energy, such as the subaqueous distributary bay and at the top of the subaqueous distributary channel. The most typical feature is the MI clay minerals that fill the intergranular pores (Figure 5a), with the clastic grains mainly in tangential, flattened contact and even concave–convex contact positions, which indicates effective compaction. Compaction and MI clay minerals filling are the main factors that will reduce porosity and permeability.

4.5.2. Pore Structure Based on High-Resolution 3D CT Scanning

The three-dimensional (3D) visualization and pore connectivity results for the UC reservoir A1(4196.54 m) showed a porosity of 12.5% and a permeability of 20.8 mD, as shown in Figure 10A. After CT scanning, the total porosity was found to be 9.44% (Table 2), the isolated porosity was 0.39% and the accessible porosity was 9.04%, indicating that the pores are well connected.

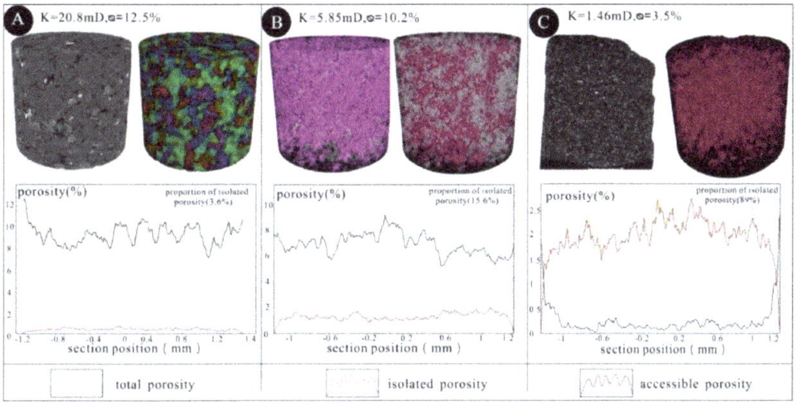

Figure 10. The three—dimensional (3D) visualization and the pore connectivity on three selected samples. (**A**) results for the UC reservoir, Well A1(4196.54 m). (**B**) results for "bottom calcium" reservoir, Well A5(4088.16 m). (**C**) results for "MI clay" reservoir, A7(4057.18 m).

Table 2. High-resolution 3D CT scanning of typical reservoirs in Pinghu Formation.

Well	Depth (m)	Average Diameter of Pore (mm)	Diameter Standard Deviation of Pore (mm)	Accessible Volume (mm^2)	Isolated Volume (mm^2)	Total Volume (mm^2)	Accessible Porosity (%)	Isolated Porosity (%)	Total Porosity (%)
A1	4196.54	0.03	0.02	1.42	0.62	1.48	9.04	0.39	9.44
A5	4088.16	0.01	0.01	0.06	0.48	0.54	0.24	1.95	2.19
A7	4057.18	0.12	0.01	0.87	0.16	1.00	7.00	1.30	8.30

Institute of Geology, Chinese Academy of Geological Sciences.

For "bottom calcium" reservoir A5(4088.16 m) was found to have a porosity of 3.5% and a permeability of 2.19 mD, as shown in Figure 10C. After CT scanning, the total porosity was found to be 2.19%, the isolated porosity was 1.95% and the accessible porosity was 0.24% (Table 2), indicating that most pores were isolated.

The "MI clay" reservoir A7(4057.18 m) was found to have a porosity of 10.2% and a permeability of 5.85 mD, as shown in Figure 10B. After CT scanning, the total porosity was 8.3%, the isolated porosity was 1.3% and the accessible porosity was 7% (Table 2). Compared with the "bottom calcium" reservoir, the "MI clay" reservoir was found to have a larger pore radius and more connected pores.

4.6. C and O Isotopic Data

As shown in Table 3 and Figure 11, the calcite $\delta^{13}C$ PDB values were from −3.3‰ to 3.3‰ and with an average value of 0.98‰; the $\delta^{18}O$ PDB values for calcite were −17.5‰ to −10.2‰ with a mean value of −13.4‰. The dolomite $\delta^{13}C$ PDB values were from 8.4‰ to 9.3‰, with an average value of 8.85%; The $\delta^{18}O$ PDB values were −11.4‰ to −11.1‰ with a mean value of −11.25‰.

Table 3. Carbon and oxygen isotopic composition for calcite and dolomite.

Number	Well	Depth (m)	Mineral	$\delta^{13}C_{V\text{-}PDB}$ (%)	$\delta^{18}O_{V\text{-}PDB}$ (%)	Z
01	A1	4202.77	calcite	−1.4	−17.5	115.72
02	A2	4183.87	dolomite	9.3	−11.1	140.82
03	A2	4183.62	dolomite	8.4	−11.4	138.83
04	A3	3446.57	calcite	−1.4	−12.6	118.16
05	A3	3447.27	calcite	2.5	−10.2	127.34
06	A5	4088.16	calcite	−0.4	−12.5	120.26
07	A5	4089.26	calcite	−3	−15.3	113.54
08	A5	4087.06	calcite	3.3	−12.6	126.15
09	A7	4348.82	calcite	−2.7	−16.1	113.75
10	A9	4220.00	calcite	−2.9	−13.4	114.69
11	A9	4610.80	calcite	−0.5	−12.3	120.15
12	A9	4612.10	calcite	−0.1	−10.6	121.79
13	A9	4615.00	calcite	−3.3	−14.1	113.52

Values per Center of Analysis, Beijing Research Institute of Uranium Geology.

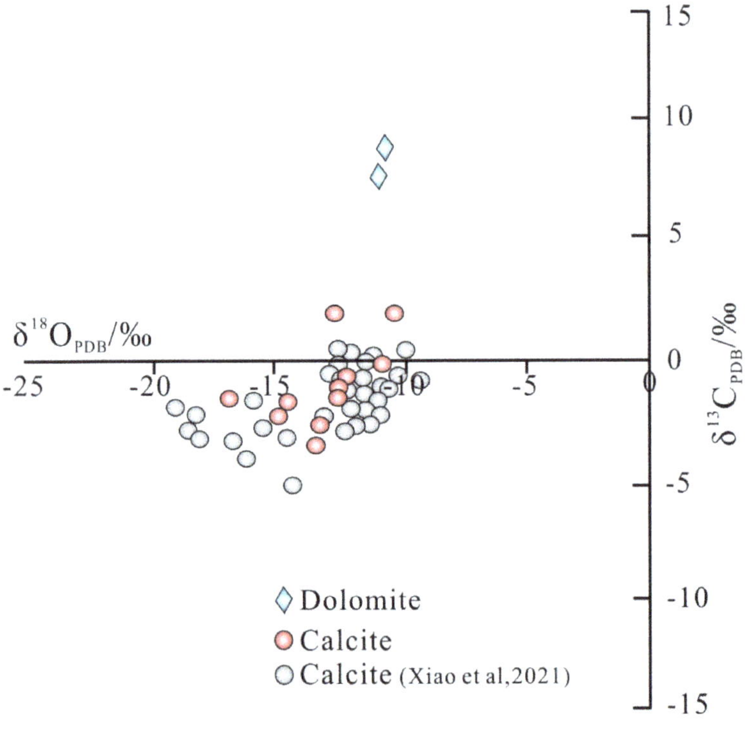

Figure 11. $\delta^{13}C$ (PDB) versus $\delta^{13}O$ (PDB) diagram of the carbonate cements (modified from [55]).

5. Discussions

5.1. Formation Mechanism of Effective Reservoir

5.1.1. The Upper Conventional Reservoir (UC Reservoir)

Calcite precipitated during eogenesis could provide a framework that may retard the adverse impacts of compaction, and the intergranular porosity keeps any potential increases during decarbonatization at greater depths [15–17]. Thus, the dissolution of early poikilotopic calcite will become a key process to improve reservoir quality through deep burial.

The petrographic examination conducted in this study suggests that poikilotopic calcite was precipitated during early eogenesis. The grains in the UC reservoir were found to be mainly point and long contacts (Figure 7g). This is significantly different from the LC reservoir, which was shown to have long to concave–convex grain contacts (Figure 7h). The main reason for this is that a large amount of calcite cement precipitated in the early diagenetic stage, occupying intergranular pores in advance, which retarded further compaction quartz overgrowth developed in the UC reservoir (Figures 5d and 7g), while no quartz overgrowth was found in the "bottom calcium" reservoir (Figures 5f,g and 7a–f). Some quartz grains were found to exhibit concave–convex contacts in the UC reservoir (Figure 7g), which is consistent with the expected continuous compaction with progressive burial. Some of the "bottom calcium" calcite cements were found to partially exhibit dull to non-luminescent characteristics (Figure 7e), suggesting that the calcite cements precipitated from a Fe-rich homogeneous fluid composition in a single event, while the late calcite presumably nucleated on earlier eogenetic calcite.

The oxygen isotopic compositions for intergranular calcite cements (−17.5‰~−10.2‰, PDB) were found to be significantly lower. It is suggested that the mixing of marine water with meteoric water might account for the similar depletion of $\delta^{18}O$ values in offshore environments [16,56–58], and positive $\delta^{18}O$ water could indicate evolved marine pore water due to interaction with silicate minerals during burial diagenesis [59]. The plot in Figure 12 gives an indication of oxygen isotope fractionation during calcite crystallization as a function of temperature and fluid composition [60,61]. The Pinghu Formation in the studied well has reached a maximum burial depth of ~4700 m, which corresponds to a temperature of approximately 150 °C by assuming a geothermal gradient of 35 °C/km. The calcite cements precipitated during mesodiagenesis would do so with the temperature being higher than 85 °C. The measured $\delta^{18}O$ of calcite (−14‰~−10‰ PDB, as the main distribution interval) and their approximate temperature of precipitation (85 °C~150 °C), imply that the $\delta^{18}O$ values of their parent diagenetic fluids were approximately between −1.5‰ and +7‰ SMOW (Figure 12), which corroborate the fluid evolution from marine and meteoric mixing water to basinal fluids. This implies that calcite cements formed in a later burial setting with a higher temperature.

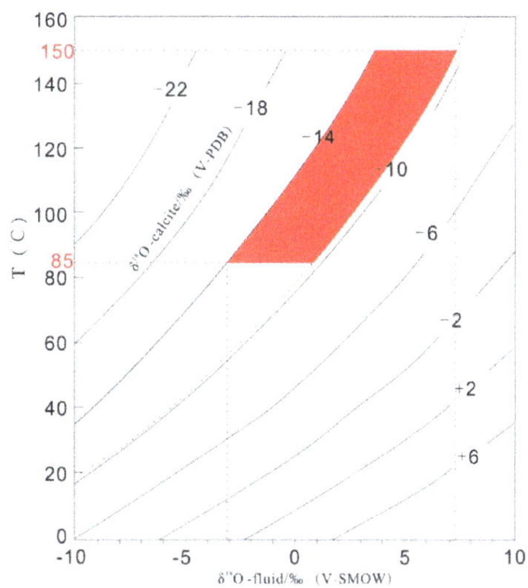

Figure 12. Oxygen isotope values of calcite plotted on the temperature—dependent, calcite—water oxygen fractionation curve ([60,61]).

The investigated late calcites $\delta^{13}C$ ranged form −3.3‰ to 3.3‰ PDB, which implies that they are likely derived from mixed sources, possibly with a contribution from marine carbonate or bioclast. The abundance of carbonate clasts and bioclasts within the sandstones provides solid evidence. In addition, calcareous siltstones adjacent to the sandstones may provide additional sources for the calcite via pore fluids expelled by compaction [12,62]. The dissolution of carbonate clasts might be induced by hot basinal fluids (e.g., hydrocarbon and coals), which brings some negative $\delta^{13}C$ signals [62–64].

The calcium involved in the formation of late diagenetic calcite can have several potential sources: the dissolution and re-precipitation of marine-derived bioclasts [65], the albitization of Ca-plagioclase [66] and the dissolution of volcano detrital fragments [67]. With the well-preserved carbonate clasts in the Pinghu Fm., it is, therefore, more likely that Ca was probably derived from the dissolution of the abundant bioclasts.

Early poikilotopic calcite normally develops in tidal channels and subaqueous distributary channels of tide-dominated delta fronts. Under the background of high-frequency fluctuations of the sea level, it was intermittently exposed to sea water. With the joint action of atmospheric fresh water and sea water, it is easy to have calcium cementation and a high content of carbonate cement. The TST tidal-dominated delta and HST tidal of sandstones have higher potential to be cemented by extensive calcite because of the abundance of carbonate bioclasts and organic matter [8,68,69]. A previous study [70] suggested that sediments were likely deposited in a TST system that was exposed to meteoric influence.

The development of secondary porosity (mostly enlarged pores) and throats by dissolution was the key process that improved the reservoir quality. The secondary porosity evolution might be related to the organic acids produced from the maturation of hydrocarbon and coals [68,71].

This early poikilotopic calcite has been also documented and studied by Hutcheon et al. (1985), Hesse and Abid (1998), Liu et al. (2014), and Xiong et al. (2016). In the Ti-3 sandstones of Flemish Pass Basin, over 75% of the initial porosity was lost through early calcite cementation, while the development of secondary porosity by later calcite dissolution improved the reservoir quality [16].

5.1.2. The Lower Conventional Reservoir (LC Reservoir)

LC reservoirs (developed in P4 and P5) are characterized by medium to coarse grains, extensive dissolution and long to concave–convex contacts. The sandstones in these areas have undergone effective compaction. LC reservoirs also develop in tidal channels and subaqueous distributary channel sandstones, with tidal channels providing better reservoirs than subaqueous distributary channels (according to 1203 data from 15 wells).

Secondary porosity (mostly enlarged, moldic pores) is extremely important in the LC reservoirs and is principally generated through the dissolution of detrital feldspar and later calcite due to the maturation of organic matter (e.g., hydrocarbon and coals). Compared with UC reservoirs, LC reservoirs have significantly higher lithic fragment contents and lower feldspar contents (Figure 9). We suggest that the sandstone variability in the quartz/feldspar ratio is related to differences in the degree of feldspar reaction and the locally associated secondary porosity, with most reactions being represented by low total feldspar contents. Secondary porosity produced by feldspar and carbonate dissolution has been noted by other authors in oversized pores, occurring locally in the Xihu depression [24,72]. However, as the content of late calcite is limited and the dissolution of substantial amounts of carbonate in pore spaces is harder to recognize, it is extremely difficult to provide unequivocal evidence for significant porosity generation by cement dissolution [73]. With the clear extensive generation of secondary porosity via feldspar dissolution, feldspar dissolution is considered to be the most important factor associated with the production of secondary porosity in LC reservoirs.

5.1.3. "Bottom Calcium" Low-Permeability Reservoir ("Bottom Calcium" Reservoir)

The clastic particles of "bottom calcium" reservoirs are mainly in point contacts, which indicates that, without early effective compaction, cementation is the main factor that reduces porosity and permeability. The connectivity between pores is poor, with most of the pores showing the isolated pores.

We consider that the source of carbon isotopes in "bottom calcium" reservoirs may differ between calcite and dolomite. The carbon sources of "bottom calcium" calcite are discussed in Section 5.1.1, and our results imply that calcite originated mostly from the diffusion of dissolved Ca^{2+} and HCO_3^- through carbonate clasts and bioclast dissolution induced by hot basinal fluids (e.g., hydrocarbon and coals). On the other hand, the dolomite cements have an isotopic composition that is more comparable with the generation of biogenic methane with the high positive values of $\delta^{13}C$ [74]. Dominant generation of biogenic methane as the origin for dolomite is therefore favored here.

Poikilotopic calcite is inferred to have been, in some cases, precipitated early in the layers of the UC sandstones and then dissolved [16,69]. In addition, late-precipitated calcites formed a barrier and sealed the underlying hydrocarbons [69].

These "bottom calcium" reservoirs normally develop at the bottom of subaqueous distributary channels and tidal channels. According to the configuration relationship between "bottom calcium" reservoirs and UC reservoirs, high-quality reservoirs can be found overlying the "bottom calcium" reservoirs.

5.1.4. "MI Clay" Low-Permeability Reservoir ("MI Clay" Reservoir)

MI clay refers to the clastic clay attached to the surface of framework clastic particles [75]. The types of textures include ridges and bridges, geopetal fabrics, loose aggregates, cutans and massive aggregates [76]. Laboratory experiments and studies on Holocene Sands [75] have shown that the basis required for the development of MI clays are a high suspended sediment concentration, a fluctuating phreatic surface and minimal sediment disturbance. The development mechanism of MI clay involves a decline in the phreatic water level and the infiltration of clay-containing water [76]. The specific process is as follows: ① when the majority of the infiltrating water accumulates and evaporates above the phreatic level, the MI clays enrichment zone can develop within the vadose zone, giving rise to geopetal fabrics; ② when the infiltrating clay-containing water concentrates near the phreatic level, MI clays can accumulate in the phreatic zone to form an enrichment zone, characterized by the texture of cutans, which appear as lamellae oriented parallel to the grain surface; ③ When clay-containing water encounters an impermeable barrier (such as bedrock, shale layer, buried soils or caliche layers) during the process of infiltration, MI clays can accumulate at the base of the permeable layer as an enrichment zone; and ④ the above three modes are all controlled by the distance from the water source of the infiltrating water, where the closer the water source is, the greater the content of MI clays will be. The migration of the subaqueous channel will form a relatively thin MI clay enrichment zone with significant lateral extension. When the channel cuts deeper, it will form a relatively thick MI clay enrichment zone. At present, it has been recognized that ① the presence of MI clays reduces the initial porosity of sandstone [77]; ② MI clays form shrinkage pores [75]; and ③ the high content of MI clays greatly reduces the permeability and forms an impermeable barrier [8,75].

"MI clays" reservoirs develop in subaqueous distributary bays and at the top of subaqueous distributary channels.

5.2. *Vertical Distribution of the "Bottom Calcium" Reservoir and the UC Reservoir*

As the pervasive calcite cement has important effects on the reservoir quality, we tried to determine the vertical distribution of "bottom calcium" reservoir and UC reservoirs.

The typical distribution rule is that UC reservoirs normally develop in tidal channels and subaqueous distributaries, with the bottom of the sand-body being extensively cemented by carbonate cements ("bottom calcium" reservoir). The thickness of the "bottom calcium" is generally less than 1 m, with an average of 0.1–0.3 m. Taking well A2 as an example (Figure 13), the tidal channel sand-body developed at a depth of 4183.5–4184 m. Under the polarized light microscope, sandstone in the bottom of the tidal channel (4148 m) was shown to display typical "bottom calcium" cementation. The cement is dolomite, which fills almost all intergranular pores; at the depths between 4183.7 m and 4184 m, the

intergranular pores are partly occupied by dolomite, and a certain number of intragranular dissolution pores and even moldic pores also occur. Therefore, the porosity in this layer can be relatively high but with few effective (connectable) pores, so the permeability is low. From 0.4 m above the bottom of the sand-body, the UC reservoir begins to develop, and the permeability rapidly increases from 10.9 mD (at 4183.6 m) to 141 mD (at 4183.5 m). The particles mainly have point and long contact characteristics. Both intergranular and intragranular pores are present, with effective connected pores accounting for a relatively high proportion of the total. The thickness of the "bottom calcium" in A2 can be estimated as 0.3 m. A similar configuration relationship was also seen at well A1 (Figure 14), with the subaqueous distributary channel developed at depths of 4190–4202 m.

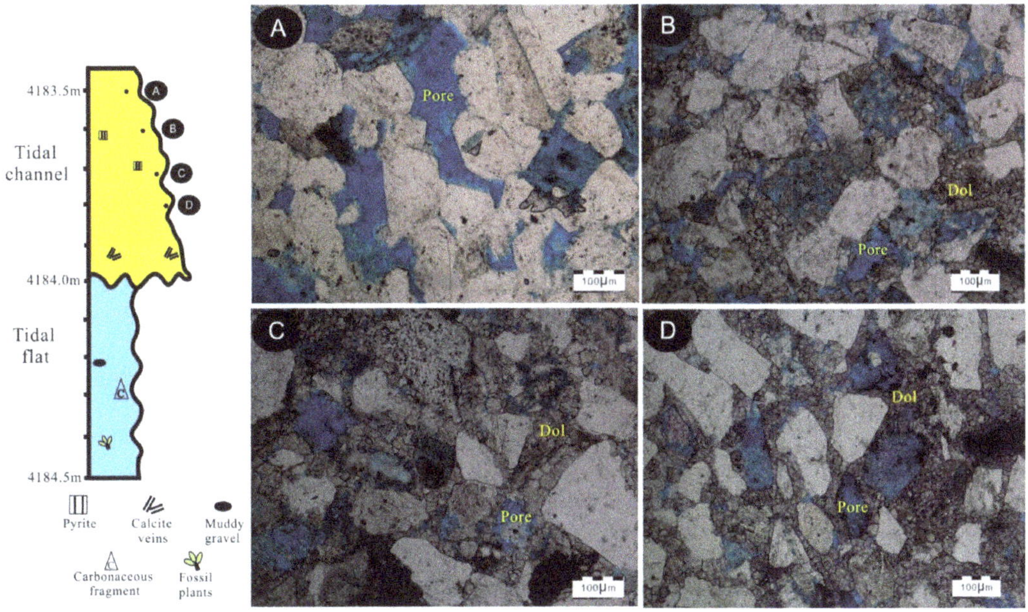

Figure 13. Typical configuration between UC reservoir and "bottom calcium" reservoir developed in tidal channel for well A2 at the depth of 4183.5–4184 m. (**A**) UC reservoir in uncemented zone with extensive secondary intergranular pores, 4183.5 m; K = 141 mD; (**B**) UC reservoir in cemented zone with secondary intergranular pores and dolomite cement, 4183.6 m, K = 10.9 mD; (**C**) "bottom calcium" reservoir cemented by dolomite with isolated pores, 4183.7 m, K = 0.267 mD; (**D**) "bottom calcium" reservoir cemented by dolomite with isolated pores, 4183.8 m, K = 0.203 mD. The thickness of the "bottom calcium" in well A2 can be estimated as 0.3 m.

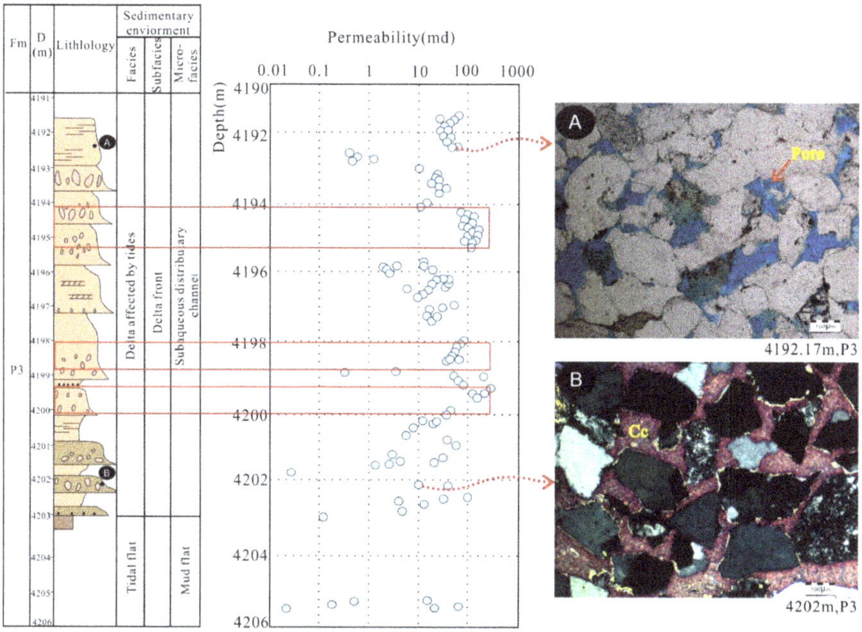

Figure 14. Typical configuration between UC reservoir and "bottom calcium" reservoir developed the subaqueous distributary channel for well A1 with the depth of 4190–4202 m. (**A**) UC reservoir in uncemented zone with extensive secondary intergranular pores, 4192.7 m; K = 85.5 mD; (**B**) "bottom calcium" reservoir with abundant calcite cements, 4202 m, K = 9.5 mD.

The above vertical distribution characteristics of "bottom calcium" reservoirs and UC reservoirs further confirms that the carbonate cements in tide-dominated delta systems play an important role in the diagenetic alterations and the reservoir quality evolution. In UC reservoirs, it is the precipitation of early poikilotopic calcite that halts mechanical compaction. Conversely, for "bottom calcium" reservoirs, the retained bottom calcium cements (include calcite, Fe-calcite and dolomite) completely fill intergranular pore areas and drastically reduce the porosity. Simultaneously, the sample density increases with the decrease in porosity (Table 4).

Table 4. Porosity, permeability and sample density of A2 well.

Depth (m)	Porosity (%)	Permeability (mD)	Grain Density	Sample Density
4183.5	20.9	141	2.63	2.08
4183.6	14.6	10.9	2.71	2.31
4183.7	11.1	0.267	2.74	2.43
4183.8	8.8	0.203	2.73	2.49
4183.9	8.2	0.154	2.73	2.50
4184	3.3	2.71	2.81	2.72

Source: CNOOC's Shanghai branch.

5.3. Summary Model for the Diagenetic Alterations and Reservoir Quality Evolution

The model used for the diagenetic alterations and reservoir quality evolution is summarized in Figure 15. The formation mechanism of effective reservoir is generally controlled by sedimentary facies and the diagenetic evolution process. In particular, the depositional facies have a considerable impact on the distribution of eogenetic and mesogenetic alterations [8].

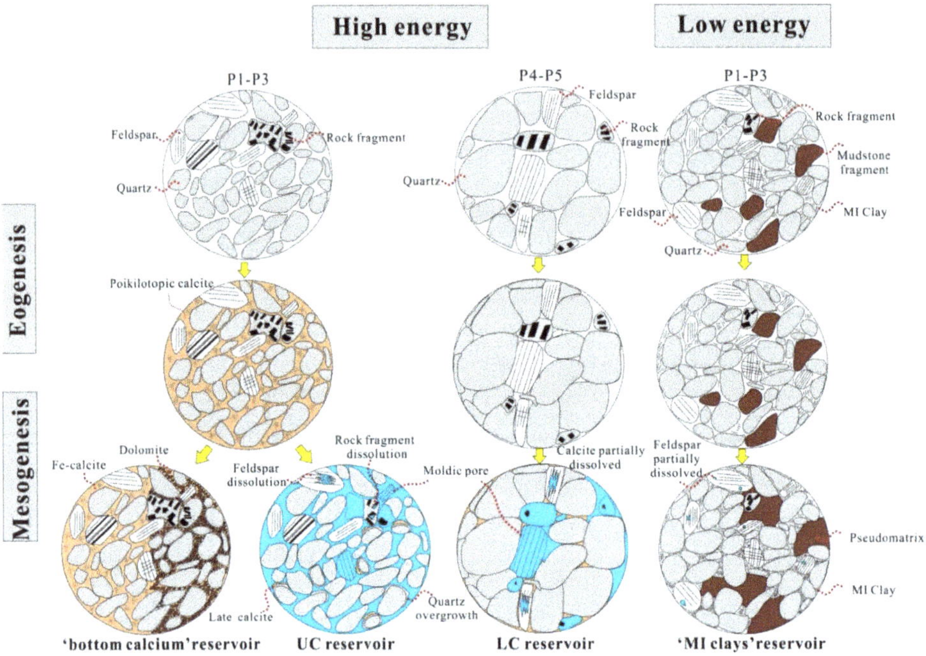

Figure 15. Schematic cartoon showing the diagenetic alterations and reservoir quality evolution.

Firstly, the clastic components controlled by sedimentary facies and provenance differ in the early stage of deposition: in high-energy environments, such as subaqueous distributary channels and tidal channels, the level of hydrodynamic energy is high, and fine-grained argillaceous sediments cannot be deposited. The initial clasts are mainly quartz, feldspar and rock debris. In the early diagenetic stage, the upper reservoirs (P1–P3) are affected by sea level fluctuations, and early poikilotopic calcite precipitates, preventing the further compaction of particles, The particles are mainly in point line contact; eogenetic alterations exert substantial control on the deep-burial, mesogenetic evolution of sandstones, because burial diagenesis is constrained by the extent of the fluid flow through the remaining pores and is influenced by the presence of earlier diagenetic products [8]. In the middle diagenetic stage, early calcite is dissolved by acid fluids, leading to the formation of UC reservoirs. Dissolved pores are also seen in low-permeability reservoirs today, but due to the filling of bottom calcium cement between particles, the pore connectivity is poor and the permeability is low. The LC reservoirs (P4-P5) did not develop poikilotopic calcite during eogenesis and are mainly characterized by long contacts or concave–convex contacts of particles. In mesodiagenesis, the dissolution of feldspar or unstable rock debris provides an opportunity for the formation of secondary pores [78]. Therefore, LC reservoirs have certain requirements in terms of the particle sizes of the initial clastic components, which are mainly medium or coarse particles.

Secondly, the relatively low-energy facies belt mainly contains pseudomatrix and MI clays, and the tide-dominated delta is characterized by MI clays (formed by the downward percolation of suspended fine-grained clay under hydrostatic or weak hydrodynamic conditions). Tidal flat sediments are characterized by pseudomatrix, which is mostly developed on the mud flats. Because the hydrodynamic energy of the overall sedimentary environment is relatively weak, ion exchange in the pore water is also weak. Without sufficient calcium sources, the early poikilotopic calcite cannot be extensively precipitated. Therefore, low-permeability reservoirs are mainly composed of long and concave–convex contacts, and the particles are filled with MI clays and pseudomatrix.

6. Conclusions

1. The petrographic results suggest that the sandstones in the study area have undergone complex diagenetic processes. The dominant eogenetic features include the pervasive precipitation of early poikilotopic calcite cements in the high-energy zone. Subsequent mesogenetic events include: (i) the extensive dissolution of feldspars and calcite due to the emplacement of organic acids; (ii) the precipitation of kaolinites and quartz cements, and (iii) the late calcite, Fe-calcite and dolomite cements.
2. UC reservoirs are characterized by high intergranular porosity, normal fine to very fine grains with point to long contacts. LC reservoirs are characterized by medium to coarse grains, extensive dissolution of feldspar and long or even concave–convex grain contacts. Compared with UC reservoirs, LC reservoirs have significantly lower quartz/feldspar ratios, which is related to differences in the number of feldspar reactions and the associated local secondary porosity. UC reservoirs normally develop in the middle of tidal channels and subaqueous distributary channels, with the base of the sand-body being extensively cemented by poikilotopic calcite cements and euhedral rhomb dolomite cements.
3. Poikilotopic calcite cements normally develop in tidal channels and subaqueous distributary channels of tide-dominated delta. They may precipitate from diagenetic fluids of evolved mixed marine and meteoric waters that became hotter with progressive burial in the basin. The carbon sources of the "bottom calcium" calcite mostly originate from the dissolution of carbonate clasts and bioclast, while dolomite is more comparable to the generation of biogenic methane with high positive $\delta^{13}C$ values.
4. The eogenesis of poikilotopic calcite provided a framework to retard the adverse impacts of compaction in the P1–P3 Formation. For the lower laryers (P4–P5), poikilotopic calcite cementation is limited, and the secondary pores that are mostly formed by the dissolution of feldspar or unstable rock fragments and LC reservoirs can only develop in coarser sandstone, which limits their regional distribution.

Author Contributions: Conceptualization, S.Z., K.W. and N.L.; experimental analysis, S.Z. and X.P.; validation, S.Z. and Y.C.; resources, K.W. and Y.C.; data curation, S.Z.; writing—original draft preparation, S.Z. and N.L.; writing—review and editing, N.L.; funding acquisition, N.L. All authors have read and agreed to the published version of the manuscript.

Funding: This work was financially supported by the National Natural Science Foundation of China (No.42272146, No.41202073). Formation mechanism and prediction of deep effective reservior in Xihu Sag (CCL2020RCPS0048ESN).

Data Availability Statement: The data is available on reasonable request from the corresponding author.

Acknowledgments: We would like to thank the geologists of CNOOC Research Institute and CNOOC's Shanghai branch district for their support of essential data.

Conflicts of Interest: The authors declare no conflict of interest.

References

1. Anovitz, L.M.; Cole, D.R.; Rother, G.; Allard, L.F.; Jackson, A.J.; Littrell, K.C. Diagenetic changes in macro-to nano-scale porosity in the St. Peter Sandstone: An (ultra) small angle neutron scattering and backscattered electron imaging analysis. *Geochem. Cosmochim. Acta* **2013**, *102*, 280–305. [CrossRef]
2. Anovitz, L.M.; Cole, D.R.; Jackson, A.J.; Rother, G.; Littrell, K.C.; Allard, L.F.; Pollington, A.D.; Wesolowski, D.J. Effect of quartz overgrowth precipitation on the multiscale porosity of sandstone: A (U) SANS and imaging analysis. *Geochem. Cosmochim. Acta* **2015**, *158*, 199–222. [CrossRef]
3. Anovitz, L.M.; Cole, D.R. Characterization and analysis of porosity and pore structures. *Rev. Mineral. Geochem.* **2015**, *80*, 61–164. [CrossRef]
4. Golab, A.N.; Knackstedt, M.A.; Averdunk, H.; Senden, T.; Butcher, A.R.; Jaime, P. 3D porosity and mineralogy characterization in tight gas sandstones. *Lead. Edge* **2010**, *29*, 1476–1483. [CrossRef]
5. Lai, J.; Wang, G.; Fan, Z.; Zhou, Z.; Chen, J.; Wang, S. Fractal analysis of tight shaly sandstones using nuclear magnetic resonance measurements. *AAPG Bull.* **2018**, *102*, 175–193. [CrossRef]

6. Oluwadebi, A.G.; Taylor, K.G.; Ma, L. A case study on 3D characterisation of pore structure in a tight sandstone gas reservoir: The Collyhurst Sandstone, East Irish Sea Basin, northern England. *J. Nat. Gas Sci. Eng.* **2019**, *68*, 102917. [CrossRef]
7. Tsakiroglou, C.D.; Payatakes, A.C. Characterization of the pore structure of reservoir rocks with the aid of serial sectioning analysis, mercury porosimetry and network simulation. *Adv. Water Resour.* **2000**, *23*, 773–789. [CrossRef]
8. Morad, S.; Al-Ramadan, K.; Ketzer, J.M.; De Ros, L.F. The impact of diagenesis on the heterogeneity of sandstone reservoirs: A review of the role of depositional facies and sequence stratigraphy. *AAPG Bull.* **2010**, *94*, 1267–1309. [CrossRef]
9. Saigal, G.C.; Bjørlykke, K. Carbonate cements in clastic reservoir rocks from offshore Norway- relationships between isotopic composition, textural development and burial depth. *Geol. Soc. Lond. Spec. Publ.* **1987**, *36*, 313–324. [CrossRef]
10. Bjørlykke, K.; Ramm, M.; Saigal, G.C. Sandstone diagenesis and porosity modification during basin evolution. *Geol. Rundsch.* **1989**, *78*, 243–268. [CrossRef]
11. Taylor, T.R. The influence of calcite dissolution on reservoir porosity in Miocene sandstones, Picaroon Field, offshore Texas Gulf Coast. *J. Sediment. Res.* **1990**, *60*, 322–334.
12. Dutton, S.P.; Willins, B.J.; White, C.D.; Bhattacharya, J.P. Outcrop characterization of reservoir quality and interwell-scale cement distribution in a tideinfluenced delta, Frontier Formation, Wyoming, USA. *Clay Miner.* **2000**, *35*, 95–105. [CrossRef]
13. Longstaffe, F.J.; Calvo, R.; Ayalon, A.; Donaldson, S. Stable isotope evidence for multiple fluid regimes during carbonate cementation of the Upper Tertiary Hazeva Formation, Dead Sea Graben, southern Israel. *J. Geochem. Explor.* **2003**, *80*, 151. [CrossRef]
14. Nyman, S.L.; Gani, M.R.; Bhattacharya, J.P.; Lee, K. Origin and Distribution of Calcite Concretions in Cretaceous Wall Creek Member, Wyoming: Reservoirquality implication for shallow-marine deltaic strata. *Cretac. Res.* **2014**, *48*, 139–152. [CrossRef]
15. Liu, S.; Huang, S.; Shen, Z.; Lü, Z.; Song, R. Diagenetic fluid evolution and water-rock interaction model of carbonate cements in sandstone: An example from the reservoir sandstone of the Fourth Member of the Xujiahe Formation of the Xiaoquan-Fenggu area, Sichuan Province, China. *Sci. China (Earth Sci.)* **2014**, *57*, 1077–1092. [CrossRef]
16. Xiong, D.; Azmy, K.; Blamey, N.J.F. Diagenesis and origin of calcite cement in the Flemish Pass Basin sand- stone reservoir (Upper Jurassic): Implications for porosity development. *Mar. Pet. Geol.* **2016**, *70*, 93–118. [CrossRef]
17. Hesse, R.; Abid, A. Carbonate cementation-the key to reservoir properties of four sandstone levels (Cretaceous) in the Hibernia Oilfield, Jeanne d'Arc Basin, Newfoundland, Canada. *Spec. Publ. Int. Assoc. Sedimentol.* **1998**, *26*, 141–3600.
18. Brown, D.M.; McAlpine, K.D.; Yole, R.W. Sedimentology and sandstone diagenesis of Hibernia Formation in Hibernia oil field, Grand Banks of Newfoundland. *AAPG Bull.* **1989**, *73*, 557–575.
19. Baker, J.C.; Havord, P.J.; Martin, K.R.; Ghori, K.A.R. Diagenesis and petrophysics of the early Permian Moogooloo sandstone, southern Carnarvon basin, western Australia. *AAPG Bull.* **2000**, *84*, 250–265.
20. McBride, E.F.; Picard, M.D.; Milliken, K.L. Calcite-cemented concretions in Cretaceous sandstone, Wyoming and Utah, USA. *J. Sediment. Res.* **2003**, *73*, 462–483. [CrossRef]
21. Odigi, M.I.; Amajor, L.C. Geochemistry of carbonate cements in Cretaceous sandstones, southeast Benue Trough, Nigeria: Implications for geochemical evolution of formation waters. *J. Afr. Earth Sci.* **2010**, *57*, 213–226. [CrossRef]
22. Azomani, E.; Azmy, K.; Blamey, N.; Brand, U.; Al-Aasm, I. Origin of Lower Ordovician dolomites in eastern Laurentia: Controls on porosity and implications from geochemistry. *Mar. Pet. Geol.* **2013**, *40*, 99–114. [CrossRef]
23. García-García, F.; Marfil, R.; De Gea, G.A.; Delgado, A.; Kobstädt, A.; Santos, A.; Mayoral, E. Reworked marine sandstone concretions: A record of highfrequency shallow burial to exhumation cycles. *Facies* **2013**, *59*, 843–861. [CrossRef]
24. Su, A.; Chen, H.H.; Chen, X.; He, C.; Liu, H.P.; Li, Q.; Wang, C.W. The characteristics of low permeability reservoirs, gas origin, generation and charge in the central and western Xihu depression, East China Sea Basin. *J. Pet. Sci. Eng.* **2018**, *53*, 94–109. [CrossRef]
25. Wang, W.G.; Lin, C.Y.; Zhang, X.G.; Dong, C.M.; Ren, L.H.; Lin, J.L. Effect of burial history on diagenetic and reservoir-forming process of the Oligocene sandstone in Xihu sag, East China Sea Basin. *Mar. Pet. Geol.* **2020**, *112*, 104034. [CrossRef]
26. Xu, F.H.; Xu, G.S.; Liu, W.; Cui, H.Y.; Wang, Y.R. Factors controlling the development of tight sandstone reser- voirs in the Huagang Formation of the central inverted structural belt in Xihu sag, East China Sea Basin. *Pet. Explor. Dev.* **2020**, *47*, 101–113. [CrossRef]
27. Wang, W.G.; Lin, C.Y.; Zhang, X.G.; Dong, C.M.; Ren, L.H.; Lin, J.L. Provenance, clastic composition and their impact on diagenesis A case study of the Oligocene sandstone in the Xihu sag, East China Sea Basin. *Mar. Pet. Geol.* **2021**, *126*, 10490. [CrossRef]
28. Zeng, F.; Dong, C.M.; Lin, C.Y.; Wu, Y.Q.; Tian, S.S.; Zhang, X.G.; Lin, J.L. Analyzing the effects of multiscale pore systems on reservoir Properties—A case study on Xihu Depression, East China Sea Shelf Basin, China. *J. Pet. Sci. Eng.* **2021**, *203*, 108609. [CrossRef]
29. Duan, D.P.; Zhang, X.G.; Liu, B.B.; Lin, J.L.; Wang, W.G. The Relationship between Chlorite and Reservoir Quality in the Huagang Formation, Xihu Depression, China. *Energies* **2022**, *15*, 3438. [CrossRef]
30. Qian, W.D.; Sun, Q.L.; Jones, S.J.; Yin, T.J.; Zhang, C.M.; Xu, G.S.; Hou, G.W.; Zhang, B. Diagenesis and controlling factors of Oligocene Huagang Formation tight sandstone reservoir in the south of Xihu sag, the East China Sea Shelf Basin. *J. Pet. Sci. Eng.* **2022**, *215*, 110579. [CrossRef]

31. Wang, Y.X.; Chen, J.F.; Pang, X.Q.; Zhang, T.; Zhu, X.J.; Liu, K.X. Hydrocarbon generation and expulsion of tertiary coaly source rocks and hydrocarbon accumulation in the Xihu Sag of the East China Sea Shelf Basin, China. *J. Asian Earth Sci.* **2022**, *229*, 105170. [CrossRef]
32. Zeng, F.; Dong, C.W.; Lin, C.Y.; Tian, S.S.; Wu, Y.Q.; Lin, J.L.; Liu, B.B.; Zhang, X.G. Pore structure characteristics of reservoirs of Xihu Sag in East China Sea Shelf Basin based on dual resolution X-ray computed tomography and their influence on permeability. *Energy* **2022**, *239*, 122386. [CrossRef]
33. Jiang, L.; Tian, H.Q.; Ma, Y.X.; Zhao, Y.S. *Petroleum System and Target Evaluation in Xihu Sag, East China Sea Shelf Basin*; Petroleum University Publishing House: Dongying, China, 2000; pp. 16–133. (In Chinese with English abstract).
34. Zhu, X.J.; Chen, J.F.; Li, W.; Pei, L.X.; Liu, K.X.; Chen, X.D.; Zhang, T.L. Hydrocarbon generation potential of Paleogene coals and organic rich mudstones in Xihu sag, East China Sea Shelf basin, offshore eastern China. *J. Pet. Sci. Eng.* **2020**, *184*, 106450. [CrossRef]
35. Wang, W.G.; Lin, C.Y.; Zhang, X.G.; Dong, C.M.; Ren, L.H.; Lin, J.L. Structural controls on sandstone compa- cation within the anticline crest and flank An example from the Xihu Sag, East China Sea Basin. *J. Pet. Sci. Eng.* **2022**, *211*, 110157. [CrossRef]
36. Li, S.X.; Shao, L.Y.; Liu, J.S.; Qin, L.Z.; Kang, S.L.; Eriksson, K.A.; Chen, X.D.; Yu, Z.K.; Liu, J.Z. Oil generation model of the liptinite-rich coals Palaeogene in the Xihu Sag, East China Sea Shelf Basin. *J. Pet. Sci. Eng.* **2022**, *209*, 109844. [CrossRef]
37. Su, A.; Chen, H.H.; Lei, M.Z.; Li, Q.; Wang, C.W. Paleo-pressure evolution and its origin in the Pinghu slope belt of the Xihu Depression, East China Sea Basin. *Mar. Pet. Geol.* **2019**, *107*, 198–213. [CrossRef]
38. Shen, Y.L.; Qin, Y.; Cui, M.; Xie, G.L.; Guo, Y.H.; Qu, Z.H.; Yang, T.Y.; Yang, L. Geochemical Characteristics and Sedimentary Control of Pinghu Formation (Eocene) Coal-bearing Source Rocks in Xihu Depression, East China Sea Basin. *Acta Geol. Sin. (Engl. Ed.)* **2021**, *95*, 91–104. [CrossRef]
39. Yang, C.Q.; Yang, Y.Q.; Yang, C.S. Tectono-sedimentary evolution of the Mesozoic in the southern East China Sea Shelf Basin and its bearing on petroleum exploration. *Mar. Geol. Quat. Geol.* **2019**, *39*, 30–40. (In Chinese with English abstract).
40. Xu, F. Characteristics of Caenozoic Structure and Tectonic Miaration of the East China Sea Shelf Basin. *J. Oil Gas Technol.* **2012**, *34*, 1–7, (In Chinese with English abstract).
41. Mi, C.Y. The Fault Characteristics and Evolution of Central Inversion Tectonic Belt and Pinghu Slope in Xihu Sag. Master's Thesis, Ocean University of China, Qingdao, China, 2015. (In Chinese with English abstract).
42. Zhou, Z.Y.; Jiang, J.Y.; Liao, Z.T. Basin inverssion in Xihu depression, East China Sea. *Gondwana Res.* **2001**, *4*, 844–845, (In Chinese with English abstract).
43. Li, C.F.; Zhou, Z.Y.; Ge, H.P. Rifting process of the Xihu depression, East China Sea Basin. *Tectonophysics* **2009**, *472*, 135–147, (In Chinese with English abstract). [CrossRef]
44. Yu, Z.K.; Ding, F.; Zhao, H. Characteristics of structural evolution and classifcation of hydrocarbon migration and accumulation units in Xihu Sag, China. *Shanghai Land Resour.* **2018**, *39*, 75–78. (In Chinese with English abstract).
45. Zhu, Y.; Yu, Y.F.; Zhang, T.; Zhang, S.I.; Tang, X.J. Geochemical characteristics of Tertiary coal- bearing source rocks in Xihu depression, East China Sea Basin. *Mar. Pet. Geol.* **2012**, *35*, 154–165. [CrossRef]
46. Su, A.; Chen, J.H.; Wang, C.W.; Li, P.J.; Zhang, H.; Xiong, W.L.; Lei, M.Z. Genesis and maturity identification of oil and gas in the Xihu Sag, East China Sea Basin. *Pet. Explor. Dev.* **2013**, *40*, 521–527. (In Chinese with English abstract). [CrossRef]
47. Abbas, A.; Zhu, H.T.; Zeng, Z.W.; Zhou, X.H. Sedimentary facies analysis using sequence stratigraphy and seismic sedimentology in the Paleogene Pinghu Formation, Xihu depression, East China Sea Shelf Basin. *Mar. Pet. Geol.* **2018**, *93*, 287–297. [CrossRef]
48. Zhu, W.; Zhong, K.; Fu, X.; Chen, C.; Zhang, M.; Gao, S. The formation and evolution of the East China Sea Shelf Basin: A new view. *Earth Sci. Rev.* **2019**, *190*, 89–111.
49. Zhao, L.N.; Chen, J.W.; Zhang, Y.G. Sedimentary characteristics of Pinghu Formation in Pinghu structural belt of Xihu depression, East China Sea. *World Geol.* **2008**, *27*, 42–47. (In Chinese with English abstract).
50. Jiang, Y.M.; Diao, H.; Zeng, W.Q. Coal source rock conditions and hydrocarbon generation model of Pinghu Formation in Xihu Depression, East China Sea Basin. *Bull. Geol. Sci. Technol.* **2020**, *39*, 30–39. (In Chinese with English abstract).
51. Liu, Y.H.; Cai, H.; Duan, D.P. The sedimentary characteristics of tidal delta and tidal flat in transgressive system tract of Pinghu Formation in Pinghu area, Xihu Sag. *Mar. Geol. Front.* **2020**, *38*, 33–40. (In Chinese with English abstract). [CrossRef]
52. Lindholm, R.C.; Finkelman, R.B. Calcite staining; semiquantitative determination of ferrous iron. *J. Sediment. Res.* **1972**, *42*, 239–242. [CrossRef]
53. McCrea, J.M. The isotopic chemistry of carbonates and a paleotemperature scale. *J. Chem. Phys.* **1950**, *18*, 849–857. [CrossRef]
54. Xie, X.J.; Xiong, L.Q.; Chen, Y.; Li, L.Z.; Zhang, C.Y.; Bai, H.Q.; Liao, J.H. Low permeability reservoir characteristics and controlling factor of "Sweet Points" of Pinghu Formation in Xihu Sag. *Sci. Technol. Eng.* **2021**, *21*, 12890–12900. (In Chinese with English abstract).
55. Xiao, X.X.; Qin, L.Z.; Zhang, W.; Jiang, X.; Xie, J.J. The origin of carbonate cements and the influence on reservoir quality of Pinghu Formation in Xihu Sag. *Chin. J. Geol.* **2021**, *56*, 1062–1076.
56. Land, L.S. Oxygen and carbon isotopic composition of Ordovician brachio pods: Implications for coeval seawater: Discussion. *Geochim. Cosmochim. Acta* **1995**, *59*, 2843–2844. [CrossRef]
57. Hutcheon, I.; Nahnybida, C.; Rouse, H.R.K. The geochemistry of carbonate cements in the Avalon sand, grand Banks of Newfoundland. *Mineral. Mag.* **1985**, *49*, 457–467. [CrossRef]

58. Scotchman, I.C. The geochemistry of concretions from the Kimmeridge Clay Formation of southern and eastern England. *Sedimentology* **1991**, *38*, 79–106. [CrossRef]
59. Denny, A.C.; Fall, A.; Orland, I.J.; Valley, J.W.; Eichhubl, P.; Laubach, S.E. A history of pore water oxygen isotope evolution in the cretaceous travis peak formation in East Texas. *Geol. Soc. Am. Bull.* **2020**, *132*, 1626–1638. [CrossRef]
60. Friedman, I.; O'Neil, J.R. *Compilation of Stable Isotope Fractionation Factors of Geochemical Interest*; US Government Printing Office: Washington, DC, USA, 1977; Volume 49.
61. Zheng, Y.F. Oxygen isotope fractionation in carbonate and sulfate minerals. *Geochem. J.* **1999**, *33*, 109–126. [CrossRef]
62. Irwin, H.; Coleman, M.; Curtis, C.D. Isotopic evidence for source of diagenetic carbonates formed during burial of organic-rich sediments. *Nature* **1977**, *269*, 209–213. [CrossRef]
63. Morad, S. *Carbonate Cementation in Sandstones: Distribution Patterns and Geochemical Evolution*; Wiley-Blackwell: Hoboken, NJ, USA, 1998; Volume 26, pp. 1–26.
64. Faure, G.; Mensing, T.M. *Isotopes: Principles and Applications*; John Wiley & Sons Inc.: Hoboken, NJ, USA, 2005.
65. Worden, R.H.; Morrall, G.T.; Kelly, S.; Mc Ardle, P.; Barshep, D.V. A renwed look at calcite cement in Marine-deltaic sandstones: The Brent Reservoir, Heather Field, Northern North Sea, UK. *Geol. Soc. Spec. Publ.* **2020**, *484*, 305–335. [CrossRef]
66. Schultz, J.L.; Boles, J.; Tilton, G.R. Tracking calcium in the San Joaquin Basin, California: A strontium isotopic study of carbonate cements at North Coles Levee. *Geochim. Cosochim. Acta* **1989**, *53*, 1991–1999. [CrossRef]
67. Mtys, S. Ferenc Molnr: The Plays and the Wives (Katalin Varge and Tams Gajd, eds). *Hungrarian Q.* **2007**, *188*, 122–129.
68. Taylor, K.G.; Gawthorpe, R.L.; Curtis, C.D.; Marshall, J.D.; Awwiller, D.N. Carbonate cementation in a sequence-stratigraphic framework: Upper Cretaceous sandstones, Book Cliffs, Utah-Colorado. *J. Sediment. Res.* **2000**, *70*, 360–372. [CrossRef]
69. Al-Ramadan, K.S.; Morad, J.N.; Proust, I.S.; Al-Aasm. Distribution of diagenetic alterations in siliciclastic shoreface deposits within a sequence stratigraphic framework: Evidence from the Upper Jurassic, Boulonnais, NW France. *J. Sediment. Res.* **2005**, *75*, 943–959. [CrossRef]
70. Haynes, S.R.; Marshall, J.; Imsland-Wathne, E.; Minielly, G.; Mortlock, E.; Walderhaug, O.; Johnson, T. Depositional interpretation and reservoir characterization of the tithonian in mizzen F-09, Flemish Pass Basin, Canada. In *Proceedings of Abstracts of Integration: CSEG, CSPG and CWLS GeoConvention*; AAPG: Calgary, AB, Canada, 2013.
71. Dutton, S.P. Calcite cement in Permian deep-water sandstones, Delaware Basin, west Texas: Origin, distribution, and effect on reservoir properties. *AAPG Bull.* **2008**, *92*, 765–787. [CrossRef]
72. Zhao, Z.X.; Dong, C.M.; Zhang, X.G.; Lin, C.Y.; Huang, X.; Duan, D.P.; Lin, J.L.; Fang, Z.; Li, D. Reservoir controlling factors of the Paleogene Oligocene Huagang Formation in the north central part of the Xihu Depression, East China Sea Basin, China. *J. Pet. Sci. Eng.* **2019**, *175*, 159–172.
73. Higgs, K.E.; Funnell, R.H.; Reyes, A.G. Changes in reservoir heterogeneity and quality as a response to high partial pressures of CO_2 in a gas reservoir, New Zealand. *Mar. Pet. Geol.* **2013**, *48*, 293–322. [CrossRef]
74. Wang, Y. Photophysical properties of fullerenes and fullerene/N, N-diethylaniline charge-transfer complexes. *J. Phys. Chem.* **1992**, *96*, 764–767. [CrossRef]
75. Matlack, K.S.; Houseknecht, D.W.; Applin, K.R. Emplacement of clay into sand by infiltration. *J. Sediment. Res.* **1989**, *59*, 77–87.
76. Moraes, M.A.S.; De Ros, L.F. Infiltrated clays in fluvial Jurassic sandstones of Recôncavo Basin, northeastern Brazil. *J. Sediment. Res.* **1990**, *60*, 809–819.
77. Wolela, A.M.; Gierlowski-Kordesch, E.H. Diagenetic history of fluvial and lacustrine sandstones of the Hartford Basin (Triassic–Jurassic), Newark Supergroup, USA. *Sediment. Geol.* **2007**, *197*, 99–126. [CrossRef]
78. Morad, S.; Ketzer, J.M.; De Ros, L.F. Spatial and temporal distribution of diagenetic alterations in siliciclastic rocks: Implications for mass transfer in sedimentary basins. *Sedimentology* **2000**, *47*, 95–120. [CrossRef]

minerals

Article

Provenance Tracing and Age Analysis of Lead–Zinc Mineralization in Qiyimuchang, Inner Mongolia, NE China

Mingxin Duan [1,2], Yunsheng Ren [3,*], Chunji Xue [1], Qun Yang [3], Yujie Hao [3] and Tao Liu [2]

1 School of Earth Sciences and Resources, China University of Geosciences (Beijing), Beijing 100083, China
2 Harbin Center for Integrated Natural Resources Survey, China Geological Survey, Harbin 150086, China
3 College of Earth Sciences, Jilin University, Changchun 130061, China
* Correspondence: author: renys@jlu.edu.cn

Abstract: The Qiyimuchang lead–zinc deposit is an important deposit in the Erguna Massif on the west slope of Daxinganling, for which the material source and age of mineralization remain unclear. The lead–zinc deposit in Qiyimuchang has been observed to occur in Jurassic volcanic strata as a vein-like orebody, and its strike is nearly perpendicular to that of the strata and the regional structures of the orogenic belt. The ore is mainly composed of sphalerite, galena, chalcopyrite, pyrite, and quartz, and hydrothermal alteration zones dominated by silicification, fluoritization, and pyritization are common within 1–5 m on both sides of the vein. The metasomatic lead–zinc mineralization is characterized by a massive vein structure. The mineralization process can be divided into three stages, pre-ore (mainly quartz, arsenopyrite and a small amount of pyrite), ore-formation (pyrite, chalcopyrite, sphalerite, galena, fluorite, and a small amount of tetrahedrite), and post-ore (quartz-calcite veinlets with a small amount of pyrite). Element and isotope geochemical studies show that the trace element compositions of the wall rocks (andesite, tuff, etc.) of Jurassic volcanic rocks in the ore bodies and surrounding rocks reflect affinity or similarity between them. The S isotopic composition of sulfide minerals in the metallogenic stage ($\delta\ ^{34}S_{V\text{-}CDT}$ = 1.6‰~4.8‰) indicate the sulfur isotopic composition of the magma. The Pb isotopic compositions of sulfide minerals (38.5–38.39, 15.55–15.62, and 18.33–18.41 for $^{208}Pb/^{204}Pb$, $^{207}Pb/^{204}Pb$, and $^{206}Pb/^{204}Pb$, respectively) in the metallogenic stage are relatively concentrated and close to the average Pb isotopic compositions of the mantle and average Pb isotopic composition of an orogenic belt, indicating that the ore-forming metals are likely to be a mixed source of crust and mantle. A new zircon U-Pb age of 150.8 ± 1.3 Ma (MSWD = 0.74) was obtained from andesitic tuff. Collectively considering the regional geology and ore deposit geological analysis, the lead–zinc mineralization in Qiyimuchang possibly occurred during the early Cretaceous. The ore-forming processes belong to Cretaceous magmatic-hydrothermal activity in extensional tectonic setting.

Keywords: Qiyimuchang lead–zinc deposit; geochemistry; S-Pb isotope; zircon U-Pb geochronology

1. Introduction

The Erguna Massif is located in the eastern section of the Xing'an-Mongolia orogenic belt in the east of the Central Asia Orogenic Belt (CAOB) (Figure 1A). Prospecting and exploration activities have revealed a large amount of lead–zinc (silver) deposits in this area (Figure 1B), with many large-size deposits, such as the Jiawula large-size lead–zinc (silver) deposit (lead and zinc resources of more than 1.3 million tons), E'rentaolegai large-size silver and manganese deposit, large-size lead–zinc (silver) deposits in Biliyagu (more than 0.7 million tons), and large-size lead–zinc deposits in Derbur (more than 1.1 million tons) [1–5]. The extensive lead–zinc mineralization in the Erguna Massif has attracted much attention [6–9]. The lead–zinc deposits in the area are distributed belt-like along the NNE, striking the Derbugan fault, and most of them exist as veins in Mesozoic volcanic rocks. Thus far, most studies on these deposits have focused on the western part of the

Derbugan fault, and the newly discovered lead–zinc deposits in the east of the fault, such as the Qiyimuchang lead–zinc deposit and Dongjun lead–zinc deposit, was poorly studied. Varied conclusions have been drawn regarding the source of the ore-forming materials, such as crust mantle mixed source [1,2,4], crust mantle mixed source dominated by mantle source [10,11], and upper crust [12,13]. Moreover, their mineralization and ore controlling elements are poorly understood.

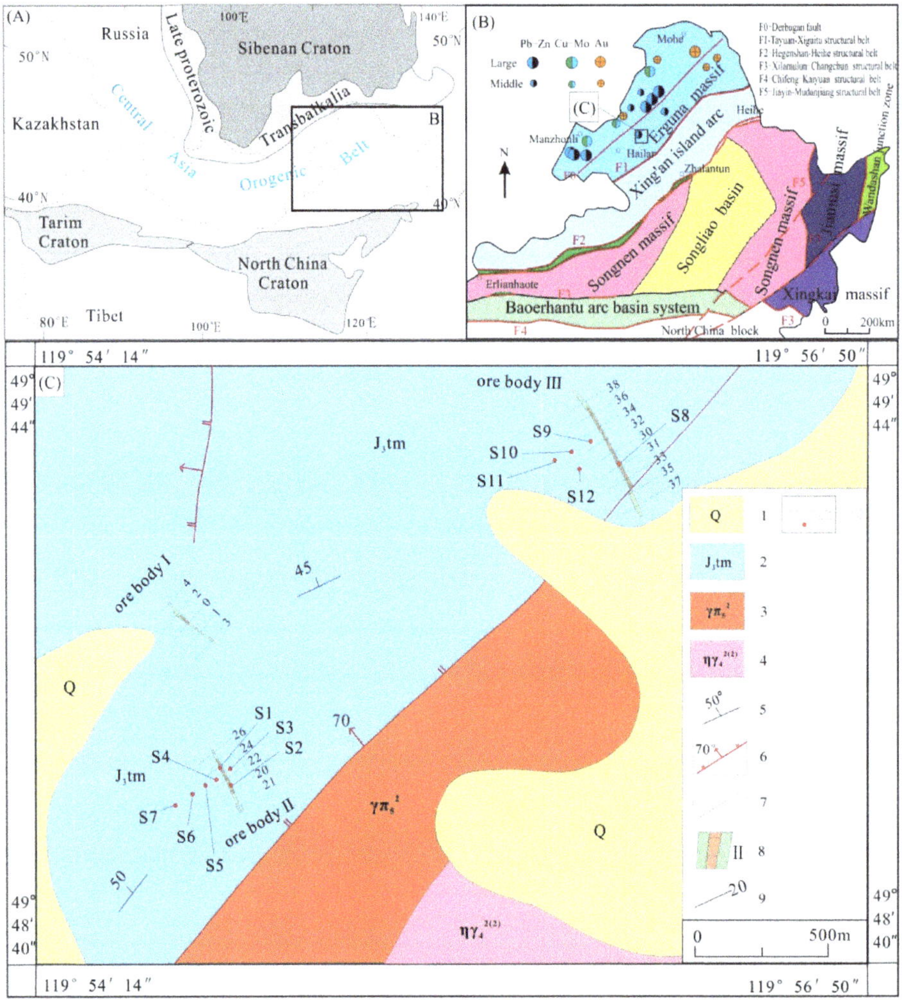

Figure 1. Geotectonic location map ((**A**) after [14]; (**B**) modified by [15]) and geological map ((**C**) after [16]) of the Qiyimuchang deposit. 1—Quaternary alluvium deposits; 2—Tamulangou formation of upper Jurassic; 3—Early Yanshanian granite porphyry; 4—Middle Variscan porphyry biotite monzogranite; 5—Strata occurrence(tuff in late Jurassic); 6—Measured normal fault and its dip angle; 7—Concealed fault; 8—Ore body and its number; 9—Prospecting line and its number; 10—Sampling location and its number; S1—13QYK1 and samples for S and Pb isotopes; S2—13QYK2 and samples for S and Pb isotopes; S3—14QY; S4—13QY6; S5—13QY25 and the sample for U-Pb Dating; S6—13QY16; S7—13QY11; S8—13QYK3 and samples for S and Pb isotopes; S9—13QY2-1; S10—13QY2-2; S11—13QY3; S12—13QY13.

In 2005, the Tenth Institute of Geology and Mineral Exploration and Development of Inner Mongolia discovered a medium-size lead–zinc deposit in Qiyimuchang. This study conducted detailed field geological surveying, petrographic and mineralogical observation, trace element analysis of volcanic rocks and ores in host rocks, and S-Pb isotopic analysis of ores to trace the source of ore-forming materials. Furthermore, the zircon U-Pb age of volcanic host rocks was determined to limit the diagenetic and metallogenic age.

2. Geological Background

The Qiyimuchang lead–zinc deposit is located in the east of CAOB, in the east of the Xing'an-Mongolia orogenic belt and in the southeast of the Erguna Massif (Figure 1B). The Erguna Massif lies adjacent to the Mongolia-Okhotsk suture zone in the northwest and the Xing'an Massif in the southeast. In the Erguna Massif, along the NE striking Derbugan fault, many lead–zinc, copper-molybdenum, and gold deposits have been found. Most lead–zinc deposits are controlled by secondary faults along regional faults, which directly control the shape and distribution of ore bodies [1–4]. Among them, faults striking NE, NNE, NW are mostly ore guiding and hosting structures of lead–zinc deposits. These lead–zinc deposits primarily occur in middle-upper Jurassic and lower Cretaceous volcanic rocks. The ore bodies are mostly vein-shaped, stockwork-shaped, and lenticular, mainly hypabyssal or ultra-hypabyssal [5–7]. The ore-forming fluid is characterized by medium low temperature and medium-low salinity. The ore body, which belongs to the epigenetic filling vein lead–zinc deposit, generally formed after the formation of the surrounding rocks (andesite, rhyolite, tuff, etc.) [17,18]. Therefore, the vein lead–zinc deposits in this area have similar metallogenic geological conditions. Unlike most lead–zinc deposits located west of the Derbugan fault, the Qiyimuchang lead–zinc deposit lies east of the Derbugan fault (Figure 1B). The Derbugan fault is a strike-slip fault located in the Erguna Massif. It is a large NE striking extensional deformation zone that was very active from the Late Jurassic to early Cretaceous [19].

The main exposed strata in the area include the lower Carboniferous Morgenhe Formation (C_1m), the lower Permian Xinyigenhe Formation (P_1x), the upper Jurassic Tamulangou Formation (J_3tm), the Manketouebo Formation (J_3mk), and Quaternary alluvium deposits (Q). The common rock combinations of each formation are presented in Figure 2. J_3tm and Q are exposed in the Erguna area (Figure 1C). With an overall strike of NE, a dip of NW, and a dip angle of 20–50°, J_3tm is mainly composed of basaltic andesite, andesite, andesitic crystal debris tuff, tuffaceous sandy conglomerate, and tuff. In particular, the lead–zinc orebodies mainly occur in the andesitic crystal debris tuff.

Two stages of plutonism masses (Variscan and Yanshanian) are exposed in the south of the Erguna area (Figure 1C). The lithology of Variscan granite is light flesh red porphyritic biotite monzogranite with medium-fine grained texture (composed of quartz, orthoclase, plagioclase, and biotite). The lithology of Yanshanian granite is light flesh red syenite granite porphyry with porphyritic texture. The matrix is mainly composed of orthoclase, albite, quartz, and biotite. This pluton is in contact with the NE fault of the Tamulangou Formation and intruded into the Variscan rock mass (Figure 1C). No lead–zinc ore body has been found in the intrusive body of the mining area.

The mining area includes NW striking and 60–70° SW dipping faults. These faults comprise the ore bearing structure of the lead–zinc orebodies. NE and near SN normal faults are developed in the mining area (Figure 1C). The NE striking fault in the mining area extends over 4 km, towards NW and dipping approximately 70°. The strike and dip are close to those (tuff, andesite, rhyolite, etc.) of the Tamulangou Formation. In the south of the mining area, the NE striking fault serves as the boundary between the Tamulangou Formation and granite porphyry, whereas in the north, it cuts through the Tamulangou Formation and the lead–zinc ore body, clearly indicating post metallogenic faulting.

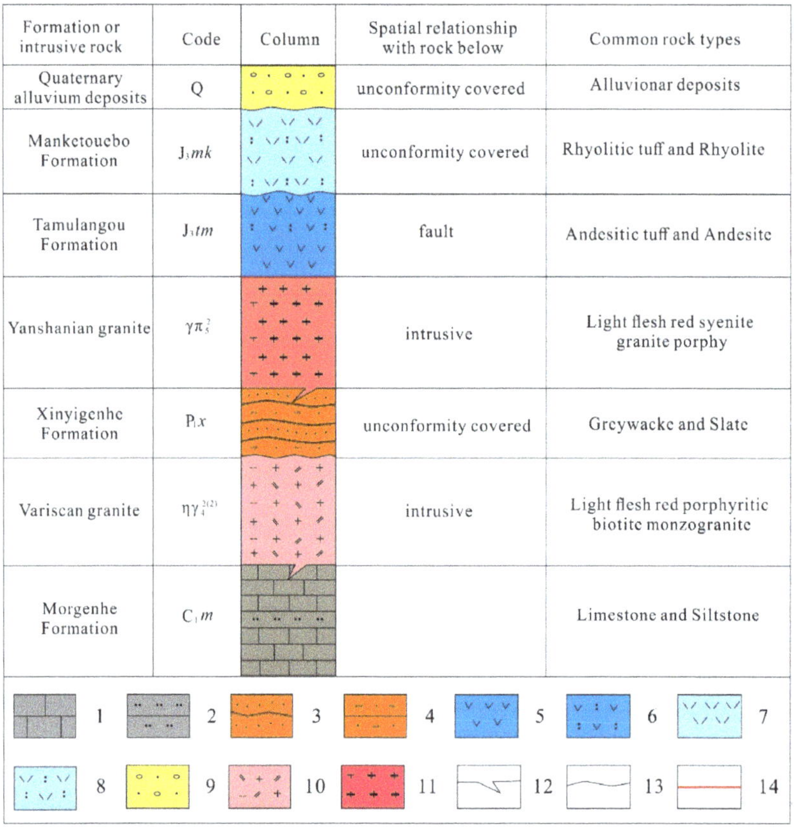

Figure 2. Common rock combinations of each formation around the Qiyimuchang deposit. 1—Limestone; 2—Siltstone; 3—Slate; 4—Greywacke; 5—Andesite; 6—Andesitic tuff; 7—Rhyolite; 8—Rhyolitic tuff; 9—Alluvium deposits; 10—Variscan light flesh red porphyritic biotite monzogranite; 11—Yanshanian light flesh red syenite granite porphyry; 12—Intrusive; 13—Unconformity covered; 14—Fault.

3. Deposit Geology

The exploration project from the Tenth Institute of Geology and Mineral Exploration and Development of Inner Mongolia delineated three lead–zinc industrial orebodies in the mining area (Figure 1C). The orebodies occur as veins (Figures 1C and 3) in the NW fracture in the Tamulangou Formation. The ore-hosting rocks are andesite, andesitic tuff, and rhyolite. The orebody trends NW (296–338°), inclining SW (Figure 3), and the dip angle varies from 62° to 77°. The exposed length of the ore body on the surface is tens to hundreds of meters, the thickness is 1–7 m, and the average ore grade is Pb 3.1% and Zn 4.5%.

Quartz and fluorite are gangue minerals of the ore also developed in the altered andesite wall rock (Figure 4A,B). Further characteristic alteration minerals are sericite, chlorite, and epidote (Figure 4C). Gangue minerals are mainly carbonates in the late metallogenic period (Figure 4D). Among them, silicification, fluoritization, and pyritization are most closely related to mineralization.

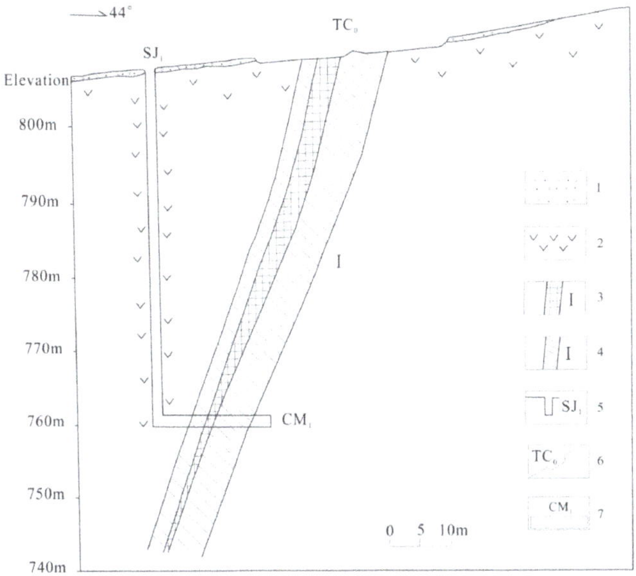

Figure 3. Cross-section of I orebody in No.0 exploration line [16]. 1—Quaternary alluvium deposits; 2—Andesite; 3—Orebody; 4—Alteration zone; 5—SJ_1:shaft and its number; 6—TC_0: exploratory trench; 7—CM_1: footrill. Hydrothermal alteration is common in the surrounding andesites of the upper and lower walls of the ore body forming an alteration zone with a width of 1–5 m.

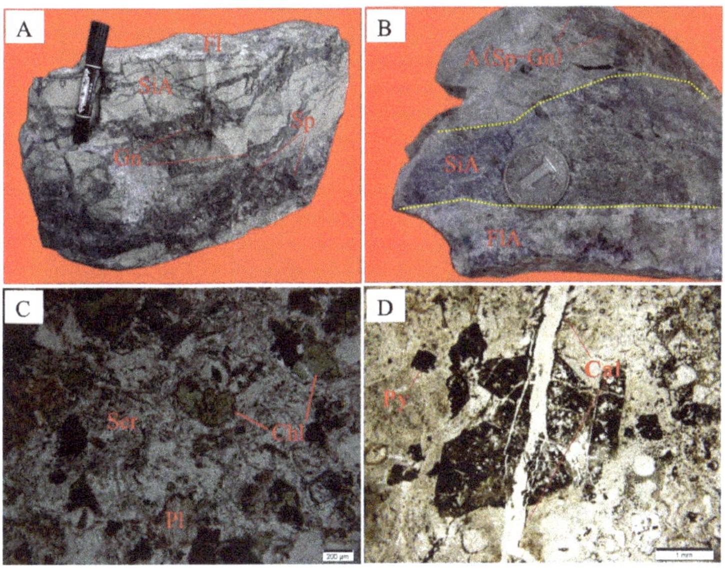

Figure 4. Alteration characteristics and micrographs of andesite and andesitic crystalline tuff of the Qiyimuchang lead zinc deposit. (**A**)—Andesite (host rock) penetrated by veins; (**B**)—Silicification and fluoritization alteration; (**C**)—Chloritization in andesitic crystalline tuff (single-polarized transmitted light); (**D**)—Later stage of carbonation, with visible carbonate veins cutting through amphibole(single polarized transmitted light); Py—pyrite; Gn—galena; Sp—sphalerite; Pl—plagioclase; Chl—chlorite; SiA—silicified andesite, Fl—fluorite, FlA—fluoritized andesite, A(Sp—Gn)—lead and zinc sulfide, Ser—sericite, Cal—carbonate.

Metal sulfide minerals in the ore mainly consist of sphalerite, galena, chalcopyrite, and pyrite, and gangue include quartz, fluorite, and calcite. Ore minerals in the ore show vein like, fine vein like, and mesh vein like structures (Figures 4A and 5A) and disseminated structures. The minerals in the ore show metasomatic dissolution structure (Figure 5B), metasomatic residual structure (Figure 5C), pseudocrystalline structure (Figure 5D), solid solution separation structure (Figure 5E), and crushing structure (Figure 5F). According to the alternation of minerals in the ore and the alteration characteristics of surrounding rocks, hydrothermal mineralization occurred through three stages, pyrite arsenopyrite (mainly quartz, arsenopyrite and a small amount of pyrite), sphalerite chalcopyrite (pyrite, chalcopyrite, sphalerite, galena, fluorite, and a small amount of tetrahedrite), and quartz calcite (quartz calcite veinlets with a small amount of pyrite).

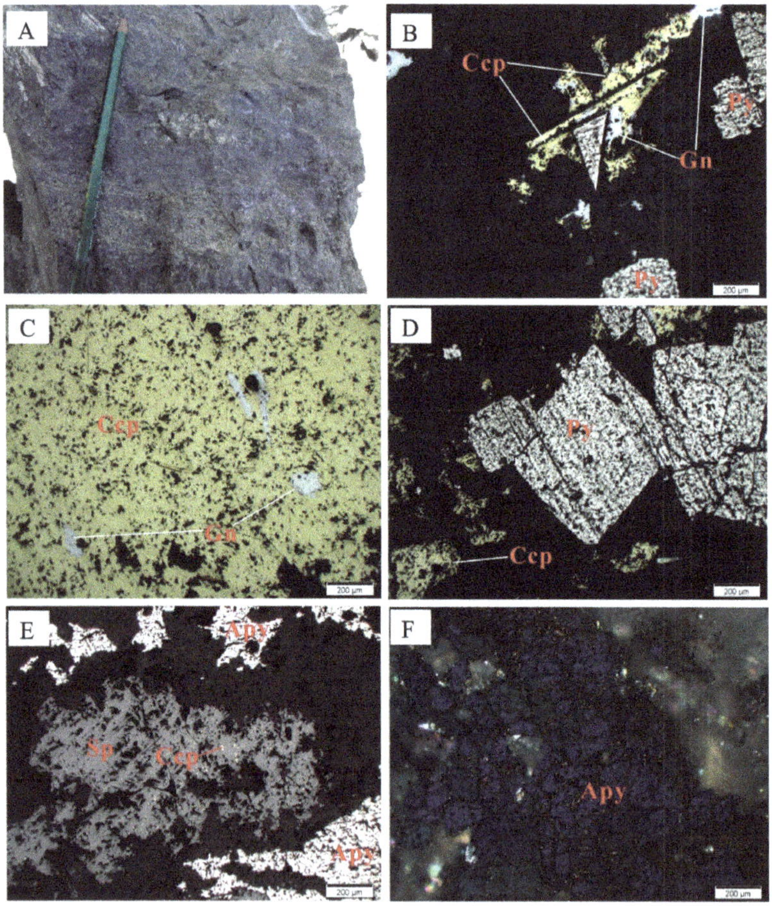

Figure 5. Ore samples and micrographs of the Qiyimuchang deposit. (**A**)—Massive ore; (**B**)—Galena irregularly distributed on the edge of pyrite, metasomatizing and dissolving pyrite(single polarized reflected light); (**C**)—Galena remaining as an island in chalcopyrite and is metasomatized by chalcopyrite to form a residual structure(single polarized reflected light); (**D**)—Pyrite presenting pseudocrystalline structure of arsenopyrite after replacing arsenopyrite(single polarized reflected light); (**E**)—Emulsion droplets of chalcopyrite in the matrix of sphalerite that show an exsolution structure (single polarized reflected light); (**F**)—Arsenopyrite in a crushed structure(orthogonally polarized reflected light).

4. Samples and Methods

4.1. Rare Earth and Trace Elements

In order to explore the relationship between host rock and mineralization of the Qiyimuchang lead–zinc deposit, samples were collected from No. 2 and No. 3 ore bodies and their contact zones for the study of ore, altered mineralized volcanic rocks, rare earth elements (REEs), and trace elements of unaltered volcanic rocks. The sample fragmentation and testing processes were completed at the Key Laboratory of Mineral Resources Evaluation in Northeast Asia of Jilin University. Samples for trace elements and REEs were melted in a Teflon melting tank. The samples is pretreated by the electrothermal plate digestion and dilution method, which dissolved by HNO_3, HF, and $HClO_4$, and diluted to 10 mL with 3% HNO_3 after constant volume with ultrapure water. Shake well and then measured by double focusing high-resolution plasma mass spectrometer ICP-MS (Finnigan MAT Company, Germany). Details of the specific sample processing are available in relevant literature [20,21].

4.2. S and Pb Isotope

In order to explore the material source of the Qiyimuchang lead–zinc deposit, samples of S and Pb isotopes of pyrite and sphalerite in the main metallogenic stage were all collected from the vein ores of No. 2 and No. 3 ore bodies. The selection of single mineral samples was completed at the Hebei Langfang Yuneng Rock and Mineral Sorting Technology Service Co., Ltd., and sample testing was completed at the Analysis and Testing Research Center of Beijing Institute of Geology of Nuclear Industry. Sulfide single mineral and cuprous oxide were ground and mixed evenly according to a certain proportion, heated under vacuum up to 2.0×10^{-2} Pa, and the reaction temperature was 980 °C. Oxidation reaction was allowed to take place in order to produce sulfur dioxide gas, which was collected using the freezing method, and the sulfur isotope composition was analyzed using the MAT 251 gas isotope mass spectrometer(Finnigan MAT Company, Germany). According to test results based on the $\delta^{34}S_{V\text{-}CDT}$, the accuracy was ±0.2‰. The Pb isotope sample was first decomposed by mixing with acid, and lead was then separated using the resin exchange method. After steaming dry, the lead isotope was measured using the IsoProbe-T hot surface ionization mass spectrometer (MicroMass Company, Britain), and the analysis accuracy is better than 0.005%. Details of the specific experimental test process are available in relevant literature [22].

4.3. Zircon U-Pb Dating

In order to define the diagenetic and metallogenic age of the Qiyimuchang lead–zinc deposit, andesitic tuff (13QY25) of the host rock was collected from the shaft of No. 2 ore body, and zircon U-Pb dating was carried out. Andesitic tuff is characterized by mottled structure and crystalline lithic tuff structure. The rock is mainly composed of lithoclasts and crystalloclasts and cemented by volcanic ash. Among them, rock debris (mainly composed of andesite) and crystalline debris (mainly plagioclase and a small amount of quartz and potassium feldspar) are angular, with particle size ranging from 0.5 to 1 mm and a content of approximately 40%. The sorting, target making, CL, and transmission and reflection light photography of zircon were completed at the Hebei Langfang Yuneng Rock and Mineral Sorting Technology Service Co., Ltd., and the testing of zircon was completed at the State Key Laboratory of Continental Dynamics of Northwest University. For detailed experimental test flow and data processing methods, please refer to relevant literature [21,23].

5. Results

5.1. Characteristics of Trace and Rare Earth Elements

The trace element test results of the ore, volcanic rocks, and altered mineralized volcanic rocks in the mining area are shown in Table 1. The trace elements in the surrounding rocks (kinds are expressed in Table 1) and ore of Qiyimuchang deposit show a relatively consistent similarity. The content of trace elements in the surrounding rocks was higher than that of the ore, and the contents of large-ion lithophile elements (Rb, Th, U) were also relatively high (up to 111.00×10^{-6}–260.00×10^{-6}, 6.46×10^{-6}–15.25×10^{-6}, 1.58×10^{-6}–4.57×10^{-6}). Among high-field-strength elements (HFSEs), Zr showed the highest content, reaching up to 233.00×10^{-6}–420.00×10^{-6}, followed by Nb (10.60×10^{-6}–17.8×10^{-6}) and Y (18.70×10^{-6}–44.50×10^{-6}), Ta showed relatively low content (0.70×10^{-6}–1.20×10^{-6}). As shown in Figure 6A, the volcanic rocks and ores of the Qiyimuchang lead–zinc deposit show relative enrichment of Th, U, and other large-ion lithophile elements (LILEs), while some HFSEs such as Ta, Nb, and Y show a trend of relative loss, reflecting the characteristics of intraplate volcanic rocks to a certain extent [24]. The relative contents of Nb and Ta elements showed weak negative anomalies, Ba, Sr, and other LILEs showed clear losses, and some LILEs such as Rb showed relatively high contents. The characteristics of these elements may indicate that crustal materials participated in the magmatic process. The distinct negative anomalies of Sr may be attributable to the residue of plagioclase and other source areas in the rock and strong alteration of the rock.

Table 1. Trace element (mg/kg) content of the representative rocks and ores in the Qiyimuchang deposit.

	13QYK1	13QYK2	13QYK3	13QY2-1	13QY6	14QY	13QY2-2	13QY25	13QY3	13QY16	13QY13	13QY11
Sample Description	Vein Ore	Massive Ore	Massive Ore	Mineralized Rhyolite	Mineralized Tuff	Altered Andesite	Rhyolite	Andesitic tuff	Basaltic Andesite	Rhyolitic Tuff	Andesite	Dacitic Crystal Tuff
Rb	100.00	36.80	1.70	260.00	276.00	182.50	246.00	192.00	149.00	131.50	229.00	111.00
Sr	8.70	3.00	0.90	39.00	13.90	16.00	50.80	13.30	41.70	250.00	190.00	338.00
Ba	122.50	36.80	20.40	368.00	261.00	235.00	482.00	293.00	286.00	825.00	629.00	516.00
Ga	12.00	6.50	3.00	19.40	19.50	22.80	20.20	22.20	23.70	18.10	23.50	18.30
Nb	5.50	1.90	0.40	13.30	13.90	17.80	13.60	15.50	14.70	11.70	15.90	10.60
Ta	0.30	0.10	0.10	1.00	0.80	1.20	0.90	0.90	0.70	0.80	0.90	0.70
Zr	122.00	35.00	2.00	319.00	333.00	420.00	324.00	359.00	355.00	245.00	380.00	233.00
Hf	2.80	0.80	0.20	8.30	8.10	10.40	8.40	8.20	7.50	6.10	8.50	6.00
Th	4.43	1.40	0.07	14.95	12.50	15.25	15.20	12.80	6.46	11.80	7.46	11.35
U	1.22	0.56	0.05	4.14	4.57	4.23	4.48	4.40	1.58	2.71	1.76	3.13
V	14.00	9.00	5.00	19.00	30.00	20.00	18.00	15.00	135.00	35.00	147.00	50.00
Cr	9.00	10.00	20.00	10.00	10.00	10.00	8.00	10.00	50.00	20.00	50.00	40.00
Sn	17.00	13.00	1.00	5.00	5.00	5.00	5.00	9.00	3.00	2.00	3.00	2.00
Cs	7.26	3.02	0.24	16.30	17.35	16.25	20.60	13.65	10.50	8.31	42.00	14.75
Y	12.00	4.10	0.50	28.10	26.10	44.50	30.80	32.90	21.60	18.70	23.20	26.80
W	5.00	4.00	3.00	2.00	16.00	2.00	2.00	19.00	9.00	2.00	7.00	2.00
La	16.40	2.70	0.50	44.30	39.80	66.70	42.10	31.00	49.20	43.10	51.80	39.40
Ce	36.20	5.90	0.70	94.90	84.60	142.50	90.80	69.90	98.20	87.90	107.50	79.50
Pr	4.17	0.69	0.06	10.70	9.83	17.05	10.30	8.01	11.80	9.72	12.50	8.99
Nd	15.80	2.80	0.30	40.00	36.20	66.00	38.20	31.50	45.70	34.00	46.80	32.50
Sm	2.62	0.53	0.16	7.13	6.49	10.90	7.33	5.36	7.74	5.70	8.00	5.75
Eu	0.40	0.09	0.03	1.38	1.66	2.38	1.32	0.76	1.72	0.95	1.92	1.19
Gd	1.99	0.52	0.08	5.86	4.95	9.40	5.19	4.34	5.16	3.91	6.01	4.87
Tb	0.35	0.09	0.01	0.79	0.81	1.29	0.85	0.76	0.76	0.53	0.82	0.75
Dy	2.11	0.57	0.08	4.51	4.53	7.17	4.81	5.17	4.02	3.14	4.24	4.34
Ho	0.41	0.17	0.01	0.96	0.89	1.53	1.16	1.21	0.75	0.70	0.84	0.92
Er	1.08	0.37	0.03	2.96	2.52	4.40	3.15	3.38	2.03	1.88	2.16	2.46
Tm	0.19	0.08	0.01	0.53	0.42	0.70	0.52	0.49	0.30	0.23	0.32	0.37
Yb	1.17	0.43	0.06	3.24	2.60	4.08	3.24	3.71	1.81	1.94	1.80	2.30
Lu	0.20	0.06	0.01	0.47	0.48	0.65	0.52	0.59	0.24	0.33	0.25	0.40
∑REE	83.09	15.00	2.04	217.73	195.78	334.75	209.49	166.18	229.43	194.03	244.96	183.74
∑LREE	75.59	12.71	1.75	198.41	178.58	305.53	190.05	146.53	214.36	181.37	228.52	167.33
∑HREE	7.50	2.29	0.29	19.32	17.20	29.22	19.44	19.65	15.07	12.66	16.44	16.41
LREE/HREE	10.08	5.55	6.03	10.27	10.38	10.46	9.78	7.46	14.22	14.33	13.90	10.20
La/Yb	9.45	4.23	5.61	9.22	10.32	11.02	8.76	5.63	18.33	14.98	19.40	11.55
Eu/Eu*	0.45	0.46	0.62	0.56	0.75	0.62	0.54	0.41	0.68	0.51	0.71	0.59
δCe	1.03	1.02	0.82	1.02	1.00	0.99	1.02	1.04	0.95	0.99	0.99	0.98

As shown in Table 1, clear differences could be observed in the total amount of REEs, light rare earth elements (LREEs), and heavy rare earth elements (HREEs) in ores and volcanic rocks in the Qiyimuchang mining area. In the volcanic rocks, the total amounts of REEs, LREEs, and HREEs were high, reaching up to 166.18×10^{-6}–334.75×10^{-6}, 167.33×10^{-6}–305.53×10^{-6}, and 12.66×10^{-6}–29.22×10^{-6}, respectively. In contrast, the total amounts of REEs, LREEs, and HREEs were lower in the ore at 2.04×10^{-6}–83.09×10^{-6}, 1.75×10^{-6}–75.59×10^{-6}, and 0.29×10^{-6}–7.5×10^{-6}, respectively, indicating that a large number of REEs were released with the ore-forming materials during mineralization, and the exchange of REEs occurred in the hydrothermal system. The LREE/HREE ratio and La/Yb ratio of all samples tested were greater than one, indicating a right dipping distribution mode. Moreover, the rare earth distribution modes of volcanic rocks and ores were very similar (Figure 6B), indicating that they may belong to the same source area and are the products of different stages of homologous magma activities. The Ce/Yb ratio of all samples ranged from 11.60 to 59.70, and the Th/Yb ratio ranged from 1.20 to 6.20, indicating that differential crystallization occurred in the process of magmatic evolution [26]. According to the trend of rare earth composition curve Eu/Eu* value, all samples showed weak negative Eu anomaly, with the Eu/Eu* value ranging from 0.41×10^{-6} to 0.76×10^{-6} (0.58×10^{-6} on average). The relative loss of HREEs indicates the presence of residual garnet in the magma source area. The negative Eu anomaly may indicate the separation of feldspar and other minerals from felsic magma during the differential crystallization of magma. It can also be said that feldspar remained in the source area during the partial melting [27].

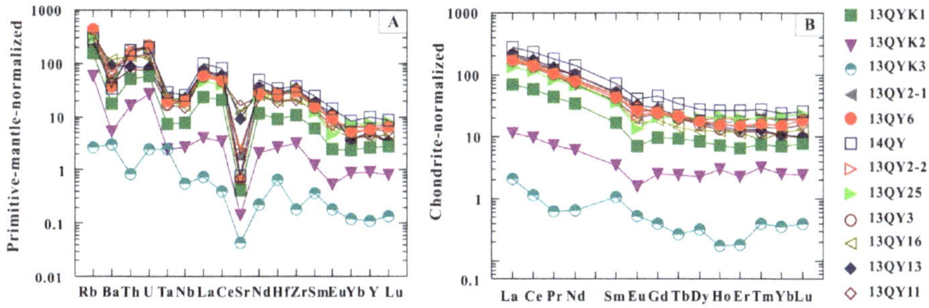

Figure 6. Primitive-mantle-normalized trace element diagrams (**A**) (normalized with data from [25]) and chondrite-normalized REE patterns (**B**) normalized with data from [25] for the representative rocks (kinds are expressed in Table 1) and ores in the Qiyimuchang deposit.

5.2. Characteristics of S and Pb Isotopic

As shown in Table 2 and Figure 7, the Qiyimuchang mining area is similar to other lead–zinc deposits in the same area. The sulfur isotopes are distributed in a tower-like manner, $\delta^{34}S$ values are positive, and the change of $\delta^{34}S$ value in a certain range is small, which indicates that the sulfur isotopic composition is relatively stable. The variation range was from +1.60‰ to +4.80‰, the average value was +4.00‰, the range was R = 3.20, and the standard deviation was S = 1.10. Among them, the pyrite $\delta^{34}S$ value was between +1.60‰ and +4.30‰, which is slightly less than that of the sphalerite $\delta^{34}S$ value (+3.80‰–+4.80‰). The reason for this phenomenon may be that the sulfur fractionation did not reach equilibrium during the evolution of the hydrothermal system; it may also be due to the loss of ^{34}S in the ore-forming fluid caused by the high oxygen fugacity in the evolution of the hydrothermal fluid [28,29].

Table 2. Sulfur isotope data of ore minerals from typical lead–zinc deposits in Erguna metallogenic belt.

Deposit	Sample	Result	Source
		$\delta^{34}S_{V\text{-}CDT}$ (‰)	
Qiyimuchang	sphalerite(13QYK1)	4.80	this paper
	pyrite (13QYK1)	4.30	
	sphalerite (13QY6)	4.70	
	sphalerite (13QYK2)	4.40	
	pyrite (13QYK2)	4.30	
	sphalerite (13QYK3)	3.80	
	pyrite (13QYK3)	1.60	

Table 2. *Cont.*

Deposit	Sample	Result $\delta^{34}S_{V\text{-}CDT}$ (‰)	Source
Jiawula	galena	3.90	[30]
	galena	4.10	
	pyrite	3.70	
	pyrite	3.20	
	pyrite	2.80	
	sphalerite	3.90	
	sphalerite	3.80	
Dongjun	galena	4.48	[31]
	pyrite	6.76	
	pyrite	6.63	
	pyrite	6.97	
	pyrite	6.89	
Derbur	pyrite	5.20	[3]
	pyrite	5.40	
	pyrite	4.90	
	pyrite	5.20	
	sphalerite	5.00	
	sphalerite	5.20	
	galena	2.90	
	galena	3.60	

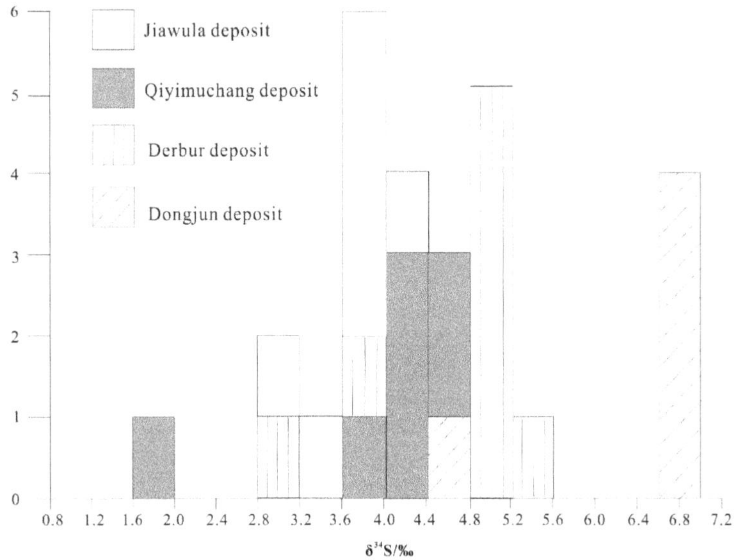

Figure 7. Sulfur isotope abundances of sulfide in typical lead zinc deposits in Erguna metallogenic belt.

The test results were consistent with those of the sulfur isotope analysis. The lead isotopic compositions of seven pyrite and sphalerite samples and other deposits in the same area are listed in Table 3. The analysis results show that the changes of $^{208}Pb/^{204}Pb$, $^{207}Pb/^{204}Pb$, and $^{206}Pb/^{204}Pb$ are relatively stable at 38.15–38.39, 15.55–15.62, and 18.33–18.41, respectively. The distribution range is relatively concentrated and the change is small, indicating that the lead in the sulfide of the Qiyimuchang deposit has the characteristics of normal lead [32].

Table 3. Lead isotope data of ore minerals from typical lead–zinc deposits in Erguna metallogenic belt.

Deposit	Sample	Isotope Ratio			μ	ω	Δβ	Δγ	Source
		$^{208}Pb/^{204}Pb$	$^{207}Pb/^{204}Pb$	$^{206}Pb/^{204}Pb$					
Qiyimuchang	Sphalerite (13QYK1)	38.26	15.58	18.35	9.44	35.92	16.83	27.05	This paper
	Pyrite (13QYK1)	38.15	15.55	18.33	9.37	35.31	14.61	24.15	
	Sphalerite (13QY6)	38.29	15.59	18.36	9.45	36.08	17.35	27.86	
	Sphalerite (13QYK2)	38.21	15.57	18.34	9.41	35.66	15.85	25.82	
	Pyrite (13QYK2)	38.29	15.59	18.36	9.45	36.09	17.41	27.94	
	Sphalerite (13QYK3)	38.38	15.62	18.38	9.50	36.58	19.17	30.27	
	Pyrite (13QYK3)	38.39	15.62	18.41	9.50	36.51	19.31	30.65	
Erdaohezi	Sphalerite	38.24	15.56	18.44					[10]
	Sphalerite	38.26	15.57	18.44					
	Sphalerite	38.22	15.56	18.44					
	Pyrite	38.19	15.55	18.13					
	Sphalerite	38.23	15.56	18.44					
	Pyrite	38.20	15.55	18.43					
	Sphalerite	38.25	15.57	18.44					
	Pyrite	38.08	15.53	18.31					
Dongjun	Pyrite	38.10	15.53	18.34					[31]
	Pyrite	38.42	15.63	18.43					
	Pyrite	38.26	15.58	18.38					
	Pyrite	38.13	15.54	18.35					
	Pyrite	38.15	15.55	18.35					
	Galena	37.98	15.49	18.32					
	Galena	38.01	15.52	18.33					
	Galena	38.20	15.56	18.37					
	Sphalerite	38.16	15.55	18.36					
	Sphalerite	38.17	15.55	18.36					
Jiawula	Galena	38.15	15.55	18.33					[30]
	Pyrite	38.08	15.54	18.28					
	Galena	38.14	15.55	18.33					
	Sphalerite	38.14	15.54	18.33					
	Pyrite	38.12	15.54	18.31					
	Pyrite	38.14	15.55	18.33					
	Sphalerite	38.14	15.55	18.34					
	Upper crust						>18	>45	
	Mixing of crust and mantle						12–40	20–45	
	Mantle						−5–15	−5–20	

Note: isotope parameters are calculated by using geokit software developed by Lu [33]; Upper crust, crust mantle mixing and mantle lead isotope parameters are cited from Li and Zhao [34]. $\mu = {}^{238}U/{}^{204}Pb$; $\omega = {}^{232}Th/{}^{204}Pb$; $\Delta\beta = [\beta/\beta M(t)-1] \times 1000$; $\Delta\gamma = [\gamma/\gamma M(t)-1] \times 1000$. β and γ are current mantle values. βM(t) and γM(t) are mantle values at "t". The "t" value is referred from the age of the Jiawula lead zinc deposit as 142 Ma.

5.3. Zircon U-Pb Ages in Andesitic Tuff

Zircons in andesitic crystal debris tuff in the Qiyimuchang mining area exhibit high transparency, medium crystal form integrity, aggregation, and octahedral shape, and the particle size is generally between 73 and 154 μm, with high automorphism and clear oscillatory ring (Figure 8). The Th and U contents of zircon were 128.00–1704.31 mg/kg and 138.07–1079.35 mg/kg, respectively, and the Th/U ratio was 0.41–2.17. All values being greater than 0.40 (Table 4) suggests that the zircons measured are of typical magmatic origin [35,36].

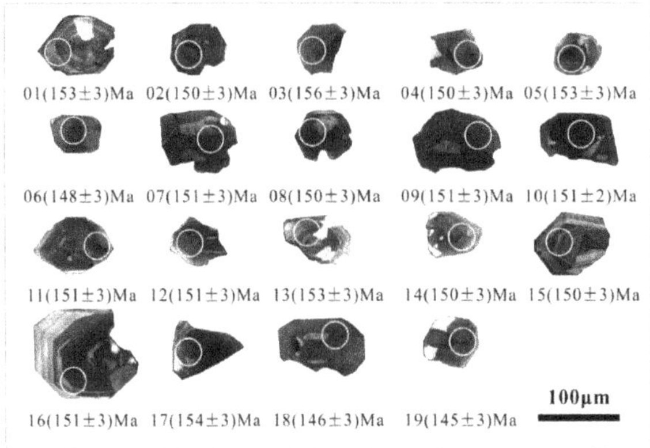

Figure 8. Zircon CL image of andesitic clastic lithic tuff in the host rock of the Qiyimuchang lead–zinc deposit.

Table 4. LA-ICP-MS zircon U-Th-Pb dating results of the andesite tuff from Qiyimuchang deposit.

Number	U (mg/kg)	Th (mg/kg)	Th/U	Isotope Ratio						Age (Ma)					
				$^{207}Pb/^{206}Pb$	1σ	$^{207}Pb/^{235}U$	1σ	$^{206}Pb/^{238}U$	1σ	$^{206}Pb/^{238}U$	1σ	$^{207}Pb/^{235}U$	1σ	$^{207}Pb/^{206}Pb$	1σ
1	138.07	127.95	0.93	0.05	0.004	0.16	0.012	0.02	0.0005	153	3	153	10	157	119
2	212.64	301.13	1.42	0.05	0.004	0.17	0.012	0.02	0.0005	150	3	156	11	236	127
3	407.5	186.43	0.46	0.05	0.002	0.16	0.007	0.02	0.0004	156	3	155	6	135	68
4	314.49	541.42	1.72	0.05	0.003	0.16	0.008	0.02	0.0004	150	3	152	7	180	87
5	266.67	579.66	2.17	0.05	0.003	0.16	0.010	0.02	0.0005	153	3	154	9	168	100
6	378.43	691.25	1.83	0.05	0.004	0.16	0.011	0.02	0.0005	148	3	152	9	213	115
7	650.57	404.52	0.62	0.05	0.002	0.16	0.007	0.02	0.0004	151	3	154	6	210	67
8	379.26	236.84	0.62	0.05	0.003	0.16	0.008	0.02	0.0004	150	3	153	7	197	79
9	322.72	132.19	0.41	0.05	0.003	0.16	0.008	0.02	0.0004	151	3	154	7	201	80
10	1079.35	1704.31	1.58	0.05	0.001	0.16	0.004	0.02	0.0004	151	2	153	4	186	30
11	239.23	354.09	1.48	0.05	0.003	0.16	0.009	0.02	0.0004	151	3	150	7	136	87
12	292.55	464.93	1.59	0.05	0.002	0.16	0.008	0.02	0.0004	151	3	152	7	172	74
13	370.54	636.61	1.72	0.05	0.004	0.17	0.012	0.02	0.0005	153	3	158	11	237	123
14	285.85	488.21	1.71	0.05	0.003	0.16	0.010	0.02	0.0005	150	3	153	9	193	102
15	450.38	303.83	0.67	0.05	0.003	0.17	0.008	0.02	0.0004	150	3	156	7	239	80
16	346.85	203.08	0.59	0.05	0.004	0.16	0.012	0.02	0.0005	151	3	153	10	183	126
17	528.17	329.13	0.62	0.05	0.002	0.16	0.006	0.02	0.0004	154	3	150	5	87	51
18	551.19	329.26	0.60	0.05	0.003	0.16	0.008	0.02	0.0004	146	3	150	7	213	75
19	305.45	477.73	1.56	0.05	0.003	0.17	0.008	0.02	0.0004	145	3	158	7	351	70

The $^{206}Pb/^{238}U$ surface ages of 19 measuring points ranged between (145 ± 3) Ma and (156 ± 3) Ma. On the zircon U-Pb age harmony diagram (Figure 9), the age data are located on and near the harmony line. The weighted mean age was (150.8 ± 1.3) Ma (MSWD = 0.74, Confidence 95%), representing the crystallization age of andesitic tuff, i.e., late Jurassic.

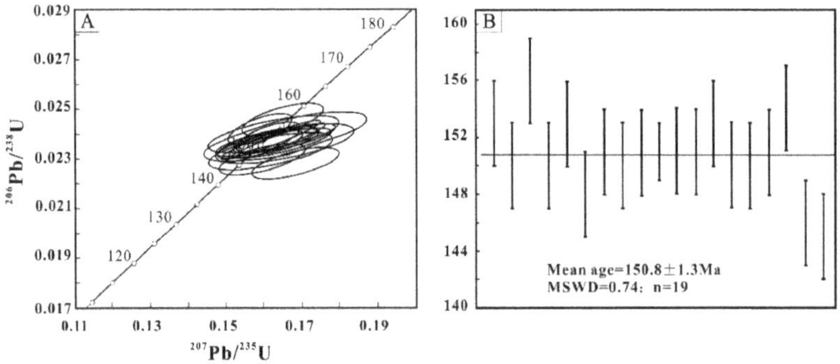

Figure 9. Zircon U-Pb concordia diagram (**A**) and weighted average diagram (**B**) for the andesitic lithic crystal tuff from the Qiyimuchang deposit.

6. Discussion

6.1. Source of Ore-Forming Materials

6.1.1. Trace and Rare Earth Element Evidence

The composition of trace elements in rocks and ores has a certain geochemical significance. The study of the geochemical characteristics of ore bodies and surrounding rocks is of great significance to the study of the whole deposit. The host rock of the Qiyimuchang lead–zinc deposit is the volcanic rock of the Tamulangou formation. According to the analysis of the trace characteristic parameters (Table 1) and primitive-mantle-normalized trace element diagrams (Figure 6A) of volcanic rocks and ores in the Qiyimuchang mining area, LILEs, such as Th and U, are relatively enriched in ores and volcanic rocks, while HFSEs, such as Ta and Nb, are relatively depleted and exhibit clear negative Sr anomalies, indicating that a large amount of Sr was lost through serious rock alteration. A distinct negative Nb anomaly could be observed, suggesting that crustal materials might have participated in the magmatic evolution [27]. The distribution pattern of incompatible elements in trace elements showed a relatively similar trend. Th, U, and Rb show clear positive anomalies, the total rake ratio is negative, and Nb and Ta show obvious negative anomalies. These suggest that the tectonic background of volcanic rocks and ores in the Qiyimuchang mining area may correspond to a continental rift environment [37]. The change trends of trace elements in volcanic rocks, mineralized altered volcanic rocks, and ores were relatively similar, but slightly different. The trace elements of the lead–zinc ores appear to not only inherit the characteristics of surrounding rocks, but also retain the characteristics of the metallogenic hydrothermal solution itself.

The content of REEs in volcanic rocks in the Qiyimuchang lead–zinc mining area was higher than that in ore, which may be due to the strong activity of REEs in the process of alteration and mineralization, resulting in the introduction and removal of REEs and metallogenic materials. Nevertheless, the differentiation relationship between REEs has not changed. Therefore, the characteristics of the REE distribution model shown in the diagram still reflect the characteristics of magmatic evolution (Figure 6B). The REE distribution model shows that the ore and volcanic rocks are relatively consistent, indicating that they may originate from homologous areas and the products of different magmatic stages.

In recent years, petrological and geochemical studies of basic volcanic rocks in the Tamulangou Formation in the Erguna Block have shown that the magma source area may be an enriched lithospheric mantle metasomatized by subduction fluid, and that the magma was clearly contaminated by crustal material during magmatic evolution and ascent [38–40]. Therefore, it can be preliminarily ascertained that the metallogenic material and surrounding rock of the Qiyimuchang lead–zinc deposit originate from the same source area and are products of different magmatic stages.

6.1.2. Sulfur and Lead Isotope Evidence

Sulfur is one of the most important metallogenic elements in lead–zinc deposits. The sulfur isotopic composition in the ore can indicate the source and genesis of sulfur in the ore [41]. Therefore, the sulfur stable isotopic composition of metal sulfide in the ore can not only reflect the initial ratio of sulfur stable isotopes, but also reveal the source of ore-forming materials in the deposit in combination with other characteristics [42,43]. Sulfur isotopes are generally believed to have three sources: mantle derived sulfur ($\delta^{34}S$ value close to 0, approximately $0 \pm 3‰$), current seawater sulfur ($\delta^{34}S$ value close to +20‰), and reduced sulfur (large negative $\delta^{34}S$ values) [44]. The $\delta^{34}S$ values of pyrite, sphalerite, and other single minerals in the Qiyimuchang mining area were close to 0, ranging from +1.6‰ to +4.8‰, with an average value of +4.0‰ and a small fluctuation range, indicating that the source of ore sulfur in the deposit is relatively singular, with the characteristics of deep source sulfur, and the ore-forming materials may be closely related to magmatic intrusion. The $\delta^{34}S$ value of pyrite was between +1.6‰ and +4.3‰, which is slightly less than that of sphalerite (from +3.8‰ to +4.8‰). The reason for this phenomenon may be that sulfur fractionation did not reach equilibrium during the evolution of the hydrothermal system, reflecting the characteristics of rapid precipitation of ore-forming materials in the process of mineralization [45,46].

In addition, the S isotopic composition of the Qiyimuchang deposit is basically consistent with that of the typical vein lead–zinc deposit in the region (Figure 7), indicating that their genesis may also be consistent.

Lead isotopes can not only be used to determine the metallogenic age of the deposit, but also has very important significance in the study of the source of metallogenic materials [47,48]. Changes in the characteristic values of lead isotopes μ can reflect the source of lead. The values of μ are 8.92 in the mantle environment, 10.87 in the orogenic belt [49], 8.69–9.24 normally [50], and 9.58 in the upper crust. The μ value of ores in the Qiyimuchang lead–zinc deposit was very high at 9.37–9.50 (9.45 on average), falling between the range of the mantle and the upper crust. The variation range of the ω value was 35.31–36.58 (36.02 on average), which is higher than that of the mantle (31.84) and lower than that of the crust (36.84) [48]. This suggests that the lead in the Qiyimuchang deposit may be of mixed source comprising the crust and mantle.

In order to eliminate the influence of time factors and better trace the source of ore-forming materials, Zhu et al. [51] proposed the expression of the three isotopes of lead as the relative deviation of the contemporary mantle Δα, Δβ, Δγ, and Δγ-Δβ genetic classification diagram to trace the source area of lead. In this study, the Δβ and Δγ values of the deposit were calculated using the GeoKit software (Table 3). The Δβ value of sulfides in the deposit ranged from 14.61 to 19.31, with an average of 17.22, and the Δγ value ranged from 2.15 to 30.65, with an average of 27.68. The Δβ and Δγ values of most samples correspond to the crust–mantle mixed source (Table 3), suggesting that the lead in the deposit may originate from a mixed source of the crust and mantle. In the Δγ-Δβ genetic classification diagram (Figure 10), all sample data points fall within the lead (magmatism) range of the subduction zone, where the upper crust and mantle are mixed.

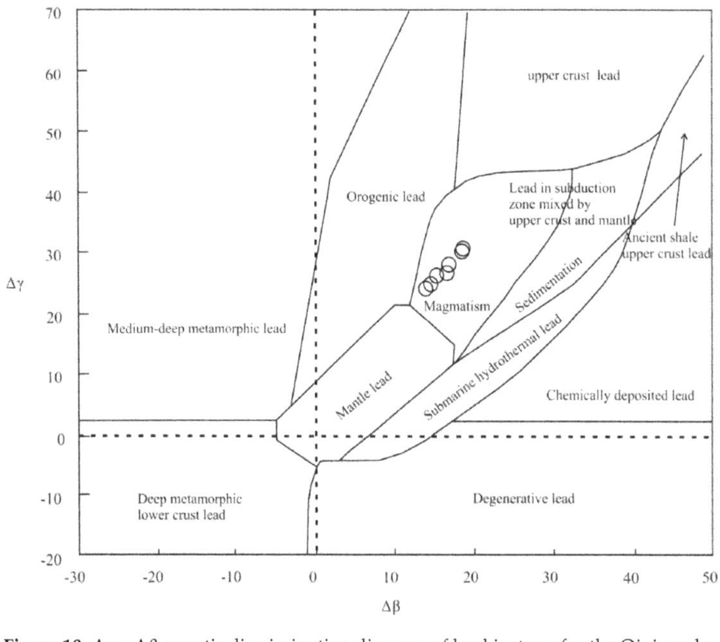

Figure 10. Δγ—Δβ genetic discrimination diagram of lead isotope for the Qiyimuchang lead–zinc deposit (base map from [49]).

Based on the above characteristics of trace elements, REEs, and sulfur and lead isotopes, the metallogenic material source of the deposit can be preliminarily considered to be a mixed source of the crust and mantle. In the period of tectonic regime transformation, during the early stage of plate subduction, the crust mantle mixture formed by the upwelling of crustal material and mantle material metasomatized by subduction fluid melted again to form magma. Under the background of the extensional structure, magma upwelling occurred along deep faults and transported a certain amount of ore-forming fluid and ore-forming materials to the shallow part. In the process of upward migration, the ore-forming material was contaminated by crust derived material.

6.2. Diagenetic and Metallogenic Age

Based on the field geology and underground investigation of the mining area, and according to the cross-cutting and co-association relationship between the lead–zinc ore body and the surrounding rock, the host rock of the Qiyimuchang lead–zinc mining area can be determined to be mainly basaltic andesite and andesitic crystal debris tuff of the Tamulangou Formation. In this study, the zircon LA-ICP-MS U-Pb dating results of andesitic crystal debris tuff reveal that the age of the upwelling, condensation, and crystallization of the diagenetic magma forming the host rock in the area should be 150.8 ± 1.3 Ma, indicating that the lower limit of mineralization of the deposit should be 150.8 ± 1.3 Ma.

Many studies have investigated the metallogenic age of vein lead–zinc silver deposits in the Erguna metallogenic belt. Li. et al. [12] carried out Rb-Sr isotopic dating of sphalerite and pyrite in the Jiawula silver-lead zinc deposit in the south of the Erguna metallogenic belt, and obtained the isochron age of 142 Ma. Li, et al. [52] obtained the sericite ^{40}Ar-^{39}Ar plateau age (138 ± 1 Ma) of the Chaganbulagan lead zinc silver deposit. In recent years, the Rb-Sr isotopic ages have also been obtained for ore minerals in large-vein lead–zinc deposits, such as Billia Valley (131.3 ± 2.4 Ma), Derbur (141 ± 1.9 Ma), and Erdaohezi (130.5 ± 3.6 Ma), which are widely distributed in the northwest of the Derbugan fault in the middle of the Erguna metallogenic belt [5,11,53]. For vein lead–zinc deposits (such as the Dongjun lead–zinc deposit) in the south and east of Derbugan fault, a more accurate Rb-Sr isochron age of sphalerite (130.2 ± 4.4 Ma) has also been obtained in recent years [54]. On the whole, the metallogenic age of vein lead–zinc silver deposits on the north and south sides of the Derbugan fault in the Erguna metallogenic belt is between 130 and 145 Ma. Therefore, the metallogenic age of the Qiyimuchang lead–zinc deposit should also be this period, i.e., the early Cretaceous. The metallogenic geological background of vein lead–zinc deposits in the Erguna Block is relatively consistent and the metallogenic age is similar, which reflects that the vein lead–zinc silver deposits in this metallogenic belt may be the output of the same metallogenic process.

6.3. Metallogenic Dynamic Background

The characteristics of trace elements in the surrounding rock and ore of the Qiyimuchang deposit show that it has the characteristics of intraplate volcanic rocks. Its metallogenic age is the same period as the vein lead–zinc deposit widely developed in the Erguna Block, i.e., the Early Cretaceous (130–140 Ma). According to the study of trace elements in the deposit and the characteristics of regional tectonic evolution, the Erguna block should be under an extensional tectonic background during this period [24,55,56]. A recent study showed that the Erguna Block has been strongly affected by the Mongolian Okhotsk tectonic domain and the Paleo-Pacific tectonic domain since the Mesozoic [57]. Under this background, it becomes unclear whether the large-scale vein lead–zinc mineralization in Erguna Block should be the product of the evolution of the Mongolia Okhotsk tectonic domain or the Paleo-Pacific tectonic domain.

According to the characteristics of volcanic rocks in Northeast China, previous studies believe that the subduction of the Paleo-Pacific Plate began in the early Jurassic [58–61]. During the late Jurassic–Early Cretaceous, the Paleo-Pacific plate was obliquely subducted, resulting in the strike slip nature of the continental margin of Northeast Asia. From the late Early Cretaceous to the Paleogene, the Paleo-Pacific Plate gradually changed from subduction to retreat [62]. Since the late last century, the "scissors" closure of the Mongolian

Okhotsk Ocean from west to east [63–65] has triggered a series of magmatism and copper molybdenum mineralization, which has been widely recognized by scholars [57,66–71]. In recent years, based on the stratigraphic evidence and paleomagnetism evidence, some scholars proposed that the Mongolia Okhotsk Ocean Basin should have been closed in the middle Jurassic. In the late Jurassic, the Erguna Block entered the regional extensional evolution history after the closure of the Mongolia Okhotsk Ocean [62]. The alkaline subalkaline transitional volcanic rock assemblages of the Tamulangou Formation and Manketou'ebo Formationdistributed in NE and NNE directions in the Erguna block all reveal the existence of a regional extensional environment [72,73]. At the same time, the vein shaped epithermal lead–zinc deposits distributed in the NNE direction in the block also indicate that the Erguna Block was in an extensional tectonic environment in this period, during which these alkaline subalkaline transitional volcanic rocks and alkaline rhyolites were absent in the east of the Songliao Basin, South Korea, and Japan [70,74]. This further shows that their formation is related to the evolution of the Mongolia Okhotsk tectonic system, but not to the Paleo-Pacific system. Therefore, from late Jurassic to early Cretaceous, the east of the Songliao Basin was mainly controlled by the subduction tectonic system of the Paleo-Pacific Plate, while the Erguna Massif and Xing'an Massif in the west of the Songliao Basin were mainly controlled by the post orogenic extension system of the Mongolia Okhotsk tectonic domain [75]. In conclusion, this paper holds that the large-scale vein lead–zinc mineralization of the Erguna Massif in the early Cretaceous should occur in a post-orogenic extensional tectonic background of the Mongolia Okhotsk tectonic domain.

7. Conclusions

1. The Qiyimuchang lead–zinc deposit is similar to the veined lead–zinc-silver deposit widely developed in the region, and their metallogenic ages are relatively consistent, i.e., the early Cretaceous. It is assumed that veined lead–zinc-silver deposits in the metallogenic belt develop according to the same metallogenic process.

2. The large-scale veined lead–zinc mineralization of the Erguna Massif in the early Cretaceous is controlled by the post orogenic extensional tectonic background of the evolution of the Mongolia Okhotsk tectonic domain.

3. The metallogenic material of the Qiyimuchang lead–zinc deposit should have a crust and mantle mixed source. In other words, the metallogenic material is the product of magma formed by the remelting of the crust and mantle mixture derived from the upwelling of the crust material and mantle material metasomatized by the subduction fluid in the early stage, which evolved under the extensional tectonic background.

Author Contributions: Conceptualization, M.D. and Y.R.; Investigation, M.D., Y.R., Y.H. and Q.Y.; Methodology, M.D., Q.Y. and Y.H., Data curation, M.D. and Y.H., Writing—original draft preparation, M.D., Writing—review and editing, Y.R., C.X. and T.L., Supervision and project administration: Y.R. and T.L. All authors have read and agreed to the published version of the manuscript.

Funding: This research was funded by the National Key Basic Research and Development Program of China (2013CB429802) and China Geological Survey Project (ZD20220103).

Data Availability Statement: The data presented in this study are available on request from the corresponding author.

Acknowledgments: The authors would like to thank all the reviewers who participated in the review and MJEditor (www.mjeditor.com) for its linguistic assistance during the preparation of this manuscript.

Conflicts of Interest: The authors declare no conflict of interest. The funders had no role in the design of the study, in the collection, analyses, or interpretation of data, in the writing of the manuscript, or in the decision to publish the results.

References

1. Sun, E.S. The metallogenic regularity of the silver deposits in the Manzhouli-Xinbaerhuyouqi metallogenic belt, Inner Mongolia. *Geol. Explor. Non-Ferr. Met.* **1995**, *4*, 23–29.
2. Zhu, H.-C.; Zhang, J.-F.; Quan, H.; Zhang, H.; Wang, H.-B. Metallogenic characteristics of nonferrous and noble metals in Ergun region. *J. Precious Met. Geol.* **1999**, *8*, 193–198.
3. Zhang, J.-F.; Wang, X.-Z.; Quan, H.; Wu, G.; Zhu, H.C. The forming conditions of nonferrous and precious metal deposits in the north of Derbugan metallogenic province. *Geol. Resour.* **2002**, *11*, 43–52.
4. Jia, B.; Liu, G.X.; Zhang, C.-H.; Yang, H.-Z.; Zhang, C.-P. Metallogenesis of the lead-zinc deposits controlled by the Mesozoic volcanic-subvolanic rocks in Daxinganling region. *Geol. Resour.* **2012**, *21*, 114–121.
5. Xu, Z.-T.; Liu, Y.; Sun, J.-G.; Liang, X.-L.; Xu, Z.-K. Nature and ore formation of the Erdaohezi Pb-Zn deposit in the Great Xing'an Range, NE China. *Ore Geol. Rev.* **2020**, *119*, 103385. [CrossRef]
6. Qin, K.Z.; Wang, D.P.; Wang, Z.T.; Sun, S. Types, geological background, metallogenic provinces and ore-forming systematics of major copper deposits in eastern China. *Miner. Depos.* **1999**, *18*, 359–371.
7. Mao, J.-W.; Xie, G.-Q.; Zhang, Z.-H.; Li, X.-F.; Wang, Y.-T.; Zhang, C.-Q.; Li, Y.-F. Mesozoic large-scale metallogenic pulses in North China and corresponding geodynamic settings. *Acta Petrol. Sin.* **2005**, *21*, 169–188.
8. Pirajno, F. *Hydrothermal Processes and Mineral Systems*; Springer: Berlin, Germany, 2009.
9. Catchpole, H.; Kouzmanov, K.; Bendezú, A.; Ovtcharova, M.; Spikings, R.; Stein, H.; Fontboté, L. Timing of porphyry (Cu-Mo) and base metal (Zn-Pb-Ag-Cu) mineralisation in a magmatic-hydrothermal system—Morococha district, Peru. *Miner. Depos.* **2015**, *50*, 895–922. [CrossRef]
10. Li, J.-W.; Liang, Y.-W.; Wang, X.-Y.; Zhang, B.; Yang, Y.-C.; She, H.-Q.; Guo, Z.-J. The origin of the Erdaohezi lead-zinc deposit, Inner Mongolia. *J. Jilin Univ. (Earth Sci. Ed.)* **2011**, *41*, 1745–1754.
11. Zhao, Y.; Lü, J.-C.; Zhang, D.-B.; Zhou, Y.-H.; Shao, J.; Wang, B. Rb-Sr isochron age of Derbur Pb-Zn-Ag deposit in Erguna massif of northeast Inner Mongolia and its geological significance. *Miner. Depos.* **2017**, *36*, 893–904.
12. Li, T.-G.; Wu, G.; Liu, J.; Hu, Y.-Q.; Zhang, Y.-F.; Luo, D.-F. Rb-Sr isochron age of the Jiawula Pb-Zn-Ag deposit in the Manzhouli area and its geological significance. *Acta Petrol. Sin.* **2014**, *30*, 257–270.
13. Cavalcante, C.; Hollanda, M.-H.; Vauchez, A.; Kawata, M. How long can the middle crust remain partially molten during orogeny? *Geology* **2018**, *46*, 839–842. [CrossRef]
14. Jahn, B.-M.; Windley, B.; Natal'in, B.; Dobretsov, N. Phanerozoic continental growth in Central Asia. *J. Asian Earth Sci.* **2004**, *5*, 599–603. [CrossRef]
15. Pan, G.; Xiao, Q.; Lu, S.; Deng, J.; Feng, Y.; Zhang, K.; Zhang, Z.; Wang, F.; Xing, G.; Hao, G. Subdivision of tectonic units in China. *Geol. China* **2009**, *36*, 1–28.
16. Zhang, Z.; Institute of Inner Mongolia Tenth Geological Mineral Exploration. *General Survey Report of Lead Zinc Deposits in I and II Ore Blocks of the Qiyimuchang Mining Area, Chenba'erhu Country, Inner Mongolia*; Inner Mongolia Geology & Mineral Technology Co.: Inner Mongolia, China, 2005.
17. Duan, M.-X.; Ren, Y.-S.; Hou, Z.-S.; Sun, D.-Y.; Sun, Z.-M.; Qin, C.-J. Ore-forming fluid and ore genesis of the Biliya valley lead-zinc (silver) deposit in Erguna region, Inner mongolia. *Mineral. Petrol.* **2014**, *2*, 60–67.
18. Duan, M.-X.; Ren, Y.-S.; Guo, H.-Y.; Wang, B.-C.; Zhao, X.-D.; Cai, Y.-L. Ore-forming fluids and genesis of Qiyimuchang lead-zinc deposit in Erguna area, Inner Mongolia. *Geol. Resour.* **2017**, *6*, 0557–0563.
19. Zheng, C.; Zhou, J.; Jin, W.; Ji, J.; Zhang, X.; Ma, Z.; Ding, X. Geochronology in the north segment of the Derbugan fault zone, Great Xing'an Range, NE China. *Acta Petrol. Sin.* **2009**, *25*, 1989–2000.
20. Zhao, W.-Z.; Zhang, Y.; Du, S.-W.; Zhang, B.-B.; Li, S.-K. Study on pretreatment of geological samples by ICP-MS method. *Chem. Eng.* **2018**, *269*, 25–28.
21. Liu, X.; Gao, S.; Yuan, H.-L.; Hattendorf, B.; Günther, D. Analysis of 42 major and trace elements in glass standard reference materials by 193 nm LA-ICPMS. *Acta Petrol. Sin.* **2002**, *18*, 408–418.
22. Liu, Z.; Shao, Y.; Zhou, X.; Zhang, Y.; Zhou, G. Hydrogen, oxygen, sulfur and lead isotope composition tracing for the ore-forming material source of Dongguashan copper (gold) deposit in Tongling, Anhui Province. *Acta Petrol. Sin.* **2014**, *30*, 199–208.
23. Li, Z.-S.; Zhang, C.-X.; Shu, L.; Shan, L.; Chi, N.-J.; Li, F.-C.; Wang, J.-L.; Yang, H. U-Pb dating of LA-MC-ICP-MS in situ of single zircon. *Shandong Land Resour.* **2022**, *38*, 71–81.
24. Xu, G.; Bian, Q.; Wang, Y. Tectonic evolution and metallization of the Erguna orogenic belt. *Chin. J. Geol.* **1998**, *33*, 84–92.
25. Sun, S.-S.; McDonough, W.F. Chemical and isotopic systematics of oceanic basalts: Implications for mantle composition and processes. *Geol. Soc. Lond. Spec. Publ.* **1989**, *42*, 313–345. [CrossRef]
26. Yin, Z.; Zhang, Y.; Yang, X.; Zhao, H.; Han, Z. Characteristics of Mesozoic volcanic rocks and magma evolution in northern Daxing'anling. *Glob. Geol.* **2006**, *25*, 120–128.
27. Rollison, H.R. *Petrogeochemistry*; China University of Science and Technology Press: Hefei, China, 2000.
28. Shan, L.; Zheng, Y.-Y.; Xu, R.-K.; Cao, L.; Zhang, L.; Lian, Y.-L.; Lian, Y.-L. Review on sulfur isotopic tracing and hydrothermal metallogenesis. *Geol. Resour.* **2009**, *18*, 197–203.
29. Li, Y.; Lü, Z.; Yan, G.; Zhen, S.; Du, Z. Isotopic characteristics of S, Pb, H and O of Jiama copper polymetallic ore deposit, Tibet and their significance. *Earth Sci. Front* **2012**, *19*, 72–81.

30. Zhai, D.; Liu, J.; Wang, J.; Yao, M.; Liu, X.; Liu, Z.; Wu, S.; Fu, C.; Wang, S.; Li, Y. A study of stable isotope geochemistry of the Jiawula large Pb–Zn–Ag ore deposit, Inner Mongolia. *Earth Sci. Front.* **2013**, *20*, 213–225.
31. Zhang, B. *The Geological Features and Genesis of the Dongjun Lead-Zinc-Silver Deposit in Inner Mongolia, China*; Chinese Academy of Geological Sciences: Beijing, China, 2011.
32. Du, Q. *Duobaoshan Porphyry Copper Deposit*; Geology Press: Beijing, China, 1988.
33. Lu, Y. GeoKit-A geochemical toolkit for Microsoft Excel. *Geochimica* **2004**, *33*, 459–464.
34. Li, Y.; Zhao, Y.-Y. Lead isotopes in the sulfide ores from the Shesuo and Lawu copper deposits, northern Xizang. *Sediment. Geol. Tethyan Geol.* **2014**, *34*, 96–105.
35. Wu, Y.; Zheng, Y. Genesis of zircon and its constraints on interpretation of U-Pb age. *Chin. Sci. Bull.* **2004**, *49*, 1554–1569. [CrossRef]
36. Li, C. A review on the minerageny and situ microanalytical dating techniques of zircons. *Geol. Surv. Res.* **2009**, *33*, 161–174.
37. Li, C.-N. *Trace Element Petrology of Igneous Rocks*; China University of Geosciences Press: Wuhan, China, 1992; Volume 195.
38. Sun, C.-Y. *Crustal Growth and Reworking Processes of the Erguna and Xing'an Massifs: Evidence from Granitoids*; Jilin University: Jilin, China, 2021.
39. Gou, J. *Geochronology, Petrogenesis and Tectonic Setting of Mesozoic Volcanic Rocks, Southern Manzhouli Area, Inner Mongolia*; Jilin University: Jilin, China, 2013.
40. Zhang, Y.T. Geochemistry of Tamulangou Formation volcanic rocks in Mohe, North Da Hinggan Mountains. *Sci. Technol. Rev.* **2013**, *31*, 25–30.
41. Yang, B.; Chen, Z.L.; Zhang, Q.; Zhou, Z.J.; Han, F.B.; Zhang, W.G.; Ma, J.; Zhang, T. Geological characteristics and sulfur and lead isotopes of the Kanling lead-zinc deposit, Southern Tianshan Mountains. *Geol. China* **2018**, *45*, 155–167.
42. Cao, G.; Gao, T.; Wu, Y. Isotope and REE geochemistry of Puziwan Gold deposit, Shanxi. *China. Geol. Geochem.* **2000**, *28*, 10–14.
43. Basuki, N.I.; Taylor, B.E.; Spooner, E.T.C. Sulfur Isotope Evidence for Thermochemical Reduction of Dissolved Sulfate in Mississippi Valley-Type Zinc-Lead Mineralization, Bongara Area, Northern Peru. *Econ. Geol.* **2008**, *103*, 783–799. [CrossRef]
44. Chaussidon, M.; Lorand, J.-P. Sulphur isotope composition of orogenic spinel lherzolite massifs from Ariege (North-Eastern Pyrenees, France): An ion microprobe study. *Geochim. Cosmochim. Acta* **1990**, *54*, 2835–2846. [CrossRef]
45. Han, Y.W.; Ma, Z.D.; Zhang, H.F.; Zhang, B.R.; Li, W.L.; Gao, S.; Bao, Z.Y. *Geochemistry*; Geology Press: Beijing, China, 2003.
46. Jiang, S.-H.; Nie, F.-J.; Liu, Y.-F.; Yun, F. Sulfur and lead isotopic compositions of Bairendaba and Weilasituo silver-polymetallic deposits, Inner Mongolia. *Miner. Depos.* **2010**, *29*, 101–112.
47. Doe, B.R.; Stacey, J.S. The Application of Lead Isotopes to the Problems of Ore Genesis and Ore Prospect Evaluation: A Review. *Econ. Geol.* **1974**, *69*, 757–776. [CrossRef]
48. Zartman, R.E.; Doe, B.R. Plumbotectonics—The model. *Tectonophysics* **1981**, *75*, 135–162. [CrossRef]
49. Doe, B.R.; Zartman, R.E. Plumbotectonics, the Phanerozoic. In *Geochemistry of Hydrothermal Ore Deposits*, 2nd ed.; Barnes, H.L., Ed.; John Wiley: London, UK, 1979; pp. 22–70.
50. Isotope Geology Laboratory of Yichang Institute of Geology and Mineral Resources. *The Basic Problem of Lead Isotope Geology*; Geological Publishing House: Beijing, China, 1979.
51. Zhu, B.-Q.; Li, S.-H.; Dai, T.-M.; Chen, Y.-W.; Fan, S.-K.; Gui, X.-T.; Wang, H.-F. *Isotopes System Theory and Its Application in Earth Science-on the Evolution of the Continental Crust and Mantle*; Science Press: Beijing, China, 1998; pp. 216–235.
52. Li, T.; Wu, G.; Liu, J.; Wang, G.; Hu, Y.; Zhang, Y.; Luo, D.; Mao, Z.; Xu, B. Geochronology, fluid inclusions and isotopic characteristics of the Chaganbulagen Pb–Zn–Ag deposit, Inner Mongolia, China. *Lithos* **2016**, *261*, 340–355. [CrossRef]
53. Xu, Z.T. *Ore Genesis and Geodynamic Background of Lead-zinc Polymetallic Deposits in Erguna Area, Inner Mongolia*; Jilin University: Jilin, China, 2020.
54. Yang, Y.-C.; Guo, M.-J.; Wang, E.-J.; She, H.-Q.; Li, J.-W.; Zhang, B. Rb-Sr dating of sphalerites from Dongjun Pb-Zn-Ag deposit, Inner Mongolia and its geological significance. *Earth Sci. Front.* **2015**, *22*, 348.
55. Ren, J.; Niu, B.; Liu, Z. Soft collision, superposition orogeny and polycyclic suturing. *Earth Sci. Front.* **1999**, *6*, 85–93.
56. Tian, S.-L.; Jin, L.-F.; Shuang, B. The relationship between vein type Ag (Pb-Zn) deposits and volcanic rocks of the Tamulangou Formation in the Ergun Ore belt. *Geol. Explor. Non-Ferr. Met.* **1995**, *4*, 334–340.
57. Wu, F.-Y.; Sun, D.-Y.; Ge, W.-C.; Zhang, Y.-B.; Grant, M.L.; Wilde, S.A.; Jahn, B.-M. Geochronology of the Phanerozoic granitoids in northeastern China. *J. Asian Earth Sci.* **2011**, *41*, 1–30. [CrossRef]
58. Yu, J.-J.; Wang, F.; Xu, W.-L.; Gao, F.-H.; Pei, F.-P. Early Jurassic mafic magmatism in the Lesser Xing' an–Zhangguangcai Range, NE China, and its tectonic implications: Constraints from zircon U–Pb chronology and geochemistry. *Lithos* **2012**, *142*, 256–266. [CrossRef]
59. Guo, P.; Xu, W.-L.; Yu, J.-J.; Wang, F.; Tang, J.; Li, Y. Geochronology and geochemistry of Late Triassic bimodal igneous rocks at the eastern margin of the Songnen–Zhangguangcai Range Massif, Northeast China: Petrogenesis and tectonic implications. *Int. Geol. Rev.* **2015**, *58*, 196–215. [CrossRef]
60. Wang, F.; Xu, Y.-G.; Xu, W.-L.; Yang, L.; Wu, W.; Sun, C.-Y. Early Jurassic calc-alkaline magmatism in northeast China: Magmatic response to subduction of the Paleo-Pacific Plate beneath the Eurasian continent. *J. Asian Earth Sci.* **2017**, *143*, 249–268. [CrossRef]
61. Wang, Z.-H.; Ge, W.-C.; Yang, H.; Bi, J.-H.; Ji, Z.; Dong, Y.; Xu, W.-L. Petrogenesis and tectonic implications of Early Jurassic volcanic rocks of the Raohe accretionary complex, NE China. *J. Asian Earth Sci.* **2017**, *134*, 262–280. [CrossRef]

62. Xu, W.; Sun, C.; Tang, J.; Luan, J.; Wang, F. Basement nature and tectonic evolution of the Xing'an-Mongolian Orogenic Belt. *Earth Sci.* **2019**, *44*, 1620–1646.
63. Zorin, Y.A. Geodynamics of the western part of the Mongolia–Okhotsk collisional belt, Trans-Baikal region (Russia) and Mongolia. *Tectonophysics* **1999**, *306*, 33–56. [CrossRef]
64. Donskaya, T.; Gladkochub, D.; Mazukabzov, A.; Ivanov, A. Late Paleozoic–Mesozoic subduction-related magmatism at the southern margin of the Siberian continent and the 150 million-year history of the Mongol-Okhotsk Ocean. *J. Asian Earth Sci.* **2013**, *62*, 79–97. [CrossRef]
65. Fritzell, E.; Bull, A.; Shephard, G. Closure of the Mongol–Okhotsk Ocean: Insights from seismic tomography and numerical modelling. *Earth Planet. Sci. Lett.* **2016**, *445*, 1–12. [CrossRef]
66. Chen, Z.; Zhang, L.; Lu, B.; Li, Z.; Wu, H.; Xiang, P.; Huang, S. Geochronology and geochemistry of the Taipingchuan copper-molybdenum deposit in Inner Mongolia, and its geological significances. *Acta Petrol. Sinca* **2010**, *26*, 1437–1449. [CrossRef]
67. Jiang, S.-H.; Nie, F.-J.; Su, Y.-J.; Bai, D.-M.; Liu, Y.-F. Geochronology and origin of the Erdenet superlarge Cu-Mo deposit in Mongolia. *Diqiu Xuebao (Acta Geosci. Sin.)* **2010**, *31*, 289–306.
68. Lü, Z.-C.; Duan, G.-Z.; Hao, L.-B.; Li, D.-C.; Dong, G.-H. The interaction system between Mesozoic tectonic, fluid, petrogenesis and mineralization in the northeast of the Northern Orogenic Belt, China. *Geotecton. Et Metallog.* **2010**, *25*, 161–170.
69. She, H.-Q.; Li, J.-W.; Xiang, A.-D.; Guan, J.-D.; Yang, Y.-C.; Zhang, D.-Q.; Tan, G.; Zhang, B. U-Pb ages of the zircons from primary rocks in middle-northern Daxinganling and its implications to geotectonic evolution. *Acta Petrol. Sin.* **2012**, *28*, 571–594.
70. Xu, W.-L.; Pei, F.-P.; Wang, F.; Meng, E.; Ji, W.-Q.; Yang, D.-B.; Wang, W. Spatial–temporal relationships of Mesozoic volcanic rocks in NE China: Constraints on tectonic overprinting and transformations between multiple tectonic regimes. *J. Asian Earth Sci.* **2013**, *74*, 167–193. [CrossRef]
71. Li, Y.; Xu, W.-L.; Wang, F.; Tang, J.; Sun, C.-Y.; Wang, Z.-J. Early–Middle Ordovician volcanism along the eastern margin of the Xing'an Massif, Northeast China: Constraints on the suture location between the Xing'an and Songnen–Zhangguangcai Range massifs. *Int. Geol. Rev.* **2018**, *60*, 2046–2062. [CrossRef]
72. Meng, E.; Xu, W.; Yang, D.; Qiu, K.; Li, C.; Zhu, H. Zircon U-Pb chronology, geochemistry of Mesozoic volcanic rocks from the Lingquan basin in Manzhouli area, and its tectonic implications. *Acta Petrol. Sin.* **2011**, *27*, 1209–1226.
73. Wang, J.; He, Z.; Xu, W. Petrogenesis of riebeckite rhyolites in the southern Da Hinggan Mts.: Geohronological and geochemical evidence. *Acta Petrol. Sin.* **2013**, *29*, 853–863.
74. Tang, J.; Xu, W.; Wang, F.; Ge, W. Subduction history of the Paleo-Pacific slab beneath Eurasian continent: Mesozoic-Paleogene magmatic records in Northeast Asia. *Sci. China Earth Sci.* **2018**, *61*, 527–559. [CrossRef]
75. Mao, A.Q. *Magmatic Rocks of Mesozoic Volcanic Basins in Central Erguna Block: Petrogenesis and Geodynamics*; Jilin University: Jilin, China, 2020.

minerals

MDPI

Article

Multi-Step Gold Refinement and Collection Using Bi-Minerals in the Laozuoshan Gold Deposit, NE China

Lin Meng [1,2], Fei Huang [1,2,*], Wenyuan Gao [1], Rongzhen Gao [1], Fude Zhao [1], Yiran Zhou [3] and Yongli Li [1]

1 School of Resources and Civil Engineering, Northeastern University, Shenyang 110819, China
2 Key Laboratory of Mineral Resources Evaluation in Northeast Asia, Ministry of Natural Resources, Changchun 130026, China
3 Analysis and Testing Center, Science and Technology Research Institute, Northeastern University, Shenyang 110819, China
* Correspondence: huangfei@mail.neu.edu.cn; Tel.: +86-24-8367-2595

Abstract: Gold and Bi-bearing parageneses are pivotal to understanding gold concentration and deposition processes. The large-scale Laozuoshan gold deposit is located in the Jiamusi Block, northeastern China, and has experienced complex mineralization processes with abundant gold and Bi-bearing minerals. However, the relationship between Bi-minerals and gold is unclear, preventing our understanding of the gold enrichment and precipitation mechanism in the Laozuoshan gold deposit. Optical microscope and SEM results show three stages of gold mineralization: pyrrhotite (Po-1) + arsenopyrite (Apy-1) + Bi-bearing minerals (Bis-1) + Au-1; arsenopyrite (Apy-2) + chlorite + Bi-bearing minerals (Bis-2) + Au-2; and arsenopyrite (Apy-3) + graphite + Bi-bearing minerals (Bis-3) + Au-3. The abundant amount of gold (Au-1~Au-3) is associated with Bi-bearing minerals (Bis-1~Bis-3), which coexist as inclusions and fill in fractures in these minerals. The mineral assemblages of arsenopyrite, Bi-minerals, and gold exhibit a clear As-Bi-Au mineralogy in the ores, and the ternary diagram of the chemical compositions of the Bi-minerals shows that Bi-minerals all fall in reducing regions, indicating that Bi-minerals are precipitated under reducing conditions. The gold compositions demonstrate a positive correlation (R^2 = 0.58) between Au and Bi. Consequently, we propose that the gold experienced the ore-forming fluids concentration and further Bi-melts scavenging for the Laozuoshan gold deposit mineralization. The Bi collector model is essential in interpreting the high-grade gold in the Laozuoshan gold deposit, indicating that the geochemical anomalies observed with bismuth may be a critical potential exploration target for the high-grade gold deposits in the Jiamusi Block.

Keywords: arsenopyrite; gold; Bi-minerals; Laozuoshan; enrichment

1. Introduction

Bismuth is widely found in various sulfide-rich gold deposits [1], including in orogenic deposits [2,3], Cu-Au porphyry-epithermal deposits [4,5], skarn Au deposits [6–10], Iron-Oxide-Copper-Gold (IOCG) deposits [11], Au-rich Volcanogenic Massive Sulfide (VMS) deposits [12], and intrusion-related gold deposits [13–16].

Gold and Bi-bearing parageneses are pivotal to understanding gold deposition and remobilization processes [1]. Bismuth acts as an effective Au scavenger in the liquid phase and dissolves ~20% of Au at 300 °C [17]. This process has been called the Bi collector model, an important mechanism for gold enrichment [10,15,17–19]. Gold scavenging using Bi-melts processes is facilitated at temperatures >271 °C, which tends to contribute to the formation of high-grade ore bodies with complex minerals [17,19]. In this context, elements such as Bi are considered as gold pathfinder elements in mineral exploration [1].

The Laozuoshan gold deposit is a large gold deposit in the north-central part of the Jiamusi Block, NE, China, and has an indicated reserve of 32 t and an average grade of

7.38 g/t [20]. Previous reports indicate that the study area experienced regional metamorphism in the early Paleozoic era, skarn mineralization (the early hydrothermal mineralization) in the late Paleozoic era (264.6 ± 2.6 Ma), and magmatic-hydrothermal superimposition (the late hydrothermal mineralization) in the Mesozoic era (104.6 ± 1.8 Ma) [21,22], resulting in multiple sulfidation stages as well as the development of complex textures and intricate mineral patterns in the Laozuoshan gold deposit [21–24]. The early hydrothermal mineralization was found to have occurred in the East and Central ore belts of the Laozuoshan gold deposit. In contrast, the late hydrothermal mineralization mainly occurred in the West ore belt, with minor occurrences in the East and Central ore belts [23,25]. C-O-S-Pb isotope data indicate that the ore-forming materials have a magmatic origin [20]. According to mineral paragenesis, there are two metallogenic periods and seven mineralization stages, and arsenopyrite and pyrrhotite are the main gold-bearing minerals [23,25,26]. We found abundant Bi-minerals coexisting with gold in the ores, occurring as inclusions in arsenopyrite or filling in arsenopyrite fissures. However, the relationship between Bi-minerals and gold is unclear, which restricts our understanding of gold enrichment and the precipitation mechanism in the Laozuoshan gold deposit.

In the present study, we focus on describing the arsenopyrite and Bi-minerals found in association with gold in the Laozuoshan gold deposit. We provide detailed information on the mineral paragenesis, chemistry, and textures of the arsenopyrite, Bi-minerals, and gold using ore microscopy, SEM-EDS (scanning electron microscopy with energy dispersive spectrometry), and EPMA (electron probe micro-analysis). These results might reveal the gold enrichment and precipitation mechanism.

2. Geological Setting

The Jiamusi Block is situated in the easternmost segment of the Central Asian Orogenic Belt (Figure 1a). The exposed geological units mainly include the Paleozoic Mashan Group, the Majiajie Group, Paleozoic–Mesozoic granitic rocks, and sedimentary rocks (Figure 1b) [27,28]. The Mashan Group is the Precambrian metamorphic basement portion of the Jiamusi Block [29]. The Mashan Group comprises graphite and phosphorus and is rich in aluminum and intercalated with marble, calcium silicate, and quartz schist that has undergone high-amphibolite–granulite facies metamorphism [30]. The Majiajie Group is mainly composed of aluminum-rich and carbon-rich sedimentary clastic rocks that have undergone contact metamorphism [31,32]. The Paleozoic–Mesozoic sedimentary strata are mainly distributed on the southern margin of the Jiamusi Block [27,33]. The early Paleozoic granitic rock is a late Pan-African magma product, as is the Mashan Group, and it constitutes the metamorphic crystalline basement of the Jiamusi Block (Figure 1b). The late Paleozoic magmatic rocks result from the southward subduction of the Paleo-Asian Ocean tectonic domains. In contrast, the Mesozoic magmatic rocks are the result of the superimposition and transformation of the Mongolia–Okhotsk Ocean's and Paleo-Pacific Ocean's tectonic domains [34–36]. The Jiamusi Block has experienced complex tectonic–magmatic activities, forming large- to medium-scale gold deposits such as the Laozuoshan, Tuanjiegou, and Malian deposits (Figure 1b) [23,37].

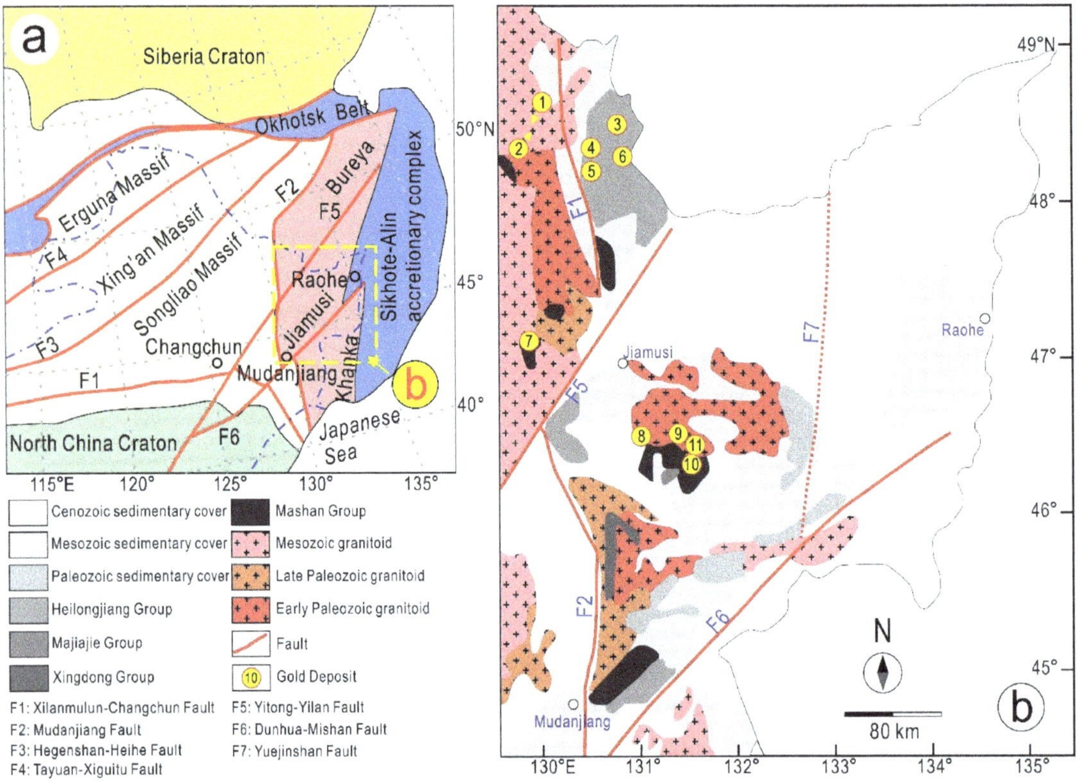

Figure 1. (**a**) Sketch tectonic map (modified according to Hao et al. [38]) and (**b**) simplified geological map of the Jiamusi Block (modified after Wilde et al. [39]). Gold deposit in (**b**) 1—Lianzhushan gold deposit; 2—Pingdingshan gold deposit; 3—Malian gold deposit; 4—Tuanjiegou gold deposit; 5—Dujiahe gold deposit; 6—Taipinggou gold deposit; 7—Dongfengshan gold deposit; 8—Yangbishan gold deposit; 9—Xinli gold deposit; 10—Jiazhuagou gold deposit; 11—Laozuoshan gold deposit.

3. Deposit Geology

The stratigraphic units of the Laozuoshan gold deposit include the Early Paleozoic Mashan Complex, the Jurassic Chengzihe Formation, and Quaternary unconsolidated sediments (Figure 2a) [27]. The Mashan Complex is exposed in the northeastern part of the mine and is dominated by biotite plagioclase granulite, biotite plagioclase gneiss, marble, and migmatite (Figure 2b). The Jurassic Chengzihe Formation consists of sandstone and carbonaceous shale exposed in the western area of the ore district (Figure 2a). The late Paleozoic gneissic granites (262 ± 3.9 Ma) are widespread in the center and western areas of the mine and make minor intrusions into the Mashan Complex and make contact with the marble in those areas to form skarn (Figure 2b) [22,40]. The magmatic activity in the late Mesozoic era resulted in the emplacement of diorite (103 ± 1 Ma), felsite (100.9 ± 1.6 Ma), granite porphyry (104.6 ± 1.8 Ma), plagiogranite, and granodiorite, as well as in widespread diorite porphyry dykes (Figure 2a) [21]. The ore-controlling structures in the ore are mainly NW-NWW- or EW-trending faults or ductile–brittle shear zones (Figure 2a,b), which provide channels for magmatism and hydrothermal fluids.

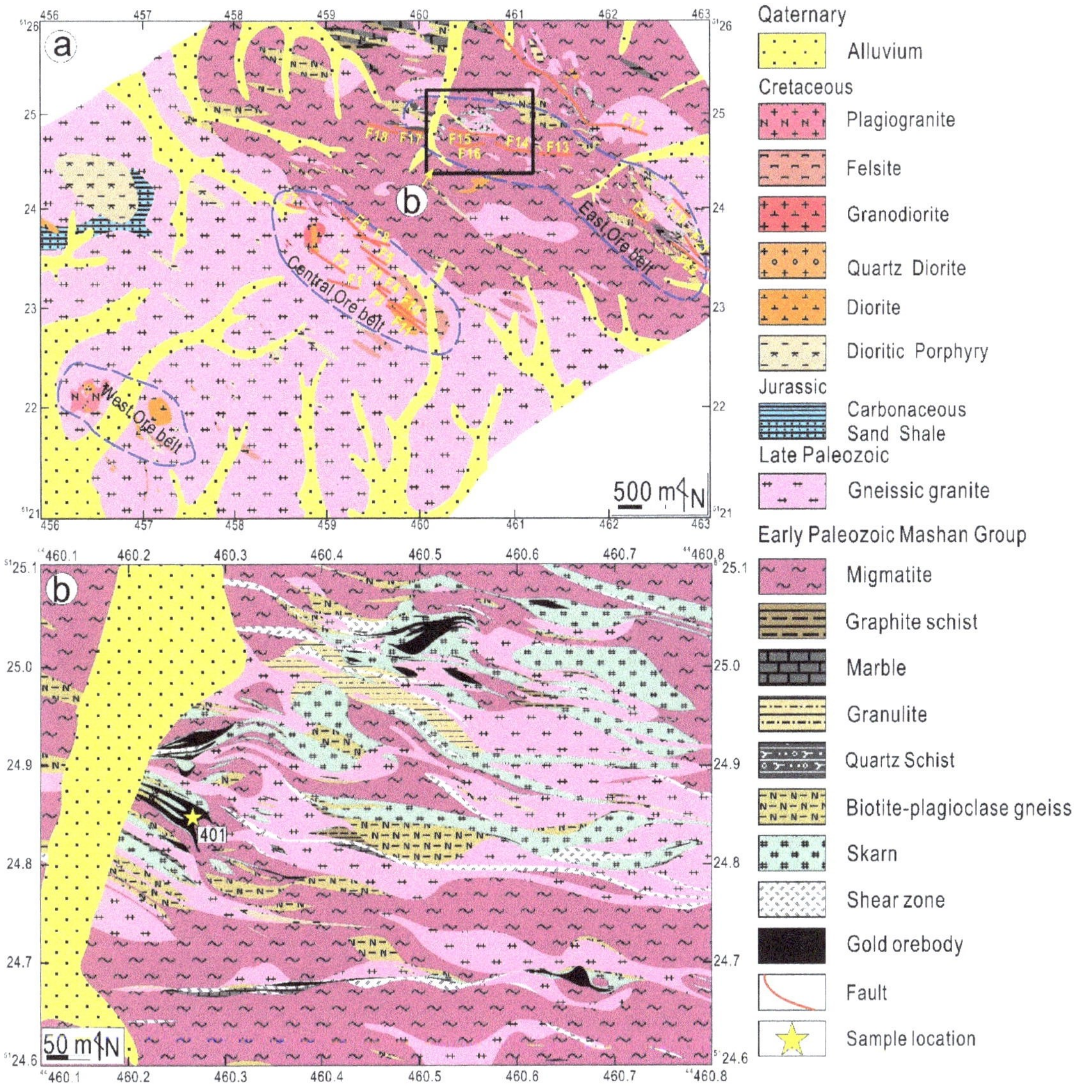

Figure 2. (**a**) Geological map of the Laozuoshan gold deposit; (**b**) and the East ore belt of the Laozuoshan gold deposit (modified after Meng et al. [23]).

The Laozuoshan gold deposit experienced two periods of metallogenesis activity (the early hydrothermal mineralization and the late hydrothermal mineralization), resulting in the formation of three ore belts (East, Central, and West) (Figure 2a). Seven stages of mineralization are recognized based on field and microscopic observations (Figure 3) [23].

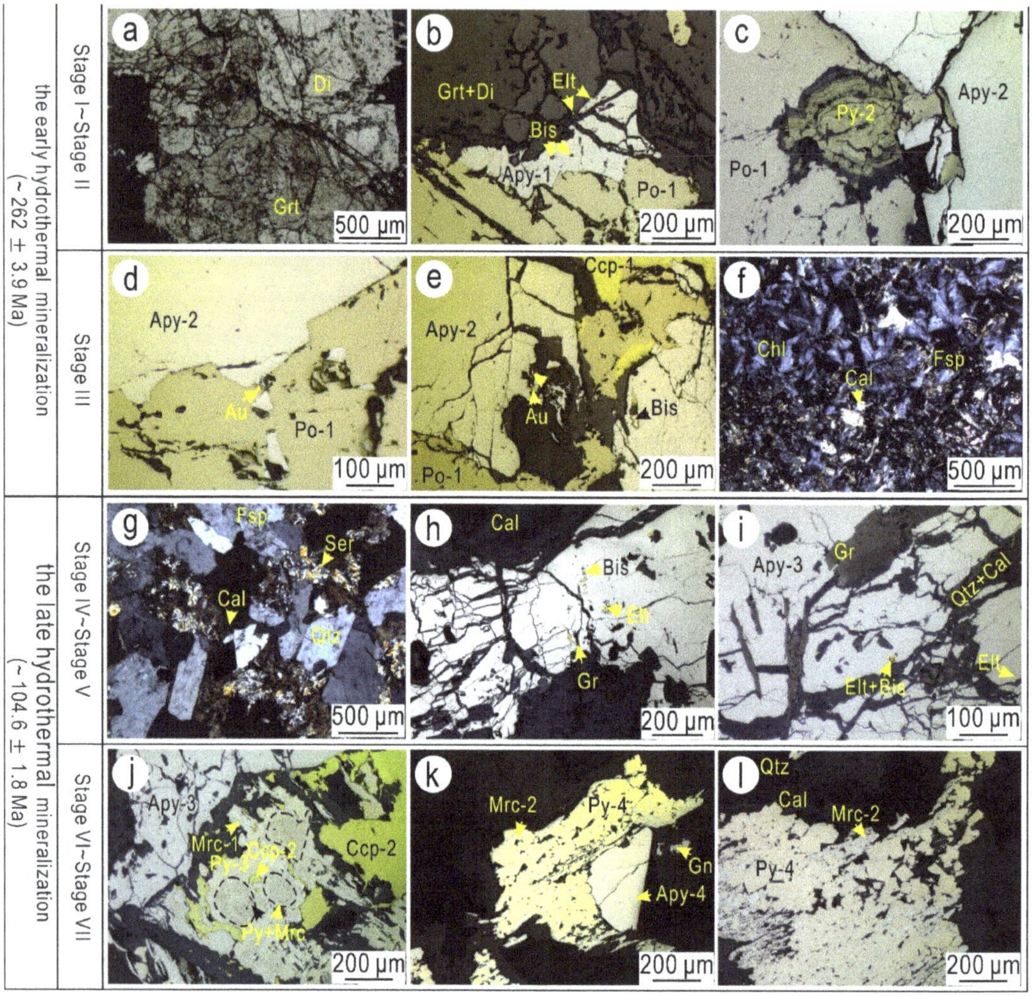

Figure 3. Photographs show the textural relationships and paragenesis features of ore samples at the different ore-forming stages in the Laozuoshan gold deposit: (**a**) Skarn alteration dominated by garnet and diopside (polarized light); (**b**) Gold-bearing arsenopyrite occurs along contact with pyrrhotite (reflected light); (**c**) Colloidal pyrite occurs along contact with pyrrhotite and then cut by arsenopyrite (reflected light); (**d**) Gold is distributed in the fissure of arsenopyrite and pyrrhotite (reflected light); (**e**) Gold and Bi-minerals occur as inclusion in arsenopyrite (reflected light); (**f**) Carbonate minerals such as calcite fill chlorite interstitial spaces (cross-polarized light); (**g**) The alteration of feldspar into sericite in alteration belt (cross-polarized light); (**h**,**i**) Bi-minerals and gold occur as the inclusion or fill in the fissure-developed arsenopyrite, and the arsenopyrite fissure is filled with graphite, quartz, and calcite (reflected light); (**j**) The fined pyrite and marcasite aggregates which containing fine chalcopyrite inclusions are surrounding by arsenopyrite and pyrite (reflected light); (**k**) Arsenopyrite occurs with pyrite and marcasite (reflected light); (**l**) Fine-grained euhedral pyrite occur in oriented banded marcasite (reflected light). Abbreviation: Apy—arsenopyrite; Au—gold; Bis—Bi-minerals; Cal—calcite; Chl—chlorite; Ccp—chalcopyrite; Di—diopside; Elt—electrum; Fsp—feldspar; Gn—Galena; Gr—graphite; Grt—garnet; Mrc—marcasite; Po—pyrrhotite; Py—pyrite; Qtz—quartz; Ser—sericite. The geochronological data come from Wu et al. [40] and Bai et al. [21].

The early hydrothermal mineralization mainly occurred in the East and Central ore belts [27], with a similar occurrence with the wall rock (Figure 2b). The contact zone and the NW-NWW- or EW-trending faults zones controlled the ore bodies (Figure 2b). These ore bodies occur as vein- and lenticular-type bodies and mostly have lengths of 13–330 m, thicknesses of 0.43–9.62 m, and downward plunge depths of 6–400 m and generally dip in the NE direction at an angle of 50°–80° [27].

Ore minerals comprise arsenopyrite, pyrrhotite, and chalcopyrite, with minor pyrite, Bi-minerals, electrum, and native gold accounting for 5%–15% of the ores (Figure 3b–e). Gangue minerals include garnet, diopside, chlorite, epidote, quartz, and calcite (Figure 3a,f). Wall rock alteration is characterized by skarnization and chloritization containing amounts of garnet, diopside, epidote, and chlorite (Figure 3a,f). Three stages (Stage I to Stage III) have been identified based on the paragenetic assemblages of minerals (Figures 3a–f and 4). Stage I (the pyrrhotite–arsenopyrite–calcite stage) is represented by massive aggregates of garnets, diopsides, and pyrrhotite, with minor amounts of arsenopyrite (Apy-1) (Figure 3a,b). Stage II (the colloidal pyrite–quartz stage) is characterized by fine- to coarse-grained aggregates of the colloidal pyrites (Figure 3c). Stage III (the coarse-grained arsenopyrite–calcite stage) is marked by abundant coarse-grained arsenopyrite (Apy-2) with anhedral to subhedral radial chlorites (Figure 3d–f).

The late hydrothermal mineralization mainly occurred in the West ore belt, with minor occurrences observed in the East and Central ore belts. The fault structure controlled the ore bodies and converged in the NE direction, spreading in the SW direction in the West ore belt. These vein-type ore bodies have lengths of 25–482 m, thicknesses of 0.19–4.96 m, and downward plunge depths of 7–390 m and generally dip in the NW direction at an angle of 45°–55° [27].

The ore minerals consist of arsenopyrite, pyrite, chalcopyrite, and marcasite with minor amounts of pyrrhotite, galena, sphalerite, Bi-minerals, electrum, and native gold within gangue minerals such as quartz, feldspar, sericite, calcite, and graphite (Figure 3g–l). K-feldspar alteration is characterized by quartz, K-feldspar, sericite, and carbonate. According to the mineral paragenetic assemblages, four stages (Stage IV to Stage VII) have been identified (Figures 3g–l and 4). Stage IV (the massive quartz stage) is characterized by a large amount of coarse-grained quartz that is largely precipitated without sulfide (Figure 3g). Stage V (the coarse-grained arsenopyrite–quartz stage) is represented by coarse-grained gold-bearing arsenopyrites (Apy-3) that are associated with graphite, quartz, and calcite (Figure 3h,i). Stage VI (the polymetallic sulfide–arsenopyrite quartz stage) is marked by polymetallic sulfide assemblages consisting of chalcopyrite, pyrite, marcasite, arsenopyrite (Apy-4), sphalerite, and galena (Figure 3j–l). Apy-4 can be euhedral to subhedral in shape, can be medium- to fine-grained, and coexists with pyrite and galena (Figure 3k). The final stage (Stage VII: the quartz–calcite stage) comprises calcite and quartz veins ranging from millimeters to centimeters in length (Figure 3l).

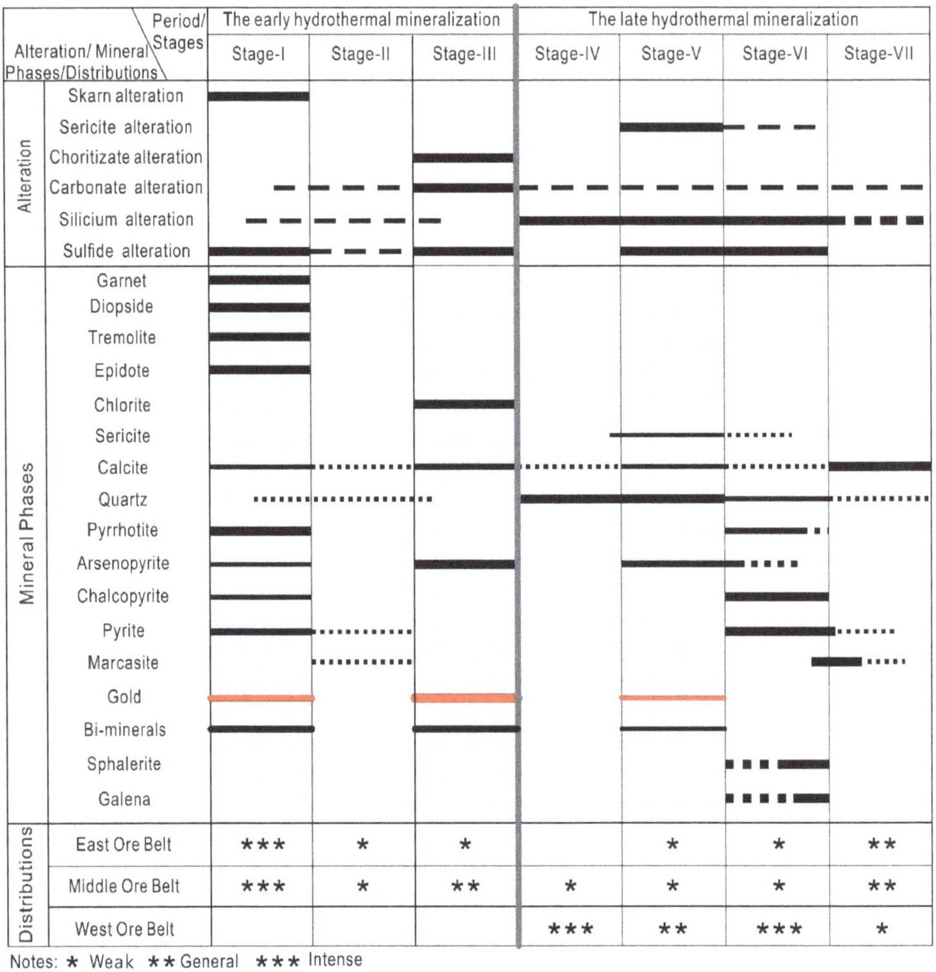

Figure 4. The paragenetic sequence of minerals in the Laozuoshan gold deposit. The shadow bar represents the gold precipitation stages (modified after He [25] and Meng et al. [23]).

4. Samples and Analytical Methods

The No. 401 ore body (which has an average grade of 13.07 g/t) and the No. 2 ore body are the main ore bodies in the East and Central ore bodies, respectively. Both of them developed skarn mineralization. In total, 41 disseminated samples were collected from the No. 401 ore body adit located in the East ore belt (Figure 2b), and 15 samples were collected from the No. 2 ore body adit in the Central ore belt. The No. 304 ore body (which has an average grade of 7.64 g/t) in the West ore belt represents the late hydrothermal mineralization. In total, 14 massive structural samples were collected from No. 304 ore body adit in the West ore belt (Figure 2a). After detailed macroscopic observations of the ore samples, the mineralogy of these samples was analyzed using transmitted and reflected light microscopy to identify silicate and ore mineral assemblages, especially those of arsenopyrite, Bi-minerals, electrum, and native gold.

Further, the microscopic textural and elemental qualitative analyses of the sulfides, Bi-minerals, electrum, and native gold were performed using an Ultra Plus field emission scanning electron microscope (FESEM) (Carl Zeiss, Berlin, Germany) at the Analytical and

Testing Center of Northeastern University, China. The device has a back-scattering electron (BSE) detector and an energy-dispersive X-ray spectrometer (EDS). The resolution was 0.8 nm, the magnification ranged from 12 to 1,000,000, and the acceleration voltage ranged from 20 V to 30 kV.

The composition data of arsenopyrite, Bi-minerals, and gold were determined using a JOEL JXA-8350F field emission electron probe microanalyzer (EPMA) (JEOL, Ltd., Tokyo, Japan) in the same laboratory. The microprobe was operated with a beam energy of 20 kV and a beam current of 20 nA with a 1 μm diameter. The counting time was 40 s for the peak and 20 s for the background of each element. The standards were pyrite–FeS_2 (Fe, S), arsenopyrite–FeAsS (As), Au metal (Au), Ag_2S (Ag), Bi metal (Bi), chalcopyrite–$CuFeS_2$ (Cu), Co metal (Co), Ni metal (Ni), Sb_2S_3 (Sb), ZnSe (Se), TeCdHg (Te), and ZnS (Zn). Analytical results were processed using the ZAF correction routines.

5. Results

5.1. Texture of the Arsenopyrite, Bi-Minerals, and Gold

The Laozuoshan gold deposit has undergone three gold mineralization stages (Au-1, Au-2, and Au-3), which occurred in Stages I and III during the early hydrothermal mineralization and in Stage V during the late hydrothermal mineralization, respectively (Figures 3 and 4). The aurous minerals (Au-1 to Au-3) are closely related to arsenopyrite (Apy-1 to Apy-3) and the Bi-minerals (Bis-1 to Bis-3) (Figures 5–7).

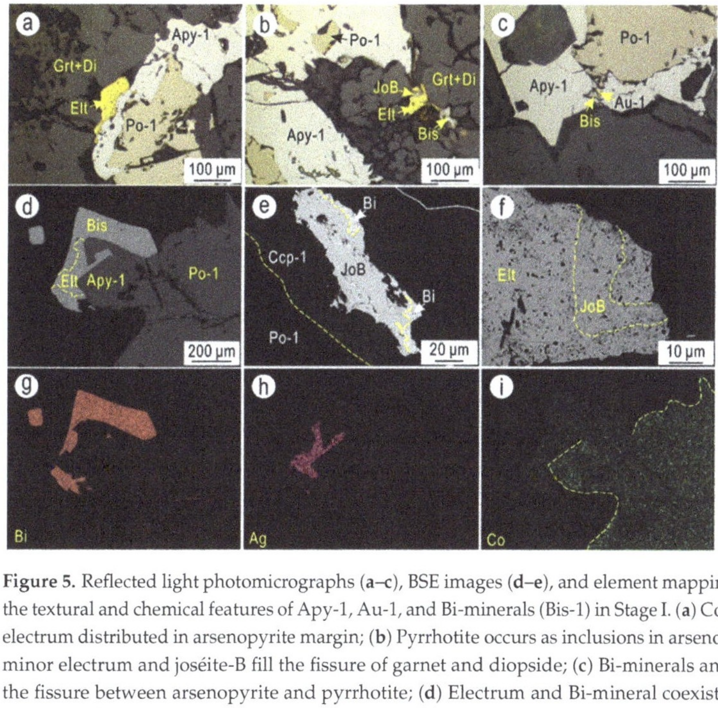

Figure 5. Reflected light photomicrographs (**a**–**c**), BSE images (**d**–**e**), and element mapping (**g**–**i**) show the textural and chemical features of Apy-1, Au-1, and Bi-minerals (Bis-1) in Stage I. (**a**) Coarse-grained electrum distributed in arsenopyrite margin; (**b**) Pyrrhotite occurs as inclusions in arsenopyrite, while minor electrum and joséite-B fill the fissure of garnet and diopside; (**c**) Bi-minerals and gold occur the fissure between arsenopyrite and pyrrhotite; (**d**) Electrum and Bi-mineral coexisted along the contact with arsenopyrite (Apy-1); (**e**) Native Bi coexisted with the joseite-B as inclusions or in the joseite-B margin, and both of them occurred as inclusion in chalcopyrite; (**f**) Coarse-grained electrum coexisted with joseite-B as the solid solution. Element mappings (**g**–**i**) are the results of (**d**); (**g**) The Bi element mapping of Bi-minerals; (**h**) The Ag element mapping of electrum; (**i**) the Co element mapping of arsenopyrite; Abbreviation: Apy—arsenopyrite; Au—gold; Bi—native Bi; Bis—Bi-minerals; Ccp—chalcopyrite; Di—diopside; Elt—electrum; Grt—garnet; JoB—joséite-B; Po—Pyrrhotite.

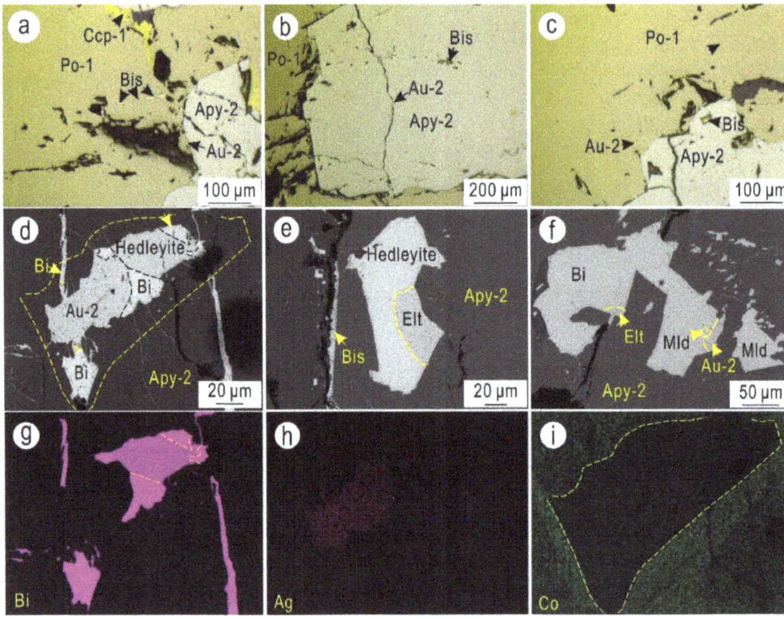

Figure 6. Reflected light photomicrographs (**a**–**c**), BSE images (**d**–**e**), and element mapping (**g**–**i**) show the textural and chemical features of Apy-2, Au-2, and Bi-minerals (Bis-2) in Stage III. (**a**) Bi-minerals occur as inclusions in pyrrhotite; (**b**) Gold fills the fissure of arsenopyrite which contains Bi-minerals inclusions; (**c**) Bi-minerals and gold occur as inclusions in arsenopyrite and pyrrhotite, respectively; (**d**) Native gold, native bismuth, and hedleyite coexisted in arsenopyrite or filled the fissure of arsenopyrite; (**e**) Electrum and hedleyite coexisted with arsenopyrite; (**f**) Maldonite occurs as inclusions in arsenopyrite or native bismuth, while native gold and electrum occur along the contact with native bismuth; Element mappings (**g**–**i**) are the results of (**d**). (**g**) the Bi element mapping of native bismuth and hedleyite; (**h**) the Ag element mapping of electrum; (**i**) the Co mapping of arsenopyrite; Abbreviation: Apy—arsenopyrite; Au—gold; Bi—native Bi; Bis—Bi-minerals; Ccp—chalcopyrite; Elt—electrum; Mld—maldonite; Po—Pyrrhotite; Py—pyrite.

5.1.1. Stage I (Pyrrhotite–Arsenopyrite–Calcite Stage)

The sulfides are disseminated in garnet and diopside (Figure 3a,b) and are distributed in the skarn ore bodies in the East and Central ore belts. Arsenopyrite (Apy-1) can be medium- to fine-grained (100 to 500 μm) and is associated with coarse-grained (>500 μm) pyrrhotite (Po-1), chalcopyrite, and pyrite (Py-1) (Figure 5a–c).

Electrum with minor amounts of native gold usually intergrowths among the Bi-minerals (Bis-1) (Figure 5). Electrum can be subhedral to anhedral shape, with particle sizes of about 40–120 μm. It occurs in inclusions in joséite-B, native bismuth, or chalcopyrite (Figure 5b,e) or at the boundary between sulfide (arsenopyrite (Apy-1), pyrrhotite (Po-1), and chalcopyrite (Figure 5c) and related gangue minerals (garnet and diopside, Figure 5b). The surface of the electrum associated with joséite-B is rough and porous (Figure 5f). The Bi-minerals (joséite-B, native bismuth, etc.) can have coarse- to fine-grain (20 to 200 μm) particles and are distributed at the margins of arsenopyrite (Apy-1) or fill in narrow fractures between pyrrhotite (Po-1) and arsenopyrite (Apy-1) or fissures in garnet and diopside (Figure 5a–d). The joséite-B also occurs as inclusions in arsenopyrite (Apy-1), chalcopyrite, and pyrrhotite (Po-1) (Figure 5e). The EDS results show that Bi, Ag, and Co are evenly distributed in Bi-minerals (Bis-1), electrum (Au-1), and arsenopyrite (Apy-1), respectively (Figure 5g–i).

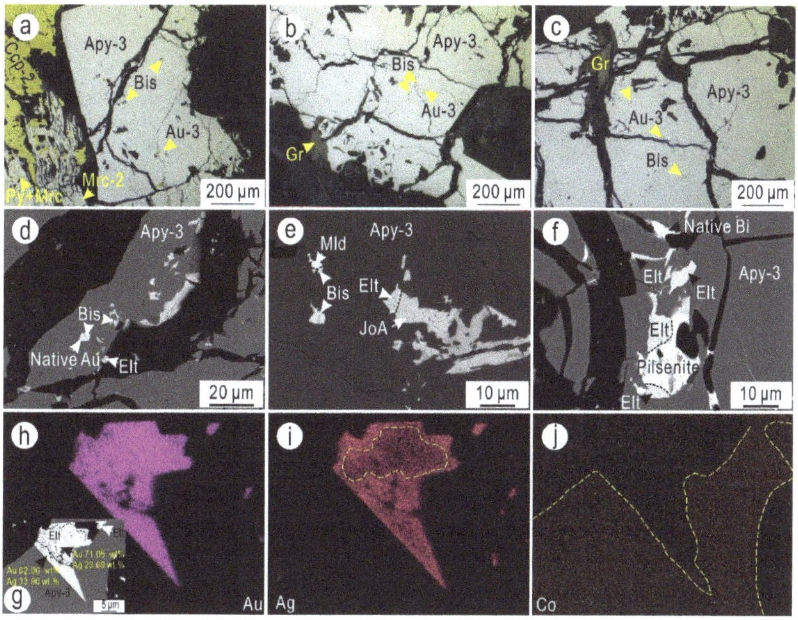

Figure 7. Reflected light photomicrographs (**a**–**c**), BSE images (**d**–**f**), and element mapping (**g**–**j**) show the textural and chemical features of Apy-3, Au-3, and Bi-minerals (Bis-3) in Stage V. (**a**) Bi-minerals and gold occur as inclusions in arsenopyrite, which surrounding by chalcopyrite, pyrite, and marcasite; (**b**,**c**) Graphite occurs in the fissure of arsenopyrite, which containing gold and Bi-minerals; (**d**) Native gold coexisted with Bi-minerals, which occurs as inclusion in arsenopyrite, while minor electrum occurs along the contact with arsenopyrite; (**e**) Electrum coexisted with maldonite and Bi-minerals such as joséite-A in arsenopyrite; (**f**) Electrum coexisted with native bismuth and pilsenite, which occur as inclusions in arsenopyrite, while minor electrum filled the fissure of arsenopyrite. Element mappings (**h**–**j**) are the results of (**g**). (**h**) the Au element mapping of electrum; (**i**) the Ag element mapping of electrum; (**j**) the Co element of arsenopyrite; Abbreviation: Apy—arsenopyrite; Au—gold; Bis—Bi-minerals; Ccp—chalcopyrite; Elt—electrum; Gr—graphite; JoA—joseite-A; Mrc—marcasite; Po—Pyrrhotite; Py—pyrite.

5.1.2. Stage III (Coarse-Grained Arsenopyrite–Calcite Stage)

Sulfides are disseminated in ore and are distributed in the skarn ore bodies in the East and Central ore belts. Arsenopyrite (Apy-2) is the diagnostic mineral of Stage III and is characterized by its coarse grain size (0.2 to 1 mm; Figure 6a–c). Apy-2 fissures are filled in with chlorites and calcites (Figure 3f). Pyrrhotite (Po-1), when replaced by Apy-2, appears to be round (Figure 3d,e). Colloidal pyrite (Py-2) is cut by Apy-2 (Figure 3c) [41].

Native gold and electrum with a minor amount of maldonite coexist with the Bi-minerals (Bis-2) (Figure 6d–f). Native gold and electrum are subhedral to anhedral shape (30 to 150 µm, Figure 6d,e). Both of them occur as inclusions in arsenopyrite (Apy-2) or the Bi-minerals (Bis-2) or fill-in arsenopyrite (Apy-2) fractures (Figure 6a–f). Minor amounts of fine electrum (<10 µm) occur as inclusions in pyrrhotite (Po-1) (Figure 6c). The Bi-minerals (native bismuth, maldonite, and hedleyite) have various particle sizes ranging from 10 to 300 µm (Figure 6d–f). Hedleyite and native bismuth occur as inclusions in arsenopyrite (Apy-2) and chalcopyrite (Figure 6a–f), with minor native bismuth filling in arsenopyrite (Apy-2) fissures (Figure 6d). EDS element mapping shows that Ag and Bi are uniformly distributed in gold, hedleyite, and native bismuth (Figure 6g,h). There are apparent Co anomalies in Apy-2 (Figure 6i). According to the EPMA data, the Co content is high in Apy-1, while minor contents are present in Apy-2 (Supplementary Materials Table S1).

Therefore, the abnormal Co in Apy-2 might indicate that Apy-1 forms Apy-2 through the dissolution–reprecipitation mechanism (Figure 6i) [42].

5.1.3. Stage V (Coarse-Grained Arsenopyrite-Quartz Stage)

Arsenopyrite (Apy-3) is the main sulfide and coexists with graphite (Figure 7a–c). It is mainly distributed in the West ore belts. Coarse-grained arsenopyrite (Apy-3) is about 2 to 3 mm in size and has developed fissures filled in with graphite, quartz, and calcite (Figure 3h,i and Figure 7a–c).

Electrum with minor native gold coexists with Bi-minerals (Bis-3) as inclusions in arsenopyrite (Apy-3) (Figure 7a–f). Additionally, minor native gold particles occur in arsenopyrite fissures and carbonate veins (Figure 7d). Compared to Au-1 and Au-2, the particle sizes of gold (electrum with minor native gold) decrease significantly and range from 5 to 30 μm (Figure 7a–f). Bi-minerals occur as inclusions in arsenopyrite (Apy-3) or at the edge of Apy-3 (Figure 7a–f). The particle sizes (5–50 μm) of Bi-minerals also decrease significantly (Figure 7d–f). The Ag in some gold particles (Au-3) shows a trend of enrichment from the center to the edge (Figure 7g–i), something that is further confirmed by EMPA data (Figure 7g and Supplementary Materials Table S2). Meanwhile, there are apparent Co anomalies in arsenopyrite (Apy-3) (Figure 7j).

5.2. *Compositions of Arsenopyrite, Gold, and Bi-Mineral Assemblages*

5.2.1. Compositions of Arsenopyrite

EPMA data show that the contents of As, S, and Fe in four generations of arsenopyrite vary significantly (Supplementary Materials Table S1 and Figure 8a). The arsenic contents in Apy-1 (average (Ave): 48.41 wt.%; $N = 13$) and Apy-3 (Ave: 49.65 wt.%, $N = 8$) are higher than those in Apy-2 (Ave: 46.39 wt.%; $N = 7$) and Apy-4 (Ave: 46.34 wt.%) (Supplementary Materials Table S1). The sulfur contents in Apy-1 (Ave: 17.90 wt.%; $N = 13$) and Apy-3 (Ave: 17.98%; $N = 8$) are lower than those in Apy-2 (Ave: 19.31 wt.%; $N = 7$) and Apy-4 (Ave: 19.67 wt.%; $N = 5$) (Supplementary Materials Table S1). The Fe contents in the four generations of arsenopyrite vary significantly, and the mean values are 33.02 wt.% (Apy-1), 35.08 wt.% (Apy-2), 29.08 wt.% (Apy-3), and 34.44 wt.% (Apy-4) (Supplementary Materials Table S1). The Co contents of the arsenopyrites also changed significantly (Supplementary Materials Table S1), e.g., Apy-3 had the highest content (Ave: 4.16 wt.%, $N = 8$), followed by Apy-1 (Ave:1.69 wt.%, $N = 13$), Apy-2 (Ave: 0.16 wt.%, $N = 7$), and Apy-4 (Ave: 0.13 wt.%, $N = 5$) (Supplementary Materials Table S1). In addition, arsenopyrite with a high arsenic content usually contains elevated Co, even though As is not significantly correlated with Co.

The scatter plot shows that the compositions of Apy-1 and Apy-3 fluctuate greatly (Figure 8a). In contrast, the composition changes observed in Ayp-2 and Apy-4 were relatively stable (Figure 8a). Arsenic was negatively correlated with S ($R^2 = 0.70$) and Fe ($R^2 = 0.43$) in arsenopyrite (Figure 8b,c). There was a strong negative correlation between Fe and Co ($R^2 = 0.88$) in arsenopyrite (Figure 8d), indicating that cobalt may replace Fe during arsenopyrite formation [42].

5.2.2. Compositions of Gold

The EPMA data show that the Au and Ag contents varied greatly from Au-1 to Au-3 (Supplementary Materials Table S2 and Figure 9). The Au content in Au-1 (electrum with minor amounts of native gold) is 68.57%–87.27%, with an average of 83.46 wt.% ($N = 14$) and a variance of 4.60 (Supplementary Materials Table S2). Au-2 is mainly composed of native gold. The Au content in gold (Au-2) is 89.87–91.42 wt.%, with an average of 90.46 wt.% ($N = 5$) and a variance of 0.58 (Supplementary Materials Table S2). Au-3 is dominated by electrum and contains minor amounts of native gold. The Au content in Au-3 varies greatly, ranging from 50.23 to 98.43 wt.%, with an average of 62.10 wt.% ($N = 10$) and variance of 15.38 (Supplementary Materials Table S2).

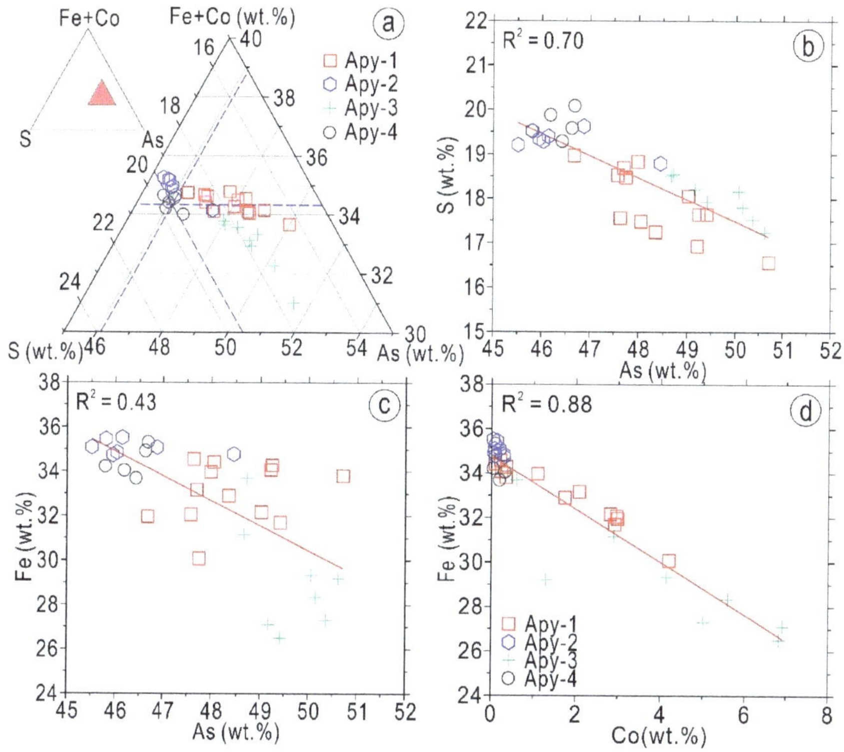

Figure 8. (**a**) Ternary plots showing compositions of the Fe, S, and As in different arsenopyrite generations; (**b**–**d**) Binary plots show compositions of selected main elements in different arsenopyrite generations. (**b**) As vs. S; (**c**) As vs. Fe; (**d**) Co vs. Fe. All data based on the results of arsenopyrite analyses by EPMA are given in Supplementary Materials Table S1.

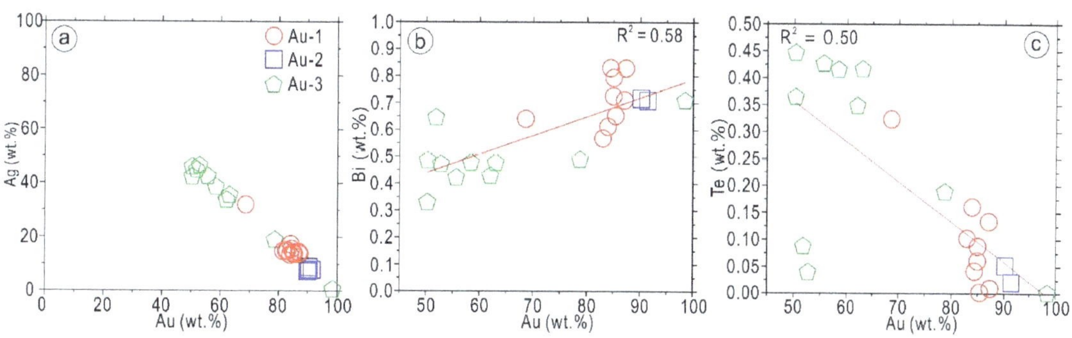

Figure 9. The binary diagrams show the concentration of selected main elements in different gold generations. (**a**) Au vs. Ag; (**b**) Au vs. Bi; (**c**) Au vs. Te. All data based on the results of gold analyses by EPMA are given in Supplementary Materials Table S2.

The above results show that the compositions of Au-1 and Au-2 are relatively homogeneous. In contrast, the compositions of Au-3 vary greatly (Figure 9a). EPMA data reflect that the Au-bearing phases (Au-1 to Au-3) also contain minor amounts of Bi (0.33–0.83 wt.%) and Te (~0.45 wt.%) (Supplementary Materials Table S2). Native gold contains more bismuth (Supplementary Materials Table S2 and Figure 9b). The amounts of

Au and Bi in gold (Au-1 to Au-3) are positively correlated (Figure 9b, R^2 = 0.58) and are negatively correlated with Te (Figure 9c, R^2 = 0.50).

5.2.3. Compositions of Bi-Minerals

The EPMA data show that the Bi-minerals (Bis-1 to Bis-3) that are associated/coexist with gold also demonstrate apparent differences (Supplementary Materials Table S3 and Figure 10). The Bi-minerals (Bis-1) mainly include joséite-B (Bi_4Te_2S: Bi 73.04~75.68 wt.%, Te 20.87~22.10 wt.%, S 2.33~2.76 wt.%), and ingodite (Bi_2TeS: Bi 63.96~69.96 wt.%, Te 21.37~30.33 wt.%, S 4.05~5.29 wt.%), with minor amounts of tsumoite (BiTe), pilsenite (Bi_4Te_3), and native bismuth (Supplementary Materials Table S3). The Bi-minerals (Bis-2) include native Bi with minor amounts of maldonite (Au_2Bi) and hedleyite (Bi_7Te_3) (Supplementary Materials Table S3). The Bi-minerals (Bis-3) include joséite-B (Bi_4Te_2S), native bismuth, and minor amounts of joséite-A (Bi_4TeS_2) as well as pilsenite (Bi_4Te_3) and maldonite (Au_2Bi) (Supplementary Materials Table S3). The Bi vs. Te binary diagram of the Bi-minerals shows relatively obvious zoning between the three generations of Bi-minerals in the Laozuoshan gold deposit (Figure 10a). The typical Bi-rich Bi-minerals of joséite-B, native Bi, with minor amounts of ingodite and joseite-A, where Bi/(Te + S) > 1 (Figure 10b), confirm that the ore-forming fluids in reducing conditions [43,44].

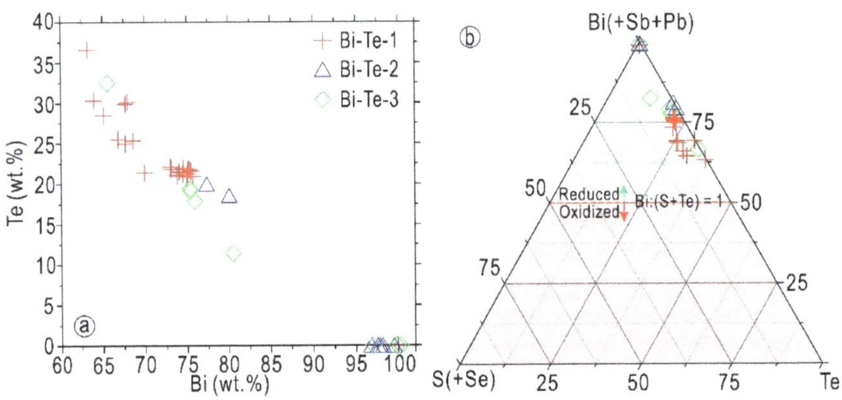

Figure 10. (**a**) The binary diagrams (Bi vs. Te) show the concentration of selected main elements in different Bi-minerals generations; (**b**) Ternary diagram (Bi + Sb + Pb − Te − S + Te shows compositions of Bi-minerals in the Laozuoshan gold deposit (modified after Cook et al. [45]; Wei et al. [46]). The Bi/Te + S = 1 line is from Ciobanu and Cook [43] and Ciobanu et al. [44], which separates the Bi-minerals formed in reduced fluids from those in oxidized fluids. Noted that the Bi-minerals from reduced fluids usually have Bi/Te + S > 1 and have an assemblage of native gold and pyrrhotite, whereas those from oxidized fluids usually have Bi/Te + S < 1 and have an assemblage of native gold and pyrite. All data based on the results of the Bi-minerals analyses by EPMA are given in Supplementary Materials Table S3.

6. Discussion

6.1. The Physical–Chemical Conditions of the Ore-Forming Fluids Indicated by Minerals Assemblages

Based on the texture and structure of the typical ores and the paragenetic association of the minerals, the Laozuoshan gold deposit can be divided into areas that experienced early hydrothermal mineralization and late hydrothermal mineralization and can be further divided into seven mineralization stages (Figures 3 and 4). Stages I and III are the main gold precipitation stage during the early hydrothermal mineralization. Stage V is the main gold precipitation stage during the late hydrothermal mineralization (Figures 5–7).

The reduced gold skarns are characterized by low garnet/pyroxene ratios and are dominated by hedenbergite and grandite [47]. There is a genetic association with reduced granitoid intrusion during the late Paleozoic era (264.6 ± 2.6 Ma) and a $Fe_2O_3/(Fe_2O_3 + FeO)$ ratio between 0.12 and 0.17 for Laozuoshan granite [21,22,40]. The earliest alteration stage is characterized by potassium hornfels. Following this alteration stage, prograde skarn formation subsequently occurred. This process is characterized by garnet in the grossular–andradite series and pyroxene in the hedenbergite–diopside series in the Laozuoshan gold deposit, similar to the reduced gold skarn deposit [20,40,47,48]. Furthermore, Bi-bearing phases and gold are often developed in later stages and accompanying base metals in reduced gold skarn deposits [47]. Usually, these assemblages postdate previous early sulfidation stages and are dominated by pyrrhotite and arsenopyrite with or without lollingite [47,49]. In the Laozuoshan gold deposits, most of the Au- and Bi-bearing phases are indeed postdated. Both the assemblages marked by coarse-grained pyrrhotite (Po-1) and arsenopyrite (Apy-1) without lollingite are from Stage I and show a distinctive Au-Bi-As ± Co geochemical association (Figure 5a–f and Supplementary Materials Table S1). Gold was also observed to coexist with the Bi-phases, and both of them occur along the contact with arsenopyrite or pyrrhotite (Figure 5a–f), which are typical mineral assemblages in reduced skarn gold deposits, suggesting their precipitation at reduced conditions (Figure 5) [47].

Arsenopyrite (Apy-1 to Apy-3) is a gold-bearing mineral found in the Laozuoshan gold deposit (Figures 5–7). The As(III) hydroxide complex $As(OH)_3$ is the most dominant As-bearing species and has a wide range of temperatures (25~500 °C) and fluid densities (0.05~1.1 g/cm^3) in the natural hydrothermal system, as indicated by experimental studies [50–52]. The decreases in fO_2 induce arsenopyrite crystallization and decrease the H_2S and arsenic concentrations in ore-forming fluid [53,54]. In the Laozuoshan gold deposit, coarse-grained arsenopyrites (Apy-2) are associated with pyrrhotite (Po-1), chlorite, and calcite in Stage III (Figures 3d–f and 6). The chlorite composition belongs to ripidolite, representing reduced conditions that favor arsenopyrite precipitation [23]. In Stage V, the abundance of graphite, quartz, and calcite filling in arsenopyrite (Apy-3) fractures suggests that arsenopyrite was precipitated under the reduced conditions (Figure 3h,i and Figure 7a–f).

After a careful examination of the mineralogy (Figures 5–7), it was possible to recognize that a wide range of disseminated Bi-minerals (Bis-1~Bis-3) coexist with the gold (Au-1~Au-3), something that is not described in Meng et al. [23]. The general forms of Bi-sulfosalts under high fS_2 conditions are associated with pyrite stability. Bismuth–chalcogenides and native bismuth usually form within the fields of pyrrhotite and arsenopyrite under low fS_2 and fO_2 conditions [10,11,55,56]. Accordingly, most of the Bi-bearing phases in the Laozuoshan gold deposit are dominated by joséite-B, ingodite, and native bismuth, with minor amounts of tsumoite, pilsenite, maldonite, hedleyite, and joséite-B and no Bi-sulfosalts (Figures 5–7), which suggest that the Bi-bearing minerals might crystallize under reduced conditions [15]. The ternary diagram depicting the chemical compositions of the Bi-minerals (Bis-1 to Bis-3) shows that the Bi-minerals all occur in the reducing regions, indicating that Bi-minerals are precipitated under reducing conditions (Figure 10b and Supplementary Materials Table S3) [44–46]. Therefore, the mineral assemblages and their chemical compositions indicate that gold mineralization occurs under the reducing conditions in the Laozuoshan gold deposit.

6.2. Gold Scavenging Using Bi-Melts and Refining Processes during the Cooling of the Mineralization System

Previous work shows that gold is commonly transported as thiosulfate and chloride complexes in hydrothermal solutions, depending on their relative fO_2 conditions and temperatures [57]. However, it might migrate by forming $[Au, As]^{2-}$ or $[Au(As, S_3)]^{2-}$ in As-rich reduced hydrothermal fluids [58,59]. In the Laozuoshan gold deposit, Bi-minerals are associated with gold and exist as inclusions or fill in arsenopyrite and pyrrhotite fractures (Figures 5–7), indicating that gold might be transported as a thiohydron complex or as an

As-bearing complex [57–59]. Gold and Bi-minerals occurred as invisible solid inclusions in arsenopyrite and pyrrhotite (Figures 5–7) and were identified using the LA-ICP-MS data of Shi et al. [22], implying that gold and bismuth were available in the fluids that were responsible for forming the earlier sulfide assemblages. The compositions and textures of the Bi-bearing assemblages related to gold where Au and Bi show a positive geochemical correlation ($R^2 \geq 0.5$) have been described in previous studies [1,5,15,60]. The EPMA data for gold (electrum, native gold) show a positive correlation ($R^2 = 0.58$) between Au and Bi (Figure 9b and Supplementary Materials Table S2), suggesting that the Bi collector model might play an important role in gold enrichment in the Laozuoshan gold deposit.

Oxidized Bi^{3+} is the predominant Bi-bearing phase of Bi in hydrothermal fluids [19,55]. If bismuth is present in the Au-bearing solution, with favorable sulfur-oxygen fugacities and temperatures (above 271 °C), Bi-melts form and effectively scavenge Au from the fluids, forming Bi-Au-melts [15,17–19,60]. The ore-forming temperature is identified according to the homogenization temperatures of the quartz and calcite fluid inclusions (366.5~410.3 °C) and of the arsenopyrite (Apy-2) geothermometer (390–420 °C) in the Laozuoshan gold deposit [23,61], all of which are above the melting point of Bi (271 °C) [17], indicating that Bi could exist as melts in the ore-forming fluids in the Laozuoshan gold deposit. Furthermore, the Bi-minerals and gold undergoing typical eutectic crystallization in the Laozuoshan gold deposit (Figures 5d–f and 7d–f) demonstrated similar textures and assemblages to those that have been previously reported in a large number of deposits worldwide [3,7,11,56,60,62]. The Bi-minerals coexist with pyrrhotite and arsenopyrite, and the Bi-bearing phases have Bi/(Te + S) ratios ≥ 1 (Figures 5, 6, 7 and 10b), which suggests that the reduced conditions might facilitate the formation of bismuth immiscible fluids [3,7].

The abundant gold and Bi-mineral particles that occur as inclusions in arsenopyrite (Apy-1~Apy-3) and pyrrhotite (Po-1) or that fill in the fractures in these sulfides (Figures 5–7) suggest the existence of Bi melts that scavenged Au, filling in the fractures of earlier sulfides. The major elements (Fe, As, Co) found in arsenopyrite varied, and the positive correlation between Co and Fe in arsenopyrite indicates that Co replaced Fe (Figure 8a,d), which might facilitate the formation of lattice defects in arsenopyrite and the mechanical incorporation of nano-micron gold and Bi-minerals inclusions [22]. Further, element mapping shows abnormal Co in arsenopyrite (Apy-2) (Figure 6i), implying that arsenopyrite (Apy-1) might experience dissolution and reprecipitation and form arsenopyrite (Apy-2) [63]. Therefore, the abundance of native gold and native bismuth particles filling in arsenopyrite (Apy-2) fractures suggests that invisible gold and Bi were remobilized from arsenopyrite during arsenopyrite recrystallization and that part of the Bi-enriched immiscible melts survived and scavenged gold after arsenopyrite crystallization until the waning stages of the hydrothermal system (Figure 6d,g,i) [1]. Similar processes have been proposed by Fougerouse et al. [64] and Morey et al. [65].

In summary, we propose that gold experienced the concentration of ore-bearing fluids and further Bi-melt scavenging during mineralization in the Laozuoshan gold deposit. During the water–rock reaction, the fO_2 of the As-rich gold-bearing fluids decreased under the reducing conditions, which led to the precipitation of base metal sulfides such as pyrrhotite and arsenopyrite [51,52,54]. The Au^+ and Bi^{3+} thiohydrogen complex in the ore-forming fluids desulfurized to form gold and bismuth nanoparticles, which were initially enriched in arsenopyrite or pyrrhotite as visible or invisible inclusions [19,55,64]. Since the melting point of bismuth is lower than the precipitation temperature of pyrrhotite and arsenopyrite [17], bismuth could exist in the ore-forming fluid as immiscible melts to further scavenge Au from the ambient fluids [18,19]. Subsequently, the gold-bearing immiscible bismuth melts precipitated and separated into gold and Bi-minerals as the temperature of the hydrothermal fluids decreased or underwent supersaturation during migration [15], which is confirmed by the gold and Bi-minerals inclusions in arsenopyrite (Apy-1~Apy-3) and pyrrhotite (Po-1) as well as by their presence in the fractures of these sulfides (Figures 5–7). The Bi collector model is essential for interpreting the high-grade gold in the Laozuoshan gold deposit, indicating that the bismuth geochemical anomaly may be a critical potential exploration target for the high-grade gold deposits in the Jiamusi Block.

7. Conclusions

1. The Laozuoshan gold deposit has experienced complex mineralization processes (the early hydrothermal mineralization and the late hydrothermal mineralization), with three gold precipitation processes occurring successively. The abundant gold (Au-1 to Au-3) and Bi-mineral (Bis-1 to Bis-3) particles are associated with pyrrhotite (Po-1) + arsenopyrite (Apy-1), chlorite + arsenopyrite (Apy-2), and graphite + arsenopyrite (Apy-3).
2. Bi-minerals (such as joséite-B, ingodite, and native bismuth without Bi-sulfosalts) are associated with gold (electrum, native gold, or minor maldonite) and coexist as inclusions or as fracture fillings in arsenopyrite (Apy-1~Apy-3) and pyrrhotite. The mineral assemblages of arsenopyrite, Bi-minerals, and gold exhibit a clear As-Bi-Au mineralogy in the ores, and the ternary diagram of the chemical compositions of the Bi-minerals shows that the Bi-minerals all fall in reducing regions, indicating that Bi-minerals are precipitated under reducing conditions.
3. The gold compositions show a positive correlation (R^2 = 0.58) between Au and Bi. The gold experienced the concentration of the ore-forming fluids and further Bi-melt scavenging during the mineralization of the Laozuoshan gold deposit. The Bi collector model is essential for interpreting the high-grade gold in the Laozuoshan gold deposit, indicating that the bismuth geochemical anomaly may be a critical potential exploration target for the high-grade gold deposits in the Jiamusi Block.

Supplementary Materials: The following supporting information can be downloaded at: https://www.mdpi.com/article/10.3390/min12091137/s1, Table S1: Compositions of the different generations of arsenopyrite in Laozuoshan gold deposit; Table S2: Compositions of gold in Laozuoshan gold deposit; Table S3: Compositions of the Bi-minerals and Bi-Au minerals in Laozuoshan gold deposit.

Author Contributions: L.M., F.H. and W.G. devised the project. L.M., F.H., R.G. and W.G. wrote the paper, assisted by Y.Z., and operated the EPMA. F.Z. and Y.L. assisted with data processing. All authors have read and agreed to the published version of the manuscript.

Funding: This research was funded by the National Natural Science Foundation of China, grant number 41272062, the National Natural Science Foundation of China, grant number 41902067, China Postdoctoral Science Foundation, grant number 2021M700713, and Key Laboratory of Mineral Resources Evaluation in Northeast Asia, Ministry of Natural Resources, grant number DBY-KF-18-04.

Institutional Review Board Statement: Not applicable.

Informed Consent Statement: Not applicable.

Data Availability Statement: All data used in this study are freely available.

Acknowledgments: The authors would like to thank Yufeng Shao for his help in the fieldwork. We also thank Degao Zhai of China University of Geosciences, Beijing, He Zhang of Nanjing University, and Jin-peng Luan of Northeastern University for their suggestions on paper revision. We appreciate the Editor for editorial help and supportive comments. We thank two anonymous reviewers for their helpful comments and suggestions to improve our paper. Thanks to MDPI for language editing services during the preparation of this manuscript.

Conflicts of Interest: The authors declare no conflict of interest.

References

1. Soares, M.B.; Alves, F.E.A.; Neto, A.V.C.; Bertolino, L.C.; Araújo, I.M.C.D.P.; Gopon, P.; Mozart, M.S. Gold refinement by the fractionation of Bi-enriched partial melts at the Quadrilatero Ferrifero, Brazil: Implications on the formation of hypozonal deposits. *Miner. Depos.* **2022**, *57*, 781–800. [CrossRef]

2. Hastie, E.C.G.; Schindler, M.; Kontak, D.J.; Lafrance, B. Transport and coarsening of gold nanoparticles in an orogenic deposit by dissolution-reprecipitation and Ostwald ripening. *Commun. Earth Environ.* **2021**, *2*, 57. [CrossRef]
3. Oberthur, T.; Weiser, T.W. Gold-bismuth-telluride-sulphide assemblages at the Viceroy Mine, Harare-Bindura-Shamva greenstone belt, Zimbabwe. *Miner. Mag.* **2008**, *72*, 953–970. [CrossRef]
4. Buzatu, A.; Damian, G.; Dill, H.G.; Buzgar, N.; Apopei, A.I. Mineralogy and geochemistry of sulfosalts from Baia Sprie ore deposit (Romania)—New bismuth minerals occurrence. *Ore Geol. Rev.* **2015**, *65*, 132–147. [CrossRef]
5. Cook, N.J.; Ciobanu, C.L. Bismuth tellurides and sulphosalts from the Larga hydrothermal system, Metaliferi Mts., Romania: Paragenesis and genetic significance. *Miner. Mag.* **2004**, *68*, 301–321. [CrossRef]
6. Cepedal, A.; Fuertes-Fuente, M.; Martin-Izard, A.; Gonzalez-Nistal, S.; Barrero, M. Gold-bearing As-rich pyrite and arsenopyrite from the El Valle gold deposit, Asturias, northwestern Spain. *Can. Miner.* **2008**, *46*, 233–247. [CrossRef]
7. Cepedal, A.; Fuertes-Fuente, M.; Martín-Izard, A.; González-Nistal, S.; Rodríguez-Pevida, L. Tellurides, selenides and Bi-mineral assemblages from the Río Narcea Gold Belt, Asturias, Spain: Genetic implications in Cu–Au and Au skarns. *Min. Pet.* **2006**, *87*, 277–304. [CrossRef]
8. Mueller, A.G. The Nevoria gold skarn deposit in Archean iron-formation, Southern Cross greenstone belt, Western Australia; I, Tectonic setting, petrography, and classification. *Econ. Geol.* **1997**, *92*, 181–209. [CrossRef]
9. Zhou, H.; Sun, X.; Wu, Z.; Yang, T.; Li, D.; Ren, Y.; Liu, Q.; Zhu, K.; Yu, H. Mineralogy of Bi-sulfosalts and tellurides from the Yaoan gold deposit, southwest China: Metallogenic implications. *Ore Geol. Rev.* **2018**, *98*, 126–140. [CrossRef]
10. Zhou, H.; Sun, X.; Cook, N.J.; Lin, H.; Fu, Y.; Zhong, R.; Brugger, J. Nano- to Micron-Scale Particulate Gold Hosted by Magnetite: A Product of Gold Scavenging by Bismuth Melts. *Econ. Geol.* **2017**, *112*, 993–1010. [CrossRef]
11. Acosta-Góngora, P.; Gleeson, S.A.; Samson, I.M.; Ootes, L.; Corriveau, L. Gold Refining by Bismuth Melts in the Iron Oxide-Dominated NICO Au-Co-Bi ($\pm$Cu$\pm$W) Deposit, NWT, Canada. *Econ. Geol.* **2015**, *110*, 291–314. [CrossRef]
12. Tormanen, T.O.; Koski, R.A. Gold enrichment and the Bi-Au association in pyrrhotite-rich massive sulfide deposits, Escanaba trough, southern Gorda Ridge. *Econ. Geol.* **2005**, *100*, 1135–1150. [CrossRef]
13. Cave, B.J.; Barnes, S.-J.; Pitcairn, I.; Sack, P.J.; Kuikka, H.; Johnson, S.; Duran, C.J. Multi-stage precipitation and redistribution of gold, and its collection by lead-bismuth and lead immiscible liquids in a reduced-intrusion related gold system (RIRGS); Dublin Gulch, western Canada. *Ore Geol. Rev.* **2019**, *106*, 28–55. [CrossRef]
14. Cepedal, A.; Martínez-Abad, I.; Fuertes-Fuente, M.; Lima, A. The presence of plumboan ingodite and a rare Bi-Pb tellurosulfide, Pb3Bi4Te4S5, in the Limarinho gold deposit, Northern Portugal. *Can. Miner.* **2013**, *51*, 643–651. [CrossRef]
15. Ciobanu, C.L.; Birch, W.D.; Cook, N.J.; Pring, A.; Grundler, P.V. Petrogenetic significance of Au–Bi–Te–S associations: The example of Maldon, Central Victorian gold province, Australia. *Lithos* **2010**, *116*, 1–17. [CrossRef]
16. McFarlane, C.R.M.; Mavrogenes, J.; Lentz, D.; King, K.; Allibone, A.; Holcombe, R. Geology and Intrusion-Related Affinity of the Morila Gold Mine, Southeast Mali. *Econ. Geol.* **2011**, *106*, 727–750. [CrossRef]
17. Douglas, N.; Mavrogenes, J.; Hack, A.; England, R. The liquid bismuth collector model: An alternative gold deposition mechanism. In Proceedings of the 15th Understanding Planet Earth, Searching for a Sustainable Future, on the Starting Blocks of the Third Millennium: Australian Geological Convention, Sydney, Australia, 3–7 July 2000; p. 135.
18. Tooth, B.; Brugger, J.; Ciobanu, C.; Liu, W. Modeling of gold scavenging by bismuth melts coexisting with hydrothermal fluids. *Geology* **2008**, *36*, 815–818. [CrossRef]
19. Tooth, B.; Ciobanu, C.L.; Green, L.; O'Neill, B.; Brugger, J. Bi-melt formation and gold scavenging from hydrothermal fluids: An experimental study. *Geochim. Cosmochim. Acta* **2011**, *75*, 5423–5443. [CrossRef]
20. Shi, K.; Ulrich, T.; Wang, K.Y.; Ma, X.L.; Li, S.D.; Wang, R. Hydrothermal evolution and ore genesis of the Laozuoshan Au skarn deposit, northeast China: Constrains from mineralogy, fluid inclusion, and O-C-S-Pb isotope geochemistry. *Ore Geol. Rev.* **2020**, *127*, 103879. [CrossRef]
21. Bai, C.L.; Sun, J.G.; Zhao, C.T.; Qin, K.Z.; Liu, Y.; Chu, X.L.; Xu, Z.K.; Li, Y.X. The multiple mineralizations and geodynamic settings of the Laozuoshan Cu-Au deposit in the Jiamusi Massif, NE China: Zircon U-Pb geochronological, elemental and Hf isotopic geochemical evidence. *Ore Geol. Rev.* **2021**, *137*, 104291. [CrossRef]
22. Shi, K.; Wang, K.; Ulrich, T.; Ma, X.; Sun, Q.; Wang, R. Two mineralization events in the Laozuoshan Au deposit, north-east China: Evidence from Re–Os geochronology and trace element geochemistry. *Geol. J.* **2020**, *56*, 1974–1986. [CrossRef]
23. Meng, L.; Huang, F.; Gao, W.; Wang, D.; Zhu, J.; Xing, C.; Tan, W.; Tang, X. Multi-stage hydrothermal processes at the Laozuoshan gold deposit in NE China: Insights from textures and compositions of sulfide assemblages. *Ore Geol. Rev.* **2020**, *117*, 103275. [CrossRef]
24. Li, M.F.; Ye, S.Q.; Yang, Y.C.; Zhang, G.B.; Hou, X.G.; Ming, T.X.; Tang, Z. Geological and geochemical characteristics of Laozuoshan gold deposit in Heilongjiang and its metallotectonic setting. *Glob. Geol.* **2014**, *33*, 543–555, (In Chinese with English Abstract).
25. He, B.L. Discussion on the origin of gold ore deposit at the eastern section of Laozuoshan gold ore field. *Gold* **2002**, *23*, 7–12, (In Chinese with English abstract).
26. Wei, C.D.; Li, X.M. The typomorphic characteristic of the main gold-bearing minerals in Laozuoshan gold deposit, Heilongjiang Province. *Gold* **2001**, *22*, 5–9, (In Chinese with English Abstract).
27. HBGMR. *Regional Geology of Heilongjiang Province*; Geological Publishing House: Beijing, China, 1993. (In Chinese)

28. Wu, F.Y.; Yang, J.H.; Lo, C.H.; Wilde, S.A.; Sun, D.Y.; Jahn, B.M. The Heilongjiang Group: A Jurassic accretionary complex in the Jiamusi Massif at the western Pacific margin of northeastern China. *Isl. Arc* **2007**, *16*, 156–172. [CrossRef]
29. Li, W.M.; Liu, Y.J.; Zhao, Y.L.; Feng, Z.Q.; Zhou, J.P.; Wen, Q.B.; Liang, C.Y.; Zhang, D. Tectonic evolution of the Jiamusi Block, NE China. *Acta Petrol. Sin.* **2020**, *36*, 665–684, (In Chinese with English Abstract).
30. Guo, X.Z.; Takasu, A.; Liu, Y.J.; Li, W.M. Zn-Rich Spinel in Association with Quartz in the Al-Rich Metapelites from the Mashan Khondalite Series, NE China. *J. Earth Sci.* **2014**, *25*, 207–223. [CrossRef]
31. Luan, J.P.; Rosenbaum, G.; Fu, J.F. Provenance and depositional history of the Mesozoic Sanjiang Basin (Northeastern China): Implications for the uplift history of the northeastern Asian continental margin. *Geol. Mag.* **2022**, *159*, 389–406. [CrossRef]
32. Luan, J.P.; Wang, F.; Xu, W.L.; Ge, W.C.; Sorokin, A.A.; Wang, Z.W.; Guo, P. Provenance, age, and tectonic implications of Neoproterozoic strata in the Jiamusi Massif: Evidence from U-Pb ages and Hf isotope compositions of detrital and magmatic zircons. *Precambr. Res* **2017**, *297*, 19–32. [CrossRef]
33. Zhou, J.B.; Pu, X.G.; Hou, H.S.; Han, W.; Gao, J.L.; Li, G.Y. The Mesozoic accretionary complex in NE China and its tectonic implications for the subduction of the Paleo-Pacific plate beneath the Eurasia. *Acta Petrol. Sin.* **2018**, *34*, 2845–2856, (In Chinese with English Abstract).
34. Bi, J.H.; Ge, W.C.; Yang, H.; Zhao, G.C.; Yu, J.J.; Zhang, Y.L.; Wang, Z.H.; Tian, D.X. Petrogenesis and tectonic implications of early Paleozoic granitic magmatism in the Jiamusi Massif, NE China: Geochronological, geochemical and Hf isotopic evidence. *J. Asian Earth Sci.* **2014**, *96*, 308–331. [CrossRef]
35. Wu, F.Y.; Sun, D.Y.; Ge, W.C.; Zhang, Y.B.; Grant, M.L.; Wilde, S.A.; Jahn, B.M. Geochronology of the Phanerozoic granitoids in northeastern China. *J. Asian Earth Sci.* **2011**, *41*, 1–30. [CrossRef]
36. Yang, H.; Ge, W.C.; Bi, J.H.; Wang, Z.H.; Tian, D.X.; Dong, Y.; Chen, H.J. The Neoproterozoic-Early Paleozoic evolution of the Jiamusi Block, NE China and its East Gondwana connection: Geochemical and zircon U-Pb-Hf isotopic constrains from the Mashan Complex. *Gondwana Res.* **2018**, *54*, 102–121. [CrossRef]
37. Sun, J.G.; Han, S.J.; Zhang, Y.; Xing, S.W.; Bai, L.A. Diagenesis and metallogenetic mechanisms of the Tuanjiegou gold deposit from the Lesser Xing'an Range, NE China: Zircon U-Pb geochronology and Lu-Hf isotopic constraints. *J. Asian Earth Sci.* **2013**, *62*, 373–388. [CrossRef]
38. Hao, Y.J.; Ren, Y.S.; Zhao, H.L.; Lai, K.; Zhao, X.; Ma, Y.P. Metallogenic Mechanism and Tectonic Setting of Tungsten Mineralization in the Yangbishan Deposit in Northeastern China. *Acta Geol. Sin.* **2018**, *92*, 241–267. [CrossRef]
39. Wilde, S.A.; Zhang, X.; Wu, F. Extension of a newly identified 500Ma metamorphic terrane in North East China: Further U–Pb SHRIMP dating of the Mashan Complex, Heilongjiang Province, China. *Tectonophysics* **2000**, *328*, 115–130. [CrossRef]
40. Wu, M.; Li, L.; Sun, J.G.; Yang, R. Geology, geochemistry, and geochronology of the Laozuoshan gold deposit, Heilongjiang Province, Northeast China: Implications for multiple gold mineralization events and geodynamic setting. *Can. J. Earth Sci.* **2018**, *55*, 604–619. [CrossRef]
41. Meng, L.; Huang, F.; Liu, J.; Xing, M.M.; Wan, Q.; Ling, J.Q.; Gao, S.; Gao, W.Y. The Genesis of Nano-Micron-Sized Spheroidal Aggregates of FeS_2 in the Laozuoshan Gold Deposit, Heilongjiang Province, NE China. *J. Nanosci. Nanotechnol.* **2017**, *17*, 6539–6548. [CrossRef]
42. Cook, N.J.; Ciobanu, C.L.; Meria, D.; Silcock, D.; Wade, B. Arsenopyrite-Pyrite Association in an Orogenic Gold Ore: Tracing Mineralization History from Textures and Trace Elements. *Econ. Geol.* **2013**, *108*, 1273–1283. [CrossRef]
43. Ciobanu, C.; Cook, N.J. Tellurides, selenides (and Bi-sulphosalts) in gold deposits. In Proceedings of the 11th Quadrennial IAGOD Symposium, Windhoek, Namibia, 22–26 July 2002.
44. Ciobanu, C.L.; Cook, N.J.; Pring, A. Bismuth tellurides as gold scavengers. In *Mineral Deposit Research: Meeting the Global Challenge*; Springer: Berlin, Germany, 2005; pp. 1383–1386. [CrossRef]
45. Cook, N.J.; Ciobanu, C.L.; Wagner, T.; Stanley, C.J. Minerals of the system Bi-Te-Se-S related to the tetradymite archetype: Review of classification and compositional variation. *Can. Miner.* **2007**, *45*, 665–708. [CrossRef]
46. Wei, B.; Wang, C.Y.; Wang, Z.C.; Cheng, H.; Xia, X.P.; Tan, W. Mantle-derived gold scavenged by bismuth-(tellurium)-rich melts: Evidence from the mesozoic wulong gold deposit in the North China Craton. *Ore Geol. Rev.* **2021**, *131*, 104047. [CrossRef]
47. Meinert, L.D.; Dipple, G.M.; Nicolescu, S. World Skarn Deposits. In *Economic Geology: One Hundredth Anniversary Volume*; Hedenquist, J.W., Thompson, J.F.H., Goldfarb, R.J., Richards, J.P., Eds.; Society of Economic Geologists: Littleton, CO, USA, 2005; pp. 299–336.
48. Meinert, L.D. Gold in Skarns Related to Epizonal Intrusions. In *Gold in 2000*; Hagemann, S.G., Brown, P.E., Eds.; Society of Economic Geologists: Littleton, CO, USA, 2000; Volume 13, pp. 347–375.
49. Zhao, H.; Wang, Q.; Groves, D.I.; Deng, J. A rare Phanerozoic amphibolite-hosted gold deposit at Danba, Yangtze Craton, China: Significance to fluid and metal sources for orogenic gold systems. *Miner. Depos.* **2019**, *54*, 133–152. [CrossRef]
50. Perfetti, E.; Pokrovski, G.S.; Ballerat-Busserolles, K.; Majer, V.; Gibert, F. Densities and heat capacities of aqueous arsenious and arsenic acid solutions to 350 °C and 300 bar, and revised thermodynamic properties of $As(OH)_3{}^{\circ}(aq)$, $AsO(OH)_3{}^{\circ}(aq)$ and iron sulfarsenide minerals. *Geochim. Cosmochim. Acta* **2008**, *72*, 713–731. [CrossRef]
51. Pokrovski, G.S.; Kara, S.; Roux, J. Stability and solubility of arsenopyrite, FeAsS, in crustal fluids. *Geochim. Cosmochim. Acta* **2002**, *66*, 2361–2378. [CrossRef]

52. Pokrovski, G.S.; Zakirov, I.V.; Roux, J.; Testemale, D.; Hazemann, J.-L.; Bychkov, A.Y.; Golikova, G.V. Experimental study of arsenic speciation in vapor phase to 500 °C: Implications for As transport and fractionation in low-density crustal fluids and volcanic gases. *Geochim. Cosmochim. Acta* **2002**, *66*, 3453–3480. [CrossRef]
53. Heinrich, C.A.; Eadington, P.J. Thermodynamic predictions of the hydrothermal chemistry of arsenic, and their significance for the paragenetic sequence of some cassiterite-arsenopyrite-base metal sulfide deposits. *Econ. Geol.* **1986**, *81*, 511–529. [CrossRef]
54. Zhang, H.; Zhu, Y. Genesis of the Mandongshan gold deposit (Xinjiang, NW China): T-P-fS_2 and phase equilibria constraints from the Au-As-Fe-S system. *Ore Geol. Rev.* **2017**, *83*, 135–151. [CrossRef]
55. Skirrow, R.G.; Walshe, J.L. Reduced and oxidized Au-Cu-Bi iron oxide deposits of the Tennant Creek inlier, Australia: An integrated geologic and chemical model. *Econ. Geol. Bull. Soc.* **2002**, *97*, 1167–1202. [CrossRef]
56. Voudouris, P.C.; Spry, P.G.; Mavrogonatos, C.; Sakellaris, G.-A.; Bristol, S.K.; Melfos, V.; Fornadel, A.P. Bismuthinite derivatives, lillianite homologues, and bismuth sulfotellurides as indicators of gold mineralization in the Stanos shear-zone related deposit, Chalkidiki, northern Greece. *Can. Miner.* **2013**, *51*, 119–142. [CrossRef]
57. Williams-Jones, A.E.; Bowell, R.J.; Migdisov, A.A. Gold in Solution. *Elements* **2009**, *5*, 281–287. [CrossRef]
58. Deditius, A.P.; Utsunomiya, S.; Renock, D.; Ewing, R.C.; Ramana, C.V.; Becker, U.; Kesler, S.E. A proposed new type of arsenian pyrite: Composition, nanostructure and geological significance. *Geochim. Cosmochim. Acta* **2008**, *72*, 2919–2933. [CrossRef]
59. Zhao, K.; Yang, L.Q.; Li, P.; Xiong, Y.Q. Morphology and chemistry composition of pyrite in the Laowangzhai gold deposit, Ailaoshan orogenic belt, SW China. *Acta Petrol. Sin.* **2013**, *29*, 3937–3948, (In Chinese with English Abstract).
60. Ciobanu, C.L.; Cook, N.J.; Damian, F.; Damian, G. Gold scavenged by bismuth melts: An example from Alpine shear-remobilizates in the Highiş Massif, Romania. *Miner. Petrol.* **2006**, *87*, 351–384. [CrossRef]
61. An, R.; Wang, K.Y.; Ma, X.L.; Liang, Y.M.; Wang, Y.C.; Wang, Z.G. Evolution mode for the ore-forming fluid system of the Laozuoshan skarn-magmatic hydrothermal vein gold deposit, Heilongjiang Province, China. *Northwest. Geol.* **2017**, *50*, 122–135, (In Chinese with English Abstract).
62. Cockerton, A.B.D.; Tomkins, A.G. Insights into the Liquid Bismuth Collector Model through Analysis of the Bi-Au Stormont Skarn Prospect, Northwest Tasmania. *Econ. Geol.* **2012**, *107*, 667–682. [CrossRef]
63. Cook, N.J.; Ciobanu, C.L.; Mao, J.W. Textural control on gold distribution in As-free pyrite from the Dongping, Huangtuliang and Hougou gold deposits, North China Craton (Hebei Province, China). *Chem. Geol.* **2009**, *264*, 101–121. [CrossRef]
64. Fougerouse, D.; Micklethwaite, S.; Tomkins, A.G.; Mei, Y.; Kilburn, M.; Guagliardo, P.; Fisher, L.A.; Halfpenny, A.; Gee, M.; Paterson, D.; et al. Gold remobilisation and formation of high grade ore shoots driven by dissolution-reprecipitation replacement and Ni substitution into auriferous arsenopyrite. *Geochim. Cosmochim. Acta* **2016**, *178*, 143–159. [CrossRef]
65. Morey, A.A.; Tomkins, A.G.; Bierlein, F.P.; Weinberg, R.F.; Davidson, G.J. Bimodal Distribution of Gold in Pyrite and Arsenopyrite: Examples from the Archean Boorara and Bardoc Shear Systems, Yilgarn Craton, Western Australia. *Econ. Geol.* **2008**, *103*, 599–614. [CrossRef]

minerals

MDPI

Article

Contrasting Sources and Related Metallogeny of the Triassic and Jurassic Granitoids in the Chifeng–Chaoyang District, Northern Margin of the North China Craton: A Review with New Data

Jian-Guo Yuan [1,2], Hua-Feng Zhang [3], Ying Tong [2,*], Yun-Yan Qu [4], Rui-Bin Liu [2] and Run-Wu Li [3]

1 Development and Research Center, China Geological Survey, Beijing 100037, China
2 Beijing SHRIMP Center, Institute of Geology, Chinese Academy of Geological Sciences, Beijing 100037, China
3 School of Earth Sciences and Resources, China University of Geosciences, 29# Xue-Yuan Road, Haidian District, Beijing 100083, China
4 China Chemical Geology and Mine Bureau, Beijing 100013, China
* Correspondence: yingtong@pku.org.cn

Abstract: Understanding of the mechanism between magma sources and metallogeny is still vague. As an important gold and molybdenum producing area, the Chifeng–Chaoyang district, located at the northern margin of the North China Craton (NCC), is a key place for this issue. New geochemical data relating to Taijiying gold-deposit-related granites are presented. These data, coupled with previous studies, are used to explore the relationship between magma sources and mineralization processes. Two major magmatic periods, the Middle Triassic (220–230 Ma) and Late Jurassic (150–160 Ma), are identified based on the compiled data. The Triassic magmatic rocks are mostly fractionated I-type and A-type granites, including monzogranite, biotite granite, and syenogranite. They have low initial $^{87}Sr/^{86}Sr$ values (0.7050), moderately enriched $\varepsilon_{Nd}(t)$–$\varepsilon_{Hf}(t)$ values (−8.5 and −5.6), and relatively young Nd–Hf model ages (T_{DM2}-T_{DM}^{C}) (1.47–1.57 Ga). These features indicate that more Archean–Paleoproterozoic mantle-derived materials were involved in their sources. On the other hand, Jurassic granites are high-K calc-alkaline of the calc-alkaline series and mainly consist of granite, monzogranite, leucogranite, and granodiorite. They have high Na_2O/K_2O, Sr/Y, and La/Yb ratios and low Y and Yb contents. The adakitic features suggest the existence of a thickened lower crust. Their significant negative $\varepsilon_{Nd}(t)$–$\varepsilon_{Hf}(t)$ values (−15.0 and −12.8) and older Nd–Hf model ages (T_{DM2}–T_{DM}^{C}) (2.17–2.11 Ga) are consistent with their derivation from thickened ancient lower crust, indicating the initial activation of NCC. It is proposed that the change in the main source resulted from the tectonic transition during the early Mesozoic initial decratonization, that is, from the post-collisional extension to the subduction of the Paleo-Pacific plate beneath the East Asia plate from the Triassic to the Jurassic. Comparative analysis suggests that the medium–large-scale gold deposits with a high grade are closely related to the Triassic granites; however, most molybdenum deposits formed in the Jurassic. The decratonization of the NCC in the early Mesozoic experienced tectonic transition and controlled the gold and molybdenum mineralizations in the different stages by the changing magma sources. This pattern is beneficial to understanding the metallogenesis in the Chifeng–Chaoyang district.

Keywords: magmatic source; metallogenesis; Sr–Nd–Hf–O isotopes; Chifeng–Chaoyang; North China Craton

1. Introduction

The distribution patterns of deposits are very important for exploration. Although many methods and parameters have been applied to decipher these rules, the feature of deep sources has been proved to be a good indicator. The transitions in the magma source indicate the changes of tectonic settings and significant mineralization events in most cases. For example, the switch in the magmatic source from enriched lithospheric mantle to depleted asthenospheric mantle was the most marked change during the NCC cratonic destruction, which promoted large-scale gold deposits. [1–3]. Therefore, the northern margin of the NCC, especially in the eastern section, is a good place to explore the relationship between the deep magmatic source and metallogenesis. This is because of the widespread Mesozoic felsic intrusive rocks and contemporaneous mineralization in this region.

The Chifeng–Chaoyang district is located in the eastern NCC, which was affected by the Central Asian Orogenic Belt (CAOB) and the Pacific plate in the Mesozoic (Figure 1a). The Paleozoic–Mesozoic magmatic rocks were emplaced, as well as many medium–large gold and molybdenum deposits (Figure 1b). Yang et al. (2017) preliminarily divided this district into four Nd isotope areas according to the model ages (T_{DM}) of the Phanerozoic granitoids [4]. In the southeastern margin of the NCC, Deng et al. (2018) proposed that the porphyry and porphyry-skarn Cu (-Au-Mo) deposits are associated with Paleoproterozoic to Mesoproterozoic reworked crustal components with minor mantle material [5]. The Jiaodong-type Au and porphyry-skarn Mo (-W-Cu) deposits are associated with Archean–Paleoproterozoic reworking crustal components. Thus, the common relationship between the crustal materials and the mineralization in this area remains unclear. The Mo mineralization is closely related to the small granitic bodies with aged continental component sources, and the Mo deposits are mainly located in the oldest terranes [6]. Some researchers proposed that the hydrated mantle wedge is the source of gold deposits through local and rapid devolatization, which suggests a juvenile source. Field works and analysis displayed high grades of Taijiying gold distributed along the granite dikes. Previous research also suggested that magmatism related to granite porphyries provided fluids and sources for the gold mineralization systems [7]. However, the kind of source controlling the Taijiying gold deposits is still unclear. In this paper, we present new petrology, U–Pb zircon ages, geochemistry, whole-rock Sr–Nd isotopes, and zircon O–Hf isotope data for the Taijiying granite porphyries (Figure 1b) and submit published data for the Triassic and Jurassic felsic magmatic rocks in the Chifeng–Chaoyang district [8]. These results can help us to better understand the sources of the granitoids and their effects on Mo and Au mineralization.

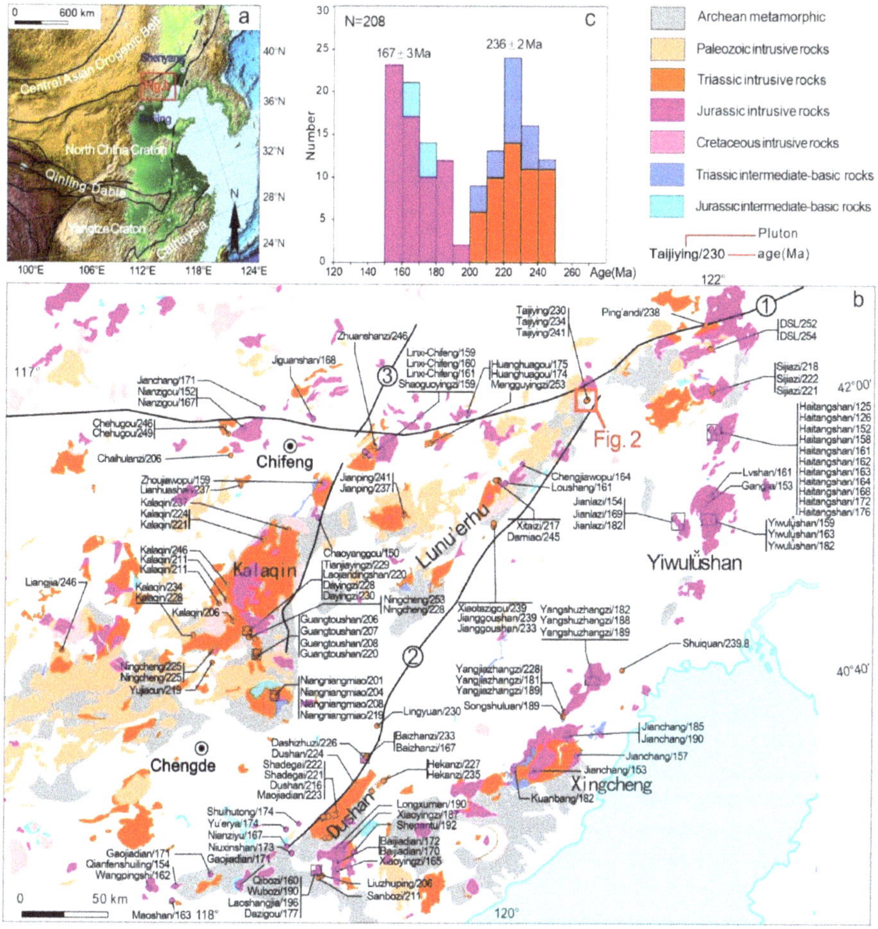

Figure 1. (**a**) Sketchmap of North China Craton (Wu et al., 2019); (**b**) outline map of Phanerozoic granitoids in Chifeng–Chaoyang and adjacent areas [8,9]. ①—Chifeng–Kaiyuan fault; ②—Lingyuan–Beipiao fault; ③—Honghan–Balihan fault; (**c**) Histogram of ages of intrusive rocks in two episodes of Chifeng–Chaoyang and adjacent areas (supplementary data are listed in the Supplementary Materials).

2. Geological Background

The triangular NCC, with an area of ca. 1,500,000 km^2, is bounded by the CAOB to the north, Sulu UHP belt to the east, Qilianshan Orogen to the west, and Qinling–Dabie orogenic belt to the south and is one of the oldest cratons in the world ([10,11]; Figure 1a). The craton had a prominent peak of crustal growth activity during the Neoarchean ca. 2.5–2.7 Ga [12]. The amalgamation of the eastern and western blocks along the NS-trending Trans-North China Orogen is generally considered to have occurred during the late Paleoproterozoic ~1.85 Ga [13,14]. In ca. 1.32 Ga, a large number of tholeiitic diabase sills or dyke swarms constituted a mid-Mesoproterozoic large igneous province in the northern NCC [15]. The multiple-stage lithospheric thinning and destruction of the NCC started from the initiation of massive circum-craton Phanerozoic subduction and collisional orogenies. In the early Paleozoic, the Qilian oceanic plate subduction and collision with the Kunlun–Qaidam Block occurred [16]. In the late Paleozoic, closure of the Paleo-Asian

Ocean and collision with the Siberia Craton happened along the northern margin of the NCC [17]. The deep subduction and collision of the Yangtze Block further destructed the NCC during the late Permian to Triassic [18]; subsequently, the Paleo-Pacific plate's westward subduction or Yangzi Craton collision started in the Early Jurassic [19]. The Pacific plate's ongoing westward subduction during the Cretaceous was the peak stage of decratonization.

The Chifeng–Chaoyang district occupies the eastern segment of the northern NCC, which is dominated by the EW-trending Chifeng–Kaiyuan fault, NE-trending Lingyuan–Beipiao fault and NNE-trending Honghan–Balihan fault (Figure 1b). The Chifeng–Kaiyuan fault divides the Hinggan–Mongolian Orogenic Belt (the middle and east part of the CAOB) from the NCC [20]. The Lingyuan–Beipiao fault, more than 1000 km in length, extends from Chengde to Beipiao city. The Honghan–Balihan fault consists of a ~3 km thick ductile and ductile-brittle shear zone passing upward into a ~100 m thick brittle chloritic breccia and a ~0.5 m thick microbreccia with fault gouges [20,21]. These compressional and shearing faults have existed since the Paleoproterozoic and have undergone multiple reactivations in the Paleozoic, Mesozoic, and especially the Yanshanian period [22]. The EW-striking faults are the earliest faults in the region, which are mostly cut by the later NE-striking faults. They divide the study area into several rhombic blocks where many gold and molybdenum deposits and granitoids have been discovered [22].

This district mainly consists of Precambrian basement rocks and Paleozoic to Mesozoic granitoids. The Precambrian basement rocks belong to the Neoarchean to Paleoproterozoic Jianping, Qianxi, and Zunhua groups, which consist of sedimentary and volcanic sequences and tonalite–trondhjemite–granodiorite (TTG) gneisses with different degrees of migmatization [22–24]. These successions were further metamorphosed to granulite phases ca. 2485 Ma and were retrograded to the greenschist phases ca. 2450–2401 Ma [23,25]. The Paleozoic rocks are distributed in the Aohanqi area, mainly belonging to a set of continental volcanic-sedimentary strata [22]. Mesozoic volcanic rocks are continental volcanic rocks which mainly consist of basalt, basaltic andesite, andesitic breccias, and tuff [26]. The Cenozoic strata are mainly residual slope and alluvial sands and loess-like sub-sandy soil.

More Triassic and Jurassic intrusions have been discovered through various dating methods (Supplementary Table S1 and Figure 1b). The intrusions in this area mainly comprise three NE-trending uplift belts, including the Kalaqin, Lunu'erhu, and Yiwulüshan metamorphic core complexes (MCCs) to the north and the Dushan composite batholiths, Xingcheng, to the south (Figure 1b). Triassic granitoids are mainly distributed in Kalaqin, Lunu'erhu, while the majority of Jurassic granites occur in the Xingcheng, Yiwulüshan MCC. The Triassic intrusions and dikes, including granodiorite, alkalic granite, and minor basic and ultrabasic rocks, are mainly distributed in the area of southern Chifeng and eastern Chengde city. The Dushan monzogranite-granodiorite, Dashizhuzi granite, Panshan monzogranite/granite, and Baizhangzi granite are well studied [27–29]. The EW-trending contemporaneous alkaline rocks belt consists of Hekanzi nepheline syenite, Guangtoushan alkaline granite, and Sungezhuang alkaline syenite [30–32]. The Jurassic plutons mainly consist of granite, granodiorite, and monzogranite exposed along regional faults, which present Qingshankou granite, Yu'erya granite, Niuxinshan granite, Jianchang–Jiumen monzogranite, Xiaojiayingzi batholiths, and Gaojiadian quartz diorite [33] (Supplementary Table S1). Generally, Triassic and Jurassic granitoids exhibit different preferences for emplacement locations [34]. Triassic granites are mostly discovered in the northern margin, close to the suture zone between the CAOB and NCC, while Jurassic granites are located in the southeast, where the area was affected by the westward subduction of the Paleo-Pacific Ocean plate.

3. Sources and Gold Mineralization of Taijiying Granite Porphyries

The granite porphyries are characterized by high-silicon dike. They mainly consist of brick-red granite porphyry, which is closely related to Taijiying gold mineralization (Figure 2a–c). Field works and analysis displayed high grades of gold distributed along the granite dikes. Previous research also suggested that magmatism related to granite porphyries provided fluids and sources in the process of gold mineralization [7]. However, the kinds of source are still unclear. The outcrop is NE-trending with a length of ca. 800 m and a width of more than 10 m (Figure 2a). This dike intruded into the Archean metamorphic rocks and Cenozoic strata with clear contact lines. The granite porphyries have a medium-to-fine-grained porphyritic texture. Thirteen samples were collected from the drilling cores and outcrops. They mainly consisted of 30%–40% phenocrysts by volume (Figure 2d–f). The phenocrysts consisted of quartz (3%–5%), perthite (5%–8%), and platy plagioclase (ca. 1%). The quartz was generally euhedral or anhedral, 0.5–1.0 mm in size, and was surrounded by recrystallized groundmass. The perthite had clear streaks with argillic alteration. The plagioclase (0.2–0.5 mm) was euhedral with strong argillic alteration and slight sericitization. The minerals in the groundmass were mainly composed of felsic spherulites (70%–75%), plagioclase (8%–10%), sericite (5%–8%), and biotite (less than 1%). The plagioclase crystal was generally the center of spherulite and surrounded by the radially fibrous corpora.

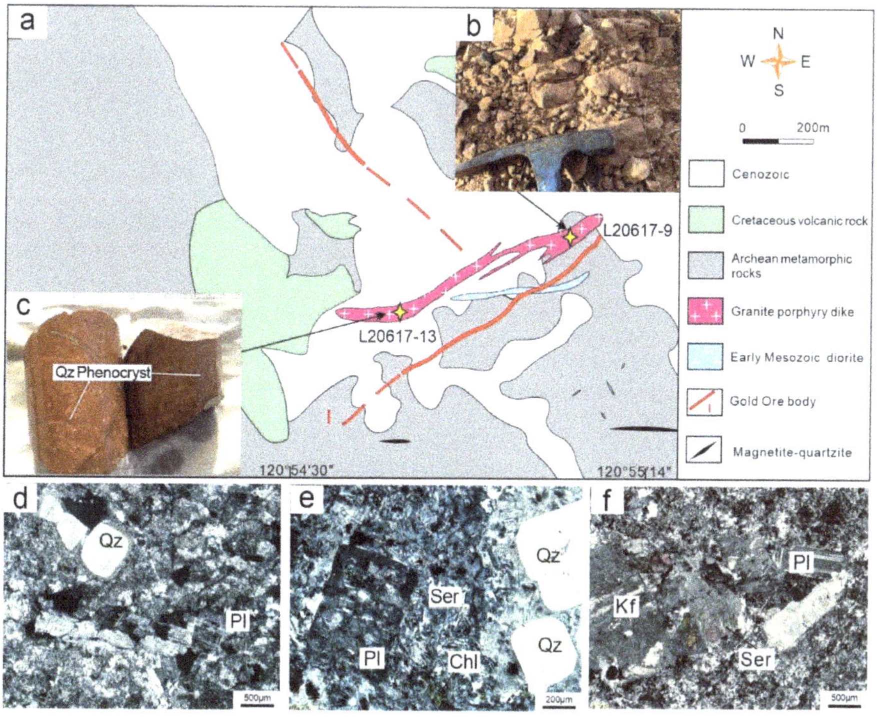

Figure 2. (**a**) Location of granite porphyry dikes in the Taijiying gold deposit; (**b**) outcrop of granite porphyry; (**c**) granite porphyry from drilling core; (**d**–**f**) photomicrographs of granite porphyries. Abbreviations: Qtz = quartz, Pl = plagioclase, Ser = sericite, Chl = chlorite, Kf = K-feldspar.

The zircon U–Pb dating results of Taijiying granite porphyries are listed in Table 1. Concordia diagrams are shown in Figure 3. Sample L20617-9 and L20617-13 exhibited zircons similar in size with lengths of 110–240 μm and widths of 80–160 μm and ratios of 1:1 to 2:1. The Th/U ratios were slightly lower, with values of 0.57 and 0.60, respectively, and exhibited a positive correlation, reflecting their magmatic origin [35,36]. For sample L20617-9, 25 spots on 25 zircon grains were analyzed by LA-ICP-MS, and the results are plotted on concordia diagrams in Figure 3. All analyses were concordant except the spot

L20617-9-02, which showed extremely low concordance for the influence of inclusion in the zircon grain. The others yielded a weighted mean $^{206}Pb/^{238}U$ age of 234 ± 1 Ma (95% confidence, MSWD = 1.1). In sample L20617-13, 27 spots on 27 zircon grains were chosen for measurement by LA-ICP-MS. The 23 analyses results were concordant and yielded a weighted mean $^{206}Pb/^{238}U$ age of 230 ± 2 Ma (95% confidence, MSWD = 1.9) which represented the emplacement age of the granite porphyry. Two granite porphyry samples had a weighted mean $^{206}Pb/^{238}U$ age ranging from ca. 230 to 234 Ma, suggesting that the emplacement of granitoid dikes occurred during the Middle Triassic period.

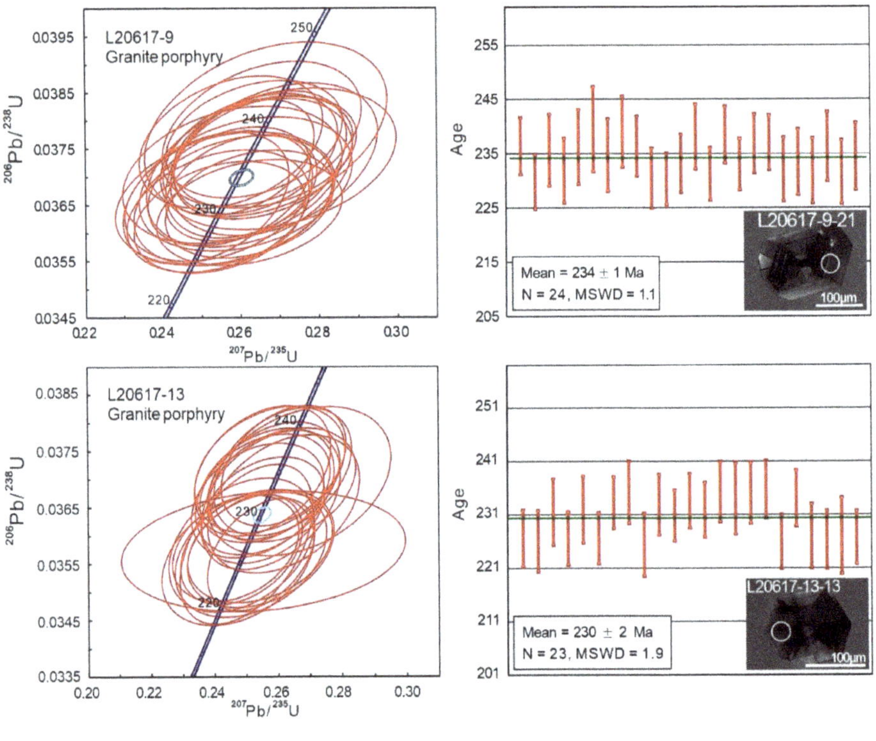

Figure 3. U–Pb concordia plots and age data histograms for samples L20617-9 and L20617-13 (granite porphyry dikes) in Taijiying gold deposit.

The geochemical results for the 14 Taijiying granite porphyry samples are presented in Table 2. The Taijiying granite porphyries had a high and narrow range of SiO_2 contents (69.0–77.6 wt%), and their mean ($K_2O + Na_2O$) value was 7.47 wt%. They had low Fe_2O_3 (0.76–1.88 wt%), MgO (0.18–0.61 wt%), CaO (0.12–1.23 wt%), and TiO_2 (0.09–0.12 wt%) contents. The samples were plotted in the high-K calc-alkaline field (Figure 4a). Al_2O_3 contents were concentrated in the range of 11.4 wt% to 12.6 wt% and were plotted in the metaluminous and peraluminous fields, with A/CNK values (molar ratios of Al_2O_3/CaO + Na_2O + K_2O) of 0.96 to 1.36, except the extremely low value of 0.58 (Figure 4b). In the Harker diagrams, the granite porphyries were plotted in a straight line (Figure 5). All samples exhibited light rare earth element (LREE) enrichment, with $(La/Yb)_N$ ratios of 7.6 to 22.0 (mean = 18.3) (Figure 6b). Several samples had high $(La/Yb)_N$ ratios of greater than 20, including L20617-7, L20617-8, and L20617-14. The samples of different ages all exhibited significant negative Eu anomalies, with δEu values of 0.10–0.40 (mean = 0.16). On the primitive, mantle-normalized trace element spider diagram (Figure 6b), all samples exhibited large ion lithophile element (LILE) (such as Rb and Th) enrichments and high field strength element (HFSE) (such as Nb, Ta) depletions with negative Ba, Sr, Eu, Nb, Ta, and P anomalies. These granite porphyries were plotted between the A-type and I-type granite fields in Figure 7. It was difficult to distinguish between fractionated I-type granites and A-type granites due to their similar features. However, based on the lower zircon saturation temperature (mean = 810 °C) and FeOT/MgO ratios (mean = 7.5), it is reasonable to conclude that the Taijiying granite porphyries are fractionated I-type granites.

Table 1. LA-ICP-MS zircon U–Pb data of the granite porphyry in Taijiying gold deposit.

Spots	Content (ppm)				Ratios						Age					
	Total Pb	^{232}Th	^{238}U	Th/U	^{207}Pb/^{206}Pb	1σ	^{207}Pb/^{235}U	1σ	^{206}Pb/^{238}U	1σ	^{207}Pb/^{206}Pb	1σ	^{207}Pb/^{235}U	1σ	^{206}Pb/^{238}U	1σ
L20617-9, granite porphyry, 24 spots, weighted mean age = 234.2 ± 1.1 Ma, MSWD = 1.07																
L20617-9-01	86.32	1218	1750	0.7	0.0499	0.0016	0.2595	0.0079	0.0374	0.0004	191	72.21	234	6.4	236	2.64
L20617-9-03	58.06	682	1236	0.55	0.0498	0.0023	0.2545	0.009	0.0363	0.0004	183	107.39	230	7.28	230	2.53
L20617-9-04	61.48	640	1289	0.5	0.0505	0.002	0.2639	0.0095	0.0372	0.0005	217	94.43	238	7.61	236	3.32
L20617-9-05	91.15	1437	1836	0.78	0.0487	0.0017	0.251	0.0087	0.0366	0.0005	200	81.47	227	7.08	232	2.97
L20617-9-06	62.56	702	1299	0.54	0.0513	0.0021	0.2684	0.0108	0.0373	0.0006	254	96.28	241	8.67	236	3.44
L20617-9-07	50.1	477	1051	0.45	0.0501	0.0026	0.2663	0.0127	0.0379	0.0006	198	124.98	240	10.18	240	3.94
L20617-9-08	64.96	771	1333	0.58	0.0505	0.0022	0.2622	0.0108	0.0371	0.0005	220	102.76	236	8.69	235	3.36
L20617-9-09	55.89	470	1186	0.4	0.0503	0.002	0.2653	0.0104	0.0378	0.0005	209	87.95	239	8.38	239	3.27
L20617-9-10	60.91	665	1250	0.53	0.0509	0.0018	0.2658	0.0094	0.0373	0.0004	235	86.1	239	7.53	236	2.77
L20617-9-11	46.02	510	1005	0.51	0.0502	0.0019	0.2541	0.0096	0.0364	0.0004	211	88.88	230	7.75	230	2.76
L20617-9-12	62.63	709	1321	0.54	0.0519	0.002	0.2621	0.0097	0.0364	0.0004	280	87.03	236	7.84	230	2.39
L20617-9-13	70.83	906	1447	0.63	0.0504	0.0017	0.257	0.0085	0.0368	0.0004	213	43.51	232	6.91	233	2.69
L20617-9-14	59.51	778	1205	0.65	0.0494	0.0018	0.2575	0.0097	0.0376	0.0005	165	87.95	233	7.8	238	2.99
L20617-9-15	80.24	928	1673	0.55	0.0518	0.0017	0.2622	0.0088	0.0365	0.0004	276	69.44	236	7.12	231	2.44
L20617-9-16	84.37	1088	1689	0.64	0.0505	0.0018	0.2635	0.0093	0.0377	0.0004	220	49.07	237	7.44	238	2.65
L20617-9-17	97.14	1446	1926	0.75	0.0509	0.0016	0.2601	0.0079	0.0368	0.0004	239	70.36	235	6.4	233	2.39
L20617-9-18	85.92	1310	1687	0.78	0.0507	0.0016	0.2638	0.0086	0.0374	0.0004	228	74.06	238	6.91	237	2.72
L20617-9-19	84.23	1161	1669	0.7	0.051	0.0018	0.2661	0.0096	0.0375	0.0004	239	83.32	240	7.69	237	2.54
L20617-9-20	55.45	609	1189	0.51	0.0498	0.0022	0.2543	0.0099	0.0367	0.0005	187	103.69	230	7.98	232	2.94
L20617-9-21	51.08	537	1093	0.49	0.0508	0.0022	0.2612	0.0112	0.0369	0.0005	232	101.84	236	9	233	3.03
L20617-9-22	92.16	1452	1840	0.79	0.0498	0.0026	0.2559	0.0115	0.0366	0.0005	183	120.36	231	9.3	232	3.01
L20617-9-23	39.61	254	882	0.29	0.0517	0.0024	0.2695	0.0123	0.0373	0.0005	333	104.62	212	9.85	236	3.19
L20617-9-24	49.55	313	1134	0.28	0.0506	0.0022	0.258	0.0104	0.0366	0.0005	233	98.14	233	8.43	232	2.94
L20617-9-25	37.27	422	804	0.52	0.0497	0.0022	0.2578	0.0115	0.037	0.0005	189	103.69	233	9.3	234	3.1
L20617-13, granite porphyry, 23 spots, weighted mean age = 231 ± 2.4 Ma, MSWD = 1.9																
L20617-13-01	64.45	792	1401	0.57	0.0509	0.0019	0.2544	0.0094	0.0358	0.0004	239	88.88	230	7.65	227	2.65
L20617-13-02	60.14	613	1322	0.46	0.0502	0.0025	0.2514	0.0098	0.0357	0.0005	211	116.65	228	7.91	226	2.87
L20617-13-03	72.1	1044	1539	0.68	0.0512	0.0019	0.2617	0.0097	0.0366	0.0005	250	89.8	236	7.82	232	3.11
L20617-13-06	86.17	1545	1787	0.87	0.0499	0.0016	0.2484	0.0078	0.0358	0.0004	191	74.06	225	6.37	227	2.5
L20617-13-07	74.82	899	1689	0.53	0.0501	0.0019	0.2576	0.009	0.0366	0.0005	198	90.73	233	7.24	232	3.13
L20617-13-08	98.93	1473	1974	0.75	0.0496	0.0017	0.2466	0.0081	0.0358	0.0004	176	79.62	224	6.62	227	2.4
L20617-13-09	76.28	1119	1561	0.72	0.0499	0.0016	0.2559	0.0084	0.0368	0.0004	191	71.29	231	6.78	233	2.44

Table 1. *Cont.*

Spots	Content (ppm)				Ratios						Age					
	Total Pb	^{232}Th	^{238}U	Th/U	^{207}Pb/^{206}Pb	1σ	^{207}Pb/^{235}U	1σ	^{206}Pb/^{238}U	1σ	^{207}Pb/^{206}Pb	1σ	^{207}Pb/^{235}U	1σ	^{206}Pb/^{238}U	1σ
L20617-13-10	78.11	1082	1617	0.67	0.0495	0.0017	0.2552	0.0088	0.0372	0.0005	169	84.25	231	7.1	235	2.93
L20617-13-11	51.49	583	1145	0.51	0.0502	0.0021	0.2472	0.0098	0.0356	0.0005	206	98.13	224	7.97	225	2.96
L20617-13-12	46.08	500	995	0.5	0.0502	0.0017	0.2558	0.0089	0.0368	0.0005	206	79.62	231	7.21	233	2.85
L20617-13-13	76.42	989	1603	0.62	0.0504	0.0015	0.2553	0.0079	0.0365	0.0004	213	72.21	231	6.37	231	2.4
L20617-13-14	35.82	321	774	0.41	0.0509	0.0018	0.2614	0.0094	0.0369	0.0004	235	81.47	236	7.53	234	2.57
L20617-13-15	53.56	520	1146	0.45	0.0497	0.002	0.2536	0.0085	0.0366	0.0004	189	94.43	229	6.9	232	2.58
L20617-13-16	61.1	635	1277	0.5	0.0498	0.0019	0.2574	0.0096	0.0372	0.0005	183	119.43	233	7.78	235	2.86
L20617-13-18	35.99	303	782	0.39	0.0519	0.0023	0.2665	0.0117	0.037	0.0005	280	106.47	240	9.39	234	3.35
L20617-13-19	77.43	930	1591	0.58	0.0506	0.0017	0.2615	0.0088	0.0371	0.0005	220	75.91	236	7.09	235	2.89
L20617-13-20	112.76	1800	2200	0.82	0.0507	0.0015	0.2618	0.008	0.0372	0.0004	233	70.36	236	6.42	236	2.71
L20617-13-22	105.11	1834	2133	0.86	0.0514	0.0014	0.2546	0.0069	0.0357	0.0004	261	61.1	230	5.6	226	2.54
L20617-13-23	53.53	562	1152	0.49	0.0505	0.0016	0.2587	0.0085	0.037	0.0004	217	80.54	234	6.83	234	2.64
L20617-13-24	31.08	308	704	0.44	0.05	0.0018	0.2484	0.0094	0.0358	0.0005	195	85.17	225	7.62	227	3.02
L20617-13-25	71.13	900	1381	0.65	0.0504	0.0033	0.2551	0.0182	0.0357	0.0004	213	149.98	231	14.69	226	2.72
L20617-13-26	96.29	1775	1937	0.92	0.0511	0.0019	0.2499	0.0085	0.0359	0.0006	256	80.54	227	6.94	227	3.6
L20617-13-27	47.77	510	1056	0.48	0.0511	0.0022	0.2542	0.0093	0.0358	0.0004	256	98.14	230	7.53	227	2.48

The granitoids had high silica contents and low MgO, FeOT contents and La/Yb, Sr/Yb ratios (Figure 8), suggesting that they were mainly derived from the partial melting of crustal materials. The isotopic data provided good constraints on the source of the Taijiying granite porphyries (Table 3). They had low initial $^{87}Sr/^{86}Sr$ ratios (0.7031 to 0.7078), concentrated $\varepsilon_{Nd}(t)$ values (−1.2 to 1.7), positive zircon $\varepsilon_{Hf}(t)$ values (+3.5 to +10.7, mean = +6.7), and relatively low $\delta^{18}O$ values (6.29 to 7.84, mean = 6.85) (Figures 9 and 10), which indicate that mantle-derived material played a critical role. The positive $\varepsilon_{Hf}(t)$ values also suggest that the Taijiying granite porphyries have not experienced extensive mixing of an ancient continental crust endmember. However, there was some reworking of ancient crustal material in the source based on the negative $\varepsilon_{Nd}(t)$ values. The Taijiying rocks had similar T_{DM2} (0.87–1.09 Ga) and $T_{DM}{}^{C}$ (0.58–1.04 Ga) ages, which indicate that the sources were derived from the depleted mantle and evolved into basaltic low crust during the Neoproterozoic. The basalts were altered by Archean seawater which increased the zircon $\delta^{18}O$ values (Figure 10). The granitic magma rapidly ascended into shallow magma chambers by partial melting and finally formed the Taijiying granite porphyries. The plagioclase surrounding quartz phenocrysts indicate rapid emplacement and cooling of the magma (Figure 2d–f).

A previous study proposed that the Taijiying gold deposit is spatially and temporally related to these Triassic granite porphyries [7]. The magmatic hydrothermal fluids evolving from the granitic magmas provided not only metals and fluids but also the external heat to promote the gold-rich fluids upward, leading to the deposition with mixed features of the magmatic and meteoric fluids.

Table 2. Whole-rock geochemical data of granite porphyry in Taijiying.

Samples	L20617-1-1	L20617-2-1	L20617-3-1	L20617-4-1	L20617-5-1	L20617-7	L20617-8	L20617-9	L20617-10	L20617-11	L20617-12	L20617-13	L20617-14	L20617-15
Major elements (wt.%)														
SiO_2	73.9	73.4	75.3	75.9	76.7	76.1	69.0	76.2	75.3	77.0	77.6	75.7	76.6	77.2
TiO_2	0.12	0.11	0.11	0.11	0.09	0.10	0.10	0.10	0.11	0.10	0.10	0.11	0.10	0.10
Al_2O_3	12.5	12.5	12.6	12.6	12.3	12.2	11.4	12.3	12.6	12.5	12.2	12.6	12.3	12.5
Fe_2O_3	0.83	0.76	1.02	0.89	0.90	1.67	1.40	1.72	1.42	1.41	1.46	1.88	1.77	1.39
MnO	0.04	0.04	0.02	0.03	0.04	0.02	0.12	0.02	0.03	0.02	0.01	0.03	0.04	0.02
MgO	0.53	0.53	0.49	0.61	0.40	0.18	0.20	0.38	0.19	0.21	0.20	0.28	0.21	0.22
CaO	1.15	1.23	0.48	0.44	0.48	0.75	5.40	0.61	0.74	0.20	0.12	0.54	0.39	0.20
Na_2O	2.90	3.27	3.04	2.65	2.84	3.89	3.52	2.72	3.48	2.35	3.18	3.24	4.12	2.36
K_2O	4.55	4.96	4.30	4.25	4.67	4.04	3.92	4.15	4.19	4.63	4.44	4.56	3.82	4.58
P_2O_5	0.02	0.01	0.01	0.03	0.02	0.02	0.02	0.02	0.02	0.02	0.02	0.02	0.02	0.02
LOI	2.09	1.80	1.50	0.96	1.07	1.12	4.93	1.58	1.32	1.19	0.83	1.28	0.91	1.28
Total	99.60	99.63	99.59	99.26	100.41	100.10	100.06	99.72	99.36	99.55	100.16	100.13	100.35	99.86
FeO	1.03	0.96	0.69	0.79	0.86	0.90	0.72	0.92	0.60	0.65	0.64	1.12	0.97	0.64
FeOT	1.78	1.64	1.61	1.59	1.67	2.41	1.98	2.46	1.88	1.92	1.95	2.81	2.56	1.89
Mg#	38.48	40.34	38.99	44.57	33.44	13.62	17.57	24.29	17.32	18.48	17.83	17.50	14.91	19.85
K_2O/Na_2O	1.57	1.52	1.41	1.60	1.64	1.04	1.11	1.53	1.20	1.97	1.40	1.41	0.93	1.94
A/CNK	1.06	0.96	1.20	1.29	1.16	1.01	0.58	1.22	1.09	1.35	1.19	1.12	1.06	1.36
Tzr	787	774	809	804	801	786	702	800	784	809	805	903	906	924
Trace elements (ppm)														
Li						3.16	3.94	12.2	4.97	8.61	4.08	3.67	1.97	3.63
Be						1.79	2.20	2.43	2.31	2.89	2.02	1.68	1.83	1.79
Sc						1.52	1.42	1.28	1.64	1.34	1.29	1.39	1.52	1.28
V	8.31	8.14	8.29	12.3	6.50	1.98	1.87	2.56	1.72	2.72	2.47	5.72	1.46	6.25
Cr	5.81	6.61	5.63	30.1	8.56	42.9	8.06	6.28	8.24	10.5	5.53	16.9	7.56	15.1
Co	0.86	0.67	0.61	2.16	0.89	0.62	0.64	0.66	0.35	0.31	0.32	0.41	0.38	0.40
Ni	4.13	4.33	4.00	14.7	5.55	2.76	1.94	12.0	1.56	1.54	1.73	2.19	1.90	2.36
Cu						4.93	3.03	4.42	2.77	3.32	2.80	3.58	3.94	3.72
Zn						29.6	22.3	25.3	26.3	38.6	28.2	36.0	32.2	36.7
Ga						16.7	16.6	17.8	19.1	20.4	16.5	17.5	17.1	17.4
Rb	167	179	178	225	246	128	137	145	164	193	160	171	142	168
Sr	43.7	57.9	51.3	50.9	41.9	44.0	127	39.0	46.3	18.2	34.9	48.5	40.5	48.4
Y	23.6	19.9	17.8	12.5	16.3	18.2	18.6	16.7	18.6	17.5	18.1	18.8	19.0	18.6
Zr	166	181	175	97.0	112	170	157	155	171	152	168	174	182	170
Nb	35.3	32.0	33.2	21.5	32.1	25.7	25.4	27.2	26.7	25.3	26.9	26.8	27.7	26.5
Sn	3.36	3.36	3.13	1.97	2.96	2.60	2.44	2.56	2.90	3.29	2.60	2.89	2.84	2.98
Cs						1.37	1.51	1.74	3.04	3.92	1.60	1.75	2.02	1.75
Ba	117.0	105.0	138.0	150.0	117.0	120	94.5	175	82.7	107	122	100.0	52.5	98.1
La	53.7	50.9	51.0	22.8	18.7	46.0	49.6	43.5	48.4	47.0	46.7	46.8	48.8	46.8
Ce	118	102	102	42.2	46.7	94.8	101	94.4	101	97.5	97.5	96.3	100	96.9
Pr	13.2	11.1	11.2	5.04	4.64	10.2	10.9	10.3	10.8	10.5	10.6	10.5	10.9	10.4

Table 2. *Cont.*

Samples	L20617-1-1	L20617-2-1	L20617-3-1	L20617-4-1	L20617-5-1	L20617-7	L20617-8	L20617-9	L20617-10	L20617-11	L20617-12	L20617-13	L20617-14	L20617-15
Nd	42.7	36.6	36.3	16.9	15.9	34.5	37.7	35.5	36.7	35.9	36.3	34.7	36.5	34.9
Sm	7.29	6.21	5.98	3.18	3.25	6.13	6.49	6.26	6.24	6.36	6.70	6.45	6.39	6.16
Eu	0.22	0.27	0.29	0.39	0.24	0.23	0.25	0.24	0.23	0.25	0.24	0.23	0.23	0.25
Gd	6.62	5.60	5.50	2.71	2.81	4.04	4.31	4.00	4.24	4.21	4.26	4.13	4.32	4.22
Tb	0.88	0.75	0.69	0.38	0.45	0.61	0.65	0.59	0.63	0.61	0.68	0.61	0.67	0.63
Dy	4.43	3.69	3.39	2.18	2.85	3.49	3.44	3.09	3.44	3.26	3.44	3.50	3.54	3.44
Ho	0.79	0.65	0.59	0.40	0.53	0.60	0.61	0.58	0.65	0.60	0.61	0.63	0.66	0.61
Er	2.48	2.06	1.94	1.26	1.64	1.75	1.77	1.54	1.80	1.59	1.69	1.78	1.75	1.67
Tm	0.37	0.30	0.28	0. 20	0.28	0.25	0.25	0.24	0.26	0.25	0.25	0.25	0.26	0.26
Yb	2.43	2.06	1.89	1.37	1.78	1.63	1.62	1.59	1.70	1.68	1.63	1.70	1.69	1.67
Lu	0.40	0.33	0.29	0.22	0.29	0.24	0.23	0.23	0.25	0.24	0.25	0.25	0.25	0.25
Hf	4.42	4.56	4.44	2.40	3.06	5.49	4.92	5.12	5.49	5.21	5.33	5.62	5.57	5.59
Ta	2.86	2.83	2.86	2.30	2.82	2.06	1.95	2.16	2.19	2.10	2.13	2.02	2.15	2.01
Tl						0.79	0.88	0.95	1.07	1.14	1.09	1.06	0.91	1.08
Pb						10.9	7.79	6.77	9.03	7.18	10.3	11.8	10.5	11.5
Th	28.0	27.7	26.8	26.8	27.2	28.2	27.4	28.8	28.9	28.4	29.7	29.6	29.2	29.2
U	4.25	8.32	6.52	3.57	3.37	6.55	5.94	4.55	4.50	3.38	6.09	6.95	6.28	6.79
Th/U	6.58	3.32	4.10	7.51	8.06	4.30	4.62	6.33	6.42	8.39	4.87	4.26	4.66	4.30
Rb/Sr	3.82	3.10	3.46	4.42	5.88	2.92	1.08	3.73	3.55	10.63	4.57	3.52	3.50	3.48
Zr/Y	7.03	9.11	9.86	7.70	6.89	9.38	8.43	9.32	9.21	8.70	9.27	9.24	9.55	9.15
Sr/Y	1.85	2.91	2.88	4.07	2.57	2.42	6.82	2.34	2.49	1.04	1.93	2.58	2.13	2.61
ΣREE	253.62	222.28	221.31	99.28	100.09	204.44	218.76	202.06	216.38	209.95	210.89	207.89	216.40	208.11
LREE/HREE	12.79	13.40	14.21	10.38	8.43	15.22	15.97	16.04	15.68	15.88	15.48	15.19	15.47	15.34
La_N/Yb_N	15.8	17.7	19.4	11.9	7.6	20.3	22.0	19.6	20.4	20.0	20.6	19.8	20.8	20.1
δEu	0.10	0.14	0.15	0.40	0.24	0.14	0.14	0.14	0.14	0.15	0.14	0.14	0.13	0.15
δCe	1.06	1.00	1.00	0.92	1.19	1.07	1.07	1.09	1.08	1.08	1.07	1.06	1.07	1.08
$^{86}Rb/^{86}Sr$						8.672	4.815	12.647	12.437			11.543	12.801	
$^{87}Sr/^{86}Sr$						0.735249	0.720037	0.747575	0.748085			0.740880	0.749636	
2σ						8	6	9	9			10	10	
$(^{87}Sr/^{86}Sr)_i$						0.70688	0.70429	0.7062	0.7074			0.70312	0.70776	
$^{147}Sm/^{144}Nd$						0.1022	0.1976	0.1530	0.1079			0.1564	0.1840	
$^{143}Nd/^{144}Nd$						0.512582	0.51258	0.512588	0.512581			0.512583	0.512582	
2σ						5	5	5	7			5	5	
$\varepsilon_{Nd}(t)$						1.7	−1.2	0.37	1.5			0.12	−0.72	
T_{DM2}						870	1089	980	886			996	1060	
$f_{Sm/Nd}$						−0.48	−0.5	−0.48	−0.45			−0.51	−0.51	

Note: Mg# = (MgO/40.31)/(MgO/40.31 + FeOT/71.85 × 0.85) × 100; FeOT = FeO + 0.8998 × Fe_2O_3.

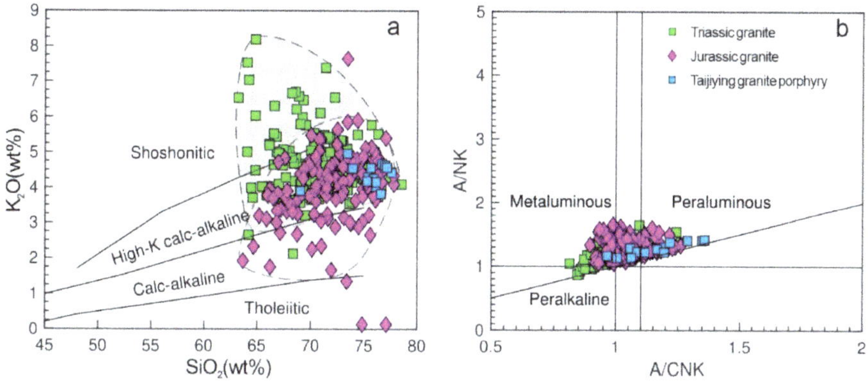

Figure 4. (**a**) Plots of K_2O vs. SiO_2 diagram; (**b**) A/NK (molar ratio $Al_2O_3/(Na_2O + K_2O)$) vs. A/CNK (molar ratio $Al_2O_3/(CaO + Na_2O + K_2O)$) diagram for granite porphyries from Chifeng–Chaoyang district ((**a**) is based on [37]).

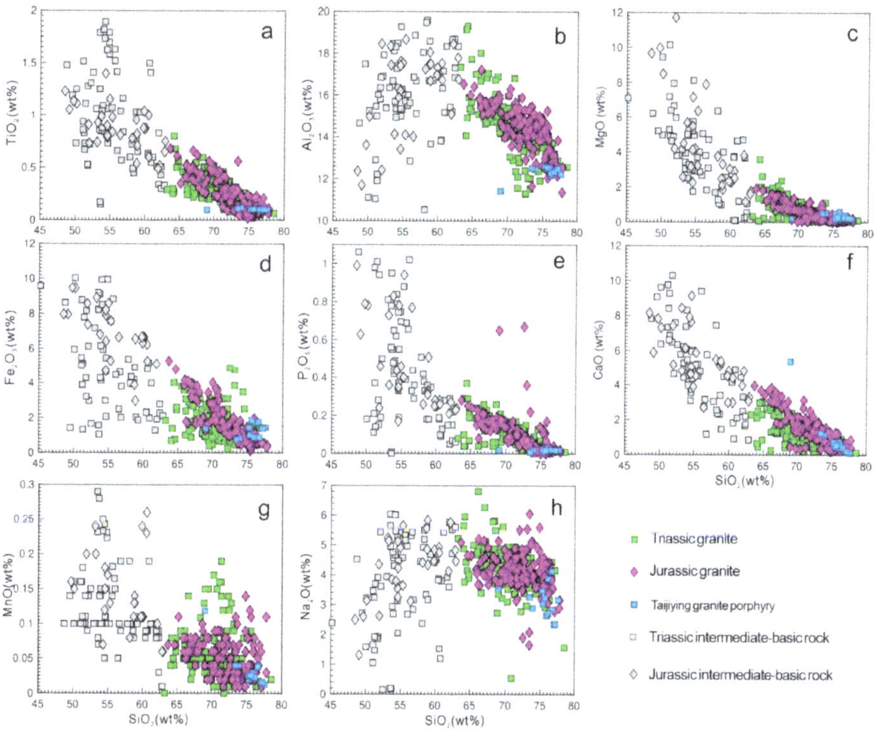

Figure 5. Harker diagrams for Chifeng–Chaoyang granites. (**a**) TiO_2 vs. SiO_2; (**b**) Al_2O_3 vs. SiO_2; (**c**) MgO vs. SiO_2; (**d**) Fe_2O_3 vs. SiO_2; (**e**) P_2O_5 vs. SiO_2; (**f**) CaO vs. SiO_2; (**g**) MnO vs. SiO_2; (**h**) Na_2O vs. SiO_2.

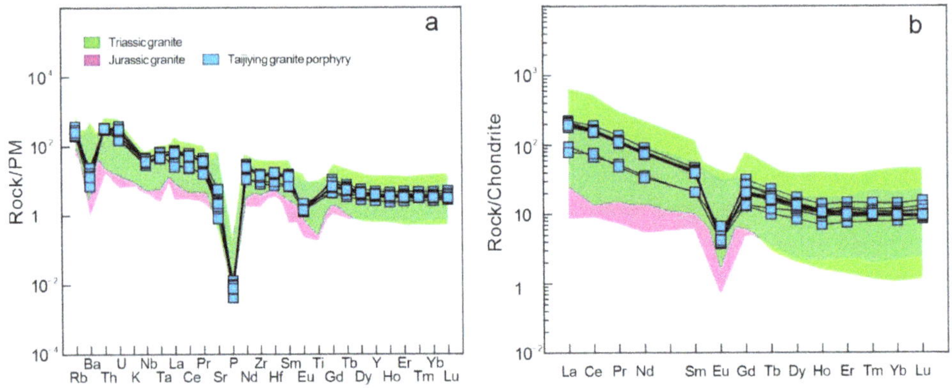

Figure 6. (**a**) Primitive, mantle-normalized trace element diagram for granite; (**b**) chondrite-normalized REE patterns diagram for granites. The chondrite values are from [38]. The primitive mantle values are from [39].

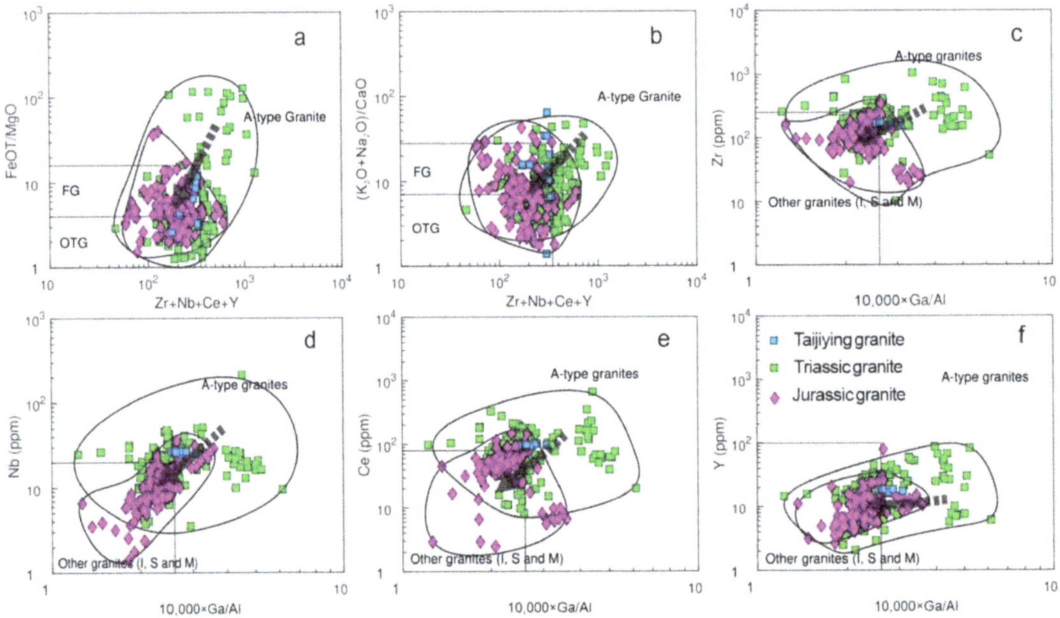

Figure 7. Classification diagrams for the Triassic and Jurassic granitic rocks. (**a**) FeOT/MgO vs. (Zr + Nb + Ce + Y), (**b**) (K_2O + Na_2O)/CaO vs. (Zr + Nb + Ce + Y), (**c**) Zr vs. 10,000 Ga/Al, (**d**) Nb vs. 10,000 Ga/Al, (**e**) Ce vs. 10,000 Ga/Al, and (**f**) Y vs. 10,000 Ga/Al classification diagrams [40], indicating that the Triassic granites are transitional between the I-, S-, M-, and A-types or highly fractionated, and Jurassic granites are mainly the I-, S-, M-, or unfractionated types, with few A-types or highly fractionated granites. A: A-type granite; I, S, and M: I-, S-, and M-type granite; FG: fractionated felsic granite; OGT: unfractionated M-, I-, and S-type granite.

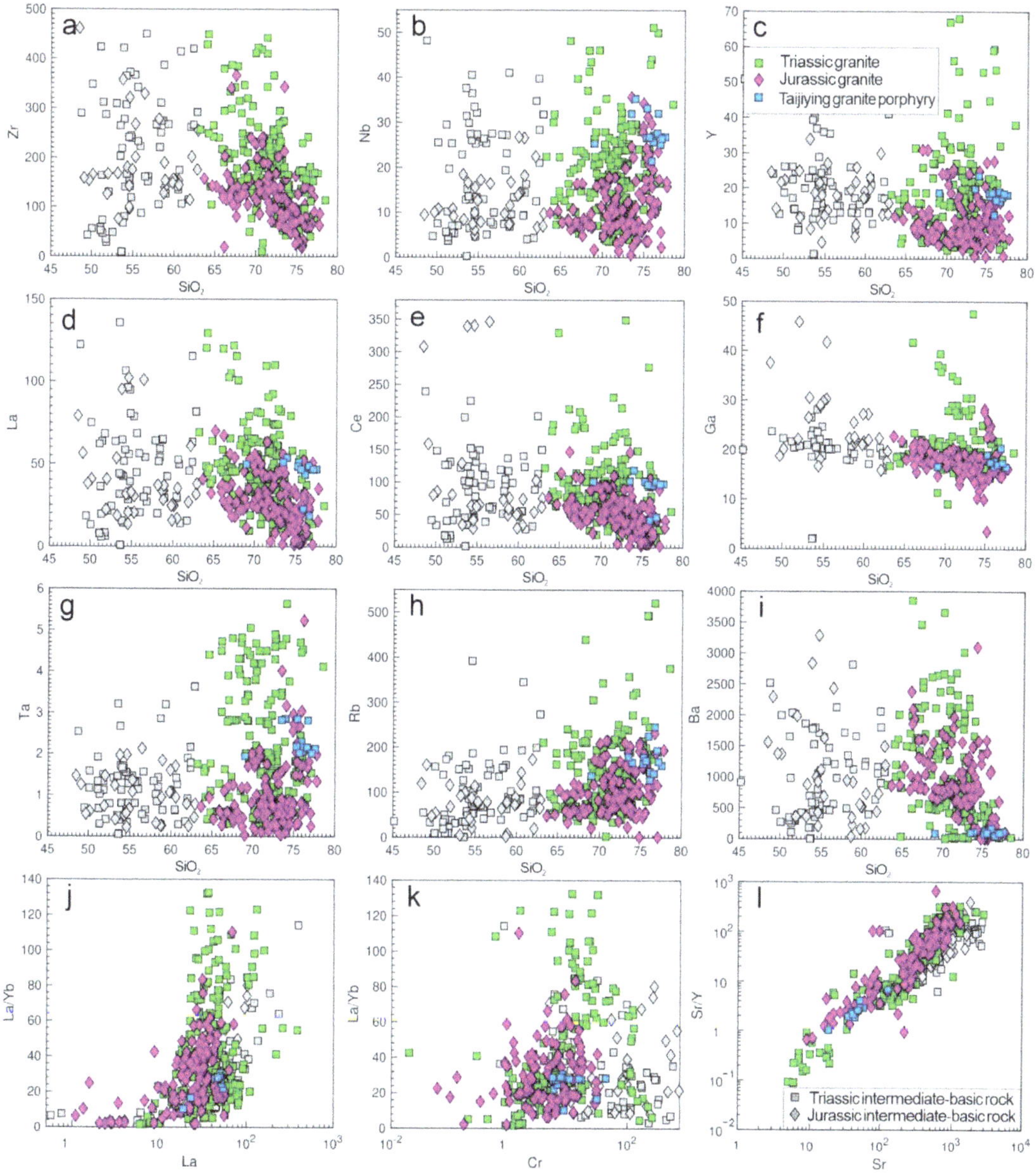

Figure 8. Elemental variation diagrams for Chifeng–Chaoyang district granites. (**a**) Zr vs. SiO_2; (**b**) Nb vs. SiO_2; (**c**) Y vs. SiO_2; (**d**) La vs. SiO_2; (**e**) Ce vs. SiO_2; (**f**) Ga vs. SiO_2; (**g**) Ta vs. SiO_2; (**h**) Rb vs. SiO_2; (**i**) Ba vs. SiO_2; (**j**) La/Yb vs. La; (**k**) La/Yb vs. Cr; (**l**) Sr/Y vs. Sr.

Table 3. Hf–O isotopic compositions of zircons from the granite porphyry samples of the Taijiying gold deposit, western Liaoning.

No.	Age (Ma)	^{176}Yb/^{177}Hf	^{176}Lu/^{177}Hf	^{176}Hf/^{177}Hf	2σ	$\varepsilon_{Hf}(0)$	$\varepsilon_{Hf}(t)$	T_{DM}	$T_{DM}{}^{c}$	$f_{Lu/Hf}$	δ^{18}O (‰)	±2σ
L20617-9-01	234	0.055540	0.001231	0.282824	0.000035	1.82	6.78	610	831	−0.96	6.69	0.15
L20617-9-02	234	0.088477	0.001983	0.282748	0.000035	−0.86	3.98	733	1011	−0.94	6.64	0.18
L20617-9-03	234	0.086536	0.001867	0.282834	0.000037	2.18	7.02	606	815	−0.94	6.73	0.13
L20617-9-04	234	0.101975	0.002190	0.282839	0.000034	2.38	7.20	603	805	−0.93	6.53	0.17
L20617-9-05	234	0.093398	0.001997	0.282854	0.000032	2.89	7.73	579	771	−0.94	6.77	0.26
L20617-9-06	234	0.068455	0.001532	0.282799	0.000034	0.94	5.86	651	891	−0.95	6.79	0.17
L20617-9-07	234	0.064702	0.001563	0.282820	0.000033	1.68	6.60	621	843	−0.95	6.76	0.19
L20617-9-08	234	0.047334	0.001183	0.282765	0.000029	−0.24	4.72	693	964	−0.96	6.96	0.07
L20617-9-09	234	0.056710	0.001461	0.282778	0.000034	0.21	5.15	679	936	−0.96	6.41	0.12
L20617-9-10	234	0.091339	0.002086	0.282878	0.000030	3.73	8.55	546	718	−0.94	6.55	0.23
L20617-9-11	234	0.074405	0.001791	0.282834	0.000035	2.20	7.09	604	813	−0.95	6.29	0.17
L20617-9-12	234	0.075216	0.001717	0.282837	0.000032	2.28	7.17	600	808	−0.95	6.87	0.14
L20617-9-13	234	0.056767	0.001338	0.282772	0.000035	−0.01	4.94	686	950	−0.96	6.44	0.20
L20617-9-14	234	0.050499	0.001185	0.282802	0.000029	1.07	6.03	640	880	−0.96	6.67	0.14
L20617-9-15	234	0.074764	0.001710	0.282915	0.000035	5.06	9.96	486	629	−0.95	6.71	0.21
L20617-13-01	230	0.074041	0.001555	0.282828	0.000030	1.97	6.79	610	828	−0.95	6.85	0.14
L20617-13-02	230	0.109095	0.002230	0.282816	0.000034	1.56	6.30	638	861	−0.93	7.49	0.12
L20617-13-03	230	0.093436	0.001965	0.282899	0.000032	4.49	9.27	513	671	−0.94	7.54	0.16
L20617-13-04	230	0.061286	0.001349	0.282798	0.000032	0.92	5.77	649	893	−0.96	6.53	0.17
L20617-13-05	230	0.112168	0.002389	0.282877	0.000030	3.71	8.42	551	724	−0.93	7.84	0.18
L20617-13-06	230	0.106823	0.002312	0.282941	0.000036	5.97	10.69	457	580	−0.93	7.51	0.20
L20617-13-07	230	0.098713	0.002224	0.282785	0.000035	0.47	5.20	683	930	−0.93	7.03	0.25
L20617-13-08	230	0.068805	0.001689	0.282810	0.000031	1.34	6.16	637	869	−0.95	6.76	0.12
L20617-13-09	230	0.059943	0.001473	0.282866	0.000032	3.32	8.17	554	741	−0.96	7.67	0.20
L20617-13-10	230	0.062086	0.001544	0.282734	0.000044	−1.35	3.47	744	1040	−0.95	6.93	0.21
L20617-13-11	230	0.078422	0.001752	0.282831	0.000034	2.07	6.86	609	824	−0.95	6.79	0.26
L20617-13-12	230	0.071436	0.001551	0.282826	0.000033	1.91	6.72	612	832	−0.95	6.59	0.16
L20617-13-13	230	0.050549	0.001240	0.282811	0.000034	1.39	6.26	628	862	−0.96	6.95	0.11
L20617-13-14	230	0.071913	0.001660	0.282803	0.000035	1.10	5.91	647	885	−0.95	6.86	0.31
L20617-13-15	230	0.058868	0.001305	0.282798	0.000033	0.93	5.80	648	893	−0.96	6.39	0.13

Note: $\varepsilon_{Hf}(t) = 10{,}000 \times \{[(^{176}Hf/^{177}Hf)_S - (^{176}Lu/^{177}Hf)_S \times (e^{\lambda t} - 1)]/[(^{176}Hf/^{177}Hf)_{CHUR,0} - (^{176}Lu/^{177}Hf)_{CHUR} \times (e^{\lambda t} - 1)] - 1\}$; $T_{DM} = 1/\lambda \times \ln\{1 + [(^{176}Hf/^{177}Hf)_S - (^{176}Hf/^{177}Hf)_{DM}]/[(^{176}Lu/^{177}Hf)_S - (^{176}Lu/^{177}Hf)_{DM}]\}$; $T^{C}_{DM} = T_{DM1} - (T_{DM1} - t) \times [(fcc - fs)/(fcc - f_{DM})]$, $f_{Lu/Hf} = (^{176}Lu/^{177}Hf)_S/(^{176}Lu/^{177}Hf)_{CHUR} - 1$, where $\lambda = 1.867 \times 10^{-11}$ year^{-1} (Soderlund et al., 2004); $(^{176}Lu/^{177}Hf)_S$ and $(^{176}Hf/^{177}Hf)_S$ are the measured values of the samples; $(^{176}Lu/^{177}Hf)_{CHUR} = 0.0332$ and $(^{176}Hf/^{177}Hf)_{CHUR,0} = 0.282772$ (Blichert-Toft and Albarede, 1997); $(^{176}Lu/^{177}Hf)_{DM} = 0.0384$ and $(^{176}Hf/^{177}Hf)_{DM} = 0.28325$ (Griffin et al., 2000); $(^{176}Lu/^{177}Hf)_{mean\ crust} = 0.015$; $fcc = [(^{176}Lu/^{177}Hf)_{mean\ crust}/(^{176}Lu/^{177}Hf)_{CHUR}] - 1$; $fs = f_{Lu/Hf}$; $f_{DM} = [(^{176}Lu/^{177}Hf)_{DM}/(^{176}Lu/^{177}Hf)_{CHUR}] - 1$; t = crystallization time of zircon.

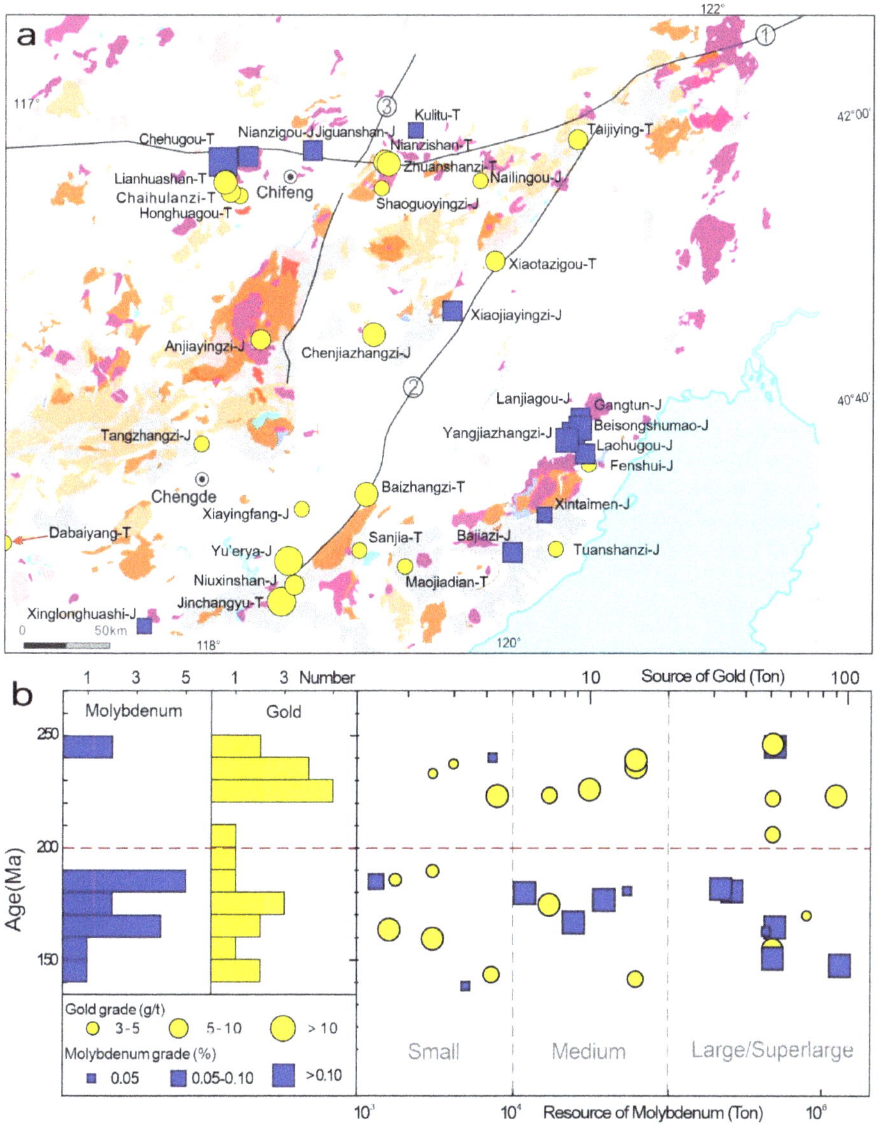

Figure 9. (**a**) Outline map of the Chifeng–Chaoyang district showing distribution of the Mesozoic gold and molybdenum deposits. ①—Chifeng–Kaiyuan fault; ②—Lingyuan–Beipiao fault; ③—Honghan–Balihan fault; (**b**) The periodical pattern of Triassic and Jurassic gold and molybdenum deposits.

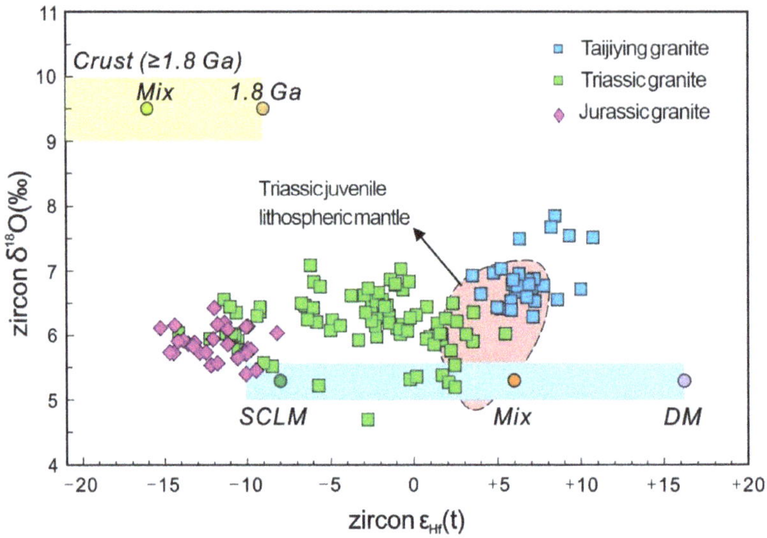

Figure 10. Zircon $\delta^{18}O$ versus $\varepsilon_{Hf}(t)$ of Triassic granite porphyries in Taijiying gold deposit. The εHf(t) values of depleted mantle, the SCLM beneath the NCC, mixed mantle, and ancient crust are +16.2, −8 [33], +6, and −16, respectively. The $\delta^{18}O$ of ancient crust and mantle are +9.5 ± 0.5% (average of 2.5 Ga and 1.8 Ga crust) and + 5.3 ± 0.3‰ [41]. Age for calculation depends on the samples. Triassic spots from [8]. Jurassic spots from [42]. Triassic juvenile lithospheric mantles are from [33].

4. Geochemical Characteristics of Triassic and Jurassic Granitoids

The Chifeng–Chaoyang district includes Kalaqin, Lunu'erhu, Haitangshan–Yiwulushan, Xingcheng, and Dushan in the northeastern NCC, which produced large volumes of Triassic and Jurassic intrusions. The metals and fluids of orogenic gold deposits and porphyry molybdenum deposits all came from felsic magma which became granites. There are close relationships between granites and deposits. This paper compiles the data from the published literature for 208 Triassic–Jurassic granitoids related to these deposits (Supplementary Table S1). The data set contains whole-rock major and trace elements for 352 samples, Sr–Nd isotopic data for 113 samples, zircon Hf isotopic data for 36 samples (709 points), and O isotopic data for 10 samples (147 points). These data are presented in Supplementary Tables S2 and S3.

4.1. Triassic Granitoids

The compiled data for the Triassic granitic rocks exhibit a wide range of major element concentrations. They have high SiO_2 (63.2–78.6 wt%) and Al_2O_3 (11.3–19.3 wt%) contents and low MgO (0.03–2.62 wt%), Fe_2O_3 (0.24–4.88 wt%), CaO (0.14–3.05 wt%), Na2O (0.54–6.80 wt%), K_2O (2.13–8.18 wt%), and Na_2O/K_2O (mean value = 0.9, mostly < 1) contents. They all have low Mg#, with a mean value of 32.26. Most samples are high-K calc-alkaline series with a small portion plotted in the shoshonitic field in the SiO_2 vs. K_2O classification diagram (Figure 4a) [43]. Most samples are metaluminous or weakly peraluminous in the A/NK vs. A/CNK diagram (Figure 4b). In the Harker diagrams, TiO_2, Al_2O_3, MgO, Fe_2O_3T, P_2O_5, and CaO decrease with increasing SiO_2, but MnO and Na_2O exhibit ambiguous correlations with SiO_2 (Figure 5).

The Triassic granites exhibit low Sr/Y ratios (mean = 53.7) and Sr contents (mean= 402 ppm) (Supplementary Table S2). They are characterized by high La/Yb ratios (1.6–132.5) and Nb (3.56–213 ppm), Cr (0.02–271 ppm), Y (2.07–87.0 ppm), Zr (10.17–1060 ppm), and Th (0.90–1233 ppm) contents. In the diagrams (Figure 8), the Sr, Zr, and Ba contents tend to decrease with increasing SiO_2, while the Nb and Y contents are not correlated with SiO_2.

The Triassic granitoids have variable total rare earth element (REE) contents, ranging from 34.0 to 1048 ppm (mean = 195). The Kalaqin batholiths monzogranite (06JH47R) has the highest REE values [8]. Except for a few samples with positive Eu anomalies (δEu = 3.37–1.00; e.g., Dushan, Nanhouding, and Hekanzi) [32,44], they mainly exhibit slight-to-strong LREE enrichment (La_N/Yb_N = 1.1–95.1) in chondrite-normalized REE patterns and have significant, negative Eu anomalies with δEu values of 0.04–0.99 (mean = 0.72) (Figure 6b). In the primitive, mantle-normalized diagrams (Figure 6a), most of the samples, including the Taijiying granite porphyry, exhibit Ba, Sr, Nb, Ta, P, and Ti depletions and Th and U enrichments.

According to their relatively high Zr, Nb, Y, La, Ce, and Ga contents (Figure 8a–f), the majority of the Triassic samples are likely A-type granites. In the FeOT/MgO and (K_2O + Na_2O)/CaO vs. (Zr + Nb + Ce + Y) diagrams (Figure 7a,b), most of the Triassic samples are plotted in the A-type granite field or near the boundary between the A-type and fractionated/unfractionated granite fields. Furthermore, the K_2O + Na_2O, molecular [K_2O + Na_2O]/Al_2O_3, Nb, and Zr vs. 10,000 Ga/Al classification diagrams also show that the samples straddle the transitional fields between A-type and M-, I-, and S-type granites (Figure 7c–f) [45]. Therefore, the Triassic granites consist of both I-type and A-type granites.

4.2. *Jurassic Granitoids*

Compared to the Triassic granitoids, the Jurassic samples exhibited similar SiO_2 contents (63.7–77.8 wt%), a wide Al_2O_3 range (11.4–17.2 wt%), high CaO contents (0.22–3.97 wt%), and low Fe_2O_3T (0.12–5.28 wt%), MgO (0.02–1.92 wt%), and Na_2O + K_2O (0.27–5.70 wt%) contents. The Na_2O/K_2O ratios ranged from 0.3 to 5.7 (mostly > 1) (Supplementary Table S2). According to Peccerillo and Taylor's (1976) classification, most of the samples were the high-K calc-alkaline series [44]. A few of the granites were calc-alkaline granites (Figure 4a), which are different from the Triassic granites. In the A/NK vs. A/CNK diagram (Figure 4b), more Jurassic samples exhibit peraluminous characteristics than Triassic samples. In the Harker diagrams (Figure 5), TiO_2, Al_2O_3, MgO, Fe_2O_3T, CaO, and P_2O_5 decrease with increasing SiO_2. The correlation between P_2O_5 and SiO_2 indicates that the Jurassic granites are not S-type granites. The Jurassic plutons also exhibited low and variable total REE contents, ranging from 6.46 to 375 ppm (mean = 122 ppm). Compared to the others, the samples from Yiwulüshan and Gaoliduntai had extremely low total REE contents, ranging from 6.46 to 20.4 ppm [40,44,46]. They had similar chondrite-normalized REE patterns and moderate LREE enrichment, with La_N/Yb_N ratios of 1.5 to 79.1 (mean = 19.7) and variable Eu anomalies (0.04–4.63, mostly = 0.80–1.00) (Figure 6b).

The Jurassic granitoids have relatively high Sr/Y ratios (0.7–665.5) and lower La/Yb ratios (2.1–110.3) (Figure 8j–l). They are characterized by low Nb (0.65–88 ppm), Ta (0.03–10.4 ppm), Zr (19.4–366 ppm), Y (0.75–225 ppm), and Cr (0.06–30.5 ppm) contents (Supplementary Table S2). In the primitive, mantle-normalized diagrams (Figure 6a), most of the samples are depleted in Ba, Nb, Sr, P, Zr, and Ti and enriched in U, Th, and Nd. In Figure 7, most samples are plotted in the I-type granite field, which indicates that they do not have the distinctive features of A-type granites. However, some of the granites emplaced in the Xingcheng, Dushan, and Haitangshan MCCs have high Na_2O/K_2O ratios (>1), Sr contents (>400 ppm), and Sr/Y (>20.0–40.0) and La/Yb (>7.6–15.0) ratios. These granitoids also have low Y (<18 ppm) and Yb (<1.9 ppm) contents and weak Eu anomalies. These geochemical features indicate that they are adakitic granites [47]. Based on the lithologies and geochemical features, most of the Jurassic granitoids belong to I-type granites with a little portion of adakitic granites [40].

5. Characteristics of Sr–Nd–Hf–O Isotopes

5.1. *Sr–Nd Isotope*

Whole-rock Sr–Nd isotopes and zircon Hf are important tools for tracing the evolution of the lithosphere and crust [48]. T_{DM2} and T_{DM}^C are used to restrict when the magmatic source is derived from the depleted mantle [49].

The Sr and Nd isotope ratios of Triassic to Jurassic granitoids in the study area are listed in Supplementary Table S2. For all collected samples from the northern NCC, the $\varepsilon_{Nd}(t)$ values exhibited a wide range from −17.2 to 0.65 (mean = −8.5), and the initial $^{87}Sr/^{86}Sr$ ratios had relatively low values (0.7010–0.7098, mostly = 0.7050), excluding the extremely low values of 0.6840 and 0.6992 of the Kalaqin batholiths [8] and Hekanzi syenite [32]. The Nd model ages (T_{DM}) plutons were 1.02–2.02 Ga with a mean value of 1.47 Ga.

The Jurassic granitoids are characterized by strongly negative $\varepsilon_{Nd}(t)$ values from −24.9 to −3.8 (mostly < −15.0) and a wide range of initial $^{87}Sr/^{86}Sr$ ratios from 0.7032 to 0.7215 (mean value = 0.7076). Although there were several plutons with a younger age (T_{DM}) in the region, such as Dayingzi (0.87–0.90 Ga) and Nianzigou (0.75–0.90 Ga) [44], contrasting with Triassic ones, these Jurassic intrusions had much older ages (T_{DM}), from 1.31 Ga to 2.96 Ga (mean age = 2.17 Ga) (Supplementary Table S2).

5.2. Zircon Hf–O Isotope

The compiled Triassic granitoids zircon Hf isotopes are characterized by relatively low $^{176}Hf/^{177}Hf$ ratios (0.281702 to 0.282933), $\varepsilon_{Hf}(t)$ values (−33.1 to 10.2, mean value = −5.6), and older Hf isotopic crustal model ages ($T_{DM}{}^{C}$), 0.60–3.32 Ga (mean = 1.57 Ga). The $\delta^{18}O$ values varied from 4.69‰ to 7.84‰ (mean = 6.44‰). The zircon Hf isotopes of the Jurassic granitoids exhibited the lowest $^{176}Hf/^{177}Hf$ ratios, from 0.281193 to 0.282801, $\varepsilon_{Hf}(t)$ values of −34.5 to 4.7 (mostly = −12.8), and $T_{DM}{}^{C}$ of 0.94 Ga to 3.36 Ga (mean = 2.11 Ga), respectively. The $\delta^{18}O$ values varied from 5.40‰ to 6.70‰ (mean = 5.96‰) (Supplementary Table S3). The $\varepsilon_{Hf}(t)$ values gradually decreased with the ages when the $T_{DM}{}^{C}$ aged from Triassic to Jurassic (Figures 10 and 11c). These features indicate that the Jurassic granitic sources had more Precambrian basement rocks involved in the process. The high $\delta^{18}O$ values of Triassic granites may be the result of basalts altered by Archean seawater [50,51].

6. Age and Types of Metallic Metallogeny

The gold and molybdenum mineralization represents the primary metallogeny in this district. More than 100 gold deposits have been discovered, which were produced in Meso-Neoarchean metamorphic rocks or Mesozoic intrusions. The large–medium gold deposits are found in Chaihulanzi, Honghuagou, Anjiayingzi, Zhuanshanzi, and Xiaotazigou, from west to east. The superlarge/large molybdenum deposits are distributed near the Xingcheng area and Chifeng City, which includes Lanjiagou, Yangjiazhangzi, Xintaimen and Chehugou [52–54]. The specific information can be seen in Supplementary Table S4.

It seems to be a pattern that gold mineralization mainly took place during the Triassic, and molybdenum deposits concentrated in the Jurassic. There were two peaks of gold mineralization in the Triassic and Jurassic: 250–230 Ma and 160–180 Ma (Figure 9). Triassic gold deposits are characterized by a higher grade and larger scale than those from the Jurassic. Twelve Triassic gold deposits exhibit high grade (ca. 5–30 g/t, mostly > 10 g/t) and large reserve scales (mostly > 5 t). For instance, the Jinchangyu orogenic gold deposit has 90 tonnes of gold reserves with a high grade of 14.6 g/t [55]. In contrast, Jurassic gold deposits are relatively small in reserve scale and lower in grade, which indicates that more intensive gold mineralization occurred in the Triassic. On the other hand, this study area is also called the Yan–Liao Mo belt and is one of the six Mo mineralization provinces in China [56]. Most large Mo deposits at the northern margins of the NCC are related to continental collision orogeny, including syn-collision to post-collision tectonism [56]. Wang et al. (2014) reported large–medium Mo polymetallic deposits in China mostly produced during 180–100 Ma and a few of deposits from the Indosinian [57]. In this area, Triassic Mo mineralization is represented by just one small-scale Kulitu and the superlarge-scale Chehugou deposits with a grade of 0.09% and 0.14%, respectively, along the north side of the Chifeng–Kaiyuan fault [58,59]. In contrast, Jurassic Mo mineralization is characterized by more deposits, larger reserve scale, and higher ore grade (Figure 9 and Supplementary Table S4). For example, the porphyry-type Lanjiagou (182 ± 7 Ma) Mo deposit has been estimated to be a reserve of molybdenum metal measuring more than 404,000 tons with an average grade of 0.13% Mo [52]. The skarn-type Yangjiazhangzi Mo deposit (181 ± 1 Ma) near the Lanjiagou deposit contains 261,800 tons of Mo with an average grade of 0.14% [54]. To the north, the Xiaojiayingzi Mo (Fe) deposit, formed in 166 ± 5 Ma, shows the features of high-grade Mo (0.29%) and is large in scale (105,000 tons) [57]. The Nianzigou Mo deposit is a quartz-vein-type Mo deposit which has 25,000 tons of Mo with an average grade of 0.31%, ranking as a middle-sized deposit [59]. The host rock porphyritic monzogranite and its associated Mo mineralization occurred at 153 ± 5 Ma [60]. The Sr–Nd isotopic compositions and Re contents in molybdenite suggest a crust-dominated source, with a minor contribution from the mantle [61]. The geochemical features of the Chehugou Mo deposit also suggest that the ore-related granitoids were derived from ancient, garnet-bearing crustal rocks related to subduction of the Palaeo-Asian Ocean and subsequent continent–continent collision between the NCC and Siberian plates. It indicates that the mineralization intensities of gold and molybdenum were different in the different periods, i.e., most gold was deposited in the Triassic, while molybdenum deposits mainly took place in the Jurassic.

7. Discussion

7.1. Transition of the Granitoid Sources from the Triassic to the Jurassic

The Triassic granitoids in the northern part of the NCC mainly consist of leucogranite, monzogranite, biotite granite, and syenogranite [52]. They belong to I-type or A-type granites, which have relatively high SiO_2, K_2O, and $K_2O + Na_2O$, low Sr contents and Mg# values, strong LREE enrichments, negative δEu values, and Ba, Nb, Ta, Sr, P, and Ti depletions. The sources are most likely to be the partial melting of crustal materials. The compiled Triassic granites had much lower negative $\varepsilon_{Nd}(t)$ and $\varepsilon_{Hf}(t)$ values and older Nd and Hf model ages (T_{DM2} and $T_{DM}{}^{C}$). In the $\varepsilon_{Nd}(t)$ vs. $(^{87}Sr/^{86}Sr)i$ diagram (Figure 11a), a portion of the granitoids overlap with or are plotted above the Ordovician kimberlites and mantle xenoliths in the eastern part of the NCC. Most of the samples plotted in the area are between the Ordovician kimberlites, the mantle xenoliths, and the Precambrian basement rocks of the NCC, but they are much closer to the former. This scenario indicates that the initial magma sources are mainly derived from the lithospheric mantle materials, mixed with a portion of ancient crustal components. Furthermore, the granitoids exhibit the same distribution pattern as the Triassic intermediate mafic rocks, which also support this conclusion. The $\varepsilon_{Nd}(t)$ and $\varepsilon_{Hf}(t)$ vs. crystallization age plots also indicate similar source features (Figure 11b,c).

The proposed mantle sources of the Triassic granitoids are also supported by many previous studies, including those on Guangtoushan alkaline granite [30], the Jianping monzogranite and syenogranite [62], the Ping'andi ferroan granitoid [15], and the Jinchanggouliang shoshonitic dikes [63]. The Triassic Jianping granites (237 Ma) exhibit $\varepsilon_{Nd}(t)$ values of −10.6 to −9.1 and higher $(^{87}Sr/^{86}Sr)i$ values, which may indicate more assimilation of crustal materials. The Late Triassic Guangtoushan alkaline–peralkaline granites (220 ± 1 Ma, $\varepsilon_{Nd}(t)$ = −9.9 to −8.6) are considered to have been derived from the enriched sub-continental lithospheric mantle [30]. The integrated geological evidence suggests that the Jinchanggouliang shoshonitic dikes were derived from the mixing of the lower crust, the lithospheric mantle of the NCC, and the ascending asthenospheric melt in a post-orogenic extensional geodynamic setting [64]. To the west, the granitoids of the Kalaqin batholith are the products of the mixing of melts that were mainly derived from three end-member sources, the ≥1.8 Ga ancient crust, the lithospheric mantle beneath the NCC, and a likely juvenile endmember formed by the mixing of melts from the lithospheric mantle and depleted mantle. The possible percentage of the crustal and mantle sources based on zircon Hf–O isotopes is 10%–40% ancient-crust-derived magma [8]. In addition, a more juvenile endmember or lithospheric mantle played a major role [30]. The Baizhangzi adakitic granites (233 Ma) were derived from the lower continental crust that was modified by basaltic underplating during the early Mesozoic, which led to the formation of juvenile crustal materials [27]. The sources of the A-type granites are mainly juvenile, mantle-derived components and a small portion of recycled ancient crust. Generally, regardless of the specific types of Triassic intrusive, the mixing of juvenile crustal melts derived from the lithospheric mantle and the depleted mantle was likely the major source. The few percentages of ancient crust materials led to ($^{87}Sr/^{86}Sr$)i values increasing in a narrow range compared to the percentages of Precambrian basement rocks (Figure 11a).

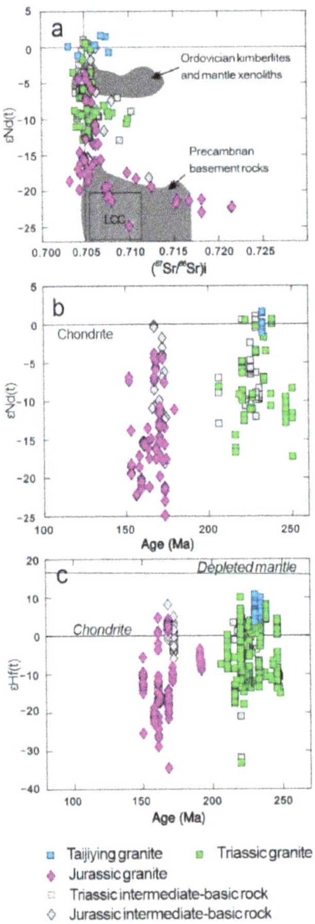

Figure 11. (**a**) Whole-rock εNd(t) vs. ($^{87}Sr/^{86}Sr$)i; (**b**) εNd(t) vs. age (Ma) diagram; (**c**) εHf(t) vs. age (Ma) diagram. Sr–Nd isotopic compositions of the Precambrian basement rocks in northern North China Craton (calculated at 310 Ma) are from [18,65], and those of the Ordovician kimberlites and mantle xenoliths (calculated at 310 Ma) in eastern North China Craton are from [63,66]. The compositions of the lower continental crust (LCC) based on [67] are also plotted.

Jurassic magmatic rocks mainly consist of granite, monzogranite with minor leucogranite, granodiorite, gabbro, lamprophyre, tonalite, and mafic microgranular enclaves (MMEs). Most of the Jurassic intrusive rocks are high-K calc-alkaline or calc-alkaline (Figure 4a). The intrusive rocks are mostly characterized by LREE enrichment, small negative or positive Eu anomalies, high Al_2O_3, Na_2O, Na_2O/K_2O, Sr, and Ba contents, and low MgO, Y, Yb, and HFSE (Nb, Ta, Zr, Hf) contents. Their adakitic features suggest that olivine and plagioclase were not residual phases in the source region, but garnet, amphibole, and pyroxene were [68]. The Jurassic plutons mostly exhibited Ba, Nb, Sr, P, Zr, and Ti depletions and U, Th, and Nd enrichments, indicating a source containing residual eclogite. The Nd/Th ratios (mean = 3.3) and Ti/Zr ratios (mean = 12.4) were closer to the values of 3 and <30 for continental crust, respectively, indicating that the genesis of granitoids also involved crustal melting. The partial melting of the thickened lower crust has been reported to be the main source of some Jurassic intrusive rocks [68–70]. The Archean–Paleoproterozoic inherited ~2.5 Ga zircons, also indicating the involvement of compositions of more ancient crustal materials, such as the Lüshan batholiths (2451 ± 12 Ma) and Shishan monzogranite (2455 ± 32 Ma) [3,66]. Based on these features, the Jurassic granitoid suit was mainly produced by the remelting of ancient mafic lower crust [24]. The presence of gabbro, lamprophyre, tonalite, and mafic microgranular enclaves (MMEs) might suggest that an enriched mantle endmember was involved.

The significant negative $\varepsilon_{Nd}(t)$ values and various initial $^{87}Sr/^{86}Sr$ ratios of the Jurassic granitoids are consistent with their derivation from ancient lower crust (Figure 11a). A portion of the samples had high negative $\varepsilon_{Nd}(t)$ values, which were similar to the values of the Ordovician kimberlites and mantle xenoliths and the coeval mantle-derived rocks (e.g., gabbro and lamprophyre). However, most of the samples were similar to the Precambrian basement rocks of the NCC. This scenario reflects that the parental magma may be more complicated, including ancient lower crust and minor juvenile lithospheric mantle components. Most of the samples had an old mean T_{DM2} age of 2.17 Ga and a mean $T_{DM}{}^{C}$ age 2.11 Ga, which required an older endmember. The Archean–Paleoproterozoic basement rocks are an optimal component. However, the crustal endmember was probably recently underplated by the Jurassic magmatism, which may, alternatively, indicate mixing with the ancient, enriched lithospheric mantle. The significant negative zircon Hf isotopic compositions also support the proposed Archean lower-crust-extracted felsic magma sources (Figures 10 and 11c). In addition, the slightly higher zircon $\delta^{18}O$ of the Jurassic Shuihutong and Baijiadian plutons is also very similar to the values of many other NCC granitoids, which are also believed to have been derived from a lower crustal source [66]. The positive $\varepsilon_{Hf}(t)$ values (Lüshan granite, LB64) may represent the involvement of the Paleo-Pacific oceanic subducted slab or underplating by the depleted mantle [68]. The basaltic melts evolved into felsic magmas with minor ancient crustal contamination and then the felsic magmas formed granites with positive $\varepsilon_{Hf}(t)$ values.

7.2. Relationship between the Sources and the Tectonic Setting

The Triassic granitoids were emplaced after the final amalgamation and suturing of the CAOB with the NCC [62,69]. In the Early Triassic, the high-K calc-alkaline granites (HKCA), including the monzogranite, K-feldspar granite and minor monzonite, were the dominant type of batholith in the Chifeng–Chaoyang district, which is one of the important characteristics of post-collisional magmatic activities [70–72]. In the final stages of orogeny, the HKCA often shift to shoshonitic or alkaline–peralkaline compositions (nepheline syenite, aegirine-augite syenite, pyroxene syenite, quartz syenite, syenite, and alkaline granite) [73,74], which is consistent with the regional lithological associations of the Middle–Late Triassic rocks (Figure 4a). The Triassic Fanshan alkaline ultramafic–syenite complex [75], Late Triassic Guangtoushan alkaline–peralkaline granites [28], and other alkaline rocks [76] support the hypothesis that the magmatic rocks were derived from the melting of an enriched mantle and post-collisional lithospheric extension along the northern margin of the NCC (Figure 12a). Late Triassic to Early Jurassic extensional deformation and large-scale gravity sliding have been identified in western Liaoning Province [76], which represent crustal uplift and rapid exhumation of the crustal rocks.

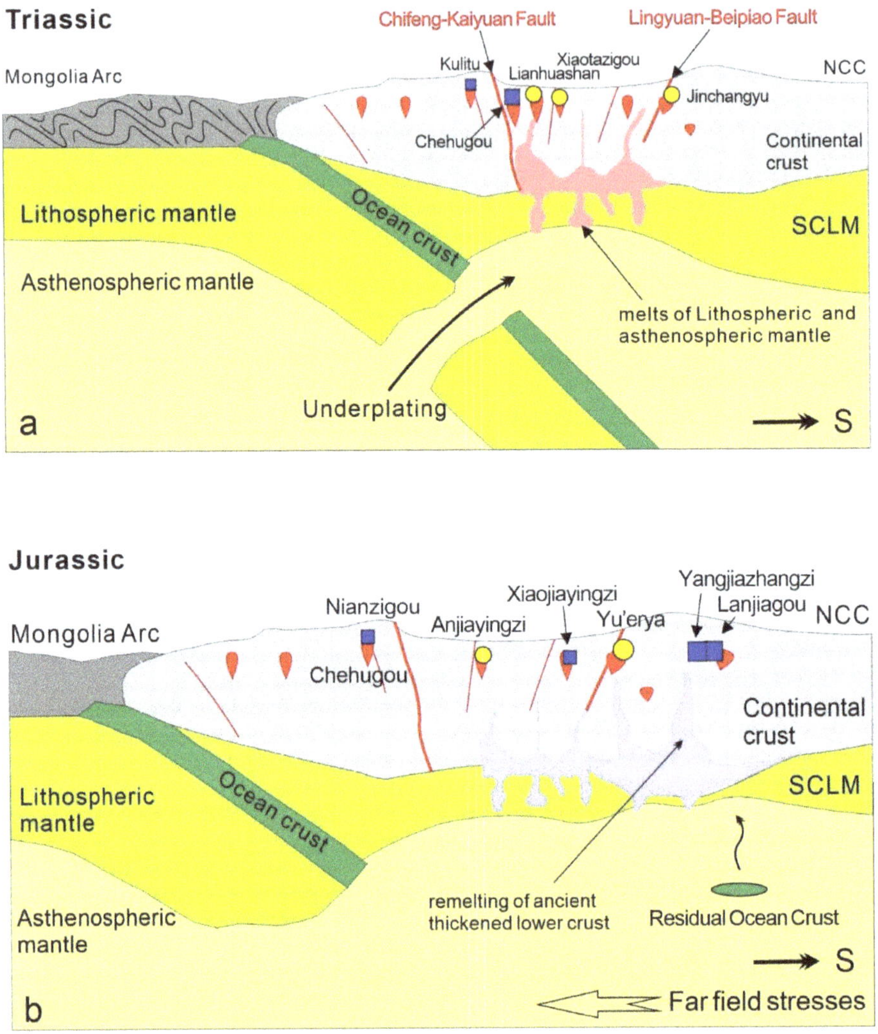

Figure 12. Outline maps of Triassic and Jurassic magmatic source and mineralization. (**a**) the metallogenic pattern and tectonic setting of study area in Triassic; (**b**) the metallogenic pattern and tectonic setting of study area in Jurassic.

The Jurassic lithological associations consist of adakitic granite (high-Sr granite), strongly peraluminous leucogranite, high-K calc-alkaline granite, and minor A-type granite (e.g., the Nianzigou pluton). The 42 ages of adakitic granite suggest that it was mainly formed during the late Indosinian to late Yanshanian (229–114 Ma), and ca. 65% of the granites were emplaced in the Jurassic. Adakitic granites are characterized by Sr enrichment, strong HREE depletion, and slightly negative δEu values. These characteristics indicate an adakitic magmatic source generated via the partial melting of crustal rocks under high-temperature (>850–900 °C) and high-pressure (≥1.5 GPa) conditions, and the original melts lacked fractional crystallization of feldspar [44]. Furthermore, high pressures indicate that the magma was produced deeper than 50 km, which represents an extreme thickened crust [77]. The existence of strongly peraluminous leucogranites (the Madi and

Lüshan plutons) is also robust evidence for a thickened crust in the Chifeng–Chaoyang district. Argillites and sandstones were brought into the deep crust and heated by ocean plate subduction and crustal thickening. The rapid crustal gravity equalization uplift caused partial melting of these heated sedimentary rocks enriched in water-bearing minerals (e.g., muscovite and biotite), producing strongly peraluminous leucogranite. These middle Mesozoic leucogranites (>160 Ma) indicate extreme crustal thickening/uplift. This thickened low crust was reactivated by the northwestward subduction of the Paleo-Pacific oceanic plate (Figure 12b).

The Triassic granitic source contained more juvenile, mantle-derived components. In contrast, the Jurassic granites were derived from the reworking of thickened ancient lower crust. It has been proposed that the most marked change during the Cretaceous cratonic destruction was the change of magmatic source from enriched lithospheric mantle to depleted asthenospheric mantle [2,3]. However, the source transition in the Triassic and Jurassic was from lithospheric mantle and depleted mantle to enriched ancient lower crust in the Chifeng–Chaoyang district. This might indicate that the Jurassic was the key period of tectonic transition from the CAOB to the Paleo-Pacific Ocean regime, and magma sources may be useful indicators to determine the tectonic settings when discrimination diagrams are invalid (Figure 13).

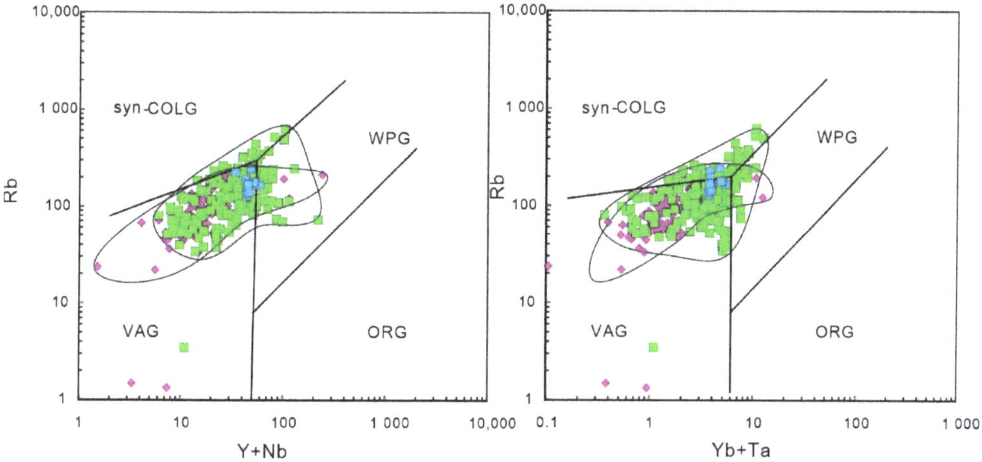

Figure 13. Discrimination diagrams of tectonic setting of granitoids from the Chifeng–Chaoyang district (based on [78]). ORG—ocean-ridge granites; syn-COLG—syn-collisional granites; VAG—volcanic arc granites; WPG—within-plate granites. Data sources and legends are the same as Figure 4.

7.3. Possible Relationship between Sources and Metallogeny

Magmatic intrusions are important lithospheric probes as they come from the deep Earth [79–81]. As powerful tools, the whole-rock Nd and zircon Hf isotopes of magmatic rocks have been broadly used to investigate crustal growth [9], lithospheric architecture, and regional metallogeny. The possible sources and ages of crustal rocks are indicated by ε_{Nd} and T_{DM}, respectively. Zircon ε_{Hf} data can be used to distinguish between juvenile mantle sources (high positive ε_{Hf} values) and ancient crustal sources (negative ε_{Hf} values) [82]. Various characteristic values can also be used to estimate the ages and components of mixed magmatic rocks, i.e., between the juvenile mantle and ancient crust endmembers. Furthermore, regional-scale Hf–Nd isotopic mapping can not only precisely identify the deep components and the temporal-spatial distributions, but also explore regional metallogenesis, which provides a new method for investigating the relationship between sources and metallogenesis [48,83].

Several studies linked isotopic architecture to the localization of mineral systems. Mole et al. (2013, 2014) consolidated and extended the Sm–Nd isotope coverage of the Yilgarn Craton and developed a comprehensive understanding of the crustal evolution of the craton and the spatial and temporal occurrence of komatiite-hosted nickel and orogenic gold [84,85]. They showed that the evolving crustal architecture of the Archean Yilgarn Craton played a key role in controlling the location of camp-scale gold, iron, and nickel mineralized systems. Komatiite-hosted nickel deposits cluster into camps localized within young, juvenile crust at the isotopic margin with older lithosphere; orogenic gold systems are typically localized along major structures within juvenile crust; and banded iron formation (BIF)-hosted iron deposits are localized at the edge of, and within, older lithospheric blocks. Based on whole-rock geochemistry and Sr–Nd–Hf isotopic data, Hou et al. (2012) proposed that Cu-bearing magmas were most probably derived from the thickened juvenile mafic lower crust beneath south Tibet. The melting of sulfide-bearing phases in the juvenile mantle components of the Tibetan lower crust probably provided portions of Cu, Au, and S for the fertile magmas. However, the Mo-bearing magmas were likely derived from the partial melting of the ancient Tibetan lower crust, while Mo was also mainly derived from the ancient crust [86]. In China, Nd–Hf isotopic mapping has been conducted in several regions, including the Da Hinggan Mountains [4], the southeastern part of the North China Craton [5], the East Qinling Orogen [6], the Sanjiang Tethyan Orogen [87], the Chinese Altai Orogen [88], and the Himalayan–Tibetan Orogen [89]. Although there is inconsistency regarding some of the specific mineral varieties, the consensus is that the distribution of the copper, gold, and copper-nickel deposits is consistent with the distribution of the juvenile, mantle-derived crust in orogenic belts and cratons, and the large-scale molybdenum deposits occurred in the reworking zone of the Paleoproterozoic ancient crust [48].

In this case, the petrological, geochemical, and Sr–Nd–O–Hf isotopic data provide important constraints on the transition of the magmatic sources from the Triassic to the Jurassic in the Chifeng–Chaoyang district. The Sr–Nd–O–Hf isotopic data indicate that the Triassic magmatism was predominantly produced by the mixing of juvenile lower crust melts derived from the lithospheric mantle and the depleted mantle, which controlled the Au(-Cu) mineralization. However, the Jurassic magmatism, with low $\varepsilon_{Nd}(t)$ and $\varepsilon_{Hf}(t)$ values, variable initial $^{87}Sr/^{86}Sr$ values, and high model age T_{DM2}–$T_{DM}{}^{C}$ values, was predominantly derived from Archean to Paleoproterozoic reworked crust and controlled the formation of the porphyry-skarn Mo mineralization.

The pattern diagrams are shown in Figure 12. Many studies on typical gold deposits suggested that asthenosphere upwelling triggered the release of gold and sulfur from an enriched and fertilized mantle lithosphere. This mechanism determined that the magma sources must be dominated by juvenile, mantle-derived components with a portion of lower crust (Figure 12a). On the contrary, ancient lower crust is responsible for providing metal Mo, which leads to more enriched isotopic signatures. The thickened lower crust and the thinning of the SCLM promoted large-scale reworking of ancient lower crust, which created favorable conditions for the release of Mo and mineralization (Figure 12b). This study may offer a new direction for studying regional metallogenic patterns.

8. Conclusions

Based on the distributions of the Triassic and Jurassic granites, as well as geochronologic, whole-rock Sr–Nd, and zircon O–Hf isotopic analyses, the primary magmatic sources changed from relatively juvenile, lithospheric, mantle-derived materials to thickened ancient lower crust. They also indicate that the tectonic setting changed from post-collisional extension to subduction of the Paleo-Pacific plate.

The medium–large-scale, high-grade Au deposits are closely related to the Triassic granitoids; however, most of the Mo deposits were formed in the Jurassic. The different magmatic sources in the deep crust controlled the particular type of mineralization.

Supplementary Materials: The following supporting information can be downloaded at: https://www.mdpi.com/article/10.3390/min12091117/s1, Table S1: The age of stocks on the northern margin of North China Craton (NCC); Table S2: Major, trace element, and Sr–Nd isotopic data for the Triassic–Jurassic intrusive rocks in the Chifeng–Chaoyang district; Table S3: Zircon O–Hf isotopic data for the Triassic–Jurassic intrusive rocks in the Chifeng–Chaoyang district; Table S4: Summary of geologic characteristics of major Au-Mo ore deposits in the Chifeng–Chaoyang district. Analytical methods.docx. [90–104].

Author Contributions: Conceptualization, H.-F.Z. and Y.T.; methodology, J.-G.Y.; software, J.-G.Y. and Y.-Y.Q.; validation, R.-W.L. and R.-B.L.; formal analysis, J.-G.Y.; investigation, J.-G.Y., R.-W.L. and R.-B.L.; resources, J.-G.Y.; data curation, J.-G.Y.; writing—original draft preparation, J.-G.Y.; writing—review and editing, J.-G.Y.; visualization, J.-G.Y. and Y.-Y.Q.; supervision, H.-F.Z. and Y.T.; project administration, Y.T.; funding acquisition, H.-F.Z. and Y.T. All authors have read and agreed to the published version of the manuscript.

Funding: The study is financially supported by National Nature Science Foundation of China (Grant No. 92162322, 41372077), the Geological Survey Program of China (Grant no. DD20190685), and the National Key Research and Development Program of China (Grant no. 2018YFC0603702, 2016YFC0600106).

Data Availability Statement: The raw data supporting the conclusions of this article will be made available by the authors without undue reservation.

Acknowledgments: We gratefully acknowledge the constructive comments of reviewers whose suggestions helped to improve the scientific quality of this paper substantially.

Conflicts of Interest: The authors declare that the research was conducted in the absence of any commercial or financial relationships that could be construed as a potential conflict of interest.

Contribution to the Field Statement: This contribution focuses on whether the characteristics of magma source affect the mineralization of gold and molybdenum deposits in the northern margin of North China Craton. This study area has been thought of as an important place for producing numerous large/superlarge deposits, e.g., the Yu'erya gold deposit and the Chehugou molybdenum deposit, as well as episodic felsic intrusions. Based on the comparative analysis of the rock assemblage and geochemical characteristics (majority and trace elements, rare earth elements and Sr–Nd–O–Hf isotopes) of the Triassic and Jurassic intrusions, it is concluded that (1) primary magmatic sources changed from juvenile lithospheric mantle to thickened ancient lower crust between the Triassic and Jurassic; (2) the tectonic setting transited from post-collisional extension to subduction of the Paleozoic-Pacific Plate beneath the East Asia Plate; and (3) different magmatic sources from the deep crust control the particular mineralization during cratonic destruction. Medium–large-scale Au deposits with high grade are closely related to Triassic magmatism; however, most Mo deposits are preferred to the Jurassic magmatic rocks. This pattern is helpful for ore exploration in this region.

References

1. Xu, Y.; Li, H.; Pang, C.; He, B. On the timing and duration of the destruction of the North China Craton. *Chin. Sci. Bull.* **2009**, *54*, 3379. [CrossRef]
2. Zhu, R.X.; Yang, J.H.; Wu, F.Y. Timing of destruction of the North China Craton. *Lithos* **2012**, *149*, 51–60. [CrossRef]
3. Wu, F.; Yang, J.; Xu, Y.; Wilde, S.; Walker, R. Destruction of the north china craton in the mesozoic. *Annu. Rev. Earth Pl. Sc.* **2019**, *47*, 173–195. [CrossRef]
4. Yang, Q.; Wang, T.; Guo, L.; Tong, Y.; Zhang, L.; Zhang, J.; Hou, Z. Nd isotopic variation of Paleozoic—Mesozoic granitoids from the Da Hinggan Mountains and adjacent areas, NE Asia: Implications for the architecture and growth of continental crust. *Lithos* **2017**, *272–273*, 164–184. [CrossRef]
5. Deng, J.; Wang, C.; Bagas, L.; Santosh, M.; Yao, E. Crustal architecture and metallogenesis in the south-eastern North China Craton. *Earth-Sci. Rev.* **2018**, *182*, 251–272. [CrossRef]
6. Wang, X.; Wang, T.; Ke, C.; Yang, Y.; Li, J.; Li, Y.; Qi, Q.; Lv, X. Nd—Hf isotopic mapping of Late Mesozoic granitoids in the East Qinling orogen, central China: Constraint on the basements of terranes and distribution of Mo mineralization. *J. Asian Earth Sci.* **2015**, *103*, 169–183. [CrossRef]
7. Yuan, J.; Zhang, H.; Tong, Y.; Gao, J.; Xiao, R. Sources of metals and fluids for the Taijiying gold deposit on the northern margin of the North China Craton. *Ore Geol. Rev.* **2021**, *139*, 104593. [CrossRef]

8. Li, R.; Yang, J.; Wang, H.; Zhu, Y.; Sun, J.; Xu, L. Triassic lithospheric modification of the northern North China Craton: Evidences from the composite Kalaqin Batholith and ultramafic-mafic Heilihe Intrusive Complex in Inner Mongolia. *Lithos* **2020**, *362–363*, 105501. [CrossRef]
9. Wang, T.; Hou, Z. Isotopic mapping and deep material probing (I): Revealing the compositional evolution of the lithosphere and crustal growth processes. *Earth Sci. Front.* **2018**, *25*, 1–19.
10. Liu, Y.; Hu, Z.; Gao, S.; Gunther, D.; Xu, J.; Gao, C.; Chen, H. In situ analysis of major and trace elements of anhydrous minerals by LA-ICP-MS without applying an internal standard. *Chem. Geol.* **2008**, *257*, 34–43. [CrossRef]
11. Zhai, M.G.; Santosh, M. The early Precambrian odyssey of the North China Craton: A synoptic overview. *Gondwana Res.* **2011**, *20*, 6–25. [CrossRef]
12. Zhai, M.; Zhou, Y. General precambrian geology in China. In *Precambrian Geology of China*; Zhai, M., Ed.; CAS: Beijing, China, 2015; pp. 3–56.
13. Zhao, G.; Cawood, P.A.; Li, S.; Wilde, S.A.; Sun, M.; Zhang, J.; He, Y.; Yin, C. Amalgamation of the North China Craton: Key issues and discussion. *Precambrian Res.* **2012**, *222*, 55–76. [CrossRef]
14. Tang, J.; Xu, W.; Wang, F.; Ge, W. Subduction history of the Paleo-Pacific slab beneath Eurasian continent: Mesozoic-Paleogene magmatic records in Northeast Asia. *Sci. China* **2018**, *61*, 527–559. [CrossRef]
15. Zhang, S.; Zhao, Y. Magmatic records of the late paleoproterozoic to neoproterozoic extensional and rifting events in the north china craton: A preliminary review. In *Main Tectonic Events and Metallogeny of the North China Craton*; Zhai, M., Zhao, Y., Zhao, T., Eds.; Springer: Singapore, 2016; pp. 359–391.
16. Song, S.; Niu, Y.; Su, L.; Xia, X. Tectonics of the North Qilian orogen, NW China. *Gondwana Res.* **2013**, *23*, 1378–1401. [CrossRef]
17. Windley, B.F.; Alexeiev, D.; Xiao, W.; Kröner, A.; Badarch, G. Tectonic models for accretion of the Central Asian Orogenic Belt. *J. Geol. Soc. London.* **2007**, *164*, 31–47. [CrossRef]
18. Wu, F.; Ge, W.; Sun, D.; Lin, Q.; Zhou, Y. The Sm-Nd, Rb-Sr isotopic ages of the Archean granites in southern Jilin Province. *Acta Petrol. Sinca.* **1997**, *13*, 499–506.
19. Zhang, K. Destruction of the North China Craton: Lithosphere folding-induced removal of lithospheric mantle? *J. Geodyn.* **2012**, *53*, 8–17. [CrossRef]
20. Tang, L.; Santosh, M. Neoarchean-Paleoproterozoic terrane assembly and Wilson cycle in the North China Craton: An overview from the central segment of the Trans-North China Orogen. *Earth-Sci. Rev.* **2018**, *182*, 1–27. [CrossRef]
21. Zeng, Q.; Liu, J.; Zhang, Z.; Zhang, W.; Chu, S.; Zhang, S.; Wang, Z.; Duan, X. Geology, fluid inclusion, and sulfur isotope studies of the chehugou porphyry molybdenum–copper deposit, xilamulun metallogenic belt, NE china. *Resour. Geol.* **2011**, *61*, 241–258. [CrossRef]
22. Wang, X.S.; Zheng, Y. $^{40}Ar/^{39}Ar$ age constraints on the ductile deformation of the detachment system of the Louzidian core complex, southern Chifeng, China. *Geol. Rev.* **2005**, *51*, 574–582, (In Chinese with English abstract).
23. LBGMR. *Regional Geology of Liaoning Province*; Geological Publishing House: Beijing, China, 1989; pp. 317–520.
24. Liu, S.; Santosh, M.; Wang, W.; Bai, X.; Yang, P. Zircon U-Pb chronology of the Jianping Complex: Implications for the Precambrian crustal evolution history of the northern margin of North China Craton. *Gondwana Res.* **2011**, *20*, 48–63. [CrossRef]
25. Liu, L.; Gu, X.; Zhang, Y.; Ouyang, X.; Wang, L.; Gao, L. Genesis of the Jinchanggouliang gold deposit, Chifeng, China: Constraints from fluid inclusions and isotopic geochemistry. *Ore Geol. Rev.* **2019**, *115*, 103180. [CrossRef]
26. Wang, W.; Liu, S.; Bai, X.; Yang, P.; Li, Q.; Zhang, L. Geochemistry and zircon U-Pb-Hf isotopic systematics of the Neoarchean Yixian—Fuxin greenstone belt, northern margin of the North China Craton: Implications for petrogenesis and tectonic setting. *Gondwana Res.* **2011**, *20*, 64–81. [CrossRef]
27. Xiong, L.; Shi, W.; Li, H.; Tian, N.; Chen, C.; Zhou, H.Z.; Zhao, S.Q.; Li, P.Y. Geochemistry, Sr-Nd-Hf isotopes and petrogenesis of Mid-Late triassic baizhangzi granitic intrusive rocks in eastern Hebei-Western liaoning province. *Earth Sci.* **2017**, *42*, 207–222, (In Chinese with English abstract).
28. Fu, L.; Wei, J.; Kusky, T.; Chen, H.; Tan, J.; Li, Y.; Kong, L.; Jiang, Y. Triassic shoshonitic dykes from the northern North China craton: Petrogenesis and geodynamic significance. *Geol. Mag.* **2012**, *149*, 39–55. [CrossRef]
29. Ma, Y.; Zeng, Q.; Song, B.; Du, J.; Yang, F.; Zhao, Y. Shrimp U-Pb dating of zircon from Panshan granitoid pluton in Yanshan orogenic belt and its tectonic implications. *Acta Petrol. Sin.* **2007**, *23*, 547–556, (In Chinese with English abstract).
30. Ye, H.; Zhang, S.; Zhao, Y.; Wu, F. Petrogenesis and emplacement deformation of the Late Triassic Dushan composite batholith in the Yanshan fold and thrust belt: Implication for the tectonic settings of the northern margin of the North China Craton during the Early Mesozoic. *Earth Sci. Front.* **2014**, *21*, 275–292, (In Chinese with English abstract).
31. Han, B.F.; Kagami, H.; Li, H.M. Age and Nd-Sr isotopic geochemistry of the Guangtoushan alkaline granite, Hebei province, China: Implications for early Mesozoic crust-mantle interaction in North China Block. *Acta Petrol. Sin.* **2004**, *20*, 1375–1388.
32. Cai, J.; Yan, G.; Mou, B.; Ren, K.; Li, F.; Yang, B. Geochronology and geochemistry of sungezhuang alkaline complex in jixian county, tianjin. *J. Jilin Univ. (Earth Sci. Ed.)* **2011**, *41*, 1901–1913.
33. Yang, J.H.; Sun, J.F.; Zhang, M.; Wu, F.Y.; Wilde, S.A. Petrogenesis of silica-saturated and silica-undersaturated syenites in the northern North China Craton related to post-collisional and intraplate extension. *Chem. Geol.* **2012**, *328*, 149–167. [CrossRef]
34. Zhang, S.H.; Zhao, Y.; Davis, G.A.; Ye, H.; Wu, F. Temporal and spatial variations of Mesozoic magmatism and deformation in the North China Craton: Implications for lithospheric thinning and decratonization. *Earth-Sci. Rev.* **2014**, *131*, 49–87. [CrossRef]

35. Elhlou, S.; Belousova, E.; Griffin, W.L.; Pearson, N.J.; O'Reilly, S.Y. Trace element and isotopic composition of GJ-red zircon standard by laser ablation. *Geochim. Et Cosmochim. Acta Suppl.* **2006**, *70*, A158. [CrossRef]
36. Belousova, E.; Griffin, W.; O'Reilly, S.Y.; Fisher, N. Igneous zircon: Trace element composition as an indicator of source rock type. *Contrib. Mineral. Petrol.* **2002**, *143*, 602–622. [CrossRef]
37. Le Maitre, R.W. *Igneous Rocks: A Classification and Glossary of Terms*, 2nd ed.; Cambridge University Press: Cambridge, UK, 2002.
38. Taylor, S.R. McLennan SM: The Continental Crust: Its Composition and Evolution. 1985. Available online: https://www.osti.gov/biblio/6582885 (accessed on 10 August 2022).
39. Sun, S.S.; McDonough, W. Chemical and isotopic systematics of oceanic basalts: Implications for mantle composition and processes. *Geol. Soc. Lond. Spec. Publ.* **1989**, *42*, 313–345. [CrossRef]
40. Whalen, J.B.; Currie, K.L.; Chappell, B.W. A-type granites: Geochemical characteristics, discrimination and petrogenesis. *Contrib. Mineral. Petrol.* **1987**, *95*, 407–419. [CrossRef]
41. Valley, J.W.; Kinny, P.D.; Schulze, D.J.; Spicuzza, M.J. Zircon megacrysts from kimberlite: Oxygen isotope variability among mantle melts. *Contrib. Mineral. Petrol.* **1998**, 1–11. [CrossRef]
42. Fan, W.; Jiang, N.; Xu, X.; Hu, J.; Zong, K. Petrogenesis of the middle Jurassic appinite and coeval granitoids in the Eastern Hebei area of North China Craton. *Lithos* **2017**, *278–281*, 331–346. [CrossRef]
43. Cui, F. *Petrogenesis of Mesozoic Granitoids and Crustal Evolution in Xingcheng Area, Western Liaoning Province*; Jilin University: Changchun, China, 2015; pp. 1–188.
44. Peccerillo, A.; Taylor, S.R. Geochemistry of eocene calc-alkaline volcanic rocks from the Kastamonu area, Northern Turkey. *Contrib. Mineral. Petrol.* **1976**, *58*, 63–81. [CrossRef]
45. Liu, H.; Zhai, M.; Liu, J.; Sun, S. The Mesozoic granitoids in the northern marginal region of North China Craton evolution from post-collisional to anorogenic settings. *Acta Petrol. Sin.* **2002**, *18*, 433–448.
46. Wu, F.Y.; Yang, J.H.; Wilde, S.A.; Zhang, X.O. Geochronology, petrogenesis and tectonic implications of Jurassic granites in the Liaodong Peninsula, NE China. *Chem. Geol.* **2005**, *221*, 127–156. [CrossRef]
47. Defant, M.J.; Drummond, M.S. Derivation of some modern arc magmas by melting of young subducted lithosphere. *Nature* **1990**, *347*, 662. [CrossRef]
48. Castillo, P.R.; Janney, P.E.; Solidum, R.U. Petrology and geochemistry of Camiguin Island, southern Philippines: Insights to the source of adakites and other lavas in a complex arc setting. *Contrib. Mineral. Petrol.* **1999**, *134*, 33–51. [CrossRef]
49. Hou, Z.; Wang, T. Isotopic mapping and deep material probing (II): Imaging crustal architecture and its control on mineral systems. *Earth Sci. Front.* **2018**, *25*, 20–41, (In Chinese with English abstract).
50. Griffin, W.L.; Wang, X.; Jackson, S.E.; Pearson, N.J.; O'Reilly, S.Y.; Xu, X.; Zhou, X. Zircon chemistry and magma mixing, SE China: In-situ analysis of Hf isotopes, Tonglu and Pingtan igneous complexes. *Lithos* **2002**, *61*, 237–269. [CrossRef]
51. Chen, Z.G.; Zhang, L.C.; Wu, H.Y.; Wan, B.; Zeng, Q.D. Geochemistry study and tectonic background of a style host granite in Nianzigou molybdenum deposit in Xilamulun molybdenum metallogenic belt, Inner Mongolia. *Acta Petrol. Sin.* **2008**, *24*, 879–889.
52. Brown, M. The generation, segregation, ascent and emplacement of granite magma: The migmatite-to-crustally-derived granite connection in thickened orogens. *Earth-Sci. Rev.* **1994**, *36*, 83–130. [CrossRef]
53. Gao, S.; Rudnick, R.; Yuan, H.; Liu, X.; Liu, Y.; Xu, W.L.; Ling, W.; Ayers, J.; Wang, X.; Wang, Q. Recycling lower continental crust in the North China Craton. *Nature* **2004**, *432*, 892–897. [CrossRef]
54. Jiang, S.; Liang, Q.; Bagas, L. Re-Os ages for molybdenum mineralization in the Fengning region of northern Hebei Province, China: New constraints on the timing of mineralization and geodynamic setting. *J. Asian Earth Sci.* **2014**, *79*, 873–883. [CrossRef]
55. Song, Y.; Jiang, S.-H.; Bagas, L.; Li, C.; Hu, J.-Z.; Zhang, Q.; Zhou, W.; Ding, H.-Y. The geology and geochemistry of Jinchangyu gold deposit, North China Craton: Implications for metallogenesis and geodynamic setting. *Ore Geol. Rev.* **2016**, *73*, 313–329. [CrossRef]
56. Yang, J.; Wu, F. Triassic magmatism and its relation to decratonization in the eastern North China Craton. *Sci. China Ser. D Earth Sci.* **2009**, *52*, 1319–1330. [CrossRef]
57. Zeng, Q.; Liu, J.; Qin, K.; Fan, H.; Chu, S.; Wang, Y.; Zhou, L. Types, characteristics, and time-space distribution of molybdenum deposits in China. *Int. Geol. Rev.* **2014**, *55*, 1311–1358. [CrossRef]
58. Wang, J.B.; Zou, T.; Wang, Y.W.; Long, L.L.; Zhang, H.Q.; Liao, Z.; Xie, H.J. Deposit associations, metallogenesis and metallogenic lineage of molydenum-polymetallic deposits in China. *Miner. Depos.* **2014**, *33*, 447–470.
59. Dai, J.; Xie, G.; Wang, R.; Ren, T.; Wang, T. Re-Os isotope dating of molybdenite seprates from the Yajishan Mo(Cu) deposit, Inner Mongolia, and its geological significance. *Geol. China* **2012**, *39*, 1614–1621, (In Chinese with English abstract).
60. Zeng, Q.; Liu, J.; Zhang, Z.; Chen, W.; Zhang, W. Geology and geochronology of the Xilamulun molybdenum metallogenic belt in eastern Inner Mongolia, China. *Int. J. Earth Sci.* **2011**, *100*, 1791–1809. [CrossRef]
61. Dai, J.; Mao, J.; Zhao, C.; Xie, G.; Yang, F.; Wang, Y. New U—Pb and Re—Os age data and the geodynamic setting of the Xiaojiayingzi Mo (Fe) deposit, western Liaoning province, Northeastern China. *Ore Geol. Rev.* **2009**, *35*, 235–244. [CrossRef]
62. Sun, Z.J.; Sun, S.G.; Yu, H.N. Chronology and geochemical characteristics of zhuanshanzi granite in chifeng and its diagenetic dynamics background. *J. Jilin Univ. (Engl. Sci. Ed.)* **2016**, *46*, 1740–1753.
63. Zhang, S.; Zhao, Y.; Liu, X.; Liu, D.; Chen, F.; Xie, L.; Chen, H.H. Late Paleozoic to Early Mesozoic mafic—Ultramafic complexes from the northern North China Block: Constraints on the composition and evolution of the lithospheric mantle. *Lithos* **2009**, *110*, 229–246. [CrossRef]

64. Fu, L.; Wei, J.; Kusky, T.M.; Chen, H.; Tan, J.; Li, Y.; Shi, W.; Chen, C.; Zhao, S. The Cretaceous Duimiangou adakite-like intrusion from the Chifeng region, northern North China Craton: Crustal contamination of basaltic magma in an intracontinental extensional environment. *Lithos* **2012**, *134*, 273–288. [CrossRef]
65. Sun, M.; Armstrong, R.L.; St, J.; Lambert, R.; Jiang, C.; Wu, J. Petrochemistry and Sr, Pb and Nd isotopic geochemistry of the paleoproterozoic kuandian complex, the eastern Liaoning province, China. *Precambrian Res.* **1993**, *62*, 171–190.
66. Wu, F.; Yang, Y.; Xie, L.; Yang, J.; Xu, P. Hf isotopic compositions of the standard zircons and baddeleyites used in U-Pb geochronology. *Chem. Geol.* **2006**, *234*, 105–126. [CrossRef]
67. Jahn, B.M.; Wu, F.Y.; Lo, C.; Tsai, C. Crust-mantle interaction induced by deep subduction of the continental crust: Geochemical and Sr–Nd isotopic evidence from post-collisional mafic–ultramafic intrusions of the northern Dabie complex, central China. *Chem. Geol.* **1999**, *157*, 119–146. [CrossRef]
68. Zhang, X.; Yuan, L.; Wilde, S.A. Crust/mantle interaction during the construction of an extensional magmatic dome: Middle to Late Jurassic plutonic complex from western Liaoning, North China Craton. *Lithos* **2014**, *205*, 185–207. [CrossRef]
69. Zhang, L.; Wu, H.; Wan, B.; Chen, Z. Ages and geodynamic settings of Xilamulun Mo-Cu metallogenic belt in the northern part of the North China Craton. *Gondwana Res.* **2009**, *16*, 243–254. [CrossRef]
70. Zhang, S.H.; Zhao, Y.; Liu, J.M.; Hu, J.M.; Song, B.; Liu, J.; Wu, H. Geochronology, geochemistry and tectonic setting of the Late Paleozoic—Early Mesozoic magmatism in the northern margin of the North China Block: A preliminary review. *Acta Petrol. Mineral.* **2010**, *29*, 824–842.
71. Black, R.; Liégeois, J.P. Cratons, mobile belts, alkaline rocks and continental lithospheric mantle: The Pan-African testimony. *J. Geol. Soc.-J. Geol. Soc.* **1993**, *150*, 89–98. [CrossRef]
72. Eby, G.N. The A-type granitoids: A review of their occurrence and chemical characteristics and speculations on their petrogenesis. *Lithos* **1990**, *26*, 115–134. [CrossRef]
73. Eby, G.N. Chemical subdivision of the A-type granitoids: Petrogenetic and tectonic implications. *Geology* **1992**, *20*, 641. [CrossRef]
74. Bonin, B. From orogenic to anorogenic settings: Evolution of granitoid suites after a major orogenesis. *Geol. J.* **1990**, *25*, 261–270. [CrossRef]
75. Liégeois, J.; Navez, J.; Hertogen, J.; Black, R. Contrasting origin of post-collisional high-K calc-alkaline and shoshonitic versus alkaline and peralkaline granitoids. The use of sliding normalization. *Lithos* **1998**, *45*, 1–28. [CrossRef]
76. Mu, B.; Shao, J.; Chu, Z.; Yan, G.; Qiao, G. Sm-Nd age and Sr-Nd isotopic characteristics of the Fanshan potassic alkaline ultramafic-syenite complex in Hebei province. *Acta Petrol. Sin.* **2001**, *03*, 358–365.
77. Peng, P.; Zhai, M.; Guo, J.; Zhang, H.; Zhang, Y. Petrogenesis of Triassic post-collisional syenite plutons in the Sino-Korean craton: An example from North Korea. *Geol. Mag.-Geol. Mag.* **2008**, *145*, 637–647. [CrossRef]
78. Pearce, J.A.; Harris, N.B.; Tindle, A.G. Trace element discrimination diagrams for the tectonic interpretation of granitic rocks. *J. Petrol.* **1984**, *25*, 956–983. [CrossRef]
79. Xiao, W.; Windley, B.F.; Sun, S.; Li, J.; Huang, B.; Han, C.; Yuan, C.; Sun, M.; Chen, H. A tale of amalgamation of three Permo-Triassic collage systems in Central Asia: Oroclines, sutures, and terminal accretion. *Annu. Rev. Earth Pl. Sc.* **2015**, *43*, 477–507. [CrossRef]
80. Milisenda, C.C.; Liew, T.C.; Hofmann, A.W.; Kröner, A. Isotopic mapping of age provinces in precambrian High-Grade terrains: Sri lanka. *J. Geol.* **1988**, *96*, 608–615. [CrossRef]
81. DePaolo, D.J.; Linn, A.M.; Schubert, G. The continental crustal age distribution: Methods of determining mantle separation ages from Sm-Nd isotopic data and application to the southwestern United States. *J. Geophys. Res. Solid Earth* **1991**, *96*, 2071–2088. [CrossRef]
82. Wang, T.; Jahn, B.; Kovach, V.P.; Tong, Y.; Hong, D.; Han, B. Nd-Sr isotopic mapping of the Chinese Altai and implications for continental growth in the Central Asian Orogenic Belt. *Lithos* **2009**, *110*, 359–372. [CrossRef]
83. Griffin, W.L.; Begg, G.C.; O'Reilly, S.Y. Continental-root control on the genesis of magmatic ore deposits. *Nat. Geosci.* **2013**, *6*, 905–910. [CrossRef]
84. Kemp, A.; Hawkesworth, C.J.; Paterson, B.A.; Kinny, P.D. Episodic growth of the Gondwana Supercontinent from hafnium and oxygen isotopes in zircon. *Nature* **2006**, *439*, 580–583. [CrossRef]
85. Mole, D.R.; Fiorentini, M.L.; Cassidy, K.F.; Kirkland, C.L.; Thebaud, N.; Mccuaig, T.C.; Doublier, M.P.; Duuring, P.; Romano, S.S.; Maas, R. Crustal evolution, intra-cratonic architecture and the metallogeny of an Archaean craton. *Geol. Soc. Lond. Spec. Publ.* **2013**, *393*, P393–P398. [CrossRef]
86. Hou, Z.; Zheng, Y.; Yang, Z.; Yang, Z. Metallogenesis of continental collision setting: Part I. Gangdese Cenozoic porphyry Cu-Mo systems in Tibet. *Miner. Depos.* **2012**, *31*, 647–670.
87. Mole, D.; Fiorentini, M.; Thébaud, N.; Cassidy, K.; McCuaig, T.; Kirkland, C.; Romano, S.; Doublier, M.; Belousova, E.; Barnes, S.; et al. Archean komatiite volcanism controlled by the evolution of early continents. *Proc. Natl. Acad. Sci. USA* **2014**, *111*, 10083–10088. [CrossRef] [PubMed]
88. Kovalenko, V.; Yarmolyuk, V.; Kovach, V.; Kotov, A.; Kozakov, I.; Salnikova, E.; Larin, A. Isotope provinces, mechanisms of generation and sources of the continental crust in the Central Asian mobile belt: Geological and isotopic evidence. *J. Asian Earth Sci.* **2004**, *23*, 605–627. [CrossRef]
89. Wang, C.M.; Bagas, L.; Lu, Y.J.; Santosh, M.; Du, B.; McCuaig, T.C. Terrane boundary and spatio-temporal distribution of ore deposits in the Sanjiang Tethyan Orogen: Insights from zircon Hf-isotopic mapping. *Earth-Sci. Rev.* **2016**, *156*, 39–65. [CrossRef]

90. Xu, X.; Jiang, N.; Fan, W.; Hu, J.; Zong, K. Petrogenesis and geological implications for the Mesozoic granites in Qinglong area, eastern Hebei Province. *Acta Pertrologica Sinica.* **2016**, *32*, 212–232, (in Chinese with English abstract).
91. Zong, K.; Klemd, R.; Yuan, Y.; He, Z.; Guo, J.; Shi, X.; Liu, Y.; Hu, Z.; Zhang, Z. The assembly of Rodinia: The correlation of early Neoproterozoic (ca. 900 Ma) high-grade metamorphism and continental arc formation in the southern Beishan Orogen, southern Central Asian Orogenic Belt (CAOB). *Precambrian Res.* **2017**, *290*, 32–48. [CrossRef]
92. Hu, Z.; Zhang, W.; Liu, Y.; Gao, S.; Li, M.; Zong, K.; Chen, H.; Hu, S. "Wave" Signal-Smoothing and Mercury-Removing Device for Laser Ablation Quadrupole and Multiple Collector ICPMS Analysis: Application to Lead Isotope Analysis. *Anal. Chem.* **2015**, *87*, 1152–1157. [CrossRef]
93. Liu, D.; Wilde, S.A.; Wan, Y.; Wu, J.; Zhou, H.; Dong, C.; Yin, X. New U-Pb and Hf Isotopic data confirm Anshan as the oldest preserved segment of the North China Craton. *Am. J. Sci.* **2008**, *308*, 200–231. [CrossRef]
94. Liu, Y.; Gao, S.; Hu, Z.; Gao, C.; Zong, K.; Wang, D. Continental and Oceanic Crust Recycling-Induced Melt Peridotite Interactions in the Trans-North China Orogen: UPb Dating, Hf Isotopes and Trace Elements in Zircons from Mantle Xenoliths. *J. Petrol.* **2010**, *51*, 537–571. [CrossRef]
95. Ludwig, K.R. ISOPLOT 3.0: A Geochronological Toolkit for Microsoft Excel. Berkeley Geochronology Center Special Publication. *US Geol. Sur. Open File Rep.* **2003**, *39*, 91–445.
96. Russell, W.A.; Papanastassiou, D.A.; Tombrello, T.A. Ca isotope fractionation on the earth and other solar system materials. *Geochim. Cosmochim. Acta* **1978**, *42*, 1075–1090. [CrossRef]
97. Lin, J.; Liu, Y.S.; Yang, Y.H.; Hu, Z.C. Calibration and correction of LA-ICP-MS and LA-MC-ICP-MS analyses for element contents and isotopic ratios. *Solid Earth Sci.* **2016**, *1*, 5–27. [CrossRef]
98. Thirlwall, M.F. Long-term reproducibility of multicollector sr and nd isotope ratio analysis. *Chem. Geol.* **1991**, *94*, 85–104. [CrossRef]
99. Li, C.F.; Li, X.H.; Li, Q.L.; Guo, J.H.; Yang, Y.H. Rapid and precise determination of Sr and Nd isotopic ratios in geological samples from the same filament loading by thermal ionization mass spectrometry employing a single-step separation scheme. *Anal. Chim. Acta* **2012**, *727*, 54–60. [CrossRef] [PubMed]
100. Tanaka, T.; Togashi, S.; Kamioka, H.; Amakawa, H.; Kagami, H.; Hamamoto, T.; Yuhara, M.; Orihashi, Y.; Yoneda, S.; Shimizu, H.; et al. Jndi-1: A neodymium isotopic reference in consistency with lajolla neodymium. *Chem. Geol.* **2000**, *168*, 279–281. [CrossRef]
101. Dong, C.Y.; Wan, Y.S.; Long, T.; Zhang, Y.H.; Liu, J.H.; Ma, M.Z.; Xie, H.Q.; Liu, D.Y. Oxygen isotopie compositions of zircons from paleoproterozoic metasedimentary rocks in the Daqingshan-Jining area, North China craton: In situ SHRIMP analysis. *Acta Petrol. Sin.* **2016**, *32*, 659–681, (in Chinese with English abstract).
102. Ickert, R.B.; Hiess, J.; Williams, I.S.; Holden, P.; Ireland, T.R.; Lanc, P.; Schram, N.; Foster, J.J.; Clement, S.W. Determining high precision, in situ, oxygen isotope ratios with a SHRIMP II: Analyses of MPI-DING silicate-glass reference materials and zircon from contrasting granites. *Chem Geol.* **2008**, *257*, 114–128. [CrossRef]
103. Black, L.P.; Sandra, L.K.; Charlotte, M.A.; Donald, W.D.; Aleinikoff, J.N.; John, N.A.; John, W.V.; Roland, M.; Ian, H.C.; Russell, J.K.; et al. Improved 206Pb/238U microprobe geochronology by the monitoring of a trace-element-related matrix effect; SHRIMP, ID–TIMS, ELA–ICP–MS and oxygen isotope documentation for a series of zircon standards. *Chem. Geol.* **2004**, *205*, 115–140. [CrossRef]
104. Hou, K.J.; Li, Y.H.; Zou, T.R.; Qu, X.M.; Xie, G.Q. Laser ablation-MC-ICP-MS technique for Hf isotope microanalysis of zircon and its geological applications. *Acta Petrol. Sin.* **2007**, *23*, 2595–2604, (In Chinese with English abstract).

minerals

MDPI

Article

Mesozoic Magmatic and Geodynamic Evolution in the Jiaodong Peninsula, China: Implications for the Gold and Polymetallic Mineralization

Bin Wang [1,2,3], Zhengjiang Ding [2,3,*], Zhongyi Bao [2,3,*], Mingchun Song [2,3], Jianbo Zhou [1], Junyang Lv [2,3], Shanshan Wang [2,3], Qibin Zhang [2,3] and Caijie Liu [2,3]

1 College of Earth Sciences, Jilin University, Changchun 130061, China
2 Shandong Provincial Engineering Laboratory of Application and Development of Big Data for Deep Gold Exploration, Weihai 264209, China
3 Shandong Provincial No. 6 Exploration Institute of Geology and Mineral Resources, Weihai 264209, China
* Correspondence: ytdzhj@126.com (Z.D.); sddzld2022@126.com (Z.B.)

Abstract: The intrusive age ranges of Linglong, Guojialing, Weideshan, and Laoshan granites in the Jiaodong Peninsula are 155–154 Ma, 131–130 Ma, 118–111 Ma, and 116 Ma, respectively. Together with the Shidao granite (227–200 Ma), five phases of magmatism can be classified by the time, all of which have different degrees of gold and polymetallic mineralization. The type of granites evolved from A–, S–type to I–, A–type from the Late Triassic to the Early Cretaceous, thus reflects the evolution of geodynamics in the Mesozoic, indicating the switch from North China Craton (NCC)–Yangtze Craton (YC) collision to subduction of the Paleo–Pacific Plate (PPP), with crustal thickening switching to lithospheric thinning and a compressional tectonic setting changing to an extensional setting. It directly leads to a series of extensional structures evolving in the Jiaodong Peninsula and demonstrates affinity for the extensive mineralization in the Early Cretaceous. The key markers of Jiaodong gold and polymetallic mineralization are magmatism, fluid activity and extensional structure. Extensive magmatic uplift and extensional structures in the Early Cretaceous formed the extensional tectonic system. During the formation process, a large proportion of crust and mantle materials exchanged and mixed, and the fluid interaction was highly active, resulting in a magmatic fluid metallogenic system, which provided favorable metallogenic conditions for gold and nonferrous metal hydrothermal deposits. Thus, a large-scale explosive mineralization occurred in Jiaodong in the middle and late Early Cretaceous.

Keywords: Mesozoic granite; geodynamic evolution; extensional structure; gold and polymetallic mineralization; Jiaodong Peninsula

1. Introduction

Jiaodong Peninsula is located in the southeast margin of the North China Craton (NCC), and the junction of the Paleo–Tethys and the Paleo–Pacific metallogenic domains. The Mesozoic collision orogeny of the Yangtze Craton (YC) and the NCC superimposed the subduction–collision of the Paleo–Pacific Plate (PPP), resulting in intense crust–mantle interaction and frequent tectonic magmatism. The gold and polymetallic metallogenic conditions in this area are extremely superior. The deposits are mainly distributed in the Sanshandao and Zhaoyuan–Pingdu fault zones in the west, the Qixia fault zone in the middle, the Taocun–Zhuwu fault zone in the northeast margin of Jiaolai Basin, and the Jinniushan, Mizhan, and Lidao fault zones in the east. Dozens of super-large, large and medium-sized deposits have been discovered here, which is known as the "Jiaodong Metallogenic Province" in China [1]. Therefore, it is an interesting area for research related to deep continental subduction, destruction of the eastern NCC, and large-scale gold and polymetallic mineralization. The Jiaodong Peninsula has been an important gold producing area since ancient times. The accumulated proven gold resources in the Jiaodong Peninsula have exceeded 5000 t, which has become the third largest gold metallogenic area in the world. Most of the deposits found in the western part of Jiaodong area are gold deposits, the central part is dominated by gold deposits with more associated metals, and the eastern part is dominated by polymetallic deposits [2,3]. Here is a natural laboratory for conducting research on the regularity of ore formation and providing demonstrations for prospecting.

The gold and polymetallic deposits are closely associated with Mesozoic granitic intrusions in Jiaodong Peninsula. Nevertheless, most previous studies focused on the Late Mesozoic gold mineralization, magmatism, and tectonism related to the Jurassic Linglong and Early Cretaceous Guojialing and Weideshan granites. Their genesis, tectonic setting, and evolution have been studied in detail [4–23]. The Jiaodong gold deposits formed between 123 and 110 Ma [24–32], much later than the formation of the Linglong granite (164–140 Ma; [6–8,10–12,18,25]) and immediately later than the formation of the Guojialing granite (135–125 Ma; [10,15,18,22]). The age of these gold deposits is consistent with previous age estimates of the Weideshan granite, which is associated with a few gold and many polymetallic deposits in the eastern Jiaodong Peninsula. It is worth noting that the only beryllium deposit in Jiaodong is produced in the Late Triassic Shidao granite [33]. Due to the small number and scale of polymetallic deposits, the research is not deep enough. However, almost every stage of magmatic activity has produced more or less polymetallic deposits since the Mesozoic. As a consequence, the relationship between the Mesozoic gold and polymetallic mineralization in Jiaodong Peninsula and tectonic evolution is unclear due to insufficient systematic research on regional magmatism, granite genesis, tectonic setting and its relationship with regional mineralization [15,34–39].

Methodical research on the origin of the above typical five-stage granites from Late Triassic to Early Cretaceous could significantly improve the general understanding of the tectonic setting and origin of gold and polymetallic mineralization. This paper is based on investigations into the composition, distribution, petrologic features, and published data of the typical five-stage granites, and presents a systematic study on morphology, geochemistry and geochronology of zircons from the granites, which elucidate associations between petrogenesis, tectonic setting, and gold and polymetallic mineralization. This study provides new insights into the relationship between magmatism and mineralization during the Mesozoic whilst also providing constraints on the tectonic setting of gold and polymetallic mineralization in the Jiaodong Peninsula.

2. Geological Background and Samples

East of the Tan–Lu fault, the study area is divided into the Sulu terrane in the southeast, the Jiaobei terrane in the northwest, and the Jiaolai basin in the middle. The two terranes are separated by the Wulian–Yantai fault in the east, which is commonly regarded as the boundary between the NCC and YC (Figure 1a). The Jiaobei terrane is bounded to the west by the Tan–Lu fault and to the east by the Wulian–Yantai fault, and contains both Precambrian metamorphic basement and Mesozoic magmatic rocks [40–45]. The Precambrian rocks in the Jiaobei terrane include the Neoarchean Jiaodong complex that consists of metamorphosed volcanic–sedimentary rocks and TTG (tonalite–trondhjemite–granodiorite) gneisses, as well as the Paleoproterozoic Jingshan and Fenzishan groups and the Neoproterozoic Penglai group, which consist of metasedimentary rocks [46]. The Paleoproterozoic Fenzishan group is the host rock of the gold and polymetallic deposits in Yantai city [47,48]. The Sulu terrane includes the Jiaonan terrane to the south and Weihai terrane to the north, and is part of the Qinling–Dabie–Sulu orogenic belt or the ultrahigh pressure (UHP) metamorphic belt of eastern central China [49–51] and represents an intercontinental collisional orogeny that evolved from the Permian to the Triassic [52–57]. The Jiaolai basin is a Cretaceous continental basin situated at the boundary between the Sulu terrane and Jiaobei terrane, and comprises volcanic–sedimentary rocks (Figure 1b).

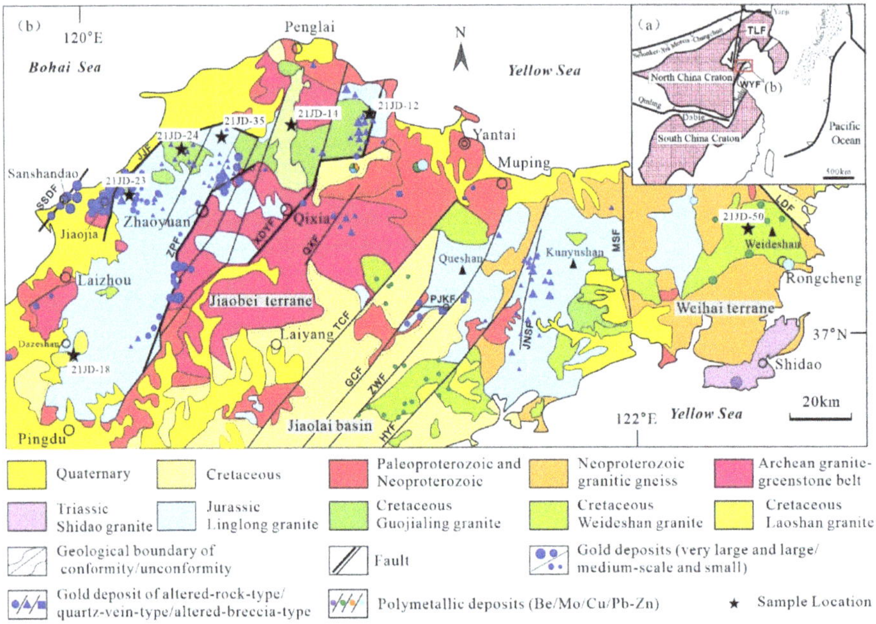

Figure 1. (**a**) Tectonic sketch map of east China, showing the location of the Jiaodong Peninsula. (**b**) Sketch map showing the regional geology and distribution of metal mineral deposits in the Jiaodong Peninsula (modified after [58,59]). Abbreviation: TLF = Tan–Lu fault zone; WYF = Wulian–Yantai fault zone; SSDF = Shanshandao fault zone; JJF = Jiaojia fault zone; ZPF = Zhaoyuan–Pingdu fault zone; XDYF = Xilin–Douya fault zone; QXF = Qixia fault zone; TCF = Taocun fault zone; GCF = Guocheng fault zone; ZWF = Zhuwu fault zone; HYF = Haiyang fault zone; PJKF = Pengjiakuang fault zone; JNSF = Jinniushan fault zone; MSF = Mishan fault zone; LDF = Lidao Fault zone.

Previous studies have shown that there is a concealed EW trending basement structure in the region, which is an anticlinorium and fault structure composed of Precambrian crystalline basement [60,61]. These basement structures are controlled by the regional tectonic stress domain of Mesozoic collision orogeny and subduction, and superimpose the fault tectonic system dominated by NE and NNE trending faults [60] (Figure 1b). Many faults are distributed on the upper wall of Tan–Lu fault from west to east, with the general trending from NE to nearly SN. These faults, which reach deep into the lower crust, are channels for the rise of deep magma, which not only affect the spatial location of regional magmatic rocks, but also control the formation and distribution of gold and polymetallic mineralization [33]. In addition, nearly EW- or NW-trending detachment faults developed at the northeast margin of Jiaolai basin, which controlled the distribution of deposits [62,63].

Besides the UHP metamorphic rocks, Mesozoic magmatic rocks are widely outcropped in the Jiaodong Peninsula (Figure 1b). They include the Triassic granite in the Shidao area, the widespread Jurassic–Cretaceous granites and the Cretaceous volcanic rocks in the Jiaolai Basin. Based on syn-magmatic evolution features, five main granitic types representing the five-stages of magmatism were defined [40]. The Jiaodong gold deposits mostly occur in the Linglong and Guojialing granites. No apparent gold mineralization exists in the Weideshan and Laoshan granites [58,64], although they have an affinity with the protolith age of the Weideshan granite [6,59,65,66]. However, all of the granites have different degrees of polymetallic mineralization.

3. Petrologic Features of the Typical Five–Stage Granites

This study focuses on the typical five-stage Mesozoic granites in the Jiaodong Peninsula. The Late Jurassic Linglong granite and Early Cretaceous Guojialing, Weideshan, and Laoshan granites are collected for analysis, and the sampling locations are shown in Figure 1b. At the same time, the published geochemical and geochronological data of the five-stage granites (including the Shidao granite) are collected for comprehensive study. Detailed geochemical data are listed in Supplementary Materials Tables S1 and S2.

3.1. The Late Triassic Shidao Granite

Late Triassic Shidao granite that intruded into the UHP granitic gneisses, covers an area of ~224 km^2, is located in the eastern UHP metamorphic belt (Figure 1b). The main lithology is composed of syenite with various structures (fine-, medium-, coarse-grained, and porphyritic), quartz syenite, and syenogranite, which together constitute the Shidao complex [40,67]. The former occupies about 40% of the total area of Shidao granite, while the latter occupy about 48% [68]. Scholars believe that syenogranite, the product of the rapid reentry after the collision between the YC and the NCC, is the source rock of Datuanliujia Be deposit, Rongcheng [2,33]. Previous studies on the chronology of the Shidao granite show that the zircon U–Pb, SHRIMP, and ^{40}Ar–^{39}Ar ages range from 227–200 Ma [69–74], suggesting that the main body of the Shidao granite was formed in the Late Triassic.

3.2. The Late Jurassic Linglong Granite

The Jurassic Linglong granite covers ~4230 km^2, and includes the Linglong pluton in the northwest Jiaodong Peninsula, as well as the Queshan and Kunyushan plutons between the Jiaobei and Weihai terranes. The main lithology is monzonitic granite. Predecessors suggested that the early intrusions were mainly gneissic garnet-bearing monzonitic granite, with later intrusions mainly being massive monzonitic granite [40,68]. The Linglong granite is the most developed granite in Jiaodong area and has always attracted much attention because of its distribution along with gold deposits. Therefore, some researchers call it one of the direct providers of ore–forming materials of Jiaodong gold deposit [75] or called it a derived source rock series [76,77]. However, the magmatic activity of this period has corresponding Mo–W polymetallic mineralization, for example, Xingjiashan Mo–W deposit is developed in Xingfushan pluton [2,78,79].

Two samples (21JD–23 and 21JD–35) are collected from Linglong pluton. Sample 21JD–23 (GPS: 37°23′27″ N, 120°10′20″ E) was collected from the abandoned quarry about 3 km east of Jiaojia gold deposit (Figure 1a). The main lithology is monzonitic granite [3]. The rock samples are light grayish white with medium- to coarse-grained texture and weak gneissic structure (Figure 2a). The main mineral composition is plagioclase (35 vol.%), K-feldspar (30 vol.%), quartz (30 vol.%) and biotite (5 vol.%). In addition, there are a few accessory minerals, such as garnet, zircon, titanite, etc. Sample 21JD–35 (GPS: 37°31′15″ N, 120°32′20″ E) is a monzonitic granite with a medium-grained texture, collected from the Longkou area, with a main mineral assemblage of K-feldspar (35 vol.%), plagioclase (30 vol.%), quartz (30 vol.%), and biotite (5 vol.%) (Figure 2b).

3.3. The Early Cretaceous Guojialing Granite

Outcrops of the Guojialing granite cover an area of ~514 km^2 in the northwest Jiaodong Peninsula. The Guojialing granite is discontinuously distributed, and the plutons, which roughly form a NEE-trending moniliform distribution (Figure 1b). The lithology is mainly porphyritic hornblende–bearing monzonitic granite, with minor granodiorite, quartz monzonite and monzodiorite [40,68]. Most of the gold deposits, especially the giant deposits, are genetically related to the Guojialing granite [2,75].

Figure 2. Field outcrop and photomicrographs (perpendicular polarized light) show textures and mineral assemblages of the Mesozoic granites in the Jiaodong Peninsula. (**a**,**b**) Monzonitic granite (Linglong granite samples 21JD–23 and 21JD–35); (**c**,**d**) porphyritic granodiorite (Guojialing granite samples 21JD–12 and 21JD–24); (**e**) porphyritic granodiorite (Weideshan granite sample 21JD–14); (**f**) porphyritic monzonitic granite (Weideshan granite sample 21JD–50); (**g**,**h**) porphyritic monzonitic granite (Laoshan granite sample 21JD–18). Abbreviation: Qtz = quartz; Pl = plagioclase; Kfs = K-feldspar; Bt = biotite; Hb = hornblende; Ser = sericite; Kln = kaoline; Mc = microcline.

Samples 21JD–12 and 21JD–24 were taken from Qujia and Guojialing plutons, respectively. Sample 21JD–12 (GPS: 37°37′38″ N, 121°03′54″ E) is a porphyritic granodiorite from the edge of the Guojialing pluton, Penglai City. It is grey in color with porphyraceous texture, and contains K-feldspar phenocrysts. The mineral assemblage comprises K-feldspar (40 vol.%), plagioclase (28 vol.%), quartz (20 vol.%), hornblende (6 vol.%), biotite

(5 vol.%), metallic minerals (1 vol.%), and minor apatite and titanite (Figure 3d). Sample 21JD–24 (GPS: 37°32′15″ N, 121°21′54″ E) is a porphyritic granodiorite from Qujia pluton, Zhaoyuan City. It is grey in color, and massive, with medium- to coarse-grained granitic texture (Figure 3c). The mineral assemblage includes K-feldspar (35 vol.%), plagioclase (25 vol.%), hornblende (20 vol.%), quartz (10 vol.%), biotite (5 vol.%), and minor secondary minerals (5 vol.% in total). Plagioclase and hornblende crystals are ~0.5–15 mm in length, with many dioritic enclaves visible in the hand specimen (Figure 2c).

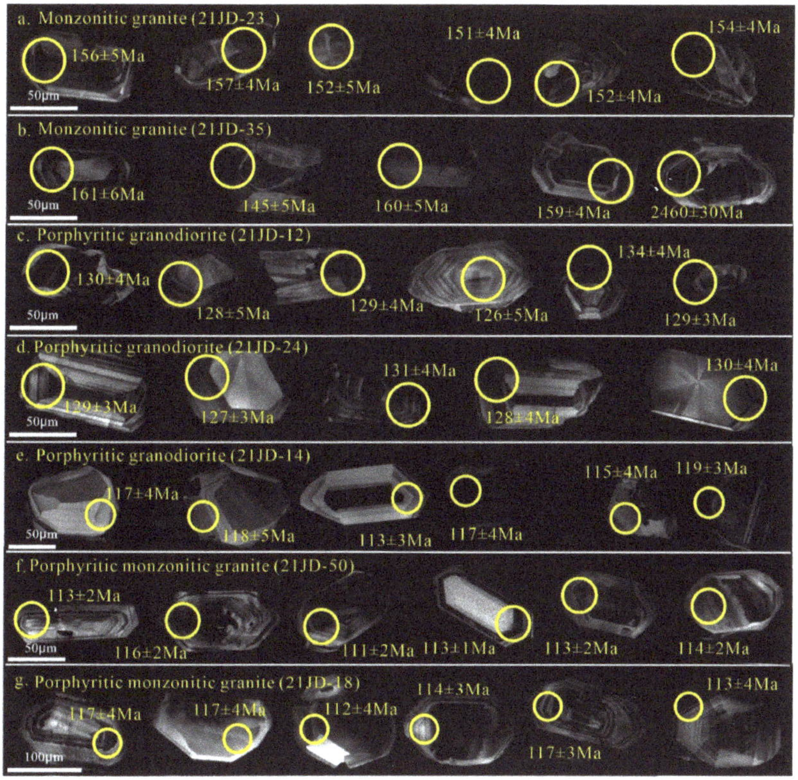

Figure 3. Representative zircon CL images for the Mesozoic granites in the Jiaodong Peninsula. Circles denote the analytical spots with LA–ICPMS zircon U–Pb ages. (**a**,**b**) Monzonitic granite (Linglong granite samples 21JD–23 and 21JD–35); (**c**,**d**) porphyritic granodiorite (Guojialing granite samples 21JD–12 and 21JD–24); (**e**) porphyritic granodiorite (Weideshan granite sample 21JD–14); (**f**) porphyritic monzonitic granite (Weideshan granite sample 21JD–50); (**g**) porphyritic monzonitic granite (Laoshan granite sample 21JD–18).

3.4. The Early Cretaceous Weideshan Granite

The Weideshan granite is dominated by quartz monzonite, monzonitic granite, and granodiorite [40]. Its outcrops cover ~2662 km^2 in the Jiaodong Peninsula, appearing as the large–scale Weideshan, Haiyang, and Jiaonan plutons in the Sulu orogenic belt, and as the smaller Aishan, Nansu, and Yashan plutons in the Jiaobei Terrane. The Weideshan granite is closely related to the regional contemporaneous Cu–Mo–Pb–Zn–Au polymetallic mineralization [2,3,65,77–79]. The magma that formed the Weideshan granite was derived from varying degrees of migmatization of mantle–crust source materials [3,6].

Sample 21JD–14 is a porphyritic granodiorite from the middle part of Aishan pluton (GPS: 37°31′10″ N, 121°44′10″ E), Penglai City. The sample is massive, with a porphyraceous texture, and contains phenocrysts of K-feldspar (10~15 vol.%) in a medium- to fine-grained granitic groundmass (Figure 2e). The mineral assemblage includes plagioclase (40 vol.%), K-feldspar (25 vol.%), quartz (25 vol.%), hornblende (5 vol.%), biotite (3 vol.%), metallic minerals (2 vol.%), and minor apatite and titanite. Dioritic enclaves (5~20 cm diameter) are visible in hand specimen (Figure 2e). Sample 21JD–50 is a porphyritic monzonitic granite from the core of the Weideshan pluton (GPS: 37°17′16″ N, 122°18′59″ E), Rongcheng City. It is grey in color with porphyraceous texture, and contains K-feldspar phenocrysts (Figure 2f). It comprises plagioclase (35 vol.%), K-feldspar (30 vol.%), quartz (20 vol.%), hornblende (10 vol.%), biotite (4 vol.%), metallic minerals (1 vol.%), and minor apatite and titanite.

3.5. The Early Cretaceous Laoshan Granite

The Laoshan granite is mainly distributed in the southeast coastal area of Jiaodong Peninsula, and also sporadically distributed in the Dazeshan area of Jiaobei terrane, with a total area of ~1327 km^2. The rock assemblage here includes monzonitic granite, syenogranite, and alkali feldspar granite [22].

Sample 21JD–18 is a porphyritic monzonitic granite from the Dazeshan pluton (GPS: 37°01′10″ N, 119°57′25″ E), Pingdu city. It is massive with medium- to coarse-grained granitic texture, and contains dense K-feldspar phenocrysts (5 mm~4 cm in length, Figure 2g). The quartz miarolitic structure is well developed. The mineral assemblage includes K-feldspar (30 vol.%), plagioclase (30 vol.%), hornblende (15 vol.%), quartz (15 vol.%), biotite (8 vol.%), metallic minerals (1 vol.%), and minor secondary minerals (~1 vol.% in total) (Figure 2h).

4. Analytical Methods

4.1. Zircon Morphology

The pretreatment, mineral separation, target fabrication and Cathodoluminescence (CL) image shooting were all carried out at the Chengxin geological Testing Co., Ltd., Langfang, China. Zircon grains were separated by conventional magnetic and density techniques to concentrate the non-magnetic, heavy fractions. After sample preparation, zircon grains, free of visible inclusions and major fractures, were handpicked and embedded in epoxy resin and then polished to expose the grain centers. CL images were obtained using a JXA–8100 electron microprobe, in order to characterize the internal structure of the zircons and to select potential target sites for U–Pb analysis. Typical CL images are presented in Figure 3, together with the U–Pb ages for the corresponding spots.

4.2. Zircon U–Pb Dating and Trace Element Analyses

Zircon U–Pb dating and trace element analyses of zircons was conducted by LA–ICP–MS at the Guangzhou Tuoyan Analytical Technology Co., Ltd., Guangzhou, China. Laser sampling was performed using a NWR 193 laser ablation system. An iCAP RQ ICP–MS instrument was used to acquire ion–signal intensities. Helium was applied as a carrier gas. Argon was used as the make-up gas and mixed with the carrier gas via a Y-connector before entering the ICP. The spot size and frequency of the laser were set to 30 μm and 6 Hz, respectively, in this study. The energy was 3.5 J/cm^2. Zircon 91,500 [80] and glass NIST610 [81] were used as external standards for U–Pb dating and trace element calibration, respectively. Each analysis incorporated a background acquisition of approximately 30 s followed by 40 s of data acquisition from the sample. An Excel-based software ICPMSDataCal was used to perform off-line selection and integration of background and analyzed signals, time-drift correction and quantitative calibration for U–Pb dating and trace element analysis. $^{206}Pb/^{238}U$ and $^{207}Pb/^{206}Pb$ ages were interpreted to denote the crystallization ages of the <1000 Ma and >1000 Ma zircon grains, respectively. The results for isotopic ratios, calculated ages and key trace element compositions are listed in Table 1.

Table 1. Summary of key REE, U–Pb isotopic data of magmatic zircons from the Mesozoic granites in the Jiaodong Peninsula, China.

No.	REE		Th/U	Isotopic Ratio						Apparent Age (Ma)					
	Eu/Eu*	Ce/Ce*		$^{207}Pb/^{206}Pb$	1σ	$^{207}Pb/^{235}U$	1σ	$^{206}Pb/^{238}U$	1σ	$^{207}Pb/^{206}Pb$	1σ	$^{207}Pb/^{235}U$	1σ	$^{206}Pb/^{238}U$	1σ
Linglong granite sample 21JD–23															
21JD-23-01	0.10	11.36	0.25	0.1261	0.0027	6.3926	0.1644	0.3678	0.0091	2044	38	2031	23	2019	43
21JD-23-02	0.24	71.24	0.52	0.0488	0.0043	0.1648	0.0143	0.0245	0.0008	136	194	155	12	156	5
21JD-23-03	0.36	42.11	0.41	0.0490	0.0024	0.1662	0.0083	0.0246	0.0007	149	111	156	7	157	4
21JD-23-04	0.44	7.49	0.87	0.0491	0.0049	0.1613	0.0158	0.0239	0.0008	151	216	152	14	152	5
21JD-23-05	0.41	42.83	0.36	0.0534	0.0062	0.1775	0.0203	0.0241	0.0009	345	244	166	18	154	5
21JD-23-06	0.17	24.11	0.32	0.0489	0.0024	0.1650	0.0081	0.0245	0.0007	143	109	155	7	156	4
21JD-23-07	0.44	69.92	0.65	0.0483	0.0132	0.1544	0.0415	0.0232	0.0015	112	545	146	36	148	10
21JD-23-08	0.32	17.11	0.62	0.0536	0.0043	0.1622	0.0129	0.0220	0.0007	352	171	153	11	140	4
21JD-23-09	0.43	146.71	0.15	0.0514	0.0029	0.1679	0.0096	0.0237	0.0007	260	124	158	8	151	4
21JD-23-10	0.31	37.52	0.57	0.0531	0.0115	0.1776	0.0376	0.0243	0.0013	332	428	166	32	155	8
21JD-23-11	0.45	8.32	0.89	0.0522	0.0058	0.1811	0.0199	0.0252	0.0009	294	236	169	17	160	6
21JD-23-12	0.50	81.46	0.59	0.0494	0.0093	0.1671	0.0308	0.0245	0.0012	167	389	157	27	156	7
21JD-23-13	0.35	36.21	0.62	0.0474	0.0057	0.1660	0.0195	0.0254	0.0009	70	262	156	17	162	6
21JD-23-14	0.13	26.62	0.45	0.1479	0.0037	8.3151	0.2364	0.4078	0.0104	2322	42	2266	26	2205	48
21JD-23-15	0.49	45.15	0.98	0.0526	0.0060	0.1733	0.0194	0.0239	0.0009	310	240	162	17	152	5
21JD-23-16	0.36	4.78	0.32	0.0493	0.0039	0.1708	0.0134	0.0251	0.0008	164	175	160	12	160	5
21JD-23-17	0.14	122.00	0.19	0.1544	0.0039	8.9175	0.2541	0.4190	0.0107	2395	42	2330	26	2256	49
21JD-23-18	0.08	144.16	0.19	0.0488	0.0027	0.1639	0.0093	0.0244	0.0007	136	126	154	8	155	4
21JD-23-19	0.29	15.17	0.33	0.0517	0.0030	0.1697	0.0099	0.0238	0.0007	271	128	159	9	152	4
21JD-23-20	0.23	69.33	0.32	0.0498	0.0035	0.1658	0.0117	0.0241	0.0007	186	157	156	10	154	4
Linglong granite sample 21JD-35															
21JD-35-01	0.79	2.28	0.17	0.0618	0.0095	0.2120	0.0316	0.0253	0.0010	667	299	195	26	161	6
21JD-35-02	0.29	11.75	0.17	0.0815	0.0068	1.7931	0.1448	0.1634	0.0047	1233	155	1043	53	976	26
21JD-35-03	0.40	57.29	0.18	0.0427	0.0098	0.1819	0.0409	0.0314	0.0015	0	305	170	35	200	10
21JD-35-04	0.34	166.97	0.18	0.1757	0.0081	8.8279	0.3896	0.3727	0.0090	2613	74	2320	40	2042	42
21JD-35-05	0.52	247.13	0.57	0.0808	0.0139	1.9159	0.1192	0.1753	0.0033	1217	306	1035	71	1056	20
21JD-35-06	0.29	252.58	0.55	0.0524	0.0069	0.1609	0.0207	0.0228	0.0007	303	276	152	18	145	5
21JD-35-07	0.50	29.19	0.88	0.0533	0.0112	0.1856	0.0382	0.0256	0.0013	343	418	173	33	163	8
21JD-35-08	0.26	6.14	0.14	0.0652	0.0157	0.2565	0.0598	0.0291	0.0018	782	438	232	48	185	11
21JD-35-09	0.51	31.98	0.14	0.1741	0.0059	10.2746	0.3306	0.4342	0.0072	2597	56	2460	30	2325	32
21JD-35-10	0.40	131.25	0.15	0.1526	0.0064	6.2519	0.2513	0.3026	0.0061	2376	70	2012	35	1704	30
21JD-35-11	0.10	1.72	0.44	0.0568	0.0068	0.1946	0.0226	0.0252	0.0007	481	245	181	19	160	5
21JD-35-12	0.71	17.95	0.33	0.0617	0.0057	0.2107	0.0190	0.0250	0.0006	665	188	194	16	159	4
21JD-35-13	0.20	26.82	0.51	0.0563	0.0096	0.1640	0.0271	0.0214	0.0008	462	338	154	24	136	5
21JD-35-14	0.81	1.91	0.09	0.0703	0.0061	1.8682	0.0733	0.1606	0.0023	936	170	915	40	968	23
21JD-35-15	0.84	1.33	0.20	0.0517	0.0096	0.1473	0.0266	0.0208	0.0008	270	377	140	24	132	5
21JD-35-16	0.49	14.86	0.79	0.0424	0.0095	0.1771	0.0390	0.0306	0.0014	0	279	166	34	194	9
21JD-35-17	0.38	137.06	0.12	0.1978	0.0286	10.6834	0.3718	0.4765	0.1838	2608	219	2341	36	2564	58

Table 1. *Cont.*

No.	REE		Th/U	Isotopic Ratio						Apparent Age (Ma)					
	Eu/Eu*	Ce/Ce*		$^{207}Pb/^{206}Pb$	1σ	$^{207}Pb/^{235}U$	1σ	$^{206}Pb/^{238}U$	1σ	$^{207}Pb/^{206}Pb$	1σ	$^{207}Pb/^{235}U$	1σ	$^{206}Pb/^{238}U$	1σ
21JD-35-18	0.56	16.61	0.79	0.0768	0.0169	0.2475	0.0526	0.0235	0.0013	1116	386	225	43	150	8
21JD-35-19	2.47	1.32	0.02	0.1181	0.0115	3.7234	0.3493	0.2295	0.0080	1928	165	1576	75	1332	42
21JD-35-20	0.12	96.34	0.04	0.2090	0.0093	16.8560	0.7084	0.5878	0.0120	2898	70	2927	40	2981	49
Guojialing granite sample 21JD-12															
21JD-12-01	0.43	6.71	0.63	0.0475	0.0100	0.1354	0.0279	0.0207	0.0011	75	435	129	25	132	7
21JD-12-02	0.41	78.84	0.53	0.0514	0.0033	0.1445	0.0094	0.0204	0.0006	259	142	137	8	130	4
21JD-12-03	0.47	49.35	0.41	0.0663	0.0050	0.2272	0.0169	0.0249	0.0008	815	149	208	14	158	5
21JD-12-04	0.62	1.32	0.62	0.0474	0.0028	0.1335	0.0078	0.0204	0.0006	68	133	127	7	130	4
21JD-12-05	0.32	47.84	0.17	0.0495	0.0057	0.1431	0.0164	0.0210	0.0007	172	250	136	15	134	5
21JD-12-06	0.50	2.07	0.65	0.0522	0.0028	0.1464	0.0080	0.0204	0.0006	293	118	139	7	130	4
21JD-12-07	0.49	9.64	0.60	0.0489	0.0073	0.1351	0.0199	0.0200	0.0008	144	317	129	18	128	5
21JD-12-08	0.46	111.40	0.52	0.1713	0.0082	10.7043	0.5267	0.4531	0.0150	2571	78	2498	46	2409	66
21JD-12-09	0.46	15.13	0.58	0.0496	0.0046	0.1366	0.0126	0.0200	0.0006	177	204	130	11	128	4
21JD-12-10	0.49	4.98	0.31	0.0487	0.0061	0.1420	0.0175	0.0212	0.0008	132	270	135	16	135	5
21JD-12-11	0.42	56.98	0.66	0.1584	0.0056	7.0552	0.2641	0.3230	0.0091	2439	58	2118	33	1804	44
21JD-12-12	0.55	1.71	0.55	0.0509	0.0058	0.1505	0.0169	0.0215	0.0008	235	243	142	15	137	5
21JD-12-13	0.72	7.01	0.53	0.1801	0.0063	10.3570	0.3889	0.4170	0.0121	2654	57	2467	35	2247	55
21JD-12-14	0.23	59.32	0.51	0.0490	0.0036	0.1361	0.0099	0.0202	0.0006	147	163	130	9	129	4
21JD-12-15	0.54	86.87	0.63	0.0489	0.0070	0.1331	0.0187	0.0198	0.0008	141	305	127	17	126	5
21JD-12-16	0.48	62.31	0.55	0.0490	0.0046	0.1416	0.0131	0.0210	0.0007	147	206	135	12	134	4
21JD-12-17	0.66	9.81	0.63	0.0489	0.0043	0.1424	0.0125	0.0211	0.0007	141	195	135	11	135	4
21JD-12-18	0.44	189.91	0.58	0.0531	0.0026	0.1485	0.0074	0.0203	0.0006	333	106	141	7	129	3
21JD-12-19	0.45	190.16	0.48	0.0491	0.0066	0.1433	0.0188	0.0212	0.0008	154	285	136	17	135	5
21JD-12-20	0.51	117.60	0.51	0.0477	0.0142	0.1271	0.0370	0.0193	0.0013	82	586	121	33	123	8
Guojialing granite sample 21JD-24															
21JD-24-01	0.61	89.13	0.64	0.0468	0.0021	0.1305	0.0061	0.0202	0.0005	37	105	125	5	129	3
21JD-24-02	0.64	354.87	0.47	0.0482	0.0025	0.1365	0.0072	0.0206	0.0006	108	118	130	6	131	3
21JD-24-03	0.53	111.08	0.62	0.0477	0.0050	0.1370	0.0143	0.0209	0.0007	82	233	130	13	133	4
21JD-24-04	0.60	93.98	0.61	0.0486	0.0024	0.1387	0.0071	0.0207	0.0006	127	114	132	6	132	4
21JD-24-05	0.50	66.09	0.78	0.0489	0.0024	0.1347	0.0068	0.0200	0.0005	144	111	128	6	127	3
21JD-24-06	0.57	242.67	0.31	0.0476	0.0080	0.1384	0.0227	0.0211	0.0009	79	356	132	20	135	6
21JD-24-07	0.56	110.01	0.71	0.0483	0.0029	0.1362	0.0083	0.0205	0.0006	112	137	130	7	131	4
21JD-24-08	0.59	253.94	0.71	0.0487	0.0027	0.1368	0.0077	0.0204	0.0006	135	125	130	7	130	4
21JD-24-09	0.50	136.14	0.58	0.0507	0.0032	0.1402	0.0088	0.0201	0.0006	226	138	133	8	128	4
21JD-24-10	0.57	77.41	0.63	0.0521	0.0059	0.1396	0.0156	0.0194	0.0007	291	240	133	14	124	4
21JD-24-11	0.58	142.38	0.89	0.0492	0.0025	0.1386	0.0072	0.0204	0.0006	157	115	132	6	130	3
21JD-24-12	0.57	122.35	0.62	0.0489	0.0027	0.1385	0.0078	0.0206	0.0006	141	126	132	7	131	4
21JD-24-13	0.56	207.60	0.57	0.0488	0.0026	0.1390	0.0074	0.0207	0.0006	139	119	132	7	132	4
21JD-24-14	0.54	57.24	0.72	0.0487	0.0098	0.1405	0.0277	0.0209	0.0011	133	415	134	25	134	7

Table 1. *Cont.*

No.	REE		Th/U	Isotopic Ratio						Apparent Age (Ma)					
	Eu/Eu*	Ce/Ce*		$^{207}Pb/^{206}Pb$	1σ	$^{207}Pb/^{235}U$	1σ	$^{206}Pb/^{238}U$	1σ	$^{207}Pb/^{206}Pb$	1σ	$^{207}Pb/^{235}U$	1σ	$^{206}Pb/^{238}U$	1σ
21JD-24-15	0.56	173.35	0.69	0.0458	0.0020	0.1292	0.0059	0.0205	0.0005	0	88	123	5	131	3
21JD-24-16	0.51	154.94	0.83	0.0507	0.0026	0.1376	0.0072	0.0197	0.0005	226	114	131	6	126	3
21JD-24-17	0.55	126.05	0.56	0.0497	0.0033	0.1394	0.0094	0.0203	0.0006	182	149	133	8	130	4
21JD-24-18	0.64	89.58	0.62	0.0490	0.0034	0.1341	0.0093	0.0199	0.0006	146	155	128	8	127	4
21JD-24-19	0.63	58.36	0.80	0.0474	0.0035	0.1366	0.0102	0.0209	0.0006	70	169	130	9	133	4
21JD-24-20	0.53	69.71	0.70	0.0502	0.0026	0.1441	0.0076	0.0208	0.0006	205	115	137	7	133	4
Weideshan granite sample 21JD-14															
21JD-14-01	0.50	82.10	1.03	0.0487	0.0094	0.1179	0.0224	0.0176	0.0009	133	401	113	20	112	5
21JD-14-02	0.57	34.24	1.12	0.0503	0.0047	0.1267	0.0118	0.0183	0.0006	210	205	121	11	117	4
21JD-14-03	0.43	43.44	1.11	0.0520	0.0067	0.1316	0.0166	0.0183	0.0007	287	270	126	15	117	4
21JD-14-04	0.61	56.34	1.36	0.0529	0.0055	0.1329	0.0136	0.0182	0.0006	323	218	127	12	117	4
21JD-14-05	0.48	2.31	0.92	0.1745	0.0058	0.5906	0.0208	0.0246	0.0007	2601	54	471	13	156	4
21JD-14-06	0.53	112.87	1.13	0.0493	0.0077	0.1255	0.0192	0.0184	0.0008	164	328	120	17	118	5
21JD-14-07	0.50	9.63	1.42	0.0492	0.0036	0.1273	0.0093	0.0188	0.0006	159	163	122	8	120	3
21JD-14-08	0.55	1.69	0.86	0.0483	0.0035	0.1180	0.0087	0.0177	0.0005	112	165	113	8	113	3
21JD-14-09	0.55	109.46	0.95	0.0510	0.0048	0.1291	0.0119	0.0184	0.0006	242	201	123	11	117	4
21JD-14-10	0.54	2.00	0.85	0.0490	0.0066	0.1208	0.0159	0.0179	0.0007	149	286	116	14	114	4
21JD-14-11	0.46	1.57	1.13	0.0527	0.0037	0.1352	0.0095	0.0186	0.0006	317	152	129	9	119	3
21JD-14-12	0.53	3.05	0.87	0.0512	0.0052	0.1275	0.0127	0.0181	0.0006	251	217	122	11	115	4
21JD-14-13	0.37	132.84	0.83	0.0485	0.0036	0.1257	0.0092	0.0188	0.0006	123	164	120	8	120	4
21JD-14-14	0.58	17.92	1.36	0.0500	0.0041	0.1296	0.0105	0.0188	0.0006	196	179	124	9	120	4
21JD-14-15	0.56	50.91	0.99	0.0500	0.0136	0.1240	0.0331	0.0180	0.0012	197	534	119	30	115	7
21JD-14-16	0.34	25.10	1.04	0.0507	0.0033	0.1316	0.0087	0.0188	0.0005	228	145	126	8	120	3
21JD-14-17	0.68	15.46	1.06	0.0518	0.0037	0.1324	0.0095	0.0186	0.0006	276	156	126	8	119	3
21JD-14-18	0.53	1.51	0.98	0.0508	0.0059	0.1395	0.0160	0.0199	0.0007	233	248	133	14	127	5
21JD-14-19	0.54	118.13	0.84	0.0488	0.0043	0.1270	0.0110	0.0189	0.0006	138	193	121	10	121	4
21JD-14-20	0.54	61.00	1.16	0.0511	0.0147	0.1347	0.0379	0.0191	0.0013	244	555	128	34	122	8
Weideshan granite sample 21JD-50															
21JD-50-01	0.51	81.79	0.69	0.0468	0.0035	0.1133	0.0083	0.0176	0.0003	39	174	109	8	113	2
21JD-50-02	0.35	112.64	0.86	0.0522	0.0032	0.1221	0.0073	0.0170	0.0002	295	139	117	7	109	1
21JD-50-03	0.46	168.41	0.68	0.0460	0.0032	0.1145	0.0081	0.0181	0.0003	110	92	110	7	116	2
21JD-50-04	0.35	196.71	0.74	0.0529	0.0036	0.1218	0.0076	0.0168	0.0003	324	156	117	7	107	2
21JD-50-05	0.35	65.87	0.89	0.0501	0.0023	0.1208	0.0053	0.0175	0.0002	211	101	116	5	112	1
21JD-50-06	0.33	162.18	0.81	0.0524	0.0027	0.1232	0.0062	0.0171	0.0002	306	117	118	6	109	1
21JD-50-07	0.33	160.43	0.87	0.0481	0.0019	0.1148	0.0046	0.0173	0.0002	106	91	110	4	110	1
21JD-50-08	0.28	127.50	0.74	0.0491	0.0037	0.1174	0.0092	0.0174	0.0003	154	170	113	8	111	2
21JD-50-09	0.46	10.59	1.42	0.0531	0.0025	0.1099	0.0047	0.0151	0.0002	345	103	106	4	97	2
21JD-50-10	0.35	2.39	0.84	0.0505	0.0023	0.1196	0.0053	0.0172	0.0002	217	106	115	5	110	1
21JD-50-11	0.58	90.86	0.71	0.0531	0.0033	0.1234	0.0073	0.0171	0.0003	345	145	118	7	109	2

Table 1. *Cont.*

No.	REE		Th/U	Isotopic Ratio						Apparent Age (Ma)					
	Eu/Eu*	Ce/Ce*		$^{207}Pb/^{206}Pb$	1σ	$^{207}Pb/^{235}U$	1σ	$^{206}Pb/^{238}U$	1σ	$^{207}Pb/^{206}Pb$	1σ	$^{207}Pb/^{235}U$	1σ	$^{206}Pb/^{238}U$	1σ
21JD-50-12	0.31	142.77	1.02	0.0470	0.0021	0.1132	0.0052	0.0174	0.0001	56	98	109	5	112	1
21JD-50-13	0.30	160.79	0.93	0.0493	0.0023	0.1204	0.0058	0.0176	0.0002	161	105	115	5	113	1
21JD-50-14	0.40	145.41	0.83	0.0500	0.0029	0.1216	0.0066	0.0177	0.0002	195	133	116	6	113	1
21JD-50-15	0.38	98.65	0.99	0.0522	0.0021	0.1268	0.0050	0.0176	0.0002	295	86	121	4	113	2
21JD-50-16	0.33	33.64	1.20	0.0486	0.0034	0.1168	0.0075	0.0177	0.0002	128	159	112	7	113	1
21JD-50-17	0.31	148.21	0.81	0.0475	0.0019	0.1125	0.0046	0.0172	0.0002	76	93	108	4	110	1
21JD-50-18	0.33	194.22	0.78	0.0462	0.0019	0.1110	0.0045	0.0175	0.0002	6	96	107	4	112	1
21JD-50-19	0.45	64.82	0.80	0.0485	0.0036	0.1180	0.0085	0.0178	0.0003	124	176	113	8	114	2
21JD-50-20	0.35	156.23	0.82	0.0470	0.0024	0.1127	0.0055	0.0175	0.0002	50	119	108	5	112	2
21JD-50-21	0.32	140.88	0.74	0.0505	0.0022	0.1219	0.0055	0.0175	0.0002	220	99	117	5	112	1
21JD-50-22	0.38	14.28	1.01	0.0511	0.0022	0.1216	0.0054	0.0172	0.0002	256	98	117	5	110	1
21JD-50-23	0.36	138.41	0.78	0.0490	0.0020	0.1165	0.0047	0.0173	0.0002	146	94	112	4	111	1
21JD-50-24	0.41	121.86	0.79	0.0508	0.0022	0.1214	0.0052	0.0174	0.0002	232	100	116	5	111	1
21JD-50-25	0.30	129.74	0.80	0.0487	0.0022	0.1169	0.0052	0.0175	0.0002	200	103	112	5	112	1
Laoshan granite sample 21JD-18															
21JD-18-01	0.50	3.69	1.02	0.0485	0.0047	0.1227	0.0117	0.0183	0.0006	125	212	118	11	117	4
21JD-18-02	0.54	15.69	1.27	0.0485	0.0030	0.1252	0.0077	0.0187	0.0005	123	137	120	7	120	3
21JD-18-03	0.53	34.34	1.01	0.0490	0.0044	0.1242	0.0110	0.0184	0.0006	149	196	119	10	117	4
21JD-18-04	0.39	3.32	0.89	0.0512	0.0042	0.1303	0.0097	0.0187	0.0003	250	186	124	9	119	2
21JD-18-05	0.47	4.01	0.91	0.0501	0.0031	0.1290	0.0083	0.0187	0.0003	198	144	123	7	115	2
21JD-18-06	0.20	11.18	0.54	0.0514	0.0027	0.1298	0.0071	0.0183	0.0005	256	118	124	6	117	3
21JD-18-07	0.48	4.06	0.61	0.0483	0.0018	0.1333	0.0054	0.0200	0.0005	115	87	127	5	128	3
21JD-18-08	0.24	72.56	0.49	0.0491	0.0049	0.1236	0.0121	0.0182	0.0006	154	216	118	11	117	4
21JD-18-09	0.39	63.42	1.02	0.0531	0.0065	0.1285	0.0155	0.0176	0.0007	332	256	123	14	112	4
21JD-18-10	0.48	26.74	1.05	0.0482	0.0038	0.1213	0.0096	0.0183	0.0006	109	177	116	9	117	4
21JD-18-11	0.43	3.32	0.83	0.0488	0.0041	0.1204	0.0100	0.0179	0.0006	138	184	116	9	114	4
21JD-18-12	0.42	7.35	0.52	0.0505	0.0047	0.1326	0.0105	0.0194	0.0005	220	200	126	9	118	3
21JD-18-13	0.41	161.66	0.71	0.0503	0.0047	0.1275	0.0109	0.0188	0.0005	209	200	122	10	120	3
21JD-18-14	0.32	75.21	0.65	0.0509	0.0036	0.1339	0.0086	0.0192	0.0003	235	163	128	8	118	2
21JD-18-15	0.54	7.56	0.70	0.0486	0.0023	0.1197	0.0059	0.0179	0.0005	126	108	115	5	114	3
21JD-18-16	0.16	5.47	0.48	0.0460	0.0047	0.1216	0.0124	0.0192	0.0006	0	227	117	11	123	4
21JD-18-17	0.39	149.88	0.98	0.0493	0.0044	0.1242	0.0111	0.0183	0.0006	160	197	119	10	117	4
21JD-18-18	0.23	2.34	0.52	0.0480	0.0038	0.1209	0.0091	0.0184	0.0003	102	178	116	8	110	2
21JD-18-19	0.41	55.26	0.68	0.0521	0.0060	0.1159	0.0132	0.0162	0.0006	288	243	111	12	103	4
21JD-18-20	0.37	92.09	0.77	0.0527	0.0058	0.1282	0.0139	0.0177	0.0006	314	233	123	13	113	4

Errors are 1-sigma. Eu/Eu*= 2 × $Eu_N/(Sm_N + Gd_N)$, represents the degree of Eu anomaly. Ce/Ce*= 2 × $Ce_N/(La_N + Pr_N)$, represents the degree of Ce anomaly.

5. Results

Seven representative samples of the Mesozoic granites including monzonitic granite (samples 21JD–23 and 21JD–35) of Linglong granite, porphyritic granodiorite (samples 21JD–12 and 21JD–24) of Guojialing granite, porphyritic granodiorite (sample 21JD–14) and porphyritic monzonitic granite (sample 21JD–50) of Weideshan granite, and porphyritic monzonitic granite (sample 21JD–18) from Laoshan granite were selected for zircon U–Pb dating. Zircon U–Pb geochronology and geochemistry data are summarized in Table 1. Errors on individual analyses are cited as 1σ, and the weighted mean $^{206}Pb/^{238}U$ ages are quoted at the 95% confidence level. CL images of the representative zircon grains with $^{206}Pb/^{238}U$ ages are shown in Figure 3.

5.1. *The Linglong Granite*

Zircons in the monzonitic granite (sample 21JD–23) are subhedral to euhedral (Figure 3a). CL imaging reveals that most grains are prismatic, colorless, transparent, and ranging from 50 to 150 μm in length, as well as show concentric oscillatory zoning (Figure 3a). In addition, the chondrite-normalized REE patterns are strongly enriched in heavy rare earth elements (HREEs) and depleted in light rare earth element (LREEs, Figure 4b). The zircons have strong negative Eu anomalies (Eu/Eu* = 0.08–0.50), high Ce/Ce* ratios (Ce/Ce* = 4.78–84.16, some reaching 146.71) and high Th/U ratios (0.15–0.98), characteristic of magmatic zircon (Table 1, [82–84]). Twenty zircon grains were analyzed for U–Pb isotopes. Except for three inherited grains yielding $^{207}Pb/^{206}Pb$ ages from 2395 $\pm$ 42 Ma to 2044 $\pm$ 38 Ma, the remaining seventeen concordant analyses yield apparent $^{206}Pb/^{238}U$ ages ranging from 162 $\pm$ 6 Ma to 140 $\pm$ 4 Ma with a weighted mean of 154 $\pm$ 2 Ma (MSWD = 1; n = 17) ($^{206}Pb/^{238}U$ and $^{207}Pb/^{206}Pb$ ages that were interpreted to denote the crystallization ages of the <1000 Ma and >1000 Ma zircon grains, respectively) (Figure 4a). The latter age can be considered as crystallization age of the granite, and is consistent with the zircon U–Pb ages of 164–140 Ma previously reported [85–87]. The Neoproterozoic zircon ages with normal magmatic zircon features indicate that the protolith rocks were not completely remelted, representing the information of magma source area, that is, derived from the Precambrian basement of NCC and subduction plate of YC.

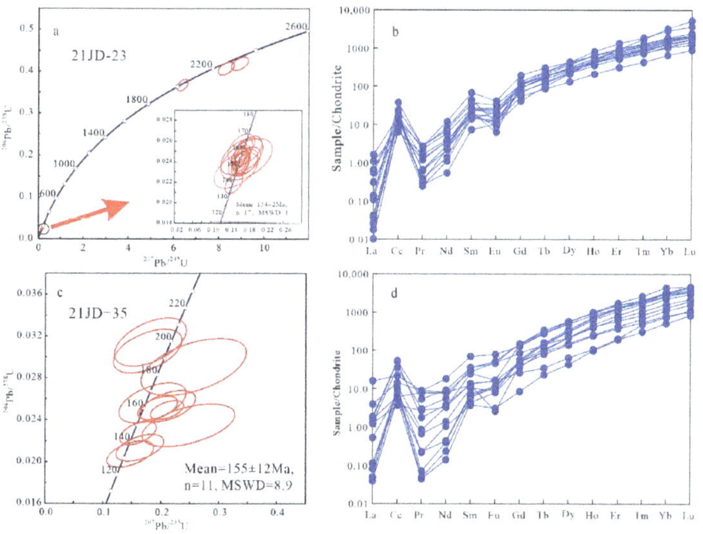

Figure 4. Samples from the Late Jurassic Linglong granite. (**a**,**c**) LA–ICP–MS magmatic zircon U-Pb concordia diagrams and dating results of samples 21JD–23 and 21JD–35, respectively. (**b**,**d**) Chondrite-normalized REE patterns of zircons from samples 21JD–23 and 21JD–35, respectively.

Zircons in the monzonitic granite (sample 21JD–35) are also euhedral, prismatic, colorless, and transparent. They range from 100 to 300 μm in size with length/width ratios of 1:1 to 3:1. CL imaging reveals that most grains have fine oscillatory growth zones (Figure 3b). All zircons have similar chondrite–normalized REE patterns, strongly enriched in HREEs and depleted in LREEs (Figure 4d), but with a slightly flatter HREE profile than the sample 21JD–35. Their strongly negative Eu anomalies (Eu/Eu* = 0.10–0.84, with one zircon grain recording a higher Eu/Eu* ratio of 2.47), high Ce/Ce* ratios (Ce/Ce* = 1.32–252.58) and high Th/U ratios (0.02–0.88) are characteristic of magmatic zircon (Table 1, [82–84]). Twenty zircon grains were analyzed for U-Pb isotopes. Eleven age-spots on zircons yield concordant $^{206}Pb/^{238}U$ ages ranging from 200 ± 10 Ma to 132 ± 5 Ma with a weighted mean of 155 ± 12 Ma (MSWD = 8.9; *n* = 11) (Figure 4c). This age can be considered as crystallization age of the granite, and is similar to the zircon U–Pb age (154 ± 2 Ma, Figure 4a) of the sample 21JD–35. Meanwhile, this sample has more Proterozoic–Archaeozoic ages than the former. It is probably that the sample location is closer to the edge of the pluton, resulting in more xenoliths and incomplete melting components of the NCC and YC.

5.2. The Guojialing Granite

Zircons in the porphyritic granodiorite (samples 21JD–12 and 21JD–24) are subhedral to euhedral and range from 100 to 300 μm in size. CL imaging reveals that most grains have fine oscillatory growth zones (Figure 3c,d). All zircons have similar chondrite-normalized REE patterns, strongly enriched in HREEs and depleted in LREEs (Figure 5b,d). Their strongly negative Eu anomalies (Eu/Eu* = 0.23–0.72), high Ce/Ce* ratios (Ce/Ce* = 1.32–354.87, most are less than 100) and high Th/U ratios (0.17–0.89) are characteristic of magmatic zircon (Table 1, [82–84]).

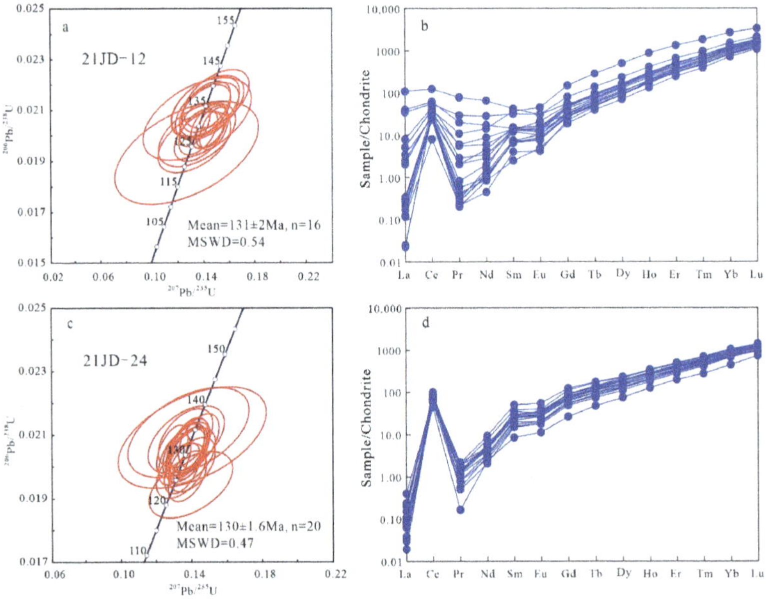

Figure 5. Samples from the Early Cretaceous Guojialing granite. (**a**,**c**) LA–ICP–MS magmatic zircon U–Pb concordia diagrams and dating results of samples 21JD–12 and 21JD–24, respectively. (**b**,**d**) Chondrite–normalized REE patterns of zircons from samples 21JD–12 and 21JD–24, respectively.

Twenty zircon grains from porphyritic granodiorite (sample 21JD–12) were analyzed for U–Pb isotopes, of which one analysis (21JD–12–03) was excluded because of high discordance caused by cracks within the zircons. Except for three inherited grains yielding

$^{207}Pb/^{206}Pb$ ages from 2439 ± 58 Ma to 2570 ± 78 Ma, the remaining sixteen age-spots on zircons yield concordant $^{206}Pb/^{238}U$ ages ranging from 137 ± 5 Ma to 123 ± 8 Ma with a weighted mean of 131 ± 2 Ma (MSWD = 0.54; n = 16) (Figure 5a). This age can be considered as crystallization age of the granite. There are also twenty zircon grains from porphyritic granodiorite (sample 21JD–24) were analyzed for U–Pb isotopes. Twenty age-spots on zircons yield concordant $^{206}Pb/^{238}U$ ages ranging from 135 ± 6 Ma to 124 ± 4 Ma with a weighted mean of 130 ± 2 Ma (MSWD = 0.47; n = 20) (Figure 5c). These ages are consistent with the zircon U–Pb ages of 135–125 Ma previously reported by [85–87].

5.3. The Weideshan Granite

Zircons in the porphyritic granodiorite (sample 21JD–14) and porphyritic monzonitic granite (sample 21JD–50) are euhedral and range from 150 to 300 μm in size. CL imaging reveals that most grains have fine oscillatory growth zones (Figure 3e,f). All zircons have similar chondrite-normalized REE patterns, strongly enriched in HREEs and depleted in LREEs (Figure 6b,d). Their strongly negative Eu anomalies (Eu/Eu* = 0.28–0.68, averaged at 0.44), high Ce/Ce* ratios (Ce/Ce* = 1.51–196.71, averaged at 83.35) and high Th/U ratios (0.68–1.42, averaged at 0.95) are characteristic of magmatic zircon (Table 1, [82–84]).

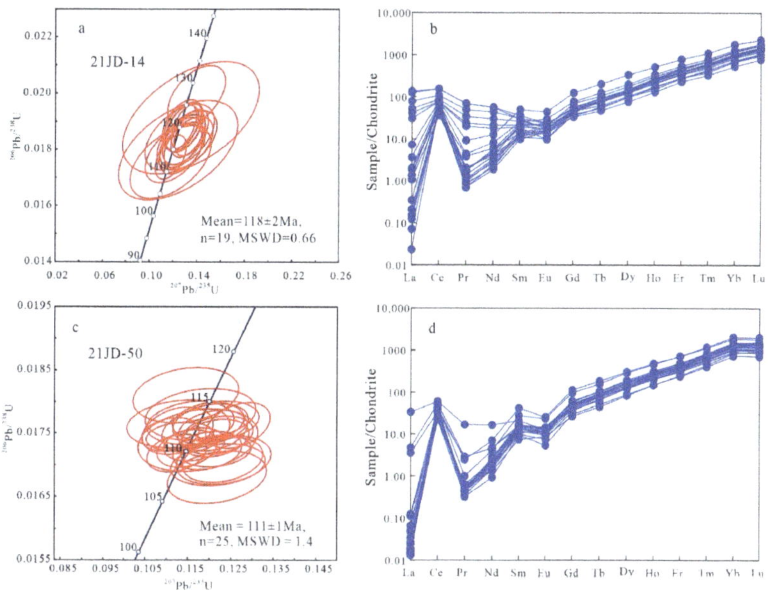

Figure 6. Samples from the Early Cretaceous Weideshan granite. (**a**,**c**) LA–ICP–MS magmatic zircon U–Pb concordia diagrams and dating results of samples 21JD–14 and 21JD–50, respectively. (**b**,**d**) Chondrite–normalized REE patterns of zircons from samples 21JD–14 and 21JD–50, respectively.

Twenty zircon grains from porphyritic granodiorite (sample 21JD–14) were analyzed for U–Pb isotopes, of which one analysis (21JD–14–05) was excluded due to being far from the concordant curve, possibly due to the presence of mineral inclusions. The remaining nineteen age–spots on zircons yield concordant $^{206}Pb/^{238}U$ ages ranging from 112 ± 5 Ma to 127 ± 5 Ma with a weighted mean of 118 ± 2 Ma (MSWD = 0.66; n = 19) (Figure 6a). This age can be considered as crystallization age of the granite. There are also twenty-five zircon grains from porphyritic monzonitic granite (sample 21JD–50) that were analyzed for U–Pb isotopes. All age-spots on zircons yield concordant $^{206}Pb/^{238}U$ ages ranging from 97 ± 2 Ma to 116 ± 2 Ma with a weighted mean of 111 ± 1 Ma (MSWD = 1.4; n = 25) (Figure 6c). These ages are consistent with the zircon U–Pb ages of 125–110 Ma previously reported by [85–87].

5.4. The Laoshan Granite

Zircons in the porphyritic monzonitic granite (sample 21JD–18) are also euhedral, prismatic, colorless, and transparent. They are granular, ranging from 80 to 300 μm in size, with aspect ratios of 1:1.5 to 4:1. CL imaging reveals that most grains have fine oscillatory growth zones (Figure 3g). All zircons have similar chondrite-normalized REE patterns, strongly enriched in HREEs and depleted in LREEs (Figure 7b). Their strongly negative Eu anomalies (Eu/Eu* = 0.16–0.54, averaged at 0.40), high Ce/Ce* ratios (Ce/Ce* = 2.34–161.66, averaged at 39.96) and high Th/U ratios (0.48–1.27, averaged at 0.78) are characteristic of magmatic zircon (Table 1, [82–84]). Of the twenty zircons from the porphyritic monzonitic granite (sample 21JD–18) that were dated by LA-ICP-MS U–Pb analysis, yield concordant $^{206}Pb/^{238}U$ ages ranging from 103 ± 4 Ma to 120 ± 3 Ma with a weighted mean of 116 ± 2 Ma (MSWD = 0.043; *n* = 20) (Figure 7a).

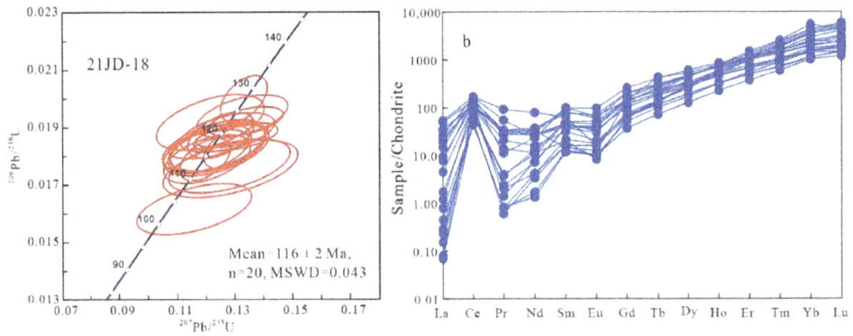

Figure 7. Samples from the Early Cretaceous Laoshan granite. (**a**) LA–ICP–MS magmatic zircon U–Pb concordia diagrams and dating results of sample 21JD–18. (**b**) Chondrite–normalized REE patterns of zircons from sample 21JD–18.

6. Discussion

6.1. Magma Sources and Petrogenesis of Granites

6.1.1. The Triassic Shidao Granite

The Shidao granite is the only one that has been reported with a Late Triassic age in the Weihai terrane. The isotopic age of Shidao granite tested by predecessors is ranging from 227 to 200 Ma [69–74].

The Shidao granite is composed of syenite, quartz syenite, and syenogranite from early to late. The only Be deposit in the Jiaodong area is found in Shidao syenogranite (205.7 ± 1.4 Ma, [72]). The Shidao granite intruded into the UHP metamorphic rocks in the Sulu orogenic belt, indicating that they were formed after the UHP metamorphic event. Many isotopic ages of eclogites in the Sulu orogenic belt have been measured by previous studies, and the results are as follows: 228–221 Ma (Sm-Nd method) [88,89], 217.1 ± 8.7 Ma (U-Pb method) [90], and 228 ± 29 Ma (zircon SHRIMP method) [91]. The zircon isotopic ages of cores and edges of UHP coesite-bearing gneiss with SHRIMP method, which were 242–209 Ma and 230–202 Ma, respectively, suggesting that 242–224 Ma were UHP metamorphic ages and 229–202 Ma were UHP retrograde metamorphic ages [92,93]. It can be seen that the formation age of Shidao granite is consistent with the retrograde metamorphism age of UHP metamorphic rocks or the roll back timing of UHP metamorphic rocks [69]. Previous studies suggest that the Shidao granite is a syn-orogenic (syn-reentry) intrusive rock that related to plate segment dissociation during the subduction of the YC and NCC [71,72,94] or post-orogenic intrusive rock [70,73,74]. In this study, we agree with the latter, that the Shidao alkaline rocks were generated under a post-collisional extensional setting. In view of the hornblende facies at the retrograde metamorphism stage of the UHP metamorphic rocks, it is considered that the Shidao granite is the product of partial melting of enriched lithospheric mantle, crystallization differentiation, and assimilation mixing of lower crust under the condition of UHP metamorphic rocks returning to hornblende facies.

6.1.2. The Jurassic–Cretaceous Granites

The Jurassic–Cretaceous intrusive rocks are well developed in the Jiaodong Peninsula, mainly including Linglong, Guojialing, Weideshan, and Laoshan granites. Previous studies have carried out a large number of isotopic tests on Late Mesozoic magmatic rocks. Some scholars have calculated the age data of Jurassic–Cretaceous granites with high accuracy by using different testing methods [95]. The diagenetic ages of the four-stage granites are approximately 164–140 Ma, 135–125 Ma, 125–110 Ma, and 120–104 Ma, respectively [6–8,10–12,15,18,22,25,30,31,65,66,72,96–106]. The data obtained in this paper are basically in line with the previous chronological data.

The Linglong granite is the main ore-hosting body of Jiaodong gold deposit. According to statistics, about 77% of Jiaodong gold deposits and a large porphyry Mo deposit (e.g., Xingjiashan, molybdenite Re–Os method, 158.7 $\pm$ 2.5 Ma) occur in the Linglong granite [107]. The Linglong granite is composed of monzogranite series intrusive rocks. The $^{87}Sr/^{86}Sr$, εNd (t), εHf(t) and Sr/Y values of Linglong granite are 0.711281~0.712418, −21.6~−19.4, −28.7~6.2, and 55.07~214.44, respectively. εHf (t) values are placed at 1.9 and 2.5 Ga crustal evolution lines [18,108]. The low εHf (t) value, high Sr/Y value and no obvious negative Eu anomaly of the rock are similar to adakite [9,18], which is similar to Jiaodong Neoarchean TTG [109]. At the same time, the Linglong granite often contains peraluminous minerals such as garnet, which indicates S–type granite feature. Magma was formed in a relatively high-pressure environment and was derived from partial melting of the thickened lower crust of North China [16]. The ages of the Linglong granite range from 164 to 140 Ma, which should be considered as the major period of magma crystallization [11,12,15,18,31,72,97,99,100,102,103]. Simultaneously, the development of a large number of inherited zircons indicates the complexity of the origin of Linglong granite, which are products of partial melting of Neoarchean crystalline basement in the NCC and remnants of deep subduction materials in the Sulu UHP metamorphic belt.

The Guojialing granite is predominantly distributed in Jiaodong gold concentration area, and is an important ore-hosting body of Jiaodong gold deposit. About 10% of the gold deposits occur in the Guojialing granite [107]. This stage of granite is closely related to Jiaodong gold deposit due to its closely metallogenic timing and spatial distribution. The Guojialing granite is an intrusive rock series composed of monzodiorite, adamellite, granodiorite, and monzonitic granite, with the I-type granite and adakite geochemical features of crust–mantle mixed sources [85,86]. The Guojialing granite is rich in Neoarchean and Paleoproterozoic inherited zircons, with a few of Jurassic, but no Neoproterozoic and early Paleozoic inherited zircons [18]. It implies that the crust source material is the Precambrian basement of NCC, and Sulu orogenic belt and YC materials are missing. At present, researchers agree on the genesis of the Guojialing granite, which is believed to be the result of the mixing of the lower crust acidic magma formed by partial melting of the Jiaodong basement metamorphic rock series and mantle-derived basic magma [85,86]. The formation of the magma is related to the subduction of the PPP beneath the NCC and the upwelling of the asthenosphere. The scholars have systematically tested the isotopic age of the Guojialing granite by various dating methods. The results show that the isotopic ages range from 132 to 125 Ma, and the host rocks are consistent with the age of the dioritic enclaves, indicating simultaneous emplacement of both acid and basic magmas [7,8,10,12,15,18,104,106].

The Weideshan granite has a certain spatial distance with gold mineralization, but they are closely related to Cu, Pb, Zn and Mo deposits [3,65,110]. The Weideshan granite is composed of rocks with an independent evolutionary history and protolith origin, different from the Linglong, Guojialing, and Laoshan granites. The Weideshan granite displays similarities in lithofacies and geochemistry with the Guojialing granite. However, the Guojialing granite has similarities in trace element composition with adakite, tonalite-trondhjemite-granodiorite, and sodic granite, while the Weideshan granite closely resembles typical island arc rocks [111,112]. The Guojialing and Weideshan granites typically contain microgranular dioritic enclaves with mantle-like geochemical characteristics [6,21,113]. The geochemical characteristics imply that the Weideshan granite is a blend of crust–mantle derived magma, and the magma is derived from partial melting of recycled continental crust material and enriched lithospheric mantle [85,86,114]. Zircon U–Pb ages of the Weideshan granite range from 120 to 107 Ma [85]. The age of dioritic enclaves is moderately younger than that of host rocks, indicating the basic magma is earlier than the emplacement of acidic magma. It is hard to see ancient inherited zircons in the rocks, demonstrating that there was no surrounding rock material blended in the process of magma emplacement.

The Laoshan granite is predominantly distributed in the UHP metamorphic belt in the east and south of the Jiaodong gold concentration area, but there is no direct spatial distribution relationship with gold mineralization and nonferrous metal mineralization. The Laoshan granite is composed of monzonitic granite, syenogranite, and alkali–feldspar granite, which transited from acidic to alkaline, with uneven distribution of miarolitic structure [22]. Geochemical characteristics show that Laoshan granite is an A-type granite formed in regional extensional tectonic background, resulting from the Mesozoic lithosphere thinning and craton destruction [114]. The magma derived from partial melting of the deep crust, which may be related to the back–arc extensional setting.

A comprehensive analysis of available data for this area reveals that from Jurassic to Early Cretaceous, the geochemical composition of four-stage granites evolved from high-potassium calc–alkaline series to peridotite series, from peraluminous to metaluminous, trace element contents change from high Ba–Sr to low Ba–Sr and from high Sr, low Y to low Sr and high Y, REE composition change from no or weak positive Eu anomaly to significant negative Eu anomaly, the granite-type evolved from S-type through I-type to A-type. It shows that the mantle beneath the Jiaodong Peninsula evolved from EM–II in the Jurassic to EM–I in the Early Cretaceous, with depleted mantle emerging in the Late Cretaceous [18,40,110,115,116]. This demonstrates that the mantle switched from enriched to depleted [86,110].

6.2. Tectonic Setting

A large number of studies have disclosed that the genetic type of granites not only reflects the nature of magma source area, but is also a discriminant mark of tectonic environment of magma formation [117–119]. A-type granites are widely regarded as important petrological markers of extensional environment, but differences in environment, scale, and depth frequently lead to different characteristics of A–type granites [120]. Consequently, further studies divided A-type granites into A_1 and A_2 types, in which A_1 type granite comprises mantle-derived melt discrete crystalline rocks, predominantly formed in anorogenic environments (continental rifts or intraplate stretching) [117]. A_2 type granites represent rocks of crustal partial melting origin, predominantly formed in post-orogenic environment [121]. A-type granites were formed in an extensional environment under the background of crustal extension and thinning, and are closely related to the post-collision orogeny [122–127]. The nature and degree of crustal extension background is one of the critical factors that inhibit the formation of A-type granites and affect the characteristics of magma and emplacement. Consequently, the determination of A-type granite has become an important petrological marker to differentiate the tectonic environment of continental crust extension, and the end time and location of orogenesis [117,128–133].

Previous in-depth studies on the Shidao granite indicated that the materials were derived from the enriched lithosphere mantle source area, and were intrusive rocks with different compositions due to different stages of melting and experiencing different degrees of crystallization differentiation and assimilation and mixing of the lower crust [74]. In the diagram of geochemical composition classification, the sample are plot in in the A-type granite area (Figure 8). In the Nb–Y–3Ga trigonometric projection map, A_2 granite area was more invested than A_1 granite area (Figure 9a). In the tectonic discrimination diagrams, the Shidao granite is mostly plot in the within-plate granite (WPG) area, and a small amount is plot in the volcanic arc granite (VAG) area (Figure 9b,c). In R_1–R_2 diagram, samples fall into the border zone of late orogenic and anorogenic (Figure 9d), shows the Shidao granite is formed in the tectonic environment transition period. The existence of UHP metamorphic rocks represented by eclogites in Sulu orogenic belt indicates that the belt underwent continental deep subduction, continental collision, and rapid exhumation of UHP metamorphic rocks in NCC and YC during the mid-late Triassic (Figure 10a). The occurrence of Shidao A-type granite marks the end of the strong collisional orogeny and the structural reentry of UHP metamorphic rocks in the Sulu orogenic belt, and the transition from the NCC–YC tectonic system to the Paleo–Pacific tectonic system in Jiaodong area [134].

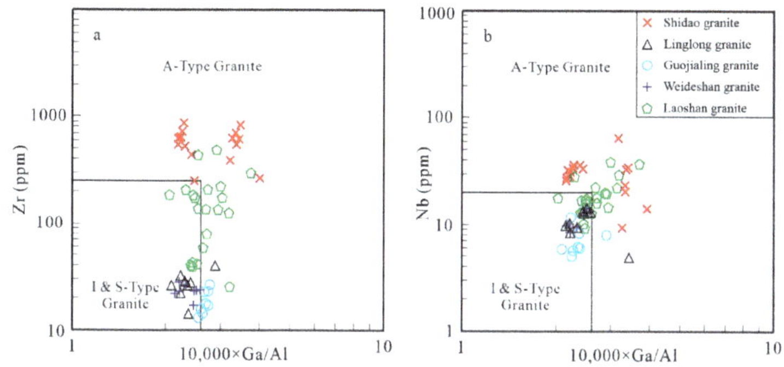

Figure 8. (**a**) Zr and (**b**) Nb versus 10,000 × Ga/Al discrimination diagrams showing the genetic types of granites (Base map after [135], data for the Linglong, Guojialing and Weideshan granites are from [86]; data for the Laoshan and Shidao granites are from [86,133] and [70,73,74], respectively). Detailed geochemical data are listed in Supplementary Materials Table S1.

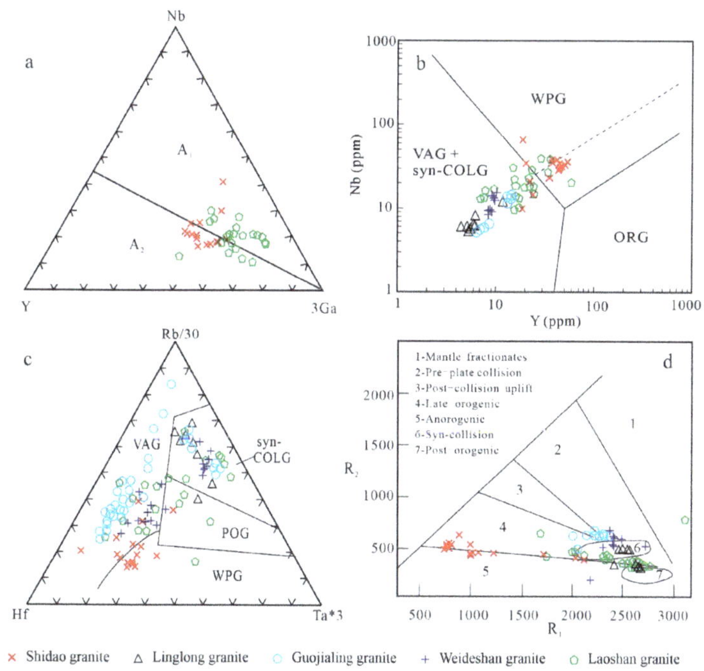

Figure 9. Tectonic discrimination diagrams for the Mesozoic granites in the Jiaodong Peninsula (Base map after [136,137]); the Shidao granite data of (**a–c**) is quoted from [70,73,74]; data of Laoshan granite in the figure are quoted from [86,133]; some data of d is from [106,112]); the Linglong, Guojialing, and Weideshan granites' data of (**b–d**) are quoted from [86,112]. Detailed geochemical data are listed in Supplementary Materials Tables S1 and S2. Abbreviations: VAG = volcanic arc granite, ORG = ocean ridge granite, WPG = within-plate granite, Syn-COLG = syn-collisional granite.

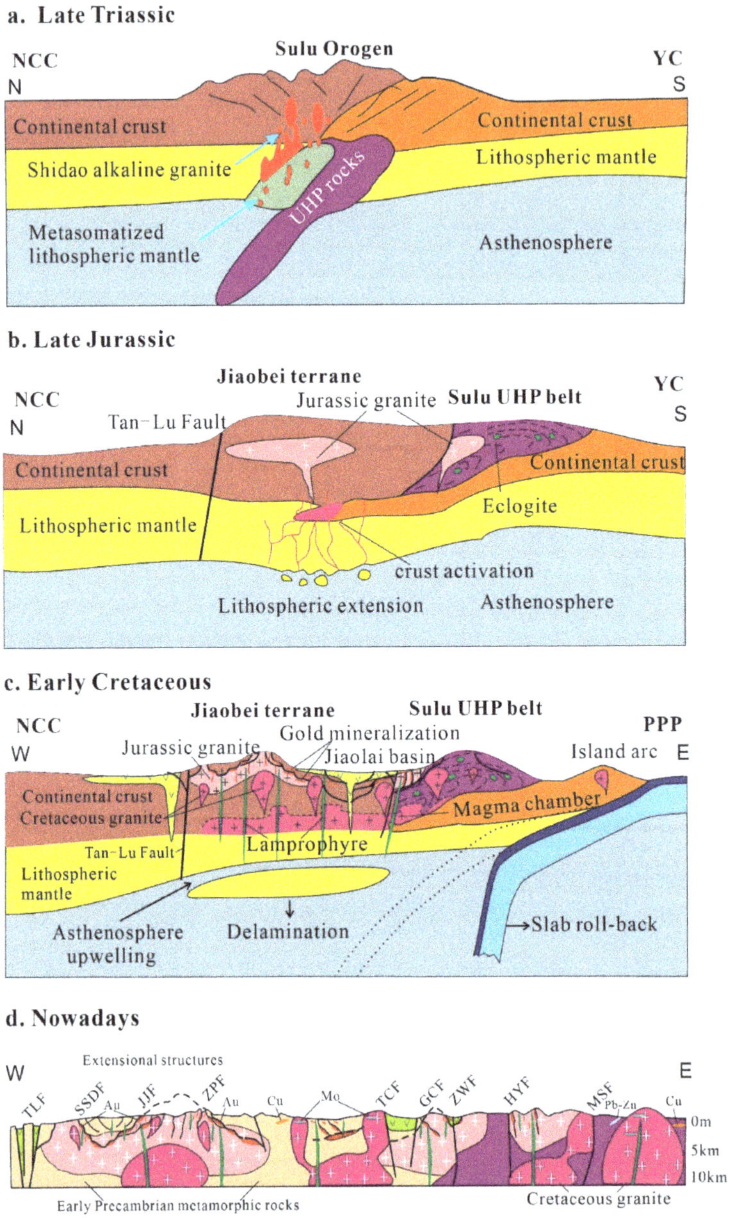

Figure 10. Schematic diagrams showing Mesozoic geodynamic evolution in the Jiaodong Peninsula, China (modified after [18,58,73,138,139]). (**a**) Partial melting of the metasomatic mantle wedge, generating the Shidao arc-like alkaline complex; (**b**) Since the Late Jurassic, the thickened crust experienced gravitational collapse under relatively high-pressure conditions, resulting in in-situ remelting of the thickened continental crust to form the crust-derived Linglong granite; (**c**) The crustal thickening event reached its peak during the early Cretaceous and the delamination of the modified lithosphere initiated a strong crust–mantle interaction, forming crust–mantle-derived mixed magma and I-type Guojialing and Weideshan granites and gold deposits; (**d**) The magma evolved along the upwelling process of the fault and formed a series of polymetallic deposits.

The tectonic setting of Linglong granite has syn- to post-collision characteristics (Figure 9c,d), and the S–type granite features and the ubiquitous gneissic phenomenon indicate that the granite was in a compressional state during diagenesis. In R_1–R_2 diagram (Figure 9d), the Linglong and Guojialing granites fall into the range of syn-collisional to post-collisional/late orogenic granite, respectively, revealing their tectonic background transformed from compression to extension. The Weideshan granite exhibits the characteristics of arc or syn-collisional granite in terms of geochemistry [86], indicating that it was formed on the volcanic arc that suffered a tectonic compression (Figure 9; [58,86]). The geological occurrence analysis indicates that the Laoshan granite is closely associated with the Weideshan granite, forming an I–A type composite granitic pluton. During the formation of the Weideshan granite, magma from the partial remelting of crustal rocks did not obviously mix with mantle-derived magma, but instead formed the A–type Laoshan granite [40,86,89]. The Laoshan granite samples are distributed in the A_1–A_2 region of the Nb–Y–3Ga triangulation map (Figure 9a). The mineral composition of Laoshan granite shows that the early unit is A_2-type granite dominated by syenogranite, and the late unit is A_1-type alkali feldspar granite that often containing the alkaline dark minerals such as sodium amphibole and aegirine. In the discriminant diagram of tectonic environment, most of the samples are plotted in the volcanic arc to collisional granite area and the volcanic arc area (Figure 9b,c), illustrating the geochemical information of the arc magmatism. In the R_1–R_2 diagram, most of the samples fall into the boundary of late-orogenic, anorogenic, and post-orogenic region (Figure 9d), but one situated in mantle fractionates indicates mantle contribution during the diagenetic process [137]. Together with related calc–alkaline granites, the Laoshan granite reveal the evolution process of late Mesozoic craton destruction and are important indicators after the peak of NCC destruction.

6.3. The Relationship between Tectonic Movement, Magmatic Activity, and Gold and Polymetallic Mineralization

Mesozoic magmatism in the Jiaodong Peninsula reflects the entire evolution of geodynamics in the Mesozoic, indicating the switch from NCC−YC collision to subduction of the PPP, with crustal thickening switching to lithospheric thinning, and a compressional tectonic setting changing to an extensional setting alternately. The superposition and transformation of the tectonic system is closely related to the formation and temporal–spatial distribution of gold and polymetallic mineralization. It can be concluded that the Late Triassic–Early Cretaceous magmatism in the Jiaodong Peninsula experienced four peaks in the Late Triassic, Late Jurassic, early and late period of the Early Cretaceous (Figure 10).

The subduction of the YC beneath the NCC resulted in continent–continent collision and HP/UHP metamorphism in the Late Triassic (Figure 10a). UHP metamorphic eclogites with ages of 240–200 Ma are widespread in the east of Jiaodong Peninsula [140]. Due to the break-off of eclogitized oceanic lithosphere at the transition zone of the continental and oceanic lithospheres, the subducted continental lithosphere ceased further descend and was rising with buoyancy. The exhumation of the deeply subducted continental crust resulted in decompression dehydration and partial melting of the UHP metamorphic slices to generate aqueous felsic melts. The felsic melts reacted with the overlying mantle wedge to cause a metasomatized mantle (Figure 10a). Partial melting of the metasomatic mantle wedge could generate post-collisional arc-like magmatism (i.e., the Shidao alkaline granite; 227–200 Ma) (Figure 10a). The alkaline magmatism experienced a successive fractional crystallization process. In the process of ascending and migrating along the fault in the late magmatic evolution, the residual magma rich in Be, as Han and Ma [141] suggested, mixed with a CO_2-rich fluid and large amount of atmospheric water. The addition of acidic fluid changed the earlier alkaline environment, destroyed the lattice balance, and precipitated a large amount of Be. Be element is wrapped in alteration minerals such as sericite in the form of bertrandite ($Be_4[Si_2O_7](OH)_2$), thus forming Be orebody, namely Datuanliujia Be deposit [33].

Since the Late Jurassic (180–170 Ma), with the rapid and low-angle subduction of the PPP along the NW direction to eastern China, the crust in the Jiaodong Peninsula has been continuously thickened [140,142,143] (Figure 10b). The peninsula experienced gravitational collapse under relatively high-pressure conditions, resulting in in-situ remelting of the thickened continental crust to form the crust-derived Linglong granite (164–140 Ma) (Figure 10b). Its sources include the orogenic belt subduction material and the NCC's lower crust. It implies that the formation processes of the granite was influenced by the NCC and YC [18,100,144]. The S-type granitic magma and the homologous magma rose along the deep fault and form the plutons and dikes. The ore-bearing gas and liquid produced in the process of magmatic condensation diffused to the surrounding rock along the fault. Extensive skarn and skarn–type Mo–W mineralization are formed in the contact zone. The hydrothermal vein W–Mo mineralization and polymetallic mineralization are mainly formed in the structural fracture zone and detachment belt far away from the contact zone, and molybdenum ore bodies that mainly formed in the later hydrothermal vein mineralization occurred in the pluton [2].

The early period of the Cretaceous was mainly influenced by the westward subduction of the PPP, with the compressional tectonic system switching to one of extension and the lithosphere responding to intense back–arc extension [38,100]. The YC merged with the NCC completely in the early Cretaceous, and the crustal thickening event reached its peak during this period. As a result, delamination of the modified lithosphere initiated a strong crust–mantle interaction, forming crust–mantle-derived mixed magma and I-type Guojialing granite and Weideshan granite (135–110 Ma; Figure 10c). The magma evolved along the upwelling process of the fault and formed the hydrothermal fluid rich in ore-forming materials, which metasomatized with the surrounding rock or at the top of the pluton in appropriate places. The ore-forming materials were further precipitated and enriched, forming a series of porphyry–skarn-type polymetallic deposits such as the Xiangkuang Cu–Pb–Zn deposit and Wangjiazhuang Cu–Zn deposit (Figure 10d).

The swift subduction of the Paleo–Pacific crust in the NWW direction and the subsequent slab roll-back resulted in strong mantle convection, intensified lithospheric delamination, and asthenospheric upwelling during the late period of the Early Cretaceous [139,145]. Underplating by rising magma caused the partial melting of the lower crust, generating large-scale crust–mantle-derived magma and forming the K-rich Weideshan granite (Figure 10). It directly leads to intense magmatism, rapid upwelling of magma, and a series of extensional structures in the Jiaodong Peninsula. Geologists have termed it "thermal upwelling–extension" [85]. The extensive gold and polymetallic mineralization in the Jiaodong Peninsula displays close affinity with these thermal upwelling–extension structures. Acting as a "heat generator", magmatic activity activated ore-bearing fluid in the surrounding rocks and lead to migration of ore-bearing fluid from the mantle to the upper crust to aid its mineralization. The fractured system generated by extension, related to magma upwelling, providing space for the Jiaodong gold deposits [146]. In the middle and late period of magmatic differentiation of Weideshan stage, a large amount of ore-bearing hydrothermal fluid gathered, and the shift of external temperature and pressure and the opening of space led to the immiscibility and boiling of fluid. Meanwhile, the mixing of different fluids and the water–rock reaction led to the shift of pH value and composition of the fluid, which led to the precipitation and mineralization of Cu and Mo minerals in the favorable parts of the structure. In the late stage of magmatic evolution, the surrounding groundwater circulates in a wide range driven by magmatic heat, and the crustal ore-forming materials are extracted, which migrate, enrich, and precipitate along the fault zone, forming the middle–low temperature hydrothermal vein type Pb–Zn–Ag–Au polymetallic mineralization in the late stage of magmatic intrusion. During this period, the porphyry Cu–Mo polymetallic deposits such as Lengjia, Shangjiazhuang, and Nantai were formed (Figure 10d).

According to the formation age of the deposit, scholars believe that there are three major ore-forming periods in Jiaodong area corresponding to the three major magma stages.

That is, the Indosinian polymetallic mineralization period corresponding to the Triassic granite; the early Yanshanian nonferrous metal mineralization period corresponding to the Jurassic granite, and the late Yanshanian gold and polymetallic mineralization period corresponding to the Cretaceous granite [65]. Further subdivision means that there were four periods of large-scale magmatic activity in the Mesozoic, namely Shidao granite in the Late Triassic, Linglong granite in the Late Jurassic, Guojialing granite in the early Cretaceous, Weideshan granite and Laoshan granite in the middle and late Early Cretaceous. Along with the large-scale magmatic activity, metal mineralization occurred. The mineralization of the Late Triassic, Late Jurassic and early Cretaceous are relatively weak. The middle and late Early Cretaceous is the "metallogenic explosion" stage [147], forming the widely distributed gold deposits in Jiaodong, accompanied by Cu, Pb, Zn and other nonferrous metal deposits.

The key markers of Jiaodong gold and polymetallic mineralization are magmatism, fluid activity and extensional structure. Extensive magmatic uplift and extensional structures in the Early Cretaceous in Jiaodong formed the extensional tectonic system [85]. During the formation of this system, a large proportion of crust and mantle materials were exchanged and mixed, and the fluid interaction was highly active [38], resulting in a magmatic fluid metallogenic system, which provided favorable metallogenic conditions for gold and nonferrous metal hydrothermal deposits. Thus, a large-scale explosive mineralization occurred in Jiaodong in the middle and late Early Cretaceous.

7. Conclusions

The intrusive age ranges of Linglong, Guojialing, Weideshan, and Laoshan granites are 155–154 Ma, 131–130 Ma, 118–111 Ma, and 116 Ma, respectively, which are basically in line with the previous age test data. Together with the Shidao granite (227–200 Ma), five phases of magmatism can be classified according to the time and geochemical features, all of which have different degrees of gold and polymetallic mineralization.

The type of granites evolved from A- and S-type to I- and A-type from the Late Triassic to the Early Cretaceous, and thus reflects the evolution of geodynamics in the Mesozoic, indicating the switch from NCC−YC collision to subduction of the PPP, with crustal thickening switching to lithospheric thinning, and a compressional tectonic setting changing to an extensional setting. The superposition and transformation of the tectonic system is closely related to the formation and temporal–spatial distribution of gold and polymetallic mineralization.

The key markers of Jiaodong gold and polymetallic mineralization are magmatism, fluid activity and extensional structure. Extensive magmatic uplift and extensional structures in the Early Cretaceous in Jiaodong formed the extensional tectonic system. During the formation process, a large proportion of crust and mantle materials exchanged and mixed, and the fluid interaction was highly active, resulting in a magmatic fluid metallogenic system, which provided favorable metallogenic conditions for gold and nonferrous metal hydrothermal deposits. Thus, a large-scale explosive mineralization occurred in Jiaodong in the middle and late Early Cretaceous.

Supplementary Materials: The following supporting information can be downloaded at: https://www.mdpi.com/article/10.3390/min12091073/s1, Table S1: Trace element compositions of the Mesozoic granite in the Jiaodong Peninsula [70,73,74,86,106,112,133], Table S2: Major element compositions of the Mesozoic granite in the Jiaodong Peninsula [70,73,74,86,133].

Author Contributions: Conceptualization, B.W., Z.D. and J.Z.; field investigation, B.W., Z.B. and J.L.; experimental analysis, C.L.; software, B.W. and Q.Z.; validation, Z.B. and S.W.; writing—original draft preparation, B.W.; writing—review and editing, B.W., Z.D. and M.S. All authors have read and agreed to the published version of the manuscript.

Funding: This work was supported by the National Natural Science Foundation of China-Shandong Joint Fund Project (No. U2006201), Youth Science and Technology Innovation Fund of Shandong Provincial No. 6 Exploration Institute of Geology and Mineral Resources (No. LY-QK-202203), High-level Scientific and technological Talents Innovation Team project of Shandong Provincial No. 6 Exploration Institute of Geology and Mineral Resources (LY-TD-202201), and Science and technology project of Shandong Provincial Bureau of Geology and Mineral Resources (No. KY202208).

Institutional Review Board Statement: Not applicable.

Informed Consent Statement: Not applicable.

Data Availability Statement: Data sharing is not applicable to this article.

Acknowledgments: The authors are grateful for the constructive comments by the anonymous reviewers.

Conflicts of Interest: The authors declare no conflict of interest.

References

1. Zhai, M.G.; Yang, J.H.; Liu, W.J. Large clusters of gold deposits and large–scale metallogenesis in the Jiaodong Peninsula, eastern China. *Sci. China Ser. D Earth Sci.* **2001**, *44*, 758–768. [CrossRef]
2. Ding, Z.J.; Sun, F.Y.; Liu, F.L.; Liu, J.H.; Peng, Q.M.; Ji, P.; Li, B.L.; Zhang, P.J. Mesozoic geodynamic evolution and metallogenic series of major metal deposits in Jiaodong Peninsula, China. *Acta Petrol. Sin.* **2015**, *31*, 3045–3080. (In Chinese with English abstract)
3. Song, M.C.; Song, Y.X.; Li, J.; Li, S.Y. Metallogenic Series of Gold and Nonferrous Metal Deposits Related to Cretaceous Granites in Eastern Shandong Peninsula, China. *Geotecton. Metallog.* **2015**, *39*, 823–843. (In Chinese with English abstract)
4. Chen, J.; Sun, F.Y.; Wang, L.; Wang, S.; Li, R.H. Zircon U–Pb geochronology and petrogeochemistry of Luanjiahe granite in Jiaodong region and their geological significance. *Glob. Geol.* **2015**, *34*, 283–295. (In Chinese with English abstract)
5. Geng, K.; Wang, R.J.; Li, H.K.; Liang, T.T.; Zhang, Y.B. Zircon SHRIMP U–Pb geochronology of Congjia granodiorite from northwest Jiaodong area. *Acta Geosci. Sin.* **2016**, *37*, 90–100. (In Chinese with English abstract)
6. Goss, C.S.; Wilde, S.A.; Wu, F.Y.; Yang, J.H. The age, isotopic signature and significance of the youngest Mesozoic granitoids in the Jiaodong Terrene, Shandong Province, North China Craton. *Lithos* **2010**, *120*, 309–326. [CrossRef]
7. Guan, K.; Luo, Z.K.; Miao, L.C.; Huang, J.Z. SHRIMP in zircon chronology for Guojialing suite granite in Jiaodong Zhaoye district. *Sci. Geol. Sin.* **1998**, *33*, 318–328, (In Chinese with English abstract).
8. Hu, S.L.; Wang, S.S.; Sang, H.Q.; Qiu, J.; Zhang, R.G. Isotopic ages of Linglong and Guojialing batholiths in Shandong Province and their geological implication. *Acta Petrol. Sin.* **1987**, *20*, 38–42. (In Chinese with English abstract)
9. Lin, B.L.; Li, B.L. Geochemistry, U–Pb dating, Lu–Hf isotopic analysis and geological significance of Linglong granite in Jiaodong Peninsula. *J. Chengdu Univ. Technol. Sci. Technol. Ed.* **2013**, *40*, 147–160. (In Chinese with English abstract)
10. Luo, X.D.; Yang, X.Y.; Duan, L.A.; Sun, W.D. Geochemical and geochronological study of the gold–related Guojialing pluton and Shangzhuang pluton in Jiaobei block. *Acta Geol. Sin.* **2014**, *88*, 1874–1888. (In Chinese with English abstract)
11. Miao, L.C.; Luo, Z.K.; Guan, K.; Huang, J.Z. The implication of the SHRIMP U–Pb age in zircon to the petrogenesis of the Linglong granite, east Shandong Province. *Acta Petrol. Sin.* **1998**, *14*, 198–206.
12. Qiu, L.G.; Ren, F.L.; Cao, Z.X.; Zhang, Y.Q. Late Mesozoic magmatic activities and their constraints on geotectonic of Jiaodong region. *Geotecton. Metallog.* **2008**, *32*, 117–123. (In Chinese with English abstract)
13. Sai, S.X.; Zhao, T.M.; Wang, Z.L.; Huang, S.Y.; Zhang, L. Petrogenesis of Linglong biotite granite: Constraints from mineralogical characteristics. *Acta Petrol. Sin.* **2016**, *32*, 2477–2493. (In Chinese with English abstract)
14. Sun, H.S.; Li, H.; Liu, L.; Chen, Q.M.; Yang, H.; Wu, P. Exhumation history of the Jiaodong and its adjacent areas since the Late Cretaceous: Constraints from low temperature thermo chronology. *Sci. China Earth Sci.* **2017**, *60*, 531–545. [CrossRef]
15. Wang, L.G.; Qiu, Y.M.; McNaughton, N.J. Constraints on crustal evolution and gold metallogeny in the northeastern Jiaodong Peninsula, China, from SHRIMP U–Pb zircon studies of granitoids. *Ore Geol. Rev.* **1998**, *13*, 275–291. [CrossRef]
16. Wang, Z.L.; Zhao, Y.X.; Zhan, Q.; Lu, H.W.; Li, J.L.; Cheng, W. Magma mixing for the high Ba–Sr Guojiling type granitoids in northwest Jiaodong Peninsula: Constraints from petrogeochemistry and Sr–Nd isotopes. *Acta Petrol. Sin.* **2014**, *30*, 2595–2608. (In Chinese with English abstract)
17. Yang, J.H.; Zhu, M.F.; Liu, W.; Zhai, M.G. Geochemistry and petrogenesis of Guojialing granodiorites from the northwestern Jiaodong Peninsula, eastern China. *Acta Petrol. Sin.* **2003**, *19*, 692–700. (In Chinese with English abstract)
18. Yang, K.F.; Fan, H.R.; Santosh, M. Reactivation of the Archean lower crust: Implications for zircon geochronology, elemental and Sr–Nd–Hf isotopic geochemistry of Late Mesozoic granitoids from northwestern Jiaodong Terrane, the North China Craton. *Lithos* **2012**, *146–147*, 112–127. [CrossRef]
19. Yang, Q.Y.; Santosh, M. Early Cretaceous magma flare–up and its implications on gold mineralization in the Jiaodong Peninsula, China. *Ore Geol. Rev.* **2015**, *65*, 626–642. [CrossRef]
20. Zhang, H.F.; Zhai, M.G.; He, Z.F.; Peng, P.; Xu, B.L. Petrogenesis and implications of the sodium–rich granites from the Kunyushan complex, eastern Shandong Province. *Acta Petrol. Sin.* **2004**, *20*, 369–380. (In Chinese with English abstract)

21. Zhang, H.F.; Zhai, M.G.; Tong, Y.; Peng, P.; Xu, B.L.; Guo, J.H. Petrogenesis of the Sanfoshan high Ba–Sr granite, Jiaodong Peninsula, eastern China. *Geol. Rev.* **2006**, *52*, 43–53. (In Chinese with English abstract)
22. Zhao, G.T.; Wang, D.Z.; Cao, Q.C. Thermal evolution and its significance of I–A type granitoid complex: The Laoshan granitoid as an example. *Sci. China Earth Sci.* **1998**, *41*, 529–536. [CrossRef]
23. Zhao, Z.F.; Zheng, Y.F. Remelting of subducted continental lithosphere: Petrogenesis of Mesozoic magmatic rocks in the Dabie–Sulu orogenic belt. *Sci. China Earth Sci.* **2009**, *52*, 1295–1318. [CrossRef]
24. Bi, S.J.; Zhao, X.F. $^{40}Ar/^{39}Ar$ dating of the Jiehe gold deposit in the Jiaodong Peninsula, eastern North China Craton: Implications for regional gold metallogeny. *Ore Geol. Rev.* **2017**, *86*, 639–651. [CrossRef]
25. Hu, F.F.; Fan, H.R.; Yang, J.H.; Wan, Y.S.; Liu, D.Y.; Zhai, M.G.; Jin, C.W. Mineralizing age of the Rushan lode gold deposit in the Jiaodong Peninsula: SHRIMP U–Pb dating on hydrothermal zircon. *Chin. Sci. Bull.* **2004**, *49*, 1629–1636. [CrossRef]
26. Li, H.M.; Mao, J.W.; Shen, Y.C.; Liu, T.B.; Zhang, L.C. Ar–Ar ages of K-feldspar and quartz from Dongji gold deposit, Northwest Jiaodong, and their significance. *Miner. Depos.* **2003**, *22*, 72–77. (In Chinese with English abstract)
27. Yang, J.H.; Zhou, X.H.; Chen, L.H. Dating of gold mineralization for super–large altered tectonite-type gold deposits in northwestern Jiaodong Peninsula and its implications for gold metellogeny. *Acta Petrol. Sin.* **2000**, *16*, 454–458. (In Chinese with English abstract)
28. Yang, J.H.; Zhou, X.H. The Rb–Sr isochron of ore and pyrite sub–samples from Linglong gold deposit, Jiaodong Peninsula, eastern China and their geological significance. *Chin. Sci. Bull.* **2000**, *45*, 2272–2276. [CrossRef]
29. Yang, J.H.; Zhou, X.H. Rb–Sr, Sm–Nd and Pb isotope systematics of pyrite: Implications for the age and genesis of lode gold deposits. *Geology* **2001**, *29*, 711–714. [CrossRef]
30. Zhang, L.C.; Shen, Y.C.; Liu, T.B.; Zeng, Q.D.; Li, G.M.; Li, H.M. $^{40}Ar/^{39}Ar$ and Rb–Sr isochron dating of the gold deposits on northern margin of the Jiaolai Basin, Shandong, China. *Sci. China Earth Sci.* **2003**, *46*, 708–718. [CrossRef]
31. Zhang, X.O.; Cawood, P.A.; Wilde, S.A. Geology and timing of mineralization at the Cangshang gold deposit, north–western Jiaodong Peninsula, China. *Miner. Depos.* **2003**, *38*, 141–153. [CrossRef]
32. Chen, B.; Deng, J.; Ji, X. Time Limit of Gold Mineralization in Muping–Rushan Belt, Eastern Jiaodong Peninsula, China: Evidence from Muscovite Ar–Ar Dating. *Minerals* **2022**, *12*, 278. [CrossRef]
33. Ding, Z.J. *Study on Metallogenic Regularity of Mesozoic Precious and Non-Ferrous Deposits in Jiaodong Peninsula*; Jilin University: Changchun, China, 2014; pp. 1–208. (In Chinese with English abstract)
34. Goldfarb, R.J.; Santosh, M. The dilemma of the Jiaodong gold deposits: Are they unique? *Geosci. Front.* **2014**, *5*, 139–153. [CrossRef]
35. Guan, K.; Luo, Z.K.; Miao, L.C.; Huang, J.Z. Petrochemical and geochemical charactristics of Guojialing suite in Zhaoyuan district and the genetic relation of gold mineralization to the granite. *Contrib. Geol. Miner. Resour. Res.* **1997**, *12*, 1–8. (In Chinese with English abstract)
36. Luo, Z.K.; Guan, K.; Miao, L.C. Discussion on relationship between lamprophyre veins and mineralization in the Linglong gold field, eastern Shandong. *Gold Geol.* **2001**, *7*, 15–21. (In Chinese with English abstract)
37. Xu, W.G.; Fan, H.R.; Yang, K.F.; Hu, F.F.; Cai, Y.C. Gold mineralizing efficiency during hydrothermal alteration of the Mesozoic granitoids in the northwest Jiaodong Peninsula: Contrasting conditions between the Guojialing and Linglong plutons. *Chem. Erde–Geochem.* **2017**, *77*, 387–398. [CrossRef]
38. Zhai, M.G.; Fan, H.R.; Yang, J.H.; Miao, L.C. Large–scale cluster of gold deposits in east Shandong: An orogenic metallogenesis. *Earth Sci. Front.* **2004**, *11*, 85–98. (In Chinese with English abstract)
39. Zhou, T.H.; Lu, G. Tectonics, granitoids and Mesozoic gold deposits in east Shandong, China. *Ore Geol. Rev.* **2000**, *16*, 71–90. [CrossRef]
40. Song, M.C.; Wang, P.C. *Regional Geology in Shandong Province*; Shandong Provincial Map Publishing House: Jinan, China, 2003; pp. 25–720. (In Chinese)
41. Tam, P.Y.; Zhao, G.C.; Sun, M.; Li, S.Z.; Iizuka, Y.; Ma, G.S.; Yin, C.Q.; He, Y.H.; Wu, M.L. Metamorphic P–T path and tectonic implications of medium–pressure pelitic granulites from the Jiaobei massif in the Jiao–Liao–Ji Belt, North China Craton. *Precambrian Res.* **2012**, *220*, 177–191. [CrossRef]
42. Tam, P.Y.; Zhao, G.C.; Zhou, X.W.; Sun, M.; Guo, J.H.; Li, S.Z.; Yin, C.Q.; Wu, M.L.; He, Y.H. Metamorphic P–T path and implications of high–pressure pelitic granulites from the Jiaobei massif in the Jiao–Liao–Ji Belt, North China Craton. *Gondwana Res.* **2012**, *22*, 104–117. [CrossRef]
43. Tang, J.; Zheng, Y.F.; Wu, Y.B.; Gong, B.; Liu, X.M. Geochronology and geochemistry of metamorphic rocks in the Jiaobei terrane: Constraints on its tectonic affinity in the Sulu orogen. *Precambrian Res.* **2007**, *152*, 48–82. [CrossRef]
44. Wan, Y.S.; Xie, H.W.; Yang, C.H.; Kröner, K.; Ma, M.Z.; Dong, C.Y.; Du, L.L.; Xie, H.Q.; Liu, D.Y. Early Neoarchean (~27 Ga) tectono–thermal events in the North China Craton: A synthesis. *Precambrian Res.* **2014**, *247*, 45–63. [CrossRef]
45. Zhou, J.B.; Wilde, S.A.; Zhao, G.C.; Zheng, C.Q.; Jin, W.; Zhang, X.Z.; Cheng, H. SHRIMP U–Pb zircon dating of the Neoproterozoic Penglai Group and Archean gneisses from the Jiaobei Terrane, North China Craton, and their tectonic implications. *Precambrian Res.* **2008**, *160*, 323–340. [CrossRef]
46. Tam, P.K.; Zhao, G.C.; Liu, F.; Zhou, X.; Sun, M.; Li, S.Z. Timing of metamorphism in the Paleoproterozoic Jiao–Liao–Ji Belt: New SHRIMP U–Pb zircon dating of granulites, gneisses and marbles of the Jiaobei massif in the North China Craton. *Gondwana Res.* **2011**, *19*, 150–162. [CrossRef]

47. Ding, Z.J.; Sun, F.Y.; Liu, D.H.; Zhang, P.J.; Yang, G.F.; Kong, Y. Metallogenic series of gold and polymetallic deposits in north Fushan area of Yantai city. *Shandong Land Resour.* **2011**, *27*, 1–6. (In Chinese with English abstract)
48. Sun, F.Y.; Ding, Z.J.; Liu, D.H.; Zhang, P.J.; Yu, X.F. Discussion on the porphyry metallogenic system in north Fushan area, Jiaodong peninsula. *Gold* **2011**, *32*, 14–19. (In Chinese with English abstract)
49. Li, S.Z.; Kusky, T.M.; Zhao, G.C.; Liu, X.C.; Zhang, G.W.; Kopp, H.; Wang, L. Two–stage Triassic exhumation of HP–UHP terranes in the western Dabie orogen of China: Constraints from structural geology. *Tectonophysics* **2010**, *490*, 267–293. [CrossRef]
50. Zhao, G.C.; Sun, M.; Wilde, S.A.; Li, S.Z. Late Archean to Paleoproterozoic evolution of the North China Craton: Key issues revisited. *Precambrian Res.* **2005**, *136*, 177–202. [CrossRef]
51. Zhao, H.G.; Sun, J.G.; Ling, H.F.; Shen, K.; Hou, S.X. Characteristics and geological significance of ore–forming fluids in the Jiaodong gold deposits. *Geol. Prospect.* **2005**, *41*, 27–33. (In Chinese with English abstract)
52. Hacker, B.R.; Ratschbacher, L.; Webb, L.; McWilliams, M.O.; Ireland, T.; Calvert, A.; Dong, S.; Wenk, H.R.; Chateigner, D. Exhumation of ultrahigh-pressure continental crust in east central China: Late Triassic-Early Jurassic tectonic unroofing. *J. Geophys. Res. Solid Earth* **2000**, *105*, 13339–13364. [CrossRef]
53. Ratschbacher, L.; Franz, L.; Enkelmann, E.; Jonckheere, R.; Porschke, A.; Hacker, B.R.; Dong, S.; Zhang, Y. The Sino–Korean–Yangtze suture, the Huwan detachment, and the Paleozoic–Tertiary exhumation of (ultra) high–pressure rocks along the Tongbai–Xinxian–Dabie Mountains. Ultrahigh–Pressure Metamorphic: Deep Continental Subduction. *Geol. Soc. Am.* **2006**, *403*, 45–77.
54. Ratschbacher, L.; Hacker, B.R.; Calvert, A.; Webb, L.E.; Grimmer, J.C.; McWilliams, M.O.; Ireland, T.; Dong, S.W.; Hu, J.M. Tectonics of the Qinling (Central China): Tectono stratigraphy, geochronology, and deformation history. *Tectonophysics* **2003**, *366*, 1–53. [CrossRef]
55. Xu, Z.; Zeng, L.; Liu, F.; Yang, J.; Zhang, Z.; McWilliams, M.; Liu, J.G. Polyphase subduction and exhumation of the Sulu High–pressure–ultrahigh–pressure metamorphic Terrane. *Geol. Soc. Am.* **2006**, *403*, 93–114.
56. Zheng, Y.F.; Fu, B.; Gong, B.; Li, L. Stable isotope geochemistry of ultrahigh pressure metamorphic rocks from the Dabie–Sulu orogen in China: Implications for geodynamics and fluid regime. *Earth Sci. Rev.* **2003**, *62*, 105–161. [CrossRef]
57. Zheng, Y.F.; Zhou, J.B.; Wu, Y.B.; Xie, Z. Low–grade metamorphic rocks in the Dabie–Sulu orogenic belt: A passive–margin accretionary wedge deformed during continent subduction. *Int. Geol. Rev.* **2005**, *47*, 851–871. [CrossRef]
58. Song, M.C.; Li, S.Z.; Santosh, M.; Zhao, S.J.; Yu, S.; Yi, P.H.; Cui, S.X.; Lv, G.X.; Xu, J.X.; Song, Y.X.; et al. Types, characteristics and metallogenesis of gold deposits in the Jiaodong Peninsula, Eastern North China Craton. *Ore Geol. Rev.* **2015**, *65*, 612–625. [CrossRef]
59. Wang, B.; Zhou, J.B.; Wilde, S.A.; Zhang, X.Z.; Ren, S.M. The timing of final closure along the Changchun–Yanji suture zone: Constraints from detrital zircon U–Pb dating of the Triassic Dajianggang Formation, NE China. *Lithos* **2016**, *261*, 216–231. [CrossRef]
60. Lv, G.X.; Kong, Q.C. *Geology of Linglong-Jiaojia Type Gold Deposit, Jiaodong*; Science Press: Beijing, China, 1993; pp. 1–266. (In Chinese)
61. Song, M.C.; Yi, P.H.; Cui, S.X.; Xu, J.X.; Zhou, M.L.; Jiang, H.L.; Huang, T.L.; Jiao, X.M.; Wan, G.P.; Cao, C.G. Thermal Uplifting Extension Ore forming Theory and Its Prospecting Significance in Jiaodong Gold Deposit. *Shandong Land Resour.* **2013**, *29*, 1–12. (In Chinese with English abstract)
62. Sun, F.Y.; Shi, Z.L.; Feng, B.Z. *Geology of Jiaodong Gold Deposit and C-H-O Fluid Differentiation and Mineralization of Mantle Source*; Jilin People's Publishing House: Changchun, China, 1995; pp. 1–170. (In Chinese)
63. Ding, Z.J.; Sun, F.Y.; Zhao, C.S.; Liu, D.H.; Zhang, P.J.; Wang, X.Q. Gold metallogenic series in northeast margin of Jiaolai Basin, Shandong Province. *Miner. Depos.* **2010**, *29*, 919–920. (In Chinese)
64. Song, M.C.; Zhang, J.J.; Zhang, P.J.; Yang, L.Q.; Liu, D.H.; Ding, Z.J.; Song, Y.X. Discovered and tectonic–magmatic background of super large gold deposit in offshore of northern Sanshandao, Shandong Peninsula, China. *Acta Geol. Sin.* **2015**, *89*, 365–383. (In Chinese with English abstract)
65. Ding, Z.J.; Sun, F.Y.; Liu, F.L.; Liu, J.H.; Liu, D.H.; Zhang, P.J.; Du, S.X.; Li, B. U–Pb dating of zircons from the Weideshan molybdenum copper polymetallic deposits in Jiaodong Peninsula, China, and its geological significance. *Acta Petrol. Sin.* **2013**, *29*, 607–618. (In Chinese with English abstract)
66. Zhou, J.B.; Zheng, Y.F.; Zhao, Z.F. Zircon U–Pb dating on Mesozoic granitoids at Wulian, Shandong Province. *Geol. J. China Univ.* **2003**, *9*, 185–194. (In Chinese with English abstract)
67. Zhu, C.Y.; Li, H.K.; Yu, X.F.; Geng, K.; Liang, T.T.; Zhang, Y.B. Structural Assemblage Classfication of Mesozoic Intrusive Rocks Associated with Gold Mineralization in Jiaodong Area. *Shandong Land Resour.* **2014**, *30*, 1–8. (In Chinese with English abstract)
68. Song, M.C.; Xun, J.X.; Wang, P.C. *Tectonic Framework and Tectonic Evolution of Shandong Province*; Geological Publishing House: Beijing, China, 2009; pp. 1–274. (In Chinese)
69. Wang, B.; Song, M.C.; Zhou, J.B.; Ding, Z.J.; Bao, Z.Y.; Lv, J.Y.; Wang, S.S. Four Stages A-type granitoids in Shandong Province and Their Implications for Tectonic Evolution. *N. China Geol.* **2022**, *in press*. (In Chinese with English abstract)
70. Chen, J.Z.; Jiang, N. Petrogenesis of the Late-Triassic alkaline magmatism in the Jiaodong area: Evidence from U-Pb age, Hf-O isotopes of zircons. *Acta Petrol. Sin.* **2011**, *27*, 3557–3574. (In Chinese with English abstract)
71. Chen, J.F.; Xie, Z.; Li, H.M.; Zhang, X.D.; Zhou, T.X.; Park, Y.S.; Ahn, K.S.; Chen, D.G.; Zhang, X. U–Pb zircon ages for a collision–related K–rich complex at Shidao in the Sulu ultrahigh pressure terrane, China. *Geochem. J.* **2003**, *37*, 35–46. [CrossRef]

72. Guo, J.H.; Chen, F.K.; Zhang, X.M.; Siebel, W.; Zhai, M.G. Evolution of syn- to post-collisional magmatism from north Sulu UHP belt, eastern China: Zircon U-Pb geochronology. *Acta Petrol. Sin.* **2005**, *21*, 1281–1301. (In Chinese with English abstract)
73. Xu, H.J.; Zhang, J.F.; Wang, Y.F.; Liu, W.L. Late Triassic alkaline complex in the Sulu UHP terrane: Implications for post–collisional magmatism and subsequent fractional crystallization. *Gondwana Res.* **2016**, *35*, 390–410. [CrossRef]
74. Yang, J.H.; Chung, S.L.; Wilde, S.A.; Wu, F.Y.; Chu, M.F.; Lo, C.H.; Fan, H.R. Petrogenesis of postorogenic syenites in the Sulu Orogenic Belt, East China: Geochronological, geochemical and Nd–Sr isotopic evidence. *Chem. Geol.* **2005**, *214*, 99–125. [CrossRef]
75. Deng, J.; Yang, L.Q.; Liu, W.; Sun, Z.S.; Li, X.Y.; Wang, Q.F. Gold origin and fluid ore-forming effect of Zhao-Ye ore deposits concentrating area in Jiaodong, Shandong, China. *Chin. J. Geol.* **2001**, *36*, 257–268. (In Chinese with English abstract)
76. Li, S.X.; Liu, C.C.; An, Y.H. *Geology of Jiaodong Gold Deposit*; Geological Publishing House: Beijing, China, 2007; pp. 1–471. (In Chinese)
77. Liu, Z.J.; Wang, J.P.; Zheng, D.W.; Liu, J.J.; Liu, J.; Fu, C. Exploration prospect and post-ore denudation in the northwestern Jiaodong Gold Province, China: Evidence from apatite fission track thermochronology. *Acta Petrol. Sin.* **2010**, *26*, 3597–3611. (In Chinese with English abstract)
78. Li, J. *Metallogenesis and Metallogenic Model of Mo-Cu-Pb-Zn Polymetallic Deposits in Jiaodong Area*; Chengdu University of Technology: Chengdu, China, 2012; pp. 1–127. (In Chinese)
79. Ding, Z.J.; Sun, F.Y.; Liu, J.H.; Liu, D.H.; Li, B.L.; Zhang, P.J.; Qian, Y.; Li, J. Re-Os dating of molybdenites from the Xingjiashan molybdenum-tungsten deposit in Jiaodong Peninsula, China and its geological significance. *Acta Petrol. Sin.* **2012**, *28*, 2721–2732. (In Chinese with English abstract)
80. Wiedenbeck, M.; Allé, P.; Corfu, F.; Griffin, W.L.; Meier, M.; Oberli, F.; Vonquadt, A.; Roddick, J.C.; Spiegel, W. Three natural zircon standards for U-Th-Pb, Lu-Hf, trace element and REE analyses. *Geostand. Newsl.* **1995**, *19*, 1–23. [CrossRef]
81. Reed, W.P. *Certificate of Analysis: Standard Reference Materials 610 and 611*; National Institute of Standards and Technology: Gaithersburg, MD, USA, 1992.
82. Hinton, R.W.; Upton, B.G.J. The chemistry of zircon: Variations within and between large crystals from syenite and alkali basalt xenoliths. *Geochim. Cosmochim. Acta* **1991**, *55*, 3287–3302. [CrossRef]
83. Hoskin, P.W.O.; Black, L.P. Metamorphic zircon formation by solid-state recrystallization of protolith igneous zircon. *J. Metamorph. Geol.* **2000**, *18*, 423–439. [CrossRef]
84. Rubatto, D. Zircon trace element geochemistry: Partitioning with garnet and the link between U–Pb ages and metamorphism. *Chem. Geol.* **2002**, *184*, 123–138. [CrossRef]
85. Song, M.C.; Li, J.; Li, S.Y.; Ding, Z.J.; Tan, X.F.; Zhang, Z.L.; Wang, S.J. Late Mesozoic thermal upwelling-extension structure in east Shandong Province and its geodynamic setting. *J. Jilin Univ. Earth Sci. Ed.* **2018**, *48*, 941–964. (In Chinese with English abstract)
86. Wang, B.; Song, M.C.; Huo, G.; Xu, Z.H.; Jiang, L.; Song, Y.X.; Li, J. Source characteristics and tectonic evolution of Late Mesozoic granites in Jiaodong and their implications for gold mineralization. *Acta Petrol. Mineral.* **2021**, *40*, 288–320. (In Chinese with English abstract)
87. Zhang, L.; Roberto, F.W.; Yang, L.Q.; Groves, D.I.; Deng, J. Mesozoic Orogenic Gold Mineralization in the Jiaodong Peninsula, China: A Focused Event at 120 ± 2 Ma During Cooling of Pregold Granite Intrusions. *Econ. Geol.* **2020**, *115*, 415–441. [CrossRef]
88. Li, S.G.; Xiao, Y.L.; Liou, D.L.; Chen, Y.Z.; Ge, N.J.; Zhang, Z.Q.; Sun, S.S.; Cong, B.L.; Zhang, R.Y.; Hart, S.R.; et al. Collision of the North China and Yangtze blocks and formation of coesite–bearing eclogites: Timing and processes. *Chem. Geol.* **1993**, *109*, 89–111. [CrossRef]
89. Li, S.G.; Wang, S.S.; Chen, Y.Z.; Liu, D.L.; Qiu, J.; Zhou, H.X.; Zhang, Z.M. Excess argon in phengite from eclogite: Evidence from dating of eclogite minerals by Sm–Nd, Rb–Sr and ^{39}Ar–^{40}Ar methods. *Chem. Geol.* **1994**, *112*, 343–350. [CrossRef]
90. Ames, L.; Zhou, G.; Xiong, B. Geochronology and geochemistry of ultrahigh–pressure metamorphism with implications for collision of the Sino–Korean and Yangtze cratons, central China. *Tectonics* **1996**, *15*, 422–489. [CrossRef]
91. Yang, J.S.; Xu, Z.Q.; Wu, C.L.; Liu, F.L.; Shi, R.D.; Wooden, J.; Maruyama, S. SHRIMP U-Pb Dating on Coesite Bearing Zircon: Evidence for Indosinian Ultrahigh-Pressure Metamorphism in Su-Lu, East China. *Acta Geol. Sin.* **2002**, *76*, 354–372. (In Chinese with English abstract)
92. Liu, F.L.; Xu, Z.Q.; Yang, J.S.; Zhang, Z.M.; Xue, H.M.; Li, T.F. Institute of Geology, Continental Dynamic Laboratory, Chinese Academy of Geological Sciences, Beijing 100037, China. Geochemical characteristics and UHP metamorphism of granitic gneisses in the main drilling hole of Chinese Continental Scientific Drilling Project and its adjacent area. *Acta Petrol. Sin.* **2004**, *20*, 9–26. (In Chinese with English abstract)
93. Liu, F.L.; Xu, Z.Q.; Song, B. Determination of UHP and Retrograde Metamorphic Ages of the Sulu Terrane: Evidence from SHRIMP U-Pb Dating on Zircons of Gneissic Rocks. *Acta Geol. Sin.* **2003**, *77*, 229–237. (In Chinese with English abstract)
94. Gao, T.S.; Chen, J.F.; Xie, Z.; Yan, J.; Qian, H. Geochemistry of Triassic igneous complex at Shidao in the Sulu UHP metamorphic belt. *Acta Petrol. Sin.* **2004**, *20*, 1025–1038. (In Chinese with English abstract)
95. Song, M.C.; Song, Y.X.; Ding, Z.J.; Wei, X.F.; Sun, S.L.; Song, G.Z.; Zhang, J.J.; Zhang, P.J.; Wang, Y.G. The discovery of The Jiaojia and the Sanshandao giant gold deposits in Jiaodong Peninsula and discussion on the relevant issues. *Geotecton. Metallog.* **2019**, *43*, 92–110. (In Chinese with English abstract)
96. Xu, H.L.; Zhang, D.Q.; Sun, G.Y. Characteristics and genesis of Kunyushan granite and its relation with gold deposits in Jiaodong. *Acta Petrol. Mineral.* **1997**, *16*, 131–143. (In Chinese with English abstract)

97. Li, J.J.; Luo, Z.K.; Liu, X.Y.; Xu, W.D.; Luo, H. Geodynamic setting for formation of large-superlarge gold deposits and Mesozoic granites in Jiaodong area. *Miner. Depos.* **2005**, *24*, 361–372. (In Chinese with English abstract)
98. Tan, J.; Wei, J.H.; Guo, L.L.; Zhang, K.Q.; Yao, C.L.; Chen, J.P.; Li, H.M. Zircon LA-ICP-MS U-Pb dating and phenocryst EPMA study of dikes in Guocheng area, Jiaodong: Implications for lithospheric evolution. *Sci. China Ser. D* **2008**, *38*, 913–929. (In Chinese with English abstract)
99. Zhang, T.; Zhang, Y.Q. Late Mesozoic tectono-magmatic evolution history of the Jiaobei uplift, Shandong Peninsula. *Acta Geol. Sin.* **2008**, *82*, 1210–1228. (In Chinese with English abstract)
100. Zhang, J.; Zhao, Z.F.; Zheng, Y.F.; Dai, M. Postcollisional magmatism: Geochemical constraints on the petrogenesis of Mesozoic granitoids in the Sulu orogen, China. *Lithos* **2010**, *119*, 512–536. [CrossRef]
101. Wang, S.J.; Wan, Y.S.; Wang, W.; Song, Z.Y.; Wang, J.G.; Dong, C.Y.; Xie, H.Q.; Liu, Q.D. Forming ages of granites in Laoshan area of Shandong Province—zircon SHRIMP U-Pb dating. *Shandong Land Resour.* **2010**, *26*, 1–6. (In Chinese with English abstract)
102. Wang, S.J.; Wan, Y.S.; Guo, R.P.; Song, Z.Y.; Wang, L.F. SHRIMP zircon dating of Linglong type (super unit) granite in eastern Shandong Province. *Shandong Land Resour.* **2011**, *27*, 1–7. (In Chinese with English abstract)
103. Ma, L.; Jiang, S.Y.; Dai, B.Z.; Jiang, Y.H.; Hou, M.L.; Pu, W.; Xu, B. Multiple sources for the origin of Late Jurassic Linglong adakitic granite in the Shandong Peninsula, Eastern China: Zircon U-Pb geochronological, geochemical and Sr-Nd-Hf isotopic evidence. *Lithos* **2013**, *162*, 251–263. [CrossRef]
104. Ma, L.; Jiang, S.Y.; Hou, M.L.; Dai, B.Z.; Jiang, Y.H.; Yang, T.; Zhao, K.D.; Pu, W.; Zhu, Z.Y.; Xu, B. Geochemistry of Early Cretaceous calc–alkaline lamprophyres in the Jiaodong Peninsula: Implication for lithospheric evolution of the eastern North China Craton. *Gondwana Res.* **2014**, *25*, 859–872. [CrossRef]
105. Dong, X.; Li, D.P.; Zhao, R.; Wang, X.R.; Wang, Q.F. Zircon U-Pb chronology and petrogenesis of Zetou pluton in the Jiaodong Peninsula: Implications for regional petrogenesis and mineralization in the Aptian to Albian. *Acta Petrol. Sin.* **2020**, *36*, 1501–1514, (In Chinese with English abstract).
106. Song, Y.X.; Yu, X.F.; Li, D.P.; Geng, K.; Yu, P.F.; Zuo, X.M.; Wang, X.F. Petrogenesis of the Beijie pluton from the northwestern Jiaodong Peninsula: Constraints from zircon U-Pb age, petrogeochemistry and Sr-Nd-Pb isotopes. *Acta Petrol. Sin.* **2020**, *36*, 1477–1500. (In Chinese with English abstract)
107. Song, M.C.; Song, Y.X.; Ding, Z.J.; Li, S.Y. Jiaodong gold deposits: Essential characteristics and major controversy. *Gold Sci. Technol.* **2018**, *26*, 406–422. (In Chinese with English abstract)
108. Huang, T.; Yang, L.Q.; Liu, X.D.; Li, H.L.; Zhang, B.L.; Wang, J.G.; Zhao, Y.F.; Zhang, N. Crustal evolution of the Jiaobei terrane Evidence from U-Pb ages, trace element compositions and Hf isotopes of inherited zircons of the Linglong biotite granite. *Acta Petrol. Sin.* **2014**, *30*, 2574–2594.
109. Wu, M.L.; Zhao, G.C.; Sun, M.; Li, S.Z. A synthesis of geochemistry and Sm–Nd isotopes of Archean granitoid gneisses in the Jiaodong Terrane: Constraints on petrogenesis and tectonic evolution of the Eastern Block, North China Craton. *Precambrian Res.* **2014**, *255*, 885–899. [CrossRef]
110. Song, M.C.; Wang, S.S.; Yang, L.X.; Li, J.; Li, S.Y.; Ding, Z.J. Metallogenic Epoch and Geological Significance of Nonferrous Metallic and Silver Deposits in Jiaodong Peninsula, China. *Acta Geol. Sin. Engl. Ed.* **2017**, *91*, 1305–1325. [CrossRef]
111. Yang, J.H.; Wu, F.Y.; Wilde, S.A. A review of the geodynamic setting of large–scale Late Mesozoic gold mineralization in the North China Craton: An association with lithospheric thinning. *Ore Geol. Rev.* **2003**, *23*, 125–152. [CrossRef]
112. Song, M.C.; Zhou, J.B.; Song, Y.X.; Wang, B.; Li, S.Y.; Li, J.; Wang, S.S. Mesozoic Weideshan granitoid suite and its relationship to large-scale gold mineralization in the Jiaodong Peninsula, China. *Geol. J.* **2020**, *55*, 5703–5724. [CrossRef]
113. Hu, F.F.; Fan, H.R.; Yang, J.H.; Zhai, M.G.; Jin, C.W.; Xie, L.W.; Yang, Y.H. Magma mixing for the origin of granodiorite: Geochemical, Sr–Nd isotopic and zircon Hf isotopic evidence of dioritic enclaves and host rocks from Changshannan granodiorite in the Jiaodong Peninsula, eastern China. *Acta Petrol. Sin.* **2005**, *21*, 569–586. (In Chinese with English abstract)
114. Yan, Q.S.; Metcalfe, I.; Shi, X.F.; Zhang, P.Y.; Li, F.C. Early Cretaceous granitic rocks from the southern Jiaodong Peninsula, eastern China: Implications for lithospheric extension. *Int. Geol. Rev.* **2018**, *61*, 821–838. [CrossRef]
115. Liu, J.M.; Zhang, H.F.; Sun, J.G.; Ye, J. Geochemical research on C–O and Sr–Nd isotopes of mantle–derived rocks from Shandong Province, China. *Sci. China Earth Sci.* **2004**, *47*, 171–180. [CrossRef]
116. Yan, J.; Chen, J.F.; Xie, Z.; Gao, T.S.; Foland, K.A.; Zhang, X.D.; Liu, M.W. Later Cretaceous basalts and mantle-derived xenoliths from eastern Shandong. *Acta Petrol. Sin.* **2005**, *21*, 99–112. (In Chinese with English abstract)
117. Eby, G.N. The A–type granitoids: A review of their occurrence and chemical characteristics and speculation on their petrogenesis. *Lithos* **1990**, *26*, 115–134. [CrossRef]
118. Picher, W.S. *Granite Type and Tectonic Environment. Mountain Building Processes*; Academic Pressa: London, UK, 1983; pp. 19–40.
119. Bedard, J. Enclaves from the A–type granite of the Meganic Complex, White Mountain magma series: Clues to granite magma genesis. *J. Geophys.* **1990**, *95*, 17797–17819. [CrossRef]
120. Chen, H.F.; He, S.Y.; Zhang, A.K.; Sun, J.L.; Yan, Z.P.; Liu, J.L.; Zhang, L.; Qian, Y. Petrogenesis and tectonic setting of Middle Silurian A-type granite in Kaerqueka area, East Kunlun. *Geol. Bull. China* **2021**, *40*, 1380–1393. (In Chinese with English abstract)
121. Whalen, J.B.; Jenner, G.A.; Longstaffe, F.J.; Robert, F.; Garipy, C. Geochemical and Isotopic (O, Nd, Pb and Sr) Constraints on A–type Granite Petrogenesis Based on the Topsails Igneous Suite, New foundland Appalachians. *J. Petrol.* **1996**, *37*, 1463–1489. [CrossRef]

122. Jia, X.H.; Wang, Q.; Tang, G.J. A type Granites: Research Progress and Implications. *Geotecton. Metallog.* **2009**, *33*, 465. (In Chinese with English abstract)
123. Zhang, Q.; Ran, H.; Li, C.D. A-type granite: What is the essence? *Acta Petrol. Sin.* **2012**, *31*, 621–626. (In Chinese with English abstract)
124. Loiselle, M.C.; Wones, D.R. *Characteristics and Origin of Anorogenic Granites*; Geological Society of America (Abstracts with Programs): Boulder, CO, USA, 1979; Volume 11, pp. 448–468.
125. Wu, S.P.; Wang, M.Y.; Qi, K.J. Present situation of researches on A-type granites: A review. *Acta Petrol. Sin.* **2007**, *26*, 57–66. (In Chinese with English abstract)
126. Sylvester, P.J. Post–collisional alkaline granites. *J. Geol.* **1989**, *97*, 261–280. [CrossRef]
127. Turner, S.; Sandiford, M.; Feden, J. Some geodynamic and compositiona1 constraints on "postorogenic" magmatism. *Geology* **1992**, *20*, 931. [CrossRef]
128. Wang, D.Z.; Zhao, G.T.; Qiu, J.S. The Tectonic Constraint on the Late Mesozoic A-type Granitoid in eastern China. *Geol. J. Univ.* **1995**, *1*, 13–21. (In Chinese with English abstract)
129. Sun, D.Y.; Wu, F.Y.; Li, H.M.; Lin, Q. Age of A-type granite after orogenic belt in northwestern Xiaoxinganling and its relationship with the eastern extension of Suolunshan-Hegenshan-Zdalaite collision zone. *Chin. Sci. Bull.* **2000**, *45*, 2217–2222. (In Chinese with English abstract)
130. Zhao, J.; Zhang, C.L.; Guo, X.J.; Liu, X.Y.; Wang, Q. Determination of the 2.4Ga A-type granite in Lvliang area of the North China Craton and its geological significance. *Acta Petrol. Sin.* **2015**, *31*, 1606–1620. (In Chinese with English abstract)
131. Sun, D.Y.; Wu, F.Y.; Gao, S.; Lu, X.P. Confirmation of two episodes of A-type granite emplacement during Late Triassic and Early Jurassic in the central Jilin Province, and their constraints on the structural pattern of Eastern Jilin Heilongjiang Area, China. *Earth Sci. Front.* **2005**, *12*, 263–275. (In Chinese with English abstract)
132. King, P.L.; White, A.J.R.; Chappell, B.W.; Allen, C.M. Characterization and Origin of Aluminous A–type Granites from the Lachlan Fold Belt, Southeastern Australia. *J. Petrol.* **1997**, *38*, 371–391. [CrossRef]
133. Zhao, G.T.; Wang, D.Z.; Cao, Q.C. The geochemistry and genesis of the Laoshan granitoids, Shandong Province. *Geol. J. China Univ.* **1997**, *3*, 1–15. (In Chinese with English abstract)
134. Chu, H.; Zhang, J.R.; Wei, C.J.; Wang, H.C.; Kang, J.L. Multiple phases of metamorphism during Paleozoic-Mesozoic and the implication for the tectonic evolution of northern North China: Research progresses and Prospects of "The Paleozoic metamorphism and tectonic dynamics in North China". *Geol. Surv. Res.* **2020**, *43*, 186–197. (In Chinese with English abstract)
135. Collins, W.J.; Beams, S.D.; White, A.J.R.; Chappell, B.W. Nature and origin of A–type granites with particular reference to southeastern Australia. *Contrib. Mineral. Petrol.* **1982**, *80*, 189–200. [CrossRef]
136. Pearce, J.A.; Harris, N.B.W.; Tindle, A.G. Trace element discrimination diagrams for the tectonic interpretation of granitic rocks. *J. Petrol.* **1984**, *25*, 956–983. [CrossRef]
137. Batchelor, R.A.; Bowden, P. Petrogenetic Interpretation of Granitoid Rock Series Using Multicationic Parameters. *Chem. Geol.* **1985**, *48*, 43–55. [CrossRef]
138. Ma, L.; Jiang, S.Y.; Hofmann, A.W.; Dai, B.Z. Lithospheric and asthenospheric sources of lamprophyres in the Jiaodong Peninsula: A consequence of rapid lithospheric thinning beneath the North China Craton? *Geochim. Cosmochim. Acta* **2014**, *124*, 250–271. [CrossRef]
139. Song, M.C. Jiaodong type gold deposit and its metallogenic tectonic magmatic background. *Miner. Depos.* **2014**, *33*, 131–132. (In Chinese with English abstract)
140. Wang, L.M.; Song, M.C.; Wang, P.C.; Wang, S.J.; Song, Z.Y.; Liu, J.W.; Li, Y.Y.; Zhang, X.D. *Structure and Evolution of Sulu ULTRAHIGH–Pressure (UHP) Metamorphic Belt*; Geological Publishing House: Beijing, China, 2005; pp. 1–126. (In Chinese)
141. Han, Y.W.; Ma, Z.D. *Geochemistry*; Geological Publishing House: Beijing, China, 2003; pp. 1–370.
142. Isozaki, Y. Jurassic accretion tectonics of Japan. *Island Arc* **1997**, *6*, 25–51. [CrossRef]
143. Maruyama, S.; Isozaki, Y.; Kimura, G.; Terabayashi, M. Paleogeographic maps of the Japanese islands: Plate tectonic systhesis from 750Ma to the resent. *Island Arc* **1997**, *6*, 121–142. [CrossRef]
144. Zhang, H.F.; Zhou, M.; Sun, M.; Zhou, X.H. The origin of Mengyin and Fuxian diamond iferous kimberlites from the north China Craton: Implication for Paleozoic subducted oceanic slab–mantle interaction. *J. Asian Earth Sci.* **2010**, *37*, 425–437. [CrossRef]
145. Ma, M.; Zhang, C.; Li, Y.; Gao, M.; Gao, J.; Li, S.; Feng, Q.; Li, J.; Tao, C.; Liu, Y. Geochronological, Geochemical and Sr-Nd-Pb-Hf Isotopic Constraints on the Petrogenesis of Pyroxene Diorites in the Sanchahe Iron-Gold Deposit, Western Shandong. *Minerals* **2022**, *12*, 601. [CrossRef]
146. Song, M.C.; Yang, L.Q.; Fan, H.R.; Yu, X.F.; Ding, Z.J.; Zhang, Y.W.; Qiu, K.F.; Li, J.; Zhang, L.; Wang, B.; et al. Current progress of metallogenic research and deep prospecting of gold deposits in the Jiaodong Peniusula during 10 years for Exploration Breakthrough Strategic Action. *Geol. Bull. China* **2022**, *41*, 903–935.
147. Hua, Z.M.; Mao, J.W. A preliminary discussion on the Mesozoic metallogenic explosion in east China. *Miner. Depos.* **1999**, *18*, 300–308. (In Chinese with English abstract)

minerals

MDPI

Article

Fluid Inclusions and C–H–O–S–Pb Isotopes of the Huoluotai Porphyry Cu (Mo) Deposit in the Northern Great Xing'an Range, NE China: Implications for Ore Genesis

Yonggang Sun [1,2], Bile Li [2,*], Xusheng Chen [3], Fanbo Meng [3], Qingfeng Ding [2], Ye Qian [2,4] and Linlin Wang [2,4]

1 School of Resources and Civil Engineering, Suzhou University, Suzhou 234000, China
2 College of Earth Sciences, Jilin University, Changchun 130061, China
3 Qiqihaer Institute of Geological Exploration, Heilongjiang 161006, China
4 Key Laboratory of Mineral Resources Evaluation in Northeast Asia, Ministry of Land and Resources, Changchun 130061, China
* Correspondence: libl@jlu.edu.cn

Abstract: The Huoluotai Cu (Mo) deposit is a recently discovered porphyry Cu deposit in the northern Great Xing'an Range, NE China. Fluid inclusion (FI) micro-thermometry results and the C–H–O–S–Pb isotope compositions of the Huoluotai Cu (Mo) deposit are presented in this study. The ore-forming process consists of the sulfide-barren quartz stage (I), the quartz + chalcopyrite ± pyrite ± molybdenite stage (II), the quartz + polymetallic sulfide stage (III), and the quartz + calcite ± pyrite ± fluorite stage (IV). Cu mineralization occurred mainly in stage II. Four types of FIs were recognized: liquid-rich two-phase FIs (L-type), vapor-rich two-phase FIs (V-type), daughter-mineral-bearing three-phase FIs (S-type), and CO_2-bearing FIs (C-type). In stage I, the ore-forming fluids belong to an $H_2O-NaCl-CO_2$ system. In stages II, III, and IV, the ore-forming fluids belong to an $H_2O-NaCl$ system. The results of the FI micro-thermometry and H–O isotope analysis show that the ore-forming fluids originated from a magmatic origin in stage I and mixed with meteoric water from stages II to IV. The S–Pb isotope results suggest that the source of the ore-forming materials has the characteristics of a crust–mantle-mixing origin. Fluid boiling occurred in stages I and II. The FI micro-thermometric data further show that Cu was mainly deposited below 400 °C in stage II, suggesting that fluid boiling occurring below 400 °C may be the primary factor for Cu precipitation in the Huoluotai Cu (Mo) deposit.

Keywords: northern Great Xing'an Range; Huoluotai Cu (Mo) deposit; fluid inclusions; C–H–O–S–Pb isotope; fluid boiling

1. Introduction

As one of the largest orogenic collages in the world, the Central Asian Orogenic Belt (CAOB) is located between the East Europe, Siberia, Tarim, and North China cratons (Figure 1A). The northern Great Xing'an Range (NGXR) is tectonically located in the eastern segment of the CAOB (Figure 1A,B) and hosts a number of epithermal and orogenic Au, hydrothermal Ag–Pb–Zn, porphyry, and skarn Pb–Zn deposits (Figure 1C). Over the past decade, many late Mesozoic porphyry deposits in the NGXR have been discovered and raised concern (Figure 1C). Examples include the Daheishan Mo (ca. 147 Ma; [1]), the Chalukou Mo (ca. 148 Ma; [2]), the Fukeshan Cu (Mo) (ca. 149 Ma; [3,4]), and the Xiaokele Cu (Mo) (ca. 150 Ma; [5]) deposits.

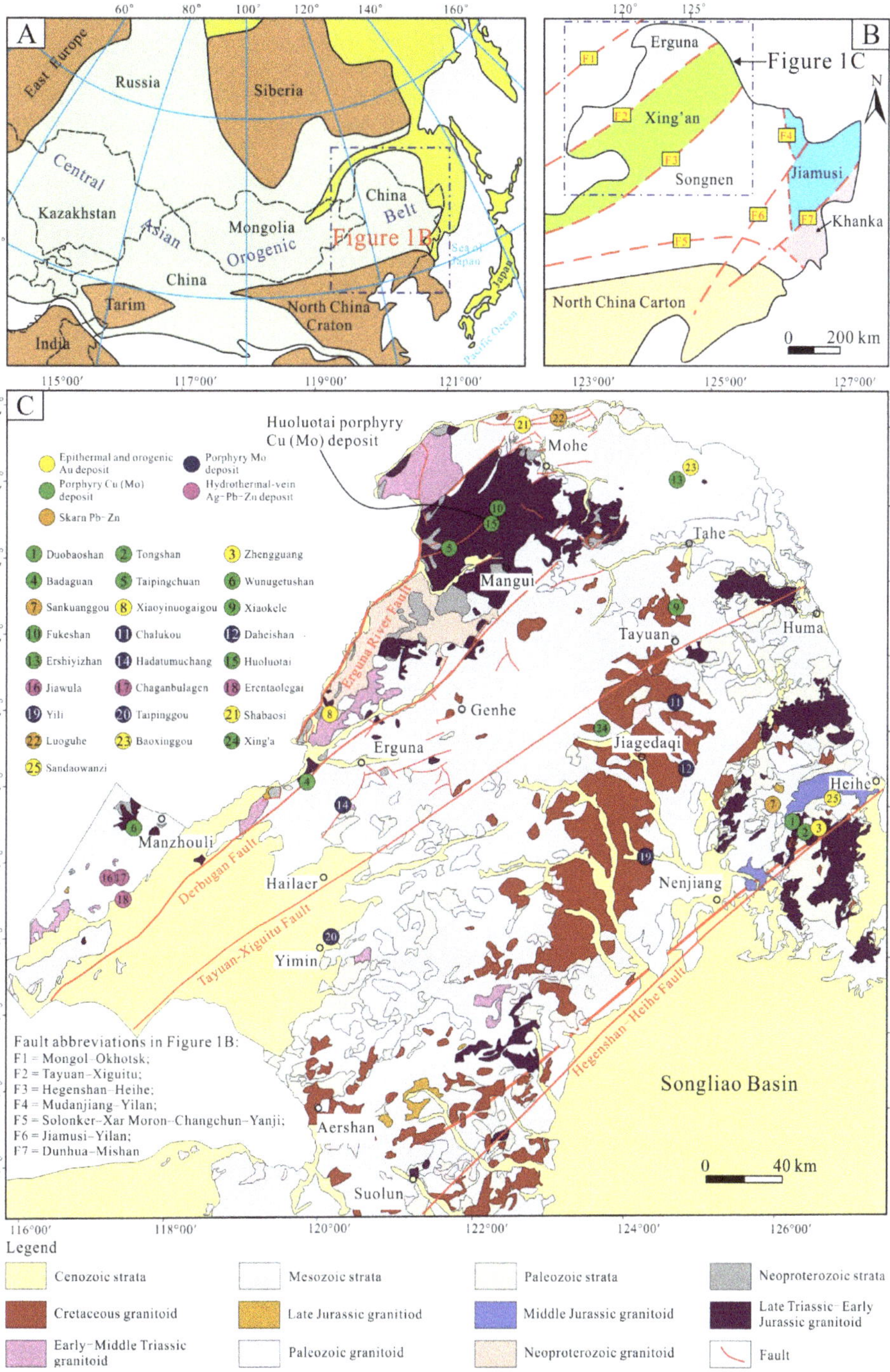

Figure 1. (**A**) Location of the CAOB [6]. (**B**) Geological map of NE China [7]. (**C**) Geological map of the NGXR (modified from [3]).

The Huoluotai Cu (Mo) deposit, which is located in the NGXR, contains estimated industrial metal resources of >4900 t Cu and >300 t Mo and is currently being explored [8]. Previous studies have focused on the geological characteristics [9], geochronology [9,10], and geochemistry of the igneous rocks in this deposit [11]. According to these studies, the Huoluotai Cu (Mo) deposit is considered to be a porphyry deposit. The magmatic activity of the Huoluotai Cu (Mo) deposit mainly occurred in the Late Jurassic–Early Cretaceous (ca. 150–140 Ma), and Cu (Mo) mineralization was closely related to the Late Jurassic granodiorite porphyry (ca. 149 Ma; [11]). Nevertheless, the lack of comprehensive research on the ore-forming fluids limits our understanding of the ore-forming conditions of the Huoluotai Cu (Mo) deposit. Three important issues have not been resolved: (1) the origin and detailed evolution of the ore-forming fluids are unknown; (2) the source of the ore-forming materials remains uncertain; and (3) the possible mineralization processes have not been determined.

To solve these problems, the present study presents micro-thermometry and laser Raman spectroscopy data of the fluid inclusions (FIs), and C–H–O–S–Pb isotope compositions. The results will provide new clues for further prospecting directions in the Huoluotai Cu (Mo) deposit, as well as the late Mesozoic porphyry Cu deposits in the NGXR.

2. Regional Geology

From west to east, NE China is divided into the Erguna, Xing'an, Songnen, Jiamusi, and Khanka blocks (Figure 1B; [12]). The NGXR is located in the western part of NE China, including the Erguna Block, the northern part of the Xing'an Block, and the northwestern part of the Songnen Block (Figure 1B; [13]). The Huoluotai Cu (Mo) deposit is located in the northern segment of the Erguna Block (Figure 1C). The Erguna Block collided with the Xing'an Block along the Tayuan–Xiguitu suture at ca. 500 Ma (Figure 1C; [14,15]). The main faults in the Erguna Block are NE-striking Erguna and Derbugan River faults (Figure 1C; [16]). The basement of the Erguna Block mainly contains Neoproterozoic granitoids and metamorphic supracrustal rocks [16–19]. The exposed strata in the Erguna Block mainly consist of Paleozoic marine sediments [20], widely distributed Mesozoic terrigenous clastic rocks, and volcanic rocks [21]. Magmatism in the NGXR can be divided into Paleozoic, Early–Middle Triassic, Late Triassic–Early Jurassic, Middle–Late Jurassic, and Cretaceous (Figure 1C; [12,22–24]).

3. Ore Deposit Geology

The Huoluotai porphyry Cu (Mo) deposit is located ~50 km southwest of Mohe City, Heilongjiang Province (Figure 1C). The NNW-striking Huoluotai River Fault is distributed in the northeastern part of the district (Figure 2A). Multiperiod igneous rocks developed in the Huoluotai ore district (Figure 2). The Early Jurassic medium–fine-grained monzogranite (ca. 180 Ma) was intruded by Late Jurassic granodiorite porphyry (ca. 149 Ma), diorite porphyry (ca. 146 Ma), and Early Cretaceous granite porphyry dikes (ca. 142 Ma) (Figure 2; [11]). The quartz diorite porphyry dikes intruded into the granodiorite porphyry (Figure 2B). Mineralization developed around the granodiorite porphyry (Figure 2B), and 66 orebodies have been identified. These orebodies are characterized by disseminations, veinlets, and stockworks and are generally 1–9 m thick, dipping toward the east at angles of 20–35°. In the dipping direction, the explored orebodies usually extend downward for 50–240 m.

Three types of alteration were identified in the Huoluotai Cu (Mo) deposit: potassic, chlorite–epidote, and phyllic alterations from early to late (Figure 3A–D). From center to edge, two alteration zones can be recognized: the potassic alteration zone and the phyllic alteration zone (Figure 2B). Hydrothermal alteration surrounds the granodiorite porphyry and is correlated strongly with the granodiorite porphyry (Figure 2B). Potassic alteration is characterized by secondary K-feldspar and biotite (Figure 3A,B). The ore minerals developed in the potassic alteration zone are mainly magnetite, chalcopyrite, hematite, and small amounts of molybdenite (Figure 3E–H). Magnetite mainly developed as disseminations and/or magnetite ± quartz ± chalcopyrite veinlets. Cu (Mo) orebodies mainly developed in the potassic alteration zone. Chalcopyrite mainly developed as disseminations, masses, or quartz + chalcopyrite veins in the potassic alteration zone (Figure 3G,H). Potassic alteration was locally overprinted by epidote and chlorite alterations. Phyllic alteration can be distinguished by secondary sericite and quartz (Figure 3C,D). The ore minerals present in the phyllic alteration zone are mainly pyrite, chalcopyrite, and molybdenite (Figure 3I). In the phyllic alteration zone, chalcopyrite occurs mainly as disseminations, while molybdenite occurs mainly as disseminations and/or quartz + molybdenite + pyrite ± chalcopyrite veins (Figure 3I).

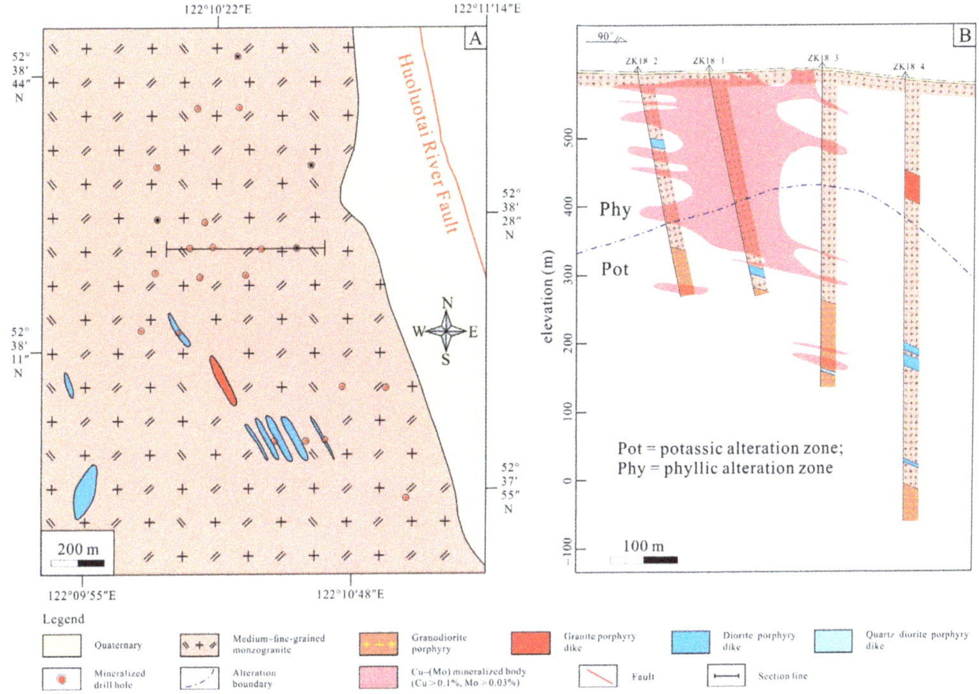

Figure 2. (**A**) Geological map of the Huoluotai Cu (Mo) deposit (modified from [8]). (**B**) Geological sections in the Huoluotai Cu (Mo) deposit (modified from [8]).

Based on the mineral assemblages and crosscutting relationships between veins in the Huoluotai Cu (Mo) deposit (Figure 3), four mineralization stages were identified: the sulfide-barren quartz stage (stage I), the quartz + chalcopyrite ± pyrite ± molybdenite stage (stage II), the quartz + polymetallic sulfide stage (stage III), and the quartz + calcite ± pyrite ± fluorite stage (stage IV) (Figure 4). Stage I is characterized by sulfide-barren quartz veins, which are rare and generally discontinuous (Figure 3E). There is no obvious cutting relationship between the veins in stage I and those in other mineralization stages. Stage I veins are mainly concentrated in the potassic alteration zone. Copper is mainly found in stage II. Stage II veins are generally continuous and irregular (Figure 3G) and mainly contain quartz, chalcopyrite, pyrite, and molybdenite. Stage II veins are distributed mainly in the potassic alteration zone. Stage III veins are generally continuous and straight (Figure 3I) and are characterized by quartz, pyrite, and molybdenite. Stage III veins are usually concentrated in the phyllic alteration zone. Stage IV veins are usually irregular and dominated by quartz and calcite (Figure 3J), with small amounts of fluorite. The sulfide observed in stage IV is only pyrite. Supergene mineralization is dominated by molybdite, azurite, and malachite (Figure 3K,L).

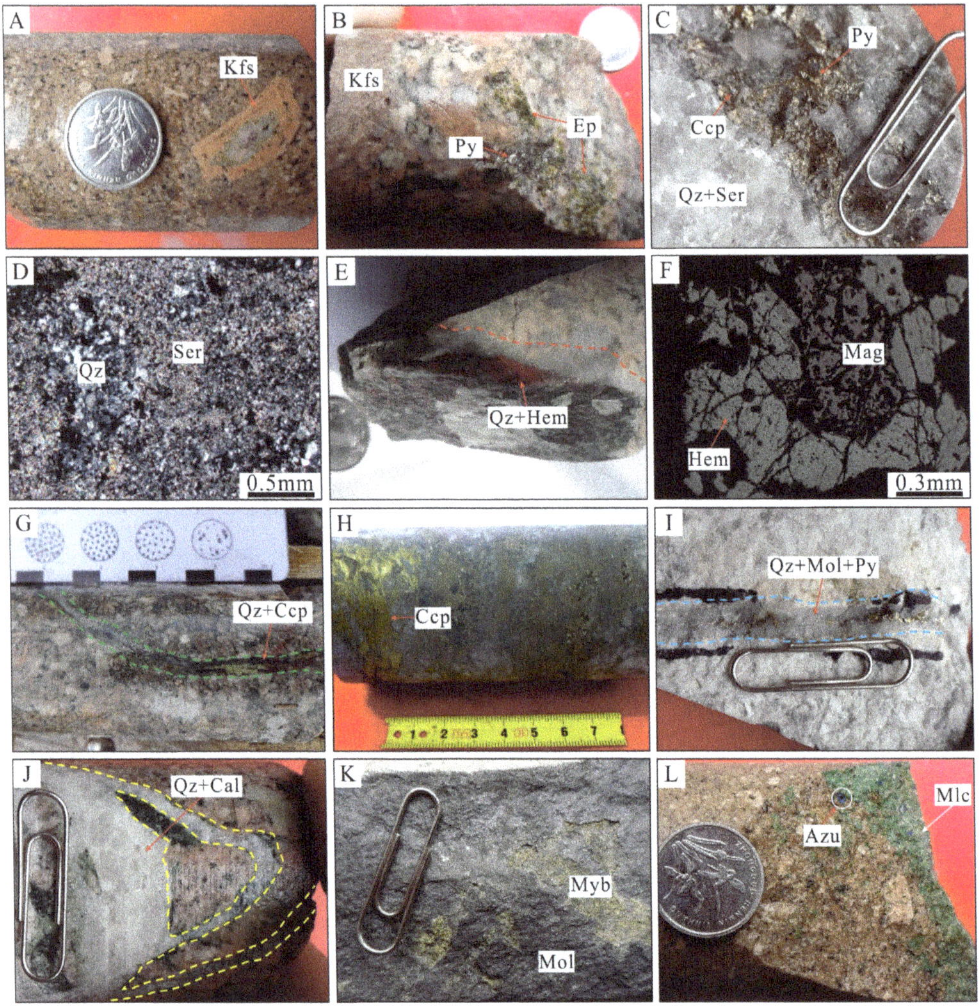

Figure 3. Photographs and photomicrographs of representative hydrothermal alterations and ore mineralization features in the Huoluotai Cu (Mo) deposit. (**A**) Potassic alteration observed in the granodiorite porphyry. (**B**) Potassic alteration overprinted by epidote alteration. (**C**) Phyllic alteration in the granodiorite porphyry. (**D**) Phyllic alteration characterized by an alteration assemblage of sericite and quartz. (**E**) Stage I quartz + hematite vein in potassic-altered granodiorite porphyry. (**F**) Anhedral hematite and magnetite in stage I. (**G**) Stage II quartz + chalcopyrite vein in potassic-altered granodiorite porphyry. (**H**) Dense disseminated chalcopyrite developed in the granodiorite porphyry. (**I**) Stage III quartz + molybdenite + pyrite vein in phyllic-altered wall rock. (**J**) Stage IV quartz + calcite vein in potassic-altered granodiorite porphyry. (**K**) Primary molybdenite oxidized to molybdite. (**L**) Primary chalcopyrite oxidized to azurite and malachite. Abbreviations: Qz = quartz; Kfs = K-feldspar; Ep = epidote; Ser = sericite; Cal = calcite; Mag = magnetite; Hem = hematite; Py = pyrite; Ccp = chalcopyrite; Mol = molybdenite; Myb = molybdite; Mlc = malachite; Azu = azurite.

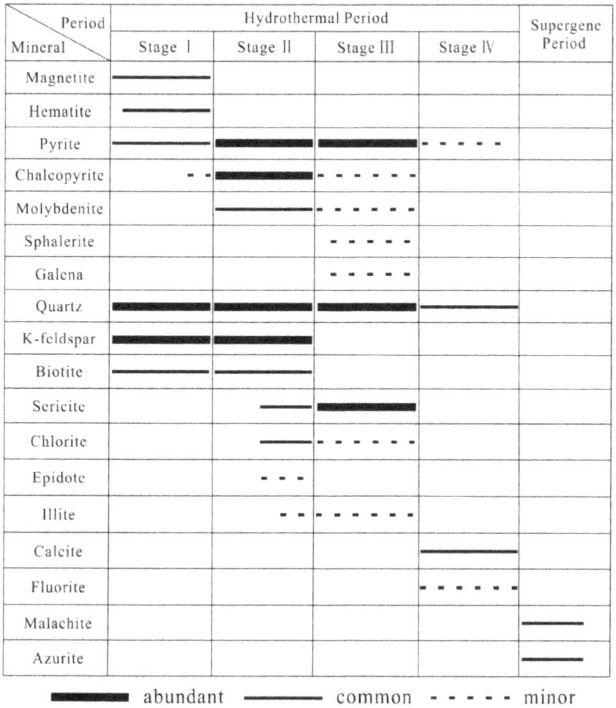

Figure 4. Paragenetic sequence of the major minerals in the Huoluotai Cu (Mo) deposit.

4. Analytical Methods

4.1. Fluid Inclusion Measurements

In total, 31 quartz samples were used to perform FI measurements at the Key Laboratory of Mineral Resources Evaluation in Northeast Asia, Ministry of Land and Resources, Changchun, China. Secondary FIs were not analyzed [25]. All types of FIs were selected for laser Raman spectra analyses using an RM-2000 Laser Raman microprobe. Microthermometric analyses were performed using a Linkam THMS-600 heating–freezing stage (Linkam Scientific Instruments Ltd., Epsom, UK).

4.2. H–O Isotope Analyses

Six quartz samples were selected for H–O isotope analysis at the Center of Analytical Laboratory in the Beijing Research Institute of Uranium Geology (BRIUG), China National Nuclear Corporation (Beijing, China). The H–O isotope compositions were analyzed by a Finnigan MAT253-EM mass spectrometer. The O and H isotope compositions were analyzed using the conventional BrF_5 method and the Zn reduction method, respectively [26,27]. The H–O isotope values were normalized with the Vienna Standard Mean Ocean Water (V-SMOW) standards. The analytical precision was better than 0.2‰ for $\delta^{18}O$ and 2‰ for δD.

4.3. C–O Isotope Analyses

Four representative calcite samples were selected for C–O isotope analysis at BRIUG. The C–O isotope analyses of calcite were performed using the 100% phosphoric acid method [28] with a MAT-251EM mass spectrometer. $\delta^{13}C$ used the Pee Dee Belemnite (PDB) standard, and $\delta^{18}O$ used the SMOW standard. The analytical precision was better than 0.1‰ for $\delta^{13}C$ and 0.2‰ for $\delta^{18}O$.

4.4. S–Pb Isotope Analyses

The S isotope compositions of seven representative sulfide samples were analyzed at BRIUG. The values of $\delta^{34}S$ for the sulfide samples were determined on SO_2 produced by the reaction between the sulfide and cuprous oxide. $\delta^{34}S$ used the Vienna Cañon Diablo Troilite (V-CDT) standard. The Pb isotope compositions of four representative sulfide samples were also analyzed at BRIUG. Sulfide samples were dissolved in a mixed solution of HF + HNO_3 at 150 °C for seven days. The Pb isotope compositions were measured using a MAT-261 thermal ionization mass spectrometer.

5. Analytical Results

5.1. Fluid Inclusions

5.1.1. Microscopy Investigations of Fluid Inclusions

According to the phases, filling degree, and combination relationship of FIs at a normal temperature (25 °C), four different FI types were identified in the quartz samples from the Huoluotai Cu (Mo) deposit, as described below.

Liquid-rich (L-type) FIs are most common in four mineralization stages, which consist of two phases with $V_{H_2O}/(V_{H_2O} + L_{H_2O}) < 50$ vol.% at ~25 °C (Figure 5A,C–E,G–I). The vapor phase accounts for 10–20 vol.% of the FI volume. The FIs range from 3 to 20 μm in size and are round or irregular in shape. Commonly, L-type FIs homogenize into liquid during heating.

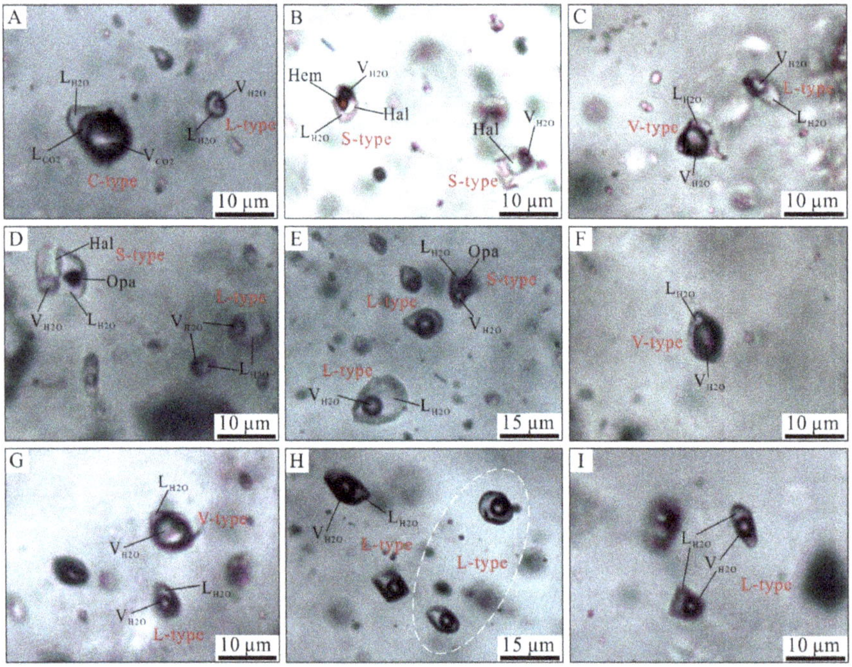

Figure 5. Photomicrographs of the representative primary FIs in four mineralization stages from the Huoluotai Cu (Mo) deposit. (**A**) The coexistence of primary L- and C-type FIs in stage I quartz. (**B**) S-type FIs with hematite and/or halite daughter minerals in stage I quartz. (**C**) The coexistence of primary L- and V-type FIs in stage I quartz. (**D**,**E**) The coexisting S- and L-type FIs in stage II quartz. (**F**) Primary V-type FIs in stage II quartz. (**G**) The coexisting L- and V-type FIs in stage III quartz. (**H**) Primary L-type FIs in stage III quartz. (**I**) Primary L-type FIs in stage IV quartz. Abbreviations: L_{H_2O} = H_2O liquid; V_{H_2O} = H_2O vapor; L_{CO_2} = CO_2 liquid; V_{CO_2} = CO_2 vapor; Hal = halite; Hem = hematite; Opa = unidentified opaque mineral.

Vapor-rich (V-type) FIs were identified in stages I, II, and III but absent in stage IV. These FIs consist of two phases with $V_{H_2O}/(V_{H_2O} + L_{H_2O}) > 50$ vol.% at ~25 °C and have 60–90 vol.% vapor bubbles (Figure 5C,F,G). These FIs are normally round or oval in shape and range in size from 5 to 20 μm. Commonly, V-type FIs homogenize into vapor during heating.

Daughter-mineral-bearing three-phase (S-type) FIs were detected in stages I and II. S-type FIs consist of an aqueous liquid, a vapor bubble, and one or more solid daughter minerals at ~25 °C (Figure 5B,D,E). These FIs are normally oval or irregular in shape and range from 5 to 20 μm in size. Additionally, these FIs may contain one, two, or three daughter minerals, which can be halite or ore minerals (Figure 5B,D,E). The halites are mainly transparent, pale white, and cubic in shape (Figure 5B,D). The ore minerals are opaque (Figure 5B,D,E). Most halites dissolved after the vapor-bubble disappeared during heating.

CO_2-bearing (C-type) FIs only occur in stage I. These FIs are normally round, oval, or irregular in shape and range from 8 to 15 μm in size. Most C-type FIs consist of three phases ($L_{H_2O} + L_{CO_2} + V_{CO_2}$), and occasionally two phases ($L_{H_2O} + L_{CO_2}$) at ~25 °C. The latter changes into three phases with the appearance of CO_2 bubbles at ~10 °C. Most C-type FIs have 50–70 vol.% CO_2 volumetric proportions $[(V_{CO_2} + L_{CO_2})/(V_{CO_2} + L_{CO_2} + L_{H_2O})]$ at ~25 °C (Figure 5A), and can be homogenized into a liquid CO_2 phase.

5.1.2. Micro-Thermometric Results

FI micro-thermometric data in this study are listed in Table 1 and shown in Figure 6. The salinities of L- and V-type FIs were estimated according to the HOKIEFLINCS_H_2O–NaCl program [29]. The salinities of S-type FIs were calculated using the equations from [30]. We calculated the salinities of C-type FIs based on the melting temperatures of CO_2 clathrate in the H_2O–CO_2–NaCl system [31].

Table 1. Micro-thermometric data and calculated parameters for FIs from the Huoluotai Cu (Mo) deposit.

Mineralized Stages	Host Minerals	FI Types	N	$T_{m\text{-}CO2}$ (°C)	$T_{m\text{-}cla}$ (°C)	$T_{h\text{-}CO2}$ (°C)	$T_{h\text{-}total}$ (°C)	$T_{m\text{-}ice}$ (°C)	$T_{h\text{-}s}$ (°C)	$T_{h\text{-}v}$ (°C)	Salinity (wt.% NaCl Equivalent)
I	Quartz	L-type	32	/	/	/	/	−8.1 to −4.5	/	406–479	7.2–11.8
		V-type	12	/	/	/	/	−7.9 to −5.8	/	427–471	8.9–11.6
		S-type	20	/	/	/	/	/	424–485	/	50.1–57.8
		C-type	8	−57.4 to −56.8	3.5–5.5	28.2–30.5	429–470	/	/	/	8.3–11.3
II	Quartz	L-type	34	/	/	/	/	−6.6 to −3.1	/	307–398	5.1–10.0
		V-type	14	/	/	/	/	−6.4 to −3.8	/	326–396	6.2–9.7
		S-type	17	/	/	/	/	/	356–414	/	43.0–48.9
III	Quartz	L-type	47	/	/	/	/	−4.5 to −0.9		223–336	1.6–7.2
		V-type	16	/	/	/	/	−3.6 to −1.8	/	265–332	3.1–5.9
IV	Quartz	L-type	40	/	/	/	/	−2.6 to −0.7		143–249	1.2–4.3

Abbreviations: $T_{h\text{-}v}$ = homogenization temperature of vapor-rich or liquid-rich fluid inclusions; $T_{h\text{-}s}$ = dissolution temperature of halite; $T_{m\text{-}ice}$ = temperature of final ice melting; $T_{h\text{-}total}$ = total homogenization temperature of CO_2 fluid inclusions; $T_{h\text{-}CO_2}$ = homogenization temperature of the CO_2 phases; $T_{m\text{-}cla}$ = final melting temperature of CO_2–H_2O clathrate; $T_{m\text{-}CO_2}$ = final melting temperature of solid CO_2.

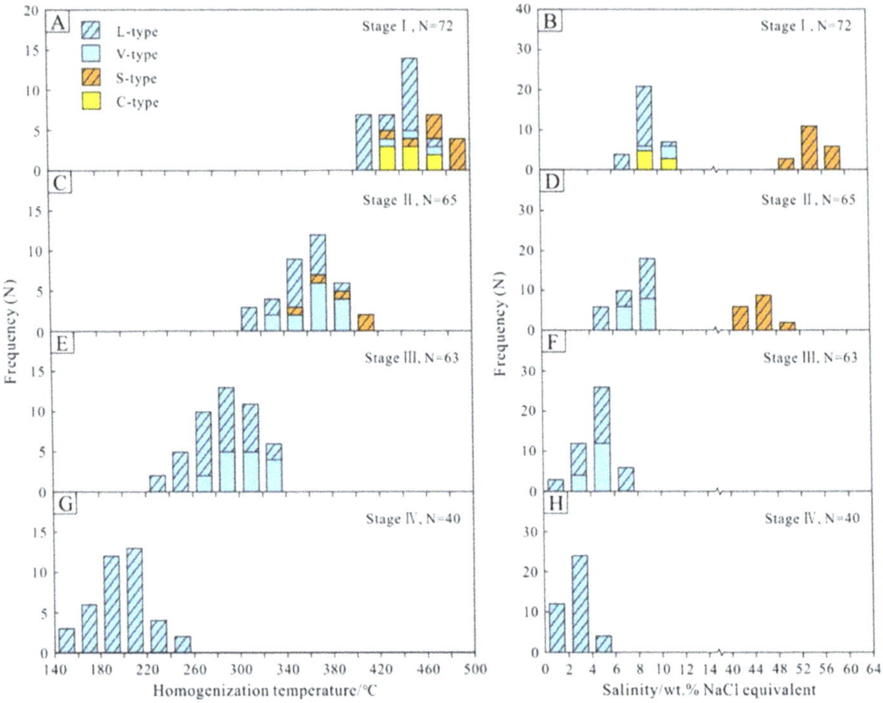

Figure 6. Frequency histograms of the total homogenization temperatures (Th) and salinities of FIs (wt.% NaCl equivalent) in four mineralization stages from the Huoluotai Cu (Mo) deposit. (**A**) Homogenization temperatures histogram for L-, V-, S-, and C-type FIs from stage I quartz; (**B**) Salinities histogram for L-, V-, S-, and C-type FIs from stage I quartz; (**C**) Homogenization temperatures histogram for L-, V-, and S-type FIs from stage II quartz; (**D**) Salinities histogram for L-, V-, and S-type FIs from stage II quartz; (**E**) Homogenization temperatures histogram for L- and V-type FIs from stage III quartz; (**F**) Salinities histogram for L- and V-type FIs from stage III quartz; (**G**) Homogenization temperatures histogram for L-type FIs from stage IV quartz; (**H**) Salinities histogram for L-type FIs from stage IV quartz.

Stage I (quartz): L-, V-, S-, and C-type FIs were observed in stage I quartz (Figure 5A–C). The L-type FIs homogenized into the aqueous liquid phase at 406–479 °C (peaking at 440–460 °C, n = 32) (Figure 6A) during heating. These FIs yielded final ice-melting temperatures ($T_{m\text{-}ice}$) of −8.1 to −4.5 °C, with calculated salinities of 7.2–11.8 wt.% NaCl equivalent (peaking at 8–10 wt.% NaCl equivalent, n = 32) (Figure 6B). The V-type FIs homogenized into the vapor phase at 427–471 °C (n = 12; Figure 6A). These V-type FIs yielded final ice-melting temperatures ($T_{m\text{-}ice}$) of −7.9 to −5.8 °C, corresponding to salinities of 8.9–11.6 wt.% NaCl equivalent (n = 12; Figure 6B). During heating, all S-type FIs commonly homogenized into the aqueous liquid phase, with halite daughter minerals dissolving at temperatures ($T_{h\text{-}s}$) varying from 424 to 485 °C (n = 20; Figure 6A), with calculated salinities of 50.1–57.8 wt.% NaCl equivalent (peaking at 52–56 wt.% NaCl equivalent, n = 20) (Figure 6B). C-type FIs were completely homogenized into liquid H_2O at temperatures ($T_{h\text{-}total}$) of 429–470 °C (n = 8) (Figure 6A). These FIs yielded initial melting temperatures ($T_{m\text{-}CO2}$) of solid CO_2 from −57.4 to −56.8 °C and clathrate melting temperatures ($T_{m\text{-}cla}$) from 3.5 to 5.5 °C, corresponding to salinities of 8.3–11.3 wt.% NaCl equivalent (n = 8) (Figure 6B). The CO_2 phase homogenized into the liquid CO_2 phase at temperatures ($T_{h\text{-}CO_2}$) of 28.2–30.5 °C.

Stage II (quartz): L-, V-, and S-type FIs were observed in stage II quartz (Figure 5D–F). The L-type FIs homogenized into the aqueous liquid phase at 307–398 °C (peaking at 340–380 °C, n = 34) (Figure 6C) during heating. These FIs yielded final ice-melting temperatures ($T_{m\text{-}ice}$) of −6.6 to −3.1 °C, with calculated salinities of 5.1–10.0 wt.% NaCl equivalent (peaking at 8–10 wt.% NaCl equivalent, n = 34) (Figure 6D). V-type FIs homogenized into the vapor phase at 326–396 °C (n = 14; Figure 6C). These V-type FIs yielded final ice-melting temperatures ($T_{m\text{-}ice}$) of −6.4 to −3.8 °C, corresponding to salinities of 6.2–9.7 wt.% NaCl equivalent (n = 14) (Figure 6D). During heating, all S-type FIs commonly homogenized into the aqueous liquid phase, with the halite daughter minerals dissolving at temperatures ($T_{h\text{-}s}$) varying from 356 to 414 °C (peaking at 360–400 °C, n = 17) (Figure 6C), with calculated salinities of 43.0–48.9 wt.% NaCl equivalent (n = 17; Figure 6D).

Stage III (quartz): L- and V-type FIs are widespread in stage III quartz (Figure 5G,H). The L-type FIs homogenized into the aqueous liquid phase at 223–336 °C (peaking at 280–320 °C, n = 47) (Figure 6E) and yielded final ice-melting temperatures ($T_{m\text{-}ice}$) of –4.5 to –0.9 °C, with calculated salinities of 1.6–7.2 wt.% NaCl equivalent (peaking at 4–6 wt.% NaCl equivalent, n = 47) (Figure 6F). The V-type FIs homogenized into the vapor phase at 265–332 °C (n = 16; Figure 6E). These V-type FIs yielded final ice-melting temperatures ($T_{m\text{-}ice}$) of −3.6 to −1.8 °C, corresponding to salinities of 3.1–5.9 wt.% NaCl equivalent (n = 16) (Figure 6F).

Stage IV (quartz): Only the L-type FIs were recognized in stage IV quartz (Figure 5I). The L-type FIs homogenized into the aqueous liquid phase at 143–249 °C (peaking at 180–220 °C, n = 40) (Figure 6G) and yielded final ice-melting temperatures ($T_{m\text{-}ice}$) of −2.6 to −0.7 °C, with calculated salinities of 1.2–4.3 wt.% NaCl equivalent (peaking at 2–4 wt.% NaCl equivalent, n = 40) (Figure 6H).

5.1.3. Laser Raman Spectroscopy

The gas phase composition of some representative FIs in the four mineralization stages were selected for laser Raman spectroscopy analysis. Representative laser Raman spectra are shown in Figure 7. C-type FIs in stage I were universally found to be rich in CO_2 (Figure 7A). Vapor phases of the V-type FIs in stages I, II, and III only contained H_2O (Figure 7B). L-type FIs in the four mineralization stages only contained H_2O (Figure 7C,D). These analytical results suggest that the ore-forming fluids of stage I represent an H_2O–CO_2–NaCl system that transforms into an H_2O–NaCl system from stages II to IV.

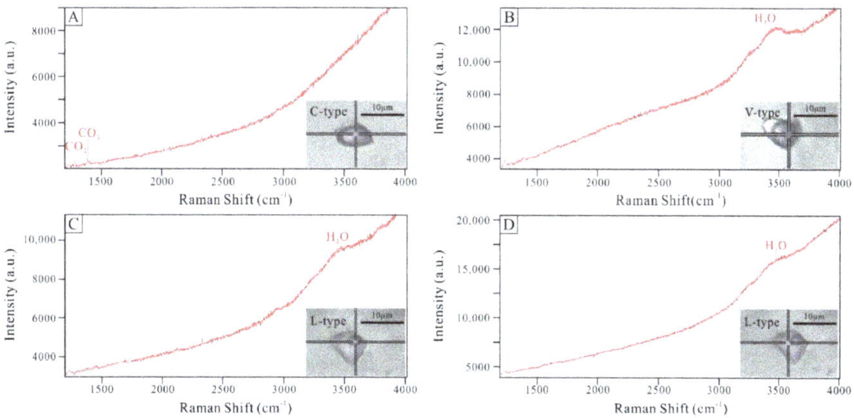

Figure 7. Representative laser Raman spectra for different types of FIs from the Huoluotai Cu (Mo) deposit. (**A**) The vapor phases of C-type FIs in stage I are rich in CO_2; (**B**) The vapor phases of V-type FIs are almost entirely H_2O; (**C**) The vapor phases of L-type FIs are almost entirely H_2O; (**D**) The liquid phases of L-type FIs are almost entirely H_2O.

5.2. H–O Isotopes

The H–O isotope results are listed in Table 2 and illustrated in Figure 8. Here, the $\delta^{18}O_{V\text{-}SMOW}$ values of the six quartz samples from the four mineralization stages range from 7.5 to 9.6‰. The $\delta^{18}O_{H_2O}$ values were calculated by the quartz–water equilibrium function [32], with the peak homogenization temperatures of FIs in the same mineralization stage. The calculated $\delta^{18}O_{H_2O}$ values for stages I, II, III, and IV are 6.5‰, 3.0‰ to 3.5‰, 0.2‰ to 1.9‰, and −2.5‰, respectively (Figure 8). The δD values for stages I, II, III, and IV are −108.1‰, −126.2‰ to −122.5‰, −132.9‰ to −130.9‰, and −139.1‰, respectively (Figure 8).

Table 2. H and O isotope data of quartz samples from the Huoluotai Cu (Mo) deposit.

Sample No.	Mineralized Stages	Mineral	$\delta^{18}O_{V\text{-}SMOW}$ (‰)	T (°C)	$\delta^{18}O_{H_2O}$ (‰)	δD (‰)
HLT-ZK18-2-HO1	I	Quartz	9.6	450	6.5	−108.1
HLT-ZK18-1-HO1	II	Quartz	8.3	370	3.5	−126.2
HLT-ZK18-1-HO2	II	Quartz	7.8	370	3.0	−122.5
HLT-ZK18-1-HO3	III	Quartz	7.5	290	0.2	−132.9
HLT-ZK18-2-HO2	III	Quartz	9.2	290	1.9	−130.9
HLT-ZK18-3-HO1	IV	Quartz	9.2	200	−2.5	−139.1

$\delta^{18}O_{H_2O}$ of water in equilibrium with quartz were calculated according to the equation of $1000\ln\alpha_{quartz\text{-}water} = 3.38 \times 10^6/T^2 - 3.4$ [32], which were defined by the peak homogenization temperatures (T) of FIs for corresponding quartz samples.

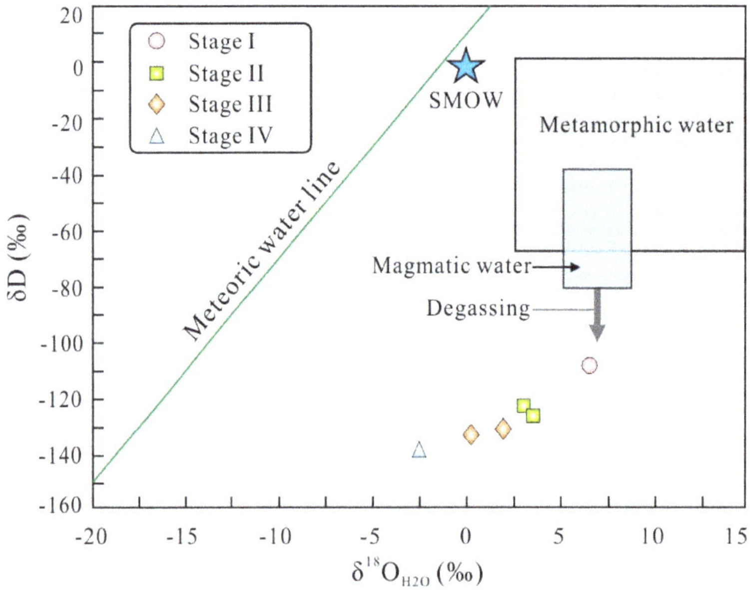

Figure 8. Plot of δD vs. $\delta^{18}O_{H_2O}$ for the ore-forming fluids in the Huoluotai Cu (Mo) deposit [33]. SMOW = Standard Mean Ocean Water.

5.3. C–O Isotopes

The C–O isotope results for the calcites in stage IV are listed in Table 3 and illustrated in Figure 9. Here, the $\delta^{13}C_{PDB}$ values of the four calcite samples in stage IV range from −3.1‰ to −0.2‰. The $\delta^{18}O_{SMOW}$ values range from 6.6‰ to 8.1‰.

Table 3. C and O isotope data of calcite samples from the Huoluotai Cu (Mo) deposit.

Sample No.	Mineral	Mineralized Stages	Sample Description	$\delta^{13}C_{PDB}$ (‰)	$\delta^{18}O_{PDB}$ (‰)	$\delta^{18}O_{SMOW}$ (‰)
HLT-ZK18-2-CO1	Calcite	IV	Qz+Cal+Py vein	−0.2	−22.1	8.1
HLT-ZK18-1-CO1	Calcite	IV	Qz+Cal+Py vein	−2.3	−22.8	7.4
HLT-ZK18-3-CO1	Calcite	IV	Qz+Cal+Py vein	−3.1	−23.5	6.6
HLT-ZK18-3-CO2	Calcite	IV	Qz+Cal+Py vein	−0.2	−23.3	6.9

$\delta^{18}O_{SMOW} = 1.03086 \times \delta^{18}O_{V\text{-}PDB} + 30.86$ [34].

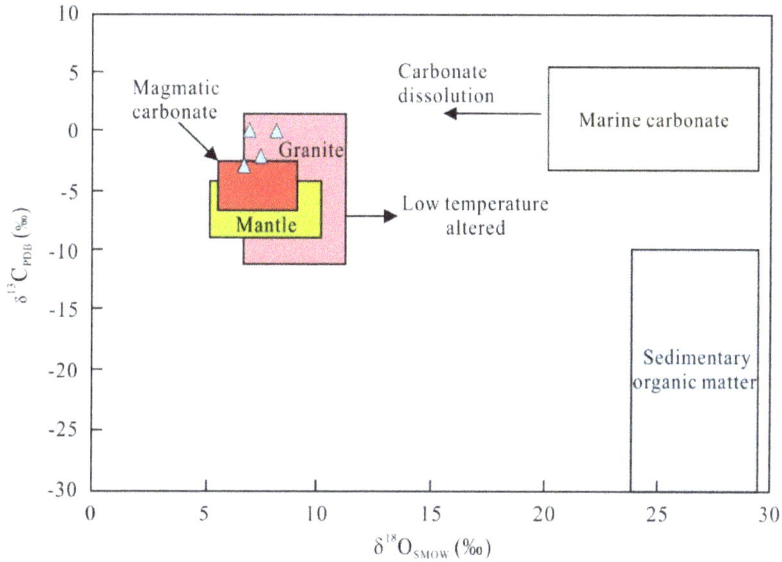

Figure 9. Plot of $\delta^{13}C_{PDB}$ vs. $\delta^{18}O_{SMOW}$ for the calcite from the Huoluotai Cu (Mo) deposit. The three main carbon sources include marine carbonate [35], sedimentary organic matter carbon [36], and magma-mantle carbonate [37].

5.4. S–Pb Isotopes

The S isotope results of the seven sulfide samples from the Huoluotai Cu (Mo) deposit are presented in Table 4 and plotted in Figure 10. The $\delta^{34}S_{V\text{–}CDT}$ values of the seven sulfides range from 0.2‰ to 3.7‰ (average = 1.9‰) (Table 4; Figure 10).

Table 4. S isotope data of sulfide samples from the Huoluotai Cu (Mo) deposit.

Sample No.	Mineral	Mineralized Stages	Sample Description	δ^{34}SV–CDT (‰)
HLT-ZK18-1-S1	Pyrite	I	Qz+Hem+Py vein	2.8
HLT-ZK18-1-S2	Pyrite	I	Qz+Kfs+Py vein	2.3
HLT-ZK18-2-S1	Chalcopyrite	II	Qz+Kfs+Ccp vein	1.8
HLT-ZK18-1-S3	Pyrite	II	Qz+Kfs+Ccp+Py vein	1.4
HLT-ZK18-1-S4	Chalcopyrite	III	Qz+Ser+Ccp+Py vein	0.2
HLT-ZK18-2-S2	Molybdenite	III	Qz+Ser+Mol vein	3.7
HLT-ZK18-3-S1	Pyrite	IV	Qz+Cal+Py vein	1.4

Abbreviations: Qz = quartz; Kfs = K-feldspar; Ser = sericite; Cal = calcite; Hem = hematite; Ccp = chalcopyrite; Mol = molybdenite; Py = pyrite.

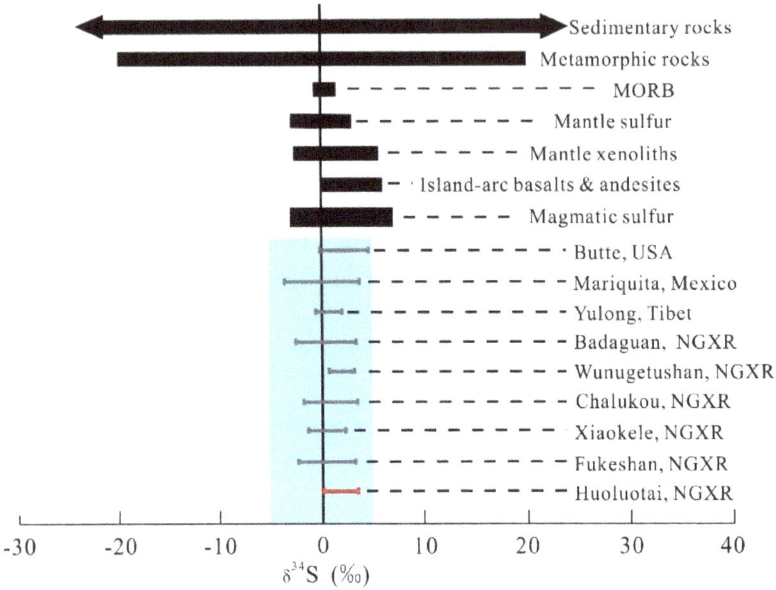

Figure 10. S isotope compositions of sulfides from the Huoluotai Cu (Mo) deposit, important S reservoirs [38–40], and other typical porphyry deposits elsewhere in the world [4,5,41–46].

The Pb isotope compositions of the four sulfide samples from the Huoluotai Cu (Mo) deposit are listed in Table 5 and illustrated in Figure 11. Here, the Pb isotope compositions of sulfides have a limited variation in $^{206}Pb/^{204}Pb$, $^{207}Pb/^{204}Pb$, and $^{208}Pb/^{204}Pb$, varying from 18.352 to 18.435, 15.552 to 15.578, and 38.212 to 38.309, respectively.

Table 5. Pb isotope data of sulfide samples from the Huoluotai Cu (Mo) deposit.

Sample No.	Mineral	Mineralized Stages	Sample Description	$^{206}Pb/^{204}Pb$	Error	$^{207}Pb/^{204}Pb$	Error	$^{208}Pb/^{204}Pb$	Error
HLT-ZK18-1-Pb1	Pyrite	I	Qz+Kfs+Py vein	18.409	0.003	15.564	0.002	38.255	0.005
HLT-ZK18-1-Pb2	Chalcopyrite	II	Qz+Ccp vein	18.352	0.002	15.552	0.002	38.212	0.005
HLT-ZK18-2-Pb1	Pyrite	III	Qz+Mol+Py vein	18.435	0.002	15.571	0.002	38.284	0.005
HLT-ZK18-1-Pb3	Pyrite	III	Qz+Mol+Py vein	18.428	0.001	15.578	0.001	38.309	0.003

Abbreviations: Qz = quartz; Kfs = K-feldspar; Ccp = chalcopyrite; Mol = molybdenite; Py = pyrite.

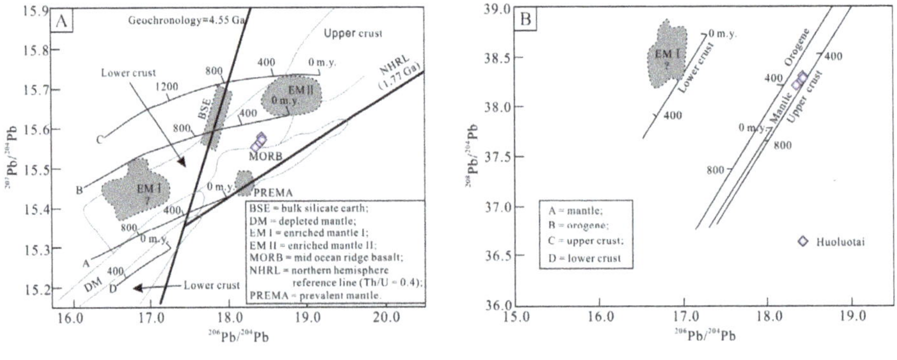

Figure 11. Pb isotope compositions of sulfides from the Huoluotai Cu (Mo) deposit [47]. (**A**) $^{207}Pb/^{204}Pb$ vs. $^{206}Pb/^{204}Pb$. (**B**) $^{208}Pb/^{204}Pb$ vs. $^{206}Pb/^{204}Pb$.

6. Discussion

6.1. Source of the Ore-forming Materials

The $\delta^{34}S_{V\text{-}CDT}$ values (0.2‰ to 3.7‰) of the seven sulfides from the Huoluotai Cu (Mo) deposit overlap those of typical porphyry deposits in the NGXR, such as the Badaguan (–2.4‰ to 3.5‰; [45]), the Chalukou (–1.9‰ to 3.6‰; [43]), the Xiaokele (–1.2‰ to 2.4‰; [5]), and the Fukeshan (–2.3‰ to 3.4‰; [4]) deposits (Figure 10). These values are also consistent with the $\delta^{34}S_{V\text{-}CDT}$ values of typical porphyry deposits elsewhere in the world (−5‰ to 5‰; [39]). The $\delta^{34}S_{V\text{-}CDT}$ values of sulfide minerals for all these deposits fall into the magmatic sulfur range and are similar to those of mantle-derived sulfur (Figure 10; [39]). The Pb isotope compositions of sulfide minerals from the Huoluotai Cu (Mo) deposit are similar and homogeneous, indicating that they share a common Pb reservoir. In the $^{207}Pb/^{204}Pb$ vs. $^{206}Pb/^{204}Pb$ diagram (Figure 11A), Pb isotope data for the sulfide samples are plotted between the mantle and orogene evolution curves, near the mid-ocean ridge basalt (MORB) region. In the $^{208}Pb/^{204}Pb$ vs. $^{206}Pb/^{204}Pb$ diagram (Figure 11B), Pb isotope data are all plotted between the orogene and upper crust evolution curves (near the mantle evolution curve), showing the characteristics of a crust-mantle-mixing origin.

6.2. Estimation of Trapping Pressure

Trapping pressure can be estimated only when the actual trapping temperature is known or if fluid boiling occurred in the hydrothermal system at the time of entrapment [48,49]. During heating, L-type FIs homogenized into the aqueous liquid phase, while the V-type FIs homogenized into the vapor phase. In the Huoluotai Cu (Mo) deposit, the coexistence of L-, V-, and S-type FI assemblages (Figure 5A–F) within the same quartz crystal in stages I and II suggests that fluid boiling occurred in those stages. The trapping pressures in stage I were estimated to range from ~300 to ~500 bar (Figure 12A), assuming a simple NaCl–H_2O system using the isobar equations given in [50]. However, the addition of CO_2 to an NaCl–H_2O system could raise the solvus and increase the entrapment pressure (Figure 12B; [51–54]). Therefore, in this study, the estimated trapping pressures in stage I represent only the minimum values. The trapping pressures in stage II were estimated to be between ~100 and ~300 bar and were mostly concentrated around ~200 bar (Figure 12A) based on the isobar equations given in [50]. There are clear decreasing trends of pressure from stages I to II (Figure 12A). Fluid boiling did not occur in stages III and IV, so the estimated trapping pressures represent only the minimum values [54]. The minimum trapping pressures in stages III and IV were estimated to range from ~30 to ~150 bar and <50 bar, respectively (Figure 12A).

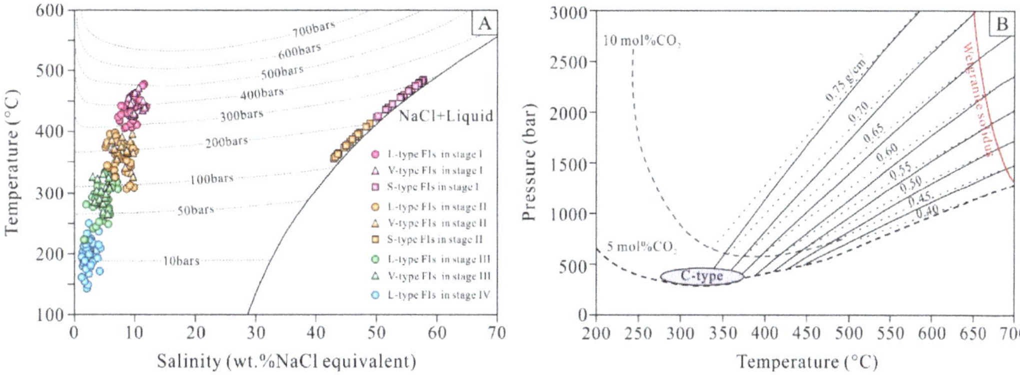

Figure 12. (**A**) Pressure estimation for FIs from the Huoluotai Cu (Mo) deposit. (**B**) Isochores for fluid of 4 wt.% NaCl, 5 mol% CO_2 (straight lines), and 10 mol% CO_2 (dashed lines). Data for the C-type FIs plot in the oval field.

6.3. Origin of the Ore-forming Fluids

The $\delta^{18}O_{H_2O}$ values for stage I (6.5‰) were close to those of magmatic water [55], indicating that the ore-forming fluids from stage I were derived from a magmatic source. The δD values for stage I were significantly lower than those for typical magmatic water (Figure 8; [56]). Previous studies have shown that the relatively depleted isotopic values of fluids could have been caused by water–rock interactions or magma degassing [57,58]. The depleted $\delta^{18}O_{H_2O}$ and δD isotopic characteristics were extensively recorded by fluids from an early stage in porphyry deposits of the NGXR [59], such as the Xiaokele Cu (Mo) (−1.2‰ to 2.4‰; [5]), the Fukeshan Cu (Mo) (−2.3‰ to 3.4‰; [4]), and the Chalukou Mo (−1.9‰ to 3.6‰; [43]) deposits, which were interpreted to be predominantly of magmatic origin. The δD and $\delta^{18}O_{H_2O}$ values for stages II, III, and IV were relatively lower than those for stage I and plotted in the region between the meteoric water line and the magmatic water field (close to the magmatic water field) (Figure 8), suggesting the involvement of some meteoric water but still dominated by magmatic water. In the $\delta^{13}C_{PDB}$ vs. $\delta^{18}O_{SMOW}$ diagram (Figure 9), the C–O isotope compositions of calcite in stage IV fell into the field of granite and magmatic carbonate, indicating that a certain amount of magmatic water was still involved in the mineralization process during the late mineralization stage.

6.4. Fluid Evolution and Mineralization Process

The occurrence of L-, V-, S-, and C-type FIs in stage I quartz indicates that the initial fluids belong to an $H_2O–CO_2–NaCl$ system. These fluids are characterized by high homogenization temperatures and high salinities (Figure 6A,B). High amounts of hematite, magnetite (Figure 3E,F), and hematite-bearing S-type FIs (Figure 5B), but few sulfides, were found in stage I, suggesting that the initial fluids were highly oxidized [60,61]. This result is supported by the detection of oxidized gases (e.g., CO_2) in FIs via laser Raman spectroscopy analysis (Figure 7A). Sulfur in the magma was mostly present in the form of sulfate due to high oxygen fugacity and CO_2 activity [41,62], but sulfide mineralization would be inevitably constrained. These kinds of initial hydrothermal fluids were observed to be particularly prone to extensive potassic alterations and quartz + K-feldspar ± magnetite ± hematite ± pyrite veins in stage I.

Cu (Mo) mineralization occurred mainly in stage II. L-, V-, and S-type FIs were found in stage II quartz crystals (Figure 5D–F), which suggests that the ore-forming fluids of stage II belong to an $H_2O–NaCl$ system. Hematite, magnetite, and C-type FIs were not found in stage II, suggesting that the CO_2 content and oxygen fugacity of the fluids were distinctly decreased compared to those in stage I. Such degassing of oxidized gas is considered to be a major process responsible for sulfate reduction [61]. This reduction in oxygen fugacity could be due to the escape of CO_2 caused by fluid boiling [63–65] and/or magnetite precipitation [66–68]. These results suggest that the fluid system only contained minor CO_2 in stage I, indicating that it is likely not a CO_2-rich system. Therefore, the immiscibility of CO_2 and water is not an important process in the Huoluotai Cu (Mo) deposit. The precipitation of magnetite was found to reduce S^{6+} to S^{2-} (i.e., $12[FeO] + H_2SO_4 = 4Fe_3O_4$ (Magnetite) + H_2S; [62,68]), so the acid in stage II ore-forming fluids became stronger with an increase in S^{2-} and H^+ activity [69]. More significantly, the homogenization temperatures of stage II fluids were obviously lower than those of the stage I fluids (Figure 6). H–O isotope data in this study indicate that mixing of hydrothermal fluids and meteoric water occurred in stage II. The addition of cooler and lower-salinity meteoric water inevitably led to a decrease in temperature and salinity of the primary hotter and high-salinity magmatic water, causing the precipitation of sulfides. Furthermore, the micro-thermometric data obtained in this study showed that Cu was mainly deposited below 400 °C in stage II, which is consistent with the temperature at which hydrothermal fluids reached Cu saturation in some micro-thermometric and thermodynamic studies of FIs [70–73]. Beyond stage I, fluid boiling also occurred in stage II. Generally, fluid boiling can cause widespread hydraulic fracturing and accelerate the influx and circulation of meteoric water, which could lead to rapid deposition of sulfides. In addition, fluid boiling can lead to the exsolution of volatile components from ore-forming fluids [51], which would also promote the enrichment of metals in the ore-forming fluids [74,75]. Thus, we believe that fluid boiling occurring below 400 °C may be the primary factor for Cu precipitation in the Huoluotai Cu (Mo) deposit.

No evidence of fluid boiling was found in stages III and IV. The ore-forming fluids in stages III and IV belong to a homogeneous H_2O–NaCl system and became more diluted and cool, as indicated by the H−O isotope data. Magmatic water mixed with meteoric water may have initiated metal sulfide precipitation in stage III [71]. The increasing meteoric water influx may have formed the quartz + calcite ± pyrite ± fluorite veins in stage IV, reflecting waning of the hydrothermal system.

7. Conclusions

(1) The Huoluotai Cu (Mo) deposit is a typical porphyry Cu deposit in the NGXR. The ore-forming process of this deposit consists of the sulfide-barren quartz stage (I), the quartz + chalcopyrite ± pyrite ± molybdenite stage (II), the quartz + polymetallic sulfide stage (III), and the quartz + calcite ± pyrite ± fluorite stage (IV). Cu mineralization occurred mainly in stage II.

(2) The results of FI micro-thermometry and H–O isotope analysis showed that the ore-forming fluids originated from a magmatic origin in stage I and then mixed with meteoric water from stages II to IV. The S–Pb isotope results suggest that the source of the ore-forming materials has the characteristics of a crust–mantle-mixing origin.

(3) Fluid boiling occurring below 400 °C may be the primary factor for Cu precipitation in the Huoluotai Cu (Mo) deposit.

Author Contributions: Conceptualization, Y.S., B.L., X.C. and Q.D.; software, F.M., Y.Q. and L.W.; investigation, Y.S., B.L., X.C. and F.M.; data curation, Y.Q. and L.W.; funding acquisition, B.L.; project administration, B.L.; writing—original draft preparation, Y.S.; and writing—review and editing, Y.S. and B.L. All authors have read and agreed to the published version of the manuscript.

Funding: This research was funded by the National Natural Science Foundation of China (41272093), the National Key R&D Program of China (2017YFC0601304), Natural Science Foundation of Jilin Province (No.20180101089JC), Key Projects of Science and Technology Development Plan of Jilin Province (No.20100445), the Shandong Provincial Natural Science Foundation of China (No. ZR2019PD017), Self-determined Foundation of Key Laboratory of Mineral Resources Evaluation in Northeast Asia, Ministry of Natural Resources (DBY-ZZ-19-04), Basic scientific research project of colleges and universities of Liaoning Provincial Department of education in 2021 (LJKZ0369), and the Heilongjiang Research Project of Land and Resources (201605 and 201704).

Data Availability Statement: Not applicable.

Acknowledgments: We would like to thank staffs of the Qiqihaer Institute of Geological Exploration, Heilongjiang, China for sample collection.

Conflicts of Interest: The authors declare no conflict of interest.

References

1. Hu, X.L.; Yao, S.Z.; He, M.C.; Ding, Z.J.; Cui, Y.B.; Shen, J.; Chen, B.; Zhu, B.P. Geochemistry, U–Pb geochronology and Hf isotope studies of the Daheishan porphyry Mo deposit in Heilongjiang Province, NE China. *Resour. Geol.* **2014**, *64*, 102–116. [CrossRef]
2. Liu, J.; Mao, J.W.; Wu, G.; Wang, F.; Luo, D.F.; Hu, Y.Q. Zircon U–Pb and molybdenite Re–Os dating of the Chalukou porphyry Mo deposit in the northern Great Xing'an Range, China and its geological significance. *J. Asian Earth Sci.* **2014**, *79*, 696–709. [CrossRef]
3. Deng, C.Z.; Sun, D.Y.; Han, J.S.; Chen, H.Y.; Li, G.H.; Xiao, B.; Li, R.C.; Feng, Y.Z.; Li, C.L.; Lu, S. Late-stage southwards subduction of the Mongol-Okhotsk oceanic slab and implications for porphyry Cu–Mo mineralization: Constraints from igneous rocks associated with the Fukeshan deposit, NE China. *Lithos* **2019**, *326–327*, 341–357. [CrossRef]
4. Sun, Y.G.; Li, B.L.; Ding, Q.F.; Qu, Y.; Wang, C.K.; Wang, L.L.; Xu, Q.L. Mineralization age and hydrothermal evolution of the Fukeshan Cu (Mo) deposit in the northern Great Xing'an Range, Northeast China: Evidence from fluid inclusions, H–O–S–Pb isotopes, and Re–Os geochronology. *Minerals* **2020**, *10*, 591. [CrossRef]
5. Sun, Y.G.; Li, B.L.; Ding, Q.F.; Meng, F.B.; Chen, X.S.; Qian, Y.; Wang, L.; Wang, L.L.; Xu, Q.L. Timing and ore formation of the Xiaokele porphyry Cu (–Mo) deposit in the northern Great Xing'an Range, NE China: Constraints from geochronology, fluid inclusions, and H–O–S–Pb isotopes. *Ore Geol. Rev.* **2022**, *143*, 104806. [CrossRef]

6. Jahn, B.M.; Wu, F.Y.; Chen, B. Granitoids of the Central Asian Orogenic Belt and continental growth in the Phanerozoic. *Earth Environ. Sci. Trans. R. Soc.* **2000**, *91*, 181–193.
7. Chen, Y.J.; Zhang, C.; Li, N.; Yang, Y.F.; Deng, K. Geology of the Mo deposits in Northeast China. *J. Jilin Univ. (Earth Sci. Ed.)* **2012**, *42*, 1223–1254. (In Chinese)
8. Sun, Y.G. Study on Metallogenesis and Prospecting Direction of Late Jurassic Porphyry Cu (Mo) Deposits in the Northern Segment of the Great Xing'an Range. Ph.D. Thesis, Jilin University, Changchun, China, 2021. (In Chinese with English Abstract).
9. Wang, L. Huoluotai, Mohe Country, Heilongjiang Province Study on Geological Characteristics and Prospecting Direction of Copper Molybdenum Deposit Summary. Master's Thesis, Jilin University, Changchun, China, 2018. (In Chinese with English Abstract).
10. Deng, C.Z.; Li, G.H. The Cu-Mo mineralization of the Late Jurassic porphyry in the Northern Great Xing'an Range: Constraints from zircon U-Pb ages of the Ore-Causative Granites. *Acta Geol. Sin-Engl.* **2019**, *93*, 236–237. [CrossRef]
11. Sun, Y.G.; Li, B.L.; Zhao, Z.H.; Sun, F.Y.; Ding, Q.F.; Chen, X.S.; Li, J.B.; Qian, Y.; Li, Y.J. Age and petrogenesis of late Mesozoic intrusions in the Huoluotai porphyry Cu-(Mo) deposit, northeast China: Implications for regional tectonic evolution. *Geosci. Front.* **2022**, *13*, 101344. [CrossRef]
12. Wu, F.Y.; Sun, D.Y.; Ge, W.C.; Zhang, Y.B.; Grant, M.L.; Wilde, S.A.; Jahn, B.M. Geochronology of the Phanerozoic granitoids in northeastern China. *J. Asian Earth Sci.* **2011**, *41*, 1–30. [CrossRef]
13. Liu, Y.J.; Li, W.M.; Feng, Z.Q.; Wen, Q.B.; Franz, N.; Liang, C.Y. A review of the Paleozoic tectonics in the eastern part of Central Asian Orogenic Belt. *Gondwana Res.* **2017**, *43*, 123–148. [CrossRef]
14. Ge, W.C.; Wu, F.Y.; Zhou, C.Y.; Rahman, A.A. Emplacement age of the Tahe granite and its constraints on the tectonic nature of the Erguna block in the northern part of the Da Hinggan Range. *Chin. Sci. Bull.* **2005**, *50*, 2097–2105. [CrossRef]
15. Zhou, J.B.; Wang, B.; Wilde, S.A.; Zhao, G.C.; Cao, J.L.; Zheng, C.Q.; Zeng, W.S. Geochemistry and U–Pb zircon dating of the Toudaoqiao blueschists in the Great Xing'an Range, northeast China, and tectonic implications. *J. Asian Earth Sci.* **2015**, *97*, 197–210. [CrossRef]
16. IMBGMR (Inner Mongolian Bureau of Geology and Mineral Resources). *Regional Geology of Inner Mongolia*; Geological Publishing House: Beijing, China, 1991; pp. 1–725. (In Chinese)
17. Sun, L.X.; Ren, B.F.; Zhao, F.Q.; Ji, S.P.; Geng, J.Z. Late Paleoproterozoic magmatic records in the Eerguna massif: Evidences from the zircon U–Pb dating of granitic gneisses. *Geol. Bull. China* **2013**, *32*, 341–352, (In Chinese with English Abstract).
18. Ge, W.C.; Chen, J.S.; Yang, H.; Zhao, G.C.; Zhang, Y.L.; Tian, D.X. Tectonic implications of new zircon U–Pb ages for the Xinghuadukou Complex, Erguna Massif, northern Great Xing'an Range, NE China. *J. Asian Earth Sci.* **2015**, *106*, 169–185. [CrossRef]
19. Sun, Y.G.; Li, B.L.; Sun, F.Y.; Ding, Q.F.; Qian, Y.; Li, L.; Xu, Q.L.; Li, Y.J. Geochronology, geochemistry, and Hf isotopic compositions of Early Permian syenogranite and diabase from the northern Great Xing'an Range, NE China: Petrogenesis and tectonic implications. *Can. J. Earth. Sci.* **2020**, *57*, 1478–1491. [CrossRef]
20. IMBGMR (Inner Mongolian Bureau of Geology Mineral Resources). *Lithostratigraphy of Inner Mongolia*; China University of Geosciences Press: Wuhan, China, 1996; pp. 1–342. (In Chinese)
21. Zhang, J.H.; Ge, W.C.; Wu, F.Y.; Wilde, S.A.; Yang, J.H.; Liu, X.M. Large-scale Early Cretaceous volcanic events in the northern Great Xing'an Range, Northeastern China. *Lithos* **2008**, *102*, 138–157. [CrossRef]
22. Tang, J.; Xu, W.L.; Wang, F.; Zhao, S.; Wang, W. Early Mesozoic southward subduction history of the Mongol-Okhotsk oceanic plate: Evidence from geochronology and geochemistry of Early Mesozoic intrusive rocks in the Erguna Massif, NE China. *Gondwana Res.* **2016**, *31*, 218–240. [CrossRef]
23. Gou, J.; Sun, D.Y.; Ren, Y.S.; Hou, X.G.; Yang, D.G. Geochemical and Hf isotopic compositions of Late Triassic–Early Jurassic intrusions of the Erguna Block, Northeast China: Petrogenesis and tectonic implications. *Int. Geol. Rev.* **2017**, *59*, 347–367. [CrossRef]
24. Sun, Y.G.; Li, B.L.; Sun, F.Y.; Ding, Q.F.; Wang, B.Y.; Li, Y.J.; Wang, K. Mineralization events in the Xiaokele porphyry Cu (–Mo) deposit, NE China: Evidence from zircon U–Pb and K–feldspar Ar–Ar geochronology and petrochemistry. *Resour. Geol.* **2020**, *70*, 254–272. [CrossRef]
25. Roedder, E. Fluid Inclusions. *Rev. Mineral. Mich. Mineral. Soc. Am.* **1984**, *12*, 644.
26. Clayton, W.M.; Mayeda, T.K. The use of bromine pent a fluoride in the extraction of oxygen from oxides and silicates for isotopic analysis. *Geochim. Cosmochim. Acta.* **1963**, *27*, 43–52. [CrossRef]
27. Friedman, I. Deuterium content of natural waters and other substances. *Geochim. Cosmochim. Acta* **1953**, *4*, 89–103. [CrossRef]
28. McCrea, J.M. On the isotopic chemistry of carbonates and a paleotemperature scale. *J. Chem. Phys.* **1950**, *18*, 849–857. [CrossRef]
29. Steele-MacInnis, M.; Lecumberri-Sanchez, P.; Bodnar, R.J. HOKIEFLINCS_H_2O-NACL: A Microsoft Excel spreadsheet for interpreting microthermometric data from fluid inclusions based on the PVTX properties of H_2O–NaCl. *Comput. Geosci.* **2012**, *49*, 334–337. [CrossRef]
30. Lecumberri-Sanchez, P.; Steele-MacInnis, M.; Bodnar, R.J. A numerical model to estimate trapping conditions of fluid inclusions that homogenize by halite disappearance. *Geochim. Cosmochim. Acta* **2012**, *92*, 14–22. [CrossRef]
31. Collins, P.L.F. Gas hydrates in CO_2-bearing fluid inclusions and use freezing data for estimation of salinity. *Econ. Geol.* **1979**, *74*, 1435–1444. [CrossRef]

32. Clayton, R.N.; O'Neil, J.R.; Mayeda, T.K. Oxygen isotope exchange between quartz and water. *J. Geophys. Res. Atmos.* **1972,** *77*, 3057–3067. [CrossRef]
33. Taylor, H.P. The application of oxygen and hydrogen isotope studies to problems of hydrothermal alteration and ore deposition. *Econ. Geol.* **1974**, *69*, 843–883. [CrossRef]
34. Friedman, I.; O'Neil, J.R. Complication of Stable Isotope Fractionation Factors of Geochemical Interest. In *Data of Geochemistry*, 6th ed.; Fleischer, M., Ed.; US Government Printing Office: Washington, DC, USA, 1977.
35. Veizer, J.; Hoefs, J. The nature of $^{18}O/^{16}O$ and $^{13}C/^{12}C$ secular trends in sedimentary carbonate rocks. *Geochim. Cosmochim. Acta* **1976**, *40*, 1387–1395. [CrossRef]
36. Hoefs, J. *Stable Isotope Geochemistry*, 6th ed.; Springer: Berlin/Heidelberg, Germany, 2009; pp. 130–135.
37. Taylor, H.P.; Frechen, J.; Degens, E.T. Oxygen and carbon isotope studies of carbonatites from the Laacher See District, West Germany and the Alnö District, Sweden. *Geochim. Cosmochim. Acta* **1967**, *31*, 407–430. [CrossRef]
38. Ohmoto, H. Systematics of sulfur and carbon isotopes in hydrothermal ore deposits. *Econ. Geol.* **1972**, *67*, 551–578. [CrossRef]
39. Ohmoto, H.; Rye, R.O. Isotopes of sulfur and carbon. In *Geochemistry of Hydrothermal Ore Deposits*; Barnes, H.L., Ed.; Wiley: New York, NY, USA, 1979; pp. 509–567.
40. Chaussidon, M.; Albarède, F.; Sheppard, S.M.F. Sulphur isotope variations in the mantle from ion microprobe analyses of micro-sulphide inclusions. *Earth Planet. Sci. Lett.* **1989**, *92*, 144–156. [CrossRef]
41. Field, C.W.; Zhang, L.; Dilles, J.H.; Rye, R.O.; Reed, M.H. Sulfur and oxygen isotopic record in sulfate and sulfide minerals of early, deep, pre-main stage porphyry Cu–Mo and late main stage base-metal mineral deposits, Butte district, Montana. *Chem. Geol.* **2005**, *215*, 61–93. [CrossRef]
42. Salas, R.D.; Ochoa-Landin, L.; Ruiz, J.; Eastoe, C.; Meza-Figueroa, D.; Zuniga-Hernandez, H.; Mendivil-Quijada, H.; Quintanar-Ruiz, F. Geology, stable isotope, and U-Pb geochronology of the Mariquita porphyry copper and Lucy Cu-Mo deposits, Cananea District, Mexico: A contribution to regional exploration. *J. Geochem. Explor.* **2013**, *124*, 140–154. [CrossRef]
43. Liu, J.; Mao, J.W.; Wu, G.; Wang, F.; Luo, D.F.; Hu, Y.Q. Fluid inclusions and H–O–S–Pb isotope systematics of the Chalukou giant porphyry Mo deposit, Heilongjiang Province, China. *Ore Geol. Rev.* **2014**, *59*, 83–96. [CrossRef]
44. Zhang, F.F.; Wang, Y.H.; Liu, J.J.; Wang, J.P.; Zhao, C.B.; Song, Z.W. Origin of the Wunugetushan porphyry Cu–Mo deposit, Inner Mongolia, NE China: Constraints from geology, geochronology, geochemistry, and isotopic compositions. *J. Asian Earth Sci.* **2016**, *117*, 208–224. [CrossRef]
45. Mi, K.F.; Liu, Z.J.; Li, C.F.; Liu, R.B.; Wang, J.P.; Peng, R.M. Origin of the Badaguan porphyry Cu-Mo deposit, Inner Mongolia, northeast China: Constraints from geology, isotope geochemistry and geochronology. *Ore Geol. Rev.* **2017**, *81*, 154–172. [CrossRef]
46. Huang, M.L.; Bi, X.W.; Gao, J.F.; Xu, L.L.; Xiao, J.F.; Liu, S.T.; Wang, X.S.; Zhou, T. Sulfur and lead isotopic variations in the giant Yulong porphyry Cu (Mo-Au) deposit from the eastern Tibetan Plateau: Implications for origins of S and Pb, and metal precipitation. *J. Geochem. Explor.* **2019**, *197*, 70–83. [CrossRef]
47. Zartman, R.; Doe, B. Plumbotectonics: The model. *Tectonophysics* **1981**, *75*, 135–162. [CrossRef]
48. Roedder, E.; Bodnar, R.J. Geologic pressure determinations from fluid inclusion studies. *Annu. Rev. Earth Planet. Sci.* **1980**, *8*, 263–301. [CrossRef]
49. Brown, P.E.; Hagemann, S.G. Macflincor and its application to fluids in Archean lode-gold deposits. *Geochim. Cosmochim. Acta* **1995**, *59*, 3943–3952. [CrossRef]
50. Driesner, T.; Heinrich, C.A. The system H_2O–NaCl. Part I: Correlation formulae for phase relations in temperature–pressure-composition space from 0 to 1000 °C, 0 to 5000 bar, and 0 to 1 XNaCl. *Geochim. Cosmochim. Acta* **2007**, *71*, 4880–4901.
51. Bowers, T.S.; Helgeson, H.C. Calculation of the thermodynamic and geochemical consequences of nonideal mixing in the system H_2O-CO_2-NaCl on phase relations in geologic systems: Equation of state for H_2O-CO_2-NaCl fluids at high pressures and temperatures. *Geochim. Cosmochim. Acta* **1983**, *47*, 1247–1275. [CrossRef]
52. Joyce, D.B.; Holloway, J.R. An experimental determination of the thermodynamic properties of H_2O-CO_2-NaCl fluids at high pressures and temperatures. *Geochim. Cosmochim. Acta* **1993**, *57*, 733–746. [CrossRef]
53. Gehrig, M.; Lentz, H.; Franck, E. The system water-carbon dioxide-sodium chloride to 773 K and 300 MPa. *Phys. Chem.* **1986**, *90*, 525–533. [CrossRef]
54. Rusk, B.G.; Reed, M.; Dilles, J.H. Fluid inclusion evidence for magmatic-hydrothermal fluid evolution in the porphyry copper-molybdenum deposit at Butte, Montana. *Econ. Geol.* **2008**, *103*, 307–332. [CrossRef]
55. Hedenquist, J.W.; Lowenstern, J.B. The role of magmas in the formation of hydrothermal ore deposits. *Nature* **1994**, *370*, 519–527. [CrossRef]
56. Barnes, H.L. Solubilities of ore minerals. In *Geochemistry of Hydrothermal Ore Deposits*; Barnes, H.L., Ed.; Wiley: New York, NY, USA, 1979; pp. 404–460.
57. Rye, R.O. The evolution of magmatic fluids in the epithermal environment: The stable isotope perspective. *Econ. Geol.* **1993**, *88*, 733–752. [CrossRef]
58. Harris, A.C.; Golding, S.D. New evidence of magmatic-fluid-related phyllic alteration: Implications for the genesis of porphyry Cu deposits. *Geology* **2002**, *30*, 335–338. [CrossRef]
59. Chen, Y.J.; Zhang, C.; Wang, P.; Pirajno, F.; Li, N. The Mo deposits of Northeast China: A powerful indicator of tectonic settings and associated evolutionary trends. *Ore Geol. Rev.* **2017**, *81*, 602–640.

60. Seedorff, E.; Dilles, J.H.; Proffett, J.M.; Einaudi, M.T.; Zurcher, L.; Stavast, W.J.A.; Johnson, D.A.; Barton, M.D. Porphyry deposits: Characteristics and origin of hypogene features. *Econ. Geol.* **2005**, *100*, 251–298.
61. Sun, W.D.; Huang, R.F.; Li, H.; Hu, Y.B.; Zhang, C.C.; Sun, S.J.; Zhang, L.P.; Ding, X.; Li, C.Y.; Zartman, R.E.; et al. Porphyry deposits and oxidized magmas. *Ore Geol. Rev.* **2015**, *65*, 97–131. [CrossRef]
62. Mungall, J.E. Roasting the mantle: Slab melting and the genesis of major Au and Au-rich Cu deposits. *Geology* **2002**, *30*, 915–918. [CrossRef]
63. Chen, Y.J.; Pirajno, F.; Li, N.; Guo, D.S.; Lai, Y. Isotope systematic and fluid inclusion studies of the Qiyugou breccias pipe-hosted gold deposit, Qinling Orogen, Henan Province, China: Implications for ore genesis. *Ore Geol. Rev.* **2009**, *35*, 245–261.
64. Li, N.; Ulrich, T.; Chen, Y.-J.; Thomsen, T.B.; Pease, V.; Pirajno, F. Fluid evolution of the Yuchiling porphyry Mo deposit, East Qinling, China. *Ore Geol. Rev.* **2012**, *48*, 442–459. [CrossRef]
65. Yang, Y.F.; Li, N.; Chen, Y.J. Fluid inclusion study of the Nannihu giant porphyry Mo–W deposit, Henan Province, China: Implications for the nature of porphyry ore-fluid systems formed in a continental collision setting. *Ore Geol. Rev.* **2012**, *46*, 83–94. [CrossRef]
66. Ulrich, T.; Guenther, D.; Heinrich, C.A. Gold concentrations of magmatic brines and the metal budget of porphyry copper deposits. *Nature* **1999**, *399*, 676–679. [CrossRef]
67. Heinrich, C.A. The physical and chemical evolution of low-salinity magmatic fluids at the porphyry to epithermal transition: A thermodynamic study. *Miner. Depos.* **2005**, *39*, 864–889.
68. Liang, H.Y.; Sun, W.D.; Su, W.C.; Zartman, R.E. Porphyry copper-gold mineralization at Yulong, China, promoted by decreasing redox potential during magnetite alteration. *Econ. Geol.* **2009**, *104*, 587–596. [CrossRef]
69. Sillitoe, R.H. Porphyry copper systems. *Econ. Geol.* **2010**, *105*, 3–41. [CrossRef]
70. Fournier, R.O. Hydrothermal process related to movement of fluid from plastic into brittle rock in the magmatic-epithermal environment. *Econ. Geol.* **1999**, *94*, 1193–1211. [CrossRef]
71. Hezarkhani, A.; William-Jones, A.E.; Gammons, C.H. Factors controlling copper solubility and chalcopyrite deposition in the Sungun porphyry copper deposit, Iran. *Miner. Dep.* **1999**, *34*, 770–783. [CrossRef]
72. Williams, P.J.; Dong, G.Y.; Ryan, C.G.; Pollard, P.J.; Rotherham, J.F.; Mernagh, T.P.; Chapman, L.H. Geochemistry of hypersaline fluid inclusions from the Starra (Fe oxide)-Au-Cu deposit, Cloncurry district, Queensland. *Econ. Geol. Bull. Soc. Econ. Geol.* **2001**, *96*, 875–883.
73. Landtwing, M.R.; Pettke, T.; Halter, W.E.; Heinrich, C.A.; Redmond, P.B.; Einaudi, M.T.; Kunze, K. Copper deposition during quartz dissolution by cooling magmatic–hydrothermal fluids: The Bingham porphyry. *Earth Planet. Sci. Lett.* **2005**, *235*, 229–243. [CrossRef]
74. Harris, A.C.; Kamenetsky, V.S.; White, N.C.; Achterbergh, E.V.; Ryan, C.G. Melt inclusions in veins: Linking magmas and porphyry Cu deposits. *Science* **2003**, *302*, 2109–2111. [CrossRef] [PubMed]
75. Zhu, J.J.; Hu, R.Z.; Richards, J.P.; Bi, X.W.; Zhong, H. Genesis and Magmatic Hydrothermal Evolution of the Yangla Skarn Cu Deposit, Southwest China. *Econ. Geol.* **2015**, *110*, 631–652. [CrossRef]

minerals

Article

Age, Genesis and Tectonic Setting of the Sayashk Tin Deposit in the East Junggar Region: Constraints from Lu–Hf Isotopes, Zircon U–Pb and Molybdenite Re–Os Dating

Zhenjun Sun [1,2], Guanghu Liu [1,2,*], Yunsheng Ren [1,2], Xi Chen [3], Xinhao Sun [4], Chengyang Wang [1,2] and Zuowu Li [5]

1 School of Earth Sciences, Institute of Disaster Prevention, Sanhe 065201, China
2 Hebei Key Laboratory of Earthquake Dynamics, Sanhe 065201, China
3 No.1 Bureau of China Metallurgical Geology Bureau, Sanhe 065201, China
4 Chengdu Center of China Geological Survey, Chengdu 610081, China
5 Xinjiang Branch of China National Geological Exploration Center of Building Materials Industry, Urumqi 830092, China
* Correspondence: liuguanghu@cidp.edu.cn

Abstract: The Sayashk tin (Sn) deposit is located within the southern part of the Eastern Junggar orogenic belt in Xinjiang Province and forms part of the Kalamaili alkaline granite belt. There are many Sn polymetallic deposits in the area. To constrain the age, genesis, and tectonic setting of the Sayashk tin deposit in the East Junggar region, we conducted a bulk-rock geochemical analysis of the granite porphyry (SR1) and medium- to fine-grained granite (SR2) hosts of the deposit, LA-ICP-MS zircon U–Pb dating and Lu–Hf c Re–OS dating and combined our results with the metallogenic conditions and other geological characteristics of the deposit. The results show that the Sayashk Sn deposit is indeed spatially, temporally, and genetically closely related to the granite porphyry and medium-fine-grained granite. Both zircon U–Pb ages are 308.2 ± 1.5 Ma and 310.9 ± 1.5 Ma, respectively. The isochron age of molybdenite is 301.4 ± 6.7 Ma, which represents the crystallization age of the granite porphyry and medium-fine-grained granite. Therefore, all of them formed in the late Carboniferous epoch. The medium-fine-grained granites and granite porphyry are characteristically rich in Si and alkali, poor in Ca and Mg, rich in high field-strength elements (HFSE, e.g., Zr, Hf) and Ce, and deficient in Ba, Sr, Eu, P, and Ti. They are typical A-type granites, showing the characteristics of a mixed crustal mantle source. The εHf(t) values of the zircon from the granite porphyry (SR1) range from 10.27 to 16.17 (average 13.71), εHf(t) values of the zircon from the medium-fine-grained granites (SR2) are between 5.72 and 9.21 (average 7.08), and the single model ages (T_{DM1}) and two-stage model ages (T_{DM2}) of the granite porphyry (SR1) fall within the ranges of 319~535 Ma and 339~644 Ma. The single model ages (T_{DM1}) and two-stage model ages (T_{DM2}) of the medium-fine-grained granites (SR2) fall within the ranges of 346~479 Ma and 309~557 Ma. There is little difference between their two-stage model ages and zircon U–Pb ages, indicating that the Sayashk granite may be the product of partial melting of juvenile crustal. Combined with previous research results, the Sayashk Sn deposit formed in a post-collision extensional tectonic setting after the late Carboniferous in the Kalamaili area.

Keywords: alkaline granite; zircon U–Pb geochronology; zircon Hf isotope compositions; molybdenite Re–Os dating; Sayashk tin deposit; east Junggar region

1. Introduction

The Beilekuduk tin (Sn) ore belt in the eastern Junggar region of northern Xinjiang is the first metallogenic belt dominated by Sn found. There are a number of Sn deposits (occurrences) distributed from the west to the east, such as Kamst, Ganliangzi, Beilekuduk,

Sayashk, Sujiquan, and Hongtujingzi in the region, which have long been studied by mineral deposit scientists locally and globally [1–5]. The eastern Junggar area is an important tin and gold metallogenic belt in northern Xinjiang. Tin mineralization is related to late Paleozoic granite [6]. Many researchers have carried out a large number of studies on petrography, geochemistry, isotope chronology, ore genesis, and ore-forming fluid properties and typical Sn deposits all over the world [1–3,7–11]. It is believed that the granites related to mineralization are mainly S-type granites originating from the partial melting of crustal materials [12–14]. However, an increasing amount of research data shows that most granites related to Sn mineralization in the world are closely related to highly differentiated granites in time and space, as well as genesis [15–19]. Generally, granites related to Sn mineralization are biotite granite, two-mica granite, potassium feldspar granite, quartz porphyry, granite porphyry, alkali feldspar granite, and other granitic rocks, and most of them belong to highly differentiated S-type, A-type, and I-type granites [20–29]. They are generally characterized by high concentrations of Si and alkali, enrichment of incompatible elements B, F, Rb, Th, U, low concentrations of elements such as Ca, Mg, Fe, Ti, Ba, Sr, Zr, and strong negative Eu anomalies, indicating that the magma has undergone a strong crystallization differentiation process [11,15].

Many researchers have studied the petrogenetic and metallogenic ages of tin deposits in the Belkuduk Sn ore belt with different dating techniques. The petrogenetic and metallogenic ages are mainly concentrated at 341–275 Ma [6,13,14,30]. The large span of petrogenetic and metallogenic ages, the diversity of testing methods, and the accuracy of testing have seriously restricted further understanding of the metallogenic regularity of tin deposits and the effect of prospecting and prediction in the eastern Junggar area. The Sabei granite is the first A-type granite discovered to develop Sn deposits in China. Similar to other rocks in the Belkuduk Sn ore belt, it has attracted extensive attention [2,8,31–33]. However, as a granite porphyry closely related to mineralization, it has not been given as much attention as the alkaline granite developed in the mining area, and there are few relevant studies. Meanwhile, a lot of researchers hold different views on its petrogenetic and metallogenic age; for example, Zhang et al. (1992) dated the single zircon U–Pb of the Sayashk amphibole granite to 290 $\pm$ 11 Ma [34], Yang (2010) obtained a weighted average age of 277 $\pm$ 11 Ma by using LA-ICP-MS zircon U–Pb [8], and Lin et al. (2007) obtained two weighted average ages of 313 $\pm$ 2 Ma and 314 $\pm$ 5 Ma by using Shrimp zircon U–Pb dating [31]. Lin et al. (2008) thought that the Sn mineralization age was 324.2 $\pm$ 3.4 Ma by using Shrimp zircon U–Pb dating of cassiterite quartz veins in the Sayashk Sn deposit [30].

To date, the metallogenic age of the Sayashk Sn deposit has not been well resolved. We carried out geochemical element analysis of granite and granite porphyry, LA-ICP-MS zircon U–Pb dating, Lu–Hf isotope composition, and molybdenite Re–Os dating to precisely determine the formation age of the Sayashk tin deposit, the types and genesis of granites related to mineralization, and the nature of the magma source area; we will use this information to discuss the geodynamic background of its formation. This study will provide an important scientific basis for further understanding the mineralization in this area.

2. Regional Geology

The eastern Junggar Orogenic Belt in northern Xinjiang is located in the northern part of the eastern Tianshan mountains (Figure 1a). It is an important part of the Central Asian orogenic belt. It is bounded by the Kalamaili fault in the south and the Erqisi fault in the north [35–37]. The strong tectonic-magmatic activity formed rich mineral resources. The Kalamaili area is located in the south of the East Junggar Orogenic Belt. A series of alkaline and ca-alkaline granite intrusions, such as the Yemaquan pluton, Laoyaquan pluton, Beilekuduk pluton, Sabei pluton, and Huangyangshan pluton, are developed on the north side of the Kalamaili deep fault and are distributed along the NW direction in the area. The Kamust Sn deposit, Ganliangzi Sn deposit, Hongtujingzi Sn deposit, Beilekuduke Sn deposit, Sayashk Sn deposit, Huangyangshan graphite deposit and Sujiquan Sn deposit (graphite deposit) are distributed in the NW direction along the Kalamaili deep fault [2,3,9,38–40].

Devonian and Carboniferous strata are mainly developed in the region (Figure 1b). The lithology is mainly tuffaceous siltstone and pyroclastic rock mixed with limestone and carbonaceous marlstone, which are exposed along both sides of the Kalamaili deep fault.

A small number of Silurian strata is sporadically exposed south of the Kalamaili deep fault. Regionally, WNW-trending faults are developed. The Kalamaili deep fault is a typical deep fault. In addition, Suji Qingshui and Kupu Kubusu deep faults roughly parallel to it are also developed [41]. They jointly control the emplacement of alkaline granitoids after the collisional orogeny, accompanied by the intrusion of a series of slightly alkaline granite porphyry and Sn mineralization [31].

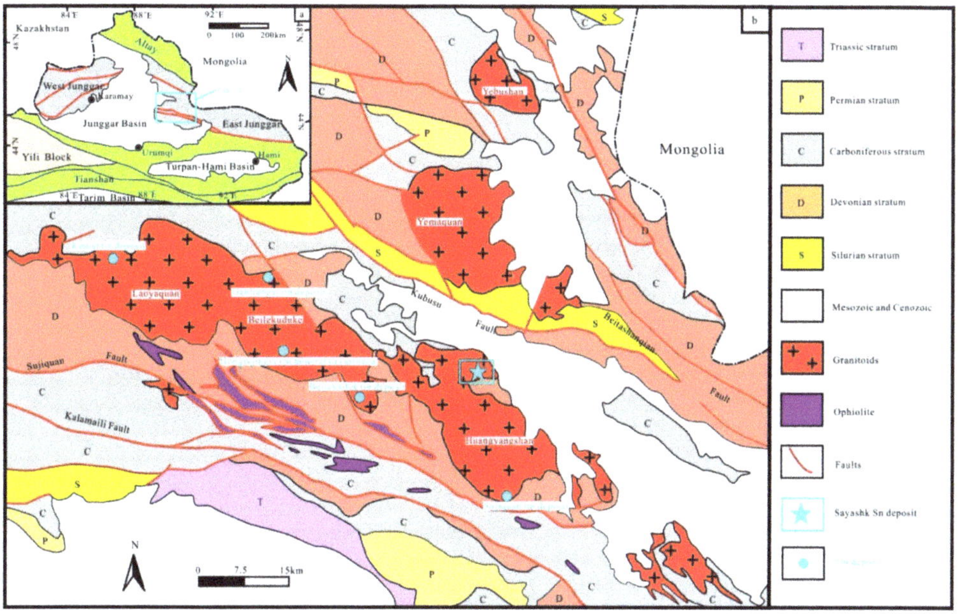

Figure 1. (**a**) Geological map of Xinjiang (modified after [42]). (**b**) Simplified geological map of the Kalamaili area, eastern Junggar (modified after [43]).

The late Carboniferous magmatic activity in the area was intense, forming large-scale and complex intrusive rocks such as Laoyaquan, Beilekuduk, Yemaquan, Sabei, and Huangyangshan, all of which occur as batholiths. Granitic magma is characterized by multicycle evolution, and its composition generally shows calc-alkaline-sub-alkaline-alkaline evolution, which is conducive to the differentiation and enrichment of Sn. The Sabei composite alkaline granite is closely related to the Sayashk Sn deposit and is located in the southeastern part of Mount Kalamaili and on the north side of the Huangyangshan pluton [9]. The Sabei composite alkaline granite is a typical rock in the Kalamali alkaline granite belt, and it intrudes into the tuffaceous sandstone and siltstone of the Beitashan Formation of the middle Devonian and the Jiangbastao Formation of the lower Carboniferous. Due to the multistage pulsating intrusion, the intrusion of the rocks from early to late is fine-grained riebeckite granite-medium fine-grained arfvedsonite granite-medium coarse-grained arfvedsonite granite-porphyritic riebeckite granite-riebeckite granite porphyry (dyke), which formed by four successive pulsation intrusions of homologous magma [32,33].

3. Geology of the Ore Deposit

The Carboniferous intrusive rocks in the Sayashk Sn deposit are widely distributed. The main lithologies are pink medium fine-grained porphyritic syenogranite and potassium feldspar granite. The medium fine-grained porphyritic syenogranite is further named medium fine-grained arfvedsonite alkaline granite. The dykes are dominated by quartz albite porphyry, granite porphyry, quartz porphyry, granite aplite, quartz vein, and other hypabyssal plutons. The strike of the dykes is NE and NNE. The fault structures in the mining area are mainly distributed in the NW, NEE or NNE direction and are generally small in scale. Many researchers generally believed that the magmatic rock closely related to mineralization was granite [2,31–33,44]. After detailed field geological surveys and exploration reports, we found that mineralization was closely related to quartz veins and granite porphyry or quartz albite porphyry veins (Figure 2). The ore body is a Sn mineralized quartz vein with obvious greisenization, silicification, epidotization, malachitization, covellite and other Cu mineralization and molybdenization. With the ore body as the center, alteration occurs to both sides, and the alteration intensity decreases with increasing distance to both sides. There is no obvious boundary between the surrounding rock and the altered rock, showing a gradual transition relationship, granite → granite porphyry → Sn-containing quartz veins.

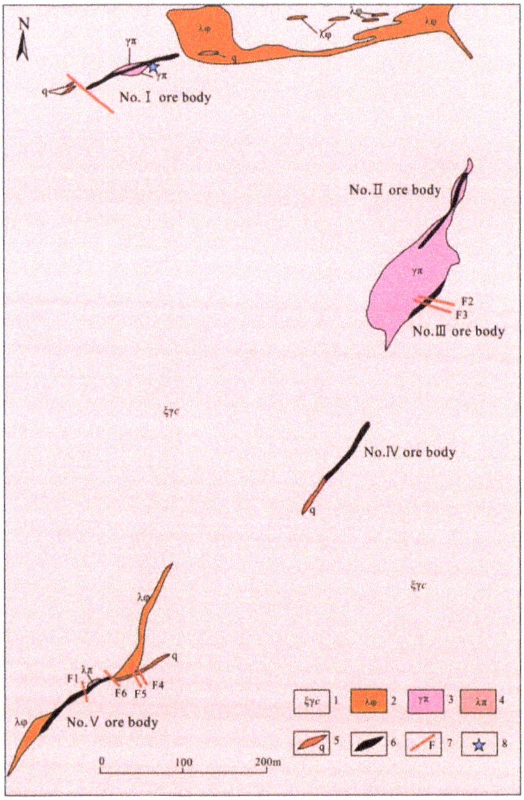

Figure 2. Geological map of Sayashk Sn mining area. 1—medium fine-grained alkaline granite; 2—quartz albite porphyry; 3—granite porphyry; 4—quartz porphyry; 5—quartz vein; 6—tin ore body; 7—fault; 8—sampling location.

Five ore bodies are preliminarily delineated in the deposit, of which the I and II ore bodies are large in scale and have strong tin mineralization.

Ore body I: the ore body is approximately 160 m long and 1–2 m wide, and the occurrence is 190° ∠ 85°. It is a quartz vein Sn ore body hosted in granite porphyry, with obvious silicification. The cassiterite and molybdenite are distributed in a dense star point, malachitization is distributed in a thin film.

Ore body II: the ore body is approximately 100 m long and 1–2 m wide, and the occurrence is 190° ∠ 85°. It occurs in granite porphyry with obvious silicification. The cassiterite and molybdenite are distributed in a dense star point, malachitization is distributed in a thin film.

Ore (mineralized) bodies III, IV, V: each ore body is approximately 50–100 m long and 0.8–1 m wide, and its occurrence is about 235° ∠ 78°. The ore bodies occur in granite porphyry or quartz albite porphyry with obvious silicification. The cassiterite and molybdenite are distributed in a dense star point, malachitization is distributed in a thin film. Sn grade is 1.0%–1.5%, Mo content is 0.001%–0.05%, and Cu content is 0.01%–0.10%.

4. Samples and Analytical Procedures

4.1. Samples Descriptions

The granite and granite porphyry samples (Figure 3) closely related to mineralization, as well as molybdenite samples, were collected from the ore body I of the Sayashk Sn deposit. The sampling coordinates are 90°21′29″ east longitude and 45°10′56″ north latitude. The weathered surface and fresh surface of the medium-fine-grained granite are gray-white, with a medium fine-grained structure. The particle size is mainly distributed between 1–3 mm, and the massive structure is mainly composed of alkali feldspar (65%), plagioclase (5%), quartz (25%), and a small number of dark minerals (biotite + amphibole accounts for approximately 5%) (Figure 3a,d,e). The weathered surface of the granite porphyry is yellowish brown, and the fresh surface is dark gray with a porphyritic structure. The phenocrysts are mainly alkali feldspar (5%) and quartz (5%), and the particle size is mainly between 2–4 mm. The matrix has cryptocrystalline and vein structures. Molybdenite mainly occurs in disseminated, micro vein, and film forms (Figure 3b,f,g).

Figure 3. Filed photos, hand specimens and photomicrographs of typical rock units and ores from the Sayashk Sn deposit. (**a**) Granite porphyry with a grayish white color and a porphyritic structure; (**b**) the medium-fine-grained granite with a grayish white color and a medium-fine-grained structure; (**c**) intrusive contact relationship between granite and granite porphyry; (**d**,**e**) granite porphyry under single polarizer and orthogonal polarizer, respectively; (**f**,**g**) medium-fine-grained granite under single polarizer and orthogonal polarizer, respectively.

4.2. Analytical Procedures

Major and trace element geochemical analyses were undertaken at Yanduzhongshi Geological Analysis Laboratories Ltd. Major oxide concentrations were measured with an X-ray fluorescence (XRF) spectrometer (Shimadzu, Kyushu, Japan) and the trace element concentrations were determined by ICP-MS techniques (Analytik Jena AG, Jena, Germany). The analytical precisions estimated through repeated analyses of standards were better than ±0.01% for XRF and ±5% for ICP-MS analysis. The detailed analytical methods and procedures are described in Liu et al. (2008) [45].

Data testing and analysis of zircon U–Pb dating were performed at Yanduzhongshi Geological Analysis Laboratories Ltd. using LA-ICP-MS. M90 ICP-MS (Shimadzu, Columbia, MD, USA)was used for the analysis with a New Wave UP213 laser ablation system (Thermo Fisher Scientific, Carlsbad, CA, USA). During the sample tests, the diameter of the laser ablation was 30 μm, SRM610 was used as the external standard, and Si was used as the internal standard to calculate the trace element contents of the zircons. Zircon standard 91,500 was used as the external standard for isotope fractionation correction, and for every 5–10 sample points analyzed, two standard sample analyses and one Plesovice analysis were conducted for monitoring. U–Pb dating was performed using the ICPMS Data Cal 4.3 program (ICPMS Data Cal 4.3, Wuhan, China) [46]. The U–Pb age calculation and concordia plots were processed using ISOPLOT 3.0 [47]. The common Pb correction process was performed according to the method proposed by Andersen [48].

Zircon Lu–Hf isotope analysis was carried out in-situ by using an NWR193 laser-ablation microprobe (Elemental Scientific Lasers LLC, Bozeman, MT, USA), attached to a Neptune multicollector ICP-MS at Yanduzhongshi Geological Analysis Laboratories Ltd. Instrumental conditions and data acquisition were comprehensively described by Wu et al. (2006) [49]. A stationary spot was used for the present analyses, with a beam diameter of 35 μm. Helium was used as carrier gas to transport the ablated sample from the laser-ablation cell to the ICP-MS torch via a mixing chamber mixed with Argon. In order to correct the isobaric interferences of ^{176}Lu and ^{176}Yb on ^{176}Hf, $^{176}Lu/^{175}Lu = 0.02658$ and $^{176}Yb/^{173}Yb = 0.796218$ ratios were determined [50]. For instrumental mass bias correction Yb isotope ratios were normalized to $^{172}Yb/^{173}Yb$ of 1.35274 [50] and Hf isotope ratios to $^{179}Hf/^{177}Hf$ of 0.7325 using an exponential law. The mass bias behavior of Lu was assumed to follow that of Yb, mass bias correction protocols details were described by Wu et al. (2006) [49]. Zircon 91,500 and Plesovice were used as the reference standards during our routine analyses, Initial $^{176}Hf/^{177}Hf$ ratios εHf (t) were calculated with reference to the chondritic reservoir (CHUR) of Blichert-Toft and Albarede (1997) [51] at the time of zircon growth from the magma. The single-stage Hf model age (TDM1) is calculated relative to the depleted mantle with present-day $^{176}Hf/^{177}Hf = 0.28325$ and $^{176}Lu/^{177}Hf = 0.0384$ [52] (e.g., Griffin et al., 2000).

Re–Os isotope analyses were completed on the separated samples at the Re–Os Laboratory of the National Research Center of Geoanalysis, Chinese Academy of Geological Sciences. Details of the chemical separation procedure are described by Shirey and Walker (1995) [53], Stein et al. (1997) [54], Du et al. (2004) [55], Xie et al. (2007) [56], and Mao et al. (1999) [57]. The Re–Os model age was calculated as $t = 1/\lambda\ [\ln(1 + ^{187}Os/^{187}Re)]$, where λ is the decay constant of ^{187}Re, and the adopted value is $\lambda = 1.666 \times 10^{-11}$ year^{-1}. Positive and inverse isochrons were computed with the ISOPLOT program [47].

5. Results

5.1. Whole-Rock Major and Trace Elements

Whole-rock major and trace element compositions of five medium-fine-grained granite samples (defined SR2) and five granite porphyry samples (defined SR1) are presented in Table 1. These samples generally have loss on ignition (LOI) values of <2 wt.%, suggesting that they have not been affected by alteration. Overall, all ten samples are characterized by high concentrations of silica and alkalis and low concentrations in oxides of Al, Mg, and Ca, and the concentration of K_2O is greater than the concentration of Na_2O. However, five medium-fine-grained granite samples (sample No. SR-2-1~SR-2-5) on average contain *w* (SiO_2) = 78.65% wt.%, *w* (Al_2O_3) = 11.10%, *w* (CaO) = 0.22%, *w* (MgO) = 0.074%, and *w* (Na_2O) + w (K_2O) = 8.25%. Five granite porphyry samples (sample No. SR-1-1~SR-1-5) contain *w* (SiO_2) = 71.27%, *w* (Al_2O_3) = 13.27%, *w* (CaO) = 0.72%, *w* (MgO) = 0.22%, and *w* (Na_2O) + w (K_2O) = 9.60%.

Table 1. Major (%) and trace (ppm) element compositions of the granite porphyry and the medium-fine-grained granite from the Qiushuwan deposit.

Sample	SR-1-1	SR-1-2	SR-1-3	SR-1-4	SR-1-5	SR-2-1	SR-2-2	SR-2-3	SR-2-4	SR-2-5
SiO_2	71.88	71.26	72.49	73.57	72.16	79.01	78.77	78.16	79.23	78.09
TiO_2	0.29	0.31	0.26	0.23	0.29	0.08	0.07	0.10	0.08	0.06
Al_2O_3	13.32	13.49	13.33	12.91	13.31	11.07	11.02	11.26	10.77	11.37
TFe_2O_3	3.05	3.15	2.85	2.71	3.12	1.20	1.23	1.42	1.27	1.12
CaO	0.85	0.89	0.80	0.71	0.84	0.18	0.19	0.20	0.20	0.31
MgO	0.23	0.23	0.20	0.21	0.23	0.07	0.08	0.08	0.07	0.07
MnO	0.06	0.06	0.07	0.03	0.04	0.02	0.02	0.02	0.02	0.01
K_2O	4.94	5.04	5.12	4.62	5.16	4.78	4.06	4.58	4.17	4.30
Na_2O	4.67	4.70	4.56	4.73	4.47	3.54	3.99	3.89	3.80	4.14
P_2O_5	0.07	0.07	0.06	0.06	0.07	0.03	0.03	0.03	0.03	0.03
LOI	0.35	0.35	0.30	0.34	0.38	0.29	0.21	0.25	0.16	0.30
Total	99.70	99.55	100.05	100.13	100.07	100.24	99.67	99.99	99.78	99.80
AR	4.87	4.78	4.65	5.55	4.43	4.40	5.92	5.24	5.50	5.87
A/CNK	0.91	0.91	0.92	0.92	0.92	0.98	0.97	0.96	0.97	0.94
A/NK	1.02	1.02	1.02	1.01	1.03	1.01	1.01	0.99	1.00	0.99
Rb	199.19	190.78	212.20	187.24	218.03	200.28	170.64	184.36	179.84	157.69
Sr	59.70	56.35	59.55	51.46	52.62	10.17	6.52	6.20	7.30	6.10
Y	68.23	69.28	73.89	67.67	72.76	52.65	51.24	56.58	52.67	58.56
Zr	633.83	654.34	671.58	612.52	625.19	293.81	315.98	280.93	312.97	289.08
Nb	14.87	14.11	14.99	15.66	14.55	11.92	12.10	15.16	12.26	14.03
Cd	0.60	0.61	0.64	0.48	0.73	0.53	0.38	0.50	0.35	0.39
In	0.17	0.17	0.17	0.17	0.18	0.15	0.15	0.15	0.14	0.11
Cs	11.14	11.12	11.51	10.00	15.15	5.88	5.98	5.27	8.90	2.98
Ba	494.93	508.19	472.93	361.18	508.23	34.19	23.57	19.77	24.70	15.55
La	39.92	43.78	51.17	26.16	43.65	36.73	37.16	35.14	39.20	32.03
Ce	84.06	100.94	108.03	56.62	100.63	74.06	72.56	71.96	71.17	69.23
Pr	10.66	12.13	13.62	7.82	12.39	9.57	9.81	9.21	10.48	8.49
Nd	45.14	51.43	56.18	32.74	52.16	38.94	40.15	37.89	42.35	33.98
Sm	11.79	12.66	13.63	9.69	12.75	10.19	10.53	10.23	10.65	9.65
Eu	0.54	0.58	0.54	0.36	0.56	0.01	0.00	0.00	0.00	0.01
Gd	12.23	12.95	13.89	10.16	13.09	10.09	9.97	9.71	9.77	9.67
Tb	2.07	2.18	2.26	1.92	2.16	1.69	1.69	1.73	1.63	1.75
Dy	12.99	12.87	13.73	12.19	13.21	9.94	9.96	10.62	9.96	10.75
Ho	2.78	2.78	2.92	2.75	2.78	2.12	2.16	2.35	2.20	2.37
Er	7.85	7.66	8.17	7.92	7.83	5.95	6.15	6.66	6.26	6.73
Tm	1.08	1.03	1.16	1.12	1.06	0.78	0.84	0.91	0.85	0.92
Yb	7.71	7.28	7.68	8.15	7.38	5.14	5.78	6.36	5.77	6.05
Lu	1.09	1.07	1.15	1.17	1.07	0.74	0.85	0.86	0.83	0.84
Hf	18.32	18.07	19.09	18.28	17.34	10.78	11.49	10.73	10.21	10.58
Ta	1.18	1.12	1.07	1.11	1.07	0.89	0.89	1.02	0.84	0.92
Tl	0.67	0.64	0.69	0.64	0.68	0.56	0.45	0.46	0.52	0.49
Th	10.74	9.93	11.23	11.40	10.23	11.76	12.01	12.17	11.87	12.23
U	3.97	3.55	4.11	4.22	3.36	3.87	3.80	5.15	3.19	3.36
ΣREE	239.91	269.34	294.13	178.77	270.72	205.95	207.61	203.63	211.12	192.48
LREE	192.11	221.52	243.17	133.39	222.14	169.50	170.21	164.43	173.85	153.39
HREE	47.80	47.82	50.96	45.38	48.58	36.45	37.40	39.20	37.27	39.08
LREE/HREE	4.02	4.63	4.77	2.94	4.57	4.65	4.55	4.19	4.66	3.92
(La/Yb_N)	3.71	4.31	4.78	2.30	4.24	5.13	4.61	3.96	4.87	3.80
Nb/Ta	38.25	45.79	52.51	29.37	48.75	43.77	45.01	37.08	50.13	15.26
Zr/Hf	34.59	36.2	35.19	33.51	36.05	27.25	27.51	26.18	30.65	27.32
δEu	0.14	0.14	0.12	0.11	0.13	0.0029	0.98	0.99	0.98	0.0021

Note: A/CNK = molar Al/(Ca + Na + K); A/NK = molar Al/(Na + K); δEu = 2 EuN/(SmN × GdN).

In the SiO_2 vs. $K_2O + Na_2O$ discrimination diagrams, all ten samples plot in the granite field, which is consistent with the identification results of hand specimens (Figure 4a). As the samples have high SiO_2 contents (>70 wt.%), we used an AR vs. SiO_2 diagram for classification (Figure 4b). The ten samples plot in the alkaline series, whereas the enclaves generally plot in the alkaline series. In the A/CNK versus A/NK diagram, all samples are metaluminous (Figure 4c).

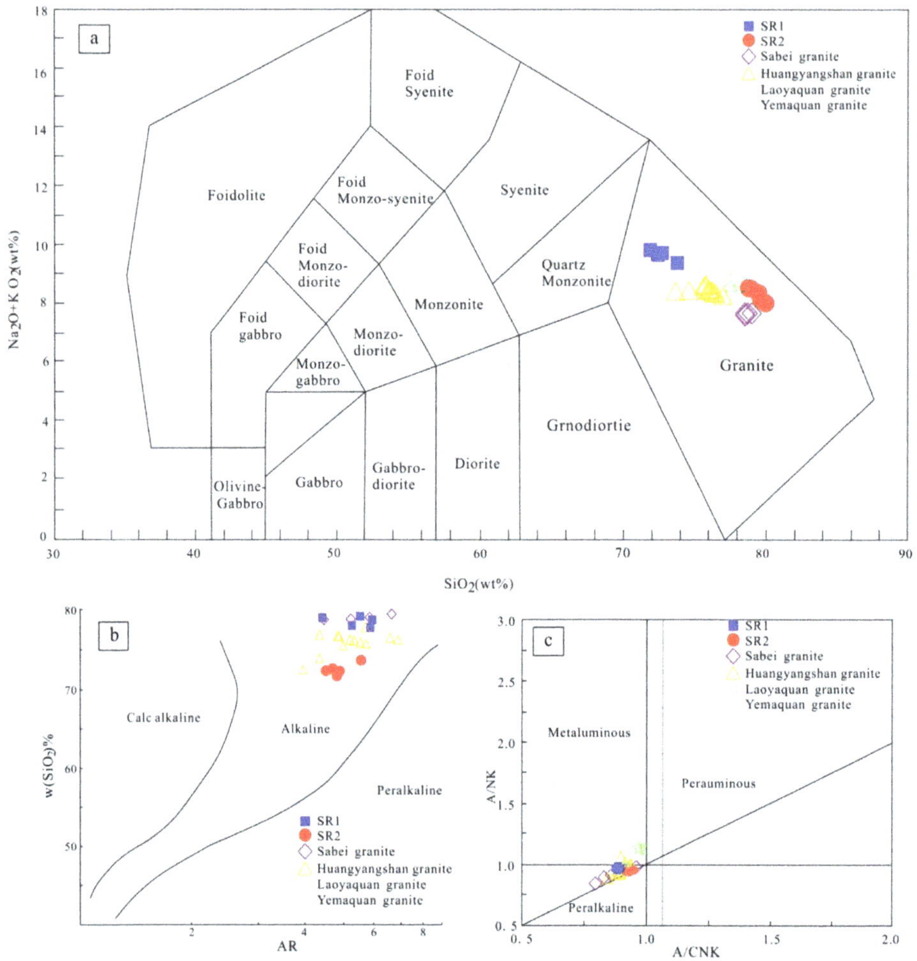

Figure 4. Geochemical characteristics of the medium-fine-grained granites and granite porphyry in the Sayashk Sn deposit. (**a**) SiO_2 vs. ($Na_2O + K_2O$) diagram. (**b**) AR vs. SiO_2 diagram. (**c**) A/CNK vs. A/NK diagram. Sabei granite after [30], Huangyangshan granite after [58], Laoyaquan granite after [9], Yemaquan granite after [59].

The total rare earth elements (REEs) of the medium-fine-grained granites and granite porphyry range from 192.48 ppm to 211.12 ppm and 178.77 ppm to 270.72 ppm, respectively. On a chondrite-normalized REE diagram (Figure 5a), the two granite distribution curves have a similar trend, showing a slightly right-leaning "seagull" shape. The fractionation of LREE and HREE is not obvious, and there is an obvious negative Eu anomaly. $(La/Yb)_N$ ratio SR2 = 3.80–5.13, $(La/Yb)_N$ ratio SR1 = 2.30–4.78.

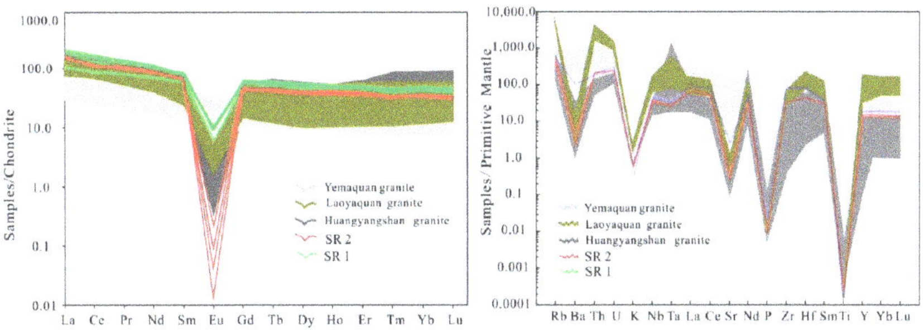

(**a**) (**b**)

Figure 5. (**a**) Primitive mantle-normalized trace element and (**b**) chondrite-normalized REE patterns of the SR2 and SR1 in the Sayashk Sn deposit. Huangyangshan granite after [58], Laoyaquan granite after [9], Yemaquan granite after [59].

The granite porphyry and granite in the Sayashk deposit are enriched in LILEs, such as Rb, Th, and Nd, and HFSEs, such as Zrand Hf, but depleted in elements such as Ba, Sr, K, P, Ti, and Eu (Figure 5b). Both are characterized by significantly negative Eu anomalies, suggesting dominant plagioclase fractionation in more evolved melts.

5.2. CL Imaging and U–Pb Geochronology

Zircon CL images of the medium-fine-grained granites and granite porphyry are shown in Figure 6. Most of the zircons in the SR1 sample are colorless and transparent, and some are dark, while most of the zircons in the SR2 sample are dark, and both zircons are long biconical. The lengths of the zircons range from 100~200 μm. Most of them are typical magmatic zircons characterized by euhedral crystals with oscillatory zoning or linear zoning, which represents the crystallization of the magma. The zircon U–Pb isotope results (Table 2) show that the Th average content of the samples SR1 is 155.00 ppm, the U average content is 346.84 ppm, the Th/U average ratio is 0.42. In sample SR2, the average value of Th is 165.01 ppm, the average value of U is 388.42 ppm, and the Th/U average ratio is 0.42. Both samples have high Th/U ratios (>0.4) and belong to typical magmatic zircons [60]. Both Th and U contents show a good positive correlation, which is consistent with the characteristics of typical magmatic zircons [61–63].

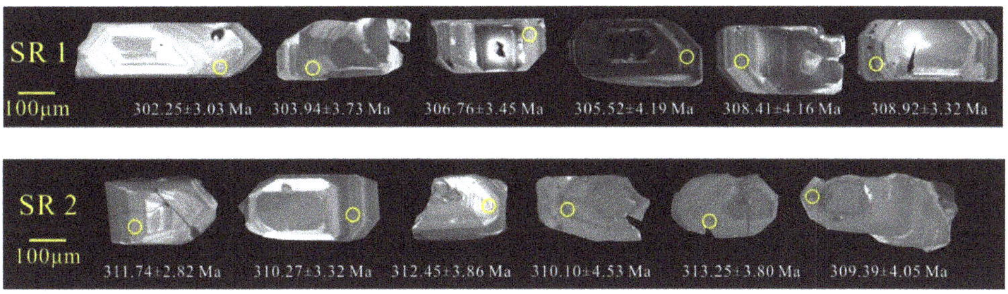

Figure 6. Part of zircon CL images showing the locations of LA-ICP-MS measurement spots and associated $^{206}Pb/^{238}U$ ages for the SR2 and SR1 in the Sayashk Sn deposit.

Table 2. Results of the zircon LA-ICP-MS U–Pb dating results for SR2 and SR1 in the Sayashk district.

Analysis Point	U (ppm)	Th (ppm)	Th/U	Isotopic Ratios						Ages (Ma)					
				$^{207}Pb/^{206}Pb$	1σ	$^{207}Pb/^{235}U$	1σ	$^{206}Pb/^{238}U$	1σ	$^{207}Pb/^{206}Pb$	1σ	$^{207}Pb/^{235}U$	1σ	$^{206}Pb/^{238}U$	1σ
SR-2-01	344	140	0.41	0.0532	0.0012	0.3619	0.0072	0.0495	0.0005	335.30	49.49	313.66	5.39	311.74	2.82
SR-2-03	380	150	0.40	0.0532	0.0013	0.3614	0.0090	0.0493	0.0005	336.80	56.31	313.23	6.70	310.27	3.32
SR-2-04	380	141	0.37	0.0541	0.0013	0.3711	0.0096	0.0497	0.0006	376.94	54.88	320.48	7.09	312.45	3.86
SR-2-05	348	148	0.43	0.0532	0.0014	0.3663	0.0109	0.0497	0.0007	338.24	58.49	316.91	8.13	312.68	4.16
SR-2-06	342	136	0.40	0.0526	0.0016	0.3578	0.0109	0.0493	0.0007	312.45	69.24	310.54	8.18	310.10	4.53
SR-2-09	306	118	0.39	0.0521	0.0011	0.3515	0.0080	0.0490	0.0005	291.54	48.94	305.85	6.00	308.17	3.07
SR-2-11	389	163	0.42	0.0535	0.0015	0.3628	0.0104	0.0491	0.0006	351.79	63.11	314.27	7.77	309.26	3.48
SR-2-12	399	145	0.36	0.0527	0.0016	0.3624	0.0114	0.0498	0.0007	314.04	68.20	313.98	8.49	313.40	4.13
SR-2-13	386	192	0.50	0.0526	0.0010	0.3588	0.0076	0.0493	0.0005	312.55	43.41	311.31	5.69	310.35	3.05
SR-2-14	410	171	0.42	0.0534	0.0013	0.3631	0.0091	0.0493	0.0007	347.15	54.79	314.56	6.75	310.42	3.99
SR-2-15	361	163	0.45	0.0529	0.0012	0.3599	0.0086	0.0493	0.0005	324.27	50.07	312.17	6.43	309.99	3.18
SR-2-16	293	99	0.34	0.0525	0.0013	0.3559	0.0093	0.0492	0.0007	305.60	55.43	309.16	6.93	309.39	4.05
SR-2-17	321	140	0.44	0.0526	0.0013	0.3622	0.0088	0.0497	0.0007	310.68	58.43	313.87	6.55	312.92	4.28
SR-2-19	690	365	0.53	0.0526	0.0010	0.3632	0.0073	0.0495	0.0006	313.47	43.87	314.61	5.41	311.75	3.91
SR-2-20	274	130	0.47	0.0519	0.0012	0.3521	0.0078	0.0490	0.0005	282.03	50.84	306.27	5.88	308.62	3.06
SR-2-23	362	156	0.43	0.0519	0.0019	0.3566	0.0109	0.0497	0.0008	279.92	82.48	309.64	8.19	312.56	5.20
SR-2-24	662	292	0.44	0.0532	0.0014	0.3651	0.0103	0.0496	0.0007	336.08	58.20	316.04	7.67	312.09	4.30
SR-2-25	331	138	0.42	0.0533	0.0011	0.3652	0.0077	0.0496	0.0005	340.20	45.72	316.10	5.71	312.00	3.16
SR-2-26	337	136	0.40	0.0528	0.0014	0.3632	0.0100	0.0498	0.0005	321.75	59.55	314.57	7.45	313.06	3.26
SR-2-27	373	141	0.38	0.0518	0.0015	0.3551	0.0096	0.0498	0.0006	277.32	67.20	308.56	7.18	313.25	3.80
SR-2-28	431	193	0.45	0.0528	0.0011	0.3584	0.0082	0.0491	0.0006	320.75	49.17	311.01	6.11	309.00	3.97
SR-2-29	425	173	0.41	0.0526	0.0017	0.3578	0.0116	0.0493	0.0006	313.00	73.37	310.54	8.67	310.16	3.92
SR-1-01	316	117	0.37	0.0529	0.0013	0.3488	0.0089	0.0480	0.0005	324.23	57.73	303.78	6.67	302.25	3.03
SR-1-02	197	73	0.37	0.0532	0.0013	0.3504	0.0089	0.0483	0.0006	337.88	56.82	305.04	6.70	303.94	3.73
SR-1-03	388	184	0.47	0.0530	0.0011	0.3563	0.0074	0.0487	0.0006	330.41	45.73	309.42	5.57	306.76	3.45
SR-1-04	435	232	0.53	0.0525	0.0011	0.3552	0.0094	0.0489	0.0006	306.14	49.76	308.59	7.04	307.91	3.51
SR-1-07	1021	493	0.48	0.0531	0.0008	0.3646	0.0076	0.0495	0.0005	333.49	34.99	315.65	5.67	311.64	3.08
SR-1-08	356	162	0.46	0.0531	0.0009	0.3637	0.0069	0.0497	0.0005	334.56	39.62	314.98	5.15	312.51	3.19
SR-1-09	322	126	0.39	0.0535	0.0014	0.3608	0.0084	0.0492	0.0007	350.53	58.94	312.83	6.29	309.64	4.03
SR-1-10	207	62	0.30	0.0543	0.0016	0.3624	0.0102	0.0485	0.0007	383.00	66.65	313.99	7.63	305.52	4.19
SR-1-11	563	286	0.51	0.0526	0.0010	0.3553	0.0065	0.0490	0.0007	310.60	41.81	308.71	4.86	308.41	4.16
SR-1-12	465	197	0.42	0.0518	0.0010	0.3481	0.0079	0.0486	0.0006	275.45	44.57	303.33	5.92	305.75	3.46
SR-1-13	680	364	0.54	0.0520	0.0008	0.3543	0.0072	0.0491	0.0004	283.71	37.31	307.96	5.39	309.28	2.70
SR-1-15	135	49	0.37	0.0521	0.0019	0.3487	0.0127	0.0488	0.0007	287.91	83.69	303.76	9.60	307.01	4.41
SR-1-16	296	132	0.45	0.0526	0.0014	0.3548	0.0090	0.0491	0.0007	311.00	59.09	308.30	6.74	308.96	4.36
SR-1-17	198	70	0.35	0.0533	0.0017	0.3595	0.0118	0.0492	0.0007	340.68	74.32	311.87	8.82	309.37	4.35
SR-1-18	166	63	0.38	0.0532	0.0023	0.3562	0.0140	0.0486	0.0008	336.45	96.75	309.39	10.50	306.19	4.84
SR-1-19	266	112	0.42	0.0524	0.0016	0.3553	0.0107	0.0492	0.0006	301.82	67.47	308.70	8.04	309.82	3.62
SR-1-20	192	67	0.35	0.0518	0.0017	0.3505	0.0116	0.0490	0.0007	275.20	75.99	305.07	8.71	308.53	4.01
SR-1-21	483	266	0.55	0.0532	0.0010	0.3608	0.0071	0.0492	0.0006	339.16	43.53	312.81	5.27	309.43	3.74
SR-1-22	117	40	0.34	0.0526	0.0022	0.3584	0.0148	0.0498	0.0008	312.52	96.59	311.01	11.07	313.10	4.67
SR-1-23	361	141	0.39	0.0515	0.0013	0.3503	0.0096	0.0493	0.0007	263.94	58.64	304.96	7.26	310.26	4.49
SR-1-25	367	175	0.48	0.0526	0.0014	0.3546	0.0100	0.0487	0.0008	312.36	60.46	308.15	7.49	306.47	5.18
SR-1-28	215	82	0.38	0.0529	0.0016	0.3540	0.0100	0.0488	0.0006	323.51	67.91	307.70	7.50	306.90	3.64
SR-1-29	177	65	0.37	0.0511	0.0022	0.3438	0.0144	0.0488	0.0008	246.47	98.83	300.06	10.86	307.17	4.66
SR-1-30	400	159	0.40	0.0530	0.0010	0.3591	0.0068	0.0491	0.0005	327.89	43.60	311.53	5.07	308.92	3.32

The results of zircons from the medium-fine-grained granites and granite porphyry yield a concordia diagram in Figure 7a,b, where the combined data points yield weighted mean $^{206}Pb/^{238}U$ ages of 308.2 ± 1.5 Ma (MSWD = 0.68, n = 24) and 310.9 ± 1.5 Ma (MSWD = 0.22, n = 22), respectively. Therefore, we interpret the mean age as the crystallization age of the medium-fine-grained granites and granite porphyry. Both of them formed in the late Carboniferous epoch.

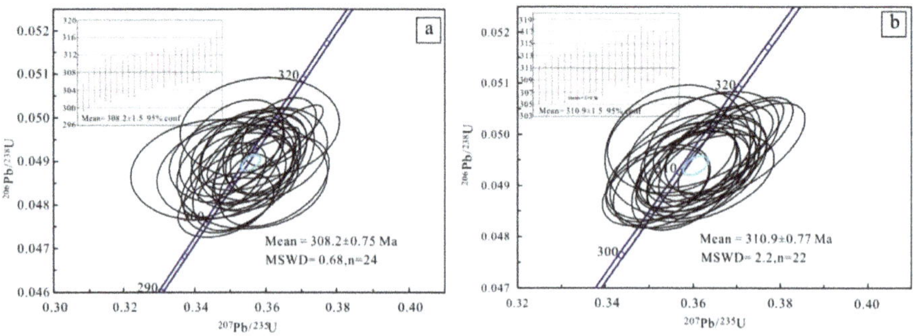

Figure 7. Zircon U–Pb concordia diagrams for (**a**) SR2, (**b**) SR1.

5.3. Zircon Hf Isotopes

The zircon Hf isotope compositions for the SR2 and SR1 are listed in Table 3. The $^{176}Lu/^{177}Hf$ ratios of all zircons are very close to or less than 0.002, indicating that after the formation of these zircons, the accumulation of radiogenic Hf was less, and the Hf isotope composition is reasonable [64]. The $^{176}Hf/^{177}Hf$ ratio can be used for geochemical tracing. The $^{176}Hf/^{177}Hf$ ratios of zircons from the SR2 vary between 0.282929 and 0.283032 (average 0.282974), with εHf(t) values ranging from 10.27 to 16.17 (average 13.71). Single model ages (T_{DM1}) and two-stage model ages (T_{DM2}) fall within the ranges of 319–535 Ma and 339–644 Ma, respectively. The $^{176}Hf/^{177}Hf$ ratios of zircons from the SR1 range from 0.282953 to 0.283033 (average 0.282972), with εHf(t) values ranging between 5.72 and 9.21 (average 7.08). Single model ages (T_{DM1}) and two-stage model ages (T_{DM2}) fall within the ranges 346~479 Ma and 309–557 Ma, respectively.

Table 3. Hf isotopic data of zircons from the SR2 and SR1 in the Sayashk Sn deposit.

Spot no.	*t* (Ma)	$^{176}Yb/^{177}Hf$	$^{176}Lu/^{177}Hf$	$^{176}Hf/^{177}Hf$	2σ	$\varepsilon_{Hf}(0)$	$\varepsilon_{Hf}(t)$	T_{DM1}	T_{DM2}	$f_{Lu/Hf}$
SR-2-001	313	0.075538	0.001940	0.282967	0.000019	6.91	13.4	414	470	−0.94
SR-2-003	291	0.086482	0.002231	0.283016	0.000017	8.63	14.61	345	374	−0.93
SR-2-004	282	0.132216	0.003455	0.282905	0.000021	4.7	10.27	526	646	−0.9
SR-2-005	320	0.075191	0.001964	0.283024	0.000017	8.9	15.53	332	339	−0.94
SR-2-006	351	0.090929	0.002342	0.282929	0.000017	5.54	12.72	475	542	−0.93
SR-2-009	305	0.078083	0.002025	0.282967	0.000020	6.88	13.19	416	478	−0.94
SR-2-011	324	0.115128	0.002893	0.282893	0.000020	4.3	10.81	535	644	−0.91
SR-2-012	312	0.098395	0.002539	0.282984	0.000017	7.49	13.84	396	441	−0.92
SR-2-013	312	0.095046	0.002361	0.282990	0.000019	7.69	14.08	386	426	−0.93
SR-2-014	336	0.074679	0.001915	0.283032	0.000019	9.19	16.17	319	309	−0.94
SR-2-015	312	0.058124	0.001507	0.282993	0.000018	7.81	14.37	372	407	−0.95
SR-2-016	347	0.074915	0.001919	0.282986	0.000017	7.58	14.79	386	408	−0.94
SR-2-017	335	0.081711	0.002086	0.282989	0.000018	7.66	14.56	384	411	−0.94
SR-2-019	313	0.116336	0.002853	0.283004	0.000018	8.19	14.49	370	400	−0.91
SR-2-020	340	0.082512	0.002134	0.282937	0.000018	5.83	12.83	461	528	−0.94
DCRH-01	334	0.055608	0.001448	0.282967	0.000018	6.91	13.93	408	451	−0.96
DCRH-03	291	0.078398	0.002021	0.282964	0.000017	6.79	12.81	419	490	−0.94
DCRH-04	324	0.126125	0.003198	0.282934	0.000020	5.72	12.15	479	557	−0.9
DCRH-06	337	0.082700	0.002123	0.282996	0.000017	7.93	14.88	374	394	−0.94
DCRH-08	383	0.144140	0.003583	0.283033	0.000020	9.21	16.75	333	309	−0.89
DCRH-09	275	0.071345	0.001849	0.282987	0.000018	7.61	13.34	384	445	−0.94
DCRH-11	336	0.054312	0.001444	0.282962	0.000018	6.71	13.8	416	462	−0.96
DCRH-12	312	0.028669	0.000773	0.282943	0.000015	6.06	12.77	435	510	−0.98
DCRH-14	330	0.083431	0.002089	0.282993	0.000017	7.84	14.66	377	403	−0.94
DCRH-15	323	0.064451	0.001673	0.282964	0.000018	6.8	13.55	415	467	−0.95
DCRH-16	287	0.066353	0.001717	0.282951	0.000016	6.35	12.33	434	517	−0.95
DCRH-17	306	0.081286	0.002078	0.282970	0.000019	6.99	13.31	412	471	−0.94
DCRH-18	310	0.046680	0.001232	0.282953	0.000021	6.4	12.98	426	494	−0.96
DCRH-23	327	0.049078	0.001291	0.282967	0.000014	6.9	13.81	407	454	−0.96
DCRH-24	310	0.043277	0.001216	0.282958	0.000018	6.59	13.15	418	482	−0.96
DCRH-26	283	0.068651	0.001867	0.283014	0.000019	8.54	14.43	346	381	−0.94

5.4. Molybdenite Re–Os Geochronology

The analytical results of molybdenite Re–Os dating are listed in Table 4. In seven molybdenite samples, Re = 70.50~462.0 ppb, Os = 0.0001~0.1658 ppb, ^{187}Re = 44.31~290.4 ppb and ^{187}Os = 0.2269~1.473 ppb. The model age of molybdenite is in the range of 306.5 Ma~298.1 Ma, with a weighted average age of 301.1 ± 3.1 Ma, MSWD = 2.0, isochron age of 301.4 ± 6.7 Ma, and MSWD = 3.4, which represents the metallogenic age of the Sayashk Sn deposit.

Table 4. Re–Os isotopic data for molybdenite from the Sayashk Sn deposit.

Sample No.	Weight (g)	Re ng/g		Os ng/g		^{187}Re ng/g		^{187}Os ng/g		Model Age (Ma)	
		Measured	2σ	Measured	2σ	Measured	2σ	Measured	2σ	Measured	2σ
SR-6-3	0.10023	246.8	1.7	0.0111	0.0018	155.1	1.1	0.7724	0.0051	298.1	4.1
SR-6-5	0.15002	168.3	1.6	0.0001	0.0008	105.8	1.0	0.5267	0.0037	298.1	4.5
SR-6-6	0.10023	208.2	1.5	0.0001	0.0012	130.8	0.9	0.6551	0.0043	299.8	4.2
SR-6-7	0.15043	134.2	1.6	0.0023	0.0008	84.37	1.00	0.4311	0.0031	305.9	5.2
SR-6-8	0.10230	281.9	3.6	0.0011	0.0007	177.2	2.3	0.8874	0.0056	299.8	5.2
SR-6-9	0.15023	462.0	4.1	0.0060	0.0005	290.4	2.6	1.473	0.009	303.7	4.4
SR-6-10	0.10056	70.50	1.05	0.1658	0.0220	44.31	0.66	0.2269	0.0024	306.5	6.4

6. Discussion

6.1. Timing of Magmatism and Mineralization in the Sayashk Sn Deposit

The granites in the East Junggar region are mainly distributed along three major tectonic faults [65]. They could be divided into two cycles, Caledonian and Variscan. The Caledonian cycle was only developed in the late Caledonian period, such as the Yemaquan biotite granite, plagioclase-rich granite, and gneiss granite, which formed when the marginal sea basin at the end of the Silurian was closed and uplifted. They are orogenic uplift-type granites [66]. The Variscan cyclic granite can be divided into three stages: early Variscan, middle-late Variscan (340–290 Ma), and late Variscan (280–240 Ma). Among them, the early Warwick granite rock mass closely coexists with the ophiolite and is the final product of basic magmatic differentiation of the ophiolite. The lithology is mainly plagioclase granite and quartz diorite, which belong to pre-orogenic oceanic granite (M-type). The granites of middle-late Warwick and late Warwick are widely developed, and the rock types are mainly calc-alkali series and alkaline series. The calcium-alkali series lithology mainly includes pyroxene diorite, diorite, quartz diorite, monzolite, quartz monzolite, granodiorite, and plagiogranite. The alkaline series lithology is mainly quartz alkaline syenite, sodium-ferrite granite, sodium-amphibole granite, porphyry sodium-amphibole granite, and alkali-feldspar granite. Their formation was closely related to the collisional orogeny of East Junggar [32,65–67].

In the petrogenetic age, the La-ICP-MS zircon U–Pb ages of the SR2 and SR1 obtained in this study are 308.2 ± 1.5 Ma (MSWD = 0.52, n = 24) and 310.9 ± 1.5 Ma (MSWD = 0.21, n = 22), respectively. Regionally, Bai Jianke et al. (2018) carried out LA-ICP-MSICP-MS zircon U–Pb dating of the ore-bearing alkaline granites of the Huangyangshan graphite deposit in the East Junggar area and concluded that the ages of the ore-bearing granites of the No. 1 and No. 2 graphite ore bodies were 303.6 ± 4.0 Ma and 304.6 ± 3.7 Ma, respectively [68]. It was considered that the petrogenetic and metallogenic ages were both in the later part of the late Carboniferous; Ai et al. (2020) reported zircon U–Pb ages of 318.6 ± 4.2 Ma, 321.4 ± 2.7 Ma, and 305.1 ± 5.4 Ma for the medium-grained arfvedsonite, amphibole, and fine-grained biotite granites, respectively [69]. Sun et al. (2021) carried out U–Pb dating of LA-ICP-MS zircon for the Huangyangshan plutonic granite, medium-fine amphibolite granite, medium-grained biotite granite, and fine-grained biotite granite, and the results showed that the weighted average ages were 322.7 ± 4.5 Ma, 303.9 ± 4.0 Ma, 303.9 ± 2.1 Ma, and 301.1 ± 3.6 Ma, respectively [39]. Gan Lin et al. (2010) carried out LA-ICP-MS zircon U–Pb dating of granodiorite, monzogranite, and alkali-feldspar granite of the Yemaquan complex rock in eastern Junggar. The results are 304 ± 3.0 Ma, 300 ± 2.0 Ma and 297 ± 6.0 Ma, respectively, indicating that the Yemaquan rock was emplaced in the late Carboniferous and belonged to the product of magmatism in the post-collision stage of the eastern Junggar area [70]. Hu Wanlong (2016) carried out LA-ICP-MS zircon U–Pb dating of porphyritic granite, coarse-grained granite, and fine-grained granite in the Laoyaquan complex granite body in the eastern Junggar, and the results were 317.6 ± 3.3 Ma and 310.3 ± 3 Ma, respectively. Ma and 308.4 ± 2.8 Ma, formed in the late Carboniferous [71]; Li, Yuechen, etal, (2007) dated the zircon SHRIMP U–Pb of the Bellekuduk syenite granite and obtained an age of 306 ± 5 Ma [72]. Nie Xiaoyong (2017) identified the early Carboniferous Qingshuidong plagioclase granite in the Eastern Junggar, with a harmonious age of 341.7 ± 1.7 Ma, quartz diorite with a harmonious age of 349.2 ± 6.7 Ma, and the Baijiigou granite has an age of 337.4 ± 5.8 Ma. In the East Junggar area, especially in the Kalamaili fault zone, magmatic activities are extensive and intense. With the characteristics of the complex rocks, it is not difficult to see from the above studies that these magmatic activities are characterized by multiple periods [73]. Han Baofu et al. (2006) believed that post-collisional plutonic magmatism occurred from 330–265 Ma and was concentrated from 330–310 Ma and 305–280 Ma [67]. Based on the extensive collection of previous research results (Table 5), this study considers that magmatic intrusions of 450–265 Ma were widely developed in this area, 420–410 Ma, 350–330 Ma, 320–300 Ma, and 300–280 Ma are four stages of relatively intense magmatic activity, especially 320–300 Ma, is the most intense period of magmatic activity (Figure 8).

Table 5. Typical intrusions and metal deposits of Kalamaili.

Name	Age (Ma)	Testing Method	Testing Object	References
Seltek pluton	314 ± 2	U-Pb	Zircon	[65]
Ertai potassium feldspar granite	319 ± 7	SHRIMP	Zircon	[67]
Xiaohongshan pluton	296 ± 4	SHRIMP	Zircon	[67]
Sujiquan pluton	295 ± 5	SHRIMP	Zircon	[67]
Kamst pluton	292 ± 7	SHRIMP	Zircon	[67]
Belage Kuduk pluton	273 ± 6	SHRIMP	Zircon	[67]
Yebushan pluton	268 ± 4	SHRIMP	Zircon	[67]
Kubu Sunan pluton	286 ± 3	LA-ICP-MS	Zircon	[8]
Hilectic Harassu pluton	381 ± 6	SHRIMP	Zircon	[74]
Harassay pluton	376 ± 10	SHRIMP	Zircon	[74]
Bieliatun granite	282 ± 5	SHRIMP, LA-ICP-MS	Zircon	[59]
Sujiquan biotite granite	304 ± 2	LA-ICP-MS	Zircon	[59]
Huangyangshan pluton	310 ± 4	LA-ICP-MS	Zircon	[75]
Huangyangshan pluton	302 ± 2	LA-ICP-MS	Zircon	[75]
Sabei granite	306 ± 3	LA-ICP-MS	Zircon	[2]
Thorange Kuduk granite	413 ± 8	SHRIMP	Zircon	[72]
Belle Kuduk granite	284 ± 5	LA-ICP-MS	Zircon	[72]
Qiongheba pluton	412.7 ± 3.3	LA-ICP-MS	Zircon	[76]
Yemaquan pluton	304 ± 3	LA-ICP-MS	Zircon	[70]
Ulungu River Chakurtu pluton	311.2 ± 2.5	SHRIMP	Zircon	[77]
Ertai pluton	279 ± 3	SHRIMP	Zircon	[77]
Sayashk granite	310.9 ± 1.5	LA-ICP-MS	Zircon	This paper
Sayashk granite porphyry	308.2 ± 1.5	LA-ICP-MS	Zircon	This paper
Mutanyao granite	349.8 ± 3.52	LA-ICP-MS	Zircon	[78]
Basque granodiorite	301 ± 2.5	LA-ICP-MS	Zircon	[58]
Basque granodiorite	310 ± 3.6	LA-ICP-MS	Zircon	[58]
Nanmingshui gold deposit	337.5 ± 3.9	LA-ICP-MS	Sericite	[73]
Shuangquan gold deposit	269 ± 9~260 ± 4	Ar-Ar	Sericite	[79]
Shuangquan gold deposit	310	Ar-Ar	Inclusions in Quartz	[80]
Qingshui No. 48 old deposit	311 ± 46	Rb-Sr	Inclusions in Quartz	[81]
Jinshuiquan gold deposit	271.7 ± 3.3	Ar-Ar	Sericite	[81]
West of the Huangyangshan gold deposit	318.4 ± 310.3	LA-ICP-MS	Zircon	[82]
Sayashk Sn deposit	301.1 ± 3.1	Re-Os	Molybdenite	This paper
Sabei Sn deposit	324.2 ± 3.4	SHRIMP	Zircon	[31]
Ganliangzi Sn deposit	314.0 ± 1	Ar-Ar	muscovite	[83]
Kamust Sn deposit	307.0 ± 1	Ar-Ar	muscovite	[83]

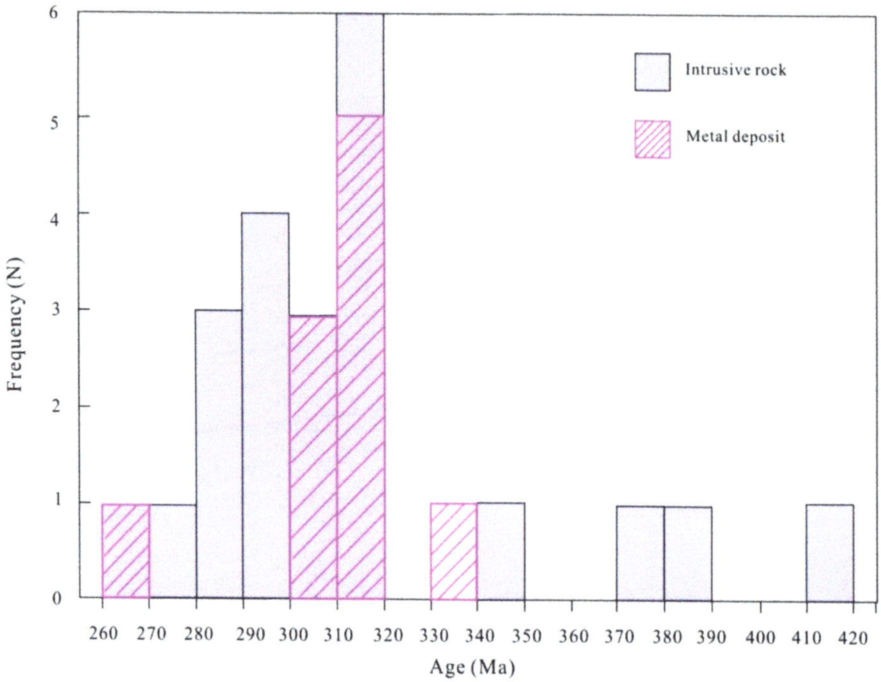

Figure 8. Age histogram of the typical intrusive rocks and some important metal deposits in the Kalamaili area.

In terms of metallogenetic epoch, a series of (tungsten) Sn deposits (points) and graphite deposits developed in the Belekuduk (tungsten) tin metallogenic belt in the East Junggar region, such as Kamster, Ganliangzi, Belekuduk, Sayashk, Sujiquan, and Hongtujingzi. In addition to the Huangyangshan and Sujiquan graphite deposits. They are mostly formed in the middle-late Carboniferous. For example, the ^{40}Ar-^{39}Ar age of muscovite in the mica-type tin ore in the Bailekuduke Sn deposit is 306.8 ± 2.4 Ma~309.7 ± 2.4 Ma [3], ^{40}Ar-^{39}Ar isotopic age of in quartz vein is 296.3 ± 2.6 Ma [6], Re–Os isotopic age of globular graphite in Huangyangshan graphite deposit is 332 ± 53 Ma [39], Rb-Sr isochron age of cassiterite-quartz vein whole rock of Ganliangzi Sn deposit is 307 ± 20 Ma and Rb-Sr isochron age of quartz fluid inclusions is 305 ± 25 Ma [41], Re–Os isochron age of molybdenite in tin ore of Sayashk Sn deposit is 307 ± 11 Ma [2], the model age of Molybdenite in the Sayashk Sn deposit is 298.1~306.5 Ma, with a weighted average age of 301.1 ± 3.1 Ma (MSWD = 2.0). The results show that the Sn deposits in this metallogenic belt were mainly concentrated between 320 Ma and 300 Ma (Figure 8). Because Sn deposits are strictly spatially controlled by alkaline granite, ore bodies of each Sn deposit occur in alkaline granite bodies and adjacent surrounding rocks, and tin metallogenesis is controlled by magma intrusion, differentiation, and gas–liquid alteration. Metallogenesis generally occurs in the late evolution of alkaline complex pluton and shows a good coupling relationship between diagenesis and mineralization in the late Carboniferous [32].

6.2. Rock Type, Petrogenesis and Nature of the Magma Source Area at the Sayashk Sn Deposit

6.2.1. Rock Type

As previously mentioned, the post-collisional plutonic magmatism in the East Junggar mainly occurred at 420–410 Ma, 350–330 Ma, 320–300 Ma, and 300–280 Ma. The late Silurian early Devonian granites are mainly found in the Jiangerkudu area and occur as batholiths or stocks. The lithology is mainly medium-grained biotite quartz monzodiorite, belonging to the quasi-aluminous high potassium calc-alkaline rocks [76,84], with no dark inclusions found. The early Carboniferous granites are sporadically exposed south of the Qingshui-Sujiquan fault, and the rock types are mainly quartz diorite, oligoclase granite, granodiorite, quartz porphyry, and quartz syenite [4]; most of them are quasi-aluminum-peraluminous high potassium calc-alkaline rock series. Late Carboniferous granites are widely distributed north of the Kalamaili fault and are the main components of the alkali-rich granite belt in Kalamaili, mainly composed of Laoyaquan, Huangyangshan, Bellekuduk, Sujiquan, and Balebanaan. They are produced as batholith or plutons [5,31,75]. The Sayashk granites are formed on the north side of the Huangshan pluton. The SR2 and SR1 are characterized by high silicon and alkali contents and belong to the per-alkali rock series. The petrogenetic age of these plutons range from 320 Ma to 305 Ma, and the rock types include biotite granite, soda amphibole granite, alkali feldspar granite, and quartz alkaline syenite. In addition, dark inclusions are developed in most plutons (Figure 6), showing the characteristics of mixed crustal mantle sources.

SR2 and SR1 are mainly composed of quartz and alkali feldspar, silicon-rich (average value of SR1 is 78.65%), alkali-rich (SR1-average 9.60%, SR2-average 8.25%), calcite-poor (SR1-average 0.82%, SR2-average 0.22%), magnesium-poor (SR1-average 0.22%, SR2-average 0.07%). They are basically consistent with the major element characteristics of typical A-type granites locally and globally, with high SiO_2 average (73.35%~73.81%), alkaline-rich (average 8.42%~8.72%), and poor CaO (average 0.75%~0.82%) [85], enriched in HFSEs (Zr and Hf) and Ce, poor in Ba, Sr, Eu, P, and Ti, and show a typical right-dipping seagull-type REE distribution pattern, which conforms to the characteristics of A-type granite. Overall, in the case of high differentiation (aluminous A-type granite), I-type, S-type, and A-type granites are difficult to distinguish [86]. They often have the same mineralogical and geochemical characteristics [85]. Therefore, they still need to be distinguished.

The distinction between A-type granite and highly differentiated S-type granite is relatively simple; highly differentiated S-type granites have a higher content of P_2O_5 (average 0.14%) [87], and the content of P_2O_5 increases with increasing differentiation degree, but A-type granite has the opposite trend. SR2 and SR1 in this paper have low contents of P_2O_5 (SR2 average 0.03%, SR1—0.06%), and high content of Na_2O (SR2 average 3.87%, SR1—4.63%). No characteristic minerals of S-type granite (such as primary muscovite, cordierite) are found, so it is impossible to be a highly differentiated S-type granite.

It is calculated that the zircon saturation temperatures of the two granites are 900 °C–908.2 °C (average 902.3 °C), and 836 °C–850.4 °C (average 843.3 °C), respectively, and the formation temperature is relatively high. The two plutons (TFeO) are 2.44–2.84 (average 2.68) and 1.01–1.27 (average 1.12), respectively. The total iron content of (TFeo) A-type granite is high, generally greater than 1%, while that of highly differentiated I-type granite is generally less than 1% [88], and it has high TFeo/(TFeo + MgO) values (average values are 0.925 and 0.938, respectively), showing the characteristics of A-type granite. Diagram of granite rock type discrimination (Nb-10,000 Ga/Al diagram and Ce-10,000 Ga/Al diagram) (Figure 9). The sample data plot in the A-type granite area. It is shown as A2-type granite in the classification diagram of the A-type granite subtype (Figure 10). Overall, the Sayashk granites are similar to the typical granite in the region from the geochemical characteristics of major elements, REE and trace elements, and so on (Figure 9) [9,30,59,70].

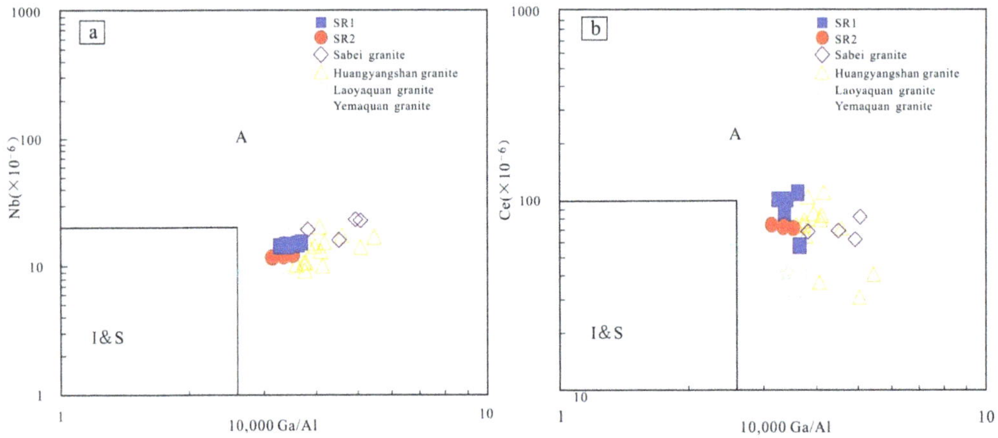

Figure 9. Nb-10,000 Ga/Al diagrams (**a**) and Ce-10,000 Ga/Al diagrams (**b**) for the medium-fine-grained granites and granite porphyry in the Sayashk Sn deposit. (After [86]). Sabei granite after [30], Huangyangshan granite after [75], Laoyaquan granite after [9], Yemaquan granite after [70].

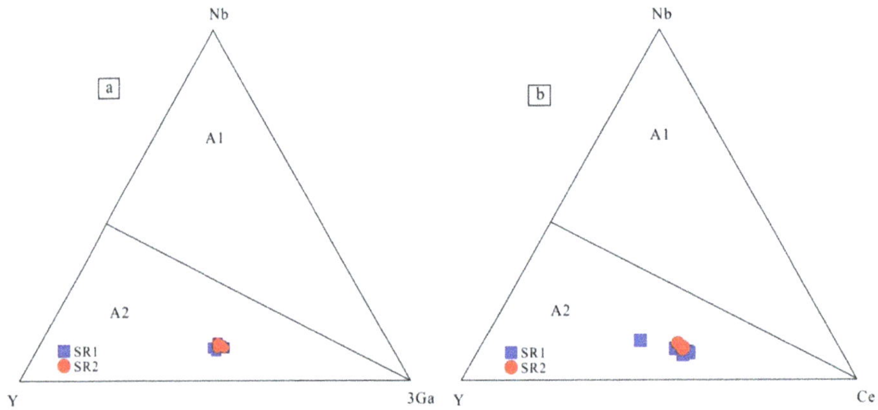

Figure 10. Y–Nb–3 Ga diagrams (**a**) and Y–Nb–Ce diagrams (**b**) for the medium-fine-grained granites and granite porphyry in the Sayashk Sn deposit (modified after [88]).

6.2.2. Petrogenesis and Nature of the Magma Source Area

At present, the main viewpoints on the petrogenetic types of A-type granite are as follows: (1) differentiation or partial melting of mantle-derived magma [89]; (2) mixing and melting of crustal mantle materials [90]; and (3) partial melting of crustal source material [91].

The acidic rocks produced by the crystal differentiation of mantle-derived alkaline basalt magmas show the characteristics of peralkaline, which is contradictory to the metalumious or weakly peraluminous characteristics of the rocks in this paper, and the mantle-derived alkaline basalt magma produced. The temporal and spatial distribution of the acidic rocks is often related to a large number of intermediate-basic magmas, which is inconsistent with the geological facts of the widely developed acidic magmatic rocks in East Junggar, especially in the Huangyangshan area. Even in the entire northern Xinjiang region, the age of basic rocks is concentrated at approximately 280 Ma, indicating that these basic magmas should have ascended and emplaced after the formation of alkaline granites in the East Junggar [92]. So, mantle-derived alkaline basalt magma is not the source rock of A-type granite in this area.

Previous studies showed that the abundance of Th in mantle-derived magma (0.05 ppm) was lower than that in a crustal source (16~21 ppm) [93]. SR2 and SR1 were 9.93~11.23 ppm and 11.76~11.23 ppm, respectively), which are larger than those in the mantle and close to those in the crust. The Nb/Ta values are 12.60~14.11 and 13.39~15.26, which are much lower than those in the mantle (Nb/Ta = 60) and closer to those in the crust (Nb/Ta = 11) [94], showing the characteristics of a mixed source in the crust and mantle.

Zircon has a strong closed system with a high Hf mass fraction and an extremely low $^{176}Lu/^{177}Hf$ value, and there is basically no radiogenic Hf accumulation after the system is closed. Therefore, zircon in situ Hf isotope research has become an important means to explain the crustal evolution and trace the magma source area.

Research shows that $\varepsilon Hf(t) > 0$ of zircon indicates that the magma came from the depleted mantle or partial melting of young crust that did not regenerate from the depleted mantle [95,96]. When ε Hf(t) is close to the current mantle value, ε Hf(t) < 0 of zircon from granite indicates that the magma originated from the ancient crust and remelted. Zircon from two granites in the Sayashk Sn deposit εHf(t) values of SR1 range from 10.27 to 16.17 (average at 13.71), εHf(t) values of SR2 are between 5.72 and 9.21 (average 7.08), and single model ages (T_{DM1}) and two-stage model ages (T_{DM2}) of SR1 fall within the ranges of 319–535 Ma and 339–644 Ma. Single model ages (T_{DM1}) and two-stage model ages (T_{DM2}) of SR2 fall within the ranges of 346~479 Ma and 309~557 Ma. There is little difference between their two-stage model ages and zircon U–Pb ages, which indicates that the Sayashk granites may have originated from the partial melting of the young crust newly accreted from the depleted mantle. In the t–εHf(t) diagram (Figure 11a), sample points plot above the Hf isotope evolution line of chondrites, characterized by enrichment. In the diagram of t–$^{176}Hf/^{177}Hf$ (Figure 11b), the sample points plot in the region between the chondrite Hf isotope evolution line and the depleted mantle, with enrichment characteristics. It is suggested that the Sayashk granites were the product of partial melting of juvenile crustal.

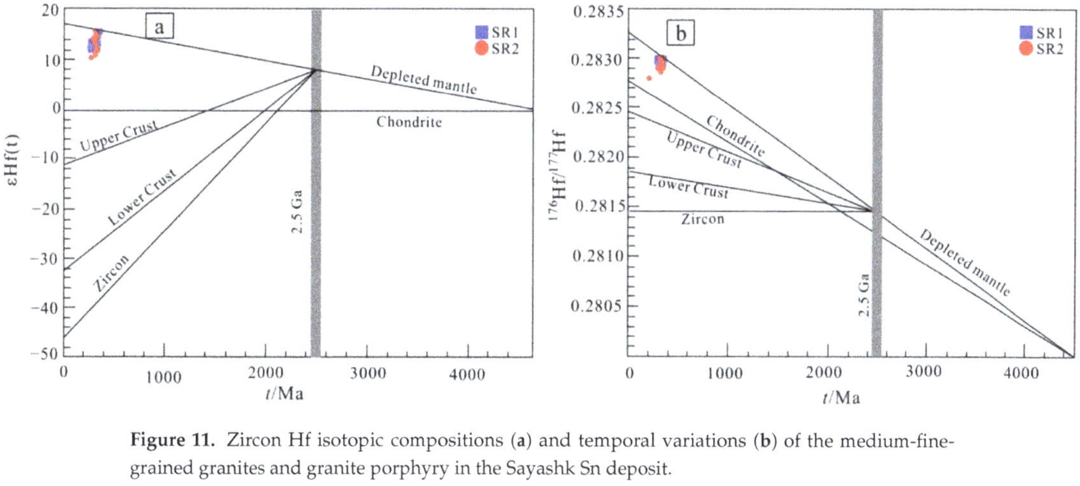

Figure 11. Zircon Hf isotopic compositions (**a**) and temporal variations (**b**) of the medium-fine-grained granites and granite porphyry in the Sayashk Sn deposit.

Previous studies have shown that A-type granites in the eastern Junggar region generally have high εNd(t) values [2,59,97], which is consistent with the conclusion of this paper.

6.3. Tectonic Setting and Metallogenic Model

The Kalamaili area experienced multiple stages of ocean basin expansion, plate subduction, collisional orogeny, post-collision, and post-orogenic processes in the Paleozoic [98,99].

Intense tectonic movement accompanied by violent magmatism provided conditions for the development of gold and copper metallogenic series related to calc-alkaline granites and tin metallogenic series related to alkaline A-type granites [2,7,31,63].

Many studies have been performed on the magmatic rocks in this area according to the evolution of the tectonic environment. Yang et al. [8], believed that the later part of the late Carboniferous Huangyangshan alkaline granite was formed in the tectonic environment of a post-collisional extensional background. Wang [100] studied the early-late Carboniferous volcanic rocks in the Kalamaili area and concluded that the rocks were formed in the transitional environment of a post-collision and post-collision tensile environment. Chen et al. [101] studied the late Carboniferous volcanic rocks in this area and pointed out that the rocks had the dual characteristics of island arc volcanic rocks and a post-collision period. At the same time, they pointed out that the rocks did not form an island arc environment but a post-collision extension environment. The island arc characteristics indicate that magma contamination led to the inheritance of early island arc materials. Luo [102] studied the early late Carboniferous magmatic rocks in this area and showed that they were formed at the end of post-collision, marking the end of orogeny in the Kalamaili area. In conclusion, it is considered that the early-late Carboniferous in the Kalamaili area entered the end of collision activity, and the middle and late part of the late Carboniferous in the Kalamaili area were in a post-collision tension environment.

The Sayashk granitoids were formed during the latter part of the late Carboniferous. In the discrimination diagram of the granite tectonic environment, the granitic rocks in Sayashk are post-collisional granites (Figure 12a,b). Simultaneous SiO_2-FeOT/FeOT+MgO (Figure 13a) and SiO_2-Al_2O_3 diagrams (Figure 13b) diagrams display that the Sayashk granitic rocks are post-orogenic granites. This is consistent with the classification diagram of the A1–A2 subtype granite. This interpretation is also consistent with the consensus that the A-type granites in northern Xinjiang were formed in a post-collisional extensional environment [92].

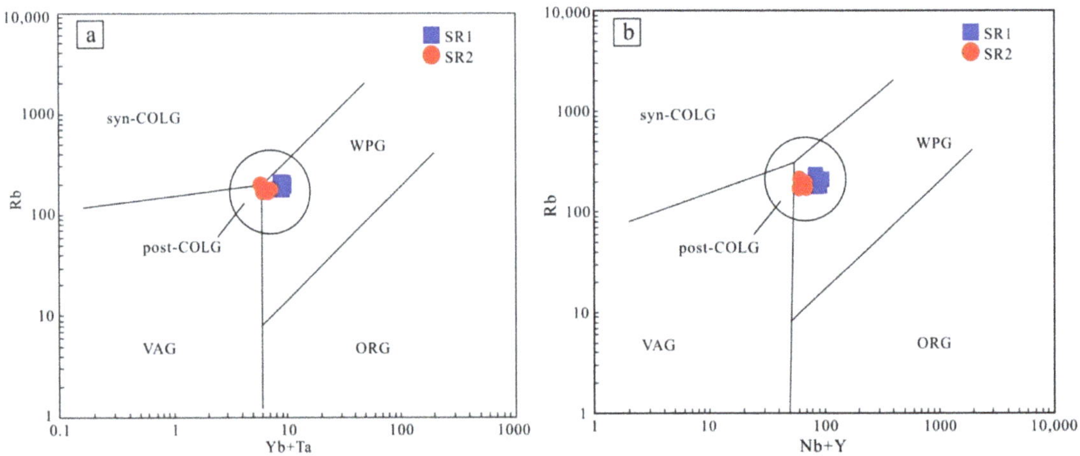

Figure 12. Tectonic environment discrimination diagram: (**a**) Yb-Ta diagram; (**b**) Y + Nb-Rb diagram (base drawing according to [103]). VAG: volcanic arc granite; ORG: ridge granite; WPG: intraplate granite; syn-COLG: collisional granite; post-COLG = post-collisional granites.

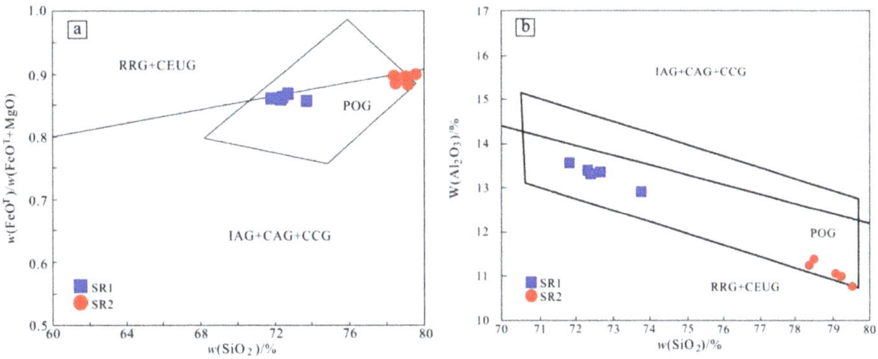

Figure 13. Tectonic environment discrimination diagram: (**a**) Yb–Ta diagram; (**b**) Y+Nb-Rb diagram (base drawing according to [104]). IAG: Island arc granite; CAG: continental arc granite; CCG: continental collision granite; POG: post-orogenic granite; RRG: rift-related granite; CEUG: continental uplift granite; OP: oceanic plagiogranite.

According to previous research results, the Paleozoic Kalamaili area can be roughly divided into five stages (Figure 14): ①: since the early Ordovician, the area has been in an environment of basin expansion [105]; ②: from the early Silurian to the late Silurian, the region formed a paleosubduction zone, the plate subducted southward, a strong fold orogeny occurred, and the uplift suffered denudation [106]; ③: in the early middle of the early Devonian, the Kalamaili ocean basin was opened, and the region was again in an extensional environment of extensional tectonics [99,106]; and ④: from the late part of the early Devonian to the early part of the late Carboniferous, the Kalamaili area changed from a continental margin facies to a mature island arc. At this stage, it was a strongly compressed island arc environment, and some magmatic rocks with a back-arc spreading environment were formed [2,74]. ⑤: After the late Carboniferous, the Kalamaili area entered the post-orogenic stage and once again changed from a compressive environment to an extensional environment [3,7,9,41]. At the same time, it is also the most frequent stage of magmatic activity in this area, forming many representative magmatic rocks, such as the Huangyangshan pluton, Sabei pluton and Kubusunan pluton.

The ore-forming material of the Kalamaili Sn ore belt comes from granite. The ore-forming hydrothermal solution is mainly late magmatic hydrothermal solution, and a small amount of atmospheric water is mixed later.

During the emplacement process of the original Sn-rich magma, with the progress of magma evolution, elements incompatible with early minerals such as Sn were gradually enriched and separated in late residual magmas and hydrothermal fluids.

Late ore-bearing (Sn) hydrothermal fluids with high contents of silicon, alkalis, and volatiles (mainly F and Cl) were gradually separated from the residual magma. The granites on both sides of the fault zone underwent alteration and alkali metasomatization (potassium feldsparization and albite). Under high-temperature conditions, mineralization alteration continues, and sand graining and silicification occur. The Sn element in the ore-bearing hydrothermal fluid continuously precipitated, and finally, a quartz vein-type tin ore body formed in a favorable position. The Sayashk Sn deposit is an Sn deposit formed in an extensional environment after the late Carboniferous.

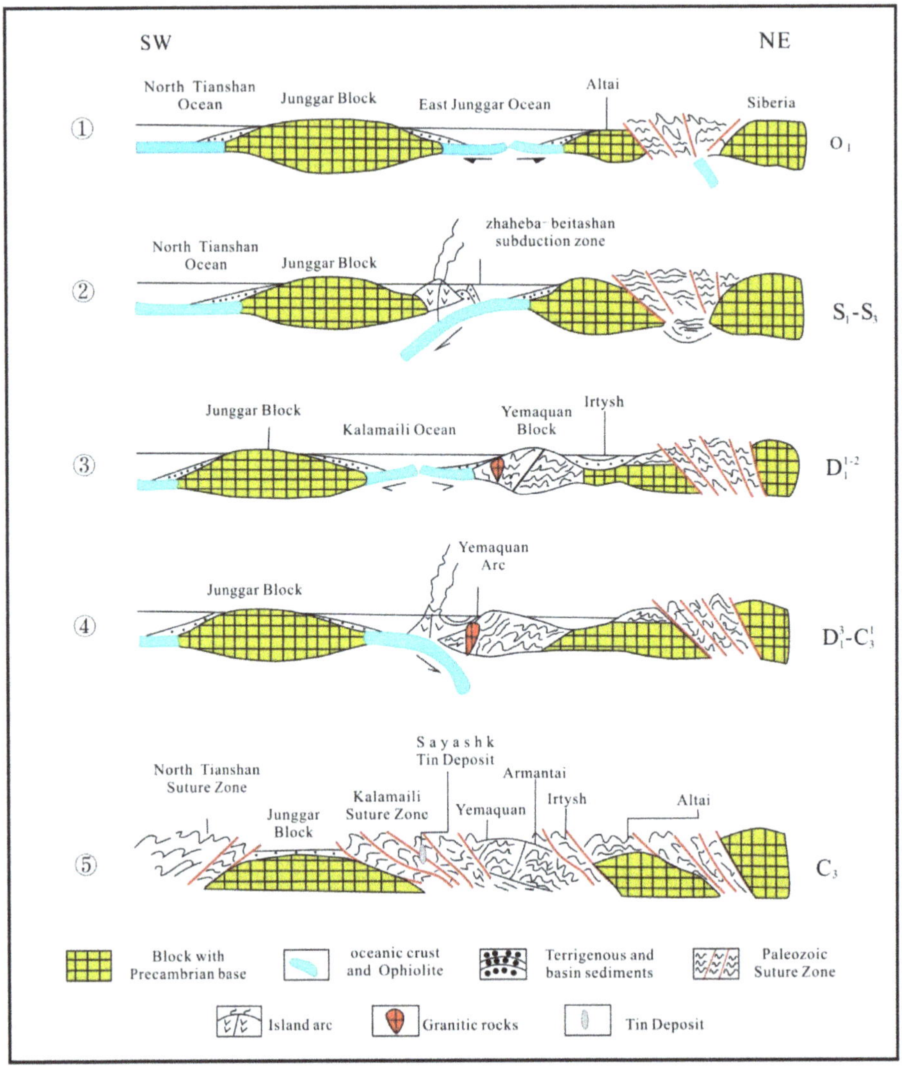

Figure 14. Diagram of Paleozoic tectonic evolution in East Junggar [106].

7. Conclusions

1. The Sayashk Sn deposit is spatially, temporally, and genetically closely related to the granite porphyry and the medium-fine-grained granite. Both zircon U–Pb ages are 308.2 ± 1.5 Ma and 310.9 ± 1.5 Ma, respectively. The molybdenite isochron age is 301.4 ± 6.7 Ma, which represents the metallogenic age of the Sayashk Sn deposit. All of them formed in the late Carboniferous epoch.
2. The medium-fine-grained granites (SR2) and granite porphyry (SR1) are rich in Si, rich alkali, poor Ca, poor Mg, enrichment HFSE (Zr, Hf) and Ce, loss Ba, Sr, Eu, P, Ti, which belongs to typical A-type granite. It shows that the mixed crustal mantle source is derived from the partial melting of juvenile crustal.

3. The Sayashk Sn deposit was formed after the late Carboniferous, and the Kalamaili area entered the post-orogenic stage, which was formed from a compressive environment to an extensional environment.

Author Contributions: Conceptualization, Z.S. and G.L.; methodology, Y.R.; software, X.C.; investigation, Z.S., C.W. and X.S. resources, Z.L.; data curation, X.S.; writing—original draft preparation, Z.S.; writing—review and editing, Y.R. All authors have read and agreed to the published version of the manuscript.

Funding: This research was funded by the Fundamental Research Funds for the Central Universities of China (Grant Number ZY20215112) and the Science and Technology Research Project of Colleges and Universities in Hebei Province (Grant Number ZD2021405).

Institutional Review Board Statement: Not applicable.

Informed Consent Statement: Not applicable.

Data Availability Statement: Not applicable.

Acknowledgments: We sincerely appreciate the detailed and constructive reviews and suggestions from two anonymous reviewers, which greatly improved this paper.

Conflicts of Interest: The authors declare no conflict of interest.

References

1. Han, B.; Wang, S.; Jahn, B.M.; Hong, D.; Kagami, H.; Sun, Y. Depleted-mantlesource for the Ulungur Biver A-type granites from Northinjiang, Chine: Geochemistry and Nd-Sr isotoic evidence, and implications for Phanerozoic crustal growth. *Chem. Geology* **1997**, *138*, 135–159. [CrossRef]
2. Tang, H.; Qu, W.; Su, Y.; Hou, G.; Du, A.; Cong, F. Genetic connection of Sareshike tin deposit with the alkaline A-type granites of Sabei body in Xinjiang: Constraint from isotopic ages. *Acta Petrol. Sinica.* **2007**, *8*, 1989–1997. (In Chinese with English Abstract)
3. Yang, F.; Mao, J.; Yan, S. Ore-forming Age and Ore-formation of the Beilekuduke tin deposit in East Junggar, Xinjiang. *Geogogical Rev.* **2008**, *54*, 626–641. (In Chinese with English Abstract)
4. Nie, X.; Liu, J.; Su, D.; Zhang, X. Zircon U–Pb age of the east Qingshui plagiogranite in Kalamaili belt of Xinjiang and its geological implications. *Geol. China* **2016**, *43*, 1729–1736. (In Chinese with English Abstract)
5. Geng, Y.; Liu, Z.; Huang, S.; He, Z. Geochronology and Geochemistry of the Kalamaili Granitic Rocks and Uranium Ore-forming Potential at the Northeastern Margin of Junggar Basin, China. *J. Earth Sci. Environ.* **2022**, *44*, 20–41. (In Chinese with English Abstract)
6. Tang, S.; Chen, Y.; Liu, Y. Isotope dating of the Beerkuduke Tin deposit in the eastren Junggar area. *Miner. Petrol.* **2006**, *26*, 71–73. (In Chinese with English Abstract)
7. Bi, C.; Shen, X.; Xu, Q.; Ming, K.; Sun, H.; Zhang, C. Geological characteristics of Stanniferous granites in the Beilekuduke tin metallogenic belt, Xinjiang. *Acta Petrol. Miner.* **1993**, *12*, 213–223. (In Chinese with English Abstract)
8. Yang, G.; Li, Y.; Si, G.; Wu, H.; Zhang, Y.; Jin, Z. Trace elements geochemistry feature and tectonic significance of the Beilekuduke aluminous A-type granite in east Junggar. *J. Earth Sci. Environ.* **2010**, *32*, 34–39. (In Chinese with English Abstract)
9. Wang, L.; Wang, J.; Wang, Y.; Long, L.; Tang, P. Study on the geology and ore-fluids of the tin deposits in Laoyaquan alkaline granites in eastern Junggar, Xinjiang. *Acta Petrol. Sin.* **2011**, *27*, 1483–1492. (In Chinese with English Abstract) [CrossRef]
10. Han, Y.; Tang, H.; Gan, L. Zircon U-Pb ages and geochemical characteristics of the Laoyaquan A-type granites in east Junggar, North Xinjiang, China. *Acta Miner. Sin.* **2012**, *32*, 193–199. (In Chinese with English Abstract)
11. Chen, Y.; Wang, Y.; Wang, J.; Wang, L.; Tang, P.; Shi, Y.; Zhao, L. Greisenized alteration-mineralization geochemistry of the tin deposit related to A-Type granite: Case study on the Kamusite and Ganliangzi deposits, Xinjiang. *Earth Sci.* **2018**, *43*, 3154–3168. (In Chinese with English Abstract)
12. Mao, J.; Zhang, S.; Rossi, P. The tin-granites and their relation to mineralization in Tengchong, Yunnan. *Acta Petrol. Sin.* **1987**, *4*, 32–43. (In Chinese with English Abstract)
13. Schwatt, M.O.; Rajah, S.S.; ASkury, A.K.; Putthapiban, P.; Djaswadi, S. The Southeast Asian tin belt. *Earth-Sci. Rev.* **1995**, *38*, 95–293.
14. Mlynarczyk, N.S.J.; Williams-Jones, A.E. The role of collisional tectonics in the metallogeny of the Central Andean tin belt. *Earth Planet. Sci. Lett.* **2005**, *240*, 656–667. [CrossRef]
15. Lehmann, B. *Metallogeny of Tin*; Springer: Berlin/Heidelberg, Germany; New York, NY, USA, 1990; pp. 1–211.
16. Hua, R.; Zhang, W.; Gu, S.; Chen, P. Comparison between REE granite and W-Sn granite in the Nanlingregion, South China, and their mineralizations. *Acta Petrol. Sin.* **2007**, *23*, 2321–2328. (In Chinese with English Abstract)
17. Gonevchuk, V.G.; Gonevchuk, G.A.; Korostelev, P.G.; Semenyak, B.I.; Seltmann, R. Tin deposits of the Sikhote-Alin and adjacent areas (Russian Far East) and their magmatic association. *Aust. J. Earth Sci.* **2010**, *57*, 777–802. [CrossRef]

18. Zhu, X.; Liu, X.; Huang, X.; Jiang, B.; Sun, Y. Characteristics of ore-controlling structures of vein-type Sn polymetallic deposit in the southern area of the Da Hinggan Mountains, Inner Mongolia. *Miner. Explor.* **2017**, *8*, 919–926. (In Chinese with English Abstract)
19. Qiu, Z.; Yan, Q.; Li, S.; Wang, H.; Tong, L.; Zhang, R.; Wei, X.; Li, P.; Wang, L.; Bu, A.; et al. Highly fractionated Early Cretaceous I-type granites and related Sn polymetallic mineralization in the Jinkeng deposit, eastern Guangdong, SE China: Constraints from geochronology, geochemistry, and Hf isotopes. *Ore Geol. Rev.* **2017**, *88*, 718–773. [CrossRef]
20. Heinrich, C.A. The chemistry of hydrothermal tin (-tungsten) ore deposition. *Econ. Geol.* **1990**, *85*, 457–481. [CrossRef]
21. Hua, R.; Chen, P.; Zhang, W.; Liu, X.; Lu, J.; Lin, J.; Yao, J.; Qi, H.; Zhang, Z.; Gu, S. Metallogenic systems related to Mesozoic and Cenozoic granitoids in South China. *Sci. China (Ser. D)* **2003**, *4*, 816–829. [CrossRef]
22. Zhou, Z.; Lv, L.; Yang, Y.; Li, T. Petrogenesis of the Early Cretaceous A type granite in the Huanggang SnFe depoislnner Mongolia: Constraints from zircon U-Pb dating and geochemistry. *Acta Petrol. Sin.* **2010**, *26*, 3521–3537. (In Chinese with English Abstract)
23. Samuel, N.W.P.; Chung, S.L.; Robb, L.J.; Searle, M.P. Petrogenesis of Malaysian Granitoids in the Southeast Asian Tin Belt: Part 1. Geochemical and Sr-Nd Isotopic Characteristics. *Geol. Soc. Am. Bull.* **2015**, *127*, 1209–1237.
24. Romer, R.L.; Kroner, U. Phanerozoic tin and tungsten mineralization: Tectonic controls on the distribution of enriched protoliths and heat sources for crustal melting. *Gondwana Res.* **2016**, *31*, 60–95. [CrossRef]
25. Yao, L.; Lv, Z.; Ye, T.; Pang, Z.; Jia, H.; Zhang, Z.; Wu, Y.; Li, R. Zircon U-Pb age, geochemical and Nd-Hfisotopic characteristics of quartz porphyry in the Baiyinchagan Sn polymetallic deposit, Inner Mongolia, southern Greatingan Range, China. *Acta Petrol. Sin.* **2017**, *33*, 3183–3199. (In Chinese with English Abstract)
26. Li, H.; Myint, A.Z.; Yonezu, K.; Watanabe, K.; Algeo, T.J.; Wu, J.H. Geochemistry and U-Pb geochronology of the Wagone and Hermyingyi A-Type granites, southern Myanmar: Implications for tectonic setting, magma evolution and Sn-W mineralization. *Ore Geol. Rev.* **2018**, *95*, 575–592. [CrossRef]
27. Yuan, S.; Williams, J.A.E.; Romer, R.L.; Zhao, P.; Mao, J. Protolith-related thermal controls on the decoupling of Sn and W in Sn-W metallogenic provinces: Sights from the Nanling Region, China. *Econ. Geol.* **2019**, *114*, 1005–1012. [CrossRef]
28. Sun, H.; Lv, Z.; Han, Z.; Du, Z.; Zhang, X.; Wang, H. Genesis and geological significance of Late Jurassichigh-B ore-bearing A-type granite in the Dayishan tin deposit, Hunan Province. *Acta Petrol. Sin.* **2021**, *37*, 1749–1764. (In Chinese with English Abstract)
29. Lehmann, B. Formation of tin ore deposits: A reassessment. *Lithos* **2021**, *402/403*, 105756. [CrossRef]
30. Lin, J.; Yu, H.; Wu, C.; Su, W.; Guo, J. Zircon SHRIMP U-Pb dating and geological implication of the Sabei Tin ore-deposit from Eastern Junggar of Xinjiang, China. *Geol. China* **2008**, *35*, 1197–1205. (In Chinese with English Abstract)
31. Lin, J.; Yu, H.; Yu, X.; Di, Y.; Tian, J. Zircon SHRIMP U-Pb Dating and geological implication of the Sabei allkall-rich sramite kromEanternJumgear ofinjiang, NW China. *Acta Petrol. Sin.* **2007**, *23*, 876–1884. (In Chinese with English Abstract)
32. Yu, H.; Xia, B.; Liu, J.; Wu, G. Collisional orogenesis and granitoids and related metallogenic series from eastern Junggar. *J. Guilin Inst. Technol.* **2000**, *3*, 213–219. (In Chinese with English Abstract)
33. Yu, H.; Xia, B.; Liu, J.; Wu, G. Crustobody tectonic evolution and tectonic—Genetic types of granitoids in eastern Junggar, Xinjiang, China. *J. Guilin Inst. Technol.* **2001**, *3*, 213–219. (In Chinese with English Abstract)
34. Zhang, Y.; Zhu, M.; Tian, H.; Ming, K.; Liu, X.; Shen, X.; Ma, W. *Geology of East Junggar and Research on Gold and Tin Minerals*; Beijing Earthquake Press: Beijing, China, 1992; pp. 162–175. (In Chinese)
35. Şengör, A.M.C.; Natal'in, B.A. Turkic-type orogenyand its role in the making of the continental crust. *Annu. Rev. Earth Planet Sci.* **1996**, *4*, 263–337. [CrossRef]
36. Wu, C.; Zhang, Z.; Zaw, R.; Dela-Pasque, I.; Tang, J.; Zheng, Y.; Wang, C.; San, J. Geochronology, geochemistry and tectonic significances of the Hongyuntan granitoids in the Qoltag area, Eastern Tianshan. *Acta Petrol. Sin.* **2006**, *22*, 1121–1134. (In Chinese with English Abstract)
37. Xiao, W.; Zhao, G. Tectonicsand metallogeny of the orogenic collagesin Centraland East Asid: Preface. *Gondwana Res.* **2017**, *43*, 1–3. [CrossRef]
38. Zhang, X.; Fan, W.; Li, Z.; Chen, Z.; Chen, J. The discovery of asuperlarge magmatic graphite deposit in Huangyangshan area, Qitai County, Xinjiang. *China Geol.* **2017**, *44*, 1033–1034. (In Chinese with English Abstract)
39. Sun, X.; Ren, Y.; Sun, Z.; Wang, C.; Li, Z. Geochronology and geochemical properties of the large-scale graphite mineralization associated with the Huangyangshan alkaline pluton, Eastern Junggar, Xinjiang, NW China. *Geochemistry* **2021**, *81*, 125820. [CrossRef]
40. Sun, X.; Ren, Y.; Sun, Z.; Li, Z.; Wang, C. Characteristics, ore sources and genesis of Huangyangshan graphite deposit in Qitai County, Xinjiang. *Acta Petrol. Sin.* **2021**, *37*, 1867–1882. (In Chinese with English Abstract)
41. Chen, F.; Li, H.; Cai, H.; Liu, H.; Chang, H. Chronology and origin of the Ganliangzi tin orefield, Xinjiang. *Miner. Depos.* **1990**, *18*, 91–97. (In Chinese with English Abstract)
42. Xu, Q.; Zhao, L.; Niu, B.; Zheng, R.; Yang, Y.; Liu, J. Early Paleozoic arc magmatism in the Kalamaili orogenic belt, Northern Xinjiang, NW China: Implications for the tectonic evolution of the East Junggar terrane. *J. Asian Earth Sci.* **2020**, *194*, 104072. [CrossRef]
43. Song, H.; Zhu, G.; Gu, C.; Zhai, M. Orogeny-related activities of Kalamaili fault zone and their indications to the orogenic processes. *Geol. Rev.* **2015**, *61*, 79–94. (In Chinese with English Abstract)
44. Chen, Y.; Wang, Y.; Wang, J.; Ding, R.; Yuan, Y.; Shi, Y. Zircon U-Pb Age, Geochemistry and Geological Implication of the 255 Ma Alkali-Rich Dykes from Ulungur Area, North Xinjiang. *J. Earth Sci.* **2013**, *24*, 519–528. [CrossRef]

45. Liu, Y.; Hu, Z.; Gao, S.; Günther, D.; Xu, J.; Gao, C.; Chen, H. Insitu analysis of major and trace elements of anhydrous minerals by LA-ICP-MS without applying an internal standard. *Chem Geol.* **2008**, *257*, 34–43. [CrossRef]
46. Liu, Y.; Hu, Z.; Zong, K.; Gao, C.; Gao, S.; Xu, J.; Chen, H. Reappraisement and refinement of zircon U-Pb isotope and trace element analyses by LA-ICP-MS. *Chin. Sci. Bull.* **2010**, *55*, 1535–1546. [CrossRef]
47. Ludwig, K.R. *User's Manual for Isoplot 3.6: A Geochronological Toolkit for Microsoft Excel*; Berkeley Geochronology Center: Berkeley CA, USA, 2003; Volume 4, pp. 1–71.
48. Andersen, T. Correction of common Pb in U–Pb analyses that do not report ^{204}Pb. *Chem. Geol.* **2002**, *192*, 59–79. [CrossRef]
49. Wu, F.Y.; Yang, Y.H.; Xie, L.W.; Yang, J.H.; Xu, P. Hf isotopic compositions of the standard zircons and baddeleyites used in U–Pb geochronology. *Chem. Geol.* **2006**, *234*, 105–126. [CrossRef]
50. Chu, N.C.; Taylor, R.N.; Chavagnac, V.; Nesbitt, R.W.; Boella, R.M.; Milton, J.A.; Germain, C.R.; Bayon, G.; Burton, K. Hf isotope ratio analysis using multi-collector inductively coupled plasma mass spectrometry: An evaluation of isobaric interference corrections. *J. Anal. At. Spectrom.* **2002**, *17*, 1567–1574. [CrossRef]
51. Blichert-Toft, J.; Albarede, F. The Lu–Hf geochemistry of chondrites and the evolution of the mantle–crust system. *Earth Planet. Sci. Lett.* **1997**, *148*, 243–258. [CrossRef]
52. Griffin, W.L.; Pearson, N.J.; Belousova, E.; Jackson, S.E.; Van Achterbergh, E.; O'Reilly, S.Y.; Shee, S.R. The Hf isotope composition of cratonic mantle: LAM-MC-ICPMS analysis of zircon megacrysts in kimberlites. *Geochim. Cosmochim. Acta.* **2000**, *64*, 133–147. [CrossRef]
53. Shirey, S.B.; Walker, R.J. Carius tube digestion for low-blank rhenium-osmium analysis. *Anal. Chem.* **1995**, *67*, 2136–2141. [CrossRef]
54. Stein, H.J.; Markey, R.J.; Morgan, M.J.; Du, A.D.; Sun, Y. Highly precise and accurate Re–Os ages for molybdenite from the East Qinling-Dabie molybdenum belt, Shaanxi province, China. *Econ. Geol.* **1997**, *92*, 827–835. [CrossRef]
55. Du, A.; Wu, S.; Sun, D.; Wang, S.; Qu, W.; Markey, R.; Stein, H.; Morgan, J.; Malinovskiy, D. Preparation and certification of Re–Os dating reference materials: Molybdenite HLP and JDC. *Geostand. Geoanalytical Res.* **2004**, *28*, 41–52. [CrossRef]
56. Xie, G.; Mao, J.; Li, R.; Qü, W.; Pirajno, F.; Du, A. Re–Os molybdenite and Ar–Ar phlogopite dating of Cu–Fe–Au–Mo (W) deposits in southeastern Hubei, China. *Miner. Petrol.* **2007**, *90*, 249–270. [CrossRef]
57. Mao, J.; Zhang, Z.; Zhang, Z.; Du, A. Re-Os isotopic dating ofmolybdenite in the Xiaoliugou W(Mo) deposit in the northern QilianMountain and its geological significance. *Geochim. Cosmochim. Acta* **1999**, *63*, 1815–1818.
58. Tao, W.; Guo, L.; Zhou, N.; Li, Y.; Wang, X.; Chen, T.; Bai, J. Zircon U-Pb ages, geochemical characterisitics of the Basikegranodiorite in East Junggar and their tectonic ignificance. *Geol. Rev.* **2002**, *68*, 488–506. (In Chinese with English Abstract)
59. Zhang, Z.; Yan, S.; Chen, B.; Zhou, G.; He, Y.; Chai, F.; He, L.; Wan, Y. SHRIMP U-Pb zircon dating of subducted granites in the northern East Junggar, Xinjiang. *Chin. Sci. Bull.* **2006**, *51*, 1565–1574. (In Chinese)
60. Claesson, S. Isotopic evidence for the Precambrian provenance and Caledonian metamorphism of high grade paragneisses from the Seve Nappes, Scandinavian Caledonides. *Contrib. Miner. Pet.* **1987**, *97*, 196–204.
61. Liu, Q.; Li, H.; Shao, Y.; Girei, M.; Jiang, W.; Yuan, H.; Zhang, X. Age, genesis, and tectonic setting of the Qiushuwan Cu-Mo deposit in East Qinling (Central China): Constraints from Sr-Nd-Hf isotopes, zircon U-Pb and molybdenite Re-Os dating. *Ore Geol. Rev.* **2021**, *132*, 103998.
62. Koschek, G. Origin and significance of the SEM cathodoluminescence from zircon. *J. Microsc.* **1993**, *171*, 223–232.
63. Hoskin, P.W.O.; Schaltegger, U. The composition of zircon and igneous and metamorphic petrogenesis. *Rev. Miner. Geochem.* **2003**, *53*, 27–62.
64. Sun, J.; Zhang, J.; Yang, J.; Yang, Y.; Chen, S. Tracing Magma Mixing and Crystal-Melt Segregation in the Genesis of Syenite with Mafic Enclaves: Evidence from in Situ Zircon Hf-O and Apatite Sr-Nd Isotopes. *Lithos* **2019**, *334/335*, 42–57. [CrossRef]
65. Liu, J.; Yuan, K. A discussion on the genesis and tectonic setting of alkali granites in the Ulunngur alkali-rich granite belt, Xinjiang. *Geol. J. China Univ.* **1996**, *2*, 257–272. (In Chinese with English Abstract)
66. Xiao, X.; Tang, Y.; Feng, Y.; Zhu, B.; Li, J.; Zhao, M. *Tectonic evolution of Northern Xinjiang and its Adjacent Areas*; Beijing Geological Press: Beijing, China, 1992; Volume 11, pp. 1–169. (In Chinese)
67. Han, B.; Ji, J.; Song, B.; Chen, L.; Zhang, L. Late Paleozoic vertical growth of continental crust around theJunggar Basin, Xinjiang, China (Part I): Timing of post-collisional plutonism. *Acta Petrol. Sin.* **2006**, *22*, 1077–1086. (In Chinese with English Abstract)
68. Bai, J.; Chen, J.; Peng, S. Geochronology and geochemistry of ore-bearing intrusions from Huangyangshan magmatic hydrothermal graphite deposit in Qitai County, Xinjiang. *Acta Petrol. Sin.* **2018**, *34*, 2327–2340. (In Chinese with English Abstract)
69. Ai, J.; Lu, X.; Li, Z.; Wu, Y. Genesis of the graphite orbicules in the Huangyangshan graphite deposit, Xinjiang, China: Evidence from geochemical, isotopic and fluid inclusion data. *Ore Geol. Rev.* **2020**, *122*, 103505. [CrossRef]
70. Gan, L.; Tang, H.; Han, Y. Geochronology and geochemical characteristics of the Yemaquan granitic pluton in East Junggar, Xinjiang. *Acta Petrol. Sin.* **2010**, *26*, 2374–2388. (In Chinese with English Abstract)
71. Hu, W. Geochemical characteristics and geological significance of late Carboniferous granites from the Laoyaquan in eastern Junggar, Xinjiang. *J. Lanzhou Univ. Nat. Sci.* **2016**, *52*, 287–294. (In Chinese with English Abstract)
72. Li, Y.; Yang, F.; Zhao, C.; Zhang, Y.; Yan, S.; Dai, J.; Xu, L. SHRIMP U-Pb zircon dating of the Beilekuduk pluton in Xinjiang and its geological implications. *Acta Petrol. Sin.* **2007**, *23*, 2483–2492. (In Chinese with English Abstract)
73. Nie, X. Carboniferous Granitic Magma Activity and Gold Mineralization in the Kalamaily Region, Northern Xinjiang. Ph.D. Thesis, China University of Geosciences, Beijing, China, 2017. (In Chinese with English Abstract)

74. Su, Y.; Yang, H.; Liu, C.; Hou, G.; Liang, L. The determination and a preliminary study of Sujiquan aluminous A-type granites in East Junggar, Xinjiang. *Acta Petrol. Miner.* **2006**, *25*, 175–184. (In Chinese with English Abstract)
75. Su, Y.; Tang, H.; Cong, F. Zircon U-Pb age and petrogenesis of the Huangyangshan alkaline granite body in East Junggar, Xinjiang. *Acta Miner.* **2008**, *2*, 117–126. (In Chinese with English Abstract)
76. Hong, T.; Hollings, P.; Gao, J.; Xu, X.; Mao, Q. Mineralization and petrogenesis of the qiongheba porphyry copper deposit in mengxi district, east Junggar, China. *Ore Geol. Rev.* **2020**, *127*, 103848. [CrossRef]
77. Tong, Y.; Wang, T.; Hong, D.; Han, B.; Zhang, J.; Shi, X.; Wang, C. Spatial and temporal distribution of the Carboniferous-Permian granitoidsin northern Xinjiang and its adjacent areas, and its tectonic significance. *Acta Petrol. Miner.* **2020**, *29*, 619–641. (In Chinese with English Abstract)
78. Zhang, Z.; Xiong, S.; Fan, X. Zircon U-Pb dating, geochemistry and geological significance of synorogenic-granit in Mutanyao area, eastern Junggar. *Geol. Rev.* **2021**, *67*, 231–241.
79. Xu, B.; Lu, Y.; Gu, X.; Zhang, W. Metallogenic epoch of the Shuangquan gold deposit in Qitai area, Xinjiang, China. *Geol. Bull. China* **2009**, *28*, 1871–1884. (In Chinese with English Abstract)
80. Lu, Y.; Zhang, D.; Fan, J.; Zhang, Y.; Zhang, W.; Zhao, J.; Ma, C.; Yang, D. Spatial-temporal coupling of rock-forming and ore-forming processes andtectonic evolution in eastern Junggar Basin, Xinjiang. *Miner. Depos.* **2008**, *27*, 33–41. (In Chinese with English Abstract)
81. Zhang, D.; Lu, Y.; Guo, X.; Fan, J.; Pan, A.; Zhang, Y.; Chao, Y. Au (Cu) mineralization associated with ductile-brittle shear zone in east Junggar area: Conversion of structural deformation, metallogenic time and geodynamics background. *Gold* **2011**, *3*, 8–15. (In Chinese with English Abstract)
82. Lu, Y.; Zhang, Y.; Pan, M.; Liu, Y.; Xu, B.; Chao, Y.; Zhang, D.; Fan, J.; Chen, X.; Pan, A. Zircon U-Pb Dating of Ore-Bearing Granite-Porphyry Vein in Western Huangyangshan Gold Deposit, Eastern Junggar and Its Geological Significance. *J. Jilin Univ. (Earth Sci. Ed.)* **2010**, *40*, 852–858. (In Chinese with English Abstract)
83. Tang, H.; Su, Y.; Qiu, H.; Han, Y. "^{40}Ar-^{39}Ar" age of tin mineralization in the Beilekuduk tin metallogenic belt, East Junggar, Xinjiang (NW China). *Acta Petrol. Sin.* **2009**, *25*, 1303–1309. (In Chinese with English Abstract)
84. Huang, G.; Niu, G.; Wang, X.; Guo, J.; Yu, F. The early Silurian arc magmatic rocks of East Junggar, Xinjiang: Evidences from geochemistry, zircon U-Pb age and Hf isotopes of the Jianggeerkuduke quartz monzobiorite. *Geoscience* **2016**, *30*, 1219–1223. (In Chinese with English Abstract)
85. Wu, F.; Li, X.; Yang, J.; Zheng, Y. Discussions on the petrogenesis of granites. *Acta Petrol. Sin.* **2007**, *23*, 1217–1238. (In Chinese with English Abstract)
86. Whalen, J.B.; Currie, K.L.; Chappell, B.W. A-type granites: Geochemical characteristics, discrimination and petrogenesis. *Contrib. Miner. Petrol.* **1987**, *95*, 407–419. [CrossRef]
87. Chappell, B.W.; White, A.J.R. Two contrasting granite types: 25 years later. *Aust. J. Earth Sci.* **2001**, *48*, 489–500. [CrossRef]
88. Eby, G.N. Chemical subdivision of the A-type granitoids: Petrogenetic and tectonic implication. *Geology* **1992**, *20*, 641–644. [CrossRef]
89. Pearce, J.A.; Harris, N.B.W.; Tindle, A.G. Trace element discrimination diagrams for the tectonic interpretation of granitic rocks. *J. Petrol.* **1984**, *25*, 956–983. [CrossRef]
90. Yang, J.; Wu, F.; Chung, S.; Simon, A.; Chu, M. A Hybrid Origin for the Qianshan A-Type Granite, Northeast China: Geochemical and Sr-Nd-Hf Isotopic Evidence. *Lithos* **2006**, *89*, 89–106. [CrossRef]
91. Collins, W.J.; Beams, S.D.; White, A.J.R.; Chappell, B.W. Nature and Origin of A-Type Granites with Particular Reference to Southeastern Australia. *Contrib. Miner. Petrol.* **1982**, *80*, 189–200. [CrossRef]
92. Guo, F.; Jiang, C.; Lu, R.; Xia, Z.; Ling, J.; Guo, N. Petrogenesis of the Huangyangshan alkali granites in Kalamaili area northern Xinjiang. *Acta Petrol. Sin.* **2010**, *26*, 2357–2373. (In Chinese with English Abstract)
93. Sun, S.S.; McDonough, W.F. Chemical and isotopic systematics of oceanic basalts: Implications for mantle composition and processes. *Geol. Soc. Lond. Spec. Publ.* **1989**, *42*, 313–345. [CrossRef]
94. Green, T.H.; Pearson, N.J. An experimental study of Nb and Ta par-titioning between Ti-rich minerals and silicate liquids at high pressure andtemperature. *Geochim. Cosmochim. Acta* **1987**, *51*, 55–62. [CrossRef]
95. Amelin, Y.; Lee, D.C.; Halliday, A.N.; Pidgeon, R.T. Nature of the Earth's earliest crust from hafnium isotopes in singledetrital zircons. *Nature* **1999**, *399*, 252–255. [CrossRef]
96. Belousova, E.A.; Griffin, W.L.; O'Reilly, S.Y.; Fisher, N.I. Igneous zircon: Trace element composition as an indicator of source rock type. *Contrib. Miner. Petrol.* **2002**, *143*, 602–622. [CrossRef]
97. Hong, D.; Wang, S.; Xie, X.; Zhang, J. Genesis of positive εHf(t) granitoids in the Dahinggan MTS.-Mongolia orogenic belt and growth continental crust. *Earth Sci. Front. Oniversity Ceasci.* **2000**, *2*, 441–456. (In Chinese with English Abstract)
98. Zhao, H.; Liao, Q.; Xiao, D.; Luo, T.; Wang, L.; Yin, T.; Liu, H. Discovery of the Early Silurian alkali basalt andits geological implications in northeastern Junggar, NW China. *Acta Petrol. Sin.* **2018**, *34*, 586–600. (In Chinese with English Abstract)
99. Yang, S.; Liu, G.; Jin, L.; Zheng, H.; Wang, S. Dating and reconstruction of protoliths of the Laojunmiao Group and its indications for basement in eastern Junggar, Xinjiang. *Geol. Bull. China* **2020**, *39*, 7–17. (In Chinese with English Abstract)
100. Wang, Y. Late Palaeozoic Tectonic Evolution of Kalamaili in Eestern Junggar, Xinjiang: Evidences from Isotopic Chronology and Geochemistry. Master's Thesis, Lanzhou University, Lanzhou, China, 2015. (In Chinese with English Abstract)

101. Chen, W.; Guo, G.; Miao, X.; Wang, J.; Hu, W.; Zhao, B.; Fan, L. Geochemical characteristics and tectonic significance of early carboniferous volcanic rocks of Ralamaili, Xinjiang. *Adv. Earth Sci.* **2016**, *31*, 180–191. (In Chinese with English Abstract)
102. Luo, T.; Chen, S.; Liao, Q.; Chen, J.; Hu, Z.; Wang, F.; Tian, J.; Wu, W. Geochronology, geochemistry and geological significance of the Late Carboniferous Bimodal volcanic rocks in the Eastern Junggar. *Earth Sci.* **2016**, *41*, 1845–1862. (In Chinese with English Abstract)
103. Pearce, J.A. Source and settings of granitic rocks. *Episodes* **1996**, *19*, 120–125. [CrossRef]
104. Maniar, P.D.; Piccoli, P.M. Tectonic discrimination of granitoids. *Geol. Soc. Am. Bull.* **1989**, *101*, 635–643. [CrossRef]
105. Li, L. Late Neoproterozoic and Paleozoic tectonic framework and evolution of eastern Xinjiang, China. *Geol. Rev.* **2004**, *50*, 304–322. (In Chinese with English Abstract)
106. Li, Y.; Li, J.; Sun, G.; Zhu, Z.; Song, B. Determination of the Early Devonian granite in East Junggar, Xinjiang, China and its geological implications. *Geol. Bull. China* **2009**, *28*, 1885–1893. (In Chinese with English Abstract)

minerals

MDPI

Article

Geochemical Data Mining by Integrated Multivariate Component Data Analysis: The Heilongjiang Duobaoshan Area (China) Case Study

Zhonghai Zhao [1,*], Kai Qiao [1], Yiwen Liu [1], Jun Chen [1] and Chenglu Li [2]

[1] College of Mining, Liaoning Technical University, Fuxin 123000, China
[2] Heilongjiang Institute of Natural Resources Survey, Harbin 150036, China
* Correspondence: zhaozhonghai@lntu.edu.cn

Abstract: The Heilongjiang Duobaoshan area is located at the confluence of the Great Xing'an Range and the Lesser Xing'an Range, and the area has undergone a complex magmatic and tectonic evolutionary history resulting in a complex and diverse geological background for mineralization. As a result of this geological complexity and the multi-period nature of mineralization, the geochemical data of the area are usually not satisfied with a single statistical distribution form, so traditional statistical methods cannot adequately explore and identify the distribution of deep-seated information in the geochemical data. Based on the above problems, this paper adopts a multivariate component data analysis method to process 14 mass fraction data elements, namely Ag, As, Au, Bi, Cu, Fe, Hg, Mn, Mo, Ni, Pb, Sb, W, and Zn, in the 1:50,000 soil geochemical data from the Duobaoshan area of Heilongjiang. The spatial distribution and internal structural characteristics of raw, logarithmic transformation and isometric logarithmic ratio (ILR) transformed data were compared using exploratory data analysis (EDA); robust principal component analysis (RPCA) was applied to obtain the PC1 and PC2 principal component combinations associated with mineralization, and a spectrum–area (S–A) fractal model was further used to decompose the geochemical anomalies of the PC1 and PC2 principal component combinations as composite anomalies. The results show the following: (i) The data transformed by the isometric logarithmic ratio (ILR) eliminate the influence of the original data closure effect, and the spatial scale of the data is more uniform; the data are approximately normally distributed, based on which RPCA can be applied to better explore the correlation between elements and the pattern of co-associated combinations. (ii) The S–A method was further used to decompose the composite anomalies of the PC1 and PC2 principal component combination in the study area. The anomalous and background fields of the screened-out PC1 and PC2 principal component combinations reflect anomalous information on mineralization dominated by Au mineralization. Moreover, the anomaly and background information after extraction were in good agreement with the known Au deposits (points), and many geochemical anomalies with prospecting potential were obtained in the periphery, providing a theoretical basis and exploration focus for the next step in the searching and exploring of the study area.

Keywords: component data; isometric logarithmic ratio transformation; robust principal component analysis; spectrum–area fractal model; Duobaoshan region

1. Introduction

Geochemical exploration methods have been dominant in mineral exploration and the quantitative prediction of mineral resources. Since the 1970s, geologists in various countries have accumulated a large amount of multi-scale and multi-element geochemical data in the process of mineral exploration [1–7], of which geochemical data processing is particularly important, indispensable and decisive in reconnaissance geochemistry [8–11]. Zuo et al. [12] pointed out how to efficiently process geochemical survey data and emphasized that mining and identifying deep information in the past, present, and future has been a hot and cutting-edge area of research in survey geochemistry. In the current research, several geochemical data processing methods were proposed for spatial pattern recognition and anomaly extraction in the geochemical data models of surveys, such as conventional statistical analysis [13], local Gap statistical methods [14], multivariate statistical analysis [15], data exploration analysis [16], geo-statistics [17], and fractal as well as multifractal methods [18], among which the fractal and multifractal theory of anomaly identification and extraction is a processing method that has been proposed and developed rapidly as well as effectively in recent years.

Mandelbrot [19] created the concept of "fractal geometry" in 1983. This concept has been applied to the analysis of complex phenomena [20,21], followed by several related studies suggesting that the spatial distribution and frequencies of geochemical elements may obey self-similarity in fractal models [22–24]. A series of fractal models has been proposed for the extraction of geochemical anomalies; common fractal and multifractal methods include the following: local singularity methods [25], the concentration–area (C–A) fractal model [26], the concentration–volume (C–V) fractal model [27], the spectrum–area (S–A) fractal model [28], the number of feature spaces–eigenvalues (N–λ) fractal model [29], and the Walsh space counterpart fractal model [30]. These methods not only consider the distribution of the deep-level information of geochemical fields, but also consider the spatial correlation, geometric patterns, and scale invariance, thus enabling the effective decomposition of complex in addition to deep background and superimposed anomalies in composite geochemistry. Cheng [25] pointed out that, by studying and quantitatively analyzing geochemical data with fractal and multifractal methods, weakly retarded geochemical anomalies that are difficult to identify can be extracted from complex geological conditions. The study and quantitative analysis of fractal and multifractal geochemical data can extract information on weakly slowed geochemical anomalies that are difficult to identify from complex geological conditions, thus enabling the understanding of the geochemical element distribution patterns. Currently, fractal and multifractal methods are widely used in exploration and geochemical data processing [15,18,31–33].

With the development of digital earth science, the requirements for the ability to identify geochemical anomalies are gradually increasing. Aitchison [34] proposed the method of log–ratio transformation to improve geochemical anomaly recognition capability, as geochemical data are typically compositional data and the geometric space of compositional data is Aitchison space [35]. In the geochemical data processing methods of surveys, as mentioned above, their geometric space is based on Euclidean space, so the use of log-ratio transformations to transform the component data into the corresponding space can assist in more accurately identifying and decomposing geochemical anomalies [28]. Many studies have shown [32,36,37] that the log-ratio transformation method can reveal the true spatial distribution patterns of elements more effectively, and the elimination of the closure effect of survey geochemical data by log-ratio transformation has gradually become an important step in survey geochemical data processing. The common log-ratio transformation methods include additive log-ratio (ALR) transformation, centered log-ratio (CLR) transformation and isometric log-ratio (ILR) transformation.

The Duobaoshan area in Heilongjiang is located at the northern end of the East Ujimqin Banner–Nenjiang polymetallic metallogenic belt, a shallow cut zone in the low to medium mountains. After recent decades of searching for minerals, several deposits (points) with a high mineralization potential have been identified in and around the area (Figure 1). Due to the geological complexity of the area and the multi-phase nature of mineralization, geochemical data are usually not satisfied with a single statistical distribution form, so traditional statistical methods are not well suited to uncover and identify the distribution of deeper information in geochemical data. Therefore, in this study, 14 elements, Ag, As, Au, Bi, Cu, Fe, Hg, Mn, Mo, Ni, Pb, Sb, W, and Zn in the 1:50,000 soil geochemical data from the Duobaoshan area of Heilongjiang were processed based on multivariate component data analysis, and the spatial distribution as well as the internal structural characteristics of the original data, logarithmic data, and ILR-transformed data were compared using exploratory data analysis (EDA). Robust principal component analysis (RPCA) was applied to obtain the PC1 and PC2 principal component combinations associated with the mineralization of the study area. The spectrum–area(S–A) fractal model was further used to decompose the composite anomalies of the combined PC1 and PC2 principal component geochemical anomalies, thus revealing the true spatial distribution pattern of the geochemical elements in the study area more effectively as well as providing ideas and directions for the further search for minerals in the area.

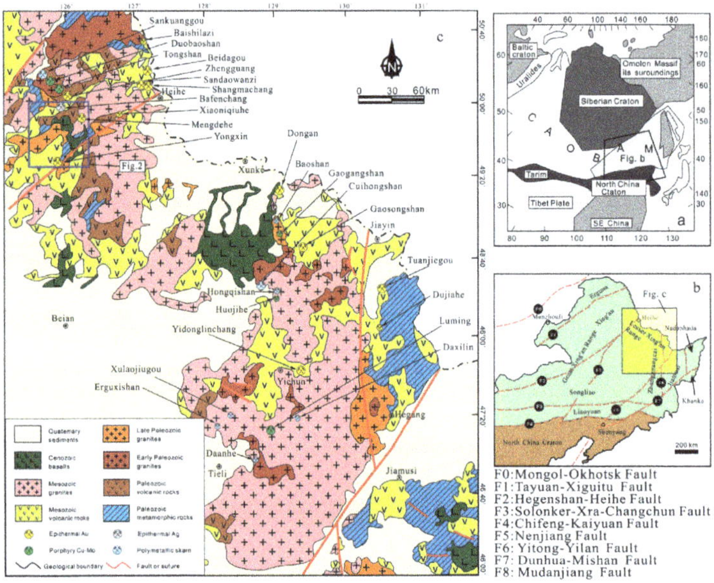

Figure 1. (**a**,**b**) Tectonic divisions of Northeast China and (**c**) a regional geological map of the Lesser Xing'an Range.

2. Geological Profile

2.1. Regional Geological Background

The study area is located at the eastern end of the Central Asian Orogenic Belt (Figure 1a), at the confluence of the Great Xing'an Range and the Lesser Xing'an Range in Heilongjiang (Figure 1b). The northeastern Great Xing'an Range is a superimposed complex tectonic zone that has undergone a long and complex magmatic and tectonic evolution, with a complex series of microplate fits between the Siberian and North China plates in addition to the tectonic evolution and eventual closure of the Paleo-Asian Ocean since the Palaeozoic [38–40]. The Lesser Xing'an Range area is characterized by the development of volcanic, metamorphic, and granitic rocks of different ages, and is the most complex and intense area of tectonic–magmatic evolution in the northeast [41–43]. In recent years, with the continuous exploration of minerals in the region, various large- and medium-sized deposits have been discovered, such as porphyry copper–molybdenum deposits in Tongshan [44] and DuobaoShan [45]; silica-type iron–copper (molybdenum) deposits, such as in Cuihongshan [46] and Xulaojiugou [47]; and shallow-forming low-temperature hydrothermal Au deposits, such as in Zhengguang [48], Sandaowanzi [49], and Tuanjiegou [50] (Figure 1c). This shows that this area has excellent metallogenic potential and a good metallogenic geological background. Up to the present, a Yongxin Au deposit, a Mengdehe Au deposit, a Bafenchang Ag-Au deposit, and many other mineralization points have been found in the study area (Figures 1 and 2).

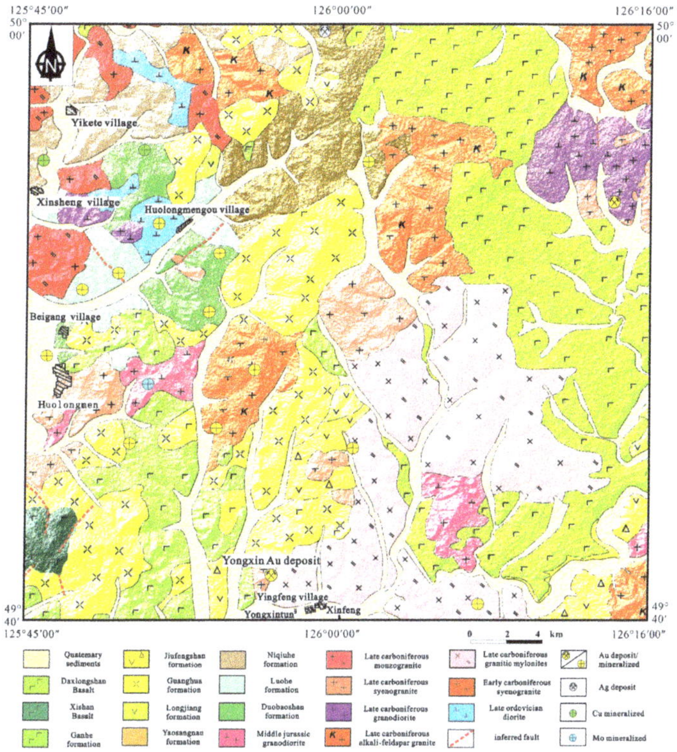

Figure 2. Geological sketch of the study area.

2.2. *Geological Background of the Study Area*

The stratigraphic units exposed in the study area are numerous and widely distributed, mainly including Palaeozoic, Mesozoic, and Cenozoic strata. The Palaeozoic strata are widely distributed in the north-western part of the area, the Mesozoic strata are concentrated in the south-western part of the area, and the Cenozoic strata are mainly distributed in the eastern part of the area. The Palaeozoic strata in the region include the Duobaoshan Formation ($O_{1\text{-}2}d$), the Luohe Formation (O_3l), the Niqiuhe Formation (S_3D_2n), and the Yaosangnan Formation (D_2y); the Mesozoic strata include the Longjiang Formation (K_1l), the Guanghua Formation (K_1gn), the Jiufengshan Formation (K_1j), and the Ganhe Formation (K_1g); and the Cenozoic strata in the region are mainly Quaternary high and low river floodplain deposits (Q*h*) (Figure 2). The intrusive rocks are widely distributed, and the rock types are complex, ranging from neutral to acidic rocks, with granites of medium-to-deep formations predominating. The formation age is in the following order: Middle Ordovician, Carboniferous, Middle Jurassic, and Early Cretaceous. The Palaeozoic Duobaoshan Formation, Mesozoic Guanghua Formation, Longjiang Formation, and Early-to-Late Carboniferous granites are the host rocks for several significant Au and polymetallic ore deposits, including the Zhengguang Au deposit, Sandaowanzi Au deposit, and Yongxin Au deposit [45,51,52]. The alteration of the surrounding rocks of the deposit mainly includes actinolitization, chloritization, epidotization, propylitization, silicification, sericitization, carbonatization, pyritization, etc. Most of the deposits (sites) are located at the intersection of NE-oriented and secondary NW-oriented fracture structures, and hydrothermal alterations such as silicification, pyritization, chloritization, and carbonatization are commonly developed.

3. Methods

3.1. Data Collection and Analysis

The geochemical data are from the 1:50,000 soil geochemical survey in the "Heilongjiang Duobaoshan Area Mineral Vision Survey Project", covering an area of 1334 km^2, with 10,314 soil samples collected. Sample collection strictly adhered to the requirements of the Geochemical Census Specification (1:50,000) (DZ/T0011-2015). The soil samples for this soil survey were collected at 333 × 333 m designed sampling points with a sampling point error of less than 15 m, and sampling was taken point-by-point by line. The sample sampling density was 9 points/km^2 of grid sampling. The sampling layer is the deposition layer (B layer), the sampling depth was 30–80 cm, mostly 40–50 cm, and the sampling media were sand, clay, and subclay. Each sample was delivered at a weight of 150 g and sieved with a −10 to +60 mesh. Samples were processed to 200 mesh by fine-grinding the collected samples according to the requirements specified by the laboratory. To prevent sample contamination, the preparation of samples using a nonpolluting sample grinder and partially by hand-grinding in an agate emulsion bowl, and the processing of the chemical probe samples, needed to be completely separate from the processing of the ore in the laboratory.

This sample was analyzed and tested by the Testing Centre of the Heilongjiang Geological Survey Research Institute and the Testing Centre of the Qiqihaer General Institute of Mineral Exploration and Development. Fourteen elements, Ag, As, Au, Bi, Cu, Fe, Hg, Mn, Mo, Ni, Pb, Sb, W, and Zn, were quantitatively analyzed and tested, and the analytical methods [53] and detection limits of the elements are shown in Table 1.

Table 1. Analysis methods and parameters.

Element	Analysis Method	Detection Limit	Precision (RSD%)
Ag	ES	0.02 mg/kg	5.32
As	AFS	0.20 mg/kg	4.98
Au	GAAS	0.30 mg/kg	4.39
Bi	AFS	0.03 mg/kg	4.71
Cu	XRF	1.00 mg/kg	4.54
Hg	AFS	0.01 mg/kg	6.72
Mn	XRF	5.60 mg/kg	2.49
Mo	ES	0.24 mg/kg	6.99
Ni	XRF	2.80 mg/kg	1.31
Pb	XRF	1.50 mg/kg	3.67
Sb	AFS	0.05 mg/kg	5.37
W	POL	0.31 mg/kg	4.92
Zn	XRF	3.00 mg/kg	2.59
Fe	XRF	0.05 mg/kg	2.22

Note: AFS—atomic fluorescence spectrometry; ES—emission spectrography; GAAS—gallium arsenide; POL—polarography; and XRF—X-ray fluorescence.

3.2. Data Processing

3.2.1. Log-Ratio Transformation and Robust Principal Component Analysis

Since the 1980s, a large number of mathematical geologists have begun to establish methods and theories for compositional data analysis [35,54]. Since geochemical data are distributed in a limited area and obey units as well as constraints, they are typical compositional data. The sum of all elemental contents in compositional data is a constant value known as the "closure effect". The closure effect can lead to pseudo-correlations between geochemical variables, making the results of data processing methods based on correlations between elements uncertain. Most compositional data do not follow the characteristics of a normal distribution; in traditional geochemical studies, log-transformation is often chosen to make them follow the characteristics of a normal distribution. However, the closure effect affects the log-transformed data and cannot obtain mineralogical element combinations with a clear indication of mineralization. Aitchison [35] proposed the ALR and CLR transformations to overcome the effects of closure in compositional data. Egozcues et al. [55] proposed the ILR transformation, which is based on the assumption that the sample space is given Euclidean geometry. The log-ratio transformation addresses the effects of closure in compositional data by transforming the original data (e.g., geochemical data) from the geometry of the compositional data into Euclidean space [7].

Compared with the non-isometric transformations, ALR and CLR, the variables in the Euclidean space will change after the transformation; the ILR transformation can ensure that the relative distance between the variables before and after the transformation of the component data remains unchanged [13], and is more suitable for the processing of component data. However, this method is asymmetric, and the correspondence between the variables will be disrupted after the ILR transformation, so the transformed variables cannot be directly interpreted. Filzmoser proposed a combination of the ILR transformation and robust principal component analysis (RPCA) [56–58], in which the ILR transforms the original data which are then analyzed using the RPCA method to obtain the principal component scores as well as loadings and then converted them into a CLR coordinate system by using a standard orthogonal basis inverse. The data are then transformed into the CLR coordinate system to establish the relationship with the original variables. The problem of mismatched variables after the ILR transformation is thus solved. Moreover, compared with the traditional principal component analysis, RPCA is based on robust statistics, which can suppress the influence of outliers in geochemical data on the results of the principal component analysis by constructing a robust covariance matrix or correlation matrix. The present study involves the ILR transformation formula as shown below:

$$CLR(x) = ln\frac{x_i}{\left\{\prod_{j=1}^{D} x_j\right\}^{\frac{1}{D}}}, i = 1,\ 2,\ \cdots,\ D$$

$$ILR(x) = \sqrt{\frac{i}{i+1}}ln\frac{\sqrt[i]{\prod_{j=1}^{i} x_j}}{x_{i+1}}, i = 1, 2, \cdots, D-1$$

3.2.2. Spectrum–Area Fractal Model

Geochemical data are generally influenced by elements such as mineralization and regional geology, giving them the advantage of complexity and diversity. Cheng et al. [59] proposed a fractal method for decomposing composite and superimposed anomalies developed based on the principle of generalized self-similarity, which is also known as the spectrum–area (S–A) fractal model or fractal filtering method. The geochemical field obeys self-similarity between indices and scales, and specific geological processes or phenomena of spatial relevance usually respond to the fraction with self-similarity. In frequency domain space, the S–A method is based on this self-similarity to construct a fractal filter and invert the fractal-filtered information back into the spatial domain using a Fourier transform transformation to obtain the decomposed background and anomaly maps. The S–A expression is shown below [28]:

$$A(\geq S) \propto S^{-\beta}$$

where S reflects the spectrum density; A(≥S) reflects the area in the spatial region where the energy spectrum density is greater than S; and β is the exponential factor. When the energy spectral density (S) is larger than the spectral density (A(≥S)) in the spatial region obeying the exponential relationship, then they are simultaneously taken logarithmically and plotted on a logarithmic graph. On the plotted lnS-lnA(≥S), different linear and fractal relationships can be reflected according to the differences between the intervals where S is taken and the differences between the straight-line segments while using the intervals. The distribution of the demarcation points is used to determine the threshold of the fractal filter [60,61].

4. Results and Discussion

4.1. Multivariate Component Data Analysis

The elemental analysis results of the 10,314 soil samples collected within the study area were analyzed using SPSS and R software. The geochemical characteristics of the 14 elements were discussed according to their means and standard deviations (Table 2), and the raw data, log-transformed data, and ILR-transformed data were statistically analyzed to identify statistical patterns between the elements. An exploratory data analysis (EDA) method was also used to visualize the data using box plots, density histograms, and double-labeled plots so that the internal structure and dispersion characteristics of the data could be obtained quickly and accurately.

Table 2. Statistics of the raw data, logarithmically transformed data, and isometric log-ratio transformation data of samples from the study area.

Element		Ag	As	Au	Bi	Cu	Hg	Mn	Mo	Ni	Pb	Sb	W	Zn	Fe
Minimum		0.04	1.00	0.10	0.05	3.20	0.01	85.00	0.25	3.10	9.00	0.04	0.27	15.00	0.59
percentiles	25%	0.07	8.30	0.70	0.30	18.90	0.03	658.00	0.80	22.40	23.20	0.49	1.76	62.70	3.57
	50%	0.08	9.80	1.00	0.34	22.50	0.03	972.00	0.96	25.70	25.60	0.57	1.96	73.50	3.96
	75%	0.10	12.10	1.50	0.38	25.90	0.04	1228.00	1.18	29.60	28.40	0.69	2.17	84.60	4.34
Maximum		3.58	151.20	1309.50	23.04	193.50	1.22	6865.00	108.90	194.20	228.60	12.89	50.86	347.00	8.67
Std		0.08	5.35	17.60	0.36	8.57	0.02	455.00	1.50	8.85	5.94	0.30	1.11	18.80	0.63
Mean		0.09	10.70	1.74	0.37	23.40	0.03	977.00	1.11	26.70	26.20	0.62	2.06	75.30	3.93
Raw	Skew	17.77	9.22	66.50	37.28	4.82	22.19	1.42	42.38	5.20	8.55	13.11	22.75	2.32	−0.27
	Kurt	547.06	169.66	4655.68	1966.62	55.15	930.34	9.08	2682.53	68.48	195.55	379.33	765.61	17.65	2.28
	MAD	0.02	2.67	0.59	0.06	5.19	0.01	418.09	0.28	5.34	3.85	0.13	0.31	16.16	0.58
log10	Skew	2.27	0.44	1.41	2.13	0.16	0.64	−0.76	2.04	−0.17	1.17	0.96	1.68	0.16	−1.58
	Kurt	9.72	6.05	7.40	19.70	4.03	5.89	0.60	13.43	5.32	9.64	7.87	19.51	2.63	7.41
	MAD	0.10	0.11	0.23	0.07	0.10	0.13	0.18	0.13	0.09	0.06	0.11	0.07	0.10	0.06
ILR	Skew	1.78	0.50	1.45	2.43	0.66	0.77	−0.73	1.86	0.67	0.12	0.94	1.87	0.41	−0.66
	Kurts	7.32	4.38	8.30	20.50	4.11	5.52	0.42	12.05	6.63	2.30	8.69	15.45	0.85	4.22
	MAD	0.24	0.26	0.49	0.15	0.19	0.28	0.44	0.28	0.17	0.18	0.22	0.16	0.23	0.15

Note: Std—standard deviation. Skew—skewness. Kurt—kurtosis. MAD—median absolute deviation. All of the element content values are expressed in an exponential form but with the exponential part (10^{-9} for Au and 10^{-6} for all the other elements) omitted from the Table for convenience.

The median absolute deviation (MAD) value reflects the median of the absolute value of the new data obtained by subtracting the median from the original data [62]. The mean and standard deviation of the data is susceptible to outliers and is less stable, whereas the MAD based on robust statistics is less affected by outliers, more stable, and more accurate in presenting the data center and the degree of dispersion of the data. In Table 2, it can be seen that the MAD values of the raw data are highly variable in the mean and standard deviation of the elemental content. This result also reflects the difference in the spatial distribution of various elements due to the influence of various geological elements in the study area.

From Table 2 and the box plot (Figure 3a), it can be seen that the skewness and kurtosis of the raw dataset are too high to meet the requirements of a normal distribution; the spatial scale of the raw dataset is large, the data are scattered, and some elements have a large number of high-value discrete points. Compared to the raw dataset, the log-transformed and ILR-transformed data (Table 2 and Figure 3b,c) show substantially less variability in the scale of the distribution of the elements, and the data for each element lie essentially at the same order of magnitude; the elements' skewness and peak state were significantly improved, with the bias and kurtosis of the elements closer to the normal distribution. The density curve (Figure 4) also shows a single-peaked distribution. In contrast, the original data do not show a corresponding density curve due to the large differences in scale; the transformed elemental data are homogeneous, and the data tend to be centered, which is more in line with the requirements of multivariate statistical analysis.

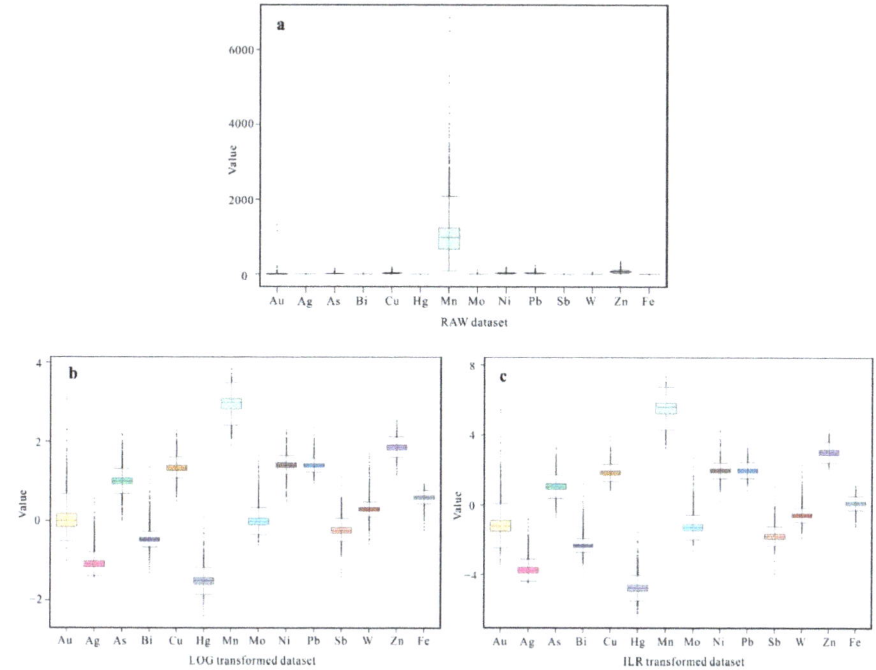

Figure 3. Box plots of (**a**) raw, (**b**) log-transformed, and (**c**) ILR-transformed datasets of metallogenetic elements.

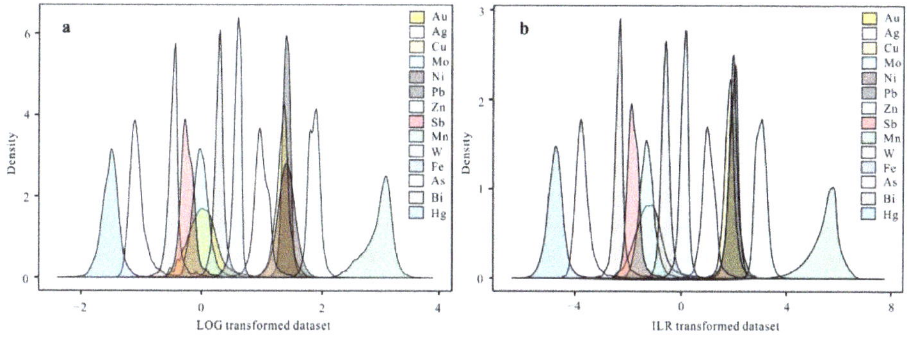

Figure 4. Density curves of (**a**) log-transformed and (**b**) ILR-transformed datasets of metallogenetic elements.

In order to better explore the correlation and co-association patterns among the elements in the study area, the principal component analysis (PCA) results are visualized in this study by combining the label plots of the EDA method. From Figure 5a,b, it can be seen that the original data and the log-transformed data have positive loadings on all of the elements in the PC1 principal component, and no information on the correlation between the elements can be derived. In contrast in the PC2 principal component, the two only show minor differences, and the results of the principal component analysis are bounded by the data closure effect. For the data based on the ILR transformation (Figure 5c,d), after PCA and RPCA, the variables were radioactive, the closure effect was obviously eliminated, and the relationship between the transformed data was much clearer. They also indicate that the principal components obtained by this method are more representative, with Au,

As, Cu, Fe, Hg, Pb, and Sb showing positive loadings in the PC1 principal component and Au, Ag, Mn, and Zn showing positive loadings in the PC2 principal component, with Au being the most discrete element as well as having high loadings in both the PC1 and PC2 principal components, indicating that Au is the main-ore forming metal, and that the PC1 and PC2 principal components may reflect a combination of elements associated with Au mineralization.

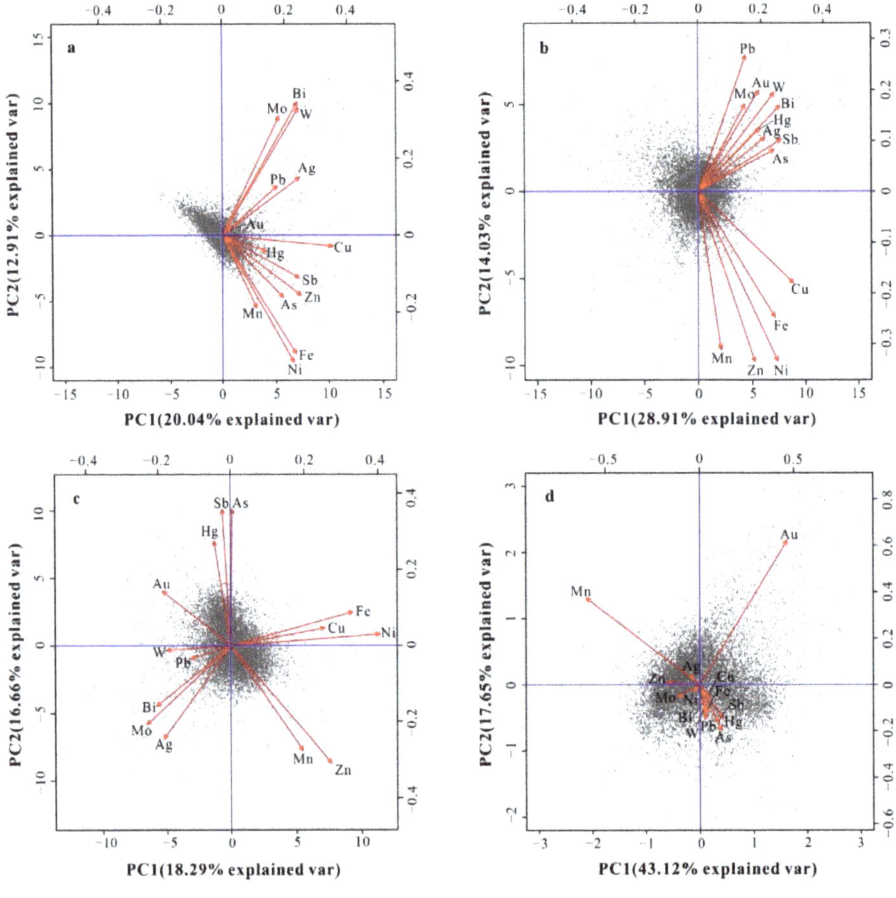

Figure 5. Biplots of the PC1 and PC2 were obtained by the raw dataset with PCA (**a**) and log-transformed dataset with PCA (**b**) and ILR-transformed dataset with PCA (**c**) and ILR-transformed dataset with RPCA (**d**).

Based on the biplot analysis, this study will plot the principal component score point plots for the PC1 and PC2 principal component s' score data obtained from the RPCA using ILR transformed data. According to the principal component score point map (Figure 6), it can be seen that the PC1 principal component scores in the western part of the study area are lower compared to the eastern part. In contrast, the higher scoring areas in the east are concentrated over the Daxiongshan Basalt, indicating that there may be a higher background of influence by the Daxiongshan Basalt in this area. At the same time, this high background generated by the eastern Daxiongshan Basalt may inhibit the identification of weak anomalies in the western part of the study area. In the PC2 principal component score, the high-scoring zones are relatively discrete, mostly overlying Early Cretaceous volcanic–subvolcanic rocks and Carboniferous granites with good coincidence and spatial correlation with known deposits (points), possibly as a result of magmatic–hydrothermal mineralization.

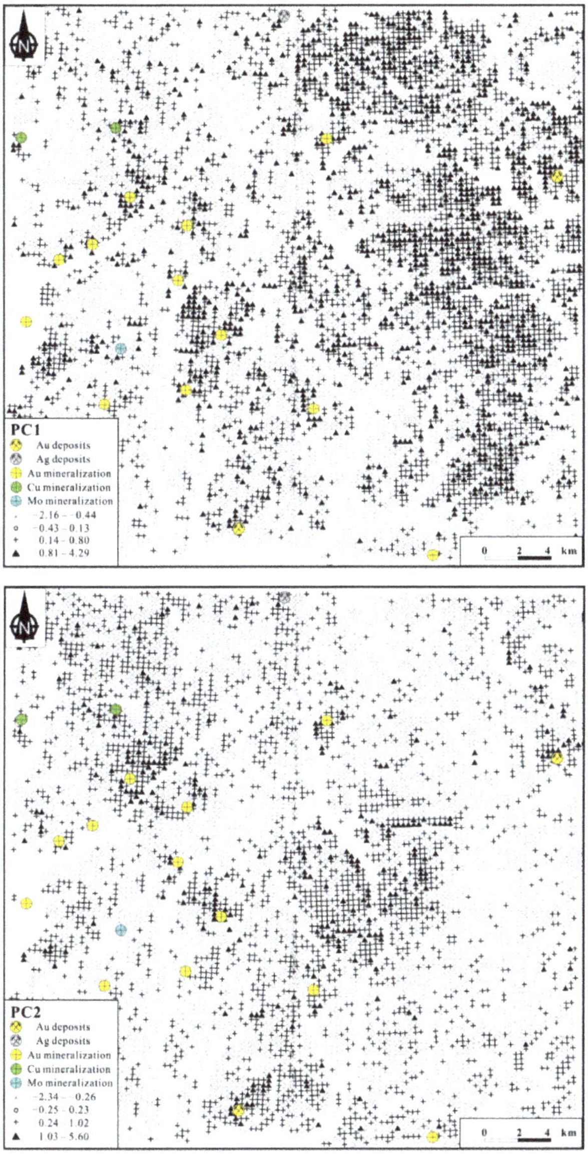

Figure 6. Maps showing the scores of the first component and second component of RPCA.

4.2. *Spectrum–Area Fractal Model Analysis*

In order to eliminate the influence of factors such as mineralization and regional geology and thus more accurately decode the geochemical anomaly information, the PC1 and PC2 principal components' data from the ILR transformation–based RPCA score will be further selected for kriging interpolation processing and S–A decomposition. In this study, the S–A model is implemented with the Geodas quantitative mineral resource prediction system developed by the China University of Geosciences. The data of the PC1 and PC2 principal components' scores were transformed into the frequency domain by Fourier transform, and the relationship between the energy spectral density (S) and the cumulative area (A) was obtained on the plotted lnS-lnA(≥S) curve. According to the variation pattern of both, a line fit based on fewest squares was performed to divide the energy spectral density values of the data into different value intervals according to the slope of the fit (number of sub-dimensions). In the lnS-lnA(≥S) plot of the PC1 principal component (Figure 7a), the line y = −2.09x + 17.10 represents the noise field, the line y = −1.69x + 14.98 represents the anomaly field, and the line y = −1.55x + 14.06 represents the background field. In the lnS-lnA(≥S) plot of the PC2 principal component (Figure 7b), the line y = −2.21x + 17.05 represents the noise field, the line y = −1.76x + 14.86 represents the anomalous field, and the line y = −2.10x + 17.05 represents the background field.

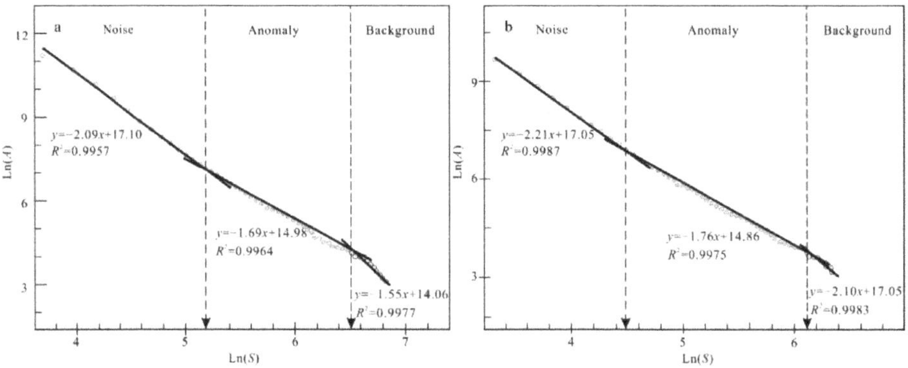

Figure 7. Log-log plots of power spectrum value S versus the area with power spectrum greater than S for principal component of log=ratio dataset of PC1 (**a**) and PC2 (**b**).

The background field obtained by S–A decomposition mainly reflects the background composition of elemental mass fractions; high-background areas may be favorable for polymetallic mineral exploration; variations in the background strength reflect the presence of elements in a favorable geological context for mineralization, and the anomaly field mainly reflects local anomalous mass molecules of elements and noise generated during data processing. Accordingly, the background and anomaly maps corresponding to the PC1 and PC2 principal components of the RPCA score were drawn based on the anomalies and background fields defined above. After decomposition, the background map of the PC1 principal component (Figure 8b) reflects the differences between the east and west of the Duobaoshan area, and, combined with the geological conditions of the study area (Figure 2), it can be seen that the high-background area is located above the Daxiongshan Basalt. In contrast, the known deposits (points) are located above the low-background area. The remaining anomalies of the PC1 principal component (Figure 8a) were obtained after the background anomalies were removed, not only reducing the extent of the anomalies in the eastern part of the study area but also increasing the strength of the local anomalies in the western part of the study area, highlighting the weak anomalous information that is hidden in the low-background area in the western part of the study area. At the same time, the known deposits (points) are located near the high-value areas of the anomalies. Combined with the geological conditions of the study area (Figure 2), the background map of the decomposed PC2 principal component (Figure 8d) shows that the background area is controlled by fractures and intrusive rocks. The high-background anomaly is located above the Late Carboniferous granitic mylonite; the decomposed anomaly (Figure 8c) is closely related to the rocks, and the distribution of the decomposed PC2 principal component anomaly is somewhat similar to that of the PC1 principal component anomaly in the middle- and high-anomaly areas; both are distributed near known Au deposits (points) and have some spatial correlation. This further illustrates the anomalous mineralization information reflected in the PC1 and PC2 principal components, which are dominated by Au mineralization. Regarding the plotted lnS-lnA(≥S), the slope of the fitted curve reflects the different self-similarity characteristics. At the same time, the energy spectrum distribution in the study area is linear, which reflects the fact that the anomalies distributed in the study area are self-similar in the frequency domain and belong to the same fractal distribution, with a high probability of being the products of the same process, further demonstrating that the anomalies and background fields extracted by the S–A method are more consistent with actual geological conditions and can more effectively indicate the location of occult deposits.

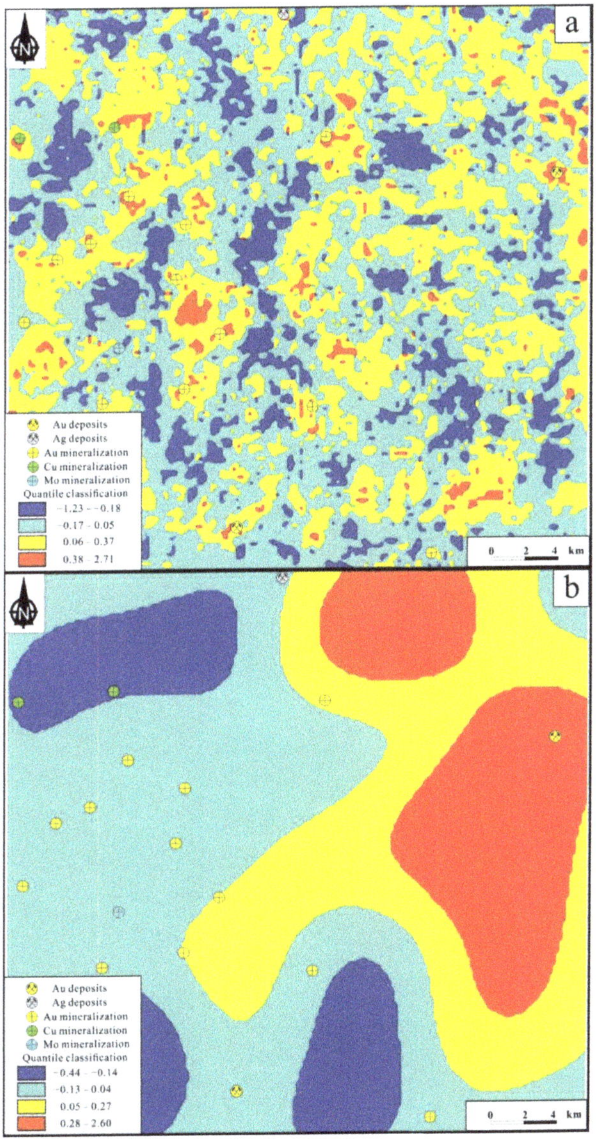

Figure 8. *Cont.*

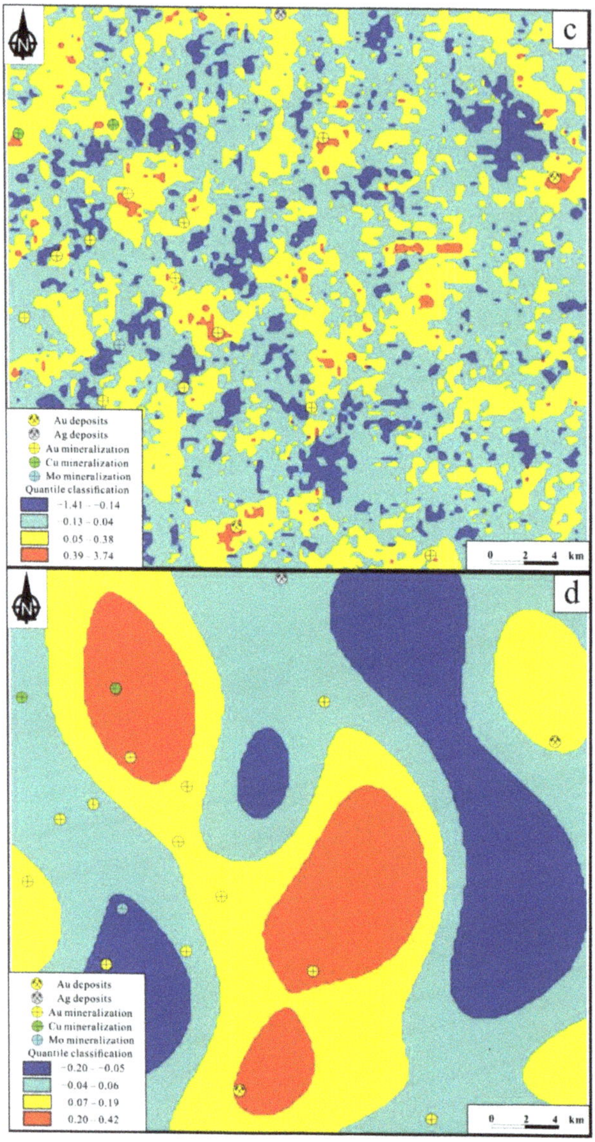

Figure 8. The decomposed maps for the anomaly map (**a**) and background component (**b**) of the PC1 component. The decomposed maps for the anomaly map (**c**) and background component (**d**) of the PC2 component.

5. Conclusions

1. Geochemical data are typical compositional data with a closure effect. Before the data can be statistically analysed, an ILR-transformed of the data is required. This method can effectively eliminate closure effects in geochemical data while revealing the true spatial distribution pattern of elements.
2. The PC1 and PC2 principal components associated with mineralization were obtained by robust principal component analysis of the ILR-transformed data from the study

area. The PC1 and PC2 principal components reflect a combination of elements associated with Au mineralization.

3. The S–A method takes into account the spatial geometry and frequency distribution of geochemical patterns. It provides an effective means for characterizing geochemical anomaly fields and decomposing diverse geochemical fields.
4. The S–A method was used to decompose the composite anomalies of the PC1 and PC2 principal component combinations in the study area, and the decomposed anomalies and background information were in good agreement with the known Au deposits (points). At the same time, a number of geochemical anomalies with prospecting potential were obtained in their periphery, which provided a theoretical basis and exploration focus for the next instance of ore prospecting and exploration in the study area.

Author Contributions: Conceptualization, K.Q. and Z.Z.; methodology, K.Q. and Z.Z.; software, K.Q. and Y.L.; validation, Z.Z., K.Q. and J.C.; formal analysis, K.Q. and Y.L.; investigation, C.L.; resources, Z.Z. and C.L.; data curation, K.Q.; writing—original draft preparation, K.Q.; writing—review and editing, K.Q. and Z.Z.; visualization, K.Q., Y.L. and J.C.; supervision, K.Q. and Z.Z.; project administration, Z.Z.; funding acquisition, Z.Z. All authors have read and agreed to the published version of the manuscript.

Funding: This work was supported by the Project of the Natural Science Foundation of Liaoning Province (2020-BS-258), and the Scientific research fund project of the educational department of Liaoning Provincial (LJ2020JCL010). The project was supported by the discipline innovation team of Liaoning Technical University (LNTU20TD-14) and the Key Research and Development Project of Heilongjiang Province (GA21A204).

Acknowledgments: Thanks to the chief editor and reviewers for their review and constructive comments, which have played a great role in the improvement of this paper.

Conflicts of Interest: The authors declare no conflict of interest.

References

1. Rugless, C.S. Lithogeochemistry of Wainaleka Cu-Zn volcanogenic deposit, Viti Levu, Fiji, and possible applications for exploration in tropical terrains. *J. Geochem. Explor.* **1983**, *19*, 563–586. [CrossRef]
2. Zhu, B.Q.; Zhang, J.M.; Zhu, L.X.; Zheng, Y.X. Mercury, arsenic, antimony, bismuth and boron as geochemical indicators for geothermal areas. *J. Geochem. Explor.* **1986**, *25*, 379–388.
3. Li, Y.G.; Cheng, H.X.; Yu, X.D.; Xu, W.S. Geochemical exploration for concealed nickel-copper deposits. *J. Geochem. Explor.* **1995**, *55*, 309–320. [CrossRef]
4. Reimann, C.; Filzmoser, P. Normal and lognormal data distribution in geochemistry: Death of a myth. Consequences for the statistical treatment of geochemical and environmental data. *Environ. Geol.* **2000**, *39*, 1001–1014. [CrossRef]
5. Cox, M.A.; Cox, T.F. Multidimensional scaling. In *Handbook of Data Visualization*; Springer: Berlin/Heidelberg, Germany, 2008; pp. 315–347.
6. Xiao, F.; Chen, J.G.; Hou, W.S.; Wang, Z.H. Identification and extraction of Ag-Au mineralization associated geochemical anomaly in Pangxitong district, southern part of the Qinzhou-Hangzhou Metallogenic Belt, China. *Acta Petrol. Sin.* **2017**, *33*, 779–790. (In Chinese with English abstract)
7. Wang, L.; Liu, B.; McKinley, J.M.; Cooper, M.R.; Li, C.; Kong, Y.; Shan, M. Compositional data analysis of regional geochemical data in the Lhasa area of Tibet, China. *Appl. Geochem.* **2021**, *135*, 105108. [CrossRef]
8. Shao, Y.; Liu, J.M. A geochemical method for the exploration of kimberlite. *J. Geochem. Explor.* **1989**, *33*, 185–194.
9. Grunsky, E.C. The interpretation of geochemical survey data. *Geochem. Explor. Environ. Anal.* **2010**, *10*, 27–74. [CrossRef]
10. Zuo, R.G.; Carranza, E.J.M.; Wang, J. Spatial analysis and visualization of exploration geochemical data. *Earth-Sci. Rev.* **2016**, *158*, 9–18. [CrossRef]
11. Zhao, P.D.; Chen, Y.Q. Digital geology and quantitative mineral exploration. *Earth Sci. Front.* **2021**, *28*, 1–5, (In Chinese with English abstract)
12. Zuo, R.G.; Wang, J.; Xiong, Y.H.; Wang, Z.Y. The processing methods of geochemical exploration data: Past, present, and future. *Appl. Geochem.* **2021**, *132*, 105072. [CrossRef]
13. Reimann, C.; Filzmoser, P.; Garrett, R.G.; Dutter, R. *Statistical Data Analysis Explained: Applied Environmental Statistics with R*; Wiley: Chichester, UK, 2008; p. 343.
14. Miesch, A.T. Estimation of the geochemical threshold and its statistical significance. *J. Geochem. Explor.* **1981**, *16*, 49–76. [CrossRef]

15. Iwamori, H.; Yoshida, K.; Nakamura, H.; Kuwatani, T.; Hamada, M.; Haraguchi, S.; Ueki, K. Classification of geochemical data based on multivariate statistical analyses: Complementary roles of cluster, principal component, and independent component analyses. *Geochem. Geophys. Geosystems* **2017**, *18*, 994–1012. [CrossRef]
16. Zheng, C.J.; Liu, P.F.; Luo, X.R.; Wen, M.L.; Huang, W.B.; Liu, G.; Wu, X.G.; Chen, Z.S.; Albanese, S. Application of compositional data analysis in geochemical exploration for concealed deposits: A case study of Ashele copper-zinc deposit, Xinjiang, China. *Appl. Geochem.* **2021**, *130*, 104997. [CrossRef]
17. Nazarpour, A.; Omran, N.R.; Paydar, G.R.; Sadeghi, B.; Matroud, F.; Nejad, A.M. Application of classical statistics, logratio transformation and multifractal approaches to delineate geochemical anomalies in the Zarshuran gold district, NW Iran. *Geochemistry* **2015**, *75*, 117–132. [CrossRef]
18. Parsa, M.; Maghsoudi, A.; Ghezelbash, R. Decomposition of anomaly patterns of multi-element geochemical signatures in Ahar area, NW Iran: A comparison of U-spatial statistics and fractal models. *Arab. J. Geosci.* **2016**, *9*, 1–16. [CrossRef]
19. Mandelbrot, B.B. *The Fractal Geometry of Nature*; WH Freeman: New York, NY, USA, 1983; pp. 17–179.
20. Bølviken, B.; Stokke, P.R.; Feder, J.; Jössang, T. The fractal nature of geochemical landscapes. *J. Geochem. Explor.* **1992**, *43*, 91–109. [CrossRef]
21. Allegre, C.J.; Lewin, E. Scaling laws and geochemical distributions. *Earth Planet. Sci. Lett.* **1995**, *132*, 1–13. [CrossRef]
22. Cheng, Q.M.; Agterberg, F.P.; Ballantyne, S.B. The separation of geochemical anomalies from background by fractal methods. *J. Geochem. Explor.* **1994**, *51*, 109–130. [CrossRef]
23. Cheng, Q.M. Multifractality and spatial statistics. *Comput. Geosci.* **1999**, *25*, 949–961. [CrossRef]
24. Xie, S.; Bao, Z. Fractal and multifractal properties of geochemical fields. *Math. Geol.* **2004**, *36*, 847–864. [CrossRef]
25. Cheng, Q.M. Mapping singularities with stream sediment geochemical data for prediction of undiscovered mineral deposits in Gejiu, Yunnan Province, China. *Ore Geol. Rev.* **2007**, *32*, 314–324. [CrossRef]
26. Ghasemzadeh, S.; Maghsoudi, A.; Yousefi, M.; Mihalasky, M.J. Stream sediment geochemical data analysis for district-scale mineral exploration targeting: Measuring the performance of the spatial U-statistic and CA fractal modeling. *Ore Geol. Rev.* **2019**, *113*, 103115. [CrossRef]
27. Afzal, P.; Alghalandis, Y.F.; Khakzad, A.; Moarefvand, P.; Omran, N.R. Delineation of mineralization zones in porphyry Cu deposits by fractal concentration–volume modeling. *J. Geochem. Explor.* **2011**, *108*, 220–232. [CrossRef]
28. Zuo, R.G.; Xia, Q.L.; Zhang, D.J. A comparison study of the C–A and S–A models with singularity analysis to identify geochemical anomalies in covered areas. *Appl. Geochem.* **2013**, *33*, 165–172. [CrossRef]
29. Cheng, Q.M. Multifractal distribution of eigenvalues and eigenvectors from 2D multiplicative cascade multifractal fields. *Math. Geol.* **2005**, *37*, 915–927. [CrossRef]
30. Chen, G.X.; Cheng, Q.M. Singularity analysis based on wavelet transform of fractal measures for identifying geochemical anomaly in mineral exploration. *Comput. Geosci.* **2016**, *87*, 56–66. [CrossRef]
31. Daya, A.A.; Afzal, P. A comparative study of concentration-area (CA) and spectrum-area (SA) fractal models for separating geochemical anomalies in Shorabhaji region, NW Iran. *Arab. J. Geosci.* **2015**, *8*, 8263–8275. [CrossRef]
32. Cicchella, D.; Ambrosino, M.; Gramazio, A.; Coraggio, F.; Musto, M.A.; Caputi, A.; Avagliano, D.; Albanese, S. Using multivariate compositional data analysis (CoDA) and clustering to establish geochemical backgrounds in stream sediments of an onshore oil deposits area. The Agri River basin (Italy) case study. *J. Geochem. Explor.* **2022**, *238*, 107012. [CrossRef]
33. Zhao, Z.H.; Chen, J.; Qiao, K.; Cui, X.M.; Liang, S.S.; Li, C.L. Remote Sensing Alteration Information and Structure Analysis Based on Fractal Theory: A Case Study of Duobaoshan Area of Heilongjiang Province. *Geoscience* **2022**, *19*, 1–16, (In Chinese with English abstract)
34. Aitchison, J. The statistical analysis of compositional data. *J. R. Stat. Soc. Ser. B (Methodol.)* **1982**, *44*, 139–160. [CrossRef]
35. Aitchison, J. *The Statistical Analysis of Compositional Data*; Chapman & Hall: London, UK, 1986; p. 416.
36. Liu, Y.; Cheng, Q.M.; Zhou, K.F.; Xia, Q.L.; Wang, X.Q. Multivariate analysis for geochemical process identification using stream sediment geochemical data: A perspective from compositional data. *Geochem. J.* **2016**, *50*, 293–314. [CrossRef]
37. Wang, Z.; Shi, W.J.; Zhou, W.; Li, X.Y.; Yue, T.X. Comparison of additive and isometric log-ratio transformations combined with machine learning and regression kriging models for mapping soil particle size fractions. *Geoderma* **2020**, *365*, 114214. [CrossRef]
38. Wu, F.Y.; Jahn, B.M.; Wilde, S.A.; Lo, C.H.; Yui, T.F.; Lin, Q.; Ge, W.C.; Sun, D.Y. Highly fractionated I-type granites in NE China (II): Isotopic geochemistry and implications for crustal growth in the Phanerozoic. *Lithos* **2003**, *67*, 191–204. [CrossRef]
39. Zheng, Y.F.; Xiao, W.J.; Zhao, G. Introduction to tectonics of China. *Gondwana Res.* **2013**, *23*, 1189–1206. [CrossRef]
40. Zhao, Z.H.; Sun, J.G.; Li, G.H.; Xu, W.X.; Lü, C.L.; Guo, Y.; Liu, J.; Zhang, X. Zircon U–Pb geochronology and Sr–Nd–Pb–Hf isotopic constraints on the timing and origin of the Early Cretaceous igneous rocks in the Yongxin gold deposit in the Lesser Xing'an Range, NE China. *Geol. J.* **2020**, *55*, 2684–2703. [CrossRef]
41. Ouyang, H.G.; Mao, J.W.; Santosh, M.; Zhou, J.; Zhou, Z.H.; Wu, Y.; Hou, L. Geodynamic setting of Mesozoic magmatism in NE China and surrounding regions: Perspectives from spatio-temporal distribution patterns of ore deposits. *J. Asian Earth Sci.* **2013**, *78*, 222–236. [CrossRef]
42. Ge, W.C.; Wu, F.Y.; Zhou, C.Y.; Zhang, J.H. Mineralization ages and geodynamic implications of porphyry Cu–Mo deposits in the east of Xingmeng orogenic belt. *Chin. Sci. Bull.* **2007**, *52*, 2407–2417. [CrossRef]

43. Zhao, Z.H.; Sun, J.G.; Li, G.H.; Xu, W.X.; Lü, C.L.; Wu, S.; Guo, Y.; Ren, L.; Hu, Z.X. Age of the Yongxin Au deposit in the Lesser Xing'an Range: Implications for an Early Cretaceous geodynamic setting for gold mineralization in NE China. *Geol. J.* **2019**, *54*, 2525–2544. [CrossRef]
44. Pang, X.Y.; Qin, K.Z.; Wang, L.; Song, G.X.; Li, G.M.; Su, S.Q.; Zhao, C. Deformation characteristics of the Tongshan fault within Tongshan porphyry copper deposit, Heilongjiang Province, and restoration of alteration zones and orebodies. *Acta Petrol. Sin.* **2017**, *33*, 398–414, (In Chinese with English abstract)
45. Zeng, Q.D.; Liu, J.M.; Chu, S.X.; Wang, Y.B.; Sun, Y.; Duan, X.X.; Zhou, L.L.; Qu, W.J. Re–Os and U–Pb geochronology of the Duobaoshan porphyry Cu–Mo–(Au) deposit, northeast China, and its geological significance. *J. Asian Earth Sci.* **2014**, *79*, 895–909. [CrossRef]
46. Hu, X.L.; Ding, Z.J.; He, M.C.; Yao, S.Z.; Zhu, B.P.; Shen, J.; Chen, B. Two epochs of magmatism and metallogeny in the Cuihongshan Fe-polymetallic deposit, Heilongjiang Province, NE China: Constrains from U–Pb and Re–Os geochronology and Lu–Hf isotopes. *J. Geochem. Explor.* **2014**, *143*, 116–126. [CrossRef]
47. Hu, X.L.; Ding, Z.J.; He, M.C.; Yao, S.Z.; Zhu, B.P.; Shen, J.; Chen, B. A porphyry-skarn metallogenic system in the Lesser Xing'an Range, NE China: Implications from U–Pb and Re–Os geochronology and Sr–Nd–Hf isotopes of the Luming Mo and Xulaojiugou Pb–Zn deposits. *J. Asian Earth Sci.* **2014**, *90*, 88–100. [CrossRef]
48. Gao, R.; Xue, C.; Lü, X.; Zhao, X.; Yang, Y.; Li, C. Genesis of the Zhengguang gold deposit in the Duobaoshan ore field, Heilongjiang Province, NE China: Constraints from geology, geochronology and S-Pb isotopic compositions. *Ore Geol. Rev.* **2017**, *84*, 202–217. [CrossRef]
49. Zhai, D.; Liu, J.; Ripley, E.M.; Wang, J. Geochronological and He–Ar–S isotopic constraints on the origin of the Sandaowanzi gold-telluride deposit, northeastern China. *Lithos* **2015**, *212*, 338–352. [CrossRef]
50. Sun, J.G.; Han, S.J.; Zhang, Y.; Xing, S.W.; Bai, L.A. Diagenesis and metallogenetic mechanisms of the Tuanjiegou gold deposit from the Lesser Xing'an Range, NE China: Zircon U–Pb geochronology and Lu–Hf isotopic constraints. *J. Asian Earth Sci.* **2013**, *62*, 373–388. [CrossRef]
51. Hao, Y.J.; Ren, Y.S.; Duan, M.X.; Tong, K.Y.; Chen, C.; Yang, Q.; Li, C. Metallogenic events and tectonic setting of the Duobaoshan ore field in Heilongjiang Province, NE China. *J. Asian Earth Sci.* **2015**, *97*, 442–458. [CrossRef]
52. Zhao, Z.H.; Sun, J.G.; Li, G.H.; Xu, W.X.; Lü, C.L.; Wu, S.; Guo, Y.; Liu, J.; Ren, L. Early Cretaceous gold mineralization in the Lesser Xing'an Range of NE China: The Yongxin example. *Int. Geol. Rev.* **2019**, *61*, 1522–1549. [CrossRef]
53. Ye, J.Y.; Jiang, B.L. Combination schemes of sample analysis methods for multitarget geochemical survey. *Geol. Bull. China* **2006**, *25*, 741–744, (In Chinese with English abstract)
54. Pawlowsky-Glahn, V.; Buccianti, A. *Compositional Data Analysis Theory and Applications*; John Wiley & Sons Ltd.: London, UK, 2021; pp. 1–372.
55. Egozcue, J.J.; Pawlowsky-Glahn, V.; Mateu-Figueras, G.; Barcelo-Vidal, C. Isometric logratio transforma-tions for compositional data analysis. *Math. Geol.* **2003**, *35*, 279–300. [CrossRef]
56. Filzmoser, P.; Hron, K.; Reimann, C. Univariate statistical analysis of environmental (compositional) data: Problems and possibilities. *Sci. Total Environ.* **2009**, *407*, 6100–6108. [CrossRef] [PubMed]
57. Filzmoser, P.; Hron, K.; Reimann, C. Principal component analysis for compositional data with outliers. *Env. Off. J. Int. Env. Soc.* **2009**, *20*, 621–632. [CrossRef]
58. Filzmoser, P.; Hron, K.; Templ, M. *Applied Compositional Data Analysis*; Springer: Cham, Switzerland, 2018; pp. 1–64.
59. Cheng, Q.M.; Xu, Y.G.; Grunsky, E. Integrated spatial and spectrum method for geochemical anomaly separation. *Nat. Resour. Res.* **2000**, *9*, 43–52. [CrossRef]
60. Zuo, R.G. Identification of geochemical anomalies associated with mineralization in the Fanshan district, Fujian, China. *J. Geochem. Explor.* **2014**, *139*, 170–176. [CrossRef]
61. Zuo, R.G.; Wang, J.L. ArcFractal: An ArcGIS add-in for processing geoscience data using fractal/multifractal models. *Nat. Resour. Res.* **2020**, *29*, 3–12. [CrossRef]
62. Malinowski, E.R. Determination of rank by median absolute deviation (DRMAD): A simple method for determining the number of principal factors responsible for a data matrix. *J. Chemom. A J. Chemom. Soc.* **2009**, *23*, 1–6. [CrossRef]

minerals

MDPI

Article

Genesis of the Dongpuzi Gold Deposit in the Liaodong Peninsula, NE China: Constraints from Geology, Fluid Inclusion, and C–H–O–S–Pb Isotopes

Cong Chen [1,2], Taotao Wu [1,*], Deming Sha [1], Dongtao Li [1], Zhongzhu Yang [1,2], Jing Zhang [1,2] and Qingqing Shang [3]

[1] Shenyang Center of Geological Survey, China Geological Survey, Shenyang 110034, China
[2] Key Laboratory of Deep Mineral Resources Exploration and Evaluation, Department of Natural Resource of Liaoning Province, Shenyang 110034, China
[3] College of Earth Sciences, Jilin University, Changchun 130061, China
* Correspondence: wutaotao@mail.cgs.gov.cn

Abstract: The Dongpuzi deposit is an epithermal gold deposit located in the southern margin of the Shaozihe volcanic fault basin in the Liaodong Peninsula. On the basis of fluid inclusion and C–H–O–S–Pb isotope data, a metallogenic model is established for the Dongpuzi deposit. The mineralization at the Dongpuzi deposit has experienced quartz–pyrite (I), quartz–sulfide (II), and quartz–calcite (III) stages. The quartz from ore stage II has liquid-dominated aqueous inclusions, which have homogenization temperatures ranging from 113 to 162 °C and salinities varying from 3.2 to 9.6 wt% NaCl equiv. The quartz from the quartz–calcite stage has decreasing homogenization temperatures (106~143 °C) and salinities (2.7~6.9 wt% NaCl equiv.). The fluid inclusion data indicate that the gold ores were precipitated from low-temperature and low-salinity solutions, with an obvious decrease in temperature and salinity from ore stages II to III. The calculated $\delta^{18}O_{water}$ values for the quartz of ore stage II range from −14.71‰ to −13.31‰, and the corresponding δD_{water} values range from −103.3‰ to −96.1‰, indicating that the ore-forming fluids could be of a meteoric origin. The calcite from ore stage III has $\delta^{13}C_{V\text{-}PDB}$ values of −4.5‰ to −4.2‰ and $\delta^{18}O_{V\text{-}SMOW}$ values of +7.0‰ to +7.4‰, indicating a mantle source for the carbon. The pyrite yielded $\delta^{34}S$ values of +4.1‰ to +6.6‰ and Pb isotopes consistent with those of the host trachyte porphyry and volcanic rocks of the Xiaoling Formation, which suggests that the S and Pb in gold ores were dominantly derived from the host trachyte porphyry and volcanic rocks of the Xiaoling Formation, with some combination of Paleoproterozoic metamorphic rocks of the Gaixian Formation. These results, together with the ore geology, indicate that the Dongpuzi deposit is a typical low-sulfidation epithermal gold deposit with important ore-forming materials input from the host trachyte porphyry, volcanic rocks of the Xiaoling Formation, and Paleoproterozoic metamorphic rocks of the Gaixian Formation. The Dongpuzi deposit was formed under an extensional setting related to the Early Cretaceous lithospheric extension and thinning of the eastern North China Craton.

Keywords: Dongpuzi gold deposit; C–H–O–S–Pb isotopes; fluid inclusions; low-sulfidation epithermal deposit; Liaodong Peninsula

1. Introduction

The Liaodong Peninsula in the eastern North China Craton (NCC) is an important gold province in China that hosts numerous medium- and large-scale gold deposits, such as the Wulong, Baiyun, Sidaogou, Maoling, Wangjiaweizi, and Xiaotongjiapuzi deposits, contributing a total gold reserve of >220 t [1]. The gold deposits are mostly hosted within the Paleoproterozoic metamorphic rocks of the Liaohe Group, structurally controlled by fault zones and associated with Mesozoic granitic intrusions [2–4]. The gold deposits have been broadly classified into the quartz vein type, represented by the large-scale Wulong deposit, and the fracture-controlled disseminated type, represented by the large-scale

Sidaogou deposit [2]. However, except for the Mesozoic intrusive magmatic-hydrothermal gold deposits, the volcanic-hosted epithermal gold ore system is poorly mentioned in the Liaodong Peninsula.

The Dongpuzi gold deposit is a gold vein deposit occurring within the Shaozi River volcanic fault basin in the Xiuyan area, the northern part of the Liaodong Peninsula. Chen [5] first described the geology and ore-forming conditions of this deposit and proposed that it belongs to an epithermal gold deposit related to the host trachyte porphyry. Zhang et al. [6] recognized the mineralizing fluids with low temperatures (90~240 °C) and low salinities (0.35~6.17 wt% NaCl equiv.) and proposed a mixed magmatic and meteoric origin based on the limited fluid inclusion studies and H–O–S isotope data. Jiang [7] proposed that the basement rocks of the Liaohe Group may provide some ore-forming components for the Dongpuzi gold mineralization. However, the previous research lacked detailed investigations on petrography, mineralogy, and paragenesis, and the conclusions were based on limited microthermometric and isotopic data. Thus, the characteristics and origins of ore-forming fluids, sources of ore-forming materials, and genetic relationships between the Dongpuzi gold mineralization and Early Cretaceous volcanic rocks remain poorly known.

Based on field observation and ore geology, this paper carries out systematic fluid inclusion studies and presents new C–H–O–S–Pb isotopes for the Dongpuzi gold deposit. These new data provide further constraints on the characteristics and sources of the ore-forming fluids and materials, which allows a better understanding of the ore genesis of the Dongpuzi gold deposit. Finally, a tectonic–metallogenic model is established for the Dongpuzi gold mineralization.

2. Geological Background

2.1. Regional Geology

The Huanghuadianzi–Dayingzi–Shaozi River region is located in the eastern part of the NCC, on the western margin of the Paleo-Pacific Plate (Figure 1a). The rocks exposed in the area mainly comprise Paleoproterozoic metamorphic rocks of the Liaohe Group and granitic and mafic intrusions, and a series of Mesozoic intrusive and volcanic rocks, as well as minor Late Triassic epimetamorphic rocks of the Cuocaogou Formation (Figure 1b). The Paleoproterozoic Liaohe Group includes the Lieryu, Gaojiayu, Dashiqiao, and Gaixian formations in the area, which have been metamorphosed to the greenschist to lower amphibolite facies and locally to the granulite facies [8]. The Lieryu and Gaojiayu formations are composed dominantly of fine-grained felsic gneiss, amphibolite, and mica–quartz schist. The Dashiqiao Formation consists of dolomitic marble intercalated with minor carbonaceous slate and mica schist [9]. The uppermost Gaixian Formation mainly comprises andalusite–cordierite–mica schist, phyllite, sillimanite–mica schist, and staurolite–mica schist, with minor marble and quartzite. The Paleoproterozoic Liaoji granitoids are widely developed in the area and mainly include deformed monzogranitic gneiss, undeformed porphyritic monzogranite, granite, and alkaline syenite [10].

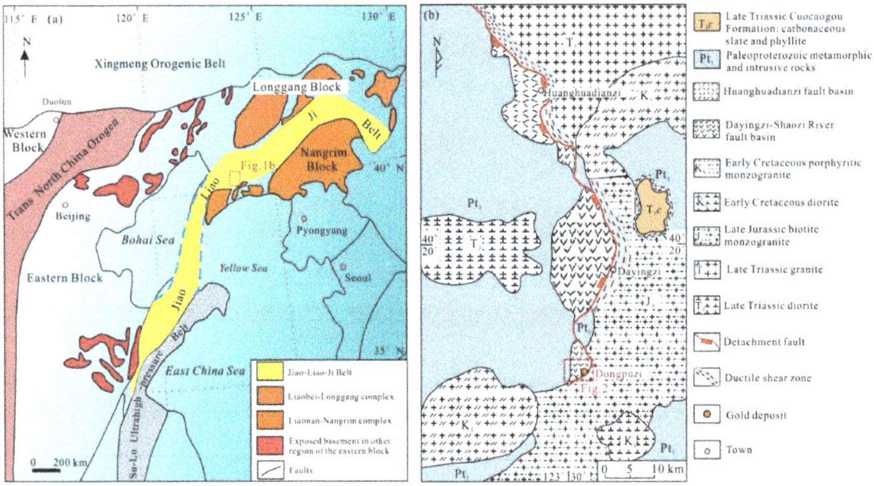

Figure 1. Geological and tectonic map of the eastern NCC (**a**), modified from [8]. A sketch map of the Huanghuadianzi–Dayingzi–Shaozi River area, Liaodong Peninsula (**b**), modified from [11]. The location of Figure 2 has been marked by a red rectangle.

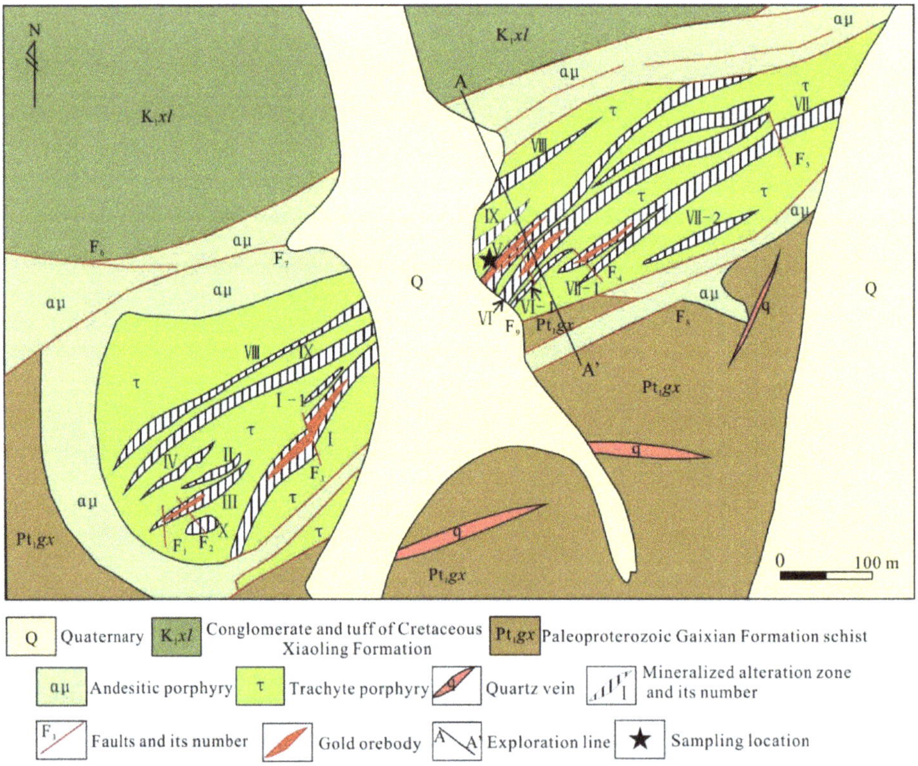

Figure 2. Geological map of the Dongpuzi gold deposit (modified from [5]).

Since the Late Triassic period, the area has experienced intensive tectonic and magmatic activities as an important part of the Circum-Pacific tectono-magmatic zone [12]. The Late Triassic Cuocaogou Formation occurs as erosion residue overlying the Paleoproterozoic metamorphosed sedimentary and volcanic successions of the Liaohe Group and includes carbonaceous slate and phyllite. Intrusive rocks in the area are dominantly Late Triassic diorite and monzogranite, Late Jurassic biotite monzogranite, and Early Cretaceous porphyritic monzogranite and diorite [11,12] (Figure 1b). The tectonic structure in the area is mainly dominated by the Huanghuadianzi–Dayingzi–Shaozi River extensional structure, which is composed of three major units: the extensional fault basins (Huanghuadianzi, Dayingzi, and Shaozi Rivers) in the hanging wall, detachment fault, and Early Cretaceous granitic plutons in the lower plate [9,13]. The extensional fault basins are mainly filled with the volcanic sedimentary rocks of the Early Cretaceous Xiaoling Formation, which include basalt, andesitic basalt, andesite, rhyolite, conglomerate, tuffceous sandstone, and tuff.

The detachment fault extends from the Huanghuadianzi in the north, through the Dayingzi, to Shaozi River in the south, with a length of about 70 km. The fault always dips SW at 24° and develops a series of brittle high-angle normal faults and ductile shear zones in the Mesozoic granitic plutons [11,13] (Figure 1b).

2.2. Geology of the Dongpuzi Orefield

The Dongpuzi gold deposit is located in the southernmost part of the Huanghuadianzi–Dayingzi–Shaozi River extensional structure, in the south margin of the Shaozi River volcanic fault basin (Figure 1b). Strata in the Dongpuzi gold field consist mostly of the Gaixian Formation of the Paleoproterozoic Liaohe Group and Early Cretaceous Xiaoling Formation (Figure 2). The Gaixian Formation is mainly composed of biotite schist, two-mica schist, and sillimanite biotite schist [5]. The Xiaoling Formation comprises limestone, coarse sandstone, and tuff. The Cenozoic stratigraphy is dominated by alluvial deposits.

The tectonic structures of the Dongpuzi orefield include a series of NE–SW- to NNE–SSW-trending faults and minor NW–SE- to NNW–SSE-trending faults (Figure 2). The NE–SW- to NNE–SSW-trending faults always dip NW at 50–70° and are characterized by multi-periodic activities. The faults control the emplacement of ore-hosting trachyte porphyry and post-ore andesitic porphyry and lead to the formation of several joints and fractures. The joints and fractures are marked by intensive mineralization and alteration and control the formation of the Dongpuzi gold-bearing quartz veins. The NW–SE- to NNW–SSE-trending faults cut the mineralized alteration zones and gold orebodies.

The intrusive rocks are mainly Early Cretaceous trachyte porphyry and andesitic porphyry, which are distributed in the middle part of the Dongpuzi orefield. The trachyte porphyry was emplaced along the NE–SW- to NNE–SSW-trending faults and hosted the mineralized alteration zones and gold orebodies. The andesitic porphyry intruded into the Gaixian Formation and trachyte porphyry, as well as the gold orebodies (Figures 2 and 3).

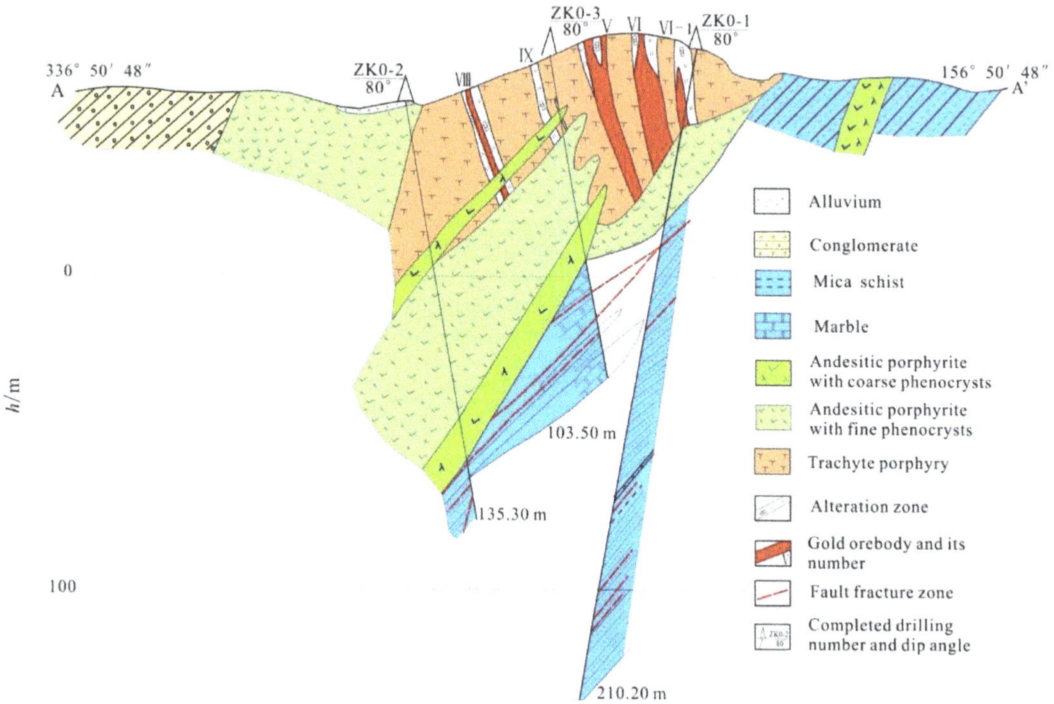

Figure 3. Geological cross section of the No.0 exploration line in the Dongpuzi gold deposit (modified after [14]; the location of the No.0 exploration line is referred as A-A′ in Figure 2).

3. Ore Deposit Geology

The Dongpuzi gold deposit consists of 13 mineralized alteration zones (No. I, I-1, II, III, V, V-1, VI, VI-1, VIII, IX, X, etc.; Figure 2), which contain 30 gold orebodies, with an average grade of 5 gt^{-1} [14]. The gold orebodies are hosted within the Early Cretaceous trachyte porphyry (Figures 2 and 3) and structurally controlled by NE–SW- to NNE–SSW-trending faults and fracture zones. Gold mineralization occurs mainly as pyrite–sericite–quartz-altered rocks and auriferous quartz–pyrite veins and veinlets. The major orebodies are No. V and No. VI, which account for more than 77% of the total metal reserves of the Dongpuzi deposit (Figure 3) [14].

The ore mineral assemblage is mainly pyrite, with minor chalcopyrite, sphalerite, tetrahedrite, native gold, and electrum (Figure 4). The nonmetallic minerals observed in the ores include quartz, sericite, calcite, kaolin, and chlorite (Figure 4). Native gold and electrum generally occur as grains in inclusions or along fractures within pyrite and quartz grains (Figure 4i). The ore minerals are characterized by subhedral to anhedral granular (Figure 4d,e), metasomatic (Figure 4e,f), poikilitic, and cataclastic textures. They occur either in veins (Figure 4a), veinlets, or stockworks (Figure 4b) or have veinlet-disseminated, disseminated, or miarolitic structures (Figure 4b).

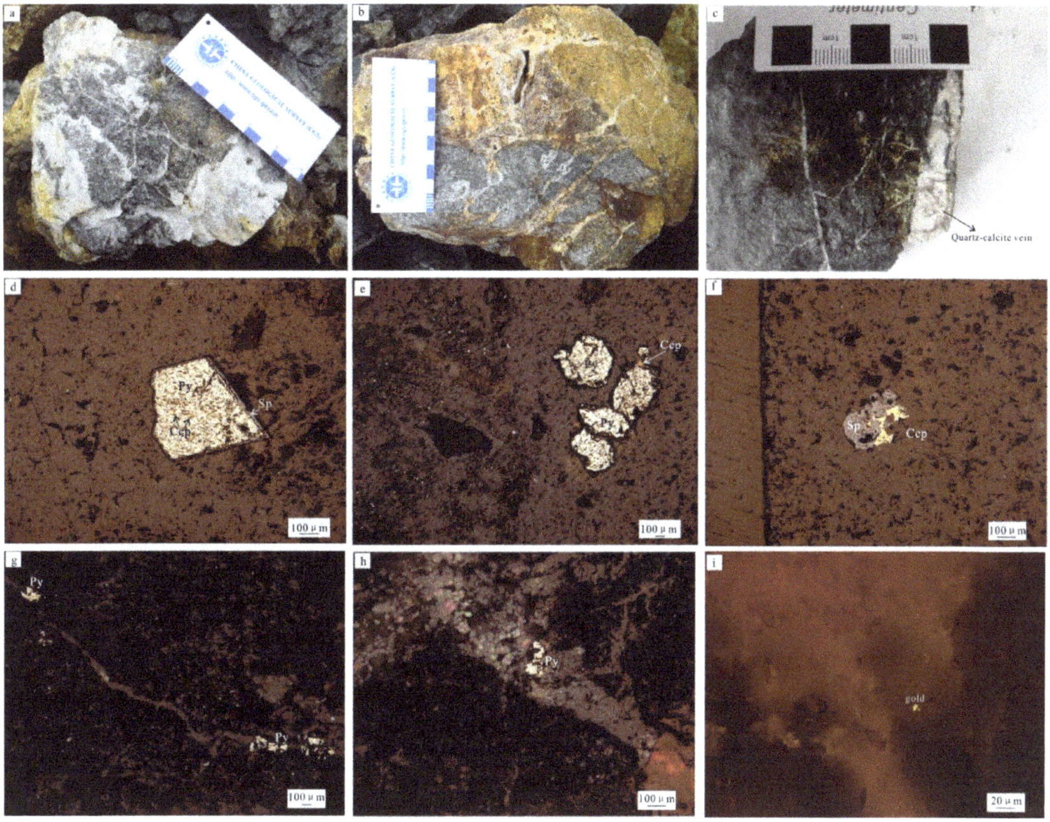

Figure 4. Photographs and microphotographs of the Dongpuzi gold ores. (**a**): The quartz–pyrite-vein-type ore. (**b**): The gold ores characterized by a miarolitic structure. (**c**): The quartz–calcite vein cut through the altered trachyte porphyry. (**d**,**e**): The subhedral pyrite was metasomatized by later chalcopyrite and sphalerite. (**f**): The sphalerite was metasomatized by later chalcopyrite. (**g**,**h**): The anhedral pyrite within the quartz veinlet. (**i**): The gold mineral within the quartz veins. Abbreviations: Ccp = chalcopyrite, Py = pyrite, Qz = quartz, and Sp = sphalerite.

The wall-rocks record various types of alterations, which include sericitization, silicification, carbonatization, kaolinization, and chloritization. Silicification usually manifests as quartz veins, veinlets, or stockwork. Carbonatization mainly occurs as quartz–calcite veins and locally cut the early-stage gold-bearing quartz veins. In addition, the wall-rock alteration displays clear spatial zonation, which can be classified into three zones from the gold orebody to wall-rocks: an inner silicification–pyritization–sericitization zone, a middle silicification–sericitization–kaolinization zone, and an external silicification–chloritization zone [15].

On the basis of the paragenetic assemblages and crosscutting relationships, gold mineralization in the Dongpuzi deposit can be divided into three hydrothermal stages: the quartz–pyrite stage (I), the quartz–sulfide stage (II), and the quartz–calcite stage (III). The quartz–pyrite stage forms large amounts of quartz and sericite and minor subhedral pyrite, with little or no gold (Figure 4a). The quartz–sulfide stage represents the main metallogenic stage of gold, which is characterized by quartz, pyrite, chalcopyrite, sphalerite, and minor native gold and electrum (Figure 4b,d–f,i). The quartz is always smoky gray and occurs as quartz veins, veinlets, or stockworks within the trachyte porphyry (Figure 4b,g,h). The eubhedral–subhedral pyrite grains are locally metasomatized by later chalcopyrite and sphalerite (Figure 4d,e). The quartz–calcite stage is characterized by quartz and calcite occurring as veins that crosscut the earlier quartz–sulfide veins (Figure 4c).

4. Sampling and Analytical Methods

The samples analyzed for fluid inclusion studies and C–H–O–S–Pb isotopes herein were collected from the No.V gold orebody derived from the underground mine. The fresh trachyte porphyry samples were collected in the mine dump for analysis of the Pb isotope. Descriptions of the samples are presented in Table 1.

Table 1. Sample characteristics of the Dongpuzi gold deposit.

Sample Number	Mineral/Sample	Mineralization Stage	Sample Location and Characteristics	Analysis
21dpl-1	Quartz	Quartz–sulfide	Quartz from intensely silicified trachyte porphyry	Fluid inclusion
21dpl-2	Quartz	Quartz–sulfide	Quartz veins within the trachyte porphyry, where sericitization and silicification occur	
21dpl-3	Quartz	Quartz–sulfide		
21dpl-5	Quartz	Quartz–calcite	Quartz–calcite veins cutting the trachyte porphyry, where sericitization and silicification occur	
21DP-1	Quartz	Quartz–calcite	Quartz–calcite vein cutting the trachyte porphyry, where sericitization and silicification occur	H–O isotopes
21DP-2	Quartz	Quartz–sulfide	Quartz from intensely silicified trachyte porphyry	
21DP-3	Quartz	Quartz–sulfide	Quartz veins within the trachyte porphyry, where sericitization and silicification occur	
21DP-4	Quartz	Quartz–sulfide		
21DP-5	Quartz	Quartz–sulfide		
21DP-6	Calcite	Quartz–calcite	Quartz–calcite vein cutting the trachyte porphyry, where sericitization and silicification occur	C–O isotopes
21DP-7	Calcite	Quartz–calcite		
21DP-8	Calcite	Quartz–calcite		
DP-2	Pyrite	Quartz–sulfide	Pyrite within intensely silicified trachyte porphyry	S–Pb isotopes
DP-3	Pyrite	Quartz–sulfide		
DP-4	Pyrite	Quartz–sulfide		
21DP-9	Pyrite	Quartz–sulfide		
21DP-10	Pyrite	Quartz–sulfide	Pyrite within the quartz veins	
DPX-1	Trachyte porphyry	-	Trachyte porphyry with little or no alteration and mineralization	Pb isotope
DPX-2	Trachyte porphyry	-		
DPX-3	Trachyte porphyry	-		
DPX-4	Trachyte porphyry	-		
DPX-5	Trachyte porphyry	-		

Note: "-" in the list of "Mineralization stage" refers to no correspondence between the trachyte porphyry and the mineralization stage.

4.1. Fluid Inclusion Petrography and Microthermometry

Quartz from intensely silicified trachyte porphyry (21dpl-1), quartz veins (21dpl-2, 3) in the gold ores from the main metallogenic stage (II), and quartz from quartz–calcite veins (21dpl-5) from the late ore stage (III) were chosen for fluid inclusion studies to determine the characteristics of the ore-forming fluids of ore stages II and III.

The petrographic observation and microthermometry of fluid inclusions were performed at the Analytical Laboratory of the Beijing Research Institute of Uranium Geology, Beijing, China. Microthermometric measurements were conducted on a LINKAM THMS-600 type heating–freezing stage, which has a temperature measurement range of −196 °C to +600 °C. During the heating and freezing process, the temperature control rate was less than

5 °C/min, and the rate decreased to less than 1 °C/min when the phase transformation was about to occur.

4.2. Isotope Analysis

Quartz grains from quartz veins (21DP-1; ore stage II), silicified trachyte porphyry (21DP-2; ore stage II), and quartz–calcite veins (21DP-3, -4, -5; ore stage III) were analyzed for H–O isotopes; and calcite grains from quartz–calcite veins (21DP-6, -7, -8; ore stage III) were analyzed for C–O isotopes to determine the source of the ore-forming fluids. Pyrite grains from intensely silicified trachyte porphyry (DP-2, -3, -4) and quartz veins (21DP-9, -10) of ore stage II were analyzed for S and Pb isotopes to constrain the source of ore-forming materials. The trachyte porphyry samples with little or no alterations that host the No.V gold orebody were analyzed for the Pb isotope to reveal the relationship between the gold ores and host rocks.

C, H, O, S, and Pb isotopes were analyzed at the Analytical Laboratory of the Beijing Research Institute of Uranium Geology, Beijing, China. The quartz, calcite, and pyrite mineral grains from ore stages II and III with a purity of >98% were ground to 40~60 mesh for analyses of C, H, and O isotopes. Furthermore, the fresh trachyte porphyry samples were crushed to 200 mesh for analysis of the Pb isotope.

H and O isotopic compositions of fluid inclusions within the quartz crystals and C and O isotopic compositions of those in calcite crystals were measured using a MAT-253 mass spectrometer. The isotopic results were standardized by Vienna-Standard Mean Ocean Water (V-SMOW) for H and O isotopes and the Vienna-Peedee Belemnite (V-PDB) standard for the C isotope. The analytical precisions were better than ±1‰ for δD, ±0.2‰ for $\delta^{18}O$, and ±0.2‰ for $\delta^{13}C$.

S isotopic compositions of five pyrite samples were determined using a Delta V Plus mass spectrometer. The S isotopic results were standardized by Vienna-Canyon Diablo Troilite (V-CDT), and analytical precision was estimated to be ±0.2‰.

Pb isotope analyses of five pyrite and five trachyte porphyry samples were conducted using an ISOPROBE-T thermal ionization mass spectrometer. The Pb isotopic ratios were corrected using NBS SRM 981 [16], and the Pb isotope uncertainties were reported as $\pm 2\sigma$. The measurement accuracy was <0.05% for $^{204}Pb/^{206}Pb$ and <0.005% for $^{208}Pb/^{206}Pb$ with 1 μg of lead. The specific procedures for C, H, O, S, and Pb isotopic analyses are described by Guo et al. [17].

5. Results

5.1. Fluid Inclusions

Fluid inclusions within the quartz crystals in gold ores formed during ore stages II and III were analyzed. Petrographic observations indicate that fluid inclusions are common in quartz crystals at room temperature and mainly include three types: (1) liquid-only aqueous inclusions (L-only), (2) vapor-only aqueous inclusions (V-only), and (3) bi-phase (liquid and vapor) aqueous inclusions dominated by liquid (L-dominated) and finally homogenized to liquid (Figure 5). The three types of fluid inclusions occur in both ore stages II and III.

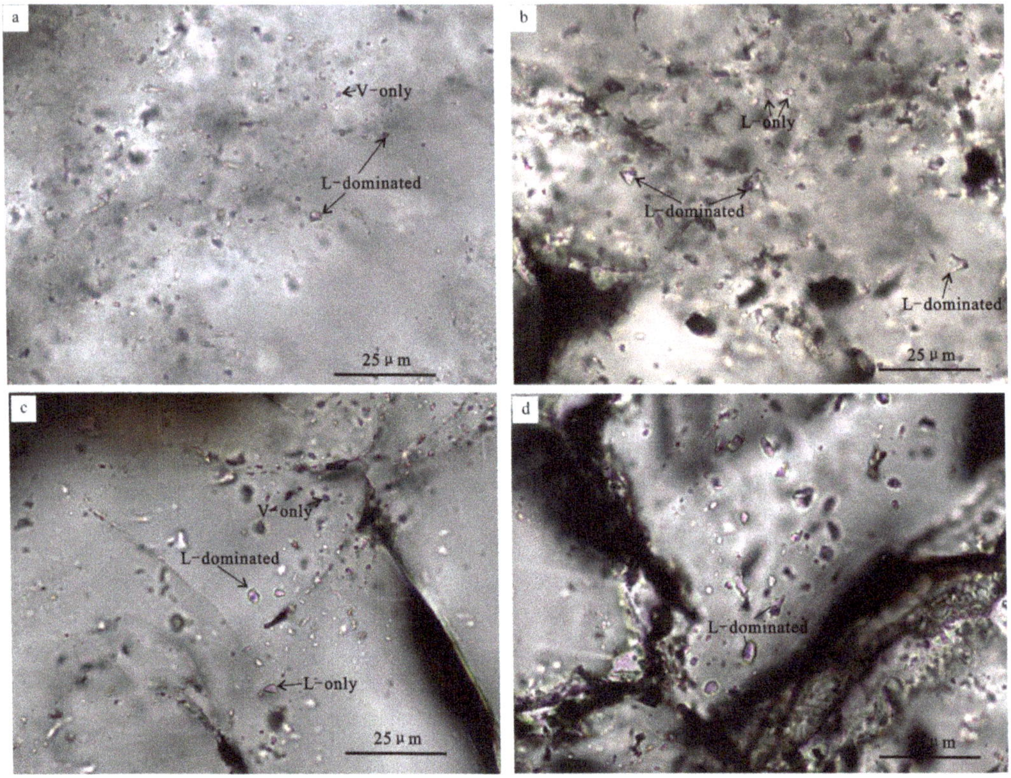

Figure 5. Photomicrographs of the fluid inclusions for the Dongpuzi gold deposit. (**a**,**b**): V-only, L-only, and L-dominated fluid inclusions within the quartz crystal from ore stage II; (**c**,**d**): V-only, L-only, and L-dominated fluid inclusions within the quartz crystal from ore stage III).

Ore stage II: L-dominated fluid inclusions are distributed in groups or scattered, and they are subrounded or irregularly polygonal in shape, with sizes of 2~7 μm (Figure 5a,b). They mostly contain a vapor phase that occupies 5~10 vol%. L-only fluid inclusions are elliptical and irregular in shape, with sizes of 2~5 μm (Figure 5b). V-only fluid inclusions generally exhibit obvious dark halos under the microscope and are dispersively distributed. They are mostly round or elliptical in shape, with sizes of 3~5 μm (Figure 5a).

Ore stage III: L-dominated fluid inclusions are mostly rounded and subrounded in shape, and their sizes ranges from 1 to 13 μm, most of which are 3~8 μm (Figure 5c,d). The volume of vapor in the L-dominated fluid inclusions of ore stage III is 5~10%. L-only and V-only fluid inclusions often coexist with the L-dominated fluid inclusions in the same quartz crystal. L-only fluid inclusions are generally scattered in the quartz crystal and are subrounded or irregular polygon in shape, with sizes of 2~8 μm (Figure 5c). V-only fluid inclusions are distributed in groups or scattered, and most of them are elliptical in shape. Their sizes usually vary from 1 to 6 μm (Figure 5c).

The primary L-dominated fluid inclusions occurring in ores of stages II and III were chosen for microthermometry analysis. Due to their poor visibility or small sizes, few ice-melting temperatures of L-dominated fluid inclusions have been measured in this study. The homogenization and ice-melting temperatures of fluid inclusions and related calculation equations are summarized in Table 2.

Table 2. Microthermometric data of the L-dominated fluid inclusions for the Dongpuzi gold deposit.

Sample Number	Host Mineral	Fluid Inclusion Types	Homogeneous Temperatures (T_h/°C)	Ice-Melting Temperatures (T_{ice}/°C)	Salinity (wt% NaCl)
21dpl-1	Quartz	L-dominated	113~151 (17)	6.3~4 (8)	6.4~9.6 (8)
21dpl-2	Quartz	L-dominated	109~162 (27)	5.3~1.9 (13)	3.2~8.3 (13)
21dpl-3	Quartz	L-dominated	110~161 (24)	6~5.9 (5)	9~9.2 (5)
21dpl-5	Quartz	L-dominated	106~143 (25)	4.3~1.6 (6)	2.7~6.9 (6)

Note: The number in the parentheses means the number of measurements; the equation for salinity calculations: $S = 0 + 1.76985 \times T_{ice} - 0.042384 \times T_{ice}^2 + 0.00052778 \times T_{ice}^3$ ([18]).

L-dominated fluid inclusions within quartz crystals of ore stage II have homogenization temperatures ranging from 113 to 162 °C, with an average of 134 °C (Figure 6a). The salinities of these inclusions vary from 3.2 to 9.6 wt% NaCl equiv. based on the final ice-melting temperatures of −6.3 to −1.9 °C.

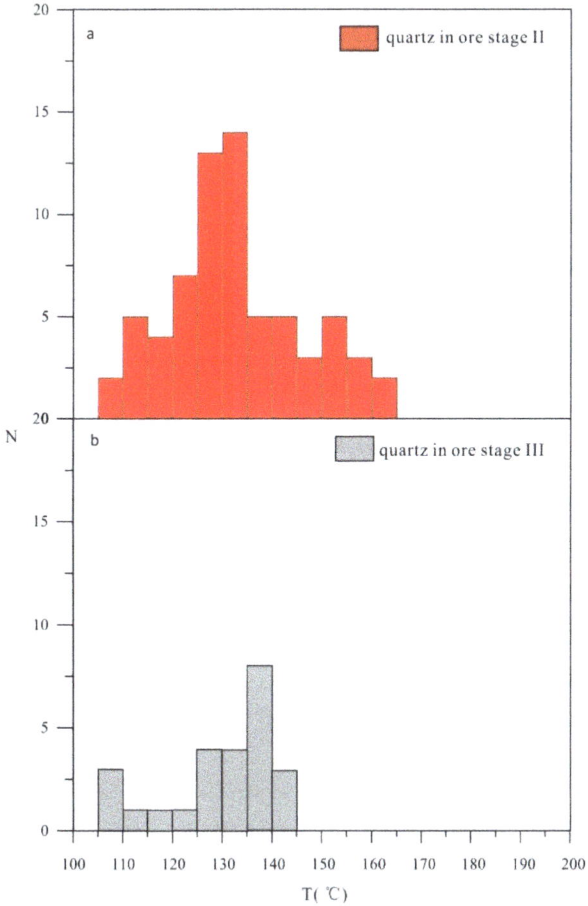

Figure 6. Histograms of homogenization temperatures of fluid inclusions of the Dongpuzi gold deposit. ((**a**): homogenization temperatures of fluid inclusions from the ore stage II; (**b**): homogenization temperatures of fluid inclusions from the ore stage III).

The homogenization temperatures of L-dominated fluid inclusions within quartz crystals of ore stage III vary from 106 to 143 °C, with an average of 129 °C (Figure 6b). The salinities of these fluid inclusions vary from 2.7 to 6.9 wt% NaCl equiv. based on the final ice-melting temperatures of −4.3 to −1.6 °C.

5.2. C–H–O Isotopic Compositions

The C–H–O isotopic data and relative fractionation equations of the Dongpuzi gold deposit are presented in Table 3 and plotted in Figure 7. Four quartz samples of ore stage II have $\delta^{18}O_{V\text{-}SMOW}$ and δD_{water} values of +2.9‰ to +4.3‰ and −103.3‰ to −96.1‰, respectively, with corresponding $\delta^{18}O_{water}$ values of −14.71‰ to −13.31‰. One quartz sample of ore stage III has $\delta^{18}O_{V\text{-}SMOW}$ and δD_{water} values of +4.4‰, and −103.9‰, respectively, with corresponding $\delta^{18}O_{water}$ values of −13.62‰. Three calcite samples from ore stage III have $\delta^{13}C_{V\text{-}PDB}$ values of −4.5‰ to −4.2‰ and $\delta^{18}O_{V\text{-}SMOW}$ values of +7.0‰ to +7.4‰.

Table 3. C–H–O isotope data of the Dongpuzi gold deposit.

Sample Number	Mineral	T (°C)	$\delta D_{V\text{-}SMOW}$ (‰)	$\delta^{13}C_{V\text{-}PDB}$ (‰)	$\delta^{18}O_{V\text{-}SMOW}$ (‰)	$\delta^{18}O_{H2O\text{-}SMOW}$ (‰)
21DP-1		129	−103.9	-	4.4	−13.62
21DP-2			−94.2	-	4.3	−13.31
21DP-3	Quartz	133	−94.9	-	4.2	−13.41
21DP-4			−103.3	-	3.7	−13.91
21DP-5			−96.1	-	2.9	−14.71
21DP-6			-	−4.2	7.4	-
21DP-7	Calcite	129	-	−4.5	7.2	-
21DP-8			-	−4.3	7	-

Note: The equilibrium temperatures for each stage are based on the fluid inclusion studies in this paper. The fractionation equation for $\delta^{18}O_{water}$ calculation: $1000\ln\alpha_{quartz-water} = 3.38 \times 10^6/T^2 - 2.9$ ([19]). The "-" in the table means no analytical results.

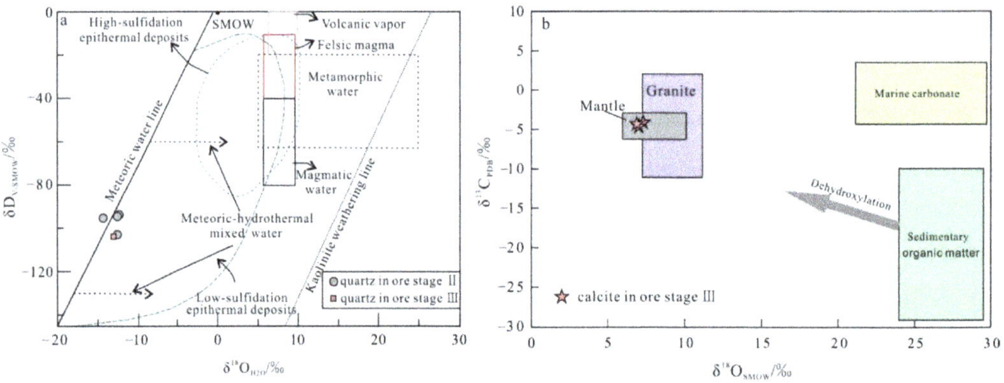

Figure 7. Diagrams of δD—$\delta^{18}O$ (**a**) and δC—$\delta^{18}O$ (**b**) ((**a**) is modified from [20,21]; the fields for the epithermal deposits and meteoric-hydrothermal mixed water are from [22] and [23], respectively; (**b**): the carbonate fields of marine carbonate, sedimentary organic carbon, and magma-mantle carbonate are from [24] and [25], respectively).

5.3. S Isotopic Compositions

The S isotopic data of the Dongpuzi gold deposit are presented in Table 4 and plotted in Figure 8a,b. Five pyrite samples yield relatively homogeneous S isotopic compositions with $\delta^{34}S_{V\text{-}CDT}$ values ranging from +4.1‰ to +6.6‰.

Table 4. Pyrite S–Pb isotopic data of the Dongpuzi gold deposit.

Sample Number	$\delta^{34}S_{V\text{-}CDT}$/‰	$^{208}Pb/^{204}Pb$	Std Err	$^{207}Pb/^{204}Pb$	Std Err	$^{206}Pb/^{204}Pb$	Std Err
DP-2	6.5	38.798	0.003	15.636	0.001	18.069	0.001
DP-3	5.5	38.827	0.004	15.638	0.002	18.039	0.002
DP-4	6.6	38.798	0.006	15.641	0.001	18.07	0.001
21DP-9	6	38.557	0.007	15.595	0.002	17.918	0.002
21DP-10	4.1	38.819	0.007	15.64	0.003	18.046	0.003

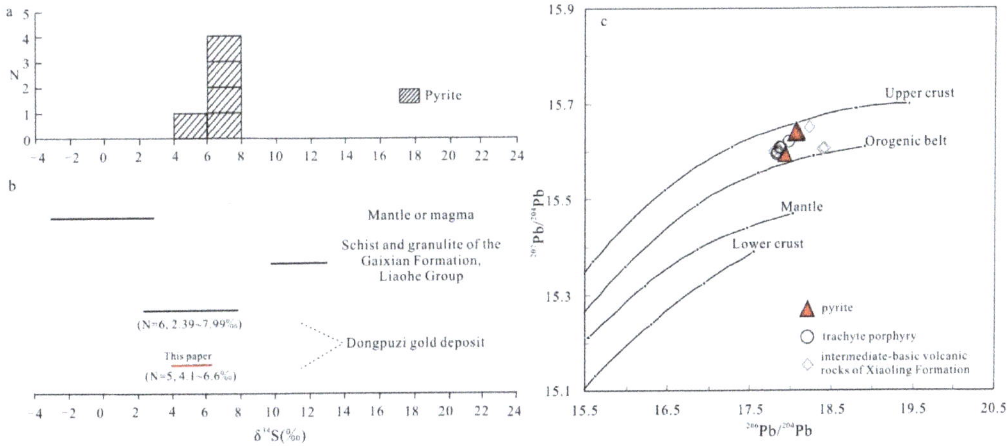

Figure 8. Histogram of the S isotope in the Dongpuzi gold deposit (**a**) and comparison of $\delta^{34}S$ values between the Dongpuzi deposit and associated geological bodies (**b**). The S isotopic data of Dongpuzi gold deposit are from this paper and [15]; data of the metamorphic rocks of the Gaixian Formation and mantle or deep magma are from [26,27], respectively. The Pb evolution curves for Dongpuzi gold ores (**c**). The data of intermediate-basic rocks of the Xiaoling Formation are from [28].

5.4. Pb Isotopic Compositions

The Pb isotopic compositions of pyrite and trachyte porphyry samples are given in Tables 4 and 5 and plotted in Figure 8c. Five pyrite samples have $^{206}Pb/^{204}Pb$ ratios of 17.918–18.07, $^{207}Pb/^{204}Pb$ ratios of 15.595–15.641, and $^{208}Pb/^{204}Pb$ ratios of 38.557–38.827 (Table 4). Five trachyte porphyry samples have $^{206}Pb/^{204}Pb$ ratios of 17.955–17.848, $^{207}Pb/^{204}Pb$ ratios of 15.595–15.621, and $^{208}Pb/^{204}Pb$ ratios of 38.963–39.116 (Table 5).

Table 5. Pb isotopic data of trachyte porphyry in the Dongpuzi gold deposit.

Sample Number	Sample	Age/Ma	$^{208}Pb/^{204}Pb$	Std Err	$^{207}Pb/^{204}Pb$	Std Err	$^{206}Pb/^{204}Pb$	Std Err	$(^{208}Pb/^{204}Pb)_i$	$(^{207}Pb/^{204}Pb)_i$	$(^{206}Pb/^{204}Pb)_i$
DPX-1	Trachyte porphyry	130	39.375	0.004	15.625	0.002	18.035	0.002	39.116	15.621	17.955
DPX-2			39.353	0.005	15.6	0.002	17.921	0.002	39.011	15.595	17.819
DPX-3			39.299	0.006	15.611	0.003	17.94	0.003	39.039	15.607	17.865
DPX-4			39.282	0.009	15.604	0.004	17.926	0.004	38.963	15.599	17.82
DPX-5			39.393	0.006	15.614	0.002	17.962	0.002	39.023	15.608	17.848

Note: The formation age of trachyte porphyry is from unpublished U–Pb data.

6. Discussion

6.1. Properties of Ore-Forming Fluids

The primary L-dominated fluid inclusions occurring in ores of stages II and III were measured in this study. In the main ore stage, the homogenization temperatures of L-dominated fluid inclusions range from 113 to 162 °C, which indicate that the main ore-forming fluids belong to the low-temperature fluid system [29]. The homogenization temperatures of L-dominated fluid inclusions in the late quartz–calcite stage decrease (106~143 °C), indicating the occurrence of fluid cooling, which is an important control on ore deposition [30]. The salinities of fluids also decrease from the main ore stage (average 7.2 wt% NaCl equiv.) to the late stage (average 6.1 wt% NaCl equiv.), which is similar to those of the low-sulfidation epithermal gold deposits worldwide (commonly <5 wt% NaCl equiv.; [22]). Previous studies have indicated that the ore-forming fluids of the Dongpuzi gold deposit are NaCl-H_2O-dominated type [3]. Therefore, the ore-forming fluids of the Dongpuzi gold deposit are a low-temperature (113 to 162 °C) and low-salinity (average 7.2 wt% NaCl equiv.) NaCl-H_2O fluid system and fluid cooling may be an important factor in the formation of this gold deposit.

6.2. Origin of Ore-Forming Fluids

The δD_{water} and calculated $\delta^{18}O_{water}$ values of the fluid trapped in quartz from the Dongpuzi ore stage II are −103.3‰ to −96.1‰ and −14.71‰ to −13.31‰, respectively, which are consistent with corresponding values of the hydrothermal fluids from ore stage III (Table 3). These data are plotted in or close to the meteoric water line in a δD—$\delta^{18}O_{H2O}$ diagram, indicating that the ore-forming fluids of the Dongpuzi deposit were predominantly derived from meteoric water, with little indication of magmatic water (Figure 7a). Furthermore, these data are compatible with those of fluids from low-sulfidation epithermal systems worldwide (Figure 7a) [22].

Three calcite samples from stage III have measured $\delta^{13}C_{V\text{-}PDB}$ values, ranging from −4.5‰ to −4.2‰, close to those of igneous- or mantle-derived CO_2 (−2‰ to −7‰; [24]) and of mixed crustal sources (average −5‰; [27]), indicating that the carbon in the Dongpuzi gold deposit was derived from either the mantle or a mixed crustal source. This is also shown in the $\delta^{13}C$—$\delta^{18}O$ diagram (Figure 7b). The variations in C–O isotope compositions are likely affected by the fluid–rock interaction during the evolution of the ore-forming fluids [31].

6.3. Sources of Sulfur and Lead

Under temperatures lower than 350 °C and reduced hydrothermal conditions where H_2S is the dominant sulfur species in the fluid, the $\delta^{34}S$ values of sulfide minerals will represent the overall S isotope composition of the hydrothermal fluid (i.e., $\delta^{34}S_{\Sigma S} \approx \delta^{34}S_{sulfide}$) [27]. The Dongpuzi gold mineralization is formed under low temperatures, as stated above. The mineralogical observations indicate that the ores are mainly composed of pyrite, chalcopyrite, and sphalerite, which are generally formed in a reducing environment, and no sulfate minerals were found. Therefore, the average $\delta^{34}S$ values of sulfide minerals approximate the sulfur isotopic compositions of hydrothermal fluid. Five pyrite samples from gold ores yield $\delta^{34}S$ values of +4.1‰ to +6.6‰ (Figure 8a and Table 4), basically consistent with the $\delta^{34}S_{V\text{-}CDT}$ values of +2.39‰ to +7.99‰ reported by Zhang et al. [15]. These values are a little higher than those of magmatic reservoirs (0‰ ± 3‰; [27]) but lower than those of the schist and granulite of the Gaixian Formation (+10.0‰ to +17.0‰; [26]; Figure 8b). Therefore, the sulfur in ores was predominantly derived from the magmatic system, with some combination of metamorphic rocks of the Gaixian Formation for the Dongpuzi deposit.

The pyrite samples from the Dongpuzi ores yield homogeneous $^{206}Pb/^{204}Pb$ ratios of 17.918~18.07, $^{207}Pb/^{204}Pb$ ratios of 15.595~15.641, and $^{208}Pb/^{204}Pb$ ratios of 38.557~38.827 (Table 4). The Pb isotope data are all plotted adjacent to the orogen line on a $^{207}Pb/^{204}Pb$–$^{206}Pb/^{204}Pb$ diagram (Figure 8b), suggesting a mixed crust–mantle source. Furthermore, the Pb isotope compositions of pyrite are consistent with those of the ore-hosting trachyte porphyry and intermediate–basic volcanic rocks of the Xiaoling Formation (Figure 8b). Previous studies have demonstrated that the Early Cretaceous intermediate–basic volcanic rocks of the Xiaoling Formation in the Xiuyan area were formed by partial melting of ancient lithospheric mantle, metasomatized by subduction fluids [28]. Therefore, the volcanic rocks of the Xiaoling Formation, especially the ore-hosting trachyte porphyry, are closely related to the Dongpuzi gold mineralization.

6.4. Ore genesis and the Metallogenic Model

As stated above, the Dongpuzi gold deposit is located in the southern margin of the Shaozihe volcanic fault basin. The gold orebodies at the Dongpuzi deposit are hosted within the trachyte porphyry and are structurally controlled by the NE–SW- to NNE–SSW-trending faults and fracture alteration zones. The ores are characterized by quartz veins, quartz veinlets, quartz stockworks, or miarolitic structures, indicating that the gold mineralization occurred on the surface or near the surface [32]. The ore mineral assemblage from the mineralogical observations is dominated by pyrite, chalcopyrite, and sphalerite, indicating a low-sulfidation state [33]. The wall-rock alterations include sericitization, silicification, carbonatization, kaolinization, and chloritization and display clear spatial zonations similar to low-sulfidation epithermal deposits [15]. Fluid inclusion studies indicate that the mineralizing fluid is characterized by a low temperature (average 134 °C, <150 °C) and salinity (average 7.2 wt% NaCl equiv.) with a meteoric source (Figure 7a). These characteristics are similar to those of typical low-sulfidation epithermal gold deposits in China and worldwide [34–36]. Therefore, the Dongpuzi gold deposit belongs to a typical low-sulfidation epithermal gold deposit in the Liaodong area.

During the Early Cretaceous period, the Paleo-Pacific Ocean Plate subducted beneath the Eurasian plate [37,38] and subsequently underwent a transformation from compression–subduction to slab rollback-extension in the east of the NCC [38,39], which triggered intense magmatism and related mineralization, as well as a series of extensional structures, such as metamorphic core complex (Liaonan–Wanfu), half graben basins (Benxi, Tongyuanpu, and Dandong), and detachment fault systems (Huanghuadianzi–Dayingzi–Shaozihe) in the Liaodong Peninsula [11,13,40–43]. The retreating of the subducted Paleo-Pacific Ocean Plate induced the underplating of the asthenospheric mantle [38], which further triggered partial melting of the ancient lithospheric mantle in the east of the NCC [28,40,44,45]. The produced mantle-derived magmas were metasomatized by subducting oceanic slab-derived fluids [28]. Through fractional crystallization, the intermediate–basic volcanic rocks of the Xiaoling Formation in the Shaozi River fault basin were thus generated in the Xiuyan area [28], accompanied by intrusion of the ore-hosting trachyte porphyry. The new hydrogen, oxygen, and carbon isotope data indicate that the ore-forming fluids were derived from meteoric water, with important input of mantle-derived carbon. The sulfur and lead isotopes suggest that the sulfur and lead were mainly leached from the ore-hosting trachyte porphyry, with some contributions of the basement rocks (metamorphic rocks of the Gaixian Formation). The trachyte porphyry supplied high heat flow, drove shallow meteoric water, and remobilized metallogenic materials (sulfur, lead, and carbon) from the volcanic rocks (mainly trachyte porphyry) and metamorphic rocks of the Gaixian Formation (Figure 9). The episodic temperature and salinity drops in the mineralizing fluid perhaps resulted in the formation of the Dongpuzi gold deposit.

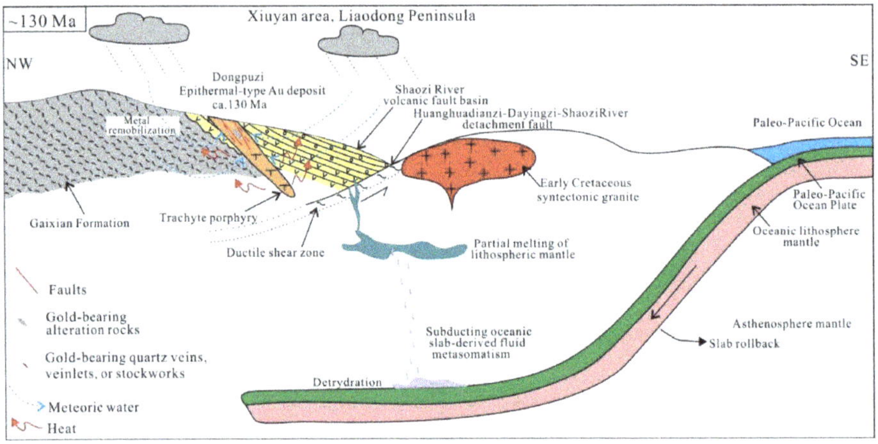

Figure 9. Schematic model explaining the formation of the Dongpuzi gold deposit (the geodynamic setting of the Early Cretaceous in the eastern NCC is after [11,13,40–43]; the diagenesis of intermediate–basic rocks of the Xiaoling Formation is after [28]; the age of ~130 Ma is taken from unpublished U–Pb data of the trachyte porphyry).

7. Conclusions

(1) The Dongpuzi gold deposit was hosted within the Early Cretaceous trachyte porphyry and controlled by the NE–SW- to NNE–SSW-trending faults and fracture zones, with three mineralization stages, i.e., quartz–pyrite, quartz–sulfide, and quartz–calcite stages.

(2) Fluid inclusion studies indicate that the Dongpuzi gold mineralization was precipitated from an epithermal fluid system with low temperatures (113 to 162 °C) and low salinity (average 7.2 wt% NaCl equiv.) and fluid cooling may be an important factor in the formation of this gold deposit.

(3) The hydrogen, oxygen, and carbon isotopes indicate that the ore-forming fluids were sourced from meteoric water and the carbon in fluids is mainly derived from the magma. The sulfur and lead isotopic compositions suggest that the ore-forming materials were mainly derived from the host trachyte porphyry and volcanic rocks of the Xiaoling Formation, with some contributions of Paleoproterozoic metamorphic rocks of the Gaixian Formation.

(4) The Dongpuzi deposit is a typical low-sulfidation epithermal gold deposit in the Liaodong area, which was formed under an extensional setting related to the Early Cretaceous lithospheric extension and thinning of the east of NCC, induced by subduction and retreat of the Paleo-Pacific Ocean Plate.

Author Contributions: Conceptualization, C.C. and T.W.; methodology, J.Z.; formal analysis, Q.S.; investigation, D.S., D.L., Z.Y. and J.Z.; writing—original draft preparation, C.C. and Q.S.; writing—review and editing, C.C., T.W. and D.S.; visualization, Q.S.; supervision, C.C. and T.W.; funding acquisition, D.S. and Z.Y. All authors have read and agreed to the published version of the manuscript.

Funding: This research was funded by the Geological Survey Project of China, grant numbers DD20221695, DD20221678, and DD20190379, and the National Key Research and Development Program of China, grant numbers 2018YFC0603804 and 2016YFC0600108.

Acknowledgments: We are grateful to the anonymous reviewers for their constructive comments, which greatly helped in improving our paper. We would like to thank Xin-hui Ma and Mu Liu from the Analytical Laboratory of the Beijing Research Institute of Uranium Geology, Beijing, China, for their help during the fluid inclusion and C–H–O–S–Pb isotope analyses.

Conflicts of Interest: The authors declare no conflict of interest.

References

1. Liu, J.; Zhang, L.J.; Wang, S.L.; Li, T.G.; Yang, Y.; Liu, F.X.; Li, S.H.; Duan, C. Formation of the Wulong gold deposit, Liaodong gold Province, NE China: Constraints from zircon U–Pb age, sericite Ar–Ar age and H–O–S–He isotopes. *Ore Geol. Rev.* **2019**, *109*, 130–143. [CrossRef]
2. Chen, C.; Li, D.T.; Wu, T.T.; Zhao, Y.; Zhao, C.Q.; Yang, J.L.; Gu, Y.C. Genesis of gold deposits in the Wulong orefield, Liaodong Peninsula, North China Craton: Constraints from ore deposit geology, REE, and C–H–O–S–Pb isotopes. *Geol. J.* **2020**, *55*, 5914–5933. [CrossRef]
3. Zhang, P.; Kou, L.L.; Zhao, Y.; Bi, Z.W.; Sha, D.M.; Han, R.P.; Li, Z.M. Genesis of the Wulong gold deposit, Liaoning Province, NE China: Constraints from noble gases, radiogenic and stable isotope studies. *Geosci. Front.* **2020**, *11*, 547–563. [CrossRef]
4. Yu, G.; Zeng, Q.D.; Frimmel, H.E.; Wang, Y.B.; Guo, W.K.; Sun, G.T.; Zhou, T.C.; Li, J.P. Genesis of the Wulong gold deposit, northeastern North China Craton: Constraints from fluid inclusions, H-O-S-Pb isotopes, and pyrite trace element concentrations. *Ore Geol. Rev.* **2018**, *102*, 313–337. [CrossRef]
5. Chen, J. Geological characteristics and ore-forming control of the Dongpuzi gold deposit in Liaoning Province. *Gold J.* **2000**, *2*, 1–5. (In Chinese)
6. Zhang, B.C.; Qin, G.J.; Wang, F.G. Fluid inclusions of Dongpuzi gold deposit in Xiuyan county, Liaoning Province. *Geoscience* **2002**, *16*, 26–31. (In Chinese)
7. Jiang, Y.Z. Geochemical characteristics of the Dongpuzi gold deposit in Liaoning Province. *Non-Ferr. Min. Metall.* **2017**, *33*, 4–8. (In Chinese)
8. Zhao, G.C.; Sun, M.; Wilde, S.A.; Li, S.Z. Late Archean to Paleoproterozoic evolution of the North China Craton: Key issues revisited. *Precambrian Res.* **2005**, *136*, 177–202. [CrossRef]
9. Liu, J.Y.; Chen, H.; Sha, D.; Wang, H. The inner zone of the Liaoji Paleorift: Its early structural styles and structural evolution. *J. Asian Earth Sci.* **1997**, *15*, 19–31. [CrossRef]
10. Li, S.Z.; Zhao, G.C. SHRIMP U-Pb zircon geochronology of the Liaoji granitoids: Constraints on the evolution of the Paleoproterozoic Jiao-Liao-Ji belt in the Eastern Block of the North China Craton. *Precambrian Res.* **2007**, *158*, 1–16. [CrossRef]
11. Shen, L.; Liu, J.L.; Hu, L.; Ji, M.; Guan, H.M. The Dayingzi detachment fault system in Liaodong Peninsula and its regional tectonic significance. *Sci. China Earth Sci.* **2011**, *41*, 437–451. (In Chinese) [CrossRef]
12. Wu, F.Y.; Lin, J.Q.; Wilde, S.A.; Zhang, X.O.; Yang, J.H. Nature and significance of the Early Cretaceous giant igneous event in eastern China. *Earth Planet. Sci. Lett.* **2005**, *233*, 103–119. [CrossRef]
13. Liu, J.L.; Ji, M.; Shen, L.; Guan, H.M. Early Cretaceous extensional structures in the Liaodong Peninsula: Structural associations, geochronological constraints and regional tectonic implications. *Sci China Earth Sci* **2011**, *54*, 823–842. (In Chinese) [CrossRef]
14. Liaoning Nonferrous Shenyang Geological Exploration Institute. Geological Survey Report of the Dongpuzi gold deposit in Xiuyan County, Liaoning Province. 1998; (Unpublished).
15. Zhang, B.C.; Li, L.; Tan, G.J. Study on gelogical feature and genesis of the Dongpuzi gold deposit, Liaoning, China. *Gold Geol.* **2002**, *8*, 29–36. (In Chinese)
16. Ozard, J.M.; Russel, R.D. Discrimination in solid source lead isotope abundance measurement. *Earth Planet. Sci. Lett.* **1970**, *8*, 331–336. [CrossRef]
17. Guo, L.N.; Goldfarb, R.J.; Wang, Z.L.; Li, R.H.; Chen, B.H.; Li, J.L. A comparison of Jiaojia- and Linglong-type gold deposit ore-forming fluids: Do they differ? *Ore Geol. Rev.* **2017**, *88*, 511–533. [CrossRef]
18. Potter, R.W.; Clynne, M.A. Freezing point depression of aqueous sodium chloride solution. *Econ. Geol.* **1978**, *73*, 284–285. [CrossRef]
19. Clayton, R.N.; O'Neil, J.R.; Mayeda, T.K. Oxygen isotope exchange between quartz and water. *Geophys. Res.* **1972**, *77*, 3057–3067. [CrossRef]
20. Taylor, H.P. The application of oxygen and hydrogen isotope studies to problems of hydrothermal alteration and ore deposit. *Econ. Geol.* **1974**, *69*, 843–883. [CrossRef]
21. Hedenquist, J.W.; Lowenstern, J.B. The role of magmas in the formation of hydrothermal ore deposits. *Nature* **1994**, *370*, 519–527. [CrossRef]
22. Simmons, S.F.; White, N.C.; John, D.A. Geological Characteristics of Epithermal Precious and Base Metal Deposits. In *Economic Geology: One Hundredth Anniversary Volume*; Hedenquist, J.W., Thompson, J.F.H., Goldfarb, R.J., Richards, J.P., Eds.; Society of Economic Geologists: Littleton, CO, USA, 2005; pp. 485–522.
23. Li, L.; Li, S.R.; Santosh, M.; Zhu, J.; Suo, X.J. Early Jurassic decratonic gold melallogenesis in the eastern North China Craton: Constraints from S-Pb-C-D-O isotopic systematics and pyrite Rb-Sr geochronology of the Guilaizhuang Te-Au deposit. *Ore Geol. Rev.* **2018**, *92*, 558–568. [CrossRef]
24. Hoefs, J. *Stable Isotope Geochemistry*, 6th ed.; Springer: Berlin/Heidelberg, Germany, 2009; pp. 130–135.
25. Ray, J.S.; Ramesh, R.; Pande, K. Carbon isotopes in Kerguelen plume derived carbonatites: Evidence for recycled inorganic carbon. *Earth Planet. Sci. Lett.* **1999**, *170*, 205–214. [CrossRef]
26. Zhao, H.Z.; Yang, S.S.; Li, H. Geologic features of Baiyun gold deposit and discussion of the genesis. *Non-Ferous Min. Metall.* **2009**, *25*, 4–8. (In Chinese)
27. Ohmoto, H.; Rye, R.O. Isotopes of sulfur and carbon. In *Geochemistry of Hydrothermal Ore Deposits*, 2nd ed.; Barnes, H.L., Ed.; John Wiley & Sons Inc.: New York, NY, USA, 1979; pp. 509–567.

28. Zhou, H.; Pei, F.P.; Xu, W.L.; Wei, J.Y.; Wang, Y.P. Petrogenesis of the Early Cretaceous volcanic rocks from the northeastern part of the North China Craton: Elemental and Sr-Nd-Pb isotopic constraints. *Lithos* **2021**, *392*, 393106149.
29. Lindgren, W. *Mineral Deposits*; McGraw-Hill Book Company, Inc.: New York, NY, USA; London, UK, 1933.
30. Fournier, R.O. Hydrothermal process related to movement of fluid from plastic into brittle rock in the magmatic–epithermal environment. *Econ. Geol.* **1999**, *94*, 1193–1211. [CrossRef]
31. Gu, F.H.; Zhang, Y.M.; Peng, Y.W.; Wang, J.L.; Liu, R.P. Geology, fluid inclusions and S-Pb-C-O isotopes of the Kuokuqueke Fe-Cu skarn deposit in Western Tianshan, China. *Ore Geol. Rev.* **2022**, *145*, 104896. [CrossRef]
32. Chen, Y.J.; Ni, P.; Fan, H.R.; Pirajno, F.; Lai, Y.; Su, W.C.; Zhang, H. Diagnostic fluid inclusions of different types hydrothermal gold deposits. *Acta Petrol. Sin.* **2007**, *23*, 2085–2108. (In Chinese)
33. Hedenquist, J.W.; Arribas, A., Jr.; Gonzalez-Urien, E. Exploration for epithermal gold deposits. *Rev. Econ. Geol.* **2000**, *13*, 245–277.
34. Ma, Y.; Jiang, S.Y.; Frimmel, H.E. Metallogeny of the Late Jurassic Qicun epithermal gold deposit in sourthern China: Constraints from geochronology, fluid inclusions, and H-O-S-Pb isotopes. *Ore Geol. Rev.* **2022**, *142*, 104688. [CrossRef]
35. Najaran, M.; Mehrabi, B.; Siani, M.G. Mineralogy, hydrothermal alteration, fluid inclusion, and O-H stable isotopes of the SiahJangal-Sar Kahno epithermal gold deposit, SE Iran. *Ore Geol. Rev.* **2020**, *125*, 103689. [CrossRef]
36. Li, S.N.; Ni, P.; Bao, T.; Wang, G.G.; Chi, Z.; Li, W.S.; Zhu, R.Z.; Dai, B.Z.; Xiang, H.L. Geological, fluid inclusion, and H–O–S–Pb isotopic studies of the Xiaban epithermal gold deposit, Fujian Province, southeast China: Implications for ore genesis and mineral exploration. *Ore Geol. Rev.* **2020**, *117*, 103280. [CrossRef]
37. Zhu, R.X.; Fan, H.R.; Li, J.W.; Meng, Q.R.; Li, S.R.; Zeng, Q.D. Decratonic gold deposits. *Sci. China Earth Sci.* **2015**, *58*, 1523–1537. [CrossRef]
38. Zheng, J.P.; Dai, H.K. Subduction and retreating of the western Pacific plate resulted in lithospheric mantle replacement and coupled basin-mountain respond in the North China Craton. *Sci. China Earth Sci.* **2018**, *61*, 406–424. [CrossRef]
39. Zheng, Y.F.; Xu, Z.; Zhao, Z.F.; Dai, L.Q. Mesozoic mafic magmatism in North China: Implications for thinning and destruction of cratonic lithosphere. *Sci. China Earth Sci.* **2018**, *61*, 353–385. [CrossRef]
40. Wu, F.Y.; Yang, J.H.; Xu, Y.G.; Wilde, S.A.; Walker, R.J. Destruction of the North China Craton in the Mesozoic. *Annu. Rev. Earth Planet. Sci.* **2019**, *47*, 173–195. [CrossRef]
41. Yang, J.H.; Wu, F.Y.; Chung, S.L.; Wilde, S.A.; Davis, G.A. Rapid exhumation and cooling of the Liaonan metamorphic core complex: Inferences from $^{40}Ar/^{39}Ar$ thermochronology and implications for Late Mesozoic extension in the eastern North China Craton. *Geol. Soc. Am. Bull.* **2007**, *119*, 1405–1414. [CrossRef]
42. Liu, J.L.; Guan, H.M.; Ji, M.; Hu, L. Late Mesozoic metamorphic core complexes: New constraints on lithosphere thinning in North China. *Prog. Nat. Sci.* **2006**, *16*, 633–638.
43. Qian, Y.; Sun, J.L.; Li, Y.J.; Yu, N.; Liu, J.L.; Li, B.L.; Sun, F.Y. Geochronology, geochemistry and metallogenic dynamics of gold-polymetallic deposits in Xicha region, the northeastern margin of the North-China Platform. *Acta Petrol. Sin.* **2020**, *36*, 1127–1150. (In Chinese)
44. Pei, F.P.; Xu, W.L.; Yang, D.B.; Yu, Y.; Wang, W.; Zhao, Q.G. Geochronology and geochemistry of Mesozoic mafic-ultramafic complexes in the southern Liaoning and southern Jilin Provinces, NE China: Constraints on the spatial extent of destruction of the North China Craton. *J. Asian Earth Sci.* **2011**, *40*, 636–650. [CrossRef]
45. Ma, Q.; Xu, Y.G.; Zheng, J.P.; Griffin, W.L.; Hong, L.B.; Ma, L. Coexisting Early Cretaceous High-Mg andesites and adakitic rocks in the North China Craton: The role of water in intraplate magmatism and cratonic destruction. *J. Petrol.* **2016**, *7*, 1279–1308. [CrossRef]

minerals

Article

The Early Cretaceous Granitoids and Microgranular Mafic Enclaves of Sanguliu Pluton, the Liaodong Peninsula: Implications for Magma Mixing and Decratonic Gold Mineralization in the Eastern North China Craton

Taotao Wu, Cong Chen *, Dongtao Li, Yan Zhao, Chunqiang Zhao and Yongheng Zhou

Shenyang Center of Geological Survey, China Geological Survey, Shenyang 110034, China
* Correspondence: chencong@mail.cgs.gov.cn

Abstract: Some Early Cretaceous granitoids characterized by abundant mafic microgranular enclaves (MMEs) formed by magma mixing have been associated with gold deposits in the eastern North China Craton (NCC). However, the genetic connection of magma mixing with gold mineralization remains unclear. The zircon U–Pb ages and in situ Lu-Hf isotopic compositions, whole-rock major- and trace-element and Sr–Nd–Pb isotopic compositions, as well as EPMA biotite compositions, were presented for the Sanguliu granodiorite and enclaves in the Liaodong Peninsula in order to obtain insights into the spatial and temporal distribution, and internal connection of magma mixing with the decratonic gold deposits in the eastern NCC. The Sanguliu granodiorite yielded coeval formation ages with the enclaves (~123 Ma), and their acicular apatites and plagioclase megacrysts suggest that the enclaves were formed by mixing between mafic and felsic magmas. Geochemically, the Sanguliu granodiorite is high-K calc-alkaline I-type granite, with an initial $^{87}Sr/^{86}Sr$ ratio of 0.70552 to 0.71470 and strongly negative $\varepsilon_{Nd}(t)$ (−11.4 to −21.3) and zircon in situ $\varepsilon_{Hf}(t)$ values (−15.1 to −25.4), indicating that the felsic magmas were ancient lower crust with the involvement of mantle-derived materials. Meanwhile, the enclaves have high MgO (4.18 to 6.17 wt.%), Cr (45.91 to 290.04 ppm), and Ni (19.65 to 88.18 ppm) contents, with high $Mg^{\#}$ values of 50 to 57 at intermediate SiO_2 contents (53.68 to 55.78 wt.%), highly negative $\varepsilon_{Nd}(t)$ values (mostly −18.42 to −22.03), and in situ zircon $\varepsilon_{Hf}(t)$ values (−18.6 to −22.7), indicating that the mafic magma was mainly derived from the enriched lithospheric mantle. Furthermore, the biotites from the Sanguliu granodiorite clustered between the MH and NNO buffers in the Fe^{2+}–Fe^{3+}–Mg diagram. This, combined with the high Ce/Ce* ratios (1.30 to 107.18) of the zircons, indicates that the primary magmas forming the Sanguliu granodiorite had a high oxygen fugacity, which is favorable for gold mineralization. These findings, together with previous studies of the Early Cretaceous granitoids with enclaves in the eastern NCC, suggest that magma mixing commonly occurred during 110–130 Ma and is temporally, spatially, and genetically related to decratonic gold systems in eastern NCC.

Keywords: mafic microgranular enclaves; magma mixing; high oxygen fugacity; enriched lithospheric mantle; decratonic gold deposit; eastern North China Craton

1. Introduction

Granites commonly contain lithic inclusions or bodies, generally referred to as enclaves, whose nature and origin are documented [1,2]. Considering their occurrences, microstructural features, and relationship with host granites, enclaves have been grouped into (1) xenoliths, as solid fragments of wall rock, (2) restite, as segregated refractory source materials left after a partial melting event, (3) autolith, as early crystallized cumulus phases or segregated mafic phases or fragments of a border rock series of a felsic pluton, (4) microgranular enclaves representing felsic, mafic, and mafic-felsic (intermediate) hybridized magmas entrained and mingled into felsic host magma, and (5) fragments of synplutonic mafic dyke intrusions [2]. In general, mafic microgranular enclaves (MMEs) are indicators of magma mingling and mixing processes and provide critical information regarding magma chamber processes and dynamics [2–4].

The eastern North China Craton (NCC) is characterized by widespread Early Cretaceous granitic intrusions. Some of these granitic intrusions contain abundant MMEs, which are considered a result of magma mixing and mingling via the intrusion of mafic magmas into felsic magma chambers [5–11]. Previous studies have mainly focused on field observations, the petrological characteristics of the host granitoids and MMEs, and the geochemistry of the host granitoids [5,6,9]. However, further MME information is limited, especially regarding their formation ages, magma sources, and mechanisms. Moreover, only the Gudaoling intrusion-enclosing MMEs in the Liaodong Peninsula have been generally researched. The intrusion is considered to have formed through a complex, multi-stage hybridization process that involved mantle- and crustal-derived magmas and several concomitant magmatic processes [7,8].

Furthermore, the mixing and mingling processes involving crustal- and mantle-derived magma may contribute to the formation of endogenetic metal deposits, especially gold deposits [10,12,13]. The Early Cretaceous granitic plutons enclosing MMEs formed by magma mixing and mingling, such as the Guojialing and Weideshan granitic intrusions, may be associated with gold deposits in the Jiaodong Peninsula in the eastern NCC [10]. However, the role of magma mixing in gold mineralization remains unclear. New field observations have identified the occurrence of MMEs in the Sanguliu intrusion, which hosts the large-scale Wulong gold deposit in the Liaodong Peninsula. The petrogenesis of the Sanguliu intrusion has not yet been investigated. The genetic relationships between the Sanguliu intrusion and the Wulong gold deposit remain controversial. Li et al. proposed that the Sanguliu intrusion is a metallogenetic intrusion of the Wulong gold deposit [14]. However, Wei et al. believed that the Sanguliu intrusion did not directly provide any ore-forming fluids or materials for Wulong gold mineralization but had similar Sr and Nd isotopic compositions to the deeper source material of the Wulong deposit [15]. The existence of deeper magma as the metallogenic intrusion has also been supported by Yu et al. [16] and Chen et al. [17]. Therefore, research on the Sanguliu intrusion may provide critical information regarding the relationship between magma mixing and gold mineralization.

On the basis of the field observation and petrography, this paper presents new zircon U-Pb ages and in situ Lu-Hf isotopes, whole-rock major and trace elements and Sr-Nd-Pb isotopic data of the MMEs and host Sanguliu granodiorite. The mineral compositions of biotite from the Sanguliu granodiorite were also analyzed. Based on the results, the petrogenesis and magma source of the Sanguliu granodiorite and monzodioritic enclaves were determined. We also discuss the relationship between the Sanguliu intrusion and the Wulong gold deposit and propose a new petrogenetic-dynamic model for these rocks. Furthermore, we review the published data for other Early Cretaceous granitoids enclosing MMEs in the Jiaodong, Liaodong, and Taihang areas in order to constrain the temporal and spatial distribution and identify any genetic relation between decratonic gold deposits and magma mixing in the eastern NCC.

2. Regional Geology

The Liaodong Peninsula is located in the eastern part of the NCC at the western margin of the Pacific Plate (Figures 1 and 2a). The exposed rocks in the area comprise metamorphosed Archean-Paleoproterozoic basement, Mesoproterozoic-Paleozoic post-cratonization cover sequences, and a series of Mesozoic intrusive and volcanic rocks (Figure 2b). The metamorphosed basement sequences include the Archean supracrustal rocks and diorite–tonalite–granodiorite suites, and the Paleoproterozoic metamorphosed sedimentary and volcanic successions of the Liaohe Group, as well as associated granitic and mafic intrusions. The Paleoproterozoic sedimentary and volcanic successions of the Liaohe Group metamorphosed from greenschist to the lower amphibolite facies and locally up to the granulite facies [18], which are conventionally divided into five formations:

Liangzishan, Li'eryu, Gaojiayu, Dashiqiao, and Gaixian Formations [19]. The lowermost Liangzishan Formation is composed of basal conglomerate-bearing quartzites, transitioning upwards to chlorite-sericite quartz schists, phyllites, garnet-bearing mica schists, minor graphite-bearing garnet-staurolite mica schists, and kyanite-bearing mica schists. It is conformably overlain by the Li'eryu and Gaojiayu Formations, which consist of boron-bearing fine-grained felsic gneiss, amphibolites, and mica quartz schist. Overlying the Gaojiayu Formation is the Dashiqiao Formation, which is composed predominantly of dolomitic marbles intercalated with minor carbonaceous slates and mica schists. The uppermost Gaixian Formation comprises phyllites, andalusite-cordierite mica schists, staurolite mica schists, and sillimanite mica schists, with minor quartzite and marble present.

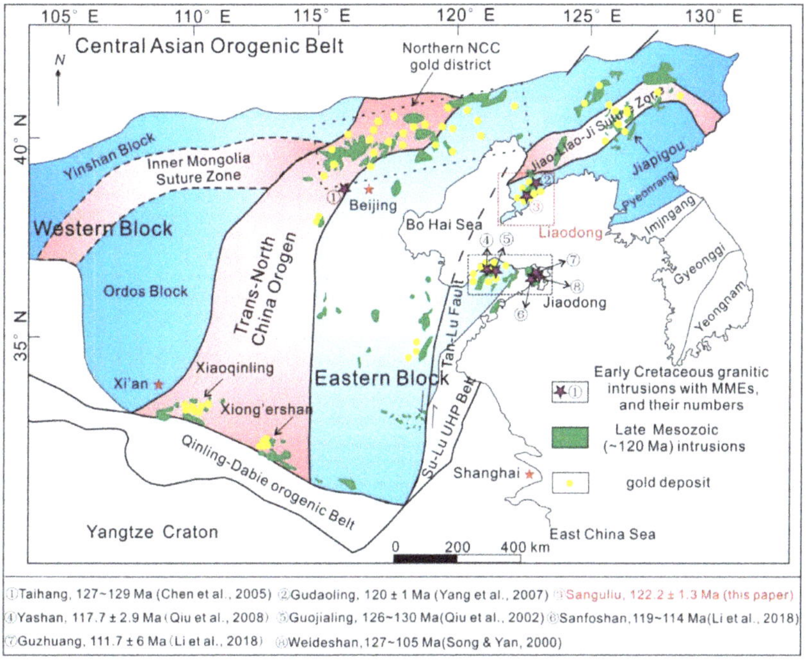

Figure 1. Tectonic location and division of the North China Craton (the zircon U–Pb ages from [6,8,11,20–23], modified from [24]). The locations of the mapped areas in Figure 2 have been marked by the red rectangle.

Voluminous Paleoproterozoic granitoid and mafic intrusions are also exposed in this area (Figure 2b). These mafic intrusions consist of gabbros and dolerites, most of which have been metamorphosed into greenschists and amphibolites, and the granitoid plutons are composed of deformed monzogranitic gneisses (traditionally called the Liaoji Granites), undeformed porphyritic monzogranites and granites, and alkaline syenites [26].

After the Paleoproterozoic rifting/orogenic events, this area was partially covered by Mesoproterozoic to Paleozoic metasediments (Figure 2b) [27]. During the Late Triassic, the area was affected by either the closure of the Paleo-Asian Ocean or the amalgamation of the Yangtze Craton (YC) and NCC to be featured by the local emplacement of Middle-Late Triassic granite (Figure 2b). Since the Late Mesozoic, the Liaodong Peninsula has become an important part of the circum-Pacific tectono-magmatic zone and is characterized by intensive tectonic, magmatic, and metallogenic events [28,29]. The Late Mesozoic intrusions are mainly divided into (1) Jurassic (180–153 Ma) tonalite, diorite, and gneissic two-mica monzogranite, which have experienced ductile deformation, and (2) Early Cretaceous (131–100 Ma) undeformed to slightly deformed diorite, granodiorite, monzogranite, and syenogranite (Figure 2b) [8,30]. Accompanied by the widespread Late Mesozoic granitoids, several episodes of volcanism, terrestrial deposition, extensional structures, and gold mineralization occurred.

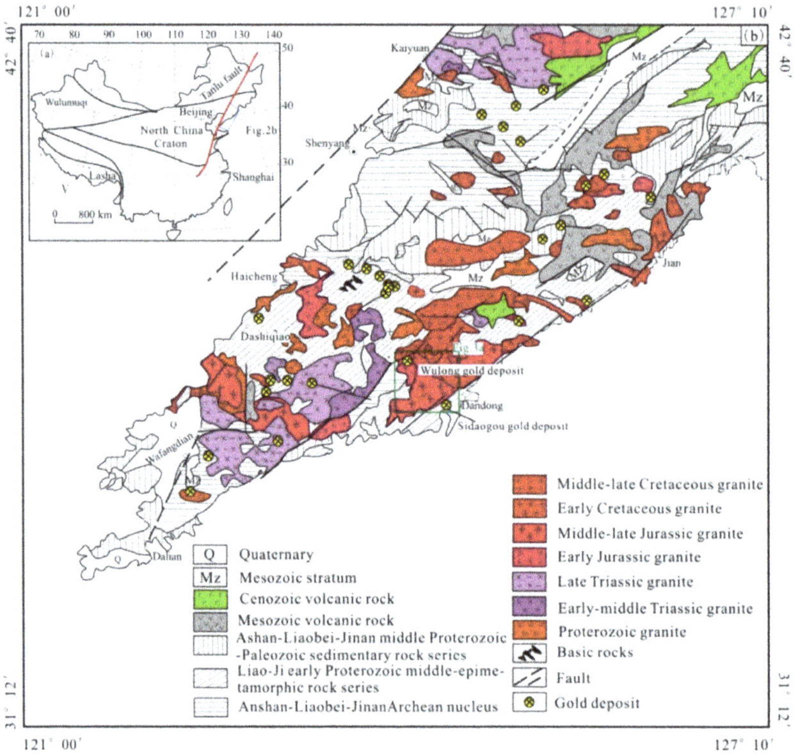

Figure 2. (**a**) Location map of the Liaodong Peninsula. (**b**) Sketch map showing the regional geology and distribution of gold deposits in the Liaodong Peninsula (modified from [25]).

3. Geology and Petrography of the Sanguliu Pluton

The Sanguliu pluton is situated in the middle of the Wulong ore field in Dandong City (Figure 3a), which contains the large tonnage of Wulong and Sidaogou gold deposits, as well as many other small gold deposits and occurrences (Figure 2a). This pluton is cut by the auriferous quartz veins of the large Wulong gold deposit (Figure 3b) and has slight sericitization and disseminated pyritization. It occurs as a stock with an outcrop of 40–50 km^2 intruding into the Paleoproterozoic metamorphic rocks of the Liaohe Group and Jurassic Wulong gneissic two-mica granite with a sensitive high-resolution ion micro-probe (SHRIMP) zircon U-Pb age of 158 ± 2 Ma [31] and is cut by a series of north-northeast-trending faults (Figure 3a). The Sanguliu intrusion yields obvious facies zoning, wherein its lithology changes from monzogranite in the center to granodiorite and quartz diorite at the border (Figure 3a).

With the exception of the xenoliths of the metamorphic rocks of the Liaohe Group, monzodioritic enclaves are ubiquitous in the Sanguliu intrusion and are oval or elliptical, ranging from five to forty centimeters in diameter (Figure 4a,b). In general, these enclaves have sharp contact with the host Sanguliu intrusion (Figure 4c). The petrography of the host Sanguliu granodiorite and its enclave studied herein are described as follows.

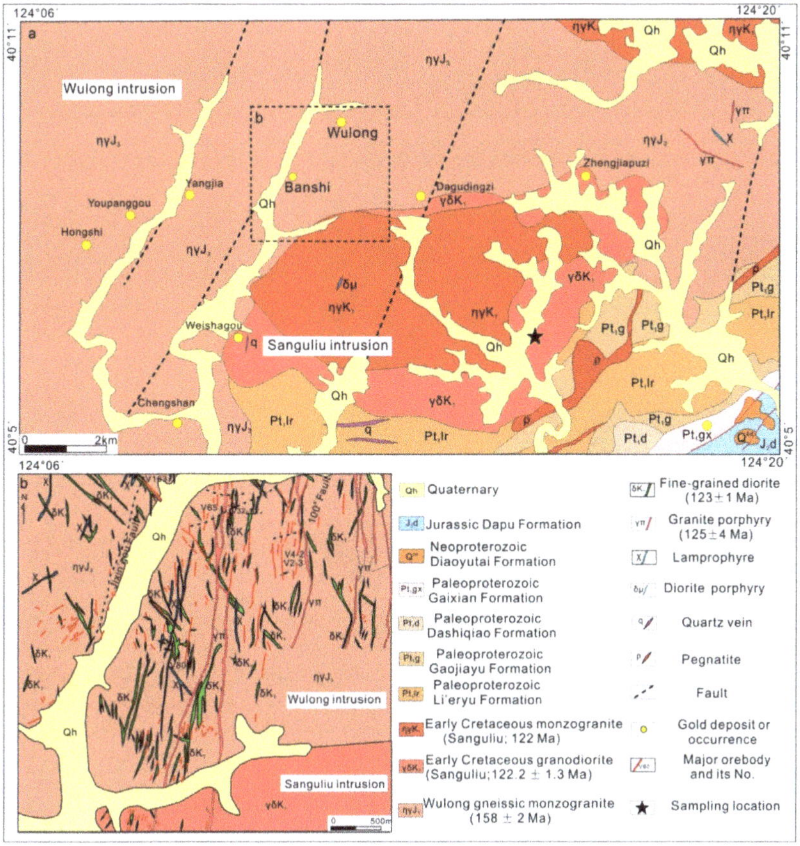

Figure 3. Geological maps of the Wulong gold orefield (**a**) and Wulong gold deposit (**b**) (modified from [16]. The geochronological data are from [30–32]).

The Sanguliu granodiorite has a medium–fine-grained granular texture and consists of quartz (20%–30%), plagioclase (35%–45%), potassium feldspar (15%–25%), biotite (10%–15%), and minor hornblende (approximately 5%), as well as accessory minerals of titanite and apatite (Figure 4c). Note that plagioclase commonly develops compositional zoning.

Meanwhile, the monzodiorite enclave has, in general, a hypautomorphic granular texture and comprises plagioclase (40%–45%), potassium feldspar (15%–20%), hornblende (approximately 20%), biotite (5%–10%), and minor quartz (<5%), as well as accessory minerals of apatite and zircon (Figure 4c–f). Furthermore, some plagioclase within the enclave occurs as an elliptic chadacryst that is trapped from the host granodiorite and develops resorption and zonal textures (Figure 4d,e). The apatite is always acicular with a length–width ratio of 10:1–20:1 (Figure 4f).

Figure 4. Photographs (**a**,**b**) and microphotographs (**c**–**f**) of the Sanguliu granodiorite and enclaves. (**a**,**b**) The monzodioritic enclaves occur in elliptic and marquise shape within the Sanguliu granodiorite; (**c**) The host granodiorite and enclave have sharp contact; (**d**) The plagioclase within the host granodiorite has zonal texture; (**e**,**f**) The monzodioritic enclave has plagioclase of zonal texture and acicular apatite. Apl—Apatite; Bt—Biotite; Hbl—Hornblende; Kfs—Potassium feldspar; Pl—Plagioclase; Qtz—Quartz.

4. Analytical Methods

4.1. Zircon LA–ICP–MS and SHRIMP U–Pb Isotopic Dating

Separation, target preparation, and cathodoluminescence (CL) images of zircon from Sanguliu granodiorite (sample SGL) were carried out by the Hongxin Geology and Exploration Technology Service Limited Company of Hebei Province, China, and laser ablation–inductively coupled plasma–mass spectrometry (LA–ICP–MS) zircon U–Pb dating was completed at Key Laboratory of Mineral Resources Evaluation in Northeast Asia, Ministry of Land and Resource.

Target preparation, CL images, and the SHRIMP U–Pb isotopic dating of zircon from the monzodiorite enclave (sample BTZ-1) were performed at the Beijing Ion Probe Center of the China Institute of Geological Sciences. Zircon SHRIMP U–Pb isotopic dating was completed using a SHRIMP II instrument. The primary O^{-2} flow intensity and beam spot diameter were maintained at 3–5 nA and 25 μm, respectively. The standard sample M257 (U = 840 $\times$ 10^{-6}, [33]) and TEM (age, 417 Ma; [34]) were used to correct the zircon U content and age, respectively. The standard zircon TEM was analyzed after investigating all three unknown samples. SQUID and ISOPLOT software were used for data processing [35]. The measured ^{204}Pb content was used to correct for ordinary lead, and $^{207}Pb/^{206}Pb$ (more than 1000 Ma) or $^{206}Pb/^{238}U$ (less than 1000 Ma) ages were adopted for the zircon age. Details of the testing methods and experimental procedures, as well as related parameters and error corrections, were identical to those applied in Williams [36].

4.2. Zircon In Situ Lu–Hf Isotopes

Zircon in situ Lu–Hf isotopic analyses were conducted at the State Key Laboratory for Endogenic Metal Deposits Research, Nanjing University, using a Neptune Laser Ablation Multiple-Receiver LA–ICP–MS with a 193 nm laser sampling system. The laser spot diameter was 60 μm, the pulse width was 15 ns, and the abrasion material carrier gas was He. Detailed analytical procedures are provided in the literature [37,38]. Measurements of the 91500 and MT zircons were conducted after every ten sample analyses. These standard samples, that is, 91500 and MT, had $^{176}Hf/^{177}Hf$ ratios of 0.282316 ± 30 and 0.282507 ± 50, respectively. In the $\varepsilon_{Hf}(t)$ and model age calculations, the ^{176}Lu decay constant was 1.867^{-11} [39], and the $^{176}Lu/^{177}Hf$ and $^{176}Hf/^{177}Hf$ ratios of the present-day chondritic and depleted mantle were 0.0332 and 0.282772 [40], and 0.0384 and 0.28325, respectively [41]. Additionally, the calculation of the crustal model ages determined an average crustal value of $^{176}Lu/^{177}Hf$ = 0.015 [42].

4.3. Whole-Rock Geochemistry

A total of 21 fresh and unaltered samples for the Sanguliu granodiorite (16) and monzodiorite enclave (5) were selected for major, trace, and rare earth element (REE) compositions analyses, which were conducted at the Northeast China Supervision and Inspection Center of Mineral Resources, Ministry of Natural Resources of China. The major element compositions were analyzed using a Philips PW2404 type X-ray fluorescence spectrometer, which has an analytical precision of better than 5%. Meanwhile, the trace element and REE compositions were determined using a nHR–ICP–MS (Element I), which has a precision of better than 5% if the contents of trace elements were more than 10 ppm, and better than 10% if their contents were less than 10 ppm.

4.4. Sr–Nd–Pb Isotopes

A total of 10 fresh and unaltered samples of the Sanguliu granodiorite (5) and monzodiorite enclave (5) were selected for Rb–Sr, Sm–Nd, and Pb isotope analyses, which were conducted at the Analytical Laboratory of Beijing Research Institute of Uranium Geology using an ISOPROBE-T type thermoelectric ionization mass spectrometer (TIMS). The mass fractionations of the Rb–Sr and Sm–Nd isotopes were corrected using $^{86}Sr/^{88}Sr$ = 0.1194 and $^{146}Nd/^{144}Nd$ = 0.7219, respectively, with corresponding standard results of NBS987 = 0.710250 ± 7 and $^{143}Nd/^{144}Nd$ = 0.512109 ± 3 (JMC). The experiment background values of Rb, Sr, Sm, and Nd were 2×10^{-10} g, 2×10^{-10} g, $<50 \times 10^{-12}$ g, and $<50 \times 10^{-12}$ g, respectively.

The Pb isotopes were identified via static acceptance and the uncorrected results of NBS981 were $^{208}Pb/^{206}Pb$ = 2.164940 ± 15, $^{207}Pb/^{206}Pb$ = 0.914338 ± 7, and $^{204}Pb/^{206}Pb$ = 0.591107 ± 2. The experimental background Pb content was $<100 \times 10^{-12}$ g.

4.5. Mineral Chemical Compositions

Mineral compositions were determined at the State Key Laboratory of Geological Processes and Mineral Resources, China University of Geosciences (Wuhan), using a JEOL JXA-8100 Electron Probe Micro Analyzer (EPMA) equipped with four wavelength-dispersive spectrometers. Prior to analysis, the samples were coated with a thin conductive carbon film. Precautions suggested by Zhang and Yang [43] were employed to minimize the differences in the carbon film thickness between samples in order to obtain a uniform coating of approximately 20 nm. During the analysis, an accelerating voltage of 15 kV, beam current of 20 nA, and 10 μm spot size were used. The data were corrected online using a modified ZAF (atomic number, absorption, fluorescence) correction procedure. The peak counting times were 10s for F, Cl, Na, Mg, Al, Si, K, Ca, and Fe, and 20s for Ti and Mn. The background counting time was half of the peak counting time for the high- and low-energy background positions. The following standards were used: Fluorite (F), Halite (Cl), Sanidine (K, Al), Almandine (Fe), Diopsode (Ca, Mg), Jadeite (Na), Rhodonite (Mn), Olivine (Si), and Rutile (Ti).

5. Results

5.1. Zircon U–Pb Dating

The LA–ICP–MS and SHRIMP zircon U–Pb dating data for the granodiorite intrusion (sample SGL) and monzodioritic enclave (sample BT1) are listed in Tables 1 and 2, respectively. The analyzed zircon grains from the granodiorite and monzodiorite are euhedral–subhedral, colorless, exhibit weak or clear oscillatory growth zoning in the CL images (Figure 5), and have Th/U ratios of 0.78 to 1.47 and 0.47 to 1.79, respectively. These zircon characteristics are indicative of magmatic origins [44].

Table 1. Zircon LA–ICP–MS U–Pb data of the host Sanguliu granodiorite.

Spot	Th (ppm)	U (ppm)	Th/U	$^{207}Pb/^{206}Pb$	1σ	$^{207}Pb/^{235}U$	1σ	$^{206}Pb/^{238}U$	1σ	$^{206}Pb/^{238}U$ Age (Ma)	1σ
SGL-1	314.54	394.93	0.80	0.04795	0.00217	0.12568	0.00579	0.01901	0.00048	121	3
SGL-2	339.58	305.64	1.11	0.05641	0.00232	0.14493	0.00611	0.01863	0.00047	119	3
SGL-3	113.06	97.69	1.16	0.0489	0.00437	0.12844	0.01139	0.01905	0.00056	122	4
SGL-4	253.45	237.22	1.07	0.04802	0.00233	0.12568	0.00617	0.01898	0.00049	121	3
SGL-5	268.08	343.25	0.78	0.04829	0.00208	0.12706	0.0056	0.01908	0.00048	122	3
SGL-6	247.74	228.56	1.08	0.05051	0.00291	0.12866	0.00742	0.01848	0.0005	118	3
SGL-7	180.27	207.67	0.87	0.04666	0.00304	0.12289	0.00798	0.0191	0.00052	122	3
SGL-8	271.5	250.81	1.08	0.04914	0.00231	0.12864	0.00614	0.01899	0.00049	121	3
SGL-9	130.22	140.73	0.93	0.05335	0.00326	0.14075	0.00859	0.01914	0.00052	122	3
SGL-10	282.68	321.85	0.88	0.04916	0.00237	0.12945	0.0063	0.0191	0.00049	122	3
SGL-11	292.1	319.06	0.92	0.04859	0.00198	0.12915	0.0054	0.01928	0.00048	123	3
SGL-12	329.77	317.11	1.04	0.05033	0.00217	0.13316	0.00586	0.01919	0.00049	123	3
SGL-13	108.91	89.63	1.22	0.07339	0.00433	0.20095	0.01176	0.01986	0.00056	127	4
SGL-14	76.28	71.69	1.06	0.04858	0.0041	0.12803	0.01073	0.01912	0.00055	122	3
SGL-15	108.26	98.94	1.09	0.05052	0.00377	0.13311	0.00986	0.01911	0.00054	122	3
SGL-16	96.35	72.88	1.32	0.05013	0.00466	0.1313	0.01209	0.019	0.00057	121	4
SGL-17	252.7	290.28	0.87	0.04837	0.00203	0.12774	0.00547	0.01915	0.00048	122	3
SGL-18	83.58	76.59	1.09	0.0665	0.00423	0.18242	0.01152	0.0199	0.00056	127	4
SGL-19	157.08	176.61	0.89	0.05092	0.0026	0.13154	0.00677	0.01874	0.00049	120	3
SGL-20	234.14	219.76	1.07	0.0481	0.00244	0.12838	0.00656	0.01936	0.0005	124	3
SGL-21	333.06	263.03	1.27	0.04828	0.00217	0.12577	0.00573	0.01889	0.00048	121	3
SGL-22	67.03	61.7	1.09	0.04893	0.00433	0.13275	0.01165	0.01968	0.00058	126	4
SGL-23	61.03	60.07	1.02	0.04927	0.00413	0.13373	0.01114	0.01968	0.00056	126	4
SGL-24	83.76	57.04	1.47	0.04748	0.00387	0.12773	0.01034	0.01951	0.00055	125	3

Table 2. Zircon SHRIMP U-Pb data of the monzodiorite enclaves.

Spot	$^{206}Pb_c$ (%)	U (ppm)	Th (ppm)	$^{232}Th/^{238}U$	^{206}Pb * (ppm)	^{207}Pb */^{206}Pb *	1σ (%)	^{207}Pb */^{235}U	1σ (%)	^{206}Pb */^{238}U	1σ (%)	Errors	$^{206}Pb/^{238}U$ Age (Ma)	$^{207}Pb/^{206}Pb$ Age (Ma)
BTZ-1-1	0.27	73	90	1.28	1.24	0.055	20	0.15	20	0.01965	2.7	0.135	124.3 ± 3.3	432 ± 440
BTZ-1-2	0.59	237	209	0.91	3.93	0.0483	8.3	0.128	8.6	0.01918	2.1	0.241	122.5 ± 2.5	115 ± 200
BTZ-1-3	0.94	121	98	0.83	2.09	0.0505	14	0.139	15	0.01991	2.3	0.159	126.8 ± 2.7	219 ± 330
BTZ-1-4	1.03	105	122	1.20	1.71	0.043	25	0.112	25	0.01864	2.6	0.104	119.8 ± 2.7	−140 ± 630
BTZ-1-5	1.55	98	94	0.99	1.7	0.0445	14	0.122	14	0.01989	2.4	0.166	127.6 ± 2.9	−80 ± 350
BTZ-1-6	0.56	371	288	0.80	6.33	0.0449	7.8	0.1223	8	0.01975	2	0.252	126.6 ± 2.5	−60 ± 190
BTZ-1-7	1.41	286	358	1.29	6.78	0.2535	3.3	0.951	3.9	0.02721	2	0.523	129.1 ± 3.1	3207 ± 52
BTZ-1-8	1.66	96	100	1.07	1.65	0.0455	16	0.123	16	0.0196	2.9	0.176	125.6 ± 3.5	−29 ± 390
BTZ-1-9	0.41	394	269	0.71	6.89	0.0475	5.6	0.1328	5.9	0.02027	2	0.338	129.6 ± 2.5	76 ± 130
BTZ-1-10	0.65	160	277	1.79	2.74	0.0462	14	0.126	14	0.0198	2.2	0.156	126.7 ± 2.7	9 ± 340
BTZ-1-11	1.51	108	172	1.65	1.79	0.0391	18	0.103	19	0.0191	2.5	0.132	123.4 ± 2.8	−413 ± 480
BTZ-1-12	0.10	179	89	0.51	45.2	0.1086	1.3	4.4	2.4	0.294	2	0.838	1649 ± 32	1776 ± 24
BTZ-1-13	0.86	158	139	0.91	2.75	0.0447	15	0.124	16	0.02005	2.3	0.145	128.6 ± 2.7	−72 ± 380
BTZ-1-14	0.95	266	204	0.79	4.49	0.0427	5.4	0.1146	5.7	0.01945	2	0.353	125.1 ± 2.5	−183 ± 130

Note: Errors = 1 σ; Pb and Pb indicate common and radiogenic portions, respectively. Common Pb corrected using measured Pb.

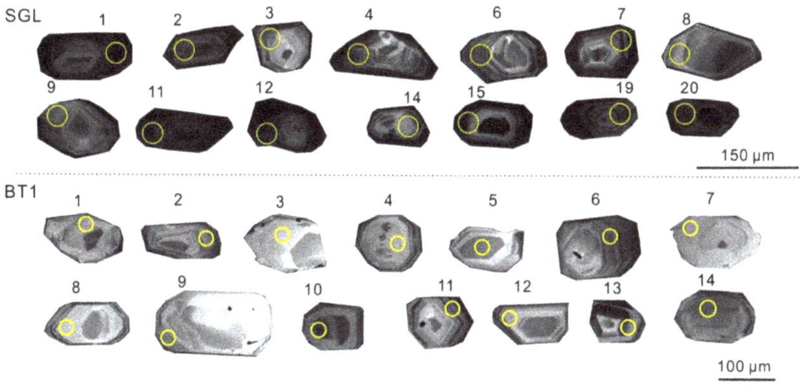

Figure 5. Cathodoluminescence (CL) images of zircons from the host Sanguliu granodiorite and monzodioritic enclaves.

Twenty-four granodiorite zircon analyses formed a relatively tight concordia cluster (Figure 6a) and their $^{206}Pb/^{238}U$ ages ranged from 118 ± 3 to 127 ± 4 Ma, with a weighted mean age of 122.2 ± 1.3 Ma (MSWD = 0.45, N = 24) (Figure 6b), which may represent the final emplacement age of the host granodiorite.

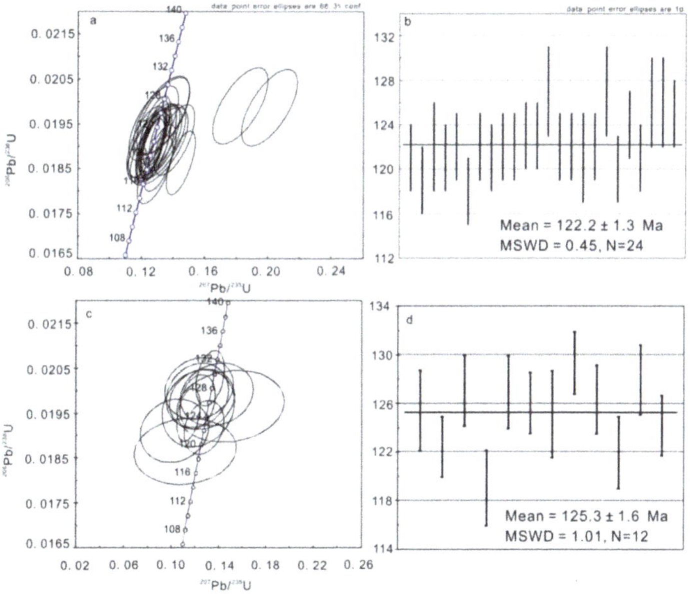

Figure 6. LA–ICP–MS and SHRIMP zircon U–Pb concordia and weighted mean age diagrams of the Sanguliu granodiorite (**a**,**b**) and monzodioritic enclave (**c**,**d**).

Similarly, thirteen zircons' analyses from the monzodioritic enclaves resulted in a tight cluster on a concordia (Figure 6c) and their $^{206}Pb/^{238}U$ ages range from 119.1 ± 3.1 to 129.4 ± 2.6 Ma, with a weighted mean age of 125.3 ± 1.6 Ma (MSWD = 1.01, N = 12) (Figure 6d), which could be considered to be the crystallization age of the monzodioritic enclave. Furthermore, one zircon spot yielded an older $^{207}Pb/^{206}Pb$ age of 1776 ± 24 Ma, which may have been inherited from the Paleoproterozoic Liaohe Group.

5.2. Zircon In Situ Lu–Hf Isotopes

Lu–Hf isotopic analyses were conducted on the 13 zircon grains with concordant SHRIMP U–Pb ages, and the results are listed in Table 3. The $^{176}Lu/^{177}Hf$ ratios for 13 spots ranged from 0.000362 to 0.00125, with $\varepsilon_{Hf}(t)$ values ranging from −18.6 to −22.7, and T_{DM2} model ages of 2404 to 2613 Ma (Figure 7).

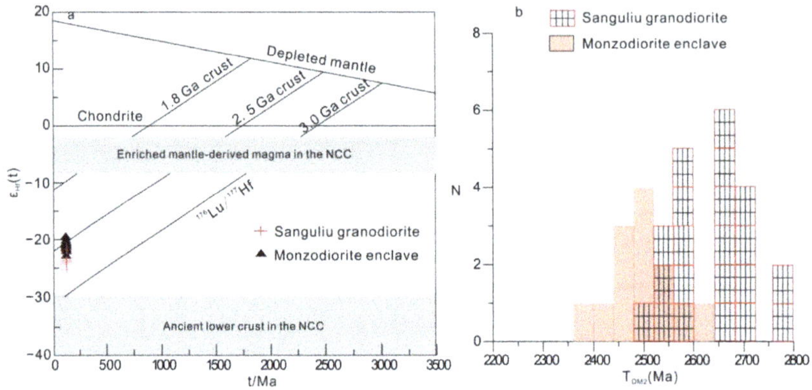

Figure 7. Evolution $\varepsilon_{Hf}(t)$ from in situ analysis of zircon Hf versus age (Ma, (**a**)) and histogram of T_{DM2} model ages (Ma, (**b**)) of the Sanguliu granodiorite and monzodioritic enclaves. The data of enriched mantle-derived magma and ancient lower crust in the NCC, the Sanguliu granodiorite, as well as enclaves are from [45–47], respectively.

5.3. Major and Trace Elements

Major and trace elemental data for the Sanguliu granodiorite and monzodioritic enclave are listed in Table 4. The Sanguliu granodiorite has medium SiO_2 (67.54 to 68.45 wt.%), medium Al_2O_3 (14.82 to 15.39 wt.%), K_2O (4.20 to 4.40 wt.%), and low MgO (1.51 to 1.72 wt.%) and Na_2O (3.22 to 3.46 wt.%) contents, with total alkalis and Na_2O/K_2O ratios ranging from 7.50 to 7.86 wt.%, and from 0.75 to 0.82, respectively. It also has medium values of aluminum saturation index (A/CNK = 0.92–1.02) and $Mg^{\#}$ values ranging from 43.73 to 47.40. Conversely, the monzodioritic enclave has low SiO_2 (54.17 to 55.78 wt.%) and Na_2O (2.15 to 4.67 wt.%), variable Al_2O_3 (12.92 to 17.07 wt.%) and K_2O (1.93 to 5.95 wt.%), and high MgO (4.18to 6.17 wt.%) and CaO (5.45 to 6.53 wt.%) contents, with total alkalis contents of 6.56 to 8.32 wt.%. The monzodioritic enclave also exhibits low values of aluminum saturation index (A/CNK = 0.62–0.84), and $Mg^{\#}$ values ranging from 49.76 to 56.62. In the total alkali–silica diagram, the Sanguliu granodiorite and monzodiorite enclave samples fell into the sub-alkalic granodiorite and alkalic monzodiorite-monzonite fields, respectively (Figure 8a), while the Sanguliu granodiorite samples plotted in the high-K calc-alkaline field in the SiO_2–K_2O diagram (Figure 8b), and all the evaluated samples plotted into the metaluminous field of the A/NK–A/CNK diagram (Figure 8c). These results indicate that the Sanguliu granodiorite is high-K calc-alkaline and metaluminous, whereas the monzodiorite is alkalic and metaluminous.

Table 3. Zircon in situ Lu–Hf isotopic data for the host Sanguliu granodiorite and monzodiorite enclaves.

Sample	SHRIMP Age (Ma)	$^{176}Yb/^{177}Hf$	2σ	$^{176}Lu/^{177}Hf$	2σ	$^{176}Hf/^{177}Hf$	2σ	$^{176}Hf/^{177}Hf_i$	$e_{Hf}(0)$	$e_{Hf}(t)$	2σ	T_{DM} (Ma)	T_{DM}^C (Ma)	$f_{Lu/Hf}$
BTZ-1-1	124.3	0.033152	0.000545	0.001250	0.000021	0.282056	0.000019	0.282053	−25.3	−22.7	0.7	1697	2614	−0.96
BTZ-1-2	122.5	0.012728	0.000112	0.000462	0.000004	0.282094	0.000019	0.282093	−24.0	−21.3	0.7	1610	2527	−0.99
BTZ-1-3	126.8	0.010377	0.000069	0.000362	0.000002	0.282114	0.000018	0.282113	−23.3	−20.5	0.6	1578	2479	−0.99
BTZ-1-4	119.8	0.012539	0.000112	0.000458	0.000004	0.282110	0.000020	0.282109	−23.4	−20.8	0.7	1587	2493	−0.99
BTZ-1-5	127.6	0.012572	0.000121	0.000452	0.000005	0.282104	0.000018	0.282103	−23.6	−20.9	0.6	1595	2501	−0.99
BTZ-1-6	126.6	0.017442	0.000028	0.000655	0.000001	0.282102	0.000016	0.282101	−23.7	−21.0	0.6	1606	2507	−0.98
BTZ-1-7	129.1	0.015908	0.000368	0.000580	0.000013	0.282140	0.000019	0.282139	−22.3	−19.6	0.7	1551	2420	−0.98
BTZ-1-8	125.6	0.010517	0.000187	0.000381	0.000006	0.282099	0.000018	0.282099	−23.8	−21.1	0.6	1599	2512	−0.99
BTZ-1-9	129.6	0.013294	0.000063	0.000485	0.000002	0.282074	0.000015	0.282072	−24.7	−21.9	0.5	1638	2568	−0.99
BTZ-1-10	126.7	0.019775	0.000241	0.000662	0.000007	0.282119	0.000019	0.282117	−23.1	−20.4	0.7	1583	2470	−0.98
BTZ-1-11	123.4	0.017998	0.000345	0.000603	0.000011	0.282150	0.000017	0.282148	−22.0	−19.4	0.6	1539	2403	−0.98
BTZ-1-13	128.6	0.014534	0.000412	0.000523	0.000016	0.282084	0.000019	0.282082	−24.3	−21.6	0.7	1626	2546	−0.98
BTZ-1-14	125.1	0.025354	0.000784	0.000968	0.000031	0.282131	0.000020	0.282128	−22.7	−20.0	0.7	1580	2446	−0.97
TW005-1.1	125.1	0.006358	0.000046	0.000274	0.000002	0.282065	0.000020	0.282042	−25.8	−23.1	0.7	1672	2642	−0.99
TW005-10.1	121.8	0.006722	0.000068	0.000280	0.000003	0.282105	0.000018	0.282082	−24.4	−21.7	0.6	1618	2555	−0.99
TW005-11.1	120.6	0.009948	0.000047	0.000420	0.000003	0.282053	0.000019	0.282030	−26.2	−23.6	0.7	1695	2672	−0.99
TW005-12.1	121.2	0.011933	0.000055	0.000500	0.000002	0.282064	0.000019	0.282041	−25.8	−23.2	0.7	1683	2647	−0.98
TW005-15.1	121.9	0.010452	0.000095	0.000439	0.000004	0.282058	0.000019	0.282035	−26.1	−23.4	0.7	1689	2660	−0.99
TW005-3.1	122.2	0.008121	0.000020	0.000340	0.000001	0.282096	0.000020	0.282073	−24.7	−22.1	0.7	1633	2576	−0.99
TW005-4.1	123.9	0.010637	0.000063	0.000442	0.000003	0.282107	0.000023	0.282084	−24.3	−21.6	0.8	1622	2550	−0.99
TW005-5.1	120.4	0.012798	0.000243	0.000519	0.000009	0.282121	0.000021	0.282098	−23.8	−21.2	0.7	1606	2521	−0.98
TW005-6.1	126.6	0.009011	0.000039	0.000371	0.000002	0.282097	0.000021	0.282074	−24.7	−21.9	0.7	1633	2571	−0.99
TW005-7.1	123.3	0.010010	0.000221	0.000411	0.000008	0.282130	0.000023	0.282107	−23.5	−20.9	0.8	1590	2501	−0.99
TW005-9.1	125.8	0.012347	0.000106	0.000492	0.000004	0.282101	0.000028	0.282078	−24.5	−21.8	1.0	1632	2563	−0.99
TW006-1.1	125.5	0.012540	0.000146	0.000524	0.000006	0.282037	0.000020	0.282014	−26.8	−24.1	0.7	1721	2704	−0.98
TW006-11.1	121.2	0.011237	0.000370	0.000467	0.000016	0.282035	0.000022	0.282012	−26.9	−24.3	0.8	1722	2713	−0.99
TW006-13.1	121.8	0.010705	0.000104	0.000441	0.000005	0.282051	0.000021	0.282028	−26.3	−23.7	0.7	1698	2675	−0.99
TW006-15.1	126.6	0.010719	0.000234	0.000445	0.000010	0.282044	0.000022	0.282021	−26.6	−23.8	0.8	1709	2689	−0.99
TW006-2.1	123.7	0.012080	0.000040	0.000510	0.000002	0.282063	0.000019	0.282041	−25.9	−23.2	0.7	1685	2648	−0.98
TW006-3.1	123.3	0.015431	0.000257	0.000642	0.000012	0.282089	0.000042	0.282066	−25.0	−22.3	1.5	1655	2591	−0.98
TW006-5.1	122.8	0.009828	0.000094	0.000401	0.000004	0.282088	0.000020	0.282065	−25.0	−22.3	0.7	1646	2593	−0.99
TW006-6.1	126.2	0.015092	0.000318	0.000627	0.000014	0.282008	0.000024	0.281985	−27.8	−25.1	0.9	1766	2769	−0.98
TW006-8.1	120.6	0.014681	0.000078	0.000609	0.000004	0.282045	0.000021	0.282022	−26.5	−23.9	0.7	1715	2691	−0.98
TW006-9.1	125.3	0.010371	0.000338	0.000426	0.000014	0.282003	0.000020	0.281980	−28.0	−25.3	0.7	1764	2781	−0.99

Note: The Hf isotopic data of the host Sanguliu granodiorite (TW005 and TW006) are from [45].

Table 4. Major (wt.%) and trace element (ppm) geochemical data for the host Sanguliu granodiorite and monzodiorite enclaves.

Sample / Element	SGL-1	SGL-2	SGL-3	SGL-4	SGL-5	BT-1	BT-2	BT-3	BT-4	BT-5	BT-6	BT-7	BT-8	BT-9	BT-10	BT-11	BT-12	BT-13	BT-14	BT-15	BT-16
Na_2O	3.45	3.46	3.46	3.22	3.43	3.87	4.24	4.37	4.21	4.41	2.43	4.46	4.61	4.67	4.63	4.39	4.37	4.60	3.02	2.39	2.15
MgO	1.59	1.64	1.51	1.53	1.72	4.82	4.91	4.60	4.41	4.62	6.17	4.87	4.40	4.82	5.54	4.31	4.18	4.20	6.03	5.95	5.91
Al_2O_3	14.82	14.82	14.84	15.39	15.09	15.90	16.54	16.91	16.77	16.91	14.10	16.70	17.07	16.70	16.50	16.86	16.89	17.03	14.45	12.92	13.50
SiO_2	68.02	67.87	68.45	67.87	67.54	55.10	54.17	55.03	55.73	54.65	52.59	54.92	55.24	54.69	53.68	55.65	55.78	55.72	54.61	55.74	54.91
P_2O_5	0.11	0.11	0.11	0.10	0.11	0.20	0.21	0.19	0.18	0.18	0.19	0.21	0.22	0.22	0.21	0.17	0.17	0.18	0.12	0.15	0.13
K_2O	4.27	4.20	4.40	4.28	4.40	3.61	2.66	2.97	3.41	2.80	5.89	2.44	2.61	2.05	1.93	3.03	3.25	2.56	4.15	5.11	5.95
CaO	3.20	3.20	2.95	2.81	3.21	5.64	6.10	5.98	5.80	6.03	5.70	5.80	5.45	6.27	6.48	5.72	5.56	5.76	6.28	6.35	6.53
TiO_2	0.48	0.52	0.48	0.47	0.49	0.89	0.94	0.88	0.88	0.90	0.90	0.94	0.98	0.92	0.85	0.82	0.82	0.83	0.65	0.90	0.81
MnO	0.06	0.06	0.06	0.06	0.06	0.15	0.15	0.14	0.14	0.14	0.18	0.14	0.13	0.15	0.15	0.14	0.13	0.14	0.18	0.18	0.18
Fe_2O_3	1.13	0.55	0.41	0.41	0.48	1.18	1.39	1.32	1.26	1.28	1.88	1.37	1.49	1.36	1.33	1.53	1.47	1.31	1.65	1.44	1.38
FeO	2.52	2.97	2.88	3.14	2.97	7.28	7.19	6.47	6.20	6.47	8.27	6.96	6.56	6.87	7.03	6.38	6.02	6.29	7.23	7.28	6.83
LOI	0.38	0.44	0.63	0.48	0.52	1.11	1.10	1.12	1.13	1.19	1.28	1.07	1.05	1.24	1.31	0.87	1.01	1.00	1.37	1.37	1.35
SUM	100.02	99.85	100.18	99.76	100.03	99.73	99.59	99.97	100.12	99.57	99.57	99.89	99.80	99.95	99.64	99.85	99.64	99.62	99.74	99.77	99.62
$Na_2O + K_2O$	7.72	7.66	7.86	7.50	7.83	7.48	6.90	7.34	7.62	7.21	8.32	6.90	7.22	6.72	6.56	7.42	7.62	7.16	7.17	7.50	8.10
Na_2O/K_2O	0.81	0.82	0.79	0.75	0.78	1.07	1.59	1.47	1.23	1.58	0.41	1.83	1.77	2.28	2.40	1.45	1.34	1.80	0.73	0.47	0.36
A/NK	1.44	1.45	1.42	1.55	1.45	1.55	1.68	1.63	1.58	1.64	1.36	1.67	1.64	1.69	1.70	1.61	1.58	1.65	1.53	1.37	1.35
A/CNK	0.92	0.92	0.94	1.02	0.93	0.78	0.79	0.80	0.79	0.80	0.68	0.81	0.84	0.78	0.77	0.81	0.81	0.82	0.69	0.62	0.62
$Mg^{\#}$	44.48	45.76	45.31	43.73	47.40	50.74	50.90	51.71	51.73	51.93	52.47	51.44	49.81	51.49	54.55	49.76	50.36	50.06	55.22	55.29	56.62
Ba	989.91	940.11	952.40	1000.00	1000.00	721.28	539.31	562.96	618.55	519.33	1600.00	567.97	565.53	531.14	483.39	566.21	634.77	481.85	987.97	1300.00	1800.00
Cr	59.04	59.20	54.60	43.02	45.92	130.54	112.59	73.85	75.83	79.05	230.02	97.34	74.06	101.45	104.43	53.61	51.77	45.91	247.62	290.04	253.74
Ga	17.27	17.87	16.87	17.96	17.26	23.34	24.99	23.60	22.41	24.31	19.43	24.86	26.14	25.89	25.81	24.75	25.31	25.45	18.96	17.84	16.06
Nb	10.11	11.20	10.84	10.46	10.51	18.93	16.26	13.52	14.23	13.69	11.36	18.10	19.09	18.12	16.65	20.80	18.41	19.04	12.97	18.05	12.96
Pb	24.08	25.83	23.93	25.30	26.34	24.12	19.48	24.79	25.06	23.75	32.35	19.57	18.52	21.60	21.31	24.63	26.20	21.90	29.05	30.00	34.32
Rb	144.85	156.41	168.43	175.55	154.60	187.28	168.28	162.14	166.57	161.04	266.68	165.08	174.25	133.00	125.83	154.96	158.27	140.98	186.87	181.20	215.11
Sr	383.88	385.99	371.63	378.42	397.14	276.09	291.76	309.92	317.66	315.11	261.95	314.28	331.28	309.25	308.86	297.53	301.49	287.52	271.62	236.82	305.79
Zr	245.26	257.54	236.68	267.02	241.48	246.84	301.37	308.43	296.56	300.95	179.14	206.22	240.36	219.86	321.37	286.55	273.99	230.20	143.61	82.61	149.12
Li	15.94	18.83	17.97	17.98	16.98	26.02	29.53	27.79	26.75	28.45	32.66	30.19	30.98	26.48	25.67	28.48	29.26	29.57	25.34	17.07	16.03
Be	2.10	2.19	2.20	2.06	2.03	3.51	3.64	3.29	3.11	3.33	1.94	3.96	3.71	3.77	3.99	3.63	3.79	3.98	2.48	2.13	1.55
Sc	9.45	9.37	7.77	6.93	9.05	30.47	30.52	27.80	25.41	28.30	21.74	25.34	20.67	28.27	30.69	26.31	26.06	26.89	26.43	33.84	29.86
Co	7.96	9.16	8.25	7.67	8.95	19.60	20.02	18.05	17.34	18.20	26.28	19.47	17.68	18.02	18.66	17.39	17.79	18.02	21.93	21.97	23.31
Cs	1.59	2.80	1.89	1.87	2.01	3.21	3.12	2.60	2.59	2.63	4.44	3.65	3.51	2.34	2.27	2.16	2.16	2.14	1.99	2.62	2.47
Ni	12.08	13.18	11.79	11.07	13.42	27.25	24.49	20.58	19.65	19.87	88.18	28.78	25.60	22.90	24.55	21.99	22.24	22.55	52.08	58.28	68.65
U	3.29	2.83	2.75	3.14	2.35	2.03	2.84	2.76	2.93	3.06	12.84	1.82	1.82	1.21	1.45	1.33	1.61	1.55	1.21	2.64	2.19
Hf	7.79	7.71	7.09	7.55	7.51	8.66	7.90	9.35	8.79	9.05	5.46	6.43	7.57	7.44	9.09	8.07	8.56	8.98	4.86	3.67	4.38
Ta	1.14	1.09	0.94	0.96	0.94	0.57	0.59	0.46	0.48	0.53	0.51	0.66	0.75	0.61	0.50	0.85	0.84	0.84	0.37	0.67	0.46
Th	29.84	25.98	25.63	26.10	24.54	6.55	8.81	7.70	9.94	8.02	14.76	4.90	6.14	3.46	4.08	4.08	4.23	3.85	2.59	3.77	2.75
La	70.71	59.78	62.02	63.90	60.76	18.71	25.97	27.23	30.92	27.71	61.80	22.02	24.69	17.57	19.30	19.10	19.02	17.74	17.43	13.79	14.45
Ce	125.43	107.05	107.71	111.22	106.73	40.78	52.01	51.64	58.75	51.42	111.33	44.01	48.29	37.16	40.04	39.49	39.48	37.17	37.51	33.98	33.56
Pr	14.10	12.30	11.85	12.47	11.96	5.36	6.38	6.16	6.96	6.23	12.90	5.63	6.18	5.08	5.35	5.20	5.09	4.86	5.33	5.28	5.09
Nd	47.10	41.43	40.00	41.45	39.89	20.95	23.31	22.51	24.97	23.22	44.96	21.86	23.07	20.56	21.70	19.55	19.16	18.18	21.95	23.53	23.05
Sm	6.88	6.12	5.70	5.94	5.67	3.89	4.13	4.05	4.32	4.16	6.71	4.02	4.30	4.21	4.35	3.65	3.46	3.35	4.57	5.69	5.57
Eu	1.26	1.22	1.24	1.27	1.25	0.95	0.92	0.91	0.98	0.92	1.29	1.02	0.96	0.86	0.87	0.93	0.91	0.88	1.05	1.22	1.44
Gd	5.98	5.38	5.10	5.14	4.95	3.51	3.59	3.55	3.76	3.67	5.87	3.51	3.66	3.60	3.83	3.16	3.07	3.02	3.81	4.80	4.54
Tb	0.79	0.72	0.68	0.69	0.65	0.58	0.56	0.54	0.56	0.57	0.78	0.57	0.56	0.60	0.62	0.51	0.49	0.49	0.65	0.84	0.80
Dy	3.97	3.69	3.40	3.50	3.25	3.14	2.91	2.88	2.87	3.06	3.90	3.00	3.01	3.26	3.37	2.75	2.69	2.68	3.60	4.76	4.38

Table 4. *Cont.*

Sample / Element	SGL-1	SGL-2	SGL-3	SGL-4	SGL-5	BT-1	BT-2	BT-3	BT-4	BT-5	BT-6	BT-7	BT-8	BT-9	BT-10	BT-11	BT-12	BT-13	BT-14	BT-15	BT-16
Ho	0.77	0.72	0.67	0.68	0.63	0.65	0.61	0.58	0.58	0.61	0.75	0.63	0.63	0.66	0.67	0.59	0.58	0.56	0.72	0.92	0.87
Er	2.17	1.97	1.86	1.90	1.74	1.90	1.79	1.62	1.69	1.70	2.02	1.75	1.72	1.80	1.88	1.72	1.70	1.64	2.00	2.40	2.22
Tm	0.34	0.32	0.30	0.31	0.28	0.33	0.31	0.28	0.28	0.28	0.32	0.30	0.30	0.31	0.31	0.31	0.31	0.30	0.32	0.38	0.34
Yb	2.14	2.05	1.91	1.95	1.79	2.34	2.21	1.97	1.91	1.89	2.09	2.00	2.00	2.02	2.11	2.27	2.19	2.19	2.16	2.42	2.14
Lu	0.31	0.29	0.27	0.29	0.26	0.41	0.38	0.33	0.32	0.32	0.32	0.34	0.32	0.35	0.36	0.39	0.38	0.38	0.35	0.40	0.32
Y	19.79	18.48	17.48	17.29	16.25	18.20	16.95	16.12	15.30	15.59	19.44	17.08	19.44	19.45	19.14	16.77	16.63	16.01	18.59	22.62	20.46
ΣREE	281.96	243.03	242.72	250.69	239.80	103.52	125.07	124.25	138.87	125.75	255.04	110.65	119.69	98.04	104.77	99.60	98.54	93.43	101.44	100.41	98.76
LREE	265.49	227.89	228.52	236.24	226.24	90.65	112.72	112.49	126.89	113.67	238.99	98.56	107.48	85.44	91.62	87.92	87.12	82.19	87.84	83.49	83.16
HREE	16.47	15.14	14.20	14.44	13.56	12.87	12.35	11.75	11.98	12.08	16.05	12.10	12.20	12.59	13.15	11.69	11.41	11.24	13.61	16.91	15.60
LREE/HREE	16.12	15.05	16.09	16.36	16.68	7.04	9.13	9.57	10.59	9.41	14.89	8.15	8.81	6.79	6.97	7.52	7.63	7.31	6.45	4.94	5.33
La_N/Yb_N	23.67	20.92	23.25	23.55	24.38	5.72	8.42	9.90	11.62	10.54	21.19	7.91	8.87	6.24	6.56	6.05	6.22	5.80	5.78	4.08	4.84
δEu	0.60	0.65	0.70	0.70	0.72	0.79	0.73	0.73	0.74	0.72	0.63	0.83	0.74	0.67	0.65	0.83	0.85	0.85	0.77	0.71	0.87
δCe	0.97	0.97	0.97	0.97	0.97	1.00	0.99	0.98	0.98	0.96	0.97	0.97	0.96	0.96	0.97	0.97	0.98	0.98	0.95	0.98	0.96

Note: A/NK = $Al_2O_3/(Na_2O + K_2O)$ (mol); A/CNK = $Al_2O_3/(CaO + Na_2O + K_2O)$ (mol); $Mg^{\#} = 100 \times Mg/(Mg + \text{total Fe})$ [48].

Table 5. Rb–Sr and Sm–Nd isotopic data for the host Sanguliu granodiorite and monzodiorite enclaves.

Sample	Rb (ppm)	Sr (ppm)	$^{87}Rb/^{86}Sr$	$^{87}Sr/^{86}Sr$	I_{Sr}	$\varepsilon_{Sr}(0)$	$\varepsilon_{Sr}(t)$	$f_{Rb/Sr}$	Sm (ppm)	Nd (ppm)	$^{147}Sm/^{144}Nd$	$^{143}Nd/^{144}Nd$	I_{Nd}	T_{DM}	T_{2DM}	T_{chur}	$\varepsilon_{Nd}(0)$	$\varepsilon_{Nd}(t)$	$f_{Sm/Nd}$
SGL−1	149	383	1.1227	0.716629	0.71467	172.2	146.4	12.58	6.65	42.8	0.094	0.511579	0.511503	1997	2467	1569	−20.66	−19.06	−0.52
SGL-2	166	389	1.2341	0.716771	0.71461	174.2	145.6	13.92	6.36	43.1	0.0893	0.511577	0.511505	1924	2465	1503	−20.7	−19.02	−0.55
SGL-3	176	369	1.377	0.717103	0.7147	178.9	146.9	15.65	5.81	38.3	0.0917	0.511584	0.51151	1953	2456	1527	−20.56	−18.92	−0.53
SGL-4	189	378	1.4503	0.717021	0.71449	177.7	143.9	16.54	6.21	42.2	0.0889	0.511559	0.511488	1940	2493	1523	−21.05	−19.35	−0.55
SGL-5	177	395	1.2954	0.716622	0.71436	172.1	142	14.66	6.06	40.7	0.09	0.511458	0.511386	2080	2653	1682	−23.02	−21.34	−0.54
BT-1	199	291	1.9851	0.718026	0.7145	192	144.1	23	3.91	19.4	0.122	0.511598	0.511498	2571	2469	2114	−20.29	−19.11	−0.38
BT-2	178	303	1.7026	0.717685	0.71466	187.2	146.3	19.59	4.15	22.5	0.1114	0.511523	0.511432	2417	2575	1986	−21.75	−20.39	−0.43
BT-3	176	321	1.5825	0.717302	0.71449	181.7	143.9	18.14	3.86	20.4	0.1143	0.510985	0.510892	3300	3423	3037	−32.24	−30.93	−0.42
BT-4	175	327	1.546	0.717299	0.71455	181.7	144.8	17.69	4.08	22.1	0.1117	0.511624	0.511533	2274	2416	1813	−19.78	−18.42	−0.43
BT-5	170	329	1.4977	0.717262	0.7146	181.2	145.5	17.11	4.11	21.5	0.1159	0.511443	0.511348	2650	2705	2245	−23.31	−22.03	−0.41

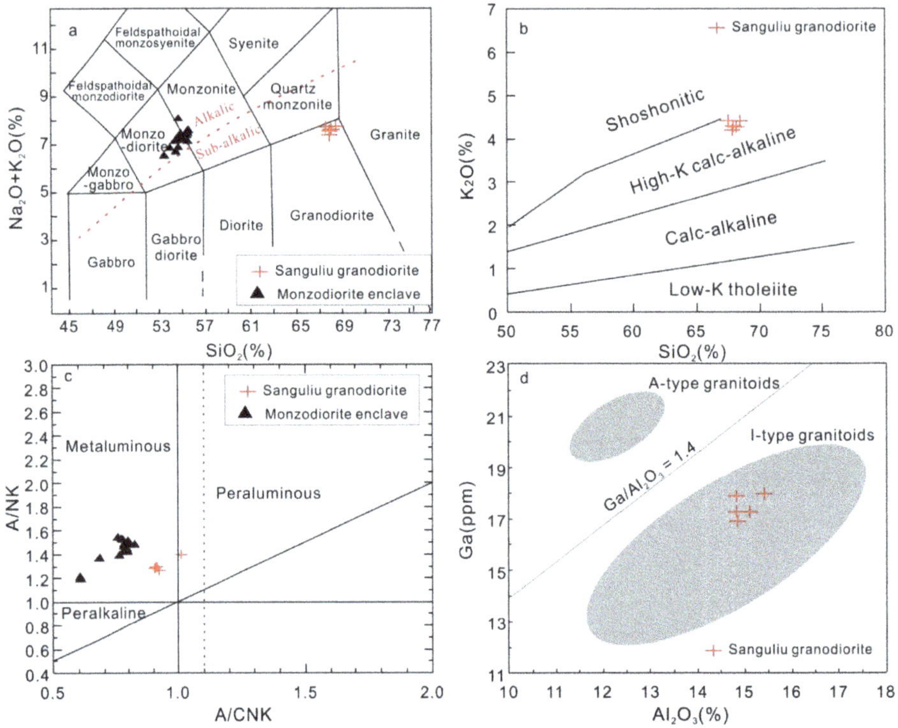

Figure 8. TAS (**a**), SiO_2 versus K_2O (**b**), A/CNK versus A/NK (**c**), and Al_2O_3 versus Ga (**d**) diagrams of the Sanguliu granodiorite and its enclaves.

The Sanguliu granodiorite samples have medium total REE contents (∑REE), ranging from 239.80 to 281.96 ppm, and exhibit strong differentiation between the light rare earth element (LREE) and heavy rare earth element (HREE), demonstrating LREE/HREE and La_N/Yb_N ratios of 15.05–16.68 and 20.92–24.38, respectively. Further, their REE patterns present an LREE-enriched pattern with medium negative Eu anomalies (δEu = 0.60–0.72) (Figure 9a). In comparison, the monzodioritic enclave samples have lower REE contents (∑REE = 93.43–138.87 ppm), a slightly flatter LREE segments pattern, and LREE/HREE amd La_N/Yb_N ratios of 4.94–10.59 and 4.08–11.62, respectively. In addition, they feature weak negative Eu anomalies (δEu = 0.65–0.85) (Figure 9a).

The granodiorite and monzodioritic enclave samples yielded similar variations of trace elemental compositions in the primitive mantle-normalized trace element spider diagram (Figure 9b). They are characterized by enrichments in large ion lithisphile elements (LILEs, i.e., Rb and U), high field strength elements (HFSEs, i.e., Zr, Hf, and Ti), and depletions in LILEs (i.e., K and Sr) and HFSEs (i.e., Nb, P, and HREE) (Figure 9b).

5.4. Sr–Nd–Pb Isotopes

The Rb–Sr and Sm–Nd isotope data for the host granodiorite and monzodioritic enclave samples are listed in Table 5 and shown in Figure 10a. The initial $^{87}Sr/^{86}Sr$ ratios and $\varepsilon_{Nd}(t)$ values were calculated at 122 and 125 Ma from the zircon U–Pb dating. All the samples exhibited similar Sr isotopic features with high $^{87}Rb/^{86}Sr$ (1.1227 to 1.9851) and initial $^{87}Sr/^{86}Sr$ ratios (0.71436 to 0.71470). In addition, both the host granodiorite and monzodiorite enclaves are characterized by strongly negative $\varepsilon_{Nd}(t)$ values of −18.92–−21.34, and

−18.42–−30.93 (mostly −18.42–−22.03), respectively. Furthermore, the enclave samples were scattered in the $\varepsilon_{Nd}(t)$–SiO_2 diagram (Figure 10b).

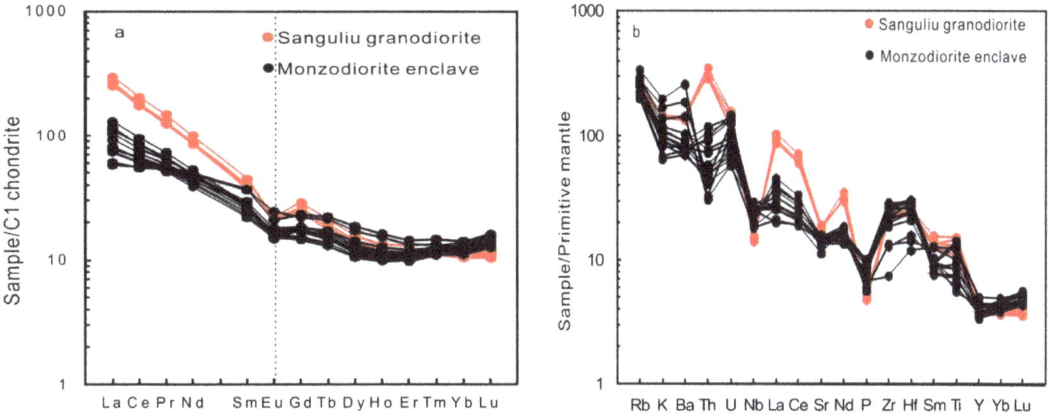

Figure 9. Chondrite-normalized REE patterns (**a**) and primitive mantle-normalized spiderdiagrams (**b**) for the Sanguliu granodiorite and monzodioritic enclaves. The chondrite and primitive mantle values are from [49].

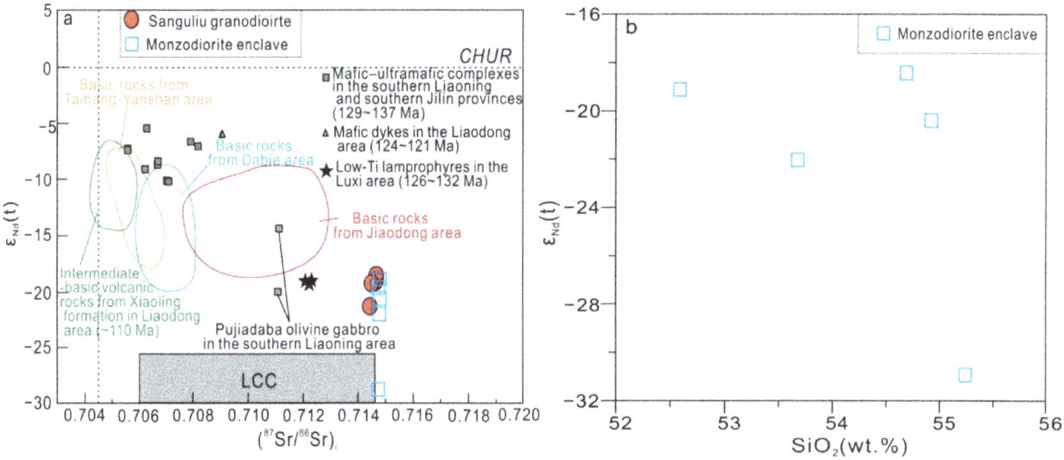

Figure 10. $\varepsilon_{Nd}(t)$ versus $(^{87}Sr/^{86}Sr)_i$ plot (**a**) and SiO_2 versus $\varepsilon_{Nd}(t)$ diagram (**b**). The data of the mafic rocks in the Taihang-Yanshan area are sourced from [50,51]; those in the Dabie area are from [52–54]; those in the Jiaodong area are from [55–57]; those in the Southern Jilin and Southern Liaoning areas are from [58]; the intermediate-basic rocks from the Liaodong area are from [59,60]; lamprophyres in the Luxi area are from [61]; LCC–lower crust of the NCC, which was taken from [62].

Moreover, the host granodiorite and monzodioritic enclave generated similar Pb isotopic ratios: $(^{206}Pb/^{204}Pb)_i$ = 17.6200–17.8360, $(^{207}Pb/^{204}Pb)_i$ = 15.5970–15.6190, and $(^{208}Pb/^{204}Pb)_i$ = 38.8410–39.9490, as summarized in Table 6 and Figure 11.

Table 6. Pb isotopic data for the host Sanguliu granodiorite and monzodiorite enclaves.

Sample	$^{206}Pb/^{204}Pb$	$^{207}Pb/^{204}Pb$	$^{208}Pb/^{204}Pb$	$(^{206}Pb/^{204}Pb)_i$	$(^{207}Pb/^{204}Pb)_i$	$(^{208}Pb/^{204}Pb)_i$
SGL-1	17.914	15.627	39.484	17.746	15.619	38.984
SGL-2	17.955	15.615	39.338	17.82	15.608	38.933
SGL-3	17.838	15.609	39.271	17.697	15.602	38.841
SGL-4	17.989	15.607	39.57	17.836	15.6	39.153
SGL-5	17.847	15.616	39.276	17.738	15.611	38.901
BT-1	17.836	15.613	38.987	17.732	15.608	38.877
BT-2	17.83	15.609	39.133	17.648	15.6	38.949
BT-3	17.82	15.617	39.138	17.681	15.61	39.011
BT-4	17.799	15.615	39.09	17.654	15.608	38.929
BT-5	17.78	15.605	38.995	17.62	15.597	38.858

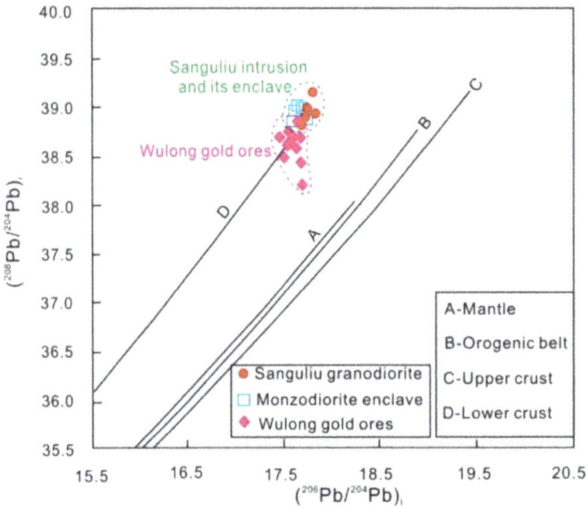

Figure 11. $(^{206}Pb/^{204}Pb)_i$ versus $(^{208}Pb/^{204}Pb)_i$ diagram (modified from [63]). The data of the Wulong gold deposit are from [17,64].

5.5. Biotite Compositions

Six biotite mineral spots on the host Sanguliu granodiorite were analyzed, and their EPMA data are listed in Table 7. The cations were initially calculated based on 22 oxygen atoms [65]. The EPMA data indicate that the biotite has high TiO_2 (2.71 to 3.71 wt.%), MgO (10.54 to 11.48 wt.%), FeO (20.10 to 21.49 wt.%), and K_2O (9.24 to 9.87 wt.%) contents, and low CaO (0.01 to 0.024 wt.%) and Na_2O (0.06 to 0.11 wt.%) contents. Further, the calculated $Fe^{2+}/(Fe^{2+} + Mg)$, $Mg/(Mg + Fe^{2+})$ ratios, and Al^{VI} values were found to be 0.46–0.50, 0.47–0.50, and 0.058–0.112, respectively. In the Mg-Al^{VI} + Fe^{3+} + Ti-Fe^{2+} + Mn diagram, the biotite from the host Sanguliu granodiorite was classified as magnesio-biotite (Figure 12a). In the Fe^{2+}-Fe^{3+}-Mg^{2+} diagram, all of the samples plotted between the NNO and MH buffers (Figure 12b).

Table 7. Electron microprobe data of the biotites from the host Sanguliu granodiorite (results in wt%).

Element \ Sample	2-bt-1	2-bt-3	2-bt-5	2-bt-6	2-bt-7	2-bt-8
SiO_2	36.453	37.332	37.391	36.849	36.95	36.949
TiO_2	3.622	3.713	2.707	3.609	3.089	2.586
Al_2O_3	13.694	13.502	13.83	13.483	13.581	13.406
FeO	21.493	20.458	20.099	21.147	20.759	21.292
MnO	0.246	0.248	0.232	0.258	0.281	0.255
MgO	10.554	10.947	11.478	10.758	10.547	11.1
CaO	0.011	0.014	0.013	0.001	0.024	0.016
Na_2O	0.077	0.055	0.129	0.108	0.105	0.09
K_2O	9.399	9.865	9.513	9.348	9.239	9.344
Si	2.8127	2.8494	2.8637	2.8350	2.8643	2.8598
Al^{IV}	1.1873	1.1506	1.1363	1.1650	1.1357	1.1402
Al^{VI}	0.0580	0.0639	0.1120	0.0576	0.1051	0.0826
Ti	0.2102	0.2132	0.1560	0.2089	0.1801	0.1506
Fe^{3+}	0.1772	0.1855	0.1695	0.1884	0.2002	0.1537
Fe^{2+}	1.2097	1.1204	1.1179	1.1722	1.1457	1.2245
Mn	0.0161	0.0160	0.0150	0.0168	0.0185	0.0167
Mg	1.2140	1.2456	1.3105	1.2339	1.2188	1.2807
Ca	0.0009	0.0011	0.0011	0.0001	0.0020	0.0013
Na	0.0115	0.0081	0.0192	0.0161	0.0158	0.0135
K	0.9252	0.9606	0.9295	0.9175	0.9137	0.9226
Total	7.8228	7.8145	7.8305	7.8116	7.7998	7.8463

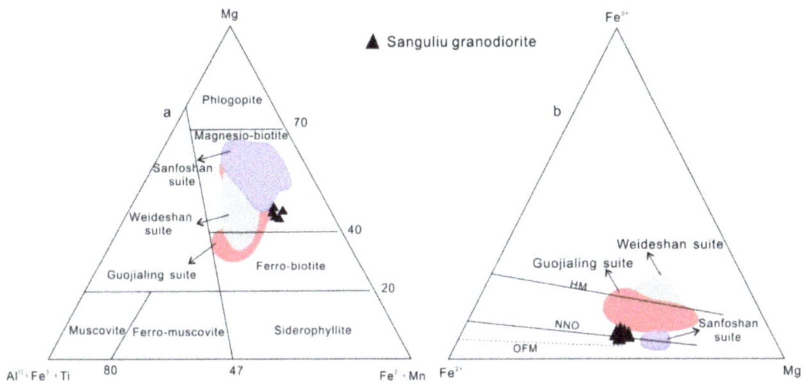

Figure 12. Diagrams of classification ((**a**), modified from [66]) and Fe^{3+}–Fe^{2+}–Mg ((**b**), modified from [67]) of the biotites from the Sanguliu, Guojialing, Weideshan, and Sanfoshan plutons. The data of EPMA biotite compositions are from [23,68] and this paper, respectively).

6. Discussion

6.1. Petrogenesis

The Sanguliu granodiorite samples show relatively low contents of SiO_2 (67.54%–68.45%) and Al_2O_3 (14.82%–15.39%), and low A/CNK values (0.92–1.02), indicating that Sanguliu granodiorite belongs to calc-alkaline I-type granite [69,70]. In addition, the granodiorite samples also exhibit low Ga/Al ratios (Figure 8d), similar to I-type granites, which are different from A-type granite (Figure 8d) [71]. This also could be supported by the petrographic features of the Sanguliu granodiorite. The Sanguliu granodiorite consists of quartz, plagioclase, potassium feldspar, biotite, and minor hornblende (Figure 4), with no muscovite. The petrographic and geochemical data indicate that the Sanguliu granodiorite belongs to typical calc-alkaline I-type granite.

The monzodioritic enclaves found in the host Sanguliu intrusion via field investigation may provide significant insights into the magma processes present [2]. The enclaves in the granitoids are generally interpreted as restites, country-rock xenoliths, autoliths, schlieren, and MMEs [2,72]. The monzodioritic enclaves in the Sanguliu intrusion are usually oval or elliptical and have typical igneous microtextures and sharp contacts with the host rocks. The monzodioritic enclaves have homogeneous U-Pb ages (125.3 ± 1.6 Ma, SHRIMP) with the host Sanguliu intrusion (122.2 ± 1.3 Ma, LA-ICP-MS), which are within the range of acceptable error. Furthermore, the monzodioritic enclaves yield similar medium to weakly negative Eu anomalies and relative Sr depletion to that of the host intrusion, as shown in Figure 9, thereby excluding the possibility of an autolith characterization [73]. These petrographic, geochemical, and geochronological features indicate that the monzodioritic enclaves in the Sanguliu intrusion were not restitic, xenoliths of wallrock, autoliths, or schlierens, but rather MMEs in origin.

MMEs usually represent remnants of a mafic component added to intermediate or felsic magma chambers [4,8,72,74,75]. The MMEs in the Sanguliu intrusion generally contain acicular apatites and megacrysts. The presence of both acicular apatites, which are attributed to rapid cooling, and megacrysts, such as K-feldspar or plagioclase, which are elliptical, implies that the enclaves are hybrids, meaning that they are related to the mingling of small volumes of hot basalt with a cooler granitic melt [3,76]. Furthermore, the MMEs are characterized by wide ranges of $\varepsilon_{Nd}(t)$ values (−18.42–−30.93), which have up to 10 epsilon units. These features suggest that magma mixing between mafic and granitic melts played an important role in the genesis of the Sanguliu granodiorite and MMEs.

The MMEs and host Sanguliu granodiorite show similar whole-rock Sr and isotopic compositions (Figure 11), indicating rapid homogenization and radiogenic isotopic equilibrium during the magma mixing process [24,77–79]. In addition to magma mixing, crystal fractionation was involved in the formation of the Sanguliu granodiorite, which is supported by the weak to medium negative Eu anomalies (Figure 9a) and the significant Sr and P depletion shown in the spidergram (Figure 9b). Therefore, the Sanguliu granodiorite and MMEs were the result of mixing between mafic and intermediate to felsic magmas, followed by apatite- and feldspar-dominated fractionation. The magma mixing process likely occurred within the primary magma chambers rather than during magma ascent or evolution, which is reflected by the lack of correlations between the $\varepsilon_{Nd}(t)$ and SiO_2 contents of the enclaves (Figure 10b) [80].

6.2. Magma Source

The host granodiorite has high SiO_2 (67.54 to 68.45 wt.%) and low MgO (1.51 to 1.72 wt.%), TiO_2 (0.47 to 0.52 wt.%) contents, and strongly negative $\varepsilon_{Nd}(t)$ values of −15.53 to −21.34, and $\varepsilon_{Hf}(t)$ values of −20.86 to −25.31, indicating a crustal derivation. The primitive magma can be inferred to have isotopic compositions with $\varepsilon_{Hf}(t)$ values less than −25.31 and ancient Hf model ages (T_{DM2}, 2501–2781 Ma, [45]). As the samples plotted close to the ancient lower crust of the NCC field in the t-$\varepsilon_{Hf}(t)$ diagram (Figure 7a), the host granodiorite may have been primarily derived from an ancient lower crust source. However, the high $Mg^{\#}$ values of 43.73–47.40 of the host granodiorite suggest the involvement of mantle-derived material as a result of magma mixing.

The MMEs have high MgO (4.18 to 6.17 wt.%), Cr (45.91 to 290.04 ppm), and Ni (19.65 to 88.18 ppm) contents, and high $Mg^{\#}$ values of 50–57 at intermediate silica contents (SiO_2 = 53.68–55.78 wt.%), making them distinct from crustal materials or crust-derived melts [81], thereby suggesting that the parental magma of the enclaves was mantle-derived basaltic magma. However, the MMEs yielded strongly negative $\varepsilon_{Nd}(t)$ values of −18.42 to −30.93, which may be the result of the partial melting of an enriched lithospheric mantle or a depleted lithospheric mantle with significant contamination by ancient crustal materials during magma evolution [80]. The voluminous incorporation of crustal components significantly modifies the concentrations of major and trace elements and $Mg^{\#}$ values [47,80]. However, the relatively low SiO_2 and high MgO contents and $Mg^{\#}$ values indicate that significant crustal contamination during magma evolution did not occur. Therefore, the enclave-forming magmas were likely derived mainly from an enriched mantle source.

The enclaves yielded lower negative $\varepsilon_{Nd}(t)$ values (-18.42 to -30.93) than those of the majority of Early Cretaceous mafic rocks in the NCC, which were derived from an enriched lithospheric mantle modified from crustal or slab melts [45,58], and the enclaves plot between the "primitive" mafic rocks from the NCC and ancient lower crust in the diagrams of t–$\varepsilon_{Hf}(t)$ and $\varepsilon_{Nd}(t)$–$(^{87}Sr/^{86}Sr)_i$ (Figures 7a and 10a). These clues suggest that the enclave-forming magmas were contaminated by lower crustal melts. This is supported by the distinct LILE and LREE enrichment and HFSE depletion in the enclaves (Figure 9) [82]. The SHRIMP zircon U-Pb age of 1776 $\pm$ 24 Ma for the enclaves also indicates the incorporation of NCC basement materials, which are characterized by late Archean and Paleoproterozoic tectonothermal events [83]. Both Pei et al. [58] and Li et al. [84] proposed that the delaminated NCC lower continental crust materials modified the lithospheric mantle in the Liaodong area. Furthermore, the enclaves have highly negative $\varepsilon_{Nd}(t)$ values (-18.42 to -30.93) that are similar to those of the Early Cretaceous low-Ti lamprophyres in the Luxi area and the Pujiadaba olivine gabbro in the southern Liaoning area (Figure 10a). These low-Ti lamprophyres were derived from the partial melting of a lithospheric mantle modified by crustal melts from both the delaminated NCC lower crust and the broken-off YC slab [45] and the Pujiadaba olivine gabbros in the southern Liaoning area were derived from the partial melting of the lithospheric mantle metasomatized by melts from the broken-off YC slab [58]. Therefore, the enclave-forming magma originated from the partial melting of lithospheric mantle modified by the delaminated NCC lower crustal and broken-off YC slab melts.

Based on the aforementioned petrography, major and trace element geochemical, and Sr, Nd, Pb, and zircon in situ Hf isotopic compositions of the MMEs and host granodiorite, a complex, multi-stage process involving magma mixing and crystal fractionation, is proposed for the generation of the Sanguliu intrusion and monzodioric enclaves. The Sanguliu granodiorite was mainly derived from the partial melting of the ancient NCC lower crust, which was mixed with mantle-derived magma. Meanwhile, the enclaves were derived from the partial melting of enriched lithospheric mantle modified by melts from the delaminated NCC lower crust and broken-off YC slab, followed by crystal fractionation. The Sr and Pb isotopic equilibrium and crystal exchange occurred during the mixing of mafic-felsic magma.

6.3. Highly Oxidized Magma

Highly oxidized magmas can incorporate significant amounts of S [12], and thus serve as an effective medium for transporting Au and other ore-forming materials [10]. The primary biotites from the Sanguliu granodiorite yield homogeneous $Fe^{2+}/(Fe^{2+} + Mg)$ ratios of 0.46–0.50, indicating that the biotites were not modified by later fluids [85]. The biotite has high $Mg/(Mg + Fe^{2+})$ ratios of 0.47–0.50, and plots in the magnesium biotite field on the Mg-Al^{VI} + Fe^{3+} + Ti-Fe^{2+} + Mn diagram (Figure 12a) indicating that the biotite is magnesium biotite. Further, biotites show high TiO_2 content (2.71 to 3.71 wt.%) and low Al^{VI} values (0.058 to 0.112) and cluster between the MH and NNO buffers (Figure 12b). Therefore, they likely formed under relatively high temperatures at high oxygen fugacities. This is supported by the high Ce/Ce* (1.30 to 107.18, average = 27.29) ratio of zircons from the Sanguliu intrusion [86], further suggesting that the parental magma of the Sanguliu intrusion has a relatively high oxygen fugacity.

The fluid inclusion Rb–Sr isochron age (120 $\pm$ 3 Ma; [15]), pyrite Rb–Sr isochron age (119 $\pm$ 1 Ma; [32]), and sericite $^{40}Ar/^{39}Ar$ plateau age (122.8 $\pm$ 0.8 Ma; [87]) date the Wulong gold mineralization. The Sanguliu granodiorite has a zircon U–Pb age of 122.2 $\pm$ 1.3 Ma, representing its emplacement age. The formation age of the Wulong gold deposit is in good agreement with this age. Further, the Sanguliu intrusion hosts the Wulong gold orebodies, which indicates that the Sanguliu intrusion may be related to gold mineralization.

The Sanguliu intrusion has been previously considered to be a metallogenetic intrusion of the Wulong gold deposit [14]. However, Wei et al. [15] proposed that the Sanguliu intrusion did not directly provide any ore-forming fluids or materials for Wulong gold mineralization but had similar Sr and Nd isotopic compositions to the deeper material source of the Wulong deposit. The Pb isotopic compositions of the Sanguliu intrusion are consistent with those of the sulfides from the Wulong deposit, both plotting near the lower crust curve (Figure 11), implying that the Pb in the Wulong gold deposit and Sanguliu intrusion was mainly derived from a similar lower crust source. Some Early Cretaceous intrusions hidden at depths below the Wulong ore field have been identified via geophysical detection and have been verified to provide important ore-forming fluids and materials for the Wulong gold mineralization [16,17]. Therefore, we suggest that the Sanguliu intrusion and the concealed metallogenic intrusion below the Wulong ore field originated from a magma chamber with a relatively high oxygen fugacity, which is favorable for gold mineralization.

6.4. Magma Mixing and Gold Mineralization

Abundant MMEs were found within the Early Cretaceous granitic plutons in the eastern NCC. Some of the host granitic plutons have been researched in terms of emplacement ages, petrogenesis, and magma sources [5–7,9,11,23,55,56,68,75,88–90]. Significant similarities were found in the Early Cretaceous plutons (Table 8), including high K_2O and Fe_2O_3 contents and $Mg^{\#}$ values [5–7,9,11,23,55,56,68,75,88–90], and the presence of magnesian biotites that were crystallized under high oxygen fugacity conditions [68,88,89].

Table 8. Characteristics of the Early Cretaceous granitoids with MMEs in the Jiaodong and Liaodong areas.

Features \ Intrusion	Sanguliu Intrusion	Gudaoling Intrusion	Guojialing Intrusion	Sanfoshan Intrusion	Weideshan Intrusion	Yashan Intrusion
Location	Liaodong Peninsula		Jiaodong Peninsula			
Occurrence	pluton	pluton	pluton	pluton	pluton	pluton
Emplacement age	122.2 ± 1.3 Ma	120 ± 1 Ma	127.9 ± 1.3 Ma	119–114 Ma	127–105 Ma	118 Ma
Rock type	granodiorite	monzogranite, and minor granodiorite, quartz diorite	granodiorite	medium-coarse- grained monzogranite, with minor syenogranite and medium-fine grained monzogranite	gneiss quartz monzodiorite, quartz monzonite, and monzogranite	granite
Enclave	monzodioritic	dioritic	dioritic	monzodioritic	dioritic	dioritic
Enclave shape	oval, irregular	angular to oval	oval, round	oval, round	oval, round, irregular	oval, round
Main minerals	quartz (20%–30%), plagioclase (35%–45%), potassium feldspar (15%–25%), biotite (10%–15%), and minor hornblende (~5%)	plagioclase (39%–43%), potassium feldspar (24%–29%), quartz (19%–35%), biotite (5%), and hornblende (1%–3%)	quartz (20%–25%), plagioclase (35%–55%), potassium feldspar (10%–30%), and minor biotite and hornblende (4%–5%)	quartz (20%–25%), plagioclase (35%–55%), potassium feldspar (30%–35%), biotite (5%–10%), and hornblende (1%–3%)	quartz (20%–30%), plagioclase (26%–33%), potassium feldspar (27%–30%), hornblende (8%–9%), and biotite (6%–7%)	quartz, plagioclase, potassium feldspar, hornblende, and biotite

Table 8. *Cont.*

Features \ Intrusion	Sanguliu Intrusion	Gudaoling Intrusion	Guojialing Intrusion	Sanfoshan Intrusion	Weideshan Intrusion	Yashan Intrusion
Accessory minerals	titanite, zircon, and apatite	apatite, zircon, titanite, and Fe–Ti oxides	magnetite, titanite, zircon, and apatite	magnetite, titanite, zircon, and apatite	magnetite, titanite, zircon, and apatite	zircon, apatite, magnetite, and titanite
Distinctive minerals	hornblende and biotite	biotite and hornblende	hornblende and biotite	biotite and hornblende	biotite and hornblende	hornblende and biotite
Relation to country rock	intrusive	intrusive	intrusive	intrusive	intrusive	intrusive
Relation to fault	cut by NE-trending faults	cut by NNE-trending faults	cut by NE- and NNE-trending faults	cut by NE- and NNE-trending faults	cut by NE-, NNE-, and NW-trending faults	cut by NE-trending faults
Relation to gold deposits	hosts Wulong gold deposit	-	hosts major gold deposits	-	hosts Mo, Cu, Pb-Zn, Ag, as well as minor Au deposits	-
ACNK	0.9–1.0	1–1.09	0.8–1.1	0.93–1.04	-	0.89–1.03
SiO_2 (wt.%)	67.54–68.45	71.44–72.56	~67	71.16–76.63	Mainly 65.29–66.65	65.5–68.82
K_2O (wt.%)	4.20–4.40	3.61–4.47	3.97–4.29	4.34–5.22	3.95–4.57	3.45–4.53
Na_2O (wt.%)	3.22–3.46	3.39–4.89	3.90–3.98	3.59–4.40	3.73–4.50	3.47–4.13
$Mg^{\#}$	47–53	Most 45–72	35–64	-	-	-
$Fe^{2+}/(Fe^{2+} + Mg)$ ratios of biotite	44–47	-	41.3	35–40	26–49	-
$\varepsilon_{Nd}(t)$	granodiorite: −18.92–−21.34; enclave: −18.42–−30.93 (mostly −18.42–−22.03)	monzogranite: −18.5–−20.9; enclave: ~−7.2	granodiorite: −11.2–−17.5	monzogranite: −18.5–−17.5; enclave: −15.1	-	granite: −17.0–−18.2; enclave: −15.9–−17.7
Origin	I type	I type	I type	I type	I type	I type
References	This paper	[7,75]	[9,22]	[11]	[6]	[90]

Zircons from the studied Sanguliu granodiorite and monzodioritic enclaves show typical zoning absorption, euhedral to subhedral shapes, and have high Th/U ratios of 0.78–1.47 and 0.47–1.79, respectively, indicating magmatic origins. Moreover, zircon U-Pb age data indicate that the Sanguliu granodiorite had coeval $^{206}Pb/^{238}U$ ages (122.2 ± 1.3 Ma, LA-ICP-MS) with the monzodiorite enclave (125.3 ± 1.6 Ma, SHRIMP) within the range of acceptable error, suggesting that the Sanguliu intrusion formed in the Early Cretaceous (125–122 Ma). This is consistent with the ages of Gudaoling monzogranite and dioritic enclaves in the Liaodong area (120 ± 1 Ma; [8]), the granitic intrusions containing enclaves in the Jiaodong area (105–130 Ma; [6,11,21,22]), and the Taihang plutons in the Taihang orogeny (127–129 Ma; [20]). Thus, the enclaves hosted within the Early Cretaceous granitic plutons were formed by magma mixing and mingling [10], with significant participation of enriched lithospheric mantle-derived magma [7]. These Early Cretaceous granitic plutons and abundant coeval enclaves indicate that magma mixing was common in the eastern NCC, even in the Taihang orogeny and mainly occurred during 110–130 Ma (Figure 1), which is consistent with the formation ages of the decratonic gold deposits (130–120 Ma) in the Jiaodong and Liaodong areas [91]. Furthermore, some gold deposits or occurrences were located near or hosted within the Early Cretaceous granitic plutons containing MMEs in the eastern NCC (Figure 3a) and are considered to be closely related to the plutons. In particular, the Guojialing granodiorite is believed to be associated with the Wang'ershan gold deposit in the Jiaodong area [92]. Although the studied Sanguliu granodiorite did not serve as a metallogenic intrusion of the Wulong gold deposit in the Liaodong area, the pluton shares a similar material source with the Wulong gold mineralization. These findings imply that magma mixing is temporally, spatially, and genetically related to the Early Cretaceous gold systems in the eastern NCC.

As previously mentioned, the enclaves within the Sanguliu granodiorite were formed by the partial melting of the mixed magma of the enriched lithospheric mantle and NCC lower crust, and the enriched lithospheric mantle was previously metasomatized by melts derived from the delaminated ancient lower crust and broken-off Yangtze slab. Zhu et al. [93] proposed that the decratonic gold deposits are closely related to the interaction between the lower crust and melts derived from the enriched lithospheric mantle, which could induce the formation of voluminous gold-bearing magma, rapidly creating giant gold accumulations. Li and Santosh [10] summarized the mineralogical, geochemical, and isotopic data of the igneous rocks, ore-forming fluids, and ore minerals from the gold deposits in the NCC and suggested that mantle fluids and metals substantially contributed to the decratonic gold systems. Therefore, the mingling and mixing of mafic magma with felsic magma not only provides insights into the magma chamber process and dynamics [3] but is also a critical process for the involvement of gold-bearing mantle components in magmatic-metallogenic systems [10,12].

During the Early Cretaceous, the Liaodong area was affected by the westward subduction of the Paleo-Pacific Plate and lithospheric thinning in the NCC [10,94], which led to magma mixing and associated large-scale gold mineralization in the NCC [95–97]. A new diagenetic and metallogenic model for the Wulong gold deposit is shown in Figure 13.

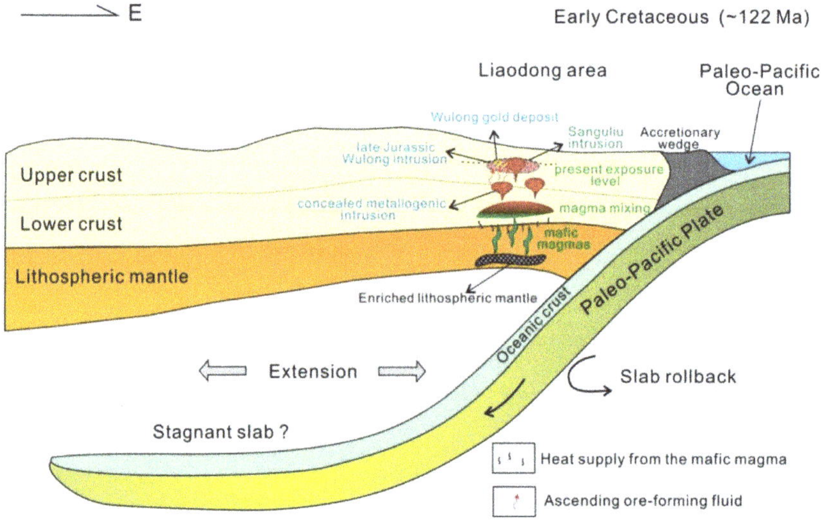

Figure 13. Schematic model for the Wulong gold mineralization and its links with the Early Cretaceous magma mixing in the Liaodong Peninsula. The tectonic setting for the Liaodong Peninsula during the Early Cretaceous is from [10,94,97].

7. Conclusions

The host Sanguliu granodiorite was emplaced at 122.2 ± 1.3 Ma and showed consistent zircon U-Pb ages with the monzodioritic enclaves (125.3 ± 1.6 Ma), as indicated by LA–ICP–MS and SHRIMP zircon U–Pb age data.

The field and petrological observations, as well as elemental, isotopic, and zircon U-Pb age data, indicate that the enclaves within the Sanguliu intrusion were mainly formed by magma mixing between mafic and felsic magmas. The felsic magmas were mainly derived from the partial melting of the ancient lower crust, and the mafic magmas were derived from the partial melting of the enriched lithospheric mantle modified by the delaminated ancient lower crust and broken-off Yangtze slab.

The primary magma forming the Sanguliu granodiorite had a relatively high oxygen fugacity, which is favorable for gold mineralization.

In general, the Early Cretaceous magma mixing occurred in the eastern NCC from 110 to 130 Ma and is temporally, spatially, and genetically related to the decratonic gold deposits in the Jiaodong and Liaodong areas.

Author Contributions: Conceptualization, T.W. and C.C.; Formal analysis, C.C.; Funding acquisition, C.C. and Y.Z. (Yongheng Zhou); Investigation, C.C., Y.Z. (Yan Zhao) and C.Z.; Methodology, D.L.; Supervision, C.C.; Visualization, C.Z.; Writing—original draft, T.W. and C.C.; Writing—review and editing, T.W. and C.C. All authors have read and agreed to the published version of the manuscript.

Funding: This research was funded by the National Key Research and Development Program of China [grant number 2016YFC0600108], and the project from China Geological Survey [grant numbers DD20160346, DD20190379, DD20221806].

Acknowledgments: We appreciate the anonymous reviewers for the critical and constructive comments and suggestions. We also deliver special thanks to Hao Yujie and Liu Xiaohe for their support during the zircon LA–ICP–MS U–Pb dating.

Conflicts of Interest: The authors declare no conflict of interest.

References

1. Kumar, S.; Rino, V.; Pal, A.B. Field evidence of magma mixing from microgranular enclaves hosted in Palaeo-proterozoic Malanjhkhand granitoids, central India. *Gondwana Res.* **2004**, *7*, 539–548. [CrossRef]
2. Kumar, S. Mafic to hybrid microgranular enclaves in the Ladakh batholiths, northwestern Himalaya: Implications on calc-alkaline magma chamber processes. *J. Geol. Soc. India* **2010**, *76*, 5–25. [CrossRef]
3. Perugini, D.; Poli, G.; Christofides, G.; Eleftheriadis, G. Magma mixing in the Sithonia Plutonic Complex, Greece: Evidence from mafic microgranular enclaves. *Miner. Petrol.* **2003**, *78*, 173–200. [CrossRef]
4. Barbarin, B. Mafic magmatic enclaves and mafic rocks associated with some granitoids of the central Sierra Nevada batholith, California: Nature, origin, and relations with the hosts. *Lithos* **2005**, *80*, 155–177. [CrossRef]
5. Qu, X.M.; Wang, H.N.; Rao, B. A study on the genesis of dioritic enclaves in Guojialing granite. *Acta Mineral. Sin.* **1997**, *17*, 302–309. (In Chinese)
6. Song, M.C.; Yan, Q.L. Characteristics of dioritic inclusions in Weideshan superunit and its magma origin in Jiaonan Area. *Shandong Geol.* **2000**, *4*, 16–22. (In Chinese)
7. Yang, J.H.; Wu, F.Y.; Chung, S.L.; Wilde, S.A.; Chu, M.F. Multiple sources for the origin of granites: Geochemical and Nd/Sr isotopic evidence from the Gudaoling granite and its mafic enclaves, northeast China. *Geochim. Cosmochim. Acta* **2004**, *68*, 4469–4483. [CrossRef]
8. Yang, J.H.; Wu, F.Y.; Wilde, S.A.; Liu, X.M. Petrogenesis of Late Triassic granitoids and their enclaves with implications for post-collisional lithospheric thinning of the Liaodong Peninsula, North China Craton. *Chem. Geol.* **2007**, *242*, 155–175. [CrossRef]
9. Chen, G.J.; Sun, F.Y.; Li, Y.C.; Liu, K. U–Pb dating, geochemical characteristics and geological significance of Guojialing granodiorite in Jiaodong Peninsula. *Glob. Geol.* **2014**, *33*, 39–48. (In Chinese)
10. Li, S.R.; Santosh, M. Geodynamics of heterogeneous gold mineralization in the North China Craton and its relationship to lithospheric destruction. *Gondwana Res.* **2017**, *50*, 267–292. [CrossRef]
11. Li, Z.D.; Yu, X.F.; Wang, Q.M.; Du, Z.Z.; Cao, Q.; Shi, M.Y.; Wang, R. Petrogenesis of Sanfoshan granite, Jiaodong: Diagenetic physical and chemical conditions, zircon U-Pb geochronology and Sr–Nd isotope constraints. *Acta Petrol. Sin.* **2018**, *34*, 447–468. (In Chinese)
12. Chai, P.; Sun, J.G.; Xing, S.W.; Men, L.J.; Han, J.L. Early Cretaceous arc magmatism and high-sulphidation epithermal porphyry Cu–Au mineralization in Yanbian area, Northeast China: The Duhuangling example. *Int. Geol. Rev.* **2015**, *57*, 1267–1293. [CrossRef]
13. Yang, X.A.; Li, C.; Hou, K.J.; Li, K.J.; Li, G.R.; Liu, J.J. Magma mixing and gold mineralization in the Maevatanana gold deposit, Madagascar. *Ore Geol. Rev.* **2017**, *91*, 1013–1024. [CrossRef]
14. Li, Z.L.; Xu, W.D.; Qin, Q.M.; Pang, W.Z. Geological characteristics and ore genesis of Wulong gold deposit, Liaoning Province. *Contrib. Geol. Miner. Resour. Res.* **1987**, *2*, 31–39. (In Chinese)
15. Wei, J.H.; Qiu, X.P.; Guo, D.Z.; Tan, W.J. Geochemistry of ore fluids and Rb-Sr isotopic dating for the Wulong gold deposit in Liaoning, China. *Acta Geol. Sin.* **2004**, *78*, 1267–1274.

16. Yu, B.; Zeng, Q.D.; Frimmel, H.E.; Wang, Y.B.; Guo, W.K.; Sun, G.T.; Zhou, T.C.; Li, J.P. Genesis of the Wulong gold deposit, northeastern North China Craton: Constraints from fluid inclusions, H–O–S–Pb isotopes, and pyrite trace element concentrations. *Ore Geol. Rev.* **2018**, *102*, 313–337. [CrossRef]
17. Chen, C.; Li, D.T.; Wu, T.T.; Zhao, Y.; Zhao, C.Q.; Yang, J.L.; Gu, Y.C. Genesis of gold deposits in the Wulong orefield, Liaodong Peninsula, North China Craton: Constraints from ore deposit geology, REE, and C–H–O–S–Pb isotopes. *Geol. J.* **2020**, *55*, 5914–5933. [CrossRef]
18. Zhao, K.D.; Jiang, S.Y.; Yang, S.Y.; Dai, B.Z.; Lu, J.J. Mineral chemistry, trace elements and Sr-Nd-Hf isotope geochemistry and petrogenesis of Cailing and Furong granites and mafic enclaves from the Qitianling batholith in the Shi-Hang zone, South China. *Gondwana Res.* **2012**, *22*, 310–324. [CrossRef]
19. Li, S.Z.; Zhao, G.C.; Sun, M.; Han, Z.Z.; Hao, D.F.; Luo, Y.; Xia, X.P. Deformation history of the Paleoproterozoic Liaohe Group in the Eastern Block of the North China Craton. *J. Asian Earth Sci.* **2005**, *24*, 659–674. [CrossRef]
20. Chen, B.; Tian, W.; Zhai, M.G.; Arakawa, Y. Zircon U–Pb geochronology and geochemistry of the Mesozoic magmatism in the Taihang Mountains and other places of the North China Craton with implications for petrogenesis and geodynamic setting. *Acta Petrol. Sin.* **2005**, *21*, 13–24. (In Chinese)
21. Qiu, L.G.; Ren, F.L.; Cao, Z.X.; Zhang, Y.Q. Late Mesozoic magmatic activities and their constraints on geotectonics of Jiaodong region. *Geotecton. Metallog.* **2008**, *32*, 117–123. (In Chinese)
22. Qiu, Y.M.; Groves, D.I.; McNaughton, N.J.; Wang, L.G.; Zhou, T.H. Nature, age, and tectonic setting of granitoid-hosted, orogenic gold deposits of the Jiaodong Peninsula, eastern North China Craton, China. *Miner. Depos.* **2002**, *37*, 283–305. [CrossRef]
23. Li, J.; Li, S.Y.; Song, Y.X.; Jiang, M.Y.; Bi, M.G.; Wang, S.S.; Jiang, L.; Sun, L.L. Study on zircon U-Pb chronology and rock geochemistry of Guzhuang diorite in Jiaodong area. *Shandong Land Resour.* **2018**, *34*, 35–42. (In Chinese)
24. Zhao, G.C.; Cawood, P.A.; Li, S.Z.; Wilde, S.A.; Sun, M.; Zhang, J.; He, Y.H.; Yin, C.Q. Amalgamation of the North China Craton: Key issues and discussion. *Precambrian Res.* **2012**, *222–223*, 55–76. [CrossRef]
25. Zhang, P.; Zhao, Y.; Kou, L.L.; Yang, H.Z.; Yang, F.C. Zircon U–Pb and molybdenite Re–Os geochronology of copper–molybdenum deposits in southeast Liaoning Province, China. *Int. Geol. Rev.* **2016**, *58*, 1481–1491. [CrossRef]
26. Li, S.Z.; Zhao, G.C. SHRIMP U–Pb zircon geochronology of the Liaoji granitoids: Constraints on the evolution of the Paleoproterozoic Jiao-Liao-Ji belt in the Eastern Block of the North China Craton. *Precambrian Res.* **2007**, *158*, 1–16. [CrossRef]
27. Wu, F.Y.; Yang, J.H.; Wilde, S.A.; Zhang, X.O. Geochronology, petrogenesis and tectonic implications of the Jurassic granites in the Liaodong Peninsula, NE China. *Chem. Geol.* **2005**, *221*, 127–156. [CrossRef]
28. Wu, F.Y.; Lin, J.Q.; Wilde, S.A.; Zhang, X.O.; Yang, J.H. Nature and significance of the Early Cretaceous giant igneous event in eastern China. *Earth Planet. Sci. Lett.* **2005**, *233*, 103–119. [CrossRef]
29. Duan, X.X.; Zeng, Q.D.; Yang, J.H.; Liu, J.M.; Wang, Y.B.; Zhou, L.L. Geochronology, geochemistry and Hf isotope of Late Triassic magmatic rocks of Qingchengzi district in Liaodong peninsula, Northeast China. *J. Asian Earth Sci.* **2014**, *91*, 107–124. [CrossRef]
30. Wu, F.Y.; Yang, J.H.; Liu, X.M. Geochronological framework of the Mesozoic granitic magmatism in the Liaodong Peninsula, Northeast China. *Geol. J. China Univ.* **2005**, *11*, 305–317. (In Chinese)
31. Gu, Y.C.; Chen, R.Y.; Li, D.T.; Yang, J.L.; Yang, F.C.; Jia, H.X.; Dick, J.; Hu, Q.H.; Ju, N.; Cao, J. Jurassic ca. 160 Ma crustal remelting and Paleoproterozoic intrusive rock residues in the Liaodong Peninsula, East China: Evidence from in situ zircon U–Pb dating and Lu–Hf isotopic analysis. *Geol. J.* **2018**, *53*, 174–188. [CrossRef]
32. Zhang, P.; Kou, L.L.; Zhao, Y.; Bi, Z.W.; Sha, D.M.; Han, R.P.; Li, Z.M. Genesis of the Wulong gold deposit, Liaoning Province, NE China: Constrains from noble gaseradiogenic and stable isotope studies. *Geosci. Front.* **2020**, *11*, 547–563. [CrossRef]
33. Nasdala, L.; Hofmeister, W.; Norberg, N.; Martinson, J.M.; Corfu, F.; Dörr, W.; Valley, J.W. Zircon M257: A homogeneous natural reference material for the ion microprobe U–Pb analysis of zircon. *Geostand. Geoanal. Res.* **2008**, *32*, 247–265. [CrossRef]
34. Black, L.P.; Kamo, S.L.; Allen, C.M.; Aleinikoff, J.N.; Davis, D.W.; Korsch, R.J.; Foudoulis, C. TEMORA 1: A new zircon standard for Phanerozoic U–Pb geochronology. *Chem. Geol.* **2003**, *200*, 155–170. [CrossRef]
35. Ludwig, K.R. Isoplot 3.0—A geochronological toolkit for Microsoft Excel. *Berkeley Geochronol. Cent. Spec. Publ.* **2003**, *4*, 1–70.
36. Williams, I.S. U–Th–Pb Geochronology by Ion Microprobe. In *Applications of Microanalytical Techniques to Understanding Mineralizing Processes*; McKibben, M.A., Shanks, W.C., III, Ridley, W.I., Eds.; Reviews in Economic Geology: Canberra, Australia, 1998; Volume 7, pp. 1–35.
37. Wu, F.Y.; Yang, Y.H.; Xie, L.W.; Yang, J.H.; Xu, P. Hf isotopic compositions of the standard zircons and baddeleyites used in U-Pb geochronology. *Chem. Geol.* **2007**, *234*, 105–126. [CrossRef]
38. Xie, L.W.; Zhang, Y.B.; Zhang, H.H.; Sun, J.F.; Wu, F.Y. In situ, simultaneous determination of trace elements, U-Pb and Lu-Hf isotopes in zircon and baddeleyite. *Chin. Sci. Bull.* **2008**, *53*, 1565–1573. [CrossRef]
39. Bouvier, A.; Vervoort, J.D.; Patchett, P.J. The Lu–Hf and Sm–Nd isotopic composition of CHUR: Constraints from unequilibrated chondrites and implications for the bulk composition of terrestrial planets. *Earth Planet. Sci. Lett.* **2008**, *273*, 48–57. [CrossRef]
40. Blichert-Toft, J.; Albarède, F. The Lu-Hf isotope geochemistry of chondrites and the evolution of the mantle-crust system. *Earth Planet. Sci. Lett.* **1997**, *148*, 243–258. [CrossRef]
41. Griffin, W.L.; Pearson, N.J.; Belousova, E.; Jackson, S.E.; Achterbergh, E.V.; O'Reilly, S.Y.; Shee, S.R. The Hf isotope composition of cratonic mantle: LAM-MC-ICPMS analysis of zircon megacrysts in kimberlites. *Geochim. Cosmochim. Acta* **2000**, *64*, 133–147. [CrossRef]

42. Griffin, W.L.; Win, T.T.; Davis, R.; Wathanakul, P.; Andrew, A.; Metcalfe, I.; Cartigny, P. Diamonds from Myanmar and Thailand: Characteristics and possible origins. *Econ. Geol. Bull. Soc. Econ. Geol.* **2001**, *96*, 159–170. [CrossRef]
43. Zhang, R.X.; Yang, S.Y. A mathematical model for determining carbon coating thickness and its application in electron probe micro analysis. *Microsc. Microanal.* **2016**, *22*, 1374–1380. [CrossRef] [PubMed]
44. Belousova, E.; Suzanne, G.W.; Fisher, Y. Igneous zircon: Trace element composition as an indicator of source rock type. *Contrib. Mineral. Petrol.* **2002**, *143*, 602–622. [CrossRef]
45. Yang, F.C.; Song, Y.H.; Yang, J.L.; Shen, X.; Gu, Y.C. SHRIMP U–Pb age, geochemical characteristics and geological significance of granite complex in Wulong-sidaogou gold deposit, eastern Liaoning. *Geotecton. Metallog.* **2018**, *42*, 1–15. (In Chinese)
46. Xu, Y.G.; Wu, X.Y.; Luo, Z.Y.; Ma, J.L.; Huang, X.L.; Xie, L.W. Zircon Hf isotope compositions of middle Jurassic-Early Cretaceous intrusions in Shandong Province and its implications. *Acta Petrol. Sin.* **2007**, *23*, 307–316. (In Chinese)
47. Chen, B.; Tian, W.; Jahn, B.M.; Chen, Z.C. Zircon SHRIMP U–Pb ages and in-situ Hf isotopic analysis for the Mesozoic intrusions from in south Taihang, north China craton: Evidence for hybridization between mantle-derived magmas and crustal components. *Lithos* **2008**, *102*, 118–137. [CrossRef]
48. Kamei, A.; Owada, M.; Nagao, T.; Shiraki, K. High-Mg diorites derived from sanukitic HMA magmas, Kyushu Island, southwest Japan arc: Evidence from clinopyroxene and whole rock compositions. *Lithos* **2004**, *75*, 359–371. [CrossRef]
49. Sun, S.S.; McDonough, W.F. Chemical and Isotopic Systematics of Oceanic Basalts: Implications for Mantle Composition and Processes. In *Magmatism in the Ocean Basins*; Saunders, A.D., Norry, M.J., Eds.; Geological Society London Special Publications; Geological Society: London, UK, 1989; pp. 313–345.
50. Chen, B.; Zhai, M.G. Geochemistry of late Mesozoic lamprophyre dykes from the Taihang Mountains, birth China, and implications for the sub-continental lithospheric mantle. *Geol. Mag.* **2003**, *140*, 87–93. [CrossRef]
51. Chen, B.; Jahn, B.M.; Zhai, M.G. Sr-Nd isotopic characteristics of the Mesozoic magmatism in the Taihang-Yanshan orogenm north China craton, and implications for Archean lithosphere thinning. *J. Geol. Soc. Lond.* **2003**, *160*, 963–970. [CrossRef]
52. Ma, C.Q.; Li, Z.; Ehlers, C.; Yang, K.; Wang, R. A post-collisional magmatic plumbing system: Mesozoic granitoid plutons from the Dabieshan high-pressure and ultrahigh-pressure metamorphic zone, east-central China. *Lithos* **1998**, *45*, 431–456. [CrossRef]
53. Jahn, B.M.; Wu, F.Y.; Lo, C.H.; Tsai, C.H. Crust-mantle interaction induced by deep subduction of the continental crust: Geochemical and Sr-Nd isotopic evidence from post-collisional mafic-ultramafic intrusions of the northern Dabie complex, central China. *Chem. Geol.* **1999**, *157*, 119–146. [CrossRef]
54. Chen, B.; Jahn, B.M.; Wei, C.J. Petrogenesis of Mesozoic granitoids in the Dabie UHP complex, central China: Trace elemant and Nd-Sr isotope evidence. *Lithos* **2002**, *60*, 67–88. [CrossRef]
55. Yang, J.H.; Chu, M.F.; Liu, W.; Zhai, M.G. Geochemistry and petrogenesis of Guojialing granodiorites form the northwestern Jiaodong Peninsula, eastern China. *Acta Petrol. Sin.* **2003**, *19*, 692–700. (In Chinese)
56. Yang, J.H.; Chung, S.L.; Zhai, M.G.; Zhou, X.H. Geochemical and Sr-Nd-Pb isotopic compositions of mafic dikes from the Jiaodong Peninsula, China: Evidence for vein-plus-peridotite melting in the lithospheric mantle. *Lithos* **2004**, *73*, 145–160. [CrossRef]
57. Guo, F.; Fan, W.M.; Wang, Y.J.; Zhang, M. Origin of early Cretaceous calc-alkaline lamprophyres from the Sulu orogen in eastern China: Implications for enrichment processes beneath continental collisional belt. *Lithos* **2004**, *78*, 291–305. [CrossRef]
58. Pei, F.P.; Xu, W.L.; Yang, D.B.; Yu, Y.; Wang, W.; Zhao, Q.G. Geochronology and geochemistry of Mesozoic mafic–ultramafic complexes in the southern Liaoning and southern Jilin provinces, NE China: Constraints on the spatial extent of destruction of the North China Craton. *J. Asian Earth Sci.* **2011**, *40*, 636–650. [CrossRef]
59. Pang, C.J. Geochronology and geochemistry of the Cretaceous mafic to intermediate volcanic rocks in the eastern North China Craton. *Guangzhou Guangzhou Insititute Geochem. Chin. Acad. Sci.* **2015**. (In Chinese)
60. Li, X.S. The Petrogenesis and Tectonic Setting of the Early Cretaceous Basic Dikes in Pulandian Area, Liaodong Peninsula. Ph.D. Thesis, China University of Geosciences, Beijing, China, 2019. (In Chinese).
61. Yang, H.T. Sr-Nd-Hf Isotopic Compositions of the Early Cretaceous Lamprophyres in Western Shandong: Constraints on the lithospheric Mentle of the Eastern North China Craton. Ph.D. Thesis, Jilin University, Changchun, China, 2018. (In Chinese).
62. Jahn, B.M.; Auvray, B.; Cornichet, J.; Bai, Y.L.; Shen, Q.H.; Liu, D.Y. 3.5 Ga old amphibolites from eastern Hebei province, China: Field occurrence, petrology, Sm–Nd isochron age and REE geochemistry. *Precambrian Res.* **1987**, *34*, 311–346. [CrossRef]
63. Zartman, R.E.; Doe, B.R. Plumbotectonics-The model. *Tectonophysics* **1981**, *75*, 135–162. [CrossRef]
64. Cheng, X.H. Integrated Research on Mesozoic Magmatism and Gold-Uranium Metallogenesis in the Liaodong Peninsula, Liaoning Province. Ph.D. Thesis, University of Science and Technology Beijing, Beijing, China, 2017. (In Chinese).
65. Lin, W.W.; Peng, L.J. The Estimation of Fe^{3+} and Fe^{2+} contents in amphibole and biotite from EPMA data. *J. Changchun Univ. Earth Sci.* **1994**, *24*, 155–162. (In Chinese)
66. Foster, M.D. Interpretation of the composition of trioctahedral Mica. United States. *Geol. Surv. Prof. Pap.* **1960**, *354*, 11–48.
67. Wones, D.R.; Eugster, H.P. Stability of biotite-experient therory and application. *Am. Mineral.* **1965**, *50*, 1228–1272.
68. Yang, Y.; Wang, X.X.; Yu, X.W.; Ke, C.H.; Wang, L.G.; Guo, R.P.; Wang, S.A.; Li, X.X. Chemicalcomposition of biotite and amphibole from Mesozoic granites in northwestern Jiaodong Peninsula, China, and their implications. *Acta Petrol. Sin.* **2017**, *33*, 3123–3136. (In Chinese)
69. Maniar, P.D.; Piccoli, P.M. Tectonic discrimination of granitoids. *Geol. Soc. Am. Bull.* **1989**, *101*, 635–643. [CrossRef]
70. Chappell, B.W.; White, A.J.R. I- and S-type granites in the Lachlan Fold Belt. *Trans. R. Soc. Edinb. Earth Sci.* **1992**, *83*, 1–26.

71. Collins, W.J.; Beams, S.D.; White, A.J.R.; Chappell, B.W. Nature and Origin of A-type Granites with Particular Reference to Southeastern Australia. *Contrib. Mineral. Petrol.* **1982**, *80*, 189–200. [CrossRef]
72. Didier, J.; Barbarin, B. Enclaves and granite petrology. *Dev. Petrol.* **1991**, *13*, 625.
73. Guan, Y.L.; Yuan, C.; Long, X.P.; Zhang, Y.Y.; Wang, X.Y.; Huang, Z.Y.; Chen, B.; Qu, S.D. Genesis of mafic enclaves from Early Paleozoic granites in the South China Block: Evidence from petrology, geochemistry and zircon U–Pb Geochronology. *Geotecton. Metallog.* **2016**, *40*, 109–124. (In Chinese)
74. Bonin, B. Do coeval mafic and felsic magmas in post-collisional to within-plate regimes necessarily imply two constrasting, mantle and crust, sources? A review. *Lithos* **2004**, *78*, 1–24. [CrossRef]
75. Yang, J.H.; Wu, F.Y.; Wilde, S.A.; Xie, L.W.; Yang, Y.H.; Liu, X.M. Tracing magma mixing in granite genesis: In situ U–Pb dating and Hf-isotope analysis of zircons. *Contrib. Mineral. Petrol.* **2007**, *153*, 177–190. [CrossRef]
76. Sparks, R.S.J.; Marshall, L. Thermal and mechanical constraints on mixing between mafic and silicic magmas. *J. Volcanol. Geotherm. Res.* **1986**, *29*, 1–124. [CrossRef]
77. Lesher, C.E. Decoupling of chemical and isotopic exchange during magma mixing. *Nature* **1990**, *344*, 235–237. [CrossRef]
78. Poli, G.; Tommasini, S.; Halliday, A.N. Trace elements and isotopic exchange during acid-basic magma interaction processes. *Earth Environ. Sci. Trans. R. Soc. Edinb.* **1996**, *87*, 225–232.
79. Zhao, K.D.; Jiang, S.Y.; Zhu, J.C.; Li, L.; Dai, B.Z.; Jiang, Y.H.; Ling, H.F. Hf isotopic composition of zircons from the Huashan-Guposhan intrusive complex and their mafic enclaves in northeastern Guangxi: Implication for petrogenesis. *Chin. Sci. Bull.* **2010**, *55*, 509–519. [CrossRef]
80. Chen, B.; Chen, Z.C.; Jahn, B.M. Origin of mafic enclaves from the Taihang Mesozoic orogen, north China craton. *Lithos* **2009**, *110*, 343–358. [CrossRef]
81. Rudnick, R.L.; Gao, S. The Composition of the Continental Crust. In *The Crust*; Rudnick, R.L., Ed.; Elsevier-Pergamon: Oxford, UK, 2003; pp. 1–64.
82. Yu, J.J.; Hou, X.G.; Ge, W.C.; Zhang, Y.L.; Liu, J.C. Magma mixing genesis of the Early Permian Liulian pluton at the northeastern margin of the Jiamusi massif in NE China: Evidences from petrography, geochronology and geochemistry. *Acta Petrol. Sin.* **2013**, *29*, 2971–2986. (In Chinese)
83. Zhai, M.G.; Guo, J.H.; Liu, W.J. Neoarchean to Paleoproterozoic continental evolution and tectonic history of the North China Craton: A review. *J. Asian Earth Sci.* **2005**, *24*, 547–561. [CrossRef]
84. Li, J.Y.; Qian, Y.; Sun, J.L.; Li, H.R.; Zhao, C.J.; Sun, F.Y.; Shen, Y.J. Early Cretaceous granitoids and gabbro in the Liaodong Peninsula: Implications for delamination of the North China Craton and Paleo-Pacific Plate subduction. *Mineral. Petrol.* **2021**, *115*, 299–322. [CrossRef]
85. Stone, D. Temperature and pressure variations in suites of Archean felsic plutonic rocks, Berens River area, northwest Superior Province, Ontario, Canada. *Can. Mineral.* **2000**, *38*, 455–470. [CrossRef]
86. Xie, J.C.; Yang, X.Y.; Sun, W.D.; Du, J.G. Early Cretaceous dioritic rocks in the Tongling region, eastern China: Implications for the tectonic settings. *Lithos* **2012**, *150*, 49–61. [CrossRef]
87. Liu, J.; Wang, S.L.; Li, T.G.; Yang, Y.; Liu, F.X.; Li, S.H.; Duan, C. Geochronology and isotopic geochemical characteristics of Wulong gold deposit in Liaoning Province. *Miner. Depos.* **2018**, *37*, 712–728. (In Chinese)
88. Li, L.; Santosh, M.; Li, S.R. The "Jiaodong type" gold deposits: Characteristics, origin and prospecting. *Ore Geol. Rev.* **2015**, *65*, 589–611. [CrossRef]
89. Li, Q.; Santosh, M.; Li, S.R.; Zhang, J.Q. Petrology, geochemistry and zircon U–Pb and Lu–Hf isotopes of the Cretaceous dykes in the central North China Craton: Implications for magma genesis and gold metallogeny. *Ore Geol. Rev.* **2015**, *67*, 57–77. [CrossRef]
90. Wang, F.F.; Han, Z.Z.; Zhang, Y.; Lai, Z.Q.; Ai, L.N. Geochemistry and petrogenesis of the Yashan granite and its enclaves, Qixia. *Contrib. Geol. Miner. Resour. Res.* **2019**, *34*, 393–405. (In Chinese)
91. Goldfarb, R.; Santosh, M. The dilemma of the Jiaodong gold deposits: Are they unique? *Geosci. Front.* **2014**, *5*, 139–153. [CrossRef]
92. Li, Y.J.; Li, S.R.; Liu, S.A.; Zhang, L.; Li, W.T.; Song, Y.X.; Wang, B.X. Zircon geochronology, geochemistry and stable isotopes of the Wang'ershan gold deposit, Jiaodong Peninsula, China. *J. Asian Earth Sci.* **2015**, *113*, 695–710. [CrossRef]
93. Zhu, R.X.; Fan, H.R.; Li, J.W.; Meng, Q.R.; Li, S.R.; Zeng, Q.D. Decratonic gold deposits. *Sci. China Earth Sci.* **2015**, *58*, 1523–1537. [CrossRef]
94. Zhu, R.X.; Yang, J.H.; Wu, F.Y. Timing of destruction of the North China Craton. *Lithos* **2012**, *149*, 51–60. [CrossRef]
95. Li, S.R.; Santosh, M.; Zhang, H.F.; Luo, J.Y.; Zhang, J.Q.; Li, C.L.; Zhang, X.B. Metallogeny in response to lithospheric thinning and craton destruction: Geochemistry and U-Pb zircon chronology of the Yixingzhai gold deposit, central North China Craton. *Ore Geol. Rev.* **2014**, *56*, 457–471. [CrossRef]
96. Tian, Y.F.; Sun, J.; Ye, H.S.; Mao, J.W.; Wang, X.X.; Bi, M.F.; Xia, X.P. Genesis of the Dianfang breccia-hosted gold deposit, western Henan Province, China–Constraints from geology, geochronology and geochemistry. *Ore Geol. Rev.* **2017**, *91*, 963–980. [CrossRef]
97. Zhu, R.X.; Zhang, H.F.; Zhu, G.; Meng, Q.R.; Fan, H.R.; Yang, J.H.; Zheng, T.Y. Craton destruction and related resources. *Int. J. Earth Sci.* **2017**, *106*, 2233–2257. [CrossRef]

minerals MDPI

Article

Ore Genesis of the Dongping Gold Deposit in the Northern Margin of North China Craton: Constraints from In-Situ Major, Trace Elemental Analysis of Magnetite and Pyrite

Chengyang Wang [1,2,3,*], Jiajia Yu [1], Yunsheng Ren [2,3], Junkang Zhao [4] and Zhenjun Sun [2,3]

[1] MNR Key Laboratory of Metallogeny and Mineral Assessment, Institute of Mineral Resources, Chinese Academy of Geological Sciences, Beijing 100037, China; yujia_2021@163.com
[2] School of Earth Sciences, Institute of Disaster Prevention, Sanhe 065201, China; rys@cidp.edu.cn (Y.R.); szj@cidp.edu.cn (Z.S.)
[3] Hebei Key Laboratory of Earthquake Dynamics, Sanhe 065201, China
[4] Institute of Geological Exploration, Zijin Mining Group Co., Ltd., Xiamen 361006, China; zhaojunkang@163.com
* Correspondence: wcy@cidp.edu.cn

Abstract: The Zhangxuan district in North China, also known as Northwestern Hebei "Golden Triangle," develops many intrusion-hosted lode-gold deposits. The Dongping gold deposit in the Zhangxuan district is well known for its unique hosting of rocks and ore mineral assemblages. Magnetite and pyrite are common minerals that widely exist in ores of the Dongping deposit. To get a better understanding of the evolution of the ore-forming fluids responsible for mineralization, we report on an integrated study on the electron microprobe analysis (EMPA) and laser ablation-inductively coupled plasma-mass spectrometry (LA-ICP-MS) analysis of magnetite and pyrite from the deposit. The major results are as follows: The magnetite grains from the Dongping deposit show a variable content of major and trace elements such as Ti, Al, Si, Fe, Mn, Cr, Na, V, and Co, and the majority of the magnetite contain low Ti contents, revealing potential properties of hydrothermal magnetite. The flat time-resolved signals of LA-ICP-MS imply that the majority of trace elements in magnetite exist in the form of isomorphism, except for some incompatible trace elements. Magnetites from the Dongping deposit have compositional characteristics of hydrothermal origins, and the genetic discriminant diagrams of Ti–V, Ti–Ni/Cr or (Ca + Al + Mn)–(Ti + V) show that they may be originated from magma differentiated hydrothermal solutions. Co, Ni in pyrite from Dongping mainly enter the lattice via isomorphism, and Cu, Zn, Ag, W, Sn, Au, Pb, and Bi are partitioned into pyrite as micro/nano- mineral inclusions. The Co, Ni content, and the Ni/Co ratios, indicated that the temperature of the ore-forming fluids has decreased from Py-1 to Py-2, and the enrichment of Au in Py-2 may be related to the cooling and boiling of the fluids.

Keywords: EPMA; LA-ICP-MS; magnetite; pyrite; Dongping deposit; northern margin of North China Craton

1. Introduction

The intrusion-hosted lode gold deposits provide China with the largest gold reserves [1,2]. The North China Craton (NCC), which is one of the three famous Precambrian cratons globally, develop a large quantity of intrusion-hosted deposits along its margins [3–5]. The Zhang-Xuan area is located along the northern margin of North China Craton, and has always been regarded as one of the most famous gold mining districts in Northern China (Figure 1). Since the Dongping gold deposit in the 1980s, several ore deposits showing similar geological characteristics have been successively discovered, evaluated, and mined in this area. Based on their distinctive ore mineral assemblages and hosting wall rocks, Chinese geologists have proposed that the Dongping and other similar deposits in the adjacent areas represent a particular type of gold mineralization, named the 'Dongping-type' deposit [6]. Among these deposits, the Dongping gold deposit is the largest deposit, with proven gold reserves of >100 tons in the Zhangxuan district. It is also the most representative deposit, which can be used as an example to explore its geological characteristics and formation mechanism [7].

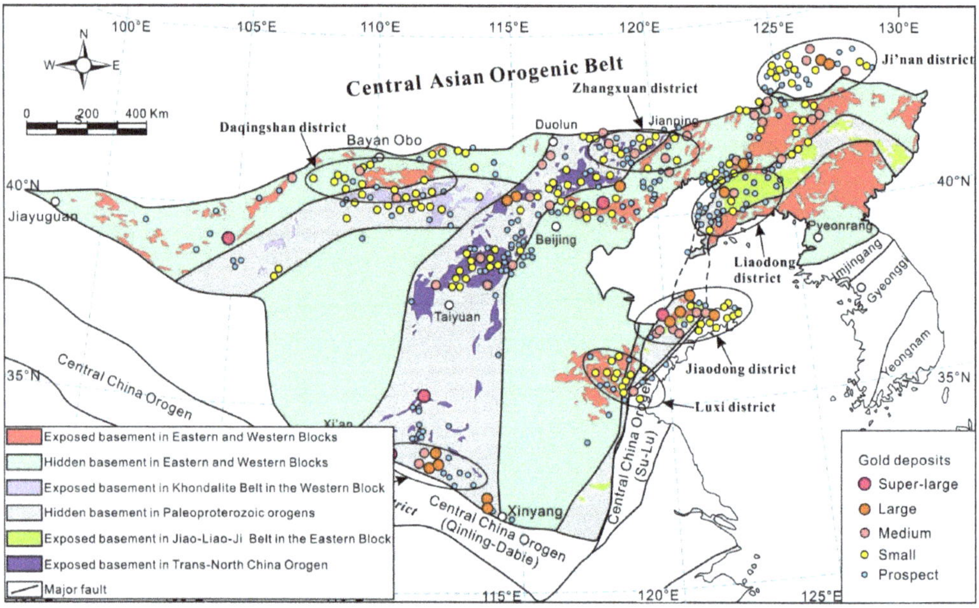

Figure 1. The distribution of gold deposits in the North China Craton (modified after Zhao et al. [12]; Deng and Wang [13]; He et al. [14]; Li and Santosh [15]).

During past decades, many studies concentrating on the genesis, fluid-rock interaction, ore-forming physicochemical conditions of the Dongping deposit have been conducted. A significant achievement obtained by previous research is that the Dongping deposit is derived from deep-derived alkaline magma [8–11]. High precise in situ chemical analysis can provide enhanced information on the metallogenic physicochemical conditions, and the nature of ore-forming fluids. Therefore, we present the first research on the in-situ element analysis for pyrite and magnetite from the deposit, through LA-ICP-MS and EPMA methods. The primary purpose of this research is to determine the genesis and evolution of the deposit.

2. Regional and Ore District Geology

Tectonically, the Zhangxuan district is located in the middle part of the northern margin of the North China Craton (Figure 2). As one of the most important gold-producing bases in China, the northern margin of the North China Craton hosts approximately 900 individual gold deposits [16]. In recent years, due to its excellent resource potential and associated Te mineralization, the gold deposits in Dongping and its adjacent areas (also known as Zhangxuan district) have attracted great attention from Chinese geologists and government officials.

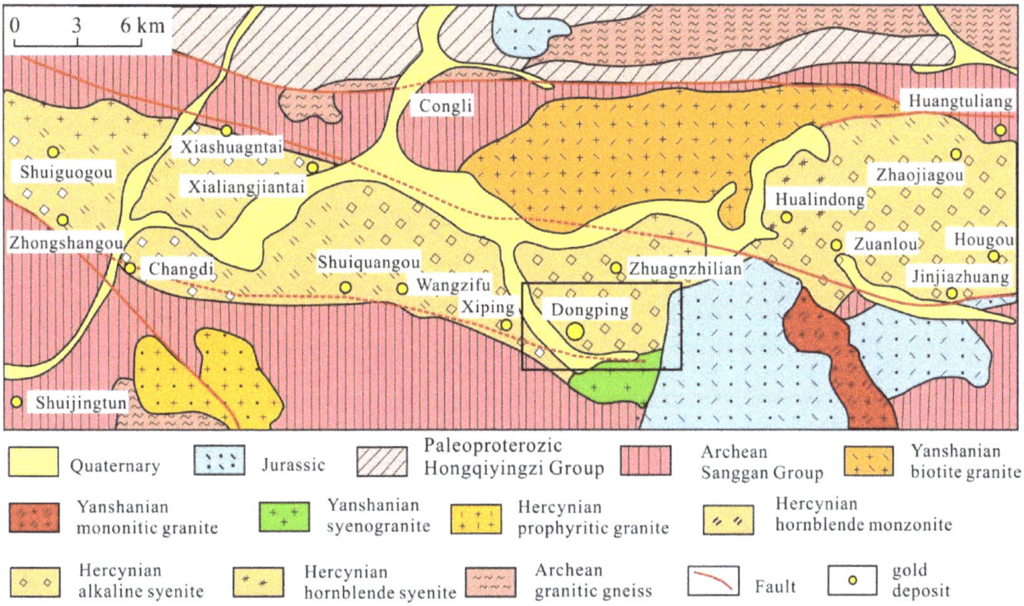

Figure 2. Geological sketch map of the area surrounding the Dongping gold deposit in the Hebei province (modified after Song and Zhao [17]; Li et al. [18]).

The outcropping successions in the regional area mainly consist of the Archean Sanggan Group and the Jurassic meta-sedimentary rocks. The Archean Shanggan Group represents the oldest crystalline basement in this area, and is composed of metamorphic rocks, including gneiss, plagio-amphibolite, and granulite. The Jurassic volcanic-sedimentary rocks, dominated by rhyolite, rhyodacite, trachyte, and pyroclastic flows, are widely distributed in this area. Three cycles of tectonic-magmatic activities have been recorded in the Dongping goldfield region: (1) the Archean-Proterozoic intrusive diorites, peridotites, and pyroxenites; (2) the Devonian alkaline intrusive rocks, mainly including the Shuiquangou igneous complex; (3) the Mesozoic alkali granite plutons, which intruded the Archaean basement and the margins of the Shuiquangou complex. Gold deposits in the regional area are mainly located within the Shuiquangou complex [19,20]. The Mesozoic plutons (formed at ca. 140 Ma) are exposed in the margins of the Shuiquangou complex, and they are likely to be related to underplate of the crust. The east-west trending Shangyi–Chongli–Chicheng fault is the most remarkable tectonic feature in the regional area, and the Shuiquangou alkaline intrusive complex was emplaced after the latest shearing along with the Shangyi–Chongli–Chicheng fault system [20]. The NE-trending Zhongshangou–Honghuabei–Shangshuiquan fault might be the main conduit for the gold-bearing hydrothermal activities. Orebodies from this area are hosted in third-order NNE- and NW- trending fractures [17].

The representative gold deposits in the Zhangxuan district are mainly composed of Dongping, Zhongshangou, Shuijingtun, Xiaoyingpan, Hougou, Huangtuliang, Guzuizi, Zhangquanzhuang, Beigangou, Dayingz, and Shiziling (Figure 2). Based on the location and occurrence of the orebodies, the above deposits can be divided into two major types, namely the Dongping type and the Xiaoyingpan type. The Dongping type deposits are distributed within the Devonian Shuiquangou alkaline complex, while the Xiaoyingpan deposits are mainly hosted in the Archean high-grade metamorphic rocks of the Sanggan Group [8]. Although numerous isotopic dating results have been published, the ages of gold mineralization in this area are still controversial.

3. Ore Deposit Geology and Mineralization Characteristics

The Dongping deposit is located in Chongli County, Northwest Hebei Province, China. The latitude and longitude of the mining area are 40°50′47″~40°53′34″ N and 115°18′17″~115°24′08″ W, respectively.

Mineralization of the deposit occurs at the contact zone between the Devonian Shuiquangou complex and the Late Archean rocks. In addition to the above two geologic units, major geological units exposed in this area include those of the Early Proterozoic Hongqiyingzi Group, Middle Proterozoic Changcheng Group, and Jurassic and Quaternary rocks. More than 70 gold veins or swarms have been discovered in the mining area, and only five of them have attained economic concentrations and been mined. Veins of the deposit are mainly hosted by Devonian intrusions and are controlled by a group of NE- and NW-trending faults (Figure 3). The lengths of ore veins ranged from several hundred to more than one thousand meters. Ores in the study area are characterized by rich telluride but low sulfide contents [17]. The identified telluride minerals mainly include calaverite, petzite, altaite, hessite, krennerite, and stutzite [9]. Quartz vein and altered syenite type ores are developed in the deposit; the Te and Au contents are quite different between the two kinds of ores. Generally, the widespread and intensive hydrothermal alteration changed the composition, structure, and texture of the wall rock, and finally formed the vein-like orebodies. In this case, the boundary between orebody and its wall rock is not well defined. In the other case, silica-rich ore-bearing hydrothermal fluids filled in along the fracture and finally formed the auriferous quartz veins, which have a distinct boundary with the wall rocks. Based on the mineral assemblages and their crosscutting relationships, previous scholars have divided the mineralization and alteration of the deposit into four main stages: (I) early silica-potassium alteration stage, (II) magnetite-hematite-pyrite–quartz stage, (III) Au bearing polymetallic sulfide–telluride quartz stage, and (IV) carbonate–quartz stage (Figures 4 and 5; [21]).

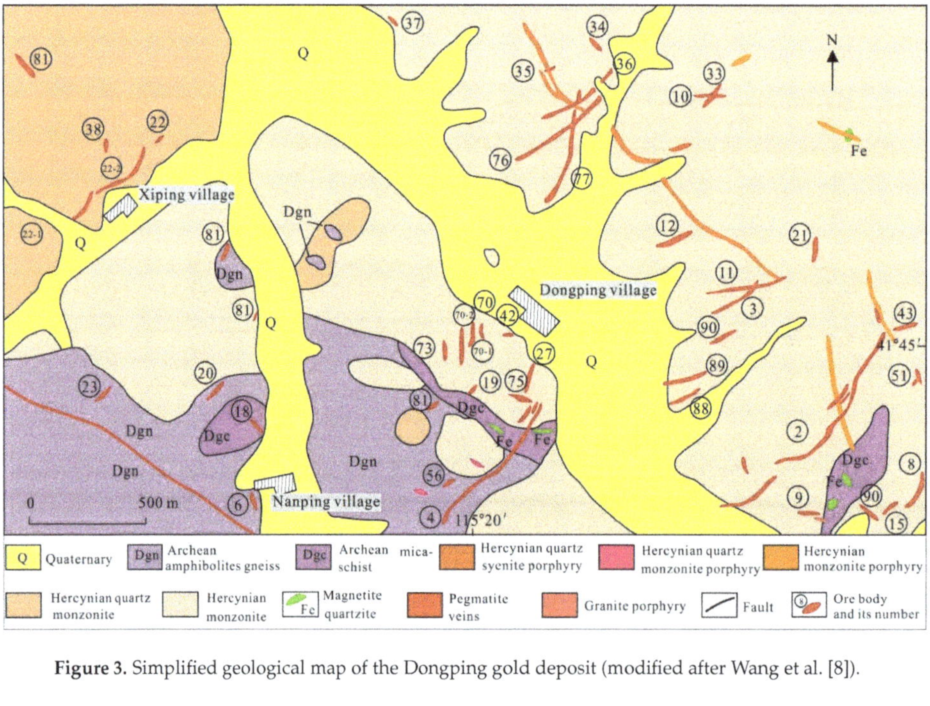

Figure 3. Simplified geological map of the Dongping gold deposit (modified after Wang et al. [8]).

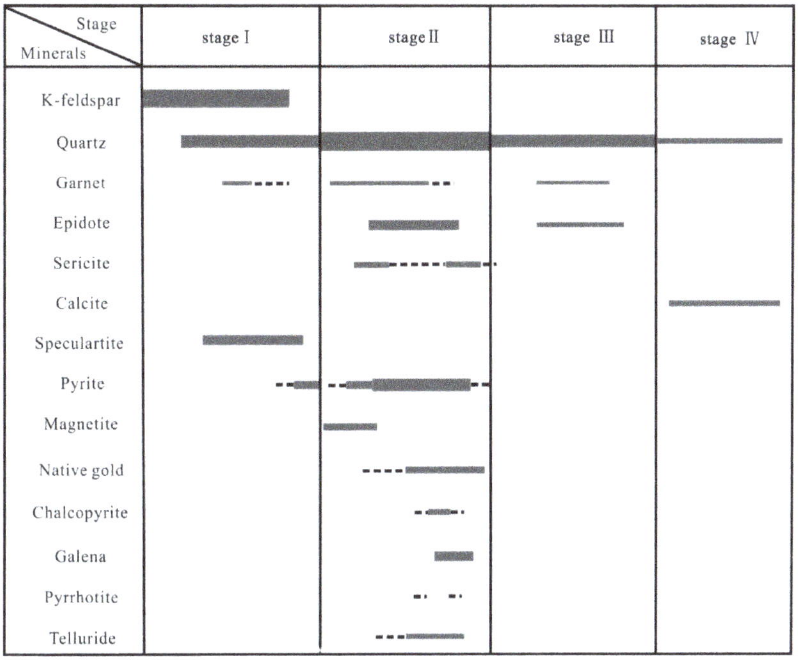

Figure 4. Paragenetic sequence for the Dongping deposit.

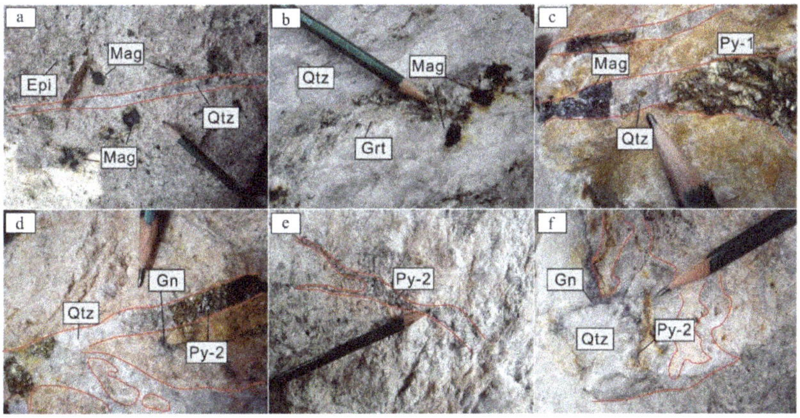

Figure 5. Representative photos of hand-specimens from the Dongping deposit: (**a**) Potassic rocks containing euhedral magnetite are crosscut by quartz vein; (**b**) Epidote, garnet, and magnetite coexist in quartz vein; (**c**) Garnet-magnetite-pyrite-quartz vein crosscut potassic rock; (**d**) Polymetallic sulfide quartz vein containing pyrite; (**e**) Quartz vein containing fine-grained pyrite aggregates; (**f**) Irregular quartz vein containing galena and pyrite. Epi—epidote; Gn—galena; Grt—garnet; Mgt—magnetite; Py—pyrite, Qtz—quartz.

4. Samples and Analytical Methods

4.1. Sample Descriptions

The iron oxide minerals within ores from the Dongping deposit are always neglected by previous studies. The magnetites from Dongping deposit have two occurrences: Mgt-1 developed in potassic igneous rocks, this type of magnetite exhibits irregular and anhedral shapes. Mgt-2 refers to magnetite occurred in the quartz veins. Compared with Mag-1, Mag-2 is favorable to form in the edge of quartz vein. Magnetite in this deposit always coexist with hematite, and show homogeneous internal textures in reflected light images. Pyrite is widespread in Dongping ores. Two pyrite types were identified within the ores, and denoted as Py-1 and Py-2. Py-1 is dominantly characterized by coarse euhedral crystals. In some cases, Py-1 coexists with magnetites in quartz veins (Figure 5). Previous studies suggest that no significant gold mineralization was associated with Py-1. Py-2 refers to the pyrite in tellurides- sulfide-quartz veins. This type of pyrite generally occurs as fine-grain aggregates in quartz veins and is replaced by other sulfide minerals. The associated sulfide minerals of Py-2 include chalcopyrite, galena, sphalerite, molybdenite, etc. (Figure 6).

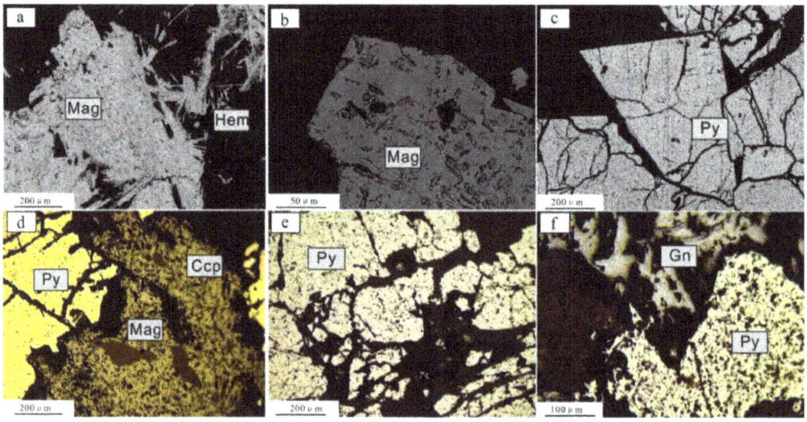

Figure 6. Photomicrographs showing textures and occurrences of magnetite/pyrite in the Dongping gold deposit: (**a**) BSE image of magnetite and hematite; (**b**) BSE image of euhedral magnetite; (**c**) BSE image of euhedral pyrite; (**d**) Reflected light image of magnetite-pyrite-chalcopyrite; (**e**) Reflected light image of pyrite aggregate; (**f**) Reflected light image of pyrite and galena. Ccp—chalcopyrite; Gn—galena; Py—pyrite; Mgt—magnetite; Hem—hematite.

4.2. Analytical Methods

Magnetite and pyrite samples of this study were mainly collected from Drilling Hole CK2, ZK1393, ZK1393, ZK301A, CK256, and open pit from the Dongping deposit. Based on the petrographic observations, 14 samples containing magnetite and pyrite were prepared for in-situ analysis. The EMPA analysis was conducted at the Wuhan Sample Solution Analytical Technology Co., Ltd., Wuhan, China. An EOL JXA-8230 electron probe micro-analyzer was used in this analysis. The accelerating voltage, current, and beam size of the EMPA are 20 kV, 5 nA and 1 μm, respectively. Peak and background counting times for Ni, Co (in pyrite), and Ca, Si (in magnetite) were 20 and 10 s, respectively. Pure FeS_2, Sb_2Te_3, ZnS, Sb_2Te_3, $(Fe, Ni)_9S_8$, Co, FeAsS, PbS, $CuFeS_2$, Bi_2Se_3, Au, and Ag were used as the standard materials for pyrite. Crystal analyzer used in this analysis included LIFH, PETH, PETJ, LIF, TAP. The test elements and their detection limits (mass fraction) are as 243×10^{-6}, Se 103×10^{-6}, Fe 138×10^{-6}, Co 125×10^{-6}, Ni 128×10^{-6}, Cu 161×10^{-6}, Zn 190×10^{-6}, Au 207×10^{-6}, S 42×10^{-6}, Pb 240×10^{-6}, Ag 85×10^{-6}, Te 94×10^{-6}, V 118×10^{-6}, Ti 91×10^{-6}. TiO_2, $CaMnSi_2O_6$, $NaAlSi_2O_6$, $(Mg, Fe)_2SiO_4$, $Mg_3Al_2Si_3O_{12}$, SiO_2, $(Fe, Ni)_9S_8$, Mn, $FeCr_2O_4$, V, $KAlSi_3O_8$, Co was used as the standard materials for magnetite. As the trace element contents of magnetite were mostly below the detection limits, EMPA results for some elements were not present in this article.

LA-ICP-MS analysis can provide enhanced resolution for trace element contents, we therefore further conducted in situ trace element research through the LA-ICP-MS method at the same laboratory as the EMPA analysis. The analysis was performed using a COMP-exPro Laser Ablation system coupled with an Agilent 7700e ICP–MS. The wavelength and maximum energy of the excimer laser were 193 nm and 200 mJ, respectively. Helium and argon were applied as the carrier and make-up gas, respectively. The spot size used during the analyses were 26 μm. Each spot analysis includes a ~20 s background acquisition and a ~60 s data acquisition. Visible mineral inclusions and possible contamination were avoided during the examination. Trace element compositions of pyrite were calibrated by reference materials NIST 610 and NIST 612, and the reference materials used for magnetite composition calibrating were NIST 610, BHVO-2G, BIR-1G, and BCR-2G. Detailed analytical procedures are available by Hu et al. [22]. Data reduction and offline data processing were performed by ICP MS DataCal and Iolite software [23].

5. Analytical Results

5.1. Magnetite

Element analysis of magnetite by EMPA included K_2O, CaO, TiO, Al_2O_3, SiO_2, FeO, MnO, Cr_2O_3, Na_2O, MgO, V_2O_3, CoO, and NiO (Table 1). The analytical results of EMPA reveal that the two types of magnetite from Dongping have similar element compositions. The Al_2O_3, Na_2O, MgO, and NiO contents of magnetite are mostly below or marginally higher than the detection limit. The maximum TiO_2 content determined by EMPA is 0.039 wt.%, revealing a characteristic of low Ti magnetite. The FeO, MnO, Cr_2O_3, and CoO contents range 88.538~92.695 wt.%, 0.080~0.195 wt.%, 0.008~0.132 wt.%, and 0.100~0.197 wt.%, respectively. The average V_2O_5 contents of Mag-1 and Mag-2 are 0.003 wt.% and 0.099 wt.%, respectively. Comparatively, the Mag-2 has higher V_2O_5 contents than Mag-1.

Table 1. Selected EPMA analyses of magnetite grains from the Dongping deposit (wt.%).

No.	Al_2O_3	SiO_2	FeO	MnO	Cr_2O_3	V_2O_3	Total
MT-1	0.003	0.007	91.78	0.24	0.115	-	92.28
MT-2	0.012	0.021	91.63	0.162	0.132	-	92.10
MT-3	-	-	91.82	0.164	0.096	0.012	92.24
MT-4	0.013	0.041	92.05	0.107	0.127	-	92.51
MT-5	0.016	-	91.866	0.102	0.088	0.007	92.25
MT-6	0.012	0.073	91.77	0.119	0.119	-	92.23
MT-7	-	0.01	92.69	0.08	0.048	0.257	93.28
MT-8	-	-	91.55	0.08	0.101	0.281	92.20
MT-9	-	0.012	91.71	0.059	0.046	0.278	92.34
MT-10	0.183	0.785	90.50	0.039	0.038	0.022	91.77
MT-11	0.027	0.346	90.79	0.025	0.008	-	91.34
MT-12	-	-	91.95	0.037	0.033	0.022	92.19
MT-13	-	0.399	90.98	0.259	0.103	0.028	91.93
MT-14	0.141	0.765	89.77	0.248	0.025	0.065	91.30
MT-15	-	0.257	90.92	0.16	0.128	0.024	91.60
MT-16	0.24	1.281	89.88	0.099	0.046	0.071	92.19
MT-17	0.302	1.676	88.54	0.182	0.008	0.068	91.35
MT-18	0.284	1.98	88.60	0.195	0.023	0.082	91.78

Notes: "-" means the content of the element is below detection limit.

The analytical results of magnetite by LA-ICP-MS can be recognized as follows (Table 2): Magnetite from the Dongping deposit has around 1300~5100 ppm Si, 100~1200 ppm Al, and 100~5000 ppm Ca, and 50~150 ppm Mg. Mag-1 has V contents of 8.62~234 ppm, Co contents of 40.8~91.4 ppm, and Ni contents of 1.48~17.5 ppm, coupled with lower Cr (with an average of 35.6 ppm), Zn(with an average of 162 ppm), Ag(<0.028 ppm), W contents of (mostly <0.4 ppm) contents. Mag-2 has average V contents of 57.6 ppm, average Cr contents of 81.23 ppm, Co contents of 10.9~66.6 ppm, Ni contents of 0.52~4.49 ppm, Cu contents of 11.5~46.7 ppm, Zn contents of 258~1787 ppm, Ag contents of 0.037~4.23 ppm, W contents of 0.091~68.2 ppm, Pb contents of 41.59~1163 ppm. The Spot 7, Spot 8, and Spot 11 have extremely higher Ca contents (6.75 wt.% for Spot 7, 2.20 wt.% for Spot 8, and 2.69 wt.% for Spot 11, respectively). Spot 7 also has a higher Mn content, suggesting that invisible carbonate mineral inclusions probably exist in magnetite.

Table 2. Selected LA-ICP-MS analyses of magnetite grains from the Dongping deposit.

Sample No.	Al_2O_3	CaO	TiO_2	V	Cr	MnO	Fe	Co	Ni	Cu	Zn	W	Pb
	wt.%	wt.%	wt%	ppm	ppm	wt.%	wt.%	ppm	ppm	ppm	ppm	ppm	ppm
DP-1-1	0.017	0.017	0.0011	12.9	127	0.18	71.78	83.0	8.35	-	104	0.40	7.90
DP-1-2	0.130	0.039	0.0049	183	11.0	0.10	71.36	83.8	7.93	1.42	66.9	60.1	24.34
DP-1-3	0.013	-	-	8.62	4.91	0.30	71.80	91.4	1.48	-	124	0.35	3.211
DP-1-4	0.043	0.0064	0.0050	59.3	21.6	0.094	71.95	40.8	17.5	110	209	0.043	194.12
DP-1-5	0.058	0.022	0.0065	173	33.1	0.095	71.50	41.7	6.67	150	276	0.16	281.76
DP-1-6	0.074	0.045	0.0092	234	40.3	0.19	71.41	42.3	15.4	69.7	196	0.085	128.49
DP-2-7	0.220	9.26	0.0019	69.2	629	0.47	64.32	10.9	3.81	42.4	258	68.2	634.52
DP-2-8	0.140	3.03	0.0018	83.5	12.9	0.10	69.11	12.5	4.31	40.1	1707	40.0	1163.33
DP-2-9	0.048	0.029	0.0008	28.0	9.35	0.032	71.77	66.6	2.31	16.5	482	0.091	41.59
DP-2-10	0.024	0.020	0.0016	44.3	6.27	0.038	71.71	20.3	4.49	11.6	217	0.72	116.57
DP-2-11	0.31	3.69	0.0014	67.4	37.4	0.32	68.25	14.1	0.52	43.5	349	18.5	641.05
DP-2-12	0.15	0.067	0.0020	75.6	40.6	0.086	71.23	26.5	3.51	120	580	4.18	585.00
DP-2-13	0.040	0.029	0.0008	21.9	13.5	0.079	71.70	28.8	3.37	15.2	374	0.24	130.87
DP-2-14	0.53	0.053	0.0009	77.6	18.9	0.18	70.16	34.3	3.69	46.7	1188	0.15	359.56
DP-2-15	0.038	0.023	0.0006	9.44	32.0	0.088	74.89	31.6	2.78	11.5	374	1.50	247.77
DP-2-16	0.20	0.036	0.0053	99.2	12.2	0.070	71.17	20.1	2.35	34.3	990	47.8	648.30

Notes: "-" means the content of the element is below detection limit.

5.2. Pyrite

Results of EMPA for pyrite from the Dongping deposit are given in Table 3. The major and minor element analysis of pyrite by EMPA mainly included Zn, Cu, Te, Sb, Se, Co, Ni, As, Au, Pb, and Ag. The average Fe, S, and Co contents are 46.05 wt.%, 53.80 wt.%, and 0.084 wt.% for Py-1; 46.10 wt.%, 53.75.05 wt.%, and 0.076 wt.% for Py-2; The contents of Ag and other trace elements (Cu, Ni, Pb, Te, Sb, Se) in pyrite were close to or below the detection limits of EMPA. Compared with Py-1, Py-2 has higher Au and Pb contents (average at 0.012 wt.%), but lower Zn and Co contents.

Table 3. Selected EPMA analyses of pyrite grains from the Dongping deposit (wt.%).

Sample No.	Zn	Cu	Te	Se	Fe	Ni	Au	S	Pb	Total
DS-1-PY-1	-	0.008	0.01	0.029	45.76	0.001	0.019	53.466	0.004	99.38
DS-1-PY-2	-	-	-	-	45.99	0.013	0.004	53.547	0.016	99.67
DS-1-PY-3	-	0.014	-	-	46.03	-	0.012	53.701	-	99.84
ZS-1-PY-1	0.015	-	0.003	-	46.06	-	-	53.996	0.005	100.17
ZS-1-PY-2	0.014	0.018	-	-	46.19	0.001	0.033	54.075	0.015	100.43
ZS-1-PY-3	0.014	-	0.027	0.032	46.26	-	0.005	54.071	0.015	100.52
ZH-2-PY-1	0.009	0.006	-	-	45.76	-	-	53.724	-	99.82
ZH-2-PY-2	0.047	-	-	-	46.26	-	-	53.827	-	100.22
ZH-2-PY-3	0.005	0.005	-	0.009	46.24	0.001	0.039	53.12	0.035	99.56
DS-2-PY-1	0.008	-	0.013	-	46.21	-	-	53.911	-	100.23
DS-2-PY-2	-	-	-	-	46.08	-	-	54.025	-	100.207
DS-2-PY-3	0.002	0.006	0.002	0.003	46.05	-	-	53.919	-	100.08

Notes: "-" means the content of the element is below detection limit.

A total of 18 pyrite LA-ICP-MS analyses were completed on the Dongping gold ores, and the representative trace elements are selected to list in Table 4. The contents of Ti, Co, Ni, Cu, Zn, Ge, and Pb are all above detection limit values in all pyrite grains, and Mg, Al, V, Mo, Ag, W, Au, Tl, and Bi are below the detection limit in some pyrite grains. Py-1 records relatively higher abundances of Co (with an average of 16.9 ppm), and Ni (average at 10.4 ppm), while Py-2 show higher contents of Ti (1.73~1637 ppm), V (0~13 ppm) and Cr (0.42~3.47 ppm). Both Py-1 and Py-2 show wide ranges of P, K, Ca, Ti, Au, and Pb contents, etc. The As, W, Mo, Sn contents determined by LA-ICP-MS are primarily below the detection limits. The relatively lower contents detected by LA-ICP-MS are consistent with the EMPA results. Nevertheless, Spot 7, 8 and 9 of Py-2 have relatively higher abundances of Cu (0.77~3.42 ppm), Zn (0.77~1.60 ppm), W (055~41.8 ppm), Au (0.25~0.84 ppm), Pb (4.4~11.5 ppm), Bi (0.31~2.27 ppm), implying possible mineral inclusions.

Table 4. Selected LA-ICP-MS analyses of pyrite grains from the Dongping deposit.

Sample No.	Ti ppm	V ppm	Cr ppm	Mn ppm	Co ppm	Ni ppm	Cu ppm	Zn ppm	As ppm	Ag ppm	W ppm	Au ppm	Pb ppm	Bi ppm
PY-2-1	2.12	0.022	0.83	0.25	1.50	0.38	0.26	0.44	-	0.023	-	0.13	0.41	-
PY-2-2	2.11	0.014	0.98	0.096	3.67	1.24	0.25	0.62	0.83	0.0073	-	-	0.12	-
PY-2-3	1.73	-	0.68	-	1.09	0.89	0.17	0.66	0.073	-	0.0092	0.0001	0.36	-
PY-1-4	0.91	0.017	0.37	0.048	0.22	0.11	1.77	0.32	0.59	0.13	0.0092	0.19	3.43	1.25
PY-1-5	1.35	-	0.28	-	0.17	0.50	0.78	0.46	0.079	0.020	-	0.032	0.91	0.10
PY-1-6	1.42	-	0.31	0.068	1.25	1.10	0.13	0.53	-	0.0060	0.0037	-	0.029	0.0034
PY-2-7	1637	13.0	3.47	0.17	2.34	6.91	3.42	1.60	0.91	0.91	41.8	0.84	11.5	2.27
PY-2-8	225	5.48	0.84	-	1.68	2.15	1.88	0.47	0.43	0.82	24.2	0.38	4.18	0.32
PY-2-9	13.6	0.15	0.42	0.11	0.99	1.76	0.77	0.77	0.054	0.094	0.55	0.25	4.40	0.31
PY-1-10	1.36	0.0048	-	-	22.43	2.43	0.76	0.78	-	0.57	-	0.13	0.52	0.0005
PY-1-11	2.19	-	0.33	0.11	74.33	84.03	0.93	0.67	-	0.11	0.0086	-	0.039	0.0012
PY-1-12	1.21	0.014	0.61	-	51.15	12.24	0.41	0.34	0.30	-	-	0.011	0.21	0.0019
PY-1-13	1.66	-	-	-	4.01	1.94	0.56	0.54	0.055	0.033	-	-	0.38	-
PY-1-14	0.87	0.019	0.20	0.064	6.41	5.77	0.61	0.49	0.18	2.17	0.0054	0.61	0.48	0.0007
PY-1-15	2.04	0.0076	1.47	0.17	18.84	10.54	0.41	0.82	-	0.13	-	0.026	0.33	-
PY-1-16	2.04	0.0012	-	-	2.77	0.93	0.25	0.55	0.0043	0.66	-	0.0094	0.63	-
PY-1-17	0.88	0.026	1.18	-	2.68	1.10	5.18	0.58	0.40	25.01	-	6.43	5.17	0.018
PY-1-18	1.98	0.0000	0.46	0.18	19.08	4.90	1.75	0.73	0.027	5.07	0.0000	1.09	0.74	0.0063

Notes: "-" means the content of the element is below detection limit.

6. Discussion

6.1. Occurrences of Elements in Magnetite and Pyrite

Minor/trace elements in magnetite and pyrite can occur in different forms: (1) invisible solid solution within the crystal lattice, (2) invisible nanoparticles of minerals, and (3) visible micron-sized mineral inclusions [24–27]. As one of the most common spinel group minerals, magnetite is generally described by a formula of AB_2O_4. Elements of Ti, V, Mn, Ni, Mg, Co, Zn, Al, Cr, or Ga can substitute into the magnetite crystal lattice via isomorphism, and these elements are thought to be compatible and detectable in magnetite [25,28]. While the incompatible elements (Al, Mg, Ca, K, and Na) probably enter magnetite structure as sub-micro-scale silicate inclusions [29]. Similar to magnetite, minor/trace elements in pyrite can substitute the Fe and S within the crystal lattice, or exist as nanoparticles and mineral inclusions of sulfides, silicates, and oxides [27]. Siderophile and chalcophile elements, including Co and Ni, are commonly entered via the lattice isomorphous replacement of Fe, while As can entered the lattice via isomorphous replacement of S. The element of Au, Ag, Cu, Pb, Zn, Te, Cr, V, Mn, Ga, Mo, Sn, and Sb generally exist as particles or nanoparticles in pyrite [30]. Mineral inclusions would significantly change the element contents detected by LA-ICP-MS, and may even result in some outliers. The following symbols can be used to identify the forms of elements in the LA-ICP-MS time-resolved depth profile: a flat time-resolved depth profiles indicate a pure mineral or the elements are incorporated into the internal structure of the mineral, while the peaks in time-resolved depth profiles are usually due to the mineral inclusions [31].

Our analyses shows that the majority of minor and trace elements in magnetite show flat patterns in the LA-ICP-MS time-resolved signal spectra; it is suggested that these elements have entered the magnetite lattice (Figure 7). Spot 7, 8, and 11 of our magnetite sample have relatively high Ca or Mn contents, indicating silicate minerals' inclusion in magnetite. Compared with magnetite, the LA-ICP-MS time-resolved signals of pyrite are more complicated. As shown in Figure 7, the signal intensities of Co and Ni are relatively stable and show similar distribution characteristics with that of Fe, inferring that Co and Ni enter pyrite as the form of isomorphism. However, the LA-ICP-MS time-resolved signals of Au, Ag, Cu, Pb, Te, and Bi are characterized by several apparent peaks. It is speculated that these elements may occur as invisible nanoparticles of sulfide, such as galena or tellurium-rich minerals.

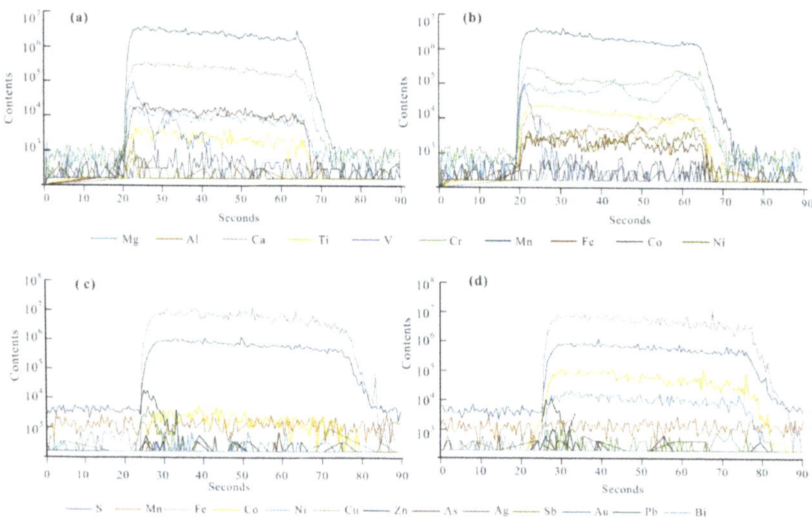

Figure 7. Representative time-resolved analytical signals of LA-ICP-MS analyses. (**a**,**b**) Magnetite; (**c**,**d**) Pyrite.

Pyrite is considered the primary mineral host for gold in the Dongping district. However, the Au contents in pyrite from both of the EMPA and LA-ICP-MS analysis are quiet low, and the Au contents of some analyzed samples are even lower than the detection limit. The measured Au contents through EMPA (between 0 and 0.039 wt.%) in this study are higher than those through LA-ICP-MS (only reaching a maximum of 6.43 ppm). This phenomenon is also found in other intrusion-posted gold deposits; an explanation is that EPMA gives the detection limit for Au, which is much higher than this one of LA-ICP-MS.

6.2. Implications for the Mineral Genesis

6.2.1. Magnetite

Magnetite is commonly developed in magmatic, hydrothermal, sedimentary, and metamorphic environments [32], and it is also widely spread in skarn, IOCG and oxidized porphyry type mineralization systems. Nevertheless, the magnetite within ores is rare in intrusion-related gold deposits. Geochemical and geological factors, including temperature, oxygen fugacity, sulfur fugacity, the composition of the fluids, and fluid-rock reaction, can influence the element compositions in magnetite [24,25]. Temperature is one of the most important factors governing compositional variations in magnetite, and the Ti, V, Mn, Al, and Ga contents in magnetite are all sensitive to temperature [25]. Elements of Cr and V occur in various valence states, and their behavior is linked to the oxygen fugacity of the hydrothermal system. For example, V can appear as V^{3+}, V4^{+}, or V^{5+} ions in magmatic or hydrothermal solutions, while only V^{3+} can readily enter the magnetite crystal lattice. The increasing oxygen fugacity would convert V^{3+} to V^{4+} or V^{5+}, resulting in the decrease of V content in magnetite [25]. W mainly replaces the trivalent ions (e.g., Al^{3+} and Fe^{3+}), and the W content in magnetite is governed by both temperature and oxygen fugacity [33,34]. The Mg, Si, Al, and Mn contents in magnetite are remarkably influenced by the degrees of fluid-rock interaction in a skarn-type metallogenic system [35]. Based on the above conclusions, a range of elemental diagrams for distinguishing the magnetite genesis have been proposed, which provides us with good baselines for the source identification of magnetite from various ore types [32].

Two alternative mechanisms can explain the magnetite genesis of Dongping ores: (1) Previous research has found that Devonian Shuiquangou syenitic complex constitutes the most important hosting rocks for ores in this area, and the magnetite commonly occur as an accessory mineral in Shuiquangou Complex [7]. Therefore, the magnetite from ores may capture, or inherit, from hosting igneous rocks. In this case, the magnetite should be formed at a higher temperature and show characteristics of magmatic origins. (2) Similar to porphyry or skarn system, magnetite of ores can form during a hydrothermal stage, and show geochemical features of hydrothermal magnetite. The magnetite compositions can discriminate the genetic type of magnetite. Mn, Ti, Cr, and Co are regarded as the most useful discriminator elements to differentiate magnetite type. Compared with hydrothermal magnetite, the igneous magnetite has, on average, higher Cr, Ti, V contents but lower Co contents [33].

In magnetite from the Dongping deposit, the Ni contents measured by LA-ICP-MS are significantly lower than the mean values of igneous magnetite; the Ni contents analyzed by EMPA are even below the detection limit. In addition, magnetite from Dongping shows features of low Ti contents. Most of the current genetic discrimination diagram of magnetite does not present the locations of the Dongping type deposit. As shown in the genetic discrimination diagram of V vs. Ti and Ti vs. Ni/Cr, the analyzed spots from Dongping are mainly located around the field of "hydrothermal" [34–36], indicating that the magnetite are formed in the hydrothermal stage. Compared with magnetite in porphyry, skarn, IOCG, and Kiruna (IOA) deposit, magnetite from the Dongping deposit shows characteristics of variable Ca, Al, and Mn contents, with wide ranges of Ti + V values (Figure 8).

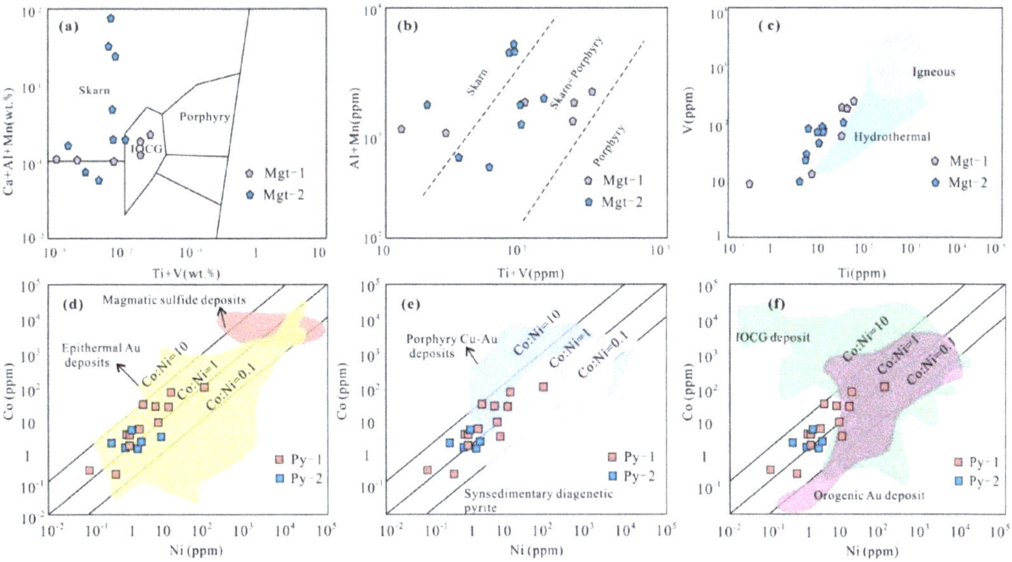

Figure 8. Genetic discrimination diagram for the magnetite and pyrite from Dongping deposit: (**a**) Ca + Al + Mn vs. Ti + V diagram for the magnetite; (**b**) Ni/(Cr + Mn) vs. Ti + V diagram for the magnetite [32]; (**c**) V vs. Ti for the magnetite [33,34]; (**d–f**) Co vs. Ni diagram for the pyrite [37]. The data used in the figure are from Tables 2 and 4.

6.2.2. Pyrite

Pyrite is the most common constituent of ore-bearing mineral assemblages in lode-type gold deposits [36,38]. As pyrites formed in different environments have distinctive compositional features, the minor and trace element contents can provide helpful information in determining the ore-forming processes. Co/Ni values of pyrite is the most sensitive indicators of the ore-forming environment. Generally, the higher Co/Ni mass ratio of pyrite (Co/Ni > 1) is attributed to magmatic hydrothermal origin, while pyrite with low Co/Ni ratio (Co/Ni < 1) is thought to be influenced by sedimentary materials [39,40].

Pyrite from the Dongping deposit has higher Co contents, but relatively lower Ni contents. The corresponding pyrite Co/Ni ratios from our EMPA analysis are significantly greater than 1.0. The LA-ICP-MS analytical results indicated that pyrites from Dongping have variable Co/Ni mass ratios ranging between 0.3 and 9.2 (Table 4), with an average of 2.3. In the Ni vs. Co plot, all of the samples from our analysis are located in the field of epithermal gold deposit, around the field of porphyry and orogenic gold deposits, which is also significantly different from magmatic sulfide deposits and IOCG deposits. Therefore, we propose that the ore-forming fluids of Dongping dominantly originate from a magmatic source, and incorporate some sedimentary materials from wall rock during the fluid migration.

6.3. Constraints on Hydrothermal Fluid Evolution

Even though the genetic type and formation age of Dongping deposit are still controversial, numerous C-H-O-S-Pb isotopic studies have also indicated that the ore-forming fluids of Dongping were dominantly magmatic hydrothermal in origin [41–43]. A remarkable feature of Dongping type mineralization is the presence of garnet in the ores, and the recent garnet U-Pb dating proved that the Dongping deposit formed at the age of 139~142 Ma, which is broadly coeval with the Early Cretaceous granitoid magmatism [44]. Our combined EMPA and LA-ICP-MS analysis on magnetite and pyrite further confirmed that the petrogenesis and metallogenesis in the Dongping district have a close temporal relationship, and the gold mineralization in this area are products of the Mesozoic tectonic-magmatic event.

In this study, two types of magnetite and two types of pyrite have been identified, providing us an excellent opportunity to trace the whole evolutionary process of ore-forming fluids. Element availability is the first-order control for minor and trace element contents in hydrothermal fluids. Spinel elements, such as Mn, Fe, Cu, Zn, As, Mo, Ag, Au, and Pb are commonly enriched in hydrothermal fluids compared to their content in igneous host rocks. Mn and Zn are widely incorporated into hydrothermal magnetite at levels above 100 ppm. In the Dongping deposit, Mn and Zn content in magnetite range 297~4348 ppm and 66.9~1707 ppm, respectively, which are significantly higher than the mean values of igneous magnetite. Contents of Cu, Au, Ag, and Mo in magnetite from Dongping are also similar to those of magmatic-hydrothermal deposits. Al and Ti are considered to be immobile during the fluid-rock reaction, and the incorporation of Al and Ti in magnetite is largely temperature-controlled. Therefore, the magnetite formed at hotter temperatures generally have higher Al and Ti contents, and the igneous magnetite has considerably higher Al and Ti contents than hydrothermal magnetite. Magnetite of this study have similar Al contents with porphyry deposit, but relatively lower Ti contents (average at 13.9 ppm), indicating that the magnetite from Dongping may be formed at a lower temperature condition than previously estimated. Considering also the fact that magnetite coexists with hematite, we can speculate that the magnetite in Dongping may have formed from relatively oxidizing fluids with lower temperatures.

The Co solubility in hydrothermal solutions is significantly influenced by temperature [45]. At Dongping, the Co contents from Py-1 to Py-2 shows a decreasing trend, indicating that the fluid temperature decreases continuously from the early to late stage. Compared with Py-1, the Cu, Zn, Ag, W, Au, Pb, and Bi, contents of Py-2 are relatively higher. The emission of CO_2 caused by the boiling of ore-forming fluids is considered to be an important mechanism for the precipitation of ore-forming materials in gold deposits. This is supported by a previous fluid inclusions study, which revealed that, the fluid heterogenization have occurred in the Dongping mineralization system [41].

7. Conclusions

1. Two magnetite sub-types, including disseminated magnetite in altered potassic igneous rocks and magnetite in quartz vein, and two pyrite sub-types, including pyrite

in oxide-ore stage (Py-1), pyrite in gold-bearing sulfide-telluride-ore stage (Py-2), are identified in the Dongping deposit.

2. A combined EMPA and LA-ICP-MS analysis on magnetite reveals that magnetite from Dongping have lower Ti and V contents, as well as wide ranges of Al, Mn, and Ca contents, indicating that they are originated from magmatic-hydrothermal fluids rather than igneous origins. These magnetites are interpreted to have resulted from low-temperature conditions of hydrothermal activity.
3. The average Au contents of pyrites from Dongping are relatively lower. The time-resolved signals of LA-ICP-MS of pyrite indicates that Co and Ni enter pyrite lattice by forming solid solutions, whereas Pb, Zn, and Cu mainly present as sulfide inclusions. From Py-1 to Py-2, Co shows a decreasing trend, indicating that the temperature of ore forming fluids has decreased. Combined with previous research of fluid inclusion, we infer that the enrichment of metal elements in Py-2 is related to the cooling or boiling of the fluids.

Author Contributions: Conceptualization, C.W. and Y.R.; methodology, Z.S.; software, C.W.; investigation, C.W.; J.Y.; J.Z.; data curation, J.Y.; writing—original draft preparation, C.W.; writing—review and editing, C.W. All authors have read and agreed to the published version of the manuscript.

Funding: This research was funded by the Youth fund of National Natural Science Foundation of China (Grant Number 42002093) and the Fundamental Research Funds for the Central Universities of China (Grant Number ZY20215134).

Acknowledgments: We sincerely appreciate the detailed and constructive reviews and suggestions from two anonymous reviewers, which greatly improved this paper. We especially thank Wuhan Sample Solution Analytical Technology Co., Ltd. for helping with the LA-ICP-MS trace element analyses.

Conflicts of Interest: The authors declare no conflict of interest.

References

1. Zhou, T.; Lü, G. Tectonics, granitoids and Mesozoic golddeposits in east Shandong, China. *Ore Geol. Rev.* **2000**, *16*, 71–90. [CrossRef]
2. Zhou, T.; Goldfarb, R.J.; Phillips, G.N. Tectonic and distribution of gold deposits in China—An overview. *Miner. Depos.* **2002**, *37*, 249–282. [CrossRef]
3. Yang, J.H.; Wu, F.Y.; Wilde, S.A. A review of the geodynamic setting of large-scale late mesozoic gold mineralization in the north china craton: An association with lithospheric thinning. *Ore Geol. Rev.* **2003**, *23*, 125–152. [CrossRef]
4. Zhai, M.G.; Santosh, M. The early Precambrian odyssey of the North China Craton: A synoptic overview. *Gondwana Res.* **2011**, *20*, 6–25. [CrossRef]
5. Zhao, G.C.; Sun, M.; Wilde, S.A.; Li, S.Z. Late Archean to Paleoproterozoic evolution of the North China Craton: Key issues revisited. *Precambr. Res.* **2005**, *136*, 177–202. [CrossRef]
6. Mao, J.W.; Li, Y.Q.; Goldfarb, R.J.; He, Y.; Zaw, K. Fluid inclusion and noble gas studies of the Dongping gold deposit, Hebei Province, China: A mantle connection for mineralization? *Econ. Geol.* **2003**, *98*, 517–534. [CrossRef]
7. Bao, Z.W.; Zhao, Z.H.; Zhang, P.H.; Wang, Y.X. Ree, Sr, Nd, and Pb isotopic evidence for the petrogenesis of the Shuiquangou syenite complex in nw hebei province. *Geol. Rev.* **2003**, *49*, 60-596.
8. Wang, D.; Liu, J.; Zhai, D.; Carranza, E.; Wang, Y.; Zhen, S.; Wang, J.; Wang, J.; Liu, Z.; Zhang, F. Mineral paragenesis and ore-forming processes of the dongping gold deposit, hebei province, China. *Resour. Geol.* **2019**, *69*, 287–313. [CrossRef]
9. Gao, S.; Xu, H.; Zhang, D.; Shao, H.; Quan, S. Ore petrography and chemistry of the tellurides from the dongping gold deposit, hebei province, China. *Ore Geol. Rev.* **2015**, *64*, 23–34. [CrossRef]
10. Bao, Z.; Sun, W.; Li, C.; Zhao, Z. U–Pb dating of hydrothermal zircon from the dongping gold deposit in north china: Constraints on the mineralization processes. *Ore Geol. Rev.* **2014**, *61*, 107–119. [CrossRef]
11. Bao, Z.W.; Li, C.J.; Zhao, Z.H. Metallogeny of the syenite-related Dongping gold deposit in the northern part of the North China Craton: A review and synthesis. *Ore Geol. Rev.* **2016**, *73*, 198–210. [CrossRef]
12. Zhao, G.C.; Cawood, P.A.; Li, S.Z.; Wilde, S.A.; Sun, M.; Zhang, J.; He, Y.H.; Yin, C.Q. Amalgamation of the North China Craton: Key issues and discussion. *Precambr. Res.* **2012**, *222–223*, 55–76. [CrossRef]
13. Deng, J.; Wang, Q.F. Gold mineralization in China: Metallogenic provinces, deposit types and tectonic framework. *Gondwana Res.* **2016**, *36*, 219–274. [CrossRef]
14. He, C.S.; Santosh, M.; Yang, Q.Y. Gold metallogeny associated with craton destruction: A geophysical perspective from the North China Craton. *Ore Geol. Rev.* **2016**, *75*, 29–41. [CrossRef]

15. Li, S.R.; Santosh, M. Geodynamics of heterogeneous gold mineralization in the North China Craton and its relationship to lithospheric destruction. *Gondwana Res.* **2017**, *50*, 267–292. [CrossRef]
16. Nie, F.J. Type and distribution of gold deposits along the northern margin of the North China Craton, People's Republic of China. *Int. Geol. Rev.* **1997**, *39*, 151–180. [CrossRef]
17. Song, G.R.; Zhao, Z.H. *Geology of Dongping Alkaline Complex-Hosted Gold Depositin Hebei Province*; Seismological Press: Beijing, China, 1996; pp. 4–109. (In Chinese with English Abstract)
18. Li, C.M.; Deng, J.F.; Su, S.G.; Li, H.M.; Liu, X.M. Two stage zircon U–Pb ages of the potash altered rock in the Dongping gold deposit, Hebei Province, and their geological implications. *Acta Geosci. Sin.* **2010**, *31*, 843–852. (In Chinese with English Abstract)
19. Zhang, Z.C.; Mao, J.W. Geology and geochemistry of the Dongping gold telluride deposit, Heibei Province, North China. *Internat. Geol. Rev.* **1995**, *37*, 1094–1108.
20. Cisse, M.; Lü, X.; Algeo, T.J.; Cao, X.; Li, H.; Wei, M.; Chen, M. Geochronology and geochemical characteristics of the Dongping ore-bearing granite, North China: Sources and implications for its tectonic setting. *Ore Geol. Rev.* **2017**, *89*, 1091–1106. [CrossRef]
21. Gao, S.; Xu, H.; Li, S.; Santosh, M.; Zhang, D.; Yang, L.; Quan, S. Hydrothermal alteration and ore-forming fluids associated with gold-tellurium mineralization in the Dongping gold deposit, China. *Ore Geol. Rev.* **2017**, *80*, 166–184. [CrossRef]
22. Hu, Z.C.; Zhang, W.; Liu, Y.S.; Gao, S.; Li, M.; Zong, K.Q.; Chen, H.H.; Hu, S.H. "Wave" Signal-Smoothing and Mercury-Removing Device for Laser Ablation Quadrupole and Multiple Collector ICPMS Analysis: Application to Lead Isotope Analysis. *Anal. Chem.* **2015**, *87*, 1152–1157. [CrossRef] [PubMed]
23. Liu, Y.S.; Hu, Z.C.; Gao, S.; Günther, D.; Xu, J.; Gao, C.G.; Chen, H.H. In situ analysis of major and trace elements of anhydrous minerals by LA-ICP-MS without applying an internal standard. *Chem. Geol.* **2008**, *257*, 34–43. [CrossRef]
24. Cook, N.J.; Ciobanu, C.L.; Mao, J.W. Textural control on gold distribution in Asfree pyrite from the Dongping, Huangtuliang and Hougou gold deposits, North China Craton (Hebei Province, China). *Chem. Geol.* **2009**, *264*, 101–121. [CrossRef]
25. Nadoll, P.; Angerer, T.; Mauk, J.L.; French, D.; Walshe, J. The chemistry of hydrothermal magnetite: A review. *Ore Geol. Rev.* **2014**, *61*, 1–32. [CrossRef]
26. Ciobanu, C.L.; Cook, N.J.; Utsunomiyas, S.; Kogagwa, M.; Green, L.; Gilbert, S.; Wade, B. Gold–telluride nanoparticles revealed in arsenic-free pyrite. *Am. Miner.* **2012**, *97*, 1515–1518. [CrossRef]
27. Thomas, H.V.; Large, R.R.; Bull, S.W.; Malennikov, V.; Berry, R.F.; Fraaser, R.; Froud, S.; Moye, R. Pyrite and pyrrhotite textures and composition in sediments, laminated quartz veins, and reefs at Bendigo gold mine, Australia: Insights for ore genesis. *Econ. Geol.* **2011**, *106*, 1–31. [CrossRef]
28. Dare, S.A.S.; Barnes, S.J.; Beaudoin, G. Variation in trace element content ofmagnetite crystallized from a fractionating sulfide liquid, Sudbury, Canada: Implications for provenance discrimination. *Geochim. Cosmochim. Acta* **2012**, *88*, 27–50. [CrossRef]
29. Xie, Q.H.; Zhang, Z.C.; Hou, T.; Jin, Z.L.; Santosh, M. Geochemistry and oxygen isotope composition of magnetite from the Zhangmatun deposit, North China Craton: Implications for the magmatic-hydrothermal evolution of Cornwall-type iron mineralization. *Ore Geol. Rev.* **2017**, *88*, 57–70. [CrossRef]
30. Belousov, I.; Large, R.R.; Meffre, S.; Danyushevsky, L.V.; Steadman, J.; Beardsmore, T. Pyrite compositions from vhms and orogenic au deposits in the yilgarn craton, western australia: Implications for gold and copper exploration. *Ore Geol. Rev.* **2016**, *79*, 474–499. [CrossRef]
31. Craig, J.R.; Vokes, F.M.; Solberg, T.N. Pyrite: Physical and chemical textures. *Miner. Depos.* **1998**, *34*, 82–101. [CrossRef]
32. Dupuis, C.; Beaudoin, G. Discriminant diagrams for iron oxide trace element fingerprinting of mineral deposit types. *Miner. Depos.* **2011**, *46*, 319–335. [CrossRef]
33. Nadoll, P.; Mauk, J.L.; Leveille, R.A.; Koenig, A.E. Geochemistry of magnetite from porphyry Cu and skarn deposits in the southwestern United States. *Miner. Depos.* **2015**, *50*, 493–515. [CrossRef]
34. Dare, S.A.S.; Barnes, S.J.; Beaudoin, G.; Méric, J.; Boutroy, E.; Potvin-Doucet, C. Trace elements in magnetite as petrogenetic indicators. *Miner. Depos.* **2014**, *49*, 785–796. [CrossRef]
35. Hu, H.; Li, J.W.; Lentz, D.; Ren, Z.; Zhao, X.F.; Deng, X.D.; Hall, D. Dissolution reprecipitation process of magnetite from the Chengchao iron deposit: Insights into ore genesis and implication for in-situ chemical analysis of magnetite. *Ore Geol. Rev.* **2014**, *57*, 393–405. [CrossRef]
36. Goldfarb, R.J.; Baker, T.; Dubé, B.; Groves, D.I.; Hart, C.J.R.; Gosselin, P. Distribution, character, and genesis of gold deposits in metamorphic terranes. In *Economic Geology 100th Anniversary Volume*; Society of Economic Geologists: Littleton, CO, USA, 2005; pp. 407–450.
37. Wang, K.; Zhai, D.; Liu, J.; Wu, H. LA-ICP-MS trace element analysis of pyrite from the Dafang gold deposit, South China: Implications for ore genesis. *Ore Geol. Rev.* **2021**, *139*, 104507. [CrossRef]
38. Koglin, N.; Frimmel, H.E.; Minter, W.; Brätz, H. Trace-element characteristics of different pyrite types in mesoarchaean to palaeoproterozoic placer deposits. *Miner. Depos.* **2010**, *45*, 259–280. [CrossRef]
39. Bralia, A.; Sabatini, G.; Troja, F. A revaluation of the Co/Ni ratio in pyrite as geochemical tool in ore genesis problems. *Miner. Depos.* **1979**, *14*, 353–374. [CrossRef]
40. Clark, C.; Grguric, B.; Mumm, A.S. Genetic implications of pyrite chemistry from the Palaeoproterozoic Olary Domain and overlying Neoproterozoic Adelaidean sequences, northeastern South Australia. *Ore Geol. Rev.* **2004**, *25*, 237–257. [CrossRef]
41. Fan, H.; Xie, Y.; Zhai, M. Ore-forming fluids in the dongping gold deposit, northwestern hebei province. *Sci. China Ser. D Earth Sci.* **2001**, *44*, 748–757. [CrossRef]

42. Zhang, G.R.; Xu, J.H.; Wei, H.; Song, G.C.; Zhang, Y.B.; Zhao, J.K.; He, B.; Chen, D.L. Structure, alteration, and fluid inclusion study on the surrounding area of the Dongping gold deposit, northern Hebei, China. *Acta Petrol. Sin.* **2012**, *28*, 637–651. (In Chinese with English Abstract)
43. Nie, F.J. Geology and origin of the Dongping alkali-type gold deposit, Northern Hebei Province, People's Republic of China. *Resour. Geol.* **1998**, *48*, 139–158. [CrossRef]
44. Fan, G.H.; Li, J.W.; Deng, X.D.; Gao, W.S.; Li, S.Y. Age and origin of the Dongping Au-Te deposit in the North China Craton revisited: Evidence from paragenesis, geochemistry, and in situ U-Pb geochronology of garnet. *Econ. Geol.* **2021**, *116*, 963–985. [CrossRef]
45. Migdisov, A.A.; Zezin, D.; Williams-Jones, A.E. An experimental study of cobalt (ii) complexation in cl and h2s-bearing hydrothermal solutions. *Geochim. Cosmochim. Acta* **2011**, *75*, 4065–4079. [CrossRef]

minerals

MDPI

Article

Paleoproterozoic Mineralization of the Lijiapuzi Gold Deposit in the Liaodong Peninsula, NE China: Constraints from ^{40}Ar-^{39}Ar Age, S-Pb Isotopes, and In Situ Analyses

Yan Zhao [1,2,*], Peng Zhang [1,2], Hongzhi Yang [1,2], Linlin Kou [1,2], Zhongwei Bi [1,2], Zhongzhu Yang [1,2] and Jiang Chen [1]

1 Shenyang Center, China Geological Survey, Shenyang 110034, China; geozhangpeng2010@163.com (P.Z.); yangxx272000@163.com (H.Y.); koulinlin@126.com (L.K.); bizhongwei@mail.cgs.gov.cn (Z.B.); yangzhongzhu@mail.cgs.gov.cn (Z.Y.); hardrock30@163.com (J.C.)

2 Key Laboratory of Deep Mineral Resources Exploration and Evaluation, Department of Natural Resource of Liaoning Province, Shenyang 110032, China

* Correspondence: zhaoyan@mail.cgs.gov.cn

Abstract: A Paleoproterozoic $^{40}Ar/^{39}Ar$ age, interpreted as the gold metallogenic epoch of the Lijiapuzi deposit in NE China, is reported in this paper. The ore body of this deposit is hosted in the marbles and schists of the Paleoproterozoic Dashiqiao Formation and exhibits parallel development to the foliation of the country rocks. Coexisting auriferous pyrite and tremolite can be observed both in the field and under a microscope, revealing the presence of hydrothermal activity in the deposit. The $\delta^{34}S$ composition of the auriferous pyrite varies from +8‰ to +12.3‰, indicating a sedimentary origin. The auriferous pyrite and host marble have similar Pb isotopic ratios, indicating the contribution of Paleoproterozoic sediments to gold mineralization. Furthermore, in situ S and trace element analyses of the auriferous pyrite reveal that the main gold mineralization stage was Py2, which coexists with radial tremolite. $^{40}Ar/^{39}Ar$ dating of these tremolite samples yields a plateau age of ~1851.23 Ma, with similar isochronal and inverse isochronal ages. The Paleoproterozoic gold metallogenesis, together with the contemporary Pb-Zn, boron, graphite, and magnetite mineralization, constitute an important mineralogenetic series during the evolution of the Jiao-Liao-Ji orogeny in the Liaodong Peninsula.

Keywords: $^{40}Ar/^{39}Ar$ dating; in situ analysis; S-Pb isotopes; gold mineralization; Jiao-Liao-Ji orogeny; the Lijiapuzi deposit

1. Introduction

The Liaodong Peninsula in the northeastern part of the North China Craton (NCC) has a long history of geological evolution. The peninsula consists of an Archean crystallized basement, Paleoproterozoic meta-sedimentary rocks, and an orogenic belt (Figure 1a) [1–4]. Multiple mineral resources, including gold, lead, zinc, boron, magnesite, and graphite, were formed during the early geological process, forming the important Liaodong-Ji'nan metallogenetic belt [5–10]. Due to the significant gold reserves in the main gold orefields, such as the Qingchengzi, Maoling, and Wulong orefields, the Liaodong Peninsula possesses one of the most important gold producing areas in China. The Qingchengzi orefield, with a total gold reserve of over 200 t [11,12], still has gold exploration potential. The Lijiapuzi gold deposit to the north and the deep gold ore bodies in the Baiyun and Linjiasandaogou deposits are newly discovered.

Many studies have focused on the metallogenesis of the Qingchengzi gold deposits; however, their conclusions remain controversial. The model of metamorphic–sedimentary processes imposed by late hydrothermal activities is preferred [14–18] since the gold ore bodies often exhibit the same attributes as the Liaohe Group country rocks, and the geochemistry of the country rocks and ore-bearing minerals are similar. Others have reported the presence of Mesozoic magmatism and some mineralized dykes within this district [19–22]. They believe that the gold deposits are magmatic–hydrothermal-related-type or decratonic-type deposits. Whether there is Paleoproterozoic gold mineralization remains one of the critical issues due to the lack of robust geochronological evidence.

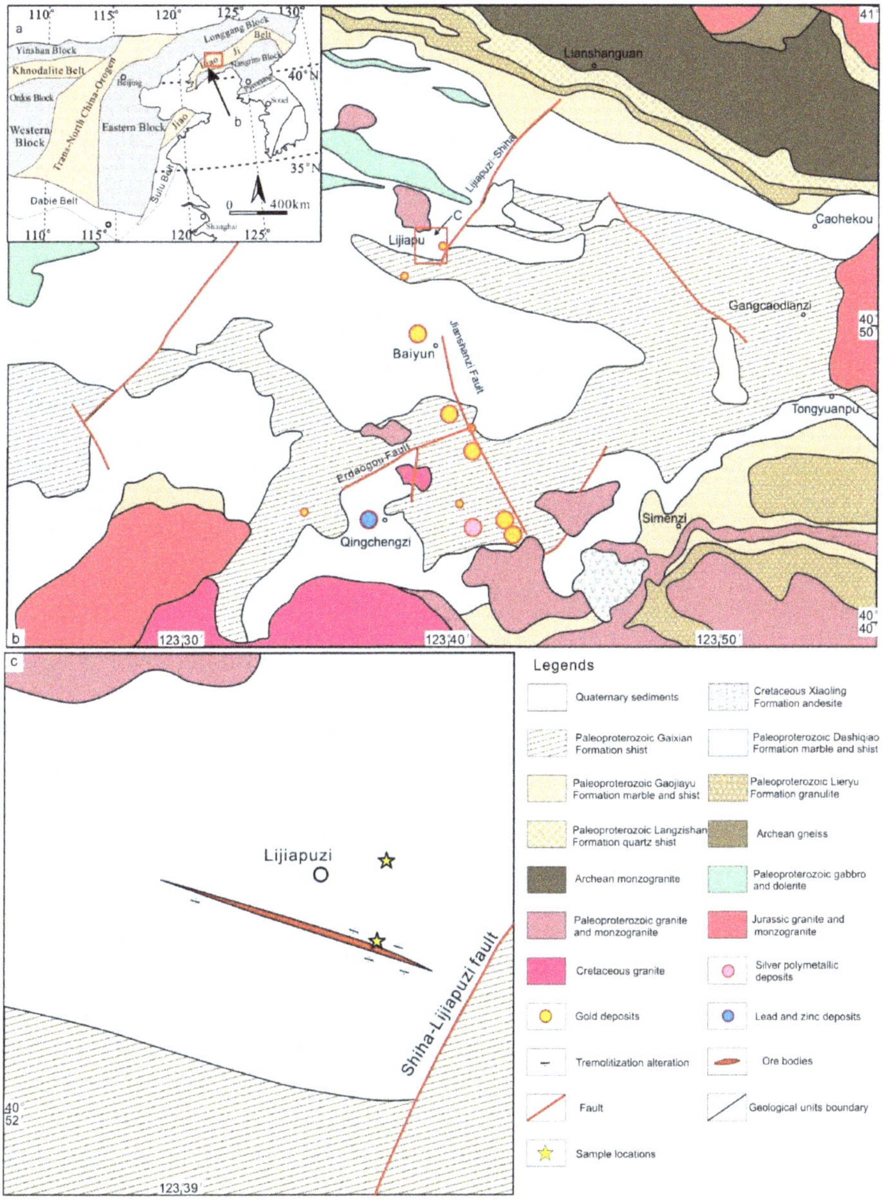

Figure 1. (**a**) Simplified tectonic map of partial North China Craton showing locations of the Liaodong Peninsula and the Jiao-Liao-Ji belt (modified after [2]). (**b**) Geological map of the Qingchengzi area showing major mineral deposits (modified after [13]). (**c**) Geological map of the Lijiapuzi gold deposit showing sample locations.

The newly discovered Lijiapuzi gold deposit was chosen as the object of this study. After detailed field and laboratory observations, important tremolitization accompanied by auriferous pyrite was identified, with euhedral radial tremolite. We conducted ^{40}Ar-^{39}Ar dating of tremolite samples to precisely constrain the metallogenic epoch of the formation of the deposit. The S and Pb isotopes of auriferous sulfides and in situ laser ablation inductively coupled mass spectrometry (LA-ICP-MS) analyses of the S and trace element contents of the auriferous minerals were also conducted. Paleoproterozoic metallogenic evidence in the Lijiapuzi deposit and the related gold mineralization and ore material sources were obtained. The results of this study also expand the Paleoproterozoic mineralogenetic series in the Liaodong Peninsula.

2. Geologic Setting of the Qingchengzi Orefield

Located in the northeastern margin of the NCC (Figure 1a), the Liaodong Peninsula consists of the Archean Longgang Block, the Archean Nangrim Block, the Paleoproterozoic Jiao-Liao-Ji Belt (JLJB), and Mesozoic volcanic and granitic intrusions [1–4,23]. The JLJB orogeny involved the deposition of voluminous Paleoproterozoic sediments, including the Liaohe Group in the Liaoning Province, the Ji'an and Laoling groups in the Ji'lin Province, and the Jingshan and Fenzishan groups in the Shandong Province [24,25], as well as igneous and metamorphic rocks [26–29]. The marble and biotite schist of the Dashiqiao Formation and the schist of the Gaixian Formation are both components of the Liaohe Group, and they comprise the country rock in the Qingchengzi polymetallic orefield and nearby gold deposits. The NCC experienced strong magmatism again in the Mesozoic, which is referred to as the decratonic process [30].

The JLJB is NW–SE-trending and consists of intensely deformed geological units. There is a giant E–W-trending anticline in the Baiyun–Qingchengzi area in the central part of the JLJB. Other subordinate synclines and anticlines, including the Xinling and Zhenzigou anticlines, have also been identified within the Qingchengzi orefield. The fault structural framework is dominated by the NW-trending Jianshanzi Fault and the NE trending Erdaogou and Lijiapuzi–Shihazhai faults, as well as their related secondary faults. Almost all of the lead–zinc deposits are located to the south of the Erdaogou Fault, and a series of gold deposits are distributed along or to the west of the Jianshanzi Fault (Figure 1b).

The Paleoproterozoic magmatism in the Qingchengzi area resulted in widespread granitoid intrusions and sporadically distributed gabbros and dolerites. The Simenzi and Fangjiaweizi monzogranite plutons in the southeast intruded before ~2.16 Ga and ~1.89 Ga [31–33]. The Lanjiagou and Lanhualing Paleoproterozoic stocks are located in the central part of the Qingchengzi orefield (Figure 1b). These granitoids are believed to have intruded in an arc setting in the Paleoproterozoic [26]. The sporadic gabbro and/or dolerite dykes have been interpreted to be related to a Paleoproterozoic back-arc basin geological setting at ~2.11 Ga [34–36]. The Triassic Shuangdinggou and Xinling granite intrusions, which have U-Pb ages of ~225 Ma [6,37], correspond with the sporadic lamprophyre dykes in this area. Jurassic diorites and monzogranites with strong deformation have also been identified [18,23]. The Mesozoic magmatism has often been interpreted as the result of the subduction of the Yangtze Craton from the south [18,21].

A series of gold, Pb–Zn, B, and magnetite metallogenies are distributed within the JLJB in the Liaodong Peninsula [7,8,15,37–39]. Among the three famous Qingchengzi, Maoling, and Wulong gold orefields, only the Qingchengzi orefield hosts giant Au–Ag–Pb–Zn polymetallic deposits. Neither the Paleoproterozoic granitoids nor the Mesozoic intrusions exhibit gold mineralization in this area [40]. The marbles and schists of the Dashiqiao Formation and the schist of the Gaixian Formation comprise the country rocks of the gold ore bodies. The ore bodies often exhibit lens, stratiform, or irregular shapes, which are controlled by the host rocks and/or faults. Silicification, sericitization, tremolitization, and pyritization are the most common alterations within the gold deposits.

3. Sampling and Analytical Methods

The Lijiapuzi gold deposit, which consists of one gold orebody, is a small-scale deposit in the northern part of the Qingchengzi orefield. The gold ore body has an NW trend, which is parallel to the foliation of the marbles of the Dashiqiao Formation (Figure 1c). Although the ore body is distributed near the NE Shiha–Lijiapuzi Fault, the gold mineralization and the ore-bearing pyrite in this ore body seem to be related to the marble country rock and the tremolite alteration therein (Figure 2). The main ore minerals include pyrite, chalcopyrite, sphalerite, and galena. Early-stage pyrite (Py1), which has a euhedral shape, is often cracked when observed under a microscope. Late-stage pyrite (Py2), which has a relatively high gold content, is the main ore phase (Figure 3). Minor chalcopyrite and sphalerite are found in association with Py2 in the main mineralization stage. Tremolitization is a common type of alteration in this deposit, along with weak silicification, sericitization, and chloritization. Tremolite often cuts through the Py1 in the cracks (Figure 2f). Paragenesis of the Py2 and radial tremolite were observed under a microscope (Figure 2c).

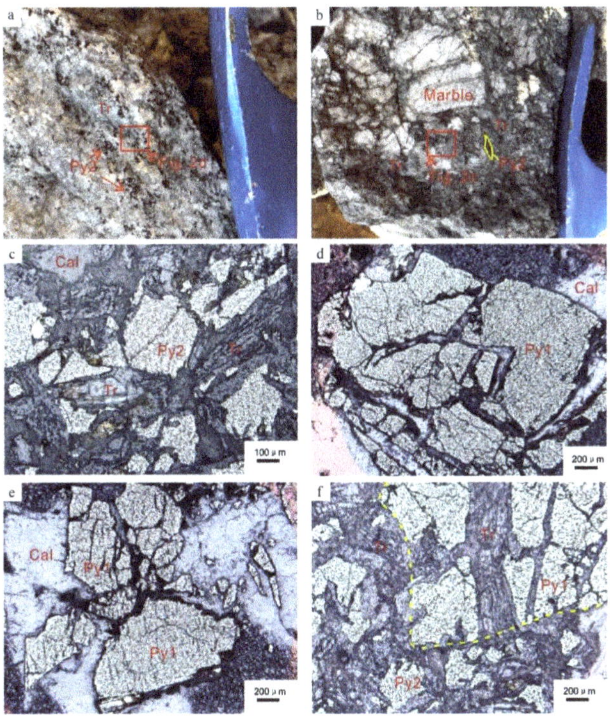

Figure 2. Representative outcrops and photomicrograph of ores from the Lijiapuzi gold deposit. (**a**,**b**) Field photos of Au-bearing pyrites and radial tremolite in country rock marbles. (**c**) Photomicrograph of ore-bearing pyrite and tremolite under a transmission ray. (**d**–**f**) Reflected light photomicrograph of different stages of ore-bearing pyrite. Cal—calcite; Py1—early-stage pyrite; Py2—late-stage pyrite; Tr—tremolite.

A tremolite sample (YP-9) that coexisted with the auriferous pyrite was collected for $^{40}Ar/^{39}Ar$ dating. Detailed S and Pb isotopic analyses were conducted on Au-bearing pyrite (Py2 type) samples (YP-1–YP-8) to study the ore-forming materials. In contrast, for determining a country rock's contribution to the mineralization process, marble samples (QC-1–QC-3) were chosen for Pb isotope analysis. In situ analysis of the S and trace element contents of seven samples was also conducted to analyze the precise textures and contents of the ore-bearing pyrite.

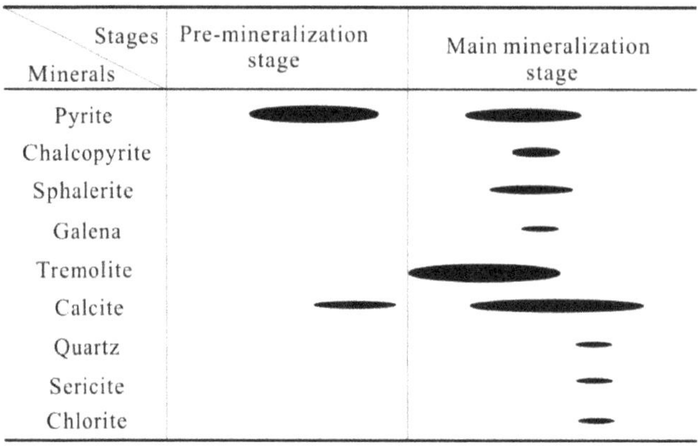

Figure 3. Paragenetic chart showing ore and gangue minerals in the Lijiapuzi gold deposit.

3.1. *Tremolite* ^{40}*Ar*-^{39}*Ar Dating*

The $^{40}Ar/^{39}Ar$ dating of the tremolite was performed at the Isotope Geology Laboratory, Institute of Geology Chinese Academy of Geological Sciences. For the tremolite analyses, we followed the procedure of Liu et al. [41]. The samples were loaded in aluminum packets and degassed at low temperatures (250–300 °C) for 20–30 min before being incrementally heated in a double-vacuum graphite furnace. The gases released during each heating step were purified using Ti and Al-Zr getters. Once cleaned, the gas was introduced into a GV Instruments HELIX-MC noble gas mass spectrometer and allowed to stabilize for 4–5 min before the static analysis was conducted. The ^{40}Ar, ^{39}Ar, ^{38}Ar, ^{37}Ar, and ^{36}Ar isotope abundances at time zero were obtained through linear extrapolation of the peak intensities. The data were corrected for system blanks, mass discrimination, Ca interference, K-derived argon isotopes, and the decay of ^{37}Ar since the time of irradiation. The decay constant used throughout the calculations was $\lambda = (5.543 \pm 0.010) \times 10^{-10}\ a^{-1}$. All of the errors are reported as 2σ.

3.2. *Pb isotope Analyses of Pyrite and Country Rock*

Pyrite and country rock marble samples were collected from the Lijiapuzi deposit and nearby for Pb isotope analysis. The pyrite grains were sorted to a purity of 99% before applying the analysis procedures described by Zhang et al. [21]. The lead isotope compositions were analyzed via thermal-ionization mass spectrometry (TIMS; MAT-261), and standard sample NBS 981 was used for the calibration. The analyses were carried out at the Beijing Research Institute of Uranium Geology. The analytical precision of the Pb isotope ratios is better than ±0.09%.

3.3. *In Situ LA-ICP-MS S and Trace Element Analyses*

In situ LA-ICP-MS trace element analysis of pyrite was conducted at the Wuhan Sample Solution Analytical Technology Co., Ltd., Wuhan, China. The laser sampling was performed using a GeolasPro laser ablation system (Coherent, Santa Clara, CA, USA), consisting of a COMPexPro 102 ArF excimer laser (wavelength of 193 nm and maximum energy of 200 mJ) and a MicroLas optical system. The spot size and frequency of the laser were set to 44 μm and 5 Hz, respectively. Detailed analysis procedures for determining the trace element compositions of sulfides were calibrated against various reference materials (NIST 610 and NIST 612) without using an internal standard [42]. Sulfide reference material MASS-1 (United States Geological Survey) was used as the unknown sample to verify the accuracy of the calibration method. Each analysis incorporated a background acquisition of approximately 20–30 s followed by 50 s of data acquisition from the sample.

4. Results

4.1. Tremolite ^{40}Ar-^{39}Ar Dating

The ^{40}Ar-^{39}Ar step-heating dating data for tremolite are presented in Table 1. Thirteen heating steps were carried out on the sample. Eight heating steps, with temperatures ranging from 1140 °C to 1400 °C, yielded a perfect plateau age of 1851.2 ± 12.8 Ma (Figure 4a). The data for the step heating from 1140 °C to 1400 °C yielded an isochronal age of 1857.88 ± 12.3 Ma, with a mean standard weighted deviation (MSWD) of 3.18 (Figure 4b), and an identical inverse isochronal age of 1858.44 ± 12.3 Ma, with an MSWD of 3.15 (Figure 4c). The plateau age of 1851.2 Ma is thus interpreted as the crystallization age of tremolite.

Table 1. ^{40}Ar-^{39}Ar data of tremolite from the Lijiapuzi gold deposit.

T	$^{36}Ar[V]$	$^{37}Ar[V]$	$^{38}Ar[V]$	$^{39}Ar[V]$	$^{40}Ar[V]$	$^{40}Ar^{R}\%$	$^{39}Ar(K)\%$	Age/Ma	2σ
900 °C	47.310	0.000	9.185	7.748	17,939.35	22.07	0.53	1697.42	52.4
980 °C	19.591	0.000	4.136	15.393	13,359.54	56.67	1.05	1655.62	12.8
1040 °C	8.685	0.000	1.981	17.744	10,505.46	75.57	1.21	1555.05	7.7
1100 °C	3.625	19.219	1.701	25.299	15,020.56	92.88	1.73	1783.26	5.8
1140 °C	4.160	1001.696	4.544	95.651	56,828.54	97.96	6.47	1854.19	4.3
1170 °C	59.297	2617.420	19.292	216.178	141,105.48	87.71	14.61	1836.72	4.4
1200 °C	10.667	7530.800	24.868	573.805	337,662.80	99.22	38.74	1860.44	4.2
1230 °C	3.437	2467.695	8.135	192.399	112,562.07	99.25	12.99	1853.74	3.8
1250 °C	2.588	379.050	1.999	36.459	21,948.31	96.64	2.47	1853.62	5.0
1300 °C	10.647	855.654	5.222	75.926	47,304.14	93.48	5.13	1855.70	4.2
1350 °C	14.091	1662.685	8.132	135.668	82,704.71	95.11	9.17	1851.42	4.0
1400 °C	24.373	896.403	7.586	83.600	55,222.62	87.07	5.65	1840.72	4.9
1450 °C	35.321	0.000	6.190	3.828	11,291.97	7.57	0.26	938.41	119.1

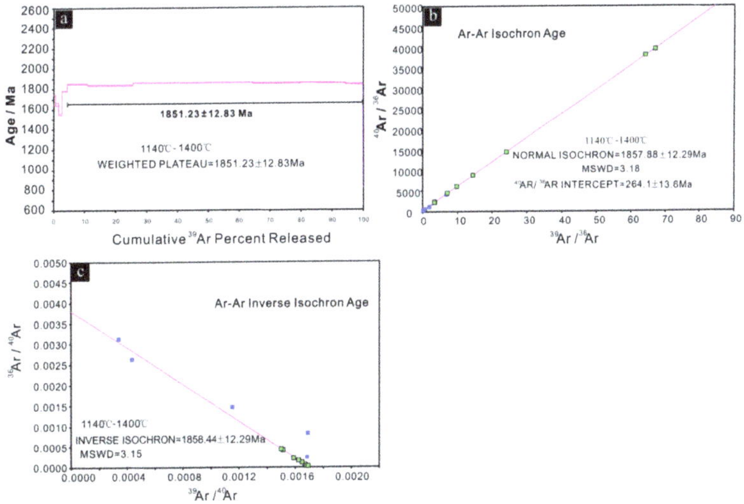

Figure 4. (**a**) ^{40}Ar-^{39}Ar-weighted mean plateau age, (**b**) ^{40}Ar-^{39}Ar isochronal line age, and (**c**) ^{40}Ar-^{39}Ar inverse isochronal line age of tremolite samples.

4.2. S Isotope Composition of Auriferous Pyrite

The S isotope compositions of the auriferous pyrite samples are presented in Table 2. The $\delta^{34}S$ values are +8.0‰ to +12.3‰, which are higher than those of the granites (Figure 5). It is believed that the sulfur source of the metallogenic deposit was probably the nearby meta-sedimentary rocks of the Liaohe Group [15,19].

Table 2. Sulfur isotopic data of ore-bearing pyrite from the Lijiapuzi gold deposit.

Sample No.	Mineral	$\delta^{34}S$/‰
YP-1	Pyrite	9.8
YP-2	Pyrite	8.8
YP-3	Pyrite	12.3
YP-4	Pyrite	9.7
YP-5	Pyrite	11.3
YP-6	Pyrite	8.6
YP-7	Pyrite	8
YP-8	Pyrite	11

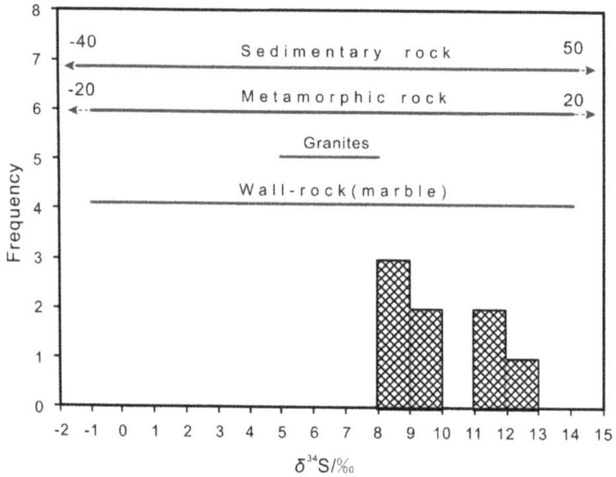

Figure 5. Sulfur isotope composition from the ore-bearing pyrite of the Lijiapuzi deposit sedimentary rocks, metamorphic rocks, granites, and wall−rock (marble) data from [43].

4.3. Pb Isotope Results

The Pb isotopic ratios of the Au-bearing pyrite and country rock marble are presented in Table 3. The pyrite exhibits $^{206}Pb/^{204}Pb$ ratios of 18.155 to 18.886, $^{207}Pb/^{204}Pb$ ratios of 15.594 to 15.687, and $^{208}Pb/^{204}Pb$ ratios of 38.26 to 39.695 (Figure 6). Meanwhile, the widespread marble in the Qingchengzi orefield area exhibits $^{206}Pb/^{204}Pb$ ratios of 21.822 to 23.035, $^{207}Pb/^{204}Pb$ ratios of 16.092 to 16.146, and $^{208}Pb/^{204}Pb$ ratios of 37.472 to 37.668 (Figure 6). The lead isotope ratios of the ore-bearing pyrite data were also plotted into the Liaohe Group country rock lead isotopic field, as previous studies have suggested [6,8,17,37].

Table 3. Lead isotope ratios of ore-bearing pyrites and marbles from the Lijiapuzi gold deposit.

No.	Target	$^{206}Pb/^{204}Pb$	σ	$^{207}Pb/^{204}Pb$	σ	$^{208}Pb/^{204}Pb$	σ
YP-1	Pyrite	18.221	0.003	15.625	0.003	38.416	0.009
YP-2	Pyrite	18.717	0.003	15.668	0.003	38.832	0.006
YP-3	Pyrite	18.886	0.003	15.684	0.002	39.695	0.008
YP-4	Pyrite	18.287	0.002	15.614	0.002	38.588	0.006
YP-5	Pyrite	18.176	0.002	15.594	0.002	38.324	0.005
YP-6	Pyrite	18.565	0.002	15.646	0.002	38.53	0.004
YP-7	Pyrite	18.249	0.003	15.687	0.002	38.568	0.006
YP-8	Pyrite	18.155	0.002	15.597	0.002	38.26	0.005
QC-1	Marble	23.035	0.006	16.146	0.004	37.668	0.011
QC-2	Marble	21.822	0.009	16.092	0.006	37.472	0.016
QC-3	Marble	22.314	0.013	16.128	0.01	37.623	0.021

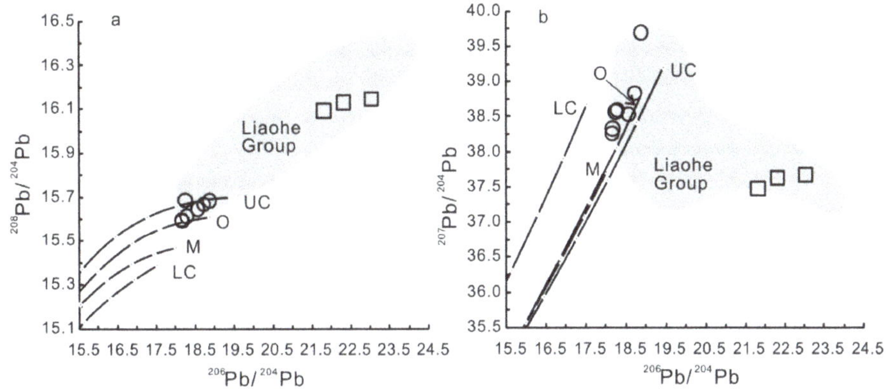

Figure 6. Lead isotope ratio from ore-bearing pyrite and marble from the Lijiapuzi deposit. Data of the Liaohe Group. (**a**) $^{208}Pb/^{204}Pb$ vs. $^{206}Pb/^{204}Pb$; and (**b**) $^{207}Pb/^{204}Pb$ vs. $^{206}Pb/^{204}Pb$. (Gaixian and Dashiqiao formations) are from [6].

4.4. S and Trace Element Compositions

A total of 37 spots on the two stages of pyrite from the Lijiapuzi gold deposit were selected for in situ LA-ICP-MS analysis of the sulfur and trace element compositions. The contents of the S and nine representative trace elements are presented in Supplementary Table S1.

Early-stage pyrite Py1, which often exhibits a euhedral cubic shape, has similar or slightly lower $\delta^{34}S$ contents than late-stage subhedral pyrite Py2 (Figure 7). The Py1 analysis spots yielded $\delta^{34}S$ values of +9.19‰ to +11.27‰, while the Py2 spots yielded $\delta^{34}S$ values of +9.49‰ to +11.49‰.

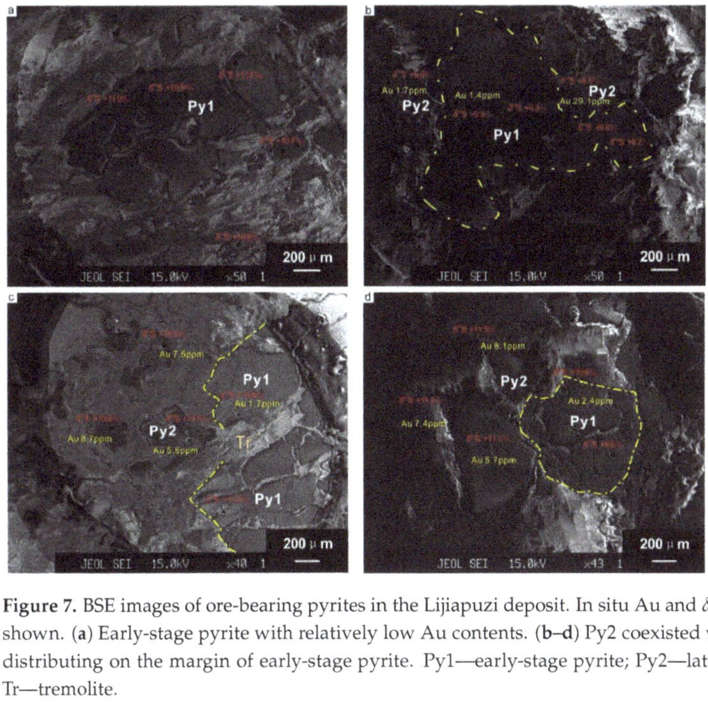

Figure 7. BSE images of ore-bearing pyrites in the Lijiapuzi deposit. In situ Au and $\delta^{34}S$ values are shown. (**a**) Early-stage pyrite with relatively low Au contents. (**b**–**d**) Py2 coexisted with tremolite distributing on the margin of early-stage pyrite. Py1—early-stage pyrite; Py2—late-stage pyrite; Tr—tremolite.

Gold and eight other gold-related trace element contents are shown in Figure 7. Compared to Py1, the Py2 shows higher Au, Sb, Cu, and Pb contents, slightly higher As and Ag contents, and similar Co, Ni, and Mo contents.

5. Discussion

5.1. Gold Metallogenic Epoch in the Lijiapuzi Deposit

The gold metallogenic epoch is usually determined through the dating of coexisting minerals, for instance, $^{40}Ar/^{39}Ar$ dating of biotite and muscovite, U−Pb zircon dating, and Re−Os and Rb−Sr dating of auriferous pyrite. Compared to biotite and muscovite, which have lower closure temperatures of ~350 °C, tremolite has an$^{40}Ar/^{39}Ar$ dating closure temperature of greater than 500 °C [42], so it can provide a much more reliable crystallization age, which is also indicative of the age of the gold metallogenic epoch.

The tremolite samples collected from the gold ore body in the Lijiapuzi gold deposit exhibit a paragenetic relationship with the auriferous pyrite (Py2) based on field and microscope observations (Figure 2). Thus, dating the tremolite could provide a robust constraint on the gold metallogenic epoch of the Lijiapuzi deposit. The $^{40}Ar/^{39}Ar$ dating of the tremolite yielded a plateau age of 1851.23 ± 12.8 Ma (Figure 4a). The similar isochronal line and inverse isochron ages of 1857.88 Ma and 1858.44 Ma, respectively, with an MSWD of 3.1, are even more persuasive (Figure 4b,c). These consistent dating results of ~1851 Ma could represent the crystallization age of tremolite, and thus, the gold metallogenic epoch of the Lijiapuzi deposit. This is the first Paleoproterozoic gold mineralization $^{40}Ar/^{39}Ar$ age reported for the Liaodong Peninsula.

5.2. Paleoproterozoic Gold Mineralization in the Lijiapuzi Deposit

Gold mineralization processes can be effectively constrained via the S-Pb isotope analysis of auriferous minerals because gold is often transported by sulfur-bearing complexes [44]. The sulfur isotope analyses of the auriferous pyrite yielded $\delta^{34}S$ values of +8.0‰ to +12.3‰, which differ from those of typical magmatic sulfur sources (Figure 5). Taking the geological setting of the deposit into consideration, the Paleoproterozoic meta-sedimentary country rocks of the Dashiqiao Formation are the most probable sulfur source of the Lijiapuzi gold deposit. The auriferous pyrite also yields lead isotopic ratios similar to those of the marbles of the Dashiqiao Formation and other Liaohe Group meta-sedimentary rocks based on previous studies (Figure 6) [6,8,38]. The country rock marbles of the Paleoproterozoic Liaohe Group provide ore-forming materials for the Lijiapuzi deposit.

In situ LA-ICP-MS sulfur and trace element analyses provide a new method for investigating the detailed mineralization processes of various mineral deposits [45–47]. Detailed observations under a microscope and EPMA studies revealed the presence of at least two stages of Au-bearing pyrite (Py1 and Py2). The Py1 grains exhibit euhedral-cubic or altered-cubic shapes, while the Py2 pyrite grains usually exhibit subhedral structures and are crystallized on the rim and/or near relics of the Py1 grains (Figure 7). Furthermore, the in situ S and trace element analyses of these types of pyrite revealed that they have relatively diverse compositions. Py1 (early-stage pyrite) has slightly lower $\delta^{34}S$ contents (Figure 5), while Py2 has higher gold, Sb, Cu, and Pb contents, slightly higher As and Ag contents, and similar Co, Ni, and Mo contents than Py1 (Figure 8). Thus, Py1 and Py2 could represent two gold mineralization stages. Py2, which coexists with tremolite, represents a main gold mineralization stage with high contents of Sb, Cu, and Pb.

5.3. Implications for Mineralization during the Jiao-Liao-Ji Orogeny

The JLJB in the NE NCC has a long and complex evolutionary history, consisting of a series of geological processes, including subduction, collision, accretion, and extension during the orogeny. The wide range of material exchange, water–rock reactions, and metamorphism were all favorable for the development of mineralizations [48,49]. In addition, the Paleoproterozoic sedimentary rocks of the Liaohe Group have relatively high gold contents of several ppm [13]. A gold mineralization model for the Paleoproterozoic metamorphosed sedimentary rocks has been proposed by economic geologists [13–17], due to the similar distributions of the gold ore bodies and the Paleoproterozoic country rocks, as well as S and Pb isotope studies on the ore minerals and country rocks. The ~1851 Ma $^{40}Ar/^{39}Ar$ age and the related isotopic and geochemical data provide direct and robust constraints on the Paleoproterozoic gold mineralization in this region.

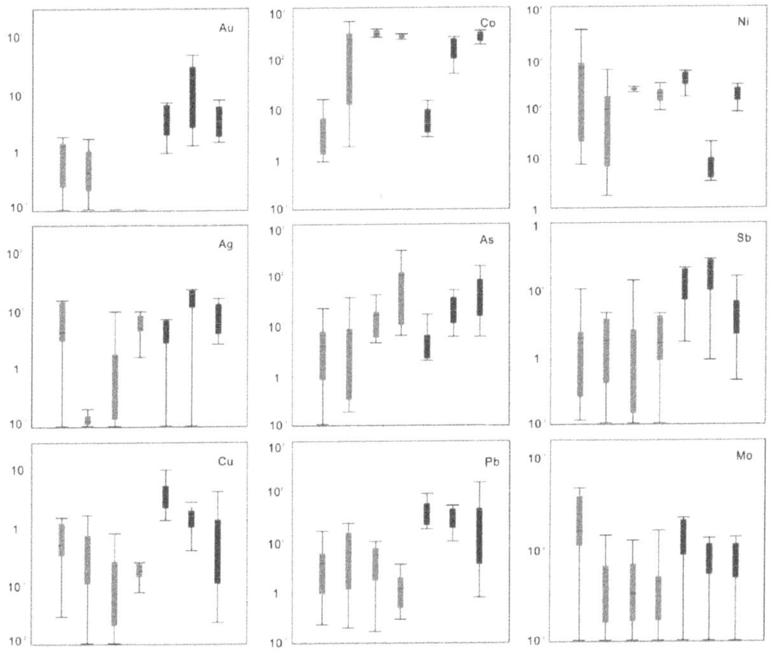

Figure 8. Box-and-whisker diagrams showing in situ Au, Co, Ni, Ag, As, Sb, Cu, Pb, and Mo from pyrite of the Lijiapuzi deposit. The red bar indicates the median value, the box encompasses the 25th to 75th percentiles of the data, and the whiskers indicate the minimum and maximum values. Light-grey bars show the Py1 spot data, while the dark-grey ones show the Py2 data.

Paleoproterozoic metallic mineralization in the Liaodong Peninsula was as remarkable as the undisputed nonmetallic boron and graphite deposition. Ma et al. [8] reported a sphalerite Rb-Sr age of 1798 ± 8 Ma from the Zhenzigou Pb−Zn deposit in the Qingchengzi orefield. Archean-banded iron formation (BIF) deposits in the Gongchangling area were also metamorphosed and experienced Fe enrichment at ~1860Ma [50,51]. Taking the gold mineralization $^{40}Ar/^{39}Ar$ age of ~1851 Ma for the Lijiapuzi deposit into consideration, a widespread Au−Pb−Zn−Fe polymetallic mineralogenetic series was formed in the Qingchengzi–Liaoyang area in the central part of the JLJB. It is also noteworthy that the ~1860 to 1850 Ma metallogenic epoch was consistent with the regional retrograde metamorphism in the JLJB [27,33,52]. Extensive metamorphic fluids related to this large-scale metamorphism may have extracted ore materials from the meta-sedimentary country rocks, and then deposited them in the available spaces due to changes in the temperature and pressure and the involvement of meteoric water.

6. Conclusions

The tremolite plateau age of ~1851 Ma implies that Paleoproterozoic gold mineralization occurred in the Lijiapuzi gold deposit. Py2, coexisting with tremolite, represents the main gold mineralization stage accompanied with high contents of Sb, Cu, and Pb. A complex Paleoproterozoic mineralogenetic series of Au−Pb−Zn−Fe−B−graphite was identified in the Liaodong Peninsula. Further studies on Paleoproterozoic mineralization and the evolutionary process of the JLJB orogeny could be a research highlight in this region.

Supplementary Materials: The following are available online at https://www.mdpi.com/article/10.3390/min12080971/s1, Table S1: In situ LA-ICP-MS S isotope and trace element compositions of pyrite from the Lijiapuzi gold deposit.

Author Contributions: Y.Z. conceived of and designed the ideas; P.Z. and L.K. tested the tremolite samples; Y.Z. and H.Y. analyzed the data; Y.Z., Z.B., Z.Y. and J.C. participated in the fieldwork; P.Z. and Y.Z. revised the manuscript. All authors have read and agreed to the published version of the manuscript.

Funding: This research was funded by National Key R&D Program of China grant number 2018YFC0603804; NSF of China, grant number 42072104; Geological Survey Project of China, grant number DD20221678, and the China Scholarship Council.

Acknowledgments: We thank the reviewers for their detailed work and constructive suggestions to improve the quality of this paper.

Conflicts of Interest: The authors declare no conflict of interest.

References

1. Zhao, G.C.; Sun, M.; Wilde, S.A.; Li, S.Z. Late Archean to Paleoproterozoic evolution of the North China Craton: Key issues revisited. *Precambr. Res.* **2005**, *136*, 177–202. [CrossRef]
2. Zhao, G.C.; Cawood, P.A.; Li, S.Z.; Wilde, S.A.; Sun, M.; Zhang, J.; He, Y.H.; Yin, C.Q. Amalgamation of the North China Craton: Key issues and discussion. *Precambr. Res.* **2012**, *222–223*, 55–76. [CrossRef]
3. Li, S.Z.; Zhao, G.C.; Sun, M.; Han, Z.Z.; Hao, D.F.; Luo, Y.; Xia, X.P. Deformation history of the Paleoproterzoic Liaohe Group in the Eastern Block of the North China Craton. *J. Asian Earth Sci.* **2005**, *24*, 659–674. [CrossRef]
4. Zhai, M.G.; Santosh, M. The early Precambrian odyssey of North China Craton: A synoptic overview. *Gondwana Res.* **2011**, *20*, 6–25. [CrossRef]
5. Xue, C.J.; Chen, Y.C.; Lu, Y.F.; Li, H.Q. Metallogenic epochs of Au and Ag deposits in Qingchengzi ore-clustered area, eastern Liaoning Province. *Miner. Depos.* **2003**, *22*, 177–184, (In Chinese with English abstract).
6. Yu, G.; Chen, J.F.; Xue, C.J.; Chen, Y.C.; Chen, F.K.; Du, X.Y. Geochronological framework and Pb, Sr isotope geochemistry of the Qingchengzi Pb–Zn–Ag–Au orefield, Northeastern China. *Ore Geol. Rev.* **2009**, *35*, 367–382. [CrossRef]
7. Zhao, Y.; Zhang, P.; Lv, J.C.; Zhang, S.; Kou, L.L.; Yang, Z.J. Characteristics of ore-formingfluids in the Gaojiapuzi Ag deposit of the Qingchengzi orefield, Liaoning Province and geological implications. *Geol. Explor.* **2015**, *51*, 441–450.
8. Ma, Y.B.; Bagas, L.; Xing, S.W.; Zhang, S.T.; Wang, R.J.; Li, N.; Zhang, Z.J.; Zou, Y.F.; Yang, X.Q.; Wang, Y.; et al. Genesis of the stratiform Zhenzigou Pb-Zn deposit in the North China Craton: Rb-Sr and C-O-S-Pb isotope constraints. *Ore Geol. Rev.* **2016**, *79*, 88–104. [CrossRef]
9. Yang, F.C.; Song, Y.H.; Zhang, P.; Chai, P.; Li, B. Forming fluid characteristics and tracing of ore-forming source materials of gold-silver deposit in the Qingchengzi ore concentration area. *Acta Geol. Sin.* **2016**, *90*, 2775–2785, (In Chinese with English abstract).
10. Zhao, Y.; Zhang, P.; Bi, Z.W. Key role of sedimentary formation played in the mineralization process of the Paleoproterozoic Yangmugan boron deposit in Liaodong Peninsula, NE China. *J. Earth Sci. Environ.* **2022**, *44*, 207–219, (In Chinese with English abstract).
11. Zeng, Q.D.; Chen, R.Y.; Yang, J.H.; Sun, G.T.; Yu, B.; Wang, Y.B.; Chen, P.W. The metallogenic characteristics and exploring ore potential of the gold deposits in eastern Liaoning Province. *Acta Petrol. Sin.* **2019**, *35*, 1939–1963.
12. Zhang, Z.C.; Wang, Y.W.; Li, D.D.; Lai, C. Lithospheric architecture and metallogenesis in Liaodong Peninsula, North China Craton: Insights from zircon Hf-Nd isotope mapping. *Minerals* **2019**, *9*, 179. [CrossRef]
13. Wang, Y.W.; Xie, H.J.; Li, D.D.; Shi, Y.; Liu, F.X.; Sun, G.Q.; Sun, Q.M.; Zhou, G.C. Prospecting prediction of ore concentration area exemplified by Qingchengzi Pb-Zn-Au-Ag ore concentration area, eastern Liaoning Province. *Miner. Depos.* **2017**, *36*, 1–24, (In Chinese with English abstract).
14. Liu, G.P.; Ai, Y.F. A discussion on some major problems of the Baiyun gold deposit, eastern Liaoning. *Miner. Depos.* **1999**, *18*, 219–225, (In Chinese with English abstract).
15. Song, J.C.; Hu, T.J.; Wang, E.D.; Jia, S.S. A comparison of metallogenic conditions between two sides of Yalu River and its inspiration to future ore-prospecting in Liaodong area. *Miner. Depos.* **2009**, *28*, 449–461, (In Chinese with English abstract).
16. Zhang, S.; Zhang, D.; Sha, D.X.; Kou, L.L.; Zhao, D.F. Metallogenetic characteristics and genesis of the gold (silver) mieralization in Linjiasandaogou–Xiaotongjiapuzi area, eatern Liaoning Province. *J. Jilin Uinv. (Earth Sci. Ed.)* **2012**, *42*, 725–732.

17. Song, Y.H.; Yang, F.C.; Yan, G.L.; Wei, M.H.; Shi, S.S. Characteristics of mineralization fluids and tracers of mineralization material sources of the Qingchengzi lead–zinc deposit in Liaoning Province. *Geol. Explor.* **2017**, *53*, 259–269, (In Chinese with English abstract).
18. Li, J.; Cai, W.Y.; Li, B.; Wang, K.Y.; Liu, H.L.; Konare, Y.; Qian, Y.; Lee, G.J.; Yoo, B.C. Paleoproterozoic SEDEX-type stratiform mineralization overprinted by Mesozoic vein-type mineralization in the Qingchengzi Pb-Zn deposit, Northeastern China. *J. Asian Earth Sci.* **2019**, *184*, 104009. [CrossRef]
19. Duan, X.X.; Zeng, Q.D.; Yang, J.; Liu, J.; Wang, Y.; Zhou, L.L. Geochronology, geochemistry and Hf isotope of Late Triassic magmatic rocks of Qingchengzi district in Liaodong Peninsula, Northeast China. *J. Asian Earth Sci.* **2014**, *91*, 107–124. [CrossRef]
20. Hao, L.B.; Zhao, X.; Zhao, Y.Y. Stabel isotope characteristics and ore genesis of the Baiyun gold deposit, Liaoning Province. *J. Jilin Univ. (Earth Sci. Ed.)* **2017**, *47*, 442–451. [CrossRef]
21. Zhang, P.; Kou, L.L.; Zhao, Y.; Bi, Z.W.; Sha, D.M.; Li, Z.M.; Han, R.P. Fluid inclusions, H-O, S, Pb, and noble gas isotope studies of the Baiyun gold deposit in the Qingchengzi orefield, NE China. *J. Geochem. Explor.* **2019**, *200*, 37–53. [CrossRef]
22. Sun, G.T.; Zeng, Q.D.; Li, T.Y.; Li, A.; Wang, E.Y.; Xiang, C.S.; Wang, Y.; Chen, P.W.; Yu, B. Ore genesis of the Baiyun gold deposit in Liaoning province, NE China: Constraints from fluid inclusions and zircon U-Pb ages. *Arab. J. Geosci.* **2019**, *12*, 299. [CrossRef]
23. Yang, J.H.; Wu, F.Y.; Xie, L.W.; Liu, X.M. Petrogenesis and tectonic implications of Kuangdonggou syenites in the Liaodong peninsula, east North China Craton: Constraints from in-situ zircon U-Pb ages and Hf isotopes. *Acta Petrol. Sin.* **2007**, *23*, 263–276, (In Chinese with English abstract).
24. Tam, P.Y.; Zhao, G.C.; Liu, F.L.; Zhou, X.W.; Sun, M.; Li, S.Z. Timing of metamorphism in the Paleoproterozoic Jiao-Liao-Ji Belt: New SHRIMP U–Pb zircon dating of granulites, gneisses and marbles of the Jiaobei massif in the North China Craton. *Gondwana Res.* **2011**, *19*, 150–162. [CrossRef]
25. Chen, J.S.; Xing, D.H.; Liu, M.; Li, B.; Yang, H.; Tian, D.X.; Yang, F.; Wang, Y. Zircon U-Pb chronology and geological significance of felsic volcanic rocks in the Liaohe Group from the Liaoyang area, Liaoning Province. *Acta Petrol. Sin.* **2017**, *33*, 2792–2810, (In Chinese with English abstract).
26. Li, Z.; Chen, B.; Wei, C.J. Is the Paleoproterozoic Jiao-Liao-Ji Belt (North China Craton) a rift? *Int. J. Earth Sci.* **2017**, *106*, 355–375. [CrossRef]
27. Li, Z.; Li, J.; Chen, B. Early Precambrian tectonic-thermal events: Coupled U-Pb-Hf of detrital zircons from Jiao-Liao-Ji Belt, North China Craton. *Arab. J. Geosci.* **2018**, *11*, 424–440. [CrossRef]
28. Liu, J.; Zhang, J.; Liu, Z.H.; Yin, C.Q.; Zhao, C.; Li, Z.; Yang, Z.J.; Dou, S.Y. Geochemical and geochronological study on the Paleoproterozoic rock assemblage of the Xiuyan region: New constraints on an integrated rift-andcollision tectonic process involving the evolution of the Jiao-Liao-Ji Belt, North China Craton. *Precamb. Res.* **2018**, *310*, 179–197. [CrossRef]
29. Zhao, Y.; Kou, L.L.; Zhang, P.; Bi, Z.W.; Li, D.T.; Chen, C. Characteristics of Geochemistry and Hf isotope from meta-gabbro in the Longchang area, eastern of North China Craton: Implications on the evolution of the Jiaoliaoji Paleoproterozoic orogeny belt. *Earth Sci.* **2019**, *44*, 3333–3345, (In Chinese with English abstract).
30. Zhu, R.X.; Fan, H.R.; Li, J.W.; Meng, Q.R.; Li, S.R.; Zeng, Q.D. Decratonic gold deposits. *Sci. China Earth Sci.* **2015**, *58*, 1523–1537. [CrossRef]
31. Song, Y.H.; Yang, F.C.; Yan, G.L.; Wei, M.H.; Shi, S.S. SHRIMP U-Pb ages and Hf isotopic compositions of Paleoproterozoic granites from the Eastern part of Liaoning Province and their tectonic significance. *Acta Geol. Sin.* **2016**, *90*, 2620–2636, (In Chinese with English abstract).
32. Wang, X.P.; Peng, P.; Wang, C.; Yang, S.Y. Nature of three episodes of Paleoproterozoic magmatism (2180 Ma, 2115 Ma and 1890 Ma) in the Liaoji belt, North China with implications for tectonic evolution. *Precambr. Res.* **2017**, *298*, 252–267. [CrossRef]
33. Zhao, Y.; Zhang, P.; Li, Y.; Li, D.T.; Chen, J.; Bi, Z.W. Geochemistry of two types of Paleoproterozoic granites, and zircon U-Pb dating, and Lu-Hf isotopic characteristics in the Kuandian area within the Jiao-Liao-Ji Belt: Implications for regional tectonic setting. *Geol. J.* **2020**, *55*, 7564–7580. [CrossRef]
34. Zhao, Y.; Lin, S.F.; Zhang, P. Geochronology and geochemical characteristics of Paleoproterozoic syn-orogenic granitoids and constraints on the geological evolution of the Jiao-Liao-Ji orogenic belt, North China Craton. *Precambr. Res.* **2021**, *365*, 106386. [CrossRef]
35. Wang, X.P.; Peng, P.; Wang, C.; Yang, S.Y. Petrogenesis of the 2115 Ma Haicheng mafic sills from the Eastern North China Craton: Implications for an intra-continental rifting. *Gondwana Res.* **2016**, *39*, 347–364. [CrossRef]
36. Xu, W.; Liu, F.L.; Tian, Z.H.; Liu, L.S.; Ji, L.; Dong, Y.S. Source and petrogenesis of Paleoproterozoic meta-mafic rocks intruding into the North Liaohe Group: Implications for back-arc extension prior to the formation of the Jiao-Liao-Ji Belt, North China Craton. *Precambr. Res.* **2018**, *207*, 66–81. [CrossRef]
37. Duan, X.X.; Zeng, Q.D.; Wang, Y.B.; Zhou, L.L. Genesis of the Pb-Zn deposits of the Qingchengzi ore field, eastern Liaoning, China: Constraints from carbonate LA-ICPMS trace element analysis and C-O-S-Pb isotopes. *Ore Geol. Rev.* **2017**, *89*, 752–771. [CrossRef]
38. Zhao, Y.; Yang, H.Z.; Yang, F.C.; Zhang, P.; Gu, Y.C.; Xu, J. Genesis of the typical gold deposits in Qingchengzi orefield, Liaodong Peninsula: Evidences from S-D-O isotopes. *Geol. Res.* **2020**, *29*, 21–28, (In Chinese with English abstract).
39. Liu, J.; Liu, F.X.; Li, S.H.; Lai, C.K. Formation of the Baiyun gold deposit, Liaodong gold province, NE China: Constraints from zircon U–Pb age, fluid inclusion, and C–H–O–Pb–He isotopes. *Ore Geol. Rev.* **2019**, *104*, 686–706. [CrossRef]

40. Zhang, Y.; Chen, W.; Chen, K.L.; Liu, X.Y. Study on the Ar-Ar age spectrum of diagenetic I/S and the mechanism of 39Ar recoil loss___Examples from the clay minerals of P-T boundary in Changxing, Zhejiang Province. *Geol. Rev.* **2006**, *52*, 556–561.
41. Liu, Y.S.; Hu, Z.C.; Gao, S.; Günther, D.; Xu, J.; Gao, C.G.; Chen, H.H. In situ analysis of major and trace elements of anhydrous minerals by LA-ICP-MS without applying an internalstandard. *Chem. Geol.* **2008**, *257*, 34–43. [CrossRef]
42. Thern, E.R.; Blereau, E.; Jourdan, F.; Nelson, D.R. Tourmaline $^{40}Ar/^{39}Ar$ geochronology and thermochronology: Example from the youngest Hadean-zircon-bearing siliciclastic metasedimentary rocks from the Yilgarn Craton. *Geochim. Cosmochim. Acta* **2020**, *27*, 285–299. [CrossRef]
43. Hoefs, J. *Stable Isotope Geochemistry*, 3rd ed.; Springer: Berlin, Germany, 1986; p. 241.
44. Zhao, H.X.; Frimmel, H.E.; Jiang, S.Y.; Dai, B.Z. LA-ICP-MS trace element analysis of pyrite from the Xiaoqinling gold district, China: Implications for ore genesis. *Ore Geol. Rev.* **2011**, *43*, 142–153. [CrossRef]
45. Zhang, J.; Deng, J.; Chen, H.Y.; Yang, L.Q.; Cooke, D.; Danyushevsky, L.; Gong, Q.J. LA-ICP-MS trace element analysis of pyrite from the Chang'an gold deposit, Sanjiang region, China: Implication for ore-forming process. *Gondwana Res.* **2014**, *26*, 557–575. [CrossRef]
46. Wen, G.; Li, J.W.; Hofstra, A.H.; Koenig, A.E.; Lowers, H.A.; Adams, D. Hydrothermal reequilibration of igneous magnetite in altered granitic plutons and its implications for magnetite classification schemes: Insights from the Handan-Xingtai iron district, North China Craton. *Geochim. Cosmochim. Acta* **2017**, *213*, 255–270. [CrossRef]
47. Yang, L.; Wang, Q.F.; Large, R.R.; Mukherjee, I.; Deng, J.; Li, H.J.; Yu, H.Z.; Wang, X. Fluid source and metal precipitation mechanism of sediment-hosted Chang'an orogenic gold deposit, SW China: Constrains from sulfide texture, trace element, S, Pb and He-Ar isotopes and calcite C-O isotopes. *Am. Mineral.* **2021**, *106*, 410–429. [CrossRef]
48. Yang, L.; Wang, Q.F.; Groves, D.I.; Lu, S.; Li, H.J.; Wang, P.; Deng, J. Multiple orogenic gold mineralization events in a collisional orogen: Insights from an extruded terrane along the southeastern margin of the Tibetan Plateau. *J. Struct. Geol.* **2021**, *147*, 104333. [CrossRef]
49. Yang, L.; Wang, Q.F.; Large, R.R.; Fougerouse, D.; Mukherjee, I.; Zhang, Q.Z.; Deng, J. Texture and geochemistry of pyrite from the Jinya, Nakuang and Gaolong gold deposits in the Youjiang Basin: Implications for basin-scale gold mineralization. *Min. Deposita* **2022**, 1–24. [CrossRef]
50. Li, H.M.; Liu, M.J.; Li, L.X.; Yang, X.Q.; Yao, L.D.; Chen, J.; Yao, T. SHRIMP U-Pb geochronology of zircons from the garnet-rich altered rocks in the mining area of the Gongchangling iron deposit: Constraints on the ages of the high-grade iron deposit. *Acta Petrol. Sin.* **2014**, *30*, 1205–1217.
51. Li, L.X.; Zi, J.W.; Li, H.M.; Rasmussen, B.; Wilde, S.A.; Sheppard, S.; Ma, Y.B.; Meng, J.; Song, Z. High-Grade magnetite mineralization at 1.86 Ga in Neoarchean banded iron formations, Gongchangling, China: In situ U-Pb geochronology of metamorphic-hydrothermal zircon and monazite. *Econ. Geol.* **2019**, *114*, 1159–1175. [CrossRef]
52. Liu, P.H.; Liu, F.L.; Yang, H.; Wang, F.; Liu, J.H. Protolith ages and timing of peak and retrograde metamorphism of the high pressure granulites in the Shandong Peninsula, eastern North China Craton. *Geosci. Front.* **2012**, *3*, 923–943. [CrossRef]

minerals

MDPI

Article

Two Epochs of Mineralization of Orogenic Gold Deposit in the East Kunlun Orogenic Belt: Constraints from Monazite U–Pb Age, In Situ Sulfide Trace Elements and Sulfur Isotopes in Wulonggou Gold Field

Zheming Zhang [1,2,3,4], Qingdong Zeng [1,3,4,*], Tong Pan [5], Hailin Xie [6], Zhanhao Wei [6], Hongrui Fan [1,3,4], Jinjian Wu [1,3,4], Kuifeng Yang [1,3,4], Xinghui Li [1,3] and Gaizhong Liang [1,3,4]

1. Key Laboratory of Mineral Resources, Institute of Geology and Geophysics, Chinese Academy of Sciences, Beijing 100029, China; zhangzheming17@mails.ucas.ac.cn (Z.Z.); fanhr@mail.iggcas.ac.cn (H.F.); wujinjian19@mails.ucas.ac.cn (J.W.); yangkuifeng@mail.iggcas.ac.cn (K.Y.); lxh@mail.iggcas.ac.cn (X.L.); lianggaizhong@mail.iggcas.ac.cn (G.L.)
2. Sinosteel Overseas Resources Limited, Sinosteel Group Corporation Limited, Beijing 100080, China
3. Innovation Academy for Earth Science, Chinese Academy of Sciences, Beijing 100029, China
4. College of Earth and Planetary Sciences, University of Chinese Academy of Sciences, Beijing 100049, China
5. Qinghai Geological and Mineral Exploration and Development Bureau, Xining 810008, China; pant66@163.com
6. The First Exploration Institute of Geology and Mineral Resources in Qinghai Province, Haidong 810600, China; dkjxhl@126.com (H.X.); qhszhw@163.com (Z.W.)

* Correspondence: zengqingdong@mail.iggcas.ac.cn

Abstract: The Wulonggou Gold Field is one of the giant gold fields in the East Kunlun Orogenic Belt, northwestern China. Previous studies mainly focused on elementary mineral isotopic studies, fluid inclusions, and geological features in the Wulonggou Gold Field. In this study, we report some research on the precise age and the specific ore-forming process of the WGF: the hydrothermal monazite U–Pb ages; the way of gold precipitation; the composition, evolution, and source of ore-forming fluids of the Wulonggou Gold Field. Finally, we demonstrate a link between two-stage hydrothermal events and sequential episodes of crust-derived magmas, with implications for gold metallogeny in the East Kunlun Orogenic Belt. There are four hydrothermal stages that are recognized: a quartz–pyrite stage (stage 1), a quartz–pyrite–arsenopyrite–chalcopyrite stage (stage 2), a quartz–galena–sphalerite–pyrite stage (stage 3) and a quartz–stibnite–carbonate stage (stage 4). The monazite U–Pb ages of the Huanglonggou and Hongqigou deposits in the Wulonggou Gold Field were 422.2 ± 2.4 Ma and 236.7 ± 3.7 Ma, respectively, which support the opinion of two epochs of mineralization. Stages 1 and 2 are the main gold mineralization stages, wherein Au and As have a close genetic relationship. The Hongqigou and Huanglonggou deposits seem to have been formed in different metallogenic events due to the contrast on the trace element compositions in pyrite. The sources of the ore-forming materials and fluids of the Hongqigou and Huanglonggou deposits show apparent characteristics of orogenic gold deposit, and the magmatic events during Paleozoic and Mesozoic have an important contribution to the formation of the gold deposits. The gold deposits in the Wulonggou Gold Field can be interpreted as an orogenic gold system related to two-epoch tectonic–magmatic events.

Keywords: orogenic gold deposit; Wulonggou Gold Field; monazite U–Pb geochronology; pyrite trace element; in situ S isotope

1. Introduction

The Wulonggou Gold Field (WGF) is located ~65 km to the east of the Golmud City and is situated in the Middle Zone of the EKOB in tectonic setting (Figure 1B). Previous studies mainly focused on elementary ore isotopic studies on bulk minerals, fluid inclusions in quartz, and geochronology research. In terms of the origin of ore-forming fluids and materials, metallogenic and pluton ages in the WGF (e.g., [1–3]), some authors suggested that the metallogenic events have a strong relationship with the Mesozoic magmatism [1–4], the majority of the authors suggested that the metallogenic events have a strong relationship with the Mesozoic magmatism [1–4], while fewer researchers have provided the evidence of Paleozoic metallogeny [2–4].

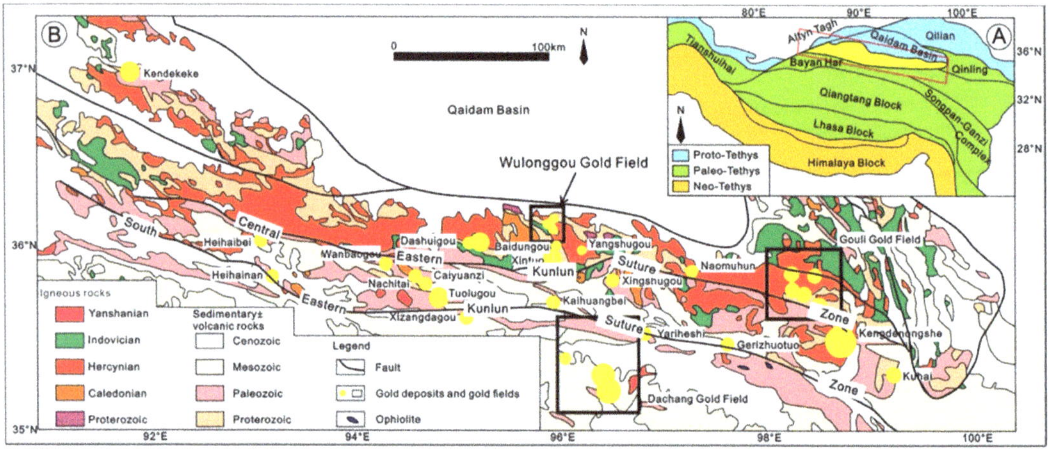

Figure 1. (**A**) The location of the East Kunlun Orogen Belt in the Qinghai–Tibetan plateau (modified after [5]). (**B**) Simplified Geological map of the East Kunlun Orogen Belt (modified after [5]).

However, some advanced research should be conducted to reveal the ore-forming process systematically and to know which tectonic setting these deposits are formed in by determining the metallogenic age by contemporaneous hydrothermal accessory minerals. In this paper, we report the results of the ore-forming age of the Wulonggou Gold Field, the evolution of the ore-forming fluids, and the sources of the fluids and metals. We studied sulfides from the Hongqigou and Huanglonggou gold deposits by using a combination of microscopy, laser ablation inductively coupled plasma mass spectrometry (LA-ICP-MS) for trace elements as well as mapping, multi-collector laser ablation inductively coupled plasma mass spectrometry (LA-MC-ICP-MS) for sulfur isotopic analysis, and backscattered electron (BSE) scanning microscopy. The LA-ICP-MS U–Pb analysis of monazite in Hongqigou and Huanglonggou deposits was carried out to determine the metallogenic ages. Moreover, we found the evidence of two epochs of mineralization and sequential episodes of crust-derived magmas in WGF through former analysis, which may provide some ideas of conducting research on gold metallogeny in the EKOB.

2. Geological Setting

The strata in the EKOB mainly include Proterozoic basement, Paleozoic sedimentary rocks interbedded with volcanic rocks, Mesozoic volcanic, and Cenozoic alluvial and pebbles (Figure 1B). The Proterozoic basement of the EKOB is composed of the Paleoproterozoic Jinshuikou, Kuhai, Wanbaogou, Baishahe groups and the Mesoproterozoic Xiaomiao Group [6–9]. The Jinshuikou Group (1.9~1.0 Ga) is characterized by granitic gneiss, migmatite, amphibolite and minor marble [8,10–12]. The Kuhai Group is composed of amphibolite, felsic gneiss, and marble [13]. The Wanbaogou Group is composed of clastic sediments, marble, and volcanic rocks, underwent greenschist face metamorphism [14]. The Baishahe Group is composed of amphibolite gneiss, mica-plagioclase gneiss, granulite, banded marble, mica schist, and mica-quartz schist [6]. The Mesoproterozoic Xiaomiao

Group is composed of quartzite, mica-quartz schist, biotite-plagioclase gneiss and marble [6]. The Paleozoic (Ordovician–Silurian) unit is related to the consumption of the Proto-Tethys and the subsequent collision between the northern and southern parts of the East Kunlun terranes [15], including the Maoniushan formation and Nachitai Group. The Ordovician–Silurian Maoniushan Formation consists of molasses, resting unconformably on pre-Ordovician–Devonian strata, which recorded the closure of the Proto-Tethys [16]. The Ordovician–Silurian Nachitai Group mainly consists of phyllite and other low-grade metavolcanic and meta-sedimentary rocks [17]. The Middle–Late Triassic Naocangjiangou and Xilikete Formations mainly consist of volcanic lava [18,19].

The East Kunlun Orogen is composed of the Northern East Kunlun Terrane and the Southern East Kunlun Terrane, which are separated by the Central East Kunlun Suture Zone. They are NWW–EW trending and north dipping and have the length of several hundred kilometers (Figure 1B). Two regional suture zones, the Central East Kunlun Suture Zone and Southern East Kunlun Suture Zone, which correspond to the evolution of the Proto-Tethys Ocean (Proterozoic–Early Paleozoic) and Paleo-Tethys Ocean (Late Paleozoic—Mesozoic), respectively, traverse the EKOB (e.g., [17]).

The mafic–ultramafic rocks mainly occur along the Central East Kunlun Suture Zone and Southern East Kunlun Suture Zone. The former was dated as Cambrian–Ordovician with ages that range from 537 Ma to 467 Ma (Figure 1B, [20–24]), whereas the latter was dated as both Cambrian–Ordovician (Figure 1B, 555–516 Ma, [25,26]) and Carboniferous ages (345–332 Ma, [26,27]). There are two epochs of granitoids range from 460–430 Ma and 290–190 Ma in the EKOB: the former recorded the Early Paleozoic terrane accretion history between Gondwana and Laurasia that was driven by the northward subduction of the Proto-Tethys oceanic crust until final closure ([28–34] and references therein); the latter records a complete evolution history from the sea-floor spreading of the Paleo-Tethys Ocean to northward subduction, then collided, and finally was the post-collisional magmatism ([29,31,33–36] and references therein).

3. Ore Deposit Geology

The WGF is structurally controlled by three synclinoria and three NW-trending sinistral ductile shear zones (Figure 2). The strata of the WGF include the Paleoproterozoic Jinshuikou Group, Mesoproterozoic Xiaomiao and Neoproterozoic Qiujidonggou formations, Early Ordovician ultramafic rocks, Late Silurian–Early Devonian granitoids and dolerite dykes, Late Permian–Late Triassic gabbro, diorite, granitoids, and metavolcanic rocks (Figure 2). The Jinshuikou Group (1927 $\pm$ 34 Ma, [6]) mainly consists of plagioclase gneiss, with minor migmatite, schist, marble, amphibolite, and granulite [2,9]. The Xiaomiao Formation (1683–1554 Ma, [37]) consists of marble, gneiss, amphibolite, granulite, schist, quartzite [2,9]. The Neoproterozoic Qiujidonggou Formation is composed of meta-conglomerate, metasandstone, schist, phyllite and marble [2,9,38]. The Ordovician Qimantage Group ultramafic rocks (zircon U–Pb age of 486.5 $\pm$ 3.4 Ma, [38]) are mainly composed of clinopyroxenites and porphyritic hornblendites intruded by porphyritic hornblende gabbro dykes and granite dykes. The Quaternary mainly distribute in ravines and valleys.

The intrusive rocks in the WGF formed during the Neoproterozoic, Early Paleozoic, Late Paleozoic and Early Mesozoic (Figure 2). The majority of intrusive rocks in the WGF formed during Early Paleozoic, which mainly include brick-red coarse-grained syenogranite with minor diorite. Their ages range from 450 to 410 Ma, and some A-type granites are about 390 Ma [1,2,39,40]. The Early Mesozoic intrusive rocks are epigenetic granite. Late Silurian-Early Devonian granitoid batholiths are composed of dominantly granodiorites and syenogranites, and the granodiorites contain abundant mafic microgranular enclaves (MMEs) (Figure 2, [39]). Late Permian–Late Triassic intrusions intruded into Precambrian metamorphic rocks or Paleozoic magmatic rocks, and comprise gabbro-diorite complexes, diorite-granodiorite batholiths, and porphyritic granodiorite stocks and porphyritic quartz diorite dykes (Figure 2). The porphyry and diorite porphyrite veins show LA-ICP-MS zircon U–Pb age of about 220 Ma [39]. Metavolcanic rocks include tuff and minor amounts of volcanic breccia and are considered to have formed in the Late Triassic according to a zircon U–Pb age of 219 ± 1 Ma [39].

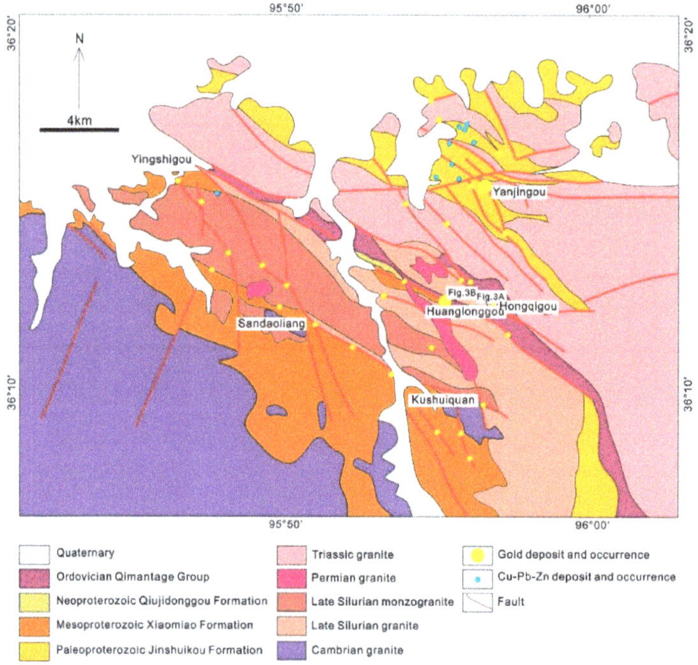

Figure 2. Geological map of the Wulonggou Gold Field (modified after [9]).

There are three ore belts of Yanjingou, Yingshigou–Hongqigou and Sandaoliang–Kushuiquan, from north to south, respectively. They are controlled by the brittle structural fracture zones hosted in the pre-existed NW-trending ductile shear zones, and they were formed during Triassic. They are 0.5–2 km in width, 10–20 km in length, and have a dip angle of ~70° NE. The majority of gold deposits in the area are distributed within or close to these three ductile shear zones (Figure 2). The Yanjinggou ore belt has ore bodies of lengths of over 5 km, widths of 0.5~1 km, and a dip to 35°~75° NE [4]. The Yingshigou–Hongqigou ore belt, located in the middle of the ore field, is the best gold mineralized ore belt in the Wulonggou area. The orebodies have lengths of over 5 km, widths of 1~2 km, and a dip to 55°~85° NE [4]. There are above 65 t gold in this belt, the gold reserve of Hongqigou is 15 t, with an average gold grade of 6.9 g/t; the gold reserve of the Huanglonggou deposit is 29.3 t, with an average gold grade of 9.39 g/t (from the work report completed by the First Exploration Institute of Geology and Mineral Resources in Qinghai Province). The Hongqigou and Huanglonggou deposits we discussed in this paper are located in this belt. The dip angle of orebodies in the Hongqigou deposit is relatively slower than that of the Huanglonggou deposit (Figure 3). Sandaoliang–Kushuiquan ore belt, located in the southmost part of the WGF, the orebodies have lengths of more than 10 km and widths of 0.5~1 km, and a dip to 45°~70° NE [4]. The Hongqigou and Huanglonggou gold deposits are located in the largest Yingshigou–Hongqigou gold belt in central WGF. Gold mineralization in these gold deposits is characterized by multi-stage quartz–sulfide veins containing abundant invisible gold.

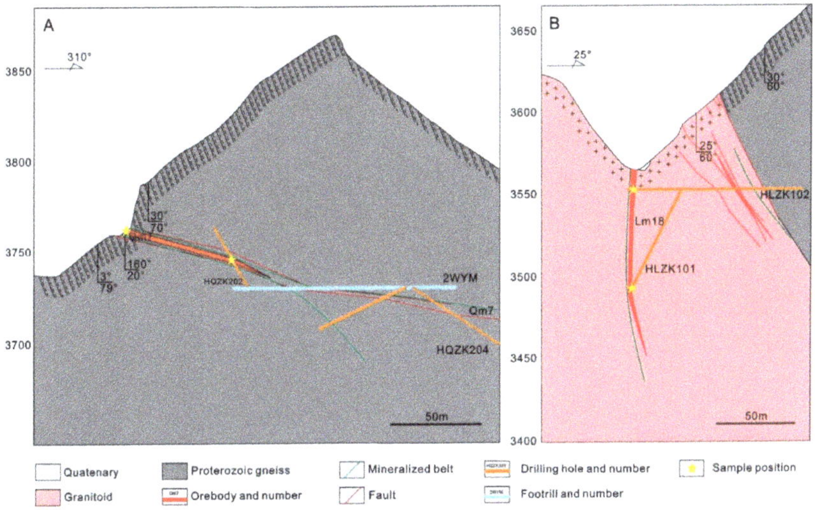

Figure 3. Profiles of the 2nd prospecting line of the Hongqigou deposit (**A**) and the 1st prospecting line of the Huanglonggou deposit (**B**) (modified after the map drawn by Qinghai Geological and Mineral Exploration and Development Bureau, 2015).

A total of 80 samples of each stage ores were collected from the tunnels of 3330 m, 3390 m, 3450 m above sea level and drilling cores in the Huanglonggou deposit, and orebodies' outcrops of 3600 m to 3800 m above sea level and drilling cores in the Hongqigou deposit.

The gold mineralization of other gold deposits in the WGF is confined in the altered and fractured zones, within which the metallic minerals are predominantly native gold, pyrite, pyrrhotite, arsenopyrite, sphalerite, galena, chalcopyrite, lollingite, and marcasite; gangue minerals are mainly quartz, sericite, chlorite, calcite, and fluorite; hydrothermal alteration includes mainly silicification, chloritization, sericitization, and carbonatization [1–4]. The gold deposits in this ore field yield age of ~240 Ma (sericite Ar-Ar, [41]; quartz Rb-Sr, [4]).

Paragenetic Sequences

Paragenetic sequences and crosscutting relationships indicate different ore-forming events in the Hongqigou and Huanglonggou deposits (Figures 4 and 5). The ore-forming process can be divided into four hydrothermal stages (Figures 4 and 5): a quartz–pyrite stage (stage 1), a quartz–pyrite–arsenopyrite–chalcopyrite stage (stage 2), a quartz–galena–sphalerite–pyrite stage (stage 3) and a quartz–stibnite–carbonate stage (stage 4). Gold mainly occurs in invisible gold in pyrite and arsenopyrite, such as cation exchange or tiny grains that are difficult to observe under a microscope and SEM; this will draw a conclusion in Section 5.3.1 through trace element mapping of pyrite. By comparing the paragenetic association of minerals, we can find that minor sphalerite but major galena is in the ores of stage 3 of the Hongqigou deposit, and that minor galena but major sphalerite is in the ores of stage 3 of the Huanglonggou deposit; the Huanglonggou deposit has more chalcopyrite in ores, while the Hongqigou deposit has more arsenopyrite in ores (Figures 4–6). Silicification and sericitization were closely related to gold mineralization.

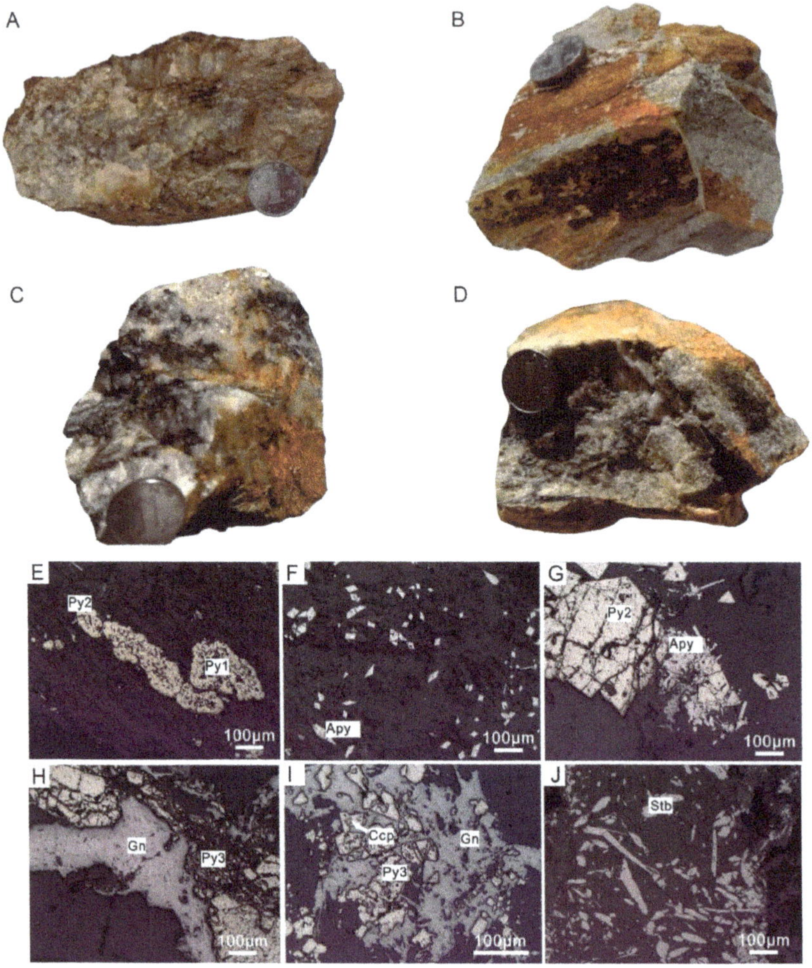

Figure 4. Photos of the ore of each stage and micrographs showing paragenesis associations in the Hongqigou deposit. (**A**) Typical barren quartz of stage 1; (**B**) Typical quartz–pyrite–arsenopyrite–chalcopyrite ore of stage 2; (**C**) Typical quartz–galena–pyrite ± arsenopyrite ore of stage 3, which represent the major lead, zinc, and silver metallogenic period; (**D**) Typical quartz–stibnite–carbonate ore of stage 4; (**E**) Rough and porous Py1 was contained by smooth Py2; (**F**) Arsenopyrite spread in stage 2 ores; (**G**) Euhedral–subhedral Py2 was crosscut by arsenopyrite in stage 2; (**H**) Subhedral Py3 was altered by galena; (**I**) Subhedral Py3 was interfilled by galena and chalcopyrite; (**J**) Stibnite spread in quartz-stibnite-carbonate ores in stage 4. Abbreviations: Apy = arsenopyrite, Ccp = chalcopyrite, Gn = galena, Py = pyrite, Stb = stibnite.

Figure 5. Photos of the ore of each stage and micrographs showing paragenesis associations in the Huanglonggou deposit. (**A**) Typical crushed altered rock type ore of stage 1; (**B**) Typical quartz-pyrite–arsenopyrite–chalcopyrite ore of stage 2; (**C**) Typical quartz–sphalerite–pyrite ± arsenopyrite ore of stage 3, which represent the major lead, zinc and silver metallogenic phase; (**D**) Typical quartz–stibnite–carbonate ore of stage 4; (**E**) Rough and porous Py1 was contained by smooth Py2; (**F**) Euhedral–subhedral Py2 was crosscut by arsenopyrite in stage 2; (**G**) Arsenopyrite and pyrite coexist in ores of stage 3; (**H**) Subhedral Py3 was interfilled by sphalerite and chalcopyrite; (**I**) Subhedral sphalerite and chalcopyrite coexist with euhedral arsenopyrite in stage 3; (**J**) Stibnite spread in quartz–stibnite–carbonate ores in stage 4. Abbreviations: Apy = arsenopyrite, Ccp = chalcopyrite, Py = pyrite, Sp = sphalerite, Stb = stibnite.

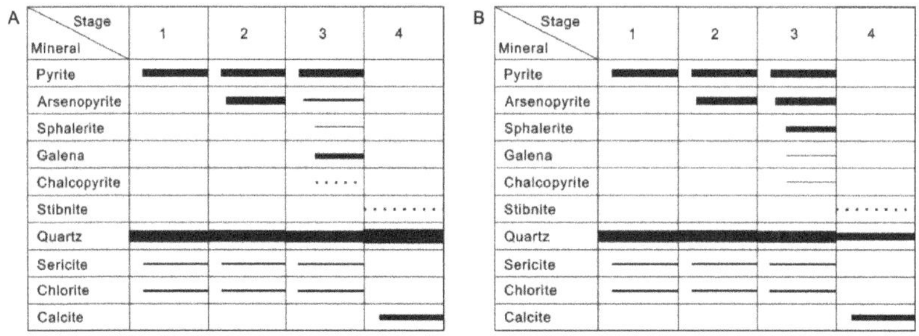

Figure 6. Paragenetic sequences for major minerals of the Hongqigou deposit (**A**) and the Huanglonggou deposit (**B**).

4. Analytical Method

4.1. Mineralogy

Doubly polished thin-sections (0.2 mm thick) were made from auriferous quartz veins for petrographic study of compositions, textures, structures, and spatial distributions of minerals. Detailed mineral identification was carried out by scanning electron microscopy (SEM) using the TM4000 plus SEM system at the Institute of Geology and Geophysics, Chinese Academy of Sciences (IGGCAS), which is equipped with a Bruker Quantax 75 energy spectrum to identify and confirm different minerals and observe special textures. Backscattered electron images (BSE) were acquired at 20 kV, and a high vacuum environment illustrated the alteration and the texture between the ore minerals and the gangue minerals. The acquisition time for each BSE image was 1 min, and the monazite used for U–Pb metallogenic age dating was observed and screened in this way. Moreover, we divided the mineralization stage, and distinguished the different types of pyrite by using the above method.

4.2. Monazite U–Pb Dating

Hydrothermal monazite, which was closely related to Au mineralization, was found by using SEM and was applied to in situ LA-ICP-MS U–Pb dating. Monazite U–Pb dating was conducted using an Agilent 7500a Q-ICP-MS equipped with a Geolas G2 193 nm excimer laser ablation system (Geolas HD, Lambda Physik, Göttingen, Germany) at IGGCAS. The analyses were conducted with a spot size of 24 μm and at a repetition rate of 8 Hz. Each analysis incorporated approximately 20 s of background acquisition and 45 s of data acquisition. The standard samples Namaqualand and Jefferson were applied to correct the U/Pb fractionation and the instrumental mass discrimination for the external standard and were measured after every six samples. All measured $^{206}Pb/^{238}U$, $^{208}Pb/^{232}Th$, $^{207}Pb/^{235}U$ and $^{207}Pb/^{206}Pb$ isotopic ratios of the Namaqualand during sample analyses were corrected and regressed afterwards [42]. We calibrated the isotopic and elemental fractionation and instrument quality deviations using Glitter 4.0 software for the calibration [43].

4.3. In Situ Sulfide Trace Element Analysis and Mapping

In situ pyrite trace element analysis and mapping were conducted using LA-ICP-MS at the State Key Laboratory of Ore Deposit Geochemistry, Institute of Geochemistry, Chinese Academy of Sciences (IGCAS), China. Laser sampling was performed by ASI RESOLution-LR-S155 laser microprobe equipped with a Coherent Compex-Pro 193 nm ArF excimer laser. We used Agilent 7700 × ICP-MS to acquire ion signal intensities. The carrier gas was He (350 mL/min). Before exiting the cell, the ablated aerosol was mixed with Ar (900 mL/min) as the transport gas. Each analysis includes approximately 30 s of background acquisition (gas blank), followed by 60 s of sample data acquisition. The analysis was performed at a pit diameter of 26 μm, a pulse frequency of 5 Hz, and a flux of 3 J/cm^2. STDGL3 was used to determine the concentration of siderophile and chalcophile elements [44]. The in-house standard pure pyrite (Py) was used to calibrate the concentrations of Fe and S; GSE-1G and GSD-1G were used for calibrating and converting the comprehensive count data of the concentrations of lithophile elements. We took the sulfide reference substance MASS-1 as an unknown sample for verifying the accuracy of the analysis. Trace elements that are relative to gold metallogeny such as ^{59}Co, ^{60}Ni, ^{65}Cu, ^{66}Zn, ^{75}As, ^{107}Ag, ^{121}Sb, ^{197}Au, ^{208}Pb, ^{209}Bi, ^{125}Te were determined and mapped.

4.4. In Situ Sulfide Sulfur Isotope Analysis

In situ pyrite sulfur isotopic analyses were performed at the Wuhan Sample Solution Analytical Technology Co., Ltd., Wuhan, China. We used a Neptune Plus MC-ICP-MS equipped with a Geolas HD excimer ArF laser ablation system. Helium was used as the carrier gas of the ablation unit in the laser ablation system, which was mixed with Ar (supplement gas) after the ablation unit. A single point ablation mode was used. A large spot size (44 μm) and slow pulse frequency (2 Hz) were applied to avoid downhole step effects, as reported by [45]. There were 90 laser pulses applied on each sample. A new signal smoothing device was applied downstream of the sample pool, which eliminated the short-term variation of the signal effectively, especially in the case of the slower pulse frequency [46]. The laser flux was stable at 5 J/cm^2. Neptune Plus came with nine Faraday cups and with a $10^{11}\Omega$ resistor. Isotopes ^{32}S and ^{34}S were collected in the Faraday cup using static mode. Newly designed X Skimmer cones and JET sample cones were used on Neptune Plus to improve signal strength. In all, 4 mL/min nitrogen was added to the central flow to reduce polyatomic interference. All measurements were made using rotational power (defined as peak-edge widths of 5%–95% of the full peak height) at a median resolution consistently >5000. The pyrite standard PPP-1 bracketing method (SSB) was applied for correcting instrumental mass fractionation. The $\delta^{34}S_{V\text{-}CDT}$ reference values of the standards have been reported by [45], and the analysis of S isotopic ratios was conducted by using the software of "Iso-Compass" [47].

5. Results

5.1. In Situ U–Pb Geochronology of Monazite

The in situ U–Pb dating analyses on monazite grains of the Huanglonggou and Hongqigou were presented in Table 1 and Figures 7 and 8. The U–Pb data were shown on the Tera–Wasserburg concordia diagram. The lower intercept age of Hongqigou was 236.7 ± 3.7 Ma (MSWD = 0.33) (Figure 7), and the lower intercept age of Huanglonggou was 422.2 ± 2.4 Ma (MSWD = 0.16) (Figure 8). Most of the analyzed monazite grains were together with pyrite grains of stage 2 (Figures 7 and 8). On the other hand, quartz and sericite are often wrapped in monazite minerals, which shows that the monazite is not the residue of the wall rocks, but the hydrothermal mineral crystalized together with quartz–pyrite–chalcopyrite vein (Figures 7 and 8). Therefore, their crystallization ages can represent the timing of gold mineralization, which is consistent with the age of granite porphyry and suggests their close genetic relationship.

Table 1. The results of the LA–ICP–MS U–Pb dating on hydrothermal monazite in the Wulonggou gold field.

Hongqigou Deposit								
Spot Number	^{207}Pb/^{206}Pb	Error	^{206}Pb/^{238}U	Error	^{207}Pb/^{235}U	Error	^{208}Pb/^{232}Th	Error
Hongqigou-1	0.42256	0.00545	0.08213	0.00126	4.79008	0.04848	0.0183	0.00025
Hongqigou-2	0.5193	0.00466	0.11783	0.00164	8.44495	0.06124	0.01918	0.00026
Hongqigou-3	0.47555	0.0068	0.09846	0.0016	6.46213	0.07121	0.01847	0.00026
Hongqigou-4	0.47919	0.00611	0.10034	0.00155	6.63648	0.0658	0.01921	0.00027
Hongqigou-5	0.49759	0.00637	0.10905	0.0017	7.48896	0.07435	0.01864	0.00026
Hongqigou-6	0.34584	0.00356	0.06569	0.00093	3.13573	0.02615	0.01761	0.00024

Table 1. *Cont.*

Hongqigou Deposit								
Spot Number	$^{207}Pb/^{206}Pb$	**Error**	$^{206}Pb/^{238}U$	**Error**	$^{207}Pb/^{235}U$	**Error**	$^{208}Pb/^{232}Th$	**Error**
Hongqigou-7	0.4282	0.0058	0.08395	0.00132	4.96137	0.05246	0.01814	0.00025
Hongqigou-8	0.46357	0.00766	0.09491	0.00164	6.07223	0.07702	0.01658	0.00024
Hongqigou-9	0.29419	0.0048	0.05799	0.00095	2.35443	0.03106	0.01898	0.00027
Hongqigou-10	0.45555	0.00944	0.0924	0.00181	5.80953	0.09164	0.01667	0.00024
Hongqigou-11	0.43754	0.00607	0.08783	0.00139	5.30415	0.05731	0.01682	0.00023
Hongqigou-12	0.56724	0.00588	0.15324	0.00223	11.99711	0.09862	0.02323	0.00032
Hongqigou-13	0.59236	0.00727	0.18292	0.00286	14.95476	0.14364	0.02494	0.00035
Huanglonggou deposit								
Spot Number	$^{207}Pb/^{206}Pb$	**Error**	$^{206}Pb/^{238}U$	**Error**	$^{207}Pb/^{235}U$	**Error**	$^{208}Pb/^{232}Th$	**Error**
Huanglonggou-1	0.07582	0.00082	0.07033	0.00091	0.73646	0.00692	0.02172	0.00029
Huanglonggou-2	0.11618	0.00135	0.07539	0.001	1.20969	0.01209	0.02214	0.0003
Huanglonggou-3	0.05336	0.00105	0.06733	0.00094	0.49625	0.00913	0.02198	0.0003
Huanglonggou-4	0.12753	0.00155	0.07672	0.00103	1.35138	0.01411	0.02182	0.00029
Huanglonggou-5	0.08696	0.001	0.07166	0.00094	0.86071	0.00867	0.02251	0.0003
Huanglonggou-6	0.07261	0.00514	0.0695	0.00197	0.69693	0.04652	0.01675	0.00023
Huanglonggou-7	0.14143	0.00165	0.07886	0.00106	1.54043	0.01526	0.02374	0.00032
Huanglonggou-8	0.07181	0.0009	0.06962	0.00092	0.6905	0.00773	0.02183	0.00029
Huanglonggou-9	0.07122	0.00074	0.06938	0.0009	0.6825	0.00608	0.02181	0.00029
Huanglonggou-10	0.06378	0.00067	0.06877	0.00089	0.60583	0.00554	0.02049	0.00027
Huanglonggou-11	0.05834	0.00075	0.06798	0.00089	0.54775	0.00634	0.02032	0.00027
Huanglonggou-12	0.07682	0.00161	0.06981	0.00103	0.74073	0.01423	0.02069	0.00029
Huanglonggou-13	0.10698	0.00144	0.07428	0.00101	1.09761	0.01294	0.01785	0.00024

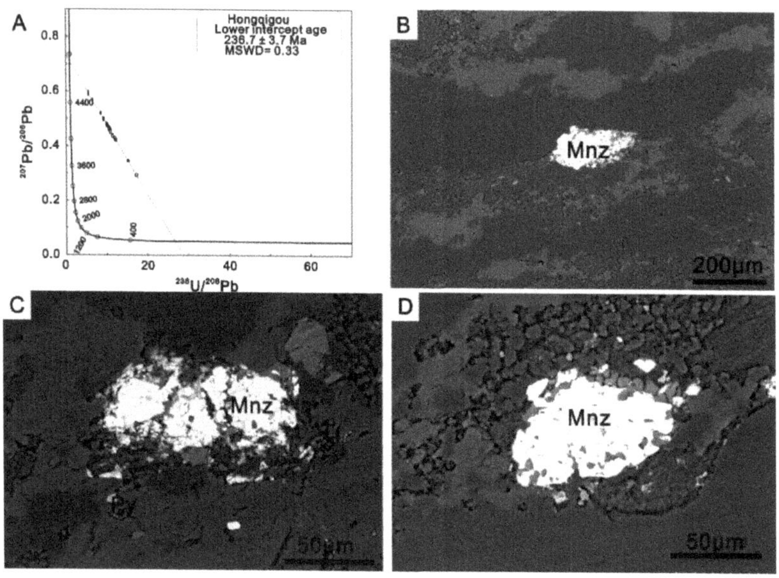

Figure 7. (**A**) The U–Pb Tera–Wasserburg Concordia diagram of monazite in the Hongqigou deposit; (**B–D**) BSE images of typical monazite grains in stage 2 ores of the Hongqigou deposit. Abbreviations: Mnz = monazite.

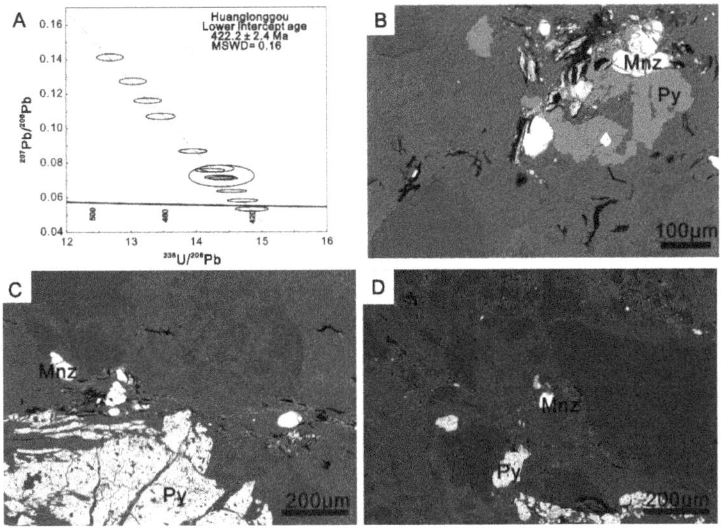

Figure 8. (**A**) The U–Pb Tera–Wasserburg Concordia diagram of monazite in the Huanglonggou deposit; (**B–D**) BSE images of typical monazite grains coexist with pyrite in stage 2 ores of the Huanglonggou deposit. Abbreviations: Mnz = monazite, Py = pyrite.

5.2. Generation and Texture of Pyrite

Based on optical characteristics and BSE observation, these kinds of pyrite and pyrrhotite were recognized: Py0 (pyrite in ultra-mylonite), Po (pyrrhotite in schist), Py1, Py2, Py3 (pyrite in ores of different stages from early to late) (Figure 9).

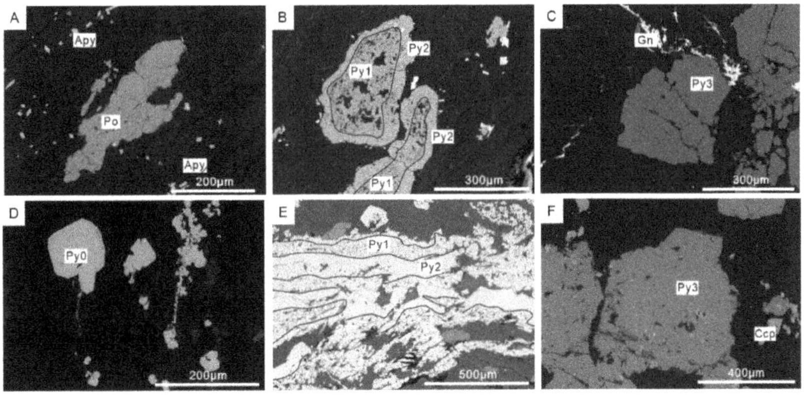

Figure 9. BSE images of sulfides of each stage. (**A**) Pyrrhotite in the schist of wall rock; (**B**) Rough and porous Py1 was contained by smooth Py2 in the Hongqigou deposit; (**C**) Py3 coexist with galena in stage 3 in the Hongqigou deposit; (**D**) Py0 in ultra-mylonite; (**E**) Rough and porous Py1 was contained by smooth Py2 in the Huanglonggou deposit; (**F**) Py3 coexist with chalcopyrite in stage 3 in the Huanglonggou deposit. Abbreviations: Apy = arsenopyrite, Ccp = chalcopyrite, Gn = galena, Po = pyrrhotite, Py = pyrite.

Py0 (pyrite in ultra-mylonite) and Po (pyrrhotite in schist) occurs as subhedral–anhedral crystals, which distribute parallelly.

Pyrite 1 (Py1) occurs as elongated anhedral crystals, which distribute parallelly, this may be a result of the deformation process. Numerous randomly oriented silicate inclusions dominated by sericite and quartz can be observed within Py1. This kind of pyrite commonly occurs in alteration zones, often being the core with Py2 and Py3 surrounding it as a rim. This suggests that Py1 is the earliest pyrite accompanying the pyrite–sericite–quartz alteration (stage 1).

Pyrite 2 (Py2), with little silicate inclusions, commonly overgrows Py1, and the pyrite may also be deformed. It occurs as coarse subhedral–anhedral grains and often has a smooth surface. Some Py1 grains without Py2 rims are accompanied with fine quartz crystals and surrounded by later coarse quartz veins, indicating Py2 precipitated during stage 2.

Pyrite 3 (Py3) occurs either as anhedral or subhedral clusters without zoning. It shows paragenetic relationships with chalcopyrite, galena, and sphalerite. These characteristics suggest that Py3 formed during stage 3.

5.3. Trace Element Distribution in Sulfide

5.3.1. Mapping of the Py1 and Py2 in the Hongqigou and Huanglonggou Deposits

The relationship between the two generations of pyrite is shown in Figure 9 where Py1 is earlier than Py2, as clearly distinguished in both optical and BSE images.

In the Hongqigou deposit, Py2 (rim) has richer As, Au, Cu and Bi and lower Co, Ni, Pb, Sb than that of Py1 (core) (Figure 10). The distribution of Pb in Py1 and the distribution of Co and Ni seem not uniform, which suggests the existence of mineral inclusions (Figure 10). We can find that the Au content in arsenopyrite is relatively higher than that of pyrite (Figure 10), suggesting that Au in the Hongqigou deposit may be carried by Au–As complexes.

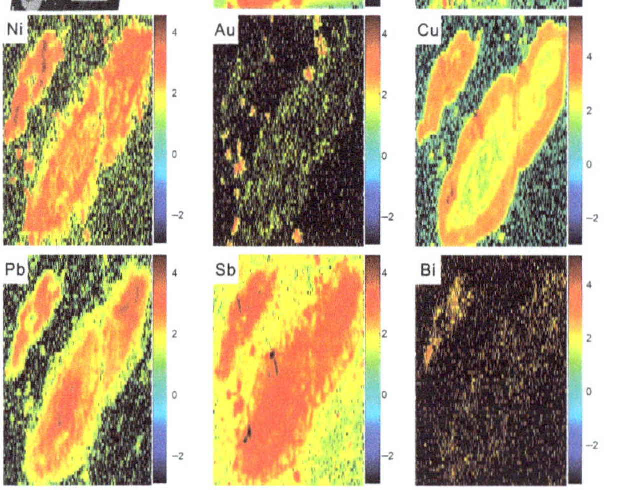

Figure 10. The LA-ICP-MS elemental mapping of Py in the Hongqigou deposit, BSE image of the selected pyrite, other pictures: element distribution of each element.

In the Huanglonggou deposit, Py2 (smooth face) has richer Sb and lower As, Co, Ni, Cu, Pb, Bi than that of Py1 (rough face) (Figure 11), and the distribution of Au is similar in these two kinds of pyrite. Some inclusions that are rich in Cu, Pb may exist in Py1 because of their nonuniform distribution (Figure 11).

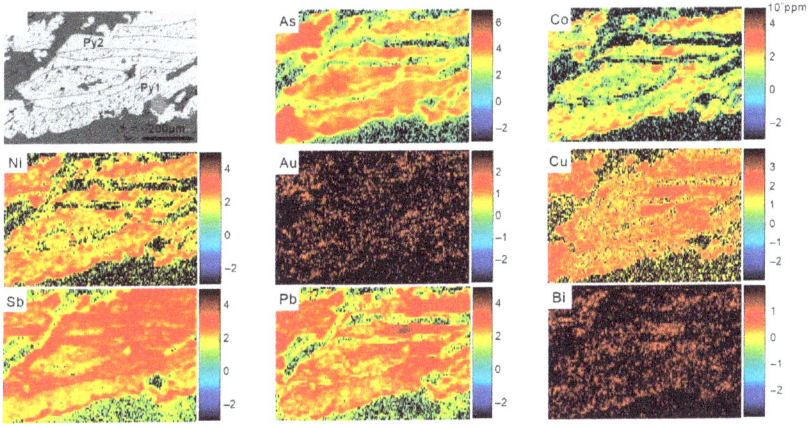

Figure 11. The LA-ICP-MS elemental mapping of Py in the Huanglonggou deposit, BSE image of the selected pyrite, other pictures: element distribution of each element.

5.3.2. In Situ Trace Element Analysis

Pyrite is one kind of important gold-bearing mineral. Hence, the Au content in different types of pyrite is crucial. We determined the trace elements in pyrites of each stage. We also determined the trace elements of sulfides in wall rock (pyrrhotite in schist and pyrite in ultra-mylonite) for contrast.

A total of 68 analyzed spots were obtained on pyrite and pyrrhotite among different stages. It is important and necessary to avoid the influence and faults carried by mineral inclusions (e.g., native gold, silicate or sulfide inclusions) when designing analysis spots, while some inclusions below the surface are difficult and unavoidable to detect, which brought some outliers. Absolute concentrations (ppm) of selected trace elements in pyrite are reported in Table 2 and Appendix A Table A1; the significant changes of trace elements are illustrated in Figures 12 and 13.

Table 2. The summarized LA-ICP-MS analytical results of trace elements for different generations of pyrite in each stage and pyrrhotite in schist of wall rock in the WGF (ppm, bdl = below detection limit).

		^{59}Co	^{60}Ni	^{65}Cu	^{66}Zn	^{75}As	^{107}Ag	^{121}Sb	^{125}Te	^{197}Au	^{208}Pb	^{209}Bi
Pyrrhotite in schist (n = 8)	Min	326.323	591.205	bdl.	bdl.	bdl.	bdl.	0.178	bdl.	bdl.	0.155	0.367
	Max	491.131	1073.904	1.949	7.117	4.594	2.240	15.199	0.461	0.124	8.720	13.182
	Median	431.729	812.913	0.773	2.840	1.818	1.093	4.338	0.077	0.052	3.069	4.630
Pyrite in ultra-mylonite (n = 8)	Min	73.199	30.788	25.344	0.937	455.063	2.355	bdl.	114.985	bdl.	bdl.	23.681
	Max	2090.737	1663.032	2124.885	8.683	6884.256	18.771	3.657	196.658	2.841	0.512	110.829
	Median	925.406	566.395	554.571	5.002	2371.613	6.766	1.829	155.343	0.494	0.138	47.744
Py1 in Hongqigou (n = 6)	Min	23.159	25.736	119.957	bdl.	21,440.451	bdl.	80.295	bdl.	0.790	17.017	0.084
	Max	55.921	103.544	360.547	3.328	56,807.293	2.491	480.888	2.123	1.691	148.769	0.256
	Median	34.156	50.335	234.019	1.568	35,862.411	1.004	234.779	0.623	1.337	65.914	0.154
Py2 in Hongqigou (n = 10)	Min	57.332	102.625	185.494	bdl.	40,115.177	bdl.	51.747	bdl.	0.263	10.809	0.068
	Max	159.267	242.640	607.798	146.296	76,993.084	5.901	491.204	1.869	3.447	121.852	0.621
	Median	90.756	157.615	423.802	16.280	54,278.866	1.303	189.854	0.338	1.023	46.533	0.358

Table 2. *Cont.*

		^{59}Co	^{60}Ni	^{65}Cu	^{66}Zn	^{75}As	^{107}Ag	^{121}Sb	^{125}Te	^{197}Au	^{208}Pb	^{209}Bi
Py3 in Hongqigou (n = 9)	Min	0.036	bdl.	0.737	bdl.	2262.859	bdl.	1.632	bdl.	bdl.	3.974	0.638
	Max	46.389	21.510	15.724	3.974	5223.606	9.880	265.700	8.308	0.751	857.226	35.190
	Median	10.758	7.550	3.483	0.981	3363.130	2.742	54.684	2.031	0.221	139.826	6.971
Py1 in Huanglonggou (n = 10)	Min	14.371	16.091	10.330	bdl.	671.534	bdl.	bdl.	99.793	bdl.	bdl.	60.713
	Max	203.787	313.391	86.059	15.131	8871.994	47.119	29.734	916.985	2.416	1.414	1210.961
	Median	88.777	136.476	46.229	5.476	4503.607	20.885	5.286	366.128	0.501	0.592	568.107
Py2 in Huanglonggou (n = 8)	Min	0.707	4.625	1.232	bdl.	250.834	bdl.	bdl.	10.803	bdl.	bdl.	4.263
	Max	135.063	147.200	54.534	79.576	13,275.196	19.131	11.649	460.266	1.229	0.592	269.836
	Median	26.475	58.107	20.032	11.410	4338.190	6.487	3.080	152.824	0.310	0.369	95.117
Py3 in Huanglonggou (n = 9)	Min	7.310	2.107	2.135	bdl.	612.995	bdl.	bdl.	3.573	bdl.	0.128	15.335
	Max	27.731	69.971	14.968	5.692	11,409.009	13.659	11.823	42.921	2.615	4.304	1853.086
	Median	11.855	19.506	7.214	1.188	5566.746	9.096	3.438	16.193	0.599	1.714	452.719

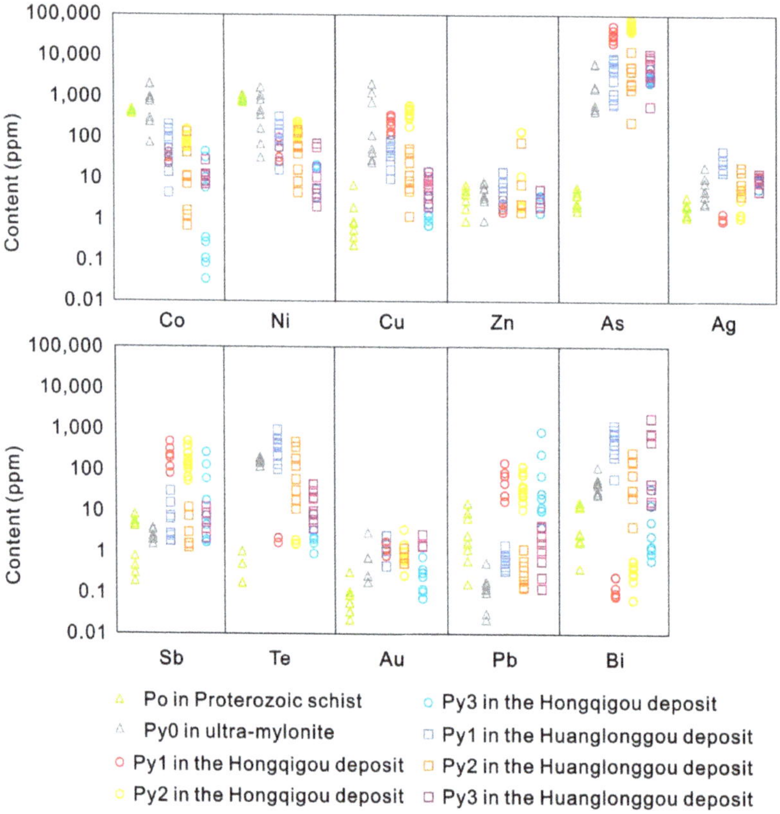

Figure 12. The variation ranges of trace elemental contents in pyrite and arsenopyrite determined by LA-ICP-MS.

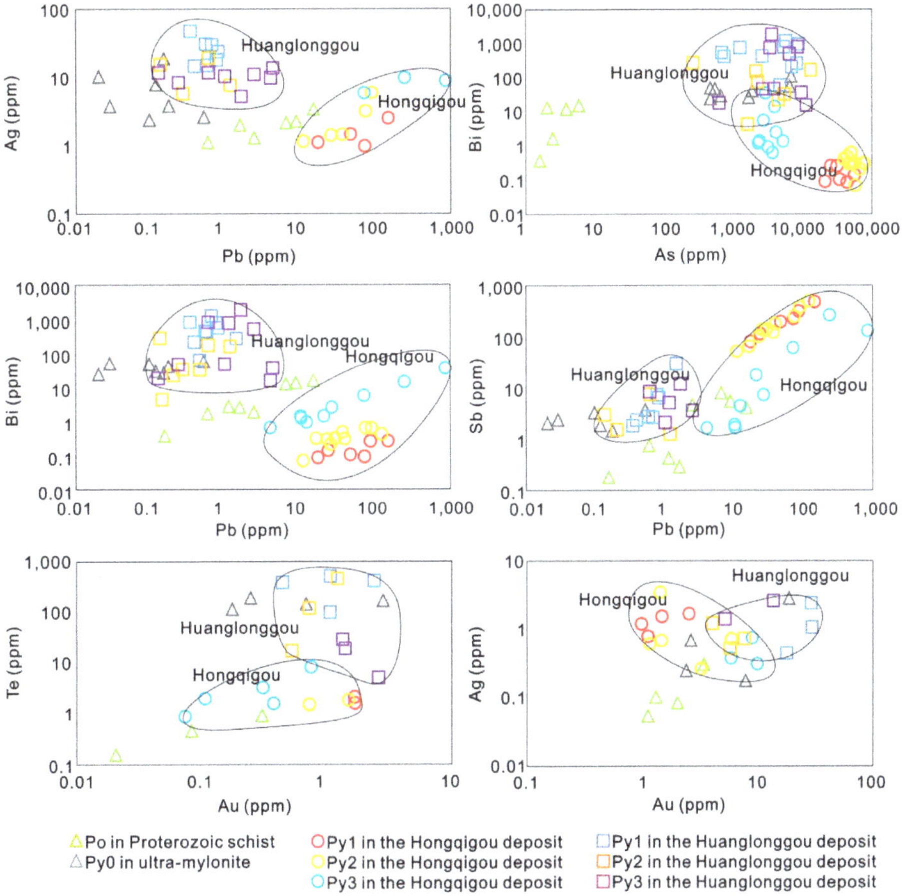

Figure 13. The selected binary plots showing the geochemical characteristics of pyrite from each stage of ores in the Hongqigou and Huanglonggou deposits, ultra-mylonite, and pyrrhotite in schist of wall rock.

5.4. Sulfur Isotopic Data on Sulfide

Sulfur isotopic data of seven types of pyrite are compiled in Table 3 and shown in Figure 14. The pyrrhotite in schist and pyrite in ultra-mylonite have $\delta^{34}S$ values and range from +6.92‰ to +9.90‰ and 2.39‰ to 3.48‰, respectively. The $\delta^{34}S$ values of Py1 Py2 and Py3 of the Hongqigou deposit range from +1.00‰ to +1.19‰, +0.95‰ to +1.22‰ and +10.91‰ to +11.66‰, respectively. The $\delta^{34}S$ values of Py1, Py2 and Py3 of the Huanglonggou deposit range from −2.94‰ to −0.68‰, +4.00‰ to +4.74‰ and +2.75‰ to +3.63‰, respectively.

Table 3. The summarized LA-ICP-MS analytical results of sulfur isotope ratios for different generations of pyrite in each stage and pyrrhotite in schist of wall rock in the WGF (ppm, bdl = below detection limit).

Sample Name	δ^{34}Sv-CDT (‰)	Delta-2SE	Sample Name	δ^{34}Sv-CDT (‰)	Delta-2SE
Schist-Po-1	9.30	0.08	Hongqigou-Py3-3	11.52	0.10
Schist-Po-2	8.95	0.11	Hongqigou-Py3-4	11.66	0.09
Schist-Po-3	7.16	0.09	Hongqigou-Py3-5	11.61	0.10
Schist-Po-4	7.00	0.08	Hongqigou-Py3-6	10.91	0.09
Schist-Po-5	6.92	0.10	Hongqigou-Py3-7	11.48	0.09
Schist-Po-6	9.90	0.09	Hongqigou-Py3-8	11.54	0.10
Schist-Po-7	7.68	0.10	Hongqigou-Py3-9	11.28	0.11
Schist-Po-8	9.67	0.08	Huanglonggou-Py1-1	−1.63	0.10
Ultramylonite-Py-1	2.70	0.22	Huanglonggou-Py1-2	−2.94	0.11
Ultramylonite-Py-2	2.92	0.19	Huanglonggou-Py1-3	−1.65	0.09
Ultramylonite-Py-3	2.39	0.27	Huanglonggou-Py1-4	−1.78	0.11
Ultramylonite-Py-4	3.30	0.20	Huanglonggou-Py1-5	−1.95	0.11
Ultramylonite-Py-5	3.48	0.19	Huanglonggou-Py1-6	−2.61	0.10
Ultramylonite-Py-6	2.78	0.20	Huanglonggou-Py1-7	−1.72	0.09
Ultramylonite-Py-7	2.68	0.17	Huanglonggou-Py1-8	−2.35	0.12
Ultramylonite-Py-8	2.87	0.16	Huanglonggou-Py1-9	−2.34	0.10
Hongqigou-Py1-1	1.07	0.10	Huanglonggou-Py1-10	−0.68	0.21
Hongqigou-Py1-2	1.09	0.09	Huanglonggou-Py2-1	4.00	0.09
Hongqigou-Py1-3	1.13	0.09	Huanglonggou-Py2-2	4.74	0.09
Hongqigou-Py1-4	1.19	0.10	Huanglonggou-Py2-3	4.67	0.07
Hongqigou-Py1-5	1.00	0.08	Huanglonggou-Py2-4	4.55	0.08
Hongqigou-Py1-6	1.01	0.09	Huanglonggou-Py2-5	4.15	0.08
Hongqigou-Py2-1	1.16	0.09	Huanglonggou-Py2-6	4.70	0.08
Hongqigou-Py2-2	1.17	0.08	Huanglonggou-Py2-7	4.17	0.08
Hongqigou-Py2-3	1.02	0.07	Huanglonggou-Py2-8	4.26	0.08
Hongqigou-Py2-4	0.95	0.08	Huanglonggou-Py3-1	3.12	0.08
Hongqigou-Py2-5	1.08	0.09	Huanglonggou-Py3-2	3.14	0.10
Hongqigou-Py2-6	1.22	0.09	Huanglonggou-Py3-3	3.06	0.11
Hongqigou-Py2-7	1.15	0.10	Huanglonggou-Py3-4	2.75	0.10
Hongqigou-Py2-8	1.20	0.08	Huanglonggou-Py3-5	3.63	0.11
Hongqigou-Py2-9	0.98	0.09	Huanglonggou-Py3-6	2.97	0.10
Hongqigou-Py2-10	1.15	0.08	Huanglonggou-Py3-7	2.81	0.12
Hongqigou-Py3-1	11.62	0.09	Huanglonggou-Py3-8	3.21	0.11
Hongqigou-Py3-2	11.43	0.10	Huanglonggou-Py3-9	2.89	0.09

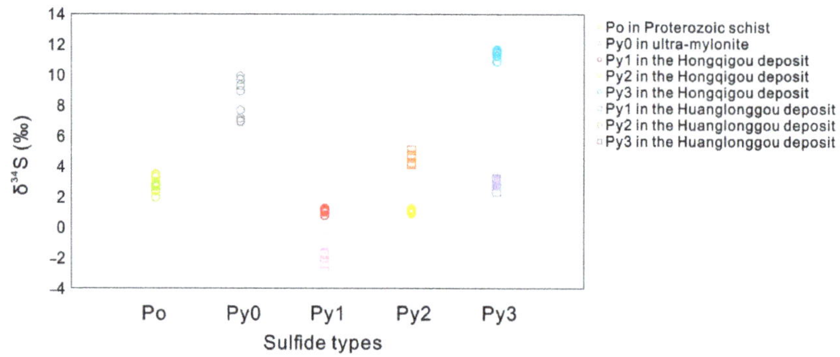

Figure 14. The δ^{34}S values of pyrite from each stage of ores in the Hongqigou and Huanglonggou deposits, ultra-mylonite, and pyrrhotite in schist of wall rock.

6. Discussion

6.1. Evolution of Pyrite Textures and Compositions

6.1.1. Mineral Paragenesis and Evolution

The classification of the different mineralization stages in this study is based on a detailed analysis of mineral assemblage, structure, and intersecting relationship. The main metallogenic stages 1 and 2 were represented by fractured altered rock type mineralization, during which there was a strong tectonic deformation, resulting in the fragmentation of the orebody and the orientation of minerals. Polymetallic sulfides were rich in stage 3, including chalcopyrite, sphalerite, and galena (Figure 9). At stage 4, stibnite, quartz, and calcite appeared. Pyrite occurs in all mineralization stages; arsenopyrite is often apparent in stages 1 and 2, which are rich in gold, while the amount of gold and arsenopyrite is negligible in stages 3 and 4. From stages 1 to 4, the complete metallogenic evolution process from early-to-late stages can be traced: the content of pyrite decreased, and their crystal morphology changed from oriented anhedral grains to euhedral—subhedral grains (Figure 9). Thus, our observations and results suggest that the metallogenic environment changed from a relatively stable environment in the early stage to a compression background, and then to a repeatedly turbulent and open environment, and finally returned to stability.

6.1.2. Relationship and Evolution of Trace Elements in Hongqigou and Huanglonggou Deposits

According to a vast literature, analysis of the trace elements (i.e., As, Au, Ag, Bi, Co, Cu, Pb, Zn, Ni, Sb, and Te) in pyrite can be used to distinguish the occurrence states of each element [48–51], know about the composition changes of ore-forming fluids [52–55], and analyse the physical and chemical changes of ore-forming fluids and growth history [50,56–59].

Some scholars suggest that gold partitioning into pyrite depends on As availability, as the Au–As coupled substitution is a key for Au precipitation (e.g., [60–64]). However, the arsenian pyrite (high As concentration) does not always record high Au values because it is obvious that the As-rich fluids are not invariably Au-bearing (e.g., [63,65]). These phenomena can be found in this paper: Au contents in pyrite from the Hongqigou and Huanglonggou deposits were similar, while the As content in Hongqigou deposit was much higher than that of Huanglonggou. Co and Ni enter the lattice via isomorphic substitution of Fe, whereas As, Te, Sb, and Bi enter the lattice substituting for S. The concentrations of Te, Sb, and Bi are extremely low and possibly caused by the high As concentration. We can find that the Te, Sb, and Bi contents of Huanglonggou were much higher than that of Hongqigou, whereas the As content represents a contrasting tendency. Hence, the compositions of the ore-forming fluids of the Hongqigou and Huanglonggou deposits seem to be different, and this may suggest that they formed in two different metallogenic events. The Au and As are rich in Py2 of the Hongqigou deposit, while in the Huanglonggou deposit, the contents of Au and As in Py1 were higher than that of Py2, which may suggest that the main Au metallogenic event in the Hongqigou happened in magmatic-hydrothermal process, while the wall-rock alteration may be the key to the mineralization in the Huanglonggou deposit. We would draw these conclusions after discussing metals ratios and sulfur isotopes. This might be further solid evidence of multiple mineralization beside ore-forming age.

Except for high Co and Ni values, other elements in Po were low (Figures 12 and 13). The contents of Co, Ni, Cu, Te, and Bi in Py0 are high, while that of Zn, As, Sb, and Pb are relatively low. In the Huanglonggou deposit, the contents of Co, Ni, Cu, Zn, Ag, and Te in Py1, Py2, and Py3 decrease successively, indicating that metal in ore-forming fluids gradually unloaded from early stage to late stage. In Hongqigou deposit, Co, Ni, Cu, and As elements increased slightly from Py1 to Py3 and then decreased sharply, while the contents of Sb and Bi gradually increased, indicating that strong precipitation occurred at stage 2. By comparing the trace elements in different stages of the Hongqigou and Huanglonggou deposits, it can be found that Cu, As, Sb, and Pb elements in the Hongqigou deposit are higher than those in the Huanglonggou deposit, especially in Stage 1 and Stage

2 (Figures 12 and 13). The Ag, Te, and Bi elements in the Hongqigou deposit are lower than those in the Huanglonggou deposit, indicating that the composition and evolution trend of the ore-forming fluids of the two deposits can be distinguished easily (Figures 12 and 13).

6.1.3. Source of Ore-Forming Materials

The ratio between different elements was generally applied upon determining the source of pyrite, and even deducing the evolution of the metallogenic stage and the genesis of the deposit. The Co/Ni ratio of pyrite has a great signature on ensuring the change of metallogenic environment, especially on the source of hydrothermal gold deposits [52,66,67]. The contents of Co and Ni in pyrite can be relatively high in the deposits that are related to mafic intrusions and contributed by degassing [68,69]. The high Co/Ni ratio of magmatic source is generally higher than that of pyrite formed in sedimentary environment [52,68,69]: magmatic–hydrothermal pyrite exhibits high Co/Ni ratios of > 1, whereas sedimentary–diagenetic pyrite has low Co/Ni ratios of < 1 [69]. The correlation between Au and As in pyrite can be applied for knowing the types of gold deposits: generally, the Au/As ratio of pyrite in orogenic gold deposits is relatively low (< 0.004, [53]), that of Carlin-type and magmatic gold deposits seems to be high [48]. The $\delta^{34}S$ values of pyrite are usually used as tracers of the sources of ore-forming materials and discussing the ore-forming processes [70,71]. Different sulfur reservoirs have their typical $\delta^{34}S$ value ranges: (1) magmatic source of sulfur (magma: $-3‰$ to $+7‰$, [72,73]; andesite, $1.0 \pm 6.1‰$, [74]; granitoids: $2.6 \pm 2.3‰$, [75]) (2) sedimentary rocks reduced sulfur (<0, [76]) and (3) marine (20‰, [76]). Generally, the source of ore-forming materials and fluids of orogenic gold deposits was commonly considered to be of a metamorphic origin with some materials from intrusive rocks [77–79]. The majority $\delta^{34}S$ values of orogenic gold deposits around the world have a range from 0 to +10‰ [80]. Hence it is difficult to determine the source of ore-forming materials and the evolution of ore deposits simply by using sulfur isotopic composition. Accordingly, we need to determine the source of ore-forming materials and processes by combining some of the trace elements and S isotopic composition of pyrite and pyrrhotite in ores of each metallogenic stages and wall rocks.

The Co/Ni ratios of Po were lower than 1 (0.457~0.613), suggesting the schist were mainly formed in sedimentary and metamorphic process; the Co/Ni ratios of Py0 have a wide range of 0.233~9.706 and have both magmatic and sedimentary characters, which represent that the ultra-mylonite formed during ductile-brittle shearing, when some magma was transported upwards. The available Au/As ratios in Po have a range of 0~0.027, suggesting a mixed origin of magma and metamorphism; and the Au/As ratios Py0 have a range of $0 \sim 1.7 \times 10^{-3}$, all of them lower than 0.004, suggesting obvious characteristics of orogenic gold deposit. The $\delta^{34}S$ values of Po and Py0 in schist and ultra-mylonite shows narrow ranges (+1.99‰ to +3.48‰ and 6.92‰ to 9.90‰, respectively). These characteristics suggest that Po originated from deep magmatic fluids and metamorphic fluids, and that Py1 mainly came from metamorphic fluids.

Hongqigou

The Co/Ni ratios of Py1 and Py2 are generally lower than 1 (0.540~1.003, 0.399~0.714, respectively), which are similar to that of Po, suggesting a similar source: sedimentary/metamorphic environment; the Co/Ni ratios of Py3 have a wide range of 0.066~2.359, suggesting a mixed source. The Au/As ratios in Py1, Py2, and Py3 have ranges of $1.8 \times 10^{-5} \sim 6.5 \times 10^{-5}$, $0.5 \times 10^{-5} \sim 5.1 \times 10^{-5}$, and $0 \sim 2.6 \times 10^{-4}$, respectively, all of which are lower than 0.004, suggesting obvious characteristics of orogenic gold deposit.

The $\delta^{34}S$ values of Py1 (+1.00‰ to +1.19‰) and Py2 (+0.95‰ to +1.22‰) in Hongqigou vary in a narrow range, suggesting a magmatic source. The $\delta^{34}S$ values of Py3 (+10.91‰ to +11.66‰) in Hongqigou seems to be higher than those of Py1 and Py2, indicating a participation of wall rock materials, because the $\delta^{34}S$ values of Py3 are quite similar to those of Po in schist.

Huanglonggou

The Co/Ni ratios of Py1, Py2, and Py3 have wide ranges of 0.223~3.352, 0.041~2.380, and 0.1771~5.826, respectively, which suggest a mixed source of sedimentary/metamorphic and magmatic origin. The Au/As ratios in Py1, Py2, and Py3 have ranges of 0~3.6×10^{-4}, 0~5.6×10^{-4}, and 0~4.1×10^{-4}, respectively, all of them lower than 0.004, suggesting obvious characteristics of orogenic gold deposit.

The $\delta^{34}S$ values of Py1 (−2.94‰ to −0.68‰), Py2 (+4.00‰ to +4.74‰), and Py3 (+2.75‰ to +3.63‰) in Huanglonggou vary in a narrow range, but there are certain differences between them, suggesting the magmatic–metallogenic process may be characterized by a multi-stage pulse, into which some wall rock materials have been mixed.

6.2. Tectonic Evolution and Mineralization

Researchers [81] promoted a long-lived subduction and accretionary tectonic model to interpret tectonic evolution of the EKOB. There were a subduction and accretionary complex and a back-arc basin during the Early Paleozoic—Triassic period. The Central Kunlun Belt was split from the Qaidam Block to an Early Paleozoic island-arc due to the spreading of the Qimantagh back-arc basin in response to the northward subduction of the Kunlun Ocean. During the Early Paleozoic—Triassic period, the EKOB evolved into a trench-arc-back-arc basin tectonic system and composite orogenic processes, which include a wide accretionary complex and long-lived northward subduction.

During ca. 460–430 Ma (Figure 15A), coevally with the ongoing northward subduction of the Kunlun oceanic slab, the oceanic crust of the Qimantage back-arc basin began to subduct towards the north beneath the Qaidam terrane [81], These processes indicate that the arc-type and subduction-related magmatic events were widespread and intense during the Ordovician–Silurian period in the EKOB. Due to northward subduction of the Kunlun oceanic slab, the island-arc of EKOB has been thickened and developed numerous gabbroic–granitic intrusions and its products [81,82].

When the Qimantage back-arc basin closed, the Central Kunlun island-arc and the Qaidam Block collided at ca. 425 Ma (Figure 15B), and the ocean crust detached from the continental crust, these processes generated syn-collisional plutons in the northern Central Kunlun Belt and the North Qimantage Belt [81]. The Huanglonggou gold deposit in this study (422.2 ± 2.4 Ma) and the Tuolugou Co–Au polymetallic deposit in the EKOB formed in this period of tectonic setting.

At ca. 400 Ma (Figure 15C), the EKOB began to delaminate from the overthickened lithosphere and a series of post-collisional plutonic magmas intruded [81,83].

During the Carboniferous–Permian period (Figure 15D), because of the delamination and post-orogenic collapse, the regional extent of the belt had transitioned from compression to extension, as well as the northward subductions of the Kunlun oceanic slab [81]. The gold deposits of Annage, Guoluolongwa in the EKOB formed during this period.

During ca. 270–250 Ma (Figure 15E), the northward subduction of the Kunlun oceanic plate had been reactivated, which resulted in the intrusions of gabbro, diorite and voluminous I-type granite in the Central Kunlun Belt [81].

During ca. 250–220 Ma (Figure 15F), slab break-off of the ongoing northward subduction of the Kunlun oceanic plate had taken place, which is evidenced by massive granitoids with MMEs in the Central Kunlun Belt; the slab break-off led to the upwelling of the mantle-derived magma from the asthenosphere, which was underplated beneath the crust of the Central Kunlun arc; this can be supported by the abundant granitic intrusions [81]. This period is the most important and largest scale metallogenic event in the EKOB. The Hongqigou gold deposit in this study (236.7 ± 3.7 Ma) and other major gold/polymetallic deposits, such as Heishigou, Kengdenongshe, Xingshugou, Baidungou, Yanjingou, Dashuigou, Asiha, Naomuhun, Kendekeke, Xizangdagou, Dachang, were all formed in this period. We prepose that its metallogenic event is related to the subduction of the EKOB, an opinion which is also supported in some previous studies [84].

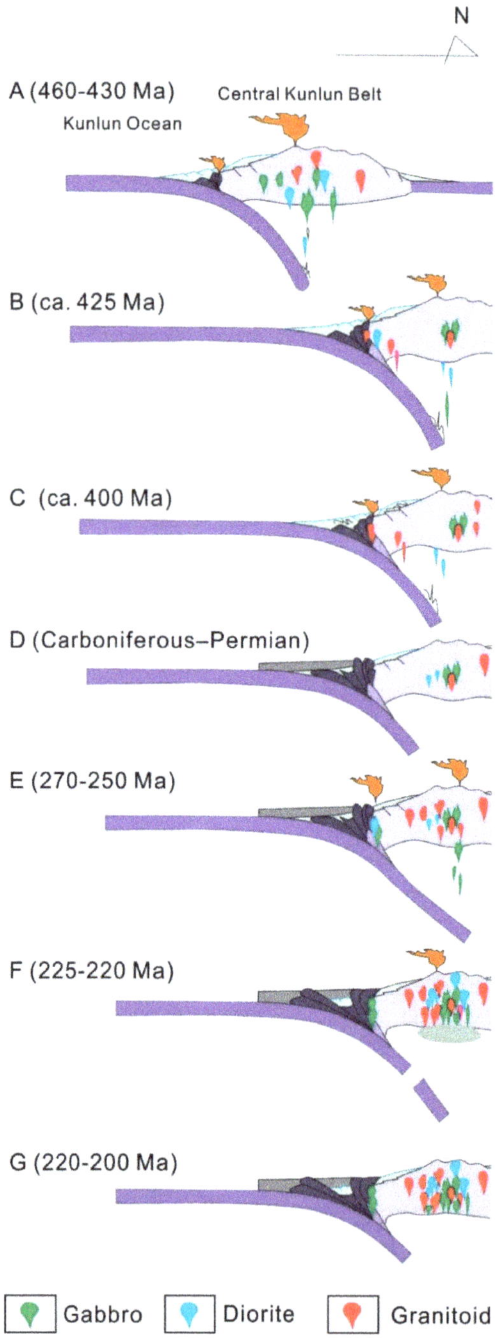

Figure 15. Schematic diagram showing the metal deposits formed in the geodynamic setting in the EKOB (modified after [81]).

During ca. 220–200 Ma (Figure 15G), the Central Kunlun arc underwent a collapse in response to underplating, and A-type granites generated at that same time, which suggests a post-collision setting [81]. Manite gold deposit in the EKOB formed during this period.

Therefore, there are two metallogenic epochs in WGF: Silurian–Devonian deposits were mainly hosted in the ductile-brittle shear zones, while the Middle–Late Triassic deposits were mainly hosted in the NW-trending brittle fractures.

6.3. Magmatism and Mineralization

In terms of the formation time, orogenic gold deposits show several prominent peaks in the late Archean to Paleoproterozoic and Phanerozoic orogenic belts [77,85]. The EKOB records two epochs of orogenesis, corresponding to the evolution of the Neoproterozoic to late Paleozoic Proto-Tethys Ocean and late Paleozoic to Triassic Paleo-Tethys Ocean [27]. In this paper, we determined the monazite U–Pb age of the Huanglonggou deposit (422.2 ± 2.4 Ma) and the Hongqigou deposit (236.7 ± 3.7 Ma), indicating that there were two epochs of magmatic—metallogenic events happening in the WGF.

The age data of the gold deposits in the EKOB have been collected (Table 4), which build up an objective geochronological framework for gold mineralization in the EKOB. There are two epochs of gold mineralization that have been recognized in the EKOB: Silurian–Devonian and Middle–Late Triassic events, and they are distinct geographically in ore fields and belts. Only a few of deposits in the EKOB reported to be formed during Silurian–Devonian period, such as the Annage, Guoluolongwa gold deposits and Tuolugou Co–Au deposit (429–374 Ma). The Middle–Late Triassic deposits are mainly located in the EKOB, such as Baidungou, Heishigou, Kengdenongshe, Xingshugou, Hongqigou, Yanjingou Dashuigou, Asiha, Naomuhun, Kendekeke, Xizangdagou, Dachang, Manite gold deposits (244–213 Ma). The granitoids in the Wulonggou gold district mostly display zircon U–Pb ages of ca. 438–390 Ma and 260–211 Ma [31,86–99]. Consequently, the main gold metallogenic event of the EKOB is Middle–Late Triassic.

Table 4. Ages of the EKOB gold deposits.

Gold Deposit	Sample	Mineral for Age Determination and Method	Age (Ma)	Reference
Annage	Pyrite	Pyrite Re–Os	383 ± 8	[100]
Asiha	Ore	Monazite U–Pb	227.1 ± 6.5	[101]
Asiha	Granite porphyry	Zircon U–Pb	1	[101]
Caiyuanzi	Granodiorite porphyry	Zircon U–Pb	210.4 ± 1.9	[102]
Dashuigou	Tonalite	Zircon U–Pb	239.5 ± 0.9	[103]
Gerizhuotuo	Granodiorite	Zircon U–Pb	213.25 ± 0.62	[86]
Guoluolongwa	Pyrite and chalcopyrite	Sulfide Re–Os	374 ± 15	[100]
Baidungou	Quartz diorite	Zircon U–Pb	241.08 ± 0.90	[3]
Hongqigou	Dolerite	Zircon U–Pb	242.8 ± 2.1	[104]
Heishigou	Dolerite	Zircon U–Pb	244.1 ± 1.6	[3]
Kendekeke	Dacite tuffs	Zircon U–Pb	227.1 ± 1.2	[87]
Kendekeke	Monzogranite	Zircon U–Pb	229.5 ± 0.5	[88]
Kendekeke	Monzogranite	Zircon U–Pb	230.5 ± 4.2	[89]
Kengdenongshe	Ore-bearing rhyolitic tuffs	Zircon U–Pb	243.3 ± 1.6	[90]
Naomuhun	Sericite	Sericite Ar–Ar	227.84 ± 1.13	[91]
Naomuhun	Quartz diorite	Zircon U–Pb	235.8 ± 0.8	[91]
Xingshugou	Quartz albite porphyry	Zircon U–Pb	243.4 ± 1.9	[105]
Yanjingou	Ore	Quartz inclusion Rb–Sr	237 ± 3	[4]
Tuolugou	Pyrite	Pyrite Re–Os	429 ± 29	[106]

As for WGF, some previous studies reported the mineralization ages: [107,108] obtained the mica Ar-Ar age of 236 Ma and 242 Ma from the WGF, respectively; [4,109,110] obtained the zircon FT age, the fluid inclusion Rb-Sr age, the monazite U–Pb age and the zircon U–Pb age of 223, 237, 239, 244 Ma from the Yanjingou gold deposit; the 236.7 ± 3.7 Ma age of Hongqigou deposit is similar with them. During that time, the Paleo-Tethys Ocean was under a transition period of flat subduction and slab rollback, which resulted in orogeny and magmatic activity, and released fluids from the subducted slab or the hydrated mantle wedge [111,112] which were conducive for generating the ore deposits in the region. However, the 422.2 ± 2.4 Ma age of the Huanglonggou deposit provide solid evidence that there was a metallogenic event during Paleozoic.

Based on the results of trace elements of pyrite, sulfur isotopic composition of sulfides and monazite U–Pb dating, we suggest that the WGF has undergone multiple-complex metallogenic processes based on multi-proxy studies, and that it can be interpreted as an orogenic gold field related to multiple tectonic–magmatic events.

7. Conclusions

(1) The monazite U–Pb ages of the Huanglonggou and Hongqigou deposits were 422.2 ± 2.4 Ma (MSWD = 0.16) and 236.7 ± 3.7 Ma (MSWD = 0.33), suggesting that there two epochs of metallogenic events that occured in the WGF.

(2) The ore-forming process can be divided into four hydrothermal stages: a quartz–pyrite stage (stage 1), a quartz–pyrite–arsenopyrite–chalcopyrite stage (stage 2), a quartz–galena–sphalerite–pyrite ± arsenopyrite ± chalcopyrite stage (stage 3) and a quartz–stibnite–carbonate stage (stage 4).

(3) Stages 1 and 2 are the main gold mineralization stages, wherein Au and As have a close genetic relationship. The Hongqigou and Huanglonggou deposits seem to be formed in different metallogenic events due to the contrast on the trace element compositions in pyrite.

(4) The source of ore-forming materials and fluids represented by δ34S and Co/Ni and Au/As ratios indicate that the Hongqigou and Huanglonggou deposits have apparent characteristics of orogenic gold deposit, and magmatic events happened during Paleozoic and Mesozoic made crucial contribution to their mineralization.

(5) We propose that the WGF went through multiple-complex metallogenic processes based on multi-proxy studies, and that this can be interpreted as an orogenic gold field related to multiple tectonic–magmatic events.

Author Contributions: Conceptualization, Z.Z. and Q.Z.; investigation, H.F., K.Y., X.L., G.L., J.W.; resources, H.X., Z.W., T.P.; writing—original draft preparation Z.Z.; writing—review and editing, Q.Z.; visualization, Q.Z.; supervision, Q.Z., H.F., K.Y.; funding acquisition, H.F., K.Y., Q.Z. All authors have read and agreed to the published version of the manuscript.

Funding: This work was funded by the Second Tibetan Plateau Scientific Expedition and Research Program (STEP) (No. 2019QZKK0801).

Institutional Review Board Statement: Not applicable.

Informed Consent Statement: Not applicable.

Data Availability Statement: Not applicable.

Acknowledgments: We are grateful to anonymous reviewers for constructive suggestions.

Conflicts of Interest: The authors declare no conflict of interest.

Appendix A

Table A1. The LA-ICP-MS analytical results of trace elements for different generations of pyrite in each stage and pyrrhotite in schist of wall rock in the Wulonggou Gold Field.

Sample	^{59}Co	^{60}Ni	^{65}Cu	^{66}Zn	^{75}As	^{107}Ag	^{121}Sb	^{125}Te	^{197}Au	^{208}Pb	^{209}Bi
Schist-Po-1	459.535	884.485	0.361	1.137	2.448	1.104	0.733	0.000	0.054	0.579	1.632
Schist-Po-2	491.131	1073.904	0.530	4.328	3.875	2.134	7.881	0.000	0.000	6.416	12.138
Schist-Po-3	436.918	746.358	0.228	1.853	0.000	1.977	0.284	0.461	0.084	1.560	2.543
Schist-Po-4	461.758	753.520	0.000	0.000	0.000	0.000	0.421	0.000	0.033	1.100	2.688
Schist-Po-5	433.093	832.637	1.453	7.117	2.040	2.240	5.391	0.000	0.000	8.720	13.182
Schist-Po-6	462.730	851.527	0.801	2.995	1.587	0.000	0.178	0.157	0.021	0.155	0.367
Schist-Po-7	326.323	591.205	1.949	5.291	4.594	0.000	15.199	0.000	0.124	3.610	2.674
Schist-Po-8	382.343	769.667	0.863	0.000	0.000	1.288	4.613	0.000	0.101	2.410	1.814
Ultramylonite-Py-1	989.711	158.880	29.096	3.412	1648.837	18.771	0.000	170.268	2.841	0.151	25.861
Ultramylonite-Py-2	73.199	66.455	43.764	3.898	6515.266	2.583	3.657	145.199	0.686	0.512	55.638
Ultramylonite-Py-3	239.699	1027.331	1290.252	8.683	456.654	10.258	2.022	161.739	0.000	0.020	23.681
Ultramylonite-Py-4	778.476	1663.032	54.432	7.999	1829.331	3.779	1.443	115.420	0.000	0.175	40.664
Ultramylonite-Py-5	2090.737	345.773	2124.885	2.923	534.603	2.355	3.283	192.794	0.248	0.096	46.837
Ultramylonite-Py-6	2045.162	798.940	753.049	5.934	455.063	3.733	2.382	196.658	0.000	0.029	49.249
Ultramylonite-Py-7	887.432	439.957	115.751	6.230	648.894	7.784	1.845	114.985	0.176	0.118	29.197
Ultramylonite-Py-8	298.831	30.788	25.344	0.937	6884.256	4.868	0.000	145.684	0.000	0.000	110.829
Hongqigou-Py1-1	32.496	55.501	256.376	1.585	34,711.167	1.453	196.130	0.000	1.545	46.914	0.102
Hongqigou-Py1-2	23.159	33.234	119.957	1.886	21,440.451	0.974	224.104	0.000	1.200	72.416	0.089
Hongqigou-Py1-3	44.046	56.473	200.034	0.000	31,916.973	0.000	313.303	2.123	1.684	86.845	0.253
Hongqigou-Py1-4	55.921	103.544	142.710	3.328	25,950.584	2.491	480.888	1.614	1.691	148.769	0.256
Hongqigou-Py1-5	23.500	27.524	360.547	2.609	56,807.293	0.000	113.957	0.000	1.115	23.523	0.136
Hongqigou-Py1-6	25.818	25.736	324.491	0.000	44,347.997	1.107	80.295	0.000	0.790	17.017	0.084
Hongqigou-Py2-1	84.478	162.908	504.067	0.000	57,046.708	0.000	64.167	0.000	0.668	16.123	0.305
Hongqigou-Py2-2	107.331	175.986	185.494	12.144	40,115.177	0.000	491.204	1.869	1.500	121.852	0.408
Hongqigou-Py2-3	159.267	242.640	366.549	2.229	50,364.569	3.160	243.525	0.000	0.263	74.326	0.621
Hongqigou-Py2-4	157.768	240.426	372.535	146.296	49,913.188	5.901	373.624	1.508	0.722	89.542	0.620
Hongqigou-Py2-5	73.299	102.625	483.004	0.000	76,993.084	0.000	138.039	0.000	0.616	29.056	0.306
Hongqigou-Py2-6	62.817	108.774	463.063	0.000	44,662.812	1.430	120.927	0.000	0.682	36.612	0.471
Hongqigou-Py2-7	81.504	204.076	607.798	0.000	57,651.922	1.140	51.747	0.000	0.606	10.809	0.068
Hongqigou-Py2-8	61.228	102.925	301.078	0.000	46,622.748	0.000	177.871	0.000	0.972	39.262	0.296
Hongqigou-Py2-9	57.332	121.853	589.064	0.000	67,080.552	1.399	135.813	0.000	3.447	25.224	0.203
Hongqigou-Py2-10	62.537	113.939	365.369	2.128	52,337.898	0.000	101.625	0.000	0.748	22.524	0.287
Hongqigou-Py3-1	0.372	5.640	4.065	3.334	3909.772	9.880	265.700	3.257	0.311	245.648	13.984
Hongqigou-Py3-2	32.846	21.510	0.737	0.000	3648.123	0.000	1.632	0.000	0.000	3.974	0.638
Hongqigou-Py3-3	46.389	19.667	1.367	0.000	4124.308	0.000	7.310	0.000	0.248	26.473	2.498
Hongqigou-Py3-4	0.036	0.000	1.285	0.000	2262.859	0.000	1.660	0.000	0.122	10.309	1.213
Hongqigou-Py3-5	0.115	0.000	0.747	0.000	2366.442	0.000	17.782	1.991	0.107	20.720	1.385
Hongqigou-Py3-6	0.088	0.000	2.524	1.517	5223.606	0.000	1.894	0.885	0.074	10.127	1.371
Hongqigou-Py3-7	0.286	0.000	0.985	0.000	3146.679	0.000	4.424	2.242	0.000	12.141	0.922
Hongqigou-Py3-8	10.640	17.532	15.724	3.974	2892.840	8.928	130.690	8.308	0.751	857.226	35.190
Hongqigou-Py3-9	6.045	3.601	3.910	0.000	2693.539	5.868	61.061	1.593	0.378	71.816	5.536
Huanglonggou-Py1-1	26.486	119.034	10.330	8.918	4662.131	0.000	0.000	99.793	1.071	0.454	60.713
Huanglonggou-Py1-2	53.938	16.091	37.423	5.330	2588.104	30.761	0.000	206.446	0.000	0.541	433.007
Huanglonggou-Py1-3	159.548	80.122	45.515	4.152	671.534	23.903	6.375	429.231	0.000	0.801	548.332
Huanglonggou-Py1-4	203.787	203.031	45.706	8.376	1233.971	17.891	7.261	390.036	0.445	0.755	761.660
Huanglonggou-Py1-5	52.195	89.540	27.779	6.579	732.027	14.941	2.718	220.313	0.000	0.568	420.303
Huanglonggou-Py1-6	14.371	58.264	16.535	0.000	6985.995	14.681	2.332	131.251	0.000	0.387	209.036
Huanglonggou-Py1-7	39.947	152.367	74.903	0.000	8270.601	0.000	29.734	916.985	0.000	1.414	264.691
Huanglonggou-Py1-8	152.760	313.391	86.059	2.255	5722.234	30.008	2.643	513.597	1.078	0.656	1210.961
Huanglonggou-Py1-9	84.559	122.308	64.932	4.021	8871.994	29.542	0.000	419.968	2.416	0.000	972.028
Huanglonggou-Py1-10	100.177	210.616	53.104	15.131	5297.483	47.119	1.799	333.662	0.000	0.342	800.335
Huanglonggou-Py2-1	7.537	61.689	6.086	0.000	5498.355	0.000	0.000	56.931	0.000	0.456	32.160
Huanglonggou-Py2-2	10.822	130.019	1.232	0.000	1575.172	0.000	0.000	10.803	0.000	0.143	4.263
Huanglonggou-Py2-3	135.063	56.744	25.632	2.440	250.834	15.419	2.932	178.409	0.000	0.135	269.836
Huanglonggou-Py2-4	1.646	39.854	12.457	0.000	2074.031	7.609	1.236	120.581	0.724	1.153	156.222
Huanglonggou-Py2-5	0.707	4.625	46.142	79.576	2193.181	4.008	11.649	460.266	1.229	0.000	76.231
Huanglonggou-Py2-6	43.340	147.200	54.534	7.673	13,275.196	19.131	7.322	348.155	0.000	0.592	167.483
Huanglonggou-Py2-7	1.268	8.446	5.126	1.594	4423.075	0.000	1.499	30.208	0.000	0.202	21.976
Huanglonggou-Py2-8	11.419	16.276	9.044	0.000	5415.672	5.730	0.000	17.236	0.527	0.271	32.765
Huanglonggou-Py3-1	8.533	4.247	9.788	5.692	3831.076	10.351	2.099	7.733	0.000	0.981	46.958
Huanglonggou-Py3-2	8.410	4.355	7.730	0.000	6464.305	11.060	3.567	20.651	0.000	2.423	503.080
Huanglonggou-Py3-3	7.393	10.818	6.513	2.156	3557.323	5.192	11.823	19.107	1.418	1.619	1853.086
Huanglonggou-Py3-4	12.274	2.107	2.553	0.000	11,409.009	9.809	0.000	3.573	0.000	4.018	15.335
Huanglonggou-Py3-5	27.731	56.888	14.968	0.000	8494.598	11.715	8.356	42.921	0.000	0.593	800.227
Huanglonggou-Py3-6	12.369	69.971	4.023	0.000	2636.416	8.345	0.000	8.120	0.000	0.238	46.526
Huanglonggou-Py3-7	7.310	18.067	2.135	0.000	612.995	11.731	0.000	9.519	0.000	0.128	17.815
Huanglonggou-Py3-8	11.240	5.630	5.456	2.840	9759.127	13.659	0.000	5.119	2.615	4.304	36.514
Huanglonggou-Py3-9	11.437	3.475	11.761	0.000	3335.870	0.000	5.096	28.996	1.354	1.118	754.925

References

1. Zhang, J.Y.; Ma, C.Q.; Li, J.W. Visible gold-forming environment evidenced by sulfide mineralogical characteristics of Shuizhadonggou-Huanglonggou gold deposit in eastern Kunlun orogen. *Miner. Depos.* **2012**, *31*, 1184–1194, (In Chinese with English abstract).
2. Zhang, J.Y.; Ma, C.Q.; Li, J.W.; Pan, Y.M. A possible genetic relationship between orogenic gold mineralization and post-collisional magmatism in the eastern Kunlun Orogen, western China. *Ore Geol. Rev.* **2017**, *81*, 342–357. [CrossRef]
3. Zhang, Y.T. *Research on Metallogenesis of Gold Deposits in the Wulonggou Ore Concentration Area, Central Segment of the East Kunlun Mountains, Qinghai Province*; Jilin University: Changchun, China, 2018; (In Chinese with English abstract).
4. Chen, B.L.; Wang, Y.; Han, Y.; Chen, J.L. Metallogenic age of Yanjingou gold deposit in Wulonggou gold orefield, eastern Kunlun Mountains. *Miner. Depos.* **2019**, *38*, 541–556, (In Chinese with English abstract).
5. Chen, J.J.; Wei, J.H.; Fu, L.B.; Li, H.; Zhou, H.Z.; Zhao, X.; Zhan, X.F.; Tan, J. Multiple sources of the Early Mesozoic Gouli batholith, Eastern Kunlun Orogenic Belt, northern Tibetan Plateau: Linking continental crustal growth with oceanic subduction. *Lithos* **2018**, *292–293*, 161–178. [CrossRef]
6. Qinghai Bureau of Geology and Mineral Resources (QBGMR). *Regional Geology of Qinghai Province*; Geological Publishing House: Beijing, China, 1991; pp. 1–662, (In Chinese with English abstract).
7. Jiang, C.F.; Yang, J.S.; Feng, B.G.; Zhu, Z.Z.; Zhao, M. *Opening-Closing Tectonics of the Kunlun Mountalns*; Geological Publishing House: Beijing, China, 1992; pp. 1–224, (In Chinese with English abstract).
8. Chen, N.S.; Wang, Q.Y.; Chen, Q.; Li, X.Y. Components and metamorphism of the basements of the Qaidam and Oulongbuluke micro-continental blocks, and a tentative interpretation of paleocontinental evolution in NW-Central China. *Earth Sci. Front.* **2007**, *14*, 43–55, (In Chinese with English abstract).
9. Ding, Q.F.; Yan, W.; Zhang, B.L. Sulfur-and lead-isotope geochemistry of the Balugou Cu-Pb-Zn skarn deposit in the Wulonggou area in the eastern Kunlun Orogen, NW China. *J. Earth Sci.* **2016**, *27*, 740–750. [CrossRef]
10. Chen, N.S.; Li, X.Y.; Zhang, K.X.; Wang, G.C.; Zhu, Y.H.; Hou, G.J.; Bai, Y.S. Lithological characteristics of the Baishahe formation to the south of Xiangride Town, Eastern Kunlun Mountains and its age constrained from zircon Pb–Pb dating. *Bull. Geol. Sci. Technol.* **2006**, *25*, 1–7, (In Chinese with English abstract).
11. Zhang, J.X.; Wan, Y.S.; Meng, F.C.; Yang, J.S.; Xu, Z.Q. Geochemistry, Sm-Nd and U-Pb isotope study of gneisses (schists) enclosing eclogites in the North Qaidam Mountains-deeply subducted Precambrian metamorphic basement? *Acta Petrol. Sin.* **2003**, *19*, 443–451, (In Chinese with English abstract).
12. Wang, G.C.; Wang, Q.H.; Jian, P.; Zhu, Y.H. Zircon SHRIMP ages of Precambrian metamorphic basement rocks and their tectonic significance in the eastern Kunlun Mountains, Qinghai Province, China. *Earth Sci. Front.* **2004**, *11*, 481–490, (In Chinese with English abstract).
13. Liu, Y.J.; Genser, J.; Neubauer, F.; Jin, W.; Ge, X.H.; Handler, R.; Takasu, A. 40Ar/39Ar mineral ages from basement rocks in the Eastern Kunlun Mountains, NW China, and their tectonic implications. *Tectonophysics* **2005**, *398*, 199–224. [CrossRef]
14. Meng, F.C.; Zhang, J.X.; Cu, M.H. Discovery of Early Paleozoic eclogite from the East Kunlun, Western China and its tectonic significance. *Gondwana Res.* **2013**, *23*, 825–836. [CrossRef]
15. Mo, X.X.; Luo, Z.H.; Deng, J.F.; Yu, X.H.; Liu, C.D.; Shen, H.W.; Yuan, W.M.; Liu, Y.H. Granitoids and crust growth in East-Kunlun Orogenic Belt. *Geol. J. China Univ.* **2007**, *13*, 403–414, (In Chinese with English abstract).
16. Zhang, B.W. *Study on Metalliferous Deposits Mineralization in Southern Qilian Orogenic Belt, Qinghai Province*; Jilin University: Changchun, China, 2010; (In Chinese with English abstract).
17. Yang, J.S.; Robinson, P.T.; Jiang, C.F.; Xu, Z.Q. Ophiolites of the Kunlun mountains, China and their tectonic implications. *Tectonophysics* **1996**, *258*, 215–231. [CrossRef]
18. Li, R.B.; Pei, X.Z.; Li, Z.C.; Liu, Z.Q.; Chen, G.C.; Chen, Y.X.; Wei, F.H.; Gao, J.M.; Liu, C.J.; Pei, L. Geological characteristics of Late Palaeozoic–Mesozoic unconformities and their response to some significant tectonic events in eastern part of Eastern Kunlun. *Earth Sci. Front.* **2012**, *19*, 244–254, (In Chinese with English abstract).
19. Li, X.W.; Huang, X.F.; Luo, M.F.; Dong, G.C.; Mo, X.X. Petrogenesis and geodynamic implications of the Mid-Triassic lavas from East Kunlun, northern Tibetan Plateau. *J. Asian Earth Sci.* **2015**, *105*, 32–47. [CrossRef]
20. Zhu, Y.H.; Pan, Y.M.; Zhang, K.X.; Chen, N.S.; Wang, G.C.; Hou, G.J. Mineralogical characteristics and petrogenesis of ophiolites in East Kunlun Orogenic Belt, Qinghai Province. *Acta Mineral. Sin.* **2000**, *20*, 128–142, (In Chinese with English abstract).
21. Bian, Q.T.; Li, D.H.; Pospelov, I.; Yin, L.M.; Li, H.S.; Zhao, D.S.; Chang, C.F.; Luo, X.Q.; Gao, S.L.; Astrakhantsev, O.; et al. Age, geochemistry and tectonic setting of Buqingshan ophiolites, North Qinghai-Tibet Plateau, China. *J. Asian Earth Sci.* **2004**, *23*, 577–596. [CrossRef]
22. Li, R.B.; Pei, X.Z.; Li, Z.C.; Sun, Y.; Feng, J.Y.; Lei, P.; Chen, G.C.; Liu, C.J.; Chen, Y.X. Geochemical features, age, and tectonic significance of the Kekekete mafic-ultramafic rocks, East Kunlun Orogen, China. *Acta Geol. Sin.* **2013**, *87*, 1319–1333.
23. Wei, B. *Study on the Geological Characteristic and Tectonic Attribute of the Ophiolite and Island-Arc-Type Igneous Rocks, Central Belt of East Kunlun (Eastern Section)*; Chang'an University: Xi'an, China, 2015; (In Chinese with English abstract).
24. Qi, X.P.; Yang, J.; Fan, X.G.; Cui, J.T.; Cai, Z.F.; Zeng, X.W.; Wei, W.; Qu, X.X.; Zai, L.M. Age, geochemical characteristics and tectonic significance of Changshishan ophiolite in central East Kunlun tectonic mélange belt along the east section of East Kunlun Mountains. *Geol. China* **2016**, *43*, 797–816, (In Chinese with English abstract).

25. Li, W.Y. *Geochronology and Geochemistry of the Ophiolites and Island-Arc-Type Igneous Rocks in the Western Qinling Orogen and the Eastern Kunlun Orogen: Implication for the Evolution of the Tethyan Ocean*; University of Science and Technology of China: Hefei, China, 2008; (In Chinese with English abstract).
26. Liu, Z.Q.; Pei, X.Z.; Li, R.B.; Li, Z.C.; Zhang, X.F.; Liu, Z.G.; Chen, G.C.; Chen, Y.X.; Ding, S.P.; Guo, J.F. LA-ICP-MS zircon U-Pb geochronology of the two suites of ophiolites at the Buqingshan area of the A'nyemaqen Orogenic Belt in the southern margin of East Kunlun and its tectonic implication. *Acta Geol. Sin.* **2011**, *85*, 185–194, (In Chinese with English abstract).
27. Chen, L.; Sun, Y.; Pei, X.Z.; Gao, M.; Feng, T. Northernmost paleo-tethyan oceanic basin in Tibet: Geochronological evidence from 40Ar/39Arage dating of Dur'ngoi ophiolite. *Chin. Sci. Bull.* **2001**, *46*, 1203–1205. [CrossRef]
28. Yin, A.; Harrison, T.M. Geologic evolution of the Himalayan-Tibetan orogen. *Annu. Rev. Earth Planet. Sci.* **2000**, *28*, 211–280. [CrossRef]
29. Roger, F.; Arnaud, N.; Gilder, S.; Tapponnier, P.; Jolivet, M.; Brunel, M.; Malavieille, J.; Xu, Z.Q.; Yang, J. Geochronological and geochemical constraints on Mesozoic suturing in east central Tibet. *Tectonics* **2003**, *22*, 1037. [CrossRef]
30. Zhang, J.Y.; Ma, C.Q.; Xiong, F.H.; Liu, B.; Li, J.W.; Pan, Y.M. Early Paleozoic high- Mg diorite-granodiorite in the eastern Kunlun Orogen, western China: Response to continental collision and slab break-off. *Lithos* **2014**, *210–211*, 129–146. [CrossRef]
31. Xiong, F.H.; Ma, C.Q.; Zhang, J.Y.; Liu, B.; Jiang, H.A. A reworking of old continental lithosphere: An important crustal evolutionmechanism in orogenic belts, as evidenced by Triassic I-type granitoids in the East Kunlun orogen, Northern Tibetan Plateau. *J. Geol. Soc.* **2014**, *171*, 847–863. [CrossRef]
32. Xiong, F.H.; Ma, C.Q.; Wu, L.; Jiang, H.A.; Liu, B. Geochemistry, zircon U-Pb ages and Sr-Nd-Hf isotopes of an Ordovician appinitic pluton in the East Kunlun orogen: New evidence for Proto-Tethyan subduction. *J. Asian Earth Sci.* **2015**, *111*, 681–697. [CrossRef]
33. Chen, X.H.; Gehrels, G.; Yin, A.; Zhou, Q.; Huang, P.H. Geochemical and Nd–Sr–Pb–O isotopic constrains on Permo–Triassic magmatism in eastern Qaidam Basin, northern Qinghai-Tibetan plateau: Implications for the evolution of the Paleo-Tethys. *J. Asian Earth Sci.* **2015**, *114*, 674–692. [CrossRef]
34. Wu, C.; Yin, A.; Zuza, A.V.; Zhang, J.Y.; Liu, W.C.; Ding, L. Pre-Cenozoic geologic history of the central and northern Tibetan plateau and the role of Wilson cycles in constructing the Tethyan orogenic system. *Lithosphere* **2016**, *8*, 254–292. [CrossRef]
35. Chen, N.S.; Zhang, L.; Sun, M.; Wang, Q.Y.; Kusky, T.M. U–Pb and Hf isotopic compositions of detrital zircons from the paragneisses of the Quanji Massif, NW China: Implications for its early tectonic evolutionary history. *J. Asian Earth Sci.* **2012**, *54–55*, 110–130. [CrossRef]
36. Dai, J.G.; Wang, C.S.; Jeremy, H.; Santosh, M. Multi-stage tectono-magmatic events of the Eastern Kunlun Range, northern Tibet: Insights from U-Pb geochronology and (U-Th)/He thermochronology. *Tectonophysics* **2013**, *599*, 97–106. [CrossRef]
37. Chen, Y.X.; Pei, X.Z.; Li, R.B.; Liu, Z.Q.; Li, Z.C.; Zhang, X.F.; Chen, G.C.; Liu, Z.G.; Ding, S.P.; Guo, J.F. Zircon U-Pb age of Xiaomiao Formation of Proterozoic in the Eastern section of the East Kunlun Orogenic Belt. *Geoscience* **2011**, *25*, 510–521, (In Chinese with English abstract).
38. Yu, C. *Characteristics of Metamorphic Rock Series and It's Fluid Inclusions from Jinshuikou in Eastern Kunlun*; Jilin University: Jilin, China, 2005; (In Chinese with English abstract).
39. Lu, L. *Study on the Ore-Controlling Structure of Wulonggou Gold Deposit in East Kunlun Mountain*; Chinese Academy of Geological Sciences: Beijing, China, 2011; (In Chinese with English abstract).
40. Liu, B.; Ma, C.Q.; Guo, P.; Zhang, J.Y.; Xiong, F.H.; Huang, J.; Jiang, H.A. Discovery of the middle Devonian A-type granite from the Eastern Kunlun Orogen and its tectonic implications. *Earth Sci. J. China Univ. Geosci.* **2013**, *38*, 947–962, (In Chinese with English abstract).
41. Zhang, D.Q.; Feng, C.Y.; Li, D.X.; Xu, W.Y.; Yan, S.H. Orogenic gold deposits in the North Qaidam and East Kunlun Orogen, West China. *Miner. Depos.* **2001**, *20*, 137–146, (In Chinese with English abstract).
42. Liu, Z.C.; Wu, F.Y.; Yang, Y.H.; Yang, J.H.; Wilde, S.A. Neodymium isotopic compositions of the standard monazites used in U-Th-Pb geochronology. *Chem. Geol.* **2012**, *334*, 221–239. [CrossRef]
43. Griffin, W.; Powell, W.; Pearson, N.; O'Reilly, S. GLITTER: Data reduction software for laser ablation ICP-MS. *Mineral. Assoc. Can.* **2008**, *40*, 308–311.
44. Danyushevsky, L.; Robinson, P.; Gilbert, S.; Norman, M.; Large, R.; McGoldrick, P.; Shelley, M. Routine quantitative multi-element analysis of sulphide minerals by laser ablation ICP-MS: Standard development and consideration of matrix effects. *Geochem. Explor. Environ. Anal.* **2011**, *11*, 51–60. [CrossRef]
45. Fu, J.; Hu, Z.; Zhang, W.; Yang, L.; Liu, Y.; Li, M.; Zong, K.; Gao, S.; Hu, S. In situ sulfur isotopes (delta (34)S and delta (33)S) analyses in sulfides and elemental sulfur using high sensitivity cones combined with the addition of nitrogen by laser ablation MC-ICP-MS. *Anal. Chim. Acta* **2016**, *911*, 14–26. [CrossRef]
46. Hu, Z.; Zhang, W.; Liu, Y.; Gao, S.; Li, M.; Zong, K.; Chen, H.; Hu, S. "Wave" signalsmoothing and mercury-removing device for laser ablation quadrupole and multiple collector ICPMS analysis: Application to lead isotope analysis. *Anal. Chem.* **2015**, *87*, 1152–1157. [CrossRef]
47. Zhang, W.; Hu, Z.; Liu, Y. Iso-Compass: New freeware software for isotopic data reduction of LA-MC-ICP-MS. *J. Anal. At. Spectrom.* **2020**, *35*, 1087–1096. [CrossRef]
48. Reich, M.; Kesler, S.E.; Utsunomiya, S.; Palenik, C.S.; Chryssoulis, S.L.; Ewing, R.C. Solubility of gold in arsenian pyrite. *Geochim. Cosmochim. Acta* **2005**, *69*, 2781–2796. [CrossRef]

49. Barker, S.L.L.; Hickey, K.A.; Cline, J.S.; Dipple, G.M.; Kilburn, M.R.; Vaughan, J.R.; Longo, A.A. Uncloaking invisible gold: Use of nano-SIMS to evaluate gold, trace elements, and sulphur isotopes in pyrite from Carlin-type gold deposits. *Econ. Geol.* **2009**, *104*, 897–904. [CrossRef]
50. Deditius, A.P.; Utsunomiya, S.; Ewing, R.C.; Chryssoulis, S.L.; Venter, D.; Kesler, S.E. Decoupled geochemical behavior of As and Cu in hydrothermal systems. *Geology* **2009**, *37*, 707–710. [CrossRef]
51. Deditius, A.P.; Utsunomiya, S.; Kesler, S.E.; Reich, M.; Ewing, R.C. Trace elements nanoparticles in pyrite. *Ore Geol. Rev.* **2011**, *42*, 32–46. [CrossRef]
52. Large, R.R.; Danyushevsky, L.; Hollit, C.; Maslennikov, V.; Meffre, S.; Gilbert, S.; Bull, S.; Scott, R.; Emsbo, P.; Thomas, H.V.; et al. Gold and trace element zonation in pyrite using a laser imaging technique: Implications for the timing of gold in orogenic and Carlin-Style sediment-hosted deposits. *Econ. Geol.* **2009**, *104*, 635–668. [CrossRef]
53. Deditius, A.P.; Reich, M.; Kesler, S.E.; Utsunomiya, S.; Chryssoulis, S.L.; Walshe, J.; Ewing, R.C. The coupled geochemistry of Au and As in pyrite from hydrothermal ore deposits. *Geochimica. Cosmochim. Acta* **2014**, *140*, 644–670. [CrossRef]
54. Gregory, D.; Meffre, S.; Large, R. Comparison of metal enrichment in pyrite framboids from a metal-enriched and metal-poor estuary. *Am. Mineral.* **2014**, *99*, 633–644. [CrossRef]
55. Tardani, D.; Reich, M.; Deditius, A.P.; Chryssoulis, S.; Sánchez-Alfaro, P.; Wrage, J.; Roberts, M.P. Copper-arsenic decoupling in an active geothermal system: A link betweenpyrite and fluid composition. *Geochim. Cosmochim. Acta* **2017**, *204*, 179–204. [CrossRef]
56. Heinrich, C.A.; Driesner, T.; Stefánsson, A.; Seward, T.M. Magmatic vapor contraction and the transport of gold from the porphyry environment to epithermal ore deposits. *Geology* **2004**, *32*, 761–764. [CrossRef]
57. Heinrich, C.A. Fluid-fluid interactions in magmatic-hydrothermal ore formation. *Rev. Mineral. Geochem.* **2007**, *65*, 363–387. [CrossRef]
58. Reich, M.; Deditius, A.; Chryssoulis, S.; Li, J.W.; Ma, C.Q.; Parada, M.A.; Barra, F.; Mittermayr, F. Pyrite as a record of hydrothermal fluid evolution in a porphyry copper system: A SIMS/EMPA trace element study. *Geochim. Cosmochim. Acta* **2013**, *104*, 42–62. [CrossRef]
59. Román, N.; Reich, M.; Leisen, M.; Morata, D.; Barra, F.; Deditius, A.P. Geochemical and micro-textural fingerprints of boiling in pyrite. *Geochim. Cosmochim. Acta* **2019**, *246*, 60–85. [CrossRef]
60. Cabri, L.J.; Chryssoulis, S.L.; de Villiers, J.P.; Laflamme, J.G.; Buseck, P.R. The nature of "invisible" gold in arsenopyrite. *Can. Miner.* **1989**, *27*, 353–362.
61. Cook, N.J.; Chryssoulis, S.L. Concentrations of invisible gold in the common sulfides. *Can. Miner.* **1990**, *28*, 1–16.
62. Morishita, Y.; Hammond, N.Q.; Momii, K.; Konagaya, R.; Sano, Y.; Takahata, N.; Ueno, H. Invisible gold in pyrite from epithermal, banded-iron-formation-hosted, and sedimentary gold deposits: Evidence of hydrothermal influence. *Minerals* **2019**, *9*, 447. [CrossRef]
63. Morishita, Y.; Shimada, N.; Shimada, K. Invisible gold in arsenian pyrite from the high-grade Hishikari gold deposit, Japan: Significance of variation and distribution of Au/As ratios in pyrite. *Ore Geol. Rev.* **2018**, *95*, 79–93. [CrossRef]
64. Kusebauch, C.; Gleeson, S.A.; Oelze, M. Coupled partitioning of Au and As into pyrite controls formation of giant Au deposits. *Sci. Adv.* **2019**, *5*, eaav5891. [CrossRef]
65. Fleet, M.E.; Wu, T.W. Volatile transport of platinumgroup elements in sulfide-chloride assemblages at 1000 °C. *Geochim. Cosmochim. Acta* **1993**, *57*, 3519–3531. [CrossRef]
66. Campbell, F.A.; Ethier, V.G. Nickel and colbalt in pyrrhotite and pyrite from the Faro and Sullivan ore bodies. *Can. Mineral.* **1984**, *22*, 503–506.
67. Bajwah, Z.U.; Seccombe, P.K.; Offler, R. Trace element distribution Co: Ni ratios and genesis of the Big Cadia iron-copper deposit, New South Wales, Australia. *Miner. Depos.* **1987**, *22*, 292–300. [CrossRef]
68. Koglin, N.; Frimmel, H.E.; Minter, W.E.L.; Brätz, H. Trace-element characteristics of different pyrite types in Mesoarchean to Paleoproterozoic placer deposits. *Miner. Depos.* **2010**, *45*, 259–280. [CrossRef]
69. Reich, M.; Simon, A.; Deditius, A.; Barra, F.; Chryssoulis, S.; Lagas, G.; Tardani, D.; Knipping, J.; Bilenker, L.; Sánchez-Alfaro, P.; et al. Trace element signature of pyrite from the Los Colorados Iron Oxide-Apatite (IOA) deposit, Chile: A missing link between Andean IOA and Iron Oxide Copper-Gold systems? *Econ. Geol.* **2016**, *111*, 743–761. [CrossRef]
70. Peterson, E.C.; Mavrogenes, J.A. Linking high-grade gold mineralization to earthquake-induced fault-valve processes in the Porgera gold deposit, Papua New Guinea. *Geology* **2014**, *42*, 383–386. [CrossRef]
71. Rye, R.O.; Ohmoto, H. Sulfur and carbon isotopes and ore genesis: A review. *Econ. Geol.* **1974**, *69*, 826–842. [CrossRef]
72. Ohmoto, H. Stable isotope geochemistry of ore deposit. *Rev. Mineral. Geochem.* **1986**, *16*, 491–559.
73. Ohmoto, H.; Goldhaber, M.B. *Sulfur and carbon isotopes, In Geochemistry of Hydrothermal ore Deposits*, 2nd ed.; Barnes, H.L., Ed.; John Wiley & Sons Inc.: New York, NY, USA, 1997; pp. 517–611.
74. Seal, R.R. Sulfur isotope geochemistry of sulfide minerals. *Rev. Mineral. Geochem.* **2006**, *61*, 633–677. [CrossRef]
75. Luhr, J.F.; Logan, M.A.V. Sulfur isotope systematics of the 1982 El Chich'on trachyandesite: An ion microprobe study. *Geochim. Cosmochim. Acta* **2002**, *66*, 3303–3316. [CrossRef]
76. Rollinson, H.R. *Using Geochemical Data: Evaluation, Presentation, Interpretation*; Routledge: London, UK, 1993; pp. 306–308.
77. Goldfarb, R.J.; Groves, D.I.; Gardoll, S. Orogenic gold and geologic time: A global synthesis. *Ore Geol. Rev.* **2001**, *18*, 1–75. [CrossRef]

78. Groves, D.I.; Santosh, M. Craton and thick lithosphere margins: The sites of giant mineral deposits and mineral provinces. *Gondwana Res.* **2021**, *100*, 195–222. [CrossRef]
79. Mernagh, T.P.; Bastrakov, E.N.; Zaw, K.; Wygralak, A.S.; Wyborn, L.A.I. Comparison of fluid inclusion data and mineralization processes for Australian orgenic gold and intrusion-related gold systems. *Acta Petrol. Sin.* **2007**, *23*, 21–32.
80. Goldfarb, R.J.; Baker, T.; Dube, B.; Groves, D.I.; Hart, C.J.R.; Gosselin, P. Distribution, character, and genesis of gold deposits in metamorphic terranes. In *100th Anniversary Volume*; Hedenquist, J.W., Thompson, J.F.H., Goldfarb, R.J., Richards, J.P., Eds.; Society of Economic Geologists: Littleton, CO, USA, 2005; pp. 407–450.
81. Dong, Y.P.; He, D.F.; Sun, S.S.; Liu, X.M.; Zhou, X.H.; Zhang, F.F.; Yang, Z.; Cheng, B.; Zhao, G.C.; Li, J.H. Subduction and accretionary tectonics of the East Kunlun orogen, western segment of the Central China Orogenic System. *Earth Sci. Rev.* **2018**, *186*, 231–261. [CrossRef]
82. Zhang, L.; Chen, R.X.; Zheng, Y.F.; Li, W.C.; Hu, Z.C.; Yang, Y.H.; Tang, H.L. The tectonic transition from oceanic subduction to continental subduction: Zirconological constraints from two types of eclogites in the North Qaidam orogen, northern Tibet. *Lithos* **2016**, *244*, 122–139. [CrossRef]
83. Wang, M.J.; Song, S.G.; Niu, Y.L.; Su, L. Post-collisional magmatism: Consequences of UHPM terrane exhumation and orogen collapse, N. Qaidam UHPM belt, NW China. *Lithos* **2014**, *210*, 181–198. [CrossRef]
84. Wu, C.; Zuza, A.V.; Chen, X.; Ding, L.; Levy, D.A.; Liu, C.; Liu, W.; Jiang, T.; Stockli, D.F. Tectonics of the Eastern Kunlun Range: Cenozoic Reactivation of a Paleozoic—Early Mesozoic Orogen. *Tectonics* **2019**, *38*, 1609–1650. [CrossRef]
85. Tomkins, A.G. On the source of orogenic gold. *Geology* **2013**, *41*, 1255–1256. [CrossRef]
86. Li, J.Q.; Wang, T.; Chen, J.; Bai, Z.H.; Lin, Y.H. Zircon U-Pb age and petrogeochemical characteristics of ore-hosting granodiorite in Manite gold deposit, East Kunlun. *J. Mineral. Petrol.* **2019**, *39*, 69–77, (In Chinese with English abstract).
87. Pan, X.P.; Li, R.S.; Yu, P.S.; Gu, P.Y.; Shi, C.; Wang, C.; Cha, X.F. The age of country rocks of the Kendekeke Fe-Co polymetalic deposit in Qimantag area and its significance. *Acta Petrol. Mineral.* **2013**, *32*, 53–62, (In Chinese with English abstract).
88. Xiao, Y.; Feng, C.Y.; Liu, J.N.; Yu, M.; Zhou, J.H.; Li, D.X.; Zhao, Y.M. LA-MC-ICP-MS zircon U-Pb dating and sulfur isotope characteristics of Kendekeke Fe-polymetallic deposit, Qinghai Province. *Miner. Depos.* **2013**, *32*, 177–186, (In Chinese with English abstract).
89. Xi, R.G.; Xiao, P.X.; Wu, Y.Z.; Dong, Z.C.; Guo, L.; Gao, X.F. The geological significances, composition and age of the monzonitic granite in Kendekeke iron mine. *Northwest. Geol.* **2010**, *43*, 195–202, (In Chinese with English abstract).
90. Liang, Y.Y.; Xia, R.; Shan, X.Y.; Ma, Y.; Zhao, E.Q.; Guo, W.H. Geochronology and Geochemistry of Ore-Hosting Rhyolitic Tuff in the Kengdenongshe Polymetallic Deposit in the Eastern Segment of the East Kunlun Orogen. *Minerals* **2019**, *9*, 589. [CrossRef]
91. Li, J.C.; Kong, H.L.; Li, Y.Z.; Namkha, N.; Jia, Q.Z.; Guo, X.Z.; Zhang, B. Ar-Ar age of altered sericite, zircon U-Pb age of quartz diorite and geochemistry of the Naomuhun gold deposit, east Kunlun. *Acta Geol. Sin.* **2017**, *91*, 979–991, (In Chinese with English abstract).
92. Song, S.G.; Niu, Y.L.; Su, L.; Xia, X.H. Tectonics of the North Qilian orogen, NW China. *Gondwana Res.* **2013**, *23*, 1378–1401. [CrossRef]
93. Xiao, Y.; Feng, C.Y.; Li, D.X.; Liu, J.N. Chronology and fluid inclusions of the Guoluolongwa gold deposit in Qinghai Province. *Acta Geol. Sin.* **2014**, *88*, 895–902, (In Chinese with English abstract).
94. Xia, R.; Wang, C.; Deng, J.; Carranza, E.J.M.; Li, W.; Qing, M. Crustal thickening prior to 220Ma in the East Kunlun Orogenic Belt: Insights from the Late Triassic granitoids in the Xiao-Nuomuhong pluton. *J. Asian Earth Sci.* **2014**, *93*, 193–210. [CrossRef]
95. Wang, W.; Yuan, W.M.; Gou, J.; Tian, C.S.; Ma, Z.Y.; Peng, W.X.; Feng, Y.L.; Cao, J.H.; Cheng, X.Q.; Hao, N.N. Zircon U-Pb ages and geochemical features of granites in Dulan area, Eastern Kunlun Mountains. *J. Chin. Electron Microsc. Soc.* **2015**, *34*, 222–233, (In Chinese with English abstract).
96. Chen, J.J.; Fu, L.B.; Wei, J.H. Geochemical characteristics of Late Ordovician granodiorite in Gouli area, Eastern Kunlun orogenic belt, Qinghai province: Implications on the evolution of Proto-Tethys Ocean. *Earth Sci.* **2015**, *41*, 1863–1882, (In Chinese with English abstract).
97. Guo, X.Z.; Jia, Q.Z.; Kong, H.L.; Li, Y.Z.; Li, J.C.; Ma, Z.Y.; Wang, Y. Zircon U-Pb geochronology and geochemistry of Harizha quartz diorite in the eastern section from East Kunlun. *Geol. Sci. Technol. Inf.* **2016**, *35*, 18–26, (In Chinese with English abstract).
98. Fan, X.Z.; Sun, F.Y.; Xu, C.H.; Xin, W.; Wang, Y.C.; Zhang, Y. Genesis of Harizha Ag-Pb-Zn deposit in the eastern Kunlun Orogen, NW China: Evidence of fluid inclusions and C-H-O-S-Pb isotopes. *Resour. Geol.* **2021**, *71*, 177–201. [CrossRef]
99. Li, J.C.; Jia, Q.Z.; Du, W.; Li, Y.Z.; Kong, H.L.; Namkha, N.; Yang, B.R. LA-ICP-MS zircon dating and geochemical characteristics of quartz diorite in Asiha gold deposit in East segment of Eastern Kunlun. *J. Jilin Univ. (Earth Sci. Edi.)* **2014**, *44*, 1188–1199, (In Chinese with English abstract).
100. Chen, J.J.; Fu, L.B.; David, S.; Wei, J.H.; Zhao, X.; Zhou, H.Z. Multiple episodes of gold mineralization in the East Kunlun Orogen, western Central Orogenic Belt, China: Constraints from Re-Os sulfide geochronology. *Ore Geol. Rev.* **2020**, *123*, 103587. [CrossRef]
101. Liang, G.Z.; Yang, K.F.; Sun, W.Q.; Fan, H.R.; Li, X.H.; Lan, T.G.; Hu, H.L.; Chen, Y.W. Multistage ore-forming processes and metal source recorded in texture and composition of pyrite from the Late Triassic Asiha gold deposit, Eastern Kunlun Orogenic Belt, western China. *J. Asian Earth Sci.* **2021**, *220*, 104920. [CrossRef]
102. Yang, X.M. *Study on the Geological Characteristics and the Genesis of Caiyuanzi Gold Deposit in Easter Kunlun Orogenic Belt, Qinghai Province*; Jilin University: Changchun, China, 2019; (In Chinese with English abstract).

103. Li, J.C.; Du, W.; Cheng, Y.S.; Kong, H.L.; Liu, J.X.; Li, Y.Z.; Jia, Q.Z.; Namkha, N.; Xia, M.Z.; Li, Y.J. Characteristics of gold deposits and ore-control factors in the East Kunlun mineralization belt, Qinghai Province. *Geol. Explor.* **2015**, *51*, 1079–1088, (In Chinese with English abstract).
104. Cheng, L. *Study on the Geological and Geochemical Characteristics and Genesis of Hongqigou Gold Deposit, Wulonggou Ore Concentration Area, East Kunlun, Qinghai Province*; Jilin University: Changchun, China, 2020; (In Chinese with English abstract).
105. Shi, C.; Li, R.S.; He, S.P.; Yu, P.S.; Wang, C.; Pan, S.J.; Zhang, H.D. A preliminary study of the Xingshugou gold mineralization spots in the eastern segment of Kunlun orogenic belt and the dating of its surrounding rocks. *Geol. Bull. China* **2012**, *31*, 1983–1990, (In Chinese with English abstract).
106. Feng, C.Y.; Qu, W.J.; Zhang, D.Q.; Dang, X.Y.; Du, A.D.; Li, D.X.; She, H.Q. Re–Os dating of pyrite from the Tuolugou stratabound Co (Au) deposit, eastern Kunlun Orogenic Belt, northwestern China. *Ore Geol. Rev.* **2009**, *36*, 213–230. [CrossRef]
107. Zhang, D.Q.; Dang, X.Y.; She, H.Q.; Li, D.X.; Feng, C.Y.; Li, J.W. Ar-Ar dating of orogenic gold deposits in northern margin of Qaidam and East Kunlun Mountains and its geological significance. *Miner. Depos.* **2005**, *24*, 87–98, (In Chinese with English abstract).
108. Kou, L.L.; Luo, M.F.; Zhong, K.H.; Zhao, C.X. 40Ar/39Ar dating of the I goldbearing shear zone on the gold ore collected belt of the Wulonggou, Qinghai, and its significance. *Xinjiang Geol.* **2010**, *28*, 330–333, (In Chinese with English Abstract).
109. Yuan, W.M.; Mo, X.X.; Zhang, A.K.; Chen, X.N.; Duan, H.W.; Li, X.; Hao, N.N.; Wang, X.M. Fission track thermochronology evidence for multiple periods of mineralization in the Wulonggou gold deposits, Eastern Kunlun Mountains, Qinghai Province. *J. Earth Sci.* **2013**, *24*, 471–478. [CrossRef]
110. Wu, J.J.; Zeng, Q.D.; Santosh, M.; Fan, H.R.; Wei, Z.H.; Yang, K.F.; Zhang, Z.M.; Li, X.H.; Liang, G.Z. Intrusion-related orogenic gold deposit in the East Kunlun belt, NW China: A multiproxy investigation. *Ore Geol. Rev.* **2021**, *139*, 104550. [CrossRef]
111. Goldfarb, R.J.; Groves, D.I. Orogenic gold: Common or evolving fluid and metal sources through time. *Lithos* **2015**, *233*, 2–26. [CrossRef]
112. Groves, D.I.; Santosh, M. The giant Jiaodong gold province: The key to a unified model for orogenic gold deposits? *Geosci. Front.* **2016**, *7*, 409–417. [CrossRef]

minerals

MDPI

Article

Ore Genesis of the Toudaochuan Gold Deposit in Central Jilin Province, NE China: Constraints from Fluid Inclusions and C–H–O–S–Pb Isotopes

Jingmou Li [1], Yunsheng Ren [1,2,3,*], Qun Yang [3] and Xinhao Sun [3]

1 School of Earth Sciences and Resources, China University of Geosciences, Beijing 100083, China; lijmmy2019@163.com
2 School of Earth Science, Institute of Disaster Prevention, Sanhe 065201, China
3 College of Earth Sciences, Jilin University, Changchun 130061, China; yangqun@jlu.edu.cn (Q.Y.); sxh610062132@126.com (X.S.)
* Correspondence: rys@cidp.edu.cn

Abstract: The Toudaochuan gold deposit is a recently discovered lode gold deposit in Central Jilin Province. Gold ore bodies are dominantly controlled by NE-trending fault. The major hydrothermal period can be further divided into the quartz–pyrite stage (stage I), quartz–gold–polymetallic sulfides stage (stage II, major gold mineralization stage), and quartz–carbonate stage (stage III). Primary fluid inclusions (FIs) identified in quartz at different hydrothermal stages include liquid-rich aqueous FIs (L-type), CO_2 FIs (C-type, including CO_2-bearing C1-type FIs and CO_2-rich C2-type FIs), and minor vapor-rich aqueous FIs (V-type). Microthermometry studies on different fluid inclusions indicate that the original ore-forming fluids belonged to the CO_2–H_2O–NaCl system characterized by a moderate–low temperature and low salinity in stages I and II, and they finally evolved into a H_2O–NaCl system characterized by low temperature and low salinity in stage III. Fluid immiscibility is considered to be the key ore-forming mechanism. The initial ore-forming fluid was originated from magmatic water and was mixed with meteoric water in the later stage. The S and Pb isotope data suggest that the ore metal materials were derived from the mixed source of mantle and crust. Based on all the above data, therefore, it can be proposed that the Toudaochuan gold deposit is a mesothermal magmatic–hydrothermal gold deposit.

Keywords: ore genesis; fluid inclusions; C–H–O–S–Pb isotopes; Toudaochuan gold deposit; Central Jilin province

1. Introduction

Central Jilin Province in Northeast (NE) China is located at the intersection of the northern margin of the North China Craton (NCC) and the eastern segments of the Central Asian Orogenic Belt (CAOB). The multi-stage and intense tectonic and magmatic activities resulted in numerous deposits of various genetic and metal types in this area, including porphyry Mo deposits represented by Daheishan and Jidetun [1,2], magmatic Cu–Ni sulfide deposits represented by Hongqiling [3,4], and mesothermal magmatic–hydrothermal vein-type Pb–Zn–(Ag) represented by Xiaohongshilazi [5], as well as the gold deposits represented by Jiapigou [6], Haigou [7], and Toudaochuan involved in this study. Previous studies on representative gold deposits (e.g., Jiapigou, Haigou, Erdaogou, Xiaobeigou, etc. [6–11]) showed that these gold deposits almost belong to magmatic–hydrothermal gold deposits. The Toudaochuan gold deposit is geographically close to these deposits, being located to the northwest of them. However, it is unknown whether its genetic type is the same as these gold deposits, thus restricting the study of regional gold metallogenic regularity and further prospecting and exploration in this area.

The Toudaochuan gold deposit is a recent discovery located about 50 km southwest of Yongji County, Jilin Province. Up to now, some properties which are important for understanding of the ore genesis of this deposit are still unclear, for instance, the ore-forming geological conditions, the source and evolution of ore-forming fluid, and the ore-forming material source. There are still some disputable views on the ore genesis of this deposit. For example, Wang et al. [12] believed that gold mineralization was closely related to Yanshanian biotite granites in the mining area periphery. Li et al. [13] proposed that the formation of the gold deposit was associated with the ore-bearing metamorphic volcanic strata, and accordingly classified the genetic type of the deposit as a medium- and low-temperature metamorphic hydrothermal quartz-vein-type gold deposit. Moreover, Wang et al. [14] also considered the deposit to be a metamorphic hydrothermal gold deposit.

Considering the above disputes, this study, on the basis of field and petrographic observation, carried out some analytical research on fluid inclusions, C–H–O, S and Pb stable isotopes of the gold ore in the Toudaochuan gold deposit. Moreover, the source of the ore-forming fluid and ore material, genetic type, gold mineralization process, and mineralization mechanism of this deposit have been discussed. The conclusions can provide a basis for knowledge on regional metallogeny and exploration of gold resource in the study area.

2. Regional Geology

The Central Jilin Province is tectonically located in the southeast part of the NE China which lies between the Siberian Craton and the North China Craton (NCC), including the Xing'an-Monglolia Orogenic Belt (XMOB), which belongs to the eastern segment of the Central Asian Orogenic Belt (CAOB) [15], and the northernmost part of the NCC (NNCC) (Figure 1a). The XMOB is composed of the Erguna, Xing'an, Songnen–Zhangguangcai Range, Jiamusi, and Khanka blocks from northwest to southeast, as well as Liaoyuan terrane in the south [16,17] (Figure 1b).

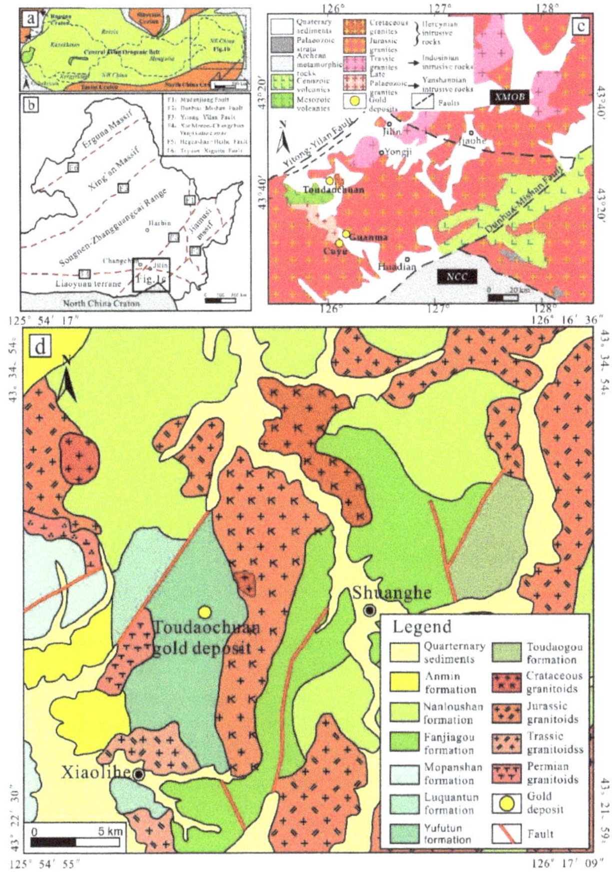

Figure 1. (**a**) Simplified tectonic map of the CAOB (after Safonova and Santosh [16]). (**b**) Tectonic divisions of the NE China (after Wu et al. [18]). (**c**) Geological map of the central Jilin Province, showing some representative gold deposits (after Yang et al. [19]). (**d**) Detailed geological map of the study area (after JBGMR [20]).

The NNCC comprises dominantly Mesoarchean to Paleoproterozoic basement [21–24] and Mesozoic igneous rocks [25–29]. This area has experienced the subduction and closure of the Paleo-Asian Ocean, the subduction of the Paleo-Pacific, and the superimposed conversion of the two tectonic systems [20,30–35], leading to extremely complex tectonic settings and large-scale magma intrusions. The interplay of tectonic and magmatic activities provided superior geological conditions for the various metallogenies and resulted in a large number of copper, gold, tungsten, molybdenum, lead, and zinc deposits.

Central Jilin Province is adjacent to Songnen–Zhangguangcai Range block in the north and NCC in the south, and bounded by Dunhua-Mishan fault and Yitong-Yilan fault to the east and west, respectively (Figure 1b,c). The outcropping strata in this area comprise the Cambrian–Ordovician Toudaogou formation composed of quartz schist and marble; the Lower Carboniferous Yufutun formation composed of intermediate–basic volcanic rocks; metamorphosed tuff and tuffaceous slate and siltstone and Luquantun formation composed of tuffaceous sandstone and rhyolitic tuff; the Middle Carboniferous Mopanshan formation composed of limestone with chert intercalations; the Lower Permian Fanjiagou formation composed of limestone, siltstone and tuffaceous sandstone; the Lower Jurassic Nanloushan formation composed of andesite, dacite, and dacite tuff,; and the

Lower Cretaceous An'min formation composed of rhyolite, rhyolitic tuff, and andesitic tuff [19] (Figure 1d). Affected by the evolution of multiphase tectonic system, many faults and folds striking E–W and NNE developed in the study area. The E–W faults were produced by N–S-trending compression in the Paleo Asian Ocean tectonic regime and reactivated by the Paleo Pacific regime in the Mesozoic. The NNE trending faults were formed under the influence of the NNW trending subduction of the Paleo Pacific plate beneath Eurasia [36,37]. Folds were represented by a large-scale synclinore striking NE in the study area. The two wings of the syncline extended to the NW and SE, formed some small-scale anticlines and thrust nappes under the influence of later tectonic activities [38]. The available geochronological data divided the widespread magmatic rocks in the study area into the following main periods: (1) Hercynian (362–250 Ma) intrusive rocks mainly consist of syenogranite, granodiorite, and monzogranite, as well as a small amount of mafic-ultramafic rocks as batholiths [39]; (2) Indosinian intrusive rocks (250–235 Ma) commonly have the characteristics of adakite, suggesting that they are related to the final closure of the paleo Asian Ocean [40–42]; (3) and Yanshanian intrusive rocks are mostly A/I-type granites, including diorite, tonalite, granodiorite, granite, etc. [20,43], which were spatially and temporally associated with the Mo mineralization in the region [1,2]. In addition, there are various dikes such as diorite porphyrite and granite porphyry intruded in the late Yanshanian period.

3. Toudaochuan Ore Deposit Geology

The lower Carboniferous Yufutun formation, composed of intermediate–basic volcanic rocks, metamorphosed tuff, and tuffaceous slate and siltstone, is developed in the mining area (Figure 2a). The intrusions in the mining area are mainly composed of granite porphyry dikes and diorite porphyrite dikes. Both the granite porphyry dikes and gold-bearing quartz vein are controlled by NE-trending faults and have a similar occurrence and close spatial relationship. The diorite porphyrite dikes intrude along NNE-trending faults and cut across ore bodies and granite porphyry (Figure 2b). Besides NE- and NNE-trending faults, an NNE-trending anticline which is subsidiary to the large-scale synclinore in Central Jilin developed in the east of the mining area. The intersection parts of the anticline and NE-trending faults provide favorable ore-hosting space for the emplacement of gold ore bodies and make mineralization more concentrated.

Fifteen gold orebodies containing about 2 t Au (average 1.01–52.36 g/t), 2.5 t Ag (average 15.64 g/t), and 5 t Cu (average 0.03%) have been identified in the Toudaochuan deposit. The orebodies consist mainly of irregular veined, stockwork, and lenticular-shape gold-bearing sulfide–quartz veins. They are intimately controlled by NE-trending faults and fractures and hosted within the Yufutun formation. The No. 1 and No. 14 are the two major orebodies in the deposit. The No. 1 orebody is about 40 m long and 6–7 m wide at most, dipping at 50°, and a maximum extension depth of about 60 m with the highest gold grade of 160.7 g/t. The No. 14 orebody is about 80 m long and 2 m wide with the highest gold grade of 19.5 g/t [44].

The mineral paragenesis and cross-cutting relationships in Toudaochuan deposit was clarified through field and specimen observations and microscopic studies. The metallic minerals in primary gold ores mainly include chalcopyrite, pyrite, sphalerite, galena, hessite, petzite, and native gold. The non-metallic minerals include quartz, potassium feldspar, chlorite, epidote, plagioclase, and calcite. Ore structures mainly include massive, veinlet, and miarolitic structures. Ore textures include euhedral–subhedral–anhedral granular texture (Figure 3d) and metasomatic textures (Figure 3e). Such alterations as silicification (Figure 3h), chloritization, epidotization, sericitization (Figure 3i), and carbonation commonly developed in orebodies and granite porphyries. Silicification and epidotization are mainly distributed around the orebodies and are closely related to mineralization. Most chloritization was formed by regional metamorphism, and a small amount is distributed around the ore bodies, which is related to mineralization. Carbonation is mainly formed in the later stage, which cut across orebodies.

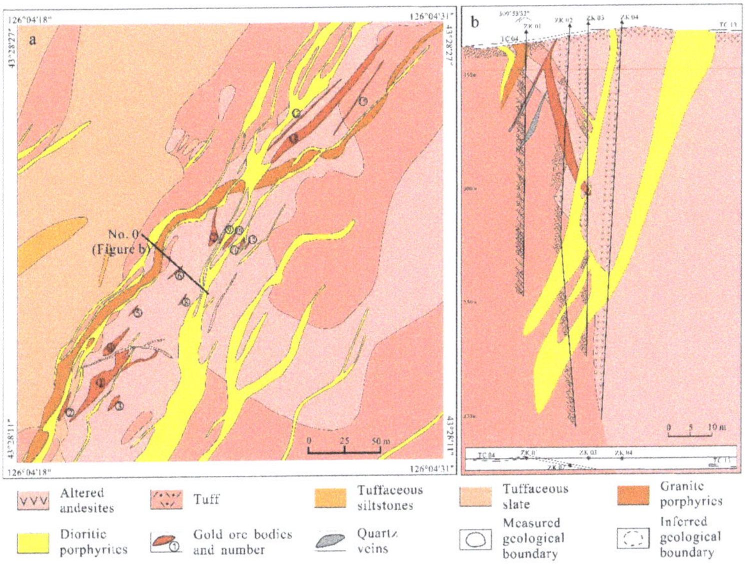

Figure 2. (**a**) Geologic map of the Toudaochuan deposit. (**b**) Geologic profile of Toudaochuan exploration line No. 0 (after JBGMR [44]).

Figure 3. Ore hand specimens, mineralization, and alteration characteristics of the Toudaochuan deposit. (**a**) sulfide-poor quartz vein from stage I; (**b**) gold-bearing polymetallic sulfide–quartz vein from in stage II (**c**) quartz–calcite veins from stage III; (**d**) euhedral–subhedral granular pyrite from stage I was replaced by the stage II sulfides (reflected light); (**e**) the sulfides assemblages in stage II (reflected light); (**f**) minor chalcopyrite distributed in sphalerite with solid-solution separation texture (reflected light); (**g**) native gold in backscatter photographs; (**h**) silicification in the surrounding rock strata near the ore bodies (crossed light); (**i**) sericitization in granite porphyry (crossed light). Abbreviations: Cal = calcite; Ccp = chalcopyrite; Cv = covellite; Dg = digenite; Gl = native gold; Py = pyrite; Qtz = quartz; Ser = sericite; Sp = sphalerite.

According to cross-cutting relationship and mineral paragenesis, mineralization can be divided into two periods: earlier hydrothermal period and later supergene period (Figure 4). The hydrothermal period can be further subdivided into three stages: (1) the quartz–pyrite stage (stage I) is characterized by the occurrence of sulfide-poor white quartz vein (Figure 3a), and the main metal mineral is pyrite as euhedral–subhedral crystal (Figure 3d); (2) the quartz–gold–polymetallic sulfides stage (stage II) is the major gold mineralization stage (Figure 3b) and mainly includes abundant chalcopyrite (Figure 3e,f), sphalerite (Figure 3e,f), native gold (Figure 3g), minor galena, hessite, and petzite, as well as minor chalcopyrite distributed in sphalerite with solid-solution separation texture (Figure 3f); and (3) the quartz–carbonate stage (stage III) is defined by the coexistence of quartz, calcite (Figure 3c), and occasionally chalcopyrite. In the supergene period, covellite (Figure 3d–f), malachite mainly developed.

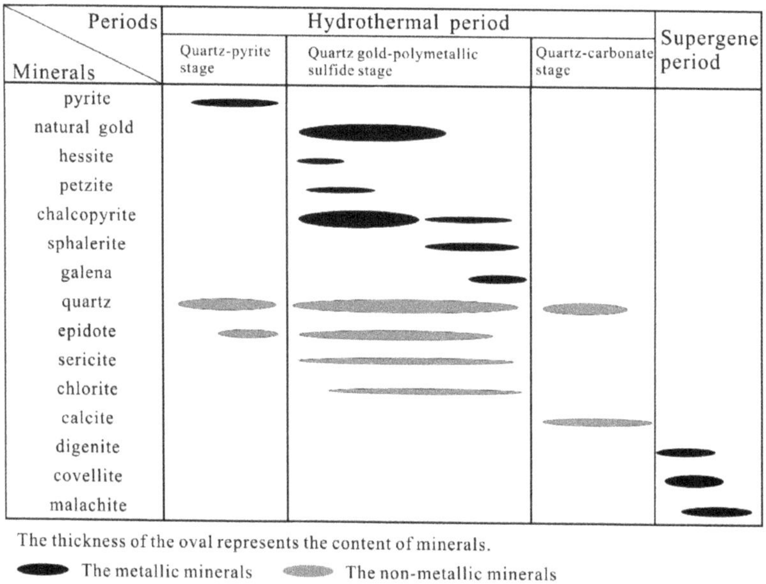

Figure 4. Paragenetic sequence of mineral deposition of the Toudaochuan gold deposit.

4. Samples Description and Analytical Methods

A total of 23 quartz samples were used for the fluid inclusion and C–H–O isotope studies. All quartz samples were selected from the exposed No. 1 orebody and open pit since No. 2–15 orebodies were inaccessible. These quartz samples were formed in different hydrothermal and mineralization stages recognized by observing intersection and overprinting relationships between various veins (veinlets), mineral assemblages, and alteration types. The average weight of each sample is about 3 to 5 kg. They were double-polished into thin sections and used for petrographically fluid inclusion studies, and 7 of them were further analyzed by microthermometry. Samples 7TDC-4–5, 8TDC-5–8, and 7TDC-6 represent the quartz–pyrite stage, quartz–gold–polymetallic sulfides stage, and quartz–carbonate stage, respectively. Four chalcopyrite samples selected from the quartz–gold–polymetallic sulfide stage were used for S and Pb measurement.

4.1. Fluid Inclusions

Microthermometry was carried out on a Linkam THMSG 600 (Linkam Scientific Instruments Ltd., Salfords, UK) heating–freezing stage within a temperature range of −196 to +600 °C at the Fluid Inclusion Laboratory of the Chinese Academy of Geological Sciences, Beijing, China. The accuracy of the stage was calibrated beforehand by using standard of synthetic FIs produced by America FLUID Inc. (Elyria, OH, USA). The estimated accuracy of the freezing/heating measurements was ±0.1 °C at temperatures of <30 °C, ±1 °C at 31–300 °C, and ±2 °C at temperatures of >300 °C. The heating/freezing rate was generally 2–5 °C/min, but less than 0.2 °C/min near the phase transition. The volatile compositions of single fluid inclusion were identified on a Renishaw RM2000 Raman microprobe (Renishaw, Wotton-under-Edge, UK), using an Ar ion laser with a surface power of 20 mW for exciting the radiation (514.53 nm); the scanning range for the spectra was set between 1000 and 4000 cm^{-1}, with an accumulation time of 60 s for each scan. The laser beam width was approximately 1 μm, and the spectral resolution is approximately 1–2 cm^{-1}.

4.2. C–H–O Isotope Analysis

The quartz samples for the C–H–O isotope analysis were collected together with those in use of fluid inclusions. Quartz grains were crushed into 40–80 mesh and further handpicked under a binocular microscope to ensure a purity of more than 99%. Analyses were measured at the Beijing Research Institute of Uranium Geology, China (BRIUG), using a MAT-253 gas isotope ratio mass spectrometer (Thermo Fisher Scientific, MA, USA). The purified quartzes were heated up to 400–500 °C in terms of an induction furnace, and a mixture of H_2O and CO_2 could be extracted, which were immediately separated by a sophisticated device. Oxygen isotopes were analyzed based on the BrF_5 extraction technique [45]. Hydrogen isotopes were measured for water released from fluid inclusions by thermal decrepitation and reaction with zinc powder. Carbon isotopes were obtained from the CO_2 extracted and separated from fluid inclusions. The δD and $\delta^{18}O$ were reported relative to the V-SMOW standard, with analytical precisions of approximately ±2‰ for δD values, and about ±0.2‰ for $\delta^{18}O$ values. The oxygen isotope compositions of the fluid ($\delta^{18}O_{fluid}$) were calculated based on the following equation: $1000\ln\alpha_{Qtz\text{-}H2O} = 3.38 \times 10^6/T^2 - 3.40$ [46]. The $\delta^{13}C$ values were reported relative to the V-PDB standard, with analytical precisions of approximately ±0.2‰.

4.3. S, Pb Isotope Analysis

The S and Pb isotope analyses were completed at the BRIUG. A MAT-251 mass spectrometer with analytical precisions of about ±0.2‰ was used for the sulfur isotopic test. The chalcopyrite grains mixed with CuO_2 were heated at 980 °C to produce SO_2, which was then analyzed by a Delta V Plus gas isotope mass spectrometer. Sulfur isotopic compositions is reported as $\delta^{34}S$ relative to the CDT standard. Lead isotope analyses were carried out by Isoprobe-T Thermal Ionization Mass Spectrometer, using the standard sample NBS 981. The samples were dissolved by acids, and lead was purified with anion exchange resin. Analytical precision was better than 0.008%. The parameters related to lead isotopes such as μ are calculated with the Geokit software (College of resources and environment, Yangtze University, Jinzhou, China) developed by Lu [47].

5. Results

5.1. Fluid Inclusions Analysis

5.1.1. Petrographic Characteristics

Since the primary fluid inclusions (FIs) are trapped during crystal growth, and the secondary fluid inclusions are trapped after the growth of the host crystal [48,49], the analytical work is focused on the primary inclusions in quartz of all three hydrothermal stages. The microscopic petrographic observation of fluid inclusions shows that the primary fluid inclusions in the quartz veins of Toudaochuan gold deposit are generally distributed as isolated individuals or as random clusters. According to the phases composition at room temperature (25 °C) and phase transitions during heating and cooling, three types of fluid inclusions can be distinguished: liquid-rich aqueous FIs (L-type), CO_2 FIs (C-type), and vapor-rich aqueous FIs (V-type).

The liquid-rich aqueous FIs (L-type) consist of liquid water (L_{H2O}) and vapor (V_{H2O}), with 10%–40% vapor–liquid ratios (Figure 5a,c,e–h) at room temperature They are abundant in quartz from all three hydrothermal stages, most of which are subcircular-, elliptical-, and strip-shaped with sizes ranging from 5 to 20 μm. During the heating process, these inclusions are completely homogenized to a liquid phase.

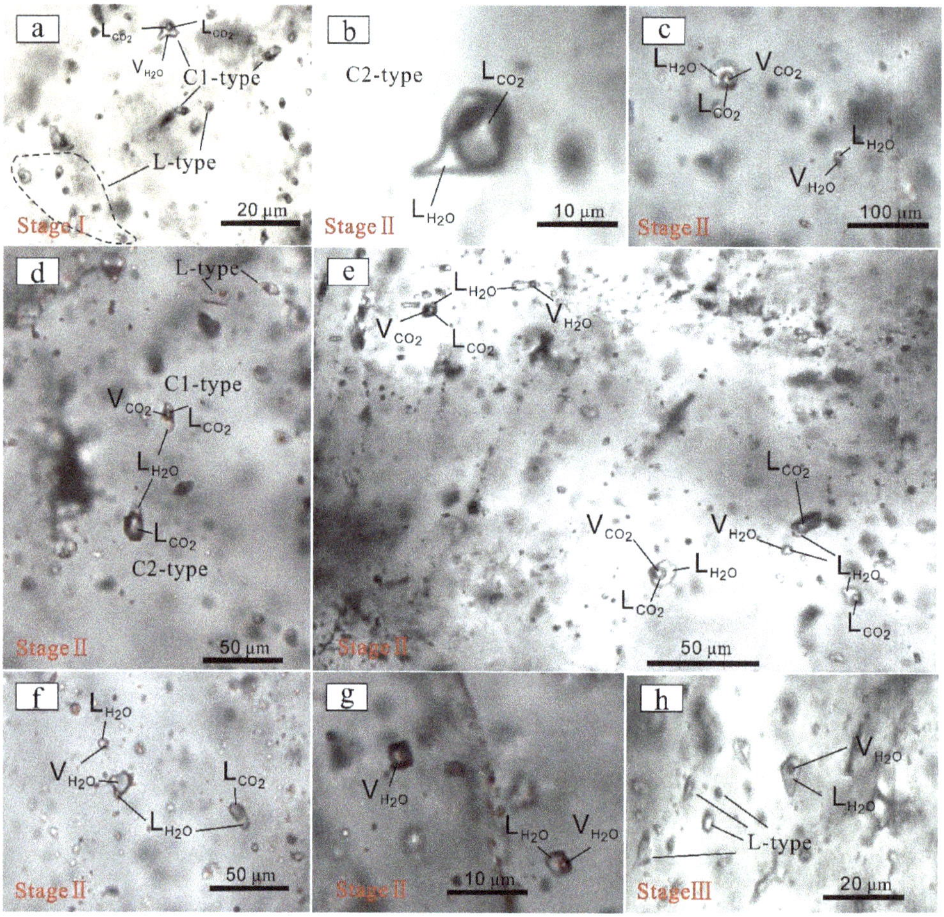

Figure 5. Photomicrographs of typical fluid inclusions in Toudaochuan deposit: (**a**) C1-type FIs coexisted with L-type FIs in stage I quartz; (**b**) C2-type FIs as two phases (liquid water + liquid CO_2) at room temperature in stage II quartz; (**c**) C1-type FIs coexisted with L-type FIs in stage II quartz; (**d**,**e**) L-type, C1-type, and C2-type FIs coexisted in stage II quartz; (**f**) L-type, V-type, and C2-type FIs coexisted in stage II quartz; (**g**) L-type and V-type FIs coexisted in stage II quartz; (**h**) L-type FIs assemblage in stage III quartz.

The vapor-rich aqueous FIs (V-type) consist of liquid water (L_{H2O}) and vapor (V_{H2O}), with >50% vapor–liquid ratios (few of them are close to pure vapor FIs) (Figure 5f,g). They have a small number and only exist in stage II, coexisting with L- and C-type FIs. They have subcircular and irregular shapes, with sizes ranging from 3 to 15 μm. During the heating process, these inclusions are completely homogenized to vapor phase.

The CO_2 FIs (C-type) consist of liquid H_2O + liquid CO_2 + vapor CO_2 (Figure 5a,c) or liquid water + liquid CO_2 phases (Figure 5b,d) at room temperature, and the latter can change into the former, with the appearance of CO_2 bubbles when the temperature drops to 10 °C or below. They have negative crystal, ellipsoidal, or irregular shapes and can be further divided into two subtypes, named C1-type and C2-type based on their CO_2-phase volumetric proportions [$(L_{CO2}$ or $V_{CO2} + L_{CO2})/(V_{CO2} + L_{CO2} + L_{H2O})$]. The C1-type is 7–20 μm in size, and their CO_2-phase volumetric proportions are mainly 30%–40% at room temperature (Figure 5a,c–e). This type of inclusion could be observed in stages I and II. They are finally homogenized to liquid H_2O phase. C2-type ones are 5–20 μm in size, and their CO_2-phase volumetric proportions are mainly 60%–85% at room temperature (Figure 5b,d,f). This type of inclusion can only be observed in stage II. They are finally homogenized to liquid CO_2 phase.

5.1.2. Microthermometry Results

The microthermometric data (e.g., $T_{m\text{-}ice}$, $T_{m\text{-}cla}$, $T_{m\text{-}CO_2}$, $T_{h\text{-}CO_2}$, and $T_{h\text{-}total}$) and calculated parameters (e.g., salinities and densities) of different types, primary FIs from the three hydrothermal stages, are listed in Table 1 and shown in Figure 6.

Table 1. Microthermometric data for fluid inclusions from the different hydrothermal stages in the Toudaochuan deposit.

Stages	Samples	Type	Quantity	$T_{m\text{-}ice}$ (°C)	$T_{m\text{-}CO2}$ (°C)	$T_{m\text{-}cla}$ (°C)	$T_{h\text{-}CO2}$ (°C)	$T_{h\text{-}total}$ (°C)	Salinity (%)
Stage I	quartz	L-type	25	−7.9–−3.2				212.6–322.8	5.2–11.6
		C1-type	8		−58.0–−56.8	4.1–6.5	25.6–29.8	235.0–335.8	6.6–10.4
Stage II	quartz	L-type	30	−5.7–−1.2				195.7–282.0	2.1–8.8
		V-type	5	−5.3–−1.1				202.5–275.4	1.9–8.3
		C1-type	24		−58.0–−57.0	6.0–9.1	26.7–30.0	210.0–298.5	1.8–7.5
		C2-type	10		−58.2–−56.9	6.1–9.4	26.8–30.0	212.5–286.2	1.2–7.3
Stage III	quartz	L-type	24	−2.7–−0.3				154.0–229.4	0.5–4.5

Quartz–pyrite stage (stage I): L-type and C1-type FIs (Figure 5a) were dominantly observed in the stage I quartz grains. The initial melting of CO_2 solid ($T_{m\text{-}CO_2}$) in C1-type fluid inclusions ranged from −58.0 to −56.8 °C, while CO_2 clathrate melting temperatures ($T_{m\text{-}cla}$) were 4.1–6.5 °C, corresponding to the salinities ranging from 6.6 to 10.4 wt.% NaCl equivalent (Figure 6b). The partial homogenization of CO_2 phase to liquid phase happened at a temperature of ($T_{h\text{-}CO_2}$) 25.6–29.8 °C. All C1-type inclusions finally homogenized to the liquid water phase at temperatures ($T_{h\text{-}total}$) between 235 and 335.8 °C (Figure 6a). The coexisting L-type fluid inclusions yielded ice-melting temperatures ($T_{m\text{-}ice}$) from −7.9 to −3.2 °C, with calculated salinities ranging from 5.2 to 11.6 wt.% NaCl equivalent (Figure 6b). They were homogenized to the liquid phase at temperatures between 212.6 and 322.8 °C (Figure 6a).

Quartz–gold–polymetallic sulfide stage (stage II): In this stage, quartz grains contained all type of FIs, including the L-type, V-type, C1-type, and C2-type FIs (Figure 5c–g). The L-type FIs homogenized to the liquid phase at temperatures 195.7–282.0 °C (mainly 220–250 °C) during heating (Figure 6c). The ice-melting temperatures ($T_{m\text{-}ice}$) ranged from −5.7 to −1.2 °C, with corresponding salinities ranging from 2.1 to 8.8 wt.% NaCl equiv. (Figure 6d). The V-type FIs homogenized to vapor phase at temperatures 202.5–275.4 °C (Figure 6c), and their ice-melting temperatures ($T_{m\text{-}ice}$) ranged from −5.3 to −1.1 °C, with corresponding salinities ranging from 1.9 to 8.3 wt.% NaCl equiv. (Figure 6d). The C1-type FIs yielded initial melting temperatures ($T_{m\text{-}CO_2}$) in the range of solid CO_2 from −58.0 to −57.0 °C, and their clathrate melting temperatures ($T_{m\text{-}cla}$) were 6.0–9.1 °C, corresponding to salinities of 1.8–7.5 wt.% NaCl equiv. (Figure 6d). The CO_2 phase homogenized to the liquid phase at a temperature of 26.7–30.0 °C ($T_{h\text{-}CO_2}$). C1-type FIs were totally homogenized to liquid H_2O at 210.0–298.5 °C ($T_{h\text{-}total}$) (Figure 6c). The C2-type FIs yielded initial melting temperatures ($T_{m\text{-}CO_2}$) of solid CO_2 from −58.2 to −56.9 °C, and their clathrate melting temperatures ($T_{m\text{-}cla}$) were 6.1–9.4 °C, corresponding to salinities of 1.2–7.3 wt.%

NaCl equiv. (Figure 6d). The partial homogenization of CO_2 phase to liquid phase happened at 26.8–30.0 °C ($T_{h\text{-}CO_2}$). The C2-type FIs were finally homogenized to liquid CO_2 at 212.0–286.2 °C ($T_{h\text{-}total}$) (Figure 6c).

Quartz–carbonate stage (stage III): Only L-type inclusions occurred in this stage quartz grains. They completely homogenized to liquid phase with the homogenization temperature of 154.0–229.4 °C (Figure 6e). Their ice-melting temperatures ($T_{m\text{-}ice}$) were −2.7 to −0.3 °C, with corresponding salinities ranging from 0.5 to 4.5 wt.% NaCl equiv. (Figure 6f).

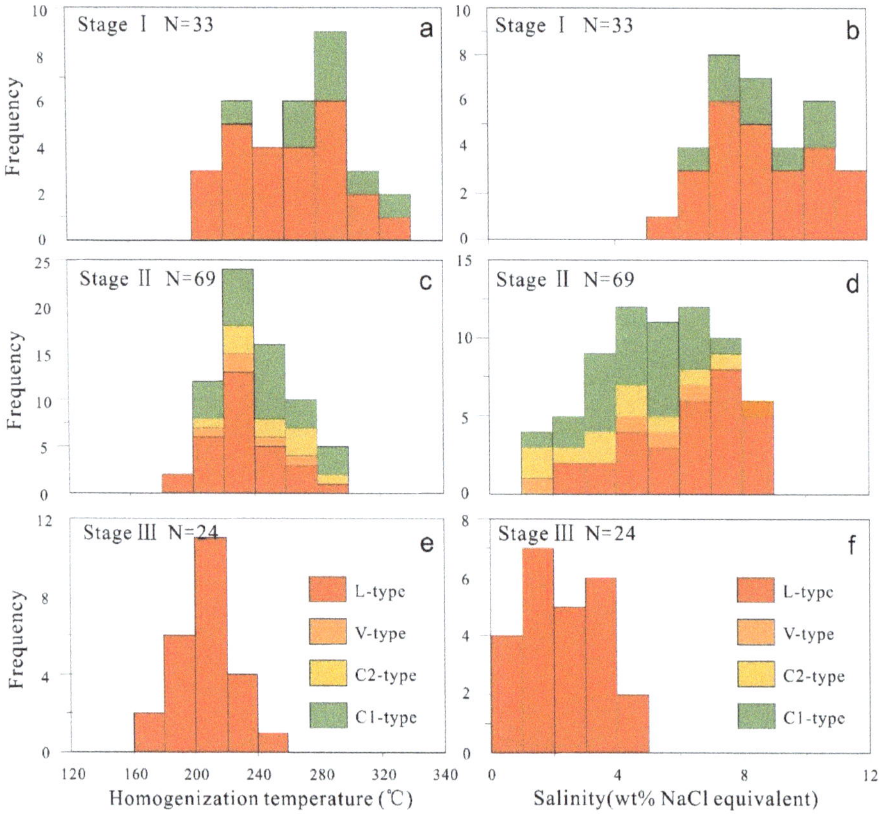

Figure 6. (**a–f**) Histograms of total homogenization temperatures (Th) and salinities of FIs from three hydrothermal mineralization stages.

5.1.3. Laser Raman Spectroscopy

Different types of representative fluid inclusions within the quartz of the first two stages were measured by using laser Raman spectroscopy to determine their gas components, and representative spectra are shown in Figure 7. The results showed that only the CO_2 component was detected in the C-type FIs (C1- and C2-type) in the both stage I and II. L-type and V-type FIs were composed of nearly pure H_2O.

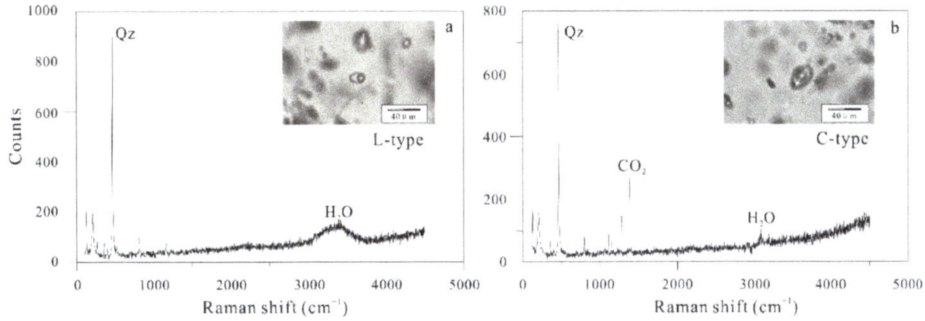

Figure 7. Representative Laser Raman spectra of fluid inclusions in the Toudaochuan deposit: (**a**) spectrum for C-type FIs and (**b**) spectrum for L-type FIs.

5.2. C–H–O Isotope Compositions

Hydrogen, oxygen, and carbon isotope data are listed in Table 2. The $\delta^{18}O_{V\text{-}SMOW}$ values in quartz samples of three hydrothermal stages were 10.9‰–11.9‰, 10.8‰–12.7‰, and 9.7‰–10.2‰, together with fluid inclusion $\delta D_{V\text{-}SMOW}$ values between −70.2 and −66.3‰, −84.7 and −80.8‰, and −89 and −84.3‰, respectively (Figure 8). The corresponding calculated $\delta^{18}O_{H2O}$ for three stages were 3.64‰–4.64‰, 1.10‰–3.00‰, and −2.67 to −2.17‰, respectively. The $\delta^{13}C$ values of the CO_2 extracted from fluid inclusions in quartz range from −8.3 to −7.7‰ (stage I), −15.5 to −8.8‰ (stage II), and −15.6 to −14.5‰ (stage III) (Figure 9).

Table 2. Hydrogen, oxygen, and carbon isotope data of inclusion fluids from the different hydrothermal stages in the Toudaochuan deposit.

Stages	Sample Description	$\delta^{13}C_{V\text{-}PDB}$ (‰)	$\delta^{18}O_{V\text{-}SMOW}$ (‰)	$\delta D_{V\text{-}SMOW}$ (‰)	Th (°C)	$\delta^{18}O_{H2O}$ (‰)
Stage I	sulfide-poor quartz veins	−8.3	11.9	−70.2	290	4.64
		−7.7	10.9	−66.3	290	3.64
Stage II	Au-bearing quartz veins	−13.1	12.7	−84.7	235	3.00
		−15.5	11.4	−80.8	235	1.70
		−8.8	10.8	−83.4	235	1.10
Stage III	Quartz veins	−15.6	10.2	−84.3	190	−2.17
		−14.5	9.7	−89	190	−2.67

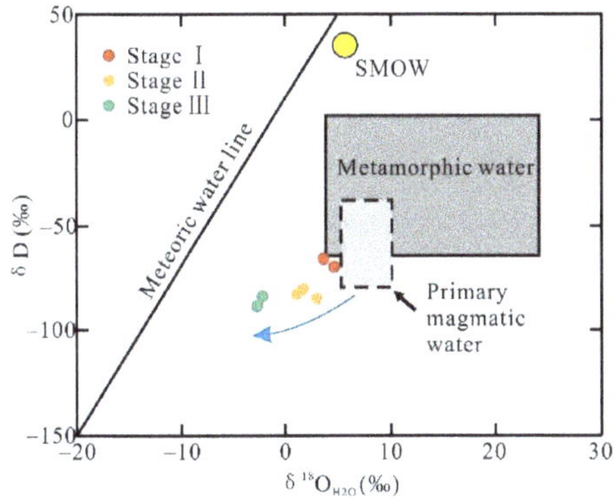

Figure 8. The diagram of hydrogen and oxygen isotopic compositions of the ore fluids for the Toudaochuan deposit. The magmatic, metamorphic water boxes and meteoric water line are from Taylor [50].

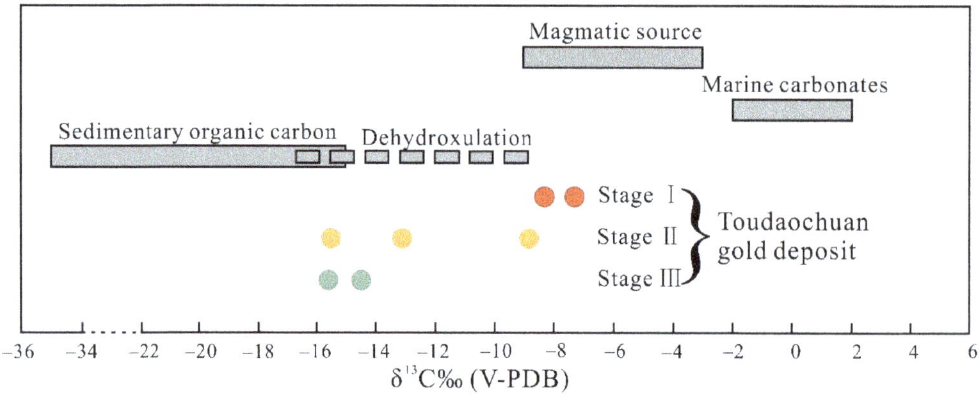

Figure 9. The diagram of carbon isotopic compositions of the ore fluids for the Toudaochuan deposit (after Niu et al. [51]).

5.3. S, Pb Isotope Compositions

The sulfur and lead isotopic analyses were performed on chalcopyrite samples coexisting with gold. Sulfur isotopic compositions are listed in Table 3 and shown in Figure 10. The δ^{34}S values of the four chalcopyrite samples were between −6.3 and −5.8‰, with an average of −6.0‰.

Table 3. The S and Pb isotope compositions of sulfides from the Toudaochuan deposit.

Sample Number	Sample Description	Samples	δ^{34}S (‰)	^{206}Pb/^{204}Pb	^{207}Pb/^{204}Pb	^{208}Pb/^{204}Pb
8TDC-5-1	massive ores	chalcopyrite	−6.3	18.370	15.574	38.295
8TDC-5-2		chalcopyrite	−5.8	18.374	15.567	38.286
8TDC-8-1		chalcopyrite	−5.8	18.388	15.582	38.324
8TDC-8-2		chalcopyrite	−6.2	18.386	15.582	38.323

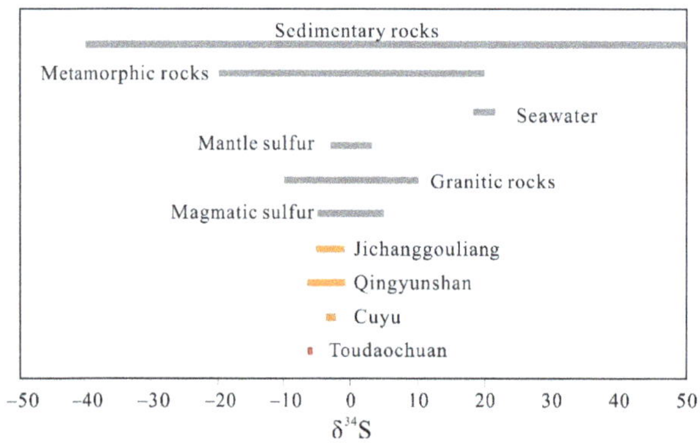

Figure 10. Ranges of the δ^{34}S values of selected geologically important sulfur reservoirs and comparison of δ^{34}S values between representative deposits and the Toudaochuan deposit (after Zhang et al. [52] and Hoefs et al. [53]).

Lead isotopic compositions are listed in Table 3 and shown in Figure 11. Four chalcopyrite samples gave Pb isotope ratios of ^{206}Pb/^{204}Pb of 18.370–18.388, ^{207}Pb/^{204}Pb of 15.567–15.582, and ^{208}Pb/^{204}Pb of 38.286–38.324.

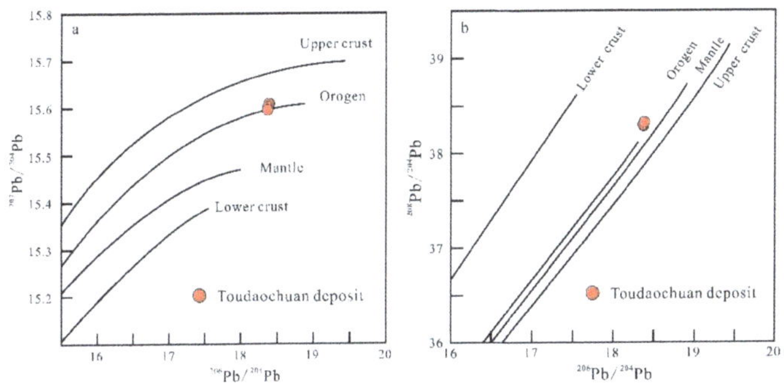

Figure 11. (**a**,**b**) Lead isotopic composition for the Toudaochuan deposit (after Zartman and Doe [54]).

6. Discussion

6.1. Sources of Ore-Forming Fluids and Materials

6.1.1. Ore-Forming Fluids

Hydrogen and oxygen isotope compositions, together with the homogenization temperatures of the corresponding FIs have been widely used to trace the source and evolution of ore-forming fluids [50,55,56]. The δD and calculated $\delta^{18}O_{H2O}$ values of quartz samples in stage I of Toudaochuan deposit are −70.2 to −66.3‰ and 3.64–4.64‰, respectively. These values are close to those of magmatic water (δD: −80 to −40‰, $\delta^{18}O_{H2O}$: 5.5‰–10‰), and different from those of metamorphic water (δD: −65 to −20‰, $\delta^{18}O_{H2O}$: 5‰–25‰) [57,58]. In the $\delta^{18}O_{H2O}$–δD diagram (Figure 8), the H–O isotopic compositions for the early stage (stage I) fluids are plotted close to the field of the primary magmatic water, indicating that the initial ore-forming fluids were primarily derived from a magmatic source. With fluid evolution, the δD and calculated $\delta^{18}O_{H2O}$ values of fluid inclusions in quartz decrease from −70.2 to −66.3‰, via −84.7 to −80.8‰, to −89 to −84.3‰, and from 3.64 to 4.64‰, via 1.10‰–3.00‰, to −2.67 to −2.17‰, respectively. In Central Jilin Province, the ore-forming fluids of numerous typical deposits (e.g., Toudaoliuhe (Au), Benqu (Au), Erdaogou (Au), Xiaobeigou (Au), Jidetun (Mo), etc.) were mixed with meteoric water during the fluid evolution. Their H-O values after the mixing were concentrated in −105 to −85‰ and −6 to 4‰, respectively [2,10,11,52,59]. The H–O values of stages II and III in Toudaochuan deposit are comparable to those of these deposits, and the sample gradually shifted toward the meteoric water evolution line in the $\delta^{18}O_{H_2O}$–δD diagram, both of which indicate that the continuous influx of meteoric water in the process of fluid evolution.

Marine carbonate source ($\delta^{13}C$ mean value of 0‰, [57]), mantle/magmatic source ($\delta^{13}C$ value of −9 to −3‰, [60]), and sedimentary organic source ($\delta^{13}C$ value mean value of −25‰, [61]) are interpreted as the three main carbon sources. The $\delta^{13}C$ values in hydrothermal quartz from the early to later stage in the Toudaochuan deposit are −8.3 to −7.7‰ (stage I), −15.5 to −8.8‰ (stage II), and −15.6 to −14.5‰ (stage III) (Figure 9), suggesting that carbon was mainly derived from a magmatic source and organic carbon source mixed in the later stage.

In general, the ore-forming fluids of the Toudaochuan gold deposit were mainly derived from magma, with continuous input of meteoric water during the evolution.

6.1.2. Ore-Forming Materials

The sulfur isotope composition of sulfide can trace the source of ore-forming materials well [57,62]. S sources are determined based on the total sulfur isotope compositions ($\delta^{34}S_{\Sigma S}$), which are impacted by parameters such as the pH value, oxygen fugacity, and temperature and not always equal to $\delta^{34}S$ values, in the ore-forming hydrothermal fluids during sulfide precipitation. However, when sulfur exists mainly in the forms of H^{S-} and S^{2-} under low oxygen fugacity and low pH circumstances [63], the $\delta^{34}S$ values are approximate to and can represent the $\delta^{34}S_{\Sigma S}$ values of the ore-forming fluids [64,65]. The sulfide assemblages of the Toudaochuan deposit are mainly chalcopyrite, pyrite, and sphalerite, and no sulfate minerals are found. Therefore, the sulfur isotope compositions of sulfide are similar to the total sulfur isotope compositions of ore-forming fluids. The following are several common sources of sulfur in ore-forming fluids: (1) mantle or magma source ($\delta^{34}S$: 0 ± 3‰ and 0 ± 5‰, respectively; [56,66,67]); (2) seawater ($\delta^{34}S$: 20‰; [56]); (3) sedimentary source ($\delta^{34}S$: −40 to 50‰; [64]); (4) metamorphic source ($\delta^{34}S$: −20 to 20‰; [64]). Four chalcopyrite samples from the Toudaochuan deposit yielded $\delta^{34}S$ values of −6.3 to −5.8‰, which are similar to Qingyunshan Cu–Au deposit ($\delta^{34}S$: −3.7 to −2.3‰; [68]), Cuyu gold deposit ($\delta^{34}S$: −6.5 to −0.9‰; [39]) and Jinchanggouliang gold deposit ($\delta^{34}S$: −5.2 to 0.9‰; [69]) (Figure 10), such sulfur isotope compositions are similar to but somewhat lower than $\delta^{34}S$ values of magmatic source, indicating that sulfur was primarily derived from magmatic source with contamination or assimilation by sulfur in the country rock strata.

Lead isotopes of sulfides can also be used to reflect the source of ore-forming metals. Zartman and Doe argued in 1979 that the lead with a μ value of 8.92 was derived from the mantle [70] and then proposed in 1981 that the lead with a μ value of more than 9.58 was derived from the upper crust [54]. The μ values of sulfides from the Toudaochuan deposit ranged from 9.40 to 9.43, indicating that lead was derived from the lower crust. In the $^{206}Pb/^{204}Pb$ versus $^{208}Pb/^{204}Pb$ diagram (Figure 11b), the Pb isotope compositions of all four sulfides samples were plotted between the lower crust and orogenic belt evolution lines but close to the latter. In the $^{206}Pb/^{204}Pb$ versus $^{207}Pb/^{204}Pb$ diagram (Figure 11a), they were projected between the upper crust and the mantle area and concentrated nearby orogenic belt evolution lines. Therefore, it is suggested that the metals in the Toudaochuan deposit were derived from a mixed source of mantle and crust.

6.2. Fluids Evolution and Gold Precipitation Mechanism

Petrography and microthermometry results of fluid inclusions showed that the L- and C1-type coexist in the quartz–pyrite stage (stage I). These FIs have similar homogenization temperatures and corresponding salinities which concentrated at 260–300 °C and 7–11 wt.% NaCl equivalent, respectively, indicating that they were captured simultaneously [49]. They have similar vapor–liquid ratios and homogenization behaviors, which both homogenized to liquid H_2O. Furthermore, laser Raman spectroscopy showed that the vapor phase components in different FIs were mainly H_2O and CO_2. Therefore, the initial ore-forming fluids belonged to the CO_2–H_2O–NaCl system characterized by moderate temperatures and moderate–low salinities.

The quartz–gold–polymetallic sulfides stage (stage II) contained all types FIs, with L-type, C1-type, and C2-type being the most developed. The peak total homogenization temperatures and corresponding salinities of all of these FIs concentrated at 220–260 °C and 3–8 wt.% NaCl equivalent, respectively. Laser Raman spectroscopy also showed that the vapor phase components in all FIs were mainly H_2O and CO_2. These characteristics revealed that the ore-forming fluid at this stage belonged to a medium–low temperature and medium–low salinity H_2O–CO_2–NaCl system.

The following petrographic and microthermometric characteristics of the FIs indicate that fluid immiscibility occurred during the quartz–gold–polymetallic sulfide stage (stage II). (1) Various types of FIs frequently coexisted in the same quartz grains, implying that they were captured at the same time [49]. (2) The C1- and C2-type FIs, as well as the L-type and a small number of V-type FIs had overlapping homogenization temperature ranges but different homogenization behaviors [48,49,71]. During heating, the C1- and C2-type FIs homogenized to liquid H_2O and liquid CO_2 phases, respectively; the L- and V-type FIs homogenized to liquid H_2O phase and vapor H_2O phase, respectively. (3) Distinct types of FIs showed comparable homogenization temperature ranges but varied salinity ranges. Previous studies of the solubility of Au in hydrothermal systems had shown that gold is most likely transported as $[Au(HS)_2]^-$, Au(HS), and $[AuCl_2]^-$ [72–74]. In the Toudaochuan deposit, the widespread occurrence of CO_2-rich FIs and the coexistence of gold and sulfides suggested that $[Au(HS)_2]^-$ was the likely agent responsible for Au transport [75,76]. During the migration and evolution of ore-forming fluids, the occurrence of fluid immiscibility made CO_2 in ore-forming fluids escape. The escape of CO_2 increased the pH values of the solution [76], destroyed the stability of Au-bisulfide complexes, and reduced the solubility of Au, resulting in the precipitation of gold and sulfides [77–82]. Gold accompanied by a large number of sulfides could be observed at stage II, while only a small amount of pyrite occurred at stage I. Therefore, fluid immiscibility is regarded as an important and plausible mechanism for gold precipitation in the Toudaochuan gold deposit.

The fluid immiscibility that occurred in stage II caused CO_2 to be eliminated from the ore-forming fluids and only L-type FIs to be developed in the quartz–carbonate stage (stage III). The FIs' homogenization temperatures dropped to 154.0–229.4 °C, and the salinities dropped to 0.5–4.5 wt.% NaCl equivalent compared to earlier stages. In addition, the H–O isotope compositions shifted dramatically toward the meteoric water line. The considerable fall in temperatures and salinities, as well as the features of H–O isotopic compositions, revealed that the ore-forming fluid had evolved into an H_2O–NaCl system with low temperatures and salinities at this stage, owing to the continual inflow of meteoric water.

6.3. Genetic Type and Ore-Forming Process

As mentioned above, there are controversies over the genetic type of the Toudaochuan gold deposit, and the arguable focus can be summarized as whether it is the metamorphic hydrothermal origin [13,14] or magmatic–hydrothermal origin [12]. This study showed that the Toudaochuan gold deposit has similar characteristics to the magmatic–hydrothermal gold deposit, as follows. The Au-bearing sulfide–quartz veins ore bodies hosted in the metamorphic volcanic rocks of the Yufutun formation were mainly controlled by the NE-trending faults and were accompanied by granite porphyry and diorite porphyrite dikes. The wall-rock alterations, such as silicification, sericitization, and chloritization, are similar to those found in mesothermal magmatic–hydrothermal deposits [83,84]. The ore-forming fluids belonged to an H_2O–CO_2–NaCl system with medium–low temperatures and medium–low salinities at stages I and II and finally evolved into a low-temperature and low-salinity H_2O–NaCl system at stage III. Such characteristics of the ore-forming fluids were also similar to those of magmatic–hydrothermal gold deposits, such as the Shajingou (China) [85], Jinchanggouliang (China) [69], Liyuan (China) [86], Berezovsk (Urals), and Linares (Spain) [87]. The C–H–O isotope data showed that the ore-forming fluids were predominantly magmatic water. The S isotopic compositions ($\delta^{34}S$: −6.3 to −5.8‰) differed significantly from those of typical metamorphic hydrothermal gold deposits (orogenic gold deposits) ($\delta^{34}S$: 0–9‰; [88,89]). The Pb isotope compositions of sulfides implied that the ore-forming materials were derived from a mantle–crust mixed source. Moreover, the geological evidence also showed that the granite porphyry dikes in the Toudaochuan gold deposit may have a close genetic relationship with gold mineralization. Firstly, no other intrusions, except for granite porphyry dikes and diorite porphyrite dikes developed in and around the mining area, and no concealed granites and ore bodies had been found. Granite porphyry dikes and ore bodies had a close spatial relationship which occurred in parallel and had similar occurrences. Secondly, previous studies have shown that the content of Au in most of the surrounding rock strata was low, and only a small amount of Au (mostly 0.01–0.1 g/t) was contained in the sections with strong alterations of the surrounding rock strata, while the content of Au in most of the granite porphyry samples was high, up to 160 g/t in some cases [44]. In addition, according to field observation, although diorite porphyrite dikes were closely related to ore bodies spatially, they cut across granite porphyry dikes and gold ore bodies and had insignificant alteration and mineralization. Therefore, the diorite porphyrite dikes should be emplaced after mineralization. In summary, based on the above features of mineralization and alteration, characteristics of the ore-forming fluids, sources of the ore-forming fluids and materials, and the comparison with some typical magmatic–hydrothermal deposits (Table 4), the Toudaochuan can be classified as the mesothermal magmatic–hydrothermal gold deposit.

Table 4. Comparison between the Toudaochuan Au deposit and typical mesothermal magmatic–hydrothermal gold deposits.

Deposits	Ore-Hosting Rocks	Orebodies Shape	Ores Structures and Textures	Metal Mineral Assembles	Wall Rock Alterations	FIs Characteristics	Source of Ore-Forming Fluids	Source of Ore-Forming Materials	Ore Genesis	References
Toudaochuan	Intermediate–basic volcanic rocks, meta-morphosed tuff and tuffaceous slate and siltstone of Yufutun formation	Irregular veined, stockwork and lens shape	Massive, veinlet, and miarolitic structures; euhedral-subhedral–anhedral granular texture and metasomatic texture	chalcopyrite, pyrite, sphalerite, galena, hessite, petzite, and native gold	Silicification, chloritization, sericitization, carbonation	Liquid-rich aqueous, vapor-rich aqueous, CO_2-rich and CO_2-bearing FIs; medium–low temperature, medium–low salinity H_2O-CO_2-NaCl fluids	Magmatic water and mixing of the meteoric water	A mixed source of mantle and crust	Mesothermal magmatic–hydrothermal deposit	This paper
Shajingou	Middle Jurassic diorite	Veined	Disseminated, vein, veinlet, taxitic and nodular structures; euhedral–subhedral granular, cataclastic, and metasomatic textures	Pyrrhotite, chalcopyrite, pyrite, native gold	Silicification, chloritization and sericitization	Liquid-rich aqueous, CO_2-pure, CO_2-rich and CO_2-bearing FIs; medium–high temperature, medium–low salinity H_2O-CO_2-NaCl fluids	Dominantly from magmatic fluid and mixed with meteoric water	-	Mesothermal magmatic–hydrothermal quartz-vein deposit	[85]
Jinchanggouliang	Xiaotazigou metamorphic rocks of the Archean Jianping Group, including biotite hornblende and plagioclase gneiss	Veined	Veinlet, banded, disseminated, brecciated and massive structures; metasomatic relict, anhedral granular, exsolution textures	Pyrite, chalcopyrite, galena, sphalerite, and native gold	Silicification, sericitization, pyritization, chloritization and carbonatization	halite-bearing inclusions, aqueous inclusions, and CO_2-H_2O inclusions; moderate–high temperature and great fluctuating salinity H_2O-CO_2-NaCl fluids	Magmatic water	derived from deep magma	Magmatic–hydrothermal deposit	[69]

Table 4. *Cont.*

Deposits	Ore-Hosting Rocks	Orebodies Shape	Ores Structures and Textures	Metal Mineral Assembles	Wall Rock Alterations	FIs Characteristics	Source of Ore-Forming Fluids	Source of Ore-Forming Materials	Ore Genesis	References
Heilongtan-Xiejiagou (China)	Middle–low grade metamorphosed volcanic rocks of Wudang group and low-grade metamorphic rocks of Yaolinghe group	Bedded, lentoid, and veined	-	Sphalerite, chalcopyrite, pyrite, galena, tetrahedrite	Potassic feldspathization, silicification, sericitization, carbonation	Aqueous, CO_2-pure and CO_2-bearing FIs; medium–low temperature, medium–low salinity H_2O-CO_2-NaCl fluids	dominantly from magmatic fluid and mixed with meteoric water	Mixture of magma and surrounding metamorphic rocks	Magmatic–hydrothermal vein-type deposit	[51]
Samhwanghak (Korea)	Precambrian metamorphic rock and granodiorite batholith	Veins	Massive and ribbon structures	Sphalerite, pyrrhotite, chalcopyrite, pyrite, galena, tetrahedrite	Silicification, chloritization, sericitization, carbonation, kaolinization, illitization	CO_2 (±CH4)-rich, aqueous FIs; Medium–high temperature, low salinity H_2O-CO_2-NaCl fluids	Magmatic water	Magmatic sulfur	Mesothermal magmatic–hydrothermal deposit	[90]

Based on the above discussion results, the ore-forming processes for the Au mineralization of the Toudaochuan deposit are proposed below and illustrated in Figure 12. For the initial ore-bearing fluid exsolved from granitic magma, NE-trending faults provided ideal migration channels. The initial ore-forming fluid was mostly constituted of magmatic water rich in Au and other metals, as well as CO_2 during the early stage of mineralization. The ore-forming fluid largely developed rich-liquid (L-type) and CO_2-bearing (C1 type) FIs and was characterized by medium temperatures and medium–low salinities. At this stage, just a limited amount of coarse euhedral–subhedral pyrite was deposited, and there was no gold mineralization. During the migration of ore-forming fluids, the CO_2 component can buffer the pH value of the solution, ensuring the steady and continuous migration of the gold–disulfide complex [76].

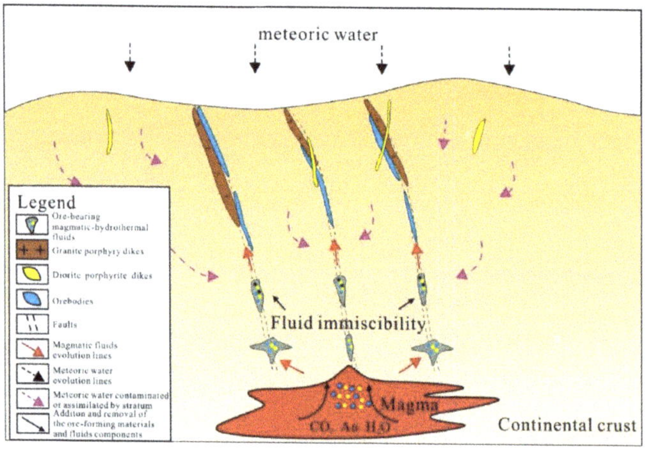

Figure 12. Simplified cartoon for the ore-forming model of the Toudaochuan deposit.

Subsequently, meteoric water carrying a small amount of organic carbon was mixed with the initial ore-forming fluid, resulting in the mixed source of carbon isotopes discussed above. Meanwhile, the inflow of meteoric water and the decreased fluid pressure induced fluid immiscibility. The immiscibility resulted in the release of CO_2 from the ore-forming fluids, thereby increasing the pH, which destroyed the stability of the gold–disulfide complex. These changes in physicochemical conditions caused the precipitation of a large number of gold and polymetallic sulfides at stage II (the main mineralization stage), leading to the formation of abundant quartz–polymetallic sulfide veins accompanied by silicification, chloritization, epidotization, and sericitization. The quartz–polymetallic sulfide veins were finally filled, enriched, and precipitated at the NE-trending structural faults or structural intersections.

With the continuous addition of meteoric water, the late ore-forming fluid gradually evolved into the NaCl–H_2O system, which is characterized by a low temperature and low salinity. At this stage, some calcite veinlets were formed and cut through the ore body, indicating that the hydrothermal mineralization was basically over.

7. Conclusions

The Toudaochuan gold deposit, in light of the received examination results, is a mesothermal magmatic–hydrothermal gold deposit. It is likely related to the granite porphyry dikes developed in the mining area. The mineralization process can be divided into the primary hydrothermal and supergene period, and the hydrothermal period can be further divided into the quartz–pyrite stage, quartz–gold–polymetallic sulfide stage, and quartz–carbonate stage, in which the gold–polymetallic sulfide stage is the major gold mineralization stage.

The primary fluid FIs in ore-bearing quartz in the Toudaochuan gold deposit mainly included liquid-rich aqueous FIs (L-type), CO_2 FIs (C-type, including CO_2-bearing C1-type FIs and CO_2-rich C2-type FIs), and minor vapor-rich aqueous FIs (V-type). The ore-forming fluids evolved from a medium-temperature, medium–low salinity CO_2–H_2O–NaCl system to a low-temperature, low-salinity H_2O–NaCl system.

The C–H–O isotope compositions suggest that ore-forming fluids were derived mainly from a magmatic source, with the inflow of meteoric water and organic carbon source in the later stage. The S isotope compositions show that sulfur was primarily derived from magmatic source with contamination or assimilation by sulfur in the country rock strata. The Pb isotope compositions indicate that the metals were derived from a mixed source of mantle and crust.

The fluid immiscibility in stage II is the key mechanism of gold mineralization. Fluid immiscibility caused CO_2 to escape from the ore-forming fluid, increased the pH value of the solution, reduced the stability and solubility of gold–sulfur complexes, and finally lead to the precipitation of gold and sulfides (e.g., chalcopyrite and sphalerite).

Author Contributions: Conceptualization, J.L. and Y.R.; formal analysis, J.L.; funding acquisition, Y.R.; investigation, J.L., Q.Y. and X.S.; project administration, Y.R.; supervision, Y.R.; visualization, J.L.; writing—original draft, J.L.; writing—review and editing, Y.R. and Q.Y. All authors have read and agreed to the published version of the manuscript.

Funding: This research was financially supported by the National Key Research and Development Program of China (2017YFC0601304), and the Opening Foundation of Key Laboratory of Mineral Resources Evaluation in Northeast Asia, Ministry of Natural Resources (No. DBY-KF-19-01).

Data Availability Statement: All data generated or used in the study are contained in the submitted article.

Acknowledgments: We thank the geologists of The Second Geologic Survey Institute of Jilin Province for their assistance during field investigations. We are also deeply grateful to anonymous reviewers for their reviews and constructive suggestions.

Conflicts of Interest: The authors declare no conflict of interest.

References

1. Zhou, L.L.; Zeng, Q.D.; Liu, J.M.; Friis, H.; Zhang, Z.L.; Duan, X.X. Ore genesis and fluid evolution of the Daheishan giant porphyry molybdenum deposit, NE China. *J. Asian Earth Sci.* **2015**, *97*, 486–505. [CrossRef]
2. Wang, Z.G.; Wang, K.Y.; Wan, D.; Konare, Y.; Liang, Y.H. Metallogenic age and hydrothermal evolution of the Jidetun Mo deposit in central Jilin Province, northeast China: Evidence from fluid inclusions, isotope systematics, and geochronology. *Ore Geol. Rev.* **2017**, *89*, 731–751. [CrossRef]
3. Lü, L.S.; Mao, J.W.; Li, H.B. Pyrrhotite Re-Os and SHRIMP zircon U-Pb dating of the Hongqiling Ni-Cu sulfide deposits in Northeast China. *Ore Geol. Rev.* **2011**, *43*, 106–119. [CrossRef]
4. Han, C.M.; Xiao, W.J.; Zhao, G.C.; Su, B.X.; Ao, S.J.; Zhang, J.E.; Wan, B.; Wang, Z.M.; Ding, J.X.; Qu, W.J.; et al. Re-Os Isotopic Age of the Hongqiling Cu-Ni Sulfide Deposit in Jilin Province, NE China and its Geological Significance. *Res. Geol.* **2014**, *64*, 247–261. [CrossRef]
5. Yang, Q.; Ren, Y.S.; Li, Y.; Hao, Y.J.; Li, J.M. Age and Tectonic Setting of Mesothermal Magmatic Hydrothermal Vein-Type Pb–Zn-(Ag) Mineralization in the Xiaohongshilazi Deposit, Central Jilin Province, Northeast China. *Res. Geol.* **2020**, *70*, 70–88. [CrossRef]
6. Miao, L.C.; Qiu, Y.M.; Fan, W.M.; Zhang, F.Q.; Zhai, M.G. Geology, geochronology, and tectonic setting of the Jiapigou gold deposits, southern Jilin Province, China. *Ore Geol. Rev.* **2005**, *26*, 137–165. [CrossRef]
7. Zhai, D.G.; Williams-Jones, A.E.; Liu, J.J.; Selby, D.; Li, C.; Huang, X.W.; Qi, L.; Cuo, D.H. Evaluating the Use of the Molybdenite Re-Os Chronometer in Dating Gold Mineralization: Evidence from the Haigou Deposit, Northeastern China. *Econ. Geol.* **2019**, *114*, 897–915. [CrossRef]
8. Zeng, Q.D.; He, H.Y.; Zhu, R.X.; Zhang, S.; Wang, Y.B.; Su, F. Origin of ore-forming fluids of the Haigou gold deposit in the eastern Central Asian Orogenic belt, NE China: Constraints from H-O-He-Ar isotopes. *J. Asian Earth Sci.* **2017**, *144*, 384–397. [CrossRef]
9. Li, L.; Sun, J.G.; Men, L.J.; Chai, P. Genesis of the Haigou gold deposit, Jilin Province, NE China: Evidence from fluid inclusions, $^{40}Ar/^{39}Ar$ geochronology and isotopes. *Geol. J.* **2017**, *52*, 992–1008. [CrossRef]
10. Li, L.; Sun, J.G.; Men, L.J.; Chai, P. Origin and evolution of the ore-forming fluids of the Erdaogou and Xiaobeigou gold deposits, Jiapigou gold province, NE China. *J. Asian Earth Sci.* **2016**, *129*, 170–190. [CrossRef]

11. Zeng, Q.D.; Wang, Z.C.; He, H.Y. Multiple isotope composition (S, Pb, H, O, He, and Ar) and genetic implications for gold deposits in the Jiapigou gold belt. Northeast China. *Miner. Depos.* **2014**, *49*, 145–164. [CrossRef]
12. Wang, Z.H.; Song, Q.H.; Yu, C. Division of gold metallogenic zonation of Jilin Province. *Jilin Geol.* **2011**, *4*, 10–14, (In Chinese with English Abstract).
13. Li, N.; Zhuang, Y.M.; Li, C.X. Formation structure and mineralization of Panshi Shizui-Guanma area, Jilin Province. *Jilin Geol.* **2012**, *31*, 15–16, 35. (In Chinese with English Abstract)
14. Wang, S.D.; Chen, H.C. Some problems on searching for gold deposits in Jilin Province. *Jilin Geol.* **1988**, *2*, 115–123, 29. (In Chinese)
15. Şengör, A.M.C.; Natal'in, B.A.; Burtman, V.S. Evolution of the Altaid tectonic collage and Palaeozoic crustal growth in Eurasia. *Nature* **1993**, *364*, 299–307. [CrossRef]
16. Safonova, I.Y.; Santosh, M. Accretionary complexes in the Asia–Pacific region: Tracing archives of ocean plate stratigraphy and tracking mantle plumes. *Gondwana Res.* **2014**, *25*, 126–158. [CrossRef]
17. Wu, F.Y.; Sun, D.Y.; Ge, W.C.; Zhang, Y.B.; Grant, M.L.; Wilde, S.A.; Jahn, B.M. Geochronology of the Phanerozoic granites in northeastern China. *J. Asian Earth Sci.* **2011**, *41*, 1–30. [CrossRef]
18. Wu, F.Y.; Zhao, G.C.; Sun, D.Y.; Wilded, S.A.; Yang, J.H. The Hulan Group: Its role in the evolution of the Central Asian Orogenic Belt of NE China. *J. Asian Earth Sci.* **2007**, *30*, 542–556. [CrossRef]
19. Yang, Q.; Ren, Y.S.; Huizenga, J.M.; Li, J.M.; Wang, E.D.; Wang, X.; Wang, B. Geological significance of Early Triassic porphyry Cu mineralization in the eastern Xar Moron–Changchun Metallogenic Belt, northeast China: A case study of the newly-discovered Guokuidingzi Cu deposit. *Ore Geol. Rev.* **2021**, *133*, 104092. [CrossRef]
20. JBGMR (Jilin Bureau of Geology and Mineral Resources). *Lithostratigraphy of Jilin Province*; China University of Geoscience Press: Wuhan, China, 1997; pp. 1–324. (In Chinese)
21. Kröner, A.; Compston, W.; Zhang, G.W.; Guo, A.L.; Todt, W. Age and tectonic setting of Late Archean greenstone–gneiss terrain in Henan Province, China, as revealed by single-grain zircon dating. *Geology* **1988**, *16*, 211–215. [CrossRef]
22. Zhao, G.C.; Wilde, S.A.; Cawood, P.A.; Sun, M. Archean blocks and their boundaries in the North China Craton: Lithological, geochemical, structural and P-T path constraints and tectonic evolution. *Precambrian Res.* **2001**, *107*, 45–73. [CrossRef]
23. Zhai, M.G.; Santosh, M. The early Precambrian odyssey of the North China Craton: A synoptic overview. *Gondwana Res.* **2011**, *20*, 6–25. [CrossRef]
24. Zhai, M.G.; Santosh, M. Metallogeny of the North China Craton: Link with secular changes in the evolving Earth. *Gondwana Res.* **2013**, *24*, 275–297. [CrossRef]
25. Zhu, R.X.; Yang, J.H.; Wu, F.Y. Timing of destruction of the North China Craton. *Lithos* **2012**, *149*, 51–60. [CrossRef]
26. Zhu, R.X.; Zhang, H.F.; Zhu, G.; Meng, Q.R.; Fan, H.R.; Yang, J.H.; Wu, F.Y.; Zhang, Z.Y.; Zheng, T.Y. Craton destruction and related resources. *Int. J. Earth Sci.* **2017**, *106*, 2233–2257. [CrossRef]
27. Yang, J.H.; Wu, F.Y.; Wilde, S.A. A review of the geodynamic setting of large-scale Late Mesozoic gold mineralization in the North China Craton: An association with lithospheric thinning. *Ore Geol. Rev.* **2003**, *23*, 125–152. [CrossRef]
28. Zhang, S.H.; Zhao, Y.; Song, B.; Hu, J.M.; Liu, S.W.; Yang, Y.H.; Chen, F.K.; Liu, X.M.; Liu, J. Contrasting Late Carboniferous and Late Permian–Middle Triassic intrusive suites from the northern margin of the North China craton: Geochronology, petrogenesis, and tectonic implications. *Geol. Soc. Am. Mem.* **2009**, *121*, 181–200. [CrossRef]
29. Zhang, X.H.; Yuan, L.L.; Xue, F.H.; Zhang, Y.B. Contrasting Triassic ferroan granitoids from northwestern Liaoning, North China: Magmatic monitor of Mesozoic decratonization and a craton–orogen boundary. *Lithos* **2012**, *144*, 12–23. [CrossRef]
30. Wu, F.Y.; Sun, D.Y.; Li, H.M.; Jahn, B.M.; Wilde, S.A. A-type granites in northeastern China: Age and geochemical constraints on their petrogenesis. *Chem. Geol.* **2002**, *187*, 143–173. [CrossRef]
31. Liu, S.; Hu, R.Z.; Gao, S.; Feng, C.X.; Feng, G.Y.; Coulson, I.M.; Li, C.; Wang, T.; Qi, Y.Q. Zircon U-Pb age and Sr–Nd–Hf isotope geochemistry of Permian granodiorite and associated gabbro in the Songliao Block, NE China and implications for growth of juvenile crust. *Lithos* **2010**, *114*, 423–436. [CrossRef]
32. Xu, W.L.; Wang, F.; Pei, F.P.; Meng, E.; Tang, J.; Xu, M.J.; Wang, W. Mesozoic tectonic regimes and regional ore-forming background in NE China: Constraints from spatial and temporal variations of Mesozoic volcanic rock associations. *Acta Petrol. Sin.* **2013**, *29*, 339–353. (In Chinese with English Abstract)
33. Eizenhöfer, P.R.; Zhao, G.C.; Zhang, J.; Sun, M. Timing of final closure of the PaleoAsian Ocean along the Solonker Suture Zone: Constraints from the provenance analysis of detrital zircons from Permian sedimentary rocks. *Tectonics* **2014**, *33*, 441–463. [CrossRef]
34. Zhou, J.B.; Li, L. The Mesozoic accretionary complex in Northeast China: Evidence for the accretion history of Paleo-Pacific subduction. *J. Asian Earth Sci.* **2017**, *145*, 91–100. [CrossRef]
35. Zhou, J.B.; Wilde, S.A. The crustal accretion history and tectonic evolution of the NE China segment of the Central Asian Orogenic Belt. *Gondwana Res.* **2013**, *23*, 1365–1377. [CrossRef]
36. Ge, X.H. Geotectonics and tectonic evolution of eastern Jilin Province. *Geoscience* **1990**, *1*, 107–113, (In Chinese with English abstract).
37. Zhao, C.J.; Peng, Y.J.; Dang, Z.X. *Tectonic Framework and Crust Evolution of Eastern Jilin and Shenyang*; Liaoning University Press: Jinzhou, China, 1996. (In Chinese)
38. Gao, N.N.; Fan, X.X.; Shan, X.L. Tectonic evolution history of Yongji basin in Central Jilin. *Gansu Sci. Technol.* **2019**, *35*, 28–31. (In Chinese)

39. Yang, Q. Late Paleozoic to Early Mesozoic Copper-Gold Polymetallic Metallogeny and Metallogenic Regularity in Central Jilin Province, Northeast China. Ph.D. Thesis, Jilin University, Changchun, China, 2020. (In Chinese with English Abstract)
40. Sun, D.Y.; Wu, F.Y.; Zhang, Y.B.; Gao, S. The final closing time of the west Lamulun River–Changchun–Yanji plate suture zone—Evidence from the Dayushan granitic pluton, Jilin Province. *J. Jilin Univ. (Earth Sci. Ed.)* **2004**, *34*, 174–181. (In Chinese with English Abstract)
41. Cao, H.H. Geochronology and Geochemistry of the Late Paleozoic–Early Mesozoic Igneous Rocks in the Eastern Segment of the Northern Margin of the North China Block. Ph.D. Thesis, Jilin University, Changchun, China, 2013. (In Chinese with English Abstract)
42. Cao, H.H.; Xu, W.L.; Pei, F.P.; Wang, Z.W.; Wang, F.; Wang, Z.J. Zircon U-Pb geochronology and petrogenesis of the Late Paleozoic-Early Mesozoic intrusive rocks in the eastern segment of the northern margin of the North China Block. *Lithos* **2013**, *170–171*, 191–207. [CrossRef]
43. Zhang, H.H.; Wang, F.; Xu, W.L.; Cao, H.H.; Pei, F.P. Petrogenesis of Early–Middle Jurassic intrusive rocks in northern Liaoning and central Jilin provinces, northeast China: Implications for the extent of spatial–temporal overprinting of the Mongol–Okhotsk and Paleo-Pacific tectonic regimes. *Lithos* **2016**, *256*, 132–147. [CrossRef]
44. JBGMR (Jilin Bureau of Geology and Mineral Resources). *General Survey and Evaluation Report on Toudaochuan Gold Deposit and its Periphery in Yongji County, Jilin Province*; JBGMR: Jilin, China, 1983. (In Chinese)
45. Clayton, R.N.; Mayeda, T.K. The use of bromine pentafluoride in the extraction of oxygen from oxides and silicates for isotopic analysis. *Geochim. Cosmochim. Acta* **1963**, *27*, 43–52. [CrossRef]
46. Clayton, R.N.; O'Neil, J.R.; Mayeda, T.K. Oxygen isotope exchange between quartz and water. *Geophys. Res. Lett.* **1972**, *77*, 3057–3067. [CrossRef]
47. Lu, Y.F. GeoKit—A geochemical toolkit for Microsoft Excel. *Geochimica* **2004**, *5*, 459–464, (In Chinese with English Abstract).
48. Roedder, E. Fluid inclusions. *Rev. Miner.* **1984**, *12*, 1–644.
49. Lu, H.Z.; Fan, H.R.; Ni, P.; Qu, G.X.; Shen, K.; Zhang, W.H. *Fluid Inclusion*; Science Press: Beijing, China, 2004; pp. 193–200. (In Chinese)
50. Taylor, H.P. The application of oxygen and hydrogen isotope studies to problems of hydrothermal alteration and ore deposition. *Econ. Geol.* **1974**, *69*, 843–883. [CrossRef]
51. Niu, P.P.; Jiang, S.Y. Fluid inclusion and stable isotope (C-H-O-S) constraints on the genesis of the Heilongtan-Xiejiagou Au deposit, northern Hubei, China. *Ore Geol. Rev.* **2022**, *144*, 104841. [CrossRef]
52. Zhang, X.T.; Sun, J.G.; Han, J.L.; Feng, Y.Y. Genesis and ore-forming process of the Benqu mesothermal gold deposit in the Jiapigou ore cluster, NE China: Constraints from geology, geochronology, fluid inclusions, and whole-rock and isotope geochemistry. *Ore Geol. Rev.* **2021**, *130*, 103956. [CrossRef]
53. Hoefs, J. *Stable Isotope Geochemistry*, 2nd ed.; Springer: Berlin/Heidelberg, Germany, 1980; pp. 1–208.
54. Zartman, R.E.; Doe, B.R. Plumbotectonics-the model. *Tectonophysics* **1981**, *75*, 135–162. [CrossRef]
55. Pirajno, F. *Hydrothermal Processes and Mineral Systems*; Springer: Berlin/Heidelberg, Germany, 2009; pp. 535–580.
56. Hoefs, J. *Stable Isotope Geochemistry*, 7th ed.; Springer International Publishing: Cham, Switzerland, 2015.
57. Ohmoto, H. Stable isotope geochemistry of ore deposits. *Rev. Mineral. Geochem.* **1986**, *16*, 491–559.
58. Sheppard, S.M.F. Characterization and isotopic variations in natural waters. *Rev. Mineral. Geochem.* **1986**, *16*, 165–183.
59. Han, J.L.; Sun, J.G.; Liu, Y.; Zhang, X.T.; He, Y.P.; Yang, F.; Chu, X.L.; Wang, L.L.; Wang, S.; Zhang, X.W.; et al. Genesis and age of the Toudaoliuhe breccia-type gold deposit in the Jiapigou mining district of Jilin Province, China: Constraints from fluid inclusions, H-O-S-Pb isotopes, and sulfide Rb–Sr dating. *Ore Geol. Rev.* **2020**, *118*, 103356. [CrossRef]
60. Hoefs, J. *Stable Isotope Geochemistry*, 4th ed.; Springer: Berlin/Heidelberg, Germany, 1997; p. 201.
61. Faure, G. *Principles of Isotope Geology*; John Wiley and Sons Inc.: New York, NY, USA, 1986.
62. Rye, R.O.; Ohmoto, H. Sulfur and carbon isotopes and ore genesis: A review. *Econ. Geol.* **1974**, *69*, 826–842. [CrossRef]
63. Ohmoto, H.; Goldhaber, M.B. Sulfur and carbon Isotopes. In *Geochemistry of Hydrothermal Ore Deposits*, 3rd ed.; Barnes, H.L., Ed.; John Wiley: New York, NY, USA, 1997; pp. 517–611.
64. Ohmoto, H. Systematics of sulfur and carbon isotopes in hydrothermal ore deposits. *Econ. Geol.* **1972**, *67*, 551–578. [CrossRef]
65. Kelly, W.C.; Rye, R.O. Geologic fluid inclusion and stable isotope studies of the tin-tungsten deposits of Panasqueira. *Portual. Econ. Geol.* **1979**, *74*, 1721–1822. [CrossRef]
66. Ohmoto, H.; Rye, R.O. Isotopes of sulfur and carbon. In *Geochemistry of Hydrothermal Ore Deposit*; Barnes, H.L., Ed.; Wiley: New York, NY, USA, 1979; pp. 509–567.
67. Ripley, E.M. Origin and concentration mechanisms of copper and nickel in Duluth Complex sulfide zones, a dilemma. *Econ. Geol.* **1986**, *81*, 974–978. [CrossRef]
68. Xiao, F.; Fan, F.P.; Xing, G.F.; Jiang, S.Y. Ore genesis of Qingyunshan Cu-Au deposit in the Dehua-Youxi area of Fujian Province, southeastern China: Constraints from U-Pb and Re-Os geochronology, fluid inclusions, and H-O-S-Pb isotope data. *Ore Geol. Rev.* **2021**, *132*, 104006. [CrossRef]
69. Liu, L.; Gu, X.X.; Zhang, Y.M.; Ouyang, X.; Wang, L.Z.; Gao, L.Y. Genesis of the Jinchanggouliang gold deposit, Chifeng, China: Constraints from Fluid inclusions and isotopic geochemistry. *Ore Geol. Rev.* **2019**, *115*, 103180. [CrossRef]
70. Doe, B.R.; Zartman, R.E. Plumbotectonics, the Phanerozoic. In *Geochemistry of Hydrothermal Ore Deposit*; Barnes, H.L., Ed.; Wiley: New York, NY, USA, 1979; pp. 22–70.

71. Lu, H.Z. Fluids immiscibility and fluids inclusions. *Acta Petrol. Sin.* **2011**, *27*, 1253–1261.
72. Pirajno, F.; Ernst, R.E.; Borisenko, A.S.; Fedoseev, G.; Naumov, E.A. Intraplate magmatism in Central Asia and China and associated metallogeny. *Ore Geol. Rev.* **2009**, *35*, 114–136. [CrossRef]
73. Stefánsson, A.; Seward, T.M. Gold(I) complexing in aqueous sulphide solutions to 500 °C at 500 bar. *Geochim. Cosmochim. Acta* **2004**, *68*, 4121–4143. [CrossRef]
74. Williams-Jones, A.E.; Bowell, R.J.; Migdisov, A.A. Gold in solution. *Elements* **2009**, *5*, 281–287. [CrossRef]
75. Benning, L.G.; Seward, T.M. Hydrosulfide complexing of Au (I) in hydrothermal solutions from 150 to 400 °C and 500–1500 bar. *Geochim. Cosmochim. Acta* **1996**, *60*, 1849–1871. [CrossRef]
76. Phillips, G.N.; Evans, K.A. Role of CO_2 in the formation of gold deposits. *Nature* **2004**, *429*, 860–863. [CrossRef] [PubMed]
77. Cole, D.R.; Drummond, S.E. The effect of transport and boiling on Ag/Au ratios in hydrothermal solutions: A preliminary assessment and possible implications for the formation of epithermal precious-metal ore deposits. *J. Geochem. Explor.* **1986**, *25*, 45–79. [CrossRef]
78. Spycher, N.F.; Reed, M.H. Evolution of a broadlands-type epithermal ore fluid along alternative PT paths; implications for the transport and deposition of base, precious, and volatile metals. *Econ. Geol.* **1989**, *84*, 328–359. [CrossRef]
79. Lowenstern, J.B. Carbon dioxide in magmas and implications for hydrothermal systems. *Miner. Depos.* **2001**, *36*, 490–502. [CrossRef]
80. Hagemann, S.G.; Lüders, V. PTX conditions of hydrothermal fluids and precipitation mechanism of stibnite-gold mineralization at the Wiluna lode-gold deposits, Western Australia: Conventional and infrared microthermometric constraints. *Miner. Depos.* **2003**, *38*, 936–952. [CrossRef]
81. Heinrich, C.A. Fluid-fluid interactions in magmatic-hydrothermal ore formation. *Rev. Mineral. Geochem.* **2007**, *65*, 363–387. [CrossRef]
82. Wagner, T.; Mlynarczyk, M.S.J.; Williams-Jones, A.E.; Boyce, A.J. Stable isotope constraints on ore formation at the San Rafael tin–copper deposit, southeast Peru. *Econ. Geol.* **2009**, *104*, 223–248. [CrossRef]
83. Thompson, J.F.H.; Sillitoe, R.H.; Baker, T. Intrusion-related gold deposits associated with tungsten-tin provinces. *Miner. Depos.* **1999**, *34*, 323–334. [CrossRef]
84. Lang, J.R.; Baker, T. Intrusion-related gold systems: The present level of understanding. *Miner. Depos.* **2001**, *36*, 477–489. [CrossRef]
85. Yang, Q.; Ren, Y.S.; Hao, Y.J.; Wang, B.; Sun, Z.M.; Li, J.M. Ore fluid, geochronology and tectonic setting of mesothermal gold metallogeny in southeastern Jilin Province, Northeast China: A case study of the Shajingou gold deposit. *Ore Geol. Rev.* **2019**, *109*, 229–252. [CrossRef]
86. Zhu, X.Q.; Wang, J.B.; Pang, Z.S.; Zhen, S.M.; Yang, F.; Xue, J.L.; Jia, H.X.; Shi, G.Y. Ore geology, fluid inclusion microthermometry and H-O-S isotopes of the Liyuan gold deposit, central Taihang Mountains, North China Craton. *Minerals* **2019**, *9*, 606. [CrossRef]
87. Cepedal, A.; Fuertes-Fuente, M.; Martín-Izard, A.; García-Nieto, J.; Boiron, M.C. An intrusion-related gold deposit (IRGD) in the NW of Spain, the Linares deposit: Igneous rocks, veins and related alterations, ore features and fluids involved. *J. Geochem. Explor.* **2013**, *124*, 101–126. [CrossRef]
88. Groves, D.I.; Goldfarb, R.J.; Gebre-Mariam, M.; Hagemann, S.G.; Robert, F. Orogenic gold deposits: A proposed classification in the context of their crustal distribution and relationship to other gold deposit types. *Ore Geol. Rev.* **1998**, *13*, 7–27. [CrossRef]
89. Goldfarb, R.J.; Groves, D.I.; Cardoll, S. Orogenic Au and geologic time: A global synthesis. *Ore Geol. Rev.* **2001**, *18*, 1–75. [CrossRef]
90. So, C.S.; Yun, S.T. Jurassic Mesothermal gold mineralization of the Samhwanghak Mine, Youngdong Area, Republic of Korea: Constraints on hydrothermal fluid geochemistry. *Econ. Geol.* **1997**, *92*, 60–80. [CrossRef]

minerals

Article

Age of the Tuchengzi Formation in Western Liaoning Province and the Jurassic–Cretaceous Boundary from the Continuous Core Records of Well YD1, Jinyang Basin

Shou-Liang Sun [1], Shu-Wang Chen [1,*], Zhong-Jie Yang [2,*], Tao Zhang [1], Yong-Fei Li [1], Ji-Chang Zhu [3], Huai-Chun Wu [4], Tian-Tian Wang [4], Yue-Juan Zheng [1] and Qiu-Hong Ding [1]

1 Shenyang Center of China Geological Surveys, Shenyang 110034, China; sunshouliang@mail.cgs.gov.cn (S.-L.S.); zhangtao@mail.cgs.gov.cn (T.Z.); liyongfei@mail.cgs.gov.cn (Y.-F.L.); zhengyuejuan@mail.cgs.gov.cn (Y.-J.Z.); dingqiuhong@mail.cgs.gov.cn (Q.-H.D.)
2 Geological and Mineral Survey Institute of Liaoning Province, Shenyang 110031, China
3 State Key Laboratory of Lithospheric Evolution, Institute of Geology and Geophysics, Chinese Academy of Sciences, Beijing 100029, China; zhujc@mail.iggcas.ac.cn
4 School of Ocean Sciences, China University of Geosciences, Beijing 100083, China; whcgeo@cugb.edu.cn (H.-C.W.); tianwang@cugb.edu.cn (T.-T.W.)
* Correspondence: chenshuwang@mail.cgs.gov.cn (S.-W.C.); yangzj870227@163.com (Z.-J.Y.); Tel.: +86-024-8600-2993 (S.-W.C.); +86-136-5403-7332 (Z.-J.Y.)

Abstract: The Tuchengzi Formation is widely distributed in western Liaoning Province with a clear top and bottom. It is the focal area for the delineation of the terrestrial Jurassic–Cretaceous boundary in China. Based on continuous core samples taken from well YD1, detailed lithostratigraphic sequences and zircon uranium–lead (U-Pb) dating WERE used to investigate the Tuchengzi Formation. The zircon U-Pb ages of the tuff samples taken from the First and Third Members of the Tuchengzi Formation ranged from 153.8 to 137.16 Ma, indicating that they were formed in the late Middle Jurassic–Early Cretaceous. Dating results from the bottom of the Second Member of the Tuchengzi Formation indicate that the sedimentary time of the stratum is no later than 145.7 ± 2.1 Ma. We concluded that the Jurassic–Cretaceous boundary of the Jinyang Basin in western Liaoning province may be located at the interface at a depth of 464 m in well YD1. This conclusion is consistent with the Jurassic–Cretaceous boundary that has been presumed by other researchers based on paleontological assemblage features found in recent years, and can provide useful geological marker beds for the future study of the terrestrial Jurassic–Cretaceous boundary. In addition, the authors also systematically sorted the potential development areas and layers of the terrestrial Jurassic–Cretaceous boundary line, which may also provide useful geological marker beds for the future study of the terrestrial Jurassic–Cretaceous boundary.

Keywords: Tuchengzi Formation; Jinyang Basin; Jurassic-Cretaceous boundary; western Liaoning region; YD1 well; zircon U-Pb dating

1. Introduction

The definition and study of the Jurassic–Cretaceous boundary are primarily based on marine sedimentary sequences, and major advances have been made in recent years with respect to these subjects. The international stratigraphic tables published by the International Union of Geological Sciences (IUGS) and the International Commission on Stratigraphy (ICS) have changed the timeframe of the Jurassic–Cretaceous boundary from 135 Ma [1] to 145 Ma [2], further reflecting the progress and dynamics of Jurassic–Cretaceous boundary studies. The Jurassic–Cretaceous stratigraphy in China is primarily terrestrial [3], and studies on its boundary are an important part of international standard stratigraphic delineation, albeit with more regional characteristics and application values.

Northeast China has the most continuous marine and continental sedimentary records of the Jurassic–Cretaceous boundary. These areas include the Northern Hebei area (Jibei), the western Liaoning area (Liaoxi), the Songliao Basin, and eastern Heilongjiang Province. It is important to note that Late Jurassic–Early Cretaceous strata in the Jibei–Liaoxi area are the most continuous and have already been studied in detail. Therefore, the Jibei–Liaoxi area has become the key area for studying the continental Jurassic–Cretaceous boundary in China [4]. However, the true age and accurate position of the Jurassic–Cretaceous boundary at these eastern Chinese terrestrial sequences are still controversial, largely due to the large depth of the target layer, poor outcrop exposure, and discontinuous drill cores.

Although there is no unified understanding of the correlation between the bottom boundary of the terrestrial Cretaceous and the international stratigraphic table (145 Ma) [5–9], researchers believe that the Jurassic–Cretaceous boundary ought to be located in the Tuchengzi Formation in northern Hebei–western Liaoning Province, and they have made this hypothesis a study focus [3,10–13]. However, the Tuchengzi Formation is mostly composed of coarse clastic conglomerate and red beds, and fossils are rare, making it difficult to correlate findings with marine bed fossils. Therefore, the age of the Tuchengzi Formation and the identification of its sedimentary hiatus are both of particular importance in determining the Jurassic–Cretaceous boundary.

Well YangD1 (YD1) was drilled into the Jinyang Basin through 2000 m of continuous volcanic–sedimentary rock in the Tuchengzi and Tiaojishan Formations. Cores taken from YD1 provide a complete and continuous record of the Late Jurassic–Early Cretaceous period, providing unique material for studying the Tuchengzi Formation and the Jurassic–Cretaceous boundary in the Jibei–Liaoxi area. In the present work, by analyzing drilling core samples from well YD1, we were able to study the age of the Tuchengzi Formation in detail. The potential position of the Jurassic–Cretaceous boundary in the Jibei–Liaoxi area was also studied. Based on a comparative study of the various potential areas and layers that might contain the Jurassic–Cretaceous boundary in northeastern China, this study provides fundamental information for its further examination and its correlation with marine stratigraphy.

2. Recent Advances in the Study of the Jurassic-Cretaceous Boundary

2.1. Advances of the Global Marine Jurassic-Cretaceous Boundary

It is widely accepted that the Berriasian Stage is located at the base of the Lower Cretaceous, and the bottom of the stage in the Tethys Zone may represent the Jurassic–Cretaceous boundary [3]. The first occurrences of the calcareous nannofossils Nannoconus kamptneri minor and N. steinmannii minor have been widely used as place markers for the Tithonian–Berriasian and Jurassic–Cretaceous boundaries [14]. The Jurassic–Cretaceous boundary is at the base of magneto zone M18r, with a recommended age of 145.0 Ma according to the international stratigraphic chart [15]. Although this is only a recommended value and lacks actual support in the geological data, an increasing number of studies have proved that this value is rational. Mahoney et al. reported a 40 Ar/39 Ar age of 144.6 $\pm$ 0.8 Ma for the bottommost reversed-polarity basalt sill cored in the lowermost Berriasian sediments on the Shatsky Rise in the northwestern Pacific [16]. The sills were bounded by both calcareous nannofossil zone NK1 and the radiolarian Pseudodictyomitra carpatica zone. Liu et al. reported a zircon uranium–lead (U-Pb) age of 142–140 Ma for the lower part of the Sangxiu Formation in southern Tibet, China, which is located in the Nannoconus steinmannii minor zone above the Berriasian Stage, suggesting that the bottom boundary age of the Berriasian Stage may be older than 142–140 Ma [17]. Vennari et al. showed that the Jurassic–Cretaceous boundary in the Andean region is located in the Nannoconus kamptneri minor zone at the bottom of the Berriasian Stage. The tuff above about 22 m of shale yielded a zircon U-Pb age of 140 Ma, suggesting that the Jurassic–Cretaceous boundary in the Andean region may also be older than 140 Ma [18].

2.2. Recent Advances in the Terrestrial Jurassic-Cretaceous Boundary in the Jibei-Liaoxi Region

In recent years, research on the terrestrial Jurassic–Cretaceous boundary in the Jibei–Liaoxi region has experienced major progress, but the definition of both the boundary and its age limit is still controversial. There are two main opinions concerning this issue.

2.2.1. The Time Limit of the Terrestrial Jurassic-Cretaceous Boundary Line Is Earlier Than 145 Ma

A number of Chinese scientists have argued that the Jurassic–Cretaceous boundary is difficult to correlate directly with international marine standards based on terrigenous stratigraphic data, and they have proposed the use of 135 or 137 Ma as its age [10]. Chen placed the Jurassic–Cretaceous boundary between the Yixian and Jiufotang Formations and recommends a boundary age of 125 Ma [19]. The All-China Commission of Stratigraphy published a specification on the regional chronostratigraphic (geological) chronology of China in 2002, which set the boundary between the Dabeigou and Yixian Formations, with an age of 137 Ma [20]. Several researchers have set the Jurassic–Cretaceous boundary between the Dabeigou and Dadianzi Formations in the Luanping Basin, northern Hebei, limiting the boundary age to 130 Ma [6,9,21,22]. Wang et al. suggested that the Jurassic–Cretaceous boundary should be in the lower Yixian Formation and recommended an age of 124 Ma [23].

2.2.2. The Terrestrial Jurassic-Cretaceous Boundary Line Is Close to 145 Ma

According to new advances in international research on the Jurassic–Cretaceous boundary, an increasing number of Chinese scientists have gradually adopted the global time limit for this boundary. At the 4th China National Stratigraphic Conference in 2013, the age of the Jurassic–Cretaceous boundary was adjusted to 145 Ma [24]. As most of the isotopic ages measured in the Tuchengzi Formation have been found to be younger than 145 Ma, the terrestrial Jurassic–Cretaceous boundary may be located within the Tuchengzi Formation [25–29]. In addition, the terrestrial sporulation assemblage of the Tuchengzi Formation also has characteristics of the Late Jurassic–Early Cretaceous [30–35]. This evidence suggests that the Chinese terrestrial Jurassic–Cretaceous boundary is located within the Tuchengzi Formation [10,13,28,29,36–40]. However, whether the boundary is located between Members 1 and 2 or Members 2 and 3 is highly controversial. Wan et al. tentatively placed the Jurassic–Cretaceous boundary in the Yanliao region between Members 2 and 3 based on foliage, mesozoans, and sporulation [41].

3. Geological Background

Northeastern China is tectonically located at the intersection of three major plates: the Siberian, Pacific, and Mongolian–North China plates. Its northern region is connected to the Siberian plate through the Mongolia–Okhotsk suture zone, and its eastern region is connected to the Pacific plate through the Sikhote–Alin tectonic zone (Figure 1). Prior to the Late Paleozoic, a series of microplates, including the Erguna–Xing'an plate, the Songliao–Xilinhot microplate, and the Jiamusi microplate, assembled and formed the Heilongjiang plate [42]. In the Late Permian, the Heilongjiang–Mongolian plate was assembled with the northern margin of the North China Plate along the Xar Moron River suture zone (Figure 1a), which formed the basement of the Songliao Basin [43]. After these events, the Jibei–Liaoxi region at the northern margin of the North China Plate and the Songliao Basin were tightly connected.

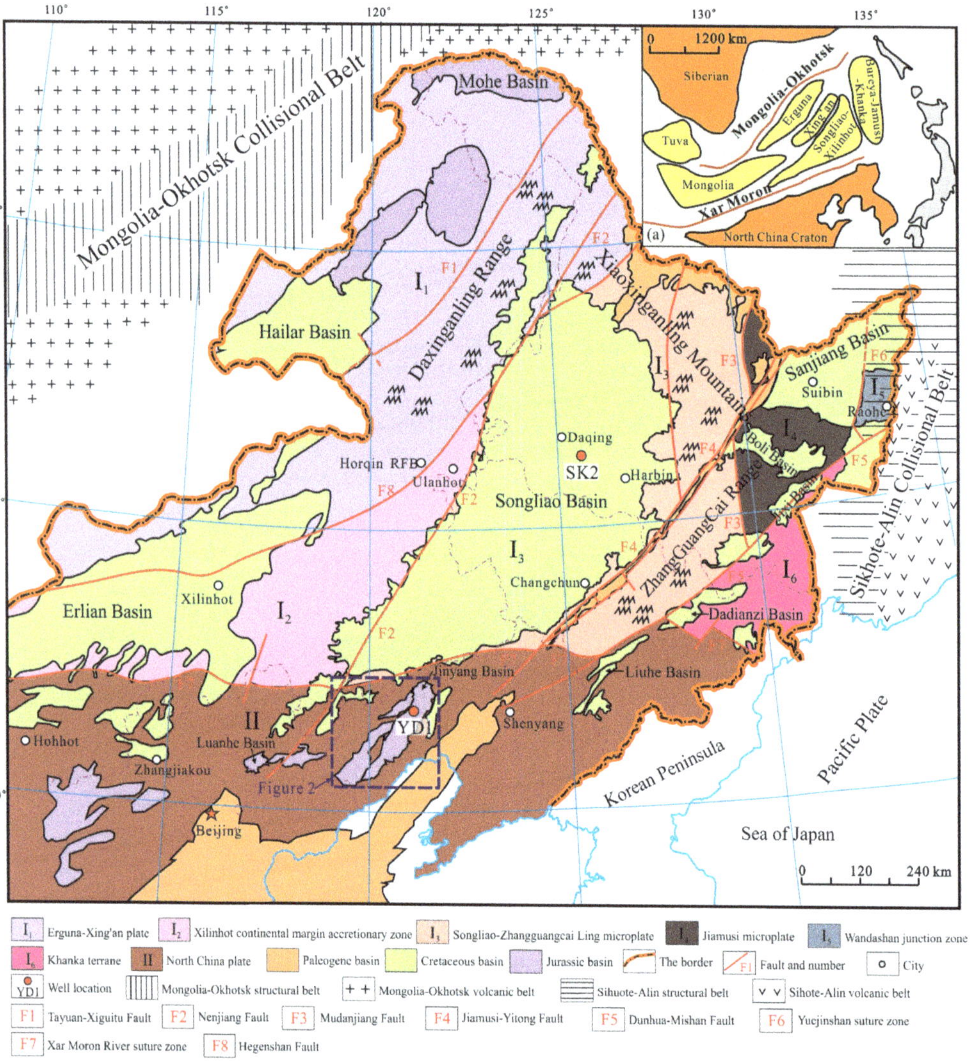

Figure 1. Simplified geological map of northeast China showing the geological tectonic units and location of the study area. (**a**) Tectonic sketch map showing the main subdivisions of eastern Asia (modified after [44]).

The Nenjiang Fault is a major fault controlling the formation of the Songliao Basin [45]. To the west of the Nenjiang fault are areas where Mesozoic Daxinganling volcanic rocks are widespread, in addition to a series of Jurassic–Cretaceous basins, including the Mohe and Hailar Basins [46]. The Jiamusi–Yitong Fault is the major basin-controlling fault of the Songliao Basin in the east. To the east of the Jiamusi–Yitong fault is the northeast-trending Dunhua–Mishan Fault, in addition to Cretaceous basins such as the Liuhe and Dadianzi Basins. The Mudanjiang Fault is north–south trending, and to the east, there is a series of Cretaceous basins, including the Sanjiang, Boli, and Jixi Basins. The Xar Moron River suture zone is on the southern boundary of the Songliao Basin, trending east–west, which belongs to the intraplate suture zone within the Mongolian–North China Plate [45]. The southern section of the belt primarily includes western Liaoning Province (Liaoxi) and the northern Hebei (Jibei) area, where a series of Mesozoic basins are located, including the Jinyang, Beipiao and Luanping Basins (Figure 1).

As a result of the Yanshanian Movement and the convergence of the East Asian multidirectional plates, the Mesozoic volcanic–sedimentary stratigraphy is most complete in western Liaoning, which is located on the eastern Yanshan orogenic belt [47,48]. The Jinyang Basin is the largest and most typical of all the basins [49–51].

Jurassic–Cretaceous strata are widespread in the Jinyang Basin, accounting for more than 90% (Figure 2b). Impacted by the Indosinian Movement and Early Yanshan orogeny, the upper basin formed in the Late Jurassic and the Tuchengzi Formation was filled with purple conglomerate, sandy conglomerate, siltstone, and argillaceous sedimentary formations. From bottom to top, the strata consist of the Xinglonggou and Beipiao Formations of the Lower Jurassic, the Haifangou and Tiaojishan Formations of the Middle Jurassic, and the Tuchengzi and Yixian Formations of the Upper Jurassic to Lower Cretaceous (Figure 2c). The contact interface between the Beipiao and Haifangou Formations is an angular unconformity, as is the interface between the Tuchengzi and Yixian Formations. However, the others are parallel unconformities. The angular unconformity between the Beipiao and Haifangou Formations and the Tuchengzi and Yixian Formations may represent Phases A1 and B of the Yanshanian Movement, respectively [52–57]. The Tuchengzi Formation was deposited during the transitional period between Phase A1 and B of the Yanshan orogeny [58,59].

The Jibei–Liaoxi area was located in a key position in the tectonic transition of East Asia during the turn of the Jurassic and Cretaceous (Figure 2a) [60]. The Yanshanian Movement triggered the Late Jurassic multidirectional extrusion and intra-land orogeny, Early Cretaceous massive extension, and lithospheric thinning. Regional tectonic and magmatic events at different times led to dramatic ecological changes [47,61], resulting in the extinction of the Yanliao Biota and the replacement of the Jehol Biota.

4. Lithologic Sections of the Tuchengzi Formation

The Tuchengzi Formation is divided into three lithologic members [62]. Based on the sedimentary environment and hiatus in this area, in addition to the core of well YD1, the boundary of Members 1 and 2 is set at the onset of gray-purple, complex-fine conglomerate, and the end of the red sand and mudstone. The boundary of Members 2 and 3 is set at the sedimentary hiatus between the green-gray lithic feldspar (intermediate-fine) sandstone and the gray-purple (complex-fine) conglomerate (Figure 2d).

At the top stratum of well YD1 is the Third Member of the Tuchengzi Formation, which is composed of gray-green aeolian sandstone. Well YD1 runs through the First and Second Members of the Tuchengzi Formation downward, and is primarily composed of sand–mudstone and polymictic conglomerates. Below the Tuchengzi Formation is the Tiaojishan Formation, which is primarily composed of neutral lava and pyroclastic. Due to the lateral differences between volcanic and sedimentary environments, the lithostratigraphy of the Tuchengzi Formation in western Liaoning Province is slightly different at each sedimentary basin. The lithostratigraphy of each member of the Tuchengzi Formation in well YD1 of the Jinyang Basin is divided as follows:

Member 1 (depth: 500.52–1207 m): The bottom is composed of gray-green, breccia-bearing, tuffaceous sandstone (medium-coarse), and gray-green andesite conglomerate (medium-coarse) with an alluvial fan facies, unconformably overlying the gray-green basaltic andesite of the Tiaojishan Formation. The middle and upper sections are brown, silty mudstone with gray-green siltstone and a thin tuff layer, which is dominated by lakeside–shallow lake subfacies.

Member 2 (depth: 50.45–500.52 m): The bottom is composed of gray-purple gravel complex conglomerate (medium-fine) with scour surface and massive bedding. The middle part is mainly gray-purple complex conglomerate. The top is gray-green lithic feldspar sandstone (medium-coarse) with grain sequence bedding, parallel bedding, and large plate crossbedding. The stratum is dominated by alluvial fan facies and conformable contact with Member 1.

Member 3 (depth: 0–50.45 m): Green-gray, lithic feldspar sandstone (medium-fine) interbedded with a thin layer of purple-red feldspar sandstone (medium-fine), with medium-large crossbedding (coarse-grained), make up Member 3. This member has an aeolian sedimentary sequence that has conformable contact with Member 2. Because the top strata of well YD1 are part of Member 3, the revealed strata in this section are incomplete. Comparing profiles near well YD1 shows that the revealed strata in this section form the bottom of the Tuchengzi Formation's Third Member.

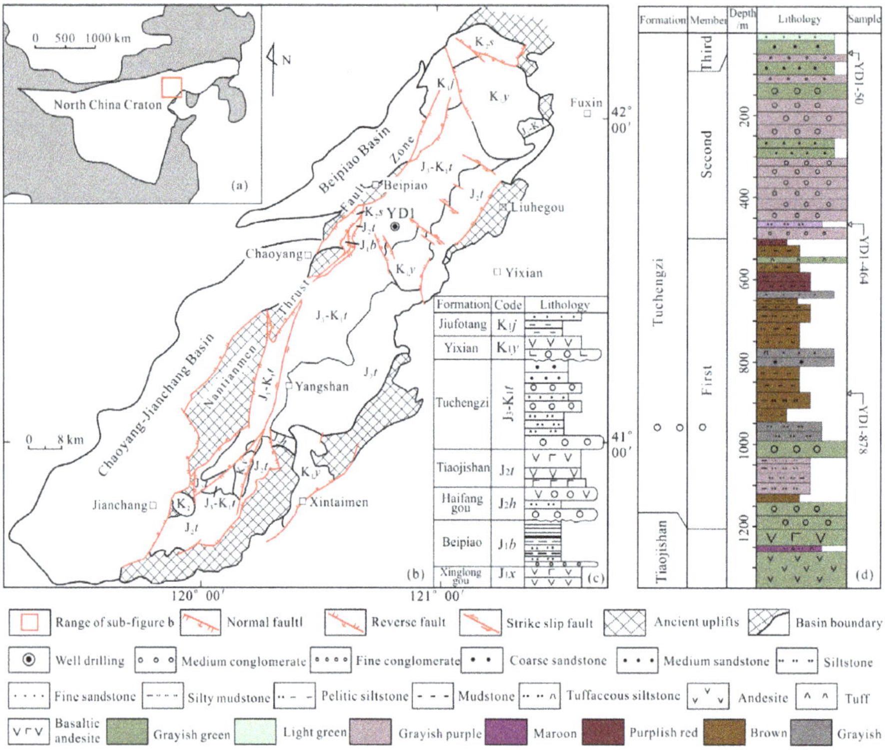

Figure 2. Geological sketch of Jinyang basin and its surrounding area and the core histogram of the Tuchengzi Formation in Well YD1. (**a**) Location of study area (modified after [63]); (**b**) geological sketch of Jinyang basin and its surrounding area modified after [64]; (**c**) filling sequence of main strata in Jinyang Basin; (**d**) the core histogram of well YD1 and sample location.

5. Analytical Methods and Sample Characteristics

5.1. Sample Characteristics

Three samples were chosen for zircon U-Pb isotopic dating. Vitreous crystalline tuff (YD1-50) at 50 m was taken from the bottom of Member 3 in well YD1. The other two samples were calcareous feldspar lithic sandstone (YD1-464) taken from a depth of 464 m at the bottom of Member 2, and rhyolitic tuff (YD1-878) at 878 m taken from the middle part of Member 1 (Figures 2d and 3).

Figure 3. Photographs and microphotographs showing sampling core and its lithology. (**a**) Glassy crystal tuff (sample YD1-50), and (**b**) under transmitted light (sample YD1-50); (**c**) calcareous feldspar lithic sandstone (sample YD1-464), and (**d**) under transmitted light (sample YD1-464); and (**e**) rhyolitic tuff (sample YD1-878), and (**f**) under transmitted light (sample YD1-878).

5.2. Analytical Methods

Zircon crystals were separated at the Mineral Separation Laboratory of Langfang Regional Geological Survey, Hebei Province. First, each sample was ground, washed, dried, and sorted. Then, magnetic separation and heavy liquid separation technology were used to retrieve the zircon crystals. Relatively complete zircons crystals were then handpicked under a binocular microscope and mounted in epoxy. To reveal the internal texture, cathodoluminescence (CL) images were obtained.

Laser Ablation Inductively Coupled Plasma Mass Spectrometry (LA-ICP-MS) zircon U-Pb analyses were performed at the State Key Laboratory of Lithospheric Evolution, Institute of Geology and Geophysics, Chinese Academy of Sciences, using an Anlyitik Jena PQMS Elite ICP-MS equipped with a 193 nm laser. Analyses were performed with a beam diameter of 25 um, a laser frequency of 10 Hz, an energy of 2.31 J/cm^2, and He as a carrier gas. LA-ICP-MS laser denudation sampling was conducted using single point denudation [65]. GJ-1 was used as an unknown to evaluate the data quality, and SRM 610 silicate glass was applied for the external standard with silicon as the internal standard [66]. ICPMSdatacal [65] and ISOPLOT 3.0 [67] were used for data reduction. In data processing, most analysis points of $^{206}Pb/^{204}Pb > 1000$ were not corrected by ordinary lead, and analysis points with abnormally high ^{204} Pb content may be affected by ordinary Pb, such as inclusions. The analysis points with abnormally high ^{204}Pb were eliminated during calculation.

6. Results of Zircon U-Pb Dating

6.1. Glassy Crystal Tuff (YD1-50)

The zircon U-Pb data derived from 58 detrital zircons taken from sample YD1-50 are listed in Table S1 and illustrated in Figures 4 and 5a,b. Zircon grain CL images showed that most were euhedral to subhedral. The zircon grains showed clear oscillatory growth zoning. This kind of zircon grain is a homologous product with a single color, a concentrated crystal group, and a similar transformation distance. These properties, together with Th/U ratios (0.48–1.69), indicate these zircon grains had a magmatic origin [68,69]. For the analyzed samples, the concordant $^{206}Pb/^{238}U$ ages ranged from 142.3 to 132.3 Ma (Table S1). A U-Pb age frequency histogram (Figure 5a) shows that the detrital zircons had three main age peaks of 137.6, 252.9, and 2511 Ma, and secondary age peaks of 314.9, 410.3, 1700, and 1912 Ma. The age of the detrital zircons ranged from 2618.5 to 123.4 Ma, which reflects the diversity and complexity of the sediment sources. The weighted average age of $^{206}Pb/^{238}U$ was 137.16 ± 0.70 Ma (MSWD = 1.3, Figure 5b), which represented the age of magma crystallization.

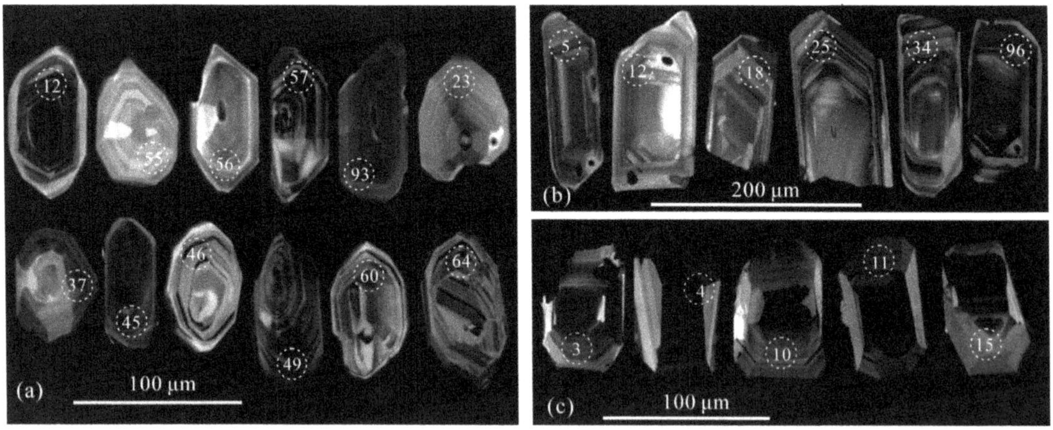

Figure 4. Cathodoluminescence (CL) images of typical detrital and magmatic zircons from the Tuchengzi Formation in well YD1. Circles indicate the location of laser ablation inductively coupled plasma source mass spectrometry, numbers in circles refer to the spot number. (**a**) Sample YD1-464; (**b**) sample YD1-50; and (**c**) sample YD1-878.

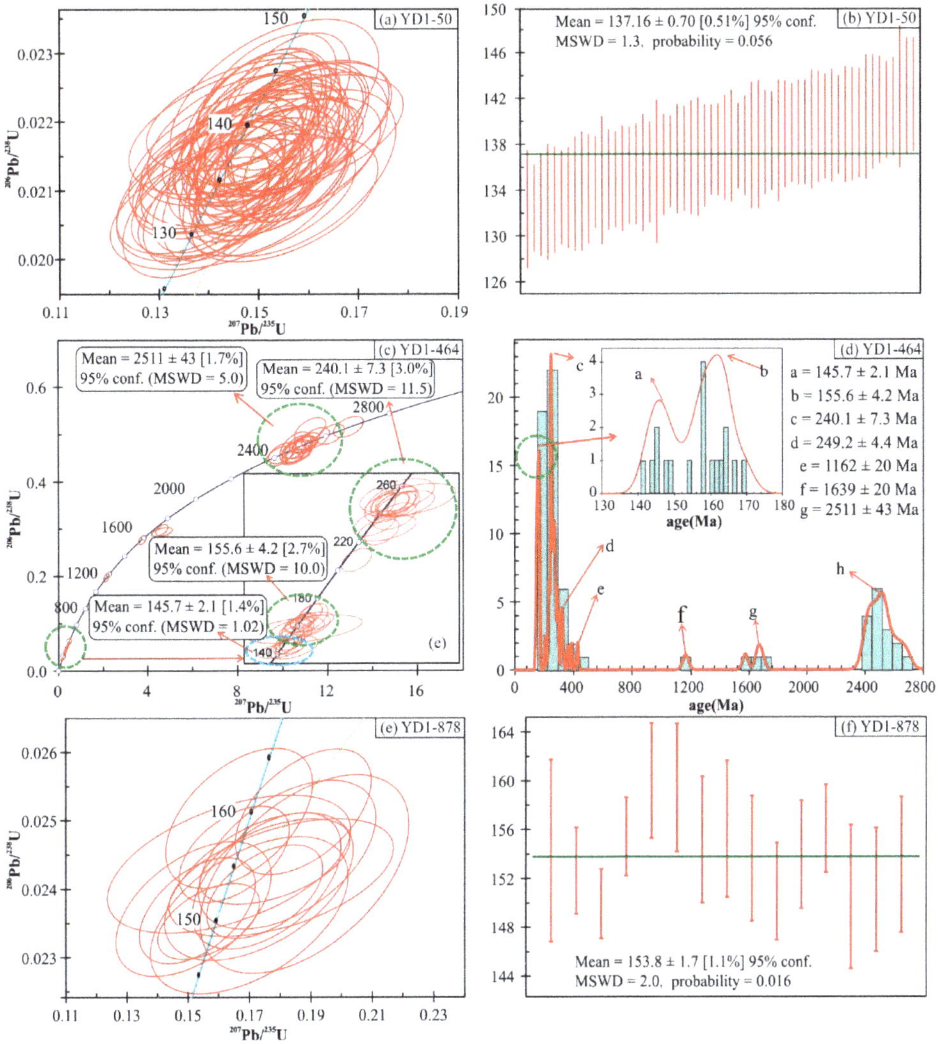

Figure 5. U-Pb Concordia diagram, frequency histogram, and weighted isochron diagram of zircon ages for samples from the Tuchengzi Formation. (**a**) U-Pb age Concordia diagram and (**b**) U-Pb weighted isochron diagram of YD1-50 magma zircon; (**c**) U-Pb age Concordia diagram and (**d**) U-Pb frequency histogram of YD1-464 detrital zircon; (**e**) U-Pb age Concordia diagram and (**f**) U-Pb weighted isochron of YD1-878 magma zircon.

6.2. Calcareous Feldspar Lithic Sandstone (YD1-464)

The zircon U-Pb data derived from 70 detrital zircons taken from sample YD1-464 are listed in Table S2 and illustrated in Figures 4 and 5c,d. Zircon grain CL images showed that most were euhedral to subhedral. The zircon grains showed clear oscillatory growth zoning. These properties, together with Th/U ratios (0.29–3.93), indicate that these zircon grains had a magmatic origin [68,69]. A total of 92 zircon grains were analyzed, and the concordant $^{206}Pb/^{238}U$ ages ranged from 2680.9 to 142.3 Ma (Figure 5c). The U-Pb age frequency histogram (Figure 5d) shows that the detrital zircons had four main age peaks of 145.7, 155.6, 240.1, and 2511 Ma and secondary age peaks of 294.2, 426.7, 1162, and

1639 Ma. The age range was large, which further reflects the diversity and complexity of the Tuchengzi Formation sediment sources.

6.3. *Rhyolitic Tuff (YD1-878)*

The zircon U-Pb data derived from 15 zircon grains taken from sample YD1-878 are listed in Table S3 and illustrated in Figures 4 and 5e,f. The zircon grains showed clear oscillatory growth zoning. These properties, together with Th/U ratios (0.59–1.81), indicate that these zircon grains had a magmatic origin [68,69]. A total of 15 zircon grains were analyzed, and the concordant $^{206}Pb/^{238}U$ ages ranged from 149.9 to 160.0 Ma (Table S3). The weighted average age of $^{206}Pb/^{238}U$ was 153.8 ± 1.7 Ma (MSWD = 2.0, Figure 5f), which represented the age of magma crystallization.

7. Discussion

7.1. *Age of the Tuchengzi Formation*

The Jurassic–Cretaceous strata in China are mostly terrestrial deposits. The high mountains formed by tectonic movements block communication between different lake ecosystems. As a result, the fauna and flora show remarkable regional differences and make it difficult to find biological markers for isochronous comparisons. This makes it difficult to reach a unified opinion on the biostratigraphic comparison of different phyla [32,33,70–77]. In terms of high precision isotopic dating, He et al. used the K-Ar and Rb-Sr methods to limit the age of the upper and lower strata of the Tuchengzi Formation to 156–145 Ma [78,79]. Davis et al. and Sun et al. limited its age to 156–139 Ma based on previous zircon U-Pb age data [26,28]. The National Stratigraphic Commission published the Manual of China's Regional Geochronology (Geochronology) table in 2014, which classified the strata of the Upper Jurassic in China as the Tuchengzi Formation. However, due to the difficulty in determining the exact attribution of the Tuchengzi Formation, they did not establish its stage [24,80].

Based on two tuff samples with zircon grains of magmatic origin, we further constrained the age of the Tuchengzi Formation in well YD1 between 137.16 ± 0.7 Ma (YD1-50) and 153.8 ± 1.7 Ma (YD1-878). However, the age of 137.16 ± 0.7 Ma in this study was the absolute age at the bottom of Member 3, where there was an aeolian dune deposition at least 1000 m above the interface [81]. Although the top age of the Tuchengzi Formation was not obtained, it can be inferred that the age of the top formation is later than 137.16 ± 0.7 Ma. Compared to the International Stratigraphic Chronology (2018), we suggest that the Tuchengzi Formation in western Liaoning province was formed between the middle–late Late Jurassic and Early Cretaceous.

7.2. *The Boundary between Jurassic and Cretaceous*

A total of 70 effective detrital zircons from sample YD-464, collected from the lower part of Member 2, were analyzed, among which, the ages of the six youngest zircon grains ranged from 142.3 to 149.5 Ma, with a weighted mean age of 145.7 ± 2.1 Ma (MSWD = 1.02, Figure 5c). The frequency histogram showed the youngest peak age at 145.0 Ma, which is consistent with the Jurassic–Cretaceous boundary of 145 Ma from the International Commission on Stratigraphy (ICS) [82]. It is also close to the minimum age (144.6 ± 0.8 Ma) of the Lower Cretaceous Berriasian Stage at the Shatsky Rise in the northwest Pacific [16]. The peak age and weighted-mean age of the samples were within the error range, which indicates that the sedimentary period of the strata was no later than 145.7 ± 2.1 Ma. Therefore, we concluded that the Jurassic–Cretaceous boundary of the Jinyang Basin in western Liaoning Province may be located at the interface at a depth of 464 m in well YD1. Based on the sedimentary facies analysis of the upper and lower strata, we believe that there may be a sedimentary gap between Members 1 and 2 and of the Tuchengzi Formation, which may be the Jurassic–Cretaceous boundary of the Mesozoic terrestrial strata in western Liaoning province (Figure 6a). This result is also consistent with Huang's opinion that the Jurassic–Cretaceous boundary is located in the upper part of the Tuchengzi Formation based on the paleontological assemblage characteristics [59,83].

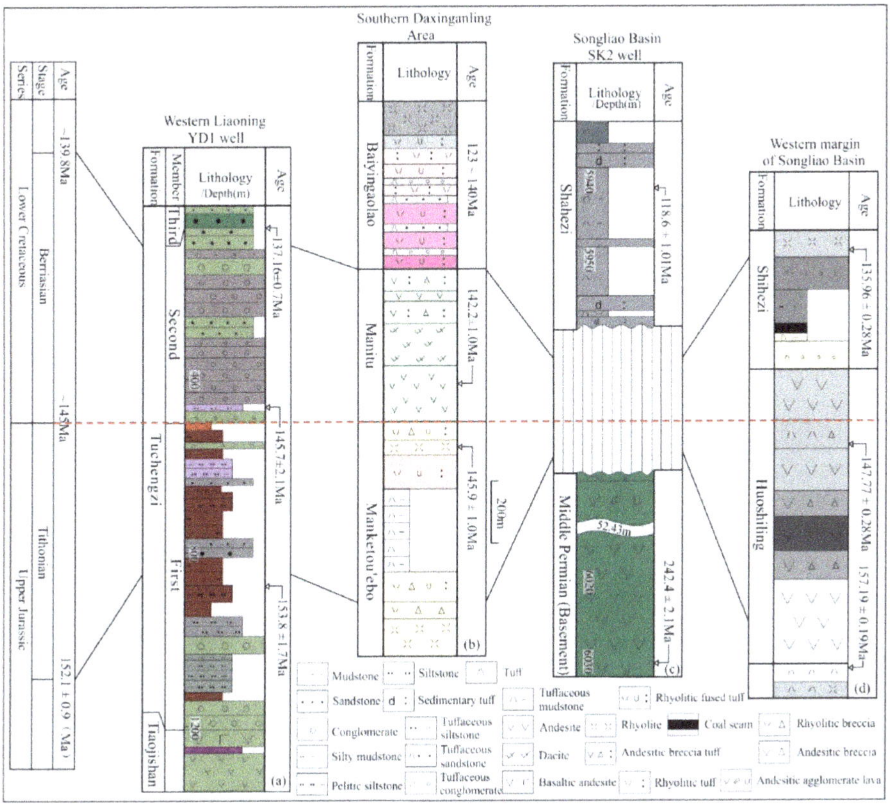

Figure 6. Potential layer comparison of terrestrial Jurassic–Cretaceous boundary in Northeast China. (**a**) Western Liaoning YD1 well; (**b**) Southern Daxinganling area; (**c**) Songliao Basin SK2 well; (**d**) Western margin of Songliao Basin.

Although a final determination of the international Jurassic–Cretaceous boundary age requires more accurate age data and the support of further research results, the latest research results indicate that 145 Ma can be used as the international Jurassic–Cretaceous boundary age. Considering that the Tuchengzi Formation was formed between 154 and 137 Ma ago, we further verified that the Jurassic–Cretaceous boundary exists within the Tuchengzi Formation in the northern Hebei area and western Liaoning Province regions. The results can provide detailed and basic geological data for the future study of the terrestrial Jurassic–Cretaceous boundary.

7.3. Potential Areas and Layers for the Development of the Terrestrial Jurassic-Cretaceous Boundary

It is generally accepted that the closure of the Mongolian–Okhotsk Ocean was the most important geotectonic event in northeast Asia during the Mesozoic, and it is characterized by a west-to-east scissor-like closure [84–86]. Its eastern closure process lasted until the Late Jurassic–Early Cretaceous [87,88].

To the west of the Nenjiang fault, i.e., the Daxinganling area, the Manketou'ebo Formation is a volcanic system that was formed by the collapse of thickened crust or delamination extension after the closure of the Mongolia–Okhotsk Ocean. Numerous isotopic dating results taken from the Manketou'ebo Formation have been reported in recent years, with age results ranging from 157 to 132 Ma [89–92]. Of note, Jia et al. performed zircon U-Pb dating on two Manketou'ebo Formation rhyolite samples from the Horqin Right Front Banner area, and the results were 145.9 ± 1.0 Ma and 146.0 ± 1.3 Ma (Figure 6b) [93]. The Manketou'ebo Formation overlies the Manitu Formation, of which the ages of volcanic rocks are between 139 and 148 Ma [94,95]. Tao et al. reported an age of 142.2 ± 1.0 Ma for the andesite of the Manitu Formation near Ulanhot in southern Daxinganling (Figure 6b) [96]. Judging from the dating results, the Manketou'ebo and Manitu formations have diachronous phenomena, which may be attributed to a younger trend from the southwest to the northeast in the spatiotemporal distribution of the Manketou'ebo Formation [93]. However, the overlying volcanic rocks of the Baiyingaolao Formation are between 123 and 140 Ma in age and belong to the Early Cretaceous (Figure 6b) [97]. Therefore, the Manketou'ebo and Manitu formations in southern Daxinganling are noteworthy formations for the study of the terrestrial Jurassic–Cretaceous boundary in these regions.

Andesite taken from a depth of 6031.9 m in well SK2, located in the Xujiaweizi fault depression in the Songliao Basin, yielded an age of 242.4 ± 2.1 Ma (Middle Triassic) [98]. The age of rhyolitic tuff approximately 88.71 m above the andesite (5943.19 m) was 118.6 ± 1.0 Ma [99]. Based on these two dating results, we suggest that, in the Xujiaweizi fault trap, where well SK2 was located, the Middle Triassic–Early Cretaceous sediments were in hiatus for around 123.8 Ma (Figure 6c). This also indicates that, after the closure of the Mongolian–Okhotsk ocean, no extensional basin was formed in the Songliao Basin. Consequently, there is no extensive Late Jurassic–Early Cretaceous continuous sedimentation in the Songliao Basin. Instead, numerous Late Jurassic volcanic rocks have been reported in the Changjiaweizi and Melisi fault depressions on the western margin of the Songliao Basin [100,101]. Qu et al. established sedimentary sequences for the Huoshiling Formation based on a detailed study in the Xujiaweizi fault depression [102,103]. The lower part is primarily composed of sedimentary rocks, whereas the upper part is composed of volcanics rocks interlayered with sedimentary sequences. Based on the established sequences, Wang et al. and Qu et al. suggested its diachronous nature (Figure 6d) [103,104]. Therefore, the Huoshiling Formation in the southern Songliao Basin is a key area in the study of the Jurassic–Cretaceous boundary.

The Jibei–Liaoxi area adjacent to the Songliao Basin is an important area for the study of the terrestrial Jurassic–Cretaceous boundary. With intensive studies and deeper stratigraphy revealed by drilling, there is an increasing amount of evidence suggesting that the terrestrial Jurassic–Cretaceous boundary in the Jibei–Liaoxi area is located in the upper portion of the Tuchengzi Formation's First Member. Zhang et al. reported a zircon U-Pb age of 142.6 ± 1.3 Ma for the welded tuff at the base of the Second Member of the Tuchengzi Formation in the Luanping Basin, Hebei [27]. Lin and Li reported a spodumene assemblage at the top of the First Member in the Luanping Basin, Hebei Province, and suggested that it is of Early Cretaceous (middle–late-stage Valanginian) origin, and the Jurassic–Cretaceous boundary in northern China is located in the upper portion of the First Member [105]. Another report suggested that the sporulation assemblage of Member 3 near Beipiao Sihetun is similar to that of the lower portion of the Yixian Formation in western Liaoning Province; no discernable differences could be found between the appearance of flora within the two formations [83]. These latest results are consistent with the conclusion that the Jurassic–Cretaceous boundary is located in the upper part of the Tuchengzi Formation's First Member. The lateral comparison of the formation in western Liaoning Province is crucial for reconstructing the sedimentary and evolutionary history of the Jurassic–Early Cretaceous in northern China.

Transgressions occurred in the Late Jurassic–Early Cretaceous Valanginian Stage in eastern Heilongjiang Province, but they were limited to the Dong'an-Sui'bin line and its northern area. Among them, the Donganzhen Formation in the Raohe and the Dongrong Formation in Suibin are two potential areas that may contain the Jurassic–Cretaceous boundary [106–109]. Marine fossil data from the interfacial strata can be compared with the international marine Jurassic–Cretaceous boundary. In contrast, the sporopollen fossil assemblage can be compared with the terrestrial Jurassic–Cretaceous boundary, thus establishing a bridge between the marine and terrestrial comparison of the Jurassic–Cretaceous boundary. Therefore, eastern Heilongjiang is a key area for making terrestrial and international marine comparisons with respect to the Jurassic–Cretaceous boundary in the Jibei–Liaoxi region. The next step of this work should be focused on the isotope chronology, biostratigraphy, magnetic stratigraphy, and cyclic stratigraphy of the Jibei–Liaoxi region and eastern Heilongjiang.

8. Conclusions

(1) The zircon U-Pb dating of two tuff samples from Members 1 and 3 of the Tuchengzi Formation in well YD1 constrained the sedimentary age range from 153.8 to 137.16 Ma, although the upper age may be much later than 137.16 Ma. According to the International Stratigraphic Chronology (2018), the age of the Tuchengzi Formation in western Liaoning province was determined to be from the middle–late Late Jurassic to the Early Cretaceous.

(2) A total of six zircon grains collected from the bottom of Member 2 of the Tuchengzi Formation (YD-464) yielded a weighted mean age of 145.7 $\pm$ 2.1 Ma, which is consistent with the peak age of the sample within the error range, and can be roughly used as the sedimentary period for this formation. Therefore, there may be a depositional gap between Members 2 and 1 of the Tuchengzi Formation, or at the top of Member 1, which can be regarded as the Jurassic–Cretaceous boundary of the Mesozoic terrestrial strata in western Liaoning Province. This conclusion is consistent with the Jurassic–Cretaceous boundary inferred by paleontological assemblage characteristics in recent years.

(3) The Tuchengzi Formation in the Jibei–Liaoxi region, the Dongrong Formation, and the Dong'anzhen Formation in eastern Heilongjiang are key layers that may contain the terrestrial Jurassic–Cretaceous boundary. The Jurassic–Cretaceous boundary may be only locally developed in the Songliao Basin, and the Huolshiling Formation in the south is a key layer in the study of the Jurassic–Cretaceous boundary.

Supplementary Materials: The following are available online at https://www.mdpi.com/article/10.3390/min12080953/s1, Table S1: LA-ICP-MS zircon U-Pb data of the glassy crystal tuff (YD1-50); Table S2: LA-ICP-MS zircon U-Pb data of the calcareous feldspar lithic sandstone (YD1-464); Table S3: LA-ICP-MS zircon U-Pb data of the rhyolitic tuff (YD1-878).

Author Contributions: Conceptualization, S.-L.S., S.-W.C. and Y.-J.Z.; Field investigation, S.-L.S., T.Z. and Y.-F.L.; Experimental analysis, J.-C.Z., S.-L.S. and T.-T.W.; Software, S.-L.S. and Z.-J.Y.; Validation, S.-L.S., H.-C.W. and Q.-H.D.; Resources, Y.-F.L. and H.-C.W.; Data Curation, S.-L.S.; Writing—Original Draft Preparation, S.-L.S.; Writing—Review and Editing, S.-L.S. and Y.-J.Z.; Visualization, S.-L.S.; Funding Acquisition, S.-L.S. and S.-W.C. All authors have read and agreed to the published version of the manuscript.

Funding: This study was financially supported by the National Natural Science Foundation of China (Grant No. 41790451), the Oil and Gas Survey Program of China Geological Survey (Grant No. DD20221664), and the National Key Research and Development Program of China (Grant No. 2019YFC0605404).

Data Availability Statement: Not applicable.

Acknowledgments: The authors would like to thank Cheng-Shan Wang of China University of Geosciences, De-He Xing of Shenyang Geological Survey Center of the China Geological Survey for their careful guidance and constructive opinions during the writing process. In addition, we appreciate the constructive and thoughtful comments by two anonymous reviewers that greatly improved the paper.

Conflicts of Interest: The authors declare no conflict of interest.

References

1. Remane, J. International stratigraphic chart, with explanatory note. In Proceedings of the 31st International Geological Congress, Rio de Janeiro, Brazil, 6–17 August 2000; pp. 1–16. (In Chinese with English Abstract)
2. Cohen, K.M.; Finney, S.C.; Gibbard, P.L.; Fan, J.X. The ICS international chronostratigraphic chart. *Phase* **2013**, *36*, 199–204. [CrossRef] [PubMed]
3. Wan, X.Q.; Gao, L.F.; Qin, Z.H.; Cui, C.; Li, W.; Xi, D.P. Jurassic-Cretaceous boundary and its terrestrial issue in Northern China. *Earth Sci. Front.* **2016**, *23*, 312–322. (In Chinese with English Abstract)
4. Zheng, Y.J.; Chen, S.W.; Ding, Q.H.; Li, Y.F.; Zhang, J.; Wang, J.; Li, X.H.; Su, F.; Gao, X.Y. Chronostratigraphy of the Jurassic Tuchengzi Formation in Northern Hebei and Western Liaoning, China. *Geol. Resour.* **2010**, *19*, 293–295. (In Chinese with English Abstract)
5. Pang, Q.Q.; Li, P.X.; Tian, S.G.; Liu, Y.Q. Discovery of ostracods in the Dabeigou and Dadianzi Formations at Zhangjiagou, Luanping County, northern Hebei Province of China and new progress in the biostratigraphic boundary study. *Geol. Bull. China* **2002**, *21*, 329–338. (In Chinese with English Abstract)
6. Liu, Y.Q.; Pang, Q.Q.; Li, P.X.; Tian, S.G.; Niu, S.W. Advances in the study of non-marine Jurassic-Cretaceous biostratigraphi cal boundary and candidate stratotype in Luanping basin,northern Hebei. *Geol. Bull. China* **2002**, *21*, 176–180. (In Chinese with English Abstract)
7. Wang, Q.F.; Chen, P.J. A brief introduction to the Cretaceous chronostratigraphic study. *J. Stratigr.* **2005**, *29*, 114–123. (In Chinese with English Abstract)
8. Tian, S.G.; Niu, S.W.; Pang, Q.Q. Redefinition of the Lower Cretaceous terrestrial Yixianian Stage and its stratotype candidate in the Luanping basin, northern Hebei, China. *Geol. Bull. China* **2008**, *27*, 739–752. (In Chinese with English Abstract)
9. Tian, S.G.; Pang, Q.Q.; Niu, S.W.; Li, P.X.; Liu, Y.Q. Terrestrial Jurassic-Cretaceous boundary stratotype candidate in Luanping basin, northern Hebei. *Geol. Bull. China* **2004**, *23*, 1170–1179. (In Chinese with English Abstract)
10. Ji, Q.; Liu, Y.Q.; Ji, S.A.; Chen, W.; Lu, J.C.; You, H.L.; Yuan, C.X. On the terrestrial Jurassic-Cretaceous Boundary in China. *Geol. Bull. China* **2006**, *25*, 336–339. (In Chinese with English Abstract)
11. Zhou, Z.H.; He, H.Y.; Wang, X.L. The continental Jurassic-Cretaceous Boundary in China. *Acta Palaeontol. Sin.* **2009**, *48*, 541–555. (In Chinese with English Abstract)
12. Xu, H.; Liu, Y.Q.; Kuang, H.W.; Jiang, X.J.; Peng, N. U-Pb SHRIMP age for the Tuchengzi Formation, northern China, and its implications for biotic evolution during the Jurassic-Cretaceous transition. *Palaeoworld* **2012**, *21*, 222–234. [CrossRef]
13. Wan, X.Q.; Li, G.; Huang, Q.H.; Xi, D.P.; Chen, P.J. Division and correlation of terrestrial cretaceous stages in China. *J. Stratigr.* **2013**, *37*, 457–471. (In Chinese with English Abstract)
14. Casellato, C.E. Calcareous nannofossil biostratigraphy of Upper Callovian—Lower Berriasian succession from the southern Alps, North Italy. *Riv. Ital. Di Paleontol. Stratigr.* **2010**, *116*, 357–404.
15. Gradstein, F.M.; Ogg, J.G.; Schmitz, M.D. *The Geologic Time Scale*; Elsevier: Amsterdam, The Netherlands, 2012; pp. 795–813.
16. Mahoney, J.J.; Duncan, R.A.; Tejada, M.L.G.; Sager, W.W.; Bralower, T.J. Jurassic-Cretaceous boundary age and mid-ocean-ridge–type mantle source for Shatsky Rise. *Geology* **2005**, *33*, 185–188. [CrossRef]
17. Liu, Y.Q.; Ji, Q.; Jiang, X.J.; Kuang, H.W.; Ji, S.; Gao, L.F.; Zhang, Z.G.; Peng, N.; Yuan, C.X.; Wang, X.R. U-Pb zircon ages of early Cretaceous volcanic rocks in the Tethyan Himalaya at Yangzuoyong Co Lake, Negarze, Southern Tibet, and implications for the Jurassic/Cretaceous boundary. *Cretac. Res.* **2013**, *40*, 90–101. [CrossRef]
18. Vennari, V.V.; Lescano, M.; Naipauer, M.; Aguirre-Urreta, B.; Concheyro, A.; Schaltegger, U.; Armstrong, R.; Pimentel, M.; Ramos, V.A. New constraints on the Jurassic-Cretaceous boundary in the High Andes using high-precision U-Pb data. *Gondwana Res.* **2014**, *26*, 374–385. [CrossRef]
19. Chen, P.J. Comments on the Classification and Correlation of Non-marine Jurassic and Cretaceous of China. *J. Stratigr.* **2000**, *24*, 114–119. (In Chinese with English Abstract)
20. All China Commission of Stratigraphy. *Specification of Regional Chronostratigraphic (Geological) Chronology in China*; Geological Press: Beijing, China, 2002; pp. 1–72. (In Chinese with English Abstract)
21. Pang, Q.Q.; Tian, S.G.; Li, P.X.; Niu, S.W.; Liu, Y.Q. Ostracod biostratigraphy of the Dabeigou and Dadianzi formations and Jurassic-Cretaceous boundary in the Luanping basin, northern Hebei, China. *Geol. Bull. China* **2006**, *25*, 348–356. (In Chinese with English Abstract)
22. Niu, S.W.; Tian, S.G.; Pang, Q.Q. Conchostracan biostratigraphy of the Dadianzi Formation in the Luanping basin, northern Hebei, China and the boundary of continental Jurassic and Cretaceous strata. *Geol. Bull. China* **2010**, *29*, 961–979. (In Chinese with English Abstract)
23. Wang, W.L.; Zhang, L.J.; Zheng, S.L.; Ren, D.; Zheng, Y.J.; Ding, Q.H.; Zhang, H.; Li, Z.T.; Yang, F.L. The age of the Yixianian Stage and the boundary of Jurassic-Cretaceous—The establishment and study of stratotypes of the Yixianian Stage. *Geol. Rev.* **2005**, *51*, 234–242. (In Chinese with English Abstract)
24. Yin, H.F. Notes on the compilation of China stratigraphic table and the revision of China stratigraphic guide. *J. Stratigr.* **2014**, *38*, 123–125. (In Chinese with English Abstract)
25. Swisher, C.C., III; Wang, Y.L.; Zhou, Z.H.; Wang, Y.Q.; Jin, F.; Zhang, J.Y.; Xue, X.; Zhang, F.C.; Wang, Y. New evidence of isotopic age of Yixian Formation and 40Ar/39Ar dating of Tuchengzi Formation. *Chin. Sci. Bull.* **2001**, *46*, 2009–2012. (In Chinese with English Abstract)

26. Davis, G.A. The Late Jurassic "Tuchengzi/Houcheng" Formation of the Yanshan fold-thrust belt: An analysis. *Earth Sci. Front.* **2005**, *12*, 331–345.
27. Zhang, H.; Yuan, H.L.; Hu, Z.C.; Liu, X.M.; Diwu, C.R. U-Pb Zircon dating of the Mesozoic volcanic strata in Luanping of North Hebei and its significance. *Earth Sci. J. China Univ. Geosci.* **2005**, *30*, 707–720. (In Chinese with English Abstract)
28. Sun, L.X.; Zhao, F.Q.; Wang, H.C.; Gu, Y.C.; Ji, S.P. Correlationship of Tuchengzi Formation and implications of the Basin Tectonic Evolution in the Intracontinental Yanshan Orogenic Belt. *Acta Geol. Sin.* **2007**, *81*, 445–453. (In Chinese with English Abstract)
29. Xu, H.; Liu, Y.Q.; Kuang, H.W.; Peng, N. Ages of the Tuchengzi Formation in northern China and the terrestrial Jurassic-Cretaceous boundary in China. *Earth Sci. Front.* **2014**, *21*, 203–215. (In Chinese with English Abstract)
30. Shen, Y.B.; Chen, P.J. Late Middle Jurassic conchostracans from the Tuchengzi Formation of W. Liaoning, NE China. Bulletin of Nanjing Institute of Geology and Palaeontology. *Chin. Acad. Sci.* **1984**, *9*, 309–326. (In Chinese with English Abstract)
31. Wang, S.N. Correlation of the continental Jurassic in The North of China to the Paralic Jurassic in Northwest Scotland, United Kingdom with a discussion of the stratigraphic subdivision and correlation of the Jurassic in the North of China. *Acta Geol. Sin.* **1998**, *72*, 11–20.
32. Wang, W.L.; Zhang, H.; Zhang, L.J.; Zheng, S.L.; Yang, F.L.; Li, Z.T.; Zheng, Y.J.; Ding, Q.H. *Standard Sections of Tuchengzi Stage and Yixian Stage and Their Stratigraphy, Palaeontology and Tectonic-Volcanic Actions*; Geological Publishing House: Beijing, China, 2004; pp. 21–91. (In Chinese with English Abstract)
33. Wang, S.E.; Li, G. New fossil clam shrimps from the Tuchengzi formation of Northern Hebei and Western Liaoning. *Acta Palaeontol. Sin.* **2008**, *47*, 319–325. (In Chinese with English Abstract)
34. Wang, S.E.; Gao, L.Z.; Wan, X.Q.; Song, B. Ages of Tuchengzi formation in western Liaoning-northern Hebei area in correlation with those of international strata. *Geol. Bull. China* **2013**, *32*, 1673–1690. (In Chinese with English Abstract)
35. Deng, S.H.; Lu, Y.Z.; Zhao, Y.; Fan, R.; Wang, Y.D.; Yang, X.J.; Li, X.; Sun, B.N. The Jurassic palaeoclimate regionalization and evolution of China. *Earth Sci. Front.* **2017**, *24*, 106–142. (In Chinese with English Abstract)
36. Li, G.; Matsuoka, A. Searching for a non-marine Jurassic/Cretaceous boundary in northeastern China. *J. Geol. Soc. Jpn.* **2015**, *121*, 109–122.
37. Li, W.; Wan, X.Q.; Atsushi, M.; Zhang, S.Q.; Qu, H.Y.; Tsubasa, T.; Kohei, Y. Jurassic-Cretaceous boundary strata of the Somanakamura Group in NE Japan and their correlation with coeval terrestrial deposits in China. *Acta Geol. Sin.* **2015**, *89*, 285–299.
38. Deng, S.H.; Wang, S.E.; Yang, Z.Y.; Lu, Y.Z.; Li, X.; Hu, Q.Y.; An, C.Z.; Xi, D.P.; Wan, X.Q. Comprehensive Study of the Middle–Upper Jurassic strata in the Junggar Basin, Xinjiang. *Acta Geosci. Sin.* **2015**, *36*, 559–574. (In Chinese with English Abstract)
39. Wang, D.N.; Wang, X.R.; Ji, Q. The Palynoflora alternation and the Paleoclimate change at the turning time between Late Jurassic and Early Cretaceous in Northern Hebei and Western Liaoning. *Acta Geosci. Sin.* **2016**, *37*, 449–459. (In Chinese with English Abstract)
40. Liu, S.F.; Lin, C.F.; Liu, X.B.; Zhuang, Q.T. Syn-tectonic sedimentation and its linkage to fold-thrusting in the region of Zhangjiakou, North Hebei, China. *Sci. China Earth Sci.* **2018**, *61*, 681–710. [CrossRef]
41. Wan, X.Q.; Sun, L.X. Terrestrial Fossil Assemblages of Tuchengzi Formation and Jurassic-Cretaceous Boundary chronostratigraphy in Yanliao Area. *Acta Palaeontol. Sin.* **2020**, *59*, 1–12. (In Chinese with English Abstract)
42. Liu, Y.J.; Zhang, X.Z.; Jin, W.; Chi, X.G.; Wang, C.W.; Ma, Z.H.; Han, G.Q.; Wen, Q.B.; Li, W.; Wang, W.D.; et al. Late Paleozoic tectonic evolution in Northeast China. *Geol. China* **2010**, *37*, 943–951. (In Chinese with English Abstract)
43. Wang, C.W.; Sun, Y.W.; Li, N.; Zhao, G.W.; Ma, X.Q. Tectonic implications of late Paleozoic stratigraphic distribution in Northeast China and adjacent region. *Sci. China Ser. D Earth Sci.* **2009**, *52*, 619–626. [CrossRef]
44. Zhou, J.B.; Wilde, S.A.; Zhang, X.Z.; Zhao, G.C.; Zheng, C.Q.; Wang, Y.J.; Zhang, X.H. The onset of Pacific margin accretion in NE China: Evidence from the Heilongjiang high-pressure metamorphic belt. *Tectonophysics* **2009**, *478*, 230–246. [CrossRef]
45. Wang, P.J.; Zhao, R.L.; Meng, Q.A.; Qu, X.J.; Zhu, D.F.; Gao, Y.F. The Cretaceous Songliao Basin: Dynamic background from volcanic rift to interior sag basin. *Earth Sci. Front.* **2015**, *22*, 99–117. (In Chinese with English Abstract)
46. Zhang, X.Z.; Guo, Y.; Zeng, Z.; Fu, Q.L.; Pu, J.B. Dynamic evolution of the Mesozoic-Cenozoic basins in the northeastern China. *Earth Sci. Front.* **2015**, *22*, 88–98. (In Chinese with English Abstract)
47. Dong, S.W.; Zhang, Y.Q.; Chen, X.H.; Long, C.X.; Wang, T.; Yang, Z.Y.; Hu, J.M. The formation and deformational characteristics of East Asia multi-direction convergent tectonic system in Late Jurassic. *Acta Geosci. Sin.* **2008**, *29*, 306–317. (In Chinese with English Abstract)
48. Dong, S.W.; Zhang, Y.Q.; Li, H.L.; Shi, W.; Xue, H.M.; Li, J.H.; Huang, S.Q.; Wang, Y.C. The Yanshan orogeny and late Mesozoic multi-plate convergence in East Asia—Commemorating 90th years of the "Yanshan Orogeny". *Sci. China Earth Sci.* **2018**, *61*, 1888–1909. [CrossRef]
49. Xing, D.H.; Sun, C.L.; Sun, Y.W.; Zhang, L.D.; Guo, S.Z.; Zhang, C.J.; Peng, Y.D.; Jia, B.; Chen, S.W.; Ding, Q.H. Stratigraphic sequence of Tuchengzi Formation in Jinlingsi-Yangshan Basin, Western Liaoning Province. *Geol. Resour.* **2006**, *15*, 166–173. (In Chinese with English Abstract)
50. Xing, D.H.; Zhang, L.D.; Guo, S.Z.; Zhang, C.J.; Peng, Y.D.; Jia, B.; Chen, S.W.; Ding, Q.H.; Zheng, Y.J. Sequence stratigraphy of the Tuchengzi Formation in the northern part of the Jinlingsi-Yangshan basin, western Liaoning, and basin evolution. *Chin. Geol.* **2001**, *28*, 5–12. (In Chinese with English Abstract)

51. He, B.; Li, M.F.; Liu, Y.X. Jinlingsi-Yangshan Basin Beipiao Formation hydrocarbon source rock characteristics. *Coal Geol. China* **2008**, *20*, 25–27. (In Chinese with English Abstract)
52. Wong, W.H. Tectonic studies on Beipiao area of Jehol. *Geol. Rep.* **1928**, *11*, 23. (In Chinese with English Abstract)
53. Wong, W.H. Crustal movements and igneous activities in eastern China since Mesozoic time. *Bull. Geol. Soc. China* **1927**, *6*, 9–37. [CrossRef]
54. Zhang, H.R.; Zhang, Y.K.; Cai, X.M.; Qu, H.J.; Li, H.L.; Wang, M. The triggering of Yanshan Movement: Yanshan Event. *Acta Geol. Sin.* **2013**, *87*, 1779–1790.
55. Huang, D.Y. Yanliao Biota and Yanshan Movement. *Acta Palaeontol. Sin.* **2015**, *54*, 501–546. (In Chinese with English Abstract)
56. Zhang, R.H. The phasing of Yanshan Movement and its several important questions. *Acta Geol. Sin.* **2016**, *90*, 2176–2180. (In Chinese with English Abstract)
57. Zhu, J.C.; Feng, Y.C.; Meng, Q.R.; Wu, F.C.; Li, H.; Liu, H.T.; Zhang, F.P.; Wang, T.Y.; Wu, G.L.; Zou, C.N.; et al. Late Mesozoic tectonostratigraphic division and correlation of Bohai Bay basin: Implications for the Yanshanian Orogeny. *Sci. China Earth Sci.* **2020**, *62*, 1783–1804. [CrossRef]
58. Xi, D.P.; Wan, X.Q.; Li, G.B.; Li, G. Cretaceous integrative stratigraphy and timescale of China. *Sci. China Earth Sci.* **2019**, *62*, 256–286. [CrossRef]
59. Huang, D.Y. Jurassic integrative stratigraphy and timescale of China. *Sci. China Earth Sci.* **2019**, *62*, 223–255. [CrossRef]
60. Xu, W.L.; Wang, F.; Pei, F.P.; Meng, E.; Tang, J.; Xu, M.J.; Wang, W. Mesozoic tectonic regimes and regional ore-forming background in NE China: Constraints from spatial and temporal variations of Mesozoic volcanic rock associations. *Acta Petrol. Sin.* **2013**, *29*, 339–353. (In Chinese with English Abstract)
61. Dong, S.W.; Zhang, Y.Q.; Long, C.X.; Yang, Z.Y.; Ji, Q.; Wang, T.; Hu, J.M.; Chen, X.H. Jurassic tectonic revolution in China and new interpretation of the Yanshan Movement. *Acta Geol. Sin.* **2007**, *81*, 1449–1461. (In Chinese with English Abstract)
62. Yang, X.D.; Li, X.Y. *Stratigraphy (Lithostratic) of Liaoning Province*; China University of Geosciences Press: Wuhan, China, 1997; pp. 1–247. (In Chinese with English Abstract)
63. Zhai, M.G.; Hu, B.; Zhao, T.P.; Peng, P.; Meng, Q.R. Late Paleoproterozoic–Neoproterozoic multi-rifting events in the North China Craton and their geological significance: A study advance and review. *Tectonophysics* **2015**, *662*, 153–166. [CrossRef]
64. Sun, S.L.; Li, Y.F.; Gao, X.Y.; Zhou, T.S.; Zhang, T.; Wang, S.H. An analysis of the genesis and geochemical characteristics of shale gas in Lower Jurassic Beipiao Formation in Jinyang Basin. *Geol. Bull. China* **2017**, *36*, 575–581. (In Chinese with English Abstract)
65. Liu, Y.S.; Hu, Z.C.; Zong, K.Q.; Gao, C.G.; Chen, H.H. Reappraisement and refinement of zircon U-Pb isotope and trace element analyses by LA-ICP-MS. *Chin. Sci. Bull.* **2010**, *55*, 1535–1546. [CrossRef]
66. Liu, Y.S.; Gao, S.; Hu, Z.C.; Gao, C.G.; Zong, K.Q.; Wang, D.B. Continental and oceanic crust recycling-induced melt-peridotite interactions in the Trans-North China Orogen:U-Pb Dating,Hf isotopes and trace elements in zircons from mantle xenoliths. *J. Petrol.* **2010**, *51*, 537–571. [CrossRef]
67. Ludwig, K.R. *ISOPLOT 3.0: A Geochronological Toolkit for Microsoft Excel*; Special Publication: Berkeley, CA, USA, 2003; p. 70.
68. Belousova, E.A.; Griffin, W.L.; O'Reilly, S.Y.; Fisher, N. Igneous zircon: Trace element composition as an indicator of source rock type. *Contrib. Mineral. Petrol.* **2002**, *143*, 602–622. [CrossRef]
69. Hoskin, P.W.O.; Ireland, T.R. Rare earth element chemistry of zircon and its use as a provenance indicator. *Geology* **2000**, *28*, 627–630. [CrossRef]
70. Pang, Q.Q. Ostracods from middle and Upper Jurassic in Yanshan area, Hebei Province and their stratigraphic significance. *J. Hebei GEO Univ.* **1982**, *Z1*, 89–110. (In Chinese with English Abstract)
71. Pu, R.G.; Wu, H.Z. Palynopollen assemblage of middle and late Jurassic in western Liaoning Province. *J. Shenyang Inst. Geol. Miner. Resour.* **1982**, *4*, 169–184. (In Chinese with English Abstract)
72. Yang, R.Q. Jurassic-Cretaceous ostracods assemblage in Hebei Province and its stratigraphic significance. *J. Hebei GEO Univ.* **1984**, *4*, 30–46. (In Chinese with English Abstract)
73. Xiao, Z.Z.; Yang, H.L.; Shan, Q.S. *Mesozoic Strata and Biota in Beijing*; Geological Press: Beijing, China, 1994; pp. 119–121. (In Chinese with English Abstract)
74. Hong, Y.C. Discovery of fossil insects from Houcheng Formation of Hebei Province and establish of Houcheng Entomofauna. *Beijing Geol.* **1997**, *1*, 1–7. (In Chinese with English Abstract)
75. Zheng, S.L. *Division, Correlation and Age of Tuchengzi Formation in Northern Hebei and Western Liaoning: Editorial Board of Theses of the Third National Stratigraphic Conference//Proceedings of the Third National Stratigraphic Conference*; Geological Press: Beijing, China, 2000; pp. 227–232. (In Chinese with English Abstract)
76. Zheng, S.L.; Zhang, W.; Ding, Q.H. Discovery of fossil plants from Middle-Upper Jurassic Tuchengzi Formation in Western Liaoning, China. *Acta Palaeontol. Sin.* **2001**, *40*, 67–85. (In Chinese with English Abstract)
77. Wang, S.E.; Ji, Q. *A Comprehensive Study of the Continental Upper Jurassic Tuchengzi Stage in China//the Third National Stratigraphic Commission: A Study Report on the Establishment of the Main Chronological Strata In China*; Geological Press: Beijing, China, 2008; pp. 105–114. (In Chinese with English Abstract)
78. He, Z.J.; Wang, Z.Q.; Ren, J.S. A preliminary research on sedimentary features and genetic mechanism of frontal basins before Jurassic large-scale nappe in the Northern Region of North China. *Chin. J. Geol.* **1999**, *34*, 186–195. (In Chinese with English Abstract)

79. He, Z.J.; Li, J.Y.; Niu, B.G.; Ren, J.H. A late Jurassic intense thrusting-uplifting event in the Yanshan—Yinshan area, Northern China, and its sedimentary response. *Geol. Rev.* **1998**, *44*, 407–418. (In Chinese with English Abstract)
80. Wang, Z.J.; Huang, Z.G.; Yao, J.X.; Ma, X.L. Characteristics and main progress oft the stratigraphic chart of China and directions. *Acta Geosci. Sin.* **2014**, *35*, 271–276. (In Chinese with English Abstract)
81. Huang, Z.A.; Pan, Y.Q.; Yang, Y.J. *Regional Geological Survey Report of 1:250,000 Zone in Jinzhou*; Liaoning Provincial Institute of Geology and Mineral Resources Investigation: Dalian, China, 2003; pp. 1–391. (In Chinese with English Abstract)
82. Fan, J.X.; Chen, D.Y.; Hou, X.D. The International Chronostratigraphic Chart (v 2016/04), Its color model and application. *J. Stratigr.* **2016**, *40*, 341–348. (In Chinese with English Abstract)
83. Lin, M.Q.; Li, J.G.; Peng, J.G. Early Cretaceous palynological assemblages from the Third Member of the Tuchengzi Formation at Sihetun, Beipiao County, Western Liaoning Province, Northeast China and their implications. *Acta Micropalaeontologica Sin.* **2016**, *33*, 261–271. (In Chinese with English Abstract)
84. Kozhevnikov, V.M.; Ruzhentsev, S.V.; Dergunov, A.B.; Filippova, I.B.; Tomurtogoo, O.; Arvisbaatar, N.; Bayasgalan, T. The South Siberia—Central Mongolia transect. *Tectonophysics* **1993**, *225*, 361–378.
85. Zorin, Y.A. Geodynamics of the western part of the Mongolia-Okhotsk collisional belt, Trans-Baikal region (Russia) and Mongolia. *Tectonophysics* **1999**, *306*, 33–56. [CrossRef]
86. Mo, S.G.; Han, M.L.; Li, J.Y. Compositions and orogenic processes of Mongolia-Okhotsk Orogen. *J. Shandong Univ. Sci. Technol.* **2005**, *24*, 50–52, 64. (In Chinese with English Abstract)
87. Kravchinsky, V.A.; Cogné, J.P.; Harbert, W.P.; Kuzmin, M.I. Evolution of the Mongol-Okhotsk Ocean as constrained by new palaeomagnetic data from the Mongol-Okhotsk suture zone, Siberia. *Geophys. J. Int.* **2002**, *148*, 34–57. [CrossRef]
88. Sorokin, A.A.; Sorokin, A.P.; Ponomarchuk, V.A.; Travin, A.V. The age and geochemistry of volcanic rocks on the eastern flank of the Umlekan-Ogodzha volcanoplutonic belt (Amur region). *Russ. Geol. Geophys.* **2010**, *51*, 369–379. [CrossRef]
89. Wang, C.Y.; Wang, D.B.; Qu, J.J. Chronology and geochemistry of rhyolites in Manketou'ebo Formation from Keyihe area,north-central Greater Xing'an Range. *J. Heilongjiang Univ. Sci. Technol.* **2017**, *27*, 46–50. (In Chinese with English Abstract)
90. Gao, L.F.; He, Z.H.; Sui, Z.M.; Lu, K.J.; Liu, F.Y. Chronology and geochemistry of volcanic rocks from Manketouebo Formation in the Suolun area, central Da Hinggan Mountains and their tectonic implications. *Geol. Bull. China* **2018**, *37*, 881–894. (In Chinese with English Abstract)
91. Liu, K.; Wu, T.T.; Liu, J.L.; Bao, Q.Z.; Du, S.Y. Geochronology and geochemistry of volcanic rocks in Manketou'ebo Formation of Tulihe area, northern Da Hinggan Mountains. *Geol. China* **2018**, *45*, 367–376. (In Chinese with English Abstract)
92. Guo, J.G.; Zhang, Y.J.; Zhang, C.; Li, W.; Wang, Y.; Wang, Q.Z.; Guo, J. Formation age and geochemical characteristics of volcanic rocks in Manketou'ebo Formation in Kundu area of Ar Horqin Banner, Inner Mongolia. *Geol. Bull. China* **2018**, *37*, 1652–1661. (In Chinese with English Abstract)
93. Jia, B.; Wei, L.Y.; Wu, H.H.; Zhang, Z.; Tang, W.D.; Liu, T.H.; Chai, C.H.; Meng, W.Y. Chronology, geochemistry and geological significance of volcanic rocks of the Manketouebo Formation in Horqin Right-Front Banner area, Inner Mongolia. *Acta Petrol. Et Mineral.* **2021**, *40*, 859–873. (In Chinese with English Abstract)
94. Si, Q.L.; Cui, T.R.; Tang, Z.; Li, W.; Wu, X.W.; Jiang, B.; Li, L.C. Chronology, geochemistry and petrogenesis of the volcanic rocks in Manitu Formation in Chaihe Area, Central Great Xing'an Range. *J. Jilin Univ. Earth Sci. Ed.* **2015**, *45*, 389–403. (In Chinese with English Abstract)
95. Zhang, X.; Zhen, S.U.; Gao, Y. Zircon U–Pb geochronology and geochemistry of the early Cretaceous volcanic rocks from the Manitu Formation in the Hongol Area, Northeastern Inner Mongolia. *Acta Geol. Sin. Engl. Ed.* **2017**, *91*, 1286–1304. [CrossRef]
96. Tao, C.Z.; Li, Z.M.; Wang, D.Q.; Shen, X.; Zhang, Y.J.; Li, B. La-icp-ms zircon U-Pb dating and geochemistry of the volcanic rocks from manitu formation in Ulanhot region of southern Daxinganling. *Geol. Resour.* **2015**, *24*, 102–109. (In Chinese with English Abstract)
97. Wang, C.Y.; Sun, D.Y.; Tian, L.H.; Bu, J.; Shang, Y.H. Latest zircon u-pb age of the baiyingaolao formation volcanic rocks in the keyihe area of north-central da hinggan mountains. *Acta Geol. Sin.* **2018**, *92*, 360–361. [CrossRef]
98. Yin, Y.K.; Gao, Y.F.; Wang, P.J.; Qu, X.J.; Liu, H.B. Discovery of Triassic volcanic-sedimentary strata in the basement of Songliao Basin. *Sci. Bull.* **2019**, *64*, 644–646. [CrossRef]
99. Liu, H.B.; Wang, P.J.; Gao, Y.F.; Hou, H.S.; Yin, Y.K.; Li, H.H.; Feng, Y.H. New data from ICDP borehole SK2 and its constraint on the beginning of the Lower Cretaceous Shahezi Formation in the Songliao Basin, NE China. *Sci. Bull.* **2020**, *66*, 411–413. [CrossRef]
100. Gao, R.Q.; Xiao, D.M. *New Progress in Oil and Gas Exploration in Songliao and Its Peripheral Basins*; Petroleum Industry Press: Beijing, China, 1995; pp. 13–15. (In Chinese with English Abstract)
101. Zhu, D.F.; Ren, Y.G.; Wu, H.Y.; Wu, G.Y.; Wan, C.B.; Wu, X.M. A preliminary study on the buried Permian and Jurassic in the Northern Songliao Basin. *Chin. J. Geol.* **2007**, *42*, 690–708, 778. (In Chinese with English Abstract)
102. Qu, X.J.; Wang, P.J.; Gao, Y.F.; Wan, X.Q. Chronostratigraphy of Huoshiling Formation in the Songliao Basin, NE China. *Earth Sci. Front.* **2014**, *21*, 234–250. (In Chinese with English Abstract)
103. Qu, X.J.; Wang, P.J.; Yao, R.S.; Gao, Y.F.; Bai, B. Stratigraphical sequence and regional correlation of Huoshiling Formation in southern Songliao Basin. *J. Cent. South Univ.* **2014**, *45*, 2716–2727. (In Chinese with English Abstract)
104. Wang, P.J.; Liu, W.Z.; Wang, S.X.; Song, W.H. 40Ar/39Ar and K/Ar dating on the volcanic rocks in the Songliao basin, NE China: Constraints on stratigraphy and basin dynamics. *Int. J. Earth Sci.* **2002**, *91*, 331–340. [CrossRef]

105. Lin, M.Q.; Li, J.G. Palynological assemblages from the Tuchengzi and Dabeigou formations in the Luanping Basin at Sanchazi Section, Northern Hebei Province and their implications. *Acta Palaeontol. Sin.* **2021**, *60*, 263–280. (In Chinese with English Abstract)
106. Sun, G.; Sha, J.G.; Wang, Y.G.; Yuan, F.S.; Xiong, X.Z.; Piao, T.Y. Jurassic-Cretaceous Boundary of Dong'anzhen, Raohe, Heilongjiang with a new knowledge of Buchia. *Acta Palaeontol. Sin.* **1989**, *28*, 579–590. (In Chinese with English Abstract)
107. Sun, G.; Zhen, S.L.; Sun, X.K.; He, C.Q.; Piao, T.Y.; Shang, Y.K.; Zhang, C.B.; Yu, Z.Z.; Zhao, Y.H. Recent advance in the research of the strata near the Jurassic-Cretaceous boundary in eastern Heilongjiang. *J. Stratigr.* **1992**, *16*, 49–54, 83. (In Chinese with English Abstract)
108. Sun, X.K.; He, C.Q. Dinoflagellates from upper Jurassic dongrong formation of Suibin, Heilongjiang. *Acta Palaeontol. Sin.* **1992**, *31*, 190–205, 267.
109. He, C.Q.; Zhu, Y.H. Dinoflagellate assemblage from the uppermost part of the Dongrong formation with discussion on Jurassic-Cretaceous boundary in Suibin area, Northeastern Heilongjiang province, Ne China. *Acta Palaeontol. Sin.* **2003**, *42*, 328–345. (In Chinese with English Abstract)

minerals MDPI

Article

Petrogenesis and Tectonic Implications of the Ore-Associated Intrusions in Bayanbaolege Ag Polymetallic Deposit, Inner Mongolia, NE China

Xi Wang [1,2], Qun Yang [1,2,*], Zhen-Ming Sun [3] and Yun-Sheng Ren [1,4,*]

1 College of Earth Sciences, Jilin University, Changchun 130061, China; xiwang@jlu.edu.cn
2 Key Laboratory of Mineral Resources Evaluation in Northeast Asia, Ministry of Natural Resources of China, Changchun 130026, China
3 Changchun Institute of Technology, Changchun 130012, China; sunzhenming.9040@163.com
4 Institute of Disaster Prevention, Sanhe 065201, China
* Correspondence: yangqun@jlu.edu.cn (Q.Y.); renys@jlu.edu.cn (Y.-S.R.); Tel.: +86-131-5437-0506 (Q.Y.)

Abstract: The large Bayanbaolege Ag polymetallic ore deposit is located in the Tuquan-Linxi Fe (Sn)-Cu-Pb-Zn-Ag-Nb (Ta) polymetallic metallogenic belt, which is an important part of the Great Xing'an Range metallogenic province, northeast China. The sulfide–quartz vein-type orebodies in the deposit are mainly hosted in the Cretaceous granodiorite porphyry and Late Permian Linxi formation. The U-Pb dating of the zircon from the post-ore diorite porphyrite yields an age of 124.8 ± 1.1 Ma, which constrains the mineralization time at the Early Cretaceous. The Sr-Nd isotope values $(^{87}Sr/^{86}Sr)_i$ = 0.708576~0.710536; ε_{Nd} (t) = −0.51~+0.69; the Hf isotope values $^{176}Hf/^{177}Hf$ = 0.2827278~0.2830095, the ε_{Hf} (t) = +3.1~+11.2, T_{DM2} = 615~1341 Ma of the metallogenic granodiorite porphyry. The Hf isotope values $^{176}Hf/^{177}Hf$ = 0.2828596~0.2829451, and the ε_{Hf} (t) = +5.7~+8.8 of the diorite porphyrite, T_{DM2} = 827~1108 Ma, indicating that the ore-forming materials were the possible involvement of heterogeneous juvenile sources including moderately depleted mantle and newly underplated lower crust. The major and trace elements (including REEs) implied that these intrusions are the I-type granite and linked intimately to the westward subduction of the Paleo-Pacific Ocean plate. From these whole-rock major and trace elements and zircon U-Pb ages, as well as Sr-Nd-Hf isotope data, we conclude that the ore-associated I-type granites in the Bayanbaolege deposit formed in an extensional tectonic setting of the Early Cretaceous, and are compactly related to the retreat of the Paleo-Pacific Ocean subducted plate linked intimately to the westward subduction of the Paleo-Pacific Ocean plate rather than the closure of the Mongol–Okhotsk Ocean. Furthermore, by integrating geological background work and previous research work, implying the mineralization age of the Bayanbaolege deposit should have been formed in the 125–130 Ma.

Keywords: LA–ICP–MS U-Pb age; Hf isotope; petrogenesis; tectonic implication; Bayanbaolege deposit

1. Introduction

The Great Xing'an Range (GXR) polymetallic metallogenic belt is located in the eastern segment of the Central Asian Orogenic Belt (CAOB), northeast (NE) China (Figure 1a,b), which hosts numerous important metal deposits, such as porphyry Sn-Cu-Ag, Cu (Mo) and Cu, and skarn polymetallic and hydrothermal vein types (Figure 1c) [1]. These deposits mainly formed in the Late Jurassic and Early Cretaceous, with a minor amount in the Triassic-Permian [2]. The formation of these deposits in this area were closely related to the contemporary tectono-magmatic events, which are considered to be the results of the evolution of Paleo-Asian, Mongol–Okhotsk, and Paleo-Pacific tectonic regimes ([3] and references therein).

The east slope of the southern GXR hosts a series of large Cu polymetallic deposits, always considered a Cu polymetallic metallogenic belt. However, several large Ag-Pb-Zn deposits have been discovered in recent years (Figure 1c), showing a considerable Ag-Pb-Zn polymetallic mineralization potential in this area [2].

The large Bayanbaolege Ag polymetallic deposit occurs in the northeastern Tuquan-Linxi Fe (Sn)-Cu-Pb-Zn-Ag-Nb (Ta) metallogenic belt in the east slope of the southern GXR, which contains proven reserves including 1440.41 t Ag, 381,826.1 t Zn and 51,493.27 t Pb, with average grades of 187.8 g/t Ag, 2.75% Zn and 1.89% Pb. A series of studies about the Bayanbaolege deposit have been reported, mainly regarding deposit geology, ore-forming conditions, Rb-Sr isotopic dating, fluid evolution, and S-Pb stable isotopes of the sulfides (pyrite and sphalerite) [4,5]. Although Wang et al. [4] have reported that the granodiorite porphyry had relationships with the polymetallic mineralization spatially, temporally, and genetically. However, the sources of the material of the granodiorite porphyry and the post-ore diorite porphyrite, as well as the relationship between mineralization and magmatism, have not been well-constrained; especially how and to what extent the mixture of the crustal or mantle materials have been involved in the forming of the deposit are still not clear, which restricts the better understanding of the metallogenetic mechanism and regularity of the deposit.

In this study, the major trace elements (including REEs), and Sr-Nd-Hf isotopic data of the ore-associated intrusive rocks and the zircon U-Pb dating of the diorite porphyrite in the Bayanbaolege deposit are combined with the Sr-Nd isotopic data of the Early Cretaceous granitoids and the age of the diagenetic and metallogenic in the adjacent regions. In combination with previous results, we present these study data and aim to discuss the petrogenesis and tectonic setting of the ore-associated intrusions in the Bayanbaolege deposit, which have important implications for better understanding the geological evolution history and metallogenic process of this region.

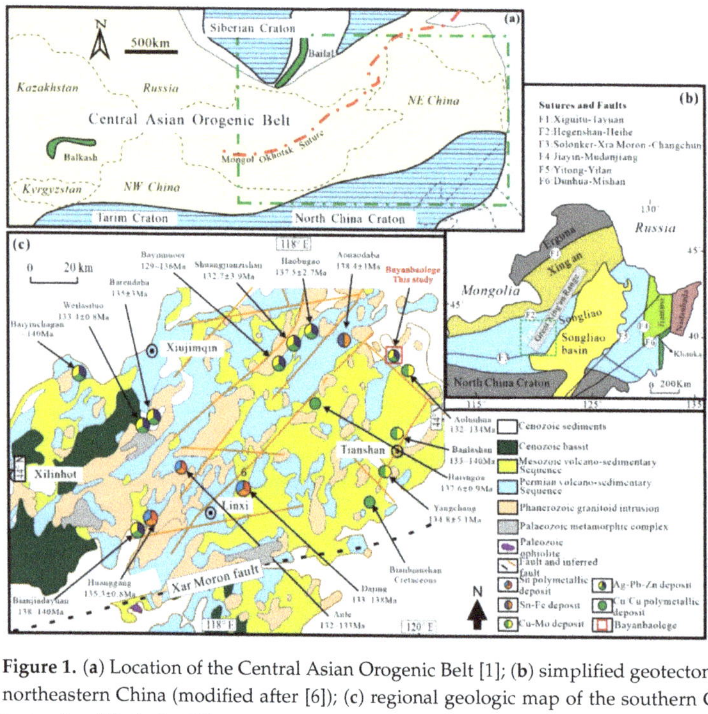

Figure 1. (**a**) Location of the Central Asian Orogenic Belt [1]; (**b**) simplified geotectonic division of northeastern China (modified after [6]); (**c**) regional geologic map of the southern Great Xing'an Range (modified after [1]).

2. Geological Setting

The large Bayanbaolege Ag polymetallic deposit is located in the Songliao block, which is considered to belong to the Xing'an-Mongolian Orogenic Belt (XMOB) in a tectonic unit of the eastern segment of the CAOB (Figure 1a,b). The XMOB is separated by the Xinlin–Xiguitu fault, Hegenshan–Heihe fault, and Solonker–Xar Moron–Changchun fault, dividing the XMOB into several elongated terranes including the Erguna, Xing'an, Songliao, and Jiamusi blocks (Figure 1b). The GXR is located in the eastern segment of the XMOB, NE China, which at least experienced the subduction closure of the Paleo-Asian Ocean, the closure of Mongol–Okhotsk Ocean, and the subduction of the Paleo-Pacific Ocean Plate, and resulted in widespread associated magmatism and metamorphism [7–9].

The regional strata are dominated by Permian and Mesozoic volcano-sedimentary rocks. The Permian strata include, from bottom to top, the Shoushangou Formation (P_1s), comprising thick-bedded limestone. The Dashizhai Formation (P_2d) is composed of slightly metamorphic intermediate-to-acidic lava, rhyolite, and volcaniclastics intercalated with some sandstone. The Zhesi Formation (P_2zs) consists of silty mudstone, tuffaceous sandstone, and tuffaceous silty sandstone intercalated with oolitic crystalline limestone lens. The Linxi Formation (P_3l) is composed of sandy conglomerate and siltstone intercalated. These formations are unconformably overlain by Mesozoic volcano-sedimentary sequences consisting of rhyolite, andesite, and felsic pyroclastic units. The Mesozoic strata, from oldest to youngest, consist of four formations, namely the Manketouebo, Manitu, Baiyingaolao, and Meiletu. The Manketouebo Formation is mainly composed of rhyoitic tuff with minor rhyolite and dacite. The Manitu Formation consists of andesite, trachy dacite, and volcaniclastic rocks. The Baiyingaolao Formation consists of rhyolite and rhyolitic tuff with intercalated sedimentary rocks. The Meiletu Formation is composed of basalt, basaltic andesite, andesite, and trachyte [10].

Faults and folds are well-developed in the GXR, which are widely distributed in this region. The faults are mainly NE- and NNE-trending and have overprinted existing EW-trending faults that dominate this area and control the Mesozoic-Paleozoic intrusive rocks. Regional magmatism includes Hercynian plutons (321–250 Ma), including quartz diorite, tonalite, granodiorite, monzogranite, gabbro, quartz monzonite, biotite granite, and K-feldspar granite [11–15]. Indosinian granitoids (250–225 Ma) consist of two-mica granite, granite porphyry [16,17], and granodiorite porphyry and Yanshanian intrusions (182–125 Ma) [1,4,9,16–18]. In addition, the W, Sn, Pb, Zn, Cu, and Ag deposits in the region are all closely related to Indosinian and Yanshanian granites, including granodiorite, quartz syenite, K-feldspar granite, hornblende monzonite, and monzogranite [2,6,14,16,19,20].

3. Deposit Geology

The Bayanbaolege deposit (44°36′ N, long. 119°57′ E) is located in Chifeng City, Inner Mongolia, NE China.

The strata in this ore district consist of the Zhesi Formation, the Linxi Formation, and the Manitu Formation (Figure 2). The Manitu Formation unconformably overlies the Zhesi Formation in the ore district and is composed of andesitic breccia clastic tuff, andesite, and dacite. The Linxi Formation consists of silty slate, sandy slate, and metamorphic fine sandstone. The Zhesi Formation consists of silty slate interbedded with a less argillaceous slate.

The NE–SW-trending F1 and F3 faults, and the NW–SE-trending F2 fault form a major tectonic framework in the ore district. The F1 and F3 faults control the distribution of intrusive and volcano-sedimentary rocks and are cut across by the F2 fault (Figure 2).

Two periods of magmatic activity have been identified in the ore district (Figures 2 and 3). The earlier period is represented by the granodiorite porphyry, which is spatial-temporal and genetically closely related to the polymetallic mineralization (Figure 4a,b) [4], and the latter is represented by diorite porphyrite (Figure 4c,d), which broke down the integrity of the ore body (Figure 4c).

The deposit consists of 86 sulfide–quartz vein-type polymetallic ore bodies, which show the characteristics of upper Ag bodies and lower Zn bodies as a whole, and distribute in the contact zone between the Late Permian strata and the Early Cretaceous granodiorite porphyry (Figure 3). The trend of these orebodies extends NE (16–67°) and dips along the NW with steep inclination (58–80°). Orebodies are 0.86–5.73 m thick, 38–1060 m in horizontal length, and 22–815 m in vertical depth. Ore structures consist of vein-type, and minor banded-type, brecciated-type and disseminated-type structures. The metallic minerals include pyrite, sphalerite, galena and minor chalcopyrite, cubanite, pyrrhotite, arsenopyrite, and loellingite. Gangue minerals include the quartz and calcite. The alterations developed such as limonitization, carbonatization, silicification and weak chloritization and epidotization.

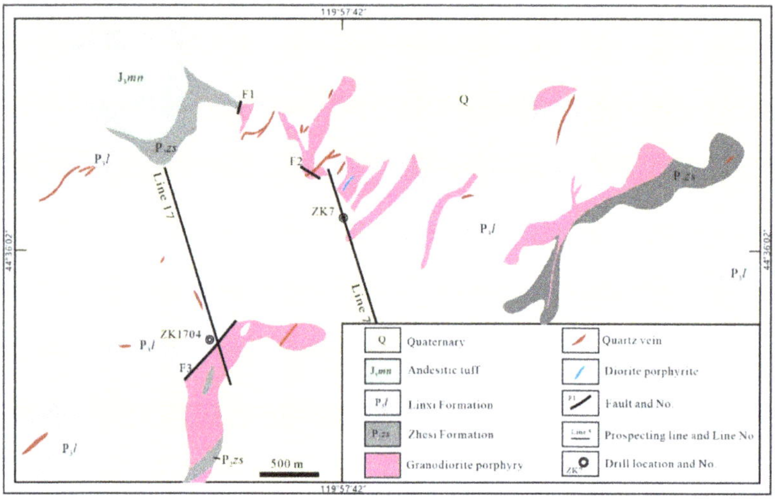

Figure 2. Geological sketch map, representative drill locations, and numbers in the Bayanbaolege deposit (modified after [4]).

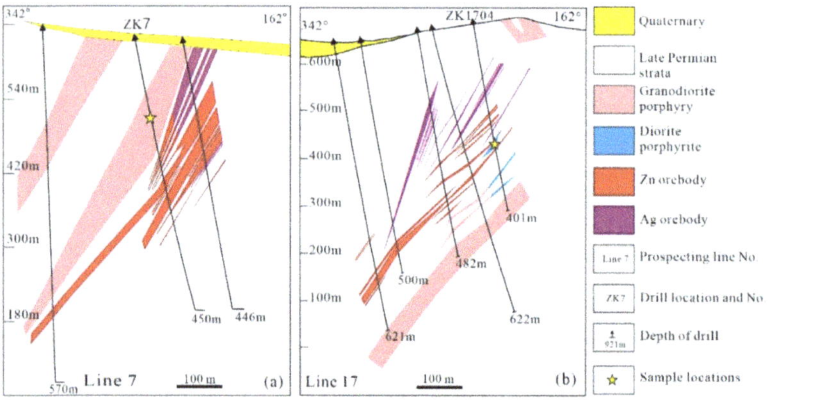

Figure 3. Geological section map of exploration lines and sampling locations in Bayanbaolege deposit (**a**,**b**).

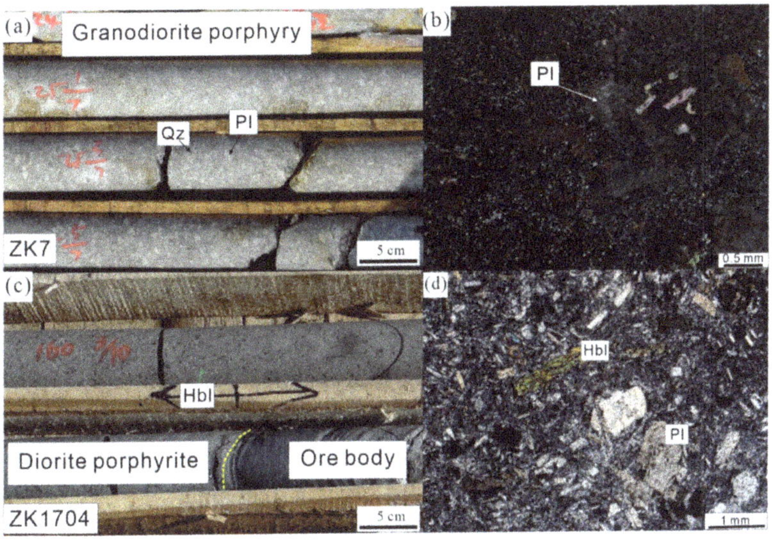

Figure 4. Representative field photographs and photomicrographs of the Bayanbaolege deposit. (**a**) Granodiorite porphyry consisting of plagioclase and minor quartz phenocrysts and matrixes. (**b**) Granodiorite porphyry showing the porphyritic texture with slight alteration of sericite. (**c**) Diorite porphyrite consisting of plagioclase and hornblende phenocrysts and microscopic matrix. (**d**) Porphyritic texture of the diorite porphyrite. Qz—quartz; Pl—plagioclase; Bt—biotite; Hbl—hornblende.

4. Sample Description and Analytical Methods

4.1. Sample Description

The granodiorite porphyry was selected (from drill hole ZK7) for Hf isotopic analyses, and five fresh samples (ZK7-1, ZK7-2, ZK7-3, ZK7-4, and ZK7-5) for whole-rock geochemistry and five samples (ZK7-T1, ZK7-T2, ZK7-T3, ZK7-T4 and ZK7-T5) of granodiorite porphyry for Sr-Nd isotopes (Figure 3a). The granodiorite porphyry is gray, with a massive structure (Figure 4a) and porphyritic texture (Figure 4b), contains phenocrysts and fine-grained groundmass with graphic texture. The principal minerals include alkali feldspar (~15%), plagioclase (~65%), quartz (~20%), biotite (~3%), and minor muscovite.

The diorite porphyrite was selected (ZK1704) for LA-ICP-MS zircon U-Pb dating and Hf isotopic analyses, and two samples (ZK1704-1 and ZK1704-2) without alterations for whole-rock geochemistry of diorite porphyrite, which obviously destroyed the integrity of the ore bodies, were selected from the ZK1704 drill core (Figure 3b). The diorite porphyrite has porphyritic texture and massive structure (Figure 4c,d), and the phenocryst minerals are dominated by plagioclase (15%–20%), alkali feldspar (5%–10%), and amphibole (<5%); the matrix minerals are composed of the plagioclase, alkali feldspar, quartz, amphibole, and biotite.

4.2. Analytical Methods

4.2.1. Ore-Forming Elements, Major and Trace Elements

Whole-rock major and trace elements of intrusive rock were analyzed at Tianjin Geological Mineral Testing Center, Tianjin, China. First, these samples were crushed in a steel jaw crusher and then powdered to 200 mesh in an agate mill, and then, trace element compositions were analyzed using the X-Series II ICP-MS (Thermo Electron Corporation, Waltham, MA, USA), following Qi et al. [21] and yielding an analytical precision better than 5%, and major element compositions were analyzed using X-ray fluorescence spectrometer (XRF), with a precision better than 2%. The selected samples were first characterized in terms of chemical composition (major oxides wt.% and trace elements ppm) as listed in Table 1.

Table 1. Major, trace, and REE contents of intrusive rocks from Bayanbaolege deposit.

Sample	ZK7-1	ZK7-2	ZK7-3	ZK7-4	ZK7-5	ZK1704-1	ZK1704-2
Type			Granodiorite Porphyry			Diorite Porphyrite	
			Major Elements (wt.%)				
Al_2O_3	12.91	12.79	13.76	12.87	12.71	16.24	16.41
CaO	1.21	1.26	1.25	1.23	1.22	3.41	2.43
TFeO	1.85	1.62	1.78	1.77	1.71	4.74	4.58
K_2O	5.26	4.04	4.73	4.85	5.16	4.3	4.35
MgO	0.43	0.32	0.41	0.32	0.31	2.71	1.86
MnO	0.07	0.06	0.07	0.08	0.07	0.43	0.38
Na_2O	3.8	4.89	4.13	3.98	4.55	3.6	3.24
P_2O_5	0.06	0.06	0.06	0.05	0.05	0.19	0.19
SiO_2	71.71	71.69	71.18	72.2	71.38	60.13	62.43
TiO_2	0.25	0.2	0.23	0.2	0.21	0.56	0.61
LOI	2.28	1.94	1.97	2	1.89	2.97	2.76
Total	99.83	98.87	99.57	99.55	99.26	99.28	99.22
$Na_2O + K_2O$	9.06	8.93	8.86	8.83	9.71	7.9	7.59
$(Na_2O + K_2O)/CaO$	7.49	7.09	7.09	7.18	7.96	2.32	3.13
A/NK	1.08	1.03	1.16	1.09	0.97	1.54	1.63
A/CNK	0.91	0.87	0.97	0.92	0.83	0.97	1.14
			Trace elements (ppm)				
Li	13.3	15.04	12.87	12.92	14.37	47.45	35.62
Be	2.05	2.1	2.13	2.03	1.98	3.25	2.86
V	12.8	11.23	16.33	11.49	11.73	131.73	140.17
Cr	16.9	15.12	13.13	13.85	10.91	67.7	66.02
Co	2.04	1.7	2.04	1.77	1.78	8.94	22.5
Ni	2.45	1.71	2.15	2.33	2.2	22	40.77
Ga	22.82	24.07	22.76	22.09	23	23.21	24.52
Rb	116.08	133.02	151.5	185.42	119.19	137.42	160.68
Sr	36.13	145.21	22.79	51.74	132.49	44.36	49.64
Zr	156.26	185.26	161.99	200.44	229.71	241.09	226.94
Nb	8.75	9.25	8.26	8.51	9.17	13.78	13.49
Ba	915.22	927.81	855.86	815.69	884.61	405.26	580.88
La	29.68	38.97	26.8	30.19	31.08	37.04	35.53
Ce	57.85	72.44	52.53	57.8	59.83	77.75	73.75
Pr	6.72	8.44	6.2	6.74	7.04	9.43	9.24
Nd	24.22	30.14	22.79	24.42	25.39	37.24	35.65
Sm	4.36	5.15	4.23	4.35	4.64	7.52	6.81
Eu	0.79	0.86	0.78	0.75	0.79	1.66	1.84
Gd	3.74	4.38	3.56	3.73	3.93	6.24	5.96
Tb	0.6	0.68	0.58	0.6	0.66	1.02	0.96
Dy	3.39	3.72	3.35	3.3	3.67	5.78	5.61
Ho	0.68	0.74	0.66	0.66	0.73	1.15	1.1
Er	2.01	2.22	1.98	2.04	2.21	3.42	3.28
Tm	0.33	0.35	0.31	0.33	0.35	0.51	0.51
Yb	2.18	2.31	2.12	2.16	2.37	3.31	3.26
Lu	0.34	0.37	0.33	0.35	0.38	0.53	0.52
Y	18.85	20.79	19	19.06	20.92	29.93	28.49
Hf	5.05	5.64	4.93	5.85	6.64	7.15	6.67
Ta	0.85	0.89	0.81	1	0.9	0.95	0.93
Th	10.42	11.94	9.7	10.27	11.19	10.12	10.17
U	3.26	2.88	3.12	2.52	2.92	2.32	2.61
∑REE	136.89	170.76	126.23	137.42	143.07	192.62	184.02
LREE/HREE	9.32	10.56	8.79	9.44	9.01	7.77	7.68
10,000 Ga/Al	2.21	2.35	2.07	2.15	2.26	1.79	1.87

4.2.2. Sr-Nd Isotopes

Whole-rock Sr-Nd isotopes of the granodiorite porphyry were analyzed at the Isotopic Laboratory, Analytical Laboratory Beijing Research Institute of Uranium Geology, China National Nuclear Corporation, Beijing, China. Sr-Nd isotopic analyses were carried out at an ISOPROBE-T thermal ionization mass spectrometer (Thermo Fisher Scientific, Waltham, MA, USA), and with the $^{86}Sr/^{88}Sr$ ratio of 0.1194 and $^{146}Nd/^{144}Nd$ ratio of 0.7219 for mass fractionation corrections. The analytical NBS987 Sr standard yields $^{87}Sr/^{86}Sr$ = 0.710250 ± 0.000007 (2σ) and the SHINESTU Nd standard yields $^{143}Nd/^{144}Nd$ = 0.512118 ± 0.000003(2σ).

4.2.3. Zircon U-Pb Dating and Hf Isotope

The zircon grain extraction and cathodoluminescence (CL) images were completed at the Institute of Regional Geology and Mineral Resources Survey, Langfang, Hebei Province, China. Zircons were extracted from the diorite porphyrite sample of No. ZK1704 and granodiorite porphyry sample of No. ZK7 using standard density and magnetic separation techniques, and cathodoluminescence (CL) images were obtained using a JSM-IT300 scanning electron microscope (JEOL, Tokyo, Japan). Zircon U-Pb isotopes were analyzed by a NEW WAVE 193 nm-FX ArF Excimer laser-ablation system (ESI Ltd., Beaverton, OR, USA) at the Isotopic Laboratory, Tianjin Institute of Geology and Mineral Resources of China Geological Survey, Tianjin, China. Samples analyses were carried out at an energy density of 11 J/cm^2, with a beam diameter of 30 μm and a repetition rate of 8 Hz. In situ zircon Hf isotopic analyses were measured on the same spot zones where U-Pb age determinations were previously analyzed, with a 50 μm laser beam diameter and a repetition rate of 11 Hz. Details of the operating conditions for the laser ablation system and the MC-ICP-MS instrument are described by Geng et al. [22]. The values of the isotopic ratios were calculated using ICPMS Data Cal 8.4 [23]. Concordia diagrams and weighted mean U-Pb ages were processed using ISOPLOT 3 [24]. Age data and Concordia plots were reported at one error and the uncertainties for weighted mean ages are given at a 95% confidence level.

5. Results

5.1. Major and Trace Elements

The granodiorite porphyry is characterized by high silica (SiO_2 = 71.18–72.20 wt.%), Al_2O_3 ranges from 12.71 to 13.76 wt.%, low MgO (0.31–0.43 wt.%) and is rich in alkali ($Na_2O + K_2O$ = 8.83–9.71 wt.%). K_2O/Na_2O ratios vary from 0.72 to 1.21, and the samples plotted in the high-k calc-alkaline field on the K_2O versus SiO_2 diagram. The molar ratio $Al_2O_3/(CaO + Na_2O + K_2O)$ (A/CNK) varies from 0.83 to 0.97, indicating the metaluminous character of all samples (Figure 5).

The diorite porphyrite samples are characterized by medium SiO_2 (SiO_2 = 60.13%–62.43%, mean = 61.28%), are rich in K_2O ($K_2O > Na_2O$), have medium CaO content (CaO = 2.43%–3.41%, mean = 2.92%), are rich in Al_2O_3 (Al_2O_3 = 16.24%–16.41%, mean = 16.33%) and have low MgO content (MgO = 1.86%–2.71%, mean = 2.29%), meaning that these samples are classified as transitional between metaluminous and peraluminous, and shoshonite series (Figure 5).

On the rare earth element (REE) chondrite-normalized spider diagram (Figure 6a), the granodiorite porphyry and diorite porphyrite rocks are characterized by the significant enrichment of light rare earth elements (LREEs) and by the variable depletion in heavy rare earth elements (HREEs). In the primitive mantle-normalized trace element spider diagrams (Figure 6b), all samples are relative enrichments in K, Rb, Th, and U, and depletions in Ba, Nb, Sr, P, and Ti.

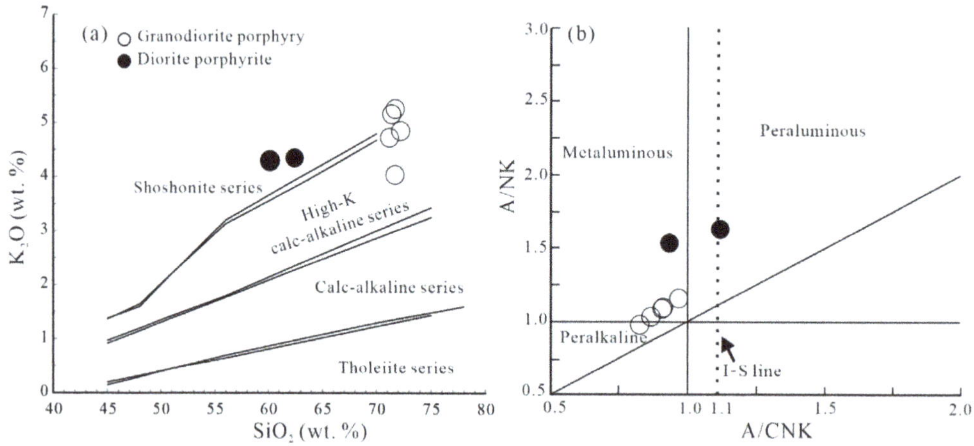

Figure 5. (**a**) K_2O versus SiO_2 [25] and (**b**) A/NK versus A/CNK diagrams [26] for the intrusions associated with the Bayanbaolege deposit.

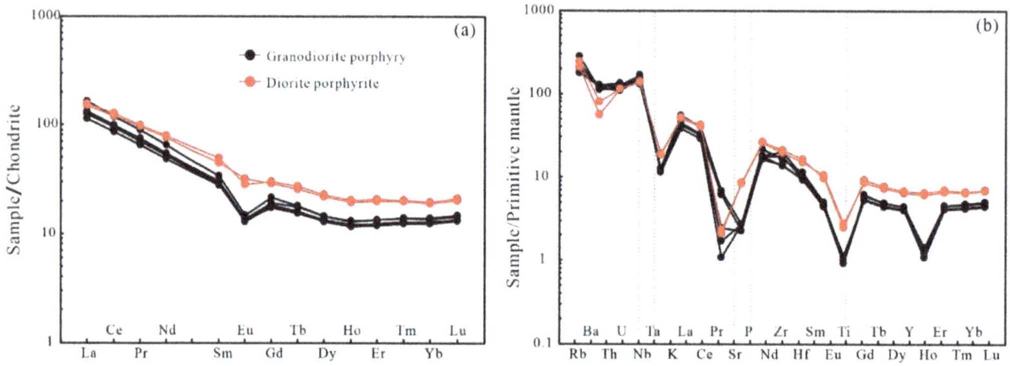

Figure 6. Chondrite-normalized REE (**a**) and primitive-mantle-normalized multi-element (**b**) diagrams for samples of the intrusions associated with the Bayanbaolege deposit (normalizing values are from [27]).

5.2. Sr-Nd Isotopes

Five samples of the granodiorite porphyry from the Bayanbaolege deposit were analyzed for Sr-Nd isotopic compositions, the aged-corrected (130 Ma) initial Sr and Nb isotope data for whole-rock samples, and the results are shown in Table 2. Initial ε_{Nd} (t) values of the samples range from −0.51 to 0.69 and $(^{87}Sr/^{86}Sr)_i$ isotope ratios vary from 0.708576 to 0.710536; all samples of the granodiorite porphyry have relatively homogeneous Nd-Sr isotopic compositions, which are consistent with the values of Mesozoic granites in NE China [28,29], and plotted within the overlapped field of the Mesozoic granites from NE China and the Mesozoic granitoids from CAOB in the ε_{Nd} (t) vs. $(^{87}Sr/^{86}Sr)_i$ diagram and the Phanerozoic granites of juvenile origin in NE China (Figure 7).

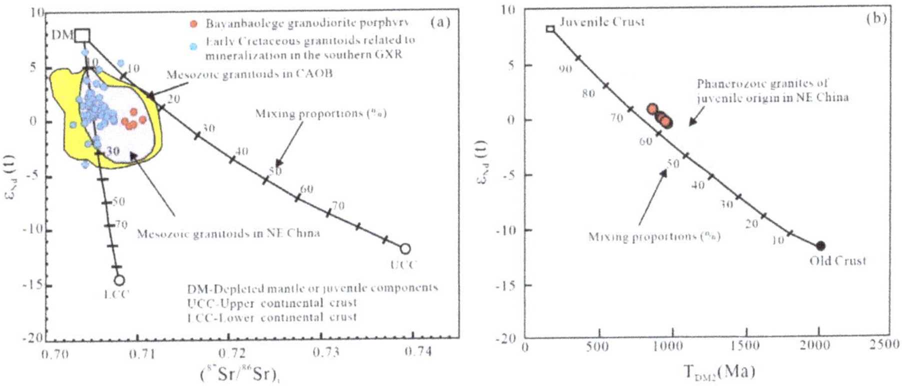

Figure 7. (**a**) ε_{Nd} (t) vs. $(^{87}Sr/^{86}Sr)_i$ isotopic ratio plot showing mixing proportions between two end members (modified after [29]). The area for Mesozoic granitoids in NE China is from [28,29]); Mesozoic granitoids in CAOB are from [30–32]; Early Cretaceous granitoids related to mineralization (data from Table 2); (**b**) ε_{Nd} (t) vs. T_{DM2} diagram [28], and the curve in this diagram approximates the proportions of mixing between juvenile and ancient crustal components. Data for Phanerozoic granites of juvenile origin in NE China are from [28].

5.3. Zircon U-Pb Dating and Lu-Hf Isotopes

Zircons grains collected from post-ore diorite porphyrite are euhedral-subhedral and obviously oscillatory growth-zoning in CL images (Figure 8a), and the value of the Th/U ratios range from 0.32 to 1.15. We interpret these zircons to be of igneous origin [32–34]. The LA-MC-ICP-MS zircon U-Pb dating results of post-ore diorite porphyrite are listed in Table 3. The analyses yielded $^{206}Pb/^{238}U$ ages ranging from 121 Ma to 129 Ma, with the weighted mean $^{206}Pb/^{238}U$ age of 124.8 ± 1.1 Ma (N = 18, MSWD = 2.6) (Figure 8b).

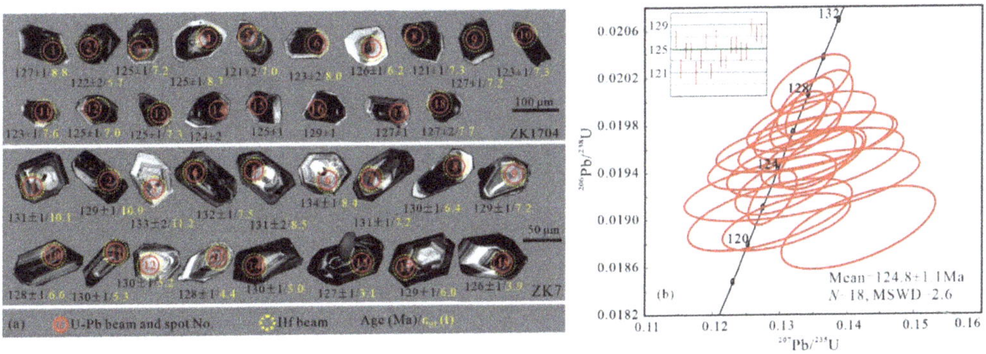

Figure 8. (**a**) Representative cathodoluminescence (CL) images of zircons from the granodiorite porphyry associated with mineralization (No. ZK7) and diorite porphyrite (No. ZK1704); (**b**) Zircon U-Pb Concordia diagrams of sample No. ZK1704.

Table 2. Early Cretaceous granitoids related to mineralization in southern Great Xing'an Range.

Deposits Name	Sample No.	Age (Ma)	Rb (ppm)	Sr (ppm)	$^{87}Rb/^{86}Sr$	$^{87}Sr/^{86}Sr$	2σ	$(^{87}Sr/^{86}Sr)_i$	Sm (ppm)	Nd (ppm)	$^{147}Sm/^{144}Nd$	$^{143}Nd/^{144}Nd$	2σ	T_{DM} (Ma)	T_{DM2} (Ma)	ε_{Nd} (t)	$f_{Sm/Nd}$	Reference
Bayanbaolege	zk7-T1	130	116.08	36.13	9.31	0.72672	18	0.70952	4.36	24.22	0.1088	0.512598	9	805	870	0.69	−0.45	This Paper
	zk7-T2	130	133.02	145.21	2.65	0.71544	14	0.71054	5.15	30.14	0.1032	0.512561	10	815	921	0.04	−0.48	
	zk7-T3	130	151.5	22.79	19.3	0.74471	17	0.70905	4.23	22.79	0.1123	0.51254	8	920	967	−0.5	−0.43	
	zk7-T4	130	185.42	51.74	10.39	0.72777	23	0.70858	4.35	24.42	0.1075	0.512556	4	855	935	−0.11	−0.45	
	zk7-T5	130	119.19	132.49	2.6	0.71432	22	0.70952	4.64	25.39	0.1104	0.512545	8	895	956	−0.38	−0.44	
Aolunhua	ALH-31	132	110.2	414.9	0.7686	0.70644	17	0.70498	3.21	19.67	0.0988	0.512631	5	689	801	1.5	−0.5	[9]
	ALH-32	132	99.81	533.4	0.5413	0.70602	18	0.70499	3.02	17.15	0.1066	0.512612	7	766	842	1	−0.46	
	ALH-44	132	70.26	383.4	0.5302	0.70619	11	0.70518	2.92	16.05	0.1099	0.512667	10	710	759	2.1	−0.44	
	ALH-54	132	102.7	448.2	0.6627	0.70609	11	0.70483	3.22	18.68	0.1043	0.512628	5	728	813	1.4	−0.47	
	ALH-56	132	65.3	203.4	0.9287	0.70665	25	0.70448	1.22	5.88	0.1251	0.512605	10	938	878	0.6	−0.36	
	ALH-63	132	79.12	519.2	0.4409	0.70596	14	0.70512	3.11	17.94	0.1047	0.512613	10	751	838	1.1	−0.47	
Haobugao	6-1	139	206	228	2.6193	0.71045	16	0.705268	7.39	33.1	0.135	0.512613	10		880	0.6		[35]
	6-2	139	191	224	2.4719	0.71009	14	0.705206	6.85	29.8	0.139	0.512627	14		862	0.82		
	6-3	139	279	74.1	10.9153	0.72764	18	0.706054	8.11	33.6	0.1459	0.512673	10		800	1.58		
	6-4	139	248	72.9	9.8622	0.7255	22	0.705996	8.04	32.4	0.15	0.512674	8		804	1.53		
	6-5	139	279	55.2	14.6526	0.73533	28	0.70636	7.29	26.7	0.165	0.512673	12		827	1.24		
Naoniushan	NNS-2-1	134			0.5855	0.70573	13	0.70458			0.1077	0.51268	9	678	737	2.4	−0.45	[36]
	NNS-2-2	134			0.4447	0.70557	13	0.70469			0.1116	0.512657	13	739	779	1.9	−0.43	
	JNNS-5a	134			0.2116	0.70538	14	0.70497			0.1155	0.512688	7	720	735	2.4	−0.41	
	JNNS-5b	134			0.2074	0.70535	13	0.70494			0.1113	0.512702	8	670	707	2.8	−0.43	
Baiyinnuoer	750-21-12	137	165	68.9	6.94	0.7193	12	0.70579	7.15	36.3	0.119	0.51269	7	743	737	2.4	−0.4	[37]
	BY13-6	137	207	142	4.24	0.71454	16	0.70632	7.36	38.2	0.1165	0.51274	9	646	654	3.4	−0.41	
	BY13-19	137	197	144	3.96	0.71363	14	0.70592	9.95	38.5	0.1562	0.512591	5	1485	948	−0.2	−0.21	
	BY13-20	137	181	124	4.23	0.71462	6	0.70639	7.29	37.1	0.1187	0.512645	6	813	808	1.5	−0.4	
	BY13-21	137	206	189	3.16	0.71087	9	0.70473	6.6	32.8	0.1215	0.512713	8	725	704	2.8	−0.38	
	BY13-49	137	193	142	3.92	0.71399	8	0.70632	8.12	41.6	0.1181	0.512608	9	867	866	0.8	−0.4	
	BY13-1	137	216	126	4.99	0.71724	9	0.70814	6.32	25.7	0.1487	0.51287	17	661	490	5.3	−0.24	
	750-21-6	129	229	40.8	16.2	0.73654	12	0.70668	5.02	26.2	0.1157	0.512622	8	824	841	1	−0.41	
	750-21-3	129	164	95.8	4.95	0.71485	14	0.70577	5.03	26	0.1172	0.521823	9	519	523	4.9	−0.4	
	750-25-9	129	199	41.9	13.8	0.7303	13	0.70505	5.3	28.7	0.1116	0.512632	6	776	819	1.3	−0.43	
	BY13-22	129	310	54	16.6	0.73527	11	0.70473	6.29	26.3	0.1446	0.512622	11	1168	879	0.5	−0.26	
	BY13-50	129	237	46.6	14.7	0.73155	12	0.70451	5.2	28.2	0.1114	0.512581	8	850	900	0.3	−0.43	
	BY13-51	129	333	66.4	14.5	0.73304	8	0.70637	4.7	25.1	0.1131	0.512567	9	886	925	0	−0.43	
Shuangjianzishan	SJS-3	137			0.3514	0.70509		0.70441			0.1134	0.512766		586	608	4	−0.42	[38]
	SJS-4	137			0.3058	0.70485		0.70425			0.111	0.512896		380	397	6.5	−0.44	

Table 2. *Cont.*

Deposits Name	Sample No.	Age (Ma)	Rb (ppm)	Sr (ppm)	^{87}Rb/^{86}Sr	^{87}Sr/^{86}Sr	2σ	(^{87}Sr/^{86}Sr)$_i$	Sm (ppm)	Nd (ppm)	^{147}Sm/^{144}Nd	^{143}Nd/^{144}Nd	2σ	T_{DM} (Ma)	T_{DM2} (Ma)	ε_{Nd} (t)	$f_{Sm/Nd}$	Reference
Haisugou	HSG01	137.6			1.01	0.70602	3	0.70403			0.1272	0.512657	5	869	799	1.6		[39]
	HSG02	137.6			0.996	0.70689	8	0.70494			0.1281	0.512633	2	921	838	1.1		
	HSG03	137.6			1.078	0.70706	4	0.70495			0.1199	0.512609	2	879	865	0.8		
	HSG09	137.6			0.939	0.70789	25	0.70605			0.1362	0.512623	2	1036	865	0.8		
	HSG41	137.6			1.106	0.70878	5	0.70661			0.132	0.512609	2	1009	882	0.6		
	HSG42	137.6			1.007	0.70728	12	0.70531			0.1264	0.512625	3	917	849	1		
	HSG43	137.6			1.029	0.70945	18	0.70743			0.1237	0.512583	2	960	911	0.2		
	HSG44	137.6			0.969	0.70723	3	0.70533			0.1354	0.512655	5	963	813	1.4		
	HSG45	137.6			1.3	0.70704	4	0.70449			0.1287	0.512619	2	952	861	0.8		
	HSG46	137.6			2.422	0.7121	16	0.70735			0.1268	0.512595	2	973	897	0.4		
	HSG47	137.6			1.083	0.70945	7	0.70733			0.1309	0.512593	11	1026	906	0.3		
	HSG48	137.6			1.624	0.70725	3	0.70406			0.1187	0.512607	3	872	866	0.8		
Baerzhe	Ba-6	125	10.1	1366	421	1.43636	7	0.68842	169.3	531.6	0.1279	0.51276	11		637	2.4	−0.35	[40]
	Ba-8	125	13.3	1364	296	1.22943	5	0.70356	172.4	510.2	0.2043	0.512761	18		724	2.24	0.04	
	Ba-14	125	9.05	1132	390	1.38568	4	0.69282	89	269.8	0.1813	0.512735	19		744	2.1	−0.08	
	Ba-16	125	10.15	920	274	1.18235	9	0.69557	64.3	214	0.1817	0.512754	10		714	2.47	−0.08	
	Ba-22	125	11.6	511	131	0.92987	6	0.69714	32.6	120.1	0.1641	0.512724	24		740	2.17	−0.17	
Yangchang	1-16	138			0.7992	0.70711		0.705546			0.0982	0.512459		914		−1.8	−0.5	[41]
	2-31	138			0.9214	0.70728		0.705479			0.1007	0.512476		911		−1.5	−0.49	
	2-32	138			0.956	0.70644		0.704573			0.1007	0.512461		931		−1.8	−0.49	
	3-3	138			0.7372	0.70691		0.705472			0.101	0.512453		944		−1.9	−0.49	
	3-4	138			0.9143	0.70733		0.705543			0.0964	0.512457		903		−1.8	−0.51	
	3-5	138			0.8772	0.70717		0.705459			0.0999	0.512457		931		−1.8	−0.49	
	4-23	138			0.7607	0.707		0.705512			0.0998	0.512447		942		−2	−0.49	
	5-8	138			0.9489	0.70738		0.705522			0.099	0.512454		927		−1.9	−0.5	
	6-7	138			0.6587	0.70692		0.705627			0.1021	0.512452		955		−2	−0.48	

Table 3. LA-MC-ICP-MS zircon U-Pb data of diorite porphyrite in the Bayanbaolege deposit.

Sample No.	Pb (ppm)	U (ppm)	Isotopic Ratios								Ages (Ma)					
			^{232}Th/^{238}U	1σ	^{207}Pb/^{235}U	1σ	^{206}Pb/^{238}U	1σ	^{207}Pb/^{206}Pb	1σ	^{206}Pb/^{238}U	1σ	^{207}Pb/^{235}U	1σ	^{207}Pb/^{206}Pb	1σ
ZK1704.1	7	338	0.7558	0.0020	0.1346	0.0078	0.0198	0.0002	0.0492	0.0028	127	1	128	7	158	135
ZK1704.2	17	810	0.9285	0.0141	0.1285	0.0099	0.0190	0.0002	0.0489	0.0034	122	2	123	9	145	163
ZK1704.3	18	874	0.6267	0.0026	0.1349	0.0063	0.0195	0.0002	0.0501	0.0024	125	1	129	6	200	113
ZK1704.4	6	306	0.9142	0.0044	0.1308	0.0094	0.0195	0.0002	0.0486	0.0034	125	1	125	9	130	165
ZK1704.5	3	159	0.4427	0.0029	0.1417	0.0094	0.0189	0.0003	0.0543	0.0035	121	2	135	9	383	147
ZK1704.6	4	197	0.6062	0.0011	0.1333	0.0143	0.0193	0.0002	0.0501	0.0054	123	2	127	14	200	250
ZK1704.7	10	500	0.6093	0.0022	0.1326	0.0053	0.0198	0.0002	0.0487	0.0019	126	1	126	5	132	90
ZK1704.8	9	422	1.0551	0.0031	0.1309	0.0077	0.0190	0.0002	0.0500	0.0029	121	1	125	7	196	136
ZK1704.9	24	1115	0.8224	0.0008	0.1327	0.0034	0.0198	0.0002	0.0485	0.0012	127	1	127	3	123	58

Table 3. *Cont.*

Sample No.	Pb (ppm)	U (ppm)	Isotopic Ratios								Ages (Ma)					
			$^{232}Th/^{238}U$	1σ	$^{207}Pb/^{235}U$	1σ	$^{206}Pb/^{238}U$	1σ	$^{207}Pb/^{206}Pb$	1σ	$^{206}Pb/^{238}U$	1σ	$^{207}Pb/^{235}U$	1σ	$^{207}Pb/^{206}Pb$	1σ
ZK1704.10	8	385	0.7000	0.0028	0.1308	0.0061	0.0193	0.0002	0.0492	0.0022	123	1	125	6	156	107
ZK1704.11	30	1309	1.1541	0.0058	0.1289	0.0029	0.0192	0.0002	0.0487	0.0011	123	1	123	3	134	51
ZK1704.12	9	429	0.7692	0.0007	0.1340	0.0110	0.0197	0.0002	0.0494	0.0040	125	1	128	10	169	189
ZK1704.13	7	318	0.6843	0.0019	0.1353	0.0094	0.0197	0.0002	0.0499	0.0034	125	1	129	9	191	160
ZK1704.14	5	261	0.4098	0.0054	0.1353	0.0107	0.0195	0.0002	0.0503	0.0040	124	2	129	10	211	184
ZK1704.15	30	1560	0.3169	0.0010	0.1355	0.0046	0.0195	0.0002	0.0504	0.0016	125	1	129	4	213	73
ZK1704.16	13	596	0.8334	0.0027	0.1345	0.0048	0.0201	0.0002	0.0484	0.0017	129	1	128	5	119	82
ZK1704.17	15	660	0.8707	0.0031	0.1337	0.0035	0.0200	0.0002	0.0486	0.0012	127	1	127	3	128	59
ZK1704.18	15	683	0.8212	0.0043	0.1357	0.0041	0.0200	0.0002	0.0493	0.0014	127	2	129	4	163	67

Table 4. Zircon Hf isotopic compositions of the granitoids from Bayanbaolege deposit.

No.	Age (Ma)	$^{176}Yb/^{177}Hf$	22σs	$^{176}Lu/^{177}Hf$	2σ	$^{176}Hf/^{177}Hf$	2σ	$^{176}Hf/^{177}Hf_i$	$e_{Hf}(0)$	$e_{Hf}(t)$	T_{DM}(Ma)	$T_{DM}{}^{C}$(Ma)	$f_{Lu/Hf}$
						Granodiorite Porphyry							
ZK7.1	131	0.048202	0.000446	0.001713	0.000005	0.282979	0.000029	0.282975	7.3	10.1	394	714	−0.95
ZK7.2	129	0.113247	0.00233	0.003574	0.000038	0.283009	0.000043	0.283001	8.4	10.9	369	635	−0.89
ZK7.3	133	0.050256	0.000219	0.001714	0.000013	0.283009	0.000036	0.283005	8.4	11.2	350	615	−0.95
ZK7.4	132	0.036693	0.00089	0.001217	0.000021	0.282904	0.000033	0.282901	4.7	7.5	496	951	−0.96
ZK7.5	131	0.031451	0.000836	0.001101	0.000019	0.282932	0.000029	0.28293	5.7	8.5	455	860	−0.97
ZK7.6	134	0.069897	0.001611	0.002781	0.000049	0.282933	0.000031	0.282926	5.7	8.4	475	869	−0.92
ZK7.7	131	0.040393	0.000108	0.001661	0.000005	0.282898	0.000027	0.282894	4.5	7.2	511	975	−0.95
ZK7.8	130	0.036409	0.000931	0.001166	0.000019	0.282874	0.000028	0.282871	3.6	6.4	539	1049	−0.96
ZK7.9	129	0.051141	0.000582	0.001867	0.000016	0.2829	0.000027	0.282896	4.5	7.2	511	971	−0.94
ZK7.10	128	0.056866	0.001234	0.002098	0.000031	0.282884	0.000029	0.282879	4	6.6	538	1026	−0.94
ZK7.11	130	0.033501	0.000676	0.001388	0.000025	0.282844	0.000026	0.28284	2.5	5.3	585	1148	−0.96
ZK7.12	130	0.047709	0.000959	0.001722	0.000022	0.282843	0.000034	0.282838	2.5	5.2	592	1155	−0.95
ZK7.13	128	0.014286	0.000608	0.000547	0.000017	0.282817	0.000027	0.282816	1.6	4.4	609	1228	−0.98
ZK7.14	130	0.041769	0.001384	0.001424	0.000038	0.282836	0.000035	0.282832	2.3	5	597	1174	−0.96
ZK7.15	127	0.080624	0.002572	0.00278	0.000081	0.282788	0.000039	0.282781	0.6	3.1	691	1341	−0.92
ZK7.17	129	0.037943	0.000651	0.001404	0.000021	0.282866	0.000031	0.282863	3.3	6	554	1078	−0.96
ZK7.18	126	0.046771	0.000295	0.001917	0.000014	0.282808	0.000026	0.282804	1.3	3.9	645	1270	−0.94
						Diorite porphyrite							
ZK1704.1	127	0.054425	0.000299	0.001397	0.000009	0.282945	0.000022	0.282942	6.1	8.8	440	827	−0.96
ZK1704.2	122	0.061272	0.000575	0.00146	0.000019	0.28286	0.00002	0.282856	3.1	5.7	564	1108	−0.96
ZK1704.3	125	0.043899	0.000121	0.001144	0.000002	0.282901	0.000018	0.282898	4.6	7.2	500	969	−0.97
ZK1704.4	125	0.023133	0.000345	0.000568	0.000006	0.282941	0.000018	0.28294	6	8.7	436	836	−0.98
ZK1704.5	121	0.028046	0.000244	0.000705	0.000004	0.282896	0.000023	0.282895	4.4	7	501	986	−0.98
ZK1704.6	123	0.053344	0.000572	0.001276	0.000009	0.282924	0.00002	0.282922	5.4	8	468	897	−0.96
ZK1704.7	126	0.031281	0.00036	0.00079	0.00001	0.282872	0.000017	0.28287	3.5	6.2	536	1058	−0.98
ZK1704.8	121	0.042613	0.000187	0.001056	0.000008	0.282905	0.000024	0.282903	4.7	7.3	493	961	−0.97
ZK1704.9	127	0.045652	0.000274	0.001133	0.000008	0.282899	0.000018	0.282896	4.5	7.2	503	974	−0.97
ZK1704.10	123	0.052782	0.000316	0.001312	0.000007	0.282904	0.000021	0.282901	4.7	7.3	497	962	−0.96
ZK1704.11	123	0.031526	0.000057	0.000801	0.000004	0.282911	0.000018	0.282909	4.9	7.6	481	936	−0.98
ZK1704.12	125	0.045432	0.000179	0.001123	0.000005	0.282895	0.00002	0.282892	4.3	7	509	989	−0.97
ZK1704.13	129	0.027104	0.000172	0.000639	0.000002	0.282901	0.000027	0.282899	4.6	7.3	494	961	−0.98
ZK1704.18	127	0.037908	0.000139	0.00091	0.000002	0.282914	0.000029	0.282912	5	7.7	478	922	−0.97

The crystallization age of the granodiorite porphyry associated genetically with the mineralization of the Bayanbaolege deposit is 130 ± 1 Ma [4]; the shapes of the zircon grain are shown in Figure 8a. In situ zircon Lu-Hf isotopic data of the granodiorite porphyry and the diorite porphyrite are shown in Table 4. The zircons of the No.ZK7 yielded initial $^{176}Hf/^{177}Hf$ ratios of 0.282788 to 0.283009 with ε_{Hf} (t) values of +3.1 to +11.2, and T_{DM2}(Hf) model ages of 615–1341 Ma. The zircons of the diorite porphyrite yielded initial $^{176}Hf/^{177}Hf$ ratios of 0.282856 to 0.282942 with ε_{Hf} (t) values of +5.7 to +8.8 and T_{DM2}(Hf) model ages of 827–1108 Ma. All analyses plot above the Chondrite uniform reservoir (CHUR) evolutionary line in a diagram of ε_{Hf} (t) vs. age (Figure 9a) and located in the region of Early Cretaceous magmatic rocks in the southern GXR (Figure 9b).

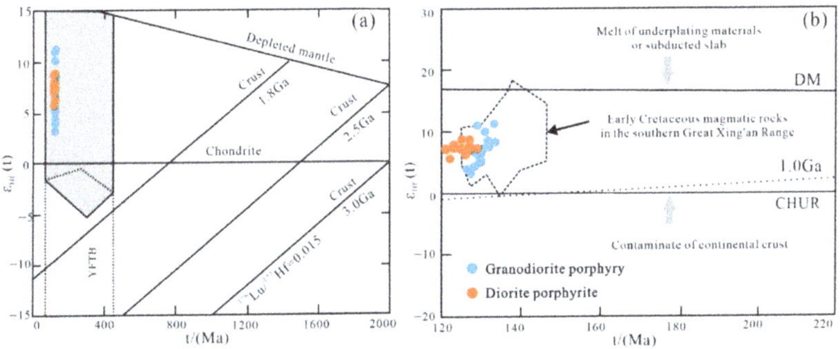

Figure 9. Correlations between ε_{Hf} (t) and ages of zircons from the intrusions. CAOB = Central Asian Orogenic Belt; YFTB = Yanshan Fold and Thrust Belt. (**a**) modified after [42,43]; (**b**) after [1,44].

6. Discussion

6.1. Petrogenesis of the Ore-Associated Intrusions

The major and trace elements of the intrusions are shown in Table 1. The granodiorite porphyry contains high concentrations of SiO_2, high K calc-alkalinity, metaluminous (A/CNK = 0.83–0.97), and sodium contents (3.80–4.89 wt.%); the CIPW normative corundum <1% and diopside of samples showed that the high-Si felsic granodiorite porphyry is I-type granites [45]. The high $(K_2O + Na_2O)/CaO$ (7.09–7.96), FeO^T/MgO ratios (4.30–5.53) and 10,000 Ga/Al ratios (2.07–2.35) are also showed in the I-type. The diorite porphyrite, A/CNK = 0.97–1.14, and A/NK = 1.54–1.63, metaluminous-peraluminous, and 10,000 Ga/Al = 1.79–1.87 samples all fall into the I-type granites field in Figure 10a,b [46].

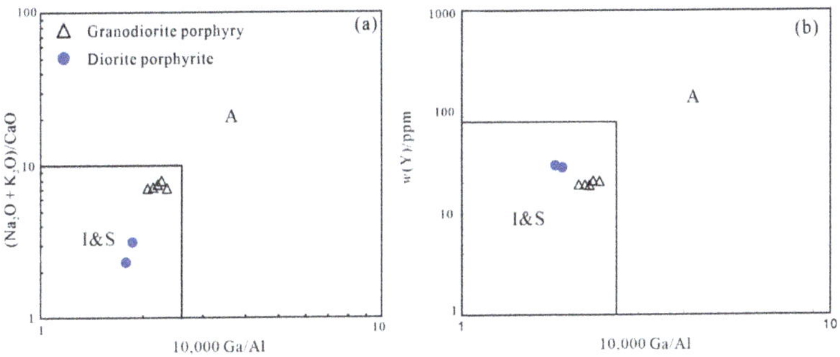

Figure 10. Discrimination diagrams of granite genetic type in the study area: (**a**) $(K_2O + Na_2O)/CaO$ vs. 10,000 Ga/Al; and (**b**) Y vs. 10,000 Ga/Al [46].

6.2. Magmatic Material Source of the Ore-Associated Intrusions

There are three possible interpretations for the source of I-type granites: (1) partial melting of metamorphic intermediate-basic igneous rocks from the crust [47,48]; (2) fractional crystallization from mantle magma and assimilation of the crust material in the late stage of magma [49,50] and (3) magma mixing of basaltic magma with felsic magmas [51].

Table 1 showed that samples of granodiorite porphyry from the Bayanbaolege deposit contain low MgO (0.31–0.43 wt.%), Ni (1.71–2.45 ppm), and Cr (10.91–16.90 ppm) element, indicating a few involvements of mantle-derived magma [52,53]. The Nb/Ta ratios for the Bayanbaolege intrusion vary between 8.47 and 10.36, significantly lower than the mantle values (17.5 $\pm$ 2) [54,55] and close to the crust values (~11) [55,56]. Combined with the high SiO_2, K_2O, low Al_2O_3 and Sr, Nb, Ta, Ti and P, and Figure 6 showed the positive Rb, Th, and K anomalies, with enrichment in LREEs and LILEs. These geochemical features are commonly matched to crust-derived rocks [57]. Hence, the granodiorite porphyry from the Bayanbaolege deposit is likely to have come from a crustal source or a mixing of the mantle- and crustal-derived magmas. The granodiorite porphyry yielded the ε_{Nd} (t) values ranging from -0.51 to 0.69 and a young two-stage Nd model age (T_{DM2} =870–967 Ma), indicating that the primary magmas were extracted from a mixed juvenile-crust and old-crust source (Figure 7).

The diorite porphyrite exhibited low SiO_2 (60.13–62.43 wt.%), high K_2O (4.30–4.35 wt.%), and Al_2O_3 (16.24–16.41 wt.%) contents, suggesting that these rocks are metaluminous-peraluminous. The trace elements diagrams in Figure 6 show that the diorite porphyrite and the granodiorite porphyry have a similar tendency. The Nb/Ta values (14.48–14.56) are higher than the crust values (11.0) and lower than those of the primitive mantle (17.5 $\pm$ 2) [55], implying these rocks are from the mixing of the mantle- and crustal-derived magmas.

These rocks of the diorite porphyrite and granodiorite porphyry have positive ε_{Hf} (t) (+5.7 to +8.8) and (+3.1 to +11.2), and T_{2DM} ages (827~1108 Ma) and (615~1341 Ma), respectively (Table 4). All analyses plot above the Chondrite uniform reservoir (CHUR) evolutionary line in a diagram of ε_{Hf} (t) vs. age (Figure 9a) located in the region of Early Cretaceous magmatic rocks in the southern GXR (Figure 9b). The similar εHf(t) values of these samples and the wider ranges and more mantle-like values suggest the possible involvement of heterogeneous juvenile sources, including moderately depleted mantle and newly underplated lower crust.

6.3. Mineralization Age of Polymetallic Mineralization

There are two main methods for determining the mineralization age of hydrothermal deposits. One is indirect, using the isotopic ages of rocks (strata or intrusions) associated with mineralization; the other directly measures the ages of ore minerals or coexistence alteration minerals with ore minerals [58]. The indirect mineralization age of the granodiorite porphyry associated genetically with mineralization reported is 130 $\pm$ 1 Ma [4] and the direct isochron age of sphalerite is 129.9 $\pm$ 2.9 Ma [5]. Hence, the mineralization of the Bayanbaolege deposit occurred in the Early Cretaceous. However, the Bayanbaolege is a large-scale polymetallic deposit including Ag, Zn, and Pb. The different metallogenic temperatures and precipitation phases of these metals have been verified by Wang et al. [4]. The mineralization of these metals should last quite a long time. However, the indirect mineralization age (130 $\pm$ 1 Ma) and the direct mineralization age (129.9 $\pm$ 2.9 Ma) are too close. To constrain the mineralization duration of the Bayanbaolege polymetallic deposit, the diorite porphyrite was measured and weighted with a mean $^{206}Pb/^{238}U$ age of 124.8 $\pm$ 1.1 Ma (Figure 8b). These results of the magmatism and mineralization mentioned above give a range of 125–130 Ma, and this result is consistent with the mineralization age (126–132 Ma) of the nearby Aolunhua porphyry Mo-Cu deposit [59]. Meanwhile, the mineralization age of the lager-scale Baiyinnuoer skarn Pb-Zn deposit (129–136 Ma) [37] and the lager-scale Dajing Sn-Cu-Pb-Zn-Ag (133–136 Ma) [60] also show that the hydrothermal mineralization is probably the product of the large-scale tectono-magmatic-mineralization events in the southern GXR in the Early Cretaceous.

6.4. Tectonic Implications

The tectonic setting of the Late Mesozoic magmatism in the southern GXR and adjacent areas has been controversial and invoked (1) post-orogenic extension after the closure of the Mongol–Okhotsk Ocean [32,61–66] and (2) the subduction of the Paleo–Pacific Ocean Plate [67–72]. Recent studies considered that the GXR was subject to an extensional tectonic regime related to the delamination of a thickened crust [73–76], which was identified by the large-area distributed synchronous A-type granitoids [77], metamorphic core complexes [78–80], diabase dyke groups, and rift basins [81,82]. Hence, the GXR was in an extensional setting during the Early Cretaceous, which resulted from post-orogenic extensional tectonics after the closure of the Mongol–Okhotsk Ocean, or the subduction of the Paleo-Pacific Ocean Plate.

The extensional environment during the 138–145 Ma was the common result of the closure of the Mongolia–Okhotsk Ocean and the subduction of the Paleo-Pacific Ocean Plate [83–85], and the active continental margin setting of the 106–133 Ma mainly related to the subduction of the Paleo-Pacific Ocean Plate [83,86]. Furthermore, there is a large crossing angle between the tendency (NE) of the Mongol–Okhotsk Ocean suture zone and the tendency (NNE) of the GXR, implying that the Mongol–Okhotsk Ocean tectonic domain cannot efficaciously control the Early Cretaceous magmatism in the GXR area [87,88].

The diorite porphyrite and granodiorite porphyry magmatism of the Bayanbaolege took place in the Early Cretaceous (125 Ma and 130 Ma), and seven samples plotted in the VAG field (Figure 11a–c) and the subduction zone field (Figure 11d) in the geochemical discrimination diagrams, indicating that the granitoids in the Bayanbaolege area have closely related to the subduction tectonic setting instead of the post Mongol–Okhotsk Ocean tectonic regime. This result is consistent with the point that the influence scope of the Paleo-Pacific plate subduction extended to the GXR during the period of the Early Cretaceous [89]. So, it can be concluded that the affected region of the westward subduction of the Paleo-Pacific Ocean plate reached the east slope of the Great Xing'an Range during the early Cretaceous. This viewpoint is also consistent with the thoughts of Zhu [90] and Tian et al. [91], who thought that the gravity gradient zones began to form in the early cretaceous and did not cross the GXR, in other words, the leading edge of the subduction Paleo-Pacific Ocean plate stagnates in the east segment of the GXR.

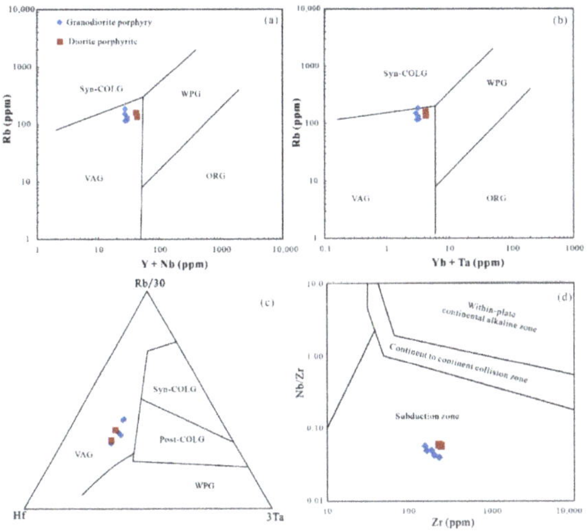

Figure 11. (**a**) Y vs. Nb diagram [92]; (**b**)Y + Nb vs. Rb diagram [92]; (**c**) Rb/30–Hf–3Ta discrimination diagram [93]; (**d**) Zr versus (Nb/Zr) diagram [94]. Abbreviations: VAG—volcanic arc granite; ORG—ocean ridge granite; WPG—within-plate granite; Syn-COLG—syncollision granite.

7. Conclusions

The magmatism in the Bayanbaolege area occurred during the Early Cretaceous (diorite porphyrite: 124.8 ± 1.1 Ma), and the mineralization age should be formed in the 125–130 Ma. The whole-rock elemental compositions of the intrusions implied that the intrusive rocks are the I-type granite. The Sr-Nd-Hf isotopes indicated that the primary magmas of the granodiorite porphyry and the diorite porphyrite were the possible involvement of heterogeneous juvenile sources including moderately depleted mantle and newly underplated lower crust. The Early Cretaceous intrusive rocks in the Bayanbaolege area were formed in an extensional tectonic setting and compactly related to the retreat of the Paleo-Pacific Ocean subducted plate rather than the closure of the Mongol–Okhotsk Ocean.

Author Contributions: Conceptualization, X.W., Q.Y. and Y.-S.R.; field investigation, X.W., Q.Y. and Y.-S.R.; experimental analysis, X.W.; software, X.W. and Q.Y.; validation, X.W., Q.Y. and Y.-S.R.; resources, Y.-S.R. and Z.-M.S.; data curation, X.W.; writing—original draft preparation, X.W.; writing—review and editing, X.W. and Y.-S.R.; visualization, X.W.; funding acquisition, X.W. and Y.-S.R. All authors have read and agreed to the published version of the manuscript.

Funding: This work was financially supported by the Self-determined Foundation of Key Laboratory of Mineral Resources Evaluation in Northeast Asia, Ministry of Natural Resources (No. DBY-ZZ-19-10) and the China Geological Survey Programs (No. DD20160048; No. 12120115031701).

Acknowledgments: We sincerely appreciate the geologists from the Inner Mongolia Chifeng Geology and Mineral Resources Exploration Development Institute for their support of our fieldwork.

Conflicts of Interest: The authors declare no conflict of interest.

References

1. Ouyang, H.G.; Mao, J.W.; Zhou, Z.H.; Su, H.M. Late Mesozoic metallogeny and intracontinental magmatism, southern Great Xing'an Range, northeastern China. *Gondwana Res.* **2015**, *27*, 1153–1172. [CrossRef]
2. Zeng, Q.D.; Liu, J.M.; Chu, S.X.; Guo, Y.P.; Gao, S.; Guo, L.X.; Zhai, Y.Y. Poly-metal mineralization and exploration potential in southern segment of the Great Xing'an range. *J. Jilin Univ.* **2016**, *46*, 1100–1123. (In Chinese with English abstract)
3. Gao, Y.; Liu, J.; Li, T.G.; Zhang, D.D.; Yang, Y.C.; Han, S.J.; Ding, Q.F.; Zhang, S. Multiple isotope (He-Ar-Zn-Sr-Nd-Pb) constraints on the genesis of the Jiawula Pb-Zn-Ag deposit, NE China. *Ore Geol. Rev.* **2021**, *134*, 104142. [CrossRef]
4. Wang, X.; Ren, Y.S.; Zhao, D.S.; Ren, X.G. Ore-forming fluids and ore genesis of the large Bayanbaolege Ag polymetallic deposit, Southern Great Xing'an Range, NE China. *Ore Geol. Rev.* **2019**, *111*, 102987. [CrossRef]
5. Wang, X.; Ren, Y.S.; Yang, Q.; Zhao, D.S. Rb–Sr dating of sulfdes and S–Pb isotopic study of the Bayanbaolege Ag polymetallic deposit, Inner Mongolia, NE China. *Geochemistry* **2021**, *81*, 125724. [CrossRef]
6. Zhang, J.H.; Gao, S.; Ge, W.C.; Wu, F.Y.; Yang, J.H.; Wilde, S.A.; Li, M. Geochronology of the Mesozoic volcanic rocks in the Great Xing'an Range, northeastern China: Implications for subduction-induced delamination. *Chem. Geol.* **2010**, *276*, 144–165. [CrossRef]
7. Jahn, B.-M.; Wu, F.Y.; Chen, B. Massive granitoid generation in central Asia: Nd isotopic evidence and implication for continental growth in the Phanerozoic. *Episodes* **2000**, *23*, 82–92. [CrossRef]
8. Wu, F.Y.; Zhao, G.C.; Sun, D.Y.; Wilde, S.A.; Zhang, G.L. The Hulan Group: Its role in the evolution of the Central Asian Orogenic Belt of NE China. *J. Asian Earth Sci.* **2007**, *30*, 542–556. [CrossRef]
9. Liu, Y.J.; Li, W.M.; Feng, Z.Q.; Wen, Q.B.; Neubauer, F.; Liang, C.Y. A review of the Paleozoic tectonics in the eastern part of Central Asian Orogenic Belt. *Gondwana Res.* **2017**, *43*, 123–148. [CrossRef]
10. Bureau of Geology and Mineral Resources of Inner Mongolia (BGMRIM). *Regional Geology of Nei Mongol (Inner Mongolia) Autonomous Region*; Geological Publishing House: Beijing, China, 1991; 725p. (In Chinese with English abstract)
11. Li, J.Y. Permian geodynamic setting of Northeast China and adjacent regions: Closure of the Paleo–Asian Ocean and subduction of the Paleo-Pacific Plate. *J. Asian Earth Sci.* **2006**, *26*, 207–224. [CrossRef]
12. Bao, Q.Z.; Zhang, C.J.; Wu, Z.L.; Wang, H.; Li, W.; Sang, J.H.; Liu, Y.S. SHRIMP U-Pb zircon geochronology of a Carboniferous quartz-diorite in Baiyingaole area, Inner Mongolia and its implications. *J. Jilin Univ.* **2007**, *37*, 15–23. (In Chinese with English abstract)
13. Zhang, J.H.; Ge, W.C.; Wu, F.Y.; Wilde, S.A.; Yang, J.H.; Liu, X.M. Large-scale Early Cretaceous volcanic events in the northern Great Xing'an Range, Northeastern China. *Lithos* **2008**, *102*, 138–157. [CrossRef]
14. Wu, F.Y.; Sun, D.Y.; Ge, W.C.; Zhang, Y.B.; Grant, M.L.; Wilde, S.A.; Jahn, B.M. Geochronology of the Phanerozoic granitoids in northeastern China. *J. Asian Earth Sci.* **2011**, *41*, 1–30. [CrossRef]

15. Xue, H.M.; Guo, L.J.; Hou, Z.Q.; Tong, Y.; Pan, X.F.; Zhou, X.W. SHRIMP zircon U-Pb ages of the middle Neopaleozoic unmetamorphosed magmatic rocks in the southwestern slope of the Da Hinggan Mountain, Inner Mongolia. *Acta Petrol. Miner.* **2010**, *29*, 811–823. (In Chinese with English abstract)
16. Ge, W.C.; Wu, F.Y.; Zhou, C.Y.; Zhang, J.H. Zircon U-Pb ages and its significance of the Mesozoic granites in the Wulanhaote region, central Da Hinggan Mountain. *Acta Petrol. Sin.* **2005**, *21*, 749–762. (In Chinese with English abstract)
17. Jiang, S.H.; Nie, F.J.; Bai, D.M.; Liu, Y.F.; Liu, Y. Geochronology evidence for Indosinian mineralization in Baiyinnuoer Pb-Zn deposit of Inner Mongolia. *Miner. Depos.* **2011**, *30*. 787–798. (In Chinese with English abstract)
18. Gu, Y.C.; Chen, R.Y.; Jia, B.; Ju, N. Zircon U-Pb Dating and Geochemistry of the Granite Porphyry from the Shuangjianzishan Silver Polymetallic Deposit in Inner Mongolia and Tectonic Implications. *Geol. Prospect.* **2017**, *53*, 495–507. (In Chinese with English abstract)
19. Zeng, Q.D.; Qin, K.Z.; Liu, J.M.; Li, G.M.; Zhai, M.G.; Chu, S.X.; Guo, Y.P. Porphyry molybdenum deposits in the Tianshan-Xingmeng orogenic belt, northern China. *Int. J. Earth Sci.* **2015**, *104*, 991–1023. [CrossRef]
20. Li, M.W. Research on Metallogenic Series and Prognosis of End Ogenetic Metallic Deposits in the Middle Section of Tianshan-Xingmeng Orogeny. Ph.D. Thesis, Chinese Academy of Geological Sciences, Beijing, China, 2006; pp. 1–172. (In Chinese with English abstract)
21. Qi, L.; Hu, J.; Gregoire, D.C. Determination of trace elements in granites by inductively coupled plasma-mass spectrometry. *Talanta* **2000**, *51*, 507–513.
22. Geng, J.Z.; Qiu, K.F.; Gou, Z.Y.; Yu, H.C. Tectonic regime switchover of Triassic Western Qinling Orogen: Constraints from LA-ICP-MS zircon U-Pb geochronology and Lu-Hf isotope of Dangchuan intrusive complex in Gansu, China. *Chem. Erde Geochem.* **2017**, *77*, 637–651. [CrossRef]
23. Liu, Y.S.; Gao, S.; Hu, Z.C.; Gao, C.G.; Zong, K.Q.; Wang, D.B. Continental and oceanic crust recycling-induced melt-peridotite interactions in the Trans-North China Orogen: U-Pb dating, Hf isotopes and trace elements in zircon s of mantle xenoliths. *J. Petrol.* **2010**, *51*, 537–571. [CrossRef]
24. Ludwig, K.R. *Isoplot v. 3.0: A Geochronological Toolkit for Microsoft Excel*; Berkeley Geochronology Center: Berkeley, CA, USA, 2003; pp. 1–70.
25. Peccerillo, A.; Taylor, S.R. Geochemistry of Eocene calc-alkaline volcanic rocks from the Kastamonu area, northern Turkey. *Contrib. Mineral. Petrol.* **1976**, *58*, 63–81. [CrossRef]
26. Maniar, P.D.; Piccoli, P.M. Tectonic discrimination of granitoids. *Geol. Soc. Am. Bull.* **1989**, *101*, 635–643. [CrossRef]
27. Sun, S.S.; McDonough, W.F. *Chemical and Isotopic Systematics of Oceanic Basalts: Implications for Mantle Composition and Processes*; Geological Society: London, UK, 1989; Volume 42, pp. 313–345.
28. Wu, F.Y.; Jahn, B.M.; Wilde, S.; Sun, D.Y. Phanerozoic crustal growth: U-Pb and Sr–Nd isotopic evidence from the granites in northeastern China. *Tectonophysics* **2000**, *328*, 89–113. [CrossRef]
29. Wu, F.Y.; Jahn, B.M.; Wilde, S.A.; Lo, C.H.; Yui, T.F.; Lin, Q.; Ge, W.C.; Sun, D.Y. Highly fractionated I-type granites in NE China (II): Isotopic geochemistry and implications for crustal growth in the Phanerozoic. *Lithos* **2003**, *67*, 191–204. [CrossRef]
30. Jian, P.; Liu, D.; Kröner, A.; Windley, B.F.; Shi, Y.; Zhang, F.; Shi, G.; Miao, L.; Zhang, W.; Zhang, Q.; et al. Time scale of an early to mid-Paleozoic orogenic cycle of the long-lived Central Asian Orogenic Belt, Inner Mongolia of China: Implications for continental growth. *Lithos* **2008**, *101*, 233–259. [CrossRef]
31. Chen, B.; Jahn, B.M.; Tian, W. Evolution of the Solonker suture zone: Constraints from zircon U-Pb ages, Hf isotopic ratios and whole-rock Nd-Sr isotope compositions of subduction and collision-related magmas and forearc sediments. *J. Asian Earth Sci.* **2009**, *34*, 245–257. [CrossRef]
32. Jahn, B.M.; Litvinovsky, B.A.; Zanvilevich, A.N.; Reichow, M. Peralkaline granitoid magmatism in the Mongolian-Transbaikalian Belt: Evolution, petrogenesis and tectonic significance. *Lithos* **2009**, *113*, 521–539. [CrossRef]
33. Pupin, J.P. Zircon and granite petrology. *Contrib. Mineral. Petrol.* **1980**, *73*, 207–220. [CrossRef]
34. Koschek, G. Origin and significance of the SEM cathodoluminescence from zircon. *J. Microsc.* **1993**, *171*, 223–232. [CrossRef]
35. Wang, X.D.; Xu, D.M.; Lv, X.B.; Wei, W.; Mei, W.; Fan, X.J.; Sun, B.K. Origin of the Haobugao skarn Fe-Zn polymetallic deposit, Southern Great xing'an range, NE China: Geochronological, geochemical, and Sr-Nd-Pb isotopic constraints. *Ore Geol. Rev.* **2018**, *94*, 58–72. [CrossRef]
36. Gu, J.F.; Ni, P.; Li, W.S.; Wang, G.G.; Pan, J.Y.; Liu, Z. The characteristics of ore-forming fluid and genesis of the Matou Mo deposit, Chizhou, Anhui province. *J. Nanjing Univ.* **2018**, *54*, 351–365. (In Chinese with English abstract)
37. Jiang, S.H.; Chen, C.L.; Bagas, L.; Liu, Y.; Han, N.; Kang, H.; Wang, Z.H. Two mineralization events in the Baiyinnuoer Zn-Pb deposit in InnerMongolia, China: Evidence from field observations, S-Pb isotopic compositions and U-Pb zircon ages. *J. Asian Earth Sci.* **2017**, *144*, 339–367. [CrossRef]
38. Wang, F.X. Magmatism and Ag-Polymetallic Mineralization in Shuangjianzishan Deposit and Its Periphery, Inner Mongolia. Ph.D. Thesis, Chinese Academy of Geological Sciences, Wuhan, China, 2017; pp. 1–189. (In Chinese with English abstract)
39. Shu, Q.H.; Lai, Y.; Wang, C.; Xu, J.J.; Sun, Y. Geochronology, geochemistry and Sr-Nd-Hf isotopes of the Haisugou porphyry Mo deposit, northeast China, and their geological significance. *J. Asian Earth Sci.* **2014**, *79*, 777–791. [CrossRef]
40. Wang, Y.X.; Zhao, Z.H. Geochemistry and origin of the Baerzhe REE-Nb-Be-Zr super-large deposit. *Geochimica* **1997**, *26*, 24–35. (In Chinese with English abstract)

41. Zeng, Q.D.; Yang, J.H.; Zhang, Z.L.; Liu, J.M.; Duan, X.X. Petrogenesis of the Yangchang Mo-bearing granite in the Xilamulun metallogenic belt: China Office of Nuclear Energy, geochemistry, zircon U–Pb ages, and Sr–Nd–Pb isotopes. *Geol. J.* **2014**, *49*, 1–14. [CrossRef]
42. Yang, J.H.; Wu, F.Y.; Wilde, S.A.; Xie, L.W.; Yang, Y.H.; Liu, X.M. Tracing magma mixing in granite genesis: In situ U–Pb dating and Hf-isotope analysis of zircons. *Contrib. Mineral. Petrol.* **2007**, *153*, 177–190. [CrossRef]
43. Tang, J.; Xu, W.L.; Wang, F.; Zhao, S.; Wang, W. Early Mesozoic southward subduction history of the Mongol-Okhotsk oceanic plate: Evidence from geochronology and geochemistry of Early Mesozoic intrusive rocks in the Erguna Massif, NE China. *Gondwana Res.* **2016**, *31*, 218–240. [CrossRef]
44. Yao, L.; Lv, Z.C.; Ye, T.Z.; Pang, Z.S.; Jia, H.X.; Zhang, Z.H.; Wu, Y.F.; Li, R.H. Zircon U-Pb age, geochemical and Nd-Hf isotopic characteristics of quartz porphyry in the Baiyinchagan Sn polymetallic deposit, Inner Mongolia,southern Great Xing'an Range, China. *Acta Petrol. Sin.* **2017**, *33*, 3183–3199. (In Chinese with English abstract)
45. Chappell, B.W.; White, A.J.R. Two contrasting granite types. *Pacific Geol.* **1974**, *8*, 173–174.
46. Whalen, J.B.; Currie, K.L.; Chappell, B.W. A-type granites: Geochemical characteristics, discrimination and petrogenesis. *Contrib. Mineral. Petrol.* **1987**, *95*, 407–419. [CrossRef]
47. Petford, N.; Atherton, M. Na-rich partial melts from newly underplated basaltic crust: The Cordillera Blanca Batholith, Peru. *J. Petrol.* **1996**, *37*, 1491–1521. [CrossRef]
48. Li, X.H.; Li, W.X.; Li, Z.X. On the genetic classification and tectonic implications of the Early Yanshanian granitoids in the Nanling Range, South China. *Chin. Sci. Bull.* **2007**, *52*, 1873–1885. (In Chinese with English abstract) [CrossRef]
49. Beard, J.S.; Lofgren, G.E. Dehydration melting and water-saturated melting of basaltic and andesitic greenstones and amphibolites at 1, 3 and 6.9 kb. *J. Petrol.* **1991**, *32*, 365–401. [CrossRef]
50. Castro, A. Tonalite-granodiorite suites as cotectic systems: A review of experimental studies with applications to granitoid petrogenesis. *Earth Sci. Rev.* **2013**, *124*, 68–95. [CrossRef]
51. Kemp, A.I.S.; Hawkesworth, C.J.; Foster, G.L.; Paterson, B.A.; Woodhead, J.D.; Hergt, J.M.; Gray, C.M.; Whitehouse, M.J. Magmatic and crustal differentiation history of granitic rocks from Hf-O isotopes in zircon. *Science* **2007**, *315*, 980–983. [CrossRef]
52. Atherton, M.P.; Petford, N. Generation of sodium-rich magmas from newly underplated basaltic crust. *Nature* **1993**, *362*, 144–146. [CrossRef]
53. Deng, J.; Mo, X.; Zhao, H.; Luo, Z.; Zhao, G. Yanshanian magma–tectonic–metallogenic belt in east China of circum-Pacific domain (I): Igneous rocks, and orogenic processes. *J. China Univ. Geosci.* **1999**, *10*, 21–24. (In Chinese with English abstract)
54. Hofmann, A.W. Chemical differentiation of the Earth: The relationship between mantle, continental crust, and oceanic crust. *Earth Planet. Sci. Lett.* **1988**, *90*, 297–314. [CrossRef]
55. Green, T.H. Significance of Nb/Ta as an indicator of geochemical processes in the crust–mantle system. *Chem. Geol.* **1995**, *120*, 347–359. [CrossRef]
56. Taylor, S.R.; McLennan, S.M. *The Continental Crust: Its Composition and Evolution*; Blackwell Scientific Publications: Hoboken, NJ, USA, 1985; p. 312.
57. Rudnick, R.L.; Fountain, D.M. Nature and composition of the continental crust: A lower crustal perspective. *Rev. Geophys.* **1995**, *33*, 267–309. [CrossRef]
58. Li, W.B.; Huang, Z.L.; Xu, D.R.; Chen, J.; Xu, C.; Guan, T. Rb-Sr isotopic method on zinc–lead ore deposits: A review. *Geotect. Metall.* **2002**, *26*, 436–441. (In Chinese with English abstract)
59. Ma, X.H.; Chen, B.; Yang, M.C. Magma mixing origin for the Aolunhua porphyryrelated to Mo-Cu mineralization, eastern Central Asian Orogenic Belt. *Gondwana Res.* **2013**, *24*, 1152–1171. [CrossRef]
60. Jiang, S.H.; Liang, Q.L.; Liu, Y.F.; Liu, Y. Zircon U-Pb ages of the magmatic rocks occurring in and around the Dajing Cu-Ag-Sn polymetallic deposit of Inner Mongolia and constrains to the ore-forming age. *Acta Petrol. Sin.* **2012**, *28*, 495–513. (In Chinese with English abstract)
61. Wang, Y.; Fan, W.; Guo, F. K–Ar dating of late Mesozoic volcanism and geochemistry of volcanic gravels in the North Huaiyang Belt, Dabie orogen: Constraints on the stratigraphic framework and exhumation of the northern Dabie orthogneiss complex. *Chin. Sci. Bull.* **2002**, *47*, 1688–1695. [CrossRef]
62. Fan, W.M.; Guo, F.; Wang, Y.J.; Lin, G. Late Mesozoic calc-alkaline volcanism of post-orogenic extension in the northern Da Hinggan Mountains, northeastern China. *J. Volcanol. Geotherm. Res.* **2003**, *121*, 115–135. [CrossRef]
63. Zhou, J.B.; Wilde, S.A.; Zhang, X.Z.; Ren, S.M.; Zheng, C.Q. Early Paleozoic metamorphic rocks of the Erguna block in the Great Xing'an Range, NE China: Evidence for the timing of magmatic and metamorphic events and their tectonic implications. *Tectonophysics* **2011**, *499*, 105–117. [CrossRef]
64. Zhou, J.B.; Wilde, S.A.; Zhang, X.Z.; Zhao, G.C.; Zheng, C.Q.; Wang, Y.J.; Zhang, X.H. The onset of Pacific margin accretion in NE China: Evidence from the Heilongjiang high-pressure metamorphic belt. *Tectonophysics* **2009**, *478*, 230–246. [CrossRef]
65. Wang, T.; Zheng, Y.D.; Zhang, J.J.; Zeng, L.S.; Donskaya, T.; Guo, L.; Li, J.B. Pattern and kinematic polarity of late Mesozoic extension in continental NE Asia: Perspectives from metamorphic core complexes. *Tectonics* **2011**, *30*, 1–27. [CrossRef]
66. Liu, J.; Mao, J.W.; Wu, G.; Wang, F.; Luo, D.F.; Hu, Y.B. Zircon U-Pb and molybdenite Re-Os dating of the Chalukou porphyry Mo deposit in the northern Great Xing'an range, China and its geological significance. *J. Asian Earth Sci.* **2014**, *79*, 696–709. [CrossRef]
67. Hilde, T.W.C.; Uyeda, S.; Kroenke, L. Evolution of the western Pacific and its margin. *Tectonophysics* **1977**, *38*, 145–165. [CrossRef]

68. Wu, F.Y.; Lin, J.Q.; Wilde, S.A.; Zhang, X.O.; Yang, J.H. Nature and significance of the Early Cretaceous giant igneous event in eastern China. *Earth Planet. Sci. Lett.* **2005**, *233*, 103–119. [CrossRef]
69. Wang, F.; Zhou, X.H.; Zhang, L.C.; Ying, J.F.; Zhang, Y.T.; Wu, F.Y.; Zhu, R.X. Late Mesozoic volcanism in the Great Xing'an Range (NE China): Timing and implications for the dynamic setting of NE Asia. *Earth Planet. Sci. Lett.* **2006**, *251*, 179–198. [CrossRef]
70. Yu, J.J.; Wang, F.; Xu, W.L.; Gao, F.H.; Pei, F.P. Early Jurassic mafic magmatism in the Lesser Xing'an-Zhangguangcai Range, NE China, and its tectonic implications: Constraints from zircon U-Pb chronology and geochemistry. *Lithos* **2012**, *142*, 256–266. [CrossRef]
71. Zhang, F.Q.; Chen, H.L.; Yu, X.; Dong, C.W.; Yang, S.F.; Pang, Y.M.; Batt, G.E. Early Cretaceous volcanism in northern Songliao Basin, NE China, and its geodynamic implication. *Gondwana Res.* **2011**, *19*, 163–176. [CrossRef]
72. Xu, M.J.; Xu, W.L.; Meng, E.; Wang, F. LA–ICP–MS zircon U–Pb chronology and geochemistry of Mesozoic volcanic rocks from the Shanghulin–Xiangyang basins in Erguna area, and its tectonic implications. *Geol. Bull. China* **2011**, *30*, 1321–1338. (In Chinese with English abstract)
73. Wei, C.S.; Zhao, Z.F.; Spicuzza, M.J. Zircon oxygen isotopic constraint on the sources of late Mesozoic A-type granites in eastern China. *Chem. Geol.* **2008**, *250*, 1–15. [CrossRef]
74. Meng, F.C.; Liu, J.Q.; Cui, Y.; Gao, J.L.; Liu, X.; Tong, Y. Mesozoic tectonic regimes transition in the Northeast China: Constriants from temporal-spatial distribution and associations of volcanic rocks. *Acta Petrol. Sin.* **2014**, *30*, 3569–3586. (In Chinese with English abstract)
75. Wang, T.; Guo, L.; Zhang, L.; Yang, Q.D.; Zhang, J.J.; Tong, Y.; Ye, K. Timing and evolution of Jurassic-Cretaceous granitoid magmatisms in the Mongol-Okhotsk belt and adjacentareas, NE Asia: Implications for transition from contractional crustal thickening to extensional thinning and geodynamic settings. *J. Asian Earth Sci.* **2015**, *97*, 365–392. [CrossRef]
76. Han, S.J.; Wang, X.; Wang, X.; Wang, Y.; Zhang, Y. Geochronology and geochemistry of late Jurassic–Early Cretaceous volcanic rocks in the southern Great Xing'an range, NE China: Constraints for late Mesozoic tectono-magmatic evolution. *Int. Geol. Rev.* **2021**, *63*, 1366–1388. [CrossRef]
77. Wu, F.Y.; Sun, D.Y.; Li, H.M.; Jahn, B.M.; Wilde, S. A-type granites in northeastern China: Age and geochemical constraints on their petrogenesis. *Chem. Geol.* **2002**, *187*, 143–173. [CrossRef]
78. Zhang, L.Q.; Shao, J.A.; Zheng, G.R. Metamorphic core complex in Ganzhuermiao Inner Mongolia. *Sci. Geol. Sin.* **1998**, *33*, 14–20. (In Chinese with English abstract)
79. Zhang, X.D.; Yu, Q.; Chen, F.H.; Wang, X.W. Structural characteristics, origin and evolution of metamorphic core complex in central basement uplift and Xujiaweizi faulted depression in Songliao basin, Northeast China. *Earth Sci. Front.* **2000**, *7*, 411–419. (In Chinese with English abstract)
80. Liu, J.L.; Davis, G.A.; Lin, Z.Y.; Wu, F.Y. The Liaonan metamorphic core complex, southeastern Liaoning Province, North China: A likely contributor to Cretaceous rotation of eastern Liaoning, Korea and contiguous areas. *Tectonophysics* **2005**, *407*, 65–80. [CrossRef]
81. Shao, J.A.; Zhang, L.Q. Mesozoic dyke swarms in the north of North China. *Acta Petrol. Sin.* **2002**, *18*, 312–318. (In Chinese with English abstract)
82. Meng, E.; Xu, W.L.; Yang, D.B.; Qiu, K.F.; Li, C.H.; Zhu, H.T. Zircon U–Pb chronology, geochemistry of Mesozoic volcanic rocks from the Lingquan basin in Manzhouli area, and its tectonic implications. *Acta Petrol. Sin.* **2011**, *27*, 1209–1226. (In Chinese with English abstract)
83. Xu, W.L.; Pei, F.P.; Wang, F.; Meng, E.; Ji, W.Q.; Yang, D.B.; Wang, W. Spatial-temporal relationships of Mesozoic volcanic rocks in NE China: Constraints on tectonic overprinting and transformations between multiple tectonic regimes. *J. Asian Earth Sci.* **2013**, *74*, 167–193. [CrossRef]
84. Xu, W.L.; Wang, F.; Pei, F.P.; Meng, E.; Tang, J.; Xu, M.J.; Wang, W. Mesozoic tectonic regimes and regional ore-forming background in NE China: Constraints from spatial and temporal variations of Mesozoic volcanic rock associations. *Acta Petrol. Sin.* **2013**, *29*, 339–353.
85. Dong, Y.; Ge, W.C.; Yang, H.; Zhao, G.C.; Wang, Q.H.; Zhang, Y.L.; Su, L. Geochronology and geochemistry of Early Cretaceous volcanic rocks from the Baiyingaolao formation in the central Great Xing'an range, NE China, and its tectonic implications. *Lithos* **2014**, *205*, 168–184. [CrossRef]
86. Yu, Y.; Xu, W.L.; Pei, F.P.; Yang, D.B.; Zhao, Q.G. Chronology and geochemistry of Mesozoic volcanic rocks in the Linjiang area, Jilin Province and their tectonic implications. *Acta Geol. Sin.* **2009**, *83*, 245–257. [CrossRef]
87. Zhu, R.X.; Xu, Y.G. The subduction of the west Pacific plate and the destruction of the North China Craton. *Sci. China Earth Sci.* **2019**, *49*, 1340–1350. (In Chinese with English abstract) [CrossRef]
88. Liu, J.L.; Ni, J.L.; Chen, X.Y.; Craddock, J.P.; Zheng, Y.Y.; Ji, L.; Hou, C.R. Early Cretaceous tectonics across the North Pacific: New insights from multiphase tectonic extension in Eastern Eurasia. *Earth Sci. Rev.* **2021**, *217*, 103552. [CrossRef]
89. Tang, J.; Xu, W.L.; Wang, F.; Ge, W.C. Subduction history of the Paleo-Pacific slab beneath Eurasian continent: Mesozoic-Paleogene magmatic records in Northeast Asia. *Sci. China Earth Sci.* **2018**, *61*, 527–559. (In Chinese with English abstract) [CrossRef]
90. Zhu, X.; Chen, H.J.; Sun, Z.C.; Zhang, Y.C. The Mesozoic-Cenozoic tectonics and petroliferous basins of China. *Acta Geol. Sin.* **1983**, *3*, 235–242. (In Chinese with English abstract)

91. Tian, Y.; Ma, J.C.; Liu, C.; Feng, Y.; Liu, T.T.; Zhu, H.X.; Yan, D.; Li, H.H. Effects of subduction of the western Pacific plate on tectonic evolution of Northeast China and geodynamic implications. *Chin. J. Geophys.* **2019**, *62*, 1071–1082. (In Chinese with English abstract)
92. Pearce, J.A.; Harris, N.B.W.; Tindle, A.G. Trace element discrimination diagrams for the tectonic interpretation of granitic rocks. *J. Petrol.* **1984**, *25*, 956–983. [CrossRef]
93. Harris, N.B.; Pearce, J.A.; Tindle, A.G. Geochemical characteristics of collision-zone magmatism. In *Collision Tectonics*; Coward, M.P., Reis, A.C., Eds.; Geological Society of London: London, UK, 1986; pp. 67–81.
94. Luchitskaya, M.V.; Morozov, O.L.; Palandzhyan, S.A. Plagiogranite magmatism in the Mesozoic island-arc structure of the Pekulney Ridge, Chukotka Peninsula, NE Russia. *Lithos* **2004**, *79*, 251–269. [CrossRef]

MDPI AG
Grosspeteranlage 5
4052 Basel
Switzerland
Tel.: +41 61 683 77 34

Minerals Editorial Office
E-mail: minerals@mdpi.com
www.mdpi.com/journal/minerals

www.ingramcontent.com/pod-product-compliance
Lightning Source LLC
Chambersburg PA
CBHW061503220926
51850CB00004B/159
9783725850990